D0854888

TOOL AND MANUFACTURING ENGINEERS HANDBOOK

VOLUME V
MANUFACTURING MANAGEMENT

SOCIETY OF MANUFACTURING ENGINEERS

OFFICERS AND DIRECTORS, 1987-1988

TOOL AND MANUFACTURING ENGINEERS HANDBOOK

FOURTH EDITION

VOLUME V
MANUFACTURING MANAGEMENT

A reference book for manufacturing engineers, managers, and technicians

Raymond F. Veilleux
Staff Editor

Dr. Louis W. Petro
Consulting Editor

Produced under the supervision of the SME Reference Publications Committee in cooperation with the SME Technical Divisions

Society of Manufacturing Engineers
One SME Drive
Dearborn, Michigan

TMEH ®

ISBN No. 0-87263-306-3

Library of Congress Catalog No. 82-60312

Society of Manufacturing Engineers (SME)

First edition published 1949 by McGraw-Hill Book Co. in cooperation with SME under earlier Society name, American Society of Tool Engineers (ASTE), and under title *Tool Engineers Handbook*. Second edition published 1959 by McGrawHill Book Co. in cooperation with SME under earlier Society name, American Society of Tool and Manufacturing Engineers (ASTME), and under title *Tool Engineers Handbook*. Third edition published 1976 by McGraw-Hill Book Co. in cooperation with SME under current Society name and under title *Tool and Manufacturing Engineers Handbook*.

Printed in the United States of America.

PREFACE

The first edition, published as the *Tool Engineers Handbook* in 1949, established a useful and authoritative editorial format that was successfully expanded and improved on in the publication of highly acclaimed subsequent editions, published in 1959 and 1976, respectively. Now, with continuing dramatic advances in manufacturing technology, increasing competitive pressure both in the United States and abroad, and a significant diversification of information needs of the modern manufacturing engineer, comes the need for further expansion of the Handbook. As succinctly stated by Editor Frank W. Wilson in the preface to the second edition: "...no 'bible' of the industry can indefinitely survive the impact of new and changed technology."

Although greatly expanded and updated to reflect the latest in manufacturing technology, the nature of coverage in this edition is deeply rooted in the heritage of previous editions, constituting a unique compilation of practical data detailing the specification and use of modern manufacturing equipment and processes. Yet the publication of this edition marks an important break with tradition in that this volume, dedicated to manufacturing management, is the fifth and final volume in the fourth edition. Volume I, *Machining*, was published in March 1983, Volume II, *Forming*, in April 1984, Volume III, *Materials, Finishing and Coating*, in July 1985, and Volume IV, *Quality Control and Assembly*, in January 1987.

The scope of this edition is multifaceted, offering a ready reference source of authoritative manufacturing information for daily use by engineers, managers, and technicians, yet providing significant coverage of the fundamentals of manufacturing processes, equipment, and tooling for study by the novice engineer or student. Uniquely, this blend of coverage has characterized the proven usefulness and reputation of SME Handbooks in previous editions and continues in this edition to provide the basis for acceptance across all segments of manufacturing.

The subject matter in this volume is divided into six sections. This division is based on a study conducted by the Manufacturing Management Council of SME. The study revealed that there were six critical issues facing manufacturing management. To regain and retain their position in the world marketplace, manufacturing managers need to understand these critical issues and then initiate the necessary changes within their organizations.

Section 1 addresses those issues that affect operations and strategic planning. Merely doing things right isn't good enough. Manufacturing companies must find the right things to do and then concentrate their efforts and resources on them. The chapters in this section discuss planning, management control, planning and analyzing manufacturing investments, and cost estimating and control.

Section 2 is designed and organized to help managers develop and upgrade their leadership skills. It begins by discussing the importance of a company's philosophy and culture and progresses into organization, leadership, and technology management.

People are a company's most valuable resource. Section 3 assists managers in the development and coordination of their workforce. It also provides managers with an overview of the legal aspects of labor relations and concerns in occupational safety and health.

The manufacturing function needs to assume a role of management partner with engineering, finance, and marketing. Section 4 discusses how manufacturing can become involved with design and standards. It also provides detailed discussions on just-in-time manufacturing, computer-integrated manufacturing, and project management.

Section 5 emphasizes the importance of using existing facilities more efficiently. Topics to assist the manufacturing manager in this responsibility include facilities planning, equipment planning, production and inventory control, and materials management.

Quality isn't something that can be mandated. It is the net result of outstanding management systems, employee dedication, and equal corporate moral commitment. To help the manager incorporate a company-wide quality policy, Section 6 addresses quality management and planning, tools and techniques for achieving quality, and quality cost and improvement.

In this and other TMEH volumes, in-depth coverage of all subjects is presented in an easy-to-read format. A comprehensive index cross-references all subjects, facilitating quick access to information. The liberal use of drawings, graphs, and tables also speeds information gathering and problem solving.

The reference material contained in this volume is the product of incalculable hours of unselfish contribution by hundreds of individuals and organizations, as listed at the beginning of each chapter. No written words of appreciation can sufficiently express the special thanks due these many forward-thinking professionals. Their work is deeply appreciated by the Society; but more important, their contributions will undoubtedly serve to advance the understanding of manufacturing management throughout industry and will certainly help spur major productivity gains in the years ahead. Industry as a whole will be the beneficiary of their dedication. In particular, special thanks is due to Hayward Thomas, consultant and retired president of Jensen Industries, for his comments on all the material in this volume as well as his contribution to several sections.

Further recognition is due the members of the SME Reference Publication Committee for their expert guidance and support as well as the many members of the SME Technical Activities Board.

The Editors

SME staff who participated in the editorial development and production of this volume include:

EDITORIAL

Thomas J. Drozda
Director of Publications

Raymond F. Veilleux
Staff Editor

Ramon Bakerjian
Staff Editor

Louis W. Petro
Consulting Editor

Ellen J. Kehoe
Technical Copy Editor

Shirley A. Barrick
Editorial Secretary

Frances Kania
Editorial Secretary

TYPESETTING

Shari L. Smith
Administrative Coordinator

Nancy Bashi
Typesetter

GRAPHICS

Kathy J. Lake
Art Director

Thomas J. Martin
Illustrator/ Keyliner

SME

The Society of Manufacturing Engineers is a professional society dedicated to advancing manufacturing through the continuing education of manufacturing managers, engineers, technicians, and other manufacturing professionals. The specific goal of the Society is to advance scientific knowledge in the field of manufacturing and to apply its resources to research, writing, publishing, and disseminating information. "The purpose of SME is to serve the professional needs of the many types of practitioners that make up the manufacturing community...The collective goal of the membership is the sharing and advancement of knowledge in the field of manufacturing for the good of humanity."

The Society was founded in 1932 as the American Society of Tool Engineers (ASTE). From 1960 to 1969 it was known as the American Society of Tool and Manufacturing Engineers (ASTME), and in January 1970 it became the Society of Manufacturing Engineers. The changes in name reflect the evolution of the manufacturing engineering profession and the growth and increasing sophistication of a technical society that has gained an international reputation for being the most knowledgeable and progressive voice in the field.

Associations of SME—The Society provides complete technical services and membership benefits through a number of associations. Each serves a special interest area. Members may join these associations in addition to SME. The associations are:

Association for Finishing Processes of SME (AFP/SME)
Computer and Automated Systems Association of SME (CASA/SME)
Machine Vision Association of SME (MVA/SME)
North American Manufacturing Research Institute of SME (NAMRI/SME)
Robotics International of SME (RI/SME)
Manufacturing Automation Protocol & Technical and Office Protocol Users Group of SME (MAP/TOP)
Composites Group of SME (CoGSME)
Electronics Manufacturing Group of SME (EM/SME)

Members and Chapters—The Society and its associations have 80,000 member in 73 countries, most of whom are affiliated with SME's 300-plus senior chapters. The Society also has some 8000 student members and more than 150 student chapters at colleges and universities.

Publications—The Society is involved in various publication activities encompassing handbooks, textbooks, videotapes, and magazines. Current periodicals include:

Manufacturing Engineering
Manufacturing Insights (a video magazine)
SME Technical Digest
SME News
Journal of Manufacturing Systems

Certification—This SME program formally recognizes manufacturing managers, engineers, and technologists based on experience and knowledge. The key certification requirement is successful completion of a two-part written examination covering (1) engineering fundamentals and (2) an area of manufacturing specialization.

Educational Programs—The Society sponsors a wide range of educational activities, including conferences, clinics, in-plant courses, expositions, publications and other educational/training media, professional certification, and the SME Manufacturing Engineering Education Foundation.

CONTENTS

VOLUME V-MANUFACTURING MANAGEMENT

SYMBOLS AND ABBREVIATIONS

The following is a list of symbols and abbreviations in general use throughout this volume. Supplementary and/or derived units, symbols, and abbreviations that are peculiar to specific subject matter are listed within chapters.

A

ANSI	American National Standards Institute
APICS	American Production and Inventory Control Society
ASQC	American Society for Quality Control

C-D-E

CAD/CAM	Computer-aided design/computer-aided manufacturing
CAPP	Computer-aided process planning
CIM	Computer-integrated manufacturing
CIM I	Computer-interfaced manufacturing
CIM II	Computer-integrative management (of the manufacturing enterprise)
CNC	Computer numerical control
DFA	Design for assembly
DFM	Design for manufacture
DNC	Direct numerical control
Eq.	Equation

F-G-H-I-J

Fig.	Figure
FGM	Fifth-generation management
FMEA	Failure mode and effects analysis
FMS	Flexible manufacturing system(s)
ft or ′	Foot
GT	Group technology
hr	Hour
Ibid.	In the same place
in. or ″	Inch
IGES	Initial Graphic Exchange Specification
ISO	International Organization for Standardization
JIT	Just-in-time

L-M-N-O

LMRA	Labor Management Relations Act
Loc. cit.	In the place cited
MAP	Manufacturing Automation Protocol
MBO	Management by objectives
MRP	Material requirements planning
MRP II	Manufacturing resources planning
NBS	National Bureau of Standards
NC	Numerical control
NIOSH	National Institute for Occupational Safety and Health
NLRA	National Labor Relations Act
NLRB	National Labor Relations Board
NSC	National Security Council
Op. cit.	In the work cited
OSHA	Occupational Safety and Health Administration (Act)

R-S-T-W

R&D	Research and development
ROI	Return on investment
SME	Society of Manufacturing Engineers
SPC	Statistical process control
TOP	Technical and Office Protocol
WIP	Work in process
α	Alpha
$\approx$	Approximately equal to
$^\circ$	Degree
$>$	Greater than
$\geq$	Greater than or equal to
$<$	Less than
$\leq$	Less than or equal to
μ	Mu
%	Percent
$\pm$	Plus or minus
Σ or σ	Sigma (summation)

OPERATIONS AND STRATEGIC PLANNING

SECTION

1

PLANNING

This chapter is based on the underlying premise that planning is important and that manufacturing managers are inherently ''managers in the middle.'' Some elements of the planning process are defined by top managers. Manufacturing managers define elements for their own subordinate units and have extensive peer communications. The new technologies demand manufacturing input to corporate strategy. SME's studies and conferences spotlight the need for manufacturing managers to aggressively push the contribution manufacturing can make, and if necessary, shoulder their way into the planning process rather than docilely accepting the role of the ''manager in the middle.''

Numerous studies in the popular press and business literature have indicated that planning is an activity that is undertaken in all organizations even though there may not be a formal plan. Further, these studies indicate that organizations that do formal planning tend to ''outperform'' organizations that ignore formal planning.[1-8]

It is important to distinguish between the plan and the planning process. General (later President) Eisenhower, the principal architect of the D-day invasion in World War II, astutely noted:

> ''A plan is often not nearly as important as the planning process itself.''

The general's words clearly reveal that both the outcome of the planning process (the plan) and the planning process itself are important. Too often, the process is evaluated by the accuracy and impact of the final plan, thereby ignoring the superior organizational understanding and flexibility that usually result from a systematic planning process.

The focus of this chapter is on issues such as planning for planning and management of the planning process. Following this brief introduction, the first section provides a review of the basics. The first segment of the section is concerned with the definition of planning, the reasons for planning, and in a general sense, the concept of planning as a system process. The section discusses some principles of planning and provides three corporate examples of planning systems. Finally, some basic management considerations in designing and managing planning systems are examined.

The second section emphasizes the fine tuning of the planning process. This section presents an advanced discussion of issues concerning the management of the planning process, the factors affecting the need for planning, the different schools of thought on strategic planning, the notion of generic strategies and ''generic'' manufacturing structures (environment), and the alternative end states of the planning process. Further, because the notion is currently popular in the literature, the influence of generic strategies on the planning process is examined. The emphasis is on how manufacturing managers can use these concepts to link manufacturing to business strategies and choices in manufacturing technologies.

In the third section, some of the paradoxes (half-truths) regarding the planning process are highlighted along with some of the pitfalls to avoid during the process. The section concludes with a brief examination of the implementation difficulties of which managers should be aware.

Finally, the chapter concludes with some remarks that summarize the middle manager's distinctive view of and role in the planning process. In addition, there are some remarks to provoke ''rethinking'' the subject of planning.

CHAPTER CONTENTS:

THE BASICS OF PLANNING

This section provides a review of the basics of planning. Included is a discussion of the following: the key objectives and capabilities of planning systems, the impact of alternate time perspectives and organizational levels on planning, some examples of planning systems, some principles of planning, and finally, issues related to managing the planning process.

***Contributors of this chapter are: Robert Ericson**, Program Manager, Digital Equipment Corp.; **Suresh Kotha**, Professor, School of Management, Rensselaer Polytechnic Institute; **Daniel Orne**, Professor, School of Management, Rensselaer Polytechnic Institute; **Dr. Richard L. Shell**, Professor and Director of Industrial Engineering, Department of Mechanical and Industrial Engineering, University of Cincinnati.*

***Reviewers of this chapter are: Frank D. Cassidy**, Senior Engineering Manager, Digital Equipment Corp.; **James C. Emery**, Professor of Decision Sciences, The Wharton School, University of Pennsylvania; **James F. Lardner**, Vice President, Tractor and Component Div., Deere & Company; **R.E. Ochs**, Vice President and General Manager, Operations Div., Aerojet Ordnance Co.; **Hayward Thomas**, Consultant, Retired—President, Jensen Industries.*

CHAPTER 1

PLANNING BASICS

WHAT IS PLANNING?

Interest in planning increased in the late 1950s when businesses, following World War I, the Great Depression, and World War II, were trying to cope with the problems of: (a) changing consumer spending patterns, (b) the development of new competitive strategies, and (c) the uncertainties of the relatively uncontrolled marketplace. In an attempt to go beyond the concerns of day-to-day operations, businesses established departments of "long-range planning" with a new philosophy—planning for future performance.[9]

The prefix "long-range" was soon discarded for "strategic," because the term "long-range planning" underemphasized the comprehensive nature of corporate planning and overemphasized the time-specific connotation. Further, the word "strategic," while eliminating the implication of a time horizon, added "the connotation of importance."[9] This changed in 1979 when Schendel and Hofer proposed the term "strategic management" as the new model for the business policy field to simulate the consideration of management focus beyond the implementation of strategic plans and integration of tactical and operational issues to strategic issues.[10] More recently, the concepts in strategic planning have been extended to include issues of strategic management and thus are often used interchangeably with strategic management to describe the same concepts.

Planning can be viewed from many different perspectives. One author provides five different and complementary views of planning (see Fig. 1-1).[11] Similarly, Mintzberg has pointed out that planning can be thought of as: (a) future thinking, (b) integrated decision making, (c) articulated procedure and articulated result, and (d) programming.[12]

Planning as future thinking refers to managerial activities that take the future into consideration. Unfortunately, in this definition, planning becomes part of all decision making with the risk of losing its own unique identity.

Planning as integrated decision making attributes a broader perspective to planning. In this view, decisions from the differ-

1. Planning as a central control system
This view has its origins in systems thinking and cybernetics, which suggests that management should have a comprehensive planning and information system covering the total enterprise. It is the most commonly held idea of planning and has proved a most useful contribution to management, but research indicated that this approach to planning is likely to fail if used alone.

2. Planning as framework for innovation
A second and complementary view of planning is that the plan should serve as a stimulus for local initiative and a process through which the staff of an organization should organize its 'self-renewal' in terms of new products, new markets, and staff development—in fact the progressive adaptation of the organization to a rapidly changing environment.

3. Planning as a social learning process
A third view of planning is that management should use the process as a means of learning about the environment and the system which they are managing. Behavioral scientists in particular have pointed out that there are no ready-made solutions to hand, and that the only means of coping with change is to move forward a step at a time accepting that mistakes will be made, looking for marginal improvements rather than comprehensive solutions. This 'consensus' model of planning is being used very effectively in hospitals, boards of education, and other fields where there are no clear lines of command.

4. Planning as a political process
Much of the early writing on planning was politically naive. In the past 5 years however governments, unions, and social action groups have challenged the legitimacy of management. In Europe, particularly, corporate planning has come to be viewed as an interorganizational process involving government and unions, e. g. through industrial democracy and 'planning agreements' between large companies and central governments. Within organizations, too, the political struggles have been more clearly identified and researched.

5. Planning as a conflict of values
Much of planning is 'instrumental', i.e. concerned with devising feasible strategies and efficient methods for accomplishing the present objectives of existing institutions.
Another 'school of thought' takes the view that planning should be more concerned with re-examining the goals and purposes of enterprises—to achieve a better match with the aspiration of employees and the expectations of society at large.
They also argue that planners in both public and private organizations should be involved increasingly in 'creatively' forecasting the future shape of major sociotechnical systems.

Fig. 1-1 Differing perspectives on planning. Source: Bernard Taylor, "New Dimensions in Corporate Planning," *Long Range Planning* (December 1976). Copyright Pergamon Journals, Ltd. Reprinted with permission

ent areas and varying time perspectives are incorporated into a larger framework.

Planning as formalized procedure and articulated result distinguishes planning from future thinking and informal approaches to decision making, because it is closer to an operational definition. Mintzberg points out that it is the orientation toward analysis—toward systematic, explicit, recoverable thought processes—that capture what is meant by planning in most of the literature on that subject.[12] In this instance, planning can be identified with two observable behaviors in organizations: (1) the use of formalized procedures such as strategic planning, capital budgeting, and management by objectives, and (2) the articulation of the result in terms of "plans"—statement of intent, such as strategic plans, personnel plans, and budgets, and statement of objectives.

Planning as programming can be viewed as delineation of intended strategy already in place as opposed to the development of intended strategy. In this view, planning is programming and is used not to conceive an intended strategy, but to elaborate the consequences of an intended strategy already conceived.

REASONS FOR PLANNING

The reasons for formal organizational planning are numerous. They range from focusing the energies and activities of the organization to providing a basis for evaluating results. Regardless of the perspective one adopts, a fundamental purpose of planning is to integrate organizational efforts in achieving a common goal. As illustrated in Fig. 1-2, the purpose is to integrate efforts in research and development, marketing, and operations (production) toward achieving organizational goals.[13] Without planning in some form, integration is unlikely.

Alternately, one can equate the reasons for planning to the commonly articulated capabilities that the "ideal" planning system should possess. These capabilities include the following:[14]

1. Ability to anticipate surprises and crises.
2. Flexibility to adapt to unanticipated changes.
3. Ability to identify new business opportunities.
4. Ability to identify key problem areas.
5. Ability to foster managerial motivation.
6. Ability to enhance the generation of new ideas.
7. Ability to communicate top management's expectations down the line.
8. Ability to foster management control.
9. Ability to foster organization learning.
10. Ability to communicate line managers' concerns to top management.
11. Ability to integrate diverse functions and operations.
12. Ability to enhance innovation.

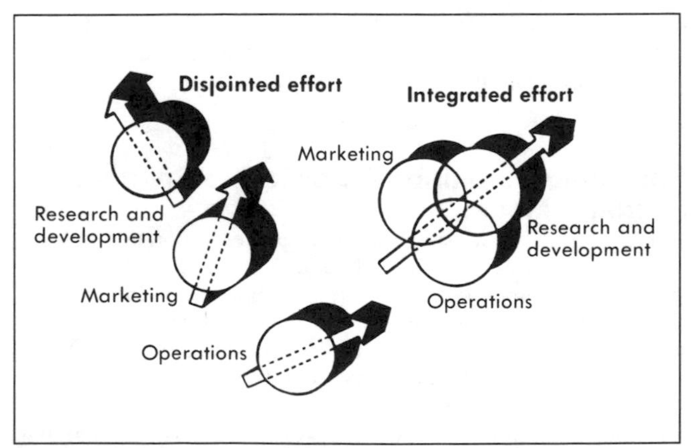

Fig. 1-2 Focus of organization effort. Source: R. Drake, "Innovative Structures for Managing Change," *Planning Review* (November 1986). Reprinted with permission

DEFINING THE PLANNING PROCESS

Given an organizational context, most generic planning and control systems consist of several key elements:[15] a plan or desired state; actual performance; and controls that compare the plan with actual performance, analyze any differences, and make necessary changes in performance or even the plan itself. A management planning and control system merely applies this to an organizational context, defining the planning function in terms of missions and purposes, goals and objectives, specific plans, and schedules. Actual performance against these desired states yields management information, resulting in control actions (see Fig. 1-3).

In later sections of this chapter, the individual elements of these general planning and control systems will be discussed more fully. Specifically, the next section considers the topic of organizational focus of the planning effort. This is followed by discussions of planning scope (or time horizon), a corporate resource planning model, a business unit-level strategic planning model, and finally a functional model for manufacturing policy determination. Chapter 2 of this volume, it should be noted, goes on to discuss more fully the concept of management control.

Unit of Organizational Focus

Before and during the planning process, it is important that the organizational unit for which the planning is undertaken is clearly defined and commonly understood. This is important because, in general, the type of goals, policies, and issues facing managers differ. This difference depends on: (a) the scope and level of organizational focus, (b) the operating level of the manager, and (c) the linkages with other organizational units and the external environment.

Although an obvious need, a fuzzy or otherwise inadequate definition of an organizational unit is a surprisingly common problem encountered in actual planning systems. A classic example is the problems encountered when an organization too narrowly defines its strategic business units and/or the industries

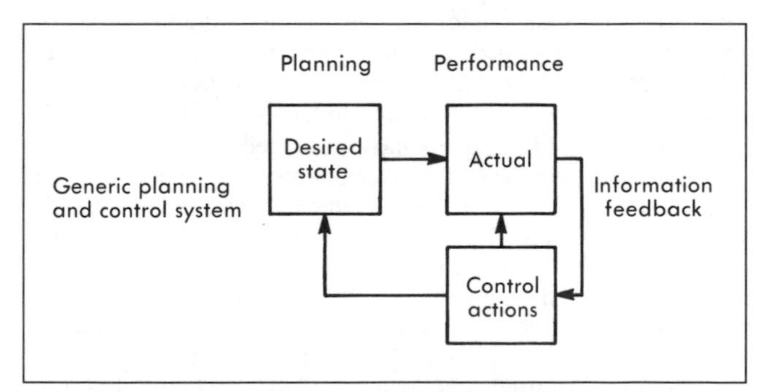

Fig. 1-3 Planning and control system relationships. Source: Lester A. Digman, *Strategic Management Concepts, Decisions, Cases* (Plano, TX: Business Publications, Inc., 1986), p. 12. Reprinted with permission

in which it is competing.[16-18] For example, both Norton and General Electric ran into this problem in the mid-1970s. To correct the problem, they inserted an organizational planning level at the group (or sector) level that specifically addressed the shared concerns between what had previously been treated as "independent" strategic business units.

To illustrate the impact of organizational level, commonly articulated levels of strategy are described. These levels are industry, corporate, business unit, and functional.[10,19]

Industry-level strategy. At the industry or government policy level, the strategic concerns revolve around broad issues such as the strategic alliances within the industry; incentives for investment; import and export trade barriers, duties, and quotas; the balance between imports and exports; inflation and the cost of capital; transportation and educational infrastructure; health and safety standards; antitrust regulations; employment levels; patent policy; and so forth.

It is commonly agreed that such industry or governmental policies can influence the manufacturing competitiveness of specific business units in specific industries. For example, Cohen and Zysman have argued that substantial elements of economic theory and policy need to be restructured, if manufacturing is to be competitive.[20] They have also stated that manufacturing managers (in the middle) have an indispensable role in setting the stage and the issues.

Corporate-level strategy. At the corporate level, the concern revolves around the definition of businesses in which the corporation wishes to participate, and the acquisition and allocation of resources to these business units.[19,21,22]

Business-level strategy. At the business level, generally referred to as strategic business unit (SBU) or strategic planning unit (SPU), three critical issues are specified: (1) the scope or boundaries of each business and the operational links with corporate strategy, (2) the basis on which the business unit will achieve and maintain a competitive advantage within its industry, and (3) the form (if any) of the strategic alliances between the business unit and other participants in the industry.

Functional-level strategy. At the functional level, strategy specifies how functional-level strategies like marketing/sales, manufacturing, research and development, and accounting/control, among others, will support the desired competitive business-level strategy. Further, the interrelations and interactions between the functions are specified to integrate the activities across the various functions.

Summary. This categorization of strategic levels has several implications for manufacturing managers in the middle. Each of the four levels described has an important and distinct role to play in achieving competitive advantage, and managers in the middle can have an impact at each level.

Many concerned managers have criticized governmental and corporate-level strategies for overlooking and underinvolving manufacturing.[23-25] The criticism is that corporate preoccupation with "diversification," "paper entrepreneurism," and "short-term profitability" has eroded the manufacturing infrastructure and, as a result, the potential for long-term profitability. Currently, planning executives and theorists have begun to address the need for a manufacturing strategy at the corporate level (a common manufacturing strategy among business units). Because the focus here is on planning and planning systems as opposed to manufacturing strategy, the reader is referred to references 26, 27, and 28 for additional information.

Planning Scope (Time Horizon)

The emphasis in this section is on the time horizon and level of detail (planning scope) addressed within the planning system. Managers have identified four common levels of planning scope in organizations: (1) strategic planning, (2) tactical planning, (3) operational planning, and (4) detailed, specific activities planning such as scheduling and dispatching. Each of the levels differs in the organizational level of responsibility, the planning issues addressed, and the planning horizon. Figure 1-4 provides a general description at each level.[29]

For example, strategic planning deals with the long-range considerations (usually 5 years), whereas scheduling and dispatching deal with issues currently under way on the factory floor. Although the activities are interrelated, there is a marked contrast between the scope of information and level of detail required at each of the levels.

Figure 1-5 provides a recent model that illustrates the type of complex planning interrelationships that can exist between:[30] (1) the previous discussion of the unit of organizational focus and (2) the current discussion of strategic, tactical, operational, and detailed planning.

It should be noted that the subtle and dynamic interrelationships between these planning activities defy their incorporation in any single model. Thus, the model presented is only a partial integration of the concepts discussed and is only applicable in a general sense to specific organizations.

If a manager is to be effective in the planning process, that manager must have a general understanding of his or her role in the overall planning system and the interrelationships in the planning activities. Too often, managers in the middle attempt to plan for only their specific activities and ignore the interrelationships just described. The result is that plans and activities are generally out of balance and have no focus. For example, managers in the middle might submit a series of independent capital appropriations requests to corporate and divisional management that substantially exceed the corporate resources that will be provided to the business unit, and they (corporate management) ignore focusing on how the business unit is attempting to generate a competitive advantage.

In the next few sections, more detailed models of some of the planning activities at the corporate, business unit, and functional levels are introduced. While some of these are intended as guides to thought and action—a pattern to follow in the planning process—others are intended as a representation of the way something actually happens. Although presented independently, they are interdependent and, in general, represent partially the dynamics and activities that encompass planning.

Corporate Planning: A Corporate Resource Planning Model

Previously, it was suggested that at the corporate level the major concerns revolved around the definition of the businesses in which the corporation wishes to participate, and the acquisition and allocation of resources to these business units. This section focuses on the latter issue. Models are often used to plan and control the allocation of corporate resources. As a result, the specific mode used affects managers in the middle who are expecting corporate resources. A key perspective is that the competitive strategy of a business unit must be consistent with the resources that will be available. Some strategies, such as cost leadership strategies, require substantial resources over a long

Increasing scope ↑

- **Strategic planning (5 years and beyond)**
  Which businesses should the firm be in?
  How should they be financed?
  How should scarce resources be allocated across business sectors?
- **Tactical planning (1-5 years)**
  What are the optimal patterns of capital investment and divestment for implementing some longer range plan?
  What decisions about facility location, expansion, or shutdown will maximize profitability?
  What products should be added to or deleted from the product line?
  What is the optimal product pricing pattern?
- **Operations planning (1-12 months)**
  What is the optimal operating plan (raw material acquisition, product sources, inventory levels, distribution system configuration, route and mode of distribution, etc.) to meet specified system objectives, consistent with some longer term plan, with existing facilities in the next planning period (for example, month, quarter, year)?
  What is the best operating plan on which to base plans for production and dispatch?
- **Scheduling and dispatching (Right now)**
  What specific operations or sequences of operations should be performed with which existing facilities, to meet specified output requirements in the next operational period (for example, hour, day, week)?

Increasing detail ↓

Fig. 1-4 A characterization of planning levels. Source: David Shirshfield, "From the Shadows," *Interfaces* (April 1983). Reprinted with permission

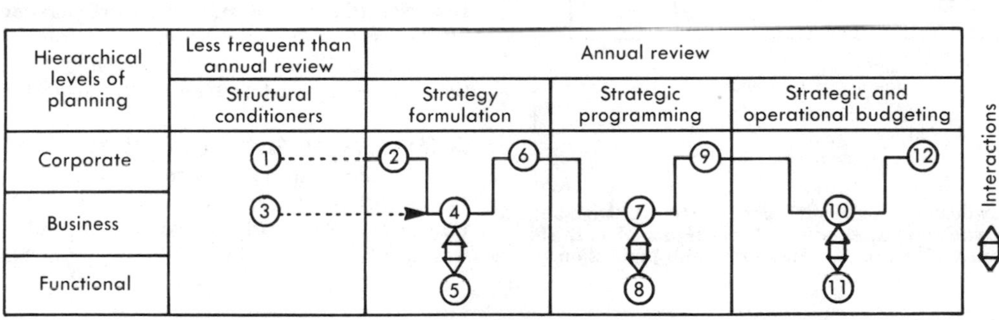

1. The vision of the firm: philosophy, mission, and identification of strategic business units (SBUs)
2. Strategic posture and planning guidelines: corproate strategic thrusts, performance objectives, and planning challenges
3. The mission of the business: business scope and identification of product market segments.
4. Formulation of business strategy and broad action programs
5. Formulation of functional strategy: participation in business planning, concurrence or nonconcurrence with business strategy proposals, broad action programs
6. Consolidation of business and functional strategies
7. Definition and evaluation of specific action programs at the business level
8. Definition and evaluation of specific action programs at the functional level
9. Resource allocation and definition of performance measurements for management control
10. Budgeting at the business level
11. Budgeting at the functional level
12. Budgeting consolidations and approval of strategic and operational funds

Fig. 1-5 Planning process model. Source: C. H. Fine and A. C. Hax, "Manufacturing Strategy: Methodology and an Illustration," *Interfaces* (November-December 1985). Reprinted with permission

period of time.[31] Without these resources, these strategies are unlikely to be executed successfully and should not be attempted.

A variety of corporate resource allocation models exist. The more popular ones have been around for years (see Figs. 1-6 and 1-7).[31-36] With slightly different techniques and effort, corporate resource allocation models generally suggest business unit capital budgets based on variables that are linked to industry growth/profitability and relative market share.

In recent years, there has been a significant controversy over the proper use of these models.[37-41] It is perceived that when these models are used with skill and sensitivity, they can be of significant help in allocating limited corporate resources to alternate business units.[42] The primary areas of controversy and sensitivity are the definition of business units and industry boundaries,[16,31] the SBU resource requirements, and the competitive dynamics within an industry.

What is often missed by managers in the middle is an understanding of how these corporate allocation models affect their units. For example, based on corporate objectives, some units are likely to be provided resources for growth, others resources to maintain and sustain their current position, and finally, some units might be forced to harvest (or cut back) their operations. Thus, manufacturing units and managers must live within their capital budgets and formulate strategies that are consistent with corporate objectives and goals.

Thus, manufacturing managers should be aware that these planning models are useful in anticipating the future need for funds by the various "independent" units. For example, a business with a strong position in a low-growth industry can usually sustain its position and generate cash. A business with a strong position in a high-growth industry will usually require additional external capital to sustain its competitive position.

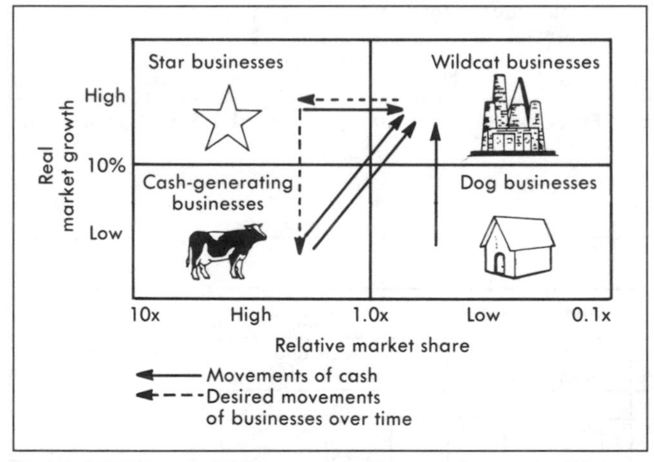

Fig. 1-6 BCG growth/share matrix. Source: Lester A. Digman, *Strategic Management Concepts, Decisions, Cases* (Plano, TX: Business Publications, Inc., 1986), p. 161. Reprinted with permission

A Strategic Management Model

The majority of critical decisions made at the business-unit level directly affect competitive positioning. At this level, the strategic management process for descriptive purposes is often broken into formulation and implementation, though in real life they are interrelated and inseparable.

Numerous models can be found in the business policy literature that attempt to describe the strategic management process for business units. Figure 1-8 is a strategic management process model that was developed by one of the most knowledgeable of the strategic management writers—the late William Glueck.[43] His process model consists of the following basic steps: (1) analyze the situation, (2) determine possible alternatives, (3) evaluate the alternatives, (4) choose the most favorable alternative, and (5) implement the chosen alternative.

As illustrated in Fig. 1-8, enterprise objectives and enterprise strategists provide the inputs to the model. After defining in general terms the objectives and mission of the organizational unit, Glueck recommends that explicit consideration be given to the values, goals, backgrounds, and skills of the enterprise strategists (the line and staff managers who develop the strategic plans) as part of the planning process. Different sets of strategists will develop different strategies, even given the same organizational context, based on their skills and abilities.[15]

A Manufacturing Policy Model

The philosophical foundation of the concept of manufacturing strategy has often been traced to the work of Wickham Skinner. He was one of the early authors who argued that manufacturing was a missing link in the strategy of business units and

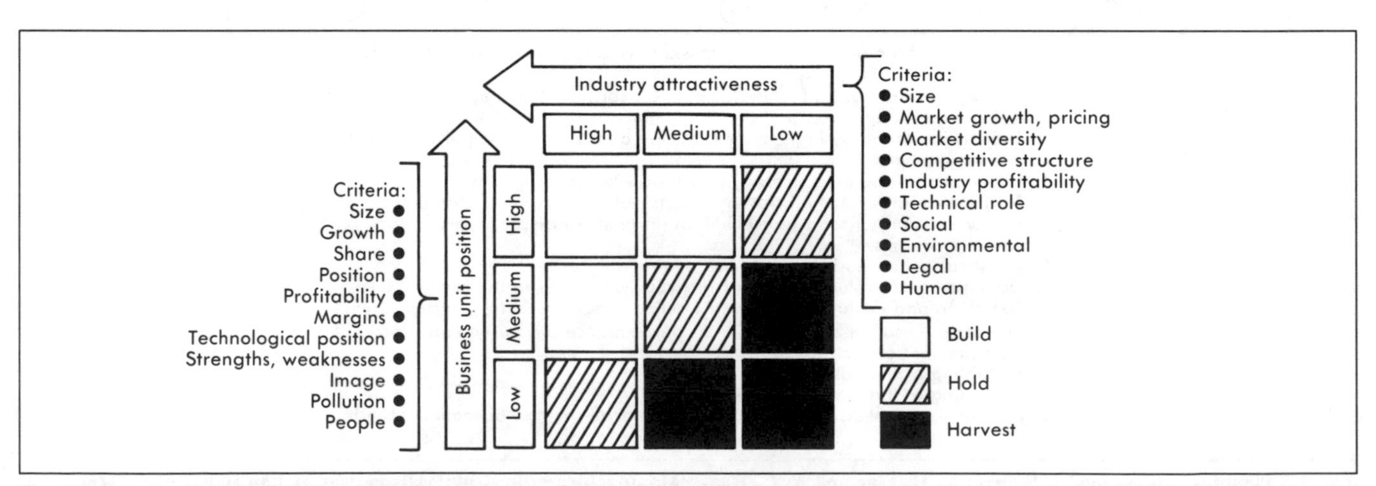

Fig. 1-7 Company position/industry attractiveness screen. Source: Michael Porter, *Competitive Strategy: Techniques for Analyzing Industries and Competitors* (New York: The Free Press, 1980), p. 365. Reprinted with permission

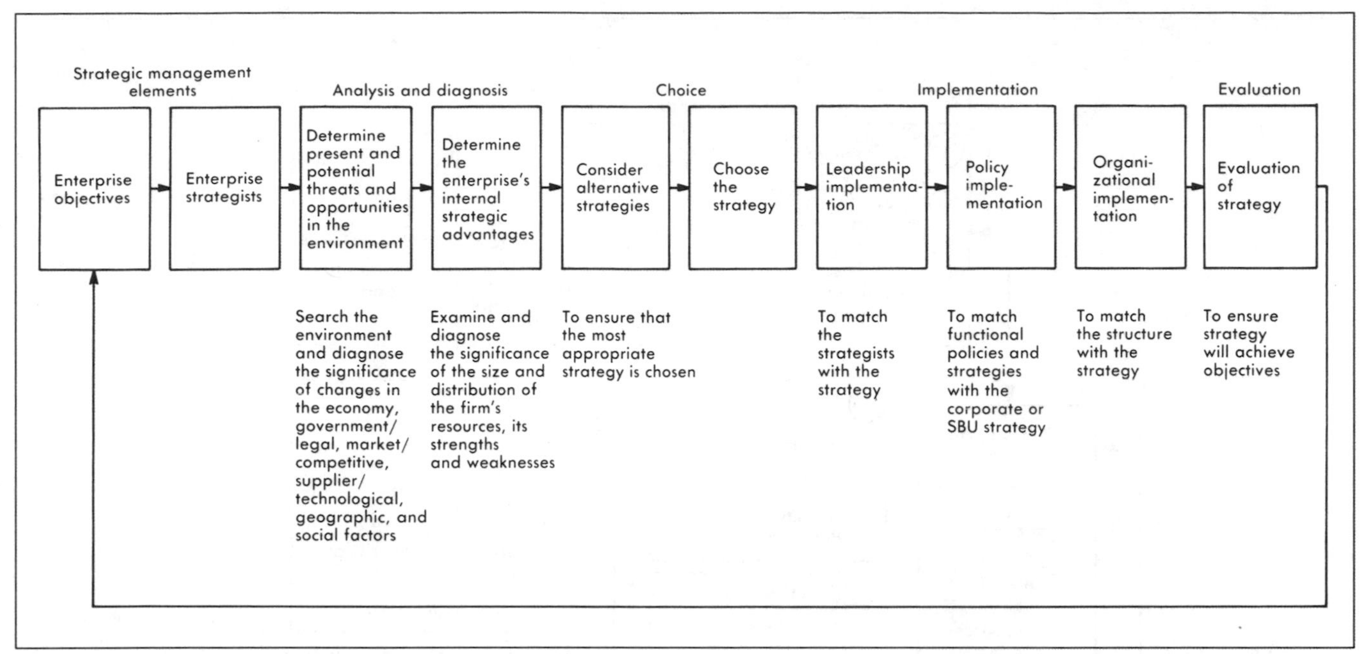

Fig. 1-8 Glueck's strategic management process model. Source: W. F. Glueck, *Business Policy and Strategic Management*, 4th ed. (New York: McGraw-Hill Book Co., 1984), p. 7. Reprinted with permission

corporations by stating that "a company's manufacturing function typically is either a competitive weapon or a corporate millstone. It is seldom neutral. The connection between manufacturing and corporate success is rarely seen as more than theachievement of high efficiency and low cost. In fact, the connection is much more critical and much more sensitive. Few top managers are aware that what appear to be routine manufacturing decisions frequently come to limit the corporation's strategic options, binding it with facilities, equipment, personnel, and basic controls and policies to a noncompetitive posture which may take years to turn around."[28]

Part of Skinner's recommendation for bridging the gap between major manufacturing decisions and the strategy of business units was a model describing the process of manufacturing policy determination (see Fig. 1-9).[44] The basic elements of the model involve an analysis of some key questions. These questions follow the steps listed in Fig. 1-9 and are as follows:

Step 1. What are competitors doing or what could they do?

Step 2. What resources does the business unit have to compete?

Step 3. What are the alternatives and chosen strategy of the SBU?

Step 4. Given the SBU strategy, what are the major tasks for the manufacturing function?

Steps 5 and 6. What are the major technologies and underlying economic factors that are likely to influence the SBU's manufacturing strategy?

Step 7. What is the assessment of the resources the SBU has or will have to compete with?

Step 8. Given all of the preceding, what are the SBU's manufacturing policies—its manufacturing strategy?

Step 9. What are the implementation requirements of the manufacturing policies?

Steps 10, 11, and 12. What are the implications for the manufacturing systems, controls, and procedures?

Step 13. How is the performance measured, and what is the performance of the manufacturing function?

Step 14. Feedback of manufacturing goals, policies, systems, and performance to subsequent SBU strategic planning.

Step 15. Feedback of manufacturing performance to modify manufacturing operations and policies.

Detailed Planning

For each functional area within a business unit there is a variety of detailed, more specific activities that require planning. Because the focus of this volume is on manufacturing and manufacturing-oriented executives, the consideration of detailed planning activities shall be limited to discussion of the manufacturing function.

Skinner and others have articulated different categories of major manufacturing management decisions that managers can focus on during the planning effort (see Fig. 1-10).[21,26,27,28,30] In each area listed, planning has an important role in structuring the activities and linking the decisions across the areas.

In subsequent chapters, the concept of planning will be periodically discussed within the context of a specific topic. The relationship of the various chapters in this volume to the decision categories in Fig. 1-10 (see p. 1-9) are listed in Table 1-1 (see p. 1-10). In addition, reference 31 discusses strategic analysis of vertical integration and capacity planning.

While each of these topics is important in isolation, it is crucial that decisions in each of these areas be linked to major manufacturing decisions, the business unit manufacturing strategy, and in a more general sense, the overall strategy of the firm.

PRINCIPLES OF PLANNING

In articles on planning, it has become somewhat common to list fundamental principles of planning. While the structure and

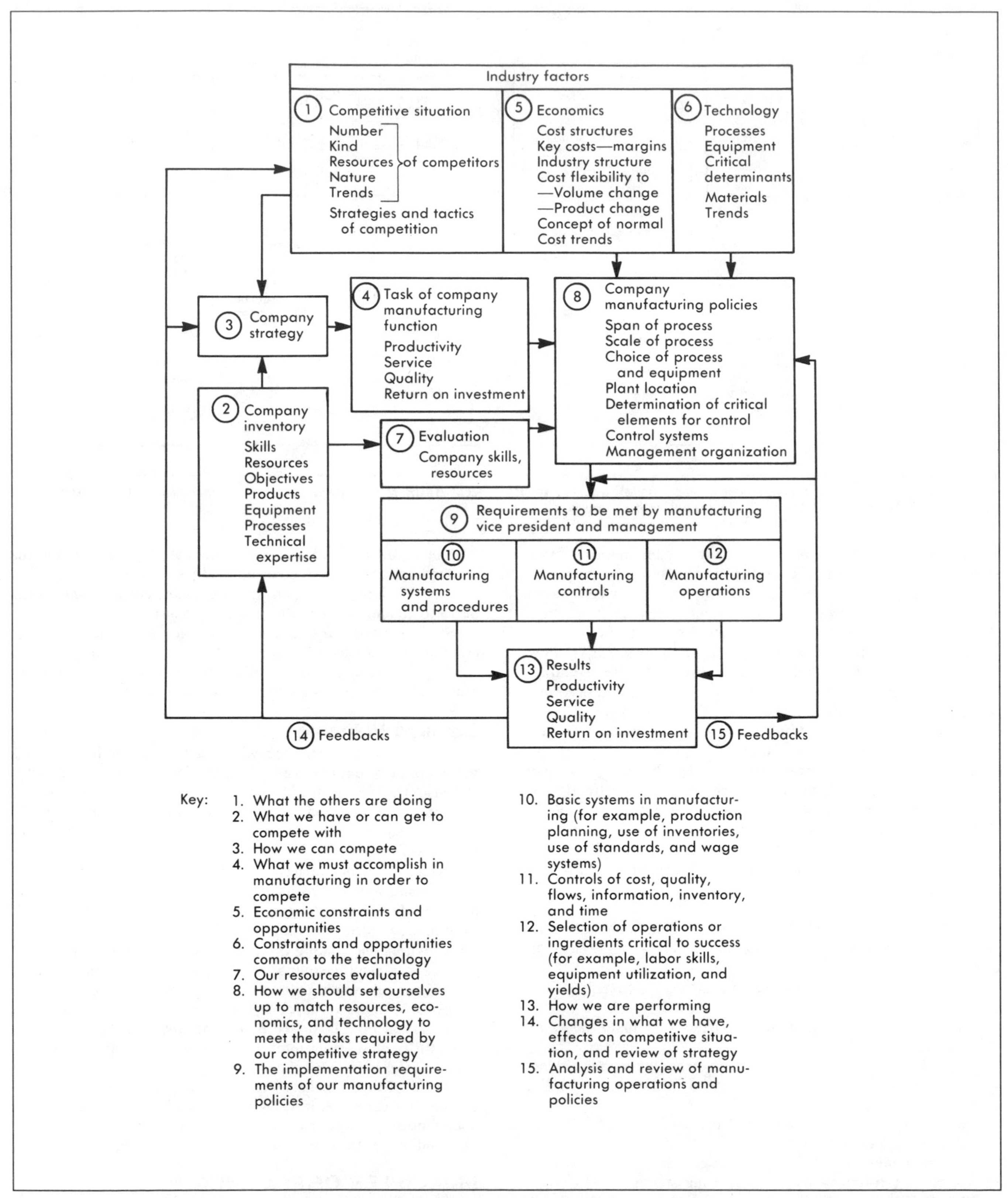

Fig. 1-9 The process of manufacturing policy determination. Source: Wickham Skinner, *Manufacturing: The Formidable Competitive Weapon* (New York: John Wiley & Sons, Inc., 1985), p. 65. Reprinted with permission

I. Hardware decisions
  a. Vertical integration (scope of the process)
  b. Capacity (scale of the process)
  c. Plant (layout, focus, location)
  d. Process and equipment technology

II. Infrastructure decisions
  a. Production and inventory control systems
  b. Purchasing systems
  c. Quality control systems
  d. Cost/information systems
  e. Workforce management
  f. Engineering and maintenance
  g. Formal organization

Fig. 1-10 Skinner's decision categories for manufacturing design. Source: Wickham Skinner, *Manufacturing in the Corporate Strategy* (New York: John Wiley & Sons, Inc., 1978). Reprinted with permission

clarity of these principles may be appealing and helpful, especially to less experienced planners, in practical terms these principles should more properly be considered as broad guidelines. In a later section of this chapter, the paradoxes and half-truths imbedded within general planning principles will be discussed. Nevertheless, the following principles by Gray will serve as a helpful reminder of some critical concepts to keep in mind while designing and managing planning systems:[45]

1. Strategic planning is a line function for which training in strategic analysis and participative skills is usually necessary.
2. The strategic business unit needs to be defined so that one executive can control the key variables essential to the execution of his or her strategic business plan.
3. A unit's concept of the business it is in must, above all, be formulated from the outside in so that it can most effectively engage the dynamics of its strategic environment.
4. Action plans for achieving business objectives are the key to implementing and monitoring strategy. They require extensive lower level participation and special leadership skills. Action plans are complete when underlying assumptions, allocation of responsibilities, time and resource requirements, risks, and likely responses have been made explicit.
5. Participative strategy development, a prerequisite for successful strategy execution, often requires cultural change at the upper levels of corporations and their business units.
6. The strategic planning system and other control systems designed to guide managerial and organizational behavior must be integrated in a consistent whole if business strategies are to be executed well.
7. Productivity improvement programs are best treated as aspects of strategic business plans because productivity takes on a significantly different meaning as the strategic balance between marketing and production shifts.
8. Well-managed organizations must be both centralized and decentralized—centralized so that strategies and control systems can be integrated, and decentralized so that units in each strategic environment can act and be treated with appropriate differentiation.
9. Over time, good strategic planning, once considered a separate activity, becomes a mind-set, a style, and a set of techniques for running a business—not something more to do, but a better way of doing what has always had to be done.

EXAMPLES

This section will present examples of a portion of the planning process in use at three separate corporations: General Electric, IBM, and Texas Instruments.

Planning at General Electric

Figure 1-11 (see p. 1-11) illustrates the annual planning cycle used by GE.[46] This annual cycle starts in December/January and continues into November and begins anew the following January. Each year, however, a new plan is developed, with increasing detail for the upcoming year, the next 5 years, and so on. The complete process occurs in four steps, described as follows:

1. The planning challenges. The planning challenges set the stage for the annual strategic planning cycle. Each year, the CEO issues a number of specific challenges that are to be addressed in the strategic plans of the strategic business units (SBUs) and the sectors.
2. Strategy development. Based on the planning challenges, the SBU planning staff and managers develop strategic alternatives and tentative plans during the next 3 months.
3. Sector reviews and evaluation. During the next 3-month cycle, June/July/August, the plans put forth by the SBU and sectors are reviewed and evaluated by corporate management and staff. Modifications, if any, are incorporated. On approval at the end of this 3-month cycle, top management decides the level of resources to be allocated to each business unit. This becomes the basis or guideline for the business unit to develop detailed operating plans and budgets. At this point, senior corporate staff executives are responsible for the objective assessment of key resources (financial, human, technology, and production) and the identification of issues affecting the company's strategic strengths, both domestic and international.
4. Resource and budget reviews. In the final 3-month cycle, the detailed operating plans and budgets are presented to top management for review. Based on the reviews, revisions are made to the plan before the final budget is approved by top management. At this point the corporate plan is updated.

Planning at IBM

At IBM, activities are grouped into a number of line operating units that have profit/loss responsibility. These units are differentiated by business area and geographic region and have the range of functional capabilities needed to conduct their assigned missions as autonomously as is practical.[47]

Operating unit management is responsible for developing and implementing its strategies and plans. Prior to implementation, these strategies and plans are approved by corporate management. Performance against the plan is then measured and controlled by unit management. The results of operations are periodically reviewed with corporate management. Further, business policies are controlled at the corporate level. These provide the broad framework within which all units function.

TABLE 1-1
Relationship of Volume V Content to Skinner's Decision Categories for Manufacturing Design

Category	Chapter Number and Title
Capacity	15. Just-in-Time Manufacturing 19. Equipment Planning 20. Production Planning and Control
Plant layout	15. Just-in-Time Manufacturing 18. Facilities Planning 19. Equipment Planning 20. Production Planning and Control
Process and equipment technology	3. Planning and Analysis of Manufacturing Investments 8. Management of Technology 16. Computer-Integrated Manufacturing 19. Equipment Planning
Production and inventory control systems	15. Just-in-Time Manufacturing 20. Production Planning and Control 21. Materials Management
Purchasing systems	15. Just-in-Time Manufacturing 21. Materials Management
Quality control systems	2. Management Control 15. Just-in-Time Manufacturing 22. Quality Management and Planning 23. Achieving Quality 24. Quality Cost and Improvement
Cost information systems	2. Management Control 3. Planning and Analysis of Manufacturing Investments 4. Cost Estimating and Control 24. Quality Cost and Improvement
Workforce management	7. Manufacturing Leadership 9. Workforce Development 10. Workforce Management 11. Legal Environment for Labor Relations 12. Management Concerns for Occupational Safety and Health
Engineering and Maintenance	13. Design for Manufacture 14. Standards and Certification 17. Project Management 19. Equipment Planning
Formal organization	5. Philosophy and Culture of Manufacturing Management 6. Organization 13. Design for Manufacture 16. Computer-Integrated Manufacturing

In assessing units' plans and performance, corporate management is assisted by corporate staff, which provides counsel and performs certain centralized services.

IBM uses two distinct, but interactive, kinds of planning that it refers to as program planning and period planning (see Fig. 1-12 on p. 1-12). These two kinds of planning are described as follows:

1. Program planning. Program planning is generally used in the planning of programs to develop a product or improve the productivity of a function. The program plan generally has a single objective, but may involve several functional elements. Its time horizon is determined by the nature of the specific program objective and the work processes required to achieve it. Its cycle for review and decision making is determined by inherent dynamics of the program. At any point in time, each operating unit has a large portfolio of product and functional programs in various stages of planning and implementation.
2. Period planning. This kind of planning complements program planning and is characterized by the following: The period plans force a balancing among the multiple program objectives to achieve the corporate targets. The plan horizons are fixed by corporate management—2 years for the commitment plan and 5 years for the strategies and strategic plans. The cycle for review and decision making is tied to the calendar to ensure the availability of an operating budget for each unit at the beginning of each year.

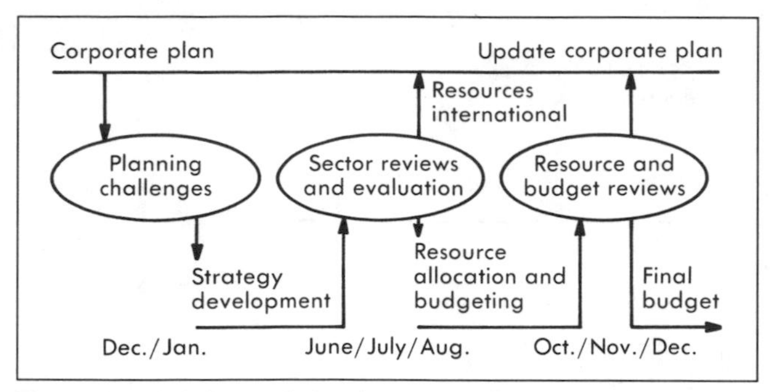

Fig. 1-11 Model of annual planning cycle at General Electric. Source: Harvard Business School Case 381-174, 1981, p. 12. Reprinted with permission

Decisions made as part of both plans affect each other, and business unit management is responsible for establishing and maintaining the balance among its objectives and resources.

Program planning generally proceeds in two distinct stages, defining the market requirements and translating them into products. Period planning, on the other hand, establishes the strategy to achieve some individual product or functional objective. Further, it forces the balancing among multiple program objectives. This balancing take place twice each year—once in the spring, as each unit sets its overall business direction, and again in the fall, as each unit commits to implementing in that direction and achieving the planned results.

Figure 1-12 illustrates the various stages in the period planning process. Strategic targets (worldwide revenues, annual growth in profit, and return on controllable assets) are established by corporate management. With these guidelines, business area operating units formulate the strategies. These are integrated by each profit center and given financial expression in a strategic plan. This is presented for review to corporate management. Management then determines the investments that best meet IBM's need for growth and profitability, informs the profit centers of its decisions, and assigns near-term targets to guide the development of the plans for implementation.

Planning at Texas Instruments—The OST System

At Texas Instruments, line management structure is divided into semiconductor, consumer electronics, materials and electrochemicals, government electronics, and geophysical exploration groups. A senior manager in charge of each of these groups reports directly to the president and is responsible for the worldwide strategic direction of the business as well as for the regular daily management functions. Each of these groups is further subdivided into divisions that are broken into product customer centers.

Texas Instruments is known for its OST system of planning. The system consists of objectives, strategies, and tactics (see Fig. 1-13 on p. 1-13). These three phases are described as follows:[48]

1. Objectives. Corporate objectives tend to project 5-10 years in a challenging fashion, with expectations for the first 2 years broken down into quarters and others expressed in annual terms. Business objectives are focused on a limited field of opportunity, its potential and market trends, and the competitive industry structure associated with it. Performance measures at this level are financial (for 5 and 10 years out). Further, the objectives are to be consistent and compatible with the corporate objectives. For each business objective, there is a business objective manager.
2. Strategies. Strategies describe in detail the environment of the business opportunity to be pursued in support of the objectives. Normally, there are several strategies supporting each objective. Each strategy is assigned to a strategy manager.
3. Tactical action plan (TAP). TAP is next in the goal hierarchy. In this phase, detailed steps of action necessary to reach the major long-range checkpoints defined by the strategies are enumerated. It is normally a short-run process, covering 6-18 months and is designated under a "responsible" individual.

OST programs are approved yearly by a growth committee, which consists of 13 permanent members, including the president, group vice presidents, and other officers reporting at the corporate level. At a strategic planning conference each year, some 500 top TI managers from around the world gather for a week of strategic planning.

MANAGEMENT CONSIDERATIONS

In this section, a series of topics related to managing the planning process are introduced and briefly discussed. These topics are: (a) viewing planning as a system process, (b) the impact of management attitudes toward planning and the activities within the planning process, and (c) managing the inevitable impact of uncertainty and luck. In a later section on fine tuning the planning process, additional topics related to managing the planning process will be discussed.

Planning as a System Process

Organizational planning is inherently a complex and multidimensional process involving numerous systems and subsystems that have causal linkages of varying strength where the direction of cause and effect between the system and subsystems is often unclear.[14] For example, it has long been accepted within strategic planning circles that there is (and should be) a close relationship between an organization's strategy and its organizational structure. Structural constraints should be explicitly considered in formulating strategy—structure dictates strategy. Conversely, changes in organizational structure are often a powerful mechanism for implementation—strategy drives structure.

Attitudes Toward Planning

In general, managers have fundamental perceptions of reality that color their attitudes toward planning. For example, some people have a "rational" world view, while others have a more "intuitive" perspective. Most managers, given the nature of their work, tend to be "action"-oriented individuals as opposed to "reflective" planners. Yet strategic management requires both elements—action and reflection.

A related concept is what organizational behavioralists call "optimal" vs. "satisficing" decisions. Optimal decisions represent the search for and the selection of the best (optimal) alternative, often with complete information about the benefits and costs of all possible outcomes. "Satisficing" decisions, on the other hand, represent the selection of the first alternative that satisfies all of the critical problem constraints, and no attempt is made to continue the search for an optimal alternative. Further, the choice is often made with incomplete information on all alternatives.

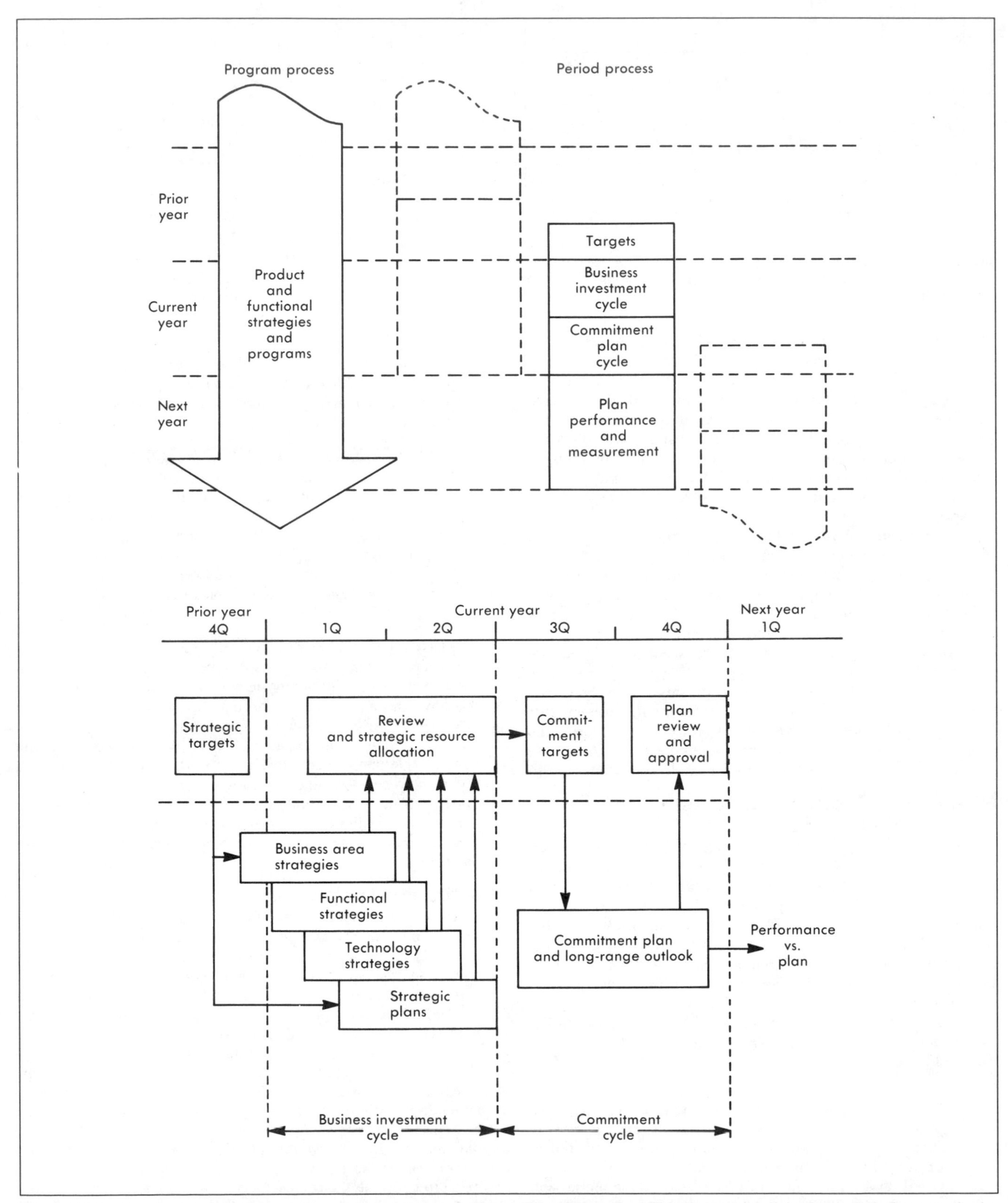

Fig. 1-12 Planning in the IBM Corporation. Source: Abraham Katz, "Planning in the IBM Corporation," in *Strategies...Success...Senior Executives Speak Out* by J. Fred McLimore and Laurie Larwood (New York: Harper & Row, 1988). Reprinted with permission

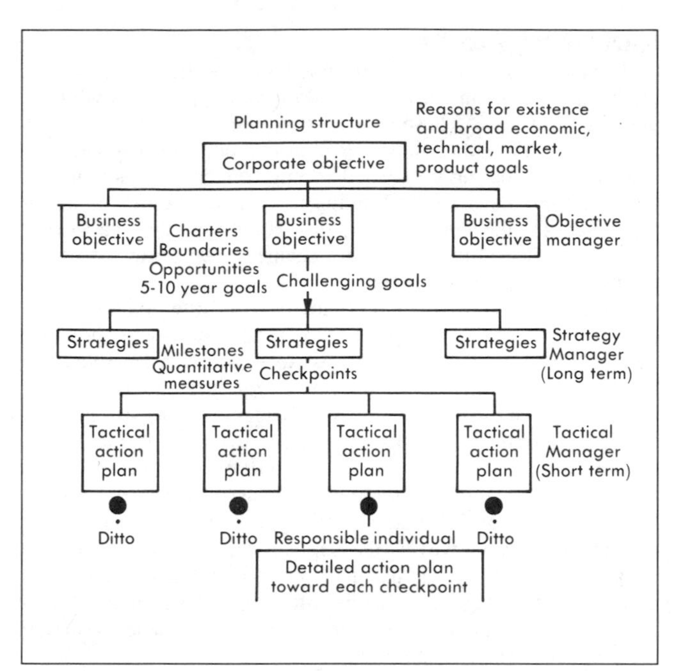

Fig. 1-13 Model of Texas Instruments' OST system.

In the context of planning, managers should be aware that: (1) they are often confronted at each stage of the planning process with the fundamental tradeoff between optimal and "satisficing" decisions; (2) given the nature of managerial work, the environment in which they work is not conducive to reflective thinking; and (3) the activity of planning is likely to appeal more to the reflective thinker. Yet the action-oriented managers have an indispensable role in both the formulation and the implementation of any plan.

Impact of Uncertainty and Luck

The inherent structure of most planning processes could lead many managers, especially those with a rational world view, to believe that it is possible to plan activities and their results in detail. However, even in the best-managed process there will remain elements of uncertainty and luck. These can dramatically influence the organization both positively and negatively. But conscious strategy does not preclude the brilliance of improvisation or the welcome consequences of good fortune.[19]

At the same time, this does not alter the fact that a business enterprise, guided by a clear sense of purpose rationally arrived at and emotionally ratified by commitment, is more likely to have a successful outcome, in terms of profit and social good, than a company whose future is left to guesswork and chance. However, the manager's role is to some degree to manage the level of risk and uncertainty he or she and the organization are willing to tolerate. A primary tool in dealing with uncertainty and luck is the organizational commitment to the planning process.

FINE TUNING THE PROCESS

Having provided planning basics in the first section of this chapter, the focus in this section is on advanced topics related to planning. This section discusses the different "schools of thought" on strategic planning, the factors affecting the need for planning, the notion of "generic" strategies, and finally, the alternative end states for the planning process itself.

FACTORS AFFECTING THE NEED FOR PLANNING

Figure 1-14 identifies the structural factors that affect the need for formal strategic planning.[49] The structural factors include demographic features of the division (business unit) and its parent company, the nature of the business, and the nature of the market and environment. These structural factors can be used systematically to select appropriate strategic planning features, as shown in Fig. 1-15.

For example, large companies generally need a planning process that is relatively elaborate, that is multilevel and multistage, and that has outputs of scientific, precise goals and business unit reorganizations. Conversely, smaller companies generally do not need these planning features unless indicated by other structural factors.

MANAGING THE PLANNING PROCESS

Earlier in this chapter, the concept of planning systems evolution was introduced. In this section, the concept will be expanded by providing two illustrated models on the subject. In addition, this section discusses the rate at which the unit of organizational focus and the planning scope are narrowed and the impact of alternate management styles on the planning process. Further, Chapter 2 of this volume extends the discussion of these topics as they relate to management control.

Planning System Evolution

Gluck indicates that formal strategic planning evolves along similar lines in various organizations, albeit at varying rates of progress (see Fig. 1-16 on p. 1-15).[50] He segments the process into four sequential phases, each marked by clear advances over its predecessor in terms of explicit formulation of issues and alternatives, quality of preparatory staff work, readiness of top management to participate in and guide the strategic decision process, and effectiveness of implementation.

Similarly, Lenz describes the evolution of planning in terms of three sequential stages of planning: introduction, consolidation, and fork in the road.[51] These stages are followed by two possible end states: (1) the self-perpetuating bureaucratic system and (2) the self-reflecting learning process (see Fig. 1-17 on p. 1-15). More importantly, he notes that on reaching the "fork-in-the-road" stage, forces are in motion that will, if not halted, drive a planning process to the end state of a self-perpetuating bureaucracy, which undermines the effectiveness of planning as a mechanism for strategic thinking and organizational learning.

FINE TUNING THE PROCESS

The important points to note on this concept are: (1) the strategic management process evolves with passage of time; (2) in general, the stages of evolution in a strategic process cannot be skipped; and (3) the process itself must be managed and shaped if it is to serve executives as a vehicle for strategic decision making.

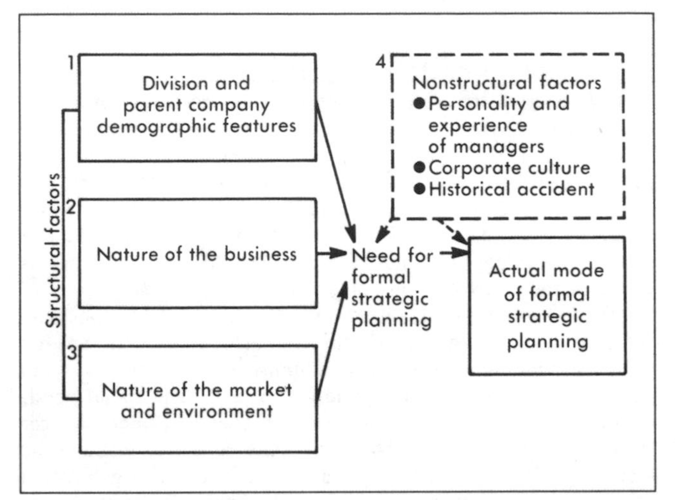

Fig. 1-14 Factors affecting need for strategic planning. Source: George S. Yip, "Who Needs Strategic Planning," *Journal of Business Strategy* (Winter 1986). Reprinted with permission

Narrowing Profile

Lorange and Vancil have provided a model that discusses the planning process in terms of a slow and rapid narrowing of planning profiles.[52] These profiles represent an explicit management decision to either slowly or rapidly reach a decision at each stage of the planning process. The key concept in this model is that managers have an opportunity to manage the rate at which "satisficing" decisions are made. How fast this narrowing occurs depends on the particular organizational setting in which the process is undertaken (see Fig. 1-18).

As a general rule, a small company with little diversity in its operation may wish to adopt an early or rapid narrowing process, because the functional and corporate executives involved are likely to be thoroughly familiar with the strategy of the few businesses in question. Functional managers can proceed directly to the development of action programs to continue the implementation of the chosen strategy. Quantitative financial linkage between the selected programs and the resulting budgets is feasible, and "tight" linkage of this type is common practice. In a large company, however, the opposite tends to be true. In this instance, the linkage is usually looser and the process more gradual. But management, as the planning system matures, can gradually accelerate the narrowing process without jeopardizing the creative aspects of planning.

Alternative Management Styles

Drake has described the characteristics of hierarchical and network organizations.[13] Figure 1-19 (see p. 1-16) compares and contrasts his descriptions. From a planning perspective, the

		Strategic planning features																					
	Value to need planning feature	Objectives:	Competitive advantage	Manage diversity	Manage turbulence	Achieve synergy	Create strategic change	Process:	Elaborate planning	Extensive data	Competitive orientation	Multilevel and multistage	Multifunctional processes	Use of outsiders	Subject matter:	Total environment	Long time horizon	Output:	Portfolio roles	Scientific, precise goals	Business unit reorganization	Market redefinition	Strategic programs
Structural factors																							
Demographic features																							
Size	Large								X			X								X	X		
Diversity	Large			X					X			X							X	X	X		
Profitability	Low		X		X																	X	X
Nature of business																							
Number of important functions	Many												X										X
Shared markets	High					X			X													X	
Number of brands	Small					X			X				X										
Complexity of market	High					X			X	X				X		X	X					X	
Nature of investment																							
Fixed assets/sales	High								X	X							X			X			
Advertising/sales	Low								X	X							X			X			
Nature of market and environment																							
Market growth rate	Low		X				X																
Competitive threat	High		X		X		X				X			X		X						X	X
Environmental threat	High				X				X	X						X	X						
Market cyclicality	High				X												X						

Fig. 1-15 Structural variables and the need for particular strategic planning features. Source: George S. Yip, "Who Needs Strategic Planning," *Journal of Business Strategy* (Winter 1986). Reprinted with permission

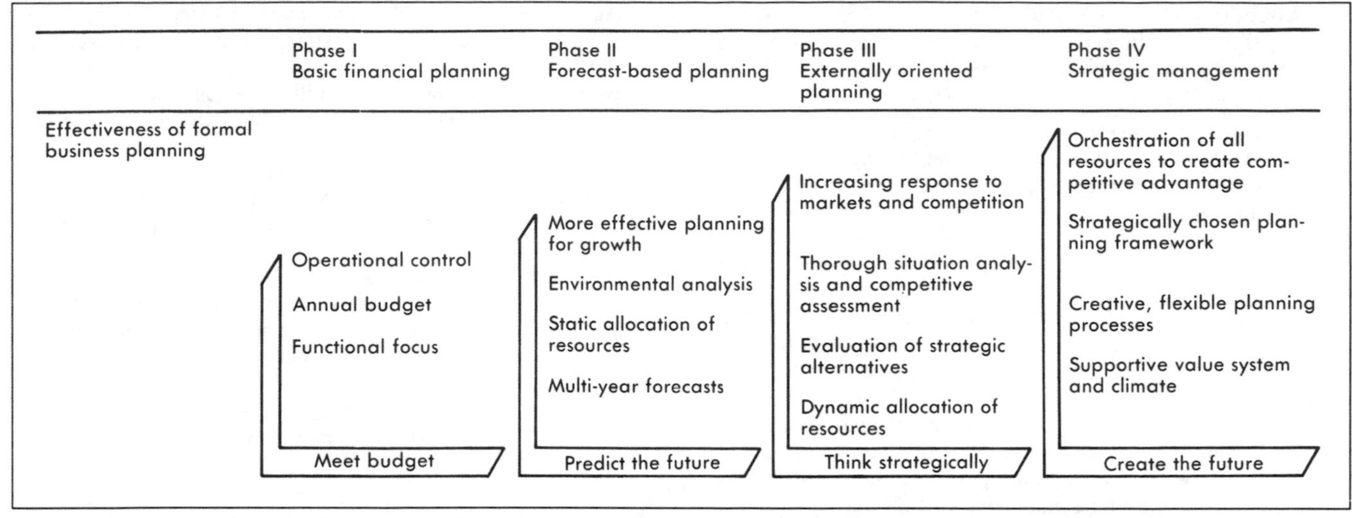

Fig. 1-16 Gluck's four phases in the evolution of planning systems. Source: Frederick W. Gluck, et al., "Strategic Management for Competitive Advantage," *Harvard Business Review* (July-August 1980). Reprinted with permission

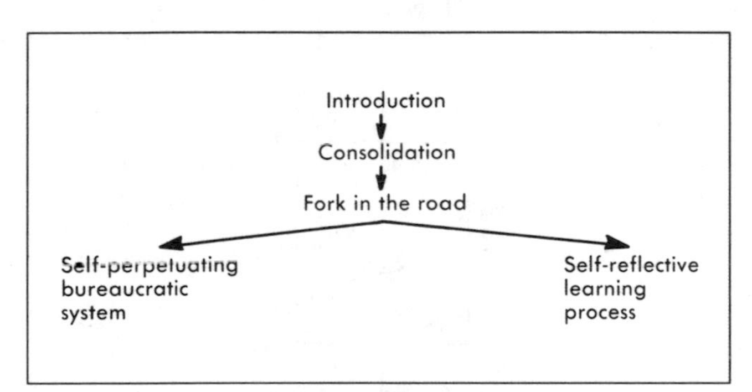

Fig. 1-17 Lenz's phases in the evolution of planning systems. Source: R. T. Lenz, "Managing the Evolution of the Strategic Planning Process," *Business Horizons* (January-February 1987). Reprinted with permission

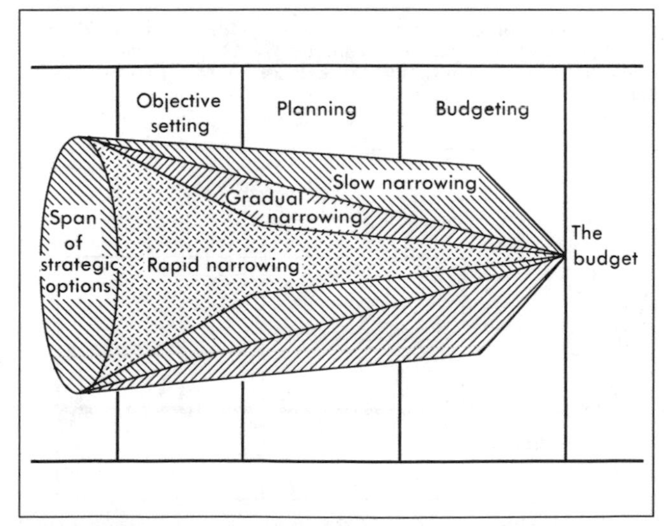

Fig 1-18 Slow vs. rapid narrowing profiles in the planning process. Source: Lorange, et al., "How to Design a Strategic Planning System," *Harvard Business Review* (September 1976). Reprinted with permission

hierarchical organization tends to "fit" a top-down orientation to planning, whereas in a network organization a bottom-up and horizontal (peer-to-peer) orientation seems more suited to the organizational culture.

Obviously, this simple framework does not adequately capture the organizational culture and management style in all organizations. However, it is important that managers tailor their planning systems to "fit" their organizations by recognizing their culture.

SCHOOLS OF THOUGHT ON PLANNING

Strategy is a very important element of planning and, in almost all cases, is crucial to achieving a sustainable competitive advantage. Further, the concept of strategic management rests squarely on the concept of strategy.[10]

Despite refinements and adjustments in strategic scanning, formulation, implementation, and management, the generalization of strategic concepts remains a source of substantial and continuing debate. This debate is reflected in the classical schools of thought, which are summarized as follows:

- Unit-specific (atomistic) approach. This method is based on the belief that strategic concepts are not generalizable because they are inherently: (1) industry and business unit specific, (2) dynamically changing over time, and (3) focused on the unexpected, not represented in past data or management practice.[19,22,53]
- Contingency theory approach. This approach is based on the belief that strategic options available to a business unit are dramatically shaped by contingency factors, such as industry type, relative market share, and product lifecycle position, and by structural factors, such as organizational size.[54-56]
- Generic strategy approach. This approach focuses on the common (generic) ways that business units generate competitive advantages across a variety of industries. These common patterns of competition represent generic strategies.[31,57-62]
- Universal (general principles) approach. In this approach, the search is for "universal laws" of strategy that hold to some extent in all settings. For example, the Boston

Consulting Group popularized the "law" of cumulative experience, calling it "almost universally observable."[63] Managers have argued that quality and flexibility are universal from a manufacturing perspective.

Each of the four levels just described has an important and distinct role to play in achieving competitive advantage. Although elements of a firm's manufacturing strategy are formulated and implemented in all of the four levels, the majority of the activities (and literature) have focused on manufacturing strategy at the functional level. See reference 18 for a more detailed discussion.

Until recently, manufacturing literature has focused implicitly on the universal principles that are applicable under all conditions such as quality, flexibility, responsiveness, and time to market. It also focuses on the system-specific approach, such as forecasting, inventory control, scheduling, shop floor control, and capacity planning. With the exception of a few articles, very little effort has been expended on articulating the attractiveness of specific manufacturing technologies in different manufacturing environments and the notion of generic strategies for manufacturing. In the section that follows, the notion of generic strategies at the business unit level will be discussed. It is hoped that this will provide some "food for thought" for applying these concepts from a manufacturing perspective.

THE IMPACT OF GENERIC STRATEGIES AND STRUCTURES

As discussed earlier, one of the key objectives of planning is to achieve a sustainable competitive advantage through strategic positioning of the business unit. Numerous frameworks have been developed by theorists to assist executives in this strategic planning activity.

Though various authors have adopted this approach, much of the thinking on competitive positioning has been influenced by Porter's comments on industry structural analysis and the interplay of five competitive forces.[31] The forces that he identified are: (1) the threat of new competitors, (2) the rivalry among existing competitors, (3) the threat of substitute products, (4) the bargaining power of buyers, and (5) the bargaining power of suppliers. It is the collective strength of these five forces that determines the ability of the firms in an industry to earn, on the average, rates of return on investment in excess of the cost of capital (see Fig. 1-20).

Bridging the gap between traditional economic thought and business literature, Porter has derived an extremely suggestive typology of strategy. It involves a simple model of generic strategies using a two-dimensional matrix (strategic advantage against strategic target). He points out that there are three generic strategies that can be used in industry: (1) overall cost leadership, (2) differentiation, and (3) focus (see Fig. 1-21).

One of the crucial problems in strategy formulation is the definition of industry boundaries. Drawing precise boundaries to estimate market share is crucial to strategic analysis. But how this can be accomplished is a source of continuing debate. Porter's framework for structural analysis of industries mitigates the importance of drawing precise boundaries for strategic analysis because it focuses on forces beyond just rivalry among existing competitors.

Hierarchy	Network
• Structure determines strategy	• Strategy determines structure
• Direct control	• Indirect control
• Rigidity	• Flexibility
• Small span of control	• Wide span of control
• Large number of organizational levels	• Small number of organizational levels
• Vertically integrated	• Not vertically integrated
• Externally disciplined	• Self-disciplined
• Homogeneous	• Diverse
• Risk averse	• Risk taking
• Change spurred by catalytic event	• Planned change
• Little involvement with outside environment	• More involvement with outside environment
• Management removed from market	• Management close to market

Fig. 1-19 Characteristics of hierarchical and network organizations. Source: R. Drake, "Innovative Structures for Managing Change," *Planning Review* (November 1986). Reprinted with permission

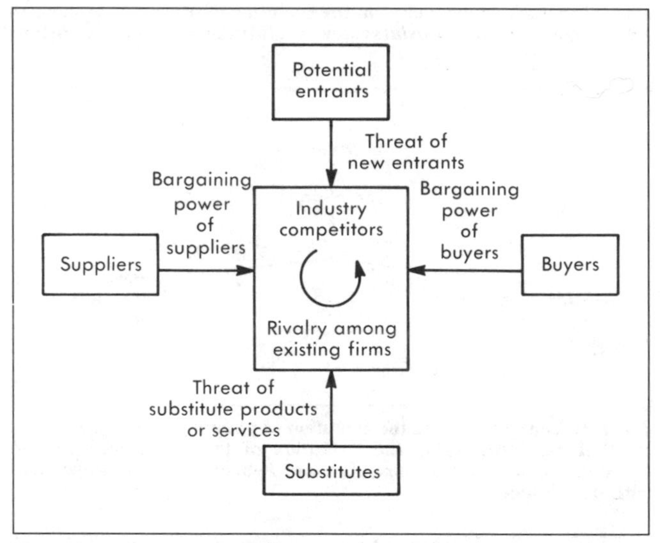

Fig. 1-20 Five forces driving industry competition. Source: Michael Porter, *Competitive Strategy: Techniques for Analyzing Industries and Competitors* (New York: The Free Press, 1980). Reprinted with permission

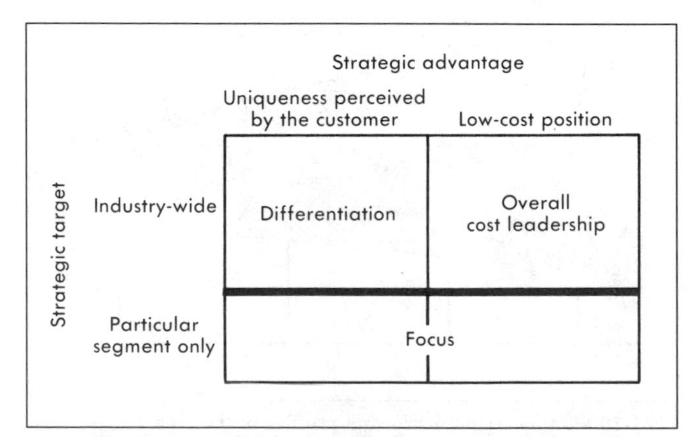

Fig. 1-21 Porter's generic strategies. Source: Michael Porter, *Competitive Strategy: Techniques for Analyzing Industries and Competitors* (New York: The Free Press, 1980). Reprinted with permission

In a similar vein, Kotha and Orne have provided a framework that explores the competitive positioning from a manufacturing perspective, especially the strategic consideration of new manufacturing technologies.[64] Their framework explores the interaction among three elements: business unit strategy, manufacturing structure, and choices in manufacturing technologies (see Fig. 1-22). They argue that for a business (firm) to be effective, it is critical that the emphasis on various manufacturing technologies be consistent with the manufacturing structure and business strategy of the firm or SBU, and vice versa. Further, they point out that the ''fit'' among choices in manufacturing structures, business strategies, and technologies has an important influence on the performance of manufacturing units.

ALTERNATIVE END STATES IN PLANNING

While discussing the stages in the evolution of planning systems as described by Lenz,[51] it was briefly mentioned that there are two possible end stages in the further evolution of the planning system: (1) the self-perpetuating bureaucratic system and (2) the self-reflective learning process. Upon reaching the ''fork-in-the-road'' stage (refer to Fig. 1-17), forces are in motion that will, if not halted, drive a planning process toward the end state of a self-perpetuating bureaucracy. This state undermines the effectiveness of planning as a mechanism for strategic thinking and organizational learning.

Figure 1-23 compares the alternate end states along four important dimensions: (1) the focal point of the planning process, (2) performance criteria, (3) modes of interaction, and (4) managerial attention.[65] Planning systems that evolve toward self-perpetuating bureaucracies tend to be mechanistic in nature. In this environment, the planning is simply an annual preparation

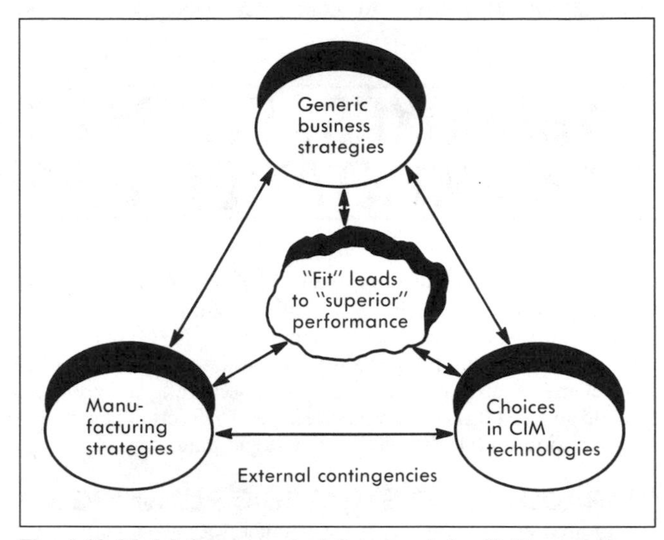

Fig. 1-22 Model for conceptual framework by Kotha and Orne. Source: Suresh Kotha and Daniel Orne, ''The Concept of Fit in Manufacturing: Implications for Investments in CIM Technology,'' Rensellaer Polytechnic Institute Working Paper, 1987. Reprinted with permission

	End state	
Points for comparison	Self-perpetuating bureaucratic system	Self-reflective learning process
Focal point of the planning process	Produce a written plan (the document itself) Control implementation through budgets and programs	Stimulate strategic thinking Foster organizational learning Coordinate executive action
Performance criteria	Historical: Current performance is compared to previous organizational performance	Strategic: Competitors' performance levels and market characteristics (for example, growth in demand, innovativeness) provide standards for assessing performance
Modes of interaction	Formalized, numbers-oriented, rigid Discussion centers on incremental adjustments to current strategy	Informal, word-oriented, self-reflective Discussion centers on exploring managers' understandings of planning assumptions, strategic issues, and the feasibility of proposed solutions
Managerial attention	Directed toward operational issues of a business Defending current competitive position	Directed toward "strategic frontiers" of current business Prospecting for new opportunities

Fig. 1-23 Lenz's alternative end states of strategic planning process. Source: R.T. Lenz, ''Strategic Capability: A Concept and Framework for Analysis,'' *Academy of Management Review*, vol. 5, no. 2 (1980). Reprinted with permission

of a document and the development of budgets for controlling implementation. Meetings are often routine, formal, and usually dominated by numbers-oriented presentations. Unfortunately, this approach deflects attention from strategic thinking toward incremental adjustments to current strategy. Self-reflective learning, in contrast, is more open and less formalized, and discussions concern a firm's overall competitive position. Words, not numbers, are the medium for explaining strategic issues and organization responses, and entrepreneurial creativity in strategy making is encouraged.

PARADOXES, PITFALLS AND IMPLEMENTATION DIFFICULTIES

No discussion on planning can be complete without reference to the paradoxes, pitfalls, and implementation difficulties that one is likely to encounter during the process. This section draws the reader's attention to these three important elements. Though the elements are covered under three separate categories, it should stressed that they are interrelated. The section begins with a discussion on the paradoxes of strategic planning.

PARADOXES OF STRATEGIC PLANNING

In articles on planning, it has become somewhat common to list fundamental principles of planning, as was done in the earlier section. While the structure and clarity of these principles are appealing and helpful, especially to less experienced practitioners, there is unfortunately a tendency for these practitioners to prematurely accept the principles as truths.[41] Alternatively, it seems more constructive to discuss some of these principles as half-truths (or loose guidelines), as done by Ramanujam and Venkatraman.[66] The paradoxes imbedded in these "principles" are summarized in the following paragraphs:

1. Effective planning should be future oriented. While it is self-evident that planning should be future oriented, in practice, however, the major cause for the failure of many planning systems is not a lack of this orientation, but an obsession with it. What is important is not the time element, but the development of an understanding of the causal forces underlying any future event. In addition, as pointed out by Hayes and other authors, long-term strategic advantage can often be the result of the skilled management of short-term operations.
2. Effective planning systems should be flexible. The underlying logic is that as assumptions or circumstances guiding the planning process change, and if it appears that the chosen plan may not succeed, the plan should be changed. In most cases, this logic cannot be faulted. Yet some stirring examples of business success reveal a streak of incredible persistence on the part of the companies or persons involved in pursuing a particular course of action.
3. Top-management involvement ensures planning success. With justification, this principle is often touted as a critical ingredient for planning success. On the other hand, planning without the cooperation and participation of lower level management usually leads to a failure of implementation. This is especially true in some manufacturing settings where top management is ill-informed about the choices with respect to key manufacturing technologies.

   In addition, top management sometimes appears underinvolved and imprecise in strategy setting. Often, based on their own insecurities, middle managers desire a greater level of involvement and direction setting than senior executives are willing to give. However, as noted by Wrapp in a provocative article, the apparent underinvolvement and imprecision may be a critical skill practiced by many successful senior executives who know how to provide a (general) sense of direction without ever publicly committing themselves to a specific set of objectives.[67] This article by Wrapp is well worth reading.
4. Effectiveness results from top-down planning with bottom-up implementation. This guideline is closely related to the previous one and is based on the logic that the overall strategy should be formulated at the top of the organization and handed down to lower levels of the organization for implementation. However, many companies possess in their lower levels a vast reservoir of creative energy that can be successfully harnessed with a planning process that includes elements of bottom-up formulation with top-down implementation.
5. Effective planning requires integration of strategic planning and issues management. Given the inability of conventional planning systems to deal with discontinuities and a wide variety of strategic issues, there is often a relentless quest among systems designers to create omnipotent systems that can tackle all problems by integrating an increasing scope of decisions. Often, the search for integration can take on a religious fervor and ignores the real benefits from deliberately not integrating certain decisions or classes of decisions.
6. Planning is a line function. This principle is partly a function of the indispensable role of line managers in the planning process and is partly a response to the relative overinvolvement of staff managers and underinvolvement of line managers during the 1960s and 1970s. However, it is a gross generalization to define planning as a line function. Some aspects of planning require the line manager's intimate knowledge of a business unit and the commitment engendered from his or her involvement in the planning process. On the other hand, other aspects of planning require the time, reflection, and special skills that are unique to staff managers.
7. Planning should be an ongoing activity. Given: (a) the scope of the planning process, ranging from long-term strategic plans to short-term operational plans, and (b) the inherent feedback loops in all planning systems, it is clearly true that planning inherently is an ongoing proc-

ess. Nevertheless, the major elements of conventional strategic planning and more recent strategic issues management, are basically periodic processes. The former is calendar driven, while the latter is event or issue driven, and neither is an ongoing process in a real sense. In general, most managerial actions are discrete interventions of one kind or another and cannot be accurately described as ongoing processes.

8. Effective planning will increase organizational performance. In the opening section of this chapter, it was suggested that organizations that do formal planning tend to outperform organizations that ignore formal planning. However, it is not true that planning is a system with tight linkages in which the impact of a change on one part of the system or another can be predicted with a certain degree of accuracy. Unfortunately, the linkages are not clear, and neither is the direction of causality. Worse, new factors are constantly being introduced because of changes in the environment, so the system itself is subject to much uncertainty and constant redefinition. Under these rapidly changing circumstances, it is only naive to argue for any sort of direct causal relationship between planning and performance.

PITFALLS TO BE AVOIDED

Each of the preceding paradoxes could be construed as a pitfall in planning. In addition, Herbert has provided a supplementary list of pitfalls to avoid during the planning process, as follows:[68]

1. Making plans in secret as an exercise in individual brilliance.
2. Being sloppy with planning assumptions. The other side is being sloppy with or not adequately questioning someone else's planning assumptions.
3. Collating functional plans and believing one now has a strategic plan.
4. Failing to consider the strategic plan's effects on operations.
5. Failing to assign appropriate new or additional resources to operational managers for the implementation of the strategic plans.

IMPLEMENTATION DIFFICULTIES

Based on a survey of business unit heads, corporate planning directors, and chief executive officers engaged in strategic planning in American multibusiness corporations, Gray has noted that more than half of his sample (59%) attribute their discontent with the strategic planning process mainly to difficulties encountered in the implementation of the plans.[45] His study also indicated that two thirds of what these managers called implementation difficulties were, on closer scrutiny, attributable to six ''preimplementation'' factors listed as follows:

1. Poor preparation of line managers.
2. Faulty definition of business units.
3. Vaguely formulated goals.
4. Inadequate information basis for action planning.
5. Badly handled reviews of business unit plans.
6. Inadequate linkage of strategic planning with other control systems.

While these difficulties cannot be totally eliminated in most organizations, they can be minimized by providing managers with timely information, planning skills training, goals/objective setting guidelines, and a clear understanding of what is expected from them in the planning process.

CONCLUSION

Inherent in the preceding discussion of planning was the understanding that both the design and the performance of the planning system should be periodically examined. In other words, is the planning process an effective and efficient use of organizational resources? In recent years, this question has come up with increasing frequency in both the academic and practitioner literature on the topic (for example, see references 3, 4, 5, 6, 7, 14, 66, and 69).

An approach to this issue would be to focus on the effectiveness of the (strategic) planning process by examining the ability of the plan to produce results that create or facilitate a sustainable competitive advantage at a superior level of profitability. Some of the issues and concerns that a plan should address include the following:[70]

1. Suitability: Is there a sustainable advantage?
2. Validity: Are the assumptions realistic?
3. Feasibility: Are the necessary skills, resources, and commitments available?
4. Consistency: Does the strategy hang together?
5. Vulnerability: What are the risks and contingencies?
6. Adaptability: Can flexibility be retained?
7. Financial desirability: How much economic value is created?

By periodically addressing these and similar questions, managers can maintain an effective sense of the overall competitive picture while simultaneously addressing increasing levels of details within the planning process.

Although presented in an entirely different context, Peck has presented some concepts that seem to have importance in summarizing the manufacturing manager's role in the planning process.[71] They are as follows:

1. Acceptance of personal responsibility. Manufacturing managers are encouraged to accept the personal responsibility for constructively participating in the planning process even in the face of less than complete participation by their superiors, subordinates, and peers. The planning process needs to be stimulated and nurtured at multiple points in the organization. The manufacturing manager has an indispensable role in the process. For example, if manufacturing managers wait for a process and a plan to be articulated by others higher in the organizational hierarchy, they may: (a) wait a very long

time and/or (b) negate their ability to influence the corporate agenda with regard to how manufacturing can and should be used to generate and sustain competitive advantages.

2. Delay of gratification. It is virtually self-evident that the planning process involves a future orientation. In addition, competitive advantages based in manufacturing require long time frames for development and implementation. Thus, corporate strength in manufacturing inherently requires, to some degree, short-term financial performance to be delayed to generate an increased level of competitiveness in the future. The worry is that without corporate delay of gratification many units will become what *Business Week* magazine has termed ''hollow corporations'' (or shells) with high levels of short-term profitability but low levels of long-term competitive strength.
3. Dedication to reality. Shared assumptions or perceptions of reality need to be periodically tested for their validity throughout the planning process. If not, blind spots may develop that offer substantial opportunity for competitors.[31]

   For example, the beer brewing industry in the United States shared the assumption that the market was not interested in low-alcohol/low-calorie beverages. Philip Morris, a relatively new entrant in the industry, as owner of Miller Brewing Co., directly challenged that assumption with its phenomenally successful introduction of Miller Lite beer. The result was a dramatic shift of industry market share positions.
4. Balancing. Throughout the planning process, there is a need to continually balance contrasting forces, issues, and perceptions. For example: (a) short-term vs. long-term time horizon, (b) top-down vs. bottom-up decision process, (c) ''satisficing'' vs. optimal decisions, (d) rational vs. intuitive decision making, and (e) broad organizational focus vs. specific, narrow focus.

As a final concluding comment for this chapter, the increasing importance of planning for manufacturing managers is partially driven by rising levels of uncertainty and rising levels of interdependence. In effect, planning is an indispensable managerial tool for coping with these powerful societal forces.

References

1. L.J. Rosenberg and C.D. Schewe, "Strategic Planning: Fulfilling the Promise," *Business Horizons* (July-August 1985).
2. "The New Breed of Strategic Planner," *Business Week* (September 17, 1984).
3. S.S. Thune and R.J. House, "Where Long Range Planning Pays Off," *Business Horizons* (August 1970).
4. L.C. Rhyne, "The Relationship of Strategic Planning to Financial Performance," *Strategic Management Journal*, vol. 7 (1986).
5. D.W. Karger and Z. Malik, "Long Range Planning and Organizational Performance," *Long Range Planning* (December 1975).
6. J.S. Bracker and J.N. Pearson, "Planning and Financial Performance of Small, Mature Firms," *Strategic Management Journal*, vol. 7 (1986), pp. 503-522.
7. L.C. Rhyne, "Contrasting Planning Systems in High, Medium and Low Performance Companies," *Journal of Management Studies* (July 1987).
8. G.A. Steiner, H. Kunin, and E. Kunin, "Formal Strategic Planning in the United States Today," *Long Range Planning*, vol. 16, no. 3 (1983).
9. Jeffrey Bracker, "The Historical Development of the Strategic Management Concept," *Academy of Management Review*, vol. 5, no. 2 (1980), pp. 219-224.
10. Charles Hofer and Dan Schendel, *Strategy Formulation: Analytical Concepts* (New York: West Publishing Co., 1978).
11. Bernard Taylor, "New Dimensions in Corporate Planning," *Long Range Planning* (December 1976).
12. Henry Mintzberg, "What is Planning Anyway?" *Strategic Management Journal*, vol. 2 (1981), pp. 319-324.
13. R. Drake, "Innovative Structures for Managing Change," *Planning Review* (November 1986).
14. N. Venkatraman and V. Ramanujam, "Planning System Success: A Conceptualization and an Operational Model," *Management Science*, vol. 33, no. 6 (June 1987).
15. Lester A. Digman, *Strategic Management Concepts, Decisions, Cases* (Plano, TX: Business Publications, Inc., 1986), p. 12.
16. George S. Day, "Strategic Market Analysis and Definition: An Integrated Approach," *Strategic Management Journal*, vol. 2 (1981), pp. 281-299.
17. Bruce D. Henderson, *Henderson on Corporate Strategy* (Boston: Boston Consulting Group, Inc., 1979).
18. Suresh B. Kotha and Daniel Orne, "Generic Manufacturing Strategies: A Conceptual Synthesis," Working Paper 37-86-P4 (Troy, NY: Rensselaer Polytechnic Institute, School of Management, 1987).
19. C.R. Christensen, et al., *Business Policy: Text and Cases*, 6th ed. (Homewood, IL: Richard D. Irwin, 1987).
20. S.S. Cohen and J. Zysman, "Why Manufacturing Matters: The Myth of the Post Industrial Economy," *California Management Review*, Vol. XXIX, No. 3 (Spring 1987).
21. Steven C. Wheelwright, "Strategy, Management, and Strategic Planning Approaches," *Interfaces*, vol. 14 (January-February 1984).
22. K.R. Andrews, *The Concept of Corporate Strategy* (New York: Dow Jones-Irwin, 1971).
23. Elwood B. Buffa, *Meeting the Competitive Challenge: Manufacturing Strategy for U.S. Companies* (Homewood, IL: Dow-Jones Irwin, 1984).
24. Daniel Orne and Leo Hanifin, "International Manufacturing Strategies and Computer Integrated Manufacturing (CIM): A Review of the Emerging Interactive Effects," forthcoming in Benjamin Lev, ed., *Production Management: Methods and Studies* (Amsterdam, Netherlands: North-Holland Publishing Co.)
25. Robert B. Reich, "The Next American Frontier," *The Atlantic Monthly* (March and April 1983).
26. Steven C. Wheelwright, "Manufacturing Strategy: Defining the Missing Link," *Strategic Management Journal*, vol. 5 (1984).
27. Wickham Skinner, *Manufacturing in the Corporate Strategy* (New York: John Wiley & Sons, Inc., 1978).
28. ——————, "Manufacturing—Missing Link in Corporate Strategy," *Harvard Business Review* (May-June 1969), p. 156.
29. David Shirshfield, "From the Shadows," *Interfaces*, 13:2 (April 1983).
30. C.H. Fine and A.C. Hax, "Manufacturing Strategy: A Methodology and an Illustration," *Interfaces* (November-December 1985).
31. Michael E. Porter, *Competitive Strategy: Techniques for Analyzing Industries and Competitors* (New York: The Free Press, 1980).
32. Arnold C. Hax and Nicolas S. Majluf, "The Use of the Growth-Share Matrix in Strategic Planning," *Interfaces*, vol. 13 (February 1983).
33. Arnold C. Hax and Nicolas S. Majluf, "The Use of the Industry Attractiveness-Business Strength Matrix in Strategic Planning," *Interfaces*, vol. 13 (April 1983).
34. D.F. Abell and J.S. Hammond, *Strategic Market Planning* (Englewood Cliffs, NJ: Prentice-Hall, Inc., 1979).
35. Bruce D. Henderson and A.J. Zakon, "Corporate Growth Strategy: How to Develop It and Implement It," *Handbook of Business Problem Solving*, K.J. Albert, ed. (New York: McGraw-Hill Book Co., 1980).
36. R.R. Osell and R.V.L. Wright, "Allocating Resources: How to Do It in Multinational Corporations," *Handbook of Business Problem Solving*, K.J. Albert, ed. (New York: McGraw-Hill Book Co., 1980).
37. P. Haspeslagh, "Portfolio Planning: Uses and Limits," *Harvard Business Review* (January-February 1982).
38. W. Kiechel III, "Corporate Strategists Under Fire," *Fortune* (December 27, 1982).

REFERENCES

39. ____________, "Oh Where, Oh Where Has My Little Dog Gone? or My Cash Cow? or My Star?" *Fortune* (October 5, 1981).
40. P. Miesing, "Drawbacks of Matrix Models as a Strategic Planning Tool," Working Paper 82-2 (SUNYA, 1982).
41. P.P. Peker, Jr., "Portfolio Analysis: A Planner's Dilemma," *Planning Review* (July 1982).
42. Bruce D. Henderson, "The Application and Misapplication of the Experience Curve," *The Journal of Business Strategy*, vol. 4, no. 3 (Winter 1984).
43. William F. Glueck, *Business Policy and Strategic Management*, 3rd ed. (New York: McGraw-Hill Book Co., 1980), p. 7.
44. Wickham Skinner, *Manufacturing: The Formidable Competitive Weapon* (New York: John Wiley and Sons, Inc., 1985), p. 65.
45. Daniel H. Gray, "Uses and Misuses of Strategic Planning," *Harvard Business Review* (January-February 1986).
46. Summarized from "General Electric, Strategic Position—1981," Harvard Business School Case 381-174.
47. Abraham Katz, "Planning in the IBM Corporation," article reprinted in *Strategies...Success...Senior Executives Speak Out*, J. Fred McLimore and Laurie Larwood (New York: Harper & Row Publishers, 1988).
48. Summarized from "Texas Instruments, Inc.: Management Systems," Harvard Business School Case 9-172-054, 1972.
49. George S. Yip, "Who Needs Strategic Planning," *Journal of Business Strategy* (Winter 1986).
50. Frederick W. Gluck, Stephen P. Kaufman, and A. Steven Walleck, "Strategic Management for Competitive Advantage," *Harvard Business Review* (July-August 1980).
51. R.T. Lenz, "Managing the Evolution of the Strategic Planning Process," *Business Horizons* (January-February 1987).
52. Peter Lorange and Richard F. Vancil, "How to Design a Strategic Planning System," *Harvard Business Review* (September 1976).
53. H.E.R. Uyterhoeven, R.W. Ackerman, and J.W. Rosenblum, *Strategy and Organization: Text and Cases in General Management* (Homewood, IL: Richard D. Irwin, 1973).
54. Donald C. Hambrick and David Lei, "Towards an Empirical Prioritization of Contingency Variables for Business Strategy," *Academy of Management Journal*, vol. 28, no. 4 (1985), pp. 763-788.
55. Ari Ginsberg and N. Venkatraman, "Contingency Perspectives of Organizational Strategy: A Critical Review of the Empirical Research," *Academy of Management Journal*, vol. 10, no. 3 (1985), pp. 421-434.
56. H.L. Tosi and J.W. Slocum, "Contingency Theory: Some Directions," *Journal of Management*, 10 (1) (1984), pp. 9-26.
57. Roderick E. White, "Generic Business Strategies, Organizational Context and Performance: An Empirical Investigation," *The Strategic Management Journal*, vol. 7 (1986), pp. 217-231.
58. Gregory G. Dess and Peter S. Davis, "Porter's (1980) Generic Strategies as Determinants of Strategic Group Membership and Organizational Performance," *Academy of Management Journal*, vol. 27, no. 3 (1984), pp. 467-488.
59. Craig Galbraith and Dan Schendel, "An Empirical Analysis of Strategy Types," *Strategic Management Journal*, vol. 4 (1983), pp. 153-173.
60. Raymond E. Miles, Charles C. Snow, Alan D. Meyer, and Henry J. Coleman, "Organizational Strategy, Structure and Process," *Academy of Management Review* (July 1978), pp. 546-562.
61. Alex Miller and Gregory G. Dess, "The Appropriateness of Porter's (180) Model of Generic Strategies as a Method of Classification and Its Implications for Business Unit Performance," Working Paper No. 212 (University of Tennessee, 1985).
62. Danny Miller, "Porter's Business Strategies in Small Firms, Environmental and Structural Correlates," Working Paper (University of Montreal, 1985.
63. Robert D. Buzzell, Bradley T. Gale, and Ralph G.M. Sultan, "Market Share a Key to Profitability," *Harvard Business Review* (January-February 1975).
64. Suresh Kotha and Daniel Orne, "The Concept of Fit in Manufacturing: Implications for Investments in CIM Technology," Working Paper (Troy, NY: School of Management, Rensselaer Polytechnic Institute, 1987).
65. R.T. Lenz, "Strategic Capability: A Concept and Framework for Analysis," *Academy of Management Review*, vol. 5, no. 2 (1980).
66. Vasudevan Ramanujam and N. Venkatraman, "Eight Half-Truths of Strategic Planning: A Fresh Look," *Planning Review* (January 1985).
67. Edward H. Wrapp, "Good Managers Don't Make Policy Decisions," *Harvard Business Review* (July-August 1984), pp. 4-11.
68. Theodore T. Herbert, "Pitfalls in the Planning Process," *Managerial Planning*, vol. 33, no. 3, pp. 42-50.
69. V. Ramanujam, N. Venkatraman, and J.C. Camillus, "Multi Objective Assessment of Effectiveness of Strategic Planning: A Discriminant Analysis Approach," *Academy of Management Journal*, vol. 29 (1986), pp. 347-372.
70. G.S. Day, "Tough Questions for Developing Strategies," *Journal of Business Strategy* (Winter 1986).
71. M.S. Peck, *The Road Less Travelled* (New York: Touchstone, Simon & Schuster, Inc., 1978).

MANAGEMENT CONTROL

This chapter of the Handbook addresses the subject of management control and, in particular, the application of management control to the manufacturing enterprise.

Management control and control systems are addressed both in the broad sense of control as an integrated part of the planning process, and in the more narrow sense of control as a feedback mechanism in the measurement of processes. The chapter defines management control as it applies to the organizational model and its evolution. It then describes control principles and systems that are important to today's manufacturing environment.

CHAPTER CONTENTS:

DEFINITION OF MANAGEMENT CONTROL

Management can be defined as the activity that allocates and utilizes resources to achieve organizational goals. Every society has fewer economic resources than it has economic needs. Labor, capital, and natural resources comprise the economic resources to be used to satisfy economic needs. Every society, from the most primitive to the most advanced, establishes certain economic institutions or organizations that perform the productive function of converting resources into usable outputs to satisfy needs. To utilize limited resources to optimal advantage, someone must "manage"—that is, allocate and utilize resources effectively. The manager is the primary factor behind the pattern of resource allocation that occurs in the many and varied profit and nonprofit organizational systems existing in society.

The search for valid, transferable management concepts has paralleled the use and growth of professional schools of business. The basic concepts of management today are planning, control, measurement, and evaluation. The goal of this section is to define management control and its relationship to the other management concepts. In so doing, it will be necessary to define and analyze control in its broad relationship to planning, measurement, and evaluation, as well as the more narrow definition of control as a feedback process.

One of the best known definitions of management control is the process by which managers ensure that resources are obtained and used effectively and efficiently in the accomplishment of the organization's objectives.[1] This definition is broad and emphasizes that the function of management control is to facilitate the accomplishment of organizational goals by implementing strategies previously identified in the planning process. The control process must be accomplishable through the use of technology and people. Three important aspects of control are pragmatism, results, and people.

Management control is not an abstract description or process. It is meant to achieve specific results within the environment in which it operates through the use of technology and people. It was noted earlier that management is the activity that allocates and utilizes resources to achieve organizational goals. In this activity, the manager is concerned with applying a system of activities called the management process to a system of resources within an environmental system.

System denotes structure, an orderly arrangement of components. Process denotes action, a moving or proceeding on a certain course. The resource system is the structure or arrangement of resources, both human and technological, that the manager uses in doing his or her job. Likewise, the environmental system is composed of the various institutional arrangements that surround the business enterprise: a legal system, social system, political system, economic system, and so on. Figure 2-1 illustrates a system model that describes management control.[2]

Relationships within the management control process can be either desired or circumstantial. Desired relationships are structured by designing the components or characteristics of a management control system to target certain results. The target is defined by the goals and objectives of the

Contributors of sections of this chapter are: Barry Holmes, *Director of Quality Consulting, Coopers & Lybrand;* ***Lyell Jennings,*** *Partner, Regional Management Consulting Services, Coopers & Lybrand;* ***Peter Malmquist,*** *Regional Director of CIM, Coopers & Lybrand.*

Reviewers of sections of this chapter are: James C. Emery, *Professor of Decision Sciences, The Wharton School, University of Pennsylvania;* ***Harold Zeschmann,*** *Director of Manufacturing, Circle Seal Controls;* ***Hayward Thomas,*** *Consultant, Retired— President, Jensen Industries.*

CHAPTER 2

DEFINITION OF MANAGEMENT CONTROL

organization when considering the situational contingencies affecting it. The components of the management control system in turn lead to performance desired by the organization, and this performance is the means to accomplish the goals and objectives of the organization.

Circumstantial relationships also result from this chain as follows:

1. Identifying and measuring performance of the entire organizational entity is extremely complex. Therefore, smaller subsets of outputs are chosen to be assessed.
2. Attributes or performance measures are then defined for the subset of outputs.
3. The performance measures then become the means for evaluating whether goals and objectives are being achieved.
4. The goals and objectives in turn influence which particular measures of the assessed outputs will be used for such evaluation.
5. Resources are allocated to ensure that the performance measures are at the desired level. This is done because rewards and sanctions are tied to the performance measures. The performance measures then come to be viewed as performance per se—that is, the measures used are believed to be true performance.

The model in Fig. 2-1 depicts the way the process of setting goals and objectives interrelates with situational contingencies to affect the characteristics of the management control system. The control system is designed to facilitate performance. Specific outputs are identified, and measures of those outputs are generated and, in turn, used within the control system. Each step

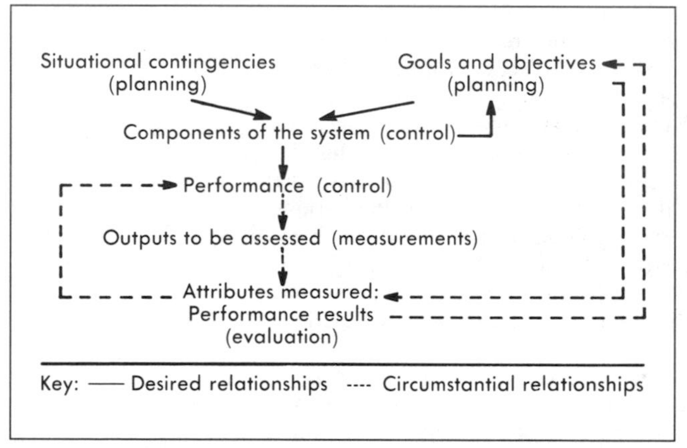

Fig. 2-1 Model of management control systems. (*From Kenneth J. Euske,* Management Control: Planning, Control, Measurement, and Evaluation, *Addison-Wesley Publishing Co., 1986; used with permission*)

depicted in the model—planning, control, measurement, and evaluation—contributes to the operation of the management control system.

In a management control system, how does planning differ from control? The plan is the means by which the manager intends to affect the future; control is the means by which the manager ensures that the plan functions. Planning is prescribing behavior. The prescription can be for a relatively short period of time and is expressed as the annual or quarterly budget, or for a longer period of time and is expressed as a program intended to occur over a 3-7 year period. *Controlling is maintaining behavior within preselected parameters.* Decision-making procedures must be designed and organized so that a plan can be carried out. The design of the procedures for anticipating and detecting variances in the system and reestablishing equilibrium is control.

Any good planning process has a control cycle. The manager cannot plan if there is no information indicating current status. On the other hand, the manager cannot control unless there is some plan that indicates the purpose of the control. The two are complementary activities.

Figure 2-2 depicts these relationships.[3] Each of the three processes requires the other two to occur as well. Some control is necessary to ensure that a plan will result from the process. The plan must be evaluated at least minimally for its reasonableness. Control and evaluation each necessitate the use of the other two processes.

The model in Fig. 2-2 describes the management control process for an organizational entity. The primary objective of this Handbook is to focus on manufacturing. Manufacturing represents a subset of the organization and as such is subject to the same management control system concepts as the organization. Figure 2-3 illustrates the production function's system for a manufacturing organization. Each of the subsystems represented is controlled by a management control system as defined by the model shown in Fig. 2-2. The total system process then is subject to the control concepts as defined by the same model.

Figure 2-3 shows the production manager drawing his or her resource inputs from a purchasing subsystem that processes fixed and current-use resources from suppliers. From a personnel subsystem, he or she draws the human resource. Together, these resources form the system that is available to the manager.

The planning/scheduling subsystems establish design specifications, product target dates, standard work methods, and route work, and they schedule personnel and machines for task performance.

Operational subsystems are the core of the production system. Manufacturing converts the resources and creates form utility. The controlling subsystem ensures that the product meets quality standards and schedules. The controlling subsystem should be integrated with the manufacturing process.

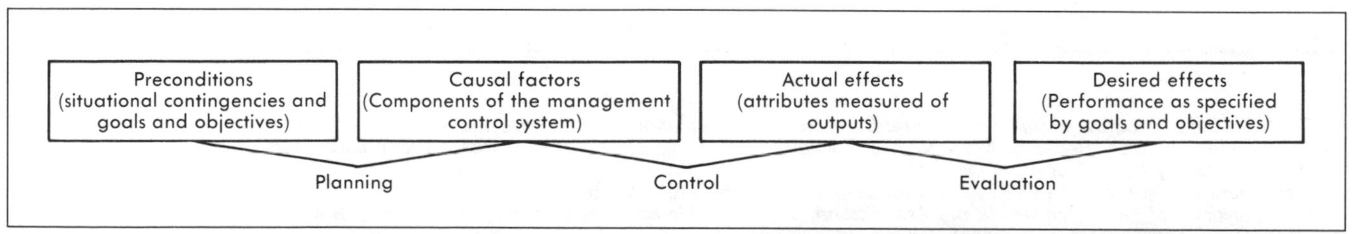

Fig. 2-2 Relationship between planning, control, and evaluation in a management control system. (*From Kenneth J. Euske,* Management Control: Planning, Control, Measurement, and Evaluation, *Addison-Wesley Publishing Co., 1986; used with permission*)

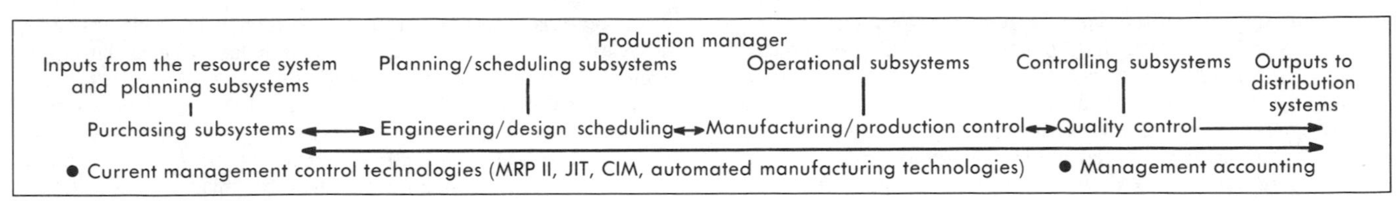

Fig. 2-3 The production system. (*Coopers & Lybrand*)

A data management (management accounting) system provides the manager with information on how well the system is functioning from a cost control perspective.

Current management control technologies that have evolved with the evolution of computer technology, such as MRP II, AMT, computer-integrated manufacturing, and philosophical techniques such as just-in-time support the entire production system. These technologies are firmly based in the fundamentals defined in the model for a management control system and include the basic concepts of planning, controlling, measuring, and evaluating.

The benefits of a management control system are myriad, but the overlying fundamental benefit is that it allows an organization and/or an individual to achieve desired results. A management control system allows an organization to record, measure, and control variability in the business process against predefined goals and objectives. This leads to the achievement of these goals and objectives and, in the case of a business entity, to profits. More detailed benefits, as related to the manufacturing function, are presented in the subsections dealing with the management control systems of management accounting and MRP II, Total Quality, CIM, and JIT.

EVOLUTION OF MANUFACTURING MANAGEMENT CONTROL SYSTEMS

Early manufacturing management control systems centered on organizational and accounting techniques.[4] According to historical records, accounting reports have been prepared dating back thousands of years. The demand to control commercial transactions has existed since people first began trading in market exchanges. But the demand for transactional management accounting within organizations is a recent phenomenon.

Manufacturing began as an owner-entrepreneur type of business. Most individuals who supplied raw materials and labor were not part of the organization. Organizational structure was of little importance. Measures of success were relatively easy to obtain. In short, money collected in sales by the owner-entrepreneur was measured against money paid out to suppliers of the production inputs, primarily labor and material. No control measurements of the production process were required. No capital requirements nor automation existed.

In the 19th century, the Industrial Revolution brought with it a number of innovations in the manufacturing enterprise and a need for increased controls and measurements. The advent of sophisticated machinery and the resultant ability to achieve economies of scale brought many changes in controls. Significant among these were the following:

- High capital investments and a shift from hiring contract workers to the hiring of permanent workforces where efficiency required measurement.
- Hierarchical organizational structures to "manage" the workforce.
- Demand to measure the "price" of output from internal operations.
- Measures to motivate management to produce.

Hierarchically managed manufacturing enterprises became the significant producers of goods. The early textile mills and the steel companies are good examples of these organizations.

The advances that occurred in the middle of the 19th century in transportation (railroads) and communication (telegraph) allowed large, hierarchical manufacturing companies to coordinate raw material acquisition and distribution of final products over larger geographical areas. These added business functions increased the requirement for measurement controls to provide performance measurements for decentralized and geographically dispersed operations. These accounting control measurements were aggregate in nature and dealt with the financial reporting aspects of the enterprise.

Controls oriented toward the efficiency of the manufacturing process itself began in the latter part of the 19th century. Scientific management engineers, such as Frederick Taylor, worked on improving the efficiency and utilization of labor and materials. They developed standards of measurement to do this that were easily integrated into the cost accounting systems approach. This integration led to the capability to aggregate a cost of finished product to support pricing decisions. In a single-product, single-process enterprise, this was a major control breakthrough.

In the early decades of the 20th century, the multiple-product, multiple-process manufacturing enterprise emerged. Organizational controls were improved in these companies in the form of decentralized operations structures representing vertically integrated divisions of the entire corporation. In this way, the enterprise was able to more effectively use controls geared to measure the performance of an enterprise with a single type of activity where only capital choices were geared to expanding the scale of one homogeneous operation.

The emergence of the multiple-activity manufacturing enterprise gave birth to a number of important operating and budgeting controls. Among these were flexible budgeting, which identifies and separates variable and fixed expenses of the manufacturing process. The variable portion of the costs is assumed to vary with direct labor activity. Modern management accountants use the "flexible budget" to compare forecast results with the results attained at actual levels (volumes) of outputs. The flexible budget distinguishes between variable and fixed costs and forecasts total costs and profits at any level of actual outputs (within a given amount of fixed capacity). The underlying principle of a flexible budget is the need for some norm of expenditure for any given volume of manufacturing business. This norm should be known beforehand to provide a guide to actual expenditures. Every business is constantly changing; it is never static. Therefore, it is erroneous to expect a manufacturing business to conform to a preconceived pattern.

Perhaps the most enduring of the operating and budgeting controls to come out of this era is the return-on-investment measure (ROI). ROI provides an overall measure of the commercial success of each operating unit and of the entire organization. ROI is used to help direct allocations of capital and to evaluate operating department performance. ROI separated into its component parts can be viewed as a combination of two of the efficiency measures—the operating ratio (return on sales) and stockturn (sales to assets)—used by single-activity organizations. Currently, both flexible budgeting and the ROI measurement are increasingly being challenged as to their effectiveness in today's globally competitive marketplace.

From the 1920s through the 1950s, manufacturing management controls remained virtually unchanged. In the last 30 years, two major occurrences have significantly affected the further development of manufacturing management controls and systems. These events are: (1) the technical evolution of computer and telecommunication technology and related information products and (2) new global competition.

The evolution of computer technology has precipitated dramatic innovations in manufacturing control systems not previously possible because of the inability to capture and analyze vast quantities of information prior to and during the manufacturing process. The most dramatic manufacturing control system innovations associated with computer technology to date have occurred in manufacturing resource planning (MRP II), process control, and robotics. The combination of these techniques have enabled many of today's manufacturing enterprises to achieve automated control over the manufacturing process, and to realize large cost savings in inventory reductions, material acquisitions, and direct labor.

Through the 1970s, there was no perceived need for the manufacturing enterprise in the United States to be concerned with further innovation of its control methodologies other than the obvious cost and productivity improvements made possible by the evolution of computer technology. There was no perceived need to be concerned with ultimate productivity, quality, and cost reduction controls; nor were existing product cost, management accounting, or performance measurement systems questioned. A combination of high inflation and a weak dollar sheltered most U.S. manufacturers from foreign competition.

Demand for U.S. products in the 1970s was high. As a result, higher costs and, occasionally, goods of substandard quality could generally be passed on to customers. This competitive position completely changed in the latter part of the 1970s and has continued through the 1980s. Foreign competitors, led by Japanese manufacturers, have made great inroads into trade and U.S. markets. These inroads are in large part the result of successful application of manufacturing philosophies and control system practices known as just-in-time (JIT), total quality control (TQC), and computer-integrated manufacturing (CIM). These control systems have resulted in significant labor reduction, high quality, improved productivity throughput, high customer satisfaction, and significantly reduced costs.

The further evolution of manufacturing control systems will be focused on the use of these new practices and technologies and on the reexamination of the value of some of the older control practices and management accounting systems.

MANAGEMENT CONTROL CONCEPTS

The first section of this chapter dealt with the definition of management control systems and the evolution of manufacturing control systems. This section deals specifically with the control aspects of a management control system and then with the aspects of controlling the manufacturing process. Emphasis is placed on the narrow definition of control as a feedback process that may or may not trigger replanning. Control techniques and guidelines currently used in manufacturing are also described.

THE ORGANIZATIONAL CONTROL PROCESS

The managerial control function allows activities in progress to be evaluated in relation to expected results. According to Kast and Rosenweig, the control function is defined as follows:[5]

> That phase of the managerial system that monitors performance and provides feedback information that can be used in adjusting both ends and means. Given certain objectives and plans for achieving them, the control function involves measuring actual conditions, comparing them to standards, and initiating feedback that can be used to coordinate organizational activity, focus it in the right direction, and facilitate the achievement of a dynamic equilibrium.

This definition is more narrow than the one provided in the first section of this chapter and emphasizes the feedback function of control. Note that the control process serves a feedback function resulting in the identification of a need for new or adjusted plans. At some point in time, every organization has to measure its progress and determine the adequacy of its performance. The control process can be applied at each level of the organization and can incorporate all enterprise activities, including, of course, manufacturing.

The control process must determine preestablished measures of performance. These standards may be quantitative (budgets) or qualitative (job instructions). Performance is measured against the preestablished standards, and a determination is made as to whether or not performance is adequate. If performance is not adequate, corrective action is taken. Not every deviation from standard requires corrective action. Only deviations that exceed accepted parameters of performance need to be corrected.

Management Control Levels

Controls may exist at various levels. These levels are described as follows:

1. High-level controls. High-level controls exist to reassure those who are not sufficiently close to day-to-day detail that intended organizational performance objectives are being met. If properly done, they will obviate investigative actions and provide top managers with proof of performance for those interested in organizational objectives. These measures have a public relations quality about them. They are measures used to demonstrate, but not prove, organizational efficiency. Profit measurement of a business enterprise is an example of such a high-level control. As is becoming more apparent, profits are rather arbitrary numbers that can be manipulated within a wide range.
2. Middle-level controls. Middle-level controls are signals to management that allow it to make decisions or act in ways that contribute to the effectiveness of the organization. Project management techniques such as PERT and critical path planning are examples of middle-level controls. These types of control allow a manager to delegate to subordinates without losing the capability to predict and make decisions to correct out-of-schedule conditions. When everyone knows the costs of failure and is committed to the same goal—as opposed to individual goals epitomized by low-level controls—there is acceptance of the need for constant review in the most critical areas.
3. Low-level controls. Low-level controls are concerned with checking and measuring detailed operations critical to the day-to-day operating efficiency of the business. These controls may be financial, operational, or technical. Some typical examples of these controls are as follows:
   - All expenditures more than $1000 have to be approved by the plant controller (financial).
   - All pharmaceuticals with decomposable ingredients must be stored in refrigerated areas (operational).
   - Batch mixing of ingredients to produce chemical products must meet certain process measurements controlled by a computer (technical).

These types of controls are well reviewed in classical scientific management, but they represent recurring problems. For instance, managers would prefer to know before something is out of control rather than after. Important decisions are often subject to before-the-fact review by a number of people. For example, in a technical program there might be many levels of concurrent sign-off prior to authorizing a design decision by an engineer. Such sign-offs are time consuming both for the overall project and for those who must review the technical details.

Control Process Characteristics

Every control process should be accurate, meaningful, appropriate, congruent, timely, understandable, and economical. Accuracy is essential because if the information received is not correct, the resulting decisions are likely to make things worse rather than better. The information measured must be meaningful in that it must be measuring performance in areas important to the manager and the organization. This measured information must be appropriate in that it fairly measures the events that it is designed to measure, and it must be congruent with the need and ability to obtain precision in measurement—for example, morale is good, not morale is 6.5. The information being fed back must be timely and forward-looking to allow management to obtain full benefits from the data, and it must be simple enough to be understandable to the people using it. Finally, it must be economical because it would be pointless for management to spend $1000 on a reporting system that only saves the company $10 a month.

In addition, a control process must be acceptable to managers. A manager does not have time to control every aspect of the operation. As a result, a control process should single out specific areas that provide comprehensive control.

Control Process Operation

A control process operates through the repetition of the following five sequential steps:

1. Establish standards of performance.
2. Measure actual performance.
3. Analyze and compare performance to standards.
4. Construct and implement a program of corrective actions.
5. Review and revise standards.

Step 1 in establishing a control process is the setting of standards. Standards are specific goals or objectives against which performance is compared. Standards as they apply to manufacturing are often set in criteria consisting of quantity, quality, timeliness, and cost.

Step 2 measures actual performance. Measurement is the assignment of numerals to attributes or objects according to rules. It should be remembered that performance is not measured, but rather some attribute of performance is measured. Euske identifies seven elements of measurement that are useful for understanding the influence of measurement and the measurement process on the control process, as follows.[6]

1. Scaling. The types of scales relate to the characteristics of the real-number series that are used in a particular set of measurements. The characteristics are: numbers are ordered, the difference between numbers is ordered (<, >, =), and the series has a unique origin, indicated by the number 0.
2. Meaningfulness. Measured information must be meaningful in that it must not exceed the inherent limitations of the data. For instance, a grading scale comparison of an A (4.0) to a C (2.0) does not indicate a level of knowledge of two times the difference, as this scale is not meant to be a ratio.
3. Specification. Specification represents the scope of the measurement. What attributes are to be measured, and under what circumstances can the measurements be made? Availability of information will also influence what is to be measured.
4. Standards. Standards provide the basis for adjusting experiences in different contexts. For example, will $40,000 per year provide the same standard of living in Los Angeles that it does in San Antonio? If standards did not exist, complete descriptions of the situation surrounding the object would have to be given so that comparisons of attributes could be made.
5. Reliability. The question of reliability deals with the degree to which repeated measurements of the same attribute vary. How much error is there in the management process?

MANAGEMENT CONTROL CONCEPTS

6. Validity. Validity is concerned with the relationship between the measures and the attributes measured. If person A measures twice as tall as person B and A is actually twice as tall as B, then the measure is valid.
7. Types. There are three types of measurements: fundamental, derived, and fiat. Fundamental measurements directly measure the extentions and attributes of objects such as a count of widgets. Derived measures are obtained by manipulating other measures such as earnings per share. Fiat measurements are simply defined to be what they are.

Step 3 in establishing a control process is evaluating performance to standards. Evaluation is applying judgment to the results of the measurement process to determine whether the attribute to be measured is in or out of control and whether or not it requires corrective action. Most control processes use one or more of the following types of evaluation:

1. Effort. Effort evaluation measures the input values as indicators of meeting objectives. For example, "number of patients" is sometimes a health care measurement.
2. Effectiveness. Effectiveness evaluation measures outputs. A measurement of meeting production goals would be how many widgets were produced.
3. Adequacy. Adequacy evaluation measures output to need. For example, did we produce the right number of widgets to meet customer demand?
4. Efficiency. Efficiency relates output to input. In other words, how much input was required to produce a given level of output.
5. Process. Process measures output as a function of input. It focuses on the mechanism by which effort is translated to output. For example, does a particular sales presentation produce a sale?

Evaluations are properly done when they are used as tools to gather data for the future improvement of a program or operational process.

Step 4 is the construction and implementation of a program of corrective action. Again, it must be emphasized that not every deviation from the plan or standard requires corrective action. Often, a range of acceptable performance or deviation from specifications is established.

Step 5 in establishing a control system is reviewing and revising standards to fit the new corrective actions implemented to bring a process back into control.

Control Tools and Techniques

The material that follows illustrates key control tools and techniques used in modern organizations. Table 2-1 summarizes these control techniques.

One of the primary ways to monitor organizational performance is through the use of financial control. Two of the major approaches to financial control are budgets and financial statement analysis.

Budgets are plans that specify results in quantitative terms and serve as a control device for feedback evaluation and follow up. All budgets have the following three common characteristics:

1. They help establish goals and provide general direction for management in terms of quantitative goals and costs.
2. They establish control points or baselines against which performance can be measured and evaluated.
3. They provide a means for coordinating organizational activity within and between components of the organization.

Most organizations, including manufacturing companies, make use of operating budgets and financial budgets. Operating budgets are used for monitoring expenses, revenues, and profits. Expense budgets are used to control organizational components such as production or marketing. Revenue budgets measure the effectiveness of sales and marketing. Financial budgets integrate the organization's financial plan with its operational plan. One question that companies ask is whether they will have sufficient money to meet their operational goals. To determine this capability, companies commonly use the following four types of financial budgets:

1. Capital expenditure budget—used for expanding physical assets, such as buildings, property, and equipment.
2. Cash budget—used to monitor flow of funds and the pattern of receipts and disbursements.
3. Financing budget—used to balance shortages in capital regardless of timing (short or long range).
4. Balance sheet budget—used to bring together all other budgets into a balance sheet projection if all budgets perform to results.

A major problem with budgets is flexibility. Variable budgets have been used by many organizations to deal with this inflexibility. Variable budgets are tied to a volume of activity in that the more output the organization produces or the more sales it generates, the greater its expenditure budget. Conversely, if the organization finds it must contract operations, the budget is decreased. The costs affected by activity changes are those directly related to production and sales (for example, materials, parts, and utilities). The major advantage of variable budgets is that they encourage the organization to examine its costs.

The financial statement analysis focuses on the balance sheet and the income statement. The balance sheet is a financial statement that shows the organization's financial position at a point in time. It consists of the following three parts:

1. Assets—things the company owns.
2. Liabilities—company debts.
3. Owner's equity—difference between assets and liabilities and representing the net worth of the company.

The income statement summarizes the company's financial performance over a given period of time, usually quarterly or yearly. Using these two financial statements, a company can conduct analyses and determine where it is doing well and where it is doing poorly. The most common way this is done is through ratio analysis. Table 2-1 includes the most common ratios.

Other than financial control, a second major form of control is operational. The tools and techniques used in operational control are designed to help the organization monitor its operations or activities. These types of controls include the modern techniques made possible by information systems. Table 2-1 summarizes some of these controls. Financial and operational controls are important at the lower and middle levels of the hierarchy. At the high level, however, managers are more interested in overall performance control. Typically, their attention is focused on five to seven key performance factors that provide the necessary performance control for the organization at large. These factors are also summarized in Table 2-1.

TABLE 2-1
Control Systems

Level	Type	Name	Control Characteristics
High level	Overall performance control	ROI	A measurement on the return on investment that analyzes cost of investment vs. payback overtime
		Market share	Percent of market share held by a company's products
		Growth	Percent of growth in revenue over time
		Profit	Percent of growth in profit over time and percent of profit to sales
		Customer satisfaction	Level of customer satisfaction
Middle and high level	Financial—statement analysis	Balance sheet	A financial statement that shows a company's financial position at a certain point in time
		Income statement	A financial statement that summarizes a company's financial performance over a given period of time
	Ratios	Liquidity ratios	Ratios designed to measure how well the organization can meet its current debt obligations
		Current ratio	Current assets—current liabilities
		Debt ratios	Ratios designed to measure the amount of financing being provided by creditors — Debt/asset ratio — Debt/equity ratio
		Operational ratios	Ratios designed to measure internal performance
		Profitability ratios	Ratios designed to measure a company's effectiveness, such as return on investment (ROI)
Low and middle level	Financial—operating budget	Expense budget	Designed to control things like production and marketing. Representing expenses vital to the operations of the organization
		Revenue budget	Used to measure marketing and sales effectiveness
	Financial—financial budget	Capital expenditure budget	Used for constructing or expanding physical assets
		Cash budget	Used to monitor the flow of funds and track cash receipts and disbursements
		Financing budget	Used to balance shortages of capital overtime
		Balance sheet budget	Used for consolidating all budgets into a projected balance sheet based on meeting all budgets
Low and middle level	Operational	Break-even analysis	Analysis used to determine the point at which the firm covers all of the costs associated with producing a particular product
		PERT/Critical path	Project management control tools used to control activities and schedules in complex projects
		MRP II	A computer-based model for planning, ordering, and controlling materials to support production
		Total quality	A concept including a number of systems to attain product quality and customer satisfaction
		Just-in-time	A philosophy and set of techniques aimed at eliminating waste in the production process
		Computer-integrated manufacturing (CIM)	An integration of all operational control systems through the use of computer technology

MANAGEMENT CONTROL CONCEPTS

The remainder of this section will focus on specific manufacturing control techniques—specifically operational process control, management accounting for a manufacturing enterprise, and the levels of control required to achieve world-class manufacturing excellence. Subsequent sections of this chapter briefly describe some of the modern manufacturing control systems in use today. These control systems are integrated with the planning step of the control system model and are not specific control techniques per se. Included in this group are MRP II, Total Quality, just-in-time, and computer-integrated manufacturing (CIM). These management systems are also described in detail in Chapters 15, 16, 20, 22, 23, and 24 of this volume.

THE MANUFACTURING CONTROL PROCESS

Manufacturing is a sequence of work processes and, as such, requires building the appropriate controls into these processes. The manufacturing process needs built-in controls with respect to flow, quality, and quantity produced in a given unit of time and with a given input of work (standards), machine maintenance and safety, and resource efficiency (machines, tools, materials, and people).

Fundamental to control of the manufacturing process is control of the work, not control of the worker. Control is a tool of the worker and must support worker flexibility and productivity, as exemplified by the success of just-in-time methodologies. Controls should never be allowed to become ends in themselves and impediments to the workers. The purpose of control is to make the process go smoothly, properly, and according to quality specifications.

A control system must maintain the process within a permissible range of deviation (variability) as presented in a recent special report in *Business Week*:

> Fixing the manufacturing system, not its products, is now seen as the primary key to better control and quality. The approach is based on a timeworn notion: an ounce of prevention is worth a pound of cure.

The objective is to tackle problems at their source by quickly spotting the cause of production defects. Removing the causes of variation is the first step in improving a process.

Manufacturing process controls have basic characteristics. They have to be preset. Performance specifications and permissible deviations must be set. As long as the process operates within the preset standards, it is under control and does not require any action. Finally, control has to be by feedback from the work done. Modern methods and automated process technologies allow for this information to be available instantaneously. Therefore, it is no longer necessary to allow an out-of-norm condition to exist through iterations of the process.

One thing that must be understood is that inspection is not control. Inspection is the control of the control system rather than the control system itself. If used as a control, it becomes cumbersome, expensive, and adversely affects the process itself.

Control has to be exercised where the problem is likely to occur. To determine what should be controlled, each manufacturing process should be systematically analyzed to establish its added value and its contribution to product value.

A critical factor to the success of a control system is the designation of the key points at which control is to be built in. This is primarily a managerial decision, not a technical decision. At what point in the system is there sufficient information to know whether control action is needed? Control must be established at points sufficient to prevent damage. The questions, "Where is preventive control needed, and where is control essentially remedial?" must be answered preliminary to the design of a control system that truly satisfies the needs of the manufacturing process.

A control system can only control the regular process. It must identify exceptions, but only to make sure that they do not hinder the process.

A later section of this chapter will deal with some of the current technologies and control systems that are now being used to achieve the goals of the manufacturer through application of the fundamentals of the control process. The next part of this section deals with the management accounting measurement process. Essentially, it deals with accounting practices in place today and whether or not they are effective in directing management toward which product lines and operations are profitable and the impact of new control practices such as just-in-time and computer-integrated manufacturing.

MANUFACTURING MANAGEMENT ACCOUNTING

Management accounting practices have existed virtually unchanged since the 1920s. Little innovation has occurred to measure the realities of the modern manufacturing environment. This is true of practices developed for cost accounts for labor, material, and overhead; budgets for cash, income, and capital; flexible budgets, sales forecasts, standard costs, variance analysis, transfer prices, and divisional performance measures. Admittedly, innovations such as cost-benefit analysis, program budgeting, and zero-based budgeting occurred in the aerospace and defense industries in the 1950s and 1960s; however, these procedures have had limited impact outside these industries.

Management accounting control systems are concerned primarily with measuring and managing costs and with measurement of profit and performance.

In the planning phase, cost accounting deals with the future. It helps management to budget the future costs of material, labor, and overhead for the manufacture of products. In the control phase, cost accounting deals with the present by measuring current results with predetermined standards and budgets.

Different manufacturing environments call for different types of cost accounting methods. A job shop or discrete manufacturing environment calls for a cost accumulation method that collects cost by discrete jobs, projects, or contracts. A process or flow manufacturing environment, such as a chemical firm, calls for a cost accumulation method that collects costs by department. The costs per unit are then calculated by dividing the total costs by the volume produced within the department. This results in an average cost rather than a unique actual cost measured by a job costing system. The following paragraphs describe in more detail the differences between job cost and process cost accounting methods.

Certain manufacturing environments produce a discrete, separately measurable unit of product. These environments are usually represented by job shops within industries such as automotive, defense, and machine tools, as well as other manufacturers whose unit of product is easily measured and can be tracked to contracts or jobs. Product cost determination may be done by job order cost methods based on contracts, work orders, or lots. A lot represents the quantity of product that can be conventionally and economically produced and costed. In job order costing, each job is an accounting unit to which material, labor, and factory overhead costs are assigned by means of a job order number. The cost of each order produced or the cost of each lot to be placed in stock is recorded and summarized on a job cost sheet. This master sheet has an order number and is designed to collect the cost of material, labor, and factory overhead applicable to a specific job.

Several jobs or orders will usually be going through the factory at the same time. Each job is given a number that is used on all material requisitions and labor associated with that job. The charges are totaled periodically (daily or weekly) and then recorded on a master summary sheet. This master cost sheet eventually becomes a summary of all costs, including factory overhead, connected with a given job. Jobs performed on a customer's specification allow the computation of a profit or loss on each order using this approach. If a job constitutes production of a specific product quantity for inventory, job costing permits computation of a unit cost for inventory costing purposes. Both actual and standard cost accounting control systems can be used with job order costing.

In manufacturing environments where it is not possible to track a unit of production because it is part of a total flow process, process costing methods must be used to control costs. Process costing control is applicable to such industries as textiles, pharmaceuticals, distilleries, and chemicals, as well as flour mills. It is also applicable to industries that use assembly lines, such as household appliances. Utility companies also use process costing controls. Because of the nature of the output, it is necessary to compute a unit cost for each process. This consists of computing an average unit cost for production by dividing the total manufacturing cost of a process by the total number of units produced over a specific period of time. A process is often identifiable with a department. Such computation is prepared on a process cost sheet or a cost of production report. A process costing system is used when products are manufactured under conditions of continuous processing or under mass production methods.

The procedures of process costing are designed to accumulate material, labor, and factory overhead costs by department. Unit costs are determined for each department. Transfer costs are computed for transfer of product from one department to another or finished goods, and costs are assigned to work still in process. The following steps normally occur in a process costing control system:

1. Costs are collected and summarized, and then the total and unit costs are computed and posted to a cost of production report.
2. Costs are charged to departments.
3. Production is accumulated and reported by departments.
4. At the end of a period, production still in process is restated in terms of completed units.
5. An average unit cost for a period is computed. This is done by dividing total costs charged to a department by total computed production of the department.
6. A cost for lost or spoiled units is computed and added to the cost of satisfactory units completed.
7. Costs of completed units of a department are transferred to the next processing department to arrive eventually at the total cost of the finished products during that period.

Cost of a period must be identified with units produced in the same period to ensure accurate unit and inventory costs.

Both actual and standard cost accounting control systems can be used in conjunction with process costing. It should also be noted that both job costing and process costing may be used by the same company to control different segments of the production environment.

The main types of cost accounting systems are standard and actual. For a standard cost system, standards are developed for material, labor, and overhead expectations. Actual costs are then collected as the job progresses, and variances against standard are measured as a means of controlling the manufacturing process. Indirect overhead costs are usually allocated to products using a cost driver such as direct labor hours or dollars. The advantage of the standard cost system over the actual cost system is the ability to measure performance variability. An actual cost system collects costs as they occur, but delays the presentation of their results until manufacturing operations have been performed. The job or process is charged with the actual quantities and costs of materials used and labor expended. Factory overhead is usually allocated on the basis of a predetermined rate. Thus, even ''actual'' cost systems are not predicated totally on actual costs. Most manufacturing environments today utilize standard costing systems.

Standard cost systems aid in controlling costs, detecting above or below performance, and in helping management better analyze probable effects on cost levels and profits.

As in any control system, planning is an integrated part of the model. The budget is the planning mechanism that integrates with standard cost control. Standards are almost indispensable in establishing a budget. When manufacturing budgets are based on standards for material, labor, and overhead, a strong capability for control and reduction of costs is created. Standard costs aid in planning by serving as building blocks for the budget. They also aid in controlling production operations by serving as benchmarks for performance evaluation, and they assist in recordkeeping by reducing the complexities of product costing.

A standard cost has two components: a standard and a cost. A standard is a norm against which production performance can be measured. In a manufacturing environment, standards are determined for material, labor, and overhead associated with each product or process of production. Standard costs are the predetermined costs of manufacturing a single unit or a number of product units during a specific period of time. Material and labor costs are generally based on normal, current conditions, allowing for alterations of prices and rates and tempered by the desired efficiency level. Factory overhead is based on normal conditions of efficiency and volume.

Once standards have been set for material, labor, and overhead by unit or process of production, costs can be accumulated and compared to standard. The differences between standard and accumulated cost are known as variances. These variances serve as performance measurements to help detect and analyze potential out-of-control costs and to help reestablish more accurate standards. The most frequently used variances in a standard cost system are as follows:

1. Materials price variance. It is often difficult to determine the price or cost to be used as a standard cost because the prices used are subject to external factors outside management's control (inflation, strikes, and so on). If the actual price paid for an item is more or less than the standard price, a price variance occurs. Price standards permit the measuring of procurement performance as well as measuring the effect of price increases or decreases on the company's profits.
2. Materials quantity variance. Quantity or usage standards are generally developed from materials specifications

prepared by manufacturing or product design engineers. The standard quantity should take into consideration allowances for acceptable levels of waste, spoilage, shrinkage, seepage, evaporation, and leakage. In such cases, the standard quantity is increased to include these factors.

The materials quantity variance is computed by comparing the actual quantity of materials used (priced at standard cost) with the standard quantity allowed (priced at standard cost). This variance helps management analyze the efficiency of the production process and to take corrective actions where required.

3. Labor rate variance. Rate standards are normally based on collective bargaining agreements by job classification in union plants. In nonunion plants, rate standards are normally set by the personnel department based on job classification.

   Any difference between standard and actual rates creates a labor rate variance. These variances can be analyzed by management to determine how the workforce is being employed and to correct rate discrepancies that could lead to employee dissatisfaction.

4. Labor efficiency variance. Labor efficiency standards are normally set by industrial engineers through the use of time and motion studies. Time standards are set for the time it should take to perform a given operation(s) in the flow of producing a finished unit. While personal factors (fatigue or working conditions) are considered a part of direct labor costs in most plants, time required for setting up machines, waiting, or breakdown are included in factory overhead instead of the direct labor standard.

   The labor efficiency variance is computed at the end of a stipulated time period by comparing standard hours allowed to actual hours worked on a given job or job operation(s). The standard hours allowed are computed by multiplying the direct labor hours predetermined to produce 1 unit times the actual number of units produced during the period.

   The labor efficiency variance allows management to determine worker efficiency and to identify potential bottlenecks in the production process.

5. Factory overhead variance. Factory overhead consists of all costs not directly associated with producing the product. These would include such costs as indirect workers, factory heat, light, maintenance, insurance, taxes, and supplies. Procedures allowed for establishing factory overhead rates can be based on a specific level of activity (fixed) or allow for variances in activity (flexible). A standard is developed to allocate these costs across departments and products based on some cost driver that is perceived to apply consistently across products.

   The standard factory overhead rate is normally a predetermined rate that uses direct labor hours as the basis for computing the rate itself and for applying factory overhead. Other cost drivers such as direct labor dollars or machine hours may also be used.

   Jobs or processes are charged with costs applicable to them on the basis of standard hours allowed multiplied by the standard factory overhead rate. At the end of every period, overhead actually incurred is compared with the expense charged into the process using the standard factory overhead rate. The difference between these two figures is called the net factory overhead variance.

   An unfavorable variance will require further analysis to guide management toward remedial action. This analysis must consider actual expenses vs. budget, idle capacity, and efficiency variances.

In summary, a standard cost control system is used for the following purposes:

1. Establishing budgets.
2. Controlling and reducing costs.
3. Motivating and measuring efficiencies.
4. Simplifying costing procedures and report preparation.
5. Assigning costs to materials, work-in-process, and finished goods inventories.
6. Forming the basis for bids and contracts by helping to determine selling prices.

Standards serve as a measurement that calls attention to cost variations. Executives and managers become cost conscious as they become aware of results. The standard cost system may be used in connection with either the process or job order cost accumulation method. It is, however, more often used in process cost accumulation because of the greater practicality of setting standards for a continuous flow of like units (repetitive) than for unique job orders.

The final point of difference in cost accounting approaches is between full costing and variable costing. A full costing system collects costs for material, labor, and fixed and variable overhead and assigns them to the product. A variable cost system assumes a fixed cost based on current operations and products and only assigns variable costs of material, labor, and overhead to new products. Full costing systems are required by the financial community for valuation of inventories and financial reporting. A survey by CAM-I has indicated that 80% of companies today use a full cost accounting approach.

A number of factors need to be addressed when designing a cost management system for today's globally competitive manufacturing environment. A well-designed cost measurement system should address allocation of costs, facilitation of process control, computation of product costs, and special studies. The considerations for each of these functions are as follows:

1. Allocate costs for periodic financial statements. This function is concerned with accumulating period expenditures and then dividing them between cost of goods sold and inventory according to generally accepted accounting practices (GAAP). This function has historically driven the practice of cost accounting.

   Unfortunately, this functional approach has been more concerned with closing the books to produce short-term financial reporting than with accurately measuring product costs in a complex multiple product environment. The result is a distortion of product cost, delayed and overly aggregated process control information, and short-term performance measures that do not reflect the increases or decreases in the organization's economic position. Common measurements such as return on sales or ROI are no longer effective yardsticks to measure the profitability of the modern manufacturer.

2. Facilitate process control. Process control must occur at the level where the process occurs. Setting standards and identifying organizational units to be measured is the first

step in the design of an effective control system. The identification of the correct cost driver is also of paramount importance. Cost drivers may include items such as direct labor, machine hours, orders received, orders shipped, engineering change orders, and many others. Unless the correct cost drivers are identified and measured, effective process control will not be achieved. Each cost center needs a clear identification of its boundaries, standards to apply to input and output, and an understanding of the cost drivers that measure variation in costs with variation in activity levels.

An effective cost management system must be able to cost effectively identify, capture, report, and measure costs by product, process, job, project, cost category (material, technology, overhead, and labor), department, workcenter, transactional activity, and element. It must also be able to identify technology costs and to capture transaction (activity) statistics.

Major advances in process cost control are possible with current production technology. Computer-controlled processes require digital information. This information can be captured to learn exactly what occurred and when in the manufacturing process, and to measure variability against standards. Cost systems need to record and process this data to produce a continual record of actual output and resource consumption. Variance reports can be produced instantaneously.

Only costs that are traceable to the cost center and measurable at the cost center level should be assigned to the cost center. The current practice of allocating indirect manufacturing costs to a cost center by some incorrect cost driver such as direct labor distorts the true product cost. This leads to management reporting of product profitability that is distorted and inaccurate. Many important management decisions regarding the profitability of product lines and ongoing product strategy are being incorrectly made because of the inability of current cost measurement systems to correctly identify process costs.

3. Compute product costs. The cost tracing process starts at the component level of the product. Systems for product costing and control should focus on components and the underlying operations. Systems for product costing control are the foundation for budgeting and variance reporting.

   The most important goal for a product cost system is to estimate the long run costs of producing each product. Emphasis on short-term costs and incorrect cost drivers can lead to incorrect decisions on pricing, product life, product mix, and make vs. buy. Short-term financial indications do not correctly factor in the costs of capital investment and long-term technology innovations.

   Current product cost systems allocate plant costs to products to place a value on the inventory and to calculate cost of goods sold to produce short-term financial indicators. This allocation usually occurs based on some oversimplified or unrepresentative cost driver such as direct labor. The primary cost drivers for manufacturing overhead are transactions involving exchange of materials or exchange of information. Reduction of these transactions is the real key to product cost reduction and control.

4. Special studies. Cost data can play an important role in special ad hoc studies. These could be studies to determine product strategies, capital expansion, equipment tradeoffs, or capacity expansion.

The performance measurement aspects of designing a cost management system are concerned with measuring the profit performance of operations and with identifying performance to standard against key cost drivers. Current performance measurement systems collect actual cost, actual transactions, and established cost factors and compare them to standards and financial plan data to produce variance reports. Comparison reports are produced on a regular or as-needed basis. Both financial and nonfinancial measures of performance are used in setting goals for a business.

Traditional performance measurements are financial and concentrate on corporate measurements and manufacturing productivity measurements. Corporate measurements are found on sales, net income, return on investment (ROI), and earnings per share. These financial measurements focus on the three key financial statements: the income statement, the balance sheet, and the cash flow statement.

Traditional manufacturing productivity measurements focus on gross profit, labor efficiency, machine utilization, materials price variance, overhead absorption, and actual vs. budget performance for plant and products.

In light of the implementation of new manufacturing control system techniques and technologies such as just-in-time (JIT) and computer-integrated manufacturing (CIM), the traditional measures of performance are being questioned. While they are effective barometers for financial reporting, they do not take into account the total impact of new technologies and automation on a business.

Application of JIT control philosophies and CIM are reshaping the physical nature of the production environment. Changes in traditional cost measurement control systems are necessary to correctly evaluate product costs and profitability in a JIT/CIM environment. Traditional cost systems measure and allocate costs to products based on direct costs, usually direct labor hours or dollars. Yet 80-90% of total manufacturing time and costs is contributed by indirect, non-value-added activities such as inspection time, move time, queue time, and storage time.

JIT control philosophies focus attention on the non-value-added transactions in the production environment as a means of cost reduction control. Direct labor usually represents only 15% or less of total production costs in the current manufacturing environment. Other cost devices, such as orders processed and engineering change orders, have been increasingly recognized as contributing a large part of the total product cost. Many of these cost drivers occur outside of the manufacturing function. These activities are non-value-adding, and their reduction through JIT/CIM control system technologies reduces the cost of the product and improves product quality. It is, therefore, necessary to design cost and performance measurement control systems that measure these transactions. Table 2-2 compares the traditional performance measurements vs. some of those that are more correct in a JIT environment.[7]

Traditional performance control measurements are ineffective in a JIT/CIM environment because they measure the wrong factors. Labor efficiency represents a small part of the total cost and focuses incorrectly on labor utilization instead of on which products should be in process. Labor productivity measurements focus on throughput at the expense of building the right products

TABLE 2-2
Performance Measures: Traditional vs. JIT

Traditional	JIT
• Direct labor — Efficiency — Utilization — Productivity • Machine utilization • Inventory turnover or months on hand • Cost variances • Individual incentives • Performance to schedule • Promotion based on seniority	• Total head count productivity — Output: Total head count (direct, indirect, administrative personnel) • Return on net assets • Days of inventory • Product cost, especially relative to competitors' costs • Group incentives • Customer service • Promotion based on increased knowledge and capability • Ideas generated • Ideas implemented • Lead time by product/product family • Set-up reduction • Number of customer complaints • Response time to customer feedback • Machine availability • Cost of quality

(*Ernest C. Huge and Alan D. Anderson, "The Spirit of Manufacturing Excellence," September 1986*)

and focusing on the quality of these products. Machine utilization encourages building inventory ahead of need and is directly opposite to JIT control principles, which simplify and reduce inventories. All of the traditional performance measurements tend to promote increasing inventory and emphasize performance to the wrong standards at the expense of quality.

There is clearly a need to reevaluate and revamp long-standing performance measurement drivers of manufacturing control systems. The results of a joint study by the National Association of Accountants and Computer Aided Manufacturing-International suggest improvements in the following broad control categories:

1. Reporting process. The respondents to this study believed that a sharper emphasis on responsibility accounting, the utilization of exception reporting, and an emphasis on variance analysis would improve the control process.

   Budget vs. actual performance reporting is still considered a valuable performance control tool. The inclusion of nonfinancial performance measure is also considered vital.
2. Long-term orientation. More than one third of the respondents believed that the performance measurement part of the management control system should take on a longer term orientation. American management has been severely criticized for its short-term orientation, and many feel it is a significant part of the reason for U.S. manufacturers' failure in international competition.
3. Context. Three contextual areas were cited as areas for potential improvement. Productivity control measurements were cited as obsolete. There is an increasing recognition of a need for a more comprehensive set of measures that effectively replace the labor productivity measure. Material and capital measurements are deemed more effective.

Cost of quality was cited as a major area of improvement. The costs of scrap, rework, inspection, consigned inventory, and material purchase reordering too often go unnoticed in controlling performance. The cost of carrying inventory was the last area of focus. Additional information on the cost of quality can be found in Chapter 24, "Quality Cost and Improvement," of this volume.

In concluding, it is emphasized that the measurement of performance is an important aspect of the design of a cost management system. Virtually any manufacturing organization can enhance its cost systems to be more supportive of the evolving JIT/CIM control environments. The first step is for financial management to adopt the primary JIT principle of continual improvement within the organization and cost management control process.

LEVELS OF MANUFACTURING CONTROL

Many levels of control exist in the pursuit of world-class manufacturing status. These levels fit into the low and mid-level controls of the management control system model previously discussed in this chapter. The goals of world-class manufacturing are continual and rapid improvement in quality, cost, lead time, customer review, and, of course, profitability of the manufacturing enterprise. According to Richard Schonberger, a full range of elements of production are affected:[8] management of quality, job classifications, labor relations, training, staff support, sourcing, supplier and customer relations, product design, plant organization, scheduling, inventory management, transport, handling, equipment selection, equipment maintenance, the product line, the accounting system, the role of the computer, automation, and others.

CHAPTER 2

MANAGEMENT CONTROL CONCEPTS

The operational levels of control in the manufacturing environment today can be summarized in the following five categories:

- Management controls.
- Technology controls.
- Quality controls.
- Cost controls.
- Delivery controls.

All of these control levels make use of the control procedures and application systems made possible by the advances of computer and communication technologies.

These levels of control are multidisciplinary and must be integrated with each other to achieve optimum performance results. They address the need for multidisciplinary planning, effective communications of objectives, effective organization structure, and the use of automation and big business systems to achieve results for the manufacturing control system. Implicit in each category is the establishment of policies and procedures that are directed toward achieving business and functional objectives. Establishing goals consistent with the business plan, education of staff, use of control procedures and systems, and feedback from performance measurement systems are inherent at each level of control.

The following manufacturing control guidelines outline the overall levels of control process that should be in effect to achieve world-class manufacturing. As can be seen, the list is extensive and addresses many types of control within each of the identified levels. This guideline can serve as a checklist to determine current effectiveness and to determine which concepts should be implemented to support a given manufacturer's environment.

A. *Management controls assessment*
  1. Management control system
    a. Organizational structure
      —Formally documented organizational charts
      —Organizational effectiveness
      —Functional staffing
      —Levels of management U.S. firm's size and goals
    b. Effectiveness of key business systems—do the following systems exist, are they effective, and are they automated?
      —Materials planning system
      —Delivery/production control system
      —Quality system
      —Cost control system
      —Audits to determine validity of financial statements
    c. Communication
      —"State-of-the-business" meetings
      —Problem-solving groups
      —Informal communication with management
      —Open-door policy
    d. Business plan
      —Existence of a formalized business plan
      —3-5 year scope
      —Comprehensive in scope:
        Market-related issues
        Financial planning
        Growth projections
        R & D projections
        Plant, facilities, and expansion issues
        Cost objectives
        Human resources development
        R & D projections
      —Competitive analysis
      —Tracking of plan conformance
      —Regular plan updates
      —Communication to employees
  2. Human resources management
    a. Training programs
      —Multidisciplined
      —Skills evaluations
      —Training records
      —Training budget
    b. Employee development
      —Documented job classification/descriptions
      —Employee participation programs
      —Formalized method of employee review
    c. Employee satisfaction
      —Appropriate benefits program
      —Union status
      —Outside activities
    d. Housekeeping
      —Neatness and cleanliness of plant facilities and grounds
      —Designated responsibility for housekeeping
    e. Health and safety
      —Conformance with state and federal regulations
      —OSHA results
      —Safety programs
  3. Management of Sources and Subcontractors
    a. Selection and assessment program
      —Competitive quoting processes
      —Comprehensive evaluation of source's business practices
    b. Performance monitoring and development
      —Monitor source performance in:
        Quality
        Delivery
        Cost reduction
  4. Responsiveness to customer's needs
    a. System for responding to customer needs
      —Tracking of big event dates
      —Tracking of customer-related issues:
        Pricing problems
        Quality problems
        Delivery problems
        Design/engineering problems
        Program timing problems

B. *Technology controls assessment*
  1. Controls for the design and development of parts and/or systems
    a. Design/development
      —Design process control:
        Performance/cost/risk tradeoff studies
        Feedback from test, production, and field support
        Use of failure mode analysis, tolerance studies, and so on
        Effective communication with customer
      —Multidisciplinary approach

—Design reviews
—Dedicated design staff
b. Product design control
—Computer audit design and analysis
c. Product testing control
—Performance testing
—Product life testing
d. Prototype support
—Integral part of business practice
—Production intent tooling and process
—Inclusion of all product parts in prototype program
—Tracking procedures to assure program benchmarks
e. Research and development program
2. Controls for assuring capabilities to manufacture parts/systems
a. Facility/equipment planning and effectiveness
—Multidisciplinary
—Process plans (work flow, automation)
—Workstation ergonomics
—Storage levels
—Labor content
—Machine/equipment selection
—Plant layout
—Material handling
b. Maintenance control
—Preventive maintenance program (rotating schedule of preventive maintenance)
c. Tool design and fabrication
—Outsourcing controls
—Tool fabrication facilities
—Tool repair capabilities
—Perishable tool tracking
—Tool inventory systems
d. Documented process instructions
—Key operation elements posted at each workstation:
Part number(s)
Operation sequence
Specifications
Process parameters
Tool requirements
Gages
Material identification and disposition
Engineering change level
Revision date
Inspection instructions
e. Productivity and operating efficiency improvements
—Identification of unscheduled machine downtime
—Documentation of machine set-up times
—Improvement programs
f. Facilities adequacy
—Suitable processes and equipment for products and volumes being run
—Appropriate automation schemes
—Programmable controls
—Equipment obsolescence

C. *Quality controls assessment*
1. Quality planning and procedural control
a. Quality systems and procedural control
—Organizational commitment to quality
—Plans for reduction of variation
—Improvement objectives based on cost of nonconformance
—Training
b. Quality function procedures
2. Process control
a. Utilization of control characteristics
—Procedures for determination of control characteristics involving customer and supplier
—Documentation of statistical control for all control characteristics
—Documented corrective action for out-of-control processes
b. Systems to assure production meets physical, visual, functional, chemical, and dimensional requirements
c. Gage capability and calibration
—Procedures for conducting gage studies and corrective action
—Records for calibration of measurement devices
—Verification of accuracy of measurement devices against recognized and traceable standards
—Procedures for inspection of gages, fixtures, jigs, and templates
d. Controls for identification and traceability of all material
—Systems to track material flow in process
—Procedures for identification, segregation, and disposition of nonconforming material
—Shipping records to assure traceability
e. Drawing and specification control
—Procedures to assure control of drawing/specification releases or changes
—Control of obsolete drawings/specifications
f. Record retention and internal audit
—Does records retention meet government regulations?
—Internal audits conducted on regular basis
3. Problem reporting and resolution systems that assure irreversible corrective action
a. Problem solving
—Procedures for resolution of nonconforming conditions
—Disposition of nonconforming material
—Causal identification
—Corrective action
—Responsiveness to customer complaints
4. Control of sourcing base to assure purchased material meets all requirements
a. Assessing, selecting, and developing supplier/subcontractor base
—Procedures to evaluate suppliers
—Conducting supplier evaluation prior to awarding business
—Reviews include evaluation of supplier quality control and SPC programs
—Periodic reevaluation of suppliers
b. Assurance of supplier's capability to assure material meets physical, chemical, visual, functional, and dimensional requirements
—Receipt of statistical data from supplier

—Receiving inspection or test data
—Verification by production process

D. *Cost assessment controls*
1. Control systems that accurately monitor all cost elements
a. Standard cost control
—Establish standard rates for labor, material, and overhead
—Efficiency standards developed by time and motion studies
—Identification of direct and indirect labor/material content
—Overhead rates based, as appropriate, on physical output, direct material cost, direct labor hours, direct labor cost, machine hours, or other
—Yearly review of standards
—Labor routings for each operation
—Generation of monthly summary reports to identify trends
—Systems in place to address variances to limits
—Distribution of overhead costs to most accurately reflect amounts used in production of a part
b. Actual costs control
—Computation of labor efficiency
—Labor overtime accounting
—Bill of materials use and accuracy
—Statistical process control for control of material usage
—Annual physical inventories, cycle count procedures
—Yearly overhead cost studies
c. Comparison of actual cost to standard
—Monthly cost comparison report of actual to standard costs—identification of significant variances
—Plant operating report comparing standard to actual costs
—Levels of comparison:
Part number
Department Commodity
Cost center
Plant
2. Reduction of costs and selling price
a. Continuous cost improvement program
—Salary and hourly groups focused on cost reductions
—Monitoring of systems and communication of results
—Scrap reduction and awareness programs
—Sharing of cost reductions with customers
b. Competitive product development and target cost achievement
—Simultaneous engineering programs
—Financial resources for product development
—Participative cost reduction programs with customers
—Control of cost-related location factors (for example, energy, cost of operations, construction costs)
—Fair negotiation of labor costs
c. Cost data sharing
—Cost data sharing with customer
—Cost data sharing with subcontractors
d. Price structure
—Base price on a combination of current material actual costs and estimated standard costs for labor and overhead
—Multidisciplinary pricing
—Profit factor application
e. Cost estimating system
—Coordinated interdepartmental participation
—Utilization of historical data for cost development
—Utilization of actual cost data when available
—Review of cost estimates to actual on a regular basis

E. *Controls to assure consistent on-time delivery*
1. Plant capacity
a. 75% (or better) of actual machine and equipment capacity
b. Burden application controls—direct machine hours preferred
c. Ability to support 10% increase in volume without expansion
d. Lead time controls for material, tooling, and operators
2. Customer schedules
a. Production control board on quick reaction to customer schedules
b. Weekly reconciliation between customer schedules and supplier records
c. Documentation to support complete customer schedule information flow to material ordering, production scheduling, and shopping functions
3. Formal production scheduling
a. Documentation correlating customer schedules and production schedules
b. Inventory tracking systems
c. Controlled scheduling factors
d. Production schedules 1 week ahead of customer requirements
e. Schedule tracking system
f. Minimize ''hot parts'' expediting
4. Shipping and shipping schedules
a. Assigned responsibility
b. Verifiable communication among production scheduling, manufacturing, and shipping functions to assure material availability
c. Computerized systems, including bar coding
5. Control of inventory
a. Systems to clearly report inventories of raw materials, work in process, and finished goods
b. Annual physical inventory, cycle count methodology
c. First in/first out control of inventory
d. Control of average inventory, inventory time
6. Scheduling communication with sources
a. 12-week material forecast to sources
b. Purchase order and blanket order release control
c. Source performance review system
7. Control of delivery performance
a. Control and monitoring of premium freight charges

b. Tracking systems for delivery performance
8. Provision of appropriate, timely data for all shipments to customers
a. Shipment notification system
—Computerized shipment notification
—External/internal communication paths for timely problem resolution
9. Control of shipping in conformance with customer requirements
a. Packaging controls
—Procedures for meeting customer specifications
—Procedures to assure returnable containers are in good condition and reviewed regularly
—Procedures for timely review and response to shipping problem reports
—Procedures to audit part counts
b. Systems to assure labeling procedures are met based on customer shipping destination requirements
—Bar code reading capability
—Part/label verification
c. Material handling control
—Adequate transfer mechanisms
—Review of damaged parts causes
d. Shipping and receiving control
—Customer specified routing
—Appropriate doors, bays, and/or wells for a company's operation

The next section of this chapter describes some of the more important evolving manufacturing control philosophies and technologies incorporated into the levels of control that have just been reviewed.

OVERVIEW OF CURRENT CONTROL SYSTEMS AND TECHNOLOGIES

The advent of computers, information systems, and automation has changed the physical nature of the manufacturing environment. These technologies have also enabled control mechanisms to be put in place that were not possible before because of the massive data capturing and processing implications. This section of the chapter describes broader based manufacturing control systems that integrate closely with planning concepts to control production scheduling, inventory costs, productivity, quality, and cost improvement. Each of these new technologies incorporates the concepts of the control system model previously reviewed in this chapter.

Four evolving major manufacturing control systems and technological advances are manufacturing resource planning (MRP II), total quality control (TQC), just-in-time (JIT), and computer-integrated manufacturing (CIM). This section will briefly review these techniques/technologies and their management control implications.

MANUFACTURING RESOURCE PLANNING

The tools for manufacturing include manufacturing resource planning (MRP II). The following paragraphs outline what MRP II is, its makeup, and the objectives that MRP II seeks to satisfy.

Manufacturing resource planning is both logic for gaining manufacturing control and computer-based application systems employed to support this logic. The MRP II conceptual model in Fig. 2-4 outlines the sequence of activities and decisions needed to provide a framework for planning, executing, controlling, and replanning production activities. Market demands for products, service, or repairs are organized in quantitative terms that are useful to manufacturing. The demands include customer orders, changes desired in final stock available for sale, distribution inventories, and forecasts of customer demand.

When properly organized, market demand portrays a production facility scheduled over relatively long periods of time, such as 6-18 months into the future. This is accomplished in production planning. The model assumes this decision is an approximate result representing management's business objectives that need to be satisfied by manufacturing.

The production plan is examined in terms of aggregate capacities of the producing facility by reviewing overloads and underloads at major producing workcenters. The master planner (a person) iterates the placement of production requirements over time until a manageable and acceptable result is produced.

Master production scheduling is the decision process for placing the production of saleable products or models with options, for example, with respect to time. The critical time is the manufacturing lead time. The master production schedule (MPS) is usually regenerated for each manufacturing lead time within the horizon time of the production plan. Here the computer systems help examine the relative stability of production activities over the MPS lead time. The "smoothing," looking through manufacturing lead time, is provided by the master scheduler.

The heart of the computer-based system is material requirements planning (MRP). Bills of materials or parts lists for products/models are "exploded" into requirements for purchases, fabrications, and so on. Their quantities and time of availability are forecasts of need calculated to support the execution of the MPS. Also, the requirements for labor and/or machine time at important production centers are calculated.

Next, the capacities of individual producing centers are analyzed to determine if large overloads are projected in the short run for the MPS-generated load. The master scheduler then simulates production activity until a reasonable level of work is available at planned workcenters on a day-to-day basis.

A series of control feedback activities ensures that work is released to producing centers in concert with the MPS and MRP. Shop activity is relayed to the master scheduler in terms of whether or not production is ahead of or behind the master production schedule (variance control). Routine adjustments to MRP and released work (jobs, runs, rates per product) are made to recognize real production achievement, thus completing the control system process.

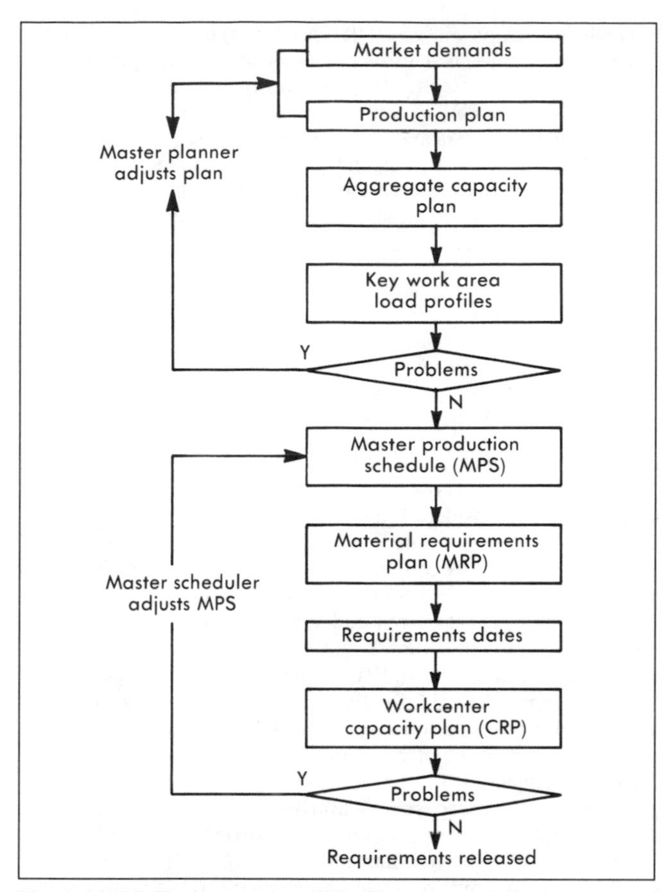

Fig. 2-4 MRP II conceptual model. ***(Coopers & Lybrand)***

The production planner forwards scheduled plant activity to meet manufacturing's policies for smoothing overloads/underloads in some sense. Master production scheduling disaggregates these results through the master scheduler.

Beginning with MRP, the computer system performs a series of necessary calculations. The master schedule is converted into a forecast of components, fabrications, and/or subassemblies using a standard product structure of ingredients that are related to MPS items.

The quantities of purchased parts and manufactured parts/subassemblies are determined with respect to time. Quantity rules for satisfying requirements include lot/run for run, replenishment, or lot-sizing algorithms.

Commitments for purchases, manufactures, and their timing are determined using standard lead times to acquire (produce) components. This is a back-scheduling technique. Back-scheduling requires availability of purchased parts and the result of fabrication to match the dates determined by the component standard lead times.

The preceding discussion provides the human scheduler with feedback concerning the practicality of executing the master production schedule.

Content of an MRP II System

MRP II computer systems provide the following information:

- Customer demand activity.
- Production plans.
- Production schedules and their execution.
- Purchasing management.
- Inventory management.
- Product cost reporting.
- Support of and financial applications of accounts receivable, accounts payable, general ledger, and payroll.

Manufacturing resource planning implies an integration of information flows and the use of modern database management techniques to improve communication among the various business functions.

The following seven areas describe the major application modules/applications in vendor-supplied MRP II application packages:

1. Engineering and product data definition. This application provides the informational base used by all functional areas, including marketing, production, and finance. Typical information includes:
   - Product structure, bill of materials.
   - Inventory data.
   - Part costs and prices.
   - Operations routings and bills of labor.
   - Operations, workcenter data.
2. Customer demand management. These applications include customer order management, inventory record keeping, and inventory management rules. Demand forecasting is not available in most vendor packages.
3. Production planning and master production scheduling. These modules employ MRP techniques usually based on product families to approximate where demands will be satisfied in the future.
4. Material planning and capacity requirements plans. Master schedule demands and other demands for components, such as their bills of materials, operations routings, and facilities data, are "exploded" through the MRP application. Their impact on near-term throughput at planned workcenters is reviewed in the capacity requirements plan (CRP) module. Production work is released to the floor based on the result of these activities.
5. Purchasing and receiving. The material plan produces requirements for purchasing. The lifecycle of purchasing activity is maintained in the purchasing application and the receiving application.
6. Shop activity reporting. Frequent monitoring of production flow and workcenter status is reported against production orders to enable the scheduler to adjust the release of work and expedite work to meet the MPS commitment. Data gathering typically includes employee labor, production and scrap, and work order location.
7. Cost and financial applications. These applications cover management accounting and general accounting needs. Product costing reports actual and standard costs by production run, as well as variances and inventory valuation. Relevant information for purchase, cost of goods, payroll, and sales accounting are also maintained. These applications ensure that the company's books are accurately maintained.

Successful employment of MRP II planning and control techniques for manufacturing control has produced improvements in several areas of manufacturing (see Table 2-3). Testimonies also include a variety of intangible benefits arising from

TABLE 2-3
MRP II Results

Benefits Area	Results
Reduced manufacturing expense	3-15% of expenses
Inventory reduction	15-40% of carrying value
Improved customer services	25-50% of delivery standard

(*Coopers & Lybrand*)

better management understanding of production activities made visible from timely information sharing.

JUST-IN-TIME

The just-in-time (JIT) manufacturing control philosophy has evolved from Japanese manufacturing techniques over the last 30 years. Manufacturers in the United States have only begun to understand the dramatic productivity and quality results available through the use of JIT techniques. In essence, Japanese manufacturers have rejected the Western obsession with complex management programs and controls, computers and information processing, and with mathematical modeling. The Japanese way is to simplify the problem. Japanese systems consist of simple procedures and techniques that do not require a particular cultural environment for implementation.

As pointed out by Schonberger in *Japanese Manufacturing Techniques*, the Japanese control system consists of two types of procedures and techniques.[9] The two types pertain to productivity and to quality. The aspect of the Japanese system pertaining to productivity is known as the just-in-time system. Japanese quality is partially addressed by just-in-time, but there are a host of other Japanese quality improvement concepts and procedures. Total quality control (TQC) covers this set of procedures and techniques. TQC encompasses some of the just-in-time techniques and improves productivity by avoiding waste.

In Japan, the workers and line managers are the focal points of implementing just-in-time procedures and techniques. There is much less emphasis on staff specialists than in the United States. While there is a growing awareness of the just-in-time philosophy, there has only been small progress made in implementing JIT in the United States. This will continue to be true as long as upper managers and their consultants and advisers remain uninformed about the power and payoffs associated with JIT.

Just-in-time is very much a misunderstood philosophy in the United States. There are many erroneous perceptions of what it is. JIT is *not*:

- An inventory program.
- An effort that involves suppliers only.
- A cultural phenomenon.
- A materials project.
- A program that displaces MRP.
- A panacea for poor management.

Rather, JIT is an enterprise-wide operating control philosophy that has as its basic objective the elimination of waste. Under JIT, waste is considered anything other than the minimum amount of equipment, materials, parts, space, and workers' time that is *absolutely essential* to add value to a product. The key operational phrase thus becomes ''adding value.'' JIT strives to identify activities that do not add value and eliminate them. JIT can be used by any manufacturer or enterprise interested in eliminating waste and simplifying the workload.

Those exposed to JIT for the first time often think that they have long followed these tenets. ''So what's so different?'' they ask. Simply put, it is not what one does as much as the way one does it. Companies in the U.S. that have implemented JIT properly have realized spectacular results as indicated by the examples from a cross-section of industries (see Table 2-4). William A. Wheeler III has identified 10 distinct steps that should be considered in any comprehensive JIT manufacturing program (see Fig. 2-5).[10] These steps identify how and where JIT can be used to control manufacturing productivity and lower costs. A detailed discussion of these 10 steps can be found in Chapter 15, ''Just-in-Time Manufacturing,'' of this volume. A summarization of these steps is as follows:

TABLE 2-4
Estimate Percent Improvements for Different Industries as a Result of JIT Implementation

Reductions	Automotive Supplier	Printer	Fashion Goods	Mechanical Equipment	Electric Components	Range
Manufacturing lead time	89	86	92	83	85	83-92
Inventory						
Raw	35	70	70	73	50	35-73
WIP	89	82	85	70	85	70-89
Finished goods	61	71	70	0	100	0-100
Changeover time	75	75	91	75	94	75-94
Labor						
Direct	19	50		5	0	0-50
Indirect	60	50	29	21	38	21-60
Exempt	?	?	22	?	?	?-22
Space	53	N/A	39	?	80(Est.)	39-80
Cost of quality	50	63	61	33	26	26-63
Purchased material (Net)	?	7	11	6	N/A	6-11
Additional capacity	N/A	36	42	N/A	0	0-42

(*Coopers & Lybrand*)

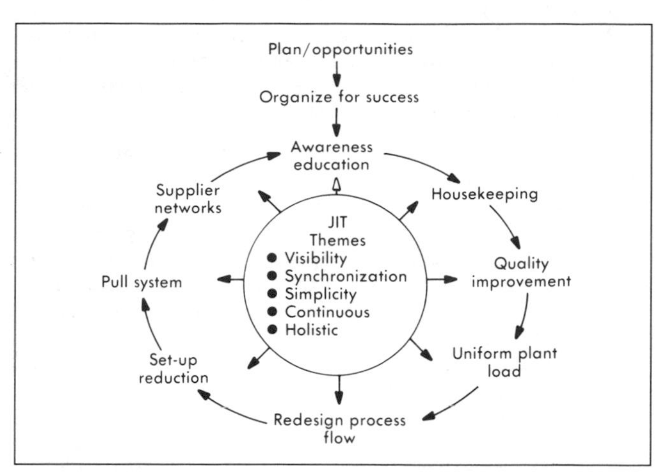

Fig. 2-5 Proposed steps in a JIT manufacturing program. (*Coopers & Lybrand*)

1. Plan/opportunities. It is very important that companies invest the appropriate time in learning JIT control principles. It is possible to invest significant time in implementation only to end up with "islands of JIT" without significant overall improvement.

   JIT shakes the very foundation of manufacturing as it is currently practiced and challenges most long-standing operating control principles. Such a profound change does not lend itself to a cherry-pick approach that is championed by an individual functional manager or supervisor. Significant change management requires a carefully conceived plan that provides a clear strategic advantage in the marketplace. Such a strategy should become manufacturing's contribution to the business plan.

   A suggested strategy would include diagnostic reviews, conceptual design, implementation planning, implementation preparation, continuous implementation, and program control monitoring.
2. Organize for success. A productivity control organization should be established to identify and implement operational improvements. This organization typically consists of the following four levels:
   - Chartered organization. Each and every person in the organization should understand where the company is going and what his or her specific contribution is to the improvement efforts.
   - Implementation steering committee. This group is charged with the execution of the JIT plan.
   - Task groups. Numerous task groups are established. These are multidisciplinary and are trained in specific JIT techniques such as variation research, set-up reduction, and procurement strategies.
   - Improvement vehicle. After a recommendation from a task group has been implemented, there is a need to continuously refine the improvement.
3. Awareness/education. A significant need exists for *all* personnel to gain both awareness of JIT technologies and "dirty education." The latter involves applying techniques introduced during the awareness portion of this activity. Everyone should be involved in trying (and usually failing at first) to control and improve operations. In the long run, it becomes the only means to ensure continuous improvement.
4. Housekeeping. Housekeeping in the JIT sense means more than just a clean workplace. It is an effort to establish an attitude that each person is responsible for his or her equipment, ensuring that necessary tools are in the right place, and that everything is clearly visible for all to see and have access to. It is not unusual to realize a 10% productivity improvement by just applying the simple principles of housekeeping.
5. Quality improvement. This activity should commence early in the JIT journey. It takes more time to realize "zero defects" than any other element of JIT. There are some differences between JIT quality programs and standard ones:
   - In JIT, the emphasis is on continuously controlling and reducing variance for improved manufacturability.
   - The operator, not the inspector, becomes responsible for "zero defects."
   - Whenever an error or problem is discovered, the operator should be allowed to stop the process and take immediate corrective action. This is an integral part of the control process.
   - Prioritization for quality problem solving leans toward the sporadic rather than the chronic defects.
6. Uniform plant load. A significant contribution from the Far East practitioners of JIT was the concept of uniform plant load (UPL). The idea is simple: If one sells daily, then make it daily. This means that every model within the product group is manufactured in small lots on a daily basis. Whenever possible, the end items are manufactured to demand and not to stock. To foster a make-to-demand environment, most companies have to dramatically reduce the manufacturing lead time.

   UPL is not the master schedule or the final assembly schedule; it is the cycle time required to meet but not exceed demand. It is not how fast a machine or process can operate; it is a production rate for *all* components and assemblies that is synchronous with the demand rate.

   The primary benefit of UPL is to eliminate indirect labor costs for managing or transporting excess inventory *and* to allow direct labor to operate multiple machines because its machines have typically been slowed down to the UPL rate.
7. Redesign process flow. To achieve rapid throughput and the productivity opportunities afforded by UPL, the process flow or functional layout usually requires rethinking. The objective is to eliminate any operations that don't add value and then group dissimilar but dedicated equipment together in a cell configuration.
8. Set-up reduction. By dedicating equipment to product groups, some set-ups or changeovers are completely eliminated. Where multiple components must be shared by the same manufacturing resource, then it becomes necessary to significantly reduce the changeover time to economically make just enough for each day's demand. There are few set-ups that cannot be reduced by 75% (reductions of 90% or more are not unusual).
9. Pull system. Once the UPL has been established and the cells put in place, it is usually time to establish the pull system. Occasionally, it is advisable to select a few

critical part numbers and institute the pull system prior to cell completion. However, the full benefits of negative feedback (self-regulating mechanisms) will not be realized until the UPL and cells are in place.

There are two types of pull systems in JIT as follows:

- Overlapped. When continuous flow has been established on a line or cell, then the empty space previously occupied by a part is the signal to make one more part.
- Linked. When parts compete for the same resource and cannot be made one at a time, or when they have to travel significant distances in a lot mode, then it is advisable to use a pull card or signal (kanban) to authorize the manufacture of the components. When a higher level workcell requires more parts, operators go to get the replacement container. When picking up the parts from the point of manufacture, a card or signal is left behind. This card becomes the authorization to make one more fixed-quantity container's worth of parts. Priority control is the sequence in which the authorizations were received. Violations of the sequence cause the entire system to collapse. Furthermore, if there are no authorizations to make, nothing is made.

The pull system causes profound changes to manufacturing resource planning (MRP II), but does not negate its overall usefulness. MRP II is still required for long-range parts and capacity planning. The pull system manages the short-term parts requirements and execution.

10. Supplier networks. The last activity in the JIT cycle is to get the supply continuum involved in delivering only when needed. Quality improvement commences at the beginning of the JIT program, but the pull system is not implemented until the flow and demand have been sufficiently smoothed out. The key criterion for specifying JIT deliveries from the supplier is a smooth demand at a commodity level, which is where co-op contracts (JIT supplier agreements) should be let (rather than at the part-number level).

The 10 steps just described deal with how and where to use just-in-time control philosophies in the manufacturing environment. A pre-production phase that can also apply JIT control techniques with dramatic benefits is design engineering. Applications of JIT in design engineering will significantly contribute to the success of JIT in the manufacturing environment.

The JIT control objectives of simplification and elimination of non-value-added activities can be effectively applied to the following aspects of design engineering:

- Control of inventory buildup in the manufacturing process can be adversely affected by poor product design. Designing products with the minimum number of parts and the fewest possible options can greatly facilitate the use of JIT in manufacturing.
- Designing tools, dies, and molds to facilitate rapid set-up will greatly enhance the capability of achieving a uniform plant load.
- Design engineering should also be aware of plant layouts and cellular design in manufacturing. Products should be designed to facilitate effectiveness through the plant.
- Elimination of bill of materials levels will simplify the manufacturing planning and build process.
- Modular building of features as plug-ins to the product can often simplify the build process and reduce costs.

A great many other opportunities will probably be identified to integrate design engineering and manufacturing just-in-time techniques. Design engineers need to become familiar with the JIT concept of manufacturing. It is important for design engineering to become a part of the JIT team. Additional information on the interrelationship between design and manufacturing can be found in Chapter 13, "Design for Manufacture," of this Handbook volume.

In summary, it can be seen that implementing JIT control techniques is not just an inventory program or only for supplies. Properly conceived, it can be a strategic tool for generating increased market participation. The advantage of making high-quality items to real demand and at a reduced cost of 10-30% becomes a powerful tool for global expansion of the market.

TOTAL QUALITY

The evolution of quality management from what was essentially a policing activity to an element of business strategy invites a rethinking of the purpose and benefits of quality systems. Quality was formerly regarded only as a control function used to identify defects in production or service. Now, the powerful effects of quality improvement on productivity, cost reduction, and customer satisfaction have led to a broader and more positive role for quality.[11] In this new environment, quality systems can be regarded in a more comprehensive control perspective as tools to identify nonproductive costs and guide efforts to eliminate these costs.

The interrelationship between productivity, cost, and quality can be understood by considering the extremes of very bad and very good quality.[12,13] Zero quality results in zero productivity because no saleable goods are produced. Perfect quality means all goods are saleable with no cost for inspection, repair, downtime, scrap, or returns. Moreover, there is no expense for quality-related design changes, retooling, renegotiated vendor prices, or lost sales due to unreliable products.

A broadly accepted operational control definition of quality is conformance to requirements.[14] This concise definition is comprehensive and very useful if "requirements" are adequately understood. Requirements must not be interpreted narrowly as engineering specifications. There are requirements for end products as well as requirements for each process and component used to achieve the end result. End product requirements are the most important, and they are established by the customer, not by engineers or managers. Process and component requirements are determined by the designs, equipment, and people involved in the full cycle of product definition, production, and delivery. Nonconformance to any of these requirements adds cost without adding value.

An integrated quality program measures conformance to requirements in each phase of the product cycle. Thus, quality systems address the costs of nonconformance not only on the production floor, but also in related activities both upstream and downstream.

Quality systems serve the following four broad purposes:

- Support product/process definition and design.
- Support production.
- Measure customer satisfaction.
- Provide information for management reports.

Earlier in this chapter, it was noted that the five levels of manufacturing control (management, technology, quality, cost, and delivery) are interdisciplinary and must be integrated to achieve optimum results. In the case of quality systems, this integration is evident from the purposes they serve and from examination of the various types of systems in use. Table 2-5 provides examples of typical quality systems. Although this list is not comprehensive, it gives an indication of the range of systems that have been found effective. A few are umbrella systems that include several of the others, such as quality function deployment and statistical process control. Implementation of many of the systems is often done manually, although software is available or is in development to support all of them. The list in Table 2-5 is arranged alphabetically and does not suggest any priority in importance or sequence of implementation for the systems.

The choice of a quality system depends on the quality awareness of the organization. The following three levels of awareness can be defined, each having a different quality objective:

1. Zero customer complaints. The focus in this level is on identifying and correcting points of customer dissatisfaction. Quality systems are used to detect and sort out nonconforming goods or services. These systems rely heavily on inspection. They are generally the least effective in achieving customer satisfaction. They are also the most costly methods; quality costs for this approach often exceed 25% of revenue.
2. Zero defects. As quality awareness increases, the objective shifts to identifying the root causes for nonconformance. Systems are intended to measure and help eliminate factors that drive processes out of statistical control. With these systems in place, costs for inspection and failure decline rapidly, and total quality costs may be in the range of 10-15% of sales.
3. Zero variability. When processes have been brought into statistical control, quality systems concentrate on identifying the causes of process variation. These systems support continuous efforts to make products and processes more resistant to forces that cause variation. Costs for quality planning increase, but are more than offset by declines in appraisal and failure costs. Total quality costs are often under 5% of revenue.

Common characteristics of quality systems that determine their effectiveness include the following:

- User friendliness. Tools and procedures for system use, including data collection and retrieval, are most effective when they are simple and fail-safe.
- Automated data analysis. Effectiveness decreases as data analysis is removed from the process in time or location. Methods that work best feature automatic data processing or forms for manual data entry that clearly reveal trends as information is being recorded.
- Results routed to users. If it is not possible to collect and analyze data at the actual process, results should be routed quickly and directly to the operator that controls the measured characteristic. The information may also be routed to management, the quality department, or a history file; but the operator is the most important user.
- Pictorial presentation. To avoid the pitfall of the "data graveyard," information should be presented in easily understood charts or graphs.[23] Examples are Pareto bar charts, scatter diagrams, trend charts, bell curves, and the quality function deployment "house of quality."
- User adaptability. Flexibility to assess new variables as they are identified by operators or analysts greatly increases usefulness. Formats used for stored data should provide extensive opportunity for sorting to examine dependent relationships that were not previously imagined.
- Use of numbers. GO/NOT-GO "measurements" and similar attribute control systems generally do not support efforts to understand process variability and usually defeat opportunities for operators to contribute to process improvement. Measurements should be made and presented numerically whenever possible. Examples are: thickness is 0.52 mm, receivables are 18.5 days past due, and packing errors occur in 3.5% of shipments.

If no tools exist to measure an important characteristic, a grading scale should be developed. For example, to evaluate a proposed new color, a satisfaction scale of 1 to 5 can be used in customer surveys to rate color density, shade, and richness.

In summary, total quality is being recognized in today's manufacturing environment as a philosophy and a set of control systems that have a dramatic affect on cost reduction and profits. At the heart of the quality approach is the use of control mechanisms to detect and correct variances or discrepancies in the total manufacturing enterprise.

COMPUTER-INTEGRATED MANUFACTURING

Computer-integrated manufacturing (CIM) frequently means very different things to different people. Chapter 16 of this Handbook also deals with CIM and adds more detail to the thoughts expressed in this section.

To define CIM requires an examination of generic manufacturing from an historical perspective. All manufacturing processes include three activities: (1) design of products and processes, (2) planning for production, and (3) producing products. Activities within each area require status and control information.

With the advent of the computer, each of the manufacturing functions individually began to use digital techniques for automation. The first use of digital technology in production was the introduction of numerically controlled machine tools in the early 1950s. The use of computers in production has been extended into robotics, programmable logic controllers, automated guided vehicle systems, and automated assembly machines.

Similarly, the computational processes used in the design function were extraordinarily enhanced by the use of computer technology. Computers replaced calculators and slide rules for a variety of engineering analysis and design simulation applications, commonly called computer-aided engineering (CAE). But even more significant to the design function was the development of the graphics technology that today is the heart of computer-aided design (CAD) systems. CAD has been advanced and enhanced to incorporate group technology coding and classification and computer-aided process planning (CAPP).

Finally, production planning and control became computerized through the development of large mainframe computers and material requirements planning (MRP) software. At the same time that MRP has been extended to manufacturing resource planning (MRP II), project management software for nonrepetitive manufacturing has been developed.

TABLE 2-5
General Descriptions of Typical Quality Systems
(Listed Alphabetically)

System	Principle	Application
Checklist	A simple yet powerful tool that aids problem solving or quality improvement. Checklists provide all of the critical characteristics of a process in a convenient format. They are developed by cross-functional groups of experts, preferably including operators and customers. The format should provide a smooth physical flow of activity when used for review or inspection	Useful at any point in the product cycle to guide a systematic evaluation; prompts the analyst to review all characteristics deemed important
Competitive ranking	Performance levels for features (requirements) most relevant to customers are rated by testing, surveys, and independent evaluations and are compared with ratings for competitors. In addition to rankings for each feature, the individual values are often weighted so that an overall comparative rank for the product can be determined	Used for design of new products or features and to establish the need for accelerated improvement of noncompetitive features
Competitor performance audits	An exhaustive evaluation of competitive products. Audits generally include comparative testing of products for durability, reliability, safety, serviceability, and performance	Through unbiased audits of competing products, it is possible to formulate realistic product design and marketing decisions
Cost of quality	A system used to capture the costs associated with nonconformance to requirements as well as the cost of efforts made to prevent nonconformance[15]	Generally used to justify and guide improvement efforts. Valuable because it expresses the impact of poor quality in terms understood at all organizational levels
Customer requirements survey	A proactive and valuable aspect of quality planning. Customer requirements are determined through market evaluations and comparisons with competition and translated into a variety of planning and employment matrixes for use in design and production Examples of this system include mail and phone surveys to new-product customers and long-term owners/users; focus groups featuring product presentation; and personal interviews with prospective buyers	Begins with new product definition and continues through development, sales, and service to determine desirable product features
Customer satisfaction survey	A method of gathering data on a product's performance after it has been designed, produced, sold, and used.[16] Often conducted with mail or telephone questionnaires, although more reliable data are gained through interviews in the field	May be needed to establish baseline performance and to correlate customer perceptions with upstream quality measurements Because it takes place late in the product cycle, this is a costly approach to detection of nonconformance. In addition, it is often difficult to obtain reliable and useful information from customers who generally describe symptoms rather than technical issues
Design review	An audit of product design that tests the actual form of a design against its intent. Considerations such as conformance to specification, manufacturability (predicted first-run capability), serviceability, materials costs, and reliability are evaluated. Conducted by a group including experts from all relevant departments	Used at critical stages in the design/development process. The need for design review diminishes if a team-based design technique such as quality function deployment is used

TABLE 2-5—*Continued*

System	Principle	Application
First-run capability (production yield or direct runner performance)	A process of measuring the percentage of first quality units that transverse the full production process without being repaired, reworked, or replaced (scrapped). An excellent measure of productivity	Measures effectiveness of production and design for manufacture. Careful exception reporting of all units that do receive repair, rework, or replacement, noting the production location and cause, permits effective quality improvement
Documentation (configuration) control	A system to ensure that the latest design level is in use in all areas, and documents reflecting obsolete designs are removed[17]	Must be an active system that simplifies control and verification. All parts or process changes require revisions of their document numbers and change dates that are automatically broadcast to all users; clear, accessible labeling of drawings, parts, and packaging is needed
Durability analysis	A proactive method of determining product life and useability. This method of quality testing is utilized to enhance overall customer satisfaction and to prevent the cost of warranty work	Normally performed on materials and components early in the design stage and on preproduction runs of the final design. Durability testing of final products often uncovers problems caused by interaction of components Also used on random samples of ongoing production to uncover changes in conformance
Failure mode and effects analysis (FMEA)	An expository technique used to analyze how key components of a product might fail.[18] Through an evaluation of the likelihood and effect of product failures, a cross-functional team may determine where design improvements are required to enhance product performance, reliability, and safety	Used early in the design stage and updated as product and process are more completely developed
Preventive maintenance (PM)	Systems that schedule preventive maintenance as part of the normal production process, and track conformance to the schedule, tend to significantly reduce costs of unscheduled downtime, process variation, and early equipment wearout	Useful to optimize productivity of machines, tools, and other equipment used for production, inspection, testing, and service
Process capability study	A system that assesses the ability of a process to consistently reproduce the intended product or service, expressed in statistical terms. The process includes the design, people, equipment, material, and environment involved in producing the product or service Measurements are made on a large continuous sample to establish consistency over time, without influencing the normal process	Results are compared to requirements (normally specifications) to determine if the process is capable of consistently delivering the required results
Quality audit	Analysis of a random sample from a product or process to evaluate conformance to requirements. Replaces mass inspection for process in statistical control	A cost-effective means of establishing performance levels at any stage in the product cycle. Most frequently used for assessing conformance to procedure or specifications. The "corporate quality audit" measures conformance of practices to established quality procedures in all departments; for example, hiring, training, customer service, machine maintenance, documentation control, and SPC feedback and corrective action[18]
Quality function deployment (QFD)	A formalized planning tool used to ensure that conformance to customer requirements is the basis for defining product design and production processes.[19] A highly visible, company-	Used for product and process design. Highlights incompatible factors that lead to delays, increased cost, and lost productivity. Requires cross-functional groups to systematically relate

(continued)

TABLE 2-5—*Continued*

System	Principle	Application
	wide system that translates both general and specific customer requirements into the appropriate technical information required at each stage of product definition, design, engineering, prototyping, production, inspection, and sales	customer needs to product specifications in a matrix format described as the "house of quality." Progressively more detailed specifications are developed without losing sight of original customer expectations Software is in development to support QFD; however, substantial manual, interactive work of interdepartmental groups is an essential aspect of this system
Quality improvement status	A system that provides structure and control to an improvement process by regularly updating a rank-ordered summary list of projects having well-defined targets, methods of measurement, and accountability	A management reporting system that clearly defines issues, responsibilities, goals, and progress, or obstacles to improvement
Quality loss function	An approach that significantly improves quality understanding by measuring losses to customers resulting from poor quality.[20,21] Traditional methods measure the producer-oriented losses of rework, scrap, and returns caused by nonconformance. The quality loss function calculates the loss to society over the entire product cycle, including the intended service life of the product Nonconformance is regarded as variation from the specified nominal value for any characteristic, irrespective tolerance limits. As variation is reduced, quality losses decline, and value to the customer and the producer increase	Used in product design and to promote continuous process improvement. Helpful in choosing between alternative concepts; a powerful tool in guiding the selection of design parameters (nominal values) and allowances (tolerances) to provide the lowest overall cost because they are robust to inevitable variations in materials, operators, processes, and use
Repair and defect tallies	Simple score cards that help to allocate problems to their cause and demonstrate results of improvement actions	Useful at any point in the product cycle to quantify problems and prevent subjective judgments about cause
Statistical process control (SPC)	SPC looks at differences in items intended to be the same: how they differ, how much, and how often. This allows timely adjustment or control of processes to prevent nonconformance[12] Many SPC systems stop at this point. The greatest benefits of SPC data, however, come from use of the information to understand and reduce process variation Most SPC applications are easily learned and used, with very user-friendly tools and software.[22] Systems requiring special mathematical knowledge represent less than 20% of applications	Powerful tools to promote improvement in all phases of the product cycle. A few examples of SPC measurement systems include: Process capability Repair tallies Production process control Production yield Warranty analysis Accounts receivable performance Design quality analysis Purchasing performance Inventory control analysiscd
Warranty analysis	A retroactive method of reviewing product quality and reliability. This process occurs once warranty work has been performed and subsequently reported	Effective warranty analysis systems provide sufficient information to allow determination of the production lot involved (production date and location, vendor detail, and so on); type of nonconformance; and use-specific facts such as climate, type of customer, and time in service Rapid reporting and extensive data sorting capability are important. It is generally very helpful to have all replaced parts returned for analysis. Because it takes place late in the product cycle, this is a costly approach to detection of nonconformance

(*Coopers & Lybrand*)

SYSTEMS AND TECHNOLOGIES OVERVIEW

As the development of computerized applications continued for each of the individual manufacturing functions, it became apparent that, for the first time in the history of manufacturing, the same technology was being used for automation of the heretofore separate manufacturing functions. Computer technology and its media for storing, transmitting, and processing data became central to the automation of the design, planning, and production processes. Because of this common basis of automation, direct communications among manufacturing functions became technically feasible. This opportunity for automated communication of information within manufacturing is the basis for computer-integrated manufacturing, or CIM.

Definition

To define CIM, it is necessary to define manufacturing and integration. Manufacturing is the manufacturing enterprise that encompasses all activities necessary to transform purchased materials into product, to deliver product to customer, and to support it in the field. By this definition, manufacturing starts with a product concept, which may exist in the marketing organization, and includes engineering where the product is designed, incorporates manufacturing planning and production, and extends through product delivery and after-sales service, which may reside in the distribution and service organizations.

Integration is the ability to provide information for each activity on a timely basis, accurately, in the format required, and without asking. Data may come directly from the source or from an intermediate database. This definition implies that the information needs for each activity must be identified and planned for in advance. This concept of integration does not necessarily include computer technology.

However, computers are the technology of choice for automation in the manufacturing enterprise and are also capable of fulfilling integration needs. Viewing the computer as the means, and using the previously stated definitions of manufacturing and integration, CIM can be defined as the use of computer technology to integrate the manufacturing enterprise and all of its planning and control systems.

Benefits and Applications

CIM has direct competitive advantages for any enterprise. The National Research Council estimates that CIM improves production productivity by 40-70%, engineering productivity by a factor of 3 to 35, and quality by a factor between 2 and 5. At the same time, CIM decreases engineering design costs by 15-30%, overall lead time by 30-60%, and work-in-process inventory by 30-60%.

These benefits of CIM are realized through automation and integration. Computerized automation is used to improve production processes and the performance of support activities. For example, in semiconductor manufacture, the use of microprocessor-based controllers on equipment such as diffusion furnaces eliminates the need for a mechanical or manual trigger to initiate each operation. The microprocessor's associated memory allows multiple sets of instructions to be entered once and recalled for subsequent executions.

Computers are also used to enhance performance and eliminate tedious, manual tasks in support activities such as planning and design. For example, in semiconductor device design, modeling programs simulate device performance allowing substantial debugging before pilot production. Computerized tools used for integrated circuit layout provide automatic checking and eliminate tedious, manual labor.

There are also benefits to be had from integration alone. By definition, integration ensures the availability of information when needed and the integrity of that information. When computers are used to integrate automated activities, the information required from one activity is available to other activities in a timely manner, and shared data is available when and where needed. Tighter links between activities are established, and immediate feedback is available, enabling the use of techniques such as adaptive control to improve the manufacturing process. In other words, process steps can be adjusted to compensate for deviations in upstream process steps without loss of product or time.

Using computers to integrate automated activities also dramatically decreases or eliminates time delays caused by waiting for information. It also eliminates non-value-added data chasing and errors due to inaccurate information. For example, integrating the microprocessor-based controllers for diffusion furnaces allows recipes from a central storage site to be downloaded to many different units. If the process is altered, the recipe needs only to be changed in one place.

These benefits translate directly to the bottom line. Quality and yield are increased as mechanisms such as adaptive control are put in place. Lead times are reduced as the time-consuming steps involved in transferring information manually are eliminated. Direct costs are reduced with automation, and indirect and overhead costs are reduced with integration. The product development cycle is shortened as fewer engineering changes are necessary, and past designs are retrievable for leveraging the design effort. Manufacturing lead times are not compromised by unnecessary waiting for parts or information.

All of this leads to a vision of the manufacturing environment of the future; this vision of CIM is a practical one.

First, manufacturing operations are rationalized through application of JIT techniques so that only activities and information flows that add value occur. Redundant and non-value-added activities are eliminated, and related activities are combined to increase efficiency.

Second, routine activities are automated using computer technologies. Third, creative activities are computer assisted. The computer does the routine part; people, the creative part.

Finally, activities are controlled and integrated—meaning that accurate information is available as needed, to people or machines. Information transfer is efficient, and the need to reconcile multiple copies of data residing in different places has been eliminated.

The result is a manufacturing environment with only value-added activities being performed. Computer technologies are used to automate routine activities and data flows and to assist creative activities. People now spend time adding value, not searching for information, expediting information, or checking each other's work. This manufacturing environment has many competitive advantages, such as lower costs, higher quality, faster throughput, and shorter product development cycles.

ISSUES FACING MANUFACTURING ENTERPRISES AND EVOLVING TECHNOLOGIES

The application of the new control practices associated with global competition, combined with the advancing computer technology discussed in this chapter, will permit greater management flexibility. Any manufacturer who fails to achieve this flexibility will inevitably find itself in a position where it is unable to compete. In the past, manufacturers have competed on economies of scale—spreading fixed factory and divisional costs over a large volume of products. In the future, manufacturers will compete on their ability to produce a wide variety of products on the same manufacturing equipment. Rapid obsolescence of products will be a fact of life. Many products will have useful lives of only a few years, some 1 year or less. Manufacturing control and productivity concepts will have to be integrated with marketing, engineering, and design when competing in the low-cost mass production of standardized products.

In early 1987, the accounting and consulting firm of Coopers & Lybrand engaged Louis Harris Associates to conduct a telephone survey to find out the American manufacturing industry's view on several key domestic and international issues facing the manufacturing industry: the competitive position of U.S. manufacturers, the future of the American economy, problems confronting U.S. manufacturers, integration of new technologies discussed in this chapter, obstacles to technology integration, and emerging concerns with attracting, training, and keeping skilled workers.

The sample for the survey consisted of two distinct groups, manufacturing executives and knowledge workers. The people conducting the survey talked to 301 senior manufacturing executives at organizations listed among the Fortune 500 industrials. The sample of knowledge workers consisted of 351 individuals randomly drawn from a list of attendees at recent professional meetings sponsored by the Society of Manufacturing Engineers. Knowledge workers were defined as full-time employees involved in manufacturing and design engineering at a for-profit company whose world headquarters is located in the U.S. and employs 500 or more people worldwide. All respondents in this group held at least a 4-year degree in engineering and spent 25% or more of their job time on the operating floor of a manufacturing facility or were primarily involved in product or process design.

Survey Observations

While different manufacturing segments may vary, overall the study revealed the following:

- Establishing and maintaining a competitive position in global markets is seen as a "very serious" problem by significant percentages of manufacturing executives and knowledge workers.
- Manufacturing executives perceive other American manufacturers to be their greatest competition in the world marketplace. While this is expected to continue in the near future, executives acknowledge a growing and longer term threat from Japan and Pacific Basin countries.
- America currently enjoys a competitive edge in several key industries, according to manufacturing executives, and could regain this edge in other industries with the "right investment, technology, and management."
- Although the need to implement technology is recognized throughout the industry, and is acknowledged to be critical in being and staying competitive, U.S. manufacturers' use of advanced systems is still relatively modest.
- While large increases in implementation are predicted over the next 5 years, approximately one half foresee only limited application of technology such as MRP II, Total Quality, and CIM.
- Manufacturing executives view lack of skilled personnel and expense as the major obstacles to wider integration of technology, while knowledge workers cite lack of management understanding as well as expense.
- Recruiting, training, and retaining systems experts and manufacturing engineers is perceived to be a "very serious" problem by both manufacturing executives and knowledge workers. Compounding this concern is the finding that while generally satisfied with their current jobs, the majority of knowledge workers feel job switching is the best way to "climb the corporate ladder."
- U.S. manufacturers are not seeking government intervention as a means to alleviate the problems facing the industry. The only actions they desire from the federal sector are an upgrade in technical training in secondary and higher education and reinstatement of the investment tax credit.

Both manufacturing executives and knowledge workers recognize the need to implement advanced technologies such as MRP II, JIT, "islands of automation" (CAD, CAM, robotics), and CIM to realize competitive benefits such as lower inventory, flexibility in response to market demand, quality improvements, and lower unit costs.

When questioned specifically about CIM, the control technology of major interest today, a substantial majority of the respondents agreed that "CIM is a breakthrough concept that will put U.S. manufacturing back on the map." More than three fourths felt that CIM implementation is likely to result in enormous cost advantages for companies willing to make the necessary investment. Additional information on CIM can be found in Chapter 16, "Computer-Integrated Manufacturing," of this volume.

Although the need to implement technology is understood and the advantages—in terms of competitive edge—documented, advanced manufacturing technologies are not widely applied in the United States. Only one quarter of manufacturing executives report very extensive use of MRP II in their industry and far fewer report extensive use of robotics, JIT, and CIM. Knowledge workers view the situation somewhat differently and more positively. Nearly one third feel that "islands of automation" such as CAD, CAM, and robotics are in very extensive use throughout their industry, while JIT, CIM, and MRP II are used to lesser degrees, but greater than those seen by executives. Both groups see exponential increases in the level of implementation of all advanced technologies in the next 5 years. However, a large percentage, approximately one half, predict only limited application of technology in that time frame.

Why are U.S. manufacturers not implementing high-level manufacturing technologies and systems in light of their recognition of the benefits they could accrue? The survey shows the existence of major obstacles to technology implementation. Both manufacturing executives and knowledge workers cite lack of skilled personnel, lack of the financial investment required, and lack of senior management understanding, and commitment as the major hurdles to integration. Knowledge workers perceive the financial commitment and lack of management understanding about technology as the biggest stumbling blocks.

Although they are aware that the benefits technology offers can strengthen their international competitive position, U.S. manufacturers have been reluctant to apply advanced technologies fully now and appear unlikely to do so in the near future. While some of this reluctance may be attributable, as knowledge workers surmise, to a lack of management understanding of such systems, it is more likely that the financial investment required is proving to be the major hurdle. This reaction represents a short-term view that may spell serious problems in the long term. It can be the result of executives emphasizing annual and quarterly profits while ignoring the investments that might ensure long-term success. This obstacle demonstrates the need to devise new cost accounting systems for a technological environment and to adopt methodologies for cost justification of advanced manufacturing systems. Justification of manufacturing expenditures is discussed in Chapter 3, ''Planning and Analysis of Manufacturing Investments,'' of this volume.

Survey Conclusions

U.S. manufacturers are—with some reservations—confident about their competitive standing in the world marketplace today and optimistic that they will remain competitive in the future. They do not fear competition from across the sea, but rather from across the street. And as a result, they are not preparing for international competition. Their plans to maintain their competitive edge include a commitment to investment, technology, and management.

However, some of their actions do not speak to this commitment. It appears that advanced technologies will not be implemented to a great extent in the near future. Although the importance of retaining manufacturing systems experts and engineers is acknowledged, steps to achieve this have not been implemented—steps such as defined career paths and improved communication with management. Additionally, manufacturers must work, both on their own and with the federal sector, to alleviate the shortage of skilled personnel to ensure the competitive standing of America's future.

The confidence of American manufacturers is reassuring. However, their work in maintaining the competitive edge is cut out for them. With the right actions, their vision of American manufacturing can be a reality.

SUMMARY

Manufacturing management control systems are in a renewed state of evolution. More traditional forms of control, such as ROI and standard cost systems, are still applicable, but face major revamping to measure product costs accurately in environments using new manufacturing control systems, technologies, and philosophies. Cost reduction and control is of critical importance to the management of a globally competitive manufacturing enterprise.

The growing role of the computer has made control technologies applicable to a more complex, multiproduct manufacturing enterprise. To compete, today's manufacturer must be able to enter product markets quickly and with cost efficiency. New control techniques such as JIT and CIM have changed the physical and production environment of manufacturing. More traditional control methodologies and cost measurement control systems must reflect the effect of these new technologies.

References

1. Robert N. Anthony, *Planning and Control Systems: A Framework for Analysis* (Boston: Harvard University, 1965).
2. Kenneth J. Euske, *Management Control: Planning, Control, Measurement, and Evaluation* (Reading, MA: Addison-Wesley Publishing Co., 1983).
3. *Ibid.*
4. H. Thomas Johnson and Robert S. Kaplan, *Relevance Lost—The Rise and Fall of Management Accounting* (Boston: Harvard Business Press, 1987).
5. Fremont E. Kast and James E. Rosenweig, *Organization and Management: A Systems Approach*, 3rd ed. (New York: McGraw-Hill Book Co., 1979).
6. Euske, *op. cit.*
7. Ernest C. Huge and Alan D. Anderson, "The Spirit of Manufacturing Excellence," September 1986.
8. Richard J. Schonberger, *World Class Manufacturing* (New York: The Free Press, 1986).
9. *Japanese Manufacturing Techniques* (New York: The Free Press, 1982).
10. William A. Wheeler III, *Straight Talk on Just in Time* (New York: Coopers & Lybrand, 1987).
11. "The Push for Quality," *Business Week* (June 8, 1987).
12. W. E. Deming, *Quality, Productivity and Competitive Position* (Boston: MIT Center for Advanced Engineering Study, 1982).
13. Wickham Skinner, "The Productivity Paradox," *Harvard Business Review* (July-August 1986).
14. P. B. Crosby, *Quality Is Free* (New York: McGraw-Hill Book Co. 1979).
15. J. T. Hagan, ed., *Principles of Quality Costs* (Milwaukee American Society for Quality Control, 1986).
16. J. W. Marr, "Letting the Customer be The Judge of Quality," *Quality Progress* (October 1986).
17. A. V. Feigenbaum, *Total Quality Control*, 3rd ed. (New York: McGraw-Hill Book Co. 1983).
18. K. Shimoyamada, "The Presidents Audit... and other related articles," *Quality Progress* (January 1987).
19. L. P. Sullivan, "Quality Function Deployment," *Quality Progress* (June 1986).
20. B. Gunter, "A Perspective on Taguchi Methods," *Quality Progress* (June 1987).
21. L. P. Sullivan, "The Power of Taguchi Methods," *Quality Progress* (June 1987).
22. J. W. Leppelheimer, "A Common-Sense Approach to SPC," *Quality Progress* (October 1987).
23. J. M. Juran, *Managerial Breakthrough* (New York: McGraw-Hill Book Co., 1964).

Bibliography

Drucker, Peter F. *Management: Tasks, Responsibilities, Practices*. New York: Harper & Row Publishers, Inc., 1974.
Hodgetts, Richard M. *Management*. New York: Academic Press, Inc., 1985.
Howell, R.A., et al. *Management Accounting in the New Manufacturing Environment*. NAA, 1987.
Made in America. New York: Coopers & Lybrand, 1987.
Matz, Adolph, and Usry, Milton. *Cost Accounting Planning and Control*. Dallas: South-Western Publishing Co., 1976.
McIlhattaw, R.D. "How Cost Systems Can Support the JIT Philosophy." *Management Accounting* (September, 1987).
Sayles, Leonard. *Organization Dynamics*. New York: AMACOM, Summer 1972.
Voich, Dan, and Wren, Daniel A. *Principles of Management: Resources and Systems*. New York: The Ronald Press Co., 1968.
Waliszewski, David A. *The JIT Starter Kit for Design Engineering*. APICS Conference Proceedings Falls Church, VA: American Production and Inventory Control Society, 1986.
World Class Manufacturing Guidelines. Detroit: General Motors Corp., 1987.

PLANNING AND ANALYSIS OF MANUFACTURING INVESTMENTS

Manufacturing firms are companies created by people to facilitate the manufacture of goods. While these firms have many social effects, they are for the most part created and managed to increase the wealth of the owner or owners. This is true whether the firm is a small garage shop owned by one individual producing a single, simple product or a worldwide conglomerate owned by thousands of shareholders producing a wide variety of different products in factories throughout the world.

The way manufacturing firms increase the wealth of their owners is first to invest in resources that are used to make products. The resources generally include raw material and machines that are purchased, as well as people who are paid salaries and wages. The resources are then used to produce products that are sold in various markets to people and firms to whom the product has a value greater than the cost of the resources used to make it. Thus, the manufacturing firm can sell the product for a price greater than the cost of the resources used and earn a profit. Ordinarily, the investments must be made and the goods produced some time before any money is received for the sale.

Because this is the main thrust of most manufacturing firms, the analysis of investments, their costs, and the benefits that are likely to accrue to firms and their owners is an important part of managing these firms.

Many changes are taking place both in the external environment where manufacturing firms operate and in the factories themselves. For example, the markets in which manufactured goods are sold have become much more competitive. At one time, manufacturing was done in only a few industrialized countries. Now manufactured goods are produced, principally for export, in many countries with low wage rates. Firms in countries with higher wage rates must find ways to compete with these goods in the marketplace.

There is growing recognition that firms must develop business strategies directed toward these competitive issues. Consequently, managers are typically devoting more of their attention to the development and implementation of business strategies. Investment plans are being driven by explicitly stated strategic objectives, the manufacturing strategies that will contribute to achieving these objectives, and the specific plans that implement the strategies.

Within the industrialized countries, the easy opportunities for improving manufacturing productivity have already been taken. Direct labor reductions in particular have been so extensive that direct labor is now only a small part of the cost of producing most goods. It has become necessary to seek other, generally more complex ways of improving the manufacturing process to stay competitive.

The opportunities that exist for improving a particular manufacturing operation at this time typically have a complex relationship with other manufacturing operations, with other nonmanufacturing functions within the firm, and with the market. These relationships and the resulting economic costs and benefits can rarely be determined by solving a single, simple equation or even a system of simultaneous equations, but require modeling of the manufacturing process.

In response to these needs, and made feasible by significant reductions in the cost of computing, methods for analyzing such improvements have been developed. Process improvements to improve productivity include elements of computer-integrated manufacturing (CIM) such as statistical process control (SPC), manufacturing resource planning (MRP II), computer-aided design (CAD), robotics, and flexible manufacturing systems. The analytical tools fall into the general category of computer programs that allow the user to design

CHAPTER CONTENTS:

Contributors of sections of this chapter are: Jay Baron*, Senior Operations Analyst, Industrial Technology Institute;* ***Michael Burstein****, Principal Scientist, Manufacturing Systems Economic Group, Center for Social and Economic Issues, Industrial Technology Institute;* ***Pearson Graham****, Consultant.*

Reviewers of sections of this chapter are: Jim Brimson*, Vice President, Business Development, CAM-I, Inc.;* ***Richard L. Engwall****, Manager, Get PRICE Program, Manufacturing Systems & Technology Center, Westinghouse Electric Corp.;* ***Chris V. Kuhner****, Manager of Manufacturing Expert Systems, Computer Systems Manufacturing Engineering & Technology Corp., Digital Equipment;* ***Steven R. Olson****, Coordinator, Research & Technical Planning, Corporate Information Systems, Eastman Kodak Co.;* ***Hamid R. Parsaei, PhD****, Assistant Professor, Department of Industrial Engineering, University of Louisville;* ***William G. Sullivan****, Professor, Department of Industrial Engineering, University of Tennessee;* ***Hayward Thomas****, Consultant, Retired—President, Jensen Industries.*

alternative manufacturing systems, as well as alternative evolutions of such systems over time, and run them on the computer. In this way, both the physical characteristics of alternative manufacturing systems and their economics can be analyzed prior to making significant investments in the systems themselves.

Further, changes are occurring at an increasing rate. Changes in markets and in both product and manufacturing technology that once required a full generation now occur in a few years. It is important for the manufacturing engineer to recognize these changes. The intensity of competitive pressures in manufacturing industries worldwide cannot be denied, and the importance of making good investment decisions is greater than ever. Because most of the simple investments have already been made, managers find it difficult to justify the more complex investments incorporating new technology. While they recognize intuitively that survival of their firms requires such investments, there have been calls to discard quantitative analyses of capital investments and return to a more intuitive approach to evaluating investment opportunities.[1]

While such an approach is clearly preferable to rejecting necessary high-technology investments because they cannot be justified by traditional, low-technology investment analysis, it is not necessary. There are ways of relating investments to the business strategy of the firm; there are new dynamic analytical models that are not only feasible, but also widely accessible; and there are many opportunities. If the manufacturing engineer of the 1950s and 1960s looked for a machine with a shorter cycle time to reduce direct labor, where should the manufacturing engineer of the 1980s and 1990s look for responses to the more complex demands being placed on manufacturing? Areas where the potential seems particularly great are as follows:

1. Better performance in the marketplace:
   - Responding to customer needs consistently, predictably, and ahead of the competition.
   - Responding to market changes with respect to product mix, product volume, and product change.
   - Offering higher quality products at competitive prices, with better customer acceptance.
2. Reduced investment:
   - Lower inventories resulting from short lead times.
   - Improved manufacturing controls, with reduced inventories.
   - Higher equipment utilization, with reduced investment.
   - Reduced tooling.
   - Less floor space.
   - Closer process control, yielding more good product per machine.
3. Lower costs:
   - Reduced direct labor.
   - Reduced set-up and changeover time, and other forms of indirect labor.
   - Higher quality products, with less scrap and rework and lower warranty and product liability costs.

Beyond these relatively tangible savings, there are many intangible benefits that can be taken into consideration, even if not as accurately as the tangible benefits.

By and large, these improvements are not confined to the single workstation where a new machine is located. The effects are felt throughout the entire manufacturing system and beyond. A machine performing a primary operation that results in a higher quality part often yields much of its savings downstream, where secondary operations can be performed more effectively. Manufacturing engineers have recognized this for years, but these savings were previously referred to as intangible savings and were for the most part ignored because there was no way of quantifying them. This has now changed with the use of dynamic models of manufacturing systems. To capture the effect of a new investment on the entire system, the baseline system can be modeled. The model can then be modified to include the proposed investment, estimate the effects on secondary operations, and see what effects a more closely controlled primary operation will have, not just on the primary operation itself, but on all elements of the system. It is then possible to link the model to a financial model and compare the economics of the baseline system with the proposed alternative.

This chapter presents not only the proven traditional methods for financial analysis of capital investments, but also a way to approach the problem of defining the needs of the manufacturing system so as to be responsive to the strategic requirements of the firm. In addition, newer methods of modeling alternative manufacturing systems to determine their operating characteristics, the means of linking the output of these models to financial analysis models, and ways of dealing with possible future events and decisions are described. These techniques are presented in such a way that, in total, the chapter provides a set of tools for use by engineers and managers, yielding greater insight into the relative advantages and disadvantages of alternative investments than either traditional analysis or an entirely intuitive approach could provide.

The first section briefly describes the business planning process, with particular emphasis on the relationship between business planning and manufacturing investment. It is the intent of this section to broaden the approach of the investment analyst from the financial analysis of specific investments as isolated opportunities for cost reduction to the analysis of investments relative to the strategic plan of the business with the objective of meeting the functional requirements demanded of the manufacturing system to respond to the strategic needs of the business. In so doing, it is necessary to relate the investment being considered to the entire manufacturing system as it exists at the time, and as it is likely to evolve in the future as products, markets, and manufacturing technologies evolve.

The next three sections of the chapter describe relatively new techniques for relating the effects of a specific investment on the manufacturing system over its lifecycle. These include dynamic modeling of manufacturing systems, translation of the output of such systems to financial results, and decision analysis. The manufacturing engineer may use such techniques to capture the benefits of investments whose major contribution is not merely the reduction of direct labor, but the improvement of the manufacturing system to better meet the functional requirements of the firm's strategy both now and as it evolves in the future.

Finally, the entire process is brought together with an illustrative analysis of alternative investments to meet the particular functional requirements for a manufacturing system that is intended to meet the strategic needs of a hypothetical evolving business.

PLANNING AND ANALYSIS OF INVESTMENTS

The four fundamental perspectives that form the basis for this chapter are as follows:

1. *The strategic perspective*. When considering an investment, one must look for benefits that assist a firm in meeting its strategic objectives.
2. *The incremental perspective*. When analyzing a specific investment, one must look for the incremental differences in both costs and benefits that occur with the investment in comparison with a realistic and reasonable baseline without the investment.
3. *The system perspective*. When seeking to define incremental costs and benefits, one must look for all those within the entire system—not only those immediately associated with the specific workstation or cell where the investment is made.
4. *The evolutionary perspective*. When examining the system for costs and benefits, one must not only seek those associated with the system in its present form, but also those associated with the evolution of the system over its life span.

The approach used in this chapter begins with a statement of the company's strategy and its business plan to achieve strategic objectives, using a strategic perspective. The manufacturing strategy is then defined in such a way that it supports the business strategy, not only with present requirements, but also as they may evolve over the life of the system. To assist in developing this evolutionary perspective, possible future chance events and decisions are explored and set forth in the form of a qualitative decision tree.

Alternative manufacturing systems, with characteristics that meet the requirements of the manufacturing strategy, are then modeled. The results are translated into economic terms. Through the system perspective, as many as possible of the costs and benefits of each system are quantified.

These costs and benefits are applied to the qualitative decision tree developed previously to estimate the economics of each alternative system over its life. These values are then compared with a realistic baseline to define the incremental economic effects of each alternative investment, thereby assisting management in its decision. The steps in this approach are summarized in Fig. 3-1.

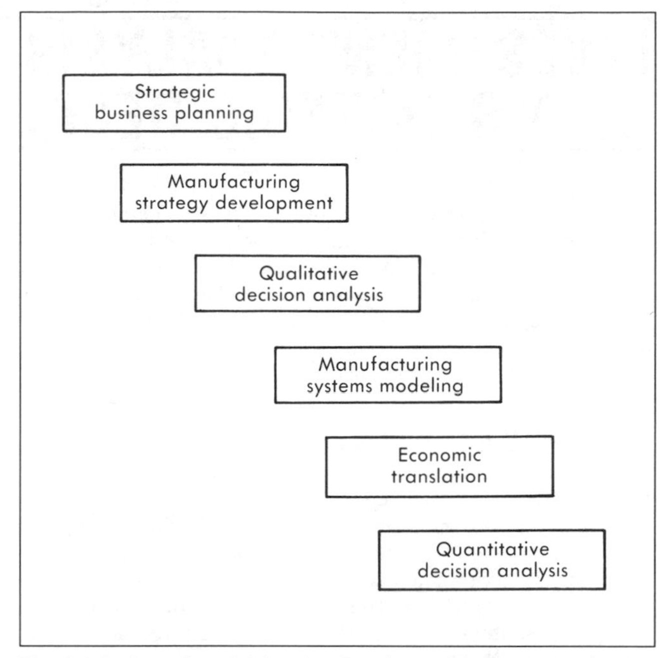

Fig. 3-1 Suggested model for the planning and analysis of manufacturing investments. (*Industrial Technology Institute*)

Certainly a full-blown approach such as the one described in this chapter is not appropriate for every investment.[2] For example, investments to implement maintenance policies would not require strategic evaluation, although design and redesign of the policies themselves would call for attention to strategy. Because the nature of strategic planning is long term, re-evaluating the strategy for every investment would defeat its purpose. There are times, however, when the strategy should be reviewed. Some possible times are as follows:

- When changing technology is expected to change the market presence of the firm through the creation of new products, new processes, or new distribution channels.
- When the investment is extremely large.
- When there is a need to communicate the business strategy of the firm to the organization.
- When it is intuitively felt that a major investment is important to the firm, and a simple analysis significantly fails to capture its benefits.
- Periodically in any event, to emphasize the necessity to consider the very best investment over the long run—not just incremental improvements.

The chapter has the words "planning and analysis" in its title rather than "justification" because of a fundamental perception of the appropriate process to use prior to investing. Justification implies making a decision on an investment, then putting together numbers to sell it to management; this is a process that, except in the hands of a genius, results in mediocrity. Planning and analysis implies assessing the needs of the firm, then seeking out the investment that, in combination with other activities, best meets those needs.

When greater understanding of a specific investment technique is required, this material is given in an appendix rather than in the body of the text. If the reader is familiar with the technique, there is no need to refer to the appendix; if not, the appendix provides adequate information to be used with the text and examples.

If the reader desires greater understanding of an investment technique than is feasible to provide in this Handbook, references are provided. These have been expanded to incorporate not only text references but software references as well.

The changes described in this chapter are still under way. Additional changes in manufacturing technology are anticipated, as are improvements in the scope, applicability, and features of the models available for simulation and analytical purposes. Therefore, the reader should view this chapter as a snapshot of the state of the art at a particular time, to be updated regularly by reading the journals noted in the bibliography.

INVESTMENT ANALYSIS AS PART OF A STRATEGIC DECISION

This section briefly describes the business planning process, with particular emphasis on the relationship between business planning and manufacturing investment. It is the intent of this section to broaden the approach of the investment analyst from the financial analysis of specific investments as isolated opportunities for cost reduction to the analysis of investments relative to the strategic plan of the business.

BUSINESS STRATEGY

Historically, capital budgets have begun in many firms as the list of capital investments requested by operations for the coming year. This list was reviewed and revised at various levels and ultimately became the capital budget for the firm. These investments in total represented an implicit strategy of some sort that might or might not have been internally consistent. Very often the implicit strategy was to continue doing whatever the firm had been doing in the past.

While this may have been adequate in a simpler and less competitive world, in recent years firms have found that their need to compete more aggressively requires an explicit statement of objectives to be achieved, a definition of the strategy intended to achieve the objectives, and a plan to put the strategy into action, usually covering a period of several years. The capital budget in these firms has become a part of this plan. The presence of a strategic plan does not eliminate or reduce the importance of analyzing capital investments. Rather, it provides guidance to the analyst as to how the firm intends to create value, and it encourages the members of the firm to work toward uniform objectives in the creation of value. As such, it places investment analysis along with new product planning, market planning, and human resources planning within the context of the strategic objectives of a firm and the strategies by which the firm intends to attain those objectives.

The typical process that has developed is one in which the firm starts with a strategic planning exercise, the object of which is to figure out what market/product position the firm should have, based on the threats and opportunities with which the firm is faced and the strengths and weaknesses of the firm that determine the feasible responses to the threats and opportunities. This is an exercise involving all functional areas—marketing, product engineering, manufacturing, finance, and others. The result is a plan that defines the products and markets in which the firm will participate, as well as the timing of such participation and the related responsibilities of each functional area.

It has been stated that strategic thinking in a firm should deal with a few basic questions, such as the following:[3]

1. Where does the firm stand now relative to the market and its competitors?
2. What major external forces are likely to have an impact on the firm and its competitors?
3. What alternatives exist for the firm and its competitors in light of their present position and the anticipated external forces?
4. Which alternative is believed by management to be appropriate for positioning the firm relative to the market and competition?
5. How will the firm arrive at this position?
6. What critical operating actions are required?

The firm's position may be defined by the industry in which it participates and how it competes within the industry. The structure of the industry largely determines the profitability of firms in the industry. Five structural forces within each industry have been defined as follows:[4]

1. The bargaining power of suppliers.
2. The bargaining power of buyers.
3. The threat of new entrants.
4. The threat of substitute products and services.
5. Rivalry among existing firms.

A firm may currently participate in an industry that is subject to increasing forces to make it less profitable; therefore, the firm might seek to leave that industry and commit its resources elsewhere with greater potential for creating value. Or it may be satisfied to remain within the industry and change or strengthen its position there. Often, different firms in a single industry compete in different ways. For example, one firm may be the low-cost producer, one may produce the highest quality product, and another may offer the fastest delivery. It is often possible to establish a matrix that displays the strategic position of various firms competing in an industry (see Table 3-1).[5]

After evaluating the attractiveness of the industry in which the firm participates, its position in the industry, its strengths and weaknesses, and the challenges and opportunities presented by external forces, management of the firm may elect to maintain its present position or to direct its efforts and resources toward repositioning itself in either the same industry or a related one.

MANUFACTURING STRATEGY

Just as managers once relied on implicit business strategies, they also tended to oversimplify their manufacturing requirements by assuming that such demands could be defined in a simple phrase such as "efficient manufacturing" or "high productivity." There is growing recognition that the diverse demands placed on manufacturing require a more thoughtful consideration of the functionality of the manufacturing system if it is to be responsive to the strategic needs of the firm. Thus, once the business strategy is established, the next task is to prepare a manufacturing strategy that broadly defines the firm's manufacturing function in a way that is consistent with and complementary to the business strategy. While the firm's business strategy is based on the characteristics of the industry and on the position of the firm in the industry, the manufacturing strategy is based on the requirements of the business strategy for manufacturing and the ability of the manufacturing system to respond. It is, in effect, the trace of the business strategy on the manufacturing plane. It deals with the following internal manufacturing issues:

TABLE 3-1
Example of a Strategic Position Matrix of Four Firms

	Firm A	Firm B	Firm C	Firm D
Specialization	Broad line	Broad line	Specials	Limited line
Brand identification	Yes	No	Yes	Yes
Distribution channel	Company stores	Private brand	Direct	Distributors
Vertical integration	Full	Manufacturing only	Full	Manufacturing and distribution
Service	125 dealers	None	Direct	Limited
Cost	High	Lo	Mid	Low
Quality	High	Low	High	Mid
Market share	35%	30%	15%	20%

- What are the possible collections of product families and variants of each family that the system should be capable of handling at a given time?
- At what rate should the system produce good product?
- How readily should the system be changed over
  —from one variant to another within a family?
  —from one current family to another?
  —from the current set of families to some range of possible future families?
- What should the firm's involvement be over the product lifecycle?
  —Engineering models.
  —Production prototypes.
  —Pilot production.
  —Production.
  —Aftermarket.
- What should be the process span of the firm?
- What is the best organization for manufacturing?
  —Multiple plant or single plant? If multiple plant, by geography, product, or process? If single plant, how to obtain focus?
  —Relationship to marketing, engineering, and other functions.
  —Ongoing employee involvement programs.

These are the internal issues that define the relationship of manufacturing to the business strategy of the firm and establish the functional requirements of the manufacturing system. While they are generally proposed by manufacturing, they relate closely to other functions, as well as to the overall management of the firm. Thus, they should have the agreement of other functional areas as well as approval and support of top management of the firm.

FUNCTIONALITY OF MANUFACTURING SYSTEMS

To define functionality of a manufacturing system, the characteristics of the products that are currently in production as well as those that the firm anticipates producing must be defined. Some characteristics of manufacturing systems are general. For example, throughput, defined as the quantity of good product produced in a given time period, is a key characteristic of every manufacturing system. It is a complex characteristic, depending on the reliability of the system as well as other characteristics that determine the amount of time spent producing product.

Other characteristics are specific to the type of production. For example, mechanical products have one set of characteristics, electronic products another, and chemical products a third. Characteristics of a manufacturing process to produce mechanical components might include shapes, quantities, sizes, surface finish, hardness, toughness, and density. Those for production of electronic components might include electronic functions as well as many mechanical characteristics. Chemical processes might include chemical composition, volume, and purity.

When these characteristics have been defined, each is considered from the following four perspectives:

1. Definition. Will the manufacturing process produce the quantity required? The shape and size required? And all the other characteristics of the product that are established by the manufacturing process?
2. Accuracy. How closely will the manufacturing process repeat the required characteristics time after time? What are the six-sigma limits of the machine tool compared with the dimensional tolerances of the parts to be produced? What is the yield of the process within the established limits of acceptability? In the chemical industry, repeatability is often an issue of optimal raw material usage rather than product quality, which is often established by grade specifications.
3. Flexibility. How easily (and quickly) can the manufacturing process be changed over from one current product to another? What is the range of shapes, sizes, and other characteristics of established products that can be accommodated? Or is the nature of the system such that it is dedicated to a single product?
4. Adaptability. How easily can the manufacturing process be adapted from the characteristics of the existing product or family of products to those of new products, perhaps not yet designed?

Only after system characteristics have been established can the important issue of cost be addressed. Of the feasible manufacturing systems in terms of product definition, accuracy, flexibility, and adaptability, it is necessary to ask which system is optimal in terms of creating value for the company over the life of the system.

EVOLUTION OF MANUFACTURING SYSTEMS

Manufacturing systems have evolved over a period of years. Consider the production of mechanical components as an example. Early in the Industrial Revolution, the manufacturing system consisted of a group of isolated machines each capable of producing a limited range of geometric shapes, such as drills producing cylindrical holes, lathes producing round shapes, and mills producing flat planes. The machines were largely under the control of the operator, requiring a high degree of skill. While they were not particularly fast, they had the advantage of being relatively easy to set up and so could easily be changed over from producing one part to another. There was little relationship between the manufacturing system and the various other elements of production.

This evolved to a set of machines producing a limited range of products—the production line—and then to highly individualized machines producing specific products with little direct labor—the automated line. As this evolution occurred, control gradually shifted from the operator to the machine itself, often using expensive and elaborate tooling. Changeover from one product to another involved changing tooling, which became expensive and time consuming.

Now, with the availability of numerical control, there is a trend to regain the flexibility that was lost when machines became more product specialized. Control of the machines is in the hands of a computer, and changeover is often little more than a software substitution, with little or no change of tooling. Typically, there is a closer relationship between manufacturing and the other elements of the production system; the most advanced systems rely on a common database for engineering, manufacturing, material control, quality, and the other elements of the production system.

A comparable evolution often occurs over the lifecycle of a product.[6,7] Engineering models, prototypes, high-volume production, and aftermarket parts all have different characteristics that require different manufacturing responses. For engineering models, the design is unstable, quantities are low, demand is irregular, and there is typically little pressure on costs. A toolroom or model shop with highly skilled operators is often used for such parts. Prototypes generally have a somewhat more stable design, quantities are low but growing, demand is becoming more regular, and cost pressure is growing. These characteristics make CNC a desirable method of manufacturing. In production, where volumes are so high as to demand dedicated machinery (as in the automotive industry), the stable design, high quantities, regular demand, and high pressure on costs call for a high level of automation. Later, when the product is being phased out, design is very stable, aftermarket quantities are low and declining, demand is becoming more irregular, and pressure on costs is declining, CNC again becomes an attractive means of manufacturing.

Often, more than one type of manufacturing system is found in a single plant. In organizing manufacturing at the plant level, the manufacturing engineer deals with the issues of organizing the plant by process vs. by product; by manual processes vs. automated; and by rigid vs. flexible automation. Even when a functional requirement is clearly defined, there might be many ways of achieving the desired effect. When the market demands flexibility, for example, this may be achieved through product design utilizing group technology, through flexible manufacturing systems, through clever tooling, through logistics, perhaps by a collection of permanently tooled lines utilized as necessary, or by using outside suppliers. As a result, sometimes one type of manufacturing system is cost effective for one application and another type for another.

QUALITATIVE DECISION ANALYSIS

At this point in the discussion, it is necessary to gain an understanding of the nature, extent, and diversity of the potential evolution of the manufacturing system. To do so, it is useful to lay out chance events that may occur in the future and decisions that must be made using a decision tree. It is not possible at this point to apply quantitative data to the tree, nor is it necessary to do so. It is necessary, however, to know what requirement may be placed on a manufacturing system before it can be modeled. Development of the qualitative decision tree, as called for at this stage of the analysis, is described subsequently in the section entitled "Decision Analysis" of this chapter.

OPERATIONAL PLANNING

At an operational planning level, it is necessary to narrow down some of the general decisions made earlier. Some of the earlier decisions may be reversed as more knowledge is gained. At this stage, there is generally more delegation to engineering, as the issues have less management content and more technical content. For example, if the decision has been made that rigidly automated lines are appropriate for the specific strategy, engineers must consider whether these should be in-line or rotary. Should they be synchronous or not? Should they be continuous or intermittent? Just as strategic planning makes investment planning no less important, but places it within a business strategy context, so manufacturing system design makes machinery selection no less important, but places it within a context of functional requirements.

DYNAMIC MODELING OF MANUFACTURING SYSTEMS

If alternative systems for manufacturing are being considered to achieve the objectives of the firm, and these alternative systems have different characteristics, require different investments, and are likely to result in different cost patterns, it is helpful to define the alternative systems in such a way that as many as possible of the important characteristics of the systems be captured prior to making the investment. If the systems are defined in terms of a dynamic modeling technique, they are likely to display the tangible effects of the investment in terms of these important characteristics. This will allow the manufacturing engineer to be more thorough in evaluating the costs and benefits of each alternative than if he or she had to consider many of them as "intangibles." In turn, thorough evaluation facilitates a better answer as to which alternative will best meet the needs of the company.

As one moves from a static model of a single workstation, as described earlier in the chapter, to a dynamic model of a cell, line, or factory, the model becomes much more complex. It also becomes impossible for all practical purposes to keep track of system activity without computer assistance.

This is not a problem, however, as long as the operation of the system can be defined mathematically, because the manufacturing engineer now has access to computers and related software for dynamic modeling. These dynamic models are capable of capturing the effects of a change, such as the introduction of a new machine, within an entire manufacturing system. At one extreme, if one defines the operation of the system mathematically and is capable of programming in Fortran, Pascal, or one of the other high-level computer languages, a mathematical model could be defined on the computer that will parallel the operation of a proposed manufacturing system. This is done, and it works very well, but it is often too expensive to be used for minor investment decisions.

As a first step in making modeling of manufacturing systems more feasible, three somewhat more specialized types of computer programs have been developed. One path has used queueing theory and modeled the manufacturing system as a network of servers with queues of work in process at each server. A second has developed general simulation languages, such as GPSS and SIMSCRIPT, that simulate the discrete events that occur in a manufacturing network on a timed basis. A third path, stochastic activity network software, has been developed permitting the modeling of communication and control systems in conjunction with manufacturing; METASAN, from the Industrial Technology Institute, is one of these packages. These three types of modeling techniques have been found to be useful and are widely used. The disadvantage has been that they require special training, and few manufacturing engineers find it feasible to maintain their professional standards in both manufacturing and programming skills. Some of these disadvantages have been overcome in the last few years by general manufacturing simulation software such as SLAM II from Pritsker & Associates and SIMAN from Systems Modeling.

Recently, easy-to-use programs have been developed that do not require extensive computer skills, but can be used by the manufacturing engineer to model manufacturing systems on a microcomputer. Examples of two such programs will be given in this section. The programs are similar in that they capture the effect of changes in manufacturing systems and are both dynamic models. They differ greatly in concept, however. The first, Manuplan, is an example of a queueing model that provides a rough-cut estimation of system performance using data readily available to the engineer. The second, XCELL, is an example of a simulation model that provides a higher resolution estimate of system performance, but requires more detailed planning of the system to provide the necessary inputs. Both types of models are useful, and modes of use are described subsequently.

ROUGH-CUT ESTIMATION OF SYSTEM PERFORMANCE

For rough-cut estimation of the performance of an existing or proposed system, available programs based on queueing network theory allow the manufacturing engineer to check the feasibility of a manufacturing system without a great deal of effort. A queueing network conceives of a manufacturing system as a collection of servers and waiting lines at each of those servers, with movement of entities from the waiting line of a server into the server and then into the waiting line of the next server in the network or to completion. In a queueing system, there are arrivals to the system, according to a particular distribution of time between arrivals, and there are operations each with a particular distribution of performance time. Parts arrive at receiving and then go to a waiting line, a station, on to another waiting line, to another station, and so forth according to a specified operation routing until they are complete and ready for shipment. Queueing models are typically used to (1) reduce the risk of expending time and effort in simulating an infeasible manufacturing system and (2) assist in establishing the design specifications for a system that will subsequently be simulated.

Several such programs are available for rough-cut estimation of the performance of an existing or proposed system. In this section, an example from Manuplan (a program offered by Network Dynamics, Inc. of Cambridge, MA, that is based on queueing network and reliability theory) will be shown. The program runs on the IBM PC AT or compatible personal computers. The following inputs are required:

- A title to identify the run.
- Minutes per day of operation.
- Days per year of operation.
- Names of equipment groups.
- Number of (identical) machines in each equipment group.
- Mean time to failure for each machine.
- Mean time to repair each machine.
- Part numbers to be run.
- Units required per year of each part.
- Lot size for each part.
- Operation names for each operation.
- Operation sequences.
- Definition of which operations are performed by which machine groups.
- Average and set-up times, along with measures of their variation.
- Where operation sequences are split, along with the proportions going to each branch (for example, to next operation or to rework).

Additionally, optional inputs include factors for demand, overtime, and operation speed to facilitate sensitivity analyses. By and large, these are the data that are commonly available to the manufacturing engineer considering an investment in a manufacturing system. Mean time to failure and repair are the most difficult to obtain, but using a mean time to repair of zero allows one to bypass this feature. However, its inclusion does provide a more valid and realistic model of the system.

Output of a Manuplan run includes the following:

- Units of each part produced per year.
- Units of each part scrapped per year.
- Average work in process in total, by part, and by equipment group.
- Average cycle time by part.
- Equipment utilization by equipment group.

The information thus obtained allows the manufacturing engineer to gain confidence in the feasibility of the system, to redefine the specifications of the system to improve its performance, or to discard the system as not feasible in less time than required to run a simulation of the system.

For purposes of this discussion, an example of a simple Manuplan run is named "SME Model." It consists of a factory producing two parts (A and B), each of which has four operations performed on four equipment groups as follows:

- MILL performs rough milling operations.
- GRIND performs finishing operations.

- DRILL performs drilling operations.
- TAP performs threading operations.

The system is initiated at DOCK. All four operations are performed on both parts. Scrap parts are all sent to SCRP; good parts are sent to DONE.

A facsimile printout of the Manuplan run is shown in Fig. 3-2. The printout shows input data, including those items previously cited, as well as production goals, run and set-up times, scrap rates, and mean time to failure and mean time to repair for each equipment group.

Output data includes utilization rates, whether or not production goals were attained, and information on scrap and work-in-process inventories.

Queueing models are useful in dealing with the following issues:

- How a product redesign may affect a manufacturing process.
- How a process may be designed so as to accommodate future increases in production requirements.
- How a process may be designed so as to accommodate future production of additional designs.
- Whether a synchronous or nonsynchronous material handling process is desirable in a particular manufacturing system.
- Where major maintenance emphasis should be placed in a manufacturing system.
- How a redesigned product and process might best be phased in.
- Where defect detection capabilities might best be placed.
- How job redesign might increase labor productivity.

However, queueing models yield only average values, and provide no information on high or low values. Work in process, for example, may yield identical averages for two systems, but be very stable for one and fluctuate greatly for the other. Thus, queueing models are rarely used for a final choice among alternative manufacturing systems. Nonetheless, their value is great, relative to the time and effort required, as a way of screening undesirable alternatives prior to performing a simulation analysis.

HIGH-RESOLUTION ESTIMATION OF SYSTEM PERFORMANCE

Simulation models are similar to queueing models in that they require input data about the parts to be produced, machine groups, and processes, and in that they yield information about the physical operation of the factory. The difference is that they move specific parts through a defined manufacturing process according to a clock instead of performing calculations based on queueing theory. Thus, a simulation model keeps track of time, and when an operation performed on a part by a machine is finished, the part moves on to the queue for the next operation. Randomness can be built into the system with specification of service times, machine failures and repairs, queue sizes, and defects in terms of statistical distributions. Because the clock moves much more quickly than a conventional clock, it is possible to gain a great deal of experience running a manufacturing system very quickly, thereby obtaining a large sample of the random events that occur during the actual operation of the factory.

The nature of a discrete event simulation is such that output data include full information on the sample of events during the run rather than just averages. Thus, if one wishes to know storage requirements for a bank of in-process inventory, simulation provides not only average work in process, but the full distribution that occurred during the run. Further, with a discrete event simulation, one is able to estimate the effect on other production variables of restricting the size of the work-in-process bank to a specified maximum. This, of course, gives better and more useful information for estimating the effects of investment in new machinery and inventories. The price that must be paid for this information is time. The amount and accuracy of data requirements are also greater; the time to obtain the data and structure the model, although greatly reduced from earlier days, is still much more than that which is needed for a queueing model.

Several such programs are available for high-resolution estimation of system performance. In this chapter, an example on XCELL (a program offered by Pritsker & Associates, Inc. of West Lafayette, IN) will be shown. The program is available in versions to run on the Hewlett-Packard HP200 microcomputer and on the IBM AT or fully compatible personal computers. It is most appropriate for the modeling of those systems with limited complexity.

XCELL operates in the following six modes:

1. Design, for model construction.
2. Run, for running the model.
3. File manager, for storage and retrieval of models.
4. Analysis, for checking out models.
5. New factory, for clearing the workspace.
6. Change display, for altering scale, position, and contents of screen.

This description will concentrate on the design mode and on the readouts available during the run mode.

To design a simulation model, one must, as with queueing models, specify the parts to be produced, the operations to be performed, and the machines at which the operations will be performed. The XCELL factory has the following five building blocks:

1. Workcenters, for performance of operations.
2. Receiving areas.
3. Shipping areas.
4. Buffers, for storing work in process.
5. Maintenance facilities, for maintaining workcenters.

Workcenters are the key elements. Each is defined in terms of the following:

- Source or sources of parts to be processed.
- Disposition of good output, whether to another workcenter, to a buffer, or to a shipping area.
- Percent of good parts yielded.
- Disposition of rejected output.
- Processing time, which may be constant or picked randomly from a specified distribution.
- Set-up time.
- Maintenance, which may be random or scheduled.
- Rules specifying how the process is started and for how long it runs.
- Number of machines.

Much of the success in using such a model to simulate an actual factory depends on how realistic the rules are defined for starting and stopping processes. These rules form the material control system of the factory. The user may specify lot sizes and

MANUPLAN INPUT TITLE LINE
SME Model

VERSION
II/1.0

Operation Unit,	Flow Time Unit,	Demand Period,	Minutes Worked Per Day	Days Worked Per Year
Minutes	Day	year	960	220

Utilization Limit	Variability Percent in Arrivals	Variability Percent in Equipment
95.0	30.0	30.0

Equipment Name	No. in Group	Reliability- mttf	(Mins) mttr	Overtime Factor	Speed-Factors Variability Detail Set-up	Run	Factor	Flag
MILL	1	9600	120	1	1	1	1	1
GRIND	1	9600	60	1	1	1	1	1
DRILL	1	9600	60	1	1	1	1	1
TAP	1	12000	30	1	1	1	1	1
DONE								

Part Name/Num	Demand Per Year	Lot Size	Demand Factor	Speed Factors Variability Detail Set-up	Run	Factor	Flag
A	10000	2	1	1	1	1	1
B	10000	2	1	1	1	1	1
DONE							

OPERATIONS FOR PARTS

Operation Assignment For Item (Part)

Operation Name	Equipment Name	Proportion Assigned	Time/Lot (Set-up)	Time/pc (Run)
RghMil	MILL	1	2	6
Finish	GRIND	1	3	7
Drill	DRILL	1	1	2
Thread	TAP	1	0	3
DONE				

Routing for Item (Part) A

From Operation	To Operation	Proportion
DOCK	RghMil	1
RghMil	Finish	0.97
RghMil	SCRP	0.03
Finish	Drill	0.99
Finish	SCRP	0.01
Drill	Thread	0.98
Drill	SCRP	0.02
Thread	STOK	0.98
Thread	SCRP	0.02
DONE		

(continued)

Fig. 3-2 A facsimile of a printout of a simple factory run using a queueing network theory model called Manuplan. (*Industrial Technology Institute*)

Operation assignment for item (part)

Operation Name	Equipment Name	Proportion Assigned	Time/Lot (Set-up)	Time/pc (Run)
RghMil	MILL	1	2	4
Finish	GRIND	1	3	6
Drill	DRILL	1	1	4
Thread	TAP	1	0	6

DONE

Routing For Item (Part) B

From Operation	To Operation	Proportion
DOCK	RghMil	1
RghMil	Finish	0.97
RghMil	SCRP	0.03
Finish	Drill	0.99
Finish	SCRP	0.01
Drill	Thread	0.98
Drill	SCRP	0.02
Thread	STOK	0.98
Thread	SCRP	0.02

DONE

DONE

MANUPLAN OUTPUT REPORT
SME Model

VERSION CODE
II/1.0

Operation Unit,	Flow Time Unit,	Demand Period,	Minutes Worked Per Day	Days Worked Per Year
Minutes	Day	Year	960	220

Utilization Limit	Variability Percent in Arrivals	Variability Percent in Equipment
95	30	30

EQUIPMENT UTILIZATION SUMMARY

	Percent of Capacity Required for SETUP	for RUN	for REPAIR	TOTAL Utilization	WORK-IN-PROCESS (in lots) at EQUIP	in QUEUE	TOTAL
MILL	10.3	51.3	0.8	62.4	0.62	0.41	1
GRIND	14.9	64.7	0.5	80.2	0.8	0.91	1.7
DRILL	4.9	29.6	0.2	34.7	0.35	0.05	0.4
TAP	0	43.5	0.1	43.6	0.43	0.08	0.5

Desired Production Can Be Achieved

	Good Production (pieces)	Scrap Production (pieces)	WORK IN PROCESS (pieces)	FLOW TIME dock-stock in Days
A	10,000	842.8	3.6	0.07
B	10,000	842.8	3.7	0.08
TOTAL PIECES:			7.3	

Fig. 3-2—*Continued*

		PART A PRODUCTION WAS ACHIEVED			PART B PRODUCTION WAS ACHIEVED		
		WORK IN PROCESS (pieces)	TIME Spent / Visit	TOTAL FLOW TIME Spent / Good Piece	WORK IN PROCESS (pieces)	TIME Spent / Visit	TOTAL FLOW TIME Spent / Good Piece
RghMil	MILL	1.1	22.1	22.1	0.9	18	18
Finish	GRIND	1.8	35.2	35.2	1.7	33.2	33.2
Drill	DRILL	0.3	6	6	0.5	10.1	10.1
Thread	TAP	0.4	7.7	7.7	0.7	13.8	13.8

		For each piece of GOOD production: number of pieces that are routed through this branch	For each piece of GOOD production: number of pieces that are routed through this branch
DOCK	RghMil	1.08	1.08
RghMil	Finish	1.05	1.05
RghMil	SCRP	0.03	0.03
Finish	Drill	1.04	1.04
Finish	SCRP	0.01	0.01
Drill	Thread	1.02	1.02
Drill	SCRP	0.02	0.02
Thread	STOK	1	1
Thread	SCRP	0.02	0.02

END_INPUT

Fig. 3-2—*Continued*

define priorities by part. Alternatively, the user may specify initiation of a process on the basis of an event within the system. For example, a "push" scheduling system may be simulated by triggering downstream production when the output buffer of an upstream process reaches a certain level. Alternatively, a "pull," kanban, or just-in-time system may be simulated by triggering upstream production when the input buffer of a downstream process declines to a certain level.

Buffers are simple—capacity is the only variable—but they are important building blocks. By varying lot size, size of buffer stock, and the rules for initiating processes, the user may significantly affect the capacity of the factory as well as the size of the investment in inventory.

By using multiple workcenters and combinations of workcenters and other elements, it is possible to create custom elements with characteristics specified by the user. For example, a conveyor holding a specified number of parts for a finite time can be modeled as a combination of a buffer with a defined capacity and a workcenter with a defined processing time. Assembly and disassembly operations may be modeled as well by processes to create discrete parts.

Receiving areas receive raw material that is supplied to workcenters, and shipping areas receive finished goods and scrap from workcenters for shipment. Maintenence areas provide repair services for workcenters on either a random or scheduled basis. By including maintenance areas within the model, maintenance becomes part of the manufacturing system instead of being a support service external to the model. Thus, differences in performance resulting from maintenance, including those resulting from idle time in a maintenance queue, can be captured within the model.

The model is run in periods of time units. The default value is 500 units. During the run mode, the user may view events during a simulation run on the screen displaying the various elements of the factory. Workcenters are shown as operating, idle because of shortages of material upstream, blocked because of excess material downstream, in set-up, or in repair. Quantity of material is shown in each buffer. Cumulative receipts and shipments are shown below receiving and shipping areas. Alternatively, during the run mode, the user may view a Gantt chart summarizing operation of the factory or concentrate on a specific buffer and view a plot of the contents of that buffer.

Output of the model is in two forms. The run may be paused, and the screen image at a particular point may be printed. This may be the trace, the Gantt chart, or the buffer plot. Alternatively, a summary chart may be printed. The information obtained from an XCELL run provides the user with the capacity of the system under specified operating conditions, work-in-process inventory requirements, and can be used to determine the sensitivity of the system to changes in conditions. While it does not directly provide the costs and benefits of alternative system proposals, it yields valuable data that can be used for such purposes, as will be shown subsequently.

An example of a simple XCELL run for this discussion is similar in many respects to the Manuplan example presented earlier. The system is named "FACTORY." The layout of FACTORY prior to running is shown in Fig. 3-3. It consists of two parts (A and B) and four operations, each of which is performed at a single workcenter. All four operations are performed on both parts. Scrap rates are specified for each operation on each part, and scrap parts are directed toward their own shipping area. Incomplete, good parts are directed toward buffers with a capacity of 10 parts; upon completion, good parts are sent to a shipping area. One maintenance center performs repair on a random basis to all four workcenters. The production line is balanced, with each operation requiring an average of one time unit, but operation times were allowed to vary in accordance with a normal distribution of specified maxima and minima. Batch size was set at 70 units for both parts A and B, and changeover required 10 time units.

The trace of FACTORY at a time when all workcenters are running part B is shown in Fig. 3-4. No machines have broken down, all buffers contain stock, and none are overfilled, although one is full. The one machinery repairperson remains in the maintenance center. Figure 3-5, on the other hand, shows a bad moment in which the drill has broken down and is under repair. The machinery repairperson is no longer in the maintenance center, having been called to the drill. The downstream buffer is empty causing the tapping operation to be idle, and both upstream buffers are full, blocking the milling and grinding operations. The only thing missing from the printout is the group of white shirts waving their arms frantically around the unfortunate machinery repairperson.

The two alternative screen displays are shown by Figs. 3-6 and 3-7. The stock of part A during one period of the run is shown to be relatively unstable, presumably principally because of the fluctuating process times on both upstream and downstream operations. The Gantt chart of the performance of all four workcenters during a segment of a run shows what part of the time each workcenter produced parts and what part each was idle, blocked, queued, or in repair (refer to Fig. 3-7).

The summary printout of this simple run is shown in Fig. 3-8, with the trace at the end of the run corresponding to the summary in Fig. 3-8 shown in Fig. 3-9. In a real-life situation, the program would not be used for a single simple line such as this, but would be used to model alternative systems for accomplishing a specified task.

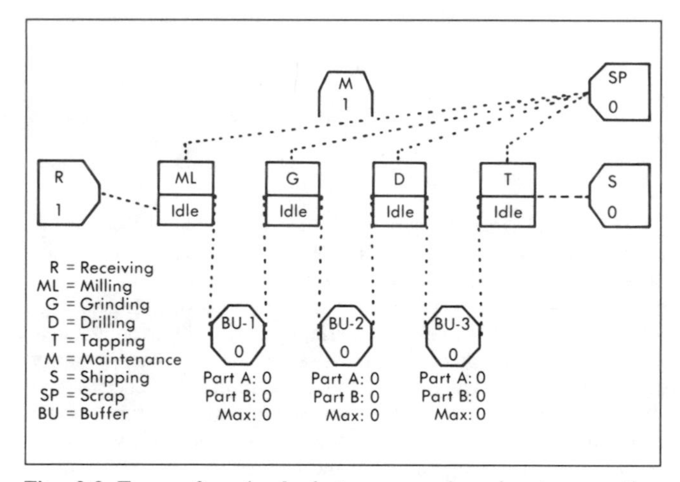

Fig. 3-3 Trace of a simple factory example prior to operation simulated on XCELL. (*Industrial Technology Institute*)

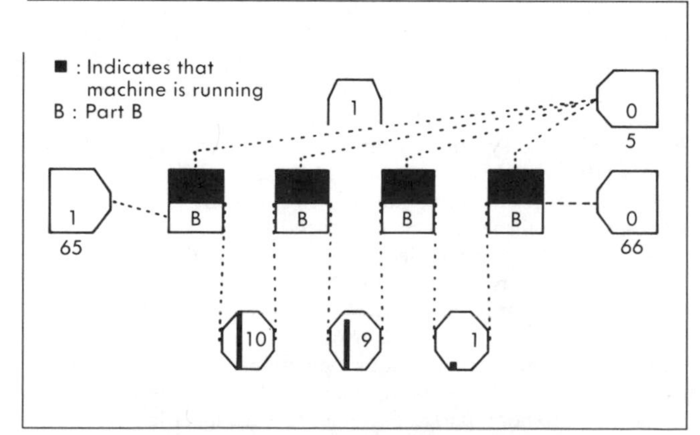

Fig. 3-4 Trace of factory operation when all workcenters are producing part B. (*Industrial Technology Institute*)

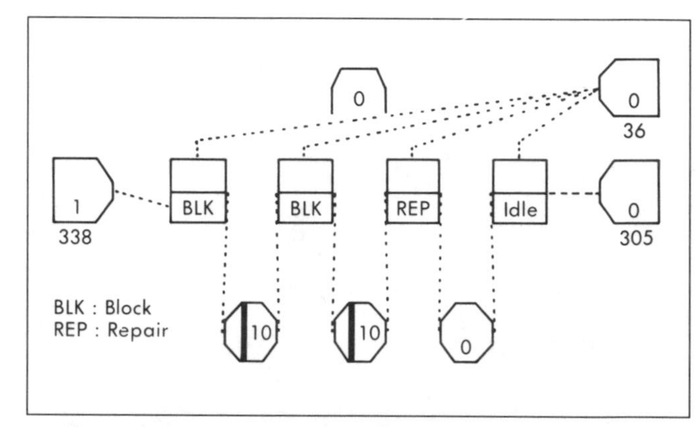

Fig. 3-5 Trace of factory operation when drilling workcenter is under repair. (*Industrial Technology Institute*)

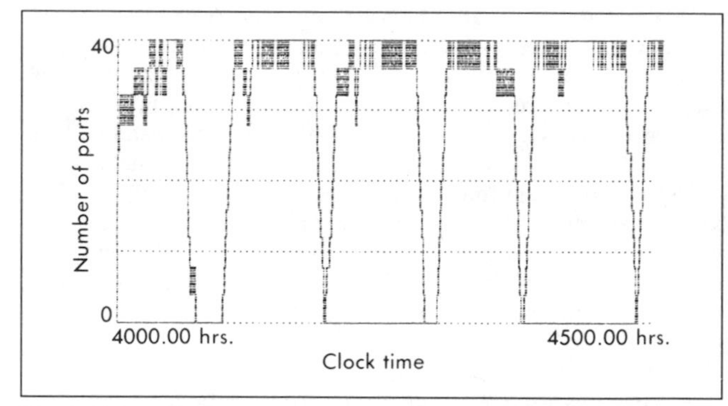

Fig. 3-6 Plot of buffer stock for part A in buffer number 2. (*Industrial Technology Institute*)

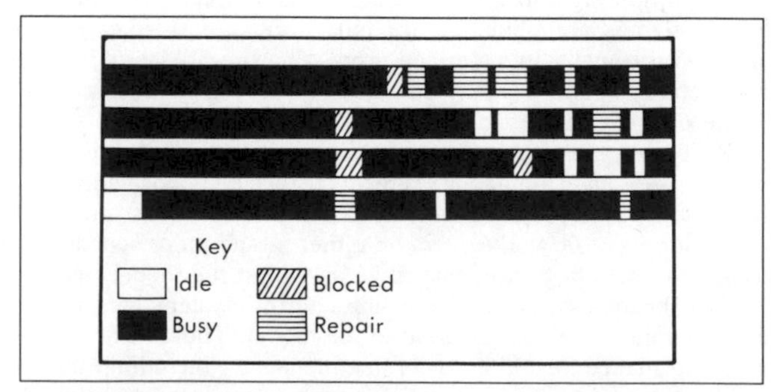

Fig. 3-7 Performance of all four workcenters during operation on a Gantt chart. (*Industrial Technology Institute*)

```
XCELL: Cellular Simulation System

Results of run of factory named "FACTORY"

Current time is:      2000.00
Period started:       1500.00

Receiving areas:
  r1
    Current period
      Units started:          366
      Units/unit time:        0.73
    Cumulative
      Units started:          1452
      Units/unit time:        0.73

Shipping areas:
  SHP
    Current period
      Units finished:         343
      Units/unit time:        0.69
    Cumulative
      Units finished:         1326
      Units/unit time:        0.66
  SCR
    Current period
      Units finished:         40
      Units/unit time:        0.08
    Cumulative
      Units finished:         139
      Units/unit time:        0.07

Workcenters:
  MIL
    Current period
      Current state
        Busy until 2000.03 on part "B"
      Percent of time busy: 73.69
      Percent of time in set-up: 10.00
    Cumulative
      Percent of time busy: 72.75
      Percent of time in set-up: 10.00
  GRD
    Current period
      Current state
        Busy until 2000.06 on part "B"
      Percent of time busy: 70.40
      Percent to time in set-up: 10.00
    Cumulative
      Percent of time busy: 69.66
      Percent of time in set-up: 10.00
  DRL
    Current period
      Current state
        Busy until 2000.15 on part "B"
      Percent of time busy: 69.61
      Percent of time in set-up: 10.00
    Cumulative
      Percent of time busy: 68.28
      Percent of time in set-up: 10.00
  TAP
    Current period
      Current state
        Being repaired
      Percent of time busy: 70.24
      Percent of time in set-up: 10.00
    Cumulative
      Percent of time busy: 67.70
      Percent of time in set-up: 10.00

Buffers:
  b1
    Current period
      Current stock:              10
      Max stock this period:      10
      Avg stock this period:      5.23
    Cumulative
      Maximum stock level:        10
      Avg stock to date:          6.36
  b2
    Current period
      Current stock:              0
      Max stock this period:      10
      Avg stock this period:      5.67
    Cumulative
      Maximum stock level:        10
      Avg stock to date:          6.41
  b3
    Current period
      Current stock:              3
      Max stock this period:      10
      Avg stock this period:      5.11
    Cumulative
      Maximum stock level:        10
      Avg stock to date:          5.31

Maintenance facilities:
  m1
    Number of repairmen:          1
    Number currently available:   0
```

Fig. 3-8 Summary printout of simple factory operation simulated on XCELL. (*Industrial Technology Institute*)

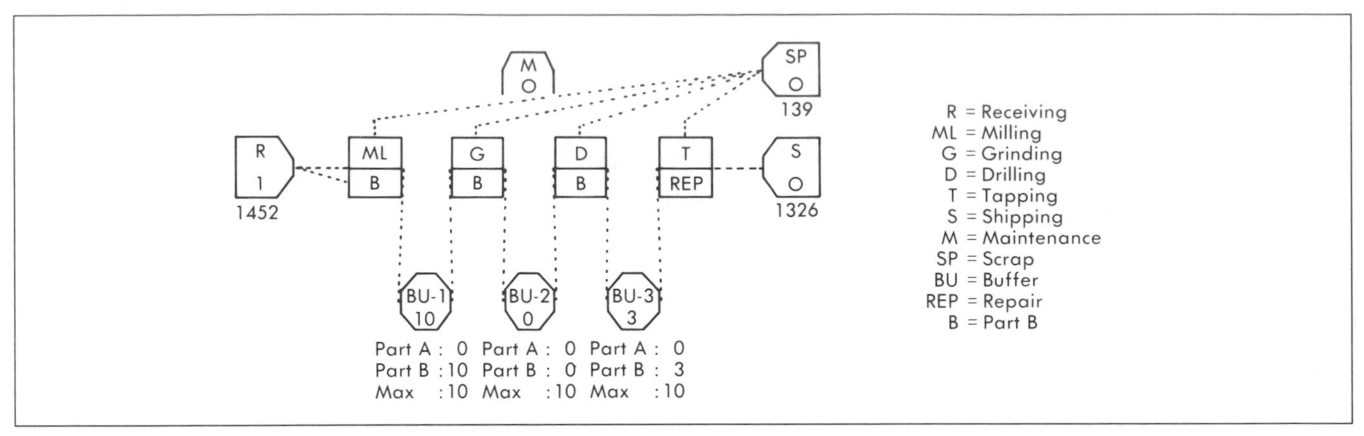

Fig. 3-9 Trace of factory operation at the end of the run. (*Industrial Technology Institute*)

ECONOMIC TRANSLATION

Although modeling results for a manufacturing system alternative on such measures as equipment utilization, inventory levels, throughput, scrap, and rework provide the designer with insights, economic translation of performance can bring the tradeoffs between sharp relief. This facilitates consideration of the following:

- Implications of product design for process configuration.
- Implications of product mix and of product routings.
- Implications of maintenance policies and equipment reliability.
- Implications of defect detection capabilities and their location.
- Layout, buffer sizes, and congestion phenomena.
- Impact of production planning and control policies.

Thus, the speedy response time of simulation and economic translation for the modeling of manufacturing systems gives manufacturing engineers the ability to use economic information in the design process rather than in rough-cut justification later.

The characteristics of specific modeling tools can be significant to the economic translation task. For example, queuing network software like Manuplan, described in the previous section, can generate a ratio for a given part type of total trips across an operation-to-operation link to total trips across that link by good parts only. This ratio can then be used to estimate the cost impact of the latter operation on the link. A simulation model, such as the XCELL program, can keep actual tallies of parts that traverse a particular process link. The implication of such differences in types of performance measures among tools for modeling manufacturing systems is that economic translation formulas must be specific to the system modeling tool. However, neither basic concerns of the manufacturing engineer about the pattern of cost accumulation in a system configuration nor the generic contributors (such as labor and materials) to this accumulation vary much by system modeling tool. These basic concerns and manufacturing-generic contributors are the focus of this section on economic translation.

Presentation of the section moves from fundamental perspectives about the incremental character of translation results to an examination of typical structures for economic spreadsheets in manufacturing. Then discussion shifts toward the calculation of spreadsheet items from projections of performance both for manufacturing system alternatives and for associated overhead such as product engineering, production planning, and control.

SELECTION OF THE BENCHMARK ALTERNATIVE

If the adoption of a new manufacturing technology by a plant would prevent the loss of market share in the mainstay product line of that plant, then it would seem advisable for the plant management to give serious consideration to the technology. However, a superficial economic evaluation of such an adoption proposal might assume stable market share regardless of technology adoption, in which case the proposal would be unlikely to find acceptance. Clearly, a close examination of the plant's economic future without the new technology, and consideration of the difference between this bleak future and life with the new technology, would make acceptance more probable. Thus, careful attention to the manufacturing context of the firm in the event of no commitments to manufacturing investments beyond those already approved within the planning horizon of interest is essential to capture the potential economic impact of newly proposed manufacturing investments. This future without approval of newly proposed manufacturing investments is called the benchmark alternative.

Three basic situations for manufacturing investment are the following: the ''brownfield'' where an existing plant is to be enhanced; the ''greenfield'' where a new plant is to be constructed for purposes of providing essential capacity; and the ''greenhouse'' where an experimental facility is to be built for purposes of acquiring valuable experience with new manufacturing technology and/or new product technology. Construction of a benchmark case is quite different from one of these situations to another. The brownfield situation was illustrated in the introductory paragraph.

A greenfield benchmark can take one of two forms. The first is the extension of existing plants to accommodate the capacity and capabilities of the greenfield opportunity; the second is the implementation of a standard plant which, in some sense, embodies current thinking about best manufacturing practice.

The latter type of benchmark for a greenfield can be used by a company if that company has no other plants or subcontracting options to absorb the work in question and can afford to develop standards for various aspects of the shop floor such as communications, the man-machine interface, and so on.

A greenhouse situation can be addressed by a benchmark alternative that comprehends the interference of manufacturing experiments at existing plants with ongoing operations. It can also consider the painfully gradual learning curves at existing plants should they adopt the greenhouse technologies without benefit of a prior greenhouse experience for the company.

INCREMENTAL COMPARISON OF ALTERNATIVES

All scenarios will be evaluated against the benchmark alternative. The procedure for deciding between two alternatives is through incremental analysis. That is, the incremental difference is evaluated to decide which alternative to choose. The economic spreadsheet reflects the difference between the benchmark and an alternative on a line-item basis over the planning horizon for the analysis (see Fig. 3-10). The example in Table 3-2 looks at just the equipment costs for two alternatives.

It is the last line, the difference, that is used in the economic comparison of the two alternatives. Each line item on the economic spreadsheet must reflect the difference between a benchmark and proposed alternative. When there is no difference between the alternatives, the line item will contain zeros. For example, if one assumes that sales will remain constant regardless of which alternative is implemented, the incremental sales revenue line will be 0.

The Economic Spreadsheet

An economic spreadsheet of incremental cash flows is provided in Fig. 3-10. This spreadsheet identifies a "shopping list" of line items that might be considered for the economic analysis of manufacturing systems. Each line item reflects the incremental economic impact of introducing the alternative vs. using the benchmark alternative. Certain line items may not be applicable to some evaluations, while others may need to be added. Therefore, the actual spreadsheet for a given analysis must be "customized" for the given technology under consideration.

The timeline across the top of the spreadsheet begins with year 0 (the time of the first expenditure for the system) and runs through the expected life of the system. Generally, the life of the system is determined by the product(s) it is expected to make. Any sequence of changes that the system goes through while producing the product should be captured within the spreadsheet. This might include, for example, technology upgrades and equipment enhancements. If the life of the benchmark is different than the alternative, then an adjustment is required so that each scenario meets the functional requirements over the entire life of the system.

Also along the top of the spreadsheet is a line showing an estimate for the annual inflation rate for each year on the spreadsheet. This line is used as an aid with some of the line items to estimate future-year cash flows. For example, new product revenue may perhaps be easily estimated for year 1. After year 1, if the level of output is expected to be constant, then one could assume that the revenue from future sales will increase with inflation. By stating an expected inflation line, values in the spreadsheet can use the line to adjust future sales revenue. There may be other types of annual adjustment assumptions to make. The following are a few commonly used adjustment factors and some line items that might use the factor:

- Inflation—sales revenue, labor rates, material costs, replacement equipment and tooling.
- Learning curve—to reflect the start-up inefficiencies of new technologies, level of output, labor hours, scrap/rework rates.
- Controlled level of capacity—to reflect the controlled level of activity in the system, level of output, labor hours, material usage.

Supporting schedules typically accompany the economic spreadsheet to show the methods and assumptions used in deriving the spreadsheet entries. Whether or not an adjustment line is incorporated directly on the spreadsheet or simply in a supporting schedule depends on the complexity of the procedure used and how much explicit detail is desired.

Explanation of Spreadsheet Categories

An advantage of performing incremental analysis is that when a particular line item is constant for all of the alternatives being evaluated, the incremental line on the spreadsheet is 0. Therefore, the value (revenue or expense) of the line item does not have to be computed explicitly for any of the alternatives. This is important to keep in mind when constructing a specific list of line items for incremental analysis; only line items with non-zero incremental entries over the life of the system are used.

The major spreadsheet categories are revenues, direct product expenses, capital equipment expenses, and indirect expenses.

Revenues. Revenues can come from several sources. Revenues from operations result from product sales and are simply the number of units sold times the selling price. If an economic analysis is being performed for a "piece" of a manufacturing system that doesn't actually produce a final product, then some other value of output must be determined (assuming the level of output of the two alternatives is different). For example, when the system produces a subcomponent, the value of that subcomponent must be determined. Its value can be determined by one of the following ways:

- Its market value if the subcomponent were sold on the open market.
- The price that would have to be paid to an outside contractor if the subcomponent were outsourced.
- A percentage of the final product's market value. The percentage could represent the portion of subcomponent cost (to produce) to the total cost to produce the final product. Therefore, the margin of the final product would be used for the margin of the subcomponent.

If the alternative being analyzed has no market effects, then revenue will be unchanged. That is, regardless of the alternative finally selected, there will be no impact on revenue. The objective of the analysis then is minimization of costs.

Other revenue line items shown in Fig. 3-10 might occur depending on the system alternatives. For example, plant sale

End of year:	0	1	2	3	4	5
Annual inflation:	0.0%	0.0%	5.0%	5.0%	5.0%	5.0%
	Jun-87	Jun-88	Jun 89	Jun-90	Jun-91	May-92
I. REVENUES						
FROM OPERATIONS						
A. Existing product sales	$0	$0	$0	$0	$0	$0
B. New product sales	$0	$20,000,000	$21,000,000	$22,050,000	$23,152,500	$24,310,125
C. Other sales	$0	$0	$0	$0	$0	$0
OTHER REVENUES						
D. Capital equipment revenue						
— Plant sale	$0	$0	$0	$0	$0	$0
— Plant lease	$0	$0	$0	$0	$0	$0
— Equipment sale	$0	$0	$0	$0	$0	$0
— Equipment lease	$0	$0	$0	$0	$0	$0
— Computer sale	$0	$0	$0	$0	$0	$0
— Computer lease	$0	$0	$0	$0	$0	$0
— Other sale	$0	$0	$0	$0	$0	$0
— Other lease	$0	$0	$0	$0	$0	$0
E. Miscellaneous revenue						
— Miscellaneous	$0	$0	$0	$0	$0	$0
TOTAL REVENUES	$0	$20,000,000	$21,000,000	$22,050,000	$23,152,500	$24,310,125
II. DIRECT PRODUCT EXPENSES						
A. Product labor & material & outside services						
Product A						
— Good production						
— Labor	$0	$2,000,000	$2,100,000	$2,205,000	$2,315,250	$2,431,013
— Material	$0	$1,000,000	$1,050,000	$1,102,500	$1,157,625	$1,215,506
— Scrap production						
— Labor	$0	$225,000	$236,250	$248,063	$260,466	$273,489
— Material	$0	$112,500	$118,125	$124.031	$130,233	$136,744
— Rework production						
— Labor	$0	$475,000	$498,750	$523,688	$549,872	$577,365
— Material	$0	$237,500	$249,375	$261,844	$274,936	$288,683
— Inventory holding	$0	$3,500	$3,675	$3,859	$4,052	$4,254
— Other production costs	$0	$0	$0	$0	$0	$0
Product B						
— Good production						
— Labor	$0	$2,500,000	$2,625,000	$2,756,250	$2,894,063	$3,038,766
— Material	$0	$1,250,000	$1,132,500	$1,378,125	$1,447,031	$1,519,383
— Scrap production						
— Labor	$0	$250,000	$262,500	$275,625	$289,406	$303,877
— Material	$0	$125,000	$131,250	$137,813	$144,703	$151,938
— Rework production						
— Labor	$0	$400,000	$420,000	$441,000	$463,050	$486,203
— Material	$0	$200,000	$210,000	$220,500	$231,525	$243,101
— Inventory holding	$0	$2,900	$3,045	$3,197	$3,357	$3,525
— Other production costs	$0	$0	$0	$0	$0	$0
Total product labor & Material & outside services	$0	$8,781,400	$9,220,470	$9,681,494	$10,165,568	$10,673,847
B. Other product material & labor & outside services	$0	$0	$0	$0	$0	$0
TOTAL DIRECT PRODUCT EXPENSE	$0	$8,781,400	$9,220,470	$9,681,494	$10,165,568	$10,673,847

Fig. 3-10 Incremental cash-flow comparison between a benchmark case and a proposed alternative. (*Industrial Technology Institute*)

End of year:	0	1	2	3	4	5
Annual inflation:	0.0%	0.0%	5.0%	5.0%	5.0%	5.0%
	Jun-87	Jun-88	Jun 89	Jun-90	Jun-91	May-92
III. CAPITAL EQUIPMENT EXPENSES						
A. Purchases						
— Plant	$2,000,000	$150,000	$0	$0	$0	$0
— Equipment	$1,000,000	$100,000	$50,000			
— Nonperishable tooling	$50,000	$5,000	$2,500	$0	$0	$0
— Computer hardware	$225,000	$22,500	$23,625	$24,806	$26,047	$27,349
— Computer software	$75,000	$10,000	$10,500	$11,025	$11,576	$12,155
— Other	$33,000	$5,000	$5,250	$5,513	$5,788	$6,078
B. Improvements						
— Plant	$0	$0	$0	$0	$0	$0
— Equipment & tooling	$0	$0	$0	$0	$0	$0
— Nonperishable tooling	$0	$0	$0	$0	$0	$0
— Computer hardware	$0	$0	$0	$0	$0	$0
— Computer software	$0	$0	$0	$0	$0	$0
— Other	$0	$0	$0	$0	$0	$0
Total capital equipment purchases & improvements	$3,383,000	$292,500	$91,875	$41,344	$43,411	$45,581
C. Leases						
— Plant	$0	$0	$0	$0	$0	$0
— Equipment	$0	$0	$0	$0	$0	$0
— Nonperishable tooing	$0	$0	$0	$0	$0	$0
— Computer hardware	$10,000	$10,000	$10,500	$11,025	$11,576	$12,155
— Computer software	$10,000	$10,000	$10,500	$11,025	$11,576	$12,155
— Other	$5,500	$5,500	$5,775	$6,064	$6,367	$6,685
Total capital equipment lease expense	$25,500	$25,500	$26,775	$28,114	$29,519	$30,995
TOTAL CAPITAL EQUIPMENT EXPENSES	$3,408,500	$318,000	$118,650	$69,458	$72,930	$76,577
D. Manufacturing supplies						
Existing product						
— Perishable tooling	$0	$0	$0	$0	$0	$0
— Fixtures	$0	$0	$0	$0	$0	$0
— Other	$0	$0	$0	$0	$0	$0
New product						
— Perishable tooling	$130,000	$260,000	$273,000	$286,650	$300,983	$316,032
— Fixtures	$150,000	$300,000	$315,000	$330,750	$347,288	$364,652
— Other	$0	$0	$0	$0	$0	$0
Total indirect manufacturing supplies	$280,000	$560,000	$588,000	$617,400	$648,270	$680,684
GENERAL & ADMINISTRATIVE						
E. Selling expenses						
— Freight out	$0	$275,000	$288,750	$303,188	$318,347	$334,264
— Sales salaries	$0	$0	$0	$0	$0	$0
— Sales commission	$0	$0	$0	$0	$0	$0
— Advertising	$0	$0	$0	$0	$0	$0
— Travel	$0	$0	$0	$0	$0	$0
— Other	$0	$0	$0	$0	$0	$0
Total selling expenses	$0	$275,000	$288,750	$303,188	$318,347	$334,264

(continued)

Fig. 3-10—*Continued*

End of year:	0	1	2	3	4	5
Annual inflation:	0.0%	0.0%	5.0%	5.0%	5.0%	5.0%
	Jun-87	Jun-88	Jun 89	Jun-90	Jun-91	May-92
F. Business Office						
— Salaries						
— Product managers	$87,500	$175,000	$183,750	$192,938	$202,584	$212,714
— Purchasing	$112,500	$225,000	$236,250	$248,063	$260,466	$273,489
— Expediting	$62,500	$125,000	$131,250	$137,813	$144,703	$151,938
— Computer support	$87,500	$175,000	$183,750	$192,938	$202,584	$212,714
— Office	$42,500	$85,000	$89,250	$93,713	$98,398	$103,318
— Other	$44,000	$88,000	$92,400	$97,020	$101,871	$106,965
— Insurance	$225,000	$225,000	$236,250	$248,063	$260,466	$273,489
— Telephone	$125,000	$125,000	$131,250	$137,813	$144,703	$151,938
— Accounting & legal	$200,000	$200,000	$210,000	$220,500	$231,525	$243,101
— Bad debts	$0	$0	$0	$0	$0	$0
— Interest	$0	$0	$0	$0	$0	$0
— Other G & A	$100,000	$1,500,00	$1,575,000	$1,653,750	$1,736,438	$1,823,259
Total business office	$1,086,500	$2,923,000	$3,069,150	$3,222,608	$3,383,738	$3,552,925
G. Miscellaneous						
— Heat & electricity	$275,000	$275,000	$288,750	$303,188	$318,347	$334,264
— Supplies	$55,000	$55,000	$57,750	$60,638	$63,669	$66,853
— Repair & maintenance parts	$400,000	$400,000	$420,000	$441,000	$463,050	$486,203
— Vehicle expense	$0	$0	$0	$0	$0	$0
— Training equipment	$120,000	$120,000	$126,000	$132,300	$138,915	$145,861
— Other	$0	$250,000	$262,500	$275,625	$289,406	$303,877
Total indirect miscellaneous	$850,000	$1,100,000	$1,155,000	$1,212,750	$1,273,388	$1,337,057
TOTAL INDIRECT EXPENSES	$3,759,000	$8,143,000	$8,550,150	$8,977,658	$9,426,540	$9,897,867
IV. INDIRECT EXPENSES MANUFACTURING						
A. Product warranty						
— Existing product	$0	$0	$0	$0	$0	$0
— New product	$0	$200,000	$210,000	$220,500	$231,525	$243,101
Total product warranty	$0	$200,000	$210,000	$220,500	$231,525	$243,101
B. Product liability						
New product						
— Insurance	$0	$0	$0	$0	$0	$0
— Expected loss	$0	$0	$0	$0	$0	$0
Existing product						
— Insurance	$0	$0	$0	$0	$0	$0
— Expected loss	$0	$0	$0	$0	$0	$0
Total product liability	$0	$0	$0	$0	$0	$0
C. Manufacturing labor						
— Engineering	$250,000	$500,000	$525,000	$551,250	$578,813	$607,753
— Custodial	$100,000	$200,000	$210,000	$220,500	$231,525	$243,101
— Maintenance	$175,000	$350,000	$367,500	$385,875	$405,169	$425,427
— Management	$100,000	$200,000	$210,000	$220,500	$231,525	$243,101
— Material handling	$0	$0	$0	$0	$0	$0
— Production control	$62,500	$125,000	$131,250	$137,813	$144,703	$151,938
— Quality control	$75,000	$150,000	$157,500	$165,375	$173,644	$182,326
— Software specialists	$50,000	$100,000	$105,000	$110,250	$115,763	$121,551
— Stores control	$50,000	$100,000	$105,000	$110,250	$115,763	$121,551
— Supervisory	$75,000	$150,000	$157,500	$165,375	$173,644	$182,326
— Tooling preparation	$62 [illegible]00	$125,000	$131,250	$137,813	$144,703	$151,938
— Trainers	$42,500	$85,000	$89,250	$93,713	$98,398	$103,318
— Other	$500,000	$1,000,000	$1,050,000	$1,102,500	$1,157,625	$1,215,506
Total indirect manufacturing labor	$1,542,500	$3,085,000	$3,239,250	$3,401,213	$3,571,273	$3,749,837

Fig. 3-10—*Continued*

TABLE 3-2
Comparison of Equipment Costs for Two Alternatives

	Year 0	Year 1	Year 2	Year 3
Alternative A:	$1000	$800	$800	$500
Benchmark:	700	700	700	700
Alternative A minus benchmark:	$ 300	$100	$100	($200)

(Industrial Technology Institute)

might occur if one alternative uses an existing facility and another one does not. Therefore, the alternative that does not use the existing facility might provide revenue through the sale of that facility, although this revenue could be offset by teardown costs. The expense of purchasing a new facility will be captured under expenses. The other revenue line items are used for similar purposes.

Direct product expenses. Direct product expenses include direct labor and material, other direct product costs (outside services), and inventory holding costs. Figure 3-10 shows these costs broken down by product for good, scrap, and rework parts; however, this detail may not be necessary for the financial analyst to whom only the total of product costs is important in justification. However, the product cost structure is important to the system designer when looking for opportune areas for cost reduction and for system improvements. The product cost structure can show the savings by eliminating or reducing scrap or rework for a particular product.

Inventory holding costs result from the raw material, work-in-process (WIP), and finished goods' inventories for each of the products produced. Each inventory cost is adjusted by an inventory holding rate to reflect the required financing to maintain that inventory. If a company's before-tax cost of capital is 20%, then the total inventory cost is multiplied by 0.20 to determine the inventory holding cost. Inventory-related labor and equipment costs, such as storage/retrieval systems and transporters, are captured separately under indirect manufacturing labor and capital equipment categories, respectively. The inventory cost is generally determined as follows:

- Raw material. Cost of raw materials (times quantity), including costs for shrinkage.
- Work-in-process. Cumulative part cost of labor and materials at every point in production (times average quantity at every point). Other related costs such as required floor space, inventory holding equipment, and even job accounting/tracking costs could be considered WIP costs, but are generally captured under other categories.
- Finished goods. Similar to work-in-process inventory, only all the products are finished.

Capital equipment expenses. Capital equipment expenses reflect investments in depreciable assets. The definition of depreciable assets varies with IRS rulings and should be checked when preparing an economic analysis. Also, the depreciation schedule for a particular asset is necessary to determine taxes and will be needed later to determine after-tax cash flows. All start-up costs, including freight, installation of the equipment, facilities modifications, and special industrial engineering studies, are added to the cost of the equipment to determine its total cost. Each start-up cost should be classified as expense or capital for cash flow calculations. Capital equipment improvements such as retrofits are sometimes depreciated and are treated like capital equipment purchases.

There are many types of leases; some treat the asset as depreciable, and some treat it as an expense, with the expense equal to the amount of the periodic payment. Expensed purchases belong in the direct expenses section of the spreadsheet (refer to Section II of Fig. 3-10). The particular lease under consideration may require the assistance of a tax specialist.

The capital equipment expenditures, regardless of taxes and depreciation, should be entered on the spreadsheet in the appropriate time column.

Indirect expenses. Quality and volume (of output) are the two key attributes affecting product warranty and liability. Therefore, if the alternatives being compared do not differ in these attributes, the incremental impact will be 0. Generally, product warranty and liability will vary proportionally with level of output. The impact of quality on warranty and liability costs requires a careful analysis as to their root cause(s) in the benchmark case. It is quite possible that improvements in quality do not change warranty or liability. A determination should be made to see if changes in quality affect sales levels or price and to see if the change affects the manufacturing system's performance through reduced downtime, less rework, and so on.

Indirect manufacturing labor includes direct manufacturing support such as floor supervision, tool crib costs, quality control, and other similar costs that are shared across multiple product lines. These costs cannot be easily traced to any one product. Manufacturing supplies are similar in that they cannot be traced to any one product and might include perishable tooling, packaging, and coolant, to name a few.

Indirect general and administrative costs are not directly tied to manufacturing. They typically do not change significantly between alternative manufacturing scenarios unless there prove to be major differences in the output or in the way the business operates.

The Financial Summary

The financial summary pulls together the incremental analysis, determines taxes, and then calculates several economic measures on the incremental cash flow analysis (see Fig. 3-11). The sequence of calculations is self-explanatory. A new category, depreciation, is necessary to determine the after-tax cash flow analysis. Because different assets have different depreciable lives as set by IRS rulings, a tax consultant should help derive the depreciation schedule. Every piece of equipment listed in the capital equipment section must have a depreciation schedule. Similarly, if the incremental analysis calls for the sale of any existing capital assets, the effect on depreciation must be considered here as well. The selection of an appropriate tax rate should be obtained from the accounting/tax department.

The measures of merit shown in Fig. 3-11 include the payback period, present value, and return-on-investment calculations. Both present value and return on investment have been calculated over two different planning horizons: 5 years and 10 years. The period that should be used depends on the firm's planning horizon and the life of the project.

Obtaining Estimates for Indirect Costs

Indirect manufacturing labor can constitute a significant portion of overall production costs. Traditional approaches have

	Jun-87	Jun-88	Jun 89	Jun-90	Jun-91	May-92
A. REVENUES	$0	$20,000,000	$21,000,000	$22,050,000	$23,152,500	$24,310,125
B. EXPENSES						
1. Total direct product	$0	$8,781,400	$9,220,470	$9,681,494	$10,165,568	$10,673,847
2. Total equipment purc. & inprov.	$3,383,000	$292,500	$91,875	$41,344	$43,411	$45,581
3. Total equipment leases	$25,500	$25,500	$26,775	$28,114	$29,519	$30,995
4. Total indirect	$3,759,000	$8,143,000	$8,550,150	$8,977,658	$9,426,540	$9,897,867
TOTAL EXPENSES	$7,167,500	$17,242,400	$17,889,270	$18,728,609	$19,665,039	$20,648,291
C. BEFORE-TAX CASH FLOW (A. LESS B.)	($7,167,500)	$2,757,600	$3,110,730	$3,321,392	$3,487,461	$3,661,834
D. TAXABLE INCOME/(LOSS) BEFORE DEPRECIATION (C. LESS B.4)	($3,784,500)	$3,050,100	$3,202,605	$3,362,735	$3,530,872	$3,707,416
E. DEPRECIATION						
3-year assets	$0	$0	$0	$0	$0	$0
5-year assets	$0	$0	$0	$0	$0	$0
10-year assets	$0	$0	$0	$0	$0	$0
F. TAXABLE INCOME/(LOSS) (D. LESS E.)	($3,784,500)	$3,050,100	$3,202,605	$3,362,735	$3,530,872	$3,707,416
G. TAXES (TAX RATE TIMES F.)	($1,324,575)	$1,067,535	$1,120,912	$1,176,957	$1,235,805	$1,297,595
H. NET AFTER-TAX CASH FLOW (C. MINUS G.)	($5,842,925)	$1,690,065	$1,989,818	$2,144,434	$2,251,656	$2,364,239
I. CUMULATIVE NET AFTER-TAX CASH FLOW	($5,842,925)	($4,152,860)	($2,163,042)	($18,608)	$2,233,048	$4,597,287

FINANCIAL MEASURES	
* Payback period (years):	3.0
* Present value (10 years/10%):	$7,593,442
* Present value (5 years/10%):	$1,777,307
* Return on investment (10 years):	34.4%
* Return on investment (5 years):	21.6%

Fig. 3-11 Financial summary of benchmark and proposed alternative comparison. (*Industrial Technology Institute*)

related indirect costs as a proportion to direct manufacturing costs. With traditional manufacturing, this approximation was adequate because indirect costs were heavily outweighed by direct costs, and small errors in estimating direct costs did not significantly affect the indirect cost estimate. New, more advanced manufacturing technologies, however, have redistributed manufacturing costs to the indirect areas. Traditional relationships have been breaking down because the indirect costs commonly exceed direct costs by several hundred percent. Therefore, estimating indirect costs can now be more important than estimating direct costs.

One approach to estimating indirect costs is through transaction analysis. A transaction is defined as some signal to do something, either initiated by the manufacturing system or by a support/indirect function external to the manufacturing system. For a particular manufacturing system, transactions can be categorized as follows:

- Logistical. Transactions that relate to the location and movement of resources such as material, jobs, equipment, and personnel in the manufacturing system.
- Balancing. Transactions that control the available amounts/volumes of resources at given periods of time.
- Quality. Transactions that pertain to process and product quality, including the operating systems necessary to carry out the transactions such as personnel, file systems, audit systems, and specifications.

When there are significant differences between alternatives in indirect cost areas, a transaction analysis can be helpful in estimating the incremental economic difference. Indirect support areas include manufacturing technical support, maintenance personnel and parts, expeditors, system auditors, and information systems. A few key occurrences (transaction sources) calling for action, and some of the support functions affected, include:

- Equipment failure—maintenance, engineering, expeditors, and schedulers.
- Production fluctuations—expeditors, scheduling, purchasing, and sales.
- Product defect detection—quality control, purchasing, and scheduling.
- Engineering change orders—engineering, expediting, scheduling, supervision, and purchasing.
- New job arrivals—engineering, purchasing, scheduling, purchasing, and auditors.

The transaction analysis in Fig. 3-12 compares the economic impact of manually generating engineering designs vs. use of a computer-aided design and manufacturing (CAD/CAM) system. This analysis lists the activities (transactions) down the left column for the occurrence of a new job arrival. Only those transactions that differ between manual and CAD/CAM processing are listed; of course, many more transactions are involved to completely process a job. The data in the table (average processing time, average labor rates, average volume of jobs, and the productivity ratios) come from a variety of sources. For example, good sources of this type of information come from the personnel involved, industrial engineering, vendors, and studies such as simulation. The resultant annual savings ($834,581) might now be adjusted to reflect a learning curve for the introduction of CAD/CAM if full realization of this benefit is not expected until a start-up period is completed. Also, this benefit reflects current-year savings that might be adjusted for changes in labor rates. Transaction analysis can be used for any area on the shop floor to ascertain the incremental economic impact in the support areas of the benchmark and proposed alternative. Although the example shows only savings, additional costs can be identified by the transactional procedure in other applications.

Obtaining Estimates for Direct Costs

Direct manufacturing costs can be directly traced to a single product. If it is a cost that is shared across two or more product lines, then some ''fair'' means of cost sharing is necessary or else the cost becomes indirect. Typically, direct cost categories include line items for labor, material, and outside services. The economic spreadsheet in Fig. 3-10 includes these categories and has also added inventory holding cost. Whether or not these costs will be direct when the manufacturing system becomes operational should not affect the economic analysis. In performing an economic comparison of alternatives, one should track costs as closely as possible with the products that incur them. Figure 3-10 goes one step further and breaks down these direct costs according to good, scrap, and rework production for each product. This detail is generally helpful if the breakdown is readily available. Detailed manufacturing models (simulation or queuing networks) are often capable of this information. If this resolution of direct costs is not available, then the total ''system'' costs should be lumped together, preferably by product.

Direct labor. Determination of direct labor requires the hourly rate (usually an average will do, and it should include wages and taxes paid by the company), the overtime rate (if

No. of Jobs Per Year: 171	Manual		Alternative A: CAD/CAM			
Activity (Transaction List)	Total Hours (per job)	Average Labor Rate, $/hr	Productivity Ratio	CAD/CAM Hours	Time Saved, hours	Savings, $
1.0 Project management						
1.1 Project planning	20	30	1.50	2213	1107	33,200
2.0 Design						
2.1 Schematics	22	25	1.40	2609	1043	26,086
2.2 Feasibility analysis	5	25	5.00	166	664	16,600
2.3 Design and layout	40	25	2.00	3320	3320	83,000
2.4 Fabrication drawing	6	25	3.00	332	664	16,600
2.5 Assembly drawing	32	25	10.00	531	4781	119,520
2.6 Artwork	40	25	40.00	166	6474	161,850
2.7 Checking	30	25	2.75	1811	3169	79,227
3.0 Development	20	30	1.25	2656	664	19,920
4.0 Manufacturing—support						
4.1 Production planning	5	30	1.25	664	166	4,980
4.2 Scheduling & expediting	5	30	1.25	664	166	4,980
4.3 Engineering	5	30	1.25	664	166	4,980
4.4 Quality control	10	25	1.25	1328	332	8,300
4.5 NC & CNC programming	10	25	3.00	553	1107	27,667
5.0 Manufacturing—operations						
5.1 Inventory reduction						
5.2 Additional capacity						
				Total:		

Fig. 3-12 Sample transaction analysis comparing the economic impact of manually generated designs to computer-aided design. (*Industrial Technology Institute*)

applicable), annual fringe benefit costs, and total hours worked. The total amount of production time (one, two, or three shifts, and the amount of overtime) is then used as a multipler for labor costs. If the amount of labor that goes into rework and scrap is available, then this breakdown can be included on the spreadsheet.

Direct material. The cost of raw materials to produce a finished product needs to be determined. Usually the purchase price including shipping is sufficient.

When a material comes into a manufacturing subsystem from another subsystem (for example, a subassembly), a transfer price must be determined. If a price has been determined by accounting, then that price represents the cost. Otherwise, the same method used for revenue determination (see the subhead entitled "Revenues" earlier) can be used for cost determination: market value, equivalent outsourcing cost, or manufacturing cost plus some markup for shared profit (margin). A balance must be maintained where the cost of a material to one system/department must represent the revenue (transfer price) to the source system/department. When both systems (the subassembly and assembly systems) are in the scope of the analysis, the subassembly is treated as work in process and is valued according to its total raw materials and direct labor. This is described subsequently.

Inventories. There are three types of inventories: raw material, work-in-process (WIP), and finished goods. It is generally acceptable to determine an average inventory level for each of these three and then to assign a cost based on the average value of that inventory. This approach may not be acceptable when inventories fluctuate widely, when costs vary significantly, or if the makeup of the inventory (distribution of part/product types) varies broadly. In these cases, a more detailed breakdown may be necessary.

Once an inventory value is derived, an inventory holding rate can be used to reflect the cost to the business for carrying the inventory. In essence, the holding rate times the inventory value represents the annual investment cost to the business. This method accounts for the investment capital the firm has tied up in inventory. The inventory holding rate can be the firm's cost of capital or, if available, the interest rate at which capital is borrowed against inventories. If the economic spreadsheet has annual cash flow summaries, then the inventory holding rate should be an annual rate; the inventory holding cost is calculated each year for the life of the system.

Raw materials are generally valued according to the material purchase price plus any additional costs for freight.

Work-in-process cost is less straightforward to estimate because it occurs throughout the manufacturing process and represents an investment of direct materials and labor. A generally accepted approach to value WIP is to accumulate direct product costs (labor, material, and services) throughout the production process for each part. Therefore, a part near the end of the production process would have a greater value than a part near the beginning of the process. The resolution of valuation that is needed depends on the volume of parts, location(s) of WIP inventories in the process, and the complexity of the process (many alternative processes require more analysis).

In general, it is helpful to identify the major WIP locations in a process (buffers) and then to assign a WIP value based on the accumulated direct costs. A distinction may have to be made between physical buffers and logical buffers. On the floor, one storage location (physical buffer) may store parts at several different levels of completion (logical buffer—one for each level of completion). An average accumulation of direct costs needs to be made for each level of completion. In general, this would occur after each process. This process is complicated if parts at the same level of completion may have gone through different series of operations, such as rework, mixture of outsourcing, and in-house processes. When there are alternative routings for a particular part, a good approach for valuation is to set up a table that accumulates the costs by operation and then weigh the accumulated WIP according to the proportion of parts through different routes.

The information in Table 3-3 shows a process sequence with process 3 going through either workcenter A or B at 40 and 60% proportions, respectively. Also, the direct material and labor vary for process 3 depending on which workcenter performs the process. The total cost column represents the total direct costs added to a part at each process. The cumulative value column adds the costs from all processes up to the current process, plus the value added at the current process. However, for process 4, the cumulative value must account for the proportion of parts that went different routes in process 3. Therefore, a weighted cumulative value is 0.4 x $4800 + 0.6 x $4900 = $4860. The average value of a WIP part waiting for process 4 is $4860.

Once this table of cumulative costs has been derived for each part in the system, the average buffer quantity, by process, can be used to estimate the average in-process value by process. Total WIP value is the sum of all buffer values and can be totaled by process or by part. The total is multiplied by the inventory holding rate to represent the company's cost of WIP inventory.

Finished goods are valued the same as raw material, based on direct costs. Finished goods, however, have direct material plus labor added. Therefore, the cost of finished goods is simply the average cumulative value of direct costs for completed product in inventory. This figure is the last value in the cumulative value column in Table 3-3. Again, the value is multiplied by the quantity and then by the inventory holding rate to represent the firm's holding cost for finished goods inventory.

Assembling Economic Spreadsheets for System Evolutions

Manufacturing systems/subsystems with significant evolution over time will have changing performance and, consequently, changing economic behavior. Averages like inventory levels, throughput, and product cost structure can change significantly with modifications to an existing system. Averages may be acceptable in one phase of the manufacturing system/subsystem, but not in the next. The effect of planned changes, such as the evolution from NC to CNC to DNC/FMS, should be estimated as part of the economic justification/planning process, even if the planned changes won't happen for several years. Different performance and economic estimates should be developed for each phase of the system. After the economic spreadsheets are developed for each phase, they should be assembled into one overall estimate with the performance measures calculated for the lifecycle of the system.

Economic Translation in Context

A manufacturing system alternative could require multiple applications of economic translation to accommodate uncertainty on the part of the system designer about the values of technological and/or market parameters at a particular time. The uncertainties relevant to one alternative evolution of the manu-

TABLE 3-3
Cumulative Costs of a Process Sequence

Process	Work-center	Proportion	Material	Labor	Services	Total Cost	Cumulative Value
1	A	100%	$2000	$800		$2800	$2800
2	A	100%	$0	$1200		$1200	$4000
3	A	40%	$500	$300		$800	$4800
	B	60%	$200	$700		$900	$4900
4	A	100%	$0	$500		$500	$5360
5	—	100%	$0	$0	$500	$500	$5860
Finish							

(*Industrial Technology Institute*)

facturing system might or might not be relevant for another alternative evolution under consideration. If essential economic information is to be gathered efficiently, a coordinated approach is called for to capture riskiness of investment and lifecycle relationships. The tools of modern decision analysis provide a framework for such coordination.

DECISION ANALYSIS

It is becoming more difficult to compete in manufacturing because of the increasing rate at which things change. What may have been an ideal manufacturing system when it was installed soon becomes obsolete for many reasons. As the product passes through its product lifecycle, not only does the market for the product change, but so do the demands placed on the manufacturing system. As the firm gains experience producing the product, the benefits of experience can be reflected in the learning curve (or experience curve), and this too affects the manufacturing system. Information on learning curves can be found in Chapter 4, "Cost Estimating and Control," of this volume. Further, technological advancements with respect to the product and process influence the manufacturing system.

Thus, production facilities, equipment, and organization being installed today will typically be forced to undergo major changes in about 3 years. Unfortunately, the future is uncertain, and one cannot predict what these changes will be. Engineers have recognized this and have intuitively favored flexible manufacturing systems. Flexible systems can respond to changing needs with less expense and disruption of production than manufacturing systems rigidly designed for a single, narrowly defined set of products, processes, and production volumes. Justification of flexible systems has presented a major problem, however. The costs of flexibility are apparent when alternative investments are analyzed, but the benefits are difficult to quantify and are typically dealt with as intangibles. Thus, it is difficult to justify an investment for flexibility even though one is aware of the benefits. Engineers would like somehow to capture those benefits, just as they are now able to capture other "intangibles" with models of manufacturing systems.

Modeling alone, however, is not sufficient to capture the benefits of flexibility. While modeling is able to quantify benefits to the entire manufacturing system and not just those at the workplace itself, the models described previously deal with a specific configuration of a system and are unable to deal with future changes to the system. It is necessary to look to the future to capture the benefits of flexibility. Engineers must think about what events may take place in the future and what decisions they may have to make in the future. Decision analysis expands the scope of analysis from the manufacturing system at one time to the evolution of the system over its lifecycle.

While one cannot foretell the future, it is possible to communicate with specialists who know what is happening in their fields. By doing so, one can often detect trends and deduce future possibilities from these trends. It is highly unlikely that a person will be able to deduce everything relevant about the future, and therefore he or she will be unable to deduce anything with certainty. Whenever one deals with the future, uncertainty is inevitable. But thinking about what might happen, what decisions the firm may make, and what risks it may be faced with are the first steps in dealing with the future and minimizing the risk of making costly errors.

In discussing future possibilities, one often finds use for some structural framework and related tools or techniques of evaluation. Jack Hirshleifer, an economist at UCLA, has given a great deal of thought to risk or uncertainty and has written extensively on it.[8] He describes two approaches to risk. One approach deals with the situation in which one knows about what will happen, but not precisely. For example, a meteorologist may have studied weather patterns and prepared a weather forecast that calls for a high temperature tomorrow of 70° (21°C). If one thinks about it, one realizes the unlikeliness of the temperature reaching precisely 70.00°—it may be 69 or 71. Likewise, if a marketing specialist has studied the market patterns and has forecast sales of 800,000 units for a particular product next year, sales will probably be about 800,000—maybe 770,000 or maybe 830,000—but probably not exactly 800,000. This is referred to as a "mean-deviation" approach to risk because the probable values have a mean; if one were to draw a probability distribution of the likely values, they would show some pattern of deviation from this mean value. The probability distribution might be in the form of the normal distribution, or it might take some other form. This approach deals with large numbers of future events, including populations, markets, and sometimes manufacturing cycle times and scrap rates.

Another approach to risk deals with the situation in which various states of the world may prevail at particular times in the future. For example, it could rain tomorrow or the sun could shine. Unlike the temperature example where the difference between a high temperature of 69 or 71° matters little, one may have different preferences for tomorrow, depending on the state of the world. If it rains, one may wish to rent a videocassette tonight and watch an old favorite movie tomorrow; if the sun shines, one may wish to reserve a tee time and be out on the golf course. Likewise, during the next 3 years firm A may or may not bring out a new and advanced product that makes firm B's product obsolete. Two possible states of the world have been defined 3 years hence—one in which firm B's product is obsolete and one in which it is not. If the product is to be obsolete in 3 years, a firm may wish to invest its available funds in designing a new product that will leap-frog the competition. If not, it may wish to invest in additional capacity to produce the existing product to maintain market share. This is called a "time-state-preference" approach to risk, in which at any given time certain states of the world might prevail, and one has a preferred position for each of these states of the world. This approach also deals with large numbers of future events, including wars, technological breakthroughs, and sometimes manufacturing cycle times and scrap rates.

How can a single measure, such as scrap rate, be an example of both mean-deviation risk and time-state-preference risk? If one's state of the world is defined as continuing to use a particular manufacturing process, the scrap rate may be somewhere between 1 and 3% with a mean of 2%. If, however, one defines an alternative state of the world in which a new manufacturing process with a higher degree of control is used, the scrap rate may be somewhere between 0.2 and 0.8% with a mean of 0.5%. The two approaches regarding risk are not mutually exclusive.

As the techniques for evaluating risk are discussed, examples will be used involving investments, the resulting cash flows, and net present value. How these investments were initiated, how the cash flows were estimated, or how net present values were calculated will not be discussed in detail, as these are covered in the appendixes of this chapter.

A MONTE CARLO TECHNIQUE TO CAPTURE DEVIATION

In the examples of capital investment analyses, a single value was given for each variable. In the investment to add capacity, for example, sales were forecast at a certain level for each year in the future. If one thought of this example in terms of the mean-deviation approach to risk, one would have considered the most likely value of probable future sales, but would have ignored the potential deviation. The resulting cash flow forecasts and net present value also represented most likely values.

Often, information about deviations from mean is at least as valuable as the values themselves. Many of the commercially available spreadsheet programs make it easy to play "what if" games, in recognition of the importance of forecasting outcomes when input variables deviate from a single most probable value. Another way of capturing this information is to define the probability distribution of input variables and use this information to calculate a probability distribution of an outcome, such as net present value. If input variables all have a normal probability distribution, it is possible, but tedious, to calculate the probability distribution of the outcome. An easier and more general approach is known as the Monte Carlo technique in which probability distributions of input variables are defined, and a sample of values is drawn at random that reflects the probability distribution of each input; then a sample of net present values is calculated with these input values. The result is a probability distribution of the net present value indicating not only the most probable value, but also an approximation of the deviations and their probabilities. The same technique can be used for internal rate of return and profitability index as well.[9,10]

A GAME THEORY TECHNIQUE FOR DEALING WITH RISK

Some firms use what is called a payoff matrix to make risky decisions. In its simplest form, the payoffs of alternative investments are considered under alternative states of the world. Referring back to the earlier example, assume that the firm may invest $1,000,000 in increased manufacturing capacity to produce the existing product or alternatively in an engineering program to design a new product. It cannot do both. Each decision is risky. If all the money is invested in designing a new product, there may not be sufficient production capacity, and the firm would lose market share. If all the money were invested in production capacity, the competitor may come out with a new product that will make the firm's existing product obsolete.

One possibility is this. Four evaluations may be made of net present value, as follows:

1. Net present value of added capacity if new competitive product: -$2 MM.
2. Net present value of added capacity if no new competitive product: $3 MM.
3. Net present value of new product if new competitive product: $2 MM.
4. Net present value of new product if no new competitive product: -$1 MM.

From this information, a "payoff matrix" may be prepared in which the two possible events are represented by two columns and the two possible actions are represented by rows. The matrix for this example is shown in Table 3-4.

A criterion is now needed on which a decision can be based. One may elect to make the decision that will lead to the largest possible gain. Adding capacity to achieve the greatest present value would be the decision of a risk-seeker firm.

A risk-averse firm would look askance at such a decision because, if the future did not turn out so well, the net present value would be negative. This firm is more likely to minimize the maximum loss by designing a new product on the basis that, at worst, the net present value would be 0.

TABLE 3-4
Example of a Payoff Matrix

	New Competitive Product	No New Competitive Product
Add capacity	-$2 MM	$4 MM
Design new product	$0 MM	$1 MM

(*Industrial Technology Institute*)

Alternatively, the firm may elect to minimize its "regret." Regret is defined as the difference between the value of an action and the value of the best action for that state of the world. Mathematically, this is equal to the difference between the value of each cell of the matrix and the highest value in that column of the matrix. The regret matrix of the payoff matrix in Table 3-4 is shown in Table 3-5. Minimal regret will occur with the decision to add capacity.

While this approach is of some use, it is limited in its application. The matrixes become complex with multiple decisions and states of the world. Also, the decision criteria may be considered mechanistic and unreasonable. The perspective is useful, however, and it leads to the concept of the decision tree, described subsequently.

EXPECTED VALUE

A somewhat different approach is to maximize expected value. For this, some estimate about the relative probabilities of the various outcomes is necessary. For example, the firm might assume equal probabilities of the two future states of the world and make the decision on the basis of expected value. This is equal to the sum of the products of each value times its probability. The expected value of adding capacity under the assumption of equal probabilities is as follows:

$$\text{Expected value} = (0.5 \times -2) + (0.5 \times 4) = -1 + 2 = 1$$

The expected value of designing a new product under the assumption of equal probabilities is:

$$\text{Expected value} = (0.5 \times 0) + (0.5 \times 1) = 0 + 0.5 = 0.5$$

Based on this decision criterion, the firm will elect to add capacity. While the concept of expected value is not usually a sufficient basis for making an investment decision, the concept is useful and, like the payoff matrix, leads to consideration of decision trees.

DECISION TREES

A decision tree is a sequenced layout of future decisions and events, their relationships, outcomes, and probabilities. It is useful for analyzing the long-term significance of near-term decisions under various future conditions. As such, it has particular value in the capture of potential costs and benefits from investments in flexible manufacturing systems.

An example of a simple decision tree is shown in Fig. 3-13. This is a printout of Arborist, a decision tree program by Texas Instruments. In this example, an initial investment of $1,000,000 is made, which Arborist has defined as function 1 (F1). It may be assumed that the investment is made in year 0 for the purpose of producing a product in subsequent years. In year 1, there is a 70% probability that the product will succeed with a positive cash flow of $400,000 (F2) and a 30% probability that it will fail with a cash flow of 0 (F5). If the product succeeds, a decision must be made whether to make a further investment of $500,000 (F3) or continue with no further investment (F5). If the investment is made, there is a 70% probability that cash flow in year 2 will double to $800,000 (F4) and a 30% probability that it will remain at $400,000 as it was in year 1 (F2). If a decision is made not to make the investment, a 100% probability that cash flow will remain $400,000 (F2) in year 2 can be assumed.

If the product fails in year 1, it is equally likely that the product will succeed in year 2 with cash flow of $400,000 (F2) or fail with cash flow of 0 (F5).

This simple example illustrates the following:

- The root node, at which the decision tree starts.
- Chance nodes (C), in which the path depends on a chance event.
- Decision nodes (D), in which the path depends on a decision.
- End nodes representing possible outcomes.
- Branches, leading from one node to another.
- Paths, leading from the root node to the various end nodes.
- Subtrees, which could be viewed as separate decision trees.

From each node, branches extend to represent all possible outcomes. Thus, the sum of the probabilities assigned to the branches emanating from each chance node is equal to 1.

For each path from the root node to a separate end node, the probability of reaching that particular end node is obtained by multiplying the probabilities assigned to each branch on the path. The sum of the probabilities of all possible paths from a chance node is equal to 1.

If the cost of capital is known, a net present value may be calculated for each path. In this case, a cost of capital of 10% is assumed, and cash flows are discounted to define the values VAL1 through VAL5. Because this simple tree does not extend far enough into the future for the investments to pay off, they all have negative values in this tree.

It is possible to calculate an expected value for a chance node by multiplying the net present value of each path from that node by its probability and summing all of these products. Expected values are useful, but are not the major reason for preparing a decision tree. The major reason for the tree is its usefulness in identifying, visualizing, and comparing the alternate decision opportunities that exist over time. While this technique is equally valuable for research, marketing, or any other near-term decision with long-term implications, the discussion in this section is concerned with investment decisions.

In addition to calculating expected value for any random event, one can use the tree to assess the riskiness of outcomes—by how much might each outcome deviate from the most likely value. This assessment may be performed in a number of ways. One could look at the distribution of possible outcomes, consider the best and worst possible outcomes, or look at a measure of the distribution of outcomes such as the standard deviation. Considering the variability of the outcome is more important than the particular measure used to do so.

Arborist, the Texas Instruments decision tree program, is designed to run on either the Texas Instruments or IBM personal computer or fully compatible machines. Inputs may be entered either directly or through a Lotus 1-2-3 format. Lotus input is

TABLE 3-5
Example of a Regret Matrix

	New Competitive Product	No New Competitive Product
Add capacity	$2 MM	$0 MM
Design new product	$0 MM	$3 MM

(*Industrial Technology Institute*)

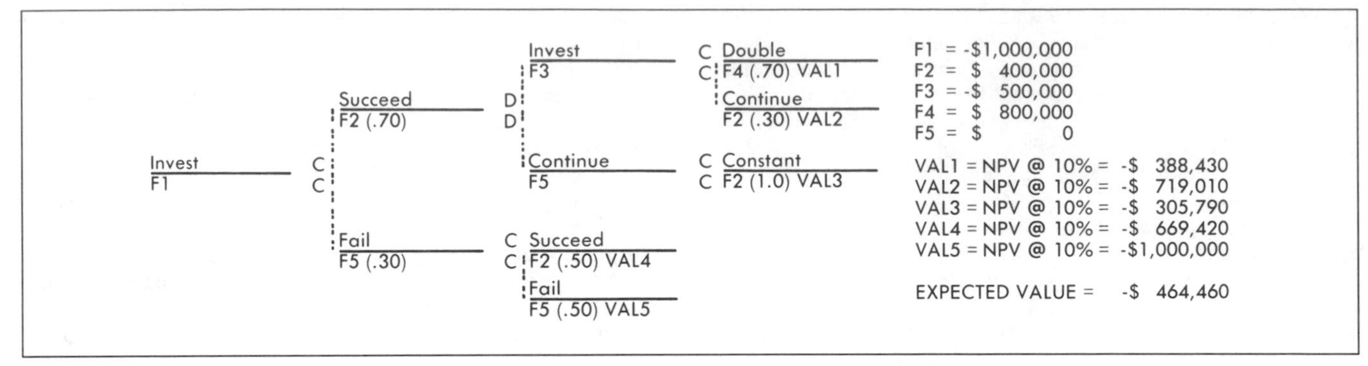

Fig. 3-13 Example of a simple decision tree printout using Arborist program. (*Industrial Technology Institute*)

somewhat easier when getting started; direct input is somewhat faster for experts. To construct the tree, one must define decision nodes, chance event nodes, and end nodes. For decision nodes, the name or description is entered, along with formulas to define quantitative variables. Chance event nodes require names, formulas to define quantitative variables, and probabilities, which may also be defined as formulas. End nodes require names and formulas for outputs—sales, cash flow, or net present value.

The program draws the decision tree and has a number of other useful features as well. In addition to calculating net present values and expected values, the program has capabilities for performing sensitivity analyses on any variable (including probabilities) and for generating probability distributions of outcomes relative to any decision within the tree.

Previously, it was mentioned that decision trees are valuable for evaluating investments in flexible manufacturing systems. As an example, two alternative investments will be analyzed with the Arborist program.

To begin, consider the decision tree of Fig. 3-14 in a qualitative way. The root of the tree is the initial investment, presumably in facilities to produce a new product. Following this investment, cash inflows are anticipated, but their magnitudes are uncertain. The broad range of possibilities in year 1 can be simplified by saying they will be either high or low, depending on chance. Then, in year 2, cash flow will either increase or decline from the year 1 level, whether it be high or low.

In year 3, the user is faced with a decision—note the D indicating a decision node. At this time, the user may elect to make an investment in retooling, in which case cash flow will remain constant in years 3 and 4, the same as in year 2. Alternatively, the user may elect not to make the additional investment, but to exit from the business. If this decision was made, cash flow will decline in each of the following 2 years.

This is clearly an abstraction from reality; the actual situation will undoubtedly be much more complicated, and one may decide to make the decision tree as complex as he or she wishes. If the tree is made too simple, it will be so divorced from reality as to be of no use. If one tries to capture every detail of possible chance events or decisions, it will be extremely complex and therefore unmanageable. The most difficult skill in using decision trees is to capture the right balance between detail and abstraction.

Having defined the structure of the decision tree, one can now take a quantitative look at it, first considering probabilities. It was estimated that there is a 67% probability of high cash flow and a 33% probability of low cash flow in year 1. In year 2, there is a 67% probability of increased cash flow and a 33% probability of decreased cash flow. Once the decision is made whether or not to retool, probabilistic estimates are not made of the outcome; a single estimate is given a probability of 100% in each case.

These estimates are known as subjective probabilities in that they have been assigned on the basis of the judgment of individuals rather than on the basis of objective tests, as one might make to find the mean time to failure of light bulbs. Subjective probabilities can be very useful, but it is important to remember that they are based on judgment.

High cash flow in year 1 was estimated to be $500,000 and low cash flow to be $300,000. The increase in year 2 was defined as a 20% jump and the decrease as a 20% fall. Thus, if cash flow was low in year 1 and increased in year 2, the cash flow in year 2 would be 20% greater than $300,000, or $360,000. If cash flow was high in year 1 and decreased in year 2, the cash flow in year 2 would be 20% less than $500,000, or $400,000.

If retooling occurred in year 3, cash flow in years 3 and 4 are assumed to be equal to cash flow in year 2. If retooling did not occur and the firm exits from the business, it is assumed that year 3 cash flow will be 50% of year 2 cash flow, and year 4 cash flow will be 50% of year 3 cash flow. Thus, if the decision was made not to retool and cash flow in year 2 was $400,000, cash flow would be $200,000 in year 3 and $100,000 in year 4.

This example assumes that there are two possible investments resulting in equal cash flows from operations, but with different investment patterns. Investment A is an investment in conventional automation. It will cost $1,000,000, will have to be completely torn out, and will require an additional $1,000,000 invested to retool in year 3. Investment B is in more flexible equipment. It will cost $1,250,000 initially, but can be retooled for only $25,000. Which investment is more cost effective?

The Arborist program will calculate net present value for each branch of the decision tree as well as the maximal expected value for the tree as a whole. The results for the two subtrees are summarized at the bottom of Fig. 3-14. Net present values are generally higher for the uppermost branches of the decision tree where cash flows are strong. Further, the maximal expected value of the tree with investment B is greater than that with investment A. It should also be noted that the decision to exit in year 3 with investment A has a higher value than the decision to retool. With investment B, the decision to retool has a higher value than the decision to exit. Unless further information is obtained, it appears that investment B will create greater value

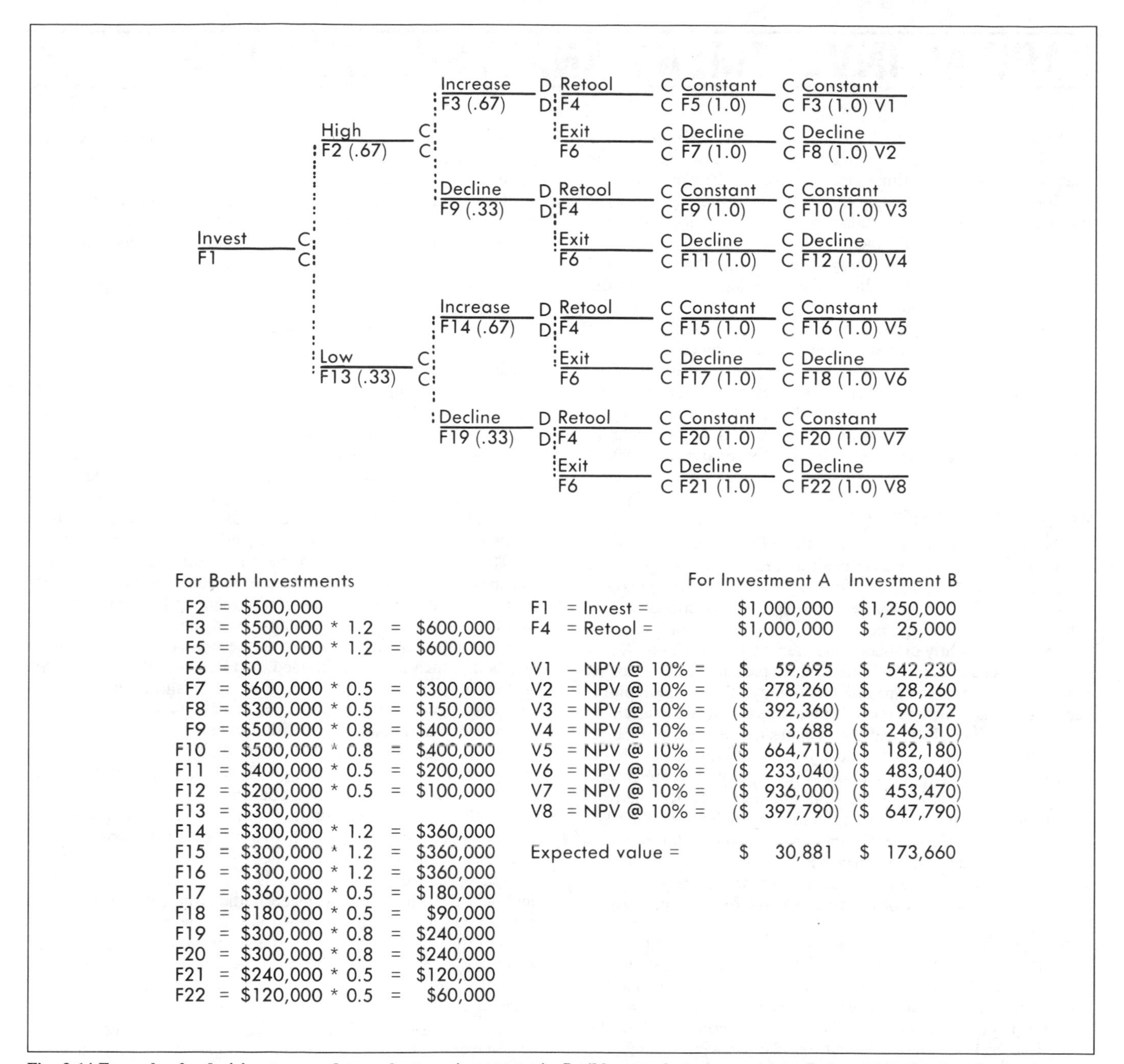

For Both Investments

F2	=	$500,000		
F3	=	$500,000 * 1.2	=	$600,000
F5	=	$500,000 * 1.2	=	$600,000
F6	=	$0		
F7	=	$600,000 * 0.5	=	$300,000
F8	=	$300,000 * 0.5	=	$150,000
F9	=	$500,000 * 0.8	=	$400,000
F10	=	$500,000 * 0.8	=	$400,000
F11	=	$400,000 * 0.5	=	$200,000
F12	=	$200,000 * 0.5	=	$100,000
F13	=	$300,000		
F14	=	$300,000 * 1.2	=	$360,000
F15	=	$300,000 * 1.2	=	$360,000
F16	=	$300,000 * 1.2	=	$360,000
F17	=	$360,000 * 0.5	=	$180,000
F18	=	$180,000 * 0.5	=	$90,000
F19	=	$300,000 * 0.8	=	$240,000
F20	=	$300,000 * 0.8	=	$240,000
F21	=	$240,000 * 0.5	=	$120,000
F22	=	$120,000 * 0.5	=	$60,000

	For Investment A	Investment B
F1 = Invest =	$1,000,000	$1,250,000
F4 = Retool =	$1,000,000	$ 25,000
V1 = NPV @ 10% =	$ 59,695	$ 542,230
V2 = NPV @ 10% =	$ 278,260	$ 28,260
V3 = NPV @ 10% =	($ 392,360)	$ 90,072
V4 = NPV @ 10% =	$ 3,688	($ 246,310)
V5 = NPV @ 10% =	($ 664,710)	($ 182,180)
V6 = NPV @ 10% =	($ 233,040)	($ 483,040)
V7 = NPV @ 10% =	($ 936,000)	($ 453,470)
V8 = NPV @ 10% =	($ 397,790)	($ 647,790)
Expected value =	$ 30,881	$ 173,660

Fig. 3-14 Example of a decision tree used to evaluate an investment in flexible manufacturing systems. (*Industrial Technology Institute*)

for the firm than investment A within the period of the analysis and has the potential for further increases in value later. Investment B will allow the firm to remain in business, while investment A will result in economic pressures for required retooling in year 3.

As noted earlier, however, the decision tree may be used in a way of greater potential value. Two possible courses of events and decisions over the next 6 years have been considered. Are they reasonable? Are there other possible courses of action suggested that should be studied as well? Could more be learned about the potential retooling in year 3? Is there an alternative way of achieving flexibility with an initial investment of less than $1,250,000? These are only a few of the questions that may be raised by preparing and studying decision trees. The greatest benefit of the decision tree is its provision of visibility for possible future events, allowing the manager to explore investment options and potentially create additional value.

Of the four techniques discussed in this section, the authors feel that the decision tree is the most valuable. It addresses the problem of justifying a greater investment in the near term to gain flexibility, thereby reducing possible future investments.

CAPITAL INVESTMENT ANALYSIS EXAMPLE

While there is no such thing as a fully comprehensive example of a capital investment analysis, this example is relatively complete. It illustrates an investment where benefits are not only spread throughout a manufacturing system as it exists at a point in time, but also throughout the lifecycle of the manufacturing system. Further, the example is of an investment in support of a business strategy in which the objective of the firm is explicitly defined. Under these circumstances, it is the responsibility of production management to devise and implement a manufacturing strategy that is part of an integrated effort by all parts of the firm to achieve the firm's objective.

It has been necessary to simplify the example in certain places and to show typical analyses, such as one year and one alternative rather than every year and each alternative, to avoid excessive length. The intention, however, is an example in sufficient detail to serve as a model for analysis of actual investment alternatives.

BUSINESS STRATEGY

The process has started with a strategic planning exercise to define the product and market position of the firm in light of its perceived threats and opportunities as well as its strengths and weaknesses. The exercise has involved all functional areas, including manufacturing, marketing, engineering, and others.

The firm presently produces and markets two products, A and B. They are engineered products, and the product design changes about every 4 years, as product technology is changing rapidly. Superior engineering and consistent quality are very important to the users of the product, and the firm has the reputation of being the technical leader in the industry and a producer of a high-quality product. The product was originally developed and introduced to the market by the firm, and until recently, the firm dominated the market. It is still the market leader with a market share of about 70%. Recently, however, a competitor emerged that is now threatening the firm's market leadership. Examination of how the competitor was able to secure market entry reveals two weaknesses of the firm relative to the competition:

- A year ago, when product B was redesigned, the competitor brought out a product which, while not as good as the new product B, was better than the old product B, and the competitor was able to get it to market 10 months before the firm's new product. Thus, for 10 months, while the firm had a better product on the drawing board, the competitor had a better product on the market. This is believed to be the primary reason for the success of the competitor, reflecting the importance of timely introduction of up-to-date engineering.
- There has been a disturbing increase in the number of customer complaints. Warranty expenses have increased as well, in spite of more stringent inspection and increased rework in the plant. It is believed that the competitor is selling a quality product and that this is playing a secondary role in its success, reflecting the importance of quality to the market.

The firm has certain strong advantages as well. The engineering department is recognized as the best in the industry. It is still the market leader, with good distribution and an excellent reputation among users of its product, in spite of the recent complaints it has received.■

The firm recognizes that completely eliminating a competitor with an initial market success is probably not feasible; but there is agreement that the firm could initiate actions which, if successful, will maintain its 70% market share or even increase it somewhat, restricting the competitor to a secondary position in the market. To do so, the firm must take certain actions, particularly with regard to getting newly engineered products on the market more quickly and restoring product quality. This will place the firm securely in the position of competing on the basis of engineering and quality once again, forcing the competitor to accept some other attribute—probably price—as its basis for competing. Management believes that this will occupy the full attention of the key people in the firm so they will not attempt to enter any new markets at this time, but will concentrate on solidifying their firm's position in the markets for A and B. This is agreed on by all departments as a feasible strategy, particularly in view of the strength of the engineering department.

The outlook for the future based on this strategy has been discussed extensively by the people working on the plan. After a great deal of discussion, general agreement has been reached, and the result is laid out qualitatively in the form of a decision tree (see Fig. 3-15). At the root of the tree is the firm's present position, which has been defined as holding a 70% share of the market with the best engineering in the business, with quality that is still good but with problems starting to show up, with an outmoded plant, and with difficulty getting new products in production and to market on a timely basis. A decision must be made whether or not the firm should make a major investment at this time. At issue is, on the one hand, whether the firm can accomplish its objective without such an investment and, on the other, whether the firm can maintain a profitable position if it goes ahead with a major investment.

The upper half of the tree deals with the chance occurrences and decisions that must be made if the firm goes ahead with a major investment at this time. In the judgment of key members of the firm, the major issue is whether or not the competitor will counter the actions of the firm with a competitive action of its own. This might be a price reduction, an increase in quality, or copying the firm's new product and getting it to market quickly. These are actions that might logically follow from the strengths of the competitor—it is believed unlikely that the competitor would be able to engineer a product superior to that of the firm. By taking one or more of the actions that are open to it, however, the competitor might continue to gain market share.

This leads to a decision about how to deal with such a competitive threat. One response would be to ignore it, to continue producing the product as planned, introducing a new product at the end of the normal 4-year product lifecycle. A more costly alternative would be to accelerate the introduction of a new product after 2 years. It is believed that an accelerated new product introduction would allow the firm either to recover or maintain market share. A normal introduction might allow the firm to maintain share, but could cause continued loss of share.

If no competitive threat was forthcoming, it is believed the firm would maintain its current market share with the new

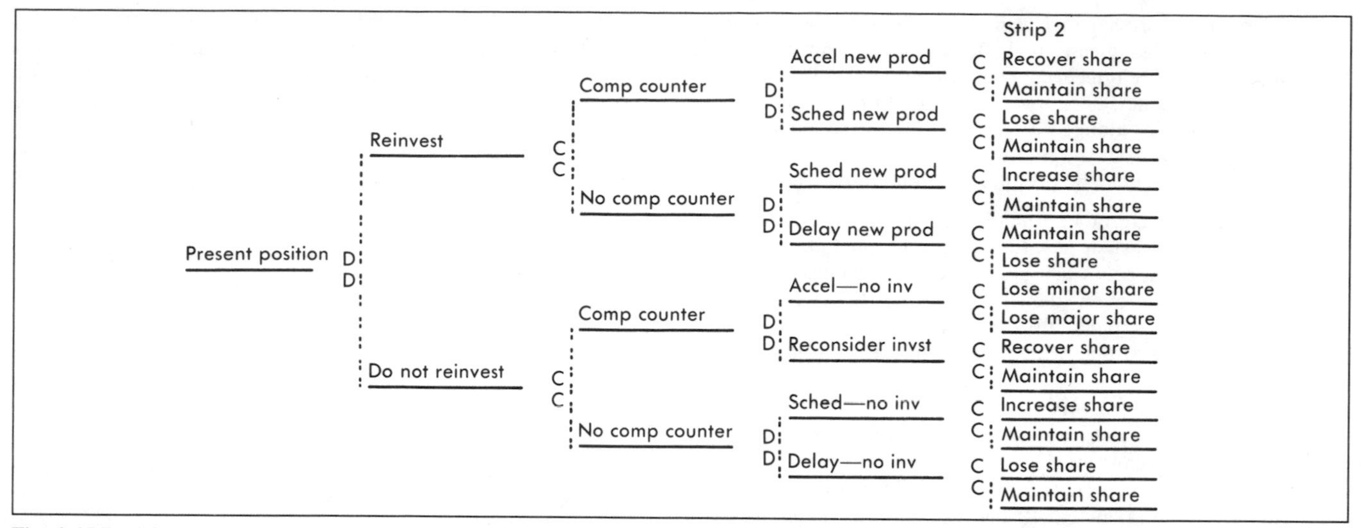

Fig. 3-15 Decision tree showing the qualitative decision of a company to solidify its position in the market. (*Industrial Technology Institute*)

product currently being introduced. It might introduce a new product as scheduled at the end of 4 years, or it might possibly extend the life of the product from 4 to 6 years, using the investment for an additional 2 years to improve cash flow. It is believed that a normal new product introduction would allow the firm either to gain or to maintain market share. A delayed new product introduction might allow the firm to maintain share, but could cause loss of share.

The lower half of the tree deals with the chance occurrences and decisions that must be made if the firm decides not to make a major investment at this time. In the judgment of key members of the firm, the major issue is the same as if the firm reinvests—whether or not the competitor will take this opportunity to initiate a competitive action to gain share. The actions open to the competitor are the same as noted previously—a price reduction, an increase in quality, or copying the firm's new product and getting it to market quickly. By taking one or more of these actions, it is believed the competitor could continue to gain market share.

The firm could deal with such a threat by accelerating the introduction of a new product after 2 years, still without a major investment. Manufacturing questioned the feasibility of such a move, but engineering insisted that it be included in the tree as a potential course of action. Alternatively, manufacturing suggested that a more feasible move would be to accelerate the introduction of the new product, at the same time reconsidering the original decision not to invest and make the major investment at that time. It is believed that an accelerated new product introduction without investment would result in a lack of responsiveness to the market that would inevitably lead to loss of market share. Whether the loss would be major or minor is an issue. Accelerating the new product introduction with a major investment might allow the firm to recover share or at least to maintain share.

If no competitive threat was forthcoming, it is believed the firm would maintain its current market share with the new product currently being introduced. It might introduce a new product as scheduled at the end of 4 years, or it might possibly extend the life of the product from 4 to 6 years. In either event, no major investment would be considered. It is believed that a normal new product introduction would allow the firm either to gain or maintain market share. A delayed new product introduction might allow the firm to maintain share, but could cause loss of share.

This entire range of decisions, chance events, and outcomes is laid out qualitatively in Fig. 3-15. At this stage, the tree is only a visual aid for the key people in the firm to help them foresee possible future decisions and chance events and to reach agreement on the most important of these potential decisions and events so they can plan for them. Later, the tree will be refined, further detail added, and quantitative data inserted.

MANUFACTURING STRATEGY

Building on this business strategy, each department is examining its own function with the objective of defining what it must do to support the business strategy and what organization, activities, tools, technologies, and investments will enable it to do so. While this example focuses principally on the investments necessary to implement the business strategy, it is important to recognize that investment alone is rarely sufficient. Successful investments to implement business strategies are usually integrated with other actions. For instance, the inability of the firm to get products to the market quickly might be partly a result of a manufacturing system that lacks adaptability and needs extensive retooling each time a new product is introduced and might be partly a result of poor organization leading to inadequate communications between engineering and manufacturing. For an investment in flexible manufacturing to be fully effective, the organization problems must be solved as well.

One of the taskforces established to implement the strategy consists of manufacturing engineers who have the responsibility of defining the functional requirements for the manufacturing system that will be able to support the strategy. After exmaining the requirements of the business strategy for manufacturing, they decide that the manufacturing system best able to respond to these requirements has the following characteristics:

- Accuracy of shape and size. The demand in the marketplace for quality, the present high level of rework, the increasing customer quality complaints, and increasing warranty costs all point to the need for improved process control.

- Adaptability of shape. The need to redesign the product every few years means that manufacturing can expect a high volume of engineering changes. Future configurations cannot be predicted, so the firm needs to be prepared for whatever comes along. Further, the loss of market share a year ago points out the importance of adapting quickly so the new product can be produced with minimal lead times.

Naturally, there are other functional requirements to which the manufacturing system must respond. It must be able to produce parts of the necessary configurations, it must produce enough of them to meet demand, it must have sufficient flexibility to be able to handle the mix of parts, and it must produce parts at a cost that allows the firm to earn a profit.

OPERATIONAL PLAN

Specific questions must be asked to come up with a viable operational plan. Examples of such questions are: What are the shapes and sizes of the parts to be produced? What manufacturing technologies are available that might produce these shapes and sizes? What is the possible range of shapes and sizes that might need to be produced? Can the range be reduced and other benefits achieved as well by introducing group technology? What quantities must be produced? What is the lot size? Might both lot size and inventories be reduced by increasing the flexibility of the system, allowing more rapid and less expensive changeover from one part to another?

In this example, both parts A and B require four operations: rough mill, grind, drill, and tap. A certain number of pieces must be reworked at each operation, and a certain number are scrapped. A diagram of the process is shown in Fig. 3-16. The firm is presently operating with an antiquated and inadequate manufacturing system. Throughput is insufficient to meet demand, estimated to be 100,000 units of each part this year. The market is expected to grow by 10% each year. Costs are high, changeover times are long, and work-in-process inventory is high. A change is clearly overdue.

A new generation of products is now being designed. They will also be identified as A and B and will require the same four operations. It is anticipated that, under normal circumstances, another generation of parts will come along 4 years from now. Management knows it cannot compete with the existing plant, so keeping the existing plant is not a viable alternative. Three alternatives are under consideration. The example does not concern itself with the physical characteristics of the three alternatives so much as it does with their performance.

The first alternative, which is considered the baseline, requires an investment of $1,000,000. It consists of selectively replacing existing conventional manufacturing equipment with similar new equipment of adequate capacity. Cycle times will remain about as at present. It is believed that with new equipment properly tooled for the next generation of parts, quality problems will become manageable. However, it is estimated that scrap will run at about 8%.

The second alternative is a highly automated system, requiring an investment of $4,250,000. It will be designed specifically for the geometry of the parts presently in design and will require extensive modification when the new parts are again redesigned. Cycle times will be substantially reduced. Changeover time will be high, resulting in a high economic lot size. The cycle time to produce parts will be very low. Part quality will be high, and it is anticipated that the scrap rate will be lowered to 2%.

The third alternative is a flexible manufacturing system costing $5,000,000. Cycle time will be shorter than with conventional equipment, but somewhat longer than with the highly automated system. However, changeover time is low, yielding a low economic batch size. It is anticipated that only a few modifications will be required by the next generation of parts in 2 years. Like the highly automated system, part quality will be high with a 2% scrap rate.

CONVENTIONAL FINANCIAL ANALYSIS

At this point, the traditional approach would be to perform an economic analysis based on direct labor savings and scrap reduction. An investment in either the hard automation manufacturing system or the flexible manufacturing system would be compared with the baseline, which is the investment in conventional equipment. The investment numbers, cycle times, scrap rates, material cost, labor rates, the firm's after-tax cost of capital, and its tax rate are shown at the top of Fig. 3-17.

These figures are then used as inputs for a conventional discounted cash flow analysis, shown at the bottom of the figure, in which incremental cash flows are calculated for the investment in hard automation and the investment in flexible manufacturing, both in comparison with conventional equipment. This analysis shows negative net present values for both hard automation and flexible manufacturing in comparison with the baseline conventional system.

This analysis is based on many assumptions, one of the most questionable being that all of the relative advantages of the

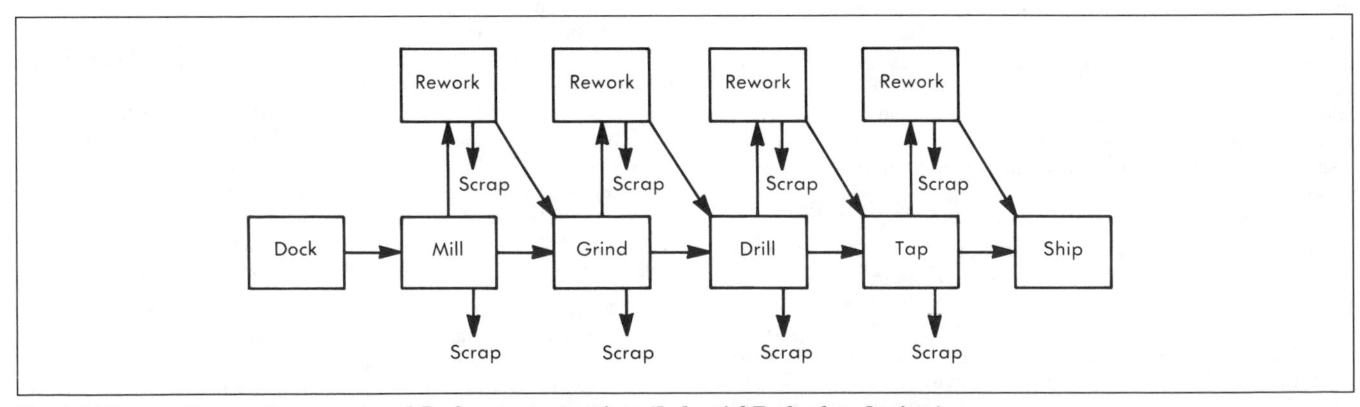

Fig. 3-16 Process diagram for parts A and B of current operation. (*Industrial Technology Institute*)

	Conventional	Hard Automation	Flexible Manufacturing
Cycle time			
Part A	18	4	8
Part B	20	5	10
Annual production			
Part A	100,000	100,000	100,000
Part B	100,000	100,000	100,000
Hours per year			
Part A	30,000	6667	13,333
Part B	33,333	8333	16,667
Total	63,333	15,000	30,000
Dollars per year	$633,333	$150,000	$300,000
Scrap rate			
Part A	8%	2%	2%
Part B	8%	2%	2%
Units scrapped per year			
Part A	8000	2000	2000
Part B	8000	2000	2000
Material cost			
Part A	$25	$25	$25
Part B	$20	$20	$20
Dollars per year			
Part A	$200,000	$50,000	$50,000
Part B	$160,000	$40,000	$40,000
Total	$360,000	$90,000	$90,000
Investment	$1,000,000	$4,250,000	$5,000,000
Annual depreciation			
Year 0	200,000	850,000	1,000,000
Year 1	320,000	1,360,000	1,600,000
Year 2	192,000	816,000	960,000
Year 3	115,200	489,600	576,000
Year 4	115,200	489,600	576,000
Year 5	57,600	244,800	288,000
Year 6	0	0	0

Year	0	1	2	3	4	5	6	NPV
HARD AUTOMATION COMPARED TO CONVENTIONAL								
Incremental investment	$3,250,000							
Labor savings		$483,333	$483,333	$483,333	$483,333	$483,333	$483,333	
Scrap material savings		$270,000	$270,000	$270,000	$270,000	$270,000	$270,000	
Tax on savings		$256,133	$256,133	$256,133	$256,133	$256,133	$256,133	
Increase in depreciation	$650,000	$1,040,000	$624,000	$374,400	$374,400	$187,200	$0	
Depreciation tax shelter	$221,000	$353,600	$212,160	$127,296	$127,296	$63,648	$0	
After-tax cash flow	($3,029,000)	$850,800	$709,360	$624,496	$624,496	$560,848	$497,200	($144,666)
FLEXIBLE MANUFACTURING COMPARED TO CONVENTIONAL								
Incremental Investment	$4,000,000							
Labor savings		$333,333	$333,333	$333,333	$333,333	$333,333	$333,333	
Scrap material savings		$270,000	$270,000	$270,000	$270,000	$270,000	$270,000	
Tax on savings		$205,133	$205,133	$205,133	$205,133	$205,133	$205,133	
Increase in depreciation	$800,000	$1,280,000	$768,000	$460,800	$460,800	$230,400	$0	
Depreciation tax shelter	$272,000	$435,200	$261,120	$156,672	$156,672	$78,336	$0	
After-tax cash flow	($3,728,000)	$833,400	$659,320	$554,872	$554,872	$476,536	$398,200	($1,108,938)

Fig. 3-17 Analysis of investments based on direct labor saving and scrap. *(Industrial Technology Institute)*

alternative investments are captured in the direct labor and scrap numbers. Other assumptions include constant labor savings over a 6-year period and no further investment requirements over this period of time.

This approach, although superficial, is used by many firms. It fails to capture the benefits of flexible manufacturing that are diffused throughout the manufacturing system at a particular point in time, as well as those that occur throughout the lifecycle of the investment. Furthermore, the analysis has failed to recognize the strategic objectives of the firm and the characteristics required of the manufacturing system to respond to these objectives. Yet many firms perform an analysis like this on proposed investments. As a result, these firms find it all but impossible to justify the purchase of equipment they know they need. To identify and quantify some of these hidden benefits and see what happens to the financial analysis, a dynamic model of the alternative manufacturing system will be made.

DYNAMIC MODELING

Before running Manuplan, XCELL, or one of the other models, one must define the manufacturing systems in greater detail than simply in terms of direct labor. The operating characteristics of the three systems are defined in Table 3-6. With the process shown in Fig. 3-16, the operating characteristics shown in Table 3-6, and a little additional data on rework outcome, it is possible to model the manufacturing system with Manuplan. The output of Manuplan indicates that all proposed systems are capable of meeting production requirements of 100,000 units of each part next year, and it provides additional information as shown at the top of Table 3-7.

Some of the advantages of flexible manufacturing are apparent in the summary. Although one cannot yet put dollars and cents on it, the work-in-process inventory will be increased somewhat with hard automation and will be reduced significantly with flexible manufacturing. In addition, the flow times suggest that the factory's response to orders will be slowed somewhat with hard automation and accelerated significantly with flexible manufacturing. Scrap rates and rework are lower for both hard automation and flexible manufacturing. This implies that costs of noncompliance will be lower for automated processes, as is usual because of the greater process control achieved.

Further, a dynamic model of the system provides a more accurate assessment of the number of stations needed for each operation to yield a given level of throughput. Not only does this allow one to estimate the investment more closely, but it also gives better direct labor figure. When the investment is made, one is able to buy either fewer machines than thought necessary with a static analysis or, more significantly, to avoid the serious problems of adding stations after the manufacturing system is in place as a result of unanticipated line shortages and blockages.

ECONOMIC TRANSLATION OF THE MODEL

To capture the benefits and costs of either hard automation or flexible manufacturing in terms useable for evaluating the alternative investments, translation of model output into economic terms is necessary. This may be done by combining the output of the model along with additional inputs, also shown in Table 3-7. The results may be calculated manually, or they may be performed on a computer. The model may include a translator as an integral part of the model, or it may be an attachment to it.

Some of the key calculations performed by the translator are shown at the bottom of Table 3-7. These, along with other figures, are used as inputs to a calculation of net present value in Table 3-8. This analysis is more complete than the analysis performed earlier in that it includes the following:

- It deals with the entire manufacturing system and all the benefits of each investment that are realized throughout the system.
- Because the output of the model calculates not only the material content of scrapped parts, but also the labor content of both scrapped and reworked parts, one is able to assess the cost savings more completely.
- An estimate has been made of the residual value of the investment—the value the investment would have to the firm at the end of the analysis period. A "going concern" approach has been used to recognize that the assets are unlikely to be sold at the end of the period, but will continue to be used for production. The residual value has been estimated in this example and henceforth as the present value of 4 additional years of cash flows that are equal to the average of the last 4 years, discounted at the firm's cost of capital of 10%.

By incorporating these benefits, both the hard automation and flexible manufacturing investments show a positive net present value in comparison with the baseline conventional system. Hard automation has higher net present value, indicating that, by this analysis, an investment in hard automation will create more value for the firm than an investment in flexible manufacturing.

The analysis still fails to capture the following key benefits:

- It fails to deal with changes in the character and overall volume of output that may occur over time.
- It fails to take into account future investments that may be required.

PUTTING QUANTITATIVE DATA ON THE DECISION TREE

To include anticipated costs and benefits over time, it is necessary to consider the future, which is uncertain. One way of dealing with an uncertain future is through the use of a decision tree, with the objective of using it as the basis for a quantitative decision analysis.

The decision tree in Fig. 3-18 is a modification of the one in Fig. 3-15 as a result of the addition of further detail and the insertion of quantitative data. It has also been modified to show the subjective probabilities of the chance occurrences. These probabilities are based on the best subjective judgment of knowledgeable people—they are not based on statistical tests, such as those used for statistical process control. Further, values have been assigned to the qualitative expressions used earlier, such as "lose market share" or "gain market share." By running these figures for the total market and the firm's share of market, along with resulting levels of production and sales, through the model and translator to determine the financial impacts of these levels of sales on investment requirements and manufacturing costs, it is possible to calculate cash flows by year for each branch of the decision tree and thus to calculate a present value for each branch. Because probabilities for each branch have already been established, the tools discussed in the earlier sections can be used to calculate expected values for each branch, as well as measures of dispersion.

To go through the calculations for each year and each branch would require many pages of calculations that are quick and simple with personal computers, but tedious to read. Figure 3-19 shows the calculations for one branch for a single year and a summary of the results of performing numerous similar calculations on all the branches.

The markets for both products A and B have been forecasted to grow at an annual rate of 10%. These figures are shown at the top of Fig. 3-19. The price for both products is forecast to remain constant, as shown. Path 1, shown as an example, is the top path on the decision tree in Fig. 3-18. It has been defined as the path in which one invests in a flexible manufacturing system, the

TABLE 3-6
Operating Characteristics of Alternative Manufacturing Systems Used as Inputs to Manuplan

	Conventional	Hard Automation	Flexible Manufacturing
Number of milling units	7 units	3 units	5 units
Mean time to failure	9600 minutes	9600 minutes	9600 minutes
Mean time to repair	120 minutes	120 minutes	120 minutes
Part A cycle time	6 minutes	1 minute	2 minutes
set-up time	120 minutes	240 minutes	0.1 minute
scrap rate	1.5 %	0.4 %	0.4 %
rework rate	1.5 %	1 %	1 %
Part B cycle time	4 minutes	1 minute	2 minutes
set-up time	120 minutes	240 minutes	0.1 minute
scrap rate	1.5 %	0.4 %	0.4 %
rework rate	1.5 %	1 %	1 %
Number of drilling units	5 units	2 units	4 units
Mean time to failure	9600 minutes	9600 minutes	9600 minutes
Mean time to repair	60 minutes	60 minutes	60 minutes
Part A cycle time	7 minutes	1 minute	2 minutes
set-up time	60 minutes	120 minutes	0.1 minutes
scrap rate	2 %	0.6 %	0.6 %
rework rate	2 %	1.5 %	1.5 %
Part B cycle time	6 minutes	1.5 minutes	3 minutes
set-up time	60 minutes	120 minutes	0.1 minute
scrap rate	2 %	0.6 %	0.6 %
rework rate	2 %	1.5 %	1.5 %
Number of grinding units	8 units	3 units	6 units
Mean time to failure	9600 minutes	9600 minutes	9600 minutes
Mean time to repair	60 minutes	60 minutes	60 minutes
Part A cycle time	2 minutes	1 minute	2 minutes
set-up time	60 minutes	120 minutes	0.1 minute
scrap rate	2 %	0.6 %	0.6 %
rework rate	1 %	0.5 %	0.5 %
Part B cycle time	4 minutes	1 minute	2 minutes
set-up time	60 minutes	120 minutes	0.1 minute
scrap rate	2 %	0.6 %	0.6 %
rework rate	1 %	0.5 %	0.5 %
Number of tapping units	5 units	3 units	5 units
Mean time to failure	12,000 minutes	12,000 minutes	12,000 minutes
Mean time to repair	30 minutes	30 minutes	30 minutes
Part A cycle time	3 minutes	1 minute	2 minutes
set-up time	30 minutes	60 minutes	0.1 minute
scrap rate	2 %	0.6 %	0.6 %
rework rate	2 %	1.5 %	1.5 %
Part B cycle time	6 minutes	1.5 minutes	3 minutes
set-up time	30 minutes	60 minutes	0.1 minute
scrap rate	2 %	0.6 %	0.6 %
rework rate	2 %	1.5 %	1.5 %

(*Industrial Technology Institute*)

competition counters with a competitive move, and one in turn counters by accelerating the introduction of the next generation of product. While the market share of 70% drops temporarily, the company's strategy is successful, and the market share is regained by year 6.

These market shares are applied to the market forecasts to calculate unit sales, which are then multiplied by unit prices to obtain revenue. Production cost is derived from the model—the figure shown for year 1 is identical with that shown for flexible manufacturing in Table 3-8, the analysis of investments based on the manufacturing system model. Investments are based on the need to reinvest only when new products are introduced, as sufficient capacity exists for normal sales growth. Other variable and other fixed costs have been estimated outside the model. In this example, identical fixed costs have been estimated for all three alternatives, with somewhat lower variable costs for

TABLE 3-7
Economic Translation of Manuplan Model

	Conventional	Hard Automation	Flexible Manufacturing
SUMMARY OF MODEL OUTPUTS FROM MANUPLAN MODEL			
Shipments (pieces)			
Product A	100000	100000	100000
Product B	100000	100000	100000
Scrap production (pieces)			
Product A	7595	2087	2087
Product B	7595	2087	2087
Average work in process (pieces)			
Product A	2620	3635	9
Product B	2806	3873	10
Equipment utilization summary (percent)			
Mill			
Set-up	8.7%	12.0%	2.0%
Run	73.9%	62.9%	67.4%
Repair	1.0%	0.9%	0.9%
Total utilization	83.6%	75.8%	70.3%
Drill			
Set-up	5.9%	11.6%	3.3%
Run	60.0%	74.1%	65.6%
Repair	0.4%	0.5%	0.4%
Total utilization	66.3%	86.2%	69.3%
Grind			
Set-up	3.8%	5.9%	2.0%
Run	83.0%	81.3%	87.0%
Repair	0.5%	0.5%	0.6%
Total utilization	87.3%	87.7%	89.6%
Tap			
Set-up	2.9%	3.9%	2.4%
Run	88.9%	73.3%	73.3%
Repair	0.2%	0.2%	0.2%
Total utilization	92.0%	77.4%	75.9%
Flow time (days)			
Part A	5.49	7.72	0.02
Part B	5.91	8.26	0.02
FINANCIAL INPUTS			
Firm's cost of capital (after-tax) (%)	10%	10%	10%
Firm's incremental tax rate (%)	34%	34%	34%
Initial investment ($)	$1,000,000	$4,250,000	$5,000,000
Tax life (yrs)	5	5	5
Material cost by product ($/unit)			
Product A	$20.00	$20.00	$20.00
Product B	$25.00	$25.00	$25.00
Labor cost by code ($/hr)			
Operation	$10.00	$10.00	$10.00
Set-up	$10.00	$10.00	$10.00
Maintenance	$10.00	$10.00	$10.00
Selling price by product			
Product A	$45.00	$45.00	$45.00
Product B	$55.00	$55.00	$55.00
Other variable costs per unit	$8.00	$5.00	$5.00
Other fixed costs per year	$100,000	$100,000	$100,000
SUMMARY OF FINANCIAL CALCULATIONS			
Revenue	$10,000,000	$10,000,000	$10,000,000
Cost of good production	$3,850,641	$3,267,308	$3,417,308
Cost of scrap	$259,401	$64,850	$72,632

TABLE 3-7—*Continued*

	Conventional	Hard Automation	Flexible Manufacturing
Cost of rework	$41,763	$23,805	$25,893
Depreciation			
Year 0	$200,000	$850,000	$1,000,000
Year 1	$320,000	$1,360,000	$1,600,000
Year 2	$192,000	$816,000	$960,000
Year 3	$115,200	$489,600	$576,000
Year 4	$115,200	$489,600	$576,000
Year 5	$57,600	$244,800	$288,000
Work-in-process inventory	$145,931	$198,997	$511

(*Industrial Technology Institute*)

TABLE 3-8
Analysis of Investments Based on Manufacturing System Model

Year	0	1	2	3	4	5	6	NPV	NPV Relative to Baseline
CONVENTIONAL									
A unit sales		100,000	100,000	100,000	100,000	100,000	100,000		
B unit sales		100,000	100,000	100,000	100,000	100,000	100,000		
Revenue		10,000,000	10,000,000	10,000,000	10,000,000	10,000,000	10,000,000		
Production cost		4,151,805	4,151,805	4,151,805	4,151,805	4,151,805	4,151,805		
Other variable		1,600,000	1,600,000	1,600,000	1,600,000	1,600,000	1,600,000		
Depreciation	200,000	320,000	192,000	115,200	115,200	57,600	0		
Other fixed		100,000	100,000	100,000	100,000	100,000	100,000		
PBT		3828195	3,956,195	4,032,995	4,032,995	4,090,595	4,148,195		
PAT		2,526,609	2,611,089		2,661,777	2,699,793	2,737,809		
Fixed invest	1,000,00								
Inventory invest		145,931	0	0	0	0	0		
Residual value							12,817,020		
Cash flow	-800,000	2,700,678	2,803,089	2,776,977	2,776,977	2,757,393	15,554,828	18,447,278	0
HARD AUTOMATION									
A unit sales		100,000	100,000	100,000	100,000	100,000	100,000		
B unit sales		100,000	100,000	100,000	100,000	100,000	100,000		
Revenue		10,000,000	10,000,000	10,000,000	10,000,000	10,000,000	10,000,000		
Production cost		3,355,963	3,355,963	3,355,963	3,355,963	3,355,963	3,355,963		
Other variable		1,000,000	1,000,000	1,000,000	1,000,000	1,000,000	1,000,000		
Depreciation	850,000	1,360,000	816,000	489,600	489,600	244,800	0		
Other fixed		100,000	100,000	100,000	100,000	100,000	100,000		
PBT		4,184,037	4,728,037	5,054,437	5,054,437	5,299,237	5,544,037		
PAT		2,761,464	3,120,504	3,335,928	3,335,928	3,497,496	3,659,064		
Fixed invest	4,250,000								
Inventory invest		198,997	0	0	0	0	0		
Residual value							17,460,804		
Cash flow	-3,400,000	3,922,467	3,936,504	3,825,528	3,825,528	3,742,296	21,119,868	23,151,538	4,704,260
FLEXIBLE MANUFACTURING									
A unit sales		100,000	100,000	100,000	100,000	100,000	100,000		
B unit sales		100,000	100,000	100,000	100,000	100,000	100,000		
Revenue		10,000,000	10,000,000	10,000,000	10,000,000	10,000,000	10000000		
Production cost		3,515,833	3,515,833	3,515,833	3,515,833	3,515,833	3,515,833		
Other variable		1,000,000	1,000,000	1,000,000	1,000,000	1,000,000	1,000,000		
Depreciation	1,000,000	1,600,000	960,000	576,000	576,000	288,000	0		
Other fixed		100,000	100,000	100,000	100,000	100,000	100,000		
PBT		3784167	4,424,167	4,808,167	4,808,167	5,096,167	5,384,167		
PAT		2,497,550	2,919,950	3,173,390	3,173,390	3,363,470	3,553,550		
Fixed invest	5000000								
Inventory invest		511	0	0	0	0	0		
Residual value							17056408		
Cash flow	-4,000,000	4,097,039	3,879,950	3,749,390	3,749,390	3,651,470	20,609,958	22,210,067	3,762,788

(*Industrial Technology Institute*)

CAPITAL INVESTMENT EXAMPLE

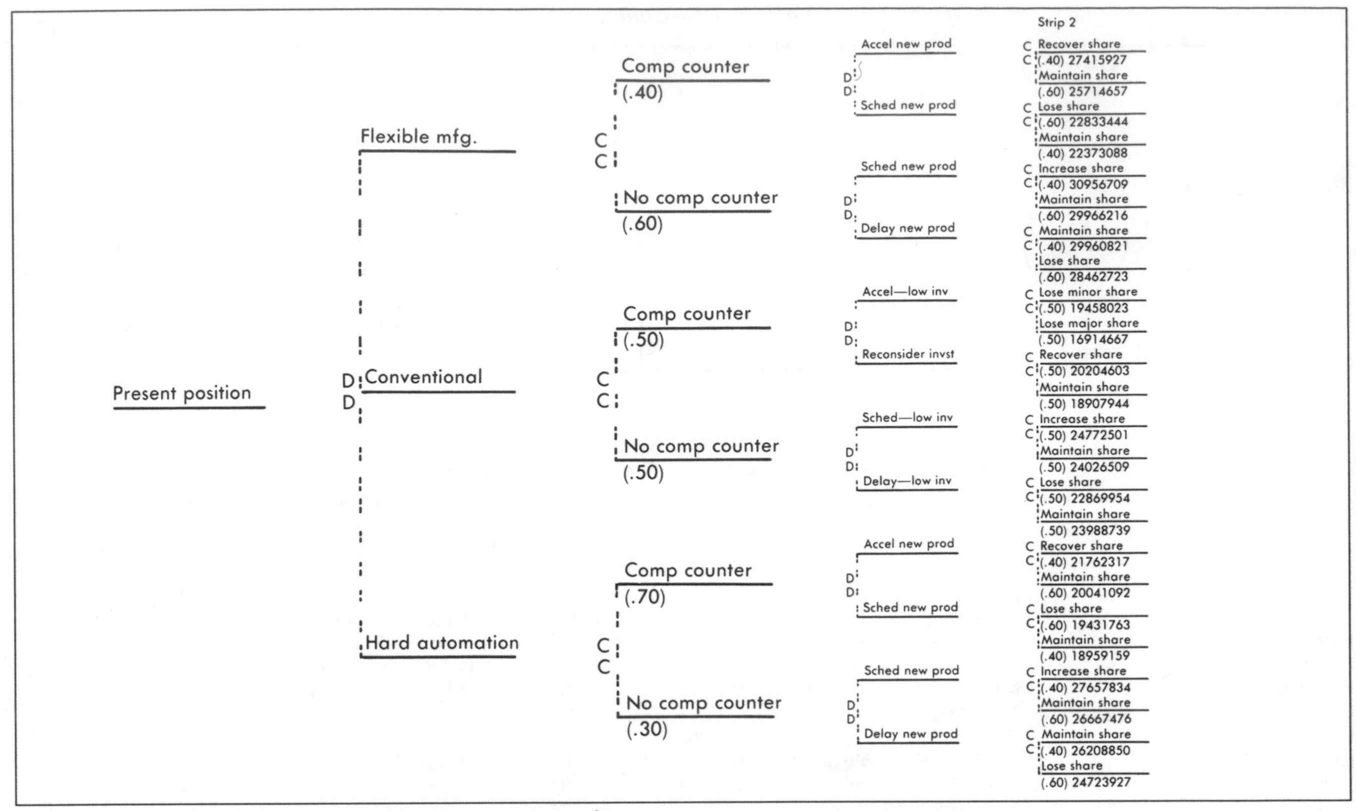

Fig. 3-18 Modified decision tree. (*Industrial Technology Institute*)

flexible manufacturing and hard automation than for the conventional system. Inventory investment is based on varying requirements for work in process in the model as volume changes with market size and the firm's share. Cash flow and net present value are calculated as defined in the appendixes.

The net present value of each branch, as determined by similar calculations, is shown along with its probability of occurrence, given that particular investment. These probabilities are calculated on the basis that the subsequent decision in response to a competitive counter is always made to the branch with the higher expected value. Based on this, the expected value is calculated for each alternative investment.

To calculate incremental costs and benefits, the investment in conventional equipment has been considered as the baseline. The baseline in this more complex example is less clear. It might be considered the expected value of the center branch of the tree; however, this value is influenced by the branch in which major investment is reconsidered.

Philosophically, the baseline is best represented by the branch that reflects a passive response to the business environment. The course of action identified as branch 14, with a net present value of $24,026,509, is probably as good as any to choose as the baseline, although it is not a perfect choice because it depends on the competition not making a competitive counter to a new product offering.

As one can see, when incremental cash flows are calculated over the life of the product, the analysis strongly favors the flexible manufacturing approach. In fact, the minimum value of the flexible manufacturing system is only slightly less than the maximum value of the hard automation system in this example. Thus, the flexible manufacturing system not only has a higher expected value, but appears to be a more robust investment in the sense that it is the only branch in which the firm has the opportunity to increase value by making good decisions even in the event of adverse chance events.

INVESTMENT DECISION

If the decision analysis had indicated a course of action not meeting some needs of the business strategy to be a better course of action than one that did, it might cast doubt on the viability of the strategy. It is likely in this event, however, that the analysis failed to capture certain categories of costs and benefits. In general, the firm would want to invest in the machinery that will help it achieve its strategic objectives. If, for example, an investment consistent with the firm's strategic needs falls short of some less suitable investment by $10,000 per year, an appropriate question is whether one has overlooked categories of benefits that will yield cash flows of $10,000 per year. If one can define such benefits convincingly, they should be added to the investment analysis. There is no need to do that with this investment example because the flexible manufacturing system that best meets the needs of the strategy also yields the highest net present value. The analysis reflects this value only because the benefits that flexibility yields have been captured, both throughout the entire manufacturing system and also throughout the lifecycle of the system. That is not to say, however, that one has necessarily captured all of its benefits. As the system runs, one might well find unexpected benefits; one might find unexpected costs as well. This is valuable information and should be retained for use when the next investment proposal is analyzed.

Year	0	1	2	3	4	5	6	NPV
Market for product A		142,857	157,143	172,857	190,143	209,157	230,073	
Market for product B		142,857	157,143	172,857	190,143	209,157	230,073	
Price for product A		45	45	45	45	45	45	
Price for product B		55	55	55	55	55	55	
Path 1								
Market share		0.7	0.6	0.6	0.625	0.65	0.7	
A unit sales		100,000	94,286	103,714	118,839	135,952	161,051	
B unit sales		100,000	94,286	103,714	118,839	135,952	161,051	
Revenue		10,000,000	9,428,562	10,371,418	11,883,916	13,595,200	16,105,083	
Production cost		3,515,833.	3,314,925	3,646,418	4,178,187.	4,779,845.	5,662,279.	
Other variable		1,000,000	94,2856	1,037,142	1,188,392	1,359,520	1,610,508	
Depreciation	1,000,000	1,600,000	97,0000	592,000	585,600	2,937,60	1,5760	
Other fixed		100,000	100,000	100,000	100,000	100,000	100,000	
PBT		3,784,167	4,100,781	4,995,859	5,831,738	7,062,075	8,716,536	
PAT		2,497,550	2,706,515	3,297,267	3,848,947	4,660,969	5,752,914	
Fixed invest	5,000,000		50,000				50,000	
Inventory invest		511	-29	48	77	87	128	
Residual value							22,036,376	
Cash flow	-4,000,000	4,097,039	3,626,544	3,889,219	4,434,470	4,954,642	27,754,922	27,415,927

SUMMARY OF OUTCOMES	NPV	Incremental NPV	Probability of Occurrence	Expected Incremental Value	Max and min Incremental Value
Flexible manufacturing					
Path 1	27,415,927	3,389,419	0.16	4,749,006	
Path 2	25,714,657	16,881,49	0 24		1,688,149
Path 3	22,833,444	-1,193,065			
Path 4	22,373,088	-1,653,420			
Path 5	30,956,709	6,930,201	0.24		6,930,201
Path 6	29,966,216	5,939,708	0.36		
Path 7	29,960,821	5,934,312			
Path 8	28,462,723	4,436,214			
Conventional:					
Path 9	19,458,023	-4,568,486		-2,048,619	
Path 10	16,914,667	-7,111,841			
Path 11	20,204,603	-3,821,906	0.25		
Path 12	18,907,944	-5,118,565	0.25		-5,118,565
Path 13	24,772,501	745,993	0.25		745,993
Path 14	24,026,509	0	0.25		
Path 15	22,869,954	-1,156,555			
Path 16	23,988,739	-37,769			
Hard automation:					
Path 17	21,762,317	-2,264,192	0.28	-1,396,715	
Path 18	20,041,092	-3,985,416	0.42		-3,985,416
Path 19	19,431,763	-4,594,746			
Path 20	18,959,159	-5,067,350			
Path 21	27,657,834	3,631,326	0.12		3,631,326
Path 22	26,667,476	2,640,968	0.18		
Path 23	26,208,850	2,182,342			
Path 24	24,723,927	697,419			

Fig. 3-19 Summary of analysis of investments based on decision tree in Fig. 3-18. (*Industrial Technology Institute*)

APPENDIX A—FINANCIAL ANALYSIS OF INVESTMENTS

Financial analysis of investments is sometimes referred to as engineering economics. By itself, it represents the traditional approach to investment evaluation. Embedded in a business planning process, this analysis is a key element in analyzing investment decisions.

THE COST OF CAPITAL

As noted in the main body of this chapter, manufacturing firms are generally formed to create wealth by producing goods. The firm hires people and purchases machinery and raw materials, produces the goods, then sells them. To do so, the firm must have sufficient money to hire the people and purchase the machinery and raw materials until such time as the goods are sold. This money is known as the firm's capital. Capital is always provided by the owners of the firm. The owners may also be the managers of the firm, as with a small job shop, or they may be separate, as with major corporations in which most of the shareholders (or owners) have no other connection with the firm. Capital may also be loaned to the firm by banks, by insurance companies, by individual investors, by other companies, and occasionally by the government.

To induce investors to purchase shares or loan money, the firm must provide compensation to them. Borrowed money requires that interest payments be made. Individuals who invest in shares may be compensated for the use of their funds by dividends, by growth in the value of the shares, or by a combination of both. This compensation provides a yield or return to the investors. Investors have a minimum required rate of return. From the perspective of the firm, this is a cost and is called the cost of capital.

The significance of the cost of capital is that if a firm earns exactly its cost of capital on its investments, it will neither create nor destroy wealth for its owners. If it earns more than its cost of capital, it will create wealth; if it earns less, it will destroy wealth. If a firm's cost of capital is 10%, it will create wealth by earning 11% on its investments; it will destroy wealth by earning 9%, even though it is operating "in the black." Thus the cost of capital is a very important figure when evaluating investments.

In most firms, the calculation of cost of capital is performed by the finance department. Even though the calculation of cost of capital is ordinarily performed by financial people rather than engineers, its significance is so great that calculation of this figure is discussed in Appendix B.

COMPOUND INTEREST

Nearly everyone is familiar with compound interest. A dollar invested at an interest rate of 10% compounded annually earns 10 cents interest the first year. The second year interest is earned not only on the dollar but also on the first year's interest of 10 cents, so the investment earns 11 cents the second year. Each year the interest compounds or is added to the principal and the interest that has already accumulated. Invested at 10% compounded annually, at the end of 10 years the dollar will have increased to \$2.59. It is useful to think of this figure of \$2.59 as the future value of a dollar invested at 10%, 10 years hence.

The compound interest equation is a simple one and is as follows:

$$FV = PV\,(1+r)^n \tag{1}$$

where:

FV = future value
PV = present value
r = interest rate
n = number of years (periods)

Thus, when the present value is \$1.00 and the interest rate is 10%, the future value calculation is as follows:

$$\begin{aligned} FV &= \$1.00\ (1 + 0.10)^{10} \\ &= \$1.00\ (1.10)^{10} \\ &= \$1.00 \times 2.59 \\ &= \$2.59 \end{aligned}$$

This calculation shows that a dollar now is worth more than a dollar at some time in the future as long as interest is earned. For example, if one can earn interest at a rate of 10% compounded annually, a dollar now would have the same value as \$2.59 in 10 years from now.

It is also possible to make the same calculations in reverse. If one knows that a transaction of some specified amount at some particular time in the future will take place and what the proper interest rate is, the future value can be discounted to obtain the present value. The equation for this calculation is as follows:

$$PV = FV/(1+r)^n \tag{2}$$

Thus, if one will receive a dollar 10 years hence, and the interest rate is 10% compounded annually, the present value is:

$$\begin{aligned} PV &= \$1.00/(1 + 0.10)^{10} \\ &= \$1.00/2.59 \\ &= \$0.39 \end{aligned}$$

Tables of compound interest and present value at various interest rates are shown in Appendix D. Compound interest tables show the future value of \$1.00 compounded periodically, and discounting tables show the present value of \$1.00 from some time in the future discounted periodically. Alternatively, it is sometimes more useful to calculate future and present values directly with a calculator or computer.

The concepts of compounding and discounting, representing the time value of money, are of particular use in evaluating investments because investments typically require an immediate cash outlay for a factory or machine, and yield cash inflows to the firm over some period in the future. By using these concepts, one can account for both the value of a dollar at present and a dollar at some time in the future.

TRADITIONAL CAPITAL INVESTMENT ANALYSIS

Years ago, there were no methods for analyzing and evaluating potential investments. Managers simply did what seemed right. In response to need, several rough techniques evolved for supplementing the manager's intuition with a more analytical approach. While these methods have value, they fail to consider whether the investment creates or destroys value.

Payback Period

Payback period is among the earliest measures that attempted to arrive at a figure of merit for investment projects. Payback period is the time required for the cash outflow for an investment to be offset by the cash inflows resulting from the investment. The concept of cash flow is discussed in some detail in Appendix C. If an investment of $100,000 resulted in cash inflows of $50,000 per year, the payback period of the investment would be 2 years. The presumption is that the investment with the shortest payback period is the best investment.

Throughout the chapter cash outflows from a firm will be referred to as negative cash flows, and cash inflows to a firm are positive cash flows; negative cash flows will be shown inparentheses. A comparison of the cash flows of two investments is shown in Table 3-9. Investment 1 has a payback period of about 3.1 years, while investment 2 has a payback period of about 2.6 years. By this criterion, investment 2 with the shorter payback period is the preferred investment.

Although payback period remains in use, it has two severe shortcomings as a criterion on which to base investments. First, it ignores cash flows in the years following the payback period, and second, it ignores the time value of money.

Accounting Rate of Return

Accounting rate of return is another traditional method for analyzing investments. Return on investment is a standard financial measure equal to net after-tax income divided by total investment. Applied to investments, an estimate is made of the increase in net after-tax income resulting from a specific investment, and this figure is divided by the value of the investment.

Typically, the effect of an investment on income varies from year to year. Further, because capital investments are depreciated over their lives, the value of the investment shown on the accounting books varies from year to year as well. This may be handled by calculating an estimated accounting rate of return for each year or by calculating a single average accounting rate of return over the life of the investment.

The accounting rate of return pattern for an investment of $20,000 depreciated on a straight-line basis over 4 years is shown in Table 3-10. Over the 4-year period, average net income is $4250, and the average value of the investment is $10,000, yielding an average accounting return of 42.5%. The difficulty of using annual accounting rate of return as an investment criterion is that the wide variation from year to year shown in Table 3-10 is typical, making it difficult to appraise the value of an investment. Using the average figure eliminates this problem, but fails to divulge the $2000 loss anticipated in year 1—a potentially important piece of information.

TABLE 3-9
Comparison of Cash Flows in Two Investments

Year	Investment 1 Cash Flow, $ Annual	Investment 1 Cash Flow, $ Cumulative	Investment 2 Cash Flow, $ Annual	Investment 2 Cash Flow, $ Cumulative
Investment	(20,000)	(20,000)	(10,000)	(10,000)
1	3000	(17,000)	2000	(8000)
2	7000	(10,000)	5000	(3000)
3	9000	(1000)	5000	2000
4	18,000	17,000	5000	7000

(*Industrial Technology Institute*)
Note: Values in parentheses are negative.

THE MAPI METHOD

In an effort to overcome the serious shortcomings of traditional investment analyses, George Terborgh of the Machinery and Allied Products Institute (MAPI) devised a method of analyzing machinery replacement directed toward selection of only those investments creating value.[11] This approach measures the annual benefits of keeping an existing machine vs. the benefits of replacing it with a new machine. The investment required to replace the asset is considered, as well as the present value of future operating costs and the potential salvage value of the new investment at the end of its life. Operating costs of the old machine are taken into account, as well as its salvage value, and the operating advantage that would result from replacing the old machine is compared with the costs of doing so.

The principal advantage of the MAPI method is its convenience; the principal disadvantage is its narrow focus on machinery replacement. To overcome this, MAPI subsequently incorporated into its methodology the investment analysis described in the following section. The resulting approach is the basis for the treatment of replacement decisions in many, if not all, current textbooks on engineering economy.

TABLE 3-10
Example of an Accounting Rate of Return Pattern for a $20,000 Investment

Year	Beginning of Year Investment, $	Annual Depreciation, $	End of Year Investment, $	Average Investment, $	Net Income, $	Accounting Rate of Return, %
1	20,000	5000	15,000	17,500	(2000)	-11
2	15,000	5000	10,000	12,500	2000	16
3	10,000	5000	5000	7500	4000	53
4	5000	5000	-0-	2500	13,000	520

Note: Values in parentheses are negative.
(*Industrial Technology Institute*)

APPENDIX A—FINANCIAL ANALYSIS

INVESTMENT ANALYSIS BASED ON CASH FLOWS AND THE COST OF CAPITAL

In recent years, managers have recognized that it is sometimes possible to use the technique of discounting cash flows, long used to value financial assets, to value potential capital investments as well. The three most widely used investment evaluation techniques using discounted cash flows are net present value, profitability index, and internal rate of return. While the details of net present value, profitability index, and internal rate of return differ, all use the same fundamental approach, as follows:

1. Define the cash flows.
2. Define the interest rate.
3. Discount the cash flows to present value.

Net Present Value

Net present value (NPV) is based on the proposition that the value created by making an investment can be calculated by setting benefits equal to the positive cash flows generated by the investment. Costs are negative cash flows associated with making the investment itself, and net present value is the present value of the benefits less the costs. If the net present value is greater than 0, the investment will create value for the firm. Mathematically, NPV can be expressed as follows:

$$NPV = \sum_{j=0}^{n} X_j/(1 + r)^j \tag{3}$$

where:

NPV = net present value, dollars
Σ = the sum of
X_j = net cash flow at the end of year j, dollars
n = number of years of cash flow
j = the year
r = the company's cost of capital

For example, a machine costing $10,000 results in positive cash flows of $3000 per year for 5 years commencing 1 year after the investment is made. The firm's cost of capital is 10%. As stated previously, cash outflows from the firm are negative, and cash inflows to the firm are positive. Compared with not making the investment, the value created by making the investment is shown in Table 3-11.

Some simplifying assumptions have been made for this calculation. In particular, the cash flows in years 1 through 5 arrive in lump payments, each conveniently arriving exactly at year-end. This is known as the "year-end convention," and it is common to use such a convention in calculating net present value. Other conventions have the cash flows occurring at the beginning of the year, at mid-year, or continuously throughout the year. Unless cash flows can be forecast with great accuracy, the differences in net present value resulting from using different conventions for cash flow patterns during the year are much smaller than the usual forecasting errors and are therefore unimportant.

Profitability Index

Some managers are accustomed to managing by using ratios or percents and would prefer to use a ratio for evaluating capital investment projects. The profitability index (PI) was designed for such managers. Its numerator is equal to the present value of future cash flows from operation with the investment where the relevant interest rate is the firm's cost of capital. The denominator is the present value of investment outlays. If the profitability index is greater than 1, the investment will create value for the firm. In calculation of the profitability index, the negative sign on the present value of investment outlays is typically ignored so that a positive ratio will result. Table 3-12 shows the calculation for an initial investment of $10,000.

Internal Rate of Return

For managers who wish to know the rate at which the investment is earning, the internal rate of return (IRR) may be calculated. The internal rate of return is a "cut-and-try" calculation in which the interest rate at which net present value is equal to 0 is found. That rate is defined as the internal rate of return. If the internal rate of return is greater than the firm's cost of capital, the investment will create value for the firm.

Once cash flows have been defined, the next step is to pick a rate. Very often, analysts pick the firm's cost of capital and calculate the net present value. If the net present value is positive, they know the internal rate of return is higher than the cost of capital. That is, future cash flows must be discounted at a higher rate to reduce the net present value to 0. In the example shown in Table 3-11, the net present value of the investment was found to be $1372 when discounted at 10%, which was the firm's cost of capital. Several tries may be required to find that the internal rate of return of this investment is 15.24% (see Tables 3-13 and 3-14).

TABLE 3-11
Example of Calculations for Determining Net Present Values

Year	Cash Flow, $	Present Value at 10%, $	
0	-10,000		= -10,000
1	3000	3000 x 1/(1.10)	= 2727
2	3000	3000 x $1/(1.10)^2$	= 2479
3	3000	3000 x $1/(1.10)^3$	= 2254
4	3000	3000 x $1/(1.10)^4$	= 2049
5	3000	3000 x $1/(1.10)^5$	= 1863
		Net present value	= 1372

(*Industrial Technology Institute*)

TABLE 3-12
Example of Calculations for Determining the Profitability Index of an Investment

Year	Cash Flow, $	Present Value at 10%, $	
1	3000	3000 x 1/(1.10)	= 2727
2	3000	3000 x $1/(1.10)^2$	= 2479
3	3000	3000 x $1/(1.10)^3$	= 2254
4	3000	3000 x $1/(1.10)^4$	= 2049
5	3000	3000 x $1/(1.10)^5$	= 1863
		Present value	= 11,372
Profitability index = 11,372 / 10,000 = 1.137			

(*Industrial Technology Institute*)

By looking at Tables 3-13 and 3-14, one can conclude that the project IRR is between 15 and 16%. Through the use of linear interpolation, the actual internal rate of return is as follows:

i	NPV
15%	57
IRR	0
16%	-178

$$IRR = 15\% + \left(\frac{57-0}{57-(-178)}\right)(16\text{-}15\%) = 15.24\%.$$

TWO EXAMPLES OF SIMPLE CAPITAL INVESTMENT PROJECTS

Prior to the mid-1970s, business strategies were typically simple, and investments were made because they reduced manufacturing costs, provided capacity to produce additional product to be sold on the market, or a combination of both. The first example will be of an investment intended to reduce manufacturing costs with no change in production volume. The second example will be an investment intended to provide manufacturing capacity needed for additional sales volume.

A Cost Reduction Investment

Several years ago, a firm was producing its product on a small, 20-ton press. It could have continued to use this equipment, but its engineers saw a potential opportunity to reduce costs by acquiring a new higher speed press with more rapid changeover capabilities and lower maintenance requirements. The new press delivered, installed, and tooled would cost $33,500. Before buying the new press, however, they were asked to analyze the investment by calculating the net present value, profitability index, and internal rate of return of the investment.

They realized that any such analysis is a comparison of two alternatives, and the alternatives were apparent in this case—continuing to use the old press vs. purchasing the new press. Thus, to make a comparison, they needed to do the following:

1. Estimate the cash flows that would occur if they continued to operate with the old press vs. those that would occur if they invested in the new press.
2. Define the interest rate.
3. Discount future cash flows to present value at the appropriate interest rate.

Their first task was to estimate cash flows. The machinery dealer informed them that their old press could not be resold and had no value and that the new press delivered and installed would cost $33,500. Naturally, if they continued to operate the old press, there was no initial outlay.

The new press would have no effect on sales (they produced and sold 12,000,000 units per year at present), the old press provided adequate capacity to continue, and the sales department saw neither a significant increase nor decrease in volume in future years. Thus, the future cash flows that would result from a new investment would be due only to operating savings.

Engineering discussed what operating savings it might expect from the new investment—a critical step in the evaluation—and decided that it could expect savings in direct labor, set-up labor, maintenance, and tooling. From existing records of standard costs and of performance to standard, the production rates and set-up times of existing equipment were found. Maintenance records yielded annual maintenance costs for the existing press, and purchasing records showed annual tooling expenditures. From financial records, it was determined that labor rates were $3.15 per hour for both press operators and set-up people, to which $1.42 per hour of fringe benefits were added for a total labor cost of $4.57 per hour.

For the new press, engineering studied the specifications and estimated both operating performance and set-up. Likewise, it was necessary for engineering to estimate maintenance and tooling costs.

A comparison of the estimated operating costs yielded by these studies is shown in Table 3-15.

Property records showed that the old press was fully depreciated. The finance department indicated that the new press would be depreciated over 5 years on a straight-line basis. As discussed in Appendix C, while depreciation is not of itself a cash flow, it affects taxes, and taxes are cash flows. The finance department indicated that the firm's tax rate was 34%, and the after-tax cost of capital (as discussed in Appendix A) was 10%.

Management indicated that there was little business risk for the next 5 years, but after the 5-year period they could not foresee what might happen. Thus, they suggested that a 5-year period would be a reasonable period to analyze cash flows. Engineering estimated that the press might be sold for around $10,000 at the end of 5 years if the firm no longer needed it.

With this information, it is possible to define initial outlay and annual cash flows in succeeding years. In this example, initial outlay is simple. There was no equipment to sell, nor were there expenses associated with the investment. Thus, the initial

TABLE 3-13
Calculation of Internal Rate of Return Using 15% as an Estimate

Year	Cash Flow, $	Present Value at 15%, $
0	-10,000	= -10,000
1	3000	$3000 \times 1/(1 + 0.15)$ = 2608
2	3000	$3000 \times 1/(1 + 0.15)^2$ = 2268
3	3000	$3000 \times 1/(1 + 0.15)^3$ = 1974
4	3000	$3000 \times 1/(1 + 0.15)^4$ = 1716
5	3000	$3000 \times 1/(1 + 0.15)^5$ = 1491
		Net present value at 15% = 57

(Industrial Technology Institute)

TABLE 3-14
Calculation of Internal Rate of Return Using 16% as an Estimate

Year	Cash Flow, $	Present Value at 16%, $
0	-10,000	= -10,000
1	3000	$3000 \times 1/(1 + 0.16)$ = 2586
2	3000	$3000 \times 1/(1 + 0.16)^2$ = 2229
3	3000	$3000 \times 1/(1 + 0.16)^3$ = 1923
4	3000	$3000 \times 1/(1 + 0.16)^4$ = 1656
5	3000	$3000 \times 1/(1 + 0.16)^5$ = 1428
		Net present value at 16% = -178

(Industrial Technology Institute)

APPENDIX A—FINANCIAL ANALYSIS

TABLE 3-15
Comparison of Operating Costs

Category	Present	Proposed
Direct labor	0.017 hours/100 pieces	0.009 hours/100 pieces
	12,000,000 pieces/year	12,000,000 pieces/year
	$4.57/hour	$4.57/hour
	$9322.80/year	$4935.60/year
Set-up	2 hours/set-up	0.5 hours/set-up
	145 set-ups/year	60 set-ups/year
	$4.57/hour	$4.57/hour
	$1325.30/year	$137.10/year
Maintenance	$1700.00/year	$300.00/year
Tooling	$33,600/year	$18,000/year

(*Industrial Technology Institute*)

outlay was equal to the investment itself, or $33,500. In keeping with the cash flow convention, cash outflow is shown as a negative number.

The calculation of annual cash flows is facilitated by a spreadsheet, as shown in Table 3-16. It is assumed that each category of annual savings in this example will occur without change through the years. This might not necessarily be true. For example, it could be assumed that direct labor savings would be less the first year because of the need for operators to learn how to use the new press. Accelerated depreciation could also be assumed, causing depreciation to be greater in earlier years and less in later years.

For each of the four savings categories—operators, set-up, maintenance, tooling—the proposed annual cost (rounded to the nearest dollar) is subtracted from the present annual cost to obtain the difference. When the difference is a reduction in cost that will result in an increase in earnings, the difference is shown as a positive amount. These four differences will flow to profit before tax, where their effect will be to increase the amount of profit. Tax must be paid on this added profit, so the tax is calculated at 34% in the next column and the effect on profit after tax in the final column.

Because expense reductions result in cash savings and the added tax will be a cash outlay, the effect on profit after tax is the same as the cash flow. This is not true of depreciation. Depreciation is not a cash expense but an accrual. It will, however, have an effect on taxes, which do affect cash flow. The increase in depreciation will have a negative effect on profit, reducing taxes by 34% of the profit reduction. This reduction in taxes will affect cash flow, so the final column in the depreciation row shows only the positive effect on cash flow resulting from the tax reduction.

By adding the various cash flow items, it is found that the annual cash flow resulting from operations will be $17,177 greater in each of the 5 years of use for the new press. There is one additional cash flow that must be taken into account. At the end of the 5-year period, the press, while fully depreciated on the books, will still have a market value estimated at $10,000. The press could be sold and an additional $10,000 of cash recovered. Taxes would have to be paid on the gain, the difference between selling price and book value; in this case, taxes would be due on the entire $10,000. If the tax rate of 34% applies, the after-tax cash flow would be + $6,600. This is often called the ''residual value'' of the asset being considered. The difference in after-tax cash flows between making the investment in the new press and continuing to use the old press is shown in Table 3-17.

It is assumed that an initial cash outlay of $33,500 at the end of year 0 is required if the new press is purchased, and no outlay is required if the old press is retained. If the new press is purchased, cash flows from operations will be $17,177 greater at the end of each year from year 1 to year 5. The new press could also be sold at the end of year 5 for a cash inflow net after taxes of $6600. Because the year-end convention is used, an assumption can be made stating that cash flows from operations in year 5 and the cash flow resulting from sale of the press can simply be added together. The last column in Table 3-17 is the end result of step 1 of the analysis and defines the cash flows that will result from making the investment.

The next step is to define the interest rate. For net present value and profitability index calculations, it is the firm's after-tax cost of capital; for the internal rate of return calculation, it is the discount rate at which net present value is equal to 0. For this example, the net present value will be calculated first. The profitability index can then be calculated by discounting the cash flows defined as the firm's cost of capital. Finally, the internal rate of return can be calculated.

The firm's after-tax cost of capital was stated to be 10% by the finance department, presumably after making an analysis such as that described in Appendix B. To calculate the net present value of the investment in the new press, the cash flows can be discounted at a rate of 10% by multiplying the after-tax cash flow each year by the discount factor for that year (see Table 3-18).

TABLE 3-16
Spreadsheet Used to Facilitate the Calculation of Cash Flows

Cost Category	Present, $	Proposed, $	Difference, $	Tax Effect, $	After-Tax Cash Flow, $
Operators	9323	4936	4387	(1492)	2895
Set-up	1325	137	1188	(404)	784
Maintenance	1700	300	1400	(476)	924
Tooling	33,600	18,000	15,600	(5304)	10,296
Depreciation	0	6700	(6700)	2278	2278
				Annual cash flows from operations =	17,177

(*Industrial Technology Institute*)

Note: Values in parentheses are negative.

TABLE 3-17
After-Tax Cash Flow Summary

Year	Initial Outlay, $	Cash Flow from Operations, $	Residual Value After Tax, $	Total After-Tax Cash Flow, $
0	(33,500)	---	---	(33,500)
1	---	17,177	---	17,177
2	---	17,177	---	17,177
3	---	17,177	---	17,177
4	---	17,177	---	17,177
5	---	17,177	6600	23,777

(*Industrial Technology Institute*)

Note: Values in parentheses are negative.

TABLE 3-18
Calculations of Net Present Value for A Cost Reduction Investment

Year	After-Tax Cash Flows, $	10% Discount Factor	Presnt Value, $
0	-33,500	1.0	-33,500
1	17,177	$1/1.10 = 0.909$	15,614
2	17,177	$1/(1.10)^2 = 0.826$	14,188
3	17,177	$1/(1.10)^3 = 0.751$	12,900
4	17,177	$1/(1.10)^4 = 0.683$	11,732
5	23,777	$1/(1.10)^5 = 0.621$	14,766
		Net present value at 10% =	35,700

(*Industrial Technology Institute*)

The net present value of $35,700 is the algebraic sum of the present values of the cash flows in all years, including the initial outlay. This indicates that investment in the new press will increase the value of the firm by $35,700 over the value the firm would have if no investment were made. Because the net present value is greater than 0, a recommendation to proceed with the investment should be made to management.

Most of the calculations have already been done for the profitability index, which is equal to the present value of operating cash flows at the firm's cost of capital, divided by the initial outlay. The initial outlay is $33,500, and the present value of cash flows in years 1 through 5 is equal to $69,200. The calculation for the profitability index is as follows:

$$\text{Profitability index} = 69{,}200/33{,}500 = 2.07$$

Because the profitability index is greater than 1.0, this calculation also leads engineering to recommend that management proceed with the investment.

Finding the interest rate at which net present value is 0 is a cut-and-try process. The internal rate of return is greater than 10%, because the net present value was positive at a 10% interest rate. It might require several tries to find that the internal rate of return is approximately 45% (see Table 3-19).

The internal rate of return is greater than the firm's cost of capital, so by this criterion as well a recommendation should be made to management that it proceed with the investment. Some pocket calculators and many spreadsheet software packages for personal computers include programs for performing all three calculations or may be readily programmed to do so.

An Investment to Add Capacity

At about the same time, the firm found that its capacity to produce another product was limited to 23,100 units, equal to sales during the current year. Sales were forecast to continue to grow as long as capacity could be added, and the sales department projected a loss of market share in coming years unless the firm added capacity to produce additional product. At present, the product sold for $100 per unit, with an after-tax margin of 12.2%, resulting in profit after tax of $281,500. Management did not want to lose market share in such a profitable product line, so the engineering, marketing, and finance departments were asked to forecast the market for the product, the investment that would be required to meet the demands of the market, and the economics of making such an investment.

The market outlook in Table 3-20 was provided by the marketing department. This outlook was based on an increase in capacity. There was general agreement that if no capacity was added, the current volume of 23,100 units would continue to be sold at the prices forecast for each year. Table 3-21 compares the volume output without an additional investment to the volume output with an investment and also shows the difference in sales. Thus, incremental sales with the investment increase from 0 in the current year to more than $1,000,000 in year 5.

The next step was to estimate the investment to produce the added sales and the incremental earnings from these added sales. This required a joint effort by the engineering and finance departments.

The engineering department found that $200,000 would buy the machinery to produce the additional product. No building

APPENDIX A—FINANCIAL ANALYSIS

TABLE 3-19
Internal Rate of Return Calculations for A Cost Reduction Investment

Year	After-Tax Cash Flows, $	44.537% Discount Factor	Presnt Value, $
0	-33.500	1.0	-33.500
1	17,177	1/1.44537 = 0.692	11,886
2	17,177	$1/(1.44537)^2$ = 0.479	8228
3	17,177	$1/(1.44537)^3$ = 0.331	5686
4	17,177	$1/(1.44537)^4$ = 0.229	3934
5	23,777	$1/(1.44537)^5$ = 0.159	3781
			Net present value at 44.537% = 15

(*Industrial Technology Institute*)

TABLE 3-20
Market Outlook for an Investment to Add Capacity Example

Year	Total Market	Firm's Market Share, %	Volume, Units	Price, $	Sales, $
0	66,000	35	23,100	100	2,310,000
1	76,000	35	26,600	104	2,766,400
2	80,000	35	28,000	108	3,024,000
3	74,000	35	29,400	112	3,292,800
4	88,000	35	30,800	117	3,603,600
5	92,000	35	32,200	122	3,928,400

(*Industrial Technology Institute*)

TABLE 3-21
Comparison of Volume Output for Example to Add Production Capacity

Year	Volume Without Investment, Units	Volume With Investment, Units	Volume Difference	Price, $	Sales Difference, $
0	23,100	23,100	0	100	0
1	23,100	26,600	3500	104	364,000
2	23,100	28,000	4900	108	529,200
3	23,100	29,400	6300	112	705,600
4	23,100	30,800	7700	117	900,900
5	23,100	32,200	9100	122	1,110,200

(*Industrial Technology Institute*)

addition would be required. Finance noted that the machinery would be depreciated on a straight-line basis over 5 years, with zero book value at the end of the period.

Engineering estimated standards, performance to standards, and labor requirements. It also took into account the changes that were likely to occur in fixed and variable costs, the costs of starting up the added machines, the effects of changes in wages, and the cost of raw materials over the years. The incremental earnings that would result from the added sales are shown in Table 3-22.

Additionally, finance estimated additional working capital requirements for inventory and accounts receivable to be equal to 18% of any additional sales (see Table 3-23).

Because the equipment to be purchased was special-purpose machinery, engineering estimated a residual value of only $10,000 at the end of 5 years. Finance informed engineering that the firm's cost of capital was 10% and that the tax rate on income was 34%.

With this information, engineering was ready to begin the analysis. The initial outlay of $200,000 for new machinery was straightforward, but with this investment, as with many investments to provide for additional sales, the investment process continued with increases in working capital throughout the entire period. Thus, the outlay is shown in Table 3-24. The cash inflow from operations is estimated to be equal to profit after tax plus depreciation and is shown in Table 3-25.

TABLE 3-22
Summary of Incremental Earnings Resulting From Added Sales

Year	Added Volume, Units	Price, $	Added Sales, $	Added Profit After Tax, $
0	0	100	0	0
1	3500	104	364,000	(35,000)
2	4900	108	529,200	26,000
3	6300	112	705,600	85,000
4	7700	117	900,900	126,000
5	9100	122	1,110,200	156,000

(*Industrial Technology Institute*)

Note: Values in parenthesis is negative.

Residual value was then considered. This includes not only the value of the machinery at the end of the analysis period, but also the value of working capital to be liquidated if the firm were to stop producing the product. The cumulative value of working capital additions due to the increased sales volume is $198,100. From the estimated $10,000 value of the machinery at the end of the period, $3400 must be subtracted for taxes that would be paid if it were sold, for a net value of $6600. Thus, the total residual value is $204,700.

The difference in after-tax cash flows between making the investment to increase capacity vs. continuing to produce and sell 23,100 units per year is shown in Table 3-26.

The next step is to define the interest rate. For net present value and profitability index calculations, it is the firm's after-tax cost of capital; for the internal rate of return calculation, it is the interest rate at which net present value is equal to 0. Net present value is calculated first, then the profitability index by discounting the cash flows defined at the firm's cost of capital. Finally, the internal rate of return can be calculated.

The firm's after-tax cost of capital was stated to be 10% by the finance department, presumably after making an analysis such as that described in Appendix B. To calculate the net present value of the investment in the new equipment, the after-tax cash flow is multiplied each year by the discount factor for that year (see Table 3-27). The net present value of $160,488 is the algebraic sum of discounted cash flows from all years, including the initial outlay. This indicates that investment in the new equipment will increase the value of the firm by $160,488 over the value the firm would have if no investment were made. Because the net present value is greater than 0, a recommendation would be made to management to proceed with the investment.

Table 3-26 contains the data for computing the profitability index. The present value of cash flows from operations at the 10% after-tax cost of capital is $387,260. A similar calculation for the investment cash flows is straightforward except that net investment in year 5 is $204,700 - $35,900. Thus, the present value for net investment cash flows over the 5-year period is -$226,850. The profitability index is calculated as follows:

$$\text{Profitability index} = 387{,}260/226{,}850 = 1.7$$

Because the profitability index is greater than 1.0, this calculation also leads to the recommendation that management proceed with the investment.

As was stated previously, the internal rate of return is a cut-and-try process. The internal rate of return is greater than 10% percent because the net present value was positive at an interest rate of 10%. Several tries may be required to find that the internal rate of return is approximately 24%. Because the internal rate of return is greater than the firm's cost of capital, a recommendation would be made to management to proceed with the investment.

Preparing a Capital Budget

The process for establishing a capital budget presented in most textbooks and generally more or less followed in practice is discussed in this section. Engineering, marketing, and operations people put together a complete list of desired capital investment opportunities and analyze them according to net

TABLE 3-23
Estimate of Additional Working Capital for Investment to Add Capacity

Year	Sales Increase, $	Working Capital, $
0	0	0
1	364,000	65,500
2	165,200	29,700
3	176,400	31,800
4	195,300	35,200
5	199,300	35,900

(*Industrial Technology Institute*)

TABLE 3-24
Investment Cash Flow Over 5-Year Period

Year	Investment Cash Flow, $
0	(200,000)
1	(65,500)
2	(29,700)
3	(31,800)
4	(35,200)
5	(35,900)

(*Industrial Technology Institute*)

Note: Values in parentheses are negative.

TABLE 3-25
Cash Inflow Over Five Years for Investment to Add Capacity Example

Year	Profit After Tax, $	Depreciation, $	Operations Cash Flow, $
0	0	0	0
1	(35,000)	40,000	5000
2	25,000	40,000	65,000
3	85,000	40,000	125,000
4	126,000	40,000	166,000
5	156,000	40,000	196,000

(*Industrial Technology Institute*)

Note: Value in parenthesis is negative.

TABLE 3-26
After-Tax Cash Flow for Investment to Add Capacity

Year	Investment Cash Flow, $	Cash Flow From Operations, $	Residual Value After Tax, $	Total After-Tax Cash Flow
0	(200,000)	0	- - -	(200,000)
1	(65,500)	5000	- - -	(60,000)
2	(29,700)	65,000	- - -	35,300
3	(31,800)	125,000	- - -	93,200
4	(35,200)	166,000	- - -	130,800
5	(35,900)	196,000	204,700	364,800

(*Industrial Technology Institute*)

Note: Values in parentheses are negative.

TABLE 3-27
Calculations of Net Present Value for an Investment to Add Capacity

Year	After-Tax Cash Flows, $	10% Discount Factor	Present Value, $
0	(200,000)	1.0	(200,000)
1	(60,000)	$1/1.10 = 0.909$	(54,540)
2	35,300	$1/(1.10)^2 = 0.826$	29,158
3	93,200	$1/(1.10)^3 = 0.751$	69,993
4	130,800	$1/(1.10)^4 = 0.683$	89,336
5	364,800	$1/(1.10)^5 = 0.621$	226,541
		Net present value at 10% =	160,488

(*Industrial Technology Institute*)

Note: Values in parentheses are negative.

present value, profitability index, or internal rate of return. These opportunities should be constructed to be independent in the sense that each of them could be implemented with any combination of the others. Total investment cost would be equal to the sum of the investments for the individual opportunities. Simultaneously, finance people analyze the availability of money for capital investment. Projects are ranked with the best investments at the top of the list. If adequate funds are available, all the investments meeting the firm's criterion for acceptance are made. Otherwise, only the best investments are made, with the less attractive investments either rejected or deferred until another time.

When following such a practice, one must choose a valid criterion for ranking potential investments. Otherwise, it is likely that some of the rejected investments will be better than those that are accepted.

The preceding section reviewed three techniques for evaluating capital investments by discounted cash flows. They are net present value, profitability index, and internal rate of return. It was noted that an investment acceptable by one criterion will be acceptable by both of the others. Thus, if the net present value of an investment project is positive, the profitability index will be greater than 1 and the internal rate of return will be greater than the firm's cost of capital. The investment projects, however, will not necessarily be ranked in the same order.

APPENDIX A—FINANCIAL ANALYSIS

If the firm ranks projects by either profitability index or internal rate of return, the cost reduction project ranks higher than the capacity increase and, if capital is constrained, has a better chance of being accepted. If the firm ranks projects by net present value, the capacity increase project ranks higher and has a better chance of being accepted. Which is best? Because the objective of the firm is to create value or wealth and net present value measures the creation of value quantitatively, it is generally accepted as the correct criterion for ranking projects. Profitability index and internal rate of return indicate whether or not a potential investment will create value, but they do not indicate how much. It is possible, as in the case of the cost reduction, to have an investment that is a very good one, as shown by the high profitability index and internal rate of return, but too small for the creation of much value. The capacity increase is not as profitable per dollar invested, but is large enough to create more value. For this reason, net present value is generally accepted as the preferred evaluation criterion. Furthermore, while most investments have an initial cash outflow followed by cash inflows resulting in one change of sign with respect to cash flow from negative to positive, cash outflows in later years result in more than one sign change. Therefore, the internal rate of return calculation yields multiple answers that are not valid.

Issues

While theoreticians agree that net present value is the best technique available for evaluating investment projects, it is not always easy to apply. To use any of the discounted cash flow methods of investment analysis, one must forecast cash flows. At times it may be impossible to do so with sufficient accuracy to make the analysis worthwhile.

It was noted earlier that cash flow forecasts may have as their source financial and management reports and engineering, marketing, or financial studies. When one seeks to use financial data as a source of cash flow forecasts, one of the greatest barriers is the problem of allocation. When an investment is made to increase capacity, one must ask what will happen to overhead expenses. Will they increase? Probably. Will they increase proportionately with direct labor? Ordinarily, there is no reason to expect that they will, but many analysts make the assumption that they will and are surprised when they do not. For investment analysis, it is best that all costs be considered direct and addressed individually. Will additional supervision be needed? Inspection? And how about tooling maintenance? In many companies, all these costs are hidden in the overhead. Thus, engineering and financial studies are needed to define the specific cost elements affected by the potential investment.

An even more difficult task is to seek out the effects on costs and revenues where customers are involved. If new machinery is acquired to maintain statistical process control, it may be difficult to determine the reductions achieved in scrap and rework. However, it is typically even more difficult to estimate what will happen to warranty costs, not to mention increases in revenue as a result of increased market share from greater consumer satisfaction.

Much of the material that follows in this chapter deals with forecasting cash flows. One section covers manufacturing system modeling as a tool for evaluating the cash flow effects of complex investments. System modeling is a great help in arriving at the economics of complex investments, but the tools currently both available and easy to use are limited in their ability to handle support activities and do not take markets into account explicitly.

A second issue closely related to defining cash flows deals with a realistic alternative to making the investment. As noted earlier, every investment analysis is a comparison of alternatives. An investment can be compared with an alternative investment or with making no investment at all. If the alternative with which an investment is compared is bad enough, any investment will look good. Every investment analyst has to deal at one time or another with the ''weeping'' manager who claims dire consequences unless a favorite investment project is approved. On the other hand, making unrealistically favorable assumptions concerning the alternative to investing can sometimes keep managers from investing when they should. Often, analysts compare new investments with the status quo, which is all right when the status quo is a viable alternative. In these times of rapidly changing markets, technologies, and international politics, however, maintaining the status quo is not always a possibility.

The difficulty of making accurate forecasts under certain circumstances has already been mentioned, but risk and uncertainty are so intimately interwoven with capital investments that they should be noted separately as the third issue. Because capital investments are inevitably made for future benefits and it is impossible to know the future, investment inevitably means taking risk.

There are numerous definitions of risk and uncertainty, but a good working definition is the possibility that things may not turn out as expected. This does not necessarily mean that things may be worse—they may be better or they may be worse. While risk is inevitable, it can sometimes be dealt with constructively. Some approaches to dealing with risk are discussed in the section on decision analysis.

APPENDIX B—THE COMPOSITE COST OF CAPITAL

In many firms, capital is provided from equity provided by the owners or shareholders leveraged with a certain amount of debt. Firms generally have a target ratio of debt to equity and try not to stray too far from this ratio.

It was stated earlier in this chapter that the yield or return to the investor must be regarded as a cost by the firm. If there were no taxes, the cost to the firm would be equal to the yield to the investor. There are taxes, however, and they must be paid before the owners of the firm can increase their wealth. Therefore, it is necessary to deal with after-tax numbers in evaluating activities, including investments, that have as their objective increasing the wealth of the owners .

Because of a peculiarity of tax laws in the United States and most other countries, the firm's income taxes are calculated on the basis of an income from which interest payments have been subtracted as a cost, but from which dividends have not. An

APPENDIX B—COMPOSITE COST OF CAPITAL

example of the profit and loss statement of a company is shown in Fig. 3-20.

As indicated in Fig. 3-20, the firm has sold goods during some period of time for $1,000,000 and spent $500,000 for the material, labor, and overhead to produce those goods. This leaves a gross profit of $500,000. Administrative costs, which generally include selling expenses as well, were $100,000, leaving net operating income of $400,000. Interest expenses on borrowings were $50,000, leaving pre-tax income of $350,000 on which federal income taxes of 34% must be paid ($119,000), leaving after-tax income of $231,000. Of this, $150,000 was paid out in dividends, and the remaining $81,000 was retained and reinvested by the firm.

COST OF DEBT

One question that a manager may ask is "where on the profit and loss statement does the firm's cost of capital appear?" The bewildering answer is that it does not. However, part of the cost of capital can be obtained from this statement. Interest expense was $50,000. If the firm had borrowed no money at all, it would have had no interest payments, in which case pre-tax earnings would have been $500,000. Taxes in this case would have been 34% ($136,000), leaving a profit after tax of $264,000. Instead of increasing profit by $50,000, it would have only increased profit by $33,000. Likewise, increasing interest payments by $50,000 will reduce profit by only $33,000. The interest payments may be said to have "sheltered" taxes by an amount equal to the interest payment multiplied by the tax rate as shown in the following equation:

$$\text{Tax shelter} = \text{Interest payment} \times \text{tax rate} \quad (4)$$

Thus, the cost to the firm of interest payments is equal to the interest payment minus the tax shelter provided as shown in the following equation:

$$\text{Cost to firm} = \text{Interest payment} - \text{tax shelter} \quad (5)$$

The after-tax cost of debt to the firm is as follows:

$$\text{After-tax cost of debt} = \text{Interest rate} \times (1\text{–tax rate}) \quad (6)$$

Sales	$1,000,000
Cost of goods sold	500,000
Gross profit	500,000
General and administrative expenses	100,000
Net operating income	400,000
Interest expense	50,000
Pre-tax income	350,000
Income taxes at 34%	119,000
Net after-tax income	231,000
Dividends	150,000
Reinvested earnings	81,000

Fig. 3-20 Example of a company's profit and loss statement. (*Industrial Technology Institute*)

Thus, if the firm is paying an interest rate of 10% and the tax rate is 34%, the after-tax cost to the firm is:

$$\begin{aligned}\text{After-tax cost of debt} &= 0.10 \times (1\text{–}0.34)\\ &= 0.10 \times 0.66\\ &= 0.066 \text{ or } 6.6\%\end{aligned}$$

COST OF EQUITY

The cost of equity is a little more difficult to compute because the concepts are a little more complex and because the data are not always readily available. There are a number of methods for calculating cost of equity, two of which are discussed.

The first approach to calculating the cost of equity is based on the observation noted earlier in Appendix A that some shareholders receive a return in the form of dividends, some in the form of increased share value, and some a combination of the two. If a firm exhibited no growth but had the same level of sales, earnings, and dividends year after year, the shareholder would receive a constant stream of dividends, the share price would be unlikely to increase, and the return to the investor (and the cost to the firm) would be equal to the annual dividend divided by the share price. The cost of equity could be calculated as follows:

$$\text{Cost of equity} = \text{Dividend / share price} \quad (7)$$

If the firm were to grow at some constant growth rate year after year, the sales, earnings, and dividends would increaseeach year. The price of the share would reflect the buyers' expectations that dividends would continue to increase each year, and it can be calculated that the cost of equity would be equal to the current dividend divided by the share price, plus the growth rate. In this instance, the cost of equity would be calculated as follows:

$$\text{Cost of equity} = (\text{Dividend / share price}) + \text{growth rate} \quad (8)$$

Thus, if a firm paid a dividend of $1.00, the shares sold for $10.00, and the market expected the firm to grow at a rate of 5% annually, the cost of equity may be calculated as follows:

$$\begin{aligned}\text{Cost of equity} &= (\$1.00 / \$10.00) + (0.05)\\ &= 0.10 + 0.05\\ &= 0.15 \text{ or } 15\%\end{aligned}$$

Because dividends are paid out of after-tax earnings, they provide the firm with no tax shelter, and the tax rate plays no role in the calculation of cost of equity.

The second approach to calculating cost of equity is based on the observation that investors demand a higher return on higher risk investments. A U.S. Treasury bill, for example, attracts investors with a relatively low rate, while securities in firms with shaky financing must pay a high rate to attract investors. The significant risk for shareholders has been defined as the fluctuation of prices in the stock markets. As the markets fluctuate, some stocks fluctuate more than others, and the extent to which they fluctuate (and hence a measure of their risk) can be calculated. The relationship between the required rate of return for a particular stock and the risk-free rate of return is known as beta. The required rate of return can be calculated as follows:

$$R_j = R_f + \mathrm{B}(R_m - R_f) \quad (9)$$

where:

R_j = return on stock j
R_f = risk-free return
R_m = return on stock market

APPENDIX B—COMPOSITE COST OF CAPITAL

Stock j is, of course, the stock of the firm in question; for the risk-free return, the U.S. Treasury Bill rate is often used; and for the return on the stock market, the yield of the Standard & Poor's 500 is often used.

If the Treasury bill rate were 7%, the market as a whole were yielding 12%, and stock j had a beta of 1.6, the cost of equity could be calculated as follows:

$$\begin{aligned} R_j &= 0.07 + 1.6\,(0.12\text{–}0.07) \\ &= 0.07 + 1.6\,(0.05) \\ &= 0.07 + 0.08 \\ &= 0.15 \end{aligned}$$

This approach is known as the capital asset pricing model, or CAPM. While CAPM and beta are widely used, in a strict sense the statistical concepts that underlie the model limit its validity to cases where cash flows are uncorrelated from period to period.[12] In practice, this is a situation rarely encountered. Because of this limitation, caution should be exercised, and CAPM should be used in conjunction with one or more other approaches to calculating cost of equity.

COMPOSITE COST OF CAPITAL

As noted in the beginning of the appendix, firms generally have a target ratio of debt to equity to which they adhere as closely as possible. The composite cost of capital is simply the cost of the two forms of financing, equity and debt, weighted by the proportions in which they are used.

The weighted cost of capital is expressed mathematically by the following equation:

$$k_b = k_d c + (1\text{-}c)k_e \qquad (10)$$

where:

k_b = composite cost of capital
k_d = after-tax cost of debt capital
c = proportion of investment financed by debt capital
$(1\text{-}c)$ = proportion of investment financed by equity capital
k_e = required return on equity (cost of equity capital)

Thus if a firm has an after-tax cost of debt of 6.6%, a cost of equity of 15%, and a target capitalization ratio of 30% debt and 70% equity, the calculation of the composite cost of capital is as follows:

	Cost	Weight	Weighted Average Cost
Debt	0.066 (k_d)	0.30 (c)	0.066 × 0.30 = 0.0198
Equity	0.16 (k_e)	0.70 (1-c)	0.15 × 0.70 = 0.105
Composite cost of capital	=		0.1248 (k_b) or about 12.5%

The firm's composite cost of capital is about 12.5%. Sometimes this approach is refined to incorporate various forms of debt, or to approximate the cost of capital to a part of a firm, particularly when the firm is in various types of businesses.

INFLATION AND THE COST OF CAPITAL

Generally speaking, if inflation is eroding the value of money by some amount each year, investors will demand a higher rate of return by an amount equal to the rate of inflation. When discounting cash flows by the firm's cost of capital, it is absolutely necesssary that inflation be included in both cash flows and the cost of capital, or that it be excluded from both.

It may be easier to remove inflation from the firm's cost of capital than to try and inflate cash flows. If the anticipated inflation rate is 4%, the firm's real (noninflated) cost of capital is calculated as follows:

$$\begin{aligned} 1 + \text{real cost} &= (1 + \text{actual cost}) \,/\, (1 + \text{inflation}) \\ &= 1.1248 \,/\, 1.04 \\ &= 1.08 \end{aligned}$$

Thus the firm's real (noninflated) cost of capital is 8%. For a discussion of the use of the real cost of capital, refer to reference 13 cited at the end of this chapter.

APPENDIX C—CASH FLOWS

The most widely used and most modern methods of evaluating capital investments involve discounting cash flows to present value. To do so, the analyst needs to know both cash flows and the firm's cost of capital. Cost of capital was discussed in Appendix B, and cash flows are discussed in this appendix. This section will deal with defining cash flows, the mechanics of dealing with them (particularly as taxes affect cash flows), and with making estimates of cash flows.

DEFINING CASH FLOWS

Most homes run on the basis of a cash budget. The paycheck comes in on payday, and the groceries, rent, and other payments are also made by cash or check. Cash flow is easy to calculate. Federal income taxes are paid on the basis of cash receipts and cash payments.

Most businesses, however, run on the basis of an accrual budget, and income includes not only cash receipts for sales, but also sales that are billed even though the cash has not yet been received. Likewise, expenses are booked as they are accrued, even though the check has not been issued.

Sometimes, firms spend money for things that will last for several years, such as capital investments. Capital investments are not charged as expenses at the time the investment is made. Instead, the entire amount of the investment is put on the books as an asset, and then, as the capital asset is used to produce goods, the wear and tear on the asset is charged or accrued as an expense over the estimated life of the asset. This accrual is called depreciation, and it is ordinarily charged as a predetermined annual amount regardless of how much or how little the asset is used in any specific year.

Thus, if one looks at the income statement of a firm, the figures are not cash figures, but a mixture of cash and accruals. To define the cash outflows and inflows that occur as the result of making an investment, the accruals must be eliminated.

Cash flow is now being recognized to a much greater extent than it was at one time. Many firms include with their financial

APPENDIX C—CASH FLOWS

statements not only a balance sheet and income statement but also a ''source and application of funds'' or ''flow of funds'' statement, which is a statement of cash received and paid out during the year.

DEALING WITH CASH FLOWS

Taxes are cash flows. As was mentioned in the introduction of this chapter, the only cash flows that count to the owners of the firm are the after-tax cash flows. While a firm is concerned with taxes, in this situation the only interest is in determining the after-tax cash flows into and out of the firm.

Sales taxes are simple. When a piece of equipment is purchased, the buyer may be charged sales tax on the value of the equipment. If so, the sales tax is a cash outflow.

Income taxes are less simple. Income taxes are charged by the federal (and often state) government on the basis of income that depends on both cash expenses and accruals, but does not depend directly on investments. Income taxes are cash outflows. Thus, to define after-tax cash flows, it is necessary to define taxes. To define taxes, the accruals that affect taxes must be defined. For most investments, the one accrual that is significant is depreciation. It is large, its effect on taxes is important, and it is specifically dependent on the capital investment being made.

With respect to investments, there are three important ways in which income taxes affect cash flow as follows:

1. If the investment reduces costs, income and therefore taxes will be increased; if it increases costs, income and therefore taxes will be reduced. Both must be taken into account.
2. If the investment increases sales, it will probably increase income. Therefore, taxes will be increased and must be taken into account.
3. If an asset is sold for an amount greater than the amount for which it is valued as an asset on the books, it results in a gain that is taxable. If sold for an amount less than the amount for which it is valued on the books, the loss will reduce taxes. For discussion purposes, the rate is assumed to be the same as the rate on income.

To explain how one deals with cash flows, an example of a cost reduction investment will be discussed. If an investment of $10,000 is made and then depreciated on a straight-line basis (the same amount each year) over 5 years while reducing direct labor by $3000 per year, there will be no effect on pre-tax earnings at the time the investment is made. In each of the next 5 years, direct labor will be reduced by $3000, and a depreciation expense will be charged against earnings in the amount of $2000. The net effect will be an increase of pre-tax earnings in the amount of $1000 per year. If the firm is paying income taxes equal to 34% of earnings, this increase of income will increase taxes by 34% of the income change. Thus, the increase in after-tax income is only 0.66 (1-0.34) x $1000, or $660. This is because the $1000 increase in income is offset by a tax increase in the amount of $340.

A question that may arise is ''What will be the effect on after-tax cash flow?'' There will be an initial outlay of $10,000 at the time the investment is made. Direct labor is not only an expense, but a cash expense. Therefore, the firm will save the $3000 per year it no longer has to pay. Depreciation in each of the following 5 years is not a cash flow—it will not have to be paid out—so it does not have to be subtracted from the direct labor saving. The tax effect, however, is as calculated in the effect on after-tax earnings. The increase in taxes will be only $340, resulting in an annual cash flow of $3000-$340 = $2660. Thus, assuming that the investment is made in year 0, the effect on earnings compared with the effect on cash flow is shown in Table 3-28.

Next, consider what will happen if the investment results in an increase in sales. Another investment of $10,000 is made that will again be depreciated on a straight-line basis over the next 5 years. This time the purpose will be to provide the capacity to increase sales by $8000 per year for the next 5 years, on which an after-tax profit of $1500 each year will be earned.

It would be possible to go through the income statement and pick out those expenses that are cash flows and those that are accruals, but as noted earlier, the single most important accrual is depreciation expense, and it is possible to approximate annual operating cash flows with the following equation:

$$\text{Cash flow} = \text{Profit after tax} + \text{depreciation} \qquad (11)$$

This is not accurate enough for the finance department to perform its cash management activities, but ordinarily it is close enough for capital investment analysis. Thus, if annual after-tax earnings are $1500 and depreciation is $2000, annual after-tax cash flows are $3500 per year. The effect on earnings compared with the effect on cash flow is shown in Table 3-29.

Finally, consider the cash flows if the initial investment of $10,000 were offset by sale of an old asset, no longer needed, for $1000. To do so, it is necessary to know the value of the asset

TABLE 3-28
Earnings Compared With Cash Flow for a Reduction Investment Example

Year	Effect on After-Tax Earnings, $	Effect on After-Tax Cash Flow, $
0	0	(10,000)
1	660	2660
2	660	2660
3	660	2660
4	660	2660
5	660	2660

(*Industrial Technology Institute*)
Note: Value in parenthesis is negative.

TABLE 3-29
Earnings Compared With Cash Flow for an Investment to Increase Capacity

Year	Effect on After-Tax Earnings, $	Effect on After-Tax Cash Flow, $
0	0	(10,000)
1	1500	3500
2	1500	3500
3	1500	3500
4	1500	3500
5	1500	3500

(*Industrial Technology Institute*)
Note: Value in parenthesis is negative.

on the books and the annual depreciation charges that would have been accrued in coming years had the old asset been kept.

First, assume the old asset had been fully depreciated and had a book value of 0. The sale of the old asset for $1000 is considered to result in a gain to the firm equal to the difference between the sale price and the book value. In this case, the full $1000 is considered a gain and is taxable. Assuming a 34% rate, the result would be a $340 increase in taxes. Thus, the initial outlay is equal to $10,000 paid for the new machine less $1000 received for the old machine plus $340 taxes paid on the gain from sale of the old machine. The initial outlay is: $10,000- $1000 + $340 = $9340. Because the old machine had been fully depreciated, the sale of the machine will have no effect on future depreciation expense, and the depreciation on the new machine of $2000 per year will be considered in the calculation of the annual cash flows.

What if, however, the old machine had a book value of $3000 and were to be depreciated at a rate of $1500 per year for the next 2 years. If the old machine is sold for $1000, a loss of $2000 would be incurred, and taxes would be reduced for the current year by 0.34 x $2000 = $680. The initial outlay is $10,000- $1000 - $680 = $8320. By incurring a loss, taxes have thus been sheltered.

By selling the old machine, the depreciation expense that would have accrued in each of the next 2 years was eliminated. The net effect on depreciation when comparing the new investment with the old is shown in Table 3-30.

RESIDUAL VALUE AS A CASH FLOW

When one forecasts future cash flows to analyze a potential investment, one cannot look for an indefinite period into the future. The further into the future one tries to forecast, the more uncertain the view. As a result, it is practical only to forecast for some finite period—perhaps as few as 3 years or perhaps as many as 20. At the end of the forecast period, whatever it is, it is likely that the assets will have some value. This is generally called the "residual value" of the assets. The firm could realize the market value of these assets by selling them. Inventories could be sold as raw material or finished goods; machinery could be sold as used machinery or scrap, and land and buildings could be sold as real estate. Alternatively, the firm could realize value by continuing to use the assets to produce goods to be sold, in which case the value of the assets would be the discounted value of future cash flows generated by the production and sale of product. In calculating the value of potential investments, this residual value needs to be considered. Once it has been estimated as the value of both inventory and fixed assets at the end of the analysis period, it is treated in the analysis as a cash inflow.

TABLE 3-30
Effect of Depreciation When Making an Investment

Year	Effect on Depreciation
1	+ $2000 - $1500 = + $500
2	+ $2000 - $1500 = + $500
3	+ $2000
4	+ $2000
5	+ $2000

(*Industrial Technology Institute*)

ESTIMATING CASH FLOWS

Estimating cash flows is, almost without exception, the most difficult and time-consuming task when evaluating potential investments. First, one must think through carefully what cash flows will be affected by the investment, and then one must seek an estimate of how great or small the cash flows will be. The importance of the first step—thinking through what cash flows will be affected by the investment—cannot be overemphasized. If something is overlooked, in effect, it is estimated to be 0. If it is other than 0, management may make an incorrect decision based on the analysis. Even a very inaccurate estimate is likely to be better than an assumption of 0.

There are three broad categories of sources for estimates of cash flow. The company's financial and management reporting system is an important source of data. It is rarely sufficient, however. For estimates of cash flows as a result of activities within the firm, such as changes in performance standards or work-in-process inventories, it is often necessary to supplement such information with engineering studies. Sometimes these are based on simulations of manufacturing systems. For estimates of cash flows of activities that reach beyond the firm and involve the responses of customers, such as changes in price, quality, or delivery, it is often necessary to use market studies.

Estimating a realistic residual value is particularly difficult. Book value is sometimes used, and while book value is often a good approximation of the residual value of inventories, it is less likely to be a good approximation of the value of fixed assets. Because accounting rules for determining book value of fixed assets do not consider market forces, book value is unlikely to represent either market value or present value of fixed assets. Difficult as it may be, it is best to estimate directly either the market value or the present value of fixed assets at the end of the analysis period. Fortunately, this value is often small relative to the other cash flows under consideration, so the results of the analysis are typically not greatly biased by an error in forecasting residual value.

APPENDIX D—TABLES

APPENDIX D—COMPOUND INTEREST AND PRESENT VALUE TABLES

Present Value of $1.00 N Periods Hence Discounted at Rate R

N R=	5%	6%	7%	8%	9%	10%	11%	12%	13%	14%	15%	16%	18%	20%	24%	28%	30%	32%	34%	36%
1	0.952	0.943	0.935	0.926	0.917	0.909	0.901	0.893	0.885	0.877	0.870	0.862	0.847	0.833	0.806	0.781	0.769	0.758	0.746	0.735
2	0.907	0.890	0.873	0.857	0.842	0.826	0.812	0.797	0.783	0.769	0.756	0.743	0.718	0.694	0.650	0.610	0.592	0.574	0.557	0.541
3	0.864	0.840	0.816	0.794	0.772	0.751	0.731	0.712	0.693	0.675	0.658	0.641	0.609	0.579	0.524	0.477	0.455	0.435	0.416	0.398
4	0.823	0.792	0.763	0.735	0.708	0.683	0.659	0.636	0.613	0.592	0.572	0.552	0.516	0.482	0.423	0.373	0.350	0.329	0.310	0.292
5	0.784	0.747	0.713	0.681	0.650	0.621	0.593	0.567	0.543	0.519	0.497	0.476	0.437	0.402	0.341	0.291	0.269	0.250	0.231	0.215
6	0.746	0.705	0.666	0.630	0.596	0.564	0.535	0.507	0.480	0.456	0.432	0.410	0.370	0.335	0.275	0.227	0.207	0.189	0.173	0.158
7	0.711	0.665	0.623	0.583	0.547	0.513	0.482	0.452	0.425	0.400	0.376	0.354	0.314	0.279	0.222	0.178	0.159	0.143	0.129	0.116
8	0.677	0.627	0.582	0.540	0.502	0.467	0.434	0.404	0.376	0.351	0.327	0.305	0.266	0.233	0.179	0.139	0.123	0.108	0.096	0.085
9	0.645	0.592	0.544	0.500	0.460	0.424	0.391	0.361	0.333	0.308	0.284	0.263	0.225	0.194	0.144	0.108	0.094	0.082	0.072	0.063
10	0.614	0.558	0.508	0.463	0.422	0.386	0.352	0.322	0.295	0.270	0.247	0.227	0.191	0.162	0.116	0.085	0.073	0.062	0.054	0.046
11	0.585	0.527	0.475	0.429	0.388	0.350	0.317	0.287	0.261	0.237	0.215	0.195	0.162	0.135	0.094	0.066	0.056	0.047	0.040	0.034
12	0.557	0.497	0.444	0.397	0.356	0.319	0.286	0.257	0.231	0.208	0.187	0.168	0.137	0.112	0.076	0.052	0.043	0.036	0.030	0.025
13	0.530	0.469	0.415	0.368	0.326	0.290	0.258	0.229	0.204	0.182	0.163	0.145	0.116	0.093	0.061	0.040	0.033	0.027	0.022	0.018
14	0.505	0.442	0.388	0.340	0.299	0.263	0.232	0.205	0.181	0.160	0.141	0.125	0.099	0.078	0.049	0.032	0.025	0.021	0.017	0.014
15	0.481	0.417	0.362	0.315	0.275	0.239	0.209	0.183	0.160	0.140	0.123	0.108	0.084	0.065	0.040	0.025	0.020	0.016	0.012	0.010
16	0.458	0.394	0.339	0.292	0.252	0.218	0.188	0.163	0.141	0.123	0.107	0.093	0.071	0.054	0.032	0.019	0.015	0.012	0.009	0.007
17	0.436	0.371	0.317	0.270	0.231	0.198	0.170	0.146	0.125	0.108	0.093	0.080	0.060	0.045	0.026	0.015	0.012	0.009	0.007	0.005
18	0.416	0.350	0.296	0.250	0.212	0.180	0.153	0.130	0.111	0.095	0.081	0.069	0.051	0.038	0.021	0.012	0.009	0.007	0.005	0.004
19	0.396	0.331	0.277	0.232	0.194	0.164	0.138	0.116	0.098	0.083	0.070	0.060	0.043	0.031	0.017	0.009	0.007	0.005	0.004	0.003
20	0.377	0.312	0.258	0.215	0.178	0.149	0.124	0.104	0.087	0.073	0.061	0.051	0.037	0.026	0.014	0.007	0.005	0.004	0.003	0.002
21	0.359	0.294	0.242	0.199	0.164	0.135	0.112	0.093	0.077	0.064	0.053	0.044	0.031	0.022	0.011	0.006	0.004	0.003	0.002	0.002
22	0.342	0.278	0.226	0.184	0.150	0.123	0.101	0.083	0.068	0.056	0.046	0.038	0.026	0.018	0.009	0.004	0.003	0.002	0.002	0.001
23	0.326	0.262	0.211	0.170	0.138	0.112	0.091	0.074	0.060	0.049	0.040	0.033	0.022	0.015	0.007	0.003	0.002	0.002	0.001	0.001
24	0.310	0.247	0.197	0.158	0.126	0.102	0.082	0.066	0.053	0.043	0.035	0.028	0.019	0.013	0.006	0.003	0.002	0.001	0.001	0.001
25	0.295	0.233	0.184	0.146	0.116	0.092	0.074	0.059	0.047	0.038	0.030	0.024	0.016	0.010	0.005	0.002	0.001	0.001	0.001	0.000
26	0.281	0.220	0.172	0.135	0.106	0.084	0.066	0.053	0.042	0.033	0.026	0.021	0.014	0.009	0.004	0.002	0.001	0.001	0.000	0.000
27	0.268	0.207	0.161	0.125	0.098	0.076	0.060	0.047	0.037	0.029	0.023	0.018	0.011	0.007	0.003	0.001	0.001	0.001	0.000	0.000
28	0.255	0.196	0.150	0.116	0.090	0.069	0.054	0.042	0.033	0.026	0.020	0.016	0.010	0.006	0.002	0.001	0.001	0.000	0.000	0.000
29	0.243	0.185	0.141	0.107	0.082	0.063	0.048	0.037	0.029	0.022	0.017	0.014	0.008	0.005	0.002	0.001	0.000	0.000	0.000	0.000
30	0.231	0.174	0.131	0.099	0.075	0.057	0.044	0.033	0.026	0.020	0.015	0.012	0.007	0.004	0.002	0.001	0.000	0.000	0.000	0.000

APPENDIX D—*Continued*

Future Value of $1.00 Compound at Rate *R* for *N* Periods

N R=	5%	6%	7%	8%	9%	10%	11%	12%	13%	14%	15%	16%	18%	20%	24%	28%	30%	32%	34%	36%
1	1.050	1.060	1.070	1.080	1.090	1.100	1.110	1.120	1.130	1.140	1.150	1.160	1.180	1.200	1.240	1.280	1.300	1.320	1.340	1.360
2	1.103	1.124	1.145	1.166	1.188	1.210	1.232	1.254	1.277	1.300	1.323	1.346	1.392	1.440	1.538	1.638	1.690	1.742	1.796	1.850
3	1.158	1.191	1.225	1.260	1.295	1.331	1.368	1.405	1.443	1.482	1.521	1.561	1.643	1.728	1.907	2.097	2.197	2.300	2.406	2.515
4	1.216	1.262	1.311	1.360	1.412	1.464	1.518	1.574	1.630	1.689	1.749	1.811	1.939	2.074	2.364	2.684	2.856	3.036	3.224	3.421
5	1.276	1.338	1.403	1.469	1.539	1.611	1.685	1.762	1.842	1.925	2.011	2.100	2.288	2.488	2.932	3.436	3.713	4.007	4.320	4.653
6	1.340	1.419	1.501	1.587	1.677	1.772	1.870	1.974	2.082	2.195	2.313	2.436	2.700	2.986	3.635	4.398	4.827	5.290	5.789	6.328
7	1.407	1.504	1.606	1.714	1.828	1.949	2.076	2.211	2.353	2.502	2.660	2.826	3.185	3.583	4.508	5.629	6.275	6.983	7.758	8.605
8	1.477	1.594	1.718	1.851	1.993	2.144	2.305	2.476	2.658	2.853	3.059	3.278	3.759	4.300	5.590	7.206	8.157	9.217	10.40	11.70
9	1.551	1.689	1.838	1.999	2.172	2.358	2.558	2.773	3.004	3.252	3.518	3.803	4.435	5.160	6.931	9.223	10.60	12.17	13.93	15.92
10	1.629	1.791	1.967	2.159	2.367	2.594	2.839	3.106	3.395	3.707	4.046	4.411	5.234	6.192	8.594	11.81	13.79	16.06	18.67	21.65
11	1.710	1.898	2.105	2.332	2.580	2.853	3.152	3.479	3.836	4.226	4.652	5.117	6.176	7.430	10.66	15.11	17.92	21.20	25.01	29.44
12	1.796	2.012	2.252	2.518	2.813	3.138	3.498	3.896	4.335	4.818	5.350	5.936	7.288	8.916	13.21	19.34	23.30	27.98	33.52	40.04
13	1.886	2.133	2.410	2.720	3.066	3.452	3.883	4.363	4.898	5.492	6.153	6.886	8.599	10.70	16.39	24.76	30.29	36.94	44.91	54.45
14	1.980	2.261	2.579	2.937	3.342	3.797	4.310	4.887	5.535	6.261	7.076	7.988	10.15	12.84	20.32	31.69	39.37	48.76	60.18	74.05
15	2.079	2.397	2.759	3.172	3.642	4.177	4.785	5.474	6.254	7.138	8.137	9.266	11.97	15.41	25.20	40.56	51.19	64.36	80.64	100.7
16	2.183	2.540	2.952	3.426	3.970	4.595	5.311	6.130	7.067	8.137	9.358	10.75	14.13	18.49	31.24	51.92	66.54	84.95	108.1	137.0
17	2.292	2.693	3.159	3.700	4.328	5.054	5.895	6.866	7.986	9.276	10.76	12.47	16.67	22.19	38.74	66.46	86.50	112.1	144.8	186.3
18	2.407	2.854	3.380	3.996	4.717	5.560	6.544	7.690	9.024	10.58	12.38	14.46	19.67	26.62	48.04	85.07	112.5	148.0	194.0	253.3
19	2.527	3.026	3.617	4.316	5.142	6.116	7.263	8.613	10.20	12.06	14.23	16.78	23.21	31.95	59.57	108.9	146.2	195.4	260.0	344.5
20	2.653	3.207	3.870	4.661	5.604	6.727	8.062	9.646	11.52	13.74	16.37	19.46	27.39	38.34	73.86	139.4	190.0	257.9	348.4	468.6
21	2.786	3.400	4.141	5.034	6.109	7.400	8.949	10.80	13.02	15.67	18.82	22.57	32.32	46.01	91.59	178.4	247.1	340.4	466.9	637.3
22	2.925	3.604	4.430	5.437	6.659	8.140	9.934	12.10	14.71	17.86	21.64	26.19	38.14	55.21	113.6	228.4	321.2	449.4	625.6	866.7
23	3.072	3.820	4.741	5.871	7.258	8.954	11.03	13.55	16.63	20.36	24.89	30.38	45.01	66.25	140.8	292.3	417.5	593.2	838.3	1179
24	3.225	4.049	5.072	6.341	7.911	9.850	12.24	15.18	18.79	23.21	28.63	35.24	53.11	79.50	174.6	374.1	542.8	783.0	1123	1603
25	3.386	4.292	5.427	6.848	8.623	10.83	13.59	17.00	21.23	26.46	32.92	40.87	62.67	95.40	216.5	478.9	705.6	1034	1505	2180
26	3.556	4.549	5.807	7.396	9.399	11.92	15.08	19.04	23.99	30.17	37.86	47.41	73.95	114.5	268.5	613.0	917.3	1364	2017	2965
27	3.733	4.822	6.214	7.988	10.25	13.11	16.74	21.32	27.11	34.39	43.54	55.00	87.26	137.4	333.0	784.6	1193	1801	2703	4032
28	3.920	5.112	6.649	8.627	11.17	14.42	18.58	23.88	30.63	39.20	50.07	63.80	103.0	164.8	412.9	1004	1550	2377	3622	5484
29	4.116	5.418	7.114	9.317	12.17	15.86	20.62	26.75	34.62	44.69	57.58	74.01	121.5	197.8	512.0	1286	2015	3138	4853	7458
30	4.322	5.743	7.612	10.06	13.27	17.45	22.89	29.96	39.12	50.95	66.21	85.85	143.4	237.4	634.8	1646	2620	4142	6503	10,143

REFERENCES

References

1. Robert Hayes and David A. Garvin, "Managing as if Tomorrow Mattered," *Harvard Business Review* (May-June 1982), pp. 70-79.
2. Robert S. Kaplan, "Must CIM Be Justified By Faith Alone?" *Harvard Business Review* (March-April 1986), pp. 87-95.
3. Ram Charan, "How to Strengthen Your Strategy Review Process," *The Journal of Business Strategy* (Winter 1982), pp. 50-60.
4. Michael E. Porter, *Competitive Strategy: Techniques for Analyzing Industries and Competitors* (New York: The Free Press, 1980).
5. *Ibid*.
6. Robert H. Hayes and Steven C. Wheelwright, "Link Manufacturing Process and Product Life Cycles," *Harvard Business Review* (January-February 1979), pp. 133-140.
7. ____________, "The Dynamics of Process-Product Life Cycles," *Harvard Business Review* (March-April 1979), pp. 127-136.
8. J. Hirshleifer, *Investment, Interest and Capital* (Englewood Cliffs, NJ: Prentice-Hall, Inc., 1970).
9. David B. Hertz, "Risk Analysis in Capital Investment," *Harvard Business Review* (January-February 1964), pp. 95-106.
10. Pearson Graham and Diron Bodenhorn, *Managerial Economics* (Reading, MA: Addison-Wesley Publishing Co., Inc., 1980), pp. 369-378.
11. George Terborgh, *Business Investment Management* (Washington, DC: Machinery and Allied Products Institute, 1967).
12. Eugene F. Fama, "Risk-Adjusted Discount Rates and Capital Budgeting Under Uncertainty," *Journal of Financial Economics* (August 1977), pp. 3-24.
13. Robert S. Kaplan, "Must CIM Be Justified By Faith Alone?" *Harvard Business Review* (March-April 1986), pp. 87-95.

Bibliography

Boer, T. "Cost Justification Is Possible." *Managing Automation* (August 1986), pp. 55-59.

Baker, J. A. "Winning Your Case for Automation." *Manufacturing Engineering*. (July 1984), pp. 72-73.

Blank, L. "The Changing Scene of Economic Analysis for the Evaluation of Manufacturing System Design and Operation." *The Engineering Economist* (Spring 1985), pp. 227-244.

Brimson, J. A., and Frescoln, L. D. "Technology Accounting—The Value-Added Approach to Capital Asset Depreciation." *CIM Review* (Fall 1986), pp. 44-52.

Canada, J. R. "Non-Traditional Method for Evaluating CIM Opportunities Assigns Weight to Intangibles." *Industrial Engineering* (March 1986), pp.66-71.

"Catalog of Simulation Software." *Simulation* (October 1987), pp. 165-181.

Huber, R. F. "Justification-Barrier to Competitive Manufacturing." *Production* (September 1985), pp. 46-51.

Shenchuk, J. "Justifying Flexible Automation." *American Machinist* (October 1984), pp. 93-96.

Snowdon, Jane L.; Ammons, Jane C.; and McGinnis, Leon F. "A Review of Queueing Network Packages for Manufacturing Systems Analysis." *1987 IIE Integrated Systems Conference Proceedings*. Norcross, GA: Institute of Industrial Engineers, 1987.

Son, Y. K., and Park, C. S. "Economic Measure of Productivity, Quality, and Flexibility in Advanced Manufacturing Systems." *Journal of Manufacturing Systems* (vol. 6, no. 3, 1987), pp. 193-207.

Varney, M. S.; Sullivan, W. G.; and Cochran, J. "Justification of Flexible Manufacturing Systems With the Analytical Hierarchy Process." *Proceedings of the Spring Annual Industrial Engineering Conference* (May 1985), pp. 181-190.

COST ESTIMATING AND CONTROL

CHAPTER CONTENTS:

Manufacturing cost estimating has been defined as the process of forecasting the "bottom line" cost totals associated with the completion of a set of manufacturing tasks.[1,2] This forecast is normally made prior to the time the sequence of tasks actually begins. In contrast, cost control has been described as the process of updating or refining prior initial cost estimates for a sequence of manufacturing tasks that is currently in process. Cost control is function-related to project performance. It is related to the problem of "living within the budget."

Engineers who complete baccalaureate programs have historically had only a limited role in the cost estimating and control process. The primary reason for this limited involvement has been the limited manufacturing experience that characterizes engineers who graduate in most disciplines. To compile accurate estimates, the estimator must be able to accurately describe the process sequence that will specify how the part(s) will be fabricated and assembled. Recent engineering graduates are often woefully unqualified to describe this sequence. Cost estimators, therefore, are typically blue collar personnel who possess a wealth of shop floor experience. These individuals often assume cost estimating responsibilities at a midpoint in their careers.

Although not actively involved with cost estimating initially, engineers of all disciplines have a variety of reasons to be interested in the cost estimating process. As the U.S. manufacturing community continues to struggle to be competitive in world markets, engineers are under increasing pressure to become more cost conscious. Many companies provide the opportunity for entry-level engineers to obtain shop floor experience through in-plant training programs. The goal is to expose engineers to a variety of manufacturing processes. Increased manufacturing literacy generally enables engineers to produce more manufacturable, yet cost effective, designs. Finally, engineers often find themselves responsible for managing or overseeing the cost estimating process as they eventually enter the supervisory ranks.

This chapter begins by describing the process of cost recovery. Types and categories of manufacturing costs are described, and components of hourly labor/overhead rates are reviewed. The budgeting process by which indirect labor, overhead, and general/administrative costs are recovered is also described. Because many manufacturers are turning to higher levels of factory automation, methods to extend the cost recovery approach to totally automated workcenters are presented.

Separate sections address the topics of the cost estimating and review processes. Parametric estimating systems are discussed. An algorithm to mathematically forecast potential job costs, throughput times, and workload levels is presented. The computer-assisted cost estimating process (CACE) is overviewed. Learning curves and their integration into the cost estimating process are also addressed.

COST RECOVERY

Each manufacturing organization has a variety of cost components that comprise its manufacturing budget. The process by which these costs are organized requires the definition of cost centers and corresponding hourly rates. It is necessary to determine the appropriate values of these rates for various workcenters, together with the different types of work activities that can occur. The supporting costs of both indirect labor and overhead are embedded in a dollar rate that is charged for each unit of work that is completed on the production floor. For conventional workcenters, this work unit is usually the direct labor hour. The supporting costs of indirect labor and overhead are charged or recovered on the basis of the number of direct labor hours worked on a given cost center in a given period of time. This cost recovery process is described more fully in the following sections.

TYPES OF COSTS

Before describing the actual cost recovery process, it is necessary to overview the types of manufacturing costs that can be incurred. Costs generally fall into the categories of fixed, variable, or semifixed costs. Among the manufacturing costs are direct labor, direct material, indirect labor, indirect manufacturing costs, general administrative costs, and tooling costs. Also described are tooling costs, fixed costs, variable costs, and semifixed costs.

Contributor of this chapter is: Eric M. Malstrom, PE, *Professor and Head, Department of Industrial Engineering, University of Arkansas.*

Reviewers of this chapter are: Willie Beatty, *Manager, Cost Control, Clark Components, North America;* ***Merrill Ebner,*** *Professor, Manufacturing Engineering Dept., Boston University;* ***Richard L. Engwall,*** *Manager—Systems Planning, Analysis and Assurance, Manufacturing Systems and Technology, Westinghouse Electric Corp.;* ***Ernest L. McGraw,*** *Staff Engineer, Materials Laboratory and Consultants Div., Naval Avionics Center;* ***Dr. Phillip Ostwald,*** *University of Colorado, Department of Mechanical Engineering-Boulder;* ***Richard L. Shell,*** *Professor and Director, Industrial Engineering, Department of Mechanical and Industrial Engineering, University of Cincinnati;* ***Hayward Thomas,*** *Consultant, Retired—President, Jensen Industries.*

COST RECOVERY

Fixed Costs

Fixed costs are those that are generally independent of the production quantity that is being built. Indirect labor and indirect manufacturing costs are generally fixed. Setup costs for machine tools are also fixed costs.

Variable Costs

Variable costs are those incurred on a per-unit basis of the quantity that is being produced. Variable costs increase with each additional unit that is produced. Per-piece direct labor and direct material costs for assembled or machined parts are examples of variable costs.

Semifixed Costs

Semifixed costs are sometimes known as step variable costs. These costs are somewhat independent of quantity and vary with specific groups of units that are produced. The cost to change cutting tools and the completion of scheduled maintenance operations after a specified number of production units are examples of semifixed costs.

Direct Labor

Direct labor is the cost of all "hands-on" effort associated with the manufacture of a specific product.[1,2] Typical direct labor activities include machining, assembly, testing, and sometimes inspection and troubleshooting. Direct labor activities are characterized by the presence of some physical contact between the worker and the workpiece. This contact usually adds value to the product being produced. Direct labor is a variable manufacturing cost.

Direct Material

Direct material is the cost of all components included in the end product being produced. To be considered as direct material, the components or raw materials must be a permanent part of the end product being manufactured. Examples of material costs that are not direct include raw material from which tooling is fabricated, test equipment, and packaging materials. Direct material is a variable manufacturing cost.

Indirect Labor

Indirect labor is the cost of all labor effort that cannot be directly associated with the manufacture of a product.[1,2] Examples of indirect labor include the salary costs of workers in the accounting, purchasing, and personnel departments, together with the salary costs of supervisors and managers. Indirect labor costs tend to increase in both absolute and percentage terms as the organization increases in size.

Indirect Manufacturing Cost

Indirect manufacturing cost (IMC) is a term often used synonymously with overhead costs. It includes all costs for rent, heat, electricity, water, and expendable factory supplies, together with the annual costs of building and equipment depreciation. Expendable factory supplies are often indirect materials that are consumed during the manufacturing process. Refer to Chapter 3 of this volume for information on equipment depreciation.

General and Administrative Costs

General and administrative (G&A) costs are those incurred at the plant or interplant level that are not easily associated with a specific workcenter or department. Examples include the costs of top executives' salaries, plant mainframe computer procurement/operation costs, and technical library facilities. Most G&A costs are fixed. The primary variable component is sales commissions.

Tooling and Test Equipment Costs

Tooling and test equipment costs are those costs incurred in the fabrication of jigs and fixtures for machining. They also include costs for the programming, generation, and checkout of NC tapes, and costs for the design and fabrication of special-purpose test equipment. Tooling and test equipment costs are generally fixed.

COST CENTERS

A cost center (CC) is a numerical way of designating different parts of an organization. Many firms use a three or four-digit numerical code to identify departments, divisions within departments, and branches within divisions. Each unit of direct labor that is charged is usually associated with a customer order number. This order number enables the labor charges to be identified with a specific customer and/or product.

Most organizations usually associate an account number with the customer order number for labor charges. The account number usually contains all or part of the organizational code of the individual performing the labor. A cost center is an identifiable accounting number that corresponds to this code. The purpose of the cost center is to identify the organizational segment performing the work on the customer order.

SHOP ORDERS

A shop order has been described as the authorization to perform a specific type of labor effort in a given cost center.[1,2] Shop orders are often identified by an alphanumeric code. The "alpha" part of the code usually refers to the type of work effort to be performed. Examples include chip turning, sheet metal operations, electronics assembly, and inspection. The numeric part of the code is a number that is traceable to the customer order number. A group of representative cost centers and related shop orders for a job shop manufacturing facility have been identified.[1,2] These cost centers and shop orders are presented in Table 4-1. The X's in the table represent unique five-digit numbers.

COMPONENTS OF LABOR/OVERHEAD RATES

Each cost center within a manufacturing organization has an hourly labor/overhead rate associated with it. The labor/overhead (LOH) rate represents the cost in dollars to the organization for each hour of direct labor that is worked on a given cost center.

The LOH rate consists of the direct labor rate, the expense rate, the IMC, and general and administrative costs. The direct labor rate is the composite average of all hourly direct labor wages on the cost center. The expense rate reflects the total indirect labor dollars to be charged per direct labor hour. It is calculated by dividing the total indirect dollars spent in a budgeting period by the total direct hours worked in a budgeting period. IMC reflects the costs of utilities, expendable supplies, and equipment/building depreciation on the cost center. General and administrative costs represent the dollar total of the plant G&A costs required to support the cost center.

Based on the assumption that the dollar values for the burden and general and administrative costs are divided by the total number of direct labor hours to be worked in a budgeting period, the labor/overhead rate for a specific cost center can be calculated using the following equation:

$$LOH_{CC} = L_D + E_R + IMC + G\&A \quad (1)$$

where:

LOH_{CC} = the labor/overhead rate for a specific cost center, dollars/hr
L_D = the direct labor rate, dollars/hr
E_R = the expense rate, dollars/hr
IMC = the indirect manufacturing cost rate, dollars/hr
$G\&A$ = general and administrative cost rate, dollars/hr

Not all cost centers within the organization have direct labor charges. Examples would include the accounting and personnel departments. All workers in these departments would be considered indirect personnel. LOH rates for these cost centers would be determined in a similar manner. The direct labor term is replaced by the average hourly wage rate on the indirect labor cost center. The expense rate term would include only those indirect labor dollars for supporting functions external to the indirect labor cost center being analyzed. The difference between direct and indirect labor cost centers becomes more apparent as the plant budgeting process is analyzed. This process is the subject of the following section.

TABLE 4-1
Representative Cost Centers and Shop Orders

Cost Center	Activity	Shop Order Code
21 Manufacturing Engineering	Process engineering	M-XXXXX
	Tool design	D-XXXXX
	Generation of NC tapes	N-XXXXX
	Packaging	K-XXXXX
22 Machining	Chip turning	C-XXXXX
	Sheet metal	S-XXXXX
	Tool fabrication and maintenance	U-XXXXX
	Painting	P-XXXXX
	Plating	L-XXXXX
	Heat treating	H-XXXXX
	Rework	R-XXXXX
23 Assembly	Assembly	A-XXXXX
	Testing	T-XXXXX
	Wire cutting	W-XXXXX
	Troubleshooting	G-XXXXX
	Rework	R-XXXXX
	Encapsulation	J-XXXXX
42 Inspection	Inspection of purchased parts	B-XXXXX
	Inspection of fabricated parts	F-XXXXX
43 Standards & Calibration	Mechanical and electronic calibration	O-XXXXX
70 Engineering	Engineering support to manufacturing	E-XXXXX

THE BUDGETING PROCESS

The goal of the cost recovery process is to capture or recover the costs of all plant operations that support direct labor activities. The recovery is accomplished by charging an hourly rate for each direct labor hour worked. This rate must be sufficient in size to fund all supporting activities and functions of the cost center. The problem becomes one of determining what rate to charge as defined by Eq. (1). This is accomplished through a budgeting process that is generally performed quarterly for all direct labor cost centers.

The budgeting process is perhaps best illustrated with an example. Consider a manufacturing plant with a total of 1000 employees. This total is distributed numerically among a variety of cost centers as shown in Table 4-2.

Suppose it is desired to determine the LOH rate for the Machining Div., cost center 22. Cost center 22 has a total of 200 employees as shown in Table 4-2. Of these, assume that 150 are direct labor employees. The remaining 50 are considered to be supervisory and clerical personnel. Suppose further that a

40-hr work week applies, with no overtime for a budgeting period of 3 months. If exactly 4 weeks are assumed for each month, the budgeting process must then consider an hourly total of the following:

40 hr/week × 4 weeks/month × 3 months = 480 hr

The 480-hr total will be worked by each of 150 direct labor employees. This yields a budgetary direct labor hour total of the following:

480 hr/employee × 150 employees = 72,000 hr

DIRECT LABOR DETERMINATION

The first step in the budgeting process is to compute the average hourly wage rate for all direct labor employees. For purposes of this example, suppose that this average is $10.00 per hour, including all benefits such as sick leave, vacation time, and insurance. This corresponds to the direct labor component, L_D, in Eq. (1).

EXPENSE RATE DETERMINATION

The next step in the budgeting process is to compute the expense rate, E_R. The first component of this term is the cost of the indirect labor employees on cost center 22. For purposes of this example, assume that all indirect labor employees are salaried and that the average annual salaries (including benefits) shown in Table 4-3 apply.

The reader should note that the salaries in Table 4-3 are only for indirect labor employees. Some of the cost centers in this table contain both direct and indirect labor employees. These are cost centers 21, 22, 23, and 42, the Manufacturing Engineering, Machining, Assembly and Inspection Divs., respectively. All other cost centers are assumed to consist only of indirect labor.

Because the budgeting process is for a fiscal quarter, the goal is to recover one fourth of the salaries shown in Table 4-3 through the budgeting procedure. On cost center (CC) 22, there exists a total of 50 indirect labor employees. From Table 4-3, the average salary, including benefits, is $28,000 per year. The amount to be recovered in a quarter becomes as follows:

$7000/employee on CC 22 × 50 employees = $350,000

Other indirect labor activities also support cost center 22. These include cost centers 30, 41, 43, 50, 60, 71, and 72. The indirect labor costs for manufacturing engineering, assembly, and inspection (CC 21, 23, and 42) are not included in this list. These costs support CC 21, 23, and 42 exclusively. They will be recovered when the budgeting process is repeated for these direct labor cost centers.

Again, because the budgeting process is only for a fiscal quarter, only one fourth of the salary levels shown in Table 4-3 need to be recovered. The process is somewhat complicated because all of the pure indirect labor cost centers support all four of the direct labor cost centers. A common way of taking this into account is to prorate the cost totals on the basis of total employees.

From Table 4-2, the Manufacturing Engineering, Machining, Assembly, and Inspection Divs. have employee totals of 100, 200, 200, and 50, respectively. The proration logic assumes that the pure indirect labor cost centers support the direct labor cost centers proportionately on the basis of people. This means that the expense rate for CC 22 must include a percentage of the indirect costs to be recovered from all indirect labor cost centers. The percentage for CC 22 would be determined as follows:

200/(100 + 200 + 200 + 50) = 200/550

Total numbers of employees on each of the four cost centers are used as opposed to totals for just the direct labor employees. This is because the other pure indirect labor cost centers support all of cost center 22, not just the direct labor activities. The total indirect costs to be recovered are tallied in Table 4-4.

The reader should note that the employee totals of Table 4-4 show only indirect labor employee totals. In some cost centers with both direct and indirect labor, the employee totals shown are less than those shown in Table 4-1. These totals are marked with an asterisk.

Table 4-4 lists all of the indirect employees and their average salaries, including benefits. Quarterly salaries are computed and totaled for each of the cost centers. The last column of the table lists the prorated cost that should be attributed to cost center 22. Note that no entries appear in column 6 for cost centers 21, 23, and 42. These indirect costs will be captured in their entirety when the budgeting process is repeated for these cost centers.

In addition, there is no entry in column 6 of Table 4-4 for cost center 22. The procedure assumes that CC 22 must pay for

TABLE 4-2
Employee Distribution by Cost Center

Cost Center	Description	Number of Employees
21	Manufacturing Engineering	100
22	Machining	200
23	Assembly	200
30	Comptroller	50
41	Quality Engineering	50
42	Inspection	50
43	Standards & Calibration	50
50	Personnel	50
60	Marketing	100
71	Design Engineering	100
72	Research & Development	50
		1000

TABLE 4-3
Average Salary Levels of Indirect Labor Employees

Cost Center	Description	Average Salary Level
21	Manufacturing Engineering	$32,000
22	Machining	$28,000
23	Assembly	$28,000
30	Comptroller	$32,000
41	Quality Engineering	$36,000
42	Inspection	$28,000
43	Standards & Calibration	$32,000
50	Personnel	$36,000
60	Marketing	$40,000
71	Design Engineering	$44,000
72	Research & Development	$48,000

TABLE 4-4
Proration of Indirect Costs for Cost Center 22

1 Cost Center	2 Average Indirect Salary	3 Column 2 x 1/4	4 Number of Indirect Employees	5 Column 3 x Column 4	6 Column 5 x 200/550
21	$32,000	$8000	50*	$400,000	
22	$28,000	$7000	50*	$350,000**	
23	$28,000	$7000	50*	$350,000	
30	$32,000	$8000	50	$400,000	$145,440
41	$36,000	$9000	50	$450,000	$163,620
42	$28,000	$7000	20*	$140,000	
43	$32,000	$8000	50	$400,000	$145,440
50	$36,000	$9000	50	$450,000	$163,620
60	$40,000	$10,000	100	$1,000,000	$363,636
71	$44,000	$11,000	100	$1,100,000	$400,000
72	$48,000	$12,000	50	$600,000	$218,160
					$1,599,916

* Number of indirect employees out of total on cost center.
** Total must be recovered in its entirety by cost center 22.

all of its own indirect costs. This is true for all other cost centers that have direct labor activities. The total of column 6 is $1,599,916. This total represents the cost to CC 22 of supporting indirect labor cost centers. The total indirect labor cost of CC 22 ($350,000) must be added to this sum. The expense rate then is determined by dividing this total by the total number of direct labor hours to be worked on CC 22 during the quarter. The expense rate, E_R, for CC 22 becomes the following:

$$E_R = \frac{\$350{,}000 + \$1{,}599{,}916}{72{,}000 \text{ hr}} = \$27.08/\text{hr}$$

INDIRECT MANUFACTURING COST

The third step in the budgeting procedure is to determine the IMC rate. Two types of IMC must be considered. The first is for CC 22 itself. Suppose the estimated cost of utilities, supplies, and equipment depreciation for the quarter for CC 22 is $500,000. A second component must be added to this term. This is the estimated cost for all IMC for the pure indirect labor cost centers. For purposes of analysis, assume that the total for the quarter is $1,000,000. The IMC totals for the other direct labor cost centers are not included because these will be recovered as the budgeting process is repeated for them.

The IMC for the indirect labor cost centers must again be prorated on the basis of employee totals between the indirect labor cost centers. The total burden to be recovered on CC 22 is as follows:

$$\$500{,}000 + \$1{,}000{,}000 \times 200/550 = \$863{,}636$$

To obtain the IMC component of the hourly rate, the total IMC to be recovered must be divided by the total direct labor hours for the quarter. The IMC rate thus becomes the following:

$$\text{IMC} = \$863{,}636/72{,}000 \text{ hr} = \$11.99/\text{hr}$$

G&A DETERMINATION

The final step in the procedure is to determine the G&A portion of the hourly rate. Assume that the total dollar value of G&A to be recovered for the budgeting quarter is $400,000. This amount is again prorated by employee totals. The total G&A to be recovered on CC 22 is as follows:

$$\$400{,}000 \times 200/550 = \$145{,}440$$

The G&A rate is obtained by dividing this total by the total direct labor hours. The G&A rate thus becomes the following:

$$\text{G\&A} = \$145{,}440/72{,}000 \text{ hr} = \$2.02/\text{hr}$$

HOURLY RATE DETERMINATION

The hourly rate for CC 22 may now be determined from Eq. (1) as follows:

$$\begin{aligned} LOH_{22} &= \$10.00/\text{hr} + \$27.08/\text{hr} + \$11.99/\text{hr} + \$2.02/\text{hr} \\ &= \$51.09/\text{hr} \end{aligned}$$

In practice this procedure would be repeated for the remaining direct labor cost centers. Adjustments would have to be made for overtime and the amount by which estimated dollar amounts are subsequently found to be in error. Resultant budgetary deficits or surpluses would be handled as fiscal adjustments in the following budgeting quarter.

BUDGETING FOR INDIRECT LABOR COST CENTERS

Once the budgeting process has been completed for all direct labor cost centers, all supporting costs are recovered through the labor/overhead rates that are determined and charged for each direct labor hour. While not necessary to recover costs, there is some value in completing a similar process for all pure indirect labor cost centers. This process has nothing to do with cost recovery. Rather, it is an exercise often performed by managers to determine the cost to the organization of each labor hour worked on an indirect labor cost center.

The procedure is basically the same as that illustrated in the preceding sections. First the average hourly wage rate on the cost center is determined. This term is analogous to the term L_D in Eq. (1). The difference is that the average hourly wage rate is for salaried indirect labor employees.

The second step is to determine an applicable expense rate for the indirect labor cost center. In this procedure, only the indirect costs for other cost centers that support the indirect labor cost center being analyzed are considered. Suppose the analysis is being completed for the Comptroller's Dept., CC 30 (refer to Table 4-2). The Personnel Dept. obviously performs supporting functions for the Comptroller by virtue of hiring and laying off workers as well as computing fringe benefits. The equivalent of the expense rate for CC 30 would be determined by summing the costs of this and all other supporting indirect cost centers. The total would be divided by the total number of hours to be worked during the quarter for the cost center being analyzed.

The third step is to determine the IMC rate. In the procedure for direct labor cost centers, the IMC for the cost center being budgeted is totaled for the quarter. The IMCs for all supporting indirect labor cost centers are also totaled and prorated based on the numbers of employees on each direct labor cost center. The total IMC is divided by the number of direct labor hours to be worked in the quarter.

In analyzing indirect labor cost centers, only the burden dollars for the quarter on the cost center being analyzed are considered. The total is again divided by the total hours in the quarter to be worked.

The final step is to determine the applicable G&A rate. The cost is prorated on the basis of numbers of employees, not only for direct labor cost centers, but for all cost centers in the organization. Recall from Table 4-2 that the Comptroller's Dept. has a total of 50 employees out of a total of 1000 for the entire organization. The G&A rate for CC 30 would be determined by taking the total G&A cost for the quarter in dollars and multiplying this sum by 50/1000. The result would be divided by the total number of hours in the quarter to be worked.

Again, the goal of this procedure for indirect labor cost centers is not cost recovery. Rather, the exercise is one of fiscal planning that enables managers to assess the cost to the organization for each hour worked on the cost center(s) for which they are responsible.

BUDGETING ON AUTOMATED WORKCENTERS

Many manufacturing organizations are increasingly turning to higher levels of factory automation. The level of automation can complicate the budgeting process described in the preceding sections. In a totally automated factory, it is conceivable that only one direct laborer might exist. This individual might operate a control console that in turn would control automatic machines, robots, and material handling systems.

The hourly cost of operating some of the automated machines might be significantly larger than that of conventional equipment. If direct labor hours continue to be used as a basis for cost recovery in this automated environment, it becomes difficult to take these cost differences into account.

A more realistic example might be to consider a semiautomatic workcell. Suppose the cell consists of one highly sophisticated automatic machine tool and two older conventional machines. Assume further that all machines in the cell are operated by only one worker.

It is clear that the cost of time on the automatic machine will be greater than either of the two conventional machines. The direct labor hour approach fails to take this into account. On a one-shift basis, the worker assigned to the cell works 40 hr each week. The costs of all three machines are often "lumped" into the overhead rate of the cost center in which the cell is located.

For highly automated equipment, the cost difference between the automated and conventional machines may be significant. Some parts routed through the cell may require the use of only conventional machines. If direct labor hours are used as the basis for cost recovery, such parts will be unfairly charged for the use of the automatic machine in the workcell. This will be true even if the parts are not routed across the automatic machine.

This problem is beginning to cause significant changes in the budgeting and cost recovery process. For totally automated workcells, some organizations are beginning to use a "machine-hour" approach to cost recovery. A standard cost rate is determined for the workcell. This rate is a function of the purchase cost of the machines, the number of direct labor workers, floor space, cutting supplies, maintenance, and indirect labor support. The cost rate is the parameter that is re-estimated each budgeting quarter. The standard cost for a part is determined by the standard time during which it resides in the workcell multiplied by the cost rate.

Corresponding actual costs are determined by computing actual cost rates at the end of the budgeting period. Actual times required by parts as they are processed through the cell are also determined. Actual costs then become the product of the actual times and the actual cost rate for the workcell.

Actual cost rates need to be adjusted for downtime and machine utilizations that are less than 100%. Workcenter efficiencies can be determined by calculating ratios of standard to actual part costs.

It is not likely that all workcells will become completely automated. Often, cells will consist of a mixture of automatic and conventional machines. Further, most plants will continue to have some workcells that consist entirely of conventional machines. These factors make necessary a system that can effectively merge two different types of cost recovery systems. Automated and mixed workcells might lend themselves toward the use of a machine-minute cost recovery approach. Conventional workcenters will probably continue to use a direct labor approach.

New cost recovery systems will have to have the capability to interface or merge both the machine-minute and direct labor hour recovery methodologies. In organizations with many different automated workcells, the number of cost centers will increase significantly. It is likely that a one-for-one correspondence between cost centers and automated or mixed workcells will exist. Future work will have to address ways to effectively merge both types of systems.

THE COST ESTIMATING PROCESS

The cost estimating process is still as much an art as a science. The accuracy of any cost estimate is a function both of the time available in which to make it and the degree of design definition present at the time the estimate is made. The cost estimating procedure relies on the use of performance standards and the separate estimation of labor and material costs. Separate LOH rates are used to convert labor hours to labor dollars on each defined cost center. The estimate total is augmented by both a contingency allowance and a profit margin. These steps are individually described in the sections that follow.

REQUIRED INPUT DATA

To make a detailed and accurate cost estimate, the design configuration of the end product to be built must be complete. A complete bill of materials, showing the assembly relationship between component parts of the final assembly, is required. For each component part, a decision must be made to purchase the part from an outside vendor or to fabricate the part in-house. This decision is often made by manufacturing engineering as opposed to design engineering. Make/purchase designations for all line items in the bill of materials must be made prior to beginning the actual estimating process.

ESTIMATING MATERIAL COSTS

The cost for each purchased part in the bill of materials is estimated separately. Computer-assisted price retrieval may assist the cost estimator in gathering price data on parts that have been previously purchased. Retrieval software can make adjustments for different purchase quantities and account for price increases over time that are due to changing technology and inflation. Computer-assisted price retrieval is usually adequate for parts with moderate unit costs and purchase quantities.

The purchase cost can be high if either the unit cost or the purchase quantity is large. Usually either of these factors necessitates obtaining an accurate price from the vendor. This can be accomplished by either obtaining a verbal quotation by telephone or by obtaining the price from a current vendor's catalog. Total costs for purchased parts are tallied by cost center. The tally usually includes an allowance for scrap and spoilage. The cost center used is that cost center where the purchased part(s) will first be used in the manufacturing process.

ESTIMATING LABOR COSTS

The estimation of labor costs is a more complicated procedure. The process is performed individually for each part in the bill of materials that is to be fabricated. The procedure relies on an established system of performance standards. The assumption is that standard times are the best prior predictor of the length of time a manufacturing task should take prior to the time the task actually begins.

The estimator begins by specifying a sketch process routing for each part. The sketch routing anticipates the process sequence for the part that will ultimately be specified by the manufacturing engineer. Generation of accurate sketch routings requires that the cost estimator have detailed knowledge of manufacturing processes and procedures. If the sequence of processes is not predicted accurately, the estimated and actual costs will be likely to differ.

The sketch routing also predicts the work content of each operation in the manufacturing sequence. The required time to complete each operation is estimated by applying performance standards. The cost center and shop order associated with each fabricated part are tallied. Completing this procedure for each part in the bill of materials to be fabricated yields an estimate of labor costs. These costs are in hours and correspond to both different shop orders and cost centers. It is common for standard hour totals to be multiplied by efficiency factors for each workcenter. These factors reflect historical ratios of actual to standard hours.

Computer-aided process planning (CAPP) can simplify the cost estimating procedure. For parts previously built, a file will exist that specifies both the process sequence and the related standard times. For parts that have not been previously fabricated, part geometries and tolerances may permit the computer generation of sketch routings and corresponding standard times.

THE COST ESTIMATE GRID

The cost estimate grid provides a systematic way to total estimated costs by shop order and cost center.[1,2] The cost estimate grid is illustrated in Fig. 4-1. The left-hand side of the grid lists all cost centers. Within each cost center are the applicable shop orders from Table 4-1. To the right is a column entitled Hours Expended to Date. This column is used in the cost review process which is described later. The next column is captioned Hours to Complete. The locations in this column are numbered 1 through 20 in Fig. 4-1. Shop order hourly totals by cost center are entered in these locations. This information is the result of the labor cost estimating procedure previously described.

The next step is to convert labor hours to labor dollars. This conversion is completed for each of the five cost centers shown. The Wage Rate column has locations numbered from 23 to 27. The direct labor rates, L_D, for each cost center are entered in these locations. This information is available from the budgeting process previously described. The Overhead Rate column has locations numbered from 28 to 32. The overhead rates consist of the sum of the expense rate, burden, and G&A for each cost center. The sum of these values, in dollars per hour, is entered for each cost center in these locations. Again, this information is available from the budgeting process previously described.

Labor hours now can be converted to labor dollars. As an example, consider the conversion process on cost center 21. The sum of all hours on this cost center is available from the Hours to Complete column. This is the total of the hourly entries in locations 1, 2, 3, and 4. This total is multiplied by the entry in the Wage Rate column, location 23. The product is entered in location 38 under the labor column. The total represents the total labor dollars estimated to be expended on cost center 21.

The sum of the hours in locations 1 through 4 is next multiplied by the Overhead Rate in location 28. The product is entered in the Overhead column, location 43. This total represents the overhead dollars to be expended on the cost center. This procedure is repeated for the remaining four cost centers.

The estimated material costs for each cost center are next entered in locations 48 through 52. This permits the total cost on each cost center to be determined. On cost center 21, the dollar entries in locations 38, 43, and 48 are totaled. The sum is entered in location 53. Repeat process for all cost centers.

The estimated production cost is the sum of all entries in the Total Cost column. The sum of the cost totals in locations 53 through 57 is entered in location 58. The next step is to enter the contingency allowance. This allowance is an entry that allows for costs that are expected to occur, but cannot be specified exactly at the time the estimate is made. Contingency allowance magnitudes vary with the degree of design definition available when the estimate is made.

COST ESTIMATING

	Cost Center	Hours Expended to Date	Hours to Complete	Wage Rate	Overhead Rate	Direct Personnel Hours	Labor	Overhead	Material	Total Cost
Mfg. Eng. CC 21	Process engineering (M—)		1	23	28	33	38	43	48	53
	Tool design (D—)		2							
	NC tapes (N—)		3							
	Packaging (K—)		4							
Mach. Div. CC 22	Chip turning (C—)		5	24	29	34	39	44	49	54
	Sheet metal (S—)		6							
	Tool fab. & maint. (U—)		7							
	Painting (P—)		8							
	Plating (L—)		9							
	Heat treating (H—)		10							
	Rework (R—)		11							
Assy. Div. CC 23	Assembly (A—)		12	25	30	35	40	45	50	55
	Testing (T—)		13							
	Wire cutting (W—)		14							
	Troubleshooting (G—)		15							
	Encapsulation (J—)		16							
	Rework (R—)		17							
Qual. Assur. CC 40	Pur. part inspection (B—)		18	26	31	36	41	46	51	56
	Fab. part inspection (F—)		19							
	Calibration (O—)		20							
CC 70	Eng. support (E—)		21	27	32	37	42	47	52	57
Subtotal (Est. production cost)			22							58
Allowance										59
Estimated total cost										60
Profit										61
Estimate total										62

Fig. 4-1 The cost estimate grid.

The allowance is calculated as a fixed percentage of the estimated production cost in location 58; percentages range from 5 to 50%. The allowance magnitude is entered in location 59. The estimated total cost is the sum of the estimated production cost and the contingency allowance. This total is entered in location 60.

The profit margin is calculated as a fixed percentage of the estimated total cost; typical percentages range from 5 to 20%. The profit value is entered in location 61. The estimate total is the sum of the estimated total cost and profit. This total is entered in location 62. Detailed information on the cost estimating procedure can be found in the cited references at the end of this chapter.[1,2]

TYPES OF COSTING SYSTEMS

The procedure described in the preceding section is meant to be performed in conjunction with a batch manufacturing process. The procedure would be completed for discrete groups of end items to be manufactured. The estimating process would be repeated for follow-on production runs of the same end item in identical or different production quantities.

Batch production usually corresponds to a job shop organizational structure. It is common for these types of organizations to use an actual costing procedure. The cost of each batch of parts is estimated separately. The unit cost is a function of the quantity to be built and the number of times (if any) that the end item has been previously manufactured. The estimated cost for each batch of parts is submitted as a bid or quotation to the customer.

A related situation is that of a flow shop. A flow shop is characterized by larger production quantities. The total number of different end items manufactured is small. It is common for an end item to be manufactured in several consecutive lots, often over a period of years. Flow shops often use a standard costing procedure. In standard costing, the unit cost of a part is accurately determined, usually after the first few production runs. This standard unit cost is then used as a basis for all pricing. The same standard cost applies, regardless of the quantity produced. Variations in actual cost and quantity cause corresponding variations in the net profit realized when the standard costing procedure is used.

THE COST REVIEW PROCESS

The cost estimating process described in the preceding section is performed for manufacturing activities that have not yet begun. In contrast, cost reviews are performed for jobs that are currently in process. The time required to make a complete cost review is significant. It takes almost as long to make a cost review as it does to make an initial cost estimate. Consequently, cost reviews are only performed on those projects that may incur significant funding deficits or surpluses.

The goal of the cost review is to compare cumulative expenditures to date with completed work activities. This information is used to predict a new cost total. These new totals may be either less or greater than those specified in the initial cost estimate. Projected funding deficits or surpluses are thus forecasted on a "per-job" basis. The cost review procedure is described in more detail in the sections that follow.

ESTABLISHING A REVIEW DATE

The first step in the cost review process is the selection of a review date. This date is essentially a "snapshot in time" of the fiscal status of the job or project being reviewed. Prior to the review date, the estimator is concerned with the fiscal history of the project and the related work that has been completed. After the review date, the estimator must estimate the work that must be completed to finish the project.

COLLECTING JOB EXPENDITURES

This task consists of collecting all labor and material expenditures that have occurred on the project from the time it began up to the review date. Hourly expenditures are recorded by both shop order and cost center. Total hours expended by shop order are entered in the Hours Expended to Date column of the cost estimate grid (refer to Fig. 4-1). The corresponding dollar amounts on each cost center reflect the actual average labor/overhead rates that have been experienced.

Funds expended for purchased parts are next recorded by cost center. These entries reflect only the costs of those parts that have been received, inspected, and accepted for use on the project. The costs do not include parts that are currently being inspected or those that are either on order or yet to be ordered. The expended costs are entered in the Material column of the cost estimate grid for each cost center.

ESTIMATING WORK YET TO BE COMPLETED

This step consists of estimating those costs yet to be completed (both labor and material) from the review date forward in time until the job is expected to be completed. Hourly labor costs are generally estimated first. For cost centers 21 and 70, each of the completion estimates is made subjectively by each shop order. Estimates are made based on expended hours and the amount of work that has been completed.

For cost centers 22, 23, and 40, a slightly different procedure is followed. Shop orders on each of these cost centers are classified into one of three different categories: (1) not yet open, (2) open, or (3) closed. A not-yet-open shop order is one with a specified amount of hours that has not yet been issued to the production floor. Work on these shop orders has not yet begun. Closed shop orders are those on which all work has been completed. Labor costs for closed shop orders are included in the expended cost totals described in the preceding section. Finally, open shop orders reflect work activities that are currently in process.

It is possible for a large project to have hundreds of separate shop orders that are either open or not yet open. The production line supervisor has the responsibility for completing these shop orders. The cost estimator generally solicits from this person estimates of hours required to complete individual work activities on per-shop-order basis. This is done for all open and not-yet-open shop orders. The supervisors make such estimates for open shop orders based on the hours already expended and the amount of work that has been completed by the review date. Often such estimates are adjusted by the amount that closed shop orders on the same project have been either over or underexpended.

This process is completed for each cost center. Within each cost center, "to complete" estimates in hours are totaled by shop order type. Subsequent totals are entered in the Hours to Complete column of the cost estimate grid (refer to Fig. 4-1). These hours are next totaled for each cost center. They are converted to labor and overhead dollars using current values of labor and overhead rates.

Outstanding or obligated material costs are estimated next. These costs are estimated for each cost center. They include the costs for parts not yet ordered, those on order, and those received but not yet accepted or paid for. On parts that are not yet ordered, the purchase cost from the preceding cost estimate is generally used. For parts received but not yet inspected or paid for, the invoice price from the vendor is used. These cost totals are entered in the Material column on the cost estimate grid for each cost center.

MERGING FISCAL INFORMATION ON THE COST ESTIMATE GRID

At this point, each cost center row in the cost estimate grid will have two sets of entries. The first will be those costs for labor and material that have already been incurred. The second will be the estimated costs for labor and material that are required to complete the job. Both sets of entries are totaled by cost center. New totals are entered in the Total Cost column of the grid (refer to Fig. 4-1). This column is then totaled by cost center to yield a revised estimated production cost.

The new value for the estimated production cost is compared with the value from the previous initial cost estimate or prior cost review. If a deficit is projected, a check is made to determine if the size of the deficit is equal to or less than the sum of the contingency allowance and profit margin from the previous cost estimate or review. If this proves to be true, the sum of the contingency allowance and profit margin is reduced by the amount of the projected deficit. The estimate total is usually left unchanged. If the projected deficit is larger than the sum of the contingency allowance and profit margin, the allowance and profit values are set equal to 0. A new estimate total is entered. This total is increased by the amount that the deficit exceeds the sum of the allowance and profit margin.

The reverse situation occurs when the estimated production cost is less than the total predicted by the previous cost estimate or review. The projected surplus is compared with the sum of the contingency allowance and profit margin. If the surplus is equal to or less than this sum, the estimate total is usually left unchanged. The new profit margin is increased by the amount of the projected surplus.

If the projected surplus is greater than the sum of the allowance and profit, a net fiscal surplus is declared. The contingency allowance is reduced by an amount that reflects the extent to which the new estimated production cost is less than its previously predicted value. This reduction also reflects the percentage of the total work effort that must still be completed. The profit margin is left unchanged. These two totals are then added to the new value of the estimated production cost yielding a new estimate total. The net fiscal surplus is the difference between the old and new values of the estimate total. This sum is made available to transfer to other project orders on which deficits are being predicted. Detailed information on the cost review process can be found in the cited references at the end of this chapter.[1,2]

LEARNING CURVES

Learning curves predict the amount by which the production time decreases as additional units are successively built. The time reduction predicted by learning curves was first observed in the aircraft industry in the 1930s. Because the obtained time reduction corresponds to reduced cost levels, learning curves are of interest to cost analysts. This section overviews learning curve theory and explains how learning curves can be integrated into the cost estimating process. For detailed information on curve types and construction, the reader should consult the references listed at the end of this chapter.[1,2,3]

CURVE PARAMETERS AND TYPES

All learning curves may be described by the following expression.

$$Y = KX^n \tag{2}$$

where:

Y = the production time expressed in either hours per unit or the cumulative average hours per unit required to build a total of X units

K = the number of hours required to build the first unit

X = the cumulative number of units built

n = a negative exponent that specifies the percent by which Y decreases each time X is doubled

Learning curves usually have a percentage associated with them. The percentage specifies the amount by which the term Y is reduced each time the quantity X is doubled. Learning curves with a common starting point and both 80 and 60% rates of reduction are illustrated in Fig. 4-2.

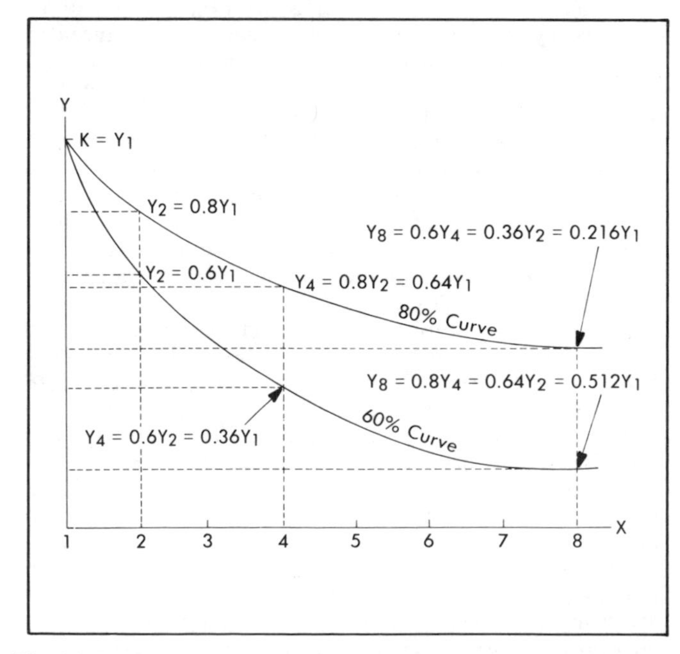

Fig. 4-2 Learning curves (80% and 60%) with a common starting value.

The exponent, n, is not the improvement curve percentage. It is a fractional value between 0 and 1 that must be derived for specific percentage reduction rates. Exponent values for typical learning curve percentages are listed in Table 4-5.

Two types of learning curves exist. The first is called a unit curve. With unit curves, the Y value represents the time in hours per unit required to build the Xth unit. The second type is called a cumulative average curve. With cumulative average curves, the Y value represents the cumulative average time in hours per unit to fabricate a total of X units. Both curve types are described by Eq. (2). The only difference is in the way the term Y is defined.

In collecting data to construct actual learning curves, different data formats apply for each curve type. Once actual curves have been constructed, it is not possible to convert a cumulative average curve percentage to a unit curve percentage. The converse is also true. Because of this conversion difficulty, it is usually best to select one of the two curve types and use it consistently for all cost analyses and estimates. The cumulative average curve is recommended for general use in manufacturing because of its wider utilization and acceptance in practice and the comparative ease with which historical time calculations can be made.

CURVE CONSTRUCTION PROCEDURE

Construction of actual learning curves is a two-step process. The first step is that of data collection. Suppose it is desired to measure the learning curve being experienced on a particular manufactured assembly. If unit curves are to be constructed, the hours expended to manufacture each assembly built must be recorded. For cumulative average curves, the procedure is simpler. As each unit is completed, the cumulative total number of units built is recorded. The cumulative total hours expended that correspond to this total is recorded as well. These totals permit the cumulative average time values to be calculated. In both cases, it is important to remember that the group of units produced must be manufactured consecutively.

The next step is to determine the actual learning curve percentage associated with the curve. If the logarithms of both sides of Eq. (2) are taken, the result is as follows:

$$\log Y = \log K = n \log X \tag{3}$$

TABLE 4-5
Exponent Values for Typical Learning Curve Percentages

Curve Percentage	n
65	-0.624
70	-0.515
75	-0.415
80	-0.322
85	-0.234
90	-0.152
95	-0.074

Eq. (3) is a first-order linear relationship. If the logarithms of the X and Y values from the data collection process are recorded, a first-order curve fit through the data may be obtained by the method of least squares. The intercept of the obtained fit will be the term log K. The obtained slope will correspond to n, from which the learning curve percentage may be obtained. This procedure is described more fully in references 1, 2, and 3.

INTEGRATING LEARNING CURVES INTO THE COST ESTIMATE

The problems associated with integrating learning curve analysis into the cost estimating procedure are perhaps best illustrated with an example. Suppose a bench assembly operation is to be performed on a quantity of 500 units. Suppose further that a sketch routing for the part has been generated. Application of standard times has yielded a setup time of 3 hr and a run time of 0.5 hr per part. Without consideration of learning, the total time required to build 500 units would be calculated as follows:

Total time = 3 hr + 500 units × 0.5 hr/unit = 253 hr

Suppose it is known that an 85% improvement rate has been historically associated with assembly operations of this type. Furthermore, analyses have revealed that the standard time for assembly operations is usually reached by the 35th unit. The procedure for determining the total hours is now changed.

From Table 4-5, $n = -0.234$ for an 85% curve. The next step is to find the time required to build the first unit, K. For this example, a cumulative average curve will be used. For this curve type, the time required to build the Xth unit is given by the following equation:

$$U = (1 + n)\,KX^{n} \qquad (4)$$

where:

U = the time in hours per unit required to build the Xth unit

If the standard time is normally reached by the 35th unit, then the time to build unit number 35 will be 0.5 hr. Substituting these values into Eq. (4) yields the following:

$$U_{35} = 0.5 \text{ hr} = (1 - 0.234)\,K\,(35)^{-0.234}$$

Rearranging Eq. (4) and solving for K is as follows:

$$K = \frac{0.5}{(1\text{-}0.234)\,(35)^{-0.234}} = 1.50 \text{ hr}$$

For a cumulative average curve, the total time required to build a total of X units is given by the following equation:

$$T = K\,X^{(1+n)} \qquad (5)$$

where:

T = the total time required to build a total of X units

Equation (5) can then be used to calculate the total time to assemble the 500 units as follows:

$$T = 1.50\,(500)^{(1\,-\,0.234)} = 175 \text{ hr}$$

As was previously stated, the set-up time for the part was 3 hr. Adding this to the above total yields a total time requirement of 178 hr. This is significantly less than the 253-hr total that was obtained without applying the learning curve analysis.

In practice, it is not always possible to integrate learning into the cost estimate by the procedure previously described. The preceding analysis required knowledge of two parameters to define the learning curve. The first was the time required to build the first unit. The second was the cumulative production quantity at which the standard time is reached. In practice, these two parameters are often not known, particularly for those production operations that have not yet begun.

Military Standard 1567A has mandated that defense contractors use performance standards in their estimating and reporting procedures. No research has been done to investigate, by product category, the point on the learning curve at which the standard time is reached. The standard time is defined as the time required by an average individual to complete a task while working at a normal pace using a prescribed method. Similarly, little is known about the time requirements for the first unit that perhaps could be defined as a multiple of the standard time, again by product type or category. Many defense contractors have circumvented this problem through the use of a procedure that relies on cost realization factors. This procedure is the subject of the following section.

REALIZATION FACTORS

The use of realization factors relies on the collection and analysis of historical data by product type or category. The effect of the learning curve is embedded in this procedure along with other factors. The procedure is one of first defining product categories on which the data collection and analysis will be performed. The categories should be defined in such a way so that the fabrication procedures for parts within a category are quite similar.

As an example, consider an organization that fabricates electronic equipment. Example categories might include the fabrication of printed circuit boards, the loading and soldering of components onto these boards, the wiring of chassis assemblies, the wiring of rack panels, and sheet metal operations required by rack panels and chassis assemblies. For an organization involved primarily with metalworking, a group technology approach might be used to define part families and corresponding workcells.

Once the categories have been defined, data collection may begin. The goal is to determine ratios of actual to standard hours as a function of both production quantity and part category. Within a category, curve-fitting procedures may be used to develop mathematical relationships that describe ratios of actual to standard hours as a function of the quantity being built. It is desirable to complete this analysis using a fairly large number of data points within a part category.

Once the curve-fitting procedures have been completed for all part categories, the procedure may be used in conjunction with the cost estimating procedure previously described. For each part, the category is first defined. The developed functional relationship is used to calculate a ratio of actual to standard hours that corresponds to the production quantity that is to be built. The standard time from the sketch routing is then multiplied by this ratio. For large production quantities, it is likely that the ratio may be less than 1. The obtained product is then entered as an hourly total in the cost estimate.

The effects of the learning curve are embedded in this new hourly total. Workcenter efficiencies are included in the total as well. These efficiencies indicate the extent to which the standard times can be achieved in practice as a function of the type of part being built.

ASSUMPTIONS AND THE EFFECTS OF FORGETTING

The learning curve process assumes that each quantity of units is consecutively built by the same worker or group of workers without interruption. This is not always true in manufacturing practice, especially in job shops. Delays and the use of multiple shifts are some practical considerations that conflict with these assumptions. Very little research has been done that has addressed the effects these practical considerations have on obtained learning curves. The effect of work-in-process inventory on learning curve construction has been described in reference 4. This particular reference also presents a procedure that defines an aggregate learning curve for a final assembly as a function of the learning curves obtained of each of its component parts.

The use of the learning curve is important because of the way in which the procedure defines a variety of nonrecurring costs associated with the early stages of production. These costs are associated with factors in addition to human learning. They have been described as:[5]

- Design improvements in the early stages of production.
- Groups of components being installed are either defective or out of tolerance.
- In the case of electronic manufacture, test operators and repair personnel become skillful enough to rapidly diagnose failure symptoms to ensure that the first attempt at repair after troubleshooting is successful.
- Correction of errors on production drawings, fabrication instructions, written test/inspection procedures, and engineering design errors and oversights.

The effect of these nonrecurring costs lessens as the production quantity increases. For large production quantities, it is possible that the actual times experienced will be less than the calculated standard times. The overall reduction in cost can be estimated with the use of the realization factor procedure that has been described.

COMPUTER-ASSISTED COST ESTIMATING

The increased use and popularity of stand-alone microcomputers has had a profound impact on all engineering fields. Spreadsheet software such as Lotus 1-2-3 and VisiCalc lends itself to automating the process of preparing cost estimates and reviews.

Spreadsheet software is only useful in automating the mathematics associated with compiling the cost estimate and cost review grids. This task is a small fraction of the total time required to complete a detailed cost estimate or review.

The computer can play a more valuable role in aiding the estimator in the process of data retrieval. Actual historical labor costs might be retrieved from a database and compared with similar parts that have not been fabricated before. Metalcutting formulas and the results of applying performance standards can be programmed. The computer can thus be used to speed up the process of manually estimating labor hours for each sketch routing operation. Tabulated data and related software for estimating such costs for metalcutting operations has been developed.[6] The tabulated data and cost estimating software may be purchased commercially.

The computer can also play an instrumental role in the forecasting of production costs. Cost forecasting and parametric estimating systems are the subject of the following section.

COST FORECASTING

There are basically two types of cost forecasting systems. The first seeks to forecast the total cost of building a group of end items prior to the time production actually begins. Parametric estimating systems are often used to complete this task. They are often used as a check on conventional estimates to determine whether the obtained estimated total cost is accurate. Parametric estimating systems partially automate the procedure of making the initial cost estimate.

The second type of forecasting system seeks to predict the potential job cost of projects that are currently in process. Expended costs are compared with completed work. Revised cost totals, together with throughput time and workload forecasts, can be obtained.

PARAMETRIC ESTIMATING SYSTEMS

Parametric estimating systems attempt to correlate product costs with a number of easily identifiable features of the part being built. For parts that require machining, some of these features include weight, volume, tolerances, material, geometry, surface finish, and the amount of material removed during the metalcutting process. For electronics products, some example part features include the number of active/passive components, number of input/output terminations, weight, volume, and power dissipation.

The best known parametric estimating system was developed by Freiman at Radio Corporation of America (RCA).[7,8] The system is called Programmed Review of Information for Costing and Evaluation (PRICE). PRICE has been used to provide post-estimate checks on the engineering and manufacturing costs for electromechanical, military, and aerospace equipment. In addition, it may be used to evaluate the financial impact of proposed engineering changes, schedule revisions, and production quantity adjustments.

PRICE uses a basic cost function that attempts to relate the total costs of engineering and production to a variety of parameters. These include a specification profile of the units to be built, the amount of work to be performed, the performance time period, and the production resources available. PRICE relies on the use of parametric relationships that relate cost to a variety of variables. The relationships are obtained through curve-fitting procedures that have been performed on a significant repository of historical cost data.

The developers of PRICE claim that the system is accurate to within 5%. This claim is impressive because the tolerance band is less than the magnitudes of many contingency allowances that are included in conventional cost estimates.

Unfortunately, none of the developed parametric cost relationships that make up the PRICE system have been published. This is also true of the validation data that substantiates the claimed accuracy of the system. The claimed accuracy can neither be endorsed nor refuted until such time as accuracy validation data, together with more specific information on the system's operating methodology, becomes available for review.

WORK PROGRESS FORECASTING ALGORITHMS

Work progress forecasting algorithms have been described in two different references.[9,10] The developed algorithms seek to forecast potential costs, completion times, and remaining workload of projects on which work has already begun.

The algorithms require two key sets of input data. The first is the repository of historical cost curve data. A cost curve relates cumulative total job expenditures in dollars to time periods for the duration of the project. The model presented in the references uses a total of 142 historical cost curves that represent a cross section of the products produced by a job shop manufacturing facility.

The second set of data is used in conjunction with a work progress function. The work progress function is used to analytically determine with considerable accuracy the percentage of the project that may be considered to be complete. The function associates one of three parameters with each operation of each process routing of each job for all active jobs. Separate parameters are defined for routing operations that are not yet started, are in process, or are already completed. These variables describe the percentage of labor effort on a project that is completed at the time a new forecast is to be made. The obtained labor completion percentage is averaged with a percentage completion on the job with respect to purchased material. The result is a detailed assessment of the percentage completed work activity on the entire project.

The process begins by forecasting the potential total job cost. The completed work is compared to the expended total costs. Optimistic and pessimistic cost bounds are defined. The new cost forecast is obtained as a midpoint between these two cost extremes.

The next step requires the use of the cost curve repository. A curve from the repository is selected on the basis of the relative sizes of the cost percentages for purchased material, machined parts, assembled parts, and inspection costs. The goal is to select the curve whose shape is most likely to accurately represent the shape of the cost curve of the job being analyzed. Data normalization and transformation procedures take into account the effects of different total costs and completion times for the cost curves in the data repository.

The selected cost curve is transformed to fit through the previously forecasted values of total cost and throughput time. The data points are transformed so that the obtained curve will also pass through the point described by the costs expended to date and the time at which the forecast is being made. The curve shape and percentage job completion are used to define a time increment that is either added or subtracted to the current throughput time forecasts. Jobs with projected overexpenditures have increased throughput time forecasts. Jobs with projected underexpenditures have decreased forecast values for throughput times.

The model has provisions to define separate forecast values for both labor and material costs. Separate cost curves for labor costs can be defined. When cost values are divided by an aggregate labor/overhead rate, a curve describing labor hour expenditures versus time can be defined. This curve can be used to define remaining per-period forecasts of the labor hours to be expended over the job's remaining duration. Readers desiring more detailed information of this cost forecasting system are urged to consult the references cited at the end of this chapter.[1,2,9,10]

SUMMARY

The estimation and control of production costs will remain a challenging task for all manufacturing personnel. The accurate estimating of costs is necessary to ensure the competitive position of the manufacturing firm. Estimated costs that are too low may result in the firm being awarded a contract that will ultimately result in a fiscal loss. Estimates that are too high often result in contracts that are awarded to competitors of the firm.

Cost control is of paramount importance for projects that are in process on the production floor. It is necessary to ensure that the completion costs of active jobs will be within the limits specified by prior cost estimates. The abundance of inexpensive direct labor in Third World countries will provide a continuing impetus for manufacturing organizations in this country to seek methods of reducing production costs. Some of these cost reductions will undoubtedly come through factory automation. Even more important than automation is the need to graduate engineers who have increased levels of manufacturing literacy. Products that can be cost effectively manufactured will be designed only by engineers who have some understanding and appreciation for how these products will be built.

References

1. Eric M. Malstrom, ed., *Manufacturing Cost Engineering Handbook* (New York: Marcel Dekker, Inc., 1984).
2. Eric M. Malstrom, *What Every Engineer Should Know About Manufacturing Cost Estimating* (New York: Marcel Dekker, Inc., 1981).
3. Eric M. Malstrom and Richard L. Shell, "A Review of Product Improvement Curves," *Manufacturing Engineering* (May 1979).
4. Eric M. Malstrom and Jack W. Fleming, "The Effects of Work-in-Process Inventory Considerations on Learning Curve Construction," *The Journal of Operations Management* (April 1982).
5. Eric M. Malstrom and Richard A. McNeely, "Empirical Predictions of Yields and Required Times in Electronic Testing,"

Proceedings of the 29th Annual Spring Conference (Norcross, GA: American Institute of Industrial Engineers, 1978).
6. Phillip F. Ostwald, *AM Cost Estimator*, 1987-1988 ed. (New York: McGraw-Hill, Inc., 1987).
7. Frank R. Freiman, "PRICE—A Parametric Cost Modeling Methodology," *Proceedings of the 26th Annual Spring Conference* (Norcross, GA: American Institute of Industrial Engineers, 1975).
8. Frank R. Freiman, *PRICE: An Overview* (Radio Corporation of America, 1974).
9. Eric M. Malstrom, "Workload Forecasting from Job Progress Measurement and Forecasts of Job Production Costs and Throughput Times," *Proceedings of the 27th Annual Spring Conference* (Norcross, GA: American Institute of Industrial Engineers, 1976).
10. Eric M. Malstrom and Colin L. Moodie, "A Work Progress Forecasting Algorithm for a Job Shop Manufacturing Facility," *The International Journal of Production Research* (January 1976).

Bibliography

Clark, F.D., and Lorenzoni, A.B. *Applied Cost Engineering*. New York: Marcel Dekker, Inc., 1978.

Clugston, R. *Estimating Manufacturing Costs*. Boston: Cahners Publishing Co., 1977.

Cochran, E.B. *Planning Production Costs: Using the Improvement Curve*. San Francisco: Chandler Publishing Co., 1968.

Hartmeyer, F.C. *Electronics Industry Cost Estimating Data*. New York: Ronald Press, 1964.

Jelen, F.C. *Cost and Optimization Engineering*. New York: McGraw-Hill, Inc., 1970.

Margets, I.G. "Estimating Labor and Burden Rates." *Manufacturing Engineering and Management* (January 1972).

Ostwald, Phillip F. *Manufacturing Cost Estimating*. Dearborn, MI: Society of Manufacturing Engineers, 1980.

__________. *Cost Estimating*, 2nd ed. Englewood Cliffs, NJ: Prentice-Hall, 1984.

Park, W.R. *Cost Engineering Analysis*. New York: John Wiley and Sons, 1973.

Vernon, Ivan R. *Realistic Cost Estimating for Manufacturing*. Dearborn, MI: Society of Manufacturing Engineers, 1968.

MANAGERIAL LEADERSHIP AND ITS FOUNDATIONS

SECTION

2

PHILOSOPHY AND CULTURE OF MANUFACTURING MANAGEMENT

One dictionary meaning of *philosophy* is "the synthesis of all learning," although one usually thinks of it as a system of motivating concepts or principles. Similarly, the word *culture* is sweeping. Indeed, it is more so for it's defined as "the totality of socially transmitted behavior patterns, arts, beliefs, institutions, and all other products of human work and thought characteristic of a community or population."

Clearly, with breadth like that involved, one can't summarize philosophy or culture as one might an article in an engineering journal. There isn't even a coherent universal view of this foundation, whether it be for a small company in a village or an international corporation that considers the world its oyster. Still, a useful part of the education of a specialist is a knowledge of the origins of his or her field that are buried in the culture.

Let's start at the beginning and unravel this subject so it can really be understood. When the cumulative store of knowledge became too great for one scholar to grasp, academic labor came to be divided into specialized disciplines. These artificial boundaries naturally tended to overlap somewhat, and this resulted in an interdisciplinary cross-fertilization that provided a wider awareness than any specialized knowledge could produce.

Also, each specialty has a limited number of legitimate topics, and sometimes a much larger lens is needed to see significant movements that may have a very strong bearing on one's actions in a chosen specialty. This is why a narrow, specialized view—however necessary for comprehension and research—can be too limiting for the making of all business judgments.

Some subject areas—like philosophy—are pervasive, but less familiar to, say, an engineer than to a philosopher. Fortunately, there is often a specific cluster of philosophical thought that deals with a specialty, or a field may deal with philosophy only slightly or in common with all studies. In a number of fields—including manufacturing engineering—detailed moral codes have been developed that exhort virtue; in others, one may find no organized guides at all. It depends on how serious the muddles about the fundamentals of morality or social responsibility tend to be in a specific pursuit. In manufacturing management, the results of such errors can be very grave indeed.

In Section II of this volume, discussion centers on the role of the manufacturing manager, and the aim of this first chapter is to provide a base for that which follows by explaining a manager's rights and responsibilities and the social justification for these. Looking back vertically, one can see the chronological development of basic philosophical concepts that underlie the role, and by looking horizontally across various cultures, the reader can gain a greater appreciation of these roots and of varying interpretations.

Through a deeper understanding, one can answer such important and practical questions as: "Where does a manager get authority?" "What is the social role of a company?" and "How can a manager distinguish between right and wrong in a controversial situation?"

For present purposes, then, the exceedingly broad subject of this chapter will be compressed into a fundamental overview of the concepts that affect manufacturing managers. Concentration will be on the following three areas:

1. Fiduciary responsibilities.
2. Social obligations.
3. Ethical considerations.

FIDUCIARY RESPONSIBILITIES

A fiduciary is a person who stands in a special relationship of trust, confidence, and responsibility for another's assets. To assure propriety, the law imposes the duty to act solely in the interest of the person(s) represented. Examples of persons with fiduciary responsibilities include officers of corporations, agents, executors, trustees, and guardians.

Contributor of this chapter is: Dr. George J. Gore, *Professor of Management, University of Cincinnati.*

Reviewers of this chapter are: Charles F. Bimmerle, *Professor, College of Business Administration, The University of North Texas;* ***Rogene A. Buchholz***, *Professor of Business and Public Policy, School of Management, The University of Texas at Dallas;* ***Chan K. Hahn***, *Professor of Management, Department of Management, Bowling Green State University;* ***George M. Kurajian***, *Professor and Chairman, Department of Mechanical Engineering, University of Michigan—Dearborn;* ***Lee E. Preston***, *Professor, College of Business and Management, Tydings Hall, University of Maryland;* ***Hayward Thomas***, *Consultant, Retired—President, Jensen Industries;* ***Richard E. Wokutch***, *Associate Professor of Management, The R.B. Pamplin College of Business, Virginia Tech;* ***Robert Wright***, *Professor of Organization and Management, Pepperdine University.*

CHAPTER 5

FIDUCIARY RESPONSIBILITIES

To better understand the justification of this managerial role and how one meets these sometimes nebulous responsibilities, developmental steps from the past will be reviewed.

ANCIENT CULTURES

In Western civilizations, business activities were conducted mainly under the heads of state in Egyptian, Babylonian, Phoenician, Greek, and Roman societies.

Egypt

Pharoahs, who were divine kings of ancient Egypt, governed by royal decree. By 2300 B.C., international trade with adjacent countries on the eastern shores of the Mediterranean Sea was developing. The crown allowed the licensing of ships for this trade, and royal provincial governors—at first the princes or companions of the king—controlled the production, distribution, and sale of agricultural products and merchandise. Each of these supervisors had a direct and regulated responsibility to the palace. Unfortunately, no written code of laws has survived, although some juridical papyri have, which include contracts and deeds. It is known, however, that state revenues were derived from royal monopolies (booty and tribute from the conquered); profits from foreign trade and crown estates; and taxation. Further, there is reason to believe that this wisdom of Egypt concerning fiduciary responsibilities slowly merged with Greek law.

While democratic forms of government were still distant, the rights of each individual were increasingly recognized, including the right of ownership.

Babylonia

Hammurabi (18th century B.C.) was a powerful king who gained control of all of Babylon. He devoted much of his reign to strengthening religion, law, and order in this confederation of city-states. Toward the end of his service, an 8′ high stone tablet was erected that included 21 columns and 282 clauses. These were essentially supplements to existing law and in total are sometimes called the "Code of Hammurabi." This granite-textured slab was discovered in 1901 and was taken to the Louvre for safekeeping because of its extreme historical importance.

A portion of the inscriptions recognized private ownership and provided for control of managerial activities. Included were mentions of the need for detailed receipts; the requirement that an agent had to return double the price of entrusted stock even if no profit accrued from the venture; moreover, if successful, the agent was entitled to half the profits, and so on. The objective was to regulate trade both inside the country and abroad and to assign brutal penalties for such violations as theft.

Phoenicia

Phoenicia, located on the northeastern shore of the Mediterranean Sea, had a turbulent history of being conquered by its neighbors, Egypt, Assyria, and Persia. Still, there were periods of independence when its basic nature—that of a seafaring and trading civilization—was allowed to flower. Through its government, the idea of market creation was instituted by establishing colonies along the entire shore of the Mediterranean Sea. These included portions of the Greek islands, the Italian peninsula, Sicily, Spain, and the entire northern coast of Africa. The heyday of this expansionism occurred during the 8th and 9th centuries B.C. Trade, and its regulation by governments, was so profitable that cities like Carthage had no taxes because the city's needs were cared for by import/export duties on the shipping trade.

The Greek and Roman empires naturally had their share of business and trade, but this was just to maintain their population. Commerce did not really flourish in these eras, because the leaders in Greece, and to some extent Rome, considered such matters rather lowly—something left to slaves and lower class citizens.

Later, in feudal times—roughly the 9th to 15th centuries—one again observes a rather arrogant regulation of production from above. Vassals owned no property, and the political-economic system of small city-states in Europe was under the direct control of a lord (baron, duke, prince, or whatever nobility was chosen to control a geographic rėgion). Farmers, artisans, and other business people had a fiduciary responsibility for any assets that had been loaned or granted them on the condition of homage and service. In this way, nobility took a yearly rent from its subjects' industry and, on the principle of noblesse oblige, provided for the citizens' welfare and protection.

EVOLVING SOCIAL SYSTEMS

The concept of tight government ownership and control gave way as Western civilizations lumbered out of the Dark Ages and into a world of expanding populations, geography, and expectations. Individuals had the right to own and bequeath cattle, housing, and land, indeed, to amass private resources into wealth to be employed as the person saw fit. In the gathering of this power beyond individual needs, widespread means were created for the capitalization of business enterprises that would fuel economic growth.

Fundamental conflict about the rights of the state versus rights of the individual had been the subject of much philosophical discourse—particularly since the times of the Greek philosopher Plato and his disciple Aristotle.

Plato (427-347 B.C.) in his greatest work, *Republic*, espoused the rights of a society, ruled by philosopher-kings, to administer a constitutional form of government. Aristotle (384-322 B.C.), whose treatises frequently centered on natural science and ethics, rejected Plato's thesis as a form of despotism. Instead, he advocated law rather than human will as the repository of the norms of a political society.

The Romans, too, noted for a civilization providing sophisticated government and military organization, also followed Aristotelian philosophy. Cicero (106-43 B.C.) was a Roman statesman and student of Greek philosophy. He stated that there were true laws, which applied to all men, and that these were unchangeable and eternal. By these laws, people could be commanded to the performance of duties and restrained from wrongdoing.

During the Renaissance period, the English *Magna Carta* in 1215 A.D. set the tone for the establishment of greater individual and collective rights over those previously enjoyed in a monarchy. Similarly, the writings of St. Thomas Aquinas (1225?-1274 A.D.) evolved into the philosophy of Thomasism that not only spelled out the position of the Roman Catholic church in society, but established a freewill doctrine that placed more emphasis on the rights and responsibilities of the individual.

A major right of the individual was the ownership of private property. At one extreme of thought, private property was condemned as an evil because it could become the major instrument for oppression of the many by the few. On the other end of the spectrum, the right to own private property can be seen as the most basic right essential for the existence and dignity of a human being. Its primary justification is that wealth is usually produced by the labor of those who "mix sweat with soil" to reap rewards.

USE OF PROPERTY

From 1100 to 1300 A.D., petty capitalism, a putting-out system, developed. Primarily involved were owners, who advanced capital goods or money in return for a promise to pay interest or dividends, and small shopkeepers or merchants, who bought goods as cheaply as they could and sold them for as much as the market would bear. It was this linkage that introduced such concepts as charging interest on loans and selling merchandise on credit.

The next 500 years—1300 to 1800 A.D.—was an age of mercantile capitalists in which the suppliers of assets did not confine themselves to local transactions. Rather, they hired agents to journey with caravans to distant lands. The merchants were usually headquartered in the great cities, where laws regulated their relations. These operations took place wherever political and religious freedoms permitted this broader and more adventurous form of business risk-taking.

INDUSTRIAL REVOLUTION

The English Industrial Revolution came about in the latter part of the 1700s as a result of several new ingredients coming together. These were: (1) large infusions of capital from French Huegonots who had migrated for religious freedom; (2) advanced managerial know-how, particularly in weaving, from these same refugees; (3) a parliamentary form of government providing more political and religious freedom; (4) a large labor pool resulting from farm laborers flocking to urban areas to improve their lot; and (5) a tremendous market for finished goods due to English colonialism and worldwide trade.

The entrepreneurs proceeded to mechanize English manufacturing through the development of systematic production tied to new power sources and machines. Prizes were offered for mechanical inventions that would speed up production and expand output. Some of the most important developments in terms of widespread social impact were Hargreave's spinning jenny, Arkwright's spinning frame, Watt's steam engine, and Cartwright's power loom.

This Industrial Revolution, with its wondrous new machines, in turn required factories to house such equipment, and the factory managers found themselves in a strong fiduciary relationship again to industrial capitalists. Still, the latter lacked the technical know-how to fully develop factory production techniques and the golden era of mercantile capitalists and owner-managers largely collapsed by 1800.

NATURAL LAWS

The tremendous upsurge in economic activity in the 18th and 19th centuries served as a catalyst to a wide-ranging development and discourse on theories designed to explain and predict social consequences of economic principles. One interchange centered on fundamental "natural laws"—such as the need to preserve liberty, property, security, and the right to resist oppression.

One of the first socioeconomic philosophers involved in these discussions was the English theorist John Locke. A major theme of his was that the origin of private property was in existence even before the formation of primitive societies. For example, a cave dwelling could be considered owned by its occupant or a seashell would be the property of its finder. It followed that property should not be taken away or annulled by either government or society because it was a natural human right.

Two other focuses on human rights by Locke concerned life and liberty—concepts that are central in the Declaration of Independence, which featured national and individualistic law ideas to protect citizens against abuses by sovereign power.

Others followed Locke's notion of government nonintervention. For example, Adam Smith's famous *Inquiry into the Nature and Causes of the Wealth of Nations*, written in 1776, was in step. His "natural liberty" economic doctrine is often associated with laissez faire. The latter is viewed today as antagonistic to regulation or interference in commerce beyond the minimum required for a free enterprise system to operate according to its own economic laws.

THE FREE ENTERPRISE SYSTEM

The system under which the United States economy operates has two basic components: (1) a free market where goods and services are exchanged and (2) a guiding philosophy or ideology that demonstrates why the free market is the ideal way to organize economic life.

An ideology is a system of fundamental beliefs that define the good life. An ideology can be social as well as personal, with an entire society tending to subscribe to a set of beliefs that shape that society's idea of what life ideally should be. The basic function of an ideology—whether for an individual, an organization, or an entire society—is to provide a view of the world and a way to organize values.

The guiding philosophy or ideology of the free enterprise system consists of the following components:

1. Individualism and freedom of choice.
2. Private property and profit (opportunity to amass wealth).
3. Equality of opportunity.
4. Competition.
5. Compensation for work expended.
6. Natural laws and limited government.

While one can see constraints or restrictions on individuals and business organizations on all sides from governments as well as other social organizations, it is within this system that one can follow the thread of fiduciary responsibility for the manufacturing manager.

OWNERSHIP AND FREEDOM

The bedrock institution on which the ideology and practice of free enterprise is founded is private property. The ownership of property allows one to control one's own destiny rather than have important decisions made by someone else. Support for this control can be seen in the work of English philosopher and economist John Stuart Mill. His treatise, *On Liberty* (1859), presented a cogent set of arguments for severe restrictions on government interference in personal as well as commercial transactions. The basic thesis can be summarized as the right of the individual to do with private possessions or property as desired so long as it does not do severe harm to others or to society as a whole.

Another supporting argument comes from the work of Max Weber, a German sociologist and economist. In *The Protestant*

Ethic and the Spirit of Capitalism (1905), Weber contended that Catholicism, through the use of the confessional, allowed the faithful to be relieved of any sense of guilt for nonperformance in the use of their wealth. Protestants, on the other hand—and Calvinists in particular, could achieve forgiveness and salvation only through hard work and being as productive as possible. Within this frame of reference, wealth and private property were to be reinvested to produce even greater wealth through hard work and risk.

Many early capitalists were guided by the widespread Protestant ethic even before its enunciation by Weber. Benjamin Franklin summed up his philosophy and strategy this way:

> The way to wealth, if you desire it, is as plain as the way to market. It depends chiefly on two words, industry and frugality; that is, waste neither time nor money, but make the best use of both. Without industry and frugality nothing will do, and with them everything.

Such accumulation of wealth and private property led to the creation of big business and the related problems of controlling that economic power through social regulation. A definitive fiduciary relationship had to exist throughout the economy.

CURRENT CONDITIONS

The United States began the 19th century as a basically agrarian society of about 5 million people. Today there is a gross national product measured in trillions of dollars, a population of almost 250 million, and a position of world leadership in economic affairs. Conditions supporting business development in this country were somewhat similar to those favoring the English Industrial Revolution in the 1700s. These were:

1. *Plentiful natural resources*. While the English needed colonialism to support growth, the U.S. had a vast continent generously irrigated by rivers, blessed by fuel, and topped by mountains fabulously rich in valuable metals. Rich soil, thick forests, and deep harbors abounded—all the ingredients for extensive business activity and trade.
2. *Favorable social and political environments*. In contrast to many European nations, the United States had no strong aristocratic class; therefore, most Americans respected and emulated business leaders. Additionally, throughout the 19th century, both government and courts pursued policies and programs supportive of business needs.
3. *Population growth and urbanism*. Tremendous population increases through immigration and favorable life conditions provided a ready labor market and a corresponding one for goods and services. Moreover, the development and expansion of cities in various regions furthered the concentration of resources for commercial growth.
4. *Transportation revolution*. The country's huge land mass virtually dictated a heavy emphasis on the development of rapid means of transport. Water, rail, and air travel accommodations were provided to move people and goods throughout the country.
5. *Innovations in communication*. The ease of information flow provides for understanding, adapting, and advancing development in any economy. Again, distance and time proved to be catalysts for the federal mail service, telegraphy, radio, television, and satellite communication systems.
6. *Technological development*. There is a tendency to think of technology in terms of mechanical objects, but as management scientist Herbert Simon has said, technology is not things, but knowledge stored in many places. The knowledge is, of course, how to do things to accomplish human goals.

This tremendous economic expansion brought with it the power of large businesses to dictate conditions for both workers and consumers. By the mid-20th century, the concept of countervailing power was advanced by Harvard's economic guru, John Kenneth Galbreath. His thesis was that big business required both big labor and big government to offset the powers of the modern corporation. By this reasoning, large international trade unions were favored by the government and given special monopoly status. Also, government involvement and regulation of business was socially justified as a protection for the consumer and society at large. The role of governments—federal, state, and local—includes fiscal and monetary policies that affect a manager's operating decisions.

GOVERNMENT OWNERSHIP PERMITS

The most direct effect of governments on the manufacturing manager's fiduciary responsibilities concerns the forms of organizations that are allowed. These fall into three general categories: the sole proprietorship, the partnership (in various forms), and the corporation (also of different types). In the last two instances, the various forms that these organizations can take are not germane to this discussion; hence, the concentration will be on just a generic basis. Additional information on organizational structures can be found in Chapter 6, "Organization," of this volume.

Sole Proprietor

The very name *sole proprietor* signifies a property situation. One individual has ownership of all assets of the organization and, ultimately, sole responsibility for the actions and outcomes of that organization. Therefore, that individual within legal limits has complete authority to decide what products or services the organization will provide, how those goals and objectives are to be achieved, and who will be granted the authority to operate certain functions of that organization. Normally, this flow of authority and responsibility is depicted by an organizational chart (see Fig. 5-1). Certainly, the fiduciary relationship between an individual responsible for a functional specialty and the owner is direct and obvious. The owner grants authority to subordinates to use assets and resources related to those activities that further organizational goals. They are trustees of these assets. Similarly, others lower down on a chart report to the heads of the various sections and they, in turn, have a fiduciary responsibility to their bosses for assets they are permitted to control.

While the sole proprietor has complete authority and control over the organization, that person also has complete liability for actions of the overall operation. When there is just one person, a proprietor, that liability is understandable and acceptable even to the point of the individual's personal assets (home, automobile, and so on). However, when the organization encompasses many people, that responsibility can weigh very heavily on the proprietor because one faulty action by an employee can doom his or her personal resources as well as wipe out those of the company. Additionally, profits go to the proprietor and are taxed as personal income.

Partnership

In the second legal form, the partnership, ownership is divided between two or more partners. They decide who will be the chief executive officer, and then the structure looks very much like that described above. Also, the flow of authority and fiduciary responsibility would be the same.

Partnerships can result in greater financial resources, pooled technical knowledge, increased management capability, and a larger overall base of operations. The disadvantages are all of those found in the sole proprietorship plus the added problem that the actions of any of the partners are binding on the remaining partners (unless specifically limited). Again, this legal responsibility attaches to personal assets as well as those of the partnership.

Corporation

While the corporate form of ownership and operation represents only about 1 firm in 8, it is the predominant type as far as capital investment and gross national product are concerned. Further, the historical concept of the modern corporation can be traced back more than 900 years. At first its use was almost exclusively for groups such as boroughs, guilds, and ecclesiastical bodies. By the 1500s, it began to acquire characteristics favorable to business in general, including the right to hold property, to sue or be sued, and to exist indefinitely so long as it is engaged in its stated legal pursuits.

Still, it wasn't until the 19th century that the business corporation as an offshoot of the joint stock trading company came into being. At that point, it provided the opportunity for gaining greater financing, an indefinite existence, a division between the owner and the manager, and liability limited to the amount of investment by any one individual. The corporation could be liable for the actions of its managers, but the individual investor's personal property was protected.

In this country, from about 1850 on, charters were available to any business that could meet the requirements of the state in which application for incorporation was made. While each state might have different standards for qualification—some very lenient and some very strict, each is a reflection of society's willingness to grant this privilege under certain conditions. Because "ownership" is involved, the concept of private property has to be translated into shares for this form.

From the standpoint of the courts, a corporation is invisible, intangible, and exists only in the contemplation of the law. As such, it possesses only those characteristics that its initiating charter grants it. (Such properties have been previously stated.) Operationally, each corporation must have a board of directors that is selected by the organization's stockholders, who are the legal owners.

Stockholders. Any person may own shares of stock in a corporation simply by transferring cash (private property) to a company for a document showing that that person is the owner of a usually small part of the corporation. Stock exchanges indicate the fluctuating price per share and therefore the legal value of that ownership. These may range in price from a few cents to hundreds of dollars per share.

While the liability of the stockholders is limited to the par or stated value of shares held, there are important rights of corporate ownership. These are the right to (1) dividends declared by the board, (2) subscribe to additional stock before it is offered to the general public, (3) sell their stockholdings at any time, (4) share in the assets of the firm at dissolution after liabilities have been met, (5) inspect the company books with sufficient cause, and (6) participate in the election of directors proportionate to the number of shares owned.

Transfer of authority. The board, elected by the stockholders, operates as a committee with a chairperson and various operating subcommittees. The directors act as trustees for the stockholders' investment and have the initial fiduciary responsibility in this structure. The board in turn selects a president, or chief executive officer, to whom authority for operation of all corporate aspects (except those specifically outlined for the board) has been granted. This person, then, has the fiduciary responsibility to the board for all the assets entrusted to that executive position. All subordinates have a fiduciary responsibility to that president, as previously stated, to use those assets in accord with the stated purpose (1) as set down in the corporate charter, (2) as understood by the stockholder when the investment was made, (3) as interpreted by the board, and (4) as passed down through corporate policies by the president.

Expanding somewhat on the organization chart presented earlier, the more extensive corporate form is shown in Fig. 5-2. While not depicted, it will be understood that the authority delegation continues down the manufacturing structure to include a superintendent, general foreman, and foreman—all levels of management. It even applies to some extent to the individual machine operator who has a fiduciary responsibility to the owners to use that piece of equipment in the fashion and manner as instructed by any forms, diagrams, blueprints, or supervision.

In Fig. 5-2, the first-line supervisor and even the rank and file are found with fiduciary relationships involving trust, confidence, and responsibility for another's assets, the "another" in the corporate form being the stockholders.

Tracing the development of business activity from ancient times, one sees the gradual emergence of laws concerning

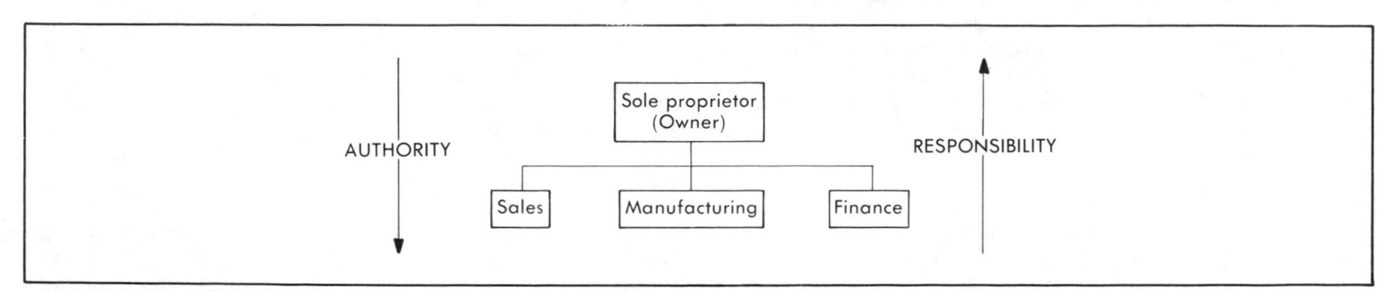

Fig. 5-1 Typical flow of authority and responsibility in a sole proprietor type of organization.

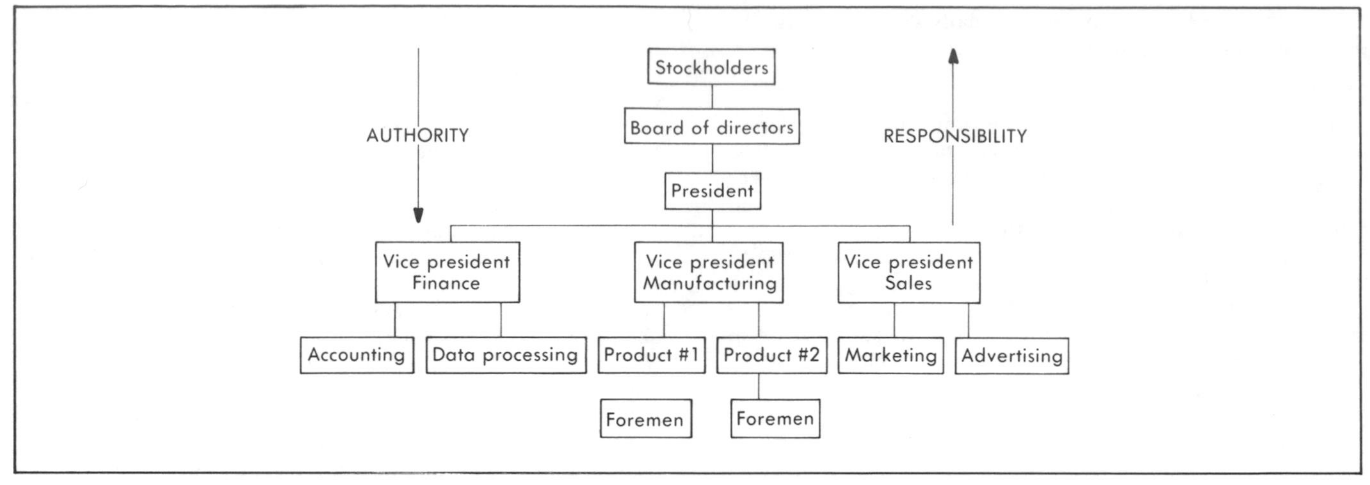

Fig. 5-2 Typical flow of authority and responsibility in a corporation.

private property. The allowable forms of legal organization serve to shape the manufacturing manager's participation in society. And within that larger framework, policies for ethical and legal behavior must be formed, not only from an individual manager's view, but also with respect to the rights of all owners and society at large.

SOCIAL OBLIGATIONS

Social responsibility for a firm essentially means acting as a "good citizen," i.e., being responsive to one's shareholders.

In management literature, systems theory describes organizations as a hierarchy of social systems that interrelate with other horizontal and vertical systems. These interfaces can be along a spectrum ranging from the highly relevant to those only slightly connected.

Each business is a social system that operates within local, state, national, and international socioeconomic frameworks, and these larger systems, in turn, are parts of the global community. Business cannot operate in a social vacuum, but is in a constant state of interaction and flux with a plethora of complex, controversial social ties both within and outside the organizational unit.

Today's social climate makes sizable legal and informal demands on business, requiring managers at all levels to be almost as occupied in responding to social needs as with more traditional concerns if they are to operate in conformance with public expectations. Fortunately for most manufacturing managers, decision-making flexibility in this area is severely limited by legal statutes, by government regulations, and by upper-level company policies. Still, it is important for all managers to have a grasp of the underlying philosophy and to know the positive and negative aspects of a firm's social responsibility.

BUSINESS AND SOCIETY

Business, to repeat, is normally conducted within a social network made up of many internal and external groups. For examples, customers are looking for fair prices, innovative designs, and high-quality products. Similarly, in the interests of social well-being, governments have their needs. Local governments levy taxes that are normally kept low enough to attract new industry, and in other ways administrations try to expand their tax base. In this way, income sources are available to be drawn on over the years and jobs are provided. Moreover, governments make numerous regulations and restrictions on the firm, and these too are a costly drain. Besides consumers and government, other groups include minorities struggling for their fair share of jobs and women seeking a greater role in economic life. Religious groups, miscellaneous in-groups, and the media all try to inform business of its societal responsibilities and also exert influence on business. In return, of course, the community provides such essentials as a labor pool, schools, consumers, police and fire protection, and other services for the firm and its employees.

Within an organization, employees may seek better conditions, more pay, meaningful tasks, savings plans, higher status, health and dental care, and assorted other fringe benefits. Sometimes ideas submitted can be foolish and inappropriate. Other desires—like providing a haven of security for workers and their families in retirement—can be prohibitively expensive.

Such networks, reaching out for money, services, protection, and opportunities, wrap themselves tightly around an enterprise. Therefore, these historically new welfare responsibilities have been accepted by business generally and are worn by managers like multiple hats.

Even the toughest-minded firm, of course, cannot escape the fact that manufacturing wastes may dangerously contaminate local drinking water or air. And a new or altered product may trigger disaster for competitors or consumers thousands of miles away. It is indeed hard to tell where business concerns end and the larger society's begin. This is why the two can be properly viewed as part of an "interactive system." Each needs the other and an action by one will inevitably affect the other. Government taxation, for example, forces managers to look for legal loopholes, deductibles, extra tax credits, and other safe strategies to soften the blow without being penalized. And, when there is disagreement about fairness, drawn-out court battles can ensue as well as fines and imprisonment for errant executives.

SOCIAL RESPONSIBILITY

Within the free enterprise system described in the first section of this chapter, government seeks optimum legal guidelines for industry and regularly examines the impact of its decisions. To some extent, this involvement is set by federal laws as in the case of safety (related to product, production, health, and pollution control). At the federal level, for example, the Occupational Safety and Health Administration (OSHA) and the National Institute for Occupational Safety and Health (NIOSH) provide standards for safety, inspection procedures, and corrective action (which includes punitive responses). Additional information on occupational safety and health concerns can be found in Chapter 12, "Management Concerns for Occupational Safety and Health," of this volume.

Pollution

While most manufacturing managers are familiar with these watchdog agencies and work cooperatively with them, Americans have yet to see the heavy prison terms levied by the Soviet Union on managers of the Chernobyl power plant in 1987 because of their violations of safety regulations. The Chernobyl incident can also serve as an example of how grave the consequences of pollution can be because produce and animal life worldwide can be affected by a single nuclear disaster. In the U.S., pollutants from the Ohio River valley power plants, landing as acid rain in the Northeast as well as the eastern provinces of Canada, also had repercussions, as did the Union Carbide Bhopal tragedy in India. Clearly, compelling laws and directives are needed.

Applicability

Social responsibility can become clouded, and too much variance between firms can occur. On the other hand, a single criterion for nonsafety factors cannot be applied to all businesses because some are marginal and struggling for survival. Insistence on the same level of social allocations would impose severe hardships resulting in business failures. Yet, some have suggested that the application of responsibilities should be equal for all companies and that those driven out perhaps should be because they are not really supportive of the whole. This thinking ignores, of course, unemployment and any liquidations that would be a waste of resources and potential.

Interactive Groups

While recognizing imprecise prescriptions, executives must at least be sensitive to what the environmental interactive elements are. The first set is all necessary, "rights-based" relationships. These are the following:

- Stockholders/funders.
- Customers.
- Employees/unions.
- Suppliers.
- Creditors.
- Wholesalers/retailers.
- Competitors.

A secondary group of interactions tends to be more troublesome and controversial. They occur beyond the usual production and marketplace theaters with groups that want a slice of the pie. Actions by the following are increasingly triggered as profits rise:

- Legal requirements of federal, state, and local governments.
- Community service groups.
- Media and public opinion.
- Business support groups and schools.
- Consumer advocates and product evaluation societies.
- Social activist groups.
- Foreign governments.

These secondary, derivative interactions do not normally work through the free market because their problems cannot be easily resolved by market action alone. In any event, when primary and secondary interactions—however positive—are combined, it can be a costly and time-consuming process to deal with pressure groups.

Impact of Pressures

In considering these forces, modern management can only conclude: (1) in making decisions, business management *shares* power with others; (2) skills are needed not only in financial and economic aspects of the business, but also in how to deal with the social and political factors involved; and (3) the acceptance of the firm—its legitimacy as an institution—depends on its successful performance in both primary and secondary spheres. The net effect is the enhancement of the quality of life while adding to the burdens and insecurities of the firm, especially those in global competition where social expectations are lower.

Business managers in "good" firms develop and express a sense of human and cultural values that generously extend beyond law and that accompany economic needs. Long term, this applied social ethic (being fair and doing that which is in the public interest) provides popular justification for the firm.

Historical Perspective

While a manager's social responsibility is hardly a new concept, an emphasis developed as the size, power, and complexity of the modern corporation took hold. Earlier capitalistic forms, discussed in the first section, tended to be quite small with few, if any, employees; hence, relations with government and other social groups were more tenuous. Of course, then as now, the proprietor of a small shop or group of artisans was always aware of the importance of pleasing the neighborhood where products or services were delivered. Other similarities also exist with ancient business organizations. Those early city and royal governments also levied taxes and imposed restrictions for the social good, and smart managers then were quite aware of (1) their own stake in a healthy society and (2) the risks involved in trying to beat the system.

Oliver Sheldon, the British management philosopher, was one pioneer concerned about the social role of business. In 1923, he eloquently deflated the Scientific Management concentration on efficiency and production so as to include the total business environment. Through this thrust, he helped shape a humanistic-production philosophy of management.

CHAPTER 5

SOCIAL OBLIGATIONS

Within the U.S. industrial system, such giants as John D. Rockefeller, Jr., as early as 1923, questioned whether industry's only role was to create wealth and jobs. He thought it should also be a major instrument in forwarding workers' health and happiness. Henry Ford was similarly concerned about social responsibility. He stressed service before profit, although the result in his firm was more profit. Both men seemed to express the concept that what was best for society as a whole was best for business.

Responsibility Arguments

Not everyone agrees with the management philosophy that espouses strong commitment to a socially responsive organization. There are, moreover, extremists on both sides of the debate, and liberal eras tend to alternate with conservative approaches to social welfare, changing what is popular in management attitudes. However, the overall tapestry of history is one of steady improvement in human life, and a brief overview follows that will describe positive opportunities for social involvement as well as some problem areas that might be foreseen.

OPPORTUNITIES FOR SOCIAL RESPONSIBILITY

Companies have a moral obligation to be responsible and help provide solutions because a business is a member of society and should have a conscience regarding its action or inactions. For example, allowing a car to be sold that is known to be defective could cost lives; hence, a moral system must be maintained. Recalls are costly in terms of repairs and reputation. So, while perhaps not saintly, American business rarely finds it advantageous in the long run to be a sinner. Rather, the posture is that of a good neighbor working with other neighbors toward a better society.

Limited Resources

Rachel Carson's *Silent Spring* presented the case well for protecting planet Earth from insecticide pollution. It awakened a new consciousness of the gravity of depleting the natural environment. Businesses have the opportunity to help protect limited natural resources so there can be a viable tomorrow. While forests and some other riches can be replaced with intelligent planning, others—like petroleum—have a limited supply life. One of many ways would be in having a sound system of material and inventory management. Applied nationwide, it would not only cut manufacturing costs, but reduce waste of raw materials throughout the nation's production system.

Better Social Environment

Business, with its controllable organizations, has the chance to aid in solving difficult social problems. One example would be in the area of overall equality of opportunity where justice may result in physically and socially handicapped individuals getting a better job. This, in turn, would lead to larger tax payments, better neighborhoods, and more family stability. This approach is also fruitful for business in that it provides a larger and more dedicated labor pool; turnover and absenteeism may be reduced; crime in the plant's neighborhood might be decreased; and, with lower frustrations, less money need be spent on insurance and direct protection of property. It is a towering challenge in which government has to be joined by business efforts if even a dent is to be made in the problem.

Long-Run Profit

Greater social good comes from an active and successful business system, and the primary goal of the firm is to remain in business through profits. This money not only provides a return to the investors, but also provides funds for reinvestment in the working organization. Evidence shows that socially responsible businesses tend to be survivors with more secure long-run profits. So, even though short-run costs may be high, long-run payouts for a balanced program are favorable due to a stronger image and wider satisfactions.

Discouragement of Regulation

Another plus in this opportunity package is that good business behavior diminishes public pressure for increased government regulation of economic activity. It is recognized that too much control can reduce precious freedom for both business and society as a whole. In the case of business, regulations tend to (1) add to economic costs, (2) restrict flexibility in decision making, (3) reduce the ability to adjust to shifting market forces, (4) add to nonproductive record keeping expenditures, and (5) decrease the willingness of people to take capital risks. Through a socially responsive business posture, the opportunity is there to influence a reduction in these effects of regulation.

Achieving a Balance

Business has vast social power because it affects so many environmental areas. The reciprocal to that power must be responsibility. Otherwise, misbehavior might result that could endanger society. But, the opportunity is there to exercise responsibility and in this way enhance the freedom of business to use its power effectively. For example, a law from outside may reach into a firm to regulate employees—such as requiring no-smoking areas or having to take health examinations before performing certain work. When a government inspection finds there are serious violations of health laws, a firm's ability to perform can be reduced or eliminated by a forced closing.

Public Image

Companies usually seek resourceful ways to enhance their public images to gain more customers, better employees, and other benefits. To do this, they don't consider people the archenemy standing in the way of profits. With a realistic perception of the world around them, there is the opportunity to use social responsiveness to adapt to current conditions.

Obstacles in today's environment are dramatically different in scope and degree from those of only 20 years ago. Many of yesterday's values are passé, so it is important for managers to stay in step with the times. Just as conditions are never perfect for employees, they are never just right for companies, so practical managers will adjust their social approaches to stay within their legal framework and maintain a favorable public image. Of course, some eras and some cultures provide more encouraging soil for business, but the opportunity to serve and yet make a profit is always there.

Prevention or Cure

The old adage that an ounce of prevention is worth a pound of cure applies to businesses dealing with social problems. Attention now, rather than later, tends to hold down the cost of problems, and a socially responsive attitude may prevent sour situations from developing. Just as a good preventive maintenance program on equipment is beneficial, so sensitive attention to social issues can prevent slowdowns or downtime for employees and management.

OPPORTUNITIES FOR THE MANUFACTURING MANAGER

As mentioned at this section's beginning, the manufacturing manager is constrained in many ways from a full decision role in societal responses. The listing of opportunities just reviewed indicates the usual full range of positive choices available to the firm. With an understanding of these, the factory manager can participate in the discussion and make suggestions or reach decisions within his or her scope of authority.

PROBLEMS IN MEETING SOCIAL RESPONSIBILITY

To fully appreciate the values coming from a socially responsive stance, business managers should be aware of the criticisms leveled at this posture as well as some of the problems in implementation.

Unrealistic Expectations

Public expectations have spiraled, and business success has often led to the pattern of providing essentially all of an employee's basic needs. But, human wants are insatiable and stretch toward a utopian existence. Expectations passed from earnings to broader social desires like more leisure time, structured recreational activities, child care facilities, educational opportunities, personal leaves, and most recently, guaranteed job security. At times, business seems to be expected to act like Santa Claus with a bottomless sack of treats. Also, what is accepted as reasonable and appropriate this year will be quickly assumed and replaced by next year's expectations.

Extraordinary Costs

This is somewhat in line with the parent who tells the child, "It's your mess. You clean it up." In many instances business actions have contributed to social problems in a very obvious way, such as industrial pollution. Sometimes the problem was unknown at the time or only of minor temporary impact. Still, one must participate responsibly in correcting each violation and not try to hide the facts from the public. Even if the firm has moved to another site, it can be held responsible for, say, contaminating the environment. Yet, fines with or without cleanup can be extraordinarily expensive.

Business Mission

Because business operates in a world of scarce resources, the economic efficiency of business is a matter of top priority and should be its primary mission. Profit maximization is classically justified because society strongly benefits when business reduces costs and improves efficiency. Thus, the proper function of business is economic, not social, and its success should be measured in terms of economic values. Managers are fiduciaries for stockholders' money, and to the extent that social responsibility reduces the return to stockholders, they are spending owners' money for purposes other than that stated.

Divided Purposes

Social involvement by business can so divide the interests of its leaders that confusion and a resultant inability to lead can result. Debating social goals can divert executive energies and attention from their main job to the detriment of all.

Hidden Costs

Many social programs are not cost effective and do not pay their own way. Often these costs are difficult to impossible to compute, and people are misled as to who will pick up the costs for these programs. Most people believe that business will pay and that the benefits will be free. In reality, such costs incurred by business are passed on to the consumer through price increases. That is partly because, if they were to be taken out of profits, the incentive for investment in those businesses would be sharply reduced; hence, they might lack growth capital. Also, employees' jobs are less secure, and eventually the nation itself may be poorer and less competitive internationally.

Business Has Enough Power

Business is one of the two or three most powerful social institutions and to give it more power over social issues would be inappropriate. Its influence already pervades society in education, the media, government, homes, and in the marketplace. Added influence, by increasing the concentration of power, could create a monolith that would tower over all other institutions and probably would reduce the effectiveness of our free society.

Lack of Accountability

There is also the problem that business has no direct lines of accountability to the people; hence, there is poor social control. Until society can develop mechanisms that establish direct lines of social accountability, business must stand clear of social issues and pursue only its profit goal.

Lack of Social Skills

The possibility always exists that a business leader lacks the perceptions and skills to work effectively on social issues. Such people are often not comfortable in the area because their tough approach—primarily concerned with economic issues—involves acts that go by different rules. Some males particularly, while not Robber Baron types, may view alertness to human interactions as "soft." Today, such executives are considered philosophically out of date, but in certain short-term competitive situations, this alternative may be required to stay afloat.

Corporate Inability to Make Moral Choices

It has also been suggested that some organizations may be inherently unable to make moral choices. Moreover, that ethical concepts apply to *people*, not organizations, and that the idea of "corporate social responsibility" is merely a smoke screen used by people with special interests or biases.

A BALANCED VIEW

Obviously, the management of producing companies must take into consideration both opportunities and problems when deciding about required and voluntary social acts. The difficulty for any manager is in achieving the optimum balance between the obvious rights of the company and the gaping needs and wants of the populace. H.L. Mencken once said the following about the awkward position of management:

> He [the businessman] is the only man above the hangman and the scavenger who is forever apologizing for his occupation. He is the only one who always seeks to make it appear, when he attains the object of his labors, i.e., the

making of a great deal of money, that it was not the object of his labors.

Why is this true? Because business recognizes its tremendous power and its stake in social well-being.

ETHICS FOR MANUFACTURING MANAGERS

Ethic is the noun used when discussing a particular system of moral principles and values—as in the "Judeo-Christian ethic." *Ethos* is also a noun, and it expresses the moral principles or values of a specific group or movement, as in "The ethos of the leadership had degenerated."

The more familiar noun, *ethics*, refers to the ideas about what is right and wrong. And, finally, the adjective *ethical* pertains, of course, to ethics; all four words come from the Greek *ethos*, which means character. While the first two are not used much in everyday communications, it is beneficial to note the distinctions in passing.

ETHICS IN BUSINESS

Ethics is not very precise or organized in business, despite the entire field of business law and a ponderous amount of business literature written on relevant moral issues. It makes a handy topic for after-dinner speeches and religious sermons because a high standard of human behavior, and the good principles that guide it, are at the core of civilized life. Further, those ethical people who follow good principles are trusted, respected, honored, and favorably viewed by people in general and especially by one's business associates.

From a more individual standpoint, a person who is unethical simply cannot be tolerated by a business organization whether the function is at the top or bottom of a firm. There are a great many money-related opportunities in commerce that tempt those with weak principles; therefore, good conduct must be immediate and instinctive. That is why several decades ago, when it was legal to do so, application forms asked for the job seeker's religious affiliation. It was believed that an employee's exposure to ethical principles on a weekly basis was ultimately meaningful and beneficial for the firm.

BUSINESS STANDARDS

Many people do not believe that the ethos of the business community is especially high, despite the great emphasis placed on this area in all business associations. This is because the top priority of business is profits. Also, the "in-house" ethos that permeates a specific firm or type of enterprise (such as printers of hard-core pornography) may be highly distasteful from an ethical standpoint; hence, these firms endanger the good name of business generally in the eyes of the public. Associations such as the Chamber of Commerce, Rotary International, and the countless business and trade associations try, through organizing, to either keep out or eliminate offenders in their ranks. The medical, legal, religious, and engineering professions also remove the privileges of those members who are proven to be a social threat and embarrassment.

RELIGION AND PHILOSOPHY

In ancient and primitive societies, religions and strict behavior codes had to be developed for reasonable social intercourse. Humans, unlike some other animals, have no special instinctive programming that limits their freedom to act in harmful ways.

Of course, from religion to religion and culture to culture one can find both interesting variances as well as universal laws—such as theft is bad and truth is good. For centuries, philosophers and religious thinkers have tried to bring some order and reason to the understanding of ethics. Buddha, Confucius, Jesus, Mohammed, Moses, and many other religious figures laid the groundwork for ethical analysis by establishing broad moral guidelines for everyday conduct.

PHILOSOPHICAL FOUNDATIONS

It is not always easy, of course, to decide what is right and wrong in a given situation. It may require weighing one social value against one or more others. That is why aspects of ethics have been elegantly debated by speculative philosophers such as St. Thomas Aquinas, Baruch Spinoza, Blaise Pascal, David Hume, Immanuel Kant, Arthur Schopenhauer, Friedrich Nietzsche, and William James. The overall theme of these cognitions is the plight of the human race without a supernatural god or ethical virtues.

Social philosophers, too, examined human capacities for harmonious relationships, and these luminaries included Aristotle, Plato, Epicurus, Epictetus, Marcus Aurelius, Confucius, Michael de Montaigne, Ralph Waldo Emerson, and John Dewey.

The civilized world was also jolted by the more hard-headed philosophers, such as Charles Darwin, Sigmund Freud, Alfred North Whitehead, and Albert Einstein, who cause one to consider the nature of things in a more scientific light and in total. Fortunately, manufacturing managers do not have to be specialists in philosophy or fully understand the history of the development of ethics.

REASONS FOR IMPORTANCE

Because so much in business is contractual and promissory in nature, the ethical qualities of the participants are central. A business must also operate through agents and get its job done through people, so this concern extends from the top of the organization, which sets overall policies, to the bottom layer. The following example will illustrate why this is true.

A flaw in the manufacture of a minor part used in the construction of a transport plane not only results in an extremely costly loss for the fleet, but in human lives and property when a crash results. Insurance rates would climb, lawsuits mount, and the airline may experience such a crushing loss in reputation that its future may be bleak. Of course, they go on to sue the parts manufacturer, which would probably have to close its doors. And, all of this occurred because one immoral employee at the bottom level chose to be unethical and used a bad part rather than be delayed in his or her production activity.

Or, consider this situation: Beneath the policy-making level in a high-tech organization, a sales director, the head of research and development, and the manufacturing manager engage in a conspiracy to steal product secrets from a competitor. Budgets are quietly manipulated and persons hired as spies to ferret out the innovative design/production/distribution methods that give the competition such an edge. Listening devices may also be employed. Such a deed is not normal business and goes well beyond the bounds of ethical behavior. So, who might society punish? While it may not seem fair, the company's owners are ultimately responsible for the ethical behavior of everyone in the organization. Of course, the company will probably discharge its overeager executives, so there is more than enough suffering to go around; small wonder that ethics is taken very seriously.

TIME VARIANCES

Attitudes can vary sharply from time to time. The philosophic thought of ancient Greek thinkers (Plato, Aristotle, Socrates, and others) shifted from religion to society in general and attempted to develop reasoned ways of clarifying various aspects of ethics.

Things can change quickly. Many readers will remember that what was considered wicked in 1950 in the U.S. might find acceptance nowadays, all because social mores (customs) have been drastically altered. The 1950s, of course, was a philosophically conservative, rational era that supported institutions and favored traditional values. Since the late 1960s, a romantic, liberal era has held sway that favors individual rights and a generous, nonauthoritarian approach to life.

INTERNATIONAL BUSINESS

To function, there must be accepted moral attitudes within a culture, and this has always been a problem in global business transactions. Traditions in a distant country can seem hopelessly corrupt while they, in turn, might be quite as stunned by another country's insensitive manners and strange values. In 1976, a special disclosure program was undertaken by the Securities and Exchange Commission (SEC) that surveyed more than 250 U.S. corporations. It was learned that many had made questionable or illegal payments to overseas persons; this led to passage, in December 1977, of the Foreign Corrupt Practices Act that prohibits attempts to influence foreign governments through gifts or bribes.

Through this legislation the government is, in effect, imposing our standards of business operations on foreign cultures. It is known that Indian, Islamic, South and Central American, as well as some European nations have quite different attitudes on such matters. In most instances, such payments are expected if one wants to conduct business in that region. This puts U.S. businesses at a distinct disadvantage because penalties for violating the act are up to $1 million in fines.

So, that which is ethical is partly what most U.S. citizens would think good social behavior right here and now. There is no way to take a quick poll, of course, so one leans on internal beliefs which, in educated people, are based on a rich heritage of thought over the centuries. For example, the Judeo-Christian ethic has been central in guiding and unifying thought in the U.S., and this cultural influence extends to shape the beliefs of atheists as well. When such a powerful, far-reaching agreement about what is right and wrong crumbles even a little, confusion, disagreements, and multiple attitudes occur.

Still, the handshake of a successful manager anywhere is sometimes sufficient for contractual reasons even today because it carries with it the executive's treasured reputation which he or she cannot function without. And, similarly, a company as a whole jealously regards its image because without it, customers, for example, would be put in a "buyer beware" situation, which is no longer tolerated by our society.

Acute international competition is, of course, having an important bearing on the ethos of American manufacturing. It is easier to act in the social interest when not in a precarious position where competition threatens survival.

Citizens in the U.S. have become accustomed to high salaries with fringe benefits and all too often of winning without really trying. But far lower wages are paid abroad, and today the nation seems to have lost its technological edge in many cases. This combination has resulted in disasters with hundreds of plant closings and lost jobs. If some operations are to continue, it would mean sharp cutbacks, quality reductions, or relocation where employee expectations are lower.

This relatively recent situation will heap on new ethical questions for business, governments, and unions. What happens after a plant closing when a community's income dips sharply and a large segment of its consumers leave town? Does a major employer, with earnings already slashed and losses high, have an obligation to retrain its former workers? Will firms turn to cheaper construction methods and materials? Is it ethical for unions to demand guaranteed job security? How about people in the community who have grown accustomed to a variety of welfare services and now find these services terminated because the national economy is slumping? And, will there be a return to more traditional maxims for managers who were trained to think another way? Does this mean that a more automated and computerized business will no longer be held to ethical behavior? Hardly. As Pope Pius XI said,

> Though economic science and moral disciplines are guided each by its own principles in its own sphere, it is false that the two orders are so distinct and alien that the former in no way depends on the latter.

CODES

An ethical problem in business is often a gray area beyond the reach of existing laws but which may, at some future date, be so formulated. One tool for dealing with gray area matters is for each company, industry, and profession to develop a code of work behavior guidelines. Of course, where employees of a specialty must abide by both professional and company codes, there can be an overlap so long as they aren't in conflict.

Moreover, a professional code may serve in two other ways. It can recommend appropriate rulings to the firm as well as help shield the professional from having to carry out inadvisable actions. A single firm might have to deal with codes of varying content and strength from a number of outside societies that can be local, regional, or national in authority. All of this is somewhat similar to dealing with different unions in which every employee involved isn't even a member of the authoring body. Further, international firms can experience even greater complications as slightly different ideologies try to merge into one.

The values themselves are normally developed formally and/or informally in work groups, and an organization's ethical norms need not exist in written form. A firm's standards are set by directors and managers as they develop rules and policies and take related actions. Research, in which a large sample size was used, indicates that about three fourths of those firms surveyed have a written code of ethics. Effectiveness, of course, varies, and enforcement of all points is often impractical or nonexistent.

ETHICS FOR MANAGERS

It will be recognized that differing conditions in separate firms would make it unrealistic to standardize a detailed code of ethics for all organizations. One strives for specificity in developing a code because, unlike platitudes, this aids in defining and guiding real-life practices. This attempt at specifics presents more problems, however.

First, the sheer number of relationships both within and outside large organizations is staggering. The code should ideally include each of these interactions because, without that degree of specificity, individual interpretation can make the code so elastic that its function can be partially destroyed.

A second problem arises from a code that provides essentially no prescription for punishment. This would, of course, erase much of the document's power to influence. Employees tend to view such a code—if they are even aware of one—as a form of moral exhibitionism that is only designed to reassure its readers. As a result, codes, so idealistically arrived at, may just gather dust, are framed, or lost, and in this forgotten state are soon too dated to be taken seriously. Despite such formidable problems, these rather antiseptic laws have become a standard way to help ensure integrity and moral sensitivity, and the very composition of them can prove very productive exercises.

These relatively compact and unifying policy statements also aid in training and in reaching judgments—including those dealing with disciplinary action. A good example of a professional code that encourages conformity was adopted by the Society of Manufacturing Engineers in April 1975 (see Fig. 5-3).

MANAGERS' ROLE IN ETHICS

Of far more importance than codes, which can be like owning but not reading a religious guidebook, is the example set by management—especially top management—as it actually goes about dealing with ethical matters. These include the following:

1. The degree of ethical/unethical behavior engaged in or allowed by management.
2. Management's swift and clear reaction to employees' and outsiders' wrong conduct.
3. The value placed on ethical behavior in training and performance evaluations.

Doing the above merely provides a foundation that follows standard planning and control procedures. Naturally, the actual implementation of ethical values can be individual and a far more complex process. The first steps in implementing an ethical program are (1) defining ethical values and drawing up specific behavioral guidelines, (2) determining the accountability and responsibility for ethical conduct, (3) devising a system of internal controls to monitor practices, (4) setting management policy for responding to callous violators, and (5) establishing procedures for reinforcing proper conduct.

Additionally, those organizations that have to be unusually watchful for some reason have sometimes set up an ombudsperson or ethics board to review policies and practices on a continuing basis. An ethics advocate is sometimes asked to identify repetitive or generic ethical questions that should be routinely asked.

Other expanded efforts have been an approach designed to raise mere issues to their full ethical level; audits of an organization's actions from the view of ethical considerations and those affected; and ethics training as described subsequently.

MANAGEMENT TRAINING

Business ethics is also taught in universities, particularly in management curricula in colleges of business administration. Course titles might be "Introduction to Organizations," "Business and Society," "Business Ethics," and "Business Policy." The specialty of business law, also found in business colleges, became a subject of academic study shortly after World War I, and it provides an important forum for debate.

Another education force of great import mushroomed in management literature in the 1970s. This, of course, was the behavioral approach to management that provided a psychological, sociological, and anthropological thrust while sublimating some of the economic values of the previous, less liberal era. All are important aspects in the training of an able practicing executive.

CHARACTER DEVELOPMENT

What is sought from all these educational inputs? Actually, it is a utilitarian sense of morality that must somehow be developed from childhood so the adult manager can make snap decisions on the production floor that won't be regretted or viewed as insensitive later. Many contemporary managers have been deprived of exposure to ethical debates and readings; indeed, in the last couple of decades there has been much cynicism and a turning upside down of what is acceptable.

Yet, great truths never die. They are just temporarily obscured, causing nice people with a character flaw or two to trip on them in rather tragic ways. The intellectual atmosphere is always in the process of change, but "right and wrong" are by and large universal laws that are remarkably stable for our species—even when unenforceable or unfashionable.

Ethical behavior is the union of the subjective (individual's conscience) with the objective (company and government laws and traditions). The subjective complexity of this is what, at the present stage of development, separates the manager from a computer in decision making. The manager may find either idealistic or potential violators above or below him in rank, or there may be a confusing combination of these who have direct influence on the executive. That is why each person has to have a distinct value set with internal guidelines for thinking through moral judgments. These are based, of course, on one's "character"—a feel for what is right based on what was absorbed in one's prior training.

CONCLUSION

Business systems, then, are a product of beliefs, mores, laws, and customs of the society in which they exist, and these can be altered by economic conditions. A firm's survival depends on social philosophies that condone and support those commercial actions needed to make a profit on funds invested.

Beliefs and value systems, concerning what is right and wrong for managers to do, have an impact on all business activities and serve as a justification for doing or not doing something in a particular way. As long as these philosophies are compatible with those of society, and the resultant business practices contribute positively to social goals, society can be expected to accept, support, and encourage the business sector. Theodore Roosevelt put it this way:

> We demand that big business give people a square deal; in return we must insist that when anyone engaged in big business honestly endeavors to do right, he shall be given a square deal.

CODE OF ETHICS OF ENGINEERS

THE FUNDAMENTAL PRINCIPLES

Engineers uphold and advance the integrity, honor and dignity of the engineering profession by:

- using their knowledge and skill for the enhancement of human welfare;
- being honest and impartial, and serving with fidelity the public, their employers and clients;
- striving to increase the competence and prestige of the engineering profession; and
- supporting the professional and technical societies of their disciplines.

THE FUNDAMENTAL CANONS

1. Engineers shall hold paramount the safety, health and welfare of the public in the performance of their professional duties.
2. Engineers shall perform services only in the ares of their competence.
3. Engineers shall issue public statements only in an objective and truthful manner.
4. Engineers shall act in professional matters for each employer or client as faithful agents or trustees, and shall avoid conflicts of interest.
5. Engineers shall build their professional reputation on the merit of their services and shall not compete unfairly with others.
6. Engineers shall associate only with reputable persons or organizations.
7. Engineers shall continue their professional development throughout their careers and shall provide opportunities for the professional development of those engineers under their supervision.

Developed by the Engineers Council for Professional Development.
Adopted by the Society of Manufacturing Engineers,
April 7, 1975.

Fig. 5-3 Professional code of ethics adopted by the Society of Manufacturing Engineers.

Bibliography

Beauchamp, Tom L., and Bowie, Norman E., eds. *Ethical Theory of Business*. Englewood Cliffs, NJ: Prentice-Hall, Inc., 1979.

Bursk, Edward C.; Clark, Donald T.; and Hidy, Ralph W. *The World of Business*. New York: Simon and Schuster, 1962.

Buchholz, Rogene A. *Business Environment and Public Policy: Implications for Management and Strategy Formulation*. Englewood Cliffs, NJ: Prentice-Hall, Inc., 1986.

Chamberlain, Neil W. *Social Strategy and Corporate Structure*. New York: MacMillan, Inc., 1982.

Clinard, Marshall B. *Corporate Ethics and Crime: The Role of Middle Management*. Beverly Hills, CA: Sage, 1983.

Davis, Keith, and Frederick, William C. *Business and Society: Management, Public Policy, Ethics*, 5th ed. New York: McGraw-Hill Book Co., 1984.

DeGeorge, Richard T. *Business Ethics*. New York: MacMillan, Inc., 1982.

Luthans, Fred; Hodgetts, Richard M.; and Thompson, Kenneth R. *Social Issues in Business: Strategy and Public Policy Perspectives*, 5th ed. New York: MacMillan, Inc., 1987.

Mintzberg, Henry. *Power In and Around Organizations*. Englewood Cliffs, NJ: Prentice-Hall, Inc., 1983.

Sharplin, Arthur. *Strategic Management*. New York: McGraw-Hill Book Co., 1985.

Sturdivant, Frederick D. *Business and Society*, 3rd ed. Homewood, IL: Irwin, 1985.

CHAPTER 6

ORGANIZATION

"We've got to get organized!" This cry is often heard in human groups; however, several different messages may be intended. In some situations, the message is that order is needed. ("The garage is a mess! When are you going to organize the trash you've collected out there?") In other situations more common to business enterprise, the message is that we need to improve on communications and/or coordination. ("But I assumed you were going to prepare the cost estimate. We've got to get organized!") Still another meaning of "getting organized" has to do with the establishment of methods and procedures. ("If we were organized, we'd have a systematic way of handling purchasing!")

In the literature of management, "getting organized" has still a different and more complex meaning. Organizing is intended to facilitate accomplishment, and the product of organizing is an organization. It is useful to recognize the relationship of planning and organizing. Planning is concerned with determining ahead of time what must be done, how it ought to be done, who is to do it, and when it should be done. Organizing is the first step in the implementation of plans and is concerned with designing and maintaining a structure of roles or positions that will allow people to work together effectively to perform the tasks and achieve the objectives set out by the plans. A close examination of organization reveals that it is a complex entity and is more organic than mechanical in character.

This chapter is intended to introduce the reader to the terms and concepts basic to an understanding of organization. Some of the topics discussed are various organizational views, formal and informal structures, and organizational subunits, structure design, and management. The reader is referred to other publications listed in the bibliography at the end of this chapter for additional information on this subject.

CHAPTER CONTENTS:

VIEWS OF THE ORGANIZATION

The organization should be viewed by managers as a complex, multidimensional entity. Complexity and multidimensionality are characteristics that develop naturally in today's organization. Managers who believe organizations are simple, easily defined structures are managers who will be regularly frustrated and who may prove dangerous to their firms.

TRADITIONAL VIEW

The traditional view suggests that the organization is prescribed by the manager. The organization is created by the manager because alone the manager cannot reach the desired goals. The activities that must be performed call for more time, effort, and talent than can be provided by the manager. Therefore, according to the traditional view, the manager groups the necessary tasks into the organizational structure or hierarchy that the manager regards as being most capable of implementing the plans. The weakness of this traditional view is that it emphasizes the managerial prescription of the organization, but does not provide a description of the organization. Other concepts and perspectives have been advanced to provide a fuller understanding of organization as an entity and a phenomenon.

ORGANIZATION AS DELEGATION

A second view suggests that organization is a structure that shows the delegation of authority. Thus viewed, organization is not only a structure of roles or positions, but also a statement of the assignment of obligations and rights. The concept of responsibility involves the assignment of "obligation to perform"; therefore, the assignment of quality control responsibilities to someone means that a designated person is obligated to perform tasks appropriate to maintaining specified quality characteristics. Similarly, the concept of authority involves the assignment of "rights to command." Thus the manager of quality control might have the authority/right to order production managers to shut down process operations when products fall below quality standards.

The concepts of authority and power are often confused and are mistakenly treated as synonyms. Power refers to the ability to command; authority refers to the legitimate right to command. Power cannot be delegated; authority can be. Power is reflected in the actions of subordinates. If a manager orders subordinates to take specific action and if the subordinates perform as directed, then that manager has power. Authority can be assigned and may be reflected in an official statement of the organization's structure. Power, however, requires the cooperation of subordinates and is manifested in subordinate responses to managerial directives.

ORGANIZATION AS A BEHAVIOR SYSTEM

A third view suggests that organization is a behavior system. Individual human beings join

***Contributor of this chapter is:* Lee Lahr**, *Professor, School of Management, Lawrence Institute of Technology.*
***Reviewers of this chapter are:* Robert E. Boughner**, *REB Associates;* **Bob F. Lahidji, PhD**, *Assistant Professor, Manufacturing Technology, College of Technology, Bowling Green State University;* **Hayward Thomas**, *Consultant, Retired—President, Jensen Industries.*

groups, assume roles, and perform according to specified patterns of behavior, and thus the organization can be viewed as a behavior network. The individuals in this behavior system choose to perform roles and tasks so that the goals and objectives of the organization might be achieved. Important to this view of organization is the motivation of the members of the system. Individuals, it is asserted, participate in the system's cooperative behavior only if and only so long as the system helps each member to achieve personal goals that could not be achieved as an independent individual. Therefore, the effective organization achieves two sets of objectives: one, the objectives of the organization and, two, the personal objectives of the individuals who participate in the organization's system of cooperative behavior. The conclusion that follows from this view is that no organization can be effective over time if only one set of objectives is satisfied. In addition, individuals do not last long in an organization if they attempt to only satisfy their own objectives.

These three views show the contrasting perspectives of managers and subordinates. Managers may see the organization only as a creature of their own making. Subordinates may see a business organization as another social group that they choose to join in the pursuit of their own personal goals; subordinates choose to cooperate with the organization only so long as the organization helps them to achieve their personal goals.

ORGANIZATION AS A MULTI-DIMENSIONAL SOCIAL ENTITY

A fourth perspective strives to integrate these views with still other views to provide a fuller, more complete understanding. In this perspective, organization is described as a multi-dimensional social entity that possesses many dynamic networks or systems of human relationships. While some systems of relationships can be altered deliberately, other systems of relationships change subtly and can only be monitored, not manipulated. The basic premise is that organizations, such as business enterprises, are communities of human beings and, therefore, subject to the same kinds of factors and conditions that beset families, clans, states, and other social organisms. Another premise is that society has changed and has become more complex and more democratic; therefore, organization is no longer a simple relationship of "boss" and undifferentiated workers who possessed similar lowly skills. Today, business organizations are social institutions, just as are schools, churches, and political communities. Today's managers must recognize and operate within the constraints imposed by government and society. Today's organization depends on the application of a variety of highly specialized skills. Therefore, today's organizational managers must induce the cooperative efforts of workers rather than coerce subordinates.

ORGANIZATIONAL STRUCTURES

Several structures, classified as formal (or official) and informal describe the system of relationships in an organization.

FORMAL STRUCTURES

Formal structures reflect the intent of top management to relate members of the organization in a hierarchy of roles and activities and to show the relationships of authority and responsibility.

Chart of Command

The best known formal structure of the organization is the official chart of command. The actual organization may differ in operation from this "official" hierarchy version because the organization is dynamic and makes adjustments in its roles and activities as operating conditions change. These adjustments should not be regarded as signs of mutiny or as compensating moves that remedy the errors made in the official structuring. Top management is well aware of the unofficial adjustments made in its formal structure. Moreover, top management condones these changes because operating conditions dictate adjustments, and updated official versions may be too slow in coming. As significant variations from the formal hierarchy of roles and tasks emerge, the top management of the organization should produce a new official chart of command.

Chart of Roles and Tasks

In addition to the chart of command that reflects the hierarchy desired by top management, a second "official" structure also exists. A variation of the official chart of command, this dynamic network pictures relationships founded on actual performance of roles and tasks rather than on titles. This performance network emerges with the knowledge and approval of top management. In time, the official chart of command may be revised to reflect the actual operating network and top management's latest views as to what constitutes the appropriate hierarchy of authority and responsibility.

Functional Network

A third formal structure can be referred to as the functional network. This structure shows the relationships that exist between members who possess special knowledge and other members of the operating organization. Thus, the manager of a manufacturing department may be forced to wait on a needed machine until a financial analyst has completed a study to justify the need for new equipment. Another example of this functional relationship appears when manufacturing managers are ordered by their manufacturing superiors to keep records in accordance with the methods and standards established by specialists.

The functional dimension of the organization has the support of top management and has become a phenomenon of management in the 20th century. It has been suggested that management is comprised of two groups: the managers who are the decision makers and the "knowledge workers" who provide the managers with the special information needed for making decisions.

In addition to providing information, knowledge workers may be given functional authority. This authority is the right delegated to individuals or groups who possess special knowledge so that they can control specific practices, policies, processes, and other activities conducted by personnel reporting to managers who are not a part of the knowledge worker group.

INFORMAL STRUCTURES

In addition to the dimensions of the organization that are formally recognized, there are other dimensions (systems of relationships) that may be appropriately regarded as the informal organization. The networks of relationships that are informal are difficult to map or trace, are dynamic and often transitory, and have irregular or unpredictable impact on the functioning of the organization.

Affiliation Network

One informal system of relationships can be called the affiliation network. These relationships have a social base. Individual members of the organization have face-to-face encounters that may develop into relationships ranging from lifelong friendships to open hostility. It may be expected that an individual who has been a member of an organization for a few years will also be part of an affiliation network, preferring to interact with some members of the organization and to avoid contacts with others.

Does the affiliation network support or undermine the formal organization? There is no clear-cut or abiding answer to that question. Managers are advised to recognize that affiliation networks emerge naturally and may have both positive and negative impacts on organizational activities. At the minimum, managers should monitor the relationships of their subordinates. Do they "get along"? Are subordinates friendly toward one another or coldly respectful? Are there indications that operations are slowed or impaired by the quality of subordinate relationships?

Influence Network

Another dimension of the informal organization may be called the influence network. Although members of the organization (who are titled managers) have the authority to make decisions, other organizational members (without managerial titles) may influence the behavior of the decision makers. Nonmanagerial members may be able to initiate interest in decision areas, or such members may provide information, estimates, and opinions that shape decision making. The importance of the influence network becomes more apparent if decision making is viewed as a multistep process rather than as a simple action of choosing from among alternatives. Few decisions, in reality, are simple. Most decisions call for the marshalling of facts and opinions; thus the significance of the influence network becomes obvious.

Decision makers seek out advice and information. Individuals who have no managerial authority and responsibility may be respected for their knowledge, experience, or judgment and, therefore, be important members of the influence network.

Within the organization are also bases of personal power, and these bases are not necessarily coincident with the official organizational structure. Individuals, often without managerial rank within the organization, are able to give commands that are obeyed. These individuals are perceived as having special knowledge or special relationships with organization members known to possess power.

Power may be found in unexpected places and in the possession of persons thought to have low status and little or no authority—an executive's secretary, a sergeant in the headquarters' company, the most experienced clerk in the purchasing department. An executive's secretary may give orders that are obeyed because it is assumed that the secretary has a special relationship with the executive. The senior clerk in the purchasing department may give commands that are followed because the clerk is perceived as having special knowledge of the procedures and policies of the department.

Communications Network

A final dimension of the organization is that of the communications network. This is not a reference to a telephone system or any other physical system for message transmission. Rather this communication network depicts the interpersonal flow of information. Again, this network need not coincide with the formal (official) organization. The office "grapevine" illustrates that communication networks emerge naturally from an organization's interpersonal relationships. What also is illustrated is that the network conveys more than factual information; the network "editorializes" in the sense that the information may be reworked by network members so as to convey feelings about the information as well as the information itself. It is important for managers to recognize that informal communications systems exist because of the significance that information transmitted by informal communication systems has to decision making, management control, relationships, and nearly all other facets of organizational life.

ORGANIZATIONAL SUBUNITS

Like most management activities, organizing has developed its own vocabulary. One of the most basic terms is *department*. This refers to a part or subunit of the organization that is to report to a designated manager who has delegated authority over that department and is held responsible for the performance of that department. In the management literature, the word *department* may be applied to any specific subunit of the organizational structure: a division, section, branch, or region.

Departmentalization by Numbers

Several approaches have been taken to structuring the organization into subunits. One of the oldest approaches is that of departmentalization by numbers. This approach is still used in some organizations; for example, all infantry companies in an army may be expected to have about the same number of soldiers.

Departmentalization by Time

Another long-established approach is departmentalization by time. Manufacturing enterprises often subdivide their organizations into departments called shifts. By doing so, the firm can utilize more fully its expensive plant facilities.

Departmentalization by Function

A widely used approach to departmentalization is to define subunits of the organization by the activities or functions performed by members. The increasing specialization of work in the last two centuries has lent support to functional departmentalization. The knowledge required by the various operations of the business has become increasingly specialized, so much so that it is not only unlikely that an engineer could substitute for an accountant, but also unlikely that an electrical engineer could substitute for a mechanical engineer or that a cost accountant could "sit in" for a tax accountant. The skills and tools of the various functional areas have also become more and more specialized.

A manufacturing enterprise, for example, might be structured on a functional basis, with major subunits performing the activities of engineering, production, marketing, personnel, and finance (see Fig. 6-1). In turn, each functional department would be further departmentalized according to specialized activities or personnel; for example, the production department could be further broken down into such subdepartments as production planning, industrial engineering, production engineering, purchasing, tooling, and general production (see Fig. 6-2).

Advantages. There are significant advantages to departmentalization by functions such as engineering, production, marketing, and personnel. Educational programs and occupational specialization by the workforce support and parallel the functional organization. A second advantage lies in the training and management development experience of this organizational form. Managers who advance in their functional departments often become highly competent in their areas. Another advantage to this organizational form is that it provides tight and direct controls at the top levels of the organization.

Disadvantages. There are also some significant disadvantages to the functional organization. Key persons in each area may be overspecialized and understand only their own functional areas. Another result is that the organization may limit the development of general management. Because personnel tend to focus on the performance of their areas, they may fail to coordinate the activities of the areas with the other functional areas. A major shortcoming of this organizational form is that only the top managers are concerned about the performance of the firm as a total entity; most managers focus on the performances of their particular departments.

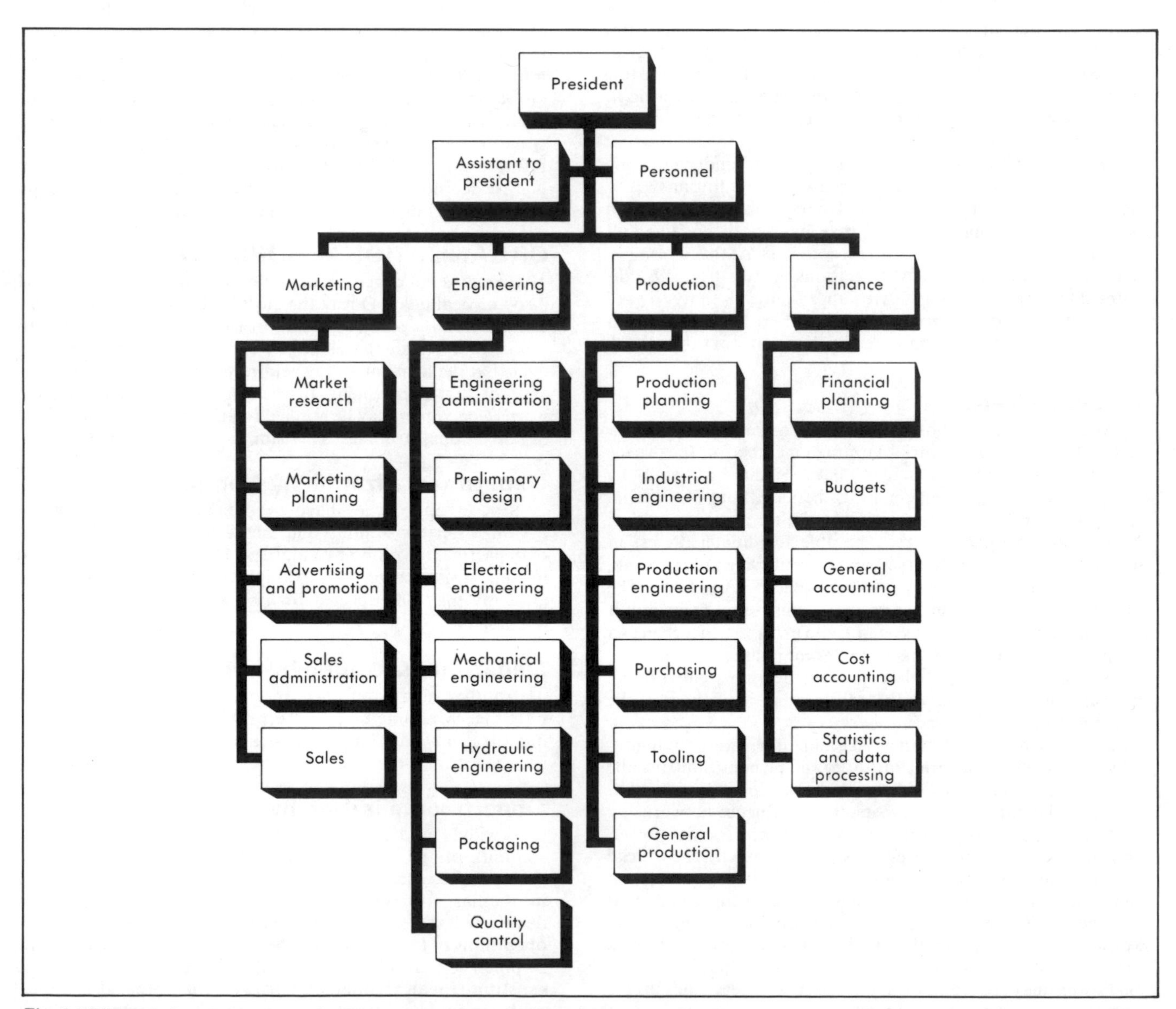

Fig. 6-1 Diagram of a functional organizational grouping for a manufacturing company.

Departmentalization by Process

A variation on this functional organization is departmentalization by process (see Fig. 6-3). This approach may be adopted because of the scale and critical nature of certain production processes to the total operation of the firm. In addition to an enterprise having functional departmentalization for sales, personnel, finance, and engineering, a metalworking firm might have departments for foundry, forging, and machining. Departmentalization by process magnifies the advantages and disadvantages of the functional organization.

Departmentalization by Product

In larger firms with several different product lines, roles and activities are often organized around the product lines (see Fig. 6-4). Top executives provide strategic direction and oversee performances of these product-oriented subunits (usually called divisions). Each division operates as an independent unit, almost as a separate enterprise. All enterprise functions (engineering, manufacturing, marketing, finance, and personnel) are performed in each division with advice and direction coming from specialists and experts who work with the top executives. The auto industry is an example of the concept of departmentalization by product.

Advantages. The advantages of this approach to departmentalization are significant. First and perhaps most important, divisionalization tends to support growth for the subunits and the firm as a whole. Second, divisionalization provides training experiences for general management. Third, coordination among the different functions tends to be improved, and responsibility for performance and profitability rests with divisional management, which can initiate and control meaningful action. Fourth, organization into divisions focuses the resources of the divisions on the respective product markets.

Disadvantages. There are disadvantages also. First, divisionalization increases the need for competent general managers. Second, many service functions (such as personnel, finance, and engineering) are often duplicated, with such services provided at the headquarters level as well as the divisional level. Also, divisionalization compounds the difficulties that top executives have of monitoring the divisional performances from the headquarters level.

Performance reporting by divisions may be slowed as divisional management requires that reports be submitted to divisional managers who then pass reports on to headquarters personnel. A greater difficulty emerges if divisional managers are allowed to approve as well as review performance reports before they go to the headquarters level.

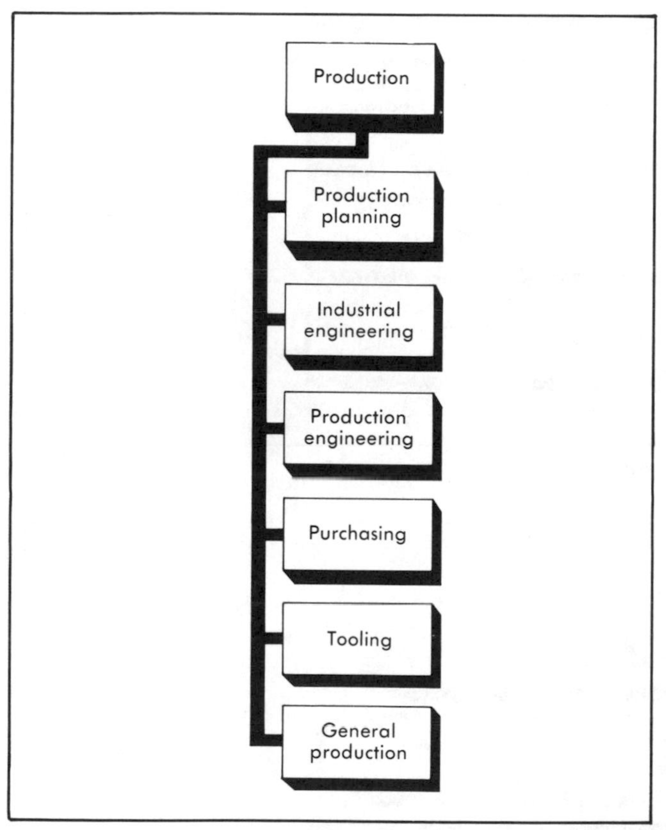

Fig. 6-2 Diagram of how a functional department in a manufacturing firm could be departmentalized.

Departmentalization by Geography

A fourth approach is to group activities by geography (see Fig. 6-5). When operations are widely dispersed physically, this departmentalization approach may be appropriate. Multinational corporations, large service organizations, and retail chain stores often organize in this fashion. Manufacturing operations that locate close to suppliers or customers for economic reasons also may divisionalize by geographic area. The advantages and disadvantages of departmentalization by geography are similar to those of departmentalization by product line.

Matrix Departmentalization

Departmentalization approaches may be combined, and variations of those mentioned above can be developed. The objective is to structure organizational roles and tasks in

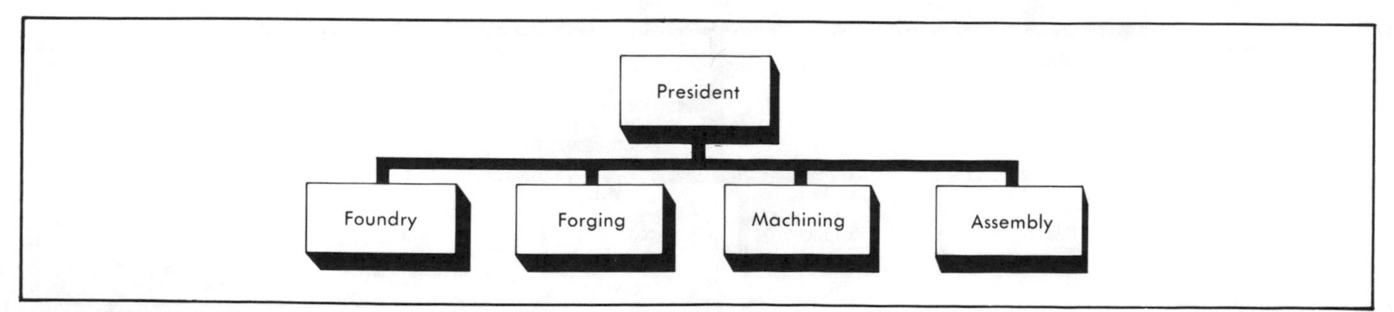

Fig. 6-3 Diagram of a departmentalization by process organizational structure.

subgroups so that organization members can work together effectively and efficiently. One variant that has received increasing use is the matrix structure (see Fig. 6-6). This hybrid organizational form calls for functional specialists to work together on specific projects until the projects are completed.

Advantages. The focus of the matrix group is on the project or program. The results orientation of the matrix group is in itself an advantage. Another advantage lies in the fact that diverse experts or specialists are brought together to work on a project. The functional expertise of each member is subordi-

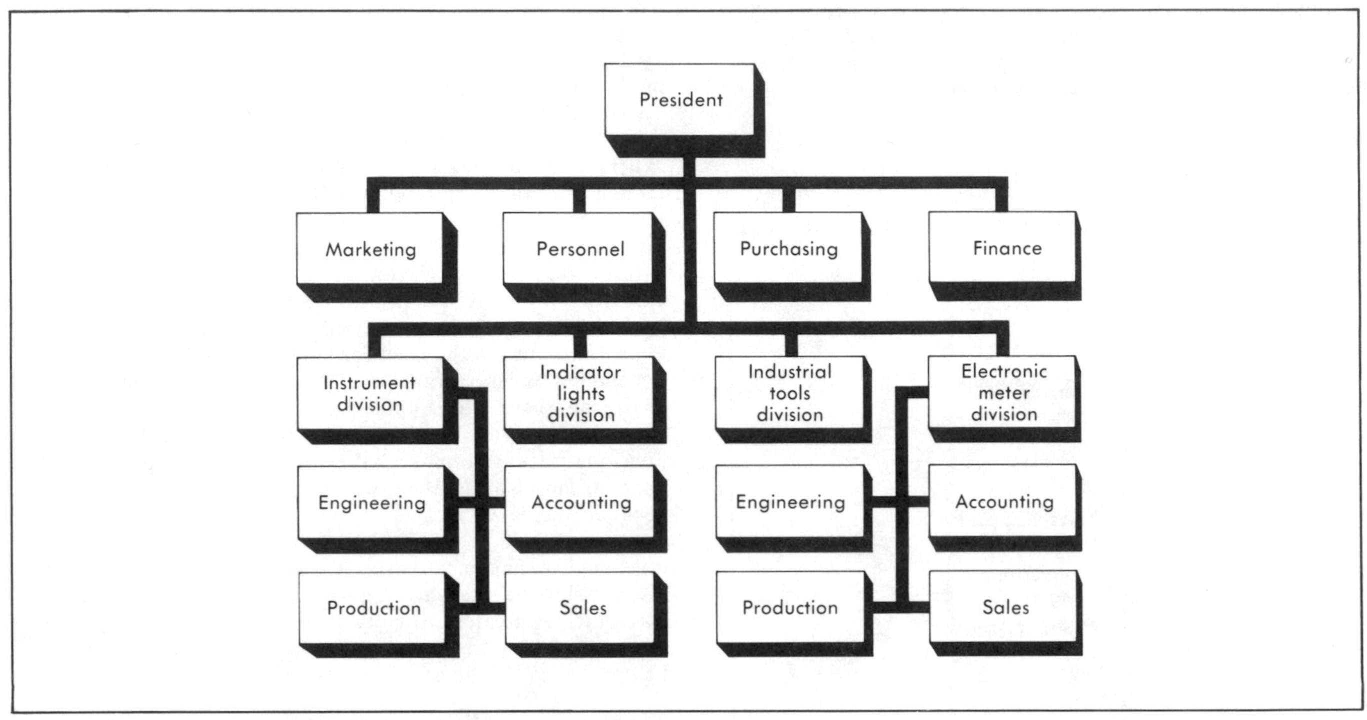

Fig. 6-4 Diagram of a product organization grouping for a manufacturing company.

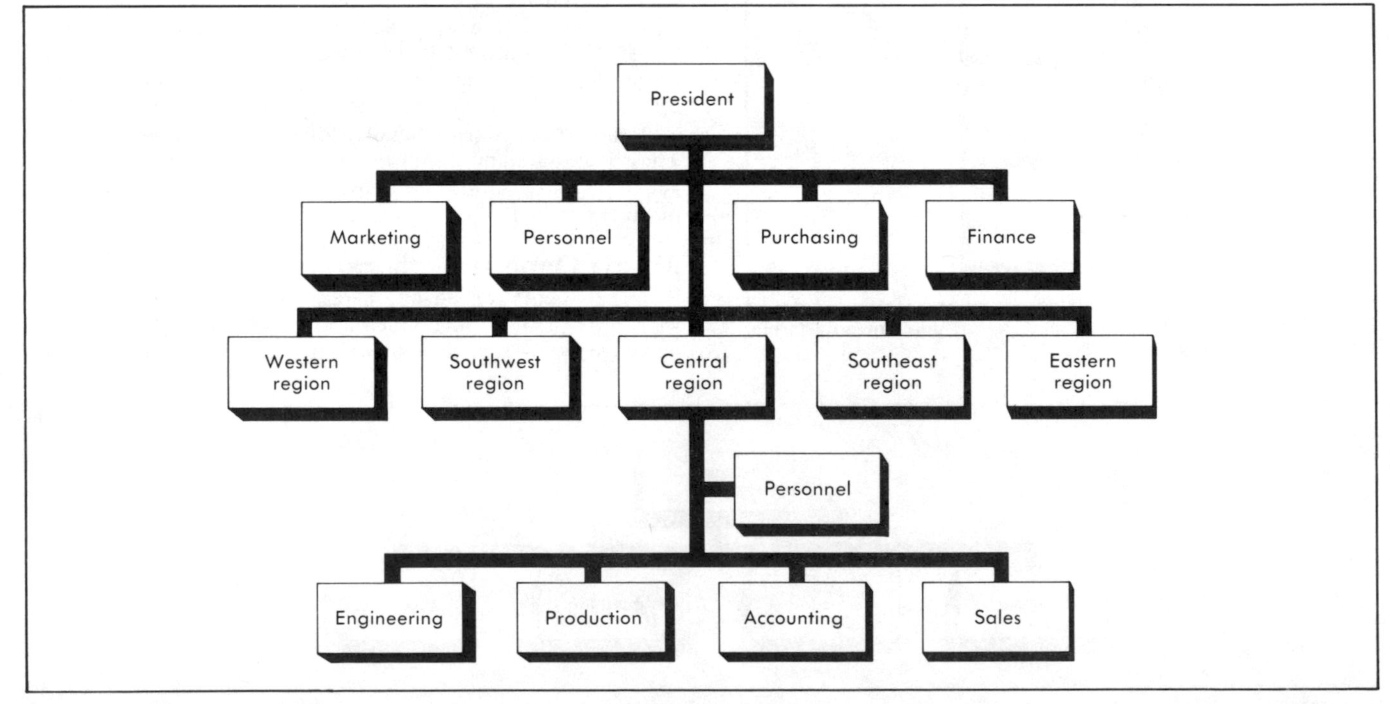

Fig. 6-5 Diagram of a territorial organization grouping for a manufacturing company.

nated to the combined effort of the group; yet the matrix group can bring a variety of skills and knowledge to bear on the problems of the project or program, thus contributing to the efficient use of human resources.

Disadvantages. Among the disadvantages of the matrix grouping is the potential for conflicts in organizational authority. Members of the project team have two superiors—the leader of the project team and their respective functional superiors. Another consequent disadvantage lies in the fact that both superiors should possess well-developed "people skills" to avoid or resolve any conflicts of authority.

Fourth and Fifth-Generation Management

According to one author, the departmentalization schemes just discussed constitute the second and third generations of management.[1] The author characterizes the small entrepreneurial business structure as the first generation, the nonmatrix departmental structures as the second generation, and the matrix approach as the third generation. The matrix approach is more or less an ad hoc, project-oriented approach cutting across traditional functional lines. The author suggests that successful organization for management requires an ongoing multifunction approach to management. In other words, it is recommended that the matrix approach be formalized in manufacturing. The fourth and fifth generations are ways to accomplish this formalization. Computer-integrated manufacturing (CIM), discussed more fully in Chapter 16, is used as a basis for the fourth and fifth-generation management approach.

CIM is a computer-based information system concept linking all of the functions relating to manufacturing. Its objective is to assure that the information necessary for manufacturing is available at the right time and in the right place. It is a way of linking all of the functional areas so that the organization's manufacturing goals can be efficiently attained.

The fourth-generation approach is merely computer-interfaced manufacturing (CIM I). It is the use of the computer to tie computer-aided design (CAD) and computer-aided manufacturing (CAM) together in a system called CAD/CAM. No change in management philosophy is involved. Only the computer linking of the design and manufacturing functions occurs.

Fifth-generation management is defined as computer-integrative management (CIM II) of the manufacturing enterprise; it is also called nodal networking. It focuses on the entire enterprise, not just the design and manufacturing functions. The key word in CIM II is integrative, not integrated. Integrative implies an ongoing integration process in which each of the enterprise functions and activities become information network nodes in an evolving information system. The network links together five threads: (1) management context, (2) business, (3) technical, (4) information architecture, and (5) production systems. Elements of the five threads are listed in Table 6-1.

The fifth-generation nodal network ties the elements of the five threads together in an information context. The formal organization structure will remain what the author calls a second-generation structure. It will be departmentalized by either function, process, product, or geography.

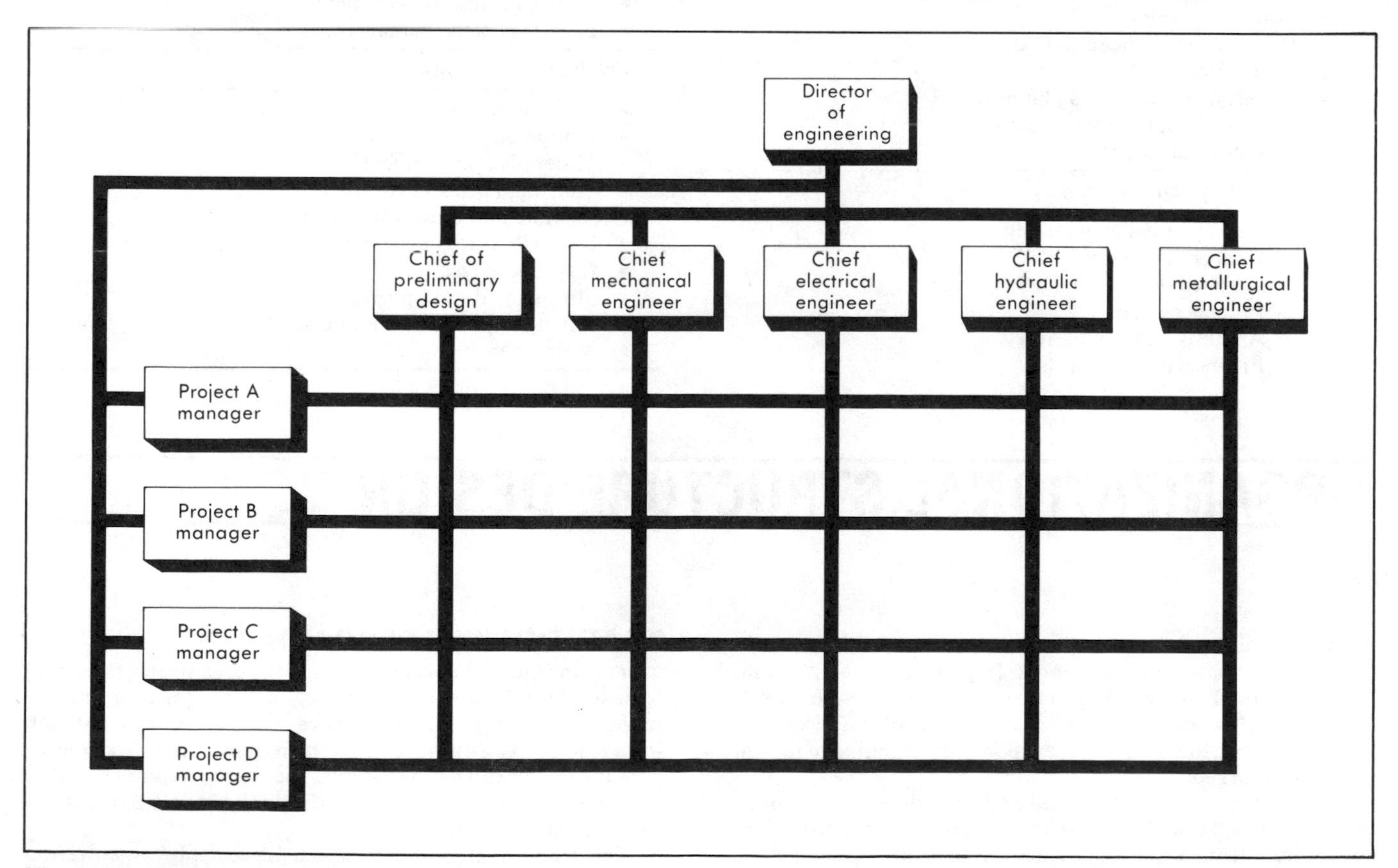

Fig. 6-6 Diagram of a matrix organizational structure.

TABLE 6-1
Five Threads in a Manufacturing Enterprise

1. Management Context Thread
   - Strategic business vision
   - Business strategy and planning
   - Nodal networking management (interactive teams)
   - Management and leadership style
   - Values and culture
   - Human resource management
     - Review and reward strategy
     - Career development
     - Hiring strategy
     - Education and training
   - Key operating norms
     - Just in time
     - Total quality
     - Design for marketability (not just design for assembly)
   - Client/vendor strategies (Interorganizational CIM)
     - Legal/contracts
   - Financial/accounting strategy
     - Cost management
     - Asset management
     - Pricing/cash flow management
   - Project management strategy
2. Business Systems Thread
   - Marketing/sales
   - Planning and scheduling
   - Procurement
   - Material requirements planning
   - Capacity requirements planning
   - Inventory management
3. Technical Systems Thread
   - Product/process engineering
     - Conceptual design
     - Detailed design simulation
     - Process design
     - Analysis and simulation
     - Engineering standards
     - Tool and fixture design
     - Group technology
     - Engineering change order management
     - Reliability engineering
   - Manufacturing engineering
   - Industrial engineering
   - Test engineering
   - Facilities and tooling engineering
   - Assembly engineering
   - Service/warranty engineering
   - Quality engineering
   - Manufacturing and assembly modeling and simulation
   - Logistics engineering (supporting JIT)
   - Standards and protocols
   - Research and development
   - Digital numerical control (DNC)
4. Information Architecture Thread
   - Neutral data reference architecture (process/product reference model)
   - Applications portfolio management
   - Decision support systems
   - Data administration/data dictionary
   - Database management system
   - Hardware selection strategy
   - Artificial intelligence
   - Telecommunications
   - Local area and wide-area networking strategy
   - Standards and protocols
   - Configuration management
   - Legacy systems management
5. Production Systems Thread
   - Shop floor control
   - Equipment
   - Materials management
   - Statistical process control
   - Flexible manufacturing system
   - DNC/PLC management
   - Inspection/test
   - Storage and transportation
   - Materials processing and assembly
   - Maintenance

ORGANIZATIONAL STRUCTURE DESIGN

Organizational design is a complex process that cannot be covered adequately in a few words. If the reader is assigned the task of actually structuring an organization, it is best to go beyond the "handbook" stage. The objective of this discussion is to help the reader to a fuller, deeper understanding of just the concept of organization.

Many approaches may be taken to the development of an organization. These various approaches may be seen to fall on a continuum from a traditional method to a situational method for designing the organizational structure.

TRADITIONAL APPROACH

The traditional approach assumes the organizational setting to be stable and that factors are unchanging in the environment of the organization. Further, tasks are well defined so that each job's role can be routinized and the required qualifications of each job holder can be spelled out. The specialization of work is presumed to provide for an efficiency-based approach to departmentalization.

The traditional designer would also apply organization principles (guidelines) to the structure.

Chain of Command

The chain of command guideline asserts that authority and responsibility should connect from the top of the hierarchy to the lowest level.

Unity of Command

The unity of command guideline asserts that each member of the organization should report directly and be directly responsible to only one superior.

Parity of Authority and Responsibility

This principle argues that managers at every level must be delegated as much authority as they need to fulfill their responsibilities.

The traditional approach to organizational design has its critics. Among the criticisms are the following:

- The assumptions are unrealistic.
- Few firms are stable.
- Some firms may not be facing as much change as others.
- The principles are guidelines that are too general.
- The traditionalist sees the organization as a static, closed, mechanical system that has no contact with the outside world.

SITUATIONAL APPROACH

At the other end of the spectrum is the situational approach to organizational design. The situational designer perceives the organization as an organic entity, in touch with its environment and capable of adjusting to changes in its environment and in itself. Situational designers begin with the enterprise mission and strategies. The basic premise is that organizational structure should be designed to support the "game plan" of the enterprise. The tasks that are critical to implementation of the firm's strategies must be identified. The interrelationships of these tasks must also be recognized so that tasks can be grouped in an appropriate manner. Control and reward subsystems must be established to provide evaluation and motivation.

EVALUATION BY EFFECTIVENESS

One yardstick applied to organizations is the effectiveness standard. The concern of this measure is the extent to which the organizational structure has contributed to the achievement of the objectives of the enterprise and the personal objectives of the members of the organization.

EVALUATION BY EFFICIENCY

A second basis for evaluation is efficiency. The concern of this measure is the extent to which the structure has contributed to the maximization of the productivity of the members of the organization and to the minimization of undesired consequences. For example, a particular type of structure may foster quality controls and, by doing so, reduce the per unit cost of production. This same structure might also lead to increased friction between production managers. Thus, while the organization has improved efficiency by way of lowered unit cost, the organizational structure has also contributed an undesired condition in the form of discord among managers.

MANAGEMENT OF THE ORGANIZATION

Some concepts essential to understanding the organizational management process are: centralization-decentralization, line and staff, and span of management.

CENTRALIZATION-DECENTRALIZATION

Centralization-decentralization is a concept concerned with an organization's tendency to assign authority. If authority is not delegated, but rather resides in one person, then that organization is regarded as highly centralized. Decentralization is a matter of the degree or of the extent to which authority is delegated.

Measurements of centralization-decentralization are qualitative rather than quantitative and can be made by looking at decision making. The more decisions made at lower levels of the organization, the more decentralized the organization; the more important the decisions made at lower ranks of the organization, the more decentralized; the broader the types of decisions made at lower levels, the more decentralized.

LINE AND STAFF

Another concept important to an understanding of organization is that of "line and staff." These terms are used in two ways in management literature. First, line and staff refer to relationships of members within an organization. Line relationships are those established by the flow of authority in the organization. Staff relationships are advisory. In this sense, the commanding general of an army would rely on staff for advice, but these staff officers would not give orders to officers in the general's line of command.

A second use of line and staff involves classification of departments or subunits within the organization. Staff departments are those units that exist primarily for the advisory services they perform. Line departments are thought to be those departments that perform operating activities important to the organization. This second usage can be confusing inasmuch as most departments in a firm perform important operating activities, but also provide important information and advice to their superiors in the organization. Despite the inadequacies of this second usage, readers should recognize that the words *line* and *staff* are widely used to describe departments rather than relationships.

SPAN OF MANAGEMENT

"Span of management" is a concept that deals with the fact that all managers are limited in the number of subordinates they can effectively supervise.

Narrow Span

A narrow span of management refers to organizational situations where a manager has a small number of subordinates reporting to him or her. A narrow span allows the manager to give close attention and supervision to subordinates. The situation also makes fast communication possible between superior and subordinates. Negative aspects are also related to narrow spans of management. Often in this situation, managers become so deeply immersed in the work of their subordinates that morale is undermined and performance suffers. Narrow spans mean that the organization must have more layers of management. The additional layers of managers translate into extra cost and increased distance between the layers.

Broad Span

Broad span of management refers to situations where managers supervise a large number of subordinates. Such situations demand that subordinates be selected carefully. Managers also must develop policies and procedures carefully and present them clearly to subordinates. Broad spans require that superiors delegate authority to their subordinates. The positive aspects of broad spans may not offset negative conditions. Some managers lose their abilities to manage their areas if the span is too broad. Managers may become overloaded and slow to respond to their subordinates.

No specific span of management is optimal. Several factors appear to determine the span that would be appropriate for an organizational setting. The quality of subordinates is an important factor. If subordinates are well trained and capable of operating with a minimum of contact with superiors, a broad span may be workable. If subordinates are performing simpler, repetitive tasks and the same tasks, a broad span can be used. The more interrelationships required within the group, the more narrow the span should be. Carefully drawn instructions generally allow a broadening of the span. If the organizational unit operates in a stable environment, then a broad span may be possible. Dynamic conditions may require a narrow span.

MANUFACTURING MANAGEMENT AND ORGANIZING

The involvement of manufacturing management in the organizing of manufacturing activities varies with the level of management. Again, it is important to recognize the relationship of organizing and planning. The top executives in the manufacturing area of the firm have the responsibility of formulating strategies that directly contribute to the accomplishment of the goals of the firm.

These manufacturing strategies determine the organizational structure of manufacturing. The structure should not be regarded as a separate, independent concern. The purpose of organizational structure is to implement organizational strategies that are courses of action intended to achieve organizational objectives. Structuring the manufacturing organization starts with the recognition of the critical tasks that must be performed if the manufacturing strategies are to be executed. These critical tasks are related by the flow of material through productive processes and by the flow of information through the managerial process. Manufacturing management must group these activities together so that they can be coordinated and must monitor these tasks to ensure that they are being performed in a manner consistent with the manufacturing strategies.

Top manufacturing management should work with the executive management in the other functional areas of the organization to see that fifth-generation (nodal network), CIM II concepts are used in the company's information system. These concepts help tie the organization together so that goals can be more efficiently attained.

Middle managers in manufacturing have the responsibility of translating the "strategic game plan" for manufacturing into actions in the various manufacturing subunits. To implement the manufacturing strategies, middle managers will develop supporting operational strategies and structure their subunits. Again, the point to organizing the subunits is to implement more effectively the support strategies formulated for the subunits. The first step in this process is also to recognize those tasks critical to the activation of the strategy.

Successful implementation of organizational strategies requires competent middle managers. These managers are concerned not only with strategic actions ("the right things to do"), but also with efficient actions ("doing things the right way").

First-line management generally has little involvement in the development of either subunit strategies or subunit organization. However, first-line management in manufacturing is most important in the execution of tasks critical to strategy implementation and achievement of organizational objectives.

A concluding, integrating view of the organization is to regard the organization as a system of systems, an organization of organizations. While the manufacturing organization is critical to the firm's survival and success, it must be remembered that the manufacturing organization exists to contribute to the firm's overall performance. The operating strategies and structure must be in harmony with and complementary to functional strategies and structures that, in turn, must be in harmony with and complementary to the strategies and structures of the larger organization, the firm. Although people may give a great deal of attention to their heart as a functioning organ, their ultimate concerns are for their overall health and their ability to perform as a human being. Similarly, the attention given to the manufacturing organization should relate to the overall well-being of the firm.

Bibliography

Chandler, A.D., Jr. *Strategy and Structure*. Garden City, NY: Anchor Books, Doubleday & Co., 1966.

Dale, E. *Planning and Developing the Company Organization Structure*. Research Report No. 20. New York: American Management Association, 1952.

Davis, S., and Lawrence, P. *Matrix*. Reading, MA: Addison-Wesley, 1977.

Etzioni, A. *A Comparative Analysis of Complex Organizations*. Glencoe, IL: The Free Press, 1961.

Huse, E.F. *Organizational Development and Change*. St. Paul: West Publishing Co., 1980.

Leavitt, Harold J.; Dill, William R.; and Eyring, Henry B. *The Organizational World*. New York: Harcourt Brace Jovanovich, 1973.

Litterer, J.A. *Analysis of Organization*, 2nd ed. New York: John Wiley & Sons, 1973.

Lorsch, J.W., and Lawrence, P.R. *Studies in Organizational Design*. Georgetown, Ontario: Irwin-Dorsey, 1970.

March, J.G., and Simon, H.A. *Organizations*. New York: John Wiley & Sons, 1958.

Miles, R., and Snow, C. *Organizational Strategy and Process*. New York: McGraw-Hill Book Co., 1978.

Mintzberg, H. *The Structuring of Organizations*. Englewood Cliffs, NJ: Prentice-Hall, Inc., 1980.

Osborn, R.N.; Hunt, J.G.; and Jauch, L.R. *Organizational Theory: An Integrated Approach*. New York: John Wiley & Sons, 1980.

Pfiffner, John M., and Sherwood, Frank P. *Administrative Organization*. Englewood Cliffs, NJ: Prentice-Hall, Inc., 1965.

Scott, William G., and Mitchell, Terence R. *Organization Theory*. Homewood, IL: Richard D. Irwin, 1972.

References

1. Charles M. Savage, *Fifth Generation Management for Fifth Generation Technology*, SME Blue Book (Dearborn, MI: Society of Manufacturing Engineers, 1987).

MANUFACTURING LEADERSHIP

There is probably no topic in the discussion of management that has generated more discussion, more research, or more suggestions than that of leadership. However, whether the focus is on leaders of state, chief executive officers, sports team coaches and managers, or supervisors of a shop, it is found to be both necessary and convenient to summarize the results of organizational performance in terms related to the quality of the leader. In 1981, one author had utilized information from almost 3000 research efforts on the topic in preparing a survey of theory and research.[1] Interestingly, there is no concluding definition of the term. It cannot be stated that "Leadership is..." in 25 words or less and expect agreement on that definition. The subtitle of one researcher's work suggests the frustration of trying to pin down what constitutes good leadership: "Another Hypothesis Shot To Hell."[2]

There are two broad approaches to leadership that can be used to provide a sense of the meaning, if not an exact definition of the term. The first approach is a broad view, labeled transformational leadership.[3] This idea suggests that leadership is a long-view concept, a visionary definition. Such a definition would have the most effective leaders be individuals with "brilliant ideas and the capacity to inspire...others."[4] Though discounting the role of leadership, the authors of *In Search of Excellence* agree that the excellent companies had strong leaders among their other positive characteristics.[5] Leaders in this "big" definition are those individuals who provide the competence and value to transform the organization. Although interesting and useful in assisting organizations in defining their culture, goals, and top-level leadership needs, the approach does not provide a great deal on a pragmatic level.

A second approach to leadership uses a transactional view. This approach to a definition is concerned with the interpersonal influence that a leader uses in face-to-face dealings with subordinates as well as superiors, peers, and others who may have an effect on the organization. This leadership is purposeful, that is, it serves to assure the accomplishment of the goals of the organization. Leadership in this sense becomes especially critical when the leader does indeed have the opportunity to use this influence—to assure productivity, motivate others, or provide individuals with a sense of direction or with needed job support. In this chapter, the transactional view of leadership will be discussed to determine some of the appropriate steps that can be taken to become an effective leader.

The essential criterion in understanding the concept of leadership is that of effectiveness, though that measure alone may be short-sighted. It could be said that if "the job gets done," there was effective leadership; unfortunately, such a statement falls short in two regards. The process of "getting the job done" may have negative consequences elsewhere in the organization or at a later time. Secondly, more must be known about the process—the "how"—if it is hoped that the transfer of knowledge to other persons in positions of leadership will take place. Therein lies much of the thrust of what has been studied about leadership, that is, what sorts of things go into good leadership and how can these ideas be repeated.

This chapter summarizes the concepts of leadership, motivation and satisfaction, and group behavior and communication. These are such broad topics that it will be necessary to focus on each separately before attempting to tie them together. Each of these concepts has been studied and researched countless times, and therefore they can be assumed to be of great importance. However, it is also true that each concept is not easily defined, nor can prescriptions be readily prepared for use in any particular setting such as manufacturing. Fortunately, because many of the studies of these ideas were conducted in industrial settings, more valid conclusions can be made.

Research studies have limited value unless recent changes in manufacturing—technology, competition, and people—are also taken into account. An important part of this chapter will be interpretations and applications of the various research concepts in the modern manufacturing environment.

Before covering each of the topics in this chapter, it is useful to know the relationship between and among them. The manufacturing environment necessarily means that managers and their organizations are dealing with how individuals interact with each other and how the collection of individuals act as one. This set of activities is referred to as group behavior. Because the way a group acts is not simply the sum of the individual behaviors, but is a separate set of activities, it is recognized and studied as such. Essential to the way a group behaves is the process by which the members of the group interact. This is the

CHAPTER CONTENTS:

***Contributor of this chapter is: Douglass V. Koch**, Associate Professor of Management, Lawrence Institute of Technology.*

***Reviewers of this chapter are: William Devaney**, President/General Manager, Stanley-Vidmar, Inc.; **James D. Holloway**, Manager, Manufacturing Engineering, Manufacturing Div., Manufacturing Engineering Dept., Pratt & Whitney UTC; **Paul Humphrey**, Project Manufacturing Engineer, Manufacturing Engineering Dept., Pratt & Whitney UTC; **George M. Kurajian**, Professor and Chair, Department of Mechanical Engineering, University of Michigan-Dearborn; **Manfred L. Spengler**, Associate Professor, Department of Industrial Engineering and Operations Research, Virginia Polytechnic Institute and State University; **Hayward Thomas**, Consultant, Retired—President, Jensen Industries.*

CHAPTER 7

LEADERSHIP THEORIES

communication process. This process is also essential to the relationship between the leaders of the organization and individuals and groups of individuals.

Another major consideration in a discussion of manufacturing leadership is motivation and the related topic of job satisfaction. Clearly, the knowledge received or assumptions made by leaders about what motivates the employees should be crucial to the decision-making process, which is the essence of leadership. Generally, it is believed that individuals are motivated, at least indirectly, to reach some level of satisfaction for themselves.

In this chapter, several theories of leadership will first be reviewed, including trait theories, leader behavior theories, and situational theories. Secondly, some of the many concepts of individual motivation will be discussed, specifically the two major categories referred to as content theories and process theories, plus two more distinct approaches to motivation, behaviorist theory and goal-setting theory. Next will be a summary of the theories of groups and group behavior, especially group decision making. In addition, there will be a brief review of the communication process. Finally, these ideas will be related with some recent ideas on manufacturing management.

THEORIES OF LEADERSHIP

Good leadership is essential to the long-term success of any organization. An obvious cliché, yes, but also true. The real issue, however, is the "why and how" of good leadership, or perhaps what it takes or what makes a good leader. After defining what the term means and what the concept involves, this section will expand on the many theories of leadership.

Leadership is the ability to influence others toward the achievement of goals. The source of this influence may be the formal role held by the leader, such as the role of manager. However, not all leaders are managers, nor are all managers leaders. In fact, a formal position of leadership is only one of many sources of leaders' influence. The ability to use that influence goes beyond the formal organization. In addition, much of this influence depends on the perception of the followers. In fact, some would say that good leadership is a function of good (or at least willing) followers. The first ideas about leadership focused on the personality or character of the leader, while later studies indicated that there were certain things that the leaders should do, such as certain behaviors that were more effective than others. Finally, it was recognized that there are a complex set of variables that affect and modify leadership effectiveness.

TRAIT THEORY

Most people would agree that there are certain personal characteristics that can be identified with good leaders. Surely there is the idea that our political and industrial leaders possess charismatic qualities that assure their success. These qualities certainly seem to assure admiration and often support the willingness to follow. Were these qualities possessed in common by famous (or perhaps infamous) leaders in both recent and past history? Can certain characteristics be identified for use in selecting potential leaders or teaching those who wish to lead? After at least 70 years of research to come up with a list of qualities, the answer to these questions is a faint "maybe."

There have been no definitive predictors of leadership success; however, there are a few qualities that correlate very slightly with the effective leader. These characteristics are intelligence, dominance, self-confidence, a high energy level, and knowledge relevant to the task. There are no particular surprises on this short list. Because the correlations are modest, organizations will not be able to alter their selection processes, nor are they likely to change in any significant way the usual sorts of leadership training progress. If the trait approach were valid, organizations would be charged with identifying the right person for the leader job. This person would presumably have been born to be an effective leader or at least would have acquired the needed characteristics by some unknown process. It is necessary to look beyond traits for an answer.

LEADER BEHAVIOR

Given the lack of meaningful conclusions about which traits can assure good leadership, it seems logical to look at what sorts of things leaders do to be successful. In other words, it should be possible to identify the behavior that effective leaders exhibit and, from this information, make decisions about the development of leadership skills for those in the designated roles.

After identifying successful behaviors of effective leaders, organizations could then plan on making leaders out of persons with those characteristics previously noted—intelligence, self-confidence, and so on—which are related to success. The emphasis shifts from uncovering just the right people to developing some of a large supply of persons through a well-ordered management training program. Organizations would have a virtually unlimited supply of leaders if proper training was implemented.

Studies begun shortly after World War II, one each at Ohio State University and the University of Michigan, showed that leader behaviors seemed to fall in one of two broad categories. The Michigan studies labeled these as employee-oriented and production-oriented, while the Ohio State research referred to these as consideration and initiating structure. Though not identical in meaning, the two sets of labels are very similar. It has been suggested that the two categories could be used to form the axes of a managerial grid that graphically portrays the two dimensions (see Fig. 7-1). The originators of this grid, Robert Blake and Jane Mouton, use the terms *concern for people* and *concern for production.*[6]

Independently, each term has value if well defined. Together they suggest an ideal set of leader behaviors. The job-centered (Michigan) leader practices close supervision so that subordinates may know specifically what is expected of them. The initiating structure of the Ohio State study suggests that the leader organizes the work, defines the relationships between group members, and tends to establish well-defined communication patterns. Each of these task-oriented approaches leads to higher productivity for the group and high ratings from superiors. The human element is not viewed in a negative light by the production-oriented leader, but rather is seen as a luxury. Perhaps it could be said that the leader is indifferent to people in

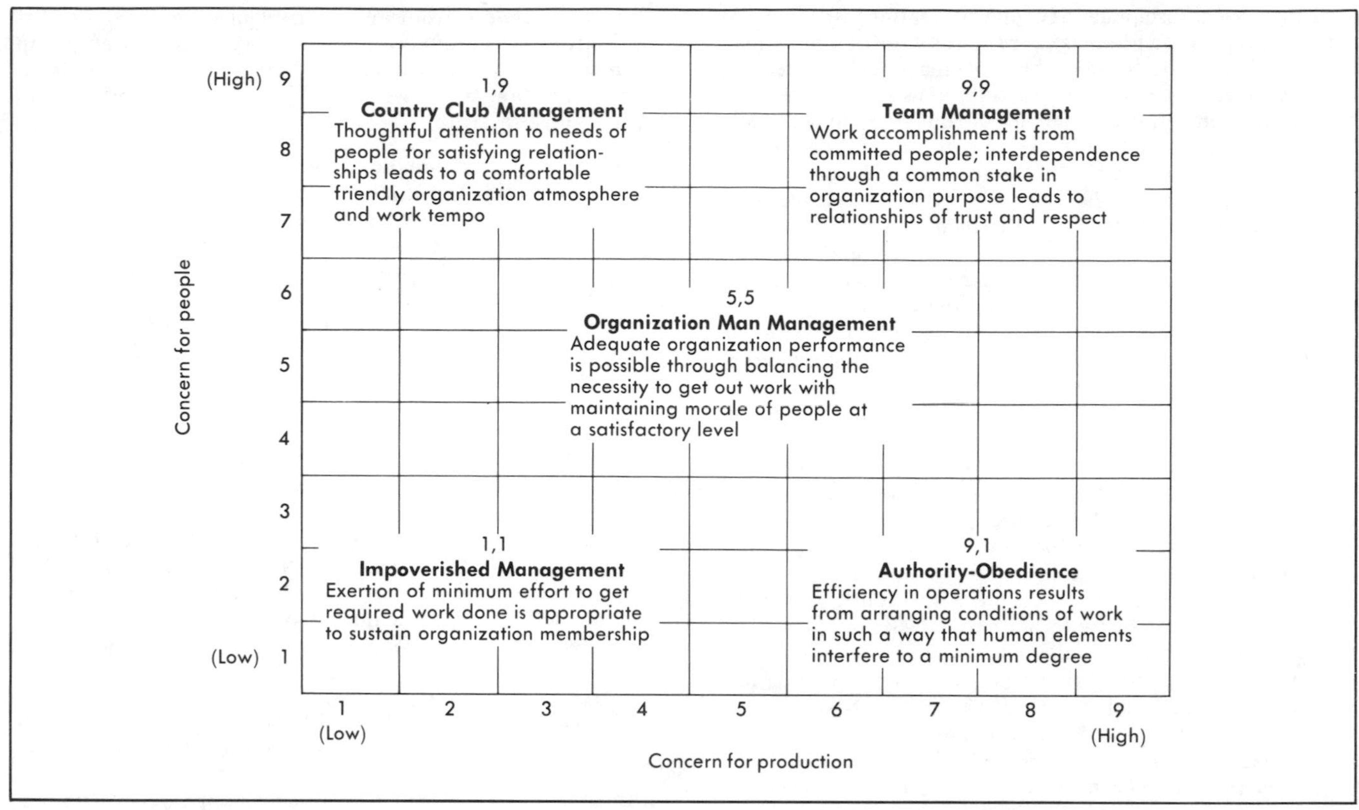

Fig. 7-1 The managerial grid.

the same sense that economists refer to an indifference curve of preferences.

The employee-centered set of behaviors includes delegating decision making, thereby creating a supportive environment that will permit subordinates to achieve and grow. Similarly, consideration means behaving so that mutual trust, respect, and rapport are developed between the leader and subordinates. These people-oriented approaches lead to more cohesive groups, greater job satisfaction, lower absenteeism, and somewhat higher group productivity. Unfortunately, the leader is sometimes seen as not being proficient by superiors. Productivity is not ignored by the leader, but rather is treated as an aspect that will take care of itself after the people concerns are addressed.

The concept of the managerial grid implies that there must be concern for both people and the task. With a dual emphasis on productivity and good relations, the leader and subordinates will form a mutually beneficial team having as its goal the well-being of each individual and the accomplishment of the group's productivity goals. This ideal model provides a framework on which to develop the organization in a broad sense, but does not specifically address what behavior will assure effective leadership. A complement to the behavioral approaches is the notion of situations. By analyzing the conditions under which the leader operates, the ability to suggest specific leader behavior is greatly enhanced.

SITUATIONAL THEORIES

The concept behind so-called situation theories is that the leader's choice of effective behavior should depend in part on the circumstances under which the leader will act. Certain variables will have an impact on a situation's favorability, other variables indicate those situations in which a more participative (less autocratic) leadership approach is more useful, and a third set of variables is used to assist the leader in determining how subordinate goals will have an effect on how the leader should behave. Each of these sets of variables is part of one of the situational theories of leadership: contingency theory, the normative or decision-making theory, and the path-goal theory.

Contingency Theory

In this set of ideas, Fiedler suggests that effective leadership will occur when there is a good match between the leader's style of interacting with subordinates and the degree to which the situation provides influence and control for the leader.[7] Fiedler essentially takes the idea of the task versus the relationship orientation that a leader might lean toward (as noted in the Michigan and Ohio State studies) and suggests that either one is appropriate given the right situation. These circumstances are defined by the following three variables:

1. The extent of the power that the leader possesses due to the role assigned by the organization.
2. The degree of structure of the task to be performed.
3. The interpersonal and psychological relationship between the leader and the subordinates.

In addition, it is important to measure which orientation the leader may have—that is, a task production concern or a person consideration point of view.

The power variable (item 1), or position power, is defined as the degree of influence that the leader possesses over such things as hiring, firing, discipline, pay, and promotions. In Fiedler's scheme, this position power is expressed as weak or strong (that is, not as a continuous variable, but as a binary one). The second variable, referred to as task structure, is used to define how routine and well defined an individual's duties might be. Again task structure is expressed in one of two ways, in this case as high or low. The third variable is simply called leader-member relations and concerns the degree of trust, confidence, and respect that subordinates have for their leader. This too is expressed as a two-sided item, good or poor. If leader-member relations are good, the job is highly structured, and the leader possesses great position power, ts said to be very favorable. If the three variables were opposite, then the situation would be very unfavorable . With other combinations of these bipolar variables, the situation falls somewhere between most favorable and least favorable. Figure 7-2 shows the eight combinations of the three binary variables and includes a favorability scale.

Given only the choices of style of task-oriented or relationship-oriented, which is appropriate in each situation? Can the leader adjust his or her style to fit the situation? In answer to the first question, it is felt that in the two extreme situations—most favorable and most unfavorable—a task-oriented leader is most effective. In the moderate situation, a relationship-oriented style of leadership is likely to be more effective. It would be difficult to envision that a leader could alter styles; therefore, a good match is obtained by altering the situation. The situation is modified by changes in the three variables, leader-member relations, task structure, and position power. Situations, however, contain many more variables than these three, and therefore, other ways have been suggested to define the situation.

Normative Theory

If leaders can be presumed to be flexible in their general style, a more useful form of situation theory is one that gives the leader a set of rules (norms) to guide his or her behavior.[8] This theory is then a decision-making theory that uses two criteria for measuring the effectiveness of a decision: quality and acceptance. *Decision quality* refers to the effect a decision may have on job performance. The quality of some decisions will have little effect on job outcome. *Acceptance* refers to the need for followers to be committed to or accepting of a decision. Some decisions can be implemented without group acceptance, while others would be unsuccessful unless the followers clearly commit to the decision.

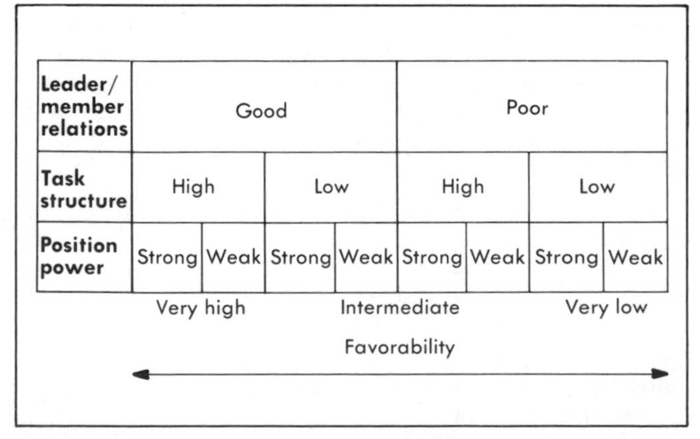

Leader/member relations	Good				Poor			
Task structure	High		Low		High		Low	
Position power	Strong	Weak	Strong	Weak	Strong	Weak	Strong	Weak

Fig. 7-2 Fiedler's contingency model, showing combinations of the three binary variables and a favorability scale.

Leaders choose from among a set of decision styles that range from a clear autocratic style to an almost entirely group decision process, with several others representing decision approaches ranging between the two ends of the autocratic-participative spectrum:

- AI—The leader makes the decision alone using whatever information is available at that time.
- AII—Subordinates provide whatever information is requested by the leader. They may or may not be aware of the nature of the problem or decision. The leader solves the problem or makes the decision.
- CI—The problem is shared by the leader with the subordinates individually. Each may provide suggestions or ideas. They do not function as a group. The leader's decision may reflect the subordinates' influence.
- CII—The problem to be solved, or decision to be made, is shared with the subordinates as a group. Ideas and suggestions are obtained, but the leader makes the decisions.
- GII—Problems are shared with the subordinates as a group. Together information, alternatives, and consequences are analyzed. An attempt is made to reach a consensus on a solution. The leader's role is primarily that of chair or facilitator of the group process. The decision made or solution chosen is one supported by the leader and the subordinates.

To select a decision style, the leader follows a set of decision rules (norms) that are simply questions with yes-no alternatives. By answering the questions in sequence, a decision tree is formed that indicates which style is most appropriate. In some cases, several styles are of equal value, and in such cases the leader selects the approach that fulfills other criteria such as the time available, the leader's preference, and the capability and willingness of the subordinates. Figure 7-3 shows the decision tree developed for this theory of leadership.[9]

Path-Goal Model

One additional theory that attempts to predict leadership effectiveness in different situations is referred to as the path-goal model.[10] The effectiveness of a leader is measured by a positive impact on the motivation of the followers, their ability to perform well, and their satisfaction. Each of these considerations has been noted in other leadership theories. In particular, the model focuses on how the leader influences the perceptions held by subordinates of their own work goals, personal goals, and the means by which they can and should reach those goals.

By considering the personal characteristics of the subordinates, especially their perception of their own abilities and experience in a given situation, and by analyzing environmental demands such as the organization's authority system, the tasks, and the nature of the work group, the leader can utilize an appropriate set of behaviors. The basis of the theory is the expectancy theory of motivation, which suggests that individuals develop perceptions about their likelihood of success and the likelihood of suitable rewards; these perceptions in turn affect their motivation.

In the path-goal model, the leader assumes a set of behaviors that tend to improve the positive expectations of the subordinates regarding their performance. At the same time, the leader works to remove environmental uncertainties surrounding their jobs. Depending on the degree of development of the subordinates, the leader attempts to clarify the path toward success and

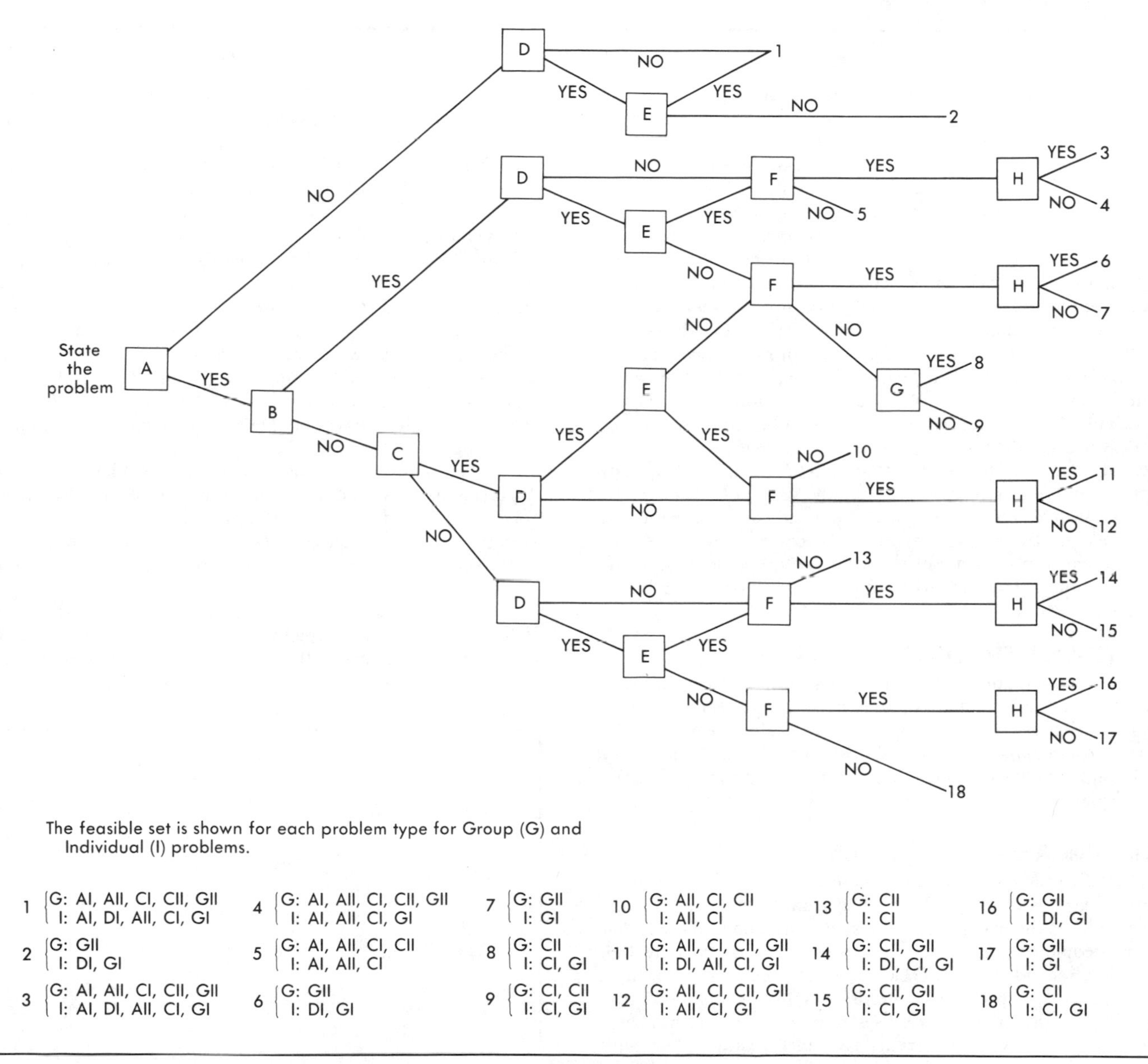

Fig.7-3 Decision-process flowchart for group problems. (*Reproduced with permission from "Decision Making as a Social Process: Normative and Descriptive Models of Leadership Behavior," by Victor H. Vroom and Arthur G. Jago published in October 1974 issue of* Decision Sciences, *p. 748.*)

positively valued rewards. The choice of leader behavior style is similar to other theories. The leader may be:

- *Directive*—The leader lets subordinates know what is expected of them.
- *Supportive*—The leader treats subordinates as equals while providing encouragement.
- *Participative*—The leader consults with subordinates and uses their suggestions and ideas.
- *Achievement-oriented*—The leader sets challenging

goals, expects subordinates to perform at a high level, and seeks continued improvement.

Though only partially developed as a model of what to do and how to do it, this last theory appears to be more comprehensive than other theories. There is also some question of which comes first, the leader style or the follower behavior, because each depends on what occurs in the other. From a management, pragmatic point-of-view, it may not be a matter of which comes first as long as there are "chickens and eggs" or an effective style and a successful performance that can be predicted and become a reality.

MOTIVATION

An essential element of the process of leadership is the determination of the forces that cause people to act in the ways they do. If one can first find out what causes action, then perhaps the leader can utilize, even harness, these forces for the benefit of the organization. It is not as if a leader can make someone behave in a certain way, but rather if the motivation is known, both the individual and the organization can benefit.

Motivation is defined simply as the force, the cause, the internal reason why people act in a certain way. Aside from automatic bodily functions, and some not-well-defined subconscious and unconscious acts, everyone chooses to behave. A person decides to eat, sleep, talk, work, or play based on some causes, including some not easily rejected ones, such as when one feels very hungry, one does eat. Motivation can also be the result of a learning experience. Certain connections are made in a person's mind between pleasant or unpleasant results and actions that one has committed. Based on these chains of actions and results, a person can choose to behave again or not.

For convenience, most of the theories of motivation are divided into two broad categories, so-called content theories and so-called process theories. After summarizing these ideas, this section will conclude with some comments on other motivation concepts.

CONTENT THEORIES

The simplest approach to an understanding of motivation is to identify the "thing," the immediate trigger for the way a person behaves. As noted earlier, hunger causes eating; in the context of manufacturing, people usually believe that, at the very least, individuals work because they wish to earn a living. Clearly, there may be hundreds of things that cause a person to act.

Maslow-Alderfer Approach

Abraham Maslow clustered these causes into five now well-known groups that he believed had a rank ordering for most people.[11] At the base level were a set of physiological needs that drive people to act. As this group of needs is generally well satisfied, individuals next seek a level of protection or security and then a group of social needs. Finally, he suggested that individuals have a need for self-esteem and a need to feel complete. Maslow's needs are arranged in a hierarchical fashion with a sense of prepotency; unless lower-order needs are essentially fulfilled, people do not seek out the next level. His needs are summarized as follows:

Highest	Self-actualization
	Self-esteem
to	Social
	Security
Lowest	Physiological

To be effective, the leader would identify the level of need of the follower and then, by providing an opportunity to fulfill that need, the leader could obtain work output from the follower. Essentially, there is, as a result, a sort of exchange process; the individual works, yielding output for the organization, and the organization, through its leaders, provides needs and satisfaction for the individual.

Although Maslow's idea seems to have a certain intuitive appeal—his groups of needs follow a certain natural order from needs relating to survival to those appearing at our highest state of humanity—there are several problems with his scheme. Great variation exists between individuals as to the magnitude and duration of these needs. From the leadership viewpoint, it would be relatively inefficient to try to provide a measure of satisfaction to each individual, if we could even be sure of the need that was operating for that individual at a point in time.

Rather than assume that the needs are hierarchical and universal, as Maslow did, another way to look at these needs is through a broader and overlapping approach suggested by Alderfer.[12] A Venn diagram (see Fig. 7-4), showing the categories of existence, relatedness, and growth (E-R-G), illustrates this idea.

Each individual has a perception of how these needs operate in terms of impact (varying the size of each circle) and in terms of their relation to each other (varying the amount of overlap). Unfortunately for the organization and its leaders, this individ-

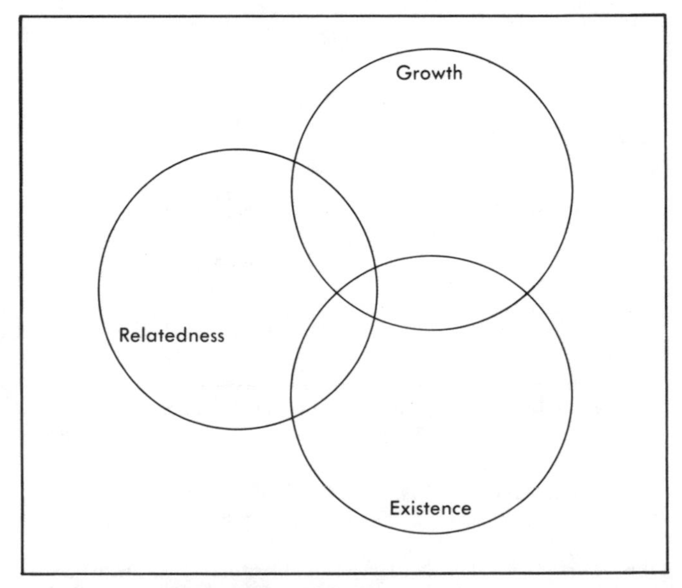

Fig. 7-4 Alderfer's approach to Maslow.

uality decreases the ability to pick a need and provide for it with any sense of effectiveness or efficiency. However, as shall be observed later, this broader approach may be useful through a change in the specifications of what the organization must do as a supplier of things that must be fulfilled.

McClelland Approach

Another content approach to motivation was developed by David McClelland.[13] He suggested a model that relates needs of leaders and followers in terms of power, achievement, and affiliation. Each individual may have a need to give or receive each of these factors to some degree. Again, by recognizing the extent of these needs in each individual, organizations can provide for need satisfaction and consequently increased productivity. The theory also suggests that more knowledge of these motivational forces in an individual being considered for a leadership role should assist the organization in doing more effective selection.

If an individual possesses a high need for power, perhaps that person would, in at least one sense, be suited for a leadership role. Of course, the individual must have a need for power and control that is not strictly a personal need, but rather an organizational one; that is, a person should not possess the need for power for primarily self-benefit (for example, the stereotypical dictator). An individual with a high need for achievement could be provided with opportunities to be successful in a job, and the individual with a high need for affiliation could be given chances to work with others in accomplishing their duties. Interesting questions arise as to the right mix of these needs in the person destined for a formal leadership position, and to the nature of these needs for those who are followers. It should be recognized that almost everyone in the organization is to some extent a follower regardless of need mix or formal position.

Herzberg Approach

The last of the content theories of motivation is that of Frederick Herzberg, who suggests in more practical ways how the leader (and the organization) can provide for fulfillment of those motivational forces.[14] He suggests a two-factor approach, with one set of factors called motivators (a straightforward label) or satisfiers and the other set called the hygiene factors (perhaps not so obvious a label) or dissatisfiers. In Herzberg's view, the organization has a normal duty to prevent dissatisfaction by providing acceptable levels of a quality work environment, reasonable supervision, adequate pay and rewards, and a host of essentially tangible job qualities. Without these, individuals become dissatisfied and less productive; however, because these things are seen as a "given," they can only cause downside behavior by their lack and hence they are potentially demotivating.

On the positive side, many intangible items provide a source of motivation. These include individual autonomy, responsibility for one's job, the work itself, and so on. The implications behind these ideas include the notion that traditional needs as usually provided by leaders will only serve to make individuals unhappy when not provided. These motivating forces, usually thought of as necessary, are also those that are easily recognized and must be weighed from a financial point of view. The less tangible items cost little, but are more difficult to understand, define, or provide. Taken to their logical conclusions, these motivators suggest that the organization should provide a certain climate and let the followers lead themselves—a, perhaps, radical idea, though not a new one.

In about 565 B.C., the Chinese philosopher Lao Tse suggested the following idea about leadership:

Leadership
A leader is best
When people barely know he exists
Not so good
When people obey and acclaim him.
Worse when they despise him.
But of a good leader
Who talks little
When his work is done
His aim fulfilled
They will say
"We did it ourselves."
Lao Tse (565 B.C.)

PROCESS THEORIES

The second major group of theories of motivation attempts to look not at what things leaders should provide to followers, but at the learning and interpreting that individuals do in the work context. By studying the process that individuals go through in coming to a decision about whether they wish to behave in a certain way, organizations can find ways to assist these individuals in behaving in productive ways. Though there are several variations of these ideas, the two most complete and practical from an applications standpoint are equity theory and expectancy theory.

Equity Theory

Equity theory looks at how individuals weigh the relationship between their input, such as effort, experience, education, and the like, and the output received, such as various rewards and returns from the organization, both tangible and intangible.[15] In addition to the ratio of output received to input given, the individual compares his or her own ratio to that of others. These others may be fellow workers in the same organization, those in other related organizations, or persons totally unrelated to the organization's immediate environment. The theory is usually expressed in the following manner:

Outcome (individual)/ Input (individual) =
Outcome (other)/ Input (other)

The thrust of the theory is that each person is motivated to maintain a balanced equation. The organization through its leader provides some of the rewards (outcome) for the individual, some sense of the required demands on the individual (input), and perhaps information used by the individual to fill in the "other" side of the equation.

In the most direct example, consider the individual who perceives that his or her outcome/ input quotient is less than another's. The theory suggests that to rebalance the equation, he or she would seek additional rewards or reduce his efforts. Either of these would tend to raise the "individual" side of the equation and therefore balance the formula. An interesting alternative in this same setting is for the individual to work harder, feeling that the organization will respond with more than sufficient reward and therefore reduce the inequity.

Contrariwise, if the individual sees that he is overrewarded or underworked, the reverse imbalance will cause an increase in efforts to "deserve" the outcome. Leaders must then provide equitable rewards, adequate definition of what is expected of the worker, and the facts in regard to the outcome/ input quotient of others. If all of these elements are provided, the individual will maintain the correct effort/motivation ratio.

Problems arise in large part with this theory because the leader cannot be sure which intangibles may exist in the individual's outcome/input ratio, or which "other" is the individual's reference, or indeed how the individual will adjust. The theory is based on individual perceptions that cannot always be well understood by the leader. Nor is it likely that efficient management would permit a distinct equity formula for each and every individual.

Most organizations would find it best to assure that each person's job is well defined and that the rewards are clearly specified. Finally, the organization must assume that the individual's reference is someone in the organization or the organization's environment—industry, competition, and so on—and that this relevant reward system is known. Of course, if the individual's reference "other" is outside the organization's sphere, then they must guess at that output/input ratio. The most essential shortcoming of the equity theory is that the perspective is the individual's, which could cause the organization to always be making estimations of the various inputs and outcomes. A more comprehensive theory of the thinking and learning process in regard to individual motivation is known as the expectancy theory.

Expectancy Theory

In this widely accepted approach (perhaps because it contains elements of other content and process theories), there are four variables of significance, and several linkages between pairs of variables are used to construct a model for the theory, which is shown in Fig. 7-5.[16,17]

Because the process is circular, or involves feedback, the individual is said to be learning whether he or she will or wishes to continue, depending on the strength of the variables and linkages. In the beginning, a person is motivated to try to accomplish something to receive a valued reward. For a given level of effort, the individual learns the likelihood (probability) of resulting success. This successful performance may or may not result in a reward. The outcome will have a certain value to the individual. If all of the variables and linkages are high enough and likely enough, then the individual will be motivated to try again (effort).

The organization has many opportunities to assist in this motivation process. Naturally, it usually provides the reward or outcomes. It sets the standard for performance or success. By providing information, tools, and training, it increases the probability of success after effort. In the administration of the reward system, it develops a strong link between success and the receiving of the reward. Only with regard to the value of given rewards to individuals does the organization have limited influence. However, the organization may establish a culture whereby certain rewards are seen to be more valuable. Overall then, the expectancy theory includes many of the concepts contained in other motivation theories, provides the mechanisms for management intervention, and allows for the interpretation of changes in either the individual's or the organization's approach to motivation.

BEHAVIORIST AND GOAL-SETTING THEORIES

Two other approaches to motivation, somewhat in contrast to each other, are the behaviorist or reinforcement theory and the goal-setting or objective-setting idea. In the latter, it is suggested that individuals with specific goals are better performers than those with no goals or with overgeneral "do-your-best" goals.[18] In the former, the leader's role is to provide the correct reinforcement at the right time for a certain (acceptable) behavior and, thereby, increase the probability that the individual will repeat the productive behavior.[19]

The behaviorist approach is not concerned with the thinking processes of an individual, but only with his or her actions. If the behavior is good, the leader rewards; if it is not, the leader is likely to not react (punishment for poor behavior is not usually imposed). This process, sometimes described as conditioning, is not actually a motivation theory, but because of its focus on behavior, it is included here.

In the goals approach, the organization, or the organization and the individual together, establishes objectives that are acceptable to both parties. Because the objectives are specific, the individual is motivated to perform so as to reach them. If the individual has the ability, tools, and data, as well as an understanding and acceptance of the goals, then the theory implies that more difficult goals yield higher levels of performance.

There are also indicators that when the individual is more actively involved in setting the goals, more effort will be put forth. Thus the participation in the establishment of the objectives is itself a motivator. If the leader unilaterally sets an easy goal, the individual is likely to accept it and be successful. As goals become harder, acceptance is less likely unless the individual helped to set this more difficult goal. The participation in the goal setting has a kind of ratchet effect on acceptance of the goal. The acceptance in turn acts as leverage on the motivation to succeed. One of the common labels for the goal-setting theory of motivation is management by objectives (MBO).

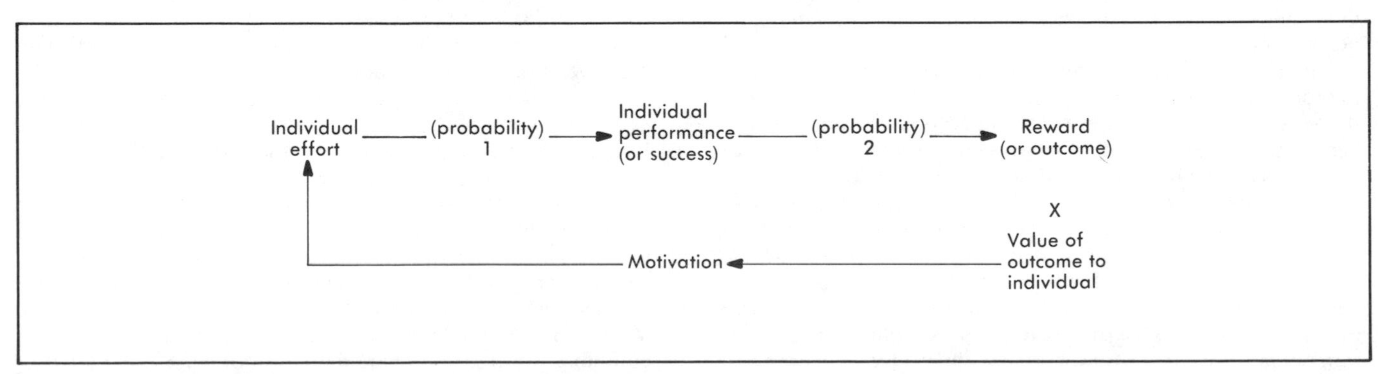

Fig. 7-5 Model of the expectancy theory.

GROUPS AND GROUP BEHAVIOR

In the manufacturing environment, as in most settings, leaders deal not only with individuals on a one-to-one basis, but with clusters of individuals in groups. How groups act is recognized not as some simple arithmetic of the addition of individual behavior, but rather as a separate noticeable behavior. Today, much is made of teamwork, committee decisions, participation, and involvement.[20,21] The value, effectiveness, and efficiency of these types of group activities can best be analyzed using a special set of concepts for groups. Leaders need to understand how membership in a group affects individuals, how groups define individual roles, how norms are established and controlled, and how decisions are made by groups.

Groups can be broadly divided into two categories, the formal and the informal. The formal group is defined by the organization when it specifies the leadership, the membership, and the expected result of the group's activities. A typical example would be that of any department or section led by a manager or supervisor with certain employees assigned certain tasks, with some particular output expected. The informal group is defined by its members and can be connected by a common bond of job duties, with the additional characteristic either being ethnic heritage, from the same neighborhood, or joined in the same social activity. The formal and informal groups may or may not be made up of the same persons. In the workplace, especially in the manufacturing environment, there is likely to be a great deal of similarity of membership between the formal and informal groups; however, the formal leader is often not a member of the informal group of employees.

This difference illustrates one important aspect of the study of groups, the need to recognize the issue of group roles and status. The organizational leader should try to come to know and understand the various roles played within the group and how status or ranking is assigned. This allows the organization's formal leader to make efforts at control that are likely to be effective. Another very significant aspect of the study of groups is that of the group's norms.

STANDARDS OF CONDUCT

Groups set standards of conduct about how one should behave in the group. The behaviors resulting from such norms may include actual work performance as well as how the members act with each other. On the negative side, group norms may call for restricting output. On the positive side, the group may call for extra effort. The choice will depend on the group's standards, but these can be influenced by how the organization and its formal leaders deal with individuals and with the group as a whole. If the group can be convinced that the organization's goals and its goals are compatible or even complementary, then everyone will benefit.

The Homan's model of group behavior, devised almost 40 years ago, can be used to analyze the behavior of groups.[22] Figure 7-6 shows a simplified version of this model.

Required behaviors are those activities, interactions between individuals, and sentiments that are defined by the organization. These are the things that the organization expects will happen when it establishes the formal structure, and through these things the organization expects a certain level of productivity. The nature of groups is such that certain other behavior will naturally begin to take place. This is the emergent behavior that takes place without the intention of the organization. These informed activities, interactions, sentiments, and a social structure are important to the level of satisfaction of the group's members. More importantly, if the group is positively reinforced by its required and emergent behavior, both productivity for the organization and satisfaction for the group member will be enhanced.

People standing around the water cooler, discussing among themselves how best to solve a work-related problem, is an example of emergent behavior. The organization does a service to all involved by not attempting to stop these behaviors and, in fact, by assisting the group in expressing this emergent behavior. In this way, the group adopts the organization's expectations for productivity while the organization accepts the group's norms for achieving satisfaction. These ideas carry over into an understanding of group decision making.

GROUP DECISION MAKING

The decision making that occurs in groups, whether the group is a formal one (committee, project team, and so on) or an informal one, is significantly different than the process used by one individual. The very process of communicating among the members of the group is likely to change not only the outcome but the sentiments of the group members about the actual decision. The principal reason usually given for having a group instead of one person (for example, "the boss") make the decision is that the additional knowledge of many should improve the quality of the decision.

However, the value of group decision making goes beyond simple volume of information; a certain synergy takes place within a well-functioning group. This concept of synergy means that the whole is greater than the sum of its parts. Each member of the group assists other members by providing ideas, which in turn trigger other thoughts, and so on. This "snowball" effect tends to increase the likelihood of a higher quality decision.

A second important and positive aspect of group decision making is that because each individual member of the group has contributed toward the solution or decision, each is likely to accept the outcome. Even if an individual disagrees with the

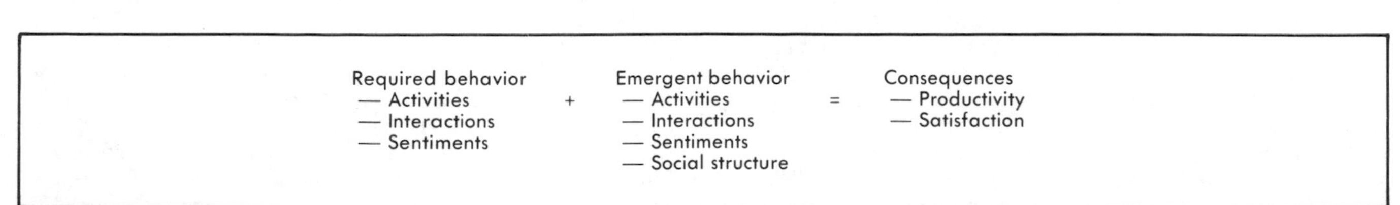

Fig. 7-6 Homan's model of human behavior.

decision, the fact that he or she was part of the decision process increases that person's acceptance of the decision. (It is presumed that there has been an equitable allowance for each person to be heard and that the final choice was made through a democratic or consensus approach.) Related to the idea of acceptance is the follow-through process or the implementation of the decision. Because acceptance is high, the group members feel an obligation to see that the decision is carried out.

This creates a second-order acceptance mode outside the group. Individuals not in the original group are affected positively by the acceptance and follow-through attitudes of the group members with whom they come in contact. This concept of "spreading the word" serves as a reinforcement to the entire organization.

There are disadvantages to group decision making, however. It should be fairly obvious that such a process involves a greater investment of the organization's resources and time than individual decision making. To assure that everyone is heard, that discussion can be held, and that choices can be explored, time and money are expended in great amounts. Therefore, group decisions, though likely to be very effective (that is, of high quality), can be low in efficiency.

However, the reader is reminded that a highly efficient individual decision made by a manager can result in a low effectiveness level. Therefore, it is necessary for a manufacturing manager to know when to effectively employ group decision making. There are situations when the only viable approach is to spend the time needed in group decision making because the organization has a need for a moderate to high quality of decision making, plus a need for a moderate to high acceptance level of that decision. Hence, the time is then well spent through a group decision making process. It is not uncommon today for organizations to successfully change their culture or work environment and improve their productivity and quality of work life through an effective group decision making process (employee involvement).

Other considerations to ponder before going ahead with an employee involvement or group decision making process is the ability and willingness of the followers, as well as the leaders to participate, to solve the problem, to design a solution, and to make a decision. Training subordinates to improve their problem solving skills and to raise their awareness level is many times the first step before actually beginning group decision making.

Another drawback of group decision making can be the tendency for some members of the group either to hold back their contributions or to conform to pressure from others to agree to a decision of lower quality. Such responses defeat the very essence of group decision making. Finally, there is some question of responsibility; because all share in the responsibility for the group's decision, perhaps no one has a responsibility for the group's decision and perhaps no one has a responsibility to actually see that the decision is carried out. This uncertainty can reduce the effectiveness of the decision-making process.

To maximize the effectiveness and efficiency of group decision making while limiting its negative aspects, the organization can take certain precautions. Whenever possible, the members of the group should be chosen to provide a contribution in terms of knowledge and expertise. The group should have a clear statement of its goals and a commitment from the organization that the decision will be implemented. Group leaders should be selected and trained to prevent dominance by anyone, including the leader, and to assist the group in maintaining progress, to provide focus and closure, and to help prevent "groupthink." This last concept is a recognized phenomenon particularly of long-standing groups.

As a result of in-group pressures, the group can exhibit some of the following symptoms:

1. Suppression of dissent; dissent is a form of disloyalty.
2. Belief that everyone agrees.
3. Feelings of superiority toward outsiders.
4. Discounting of warnings about risk in certain decisions.
5. Willingness to take excessive risks.
6. Existence of "mindguards," individuals who protect the group against negative information from inside and outside the group.

The prevention of "groupthink" generally involves being alert to its symptoms and adding a group role, probably a rotating assignment, of devil's advocate.

COMMUNICATION

It was a "failure to communicate." Understanding the problem is "halfway to a solution." "He just doesn't understand." These are familiar expressions, each concerning the communication process. At least half of all of a person's time is spent in the business of speaking and listening, writing and reading. The issue is not the fact of communication, but the way to do it well. Effective communication is what is sought. In addition, it is important to recognize that the mechanics are not so nearly as important as the understanding, and this occurs for each individual. Sometimes, therefore, the leader may be able to communicate with all of the followers, but there may be a need for a one-on-one process.

The communication process itself is simple enough in its form. Figure 7-7 is a model of the mechanics of the process.

In the organizational setting, the communicator is any employee, whether leader or follower, with an idea or information, that is, his or her message. The individual has a purpose for transmitting. Encoding is simply the means for translating the idea into some symbolic form, a language. The result of this encoding is the message. Included in the message is the purpose. The message may take verbal and nonverbal form. In addition, however, there may be unintended messages, perhaps unconscious ones sent by the communicator.

The medium is the carrier of the message whether it be written, oral, face-to-face, or in the form of memos, announcements, schedules, forecasts, evaluations, and so on. Next is the decoding, the receiving done by the receiver. This is essentially the thought process of the receiver. Decoding is the interpretation of the message and depends in part on the role of the individual in the organization. The effectiveness of the communication process can be evaluated by the degree to which the receiver obtains the message as intended by the sender. In fact,

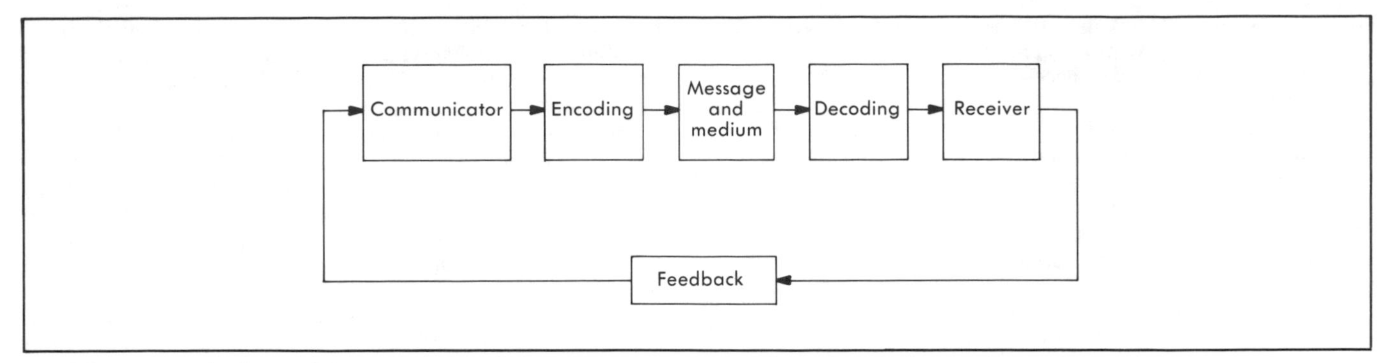

Fig. 7-7 A communication model.

the last element of the process—feedback—is the means whereby the receiver informs the sender of the reception of the message. This feedback, of course, begins another communication.

There are, however, many barriers to effective communication. These barriers take many forms and can occur during any of the stages of the communication process. Probably the most important of these barriers is the frame of reference of the sender or receiver. Each has had different experiences. Each holds a different position in the organization. Each has a different interest in the subject or outcome of the communication. If the purpose of communication is to develop a common reference for the sender and receiver on the subject of the communication, then the degree to which each individual differs initially will surely alter the effectiveness of the communication. Too often the leader and the subordinate fail to recognize in a meaningful way the other's point of view. Expectations, needs, values, and attitudes all can modify both the frame of reference and the individual's ability to perceive the other's perspective.

Other barriers to communication include selective listening, jargon and other types of specialized language, value judgments, perceived credibility of the source, filtering, and noise. Selective listening is a form of decoding and receiving in which the receiver only hears or believes that portion of the message that is acceptable to the receiver.

Jargon and other specialized language causes the same sort of difficulties that the use of a foreign language might cause. Word meanings are often totally different for sender and receiver. Source credibility and value judgments are similar barriers. In the first, the receiver has a certain level of confidence and trust in the sender, and this affects how the receiver reacts to and is willing to understand the message. In the latter, the receiver sets an overall worth to the message depending on the importance of the message, or perhaps on previous experience with that communicator or others from that department.

Finally, noise and filtering can alter the content of the message. Noise may be some sort of distraction or additional message not intended for transmission that causes the actual message to be changed. Filtering occurs most often as messages pass through the organization and usually takes place as information goes upward in the organization as successive receivers/senders change the message to suit their own particular agenda; for example, middle levels of management often avoid passing bad news upward to higher levels.

Two additional barriers to effective communication are those having to do with time. Time pressures may cause individuals to send too little or may cause receivers to listen too little. Consequently, the entire message is not transmitted or received. Secondly, communication overload occurs when the receiver is obtaining more information than can be processed.

How can the communication process be improved? Leaders (and, indeed, everyone in the organization) can improve the process by considering the points in the model where a breakdown might occur. These occur as noted above either in the process itself or in the two parties to the process. By using a common vocabulary and simplifying the language, the sender reduces some problems. By repeating the message, perhaps using different forms of the message or different media, by timing the release of the information, and by reducing the amount of information, better transmission and reception is likely. In addition, it is important for follow-up communication to determine if the message was received. This is related to assuring that feedback takes place, which indicates that each sender must be a good receiver of this feedback. Both sides of the process must learn to listen. On the psychological level, mutual trust and empathy must be developed.

CONCLUSIONS

Linking these many ideas has always been a difficult thing to do. Perhaps that is what leadership is about. Complicating these attempts are recent data that suggests that the best way to lead is with the methods used by Japanese management. Such statements have some validity to them, but are no means a certainty.

Theory Z, a term coined by William Ouchi, an American researcher, is used to represent the U.S. version of a style of management that appears often in other countries as well. Thomas J. Peters and Robert Waterman, Jr., point to certain variables that seem to characterize effective management and which coincidentally occur throughout the world.[25] Some spe-

CONCLUSIONS

cific applications at the shop floor level, supervisory level, and operating level are just now beginning to appear with some frequency. It is not always necessary to look elsewhere, but rather to look around.

In the context of the subject of leadership, there are three essential components: the leader, the follower, and the situation. These should be considered in broader ways than proposed earlier. The leader must consider personal preference, training, and style. In addition, the situation is defined by the organization's structure, its internal communication system, its goals—in the short and long run, its adaptability and willingness to change, and its environment. The most significant change from past years is the nature of the followers. They are probably more technically able, very likely to be more demanding in terms of rewards, and surely have greater expectations about what it is that they might contribute.

Effective leadership includes the recognition of the state of these variables and a willingness to adapt to their differences. If it is true that any production worker can stop the line for good reason, then leaders must plan for and accept such an event. The tendency of current research is to infer that more involvement by subordinates is likely, expected, and should be capitalized on. Such concepts as quality circles and employee involvement are increasingly common.

Such attempts to increase participation create a bias toward less autocratic behavior on the part of the leader, more information from subordinates, greater acceptance of the decisions, increased motivation for the subordinates, greater reliance on group effort, and overall a more complete commitment by leader and follower. Of course, if any of the parties involved will not or cannot behave in these ways, then effectiveness is diminished. Determining whether or not the ability and willingness are present requires that leaders listen for the signals. Peters and Waterman note that several successful organizations encourage their leaders to get out of their offices and learn to "manage by wandering around." In this way, the leader obtains the needed information to determine what the situation is and what the followers are doing or will do. With that information, the leader can decide to lead actively in the traditional way or to lead in a facilitating manner.

References

1. Barnard M. Bass, *Stogdill's Handbook of Leadership* (New York: The Free Press, 1981).
2. Fred E. Fiedler, "Leadership Experience and Leader Performance—Another Hypothesis Shot to Hell," *Organizational Behavior and Human Performance* (January 1970), pp. 1-14.
3. James MacGregor Burns, *Leadership* (New York: Harper & Row Publishers, Inc., 1978).
4. Abraham Zaleznik, *Human Dilemma of Leadership* (New York: Harper & Row Publishers, Inc., 1966).
5. Thomas J. Peters and Robert H. Waterman, Jr., *In Search of Excellence: Lessons from America's Best Run Companies* (New York: Harper & Row Publishers, Inc., 1982), p. 26.
6. Robert R. Blake and Jane S. Mouton, *The Managerial Grid III* (Houston: Gulf Publishing, 1985).
7. Fred E. Fiedler and Martin M. Chemers, *Leadership and Effective Management* (Glenview, IL: Scott, Foresman & Co., 1974).
8. Victor H. Vroom, "A Look at Managerial Decision Making," *Organizational Dynamics* (Spring 1973).
9. Victor H. Vroom and Arthur G. Jago, "Decision Making as a Social Process: Normative and Descriptive Models of Leadership Behavior," *Decision Sciences* (October 1974) p. 748.
10. Robert J. House, and Terence R. Mitchell, "Path-Goal Theory of Leadership," *Journal of Contemporary Business* (Autumn 1974), pp. 81-98.
11. Abraham H. Maslow, *Motivation and Personality* (New York: Harper & Row Publishers, Inc., 1954).
12. Clayton P. Alderfer, *Existence, Relatedness, and Growth: Human Needs in Organizational Settings* (New York: The Free Press, 1972).
13. D.C. McClelland and D.H. Burnham, "Power is the Great Motivator," *Harvard Business Review* (March/April 1976), pp. 100-110.
14. F. Herzberg, B. Mausner, and B. Snyderman, *The Motivation to Work* (New York: John Wiley and Sons, 1959).
15. J. Stacy Adams, "Inequity in Social Exchanges," *Advances in Experimental Social Psychology* (Orlando, FL: Academic Press, Inc., 1965).
16. Victor H. Vroom, *Work and Motivation* (New York: John Wiley and Sons, 1964).
17. L.W. Porter and E.E. Lawler, *Managerial Attitudes and Performance* (Homewood, IL: Richard D. Irwin, Inc. 1968).
18. E.A. Locke and G.P. Latham, *Goal Setting: A Motivational Technique That Works* (Englewood Cliffs, NJ: Prentice-Hall, Inc., 1984).
19. Fred Luthans and Robert Kreitner, *Organizational Behavior Modifications and Beyond*, 2nd ed. (Glenview, IL: Scott, Foresman & Co., 1984).
20. R.J. Pascale and A.G. Athos, "The Art of Japanese Management," *Academy of Management Review* (April 1984).
21. T.R. Miller, "The Japanese Management Theory Jungle," *Academy of Management Review* (April 1984).
22. George C. Homans, *The Human Group* (San Diego, CA: Harcourt Brace Jovanovich, Inc., 1950).
23. William G. Ouchi, *Theory Z: How American Business Can Meet the Japanese Challenge* (Reading, MA: Addison-Wesley Publishing Co., Inc., 1981).
24. Peters and Waterman, Jr., *op. cit.*

Bibliography

Blake, Robert R., and Mouton, Jane S. *The Versatile Manager: A Grid Profile*. Homewood, IL: Richard D. Irwin, Inc., 1982.

Burns, James MacGregor. *Leadership*. New York: Harper & Row Publishers, Inc., 1978.

Fiedler, Fred E. *A Theory of Leadership Effectiveness*. New York: McGraw-Hill Book Co., 1967.

Hampton, David R. *Management*, 3rd ed. New York: McGraw-Hill Book Co., 1986.

Herzberg, Frederick. "One More Time; How Do You Motivate Employees?" *Harvard Business Review* (Jan.-Feb. 1968).

Ivancevich, John M., and Matteson, Michael T. *Organizational Behavior and Management*. Plano, TX: Business Publications, Inc., 1987.

McGregor, Douglas. *The Human Side of Enterprise*. New York: McGraw-Hill Book Co., 1960.

Naisbitt, John. *Megatrends*. New York: Warner Books, Inc, 1984.

Robbins, Stephen P. *Essentials of Organizational Behavior*. Englewood Cliffs, NJ: Prentice-Hall, Inc., 1984.

Stoner, James A.F., and Wankel, Charles. *Management*, 3rd ed. Englewood Cliffs, NJ: Prentice-Hall, 1986.

Vroom, Victor H., and Yetton, Phillip W. *Leadership and Decision-Making*. Pittsburgh: University of Pittsburgh Press, 1973.

MANAGEMENT OF TECHNOLOGY

Management is increasingly aware of the opportunities that new and emerging manufacturing technology offers to their business. These opportunities range from increased flexibility to serve marketplace needs to entire new business strategies that protect and promote market share growth in an increasingly competitive international marketplace. However, the very complexity and rate of introduction of technology that makes these opportunities possible has created a major problem for most business managers. Technology jargon and a confusing array of overlapping vendor product claims has fostered management inaction in businesses where new manufacturing approaches are essential to survival. Experience indicates that only those companies with strong management and direct management involvement will succeed in applying today's flexible manufacturing technologies for the strategic benefit of the overall business. For specific information on strategic planning, the reader should refer to Chapter 1, "Planning," of this volume.

The following sections describe a comprehensive approach to the management of technology in manufacturing. In practice, the complexities of specifying, designing, and implementing individual facilities and processes will vary widely and often overlap in time as multiple technology planning, selection, and application projects are started and terminated. The guidelines presented in this chapter are intended to help management sort through these complexities and then tailor a successful, economically viable approach that is appropriate to a particular business environment.

CHAPTER CONTENTS:

CONCEPTS

The methods described in this chapter can help management evaluate the bewildering array of technology options available today and to define and execute a strategy appropriate to each manufacturing unit. These methods are based on applying three concepts: (1) a technology management lifecycle, (2) a technology selection model, and (3) a management action model. These concepts are interdependent on each other but independent of specific manufacturing needs or technologies and should be applicable in a wide variety of business situations.

TECHNOLOGY MANAGEMENT LIFECYCLE

Technology management involves three distinct phases that must be repeated over time as both technology and the business situation change: technology planning, technology selection, and technology application (see Fig. 8-1). These phases may be applied at the corporate level to all manufacturing units, if sufficient product and process commonality can be identified, or at the manufacturing unit level for specific applications.

The first step of the lifecycle is technology planning. This planning starts with both a business strategy that depends on manufacturing contributions, and one or more established manufacturing (AS IS) environments that will be the focus of the plan. The technology planning process, described in the next section, identifies new technology opportunities to meet strategic objectives, assesses the benefits that would result from realizing the opportunities, determines the feasibility (cost, time) of accomplishing the opportunities within the AS IS manufacturing environment, and produces a time-phased action plan and associated requirements for selecting technology approaches. This step results in a common understanding by all management (manufacturing, marketing, finance, and so on) of both the short-range (1-4 years) and long-range (5-10 years) manufacturing technology program and the impact of the programs on the business.

The second step of the lifecycle is technology selection. Guided by business-based requirements and planned resource commitments over time, relevant current and emerging technology is reviewed to select the best approaches to achieve planned opportunities. Compatible technology approaches are incorporated into a standard manufacturing system architecture or platform that becomes the basis for software and equipment purchase decisions and implementation constraints. New hardware and software technology beyond that which already exists in the AS IS environment is specified, and backup plans are formulated if the required technology is unavailable, too costly, or too risky. A communication and integration approach is specified to ensure

Contributors of this chapter are: ***Christopher S. Fuselier****, Manager—CIM Applications, GE Fanuc North America, Inc., Factory Automation Systems;* ***Gerald F. Roberts****, Project Manager, GE Fanuc North America, Inc., Factory Automation Systems.*

Reviewers of this chapter are: ***Joel D. Goldhar****, Dean—School of Business, Illinois Institute of Technology;* ***Dundar F. Kocaoglu, Ph.D.****, Director—Engineering Management Program, School of Engineering, University of Pittsburgh;* ***Robert L. McMahon****, Manager—Manufacturing Systems, Fort Worth Div., General Dynamics;* ***Ronald S. Petersen****, Director of Manufacturing, Astronautics Div., Lockheed Missiles & Space Co.;* ***Hayward Thomas****, Consultant, Retired—President, Jensen Industries;* ***Peter C. Van Hull****, Director of North American Automotive Consulting Practice, Arthur Andersen & Co.*

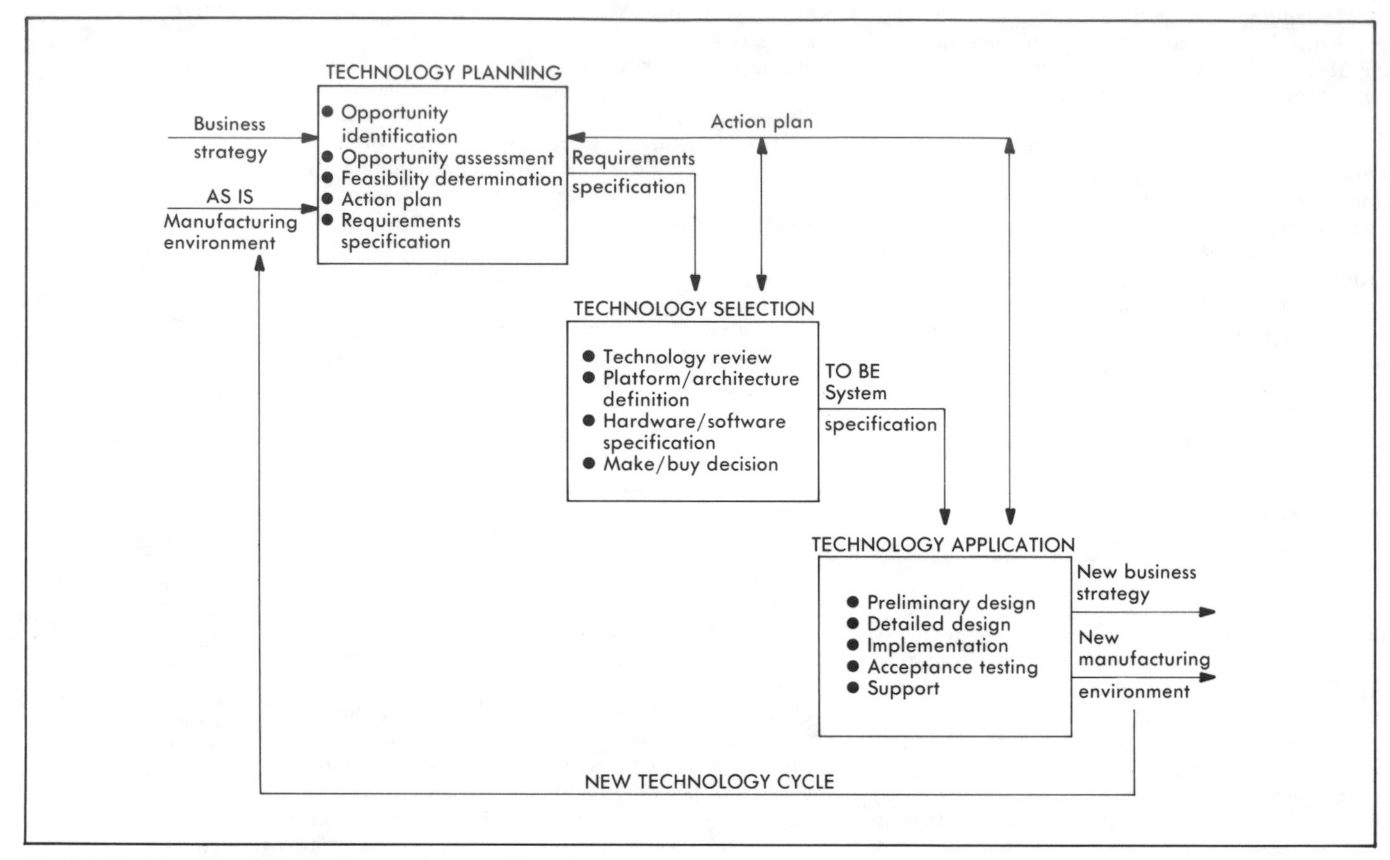

Fig. 8-1 Diagram depicting technology management lifecycle.

technology elements interconnect and can be replaced with improved technologies in the future without disrupting the overall manufacturing system. Interfaces with engineering, manufacturing, finance, and support functions are also specified to fit or eliminate AS IS manufacturing environment constraints. A system specification is developed to guide implementation and training, make and buy decisions are made, procurement specifications are developed, and the action plan is updated. This step results in general system technology guidelines and system specifications to achieve the desired (TO BE) manufacturing environment.

The final step of the lifecycle is technology application. Technology application consists of preliminary design, detail design, build, and test activities. Guided by a detailed action plan and the system specification, new technology is applied to the (AS IS) manufacturing environment. This application is defined by a preliminary design process that establishes specific subsystem and user interface requirements and by a detailed design process that identifies how each subsystem will be built or procured. The subsystems are then built and interfaced with other subsystems procured from external vendors. A series of tests is conducted to ensure that subsystem, system, and business requirements are met. In parallel, support activities such as user training are conducted. Finally, a maintenance program is established to ensure continued operation of the new or modified manufacturing technology. A key to success, where technologies are procured, is to establish a "partnership" relationship with the supplier with clearly defined objectives and incentives for success. The result of this step is a new manufacturing environment and often, over time, a new business strategy as manufacturing flexibility increases and costs decline.

As technology evolves, as the business environment changes, or as the scope of technology application increases, the entire three-phase lifecycle will be repeated. In any case, the technology plan and technology selection guidelines should be reviewed on a yearly basis to ensure that new opportunities are recognized and acted on.

TECHNOLOGY SELECTION MODEL

The large number of technology alternatives for a business may be reduced by categorizing the technologies according to the desired manufacturing automation strategy. By determining where the AS IS manufacturing environment fits within this categorization, and by choosing a future direction for manufacturing automation, technology considerations can be integrated with the business strategy rather than narrowly focused on short-term cost improvements.

The model in Fig. 8-2 divides manufacturing technologies along three primary dimensions (process mechanization, materials flow and control, and information management and control) and three essential interfaces (engineering, business, and support functions such as quality and maintenance). Each dimension is subdivided by the automation strategy appropriate to the business (manual, semiautomatic, and automatic), with each automation strategy or interface employing a range of technology options to achieve a planned result. The appropriate technology strategy results from a planned, phased application of manufacturing technology over time.

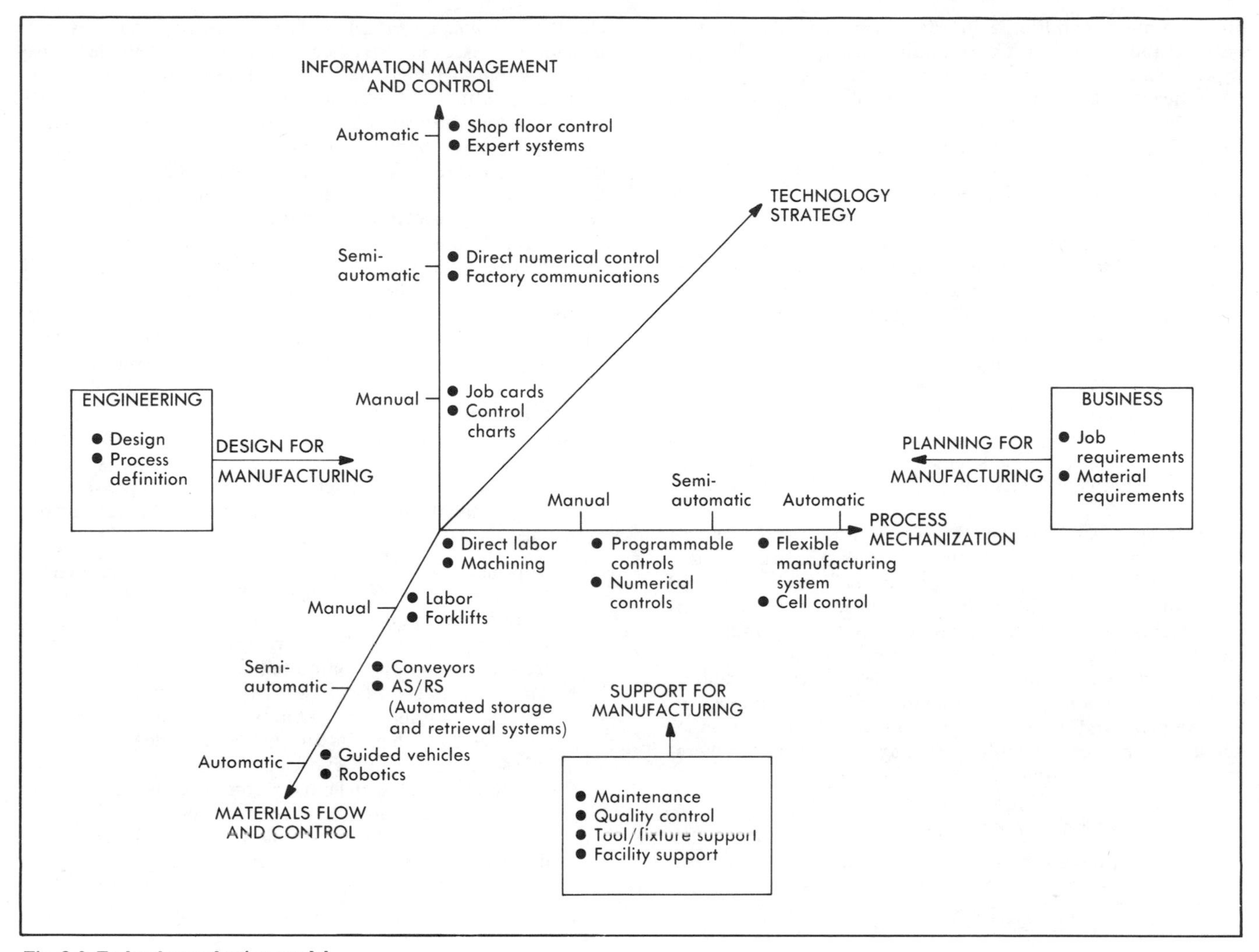

Fig. 8-2 Technology selection model.

The primary dimensions of the technology selection model correspond to typical organizational and vendor divisions of technology responsibility. The process mechanization dimension includes the personnel, machine, process, tool, fixture, sensor, and control technologies used to formulate, fabricate, and assemble a manufactured product. The materials flow and control dimension includes the handling, storage, transfer, and tracking technologies used to supply processes, hold inventory, or remove waste. The information management and control dimension includes the planning, scheduling, supervising, monitoring, coordinating, analyzing, and reporting technologies used to control processes, direct material flow, report status, and analyze manufacturing performance.

Most businesses tend to focus on a single dimension when making technology decisions because of tradition, organizational charters, or outside vendor limitations. However, to have a quantum impact on the business, all dimensions must be considered together and an appropriate balance struck to avoid isolated, irreconcilable "islands of automation" approaches that miss significant business opportunities. The technology management lifecycle avoids this problem by depending on top-level management champions to reduce organizational biases and on multifunctional teams to ensure all technology interests are considered.

The essential interfaces of the technology selection model correspond to functional organizations outside manufacturing that have a critical impact on the manufacturing process. The engineering interface is the source of product and process definition. By designing products for flexible manufacturing (common parts, minimal tolerance stackup, easily positioned parts, use of existing tools/fixtures) and by implementing compatible engineering information interfaces (geometry, dimensions, tolerances, materials, and components), product release times and manufacturing costs/risks can be reduced. The business interface provides job and material requirements. By implementing on-line manufacturing interfaces to business data, a tool is provided to assist management in product cycle time and inventory level reduction. The support interface is the source of maintenance, quality control, tool/fixture management, and facility support. By implementing on-line interfaces to support functions, process uptime can be improved while rework and indirect costs are reduced. Although three interfaces are described, additional interfaces should be considered if essential to manufacturing technology selection and application. Examples of additional interfaces are suppliers, corporate systems, and customers. The technology management lifecycle approach addresses interfaces to manufacturing technology as a significant business opportunity by involving representatives of external functional organizations in the technology planning, selection, and application phases.

The automation categories of the technology selection model correspond to major technology application strategies for each of the manufacturing dimensions. Although the strategies form a continuum, three classifications simplify the use of the model. Manual technology approaches (such as manual assembly and stand-alone machine tools) rely primarily on personnel to perform the process, move materials, and prepare and record information. Semiautomated technology approaches replace well-defined, repetitive manual procedures with automated controls, sensors, and communications. Fully automated technology approaches replace supervisory and more complex, infrequent manual procedures with adaptive controls, actuators, feedback loops, and paperless information transfer. Examples of technology for each classification are shown in Fig. 8-2.

Most businesses begin with a manual base and focus on increasing process mechanization technology. Although this strategy will provide paybacks, it may become a barrier to achieving major overall benefits to the business. Multidimensional technology strategy can significantly reduce manufacturing costs and bottlenecks and improve manufacturing flexibility. The technology management lifecycle begins with a calibration of the technology and automation starting point (AS IS environment) and, based on opportunity identification, maps out a time-phased technology strategy that provides incremental financial returns.

MANAGEMENT ACTION MODEL

Figure 8-3 indicates the primary areas of management responsibility throughout the technology management lifecycle. *Payoff management* selects technology opportunities that provide strategic business benefits within acceptable costs. *Planning management* defines manufacturing technology requirements that are feasible within the manufacturing environment and the available time. *Process management* identifies, applies, and supports technology approaches to implement defined requirements. *People management* builds a consensus in the organization on the technology strategy, trains affected personnel to work with the new technologies, and communicates the status and impacts of technology in a timely and open fashion. Management must act on all four areas of responsibility at each phase of the technology management lifecycle for each selection from the technology selection model to ensure that proper tradeoffs are made and actions taken to meet previously agreed on manufacturing commitments to the business.

GUIDELINES

The following sections on technology planning, selection, and application illustrate how to apply the technology selection model and the management action model for each phase of the technology management lifecycle. However, some general guidelines include the following:

- Top management should champion an ongoing manufacturing technology lifecycle process involving all manufacturing operations. Planning must start from the top with a sound business strategy and be implemented from the bottom by committed personnel.
- Management should start with applications of new technology with limited scope but with a broad enough base to demonstrate benefits attainable, provide positive visible success, and gain experience and confidence. These initial successes will help to convince the rest of the organization to adopt new technologies.
- Over time, management should begin establishing standard technology "platforms" for all manufacturing operations. The tendency is to want to lift a successful applica-

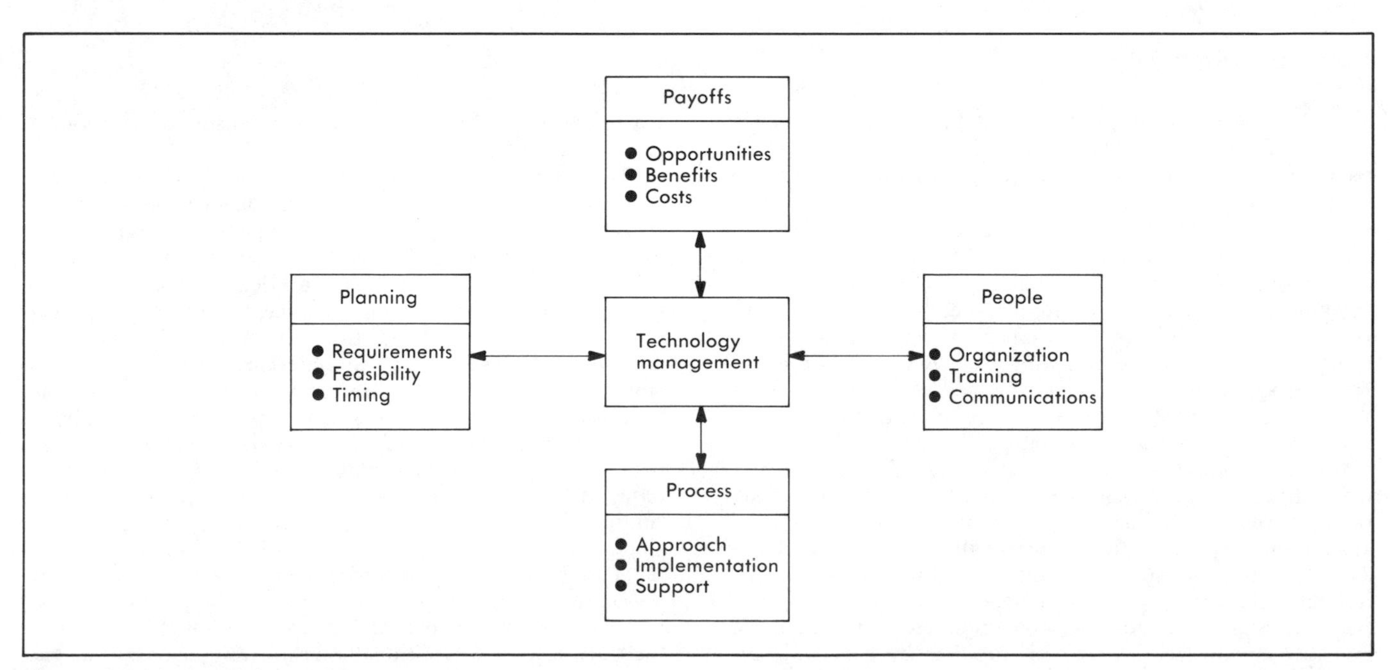

Fig. 8-3 Management action model.

tion in its entirety and try to transplant it as is. This rarely succeeds unless the environments are identical (for example, people skills, process facilities). What is possible is to select fundamental elements of the technology and assemble them into platforms for other applications. These standard selections will reduce both costs and risks if the platforms are sufficiently flexible to handle changing business needs.

TECHNOLOGY PLANNING

The objectives of technology planning are to select manufacturing technology opportunities that will support the business goals, objectives, and needs to fulfill voids in the AS IS manufacturing environment; specify those opportunities to be implemented; establish a time-phased cost/benefit plan for execution; and build an organizational consensus in support of the plan.

Effective technology planning demands a champion, backed by top management, to direct the planning process, support and promote evolving planning recommendations, and defend the plan against advocates of the status quo. This champion must also understand overall business goals and the role that manufacturing is to play in meeting these goals.

Technology planning also requires a team of the highest qualified, motivated, and committed personnel available throughout the planning process—personnel with credibility within both manufacturing and management who can effectively "sell" their recommendations. The length of the initial planning period varies from 2 months to 8 or more months depending on: product complexity, market predictability and required response time, organization complexity, scope of manufacturing plan (single product/process, entire plant, multiple plants, or all manufacturing units), manufacturing constraints (add technology to existing operation environment versus "green-field" startup), degree of integration (suppliers, vendors, other plants), and the "newness" of proposed technologies. After the planning process is completed, periodic updates will be required to maintain the plan. Care must be taken not to make the planning period too long. It is a very intense, fast-moving period, with the best skills in the organization involved. Extending the process beyond the 8 months will result in loss of team and organizational interest and deterioration of the program.

Because it is important to build commitment and interest quickly, the planning process is broken into opportunity identification and opportunity assessment activities (see Fig. 8-4). A front-end opportunity identification activity, which can be completed within 1 to 2 months, ensures that business goals and associated manufacturing objectives are clarified and documented. The objectives are then transformed into technology opportunities that provide the means for achieving the objec-

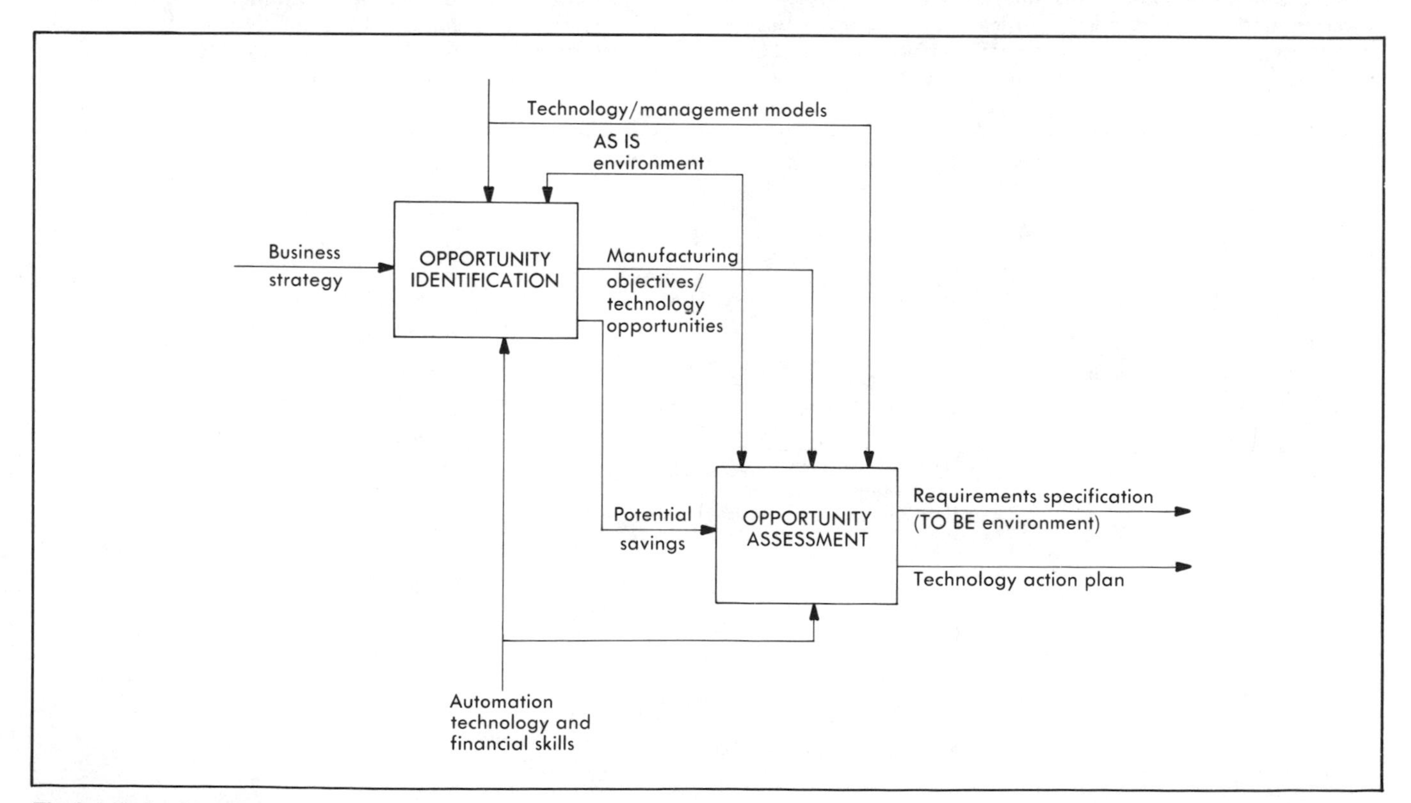

Fig. 8-4 Technology planning activities model.

tives within the constraints of the AS IS environment. Finally, the technology opportunities are prioritized by business payoff to determine short-term and longer range improvements. Short-term improvements should be evaluated quickly for overall compatibility with the plan and be implemented immediately to begin reaping benefits.

A 2 month to 8 month or more opportunity assessment activity includes a detailed technology feasibility analysis to determine benefits costs, time requirements, and risk factors; requirements specification preparation for feasible technologies; and action plan preparation including tasks, milestones, measures, resource requirements, and organizational requirements necessary to implement the plan.

The technology selection model is used to establish the technology starting point of the opportunity identification activity and to identify the technology direction of the action plan. The management action model provides a checklist for the four critical areas that must be addressed during opportunity assessment to ensure both complete requirements and a feasible action plan.

In summary, technology planning provides a manufacturing organization with the ability to dismantle organizational barriers by developing a consensus-based, high-payback, phased implementation program focused on common objectives. The major objective is to convert materials into quality products that meet changes in customer demands in minimum time with optimal utilization of resources.

OPPORTUNITY IDENTIFICATION

Opportunity identification is performed by a multifunctional team representing the three manufacturing dimensions and the critical manufacturing interfaces of the technology selection model. Opportunity identification begins with a business strategy and a manufacturing environment (current and planned) focus such as a single product line/process, a plant, related plants, or all manufacturing operations. These elements are defined initially if not available. Automation engineering skills are required to identify opportunities for process modernization/retrofit, automated material handling, and revised factory layout. Technology skills are required to identify opportunities for computer applications, factory communications, and automated control. Functional skills are required to identify opportunities for product redesign, process redesign, business system integration, and on-line support services. Financial skills are required to identify direct/indirect benefits and associated costs. A program manager or management-designated facilitator is required to organize and guide the identification process and to focus and balance results against both short-term and longer range objectives. Personnel from other plants, corporate services, vendors, and outside consultants can be invaluable as participants or facilitators to provide new ideas and insights.

A recommended approach is to divide the team into working groups of no larger than 10 individuals with an assigned facilitator. The program manager/facilitator can use a structured brainstorming technique such as "storyboarding," which is discussed subsequently, with selected data gathering to perform the following tasks (see Fig. 8-5):

- Review the business strategy and associated financial targets to establish objectives for the manufacturing plan. Selected management interviews across critical business functions can provide this information if a formal plan is unavailable. Storyboarding transforms the information into a set of business goals and manufacturing objectives that the team agrees on.
- Review the current manufacturing environment and committed plans to define constraints and to establish the AS IS starting point of the technology plan. Plant walk-throughs and selected functional interviews provide critical information. Storyboarding or a more formal

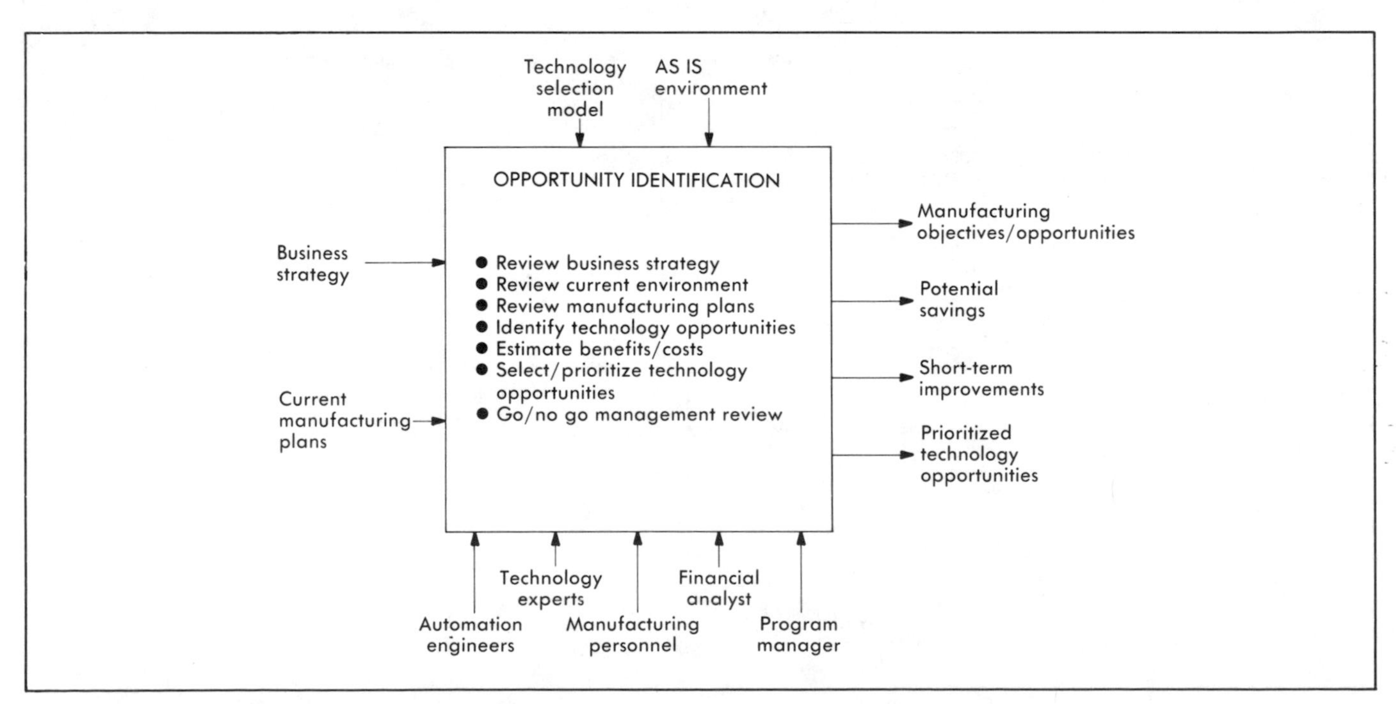

Fig. 8-5 Tasks involved in identifying technology opportunities.

modeling process can be used to build a consensus on this information.

- Define technology opportunities along each dimension and for each interface using the technology selection model as a catalyst. Industry expertise is critical to this process, and outside corporate, consultant, or vendor experts should be involved by interviews or by team participation. Industry technology baselines are also useful if available.
- Group related opportunities together; list and estimate benefits and costs for each opportunity. If this is not possible, categorize benefits and costs (high, medium, or low). The financial analyst plays a critical role in this step and helps to smooth future financial and management approval.
- Select opportunities to be pursued and categorize them as short-term improvements that can be acted on immediately or as opportunities that require further assessment and planning.

The final task in the opportunity identification process is to review results with the management champion and other key decision makers. If storyboarding was used, the storyboard "war room" provides an excellent place for progress reviews. Additional information on storyboarding can be found in the section entitled "Management Planning Tools." An example of an opportunity storyboard is shown in Fig. 8-6.

The opportunity identification process can often be completed within several weeks and results in the following:

- Planning storyboards (or other forms) documenting business goals, manufacturing objectives, and technology opportunities.
- Preliminary budgetary estimates of potential dollar benefits.
- Identification of highly visible short-term improvements where immediate positive actions can result in quick paybacks.
- A prioritized list of longer range, more complex opportunities to be reviewed in greater detail in the opportunity assessment activity.

The net result from this process is a common understanding throughout the organization and supplier community of business goals and manufacturing objectives; a set of opportunities with potential savings to focus the planning process; short-term improvements that will provide immediate payback; a management planning focus within the business that cuts through the conventional organization structure; and management confidence that new manufacturing technology will have tangible business payoff.

BUSINESS GOALS	MANUFACTURING OBJECTIVES	TECHNOLOGY OPPORTUNITIES	POTENTIAL SAVINGS	SHORT-TERM IMPROVEMENTS	PRIORITIZED OPPORTUNITIES
Shorten product introduction cycle	Decrease product startup time	Automated process planning	50% reduction in product introduction/ change time	Apply manual control charting	Automate process planning
Reduce product cost	Reduce product flow time	Design for manufacturability	80% reduction in product flow time	Apply design for manufacturing guide lines on new products	Automate material transfer
Improve customer image	Increase equipment utilization	Automate material transfer	20% increase in equipment utilization	Automate maintenance management	Apply in-line inspection
	Increase product mix flexibility	Apply in-line inspection	30% reduction in economic batch size	Download manufacturing instructions	Automate tool changing
	Increase inventory turns	Download manufacturing instructions	400% increase in inventory turns		Use universal fixtures
	Reduce labor costs	Automate tool changing	30% decrease in indirect labor costs		
	Increase response to supported plants	Use universal fixtures	80% reduction in scrap/ rework		
	Increase quality	Use just-in-time scheduling			
		Automate maintenance			
		Apply control charting			
		Automate feedback to manufacturing resource planning			

Fig. 8-6 Opportunity identification storyboard.

CHAPTER 8

TECHNOLOGY PLANNING

OPPORTUNITY ASSESSMENT

Opportunity assessment begins with the prioritized technology opportunities and is guided by the management action model, management objectives, and potential savings. Assessment is performed by a team with the same skills used during opportunity identification, but more effort and time is required to produce the detailed requirements and financial assessments necessary for a successful, multiyear technology plan. Small working groups guided by a facilitator using storyboarding or a similar technique is the primary method used to gather, process, and organize information into a consensus plan of action. However, other tools such as simulation, spreadsheet, and financial models are used to formalize assessment results for both the technology requirements specification and the action plan.

The objective of the selected assessment team is to systematically transform the prioritized technology opportunities into a series of justified manufacturing TO BE scenarios that are feasible relative to the payoff, planning, people, and process considerations of the management action model. In this process, the impact of technology on the AS IS manufacturing environment is reduced to the specific changes to be made, planned timing of the changes, budgetary costs of the changes, and the benefits that will result. The resultant plan is organized to use the savings of the initial technology implementation phases to help fund future phases.

The assessment team produces the requirements specification and technology action plan by performing the following tasks (see Fig. 8-7):

- Define the AS IS manufacturing environment in detail for those areas impacted by priority opportunities. The detailed AS IS model should indicate current facility layout, functions currently performed, processes and methods used, assigned resources, and critical interfaces. The information definition (IDEF) modeling technique is useful for capturing AS IS functions and approaches. This technique was developed during the U.S. Air Force integrated computer-aided manufacturing (ICAM) program to formally communicate manufacturing functions and information flow ($IDEF_0$) and information structure ($IDEF_1$) relationships. Figure 8-8 provides the basics of this nomenclature, and Fig. 8-9 provides an example. Although somewhat laborious, clearly understanding the AS IS environment is essential to determine the feasibility and benefits of technology change.
- Define a scenario of the desired TO BE environment that could evolve out of the existing AS IS environment for each opportunity. The scenario should define new functions, processes, and methods as well as the benefits that would result if implemented. Figure 8-10 provides an example of a scenario for process planning. Other typical scenarios that could be developed and the associated benefits are listed in Table 8-1.
- Estimate the benefit that would result from scenario implementation for each scenario. Financial analyst support using approved business tools and techniques is critical in establishing quantified, justifiable benefits. It is important to consider such indirect benefits as inventory costs or warranty costs during this process to ensure that strategic opportunities are not missed (experience indicates that significant unplanned benefits from increased sales or early product introductions invariably accompany an aggressive technology plan).
- The impacts on personnel and processes should be detailed for each scenario. Personnel training require-

Management action models
Manufacturing objectives
Potential savings
OPPORTUNITY ASSESSMENT
● Define AS IS manufacturing environment
● Define TO BE scenarios
● Estimate benefits/impacts/costs
● Determine go-no go
● Specify requirements and prepare plan
Prioritized technology opportunities
AS IS model
TO BE scenarios
Cost versus benefit analysis
Requirements specification
Action plan
Technology experts
Financial analyst
Automation engineers
Manufacturing personnel
Program manager

Fig. 8-7 Tasks involved in assessing technology opportunities.

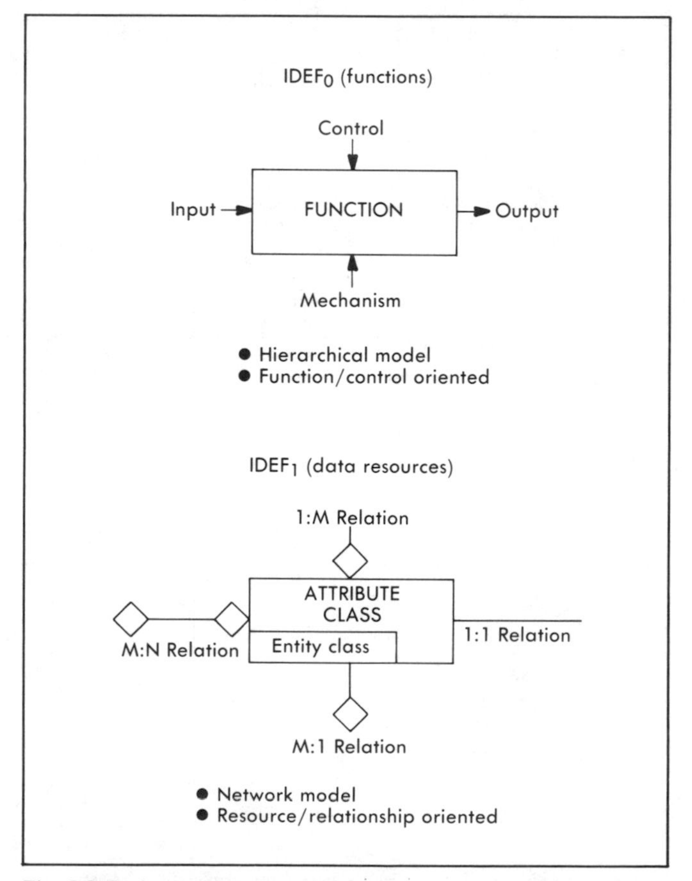

Fig. 8-8 Basic nomenclature used for the integrated computer-aided manufacturing definition (IDEF) model.

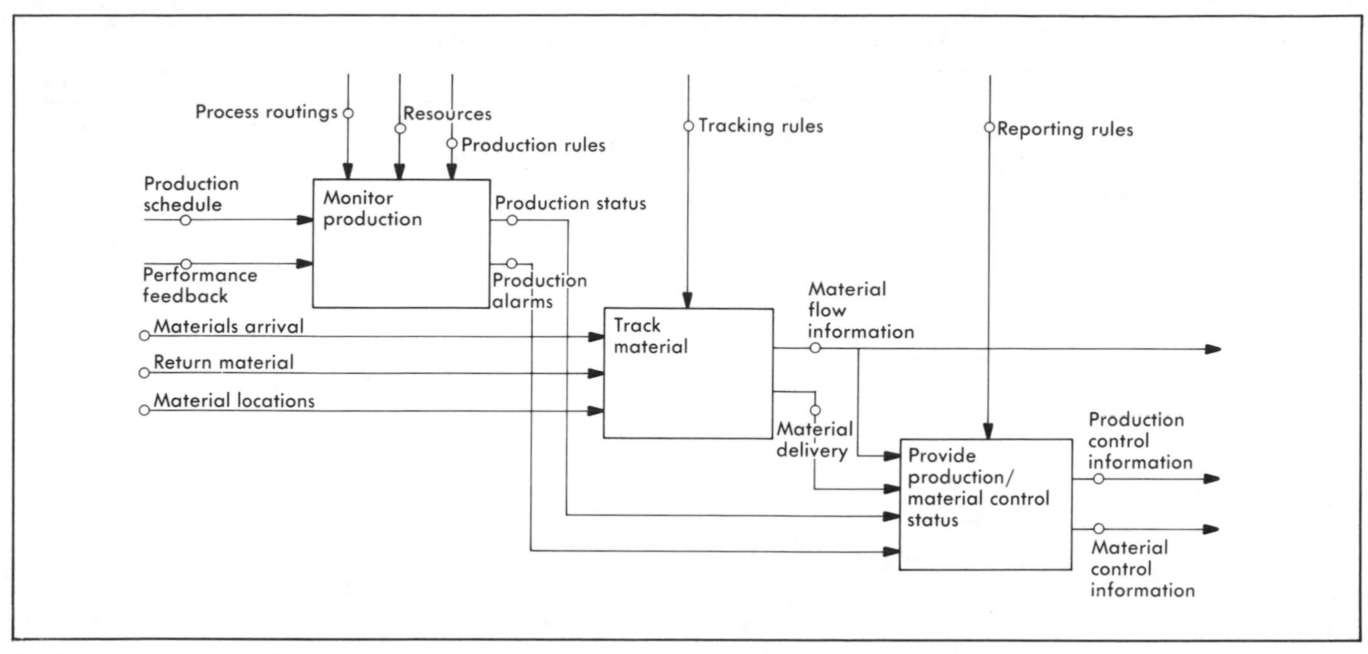

Fig. 8-9 Example of AS IS IDEF environment model.

PROCESS PLANNING

The process planning function translates the part or assembly design into the production instructions used in manufacturing or assembly. Work order routings or production routings are created which contain the part/assembly operation sequence and description. The information contained in the operation description includes the machine identification and location, the number of men and machines required, the operation setup and run time, the man and machine burden, the required tooling and tool documentation, the required assembly parts/assemblies, and the operational text details.

The production routing contains information assimilated by several manufacturing planning functions. As each function completes its planning task, the planning documents are signed-off and passed to the next manufacturing planning function. Currently, the production routings are created and signed-off manually.

A need exists for Computer Aided Process Planning (CAPP). A Group Technology based CAPP system is envisioned which will provide the capability to create and modify the production routings on an on-line basis. The Group Technology classification and coding feature of the system will enable the process engineers to identify and retrieve production routings for similar previously planned parts which can be copied and modified for use as the production routing for a new part.

The database nature of the system will provide the capability to identify production routings containing specific part and assembly information, and to retrieve information from the routings which may be needed by other CIM functions.

POTENTIAL BENEFITS

This capability provides the following potential benefits:

- Reduction in the process planning cycle. The information retrieval capability of the system will provide a means to quickly identify similar process plans which can be copied and modified for use as new process plans. The number of completely new plans will be reduced.
- Reduction in the amount of scrap. By providing standardized shop instructions, the number of incorrectly produced parts is reduced. This factor is impacted by several areas. As plans become more standardized and operators become familiar with the consistent planning, the number of scrapped parts due to operator errors is reduced. In addition, as routings become standardized, the number of new routings is reduced. This reduces the number of potential setup errors and the number of parts processed on incorrect setups. This benefit will not be accrued on permanent setup equipment.
- Reduction in the direct labor effort. The number of direct labor hours will be reduced. This factor is impacted by several areas. The reduction in scrap will directly impact the number of direct labor hours required. As less scrap is produced, the number of direct labor hours required to replace scrapped parts is reduced. The amount of rework required is reduced. By using standardized plans or obtaining standardization through modifying existing plans, the number of incorrectly made parts will be reduced.

ESTIMATED BENEFIT VALUE

Opportunity Area	Annual Savings ($000)
Direct Labor	
Indirect Labor	
Weekly Labor	
Management Labor	
Scrap	
Perishable Tools	
Optimum Use of Drop Off	
Freight Special Handling	
Total	

Fig. 8-10 Example of a process planning scenario.

TABLE 8-1
Potential Technology Scenarios and Associated Benefits

Example Scenarios	Associated Benefits						
	Increase Product Mix	Increase Quality	Reduce Inventory	Reduce Product Flow Time	Increase Responsiveness to Supported Plants	Reduce Labor Costs	Increase Equipment Utilization
Provide production feedback		X	X	X	X	X	X
Provide material movement			X	X			
Provide process monitoring		X		X			
Reduce rejects		X	X	X	X		
Provide link between design and manufacturing		X		X	X	X	
Reduce product changeover time	X		X	X	X		X
Provide maintenance management		X		X	X	X	X
Provide quality management		X		X	X	X	X
Provide schedule validation			X	X	X	X	X
Provide engineering change control	X	X	X		X		
Provide control of rework			X	X	X	X	
Provide instructions		X		X		X	X
Provide programmable device library support	X	X				X	X
Provide process plans		X		X		X	
Provide manufacturing resource planning	X		X	X	X	X	X
Provide production schedules	X		X	X	X	X	X

ments are assessed. Individuals within the organization critical to acceptance and success of the plan are identified. Requirements for equipment rearrangement are specified. New or upgraded maintenance procedures should be identified for each scenario. These impacts can dramatically influence feasibility assessments.

- Estimate the budgetary cost and time to implement each scenario. Because detailed estimates are not possible at this time, experience in past projects, similar industry/vendor experiences, or consensus "guesses" are the best that can be expected. The risk of failure, however, increases significantly if this is a first-time industry-wide implementation. Multidisciplinary storyboarding with outside participation is useful in quantifying both costs and risks of identified opportunities. Several "what if" scenarios should be analyzed for the risks of not being able to achieve specified benefits or if missing additional costs are identified later.
- Decide which technology opportunities to proceed with based on benefits versus costs and risks. If possible, time-phase the application of technology opportunities to provide early savings to partially offset costs. An example of a phased cost impact is shown in Fig. 8-11.
- Prepare a requirements specification covering each selected technology opportunity using AS IS information and TO BE scenarios. The requirements specification should begin with a description of the AS IS environment, including primary interfaces and organization structures. The needs of the AS IS environment that led to technology opportunity identification are documented next. Finally, the requirements to satisfy these needs are defined. Performance requirements are essential because they will ultimately determine system acceptability. Special requirements should also be described to guide the technology selection and technology implementation steps (see Fig. 8-12).
- Prepare a technology action plan using benefit, budgetary cost, and time-phasing data (see Fig. 8-13). The plan describes time-phased projects to satisfy identified requirements. Estimated costs and expected benefits are also related to each planned project. Associated benefits contained in Table 8-1 have now been quantified. Storyboarding is useful for organizing and for making the plan visible.
- A part of the preparation for and execution of the technology assessment phase is to become aware of and conduct preliminary assessments of supplier offerings and capabilities.

This may require a search for available technologies through technology databases, catalogs, libraries, trade journals, sup-

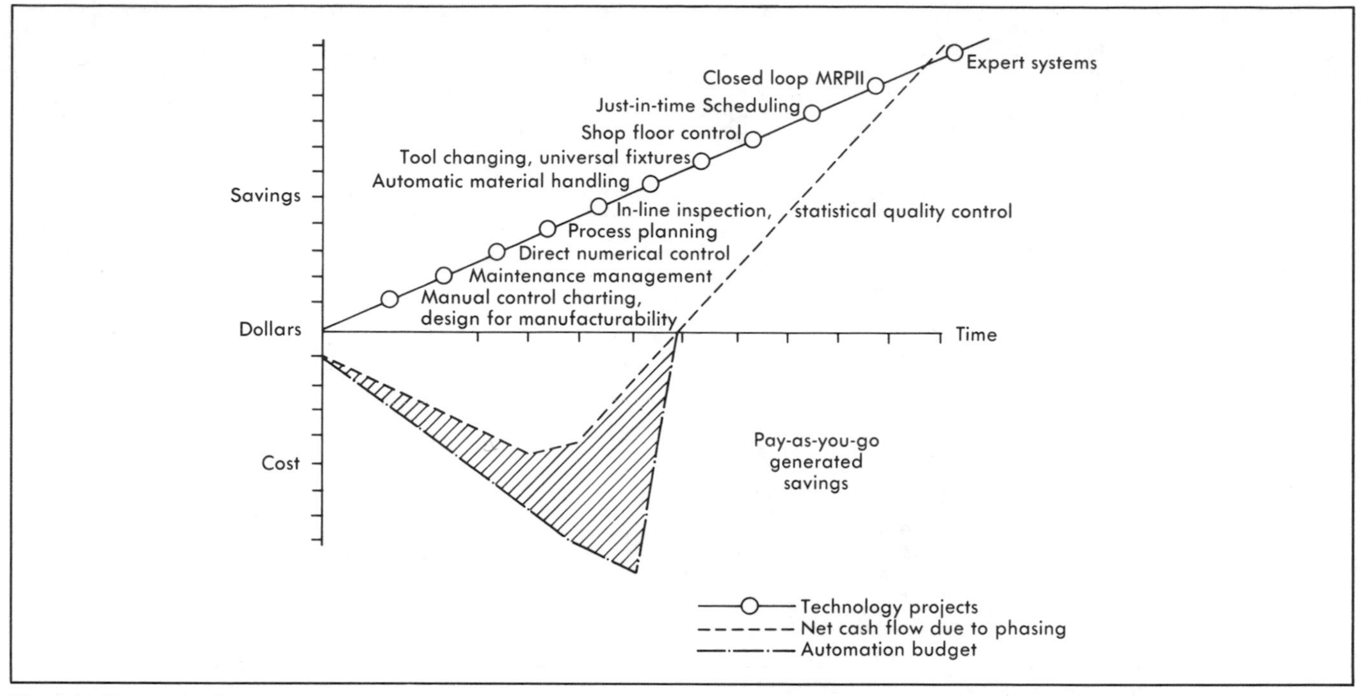

Fig. 8-11 Phased cost impact.

1. Introduction
   1.1. Scope/objectives
   1.2. Applicable documents
   1.3. Definitions
2. AS IS overview
   2.1. AS IS facility/hardware environment
   2.2. AS IS software environment
   2.3. AS IS interfaces
   2.4. AS IS organizational structure
3. Needs identification
   3.1. Facility/hardware needs
   3.2. Software needs
   3.3. Interface needs
   3.4. Organizational needs (resources, skills)
4. TO BE requirements summary
   4.1. Facility/hardware functional requirements
   4.2. Software functional requirements
   4.3. Interface requirements
   4.4. Organization requirements (training)
   4.5. Performance requirements
      4.5.1. Response time
      4.5.2. Throughput
      4.5.3. Frequency
      4.5.4. Stress
   4.6. Special requirements (security, backup/recovery, tolerance to failure, OSHA)

Fig. 8-12 Items to be included in a requirements specification.

plier product briefs and interviews, or through independent consulting firms.

The scenarios from the assessment phase and ultimately the specifications from the technology selection phase become the basis (checklists) for competitive evaluation. The assessment phase establishes a "budgetary cost" and supplier list.

The final task in the opportunity assessment process is a series of high-level management reviews. The cost/benefit data and action plan should provide required justification for proceeding with the technology management lifecycle. Also, management now has a technology plan that supports the overall business strategy, that addresses real needs versus unjustifiable "wants," and that is understood and accepted by the whole organization.

MANAGEMENT PLANNING TOOLS

Several tools that are useful during the opportunity identification and opportunity assessment processes are storyboarding, baselines, and scenarios.

Storyboarding is a structured brainstorming technique that captures ideas on cards and pins them to a corkboard wall (usually in a "war room" assigned to the planning team) for review by all storyboard participants. The storyboard session is led by a trained facilitator (formal training courses are available) who first conducts a brainstorming session in which all ideas are pinned to the wall in an organized fashion (purpose of storyboard, major topics, items of interest). Each brainstorming session is followed by a critiquing session in which all ideas that are not accepted by the entire team are dropped. The resulting ideas and information on the wall constitutes the storyboard (refer to Fig. 8-6). Storyboarding can be used for information gathering, planning, scheduling, organizing, and communicating. Although labor intensive, a properly facilitated storyboard is very productive in eliciting new approaches and has the significant benefit of building a consensus.

Baselines are structured definitions of the state of technology across an industry or a set of industries (see Fig. 8-14). Baselines may be purchased externally or produced by the business based on industry and vendor surveys. To be effective, baselines should define the state of application experience in

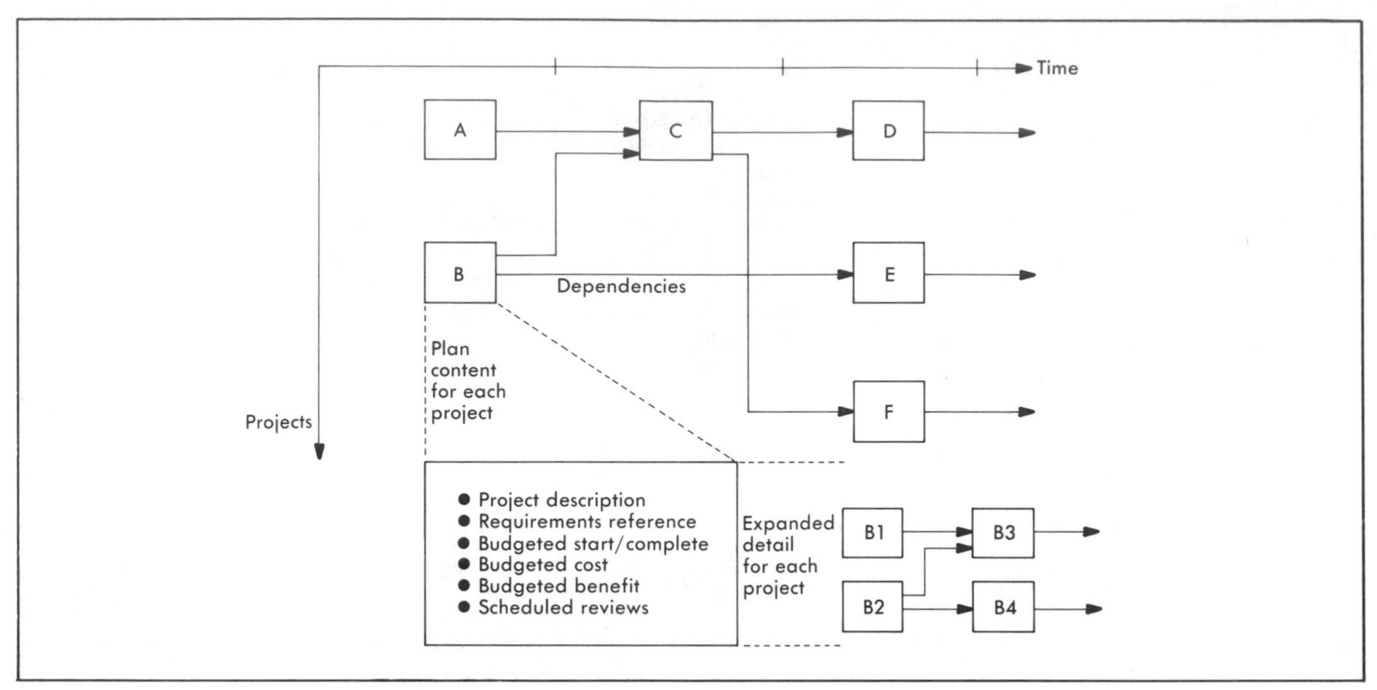

Fig. 8-13 Content of a technology action plan.

Fig. 8-14 Baselines used in management planning.

other businesses, such as different approaches to material control, and the associated benefits. The baseline should also provide a self-appraisal method for comparing the AS IS environment to the state-of-application baseline and then quantifying the resulting opportunity. When available, baselines provide a ready source of technology opportunity ideas, associated benefit data, and industry contacts to discuss costs and risks.

Scenarios are prose descriptions of what can be accomplished by applying manufacturing technology to the manufacturing environment and interfaces (refer to Fig. 8-10 and Table 8-1). Predefined scenarios are often available from outside vendors for each dimension and interface of the technology selection model. However, it is often necessary to consolidate and adapt multiple vendor scenarios into an approach that fits the manufacturing environment. Scenarios can usually be reapplied to similar environments in other manufacturing units and provide an excellent basis for establishing a standard set of manufacturing requirements across the entire business.

TECHNOLOGY SELECTION

Technology selection (refer to Fig. 8-1) is the process of specifying a new TO BE manufacturing system that will satisfy the technology planning requirements consistent with the action plan. The new manufacturing system results from a combination of existing AS IS technology (hardware, software, methods, and processes), new technology, existing and new or modified facilities, and existing and new personnel with new skills. System specification is a very creative process that depends on the skills of experienced system engineers, the constraints of the AS IS environment, the state of technology, and the degree of familiarity with new technology. Although this process is unique to each business, the tasks to be performed are common and are listed as follows (see Fig. 8-15):

- Functionally define a TO BE system that satisfies the technology planning requirements and scenarios.
- Review and select technologies to be used in the TO BE system.
- Specify a physical system architecture (hardware, software, facilities, and personnel) in terms of the technolo-

gies. Allocate the TO BE functions to this system architecture.

- Verify that the system specification satisfies each of the technology planning requirements and scenarios.
- Decide whether to make or buy each element of the system. Produce necessary procurement specifications.
- Revise the action plan to reflect tradeoffs between requirements, benefits, costs, and timing that were made during the technology selection process.
- In those cases where an emerging or undeveloped technology would be extremely beneficial, the organization might elect to fund a university or nonprofit organization project to obtain that technology.

The following sections provide management guidelines on how to accomplish these tasks using the technology selection model, management action model, and tools previously described.

"TO BE" SYSTEM DEFINITION

The TO BE system definition begins with the AS IS model, TO BE scenarios, and the requirements specification produced in the technology planning step. These are transformed into a structured set of related functions, interfaces, and data relationships that would satisfy the requirements, scenarios, and AS IS environment constraints if implemented. This task is performed by a team with the following knowledge and skills:

- Knowledge of the AS IS environment, critical interfaces, TO BE scenarios, and requirements.
- Skill in defining and modeling a system of the type required. Specific skills in hardware, software, facility, and organization definition may be required based on system complexity. Outside support (consultants or vendors) may also be required if this is a first-time or significantly different system project.

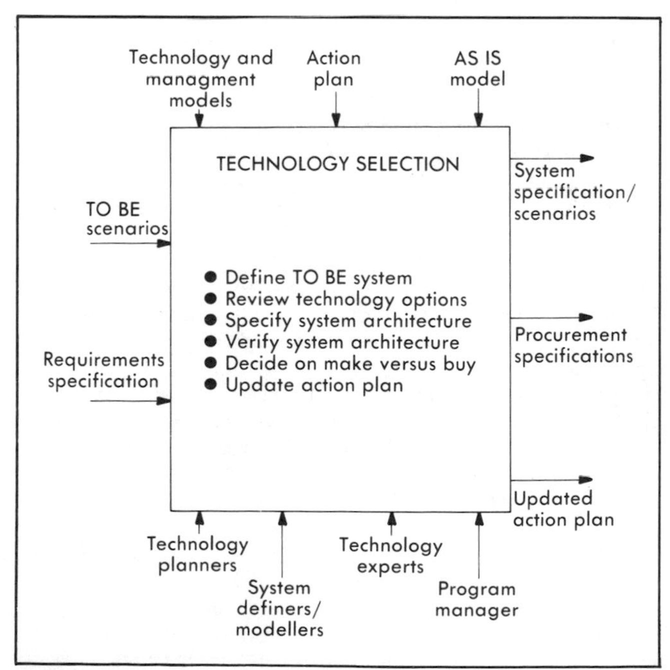

Fig. 8-15 Technology selection procedure.

- Skill in coordinating and directing the definition process.

The IDEF modeling technique is also useful during a large system definition. This tool is increasingly being used for computer-integrated manufacturing system specification. Outside training and vendor support is available. TO BE $IDEF_0$ models allow system functions and interfaces to be clearly defined and understood by all trained personnel (see Fig. 8-16). Similarly, TO BE $IDEF_1$ models allow all resource and data interrelationships to be detailed (see Fig. 8-17). Together, these models cover the structural aspects of the system. Numerous physical layout tools are also available, as are specialized tools for defining complex system control sequences.

Whatever set of tools is selected, the TO BE system definition should cover the following:

- All elements of the AS IS environment (hardware, software, data, procedures, methods, and resources) that are to be preserved in the new system.
- New requirements (hardware, software, data, procedures, methods, and resources) and their relationships to the AS IS environment.
- Physical layouts of equipment or facility rearrangements and additions.
- Descriptions of how the new system is to functionally operate (control diagrams, functional scenarios, and process descriptions) and perform (throughput, cycle time, and availability).
- Related organization functions and skills including user procedures and interfaces.

Usually a two or three-level block diagram description is sufficient because details will be added during the technology application step. The process is complete when all specified requirements are assigned to system functions and when all TO BE scenarios can be described by a sequence of system functions. Although this process is time consuming, it is the prerequisite for successful technology selection and implementation.

TECHNOLOGY REVIEW

After TO BE system functions and interfaces have been defined, technology is selected for implementing each related set of system functions. Because this technology is dependent on both the application and current state of technology, specific technology alternatives must be investigated.

Sources of current technology information include the following:

- *In-house experience and research.* Each manufacturing operation should have an active program of manufacturing technology evaluation that investigates and prototypes promising new processes, material movement techniques, and information system techniques. First-hand experience on a trial basis prior to system implementation will significantly reduce technology implementation risks.
- *Vendor and consultant recommendations* (including technology selection participation). The TO BE system definition provides a focused set of requirements for external assistance and will help eliminate the confusion of overlapping claims.
- *Trade journals, reports, studies, and seminars.* Although of limited use for specific projects, each manufac-

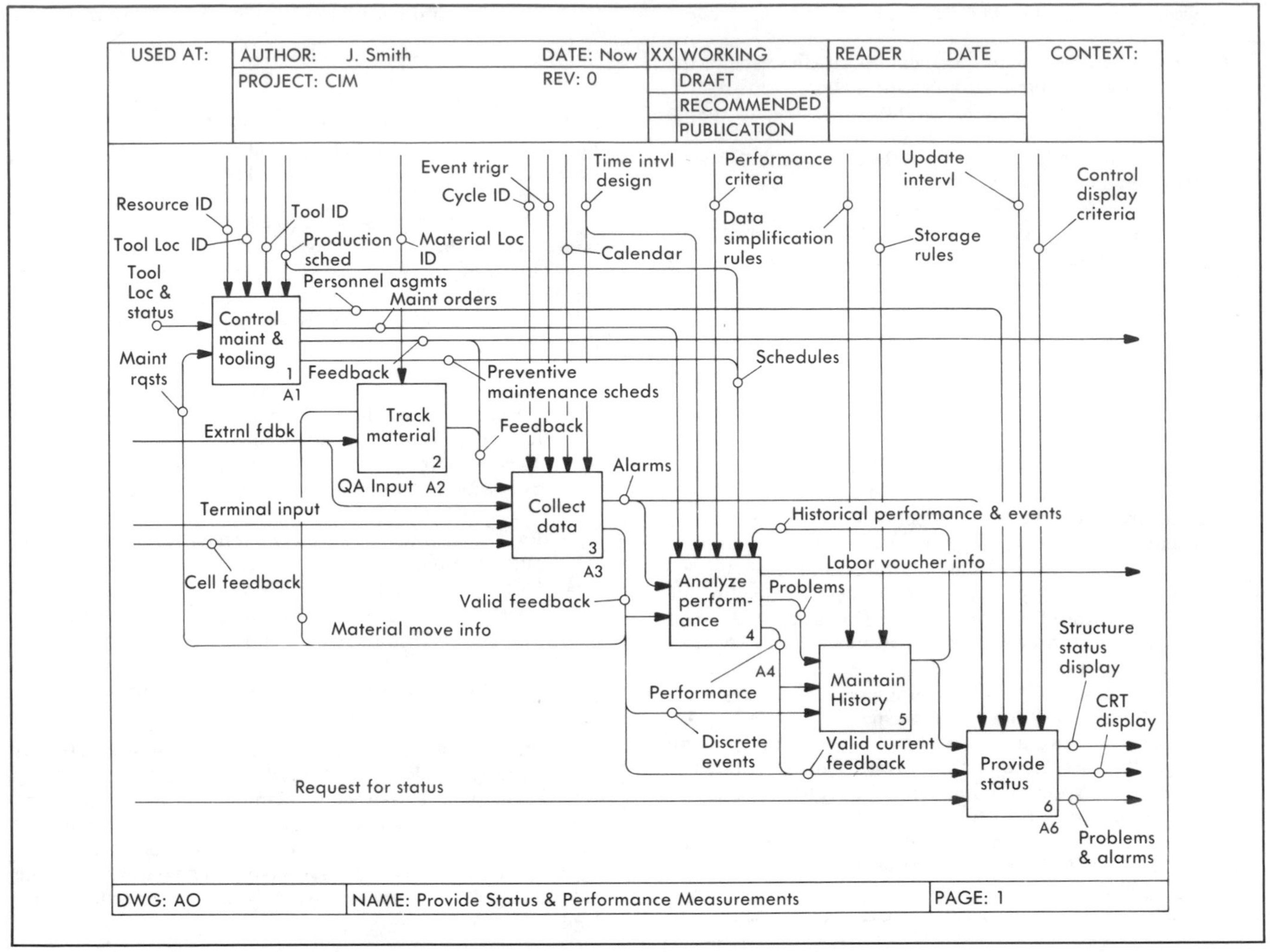

Fig. 8-16 Example of a TO BE IDEF$_0$ model.

turing operation should maintain up-to-date technical knowledge.

- *Visits to other plants* that have addressed similar requirements or are manufacturing technology leaders. The ability to see working systems and discuss implementation costs, benefits, and issues is one of the best approaches to selecting new technology and reducing implementation risks.
- *Participation in industry standards activities.* Technology standards are reducing both the costs and risks of new technology applications. A well-known standards activity is the effort to establish a common manufacturing automation protocol (MAP) for manufacturing data communications. This activity is reducing the complexity of interconnecting dissimilar equipment, computers, and terminals on the factory floor to a set of standard hardware and software communication options available from many vendors. Usually selection of a proven, industry standard technology should be the first choice among alternatives. Additional information on MAP can be found in Chapter 16, "Computer-Integrated Manufacturing," of this volume.

The review is conducted by a team familiar with the TO BE system functions and with technology alternatives. Technology review is guided by the technology selection model, which requires that choices be made along each manufacturing dimension and for each critical interface.

The review process is complete when technologies have been chosen for each element of the TO BE system description. Sometimes a technology selection is not possible (not yet available). These situations are managed by identifying contingency approaches or, if the technology is not required in the near term, the system implementation proceeds with a "black box" function identified in the action plan. These black boxes provide a focus for in-house research and vendor requirements.

SYSTEM SPECIFICATION

A system specification describes the specific technology and associated functions that are to be implemented. The specification includes prose and diagramatic descriptions of existing and new hardware, existing and new software, communication approaches, interfaces between system elements and the AS IS environment, TO BE functions assigned to each system component, data requirements of each system component, perform-

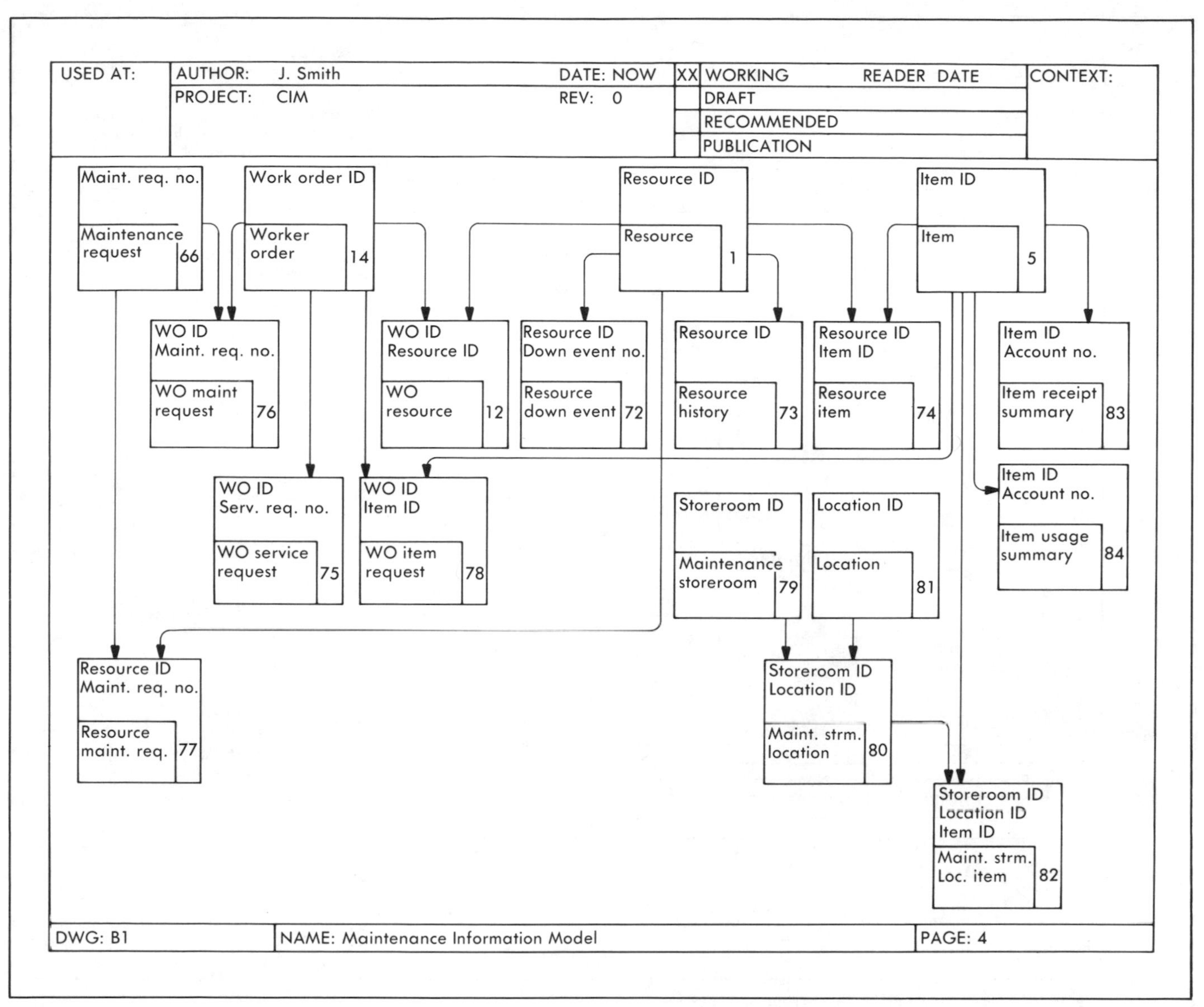

Fig. 8-17 Example of a TO BE $IDEF_1$ model.

ance requirements of each element, system control scenarios, physical layouts, and organizational relationships. The specification results from a knowledgeable team constructing a system architecture based on selected technology.

Figure 8-18 illustrates one possible architecture for a flexible manufacturing system project. Defined TO BE functions are mapped into an architecture consisting of new computers, software, programmable controllers, numerical controllers, material handling equipment, and MAP standard local area network communication subsystems. Interfaces are defined to AS IS processes, gages, machine tools, engineering systems, and business planning systems. This architecture, along with the associated detailed description, physical layout, data structures, control scenarios, and organizational relationships, are included in the system specification. For some multiplant or large-scale applications, it is more effective to define a standard system architecture or platform. The platform consists of standard system building blocks and conventions that provide specified functions through well-defined interfaces. Following are some examples:

- Guidelines should be established for developing common subsystems such as statistical process control, maintenance management, and data acquisition so that they can be integrated with other subsystems and do not themselves become inseparably locked into larger subsystems.
- Common interface software for integrating existing systems.
- Common man/machine interface conventions (forms management software) so that operators, equipment attendants, inspection, test, and maintenance personnel moving about the factory can use the same procedure to interact with any subsystem on any terminal.
- Standard database management system.
- Selection of common computer equipment and terminals for text and graphics information display.

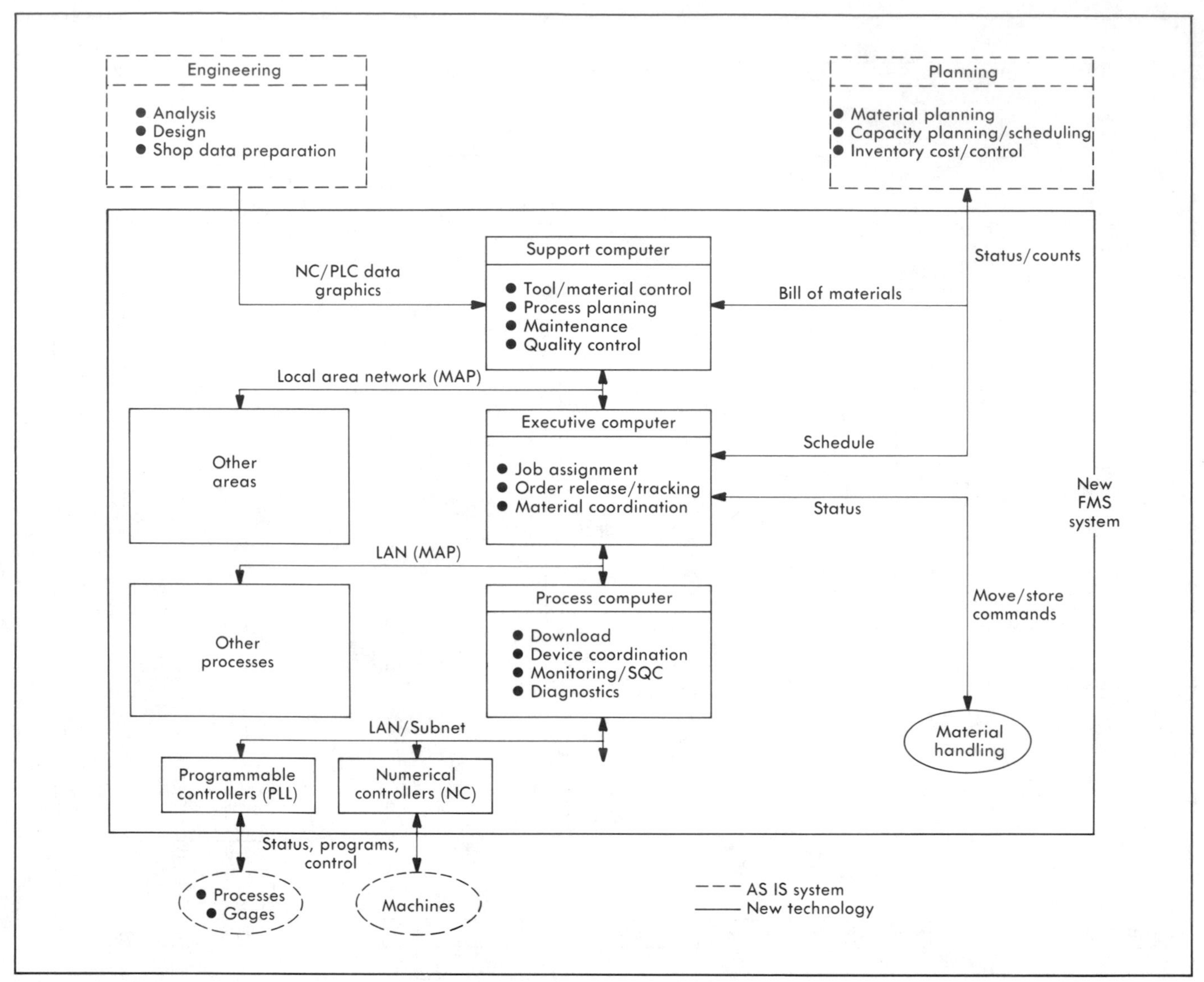

Fig. 8-18 Diagram of a flexible manufacturing system specification.

- Equipment control programming specifications to suppliers should contain standards for incorporating communications interfaces.
- Standard control equipment for ease of application and use as well as component repair, part stocking, and maintenance procedures.

The actual implementation approach of the building block is of secondary importance as long as functional, interface, and performance specifications are met. Such a building block approach allows the business to do the following:

- Plug in new technology building blocks without disrupting the overall system (technology transparency).
- Procure equipment from any vendor capable of meeting building block specifications.
- Incrementally expand a system implementation in both functionality and scope.

Several typical building blocks are illustrated in Fig. 8-19.

SYSTEM VERIFICATION

The verification process ensures that the specified system meets all the technology requirements and that side effects of the system architecture are understood. Verification is performed by the specification team or, if possible, by an independent team that understands the system requirements and the specified architecture. The team performs the following tasks:

- Collectively "walks through" each scenario developed during the planning phase to determine if the specified system can perform the scenario.
- Runs simulations to verify physical layouts and interactions. Simulations with usual graphical output should be used to allow management and manufacturing personnel to view results and recommend changes.
- Reviews the requirements specification to ensure that the "paper" system could satisfy the requirements if implemented.
- Determines any additional results, or conditions, that

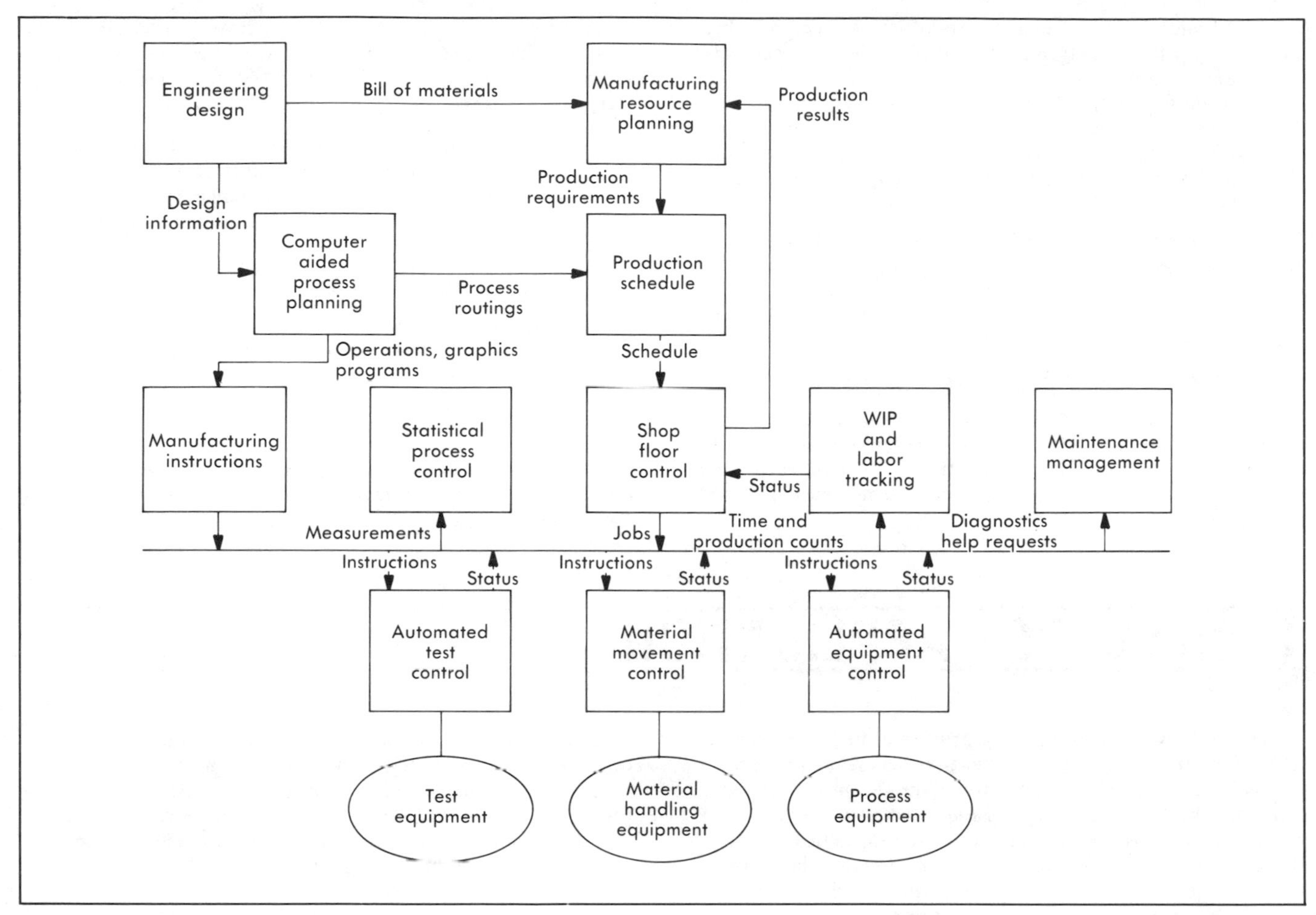

Fig. 8-19 Typical system building blocks.

could occur during system operation and how the system should respond (for example, what happens if a communication network fails, what happens if a tool breaks). These are defined in revised system scenarios that are used during technology application to verify system operation.

- Updates the system specification to correct any problems detected during the verification process.

The verification process is often overlooked with costly consequences during implementation or after system startup. At this time, specification deficiencies are easy to correct and associated costs can be assessed and managed.

MAKE/BUY DECISION

With a firm system specification in hand, management must decide what will be made or developed in-house and what will be bought externally. Some of the items to consider when making a decision are as follows:

- *MAKE*
  Resource/technical skill availability.
  Need for new technology or approach.
  In-house support costs and resource commitments.
  Benefits of in-house learning and need for intimacy with the system.
- *BUY*
  Solution availability and speed of delivery.
  Vendor capability and relevant experience.
  Lifecycle support costs.

For subsystems that are procured externally, a procurement specification is required. This specification should be extracted from the system specification with appropriate terms and conditions added. Some of the terms and conditions to include in a procurement specification are as follows:

- Ensure that the vendor clearly understands interface standards (electrical, communication, data, and function), performance criteria, and system scenarios. Clear vendor understanding of the system requirements often results, in suggested changes that reduce risk and costs.
- Have the vendor demonstrate full functionality against system requirements and scenarios prior to deliveries.
- Build in interim reviews to ensure adequate progress. Anticipate that system hardware will have to go through several design iterations if used differently than in stand-alone applications. System-compatible staging can reduce surprises, but the schedule must anticipate a certain amount of redesign time.
- Require software testing against system parameters and, if possible, interfaced with planned system hardware.

Off-site interface testing can often be accomplished using test/simulation tools and phone line interfaces to remote equipment.

- Procure system software early so that it is available as a test tool during system installation. For example, a software data collection and quality analysis package can be used to verify the performance of process equipment during installation.

Sometimes the make versus buy decision is delayed until after preliminary design in the technology application step. This may delay project completion, but provides the vendor with better definition of how the procured subsystem will fit in the overall system.

TECHNOLOGY PLANNING UPDATE

The final step of technology selection is to update the technology action plan to reflect all changes and tradeoffs and to expand planning detail (projects identified, costs allocated, and resources assigned) to encompass system elements and planned procurements.

All aspects of the management action model should be reviewed against the revised action plan to ensure project viability. Some of the questions that should be answered during this step are as follows:

- Have the impacts on the organization been considered? Have system-related training requirements (new computers, new machines, new maintenance) been considered and planned? Does the organization understand the system concept?
- Have the impacts on the process been considered? How will the system change the process? Is adequate support planned for the new system?
- How have benefits and costs changed? Are the payoffs still acceptable?
- What requirements have changed? How has timing changed? Is the project still feasible?

The technology selection process concludes with management review and acceptance of the system concept, revised costs and benefits, and revised action plan.

TECHNOLOGY APPLICATION

Technology application is the process of implementing the TO BE manufacturing system over time according to the projects defined in the technology action plan. This process involves managing both in-house projects and external vendor activities to ensure that subsystem designs conform to specifications, that designs fit together into a working system, and that the resultant system is supportable and satisfies manufacturing objectives. Although technology applications require a variety of tools to accomplish and manage (for example, a machine retrofit is handled differently than implementation of a software program), the tasks to be performed are common (see Fig. 8-20). In general, the tasks are as follows:

- Involve the ultimate user in the program as early as possible, such as when the user view is sufficiently stable so as not to cause confusion.
- Partition the system specification into a manageable set of subsystems and specify what each preliminary subsystem design must accomplish and how it will be used.
- Define in detail how each subsystem will accomplish its functions.
- Implement each subsystem and verify that assigned functions are accomplished. Link the subsystems and verify that they operate together. Produce "as-built" system documentation.
- Test the overall system to verify that it meets the specification and action plan milestones. Phase in the system application through a series of pilot runs and acceptance tests.
- Establish required system support functions. Train all required personnel to operate, maintain, and support the system.

Throughout this process, continue to revise and add to the action plan to cover all projects and activities. Use the management action model (refer to Fig. 8-3) as a checklist to ensure that costs and benefits are tracked against budget, that the organization is aware of progress and is being properly trained, that the new process can be cut over smoothly into the existing process environment with all necessary support in place, and that the planned projects or contingency plans continue to meet manufacturing objectives and timing.

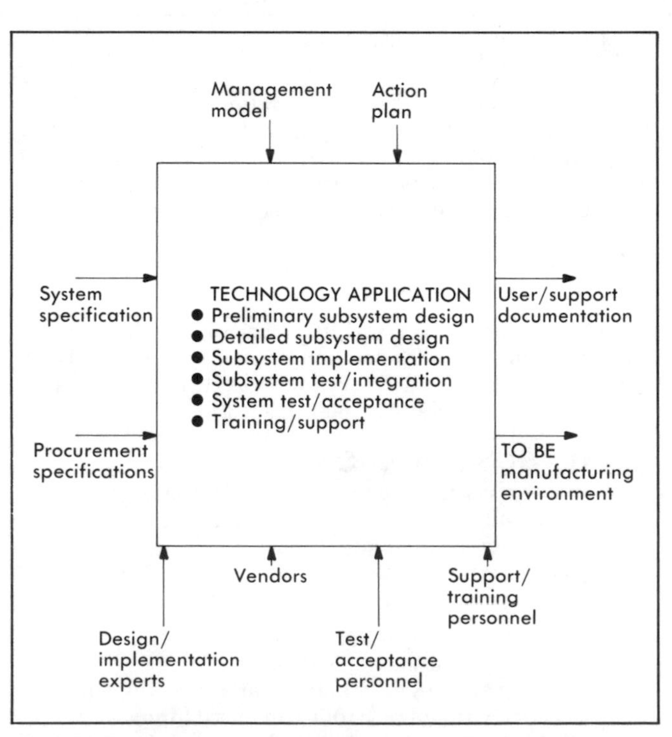

Fig. 8-20 Tasks performed in the technology application process.

TECHNOLOGY APPLICATION

PRELIMINARY DESIGN

The preliminary design process begins with the system specification, which provides a high-level description of what technology is to be employed and what functions are to be associated with that technology. Preliminary design is performed by experienced system engineers who break the system up into manageable subsystems that can be assigned to project teams for design and implementation. Using techniques such as IDEF modeling, preliminary design develops functional, informational, and performance requirements for each subsystem; interfaces to other subsystems and the AS IS environment; user interface requirements and a preliminary users manual; technology constraints such as type of hardware and standards to be used; information on how the subsystem will operate; and a plan for implementing the subsystem, including time, resources, and costs.

Preliminary design is completed when all system specifications have been assigned to subsystems and the action plan is revised. The verification process should be repeated to ensure complete coverage and to provide updated system scenarios.

As an example of the scope of preliminary design, Fig. 8-21 breaks the process level of the system shown in Fig. 8-18 into subsystems. Requirements would be specified for each subsystem and the subsystem assigned to in-house or vendor teams.

Preliminary design guidelines are as follows:

- Specify subsystems that are functionally complete with minimal interfaces to other subsystems and the AS IS environment. This will simplify integration testing during the implementation activity.
- Establish comprehensive performance and acceptance test criteria for both internally developed and procured subsystems. Define a test plan for each subsystem and add to the overall action plan.
- Standardize subsystem interfaces wherever possible; for example, standard user controls and displays, standard control register assignments, and standard software code/control sequences. This will reduce training requirements and facilitate early testing.
- Develop a comprehensive user manual on how the subsystems are supposed to operate. Involve personnel who will be using the system to contribute to and review these manuals. For computer-based systems, mock up the user interfaces as "dummy" screens to allow early user training and feedback. Other forms of "rapid prototypes" and mockups are also useful to ensure that specifications are understood.
- For software systems, allocate system information creation and access responsibilities to each subsystem. Standardize access methods. If possible, analyze and simulate critical performance areas because it is much easier to change the system information prior to subsystem design and implementation.

DETAILED DESIGN

The detailed design process results in a detailed description of how the associated subsystem will be implemented. This task is performed by one or more designers knowledgeable in the specific technologies that are selected. For example, the download/upload subsystem of Fig. 8-21 would require software

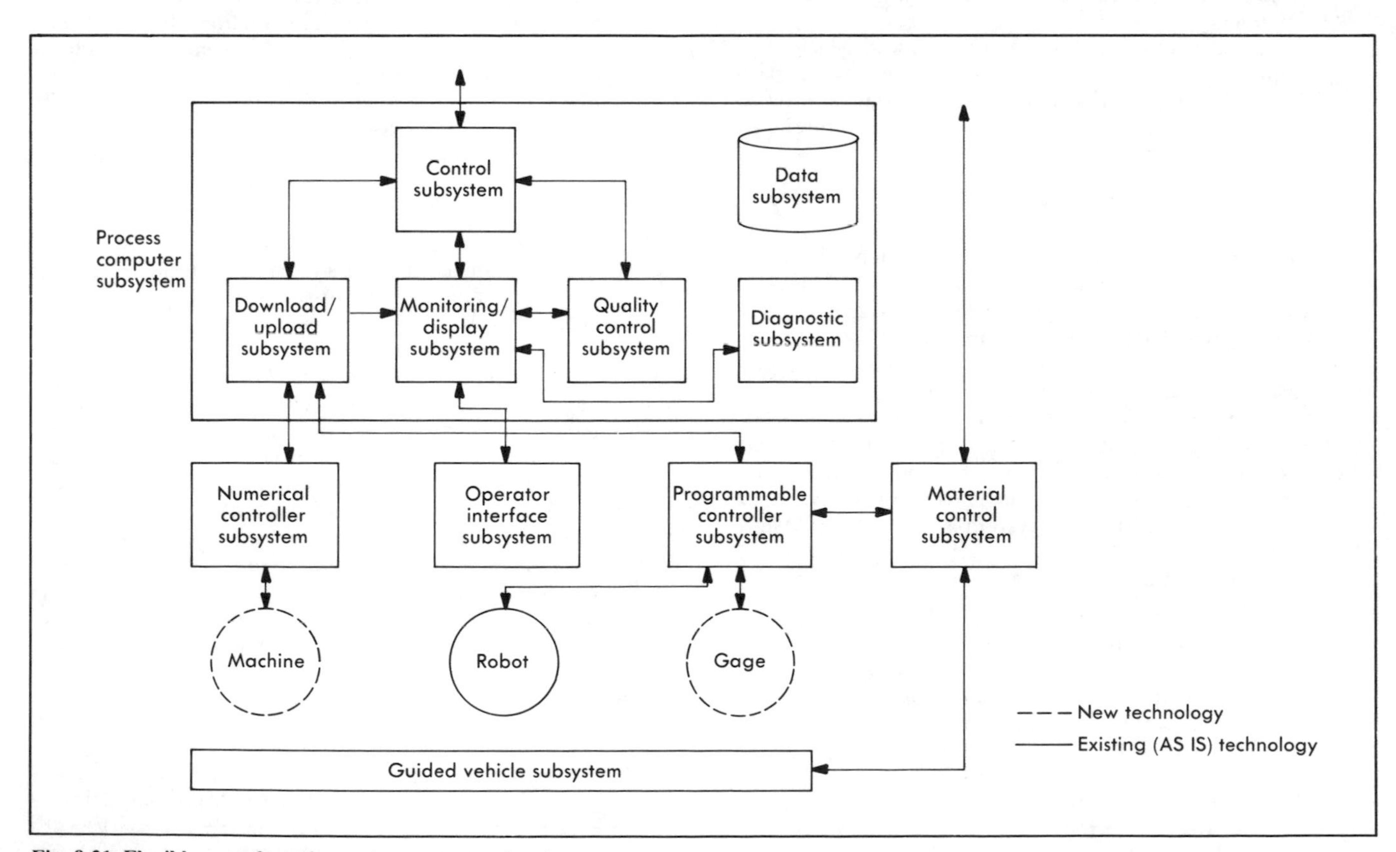

Fig. 8-21 Flexible manufacturing system process subsystems.

design skills, knowledge of the selected controller communication protocols, and some knowledge of controller program formats and operations. Similarly, retrofitting the AS IS machine of Fig. 8-21 with a new controller would require machine tool knowledge, electrical skills, and numerical programming skills. Management must ensure that proper skills are available when the subsystem project is assigned.

The actual detailed design process and associated tools are highly dependent on the specific subsystem. For software subsystems, structure charts, program description languages (PDL), and database definition languages (DDL) could be used to formally capture the design. For hardware subsystems, facility layouts, design drawings, and control charts might be used. In both cases, prototypes or models would help to ensure design completeness.

The following guidelines will facilitate a successful detailed design process:

- Utilize a computer-based project tracking system to track in-house and vendor project milestones and costs. This becomes critical when multiple projects are progressing in parallel.
- Conduct periodic reviews of both in-house and vendor designs (independent review is desirable) to ensure that preliminary design specifications are met. System scenarios should be revised and new scenarios created to cover design side effects.
- Update user manuals and use them, along with models and prototypes, to train and get feedback from users. End-user involvement is critical as the design progresses to ensure acceptance and smooth transition from existing approaches.
- Keep designs as simple as possible to meet the specified requirements. If a design is becoming too complex, revisit and revise the requirements. Complex designs invariably lead to cost and schedule overruns and unreliable implementations.

IMPLEMENTATION

The implementation process results in subsystems that meet specified functional, performance, and interface requirements. This task is either performed by the detailed designers or personnel with appropriate implementation skills. For example, software implementation would utilize programmers experienced in coding and testing on the selected computer system. Hardware implementation/modification would utilize electrical, mechanical, and facility engineers; technicians; and skilled trades workers familiar with the hardware and interfaces. The implementation team must work closely with the designers to ensure that the resulting subsystem satisfies the design objectives.

The following guidelines will facilitate a successful implementation process:

- Plan for several implementation modification cycles. Shorten these cycles by obtaining a partial implementation early that provides key functions and interfaces. Although difficult to do with hardware, this will simplify software projects by allowing system integration to proceed in parallel with subsystem implementation.
- Address higher risk implementation activities as early as possible. Create alternative approaches to reduce risk. Be prepared to redefine the project if the risk is too great. For example, work with engineering to redesign a difficult-to-orient part instead of using a complex vision/robot subsystem to recognize and reorient the part.
- Create a test bed in parallel with the implementation and use it to verify subsystem functionality and performance. Evaluate external interfaces by simulations or early connections to other subsystems. Use the preliminary design test plan and system scenarios to ensure proper operation after system integration.
- Use the test bed to train and gain acceptance from end users.

ACCEPTANCE TESTING

The acceptance testing process is used to integrate all subsystems into a working system, to verify that the overall system meets specifications, and to phase the system into operation. Although subsystem designers and implementers can be used to perform this activity, an independent team skilled in bringing up a manufacturing system is desirable. If the preceding lifecycle steps have been followed, the system should come up on schedule with minimal surprises. However, unrealistic schedules, overly complex technology applications, and unanticipated system side effects often combine to make this the most demanding management activity. By this time, however, almost all recovery plans are costly in resources, time, and missed opportunities. Management guidelines include the following:

- Perform integration testing one subsystem at a time. Minimize the number of new variables to allow rapid problem resolution.
- Use system software as a test tool to verify hardware performance and subsystem interactions. Properly developed, this software can provide system-wide visibility that is costly to obtain using standard test tools.
- Begin overall acceptance testing at a low process rate and increase the rate as confidence increases. Use the acceptance test period to train users on the actual system.

SUPPORT

The system support process ensures that all requirements are in place to sustain continued system operation, including trained users and support personnel, a maintenance program, and compatible organization structures.

Support requirements vary widely based on the technology application, but certain elements such as training, maintenance, and change are common. All users, support personnel, managers, and external organization personnel affected by the new technology must be trained. If the product has changed, customers and field support personnel may also require training. This training should be an ongoing process beginning when the system is specified and concluding prior to or during system acceptance. As indicated, the training ranges from classroom training to hands-on use of prototypes and final equipment. If possible, a separate training area should be constructed in parallel with the system. This can be revised later as a customer showcase.

Plans should be made to retain system development personnel in a support role until the user expertise and operating environment have become stable.

A maintenance program should be planned during the system specification process and be in place at the completion of acceptance testing. This plan should include necessary skills, tools, spare parts, outside vendor support, scheduling of preventive maintenance, and diagnostic procedures. Because downtime is usually one of the greatest operating costs, the maintenance plan should consider all scenarios.

The manufacturing organization must often change to be compatible with the new manufacturing system capabilities. New procedures and processes should reduce indirect support requirements. New organization structures may be required to support computer operations. System flexibility increases will change the manufacturing planning approach with ripple effects on engineering, marketing, and sales.

Top management involvement is required early in the technology application step to resolve these issues. Finally, management must continue to evaluate the actual results (costs, benefits) of the completed projects while beginning the next iteration of technology identification, selection, and application projects.

Bibliography

Allmendinger, Glen. "Management Goals for Manufacturing Technology." *Manufacturing Engineering* (November 1985).

Clark, Kim B., and Hayes, Robert H. "Why Some Factories are More Productive Than Others." *Harvard Business Review* (Sept./Oct. 1986).

Function Modeling Manual (IDEF$_0$). Manual AFWAL-TR-81-4023. Wright-Patterson Air Force Base, OH: U.S. Air Force Materials Laboratory.

Hales, H. Lee. *CIMPLAN, The Systematic Approach to Factory Automation.* Boston: Cahners Publishing Co.

Information Modeling Manual (IDEF$_1$). Manual AFWAL-TR-81-4023. Wright-Patterson Air Force Base, OH: U.S. Air Force Materials Laboratory.

Introducing the CASA/SME CIM Enterprise Wheel. Dearborn, MI: Technical Council of the Computer and Automated Systems Association of SME. Copyright 1985, Society of Manufacturing Engineers.

Kaplan, Robert S. "Must CIM Be Justified by Faith Alone." *Harvard Business Review* (March/April 1986).

Martin, James. *Design of Real Time Computer Systems.* Englewood Cliffs, NJ: Prentice-Hall, Inc.

McGroarty, J. Stanton. "Functional Costing: Understanding Where the Money Goes." *Manufacturing Engineering* (January 1987).

Rhudy, Oscar G. *Software Package Selection and Integration.* SME Technical Paper MS84-714. AUTOFACT Conference. Held October 1984, Detroit. Dearborn, MI: Society of Manufacturing Engineers, 1984.

Schonberger, Richard J. *Japanese Manufacturing Techniques.* Free Press, 1982.

Schweikert, C.L. *A Software Development Life Cycle.* Report TIS86CM0101. Bridgeport, CT: General Electric Co., Corporate Information Systems, January 1986.

Smith, Frank O. "FMS at NBS." *Manufacturing Systems* (November 1986).

Smithers, Larry E. *The Storyboard.* Hudson, OH: The Creative Thinking Center, Inc.

Yourdon, Edward, and Constantine, Larry L. *Structured Design Fundamentals of a Discipline of Computer Program and Systems Design.* New York: Yourdon Press, 1978.

HUMAN RESOURCES

SECTION

3

WORKFORCE DEVELOPMENT

Ongoing workforce development plays a key role in ensuring the success of an organization. The ability to plan workforce change to address current and future organization and work group performance issues is critical to developing and implementing effective strategic plans. Management and economic literature presents evidence that improved organizational performance occurs in enterprises that use long-range planning. This literature also suggests that effective planning and utilization of human resources contributes significantly to achievement of the organization's mission.

Managers face difficult decisions and challenges when dealing with the worker side of the enterprise. They must determine work group skill and knowledge requirements, obtain and/or create appropriate work groups, develop and/or improve work group skill and knowledge, and efficiently and effectively apply the proper quantity and quality of labor to create the organization's products and services.

The purpose of this chapter is to identify some of the issues related to workforce development. In addition, this chapter will suggest specific techniques to support a variety of aspects of workforce development. Each of the areas discussed have volumes of literature and research written about them. In keeping with the intent of this Handbook volume, the chapter provides a synthesis of the literature and experiences of professionals working within management and human resource development regarding the topics being discussed.

This chapter is organized around the human resource strategic planning issues identified previously. The issue of determining skill and knowledge requirements of the work group is dealt with in the section entitled "Job Analysis." The importance of job analysis is explored, and specific techniques are presented for guiding this important task. The strategic role of job analysis is described, and the roles of the work expert and the manager in the process are defined.

Obtaining and creating an appropriate work group is explored in the section entitled "Recruitment and Selection." The issues of internal vs. external recruitment as well as interview techniques are discussed.

Applying the work group and its members to the tasks required is dealt with in two sections. It is discussed from the perspective that for individuals or groups to focus on tasks, there are several critical interpersonal and intragroup dynamics that must be managed. The first section, "New Employee Orientation," deals with the issues and problems of introducing new members into the organization in an effective and efficient manner. The second section, "Team Development," focuses on introducing individuals into existing groups. In addition, this section discusses how to develop and manage the development and growth of groups. Issues of group value development, goal development, behaviors of different members with different roles, and interpersonal relations are also explored.

Developing and improving worker performance is addressed in two sections of the chapter. The first section, "Training and Retraining," explores training as it relates to the needs assessment outcomes. This section discusses crisis vs. the strategic planning issues of various kinds of training strategies and resources. The section entitled "Continuing and Adult Education" summarizes these models for adult development as well as some nontraditional alternatives of higher education.

CHAPTER CONTENTS:

JOB ANALYSIS

Historically, it was possible for a novice to watch an expert at work, copy the behavior, and produce outcomes of acceptable quantity and quality. This process did not require the learner to understand the nature or content of the work or how elements of the work activity interrelated to produce the desired outcome. Likewise, personnel professionals and managers have applied this same

***Contributors of sections of this chapter are:* Vincent W. Howell, CMfgE, CSP**, *Section Manager, Industrial Engineering, Computer Consoles, Inc., Communications Systems Div.;* **John T. Clifford**, *Director, Consulting Services, The General Systems Consulting Group;* **Dr. Gary D. Geroy**, *Assistant Professor and Director, Institute for Research in Training & Development.*

***Reviewers of sections of this chapter are:* Paul Humphrey**, *Project Manufacturing Engineer, Pratt & Whitney Aircraft;* **Ronald L. Jacobs, PhD**, *Assistant Professor, Department of Educational Studies, Ohio State University;* **David LaBeau**, *Manager Human Resources, Society of Manufacturing Engineers;* **Robert T. Lund**, *Professor, Center for Technology and Policy, Boston University;* **Dr. Gene D. Minton**, *CIM.TRAIN, Inc.;* **James L. Sheedy**, *Manager Human Resources, Ingersoll-Rand Co.;* **Dr. Gene R. Simons, PhD**, *Director, Graduate Programs in Manufacturing Systems Engineering;* **Hayward Thomas**, *Consultant, Retired—President, Jensen Industries.*

method to develop documents describing work behavior to be used as a template to measure qualifications for promotion, hiring, or job assignment.

The differences between the nature of work in the past and the nature of work today revolve around two factors: the growing technical complexity of the work and the rapidity of change in technology that causes changes in the knowledge and skill needs of workers. Both factors cause constant change in the knowledge and skill requirements of workers. The workers affected by these changes may be within the skilled as well as the management groups. Within this environment, the manager must manage the application of worker skills and knowledge in an effective and efficient manner. The key to managing worker performance is understanding what workers do. Performance is the key to the success of any organization, and understanding what enables desired performance to occur is a key to successful management. One key to understanding what workers do is to analyze their behavior. Likewise, when managers understand the knowledge and skill requirements of a job, they are more successful in matching existing worker skills, planning training to provide workers with needed skills, and selecting workers with desired skills. Success in these activities contributes to both the development as well as the implementation of the strategic plans of the organizational unit.

The process for determining what workers do (will do) and what skills and knowledge they use (will use) to achieve desired job performance is the job analysis and description process. It consists of three analysis activities: (1) job description development, (2) task inventory identification, and (3) task analysis.

PURPOSES FOR JOB ANALYSIS AND DESCRIPTION

There are four primary purposes for doing a job analysis. The most common purpose is to provide an information base from which to develop training. Another purpose is to provide information to plan the reassignment of existing workers to best match existing skills with activities requiring specific skills and knowledge. The information gathered about skill and knowledge requirements for jobs also provides a basis for selection and screening of new workers. A final purpose for a job analysis is to aid in job redesign. In one organization in north central Pennsylvania, this analysis was critical to developing a strategic plan for merging 70 job classifications into 13 and shifting the production system from a linear model to a cell model.

WHY JOB ANALYSIS IS IMPORTANT

Job analysis helps managers understand the strengths and weaknesses of a work group. When managers understand where and what kinds of skills and knowledge exist in a group, they are able to match the work group's strengths and weaknesses to the tasks that must be performed. It is common for managers, supervisors, and peer workers to have a perception of what a person does or what skills he or she possesses that is quite different from what the worker actually does. This usually occurs as a result of change in some aspect of the job. This change can be rapid or slow and subtle over a period of time. This change can occur for a variety of reasons. The most common causes are as follows:

- New system, same job.
- New procedure for old system.
- New responsibilities, same job and system.

Frequently, the worker makes adjustments by acquiring new skills and knowledge by trial and error, emulating more experienced workers, and/or applying old skills in new ways. However, how a worker responds to a new job task raises issues of efficiency and cost effectiveness. Can an organization afford trial-and-error, unstructured on-the-job, or apprenticeship programs to develop skills needed to achieve desired performance? Until the expert work behavior is analyzed for its skill and knowledge requirements, decisions about training needs or who should perform certain tasks cannot be effectively and efficiently determined.

JOB ANALYSIS: TOOLS AND TECHNIQUES

Most work today involves systems and complex work relationships. The systems and the relationships have one common component—the worker. Workers' performance can be described as resulting from the output of several systems and various relationships in which the workers are involved. Workers relate to their work through three primary relationships: worker to machine or thing, worker to process, and worker to worker or abstract ideas. The worker-to-machine relationship can be observed. The worker controls or manipulates the machine, some physical part of the machine, or some physical item. Examples include operating a multiple-head drill press, CNC lathe, and completing forms. Worker-to-process relationships may be partially observable and involve troubleshooting, diagnosing, or making adjustments to systems. Examples include a worker determining why a computer system is not storing data and making appropriate adjustments to the computer system. Worker-to-worker relationships are not observable and are structured around the interaction and communication of abstract ideas and concepts. An example would be a manager of marketing developing product advertising with his or her sales force.

As discussed earlier, effective and efficient training, appropriate employee selection, efficient and effective workforce development, and valid job design are predicated on knowing what an expert does and not what managers think they do. The approach taken to analyze the job is one that involves systematically moving from the most macro to the most micro view. The goal is to identify the expert's skills and knowledge that are used to successfully complete the job. The identified skills and knowledge are organized around three major categories and gathered through three analysis activities. These analysis activities include: job description development, task inventory identification, and task analysis. There are several books currently available on the various techniques and strategies for doing these analysis activities. A suggested resource list is provided at the end of this chapter. Presented next are guidelines and suggested procedures for carrying out a job analysis.

Good job analysis involves three elements. The first is a real setting. The best analyses are conducted on the job site where the expert is working. The second element is the analyst. The analyst must be objective. This requires that he or she set aside any knowledge, experience, or bias that he or she has about the job being analyzed. The third element is the expert. If one wants to know what a job entails, how it is carried out, and what kinds of skills and knowledge are required, the key is the person who is considered the expert at that job.

Initially, it is important to say something about experts. Experts may or may not make effective trainers. Experts are good sources for information about a job's skill and knowledge requirement. However, if left to analyze their own job or to develop training, experts may leave out much of the information needed from the analysis. Why? Because they are experts. Experts many times have so assimilated the skills and knowledge that they draw on that unless someone without that knowledge causes them consciously to recall and acknowledge it, they will forget they possess or use it. Therefore, it is always a good practice not to have the expert in the job being analyzed do the analysis. They should be assigned the role of being the expert, being the source of needed information, and being the object of the analysis activity.

Job Description Development

The job description is a picture of the total job. Writing a job description is like painting a house. As one paints a picture of the job, one uses broad general terms—just as a painter would use a large brush to paint a house. The job description should contain the following elements: job title, effective date, department or work group where the job is located, location of department or work group, and the job description itself.

The job title should be a succinct combination of words that reflect the job. Consider the example of a worker who is employed by an organization that makes a variety of wood laminated products. The worker's job is to operate a laminating gluing machine to make plywood panels. The machine he or she operates is one of several types of specialized glue machines. A valid job title would be "wood panel laminator machine operator" rather than a title of "glue machine operator." This sets the job apart from the other types of glue machines. On the other hand, if the worker operates a variety of glue machines, the most appropriate job title might be "wood laminating glue machine operator." This larger picture takes into account and defines the larger responsibility area of the worker.

The effective date of the job description is an important administrative organizer. Jobs change over time and for a variety of reasons. The purpose of doing job descriptions and analyses is to keep current on job content and responsibilities as well as skill and knowledge requirements. This piece of information will ensure that whoever reviews the document will be able to locate the most current and appropriate version.

The department or work group identification for which the job description is appropriate is a small but important piece of information. It provides both the organizational context in which the job occurs and designates the work group in which the worker performs the job. The identification of this item helps organize and identify the subtle or not-so-subtle differences in job activities performed by workers in the same job classification throughout an organization. For example, a budget clerk working for a production manager may have different job activities and responsibilities than a budget clerk working in payroll.

As with identifying the department or work group, locating the department or work group is important to ensure that the job description is linked to the appropriate setting. It is very common for workers in a particular job classification to have different job activities and content in different locations. For example, a budget clerk working in a St. Louis plant for the production manager may have different job activities and responsibilities than a budget clerk working for the production manager of the Pittsburgh plant.

The job description itself is a brief statement of what the worker does. For the analyst, this perspective is important to maintain. The purpose of doing a job analysis is to determine what a worker does and the skills and knowledge that are required to carry out the job. After this has been determined, it is possible to use that information to design or redesign a job, to establish criteria against which to screen candidates, or to deploy existing workforce resources. The scope of the job description should be large enough to include all activities performed by the worker. This is best presented by writing a short paragraph that describes the large clusters of activities rather than listing each individual activity.

Let us look at a worker who secures stock, blueprints, work orders, sets up a lathe, and runs production for prototype parts as part of his or her job. These are separate activities that together constitute a cluster of activities that could be described as "prototype production." At the other extreme, a job description is not an editorial about value or relevance of the worker and his or her job to the organization or the work group. Figure 9-1 is an example of a good job description; Figure 9-2 is a poor example of the same job description.

Task Inventory Identification

Developing the task inventory is the second phase of information gathering that focuses on identifying what an expert worker does. The outcome of this activity is a list of discrete units of activities which, in total, constitute the job of the expert. The obvious question is how a worker does his or her job as described in the job description and what knowledge and skills are required to do it. Why is the task inventory needed? The answer is that the skills and knowledge required to carry out a job are often complex and not used equally in all parts of the job activity. What is needed is an intermediate organizer of expert work behavior.

One of the characteristics of this list is that each item (task) on the list can be completed by the worker independently of the other tasks in the job. For example, a task list for an operator of a Seiki 3NE CNC lathe may include: set up, operate, check parts to prints, load tape, program controls, and secure maintenance.

Job title: Millworker, Rough Stock
Location: Fancy Wood, Our Town and State
Department: Wood Stock Layout
Analyst: J.Q. Detailfinder

Effective date: Mo./Day/Yr.
Reviewed: ______________
J.B. Expert
Approved: ______________
M.I. Boss

Job description: The Rough Stock Millworker receives and grades mill lumber, cuts lumber to size, and glues sized lumber into custom panels for commercial sale. During the winter production months, the millworker operates the incinerator/heating systems.

Fig. 9-1 Example of a good job description. ***(Penn State University)***

Job title: Millworker
Location: Plant 5
Department: Layout
Analyst: J.Q. Detailfinder

Job description: The millworker checks wood that is received and stores it. The millworker must be a high school graduate, punctual, neatly dressed, and able to operate a variety of woodcutting machines. In addition to operating equipment, the millworker will perform other duties as assigned.

Fig. 9-2 Example of a poor job description. (*Penn State University*)

The worker is able to begin, carry out, and complete each of these tasks independently of the others listed. Yet collectively, these tasks constitute the job of "CNC lathe operator." When analyzing expert work behavior for tasks, the guideline is that each task has a beginning point, an activity profile (simple or complex), and an end.

The tasks listed on the task inventory serve a variety of management and training purposes. First, they become the organizers around which detailed analyses of skills and knowledge used by the expert worker are carried out. It will be the outcome of these analyses that provides content for training programs and guides selection of education programs and activities.

Additionally, this list serves to support a variety of important management decision-making activities regarding hiring and workforce deployment and organization. Consider the Pennsylvania manufacturing plant. One of the major tasks facing management was reconfiguring the jobs of workers from 70 current job classifications into 13. This was compounded by physical and process reconfiguration of how the firm produced its products. In this case, the only constant was the tasks in the firm. Machining cells were created around products and functional relationships of equipment. The task profile for individual workers and job classifications supporting cell production were redesigned. A worker could find himself or herself doing some tasks that he or she: (1) had been doing, (2) was vaguely familiar with, and (3) had no familiarity with. One solution was to look at the task familiarity of members of the work group and reassign individuals accordingly. Another solution was to hire workers with a desired task familiarity profile. A third and integrated solution was to provide training for the tasks not familiar to existing and new workers.

Task Analysis

The most detailed level of job analysis is frequently referred to as task analysis[1] or specific work behavior analysis.[2] This involves the determination of specific knowledge and skills used by the expert to complete each task. The analysis activity focuses on one task at a time. The types of knowledge and skills can be organized into three general categories. The first category is knowledge about required steps to complete each task.

The second category of knowledge is the knowledge that a worker needs to troubleshoot or solve problem situations and make appropriate adjustments when completing tasks. Swanson and Gradous (1986) suggest that the greatest likelihood for the need of this type of knowledge is in tasks where a worker is required to interact with some type of system. The types of knowledge within this category are: knowledge about how the system is put together, the systems' parts and their purposes, variables or conditions within the systems that can change or be changed, specifications for these conditions or variables, the controls that change these conditions or variables, the systems' outcomes (resulting problems) if these conditions or variables are allowed to exceed or fall short of the specifications, and specifications for what to do or how to correct problems.

The third category of knowledge is abstract knowledge that may be technical or general in nature and that is required to understand processes and make decisions necessary to complete tasks within the job. This type of knowledge is generally used in people-to-people or people-to-idea interactions and processes. The technical knowledge in this category is considered abstract, but specific to a process that is unique to an operation, activity, or organization. Frequently, technical knowledge is focused on a proprietary process or element in an organization. The general knowledge in this category can be generalized and is nonproprietary knowledge that a worker (expert) possesses and that allows him or her to successfully complete the tasks in his or her job. This knowledge is frequently described in terms of the basic skills (reading, writing, math) and other generalized subject knowledge such as chemistry, welding, statistical quality control, or blueprint reading.

The strategies used to gather information during task or specific work analyses include: interviews (individual or group), questionnaires, observations (covert and overt), and literature reviews. However, it is necessary to allow expert workers to review and validate the final analyses outcomes to ensure that all relevant information has been gathered. This ensures valid training content, appropriate educational experiences, and effective management decisions about workforce deployment, job design, and hiring guidelines.

SUMMARY

Job analysis is a linear systematic process. It begins with an understanding of the broadest view of an expert worker's activities. It ends with a very detailed summary of the specific skills and knowledge needed to carry out each task of the job.

The job analysis outcome can be used to support a variety of management planning and decision-making processes that include organization and job-specific training design, education program planning, workforce deployment, job design, and development of screening criteria for new hires. This linear process is presented in Fig. 9-3.

A final word of caution on job analysis is that some jobs are extremely complex or may cycle through time frames such as annual or semiannual maintenance, audits, and report generation. Therefore, a single view of the expert's activities at one point in time may not be sufficient.

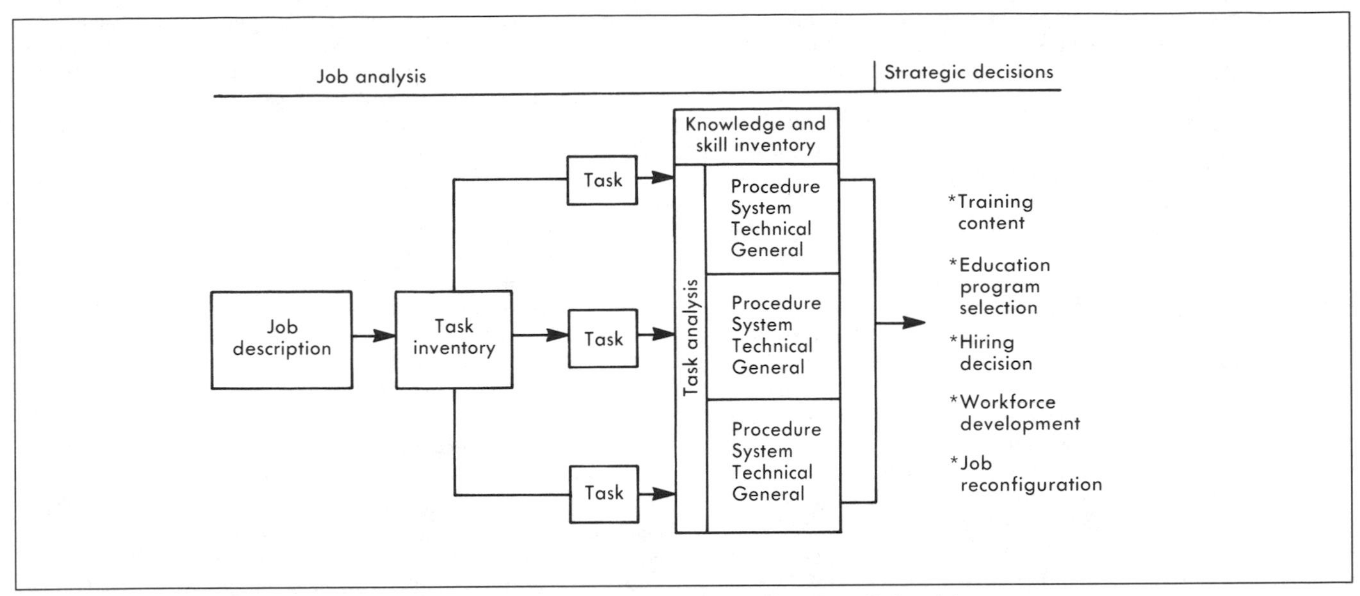

Fig. 9-3 Linear model of job analysis and resulting strategic decision making. (*Penn State University*)

RECRUITMENT AND SELECTION

A critical need that arises in many manufacturing organizations is the need to add staff who are involved with new businesses, new ideas and concepts for product lines, or new equipment in the company. The fulfillment of this need is generally through the recruitment and selection process.

RECRUITMENT

Recruitment is the process of finding qualified people and encouraging them to apply for the position. Before recruiting a candidate for a particular position, it is important to have an accurate job description for the position. This will help to ensure that the potential candidates have the necessary skills and qualifications for the position.

In most companies, the recruitment process is handled by the human resources department. It is responsible for designing and implementing a recruitment program to meet the company's personnel needs while complying with all legal requirements. This responsibility includes finding sources of applicants; writing and placing advertisements; contacting schools, agencies, and labor unions; establishing procedures to guarantee equal employment opportunity; and administering the funds budgeted for recruitment.

But these goals probably would not be attained without the cooperation of other managers, who are in the best position to predict the needs of their own departments. They are responsible for deciding how tasks should be accomplished and what kinds of people are needed to fill each type of position. They can often anticipate retirements, resignations, and other kinds of vacancies and can determine whether any of their current staff members are ready for promotion. Typically, when a vacancy occurs, the supervisor or manager completes a personnel requisition form, which usually requires higher management approval.

Because the intent of this volume as well as this chapter is to assist manufacturing managers in carrying out their responsibilities, recruitment techniques will not be discussed in detail. That type of information is more appropriate in a publication with a personnel or human resources focus. However, it is beneficial for the manufacturing manager to have a general idea of the techniques used, especially when he or she is operating under the constraints of a union contract or a newly implemented affirmative action program. For this discussion, these techniques will be divided into internal and external recruiting methods.

Internal Recruiting Methods[3]

Finding qualified applicants within the organization is the main goal of the internal recruiting effort. There are several methods for locating these applicants. Among the most common are job posting, referrals, and skills inventories.

Job posting. Perhaps the most common method, job posting involves announcing job openings to all current employees. Bulletin board notices or printed bulletins can be used for this purpose. In some companies, the personnel or human resources office publishes a monthly newsletter that lists the positions available. The announcements carry information about the nature of the position and the qualifications needed, and any employee who is interested may bid on the job—that is, enter the competition for it. Job posting can help ensure that minority workers and other disadvantaged groups become aware of opportunities to move up in the organization. For this reason, legal mandates and federal agencies require job posting and bidding as part of the settlement of discrimination cases.[4,5]

Employee referrals. Another way to find applicants within the organization is through *employee referrals* by other departments. Informal communications among managers can lead to the discovery that the best candidate for a job is already working in a different section of the firm. In some cases, referrals are made through "support networks" established by certain groups of employees; in recent years, women's groups have had a noticeable influence in this area.

Skills inventories. Many firms have developed computerized *skills inventories* of their employees. Information on every employee's skills, educational background, work history, and other important factors is stored in a database, which can then be used to identify employees with the attributes needed for a particular job.

External Recruiting Methods[6]

Finding qualified applicants from outside the organization is the most difficult part of recruitment. The success of an expanding company or one with many positions demanding specialized skills often depends on the effectiveness of the organization's recruitment program. Typically, the external recruitment process makes use of a variety of methods.

As in internal recruiting, employee referrals can be an important source of applicants. To encourage referrals, many firms offer incentive programs that may include cash bonuses. In addition, labor unions can be a key source for applicants skilled in a particular craft. Under some contracts, the company may be required to give the union first notice of job openings; even when this is not required, it may help the company maintain good labor-management relations.

Often, however, a firm must take a more active recruitment role, mounting advertising campaigns, contacting a variety of placement agencies, or sending recruiters into the field to search out qualified candidates.

SELECTION

Assuming a successful recruiting effort, there will be several candidates to select from for the open position. The actual selection process involves several steps. The number of steps and their sequence varies with the organization as well as with the type and level of the position being filled. Figure 9-4 shows the typical steps comprising the selection process. It is important to recognize that not all applicants will go through all these steps. For example, some may be rejected after the preliminary interview, others after taking tests, and so on.

In most organizations, the human resources department designs the selection system and manages its everyday operation.

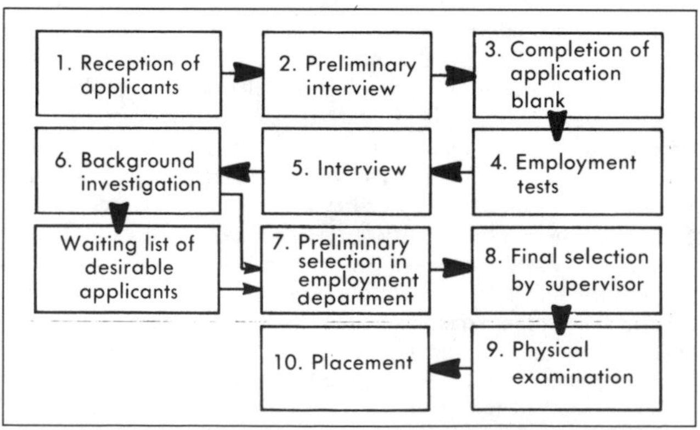

Fig. 9-4 Steps in the selection process. ***(Source: Herbert J. Chruden and Arthur W. Sherman, Jr.,*** **Managing Human Resources, 7th ed.** ***(Cincinnati: South-Western Publishing Co., 1984), p. 129), reprinted with permission, all rights reserved)***

It also makes sure that the legal requirements are consistent with the company's hiring practices. Managers should have at least a working knowledge of the federal antidiscrimination statutes to avoid inadvertent violation. A summary of the major federal statutes on hiring is given in Table 9-1. Information regarding fair employment practices as well as other legal issues can be found in Chapter 11, "Legal Environment for Labor Relations," of this volume.

The eventual hiring decision usually is made by the supervisor or manager in whose department there is an opening after the various candidates have been interviewed. Because a manager's primary responsibility is interviewing and evaluating the perspective candidates, this section will discuss the types of interviews, interview techniques, problems in interviewing, interview guidelines, and decision guidelines.

Types of Interviews[7]

Several types of interviews are used throughout industry. The most common is the one-on-one interview, in which the candidate meets privately with a single interviewer. Often, a well-qualified candidate will pass through a series of such interviews, first with a member of the human resources department, then with the manager in whose unit there is a job opening, and finally perhaps with the manager's superior.

Another type of interview is referred to as the panel interview. During this interview, one candidate meets with a panel of two or more representatives of the firm. One of the panelists may act as a chairperson, but each of the firm's representatives takes part in the questioning and discussion. This format allows the interviewers to coordinate their efforts and follow up on each other's questions.

A third type of interview is the group interview, in which a number of candidates are interviewed at once. Generally, they are allowed to discuss job-related matters among themselves while one or more observers rate their performance. This type of interview is usually considered most appropriate in the selection of managers; it can also be used with groups of current employees to evaluate their potential for supervisory roles.

Interview Techniques[8]

The three basic types of interviews lend themselves to a variety of specific interviewing techniques. The most common are the structured interview, the nondirective interview, and the situational-problem interview.

In the structured or patterned interview, the interviewer follows a standard list of questions to be asked of all applicants. This method produces uniformity of data from one interview to the next, it ensures that no important questions will be forgotten, and it helps to guarantee that all the applicants have been treated in the same way. If the interview is too rigidly structured, however, the interviewer may neglect chances for follow-up questions, and the candidate is unlikely to provide any information spontaneously.

The nondirective interview, as its name implies, takes the opposite approach. The interviewer's questions are held to a minimum, and they are open ended. Rather than asking about specific details of the candidate's last job, the interviewer may say, "Tell me about your work in this field." The aim is to follow the applicant's own lead, to let him or her express thoughts and feelings that might be relevant to the job. Instead of filling silences with new questions, the interviewer may simply nod to encourage the applicant to continue.

TABLE 9-1
The Major Federal Statutes On Hiring Practices

Act or Law	Prohibited Practices	Applicable To	Enforced By
Title VII of the Civil Rights Act of 1964 (as amended by the Equal Employment Opportunity Act of 1972)	Discrimination based on color, race, creed, national origin	Most employers of 15 or more employees, public and private employment agencies, labor unions with 15 or more members, and joint labor-management training programs. (Exemptions: Indian tribes and religious institutions with respect to religious activities)	Equal Employment Opportunity Commission (EEOC), state and local agencies
Age Discrimination in Employment Act	Child labor: i.e., hiring a minor under 16 years of age without a permit	All public employers, private employers of 20 or more employees, employment agencies serving covered employers, and labor unions with more than 25 members.	Wage and Hour Division (WHD) of the U.S. Dept. of Labor
Executive Order 11246	Improper representation of minorities in companies participating in government contracts of $10,000 or more	All facilities of federal contractors or subcontractors and state and local agencies participating in government contracts	Office of Federal Contract Compliance Programs (OFCCP) of the U.S. Dept. of Labor
Equal Pay Act of 1963 (Amendment of Fair Labor Standards Act)	Misuse of payment under minimum wage requirements	All employers covered by the Fair Labor Standards Act	WHD
Rehabilitation Act	Discrimination against rehabilitated employees who meet job requirements	Federal contractors and subcontractors whose contracts are in excess of $2500	OFCCP
Pregnancy Discrimination Act	Discriminatory hiring practice against pregnant women who meet job requirements	All Title VII employers	EEOC
Vietnam Era Veterans Readjustment Assistance Act of 1974	Discrimination against Vietnam veterans because of their veteran status by companies participating in government contracts of $10,000 or more	Government contractors and subcontractors	OFCCP

(*Modern Machine Shop*)

The nondirective technique can reveal information that would never have come up in a structured interview. It requires substantial time, however, and it may fail to touch on important aspects of the candidate's qualifications. For these reasons, few organizations use it in its pure form. Typically, the interviewer tries to mix the nondirective technique with a structured approach, encouraging the applicant to expand on his or her ideas, but making certain that standard topics are covered.

In the situational-problem interview, the candidate is given a specific problem to solve or project to complete. Often this technique is used in the group interview scenario; while the group discusses a problem and works out an answer, the interviewers rate each candidate on, for example, quality of ideas, leadership capacity, and ability to work with others.

Problems in Interviewing[9]

Whatever their format, interviews must overcome a number of typical problems. One is the personal bias of the interviewer. What if the interviewer likes blondes or dislikes green eyes or thinks that striped ties look foolish? Those who conduct interviews must learn to ignore their personal preferences. In some cases, however, a belief about a certain category of people may be so deeply ingrained that the interviewer is unaware of its existence. Such stereotypes are a principal factor in discrimination against minorities and women, and they can influence attitudes toward other groups as well. For example, an interviewer may tend to assume, without realizing it, that overweight people are lazy; if so, any candidate who is overweight will suffer from this stereotype.

If an applicant looks very impressive in one particular area, the interviewer may concentrate on that to the exclusion of other matters. Weaknesses in the candidate's background may be overlooked, or their significance may be discounted. This tendency is known as the *halo effect*. The opposite tendency, known as the *horn effect*, is to turn one negative characteristic into a conclusion that the candidate is weak on all fronts.

The proper phrasing and timing of questions is often a difficult problem to solve. In a structured or semistructured interview, questions should be specific enough to draw out the necessary information, but they should not prematurely reveal what answer the interviewer would like to hear. For instance, the question ''have you had experience running a Model 2000B?'' suggests that the operation of a 2000B is a job requirement, and the candidate may be tempted to inflate his or her experience. A better initial question might be ''what kinds of machinery have you handled?'' Then a later question could zero in on the 2000B.

Interview Guidelines[10]

Given the importance most organizations attach to interviews, it seems advisable for larger firms, at least, to undertake research and training programs to increase the effectiveness of their interviewing. In many cases, these programs may be needed to satisfy antidiscrimination requirements. For example, certain questions should not be asked in an interview because they are potentially discriminatory (see Fig. 9-5). Even in smaller firms, interviewers should be selected carefully and trained as much as possible.

Although the needs of different organizations will vary, it is possible to suggest the elements that an effective interview generally includes. The interviewer begins by studying all the materials already available on the candidate. By planning every session in advance, the interviewer knows which topics are critical for each individual applicant.

The candidate is shown to a quiet room, free from distractions. No phones are ringing; no one interrupts with messages for the interviewer. To begin, the interviewer engages in general talk about neutral subjects to put the applicant at ease and establish rapport. Soon, however, the interviewer turns the discussion to the matter at hand.

Whether the format is structured or nondirective, the interviewer listens closely to everything the applicant has to say, avoiding snap judgments and categorizations. One goal is to understand the applicant's own outlook. If the format allows, the interviewer pursues hints and follows up important leads, but remains careful not to turn the session into an interrogation. Encouraged to ask questions, the applicant understands that his or her needs are being considered. The interviewer avoids not only verbal threats, but also gestures and other nonverbal signals—such as glancing impatiently at the clock—that might increase the applicant's nervousness. If the applicant tends to be aggressive or abrupt, the interviewer remains unruffled.

Throughout the discussion, the interviewer focuses on both technical qualifications for the job and the intangible qualifications, such as motivation, energy, and enthusiasm. In evaluating the intangibles, the interviewer considers not just the particular job opening, but also the entire staff of that department, the characteristics of the coworkers and superiors, and the organizational climate and culture. The interviewer takes notes on the candidate's answers, but does so discreetly, so as not to hinder the discussion.

Before the session ends, the interviewer makes certain that the applicant understands the exact nature of the job. It is important for both the company and the candidate that any misconceptions be dispelled. Sometimes an interviewer will need to ''sell'' the firm—that is, emphasize its attractiveness to persuade a good candidate to accept the job—but the persuasion should not encourage unreasonable expectations.

The applicant is given a date by which he or she will learn the firm's decision; or if a next step is already planned, the candidate is told exactly what it will entail. When the interview is over, the interviewer personally conducts the candidate back to the reception area.

Don't ask if an applicant is single, married, divorced, or separated. Unless this information is job related (and it is very unlikely to be), the firm has no legitimate use for it.

Don't ask the ages of an applicant's children. This question might lead to charges of sex discrimination if the organization declined to hire a woman who had small children at home.

Don't ask about pregnancy or plans for a family. An interviewer may ask how long the applicant plans to stay on the job and whether any absences are anticipated—*if* the same questions are asked of males and females. The firm may also refuse to hire a pregnant woman if the job might endanger her health.

Don't ask someone's age or date of birth. This might lead to age discrimination. The firm may ask if the applicant is between the ages of eighteen and seventy, and if he or she is not, it is permissible to inquire the exact age.

Don't ask an older applicant how much longer he or she plans to work before retiring. Since this question would not be asked of a thirty-year-old, it should not be asked of a sixty-year-old.

Don't ask if the applicant has ever been arrested. The company may ask if the applicant has been convicted of a felony, but only if this is clearly job related.

Don't ask about a person's birthplace, ancestry, citizenship, or "native tongue." The answers might indicate national origin. The firm may inquire whether the applicant is a U.S. citizen and what foreign languages he or she speaks.

Don't ask for a maiden name or the name of the next of kin. Again the matter of national origin is involved. But it is legal to ask for the names of relatives who already work for the organization or for competing firms.

Don't ask if the applicant has a handicap or disability or has ever been treated for certain diseases. The firm may legally ask if the applicant has any health problems that might affect performance on the job.

Fig. 9-5 An example of questions that shouldn't be asked during an interview. ***(For more information, see Richard D. Arvey,*** **Fairness in Selecting Employees** ***(Reading, MA: Addison-Wesley Publishing Co., Inc., 1979), pp. 164-181)***

Finally, the interviewer writes up the notes from the interview, often on a standardized form. In evaluating the candidate, the interviewer tries to allow for subjective factors that might influence his or her judgment. Was there anything personally objectionable about the applicant? If so, is it related to the job requirements? Do the judgments entered on the interview report have a rational basis?

When the interview follows these guidelines, the firm should have a solid body of information on which to base the last phase of the selection process, the selection decision.

Decision Guidelines[11]

When all of the candidates have been tested and interviewed and their references have been checked, the final decision generally rests with the supervisor or head of the department in which the job opening exists and who will select the one candidate most qualified for the job. The human resources department, however, regularly must approve the salary and benefit package to ensure consistency of pay scales throughout the firm.

The job offer itself can be made by a supervisor, or manager, but in large firms it is frequently handled by the human resources staff. The offer is sometimes extended by phone or in a letter. In other cases, the candidate is called in for a final interview, and the offer is made in person. At this time, the salary and benefits are stated precisely; the prospective employee is told of any further conditions that must be met, such as passing a physical examination; a starting time is established; and, if the candidate needs time to think the matter over, a date should be set for notifying the firm of acceptance or rejection. Aspects of the position such as salary and benefits should have been discussed earlier in general terms so that the candidate is not surprised by any of the particulars.

In filling a job opening, the human resources staff should not forget the applicants who failed to get the job. All those who participated in tests or interviews should be quickly notified of the firm's decision, and those who progressed through the selection process should receive personal letters. Consideration for these people is important for the organization's reputation. Moreover, the "near misses" may be candidates for future job openings.

The success of a selection decision depends on all of the previous steps in the staffing process: the initial planning, the recruitment of qualified candidates, and the various selection techniques used to find the best candidate for the job. Only by careful attention to all of these stages can an organization assure itself of hiring capable people.

NEW EMPLOYEE ORIENTATION

On the first day on a new job, an employee is faced with an unfamiliar situation to which he or she must somehow adjust.[12] New surroundings, new coworkers, and new job procedures can make even the calmest and most competent workers feel anxious and insecure. Therefore, most organizations offer some kind of program to help the new employee get acquainted with the company and make a productive beginning on the job. This is normally referred to as the orientation program.

The orientation procedures vary in their usefulness to the new employee, depending on whether the program is haphazardly or systematically designed. It is important to periodically update the program to reflect changes in the company. While the focus in this section is on new employees, current employees should be kept up to date on any new changes.

In general, research and experience indicate that orientation procedures should be thoroughly planned and that those conducting the programs should address specific problems faced by new employees. No one orientation system will be best in all circumstances, but it is clear that participative approaches and genuine human warmth and concern for each individual are vital. The human resources department should play a key role in planning and coordinating the orientation program in collaboration with managers and supervisors. Managers and supervisors should also receive the necessary training so that they can effectively carry out their orientation responsibilities.

A poorly planned or nonexistent orientation program can have unfortunate results. Consider the case of a new employee who is assigned to sweeping the warehouse three times a day when once would have sufficed, or the new worker who is sent on foolish errands by other employees. This behavior is known as hazing, or harassing with unnecessary tasks or practical jokes. Hazing is damaging to the newcomer's morale and also lengthens the time it takes for the new employee to become productive in the new working environment.

The type and amount of information that employees will need will vary with the job.[13] However, it is customary to initially provide information about matters of immediate concern to them, such as working hours, pay, introduction to coworkers, a tour of the facility, and parking facilities. New employees should also have a clear understanding of work rules, safety rules, security requirements, and any other important matters. Later on, attention may be devoted to informing them about those areas that have a lower priority and/or that require more time for presentation and comprehension. Orientation sessions should be supplemented with a packet of materials that new employees can read at their leisure. Some materials that a packet might include are shown in Fig. 9-6. Instructions to the employee on how to use the packet are recommended.

A systematic orientation program may last only a few hours or may extend over several weeks. Information may be given through interviews, group meetings and discussion, handbooks, films, tours, or combinations of these and other methods. To avoid overlooking items that are important to employees, many organizations devise checklists for use by those responsible for conducting some phase of orientation. The use of a checklist in the initial orientation of new employees compels a supervisor to pay more attention to each new employee at a time when personal attentiveness is critical to building a long-term relationship. An example of an orientation checklist is shown in Fig. 9-7 on p. 9-11. Many programs include follow-up interviews at the end of 3 or 6 months' employment to determine how well the new employee is getting along.

Items for an orientation packet:

- ☐ A company organization chart
- ☐ A projected company organization chart
- ☐ Map of the facility
- ☐ Key terms unique to the industry, company and/or job
- ☐ Copy of policy handbook
- ☐ Copy of union contract
- ☐ Copy of specific job goals and descriptions
- ☐ List of company holidays
- ☐ List of fringe benefits
- ☐ Copies of performance evaluation forms, dates, and procedures
- ☐ Copies of other required forms (e.g., supply requisition and expense reimbursement)
- ☐ List of on-the-job training opportunities
- ☐ Sources of information
- ☐ Detailed outline of emergency and accident-prevention procedures
- ☐ Sample copy of each important company publication
- ☐ Telephone numbers and locations of key personnel and operations
- ☐ Copies of insurance plans

Fig. 9-6 Items for an orientation pack. ***(Source: Walter D. St. John, "The Complete Employee Orientation Program,"*** **Personnel Journal** ***(Costa Mesa, CA: May 1980), reprinted with permission, all rights reserved)***

A number of potential problems are associated with the orientation procedure for new employees. For example, giving too much information in an orientation session can be as much of a problem as providing too little. If a great deal of information is given to the employee all at once, he or she may feel overwhelmed and is not likely to retain much. Those who design the orientation program should be sensitive to such matters as how much information to provide at a given session, how to sequence the various parts of the program, and how well the new employee is assimilating the information. One way to help a new employee retain all of the information is to develop an orientation booklet. Providing plenty of opportunity for questions and discussion is an effective way to clarify the presentation. The new employee should also be assigned to a specific person until he or she is comfortable in the new job and to whom he or she could go to for answers to questions that may arise.

TEAM DEVELOPMENT

Being with people is a lifetime experience. At work, at worship, at recreation, and at home, one must interface with others. The story of civilization is how people have worked together in increasingly complex ways. This section of the chapter examines how a *group* of people can work together so as to grow and develop into a *team*.

"Group," "task force," and "committee" are words to describe a collection of people who work together. How effectively they work together depends on the degree to which they are a team. A team is a configuration of people with special characteristics. Individuals can be energized by being on a team. They tend to perform better and with more creativity together than alone. They usually are comfortable with each other, and being on the team fulfills their personal and professional needs. Fundamentally, team members feel they belong to a dynamic and productive unit in which the whole is greater than its constituent parts.

While it is known that there is a special magic about a good team, the question is how does one create it. What does one have to do either as leader or member to make the group move from independence to interdependence?

All teams start as a collection of individuals. They start to work together, and something happens to weld them into a functioning team. The complex pattern of their interactions help the team grow and develop. Fortunately for those who want to build teams, much is known about how this happens.

All groups go through certain predictable stages as they mature. Knowing these stages gives the leaders and members of groups control over the process and speeds the building of the team. This section reviews these stages and points out what must be done at each stage to help the team grow. Team leaders learn what behaviors to look for from the members and what he or she should do with the team at each stage of its growth.

All successful teams go through five stages of growth. Each stage has its own specific behavioral characteristics. Unsuccessful teams usually get stuck at one particular stage and develop no further. They are locked into infancy or adolescence. The five stages are best remembered as forming, norming, storming, producing, and ending. Each stage has its own issues and its own predictable member behavior. At each stage, there are specific things the leader must do to ensure maturation of the team.

Employee's Name:	Discussion Completed (check each item)
I. Word of welcome	
II. Explain overall departmental organization and its relationship to other activities of the company	
III. Explain employee's individual contribution to the objectives of the department and his [or her] starting assignment in broad terms	
IV. Discuss job content with employee and give him [or her] a copy of job description (if available)	
V. Explain department training program(s) and salary increase practices and procedures	
VI. Discuss where the employee lives and transportation facilities	
VII. Explain working conditions: a. Hours of work, time sheets b. Use of employee entrance and elevators c. Lunch hours d. Coffee breaks, rest periods e. Personal telephone calls and mail f. Overtime policy and requirements g. Paydays and procedure for being paid h. Lockers i. Other ______	
VIII. Requirements for continuance of employment—explain company standards as to: a. Performance of duties b. Attendance and punctuality c. Handling confidential information d. Behavior e. General appearance f. Wearing of uniforms	
IX. Introduce new staff member to manager(s) and other supervisors. Special attention should be paid to the person to whom the new employee will be assigned	
X. Release employee to immediate supervisor who will: a. Introduce new staff member to fellow workers b. Familiarize the employee with his [or her] work place c. Begin on-the-job training	

If not applicable, insert N/A in space provided.

Employee's Signature	Supervisor's Signature
Date	Division

Form examined for filing:

Date	Personnel Department

Fig. 9-7 Example of an orientation checklist. *(Source: Joseph Famularo, ed.,* **Handbook of Modern Personnel Administration,** *Chapter 23, "Orientation of New Employees," by Joan Holland and Theodore Curtis (New York: McGraw-Hill Book Co., Inc., 1972), reproduced by permission)*

The leader has to get the team to work well at its tasks. The team must be productive, hard working, and quality minded. Unfortunately, many leaders see this as their only concern. They push and drive the team forward, and discussion is focused on task issues only. Anything else is considered time wasting.

There is another dimension to consider, however. The leader must monitor the interpersonal transactions that take place in the group. If a leader is a highly task-oriented person with little sensitivity for interpersonal relations, the group struggles and fails to become a cohesive unit. It is ironic, but a 100% concentration on task to the exclusion of interpersonal issues is counterproductive.

The leader's role is to carefully balance out both the task and relationship dimensions as required. There are times when the team needs to totally concentrate on task. There are also times when the leader has to take time out from task issues to work on the interpersonal relationship issues. Then there are times when task and relationship must be worked on simultaneously. At each stage of development, the balance is different. Figure 9-8 summarizes the five stages in terms of the major issues, member behaviors, and leadership tasks typical of each stage.

FORMING

Stage 1, forming, is best typified as the time when there are many unspoken issues worrying the group. The members' behavior is at its most formal at this time, so the very necessary questions about goals, operating structure, and role relations never get asked. Instead, the members of a group in stage 1 secretly monitor each other very carefully for clues and hints on how to act in the group.

The members in this stage cannot be regarded as a team in any real sense of the word. They are a group of individuals. They feel no sense of group identity. They are not sure about whom they can trust. They are uncertain about what they are there for, what is supposed to happen, and what is in it for them. They are uncertain about how they will be treated. They are fractured.

It is possible for a group to remain in this state for a very long time. Some groups exhibit the above characteristics indefinitely. Individuals representing widely differing skills or experience, brought together for purposes that are not clearly stated and who

	Stage 1—Forming	Stage 2—Norming	Stage 3—Storming	Stage 4—Producing	Stage 5—Ending
Task functions + major issues	Trust, orientation, and answers to these questions: — What is going to happen? — What will the experience be like? — Who are the other people? — Where do I fit with the other people? — How will I be treated?	Primary norms established are: — Team responsibility — Cooperation plus participation — Decision making — Confronting problems — Leadership — Quality plus excellence — Interaction analysis	Increased conflict results from: — Openly confronting problems — Increased participation — Testing the norms — Increasing independence from leader	Productive Groups: — Use a range of task and process behaviors — Monitor their own accomplishments — Work on the accomplishment of goals while attending to the interpersonal needs of the members	Any separation from, and new member additions to, the team: — Creates anxiety — Reduces productivity — Creates conflicts
Interpersonal relations + member behavior	Characterized by: — Anxiety — Search for structure — Silence — Reactive to leader and other members	Characterized by: — Power struggles — Attempts to keep discussions leader focused — Silence — Hostility — Physical or psychological withdrawal — Frequent topic changes — Aggression — A failure to commit to action plans	Characterized by: — Attacks on the leader — Polarization of the team members — Increased testing of the group norms — Flight or fight behaviors	Characterized by: — Increased interpersonal cohesiveness — Work in both the task and process areas — Increased closeness with the leader — Active listening — Active resolution of conflict — Active participation — Discussions which focus on here and now information	Characterized by: — Increased conflict — A breakdown in group skills — Lethargy — Frantic attempts to work well — Anger directed at the leader — Withdrawal — Anger directed at other members — Denial that change has occurred
Leadership behavior	Reduce the uncertainty by: — Explain goals and purposes — Provide time for questions — Provide time for them to get to know each other — Model expected behaviors — Be aware and attend to your own non-verbal behaviors	Encourage norm development by: — Redirecting questions — Reinforcing positive listening — Modeling positive listening — Fostering the development of mutual goals — Teaching and enforcing consensus — Providing team-centered learning	Legitimize conflict by: — Examining your own response to conflict — Reinforcing positive conflict resolution efforts — Acknowledging it as a condition for change — DO NOT become more authoritarian	Help the team maintain its skills by: — Being prepared for temporary regressions — Reinforcing work in both the task and process areas — Providing feedback on the effectiveness of the group — Assisting in gaining more meaning from the meetings	Legitimize termination anxiety by: — Acknowledging that the group is really ending or a member leaving — Encouraging an expression of feelings — Helping members review experience — Developing action plans for outside the group — Preparing for Stage 1 Major Issues and Member Behavior

Fig. 9-8 Five stages of team development. (*The General Systems Consulting Group*)

have no opportunity to interact outside of the group, may find themselves in this situation. Unless someone takes time away from the main business of the meeting to deal with the individual uncertainties and questions about structure, it becomes progressively harder for the team to develop any further.

Relationship Issues

A major relationship issue for the group at this stage is trust. A high level of mutual trust is the one absolutely essential ingredient of a productive team. The evolution of the team is closely linked to the growth of trust within the team. It is the barometer of the team's progress toward maturity.

How does one gage the level of trust in the team? It is helpful if the team identifies the following behaviors and qualities:

- Sharing—passing information about some personal matters, feelings, and so on.
- Vulnerability—acknowledging that to err is human.
- Loyalty—committing to the goals of the team.
- Accepting—tolerance for the differences between people.
- Involving—using the team for ideas and decisions.
- Valuing—willingness to share beliefs and ideals with others.
- Awareness—sensitivity to needs of others.
- Communicating—being clear in spoken and written dialogue.
- Openness—willingness to explore new experiences.
- Honesty—avoiding deceit.

The behaviors just described identify the level of trust within the team. If each and every member is convinced that these 10 qualities are present to a large extent within the team, then trust levels are high.

Task Issues

A major task issue for a team at this stage is orientation. People in a stage 1 team usually have a set of questions that need answering. Sometimes these questions are well formed, such as "what is the agenda, what is this guy going to be like, and when will this be over?" Sometimes the questions are unformed and reflect little more than a sense of discomfort and apprehension. Until these individual issues are addressed, the members are probably reticent and guarded in their interactions. In short, they do not move toward team cohesiveness until these initial orientation issues are dealt with honestly.

Leader Behavior

Stage 1 can be a tough time for those leaders who want the team to be truly participative. On the other hand, for those leaders who want to dominate and make most of the decisions themselves, stage 1 is perfect. This is the stage when the members want to be steered, given cues, and given decisions. They have no feeling of responsibility for the work of the team. They do not clearly understand or accept the goals of the group. They are reactive to the leader.

If the leader takes up the role of the front runner, accepting total responsibility for the work of the group and not fully sharing all of the relevant information, then the group will not develop any further. As with all the other stages of team development, the role of the leader is crucial. The correct actions taken now will move the team toward further development.

The major question to be answered is what can a leader do at this stage to ensure the development of the group. At this stage, the major answer must be to reduce the uncertainty. This uncertainty can be reduced through by task and relationship actions.

Task Actions

At this point in the development process, team members have a great need for structure. The leader should continually remind the group of its goals. "Tell 'em, tell 'em, and finally, tell 'em again," is a worthwhile motto at this stage. The leader should look for opportunities to relate the current discussion to the goals of the team. Constantly referencing the goals of the group eventually removes the sense of uncertainty being felt by all the members at this stage.

There is often a temptation by the leader of a stage 1 team to rush ahead with the task. Because the predominant response to the leader is silence at this stage, this temptation is natural. As has already been emphasized, the silence should not be taken as meaning that there are no questions. The leader has to devise strategies for exposing the questions and underlying issues that are holding back the team's development. This may just take the form of asking and waiting in silence until the questions flow. Alternatively, the leader might hand out paper and ask for anonymous questions to be written down.

Relationship Actions

An early issue for people in this stage is the desire to know more about the other team members. They want to know what they think and what they are like. A prudent team leader provides opportunity for the group to meet socially outside of the business setting. Sports coaches take their teams to training camp. The reasons for doing this are only partly athletic. The camp also provides the opportunity for developing the social togetherness that welds a group of skilled individuals into a team.

The effective team leader should have a clear idea of the sort of member behavior he or she wants. Frankness, honesty, an ability to question assumptions, flexibility, and energy are all desirable features. The leader needs to be a role model for the team in all of these areas at this time. The team is very reactive to the leader, and each member closely studies the leadership style for clues on how to behave.

Knowing this, the leader carefully monitors his or her nonverbal behavior also. If the leader is talking about participative decision making, then his or her whole demeanor should indicate a willingness to accept input from others. Looking intensely bored, fatigued, or angry does nothing for the development of the team at this stage, and cutting people off in the middle of a sentence and having side conversations will send eloquent, but disruptive, messages to every member of the group.

In summary, the leader must shape the team from the front. The members need knowledge on a variety of issues. This knowledge has to be provided by the leader by any means possible. This means the leader has to provide opportunities, strategies, and role modeling to get the team to fully understand each other, the team goals, and the team's mode of operation.

NORMING

Stage 2, norming, is best typified as the time when the group develops both formal and informal procedures by which it will operate. Initially, a degree of trust has been developed, and the members have begun to feel a part of the group. The members now have to work out their decision-making processes, understand the leader's style, and develop their interaction patterns.

TEAM DEVELOPMENT

At this stage, a work team begins to adopt and develop a set of observable behaviors called operating norms. The group does not become effective until productive norms are in place. The task of the group in stage 2, therefore, is to concentrate on developing sound operating practices in such areas as decision making, group responsibility, leadership style, and the review and assessment of team performance. These are important task areas that ensure that the team effectively confronts problems, is committed to quality and excellence, and gives full cooperation and participation. Until this is done, group effectiveness is compromised.

As the task norms develop, certain relationship issues often surface. Power struggles and some hostility to the group leadership are common. The team might find it hard to commit to any action plans and constantly ask for more time to revisit issues that had been decided earlier. It is important at this time for the leader to steer the group toward productive norms. This is best done by setting aside time for the members to explore and develop guidelines for their operating behavior in the team. The main benefit of doing this is to give each member a mental checklist for observing the team behaviors.

The leader of the group, once the operating norms have been identified, must constantly reinforce those norms by his or her own behavior. Having agreed-on norms enables the leader to refer to them at appropriate moments for maximum training effect. They are also useful for group critique sessions as they provide the team with a set of operating standards for assessing the team's performance.

Task Norms

The task norms that all productive groups need to develop are decision making, confronting problems, leadership, and quality and excellence.

Decision making. All groups make decisions. It is important that the group develop the manner in which the decisions are to be made. A productive group has a range of decision-making strategies suitable for all occasions.

Confront problems. Confronting problems means that the group "owns its problems" and takes time to discuss them openly. It does not try to shirk its responsibilities. All groups have difficulties that may arise out of the nature of their work or out of the personality mix in the team. A very productive team has norms that facilitate addressing these difficulties. The assumption is that by confronting the problem issues the team will grow in levels of trust and cooperation.

Leadership. Any group will have certain assumptions about the manner in which the leadership of the group is to be managed. It is best if these assumptions are initially clarified so that the group can be certain of the style to be expected. A productive norm in this area is for the team to actively pursue the concept of shared leadership whenever appropriate. This means that members take responsibility for guiding the discussions, clarifying issues, and monitoring the progress of the group.

Quality and excellence. Good teams always strive for being the best and achieving the best performance of which they are capable. All teams experience periodic cycles of low energy. A commitment to quality and excellence keeps the group from taking shortcuts or making hasty decisions at such times.

Relationship Norms

The responsibility norms that all productive groups need to develop are team responsibility, cooperation and participation, and interaction analysis.

Team responsibility. The team responsibility norm is observed when group members begin to direct their attention to group goals rather than their own individual issues. In other words, each team member gets their personal needs met through the achievements of the group. This shows itself by members sharing ideas, changing their minds, being flexible, actively listening, and supporting the efforts of other team members. This norm develops an awareness of interdependence.

Cooperation and participation. A cooperation and participation norm is allied to team responsibility. A member who actively abides by this norm is seen to be very involved in the work of the team and contributes a united effort. Participation means sharing both ideas and doubts. It means not only providing information and insights, but also soliciting input.

Interaction analysis. In most cultures, it is not considered "good form" to critique the behavior or performance of individuals in front of others. Yet, if a team is to successfully develop any of the operating norms previously outlined, there must be an agreement to openly discuss the group's process in the group setting. This is a key norm, therefore, because without it the team will not continue to improve its performance.

Roles

It is at this stage that the team members start to develop characteristic roles within the team. Some of these roles are productive, some counterproductive. It is convenient for a team leader to break the roles into those that contribute to pushing the team on with their task, and those that help develop the working relationships within the group. And, of course, identify those counterproductive or blocking roles.

Some of the task facilitating roles are as follows:

- Direction giving—providing strategies for moving the team forward.
- Information seeking—asking those important questions that clarify the thinking of the team.
- Information giving—voicing ideas, facts, and judgments.
- Elaborating—developing and building ideas based on the information being shared.
- Coordinating—putting various ideas together in novel ways to create new concepts.
- Reality testing—preventing "group think." Asking the important question of "will it work?"

Some of the relationship building roles that the team leader should begin to develop are as follows:

- Supporting—acknowledging the good work of others with positive feedback.
- Harmonizing—finding common ground for differing opinions.
- Tension relieving—bringing humor into tense situations.
- Norming—drawing attention to unproductive behaviors.
- Energizing—taking time to revive the team's enthusiasm.

Some of the roles that are counterproductive to a team's working relationship are as follows:

- Fault finding—being negative about all new ideas.
- Overgeneralizing—getting obsessive about one idea as the answer to everything.
- Overanalyzing—excessive nit-picking; paralysis by analysis.
- Premature decision making—not analyzing enough; rushing into a decision.

- Presenting opinions as facts—claiming factual knowledge without real evidence.
- Rejecting—putting a low value on ideas because of the personality of the author.
- Pulling rank—overriding the team through the use of status.
- Dominating—cutting people off, talking over people.

STORMING

Growth implies struggle, a pushing against boundaries and limits. Physical, or athletic, development demands muscular stress and strain. A distance runner will lose the fine edge of performance if there is not a constant regimen of running.

Teams are not an exception to this general pattern of development. A time comes when the team has reached a level of understanding of the norms and personalities within the group. Now comes the period of trial. The team's health and fitness require it.

A developing team has to this point understood the goals of the group. Emphasis has been placed on participation and shared responsibility. Attempts should have been made by the leader to build trust among all members so that they can freely contribute their ideas to the discussions. The leader has been trying to take a less prominent role in the work of the team.

In a sense, everything is now set up for conflict. The group members feel free to speak openly about what is on their minds. If they are irritated with each other, they will tend to express their feelings. They voice their reservations about the ideas of others in the group. They express their differences. This is the time that they try to reshape the group norms or the team goals. The leader may be ignored. Emotionality may creep into the interactions.

What has to be understood at this point is that stage 3 is a situation of uncontrolled conflict for a newly developing group. If it learns how to handle or control this first conflict situation, then future conflicts are brief, less painful, and lead to a productive resolution of the issues. In other words, storming is not a once-and-for-all occurrence; it is cyclical. The major difference between a mature group and a forming group lies in the productive way the conflict is resolved.

A mature group listens to the reasons for the meeting, gives its input, and analyzes its differences. Disagreements are expressed and explored. Alternative suggestions are made, and final agreements are reached. Conflicts are certainly present, but they are contained, handled well, and produce a quality outcome from the team's interaction. The main reason for developing a team is to use the creative input of its members. A group of clones produces nothing more than one member could have produced alone. Diversity of experience and differences of opinion will produce the creativity and insights essential for good team outputs. In short, out of conflict comes productivity.

Unfortunately, conflict can also tear a team apart. A leader needs to know what identified the conflict stage, what are typical member behaviors at this stage, and finally, what he or she should do at this time to assist the team in building on its experiences.

Conflict

One of the most important things for a leader to determine is whether or not the team is in conflict. This may seem like a foolish question because surely it is very obvious when a team is in conflict. Yes, in some ways it is; however, there are times when the behavior is so low key that a leader or member of the group might be misled into thinking that all is well. Most people tend to think of conflict as being noisy. Red faces, bulging eyes, and table thumping are the typical images. There are other less obvious behaviors, however, to the skilled observer.

Attacks on the leader. In the earlier stages of development, the membership of the team is very reactive to the leader. The leader is held in some respect, and the members look for a great deal of guidance from their leader. A team in stage 3, however, starts to turn on the leader. Very frequently, the members blame the latter for any lack of progress. This may be a low-key criticism, but the implication usually is that "we were misled" or "you told us differently last time."

It is easier at this stage for members to vent their frustrations on the leader than on each other, even though the real cause for the conflict may be coming from the behaviors of the other members. When a leader starts to get suggestions about how they might do things differently or what they have done wrong in the past, then he or she should realize that they are observing storming behavior.

Polarization of team members. A team in stage 1 is a fragmented group of individuals. They build cohesiveness as they work together. There is a point, however, when the group starts to pull into subgroups or cliques. Sometimes the cliques are formed around strong personalities in the team, sometimes around interdepartmental structures, or sometimes around positions on critical issues. Whatever the reason, a seasoned observer will note that instead of a cohesive team unified by a common purpose, there are several small groups each committed to its own agenda. Again, this may not be evidenced by a strident drawing up of battle lines. The symptoms may be relatively peaceful. A team, however, cannot reach true productivity while having several factions existing within the group.

The goal of any team should be to have its members sharing no allegiance other than to the successful completion of the team task. If any of the members cannot give this commitment, the team will always be operating in stage 3 and never reaching true 100% team productivity.

Testing of group norms. An established group has developed understandings about how to work together. Some of these will be stated rules and procedures. Others are more informal understandings about the approved ways of behaving, dressing, listening, and speaking. Up to this point, these formalities and expectations (norms) have been useful and helped the team members find the structure necessary for their own comfort within the group.

Team members find, however, that when they reach a certain level of comfort within the group, they need to push a little against the constraints that teamwork necessarily implies. In many ways, this is a parallel of adolescent behavior where a developed knowledge of the operating standards of society is a trigger for testing behavior. Once a level of ease has been achieved with those standards, there develops a need to push and shove against what are now seen as annoying constraints.

This testing of the group's norms may be overt. There may be dominance exerted where none was exhibited earlier. Attempts to talk individuals "down" may occur. Side conversations may erupt where once people listened to one speaker at a time. Sometimes the behavior is less obvious and on the surface very passive. Some members indicate conflict by such behavior as arriving late to meetings or even missing meetings. Good excuses may be offered, but the truth is that the meeting or the group task now takes a lower priority in that person's mind than it once did. These characteristics indicate a team in stage 3.

Leader Behavior

The storming stage is a tough time for team leaders. In the earlier stages, the team was leader dependent. This might bring its own problems, but at least the group was somewhat pliant and responsive to the leader's direction. In a productive team, the members fully share the responsibilities for the team's work and understand when the leader needs to make command decisions. This in-between stage of storming tests the leader's skills.

The fundamental strategy the leader must adopt is to get the team to understand, and admit, that they are currently in stage 3. In addition, they should accept that being in this condition is a legitimate part of teamwork and to be expected. As long as the conflict is kept submerged, it cannot be resolved. Making it an open agenda item is important for the team's growth.

Some of the things that a leader might do and some of the issues he or she must be aware of to bring the team to stage 4 are to examine personal response to conflict, provide positive reinforcement, acknowledge conflict, and control authority.

Examine personal response to conflict. The point of understanding team dynamics is to give the leader and the members better control over what happens in the team and to reach high levels of productivity in a short time. The leader at stage 3 needs to focus on the word "control." The leader needs to ensure his or her inner conflicts do not sway them or divert them from the team's goals. A leader's task is to struggle with their team's "knee-jerk" responses to situations and to try and take a rational and controlled approach to the issues.

Positive reinforcement. The leader may be able to resolve, single-handedly, all of the conflicts that arise in a team. But it is certainly easier if some of the membership helps too. In stage 2, positive task and relationship roles that people can play in the team were discussed. Whenever the leader sees someone "relieving tension," "harmonizing," or any other positive behaviors, he or she should find ways to reinforce the effort. A quiet comment after the meeting, or even within the meeting, models the sort of behavior that the leader requires.

Acknowledging conflicts. Change means moving into the unknown. It means going from comfort to risk and uncertainty. It is hard to imagine change without some degree of anxiety and conflict. If the purpose is placid continuity of the status quo, conflict is unhealthy. If change and decisions have to be made, then conflict is healthy and, indeed, necessary. Conflict never feels that way, of course. The leader's task at this stage is to keep emphasizing that out of conflict and stress come creativity and quality decision making.

Authority. An earlier statement in the section on forming said, "The evolution of the team is, in reality, linked to the growth of trust within the team." Up to now, the team has been growing in trust and understanding of each other. The very fact that they are in conflict is testimony to the growth of trust and confidence within the group. They feel free to speak out, they feel free to challenge. The leader does great damage to the group if, at this stage, he or she starts to assert authority. This certainly submerges the conflict, but instead of moving the team forward, the group regresses back to stage 2 or even stage 1. The group starts exhibiting all of the earlier characteristics of being reactive to the leader, such as silence, some hostility, and so on.

On occasion, a leader has to exert authority. For example, a member may be blocking the team so badly that little else remains to be done but to remove that member from the team. This is the last resort, of course. The leader should expect, however, a temporary regression of the group. The hope, of course, is that the team will speedily move back through the stages and become productive. This can be risky. It must be clear and accepted by the team that the person has left not because his or her personality did not suit the group, but because there was an insurmountable block to progress.

PRODUCING

Stage 4 in team development is what the leader has been trying to achieve—a dynamic, creative, and hard-working team of people. The leader will recognize that the group is a skilled and mature team by the way it balances the task and relationship dimensions of its work. A truly productive team keeps a nice balance between working hard on the assigned tasks and taking care of the working relationships within the team. The team members share in the leadership of the team, freely making suggestions and exploring all of the issues in a positive manner. Any disagreements are openly dealt with in a rational and controlled manner, and there are no hidden agendas.

Task Skills

A productive team provides itself with as much feedback as possible. Constant monitoring of its achievements is important for a team at this stage. Every member needs to feel that the team is accomplishing its tasks successfully. A good team has confidence in its history and in its abilities. The members constantly reinforce the team's work by monitoring its progress.

Perhaps the key feature that marks out a productive team is its sense of oneness and common purpose. A team without common goals is fractured, mistrustful, and prone to destructive conflict. Unity of purpose provides the bond for a team. There is a shared responsibility for the work of the team.

Interpersonal Relations

In a producing team, there is a great deal of personal cohesiveness. The members feel close to the leader and each other. They actively listen to each other and are very skilled at asking the right questions at the right time. They do not shy away from conflict, but when it arises they handle it effectively.

The productive team has focus. Its members are absorbed in the work, and there is a high level of participation. Everyone contributes to the task, and the discussions seldom stray off the point. In summary, it feels good to be on such a team. The members are positive about being on the team and feel effective.

Leader's Actions

The leader's function in this stage is mainly that of maintenance. The leader has to reinforce the team's good work and monitor both the task and the relationship activities of the members. Periodically, the team will regress. Conflict becomes uncontrolled, and the team's operating norms are violated. The leader then has to bring the team back to stage 4.

ENDING

The leader of a productive team should be aware that if there is any change in the team's membership, the team regresses. An old member leaving breaks up the team's cohesiveness. The unity is fractured, and there will likely be noticeable behavior changes. The relationship and task skills of the team tend to break down. Many of the behaviors typical of stage 1 and stage 2 are evident.

This kind of reaction occurs when members leave and particularly when the leader leaves. Any new leader taking over a productive team should not expect to slide into his or her seat without some breakdown in the team's cohesiveness. Levels of trust have to be reasserted. New norms may have to be learned. With skill and sensitivity, however, the new leader should be able to bring the team back to its previous levels of productivity.

Finally, when a successful and productive team finishes a major project, there is also a breakdown of team spirit and team skills. The leader needs to be ready for the ''ending'' symptoms of lethargy, withdrawal, and even some anger that a ''good thing'' is coming to a close.

The leader, when faced with a significant ending for the team, should help the team acknowledge its feelings and structure some activity that helps people express their thoughts on the matter. It is helpful to review the team's accomplishments at this time and to develop some action plans for what might be done next. Put at its most basic element, the leader should make the change, or ending, an event. Allowing the change to take place without comment is counterproductive at this stage. The team members need to ''celebrate'' the passing of an old member, leader, or successful project. The wise leader stages that event.

SUMMARY

A good team is something that has grown and developed. It does not happen by chance. The skilled leader can build a good team by taking it through the first three stages of growth, by helping it deal with membership and task changes, and by maintaining it at a high level of performance.

Team development is not a linear process continually moving forward from stage 1 through to stage 5. It will be cyclical, moving forward and then slipping backward as the team struggles with its dynamics. The skilled leader is not depressed by the continual regressions. He or she will use them as learning experiences for the team.

TRAINING AND RETRAINING

The issues of training and retraining arise each time there are changes in work processes, technology, or organizational systems. The dimension of the need depends on the extent of the desired change in organization, group, or individual performance. The training effort is focused on those elements of change that require application of new knowledge and skills or the application in news ways of existing knowledge and skills. The determination of the difference between existing and desired performance becomes the key to identifying the content for training and retraining programs. In addition, it may become the basis for the selection of various educational programs or even technologically driven alternatives to meet the performance need. This process is referred to as needs assessment.

Needs assessments are carried out in response to several situations involving organizational, group, or individual performance. The first situation is where there is a currently perceived performance problem. It is a ''now'' situation. What is desired now does not exist, and a determination must be made of what solutions can be implemented to resolve the difference.

The second situation is where a process of planned change is about to occur and a determination of what needs to be done to bring about this planned change must be made. This also is a ''now'' situation, but unlike the first situation, it is not reaction based. Rather, as a needs assessment process, it is proactive.

The third and final situation is where a process of change is planned as part of a long-term strategy. This is a future situation. The needs assessment process is a proactive one that focuses on future performance needs and the identification of what preparation must be made to ensure successful implementation.

In these three situations, a needs assessment may identify a need for some combination of technological change, system or process change, compensation or benefit change, knowledge or skill acquisition or application, or some other type of change. The causes of the performance problem that are identified as a result of the needs assessment can be classified into four broad categories: lack of skills and knowledge, environmental, attitude/motivation, and aptitude.

Training and retraining activities are designed around the outcomes of these needs assessment activities. However, there are some best and most appropriate applications for training and retraining strategies. First, training and retraining strategies are most effective when focused on those items identified as performance problems that have skills and knowledge acquisition and application as part of the solution. The first mandate of a training or retraining strategy decision is to determine if a lack of skill or knowledge contributes to the cause of the performance problem. Second, training generally is not effective when dealing with problems caused by environmental factors such as working conditions or union-management relationships. Third, training cannot deal with problems caused by motivation or attitude factors such as pay, benefits, or the organization's personnel policies. However, training and retraining may be effective in those instances where additional knowledge or skills may influence attitude or motivation.

TRAINING VS. EDUCATION

The section on job analysis identified categories of knowledge and skills that may be used by an expert. Included in that profile of knowledge was generalizable knowledge, as well as knowledge that was specific to a job, organization, or process. In planning for training or retraining of a work group, it is helpful to determine the skill and knowledge needs that are the most appropriate for training to provide. To do this, it is beneficial to understand the difference between training and education.

Education explores the various views, perspectives, and dimensions associated with any topic, concept, or skill application. In addition, knowledge derived from an educational experience is considered to have a high degree of transferability. Training limits the scope of knowledge and skills it provides to specific content and applications to achieve desired expert worker behavior and in response to the needs of the organization. As such, there are several key words that differentiate training from education. However, the definition of training contains key words that embrace the quality criteria of education.

CHAPTER 9

TRAINING AND RETRAINING

Organizations may choose to develop, acquire, and deliver educational programs. When this is done, it is frequently called training. Educational institutions frequently deliver educational curriculums and programs at an organization's sites and call this training. Recognizing that the term ''training'' is used by many individuals and organizations to cover a wide variety of activities allows managers to think more clearly and strategically about training and retraining issues.

ECONOMIC DIMENSIONS OF TRAINING AND RETRAINING

Organizations have limited economic options when faced with the dilemma of developing new knowledge and skill bases within their workforces to remain competitive or to carry out goals. The first option is to fire the elements of the existing workforce that lack the required knowledge and skill base to carry out new tasks and to hire new workers who possess the desired skills and knowledge. The second option is to retain and retrain the existing work group. The third option is to foresee the economic, organizational, legal, political, or social implications of either of the first two options and to quit the marketplace or flee to another environment that has more acceptable alternatives to these issues. One outcome of exercising this latter option has been a shift of industrially skilled and semiskilled labor-intensive production from economically and industrially advanced Western nations to less developed ones.

TRAINING STRATEGIES

Training can take place through a variety of strategies. The most commonly used strategy has been unstructured on-the-job training. Unfortunately, most on-the-job training programs are frequently not planned and thus do not work well. This process pairs new workers with older, experienced ones to ''watch and help out.'' After some period of time, this shadowing activity results in a worker who is able to perform most of the tasks required to acceptable standards. However, it is possible for on-the-job training to be very structured. In this scenario, the learner spends time with specific experienced workers learning desired skills and obtaining knowledge in a planned and structured sequence of experiences and activities. Still, on-the-job training is frequently considered the most economical because of the apparent lack of training fees and other instructional costs. However, much research in recent years suggests that this is not accurate. During periods of on-the-job training, trainees perform at less than desired levels. There is a severe cost associated with this less than desired performance.

Unstructured training is most effective when there are very small numbers of workers to be trained and when the tasks being learned are of a simple procedural nature. Studies, however, have shown that workers who go through unstructured training take longer to achieve desired levels of job performance and are less able to troubleshoot, diagnose, or correct malfunctions as they carry out tasks than those who go through structured training programs.

The second strategy for training is to send workers through a formalized (structured) training experience. This may take the form of classes that may be on or off site. The instructors in these types of training programs may be trainers, expert workers from the organization, or vendors who are expert at the content to be delivered. Structured training may also take place on the job. When structured training involves on-the-job activities, the worker's activities and experiences must be planned and structured to ensure total exposure to all skill and knowledge applications required for all possible aspects of the job and not just those that happen to occur during the training period.

Structured training is most effective and efficient when there are more than a few workers to be trained or when the content is complex or requires the acquisition of abstract knowledge, diagnostic skills, and troubleshooting skills. Structured training can also take many forms. As indicated previously, structured training can take place on the job. However, it takes detailed planning on the part of managers to ensure that the worker is provided an opportunity to observe, practice under expert guidance, and carry out under qualified supervision all aspects of the job. Some strategies for structured on-the-job training include coaching, special assignments, and task detailing.

Coaching offers an individualized approach to the training needs of the worker. The one-to-one communications provide an opportunity for immediate feedback and constant reappraisal of the progress toward the learning objectives. In addition, the interaction between the learner and the trainer takes place at a point in the organization where there are immediate and directly observable payoffs.

Special assignments are especially useful for specialists who need information about practices and procedures in other organizational areas affecting their performance. The task detailing strategy is useful for workers who need to correct one deficient characteristic of their work behavior. This process requires the worker to carefully document each step, phase, and decision-making activity that he or she completes as they perform job tasks. This documentation is then compared to required procedures or recommended decision-making processes to help identify and correct performance problems.

Other forms of structured training include self-instructional courses as well as formalized classroom and laboratory learning experiences. Self-instructional courses may be designed around the use of computer-aided instruction (CAI), videotapes, correspondence courses, and/or self-paced programmed instructional (PI) materials.

Computer-aided instruction and programmed instructional materials are easily packaged and then sent to training centers, and they allow individuals to proceed at a pace most comfortable for their ability level. The major limitation of programmed and computer-aided instruction is the expense and extensive preparation necessary. Another disadvantage of PI is that it focuses primarily on factual material and limits learning of value-based skills and supporting principles. People-to-people skills by using knowledge gained from CAI and PI need to be taught by using other strategies.

The formalized classroom experience with supporting laboratory training activities is the most frequently considered form of structured training. Lectures, supplemented by structured activities and audiovisuals, are the most common strategies in this situation. The lecture method as a stand-alone strategy for training is generally considered economical. However, it may not be a good strategy when complex responses (for example, motor skills) are required from a large number of trainees during training. It is a good strategy, however, when the acquisition of abstract knowledge is the goal. Identifying and understanding the training objectives will help determine the appropriateness of the lecture option.

The laboratory experience provides an opportunity to practice skills and apply knowledge in a controlled environment. This is frequently referred to by trainers as simulation. It allows the trainer to control parameters of the real-world setting that might make the learning difficult. A well-designed laboratory experience allows the trainees to have expanded time or repeat experiences on tasks to meet their individual learning needs. In some cases, laboratory experiences result in usable products or services that help offset associated costs.

TRAINING RESOURCES

When it is determined that a need exists, the search to address the skill and knowledge deficiency begins. There are two general resources that can be used to meet training needs. The first and most obvious place to look for resources is internal to the organization. Can the behavioral or learning needs be met in whole or in part by already existing programs in the organization? Frequently, there is not a 100% match. Rather, distributed throughout several training programs may be modules dealing with the topics or issues of the training need. What is needed is simply to reconfigure several of these modules into a focused training program. Another internal process is to develop "in-house" the needed program from scratch. As discussed earlier, this will require the determination of skills and knowledge that expert workers use in the performance of their job, determination and selection of the most cost-effective and efficient training strategy, and the implementation of the training program.

The second resource is an outside vendor who can develop and deliver the desired training. As with the internal resource, the training program may have to be developed entirely from scratch or may be put together from modules or components of the vendor's existing programs.

Another approach is to combine the first two resources. In this arrangement, an organization's trainers and the vendors collaborate to develop and deliver the required training. This particular alternative ensures access to expertise for all aspects of the project and still allows the best compromise to control costs.

External vendor sources include professional training organizations, consultants, educational institutions, and suppliers of the technology around which the training needs revolve. Professional training organizations typically specialize in some subject matter or area of training, such as management, electronics, computers, and others. Consultants tend to specialize in some subject matter area as well as some element of the process associated with the development and delivery of training. One example would be a consultant who may be an expert at needs assessment, training design, or evaluation and may also be a subject matter expert in management communications.

Suppliers of technology frequently provide training for the technology or systems they sell. Generally, this is part of a sale of the system or technology, although several vendors offer the training component as a separate optional cost item in the sale. As with the case of any training, the suppliers may develop and deliver the training using internal resources, or they may have exclusive contracts with vendors to deliver required systems support training.

The final major source of training is educational institutions. These may include vocational and technical schools, community colleges, and universities. Frequently, organizations that receive public funds to aid in workforce development may be required to obtain training support from these public institutions as part of the terms of the funding. An example of this is the Ben Franklin Partnership Act in the Commonwealth of Pennsylvania. Through this act, funds are made available on a matching basis to organizations to meet workforce training needs. However, the training must be delivered or coordinated through an educational institution in the geographic location of the enterprise. Several states currently have programs similar to this with many of the same constraints and guidelines.

CONTINUING AND ADULT EDUCATION

Several distinct categories of knowledge used by expert workers were identified previously in this chapter. One major characteristic of the knowledge categories discussed was that some of the knowledge was very job and organization focused, while other knowledge was of a very generalizable nature. Training focuses on the delivery of knowledge and skills that are job and organization specific. Education focuses on knowledge that is highly generalizable from organization to organization and even industry to industry. For the adult learner, the formal process of education tends to be organized around degree or nondegree programs. The latter includes certificate earning programs, special-interest workshops, and topical courses.

Within this framework, continuing education and adult education are treated as having shared commonalities and distinct differences. Colleges and universities, as well as many professional organizations, generally refer to their adult education activities as continuing education. Within this context, continuing education is most often associated with nondegree, certificate-earning, and licensing-maintenance education activities. Course content is organized around defined bodies of knowledge that represent minimum requirements associated with recognition of some level of professional expertise. For example, the required completion of minimum continuing education credits from defined course lists accompany the issuance of long-term licensure in many states. Adults may earn professional certification by completing prescribed continuing education courses. For example, it is possible to earn an accounting certificate from several universities after the completion of 20 credits of prescribed continuing education coursework from the institution.

Adult education is frequently separated from continuing education based on the formality and loci of content control of the learning experience. This suggests that adult education is a learning process that takes into account the following elements: natural and unplanned learning, formal learning experiences where the learner is primarily responsible for the design and conduct of his or her own learning activities, and formal learning experiences where some other educational agent is primarily responsible for the management of the learning. Implicit in this perspective of adult education is the concept that adult education is not concerned with preparing people for life, but with helping people to live more successfully. The overarching function of adult education is to assist adults to increase their competence, to negotiate transitions in their social roles, and to better prepare them to solve personal and community problems.[14]

DELIVERY SYSTEMS FOR ADULT AND CONTINUING EDUCATION

The delivery systems available for adult and continuing education include a variety of traditional and nontraditional educational strategies. Several traditional strategies include formal coursework for adult learners at vocational education institutions, community colleges, and four-year universities and colleges. Frequently, these courses are offered through special outreach programs that may involve the instructor traveling to remote sites to offer the course in a sequence of special all-day modules. Another strategy is to offer coursework through special closed-circuit television links to several designated sites simultaneously. Depending on the available technology, this may be a one-way presentation or the activity may have two-way television and sound linkage. Another variation on this is the offering of coursework through the public broadcasting systems.

Another less traditional continuing education delivery system is teleconferencing. Although this system requires a great deal of planning and preparation to have visual support materials available to follow along with the lecture, it allows relatively inexpensive and simultaneous access to several sites as well as feedback and question capabilities. Radio also provides a vehicle for continuing and adult education courses. Special interest group workshops, correspondence courses, cassette courses, video courses on special topics, and computer-based courses complete the list of frequently used methods for delivering continuing and adult education programs.

Frequently, the unit of recognition for successful completion of a formal adult education or continuing education course is the "continuing education unit." This widely adopted method for quantifying the value of noncredit continuing education activities was approved in 1971 by the Southern Association of Colleges and Schools. Since its implementation, it has been commonly used as a barometer for advancement and relicensing of professionals in a variety of occupational areas.

NONTRADITIONAL ALTERNATIVES TO HIGHER EDUCATION

During the 1970s, a variety of nontraditional degree-granting programs evolved to serve a need by adult learners for formal recognition of lifelong learning experiences and acquired knowledge and skills through nonformal structured learning experiences. These nontraditional programs tend to possess one or more of the following characteristics: open admissions, granting of credit for knowledge obtained outside of academia, qualification for degree solely by examination, utilization of part-time nonfaculty teachers and mentors, development of individualized learning programs by students with or without the assistance of mentors, fulfillment of learning objectives through contracts, utilization of unconventional methods of learning and teaching, and the inclusion of noneducational institutions in a degree or certification-granting consortium.

An emerging alternative to degree-directed studies at institutions of higher education are the corporate colleges. These programs often work with local colleges and universities to offer courses that can be applied to a degree.

IMPLICATIONS TO MANAGERS

Continuing and adult education programs provide choices for managers. As discussed, there are two generally defined groups of knowledge associated with expert work behavior: job and organization focused and highly generalizable. From a management perspective, supporting worker involvement in continuing or adult education activities may be viewed as a costly benevolant (fringe benefit) activity, as an investment for the organization, or as a contribution to society. At any given time, managers will have to analyze these choices within the economic, political, and social climates of the organization and its environment. The short-term economic implications of the investment or noninvestment decision may be difficult to measure. However, the literature and current world economic and social trends may provide some insight into the long-term return.

SUMMARY

Workforce development is the part of a strategic management process concerned with determining the direction of the organization and effecting decisions aimed at applying appropriate quality and quantity of labor to meet the organization's short and long-term objectives. Workforce development is a strategic planning process that is the responsibility of all managers. While the overall strategic plan for an organization may be evolved at the highest enterprise level, decisions regarding workforce redeployment, creation, and development to support the organization's goals and objectives are made by managers throughout the organization.

As has been brought out in the various sections of this chapter, this can be a complex, potentially cumbersome, but nonetheless critical task. However, managers who wish to be successful in the arena of workforce development can achieve this goal by considering the strategies, concepts, and techniques offered in this chapter. Managers are encouraged to pursue the further suggested readings and to contact individual researchers and practitioners whose names appear in the chapter and in the list of cited literature.

References

1. R. Zemke and T. Kramlinger, *Figuring Things Out: A Trainer's Guide to Needs and Task Analysis* (Reading, MA: Addison-Wesley Publishing Co., Inc., 1982).
2. R.A. Swanson and D. Gradous, *Performance at Work: A Systematic Program for Analyzing Work Behavior* (New York: John Wiley & Sons, Inc., 1986).
3. Wendell L. French, *Human Resources Management* (Boston: Houghton Mifflin Co., 1986), pp. 242-243.
4. Mary T. Matthies, "The Developing Law of Equal Opportunity," *Journal of Contemporary Business* (Winter 1976).
5. Ruth Gilbert Schaeffer and Edith F. Lynton, *Corporate Experiences in Improving Women's Job Opportunities* (New York: The Conference Board, 1979), pp. 36-39.

6. French, *op. cit.*, p. 244.
7. *Ibid.*, pp. 263-264.
8. *Ibid.*, pp. 264-265.
9. *Ibid.*, p. 265.
10. *Ibid.*, pp. 266-268.
11. *Ibid.*, pp. 268-269.
12. *Ibid.*, pp. 278-283.
13. Herbert J. Chruden and Arthur W. Sherman, Jr., *Managing Human Resources*, 7th ed. (Cincinnati: South-Western Publishing Co., 1984), pp. 181-185.
14. Gordon G. Darkenwald and Sharan B. Merriam, *Adult Education: Foundations and Practices* (New York: Harper & Row Publishers, Inc., 1982).

Bibliography

Gael, S. *Job Analysis* (San Francisco: Jossey-Bass, Inc., Publishers, 1975).

Goldstein, Irwin L. *Training in Organizations: Needs Assessment, Development and Evaluation*, 2nd ed. (Monterey, CA: Brooks Cole Publishing Co., 1986).

Gross, Ronald, ed., *Invitation to Lifelong Learning* (Chicago: Follet Corp., 1982).

Harless, J.H. *An Ounce of Analysis (Is Worth a Pound of Objectives)* (Boulder, CO: Marlin Press).

Heinrich, R.; Molenda, M.; and Russell, J.D. *Instructional Media and the New Technologies of Instruction* (New York: John Wiley & Sons, Inc., 1982).

Houle, Cyril O. *The Design of Education* (San Francisco: Jossey-Bass Inc., Publishers, 1972).

Kearsley, G. *Computer Based Training* (Reading, MA: Addison-Wesley Publishing Co., Inc., 1983).

Knowles, M. *The Adult Learner* (Houston: Gulf Publishing Co., 1973).

Laird, D. *Approaches to Training and Development*, rev. ed. (Reading, MA: Addison-Wesley Publishing Co., Inc., 1985).

McCormick, E.J. *Job Analysis: Methods and Applications* (New York: American Management Association, 1979).

McGregor, D. *The Human Side of Enterprise* (New York: McGraw-Hill Book Co., 1985).

John M. Peters and Associates. *Building an Effective Adult Education Enterprise* (San Francisco: Jossey-Bass Inc., Publishers, 1980).

Walton, R.E., and Lawrence, P.M., eds. *HRM Trends and Management* (Boston: Harvard Business School Press, 1985).

Warren, M.W. *Training for Results: A Systems Approach to the Development of Human Resources in Industry* (Reading, MA: Addison-Wesley Publishing Co., Inc., 1979).

WORKFORCE MANAGEMENT

It is the responsibility of the managers—line and staff—in a manufacturing operation to build a smooth working team. Sloppy organization, poor communications, bureaucratic "empire building," and adversarial employee relations won't cut it into today's world of fast-paced technological change and intense domestic and international competition.

In preceding chapters, philosophy and culture, organization, manufacturing leadership, management of technology, and workforce development were discussed. Now it is time to apply theory and make things happen where the action is—on the factory floor. This chapter will discuss some basics of workforce management—programs that are being used throughout industry that may be applicable to a particular situation or provide a frame of reference for improvement. At the end of the chapter, the essential elements for developing a workforce that can use computer-integrated manufacturing (CIM) and the new technologies to survive and prosper in today's dynamic world will be summarized.

CHAPTER CONTENTS:

COMPENSATION

Compensation can take many forms. It is generally related directly to work performed and is most often paid in the form of wages, salaries, overtime pay, or piecework pay. The term "wages" is frequently used to refer to the cash compensation given to hourly paid workers who are employed in nonsupervisory positions. Salaries, on the other hand, are considered the sums paid to nonhourly rated employees.

It is important for every company to develop its own compensation policy. Depending on the size of the company, this responsibility is carried out by a staff department of wage and salary administration (large company) or by the controller, treasurer, or administrator whose responsibilities are in other fields (small company). Although it is difficult to state a policy that will have universal application to all companies, the following factors are usually considered important:[1]

1. The company's compensation policy should be consistent with its stated personnel policy. The company's wage and salary levels should be at least equal to those prevailing for similar jobs in the labor market.
2. A consistent basis for determining the relative worth of the jobs within the company should be established and adhered to. As a part of this, the job content should be reviewed periodically.
3. Provision should be made for periodic review of wages and salaries and a sound basis for merit increases established.
4. Consideration should be given to the establishment of individual or group incentives wherever applicable to reward additional effort of the employees.
5. Standards of performance should be established at reasonable levels consistent with the policy of a "fair day's pay for a fair day's work," and these standards should be maintained at this level.
6. Complaints arising from any phase of wage and salary administration should be given prompt attention and settled in accordance with the facts of the case.
7. Complete information on all phases of the program should be expressed in writing and distributed to all employees, and management should be willing at any time to discuss any phase of the program.

The purpose of this section is to examine various compensation techniques, the settings in which they have or have not been effective, and where they would most probably be useful. For the purpose of this discussion, the techniques have been broken into two categories: hourly rate programs and salary programs.

***Contributors of sections of this chapter are:* Jessie Bernstein**, *President, Employee Assistance Associates, Inc.;* **Derek B. Cross**, *Principal, Hay Management Consultants;* **Robert J. Giovannetti**, *Partner, Jackson, Lewis, Schnitzler, and Krupman;* **Frank Petrock**, *President, General Systems Consulting Group;* **Hayward Thomas**, *Consultant, Retired—President, Jensen Industries;* **Peter M. Tobia**, *Vice President, Kepner Tregoe, Inc.;* **Benjamin B. Tregoe**, *Chairman, Kepner Tregoe, Inc.*

***Reviewers of sections of this chapter are:* Thomas E. Wood**, *Retired Partner, Hewitt Associates;* **James R. Bowers**, *Senior Principal, Hay Management Consultants;* **Paul F. Geene**, *Corporate Personnel Manager, U.S. Manufacturing Corp.;* **David LaBeau**, *Manager Human Resources, Society of Manufacturing Engineers;* **Lawrence A. McKay**, *Senior Vice President, Theodore Barry & Associates;* **Dr. H. E. Trucks**, *Consultant.*

CHAPTER 10

COMPENSATION

FACTORS INFLUENCING COMPENSATION

Before discussing the various compensation techniques, it is important to set the stage by providing a brief overview of the factors influencing the type and amount of compensation given to each employee. Because many complex factors enter into the picture, it isn't as simple as determining that each employee be paid a certain amount of money per hour or per piece. Workers also receive paid sick leave, paid vacations and holidays, and premium pay for working overtime and on holidays. These and other forms of indirect renumeration are commonly referred to as fringe benefits; benefits and services are discussed in a separate section of this chapter.

Traditional compensation practice has been to pay employees a straight hourly, weekly, or annual base rate and add a package of benefits worth up to 40% of that base. Certain limited groups such as executives and salespeople have traditionally had some long or short-term incentive arrangements. Management positions frequently include some bonus eligibility.

These traditional practices are under increasingly critical scrutiny. Companies, under the stress of intense competition, are seeking ways to make their people more cost effective and responsive to organizational needs.

There is a strong desire to make compensation a more variable expense, by sharing some of the business risks and rewards with employees. At the same time, the advent of flexible and easily accessible computer power has made possible the administration of complex, highly individualized programs.

Many of the techniques that are "nontraditional" in this context have long but limited histories of successful application. In the proper set of circumstances, they can be adapted and brought into the mainstream compensation practices of many organizations.

In addition to the intense pressure that manufacturing organizations in North America are experiencing because of foreign competition, the factors influencing compensation include labor legislation, collective bargaining, the labor market, geographic location, cost of living, comparable worth, and ability to pay.[2]

Labor Legislation

Several federal laws have had a marked influence on minimum wages, working hours, and payroll withholding. Chief among these are the Davis-Bacon Act of 1931, the Social Security Act of 1935, the Walsh-Healey Public Contracts Act of 1936, the Fair Labor Standards Act of 1938, and the Equal Pay Act of 1963. The principal provisions of these five laws are summarized in Table 10-1. Additional information on the legal aspects of wage and hour legislation can be found in Chapter 11, "Legal Environment for Labor Relations," of this volume.

Collective Bargaining

For those employees who are members of unions, one of the major factors influencing their wages will be the union scale, which results from collective bargaining. The labor agreement usually provides a pay scale for various job classes that will be in effect during the life of the agreement. It may also stipulate the minimum earning potential under incentive systems.

The effect of collective bargaining in a labor market will be felt even by nonunionized companies. Some employers may want to equal or exceed the union scale to avoid the unionization of their employees.

Over the years, the collective bargaining in several major industries has had considerable influence on other companies. There has been a tendency to use one large company as a lever against another company, and this has had tremendous impact on the wage picture in both large and small firms. Also, the union's demands for a guaranteed employment plan and other fringe benefits have added considerable indirect payment to the workers.

One of the goals of management in many companies is to offset wage increases with increases in productivity. As a move in this direction, they have asked for changes in work rules that have led to rigid job classifications and inflated staff levels.

Another management strategy is to reduce the number of full-time workers with seniority who do all of the jobs. Companies are relying increasingly on part-time and temporary workers, who get lower wages and fewer benefits than full-time employees. Among the industries most aggressively moving in this direction are airlines, electronics, trucking, and retail food.

The Labor Market

In a labor market, the supply and demand for a particular type of labor will tend to seek a balance. Seldom, however, will it attain an absolute balance, and wide differences will usually prevail in the hourly rates paid by various employers for equivalent labor on similar jobs. Owing to the immobility of labor, surpluses of some skills may exist in one area while shortages exist in others.

The shortage of qualified workers for certain jobs may have considerable influence on the wages demanded by persons possessing those skills. From time to time, this has been the situation for engineers. The frequent shortage of engineers has had considerable influence on young engineers' starting pay.

Geographic Location

Wage differentials have always existed among different geographical areas. Some jobs are rated higher, or paid more, in a large city than in a small community, or in the North than in the South (although the South is catching up). The common explanation for this difference is that the cost of living is less in one area than in another.

Cost of Living

During periods of rising prices, the cost of living becomes an item of major importance in wage determination. It frequently may be one of the foremost items posed by the unions in collective bargaining sessions and has led to "escalator" clauses being incorporated into many union agreements. Based on the provisions of these clauses, the wages of the employees are adjusted upward or downward periodically, depending on the rise or fall of a cost-of-living index (the one prepared by the Bureau of Labor Statistics, United States Department of Labor).

Tying wages into the cost of living may prove to be a troublesome consideration for management because sales, income, and profits may not coincide with the retail prices used in calculating the cost-of-living index.

Comparable Worth

One of the questions that a number of women's organization have raised in recent years is "Why should there be a wage gap between men and women, and why should women earn only about 72 cents for every dollar earned by men?" This has led to much dispute (including strikes) and to a concept referred to as comparable worth. Comparable worth is a controversial theory that jobs have an intrinsic, measurable value that should dictate the wages paid. Federal law states clearly that workers perform-

TABLE 10-1
Principal Provisions of Federal Labor Legislation Affecting Wages and Salaries

Name of Law	*Provisions Pertaining to Wages and Hours*
Davis-Bacon Act of 1931	Provided for payment of prevailing wage rates (for similar job classifications on similar projects) on Federal construction projects in excess of $2000 and established an eight-hour working day, except for certain emergency contracts
Social Security Act of 1935	Provided for payroll withholding from employees and for employer contributions to cover a system of old-age and survivors insurance and unemployment insurance
Walsh-Healey Public contracts Act of 1936	Established certain standards of work done on U.S. Government contracts exceeding $10,000, specifically, payment of time and one-half for all hours worked in excess of eight in one day or 40 in one week. It also permitted the fixing of prevailing minimum wages in those industries performing contracts
Fair Labor Standards Act (Wage and Hour Law) of 1938	Provided for minimum hourly wages and time and one-half the employee's regular rate for overtime over 40 hours per week for employees engaged in interstate commerce, including those in any closely related process or occupation directly essential to such production. The statute set forth what was meant by minimum wages, stated conditions under which overtime provisions might be exceeded, exempted certain employees—executives, administrators, professionals, and outside salesmen—from coverage, and prohibited the employment of children under 16 (18 for hazardous occupations) in commerce or in the production of goods for commerce. The Act has been amended several times, and the minimum rate has steadily increased.
Equal Pay Act of 1963	Amended the Fair Labor Standards Act to provide that both sexes should receive equal pay for work demanding equal skill, effort, and responsibility. Made provision for wage differentials based on merit, seniority, and wage incentives

ing the same job cannot be paid differently because of their sex or race. Nonetheless, the advocates of comparable worth would take these guarantees a step further: employees in such traditionally male jobs as accountant, construction worker, and truck driver would no longer make 30-40% more than the holders of traditionally female jobs like nurse and secretary.

An analysis of the wage gap between men and women reveals that it is a complex phenomenon. In general, women work at different jobs from men, with fully half of all women concentrated in three occupations: clerical, sales, and professional. These occupations command salaries in the labor market commensurate with the supply and demand of people able and willing to do the work. Thus, many people question the need to make comparable worth the law of the land.

Ability to Pay

"Ability to pay" has been used frequently by employers to counter a demand for wages that they feel are excessive. One of the best illustrations of this usage arose in the General Motors negotiations several years ago where both the company and the union made claims and counterclaims as to the company's break-even point, costs, and profit picture. In other cases, the employer has used this argument to request a reduction in wages on the grounds that the present wage level would jeopardize the future of the company. For the most part, however, employers have argued that the ability to pay has nothing to do with wages.

Over the long run, a company's wage policy must reflect its ability to pay. The company cannot afford to pay wages at an exorbitant level at the expense of its customers or owners. At the same time, it must pay wages at the level necessary to maintain the desired caliber of the workforce. If wages were cut because the company was operating at a loss, some of the employees would leave to work elsewhere.

HOURLY RATE PAY PROGRAMS

The term "hourly rate employee" generally refers to blue collar factory workers and the less skilled office and clerical workers who are paid for the hours they work, or with incentive plans for the pieces produced. Hourly workers are usually hired at a starting rate from which they may move to the top rate for their particular classification through a series of either automatic progressions in rate, usually made at monthly or quarterly intervals and/or a series of periodic merit reviews. Once at the top rate for a specific classification, an hourly rate worker's pay will be increased only if he or she is promoted to a higher rated classification or if there is a general increase for all hourly rated employees. Annual general increases were the norm for many manufacturing companies before the decade of the 1980s. Now a number of companies, facing stiff foreign and domestic competition, have negotiated zero-increase or wage reduction contracts with their unions or have implemented similar programs in nonunion shops. A partial or full alternative to the annual increase that is gaining some popularity is an annual or periodic bonus that is tied to profitability and that does not increase the base rate and the employee benefits that go with it.

Employee benefits, which are covered in a following section, may add as much as 40% to the cost of an hour of labor. The combination of hourly pay and these benefits have made the American factory worker the most highly paid in the world. Averages can be deceptive, however, There are rural areas, not limited to the South, and areas within cities, often those with a large immigrant population, where hourly pay scales are relatively close to the minimum wage and where employee benefits may be limited. Such pockets of lower wages can be found in many major industrial cities. These wages are still several times higher than the pay scales in emerging countries. For example, workers in the Mexican Maquiladores Program doing work just across the border for American firms may be taking home less than $1.00 per hour, and the total cost to the employer including fringes may be in the area of $1.50-$3.00 per hour. The point is that an hour of labor in an American factory is an expensive commodity, and it must be used effectively for the factory to be competitive; the manager's challenge is to use this hour wisely. Some of the hourly rate pay programs in general use today are the measured day work and fair day's work.

Measured Day Work

Most volume manufacturers have labor standards that are used for costing products, calculating plant capacity, scheduling work, and controlling productivity. As the term implies, in a measured day workshop output per worker or group of workers (like on a paint line or assembly line) is compared to the standard to calculate efficiency or productivity, terms which will be discussed in more detail under measuring performance. A measured day workshop is considered to be doing very well with 100% performance against a methods time measurement (MTM) base standard or equivalent. Average factories are in the 60-70% range. Standards may be known to the worker and may be the subject of considerable debate in both union and nonunion shops.

Fair Day's Work

A fair day's work program is a variation of measured day work in which the company policy can be summarized as follows: "We pay a fair wage based on the work that you are doing and pay scales for other employers in this area. In return, we ask that you come prepared to work using our tools and our methods, stay on the job, and apply yourself during the work day. If you maintain your side of the bargain we will be satisfied with your production." While labor standards are usually available as a guide to supervision, specific output numbers are not discussed with employees, particularly when there is a problem with the employee's output. This tends to reduce "speed-up" complaints or grievances. If proper tools and methods are specified, as they should be in any circumstance, good productivity can be achieved. When there are problems, the manufacturing, methods, or industrial engineer can concentrate on assuring management that proper methods are followed. If this kind of an approach is taken, most employees will respond favorably. When they do not, discipline can be administered for failing to follow instructions rather than for failing to produce the required number of pieces. This approach appeals to most people's sense of fairness. It is used to varying degrees in many shops and departments by good supervisors, even though there may be no formal company policy.

Pay for Hours Worked

There are still many factories where standards are not available; a large number of these are ripe for productivity improvement. A manager in this situation should give priority to developing some kind of output objective or target against which to measure the workers. In a job shop, the job cost estimate may provide a basis for determining satisfactory output. Work sampling or ratio delay studies can also provide the manager or supervisor with a reasonably accurate picture of how fully occupied his or her people are. One technique that can be used is to calculate the number of people working and those idle or off the job at random times and then calculate the percentage working. If less than an average of 90% of the people are working, there is a good potential for significant improvement. A good manager or supervisor can usually develop a feel for how good the work pace is when people are working. Visiting other factories and taking MTM and/or methods training can hone this capability.

With these perceptions, measuring the current output and asking for enough more to fill out the day can produce surprising results, particularly if there is a good supervisory relationship with the people. Americans generally like to play on a winning team and have a sense of fair play. They usually can do a great deal more than is expected of them and often will when approached in a positive manner.

Production Incentive Plans

The most basic production incentive plans are built on the observation that workers paid by the day or hour are usually not exerting as much effort as they could be. They could literally be twice as productive.

Manufacturing engineers use the term "low task" to refer to the basic state of moderate effort and minimal structuring of the work. Output at moderate effort can be improved by methods analysis and standardization to attain medium task. Then effort can be increased by proper incentives to attain high task output.

Motivational psychology states that an incentive can stimulate greater raw effort. The management challenge is to accurately direct that effort in areas most profitable to the company.

Factors that have varying degrees of incentive value are improvement in present lifestyle (money, perks), rewards testifying to one's value to the company (money, perks, awards-as-scorekeeping), and less tangible daily strokes for being a valued member of a group. The simplest incentive uses money leveraged to individual productivity on a one-to-one basis.

One of the inescapable limitations of piecework and standard hour plans is the lack of any incentive to reward people for working together. For example, a worker who discovers a trick to increase his or her own productivity (and earnings) will keep it a secret. If the others use it, their production will go up also, and management will raise the standard. The logical solution is to implement a group-type production incentive that will reward cooperation.

Straight piecework. The oldest incentive system is straight piecework. The worker is paid so much per piece produced. In a pure straight piecework plan, the labor price of output is a constant, regardless of the amount of production. The setting of a production standard is not terribly important, except for weeding out the very least productive.

Selective piecework. Selective piecework plans offer low rates for the first units produced and then a higher rate for production above the standard or some percentage of standard (see Fig. 10-1).

Basic piecework exposes employees to heavy loss of earnings anytime anything interrupts or delays production. The addition of a "Manchester" guarantee pays the employee a base wage if he or she does not make minimum incentive for the day because of material shortages or equipment problems (see Fig. 10-2).

The more modern piecework plans give a guaranteed minimum up to a point between 65-80% of standard and add a piece incentive above that threshold (see Fig. 10-3).

Selective piecework plans are designed to provide some security, but to reserve most of the reward for the highest producers. Addition of the incentive threshold requires the setting of a production standard for each product.

Best candidates. Organizations that are good candidates for selective piecework plans have the following characteristics:

- Individual producers are self-paced and not dependent on anything but availability of materials.
- There is sufficient manufacturing engineering (time/motion study) capability for calculating standards.
- There are frequent changes of production, as in job shops. (This gives the standard little time to deteriorate.)

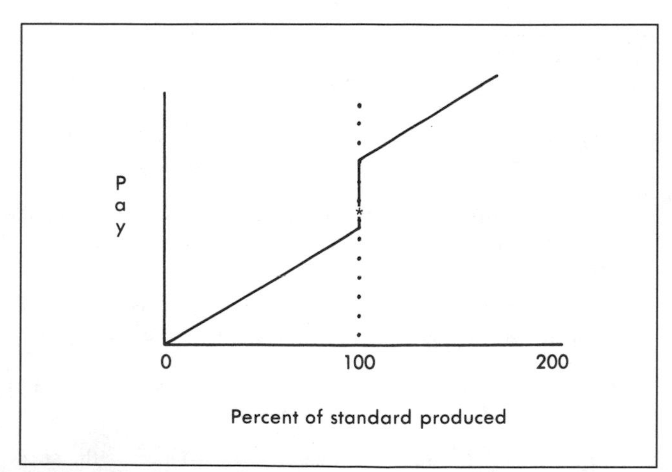

Fig. 10-1 Typical selective piecework payout/production plan.

Drawbacks. Two of the main drawbacks to selective piecework plans are as follows:

1. The definition of the standard is a continuous source of contention and manipulation by both company and employees.
2. The quest for speed/volume can lead to poor quality, excessive inspection requirement, and/or excessive scrap.

Standard hour plans. Standard hour plans calculate the standard in terms of units per hour at task. These are particularly appropriate when it takes a long time to produce a single unit.

If the standard is 0.5 units per hour, then the employee would be expected to produce 4 in an 8 hr day. In some instances, the employee would be quite capable of producing 5 units in the same time. An employee that produced 5 units would achieve 125% productivity. His or her wages would then be 125% of base pay.

In the absence of a minimum guarantee, the standard hour is as linear as the straight piecework plan. This rewards performance and strongly discourages those consistently unable to

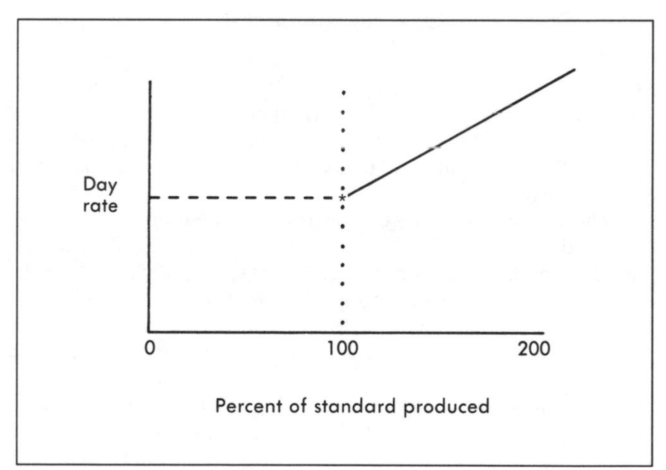

Fig. 10-2 Typical minimum or "Manchester" guarantee/production plan.

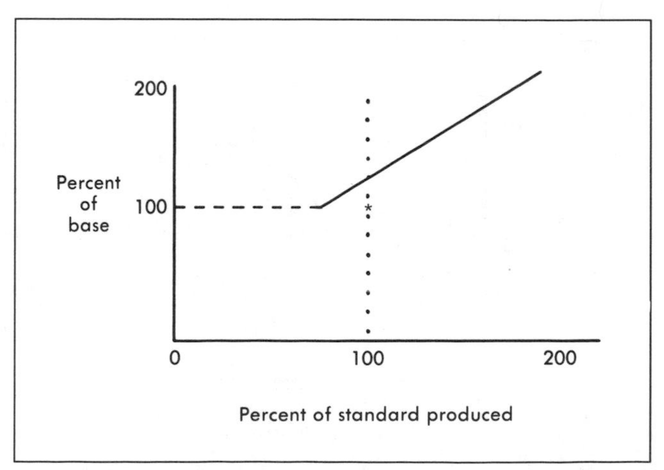

Fig. 10-3 Typical guaranteed minimum plus incentive/production plan.

meet standard. The addition of a minimum guarantee is not uncommon.

Best candidates. Organizations that are good candidates for implementing standard hour plans have the following characteristics:

- Stable, standardized, and repetitive tasks, especially those with long cycles.
- Individual producers are self-paced and not dependent on anything but availability of materials.
- There is sufficient manufacturing engineering (time/motion study) capability for calculating standards.

Drawbacks. Three of the drawbacks associated with standard hour plans are as follows:

1. Require thorough standardization of work and accurate calculation of standard (based on a percentage of "high task" output).
2. Severe fluctuation in earnings will occur if the standard is unrealistically high or if there are material/method disruptions.
3. Frequent changes in production, methods, or materials cause earning loss during "learning curve" and increase likelihood of problems/grievances.

Piece rate sharing. When conditions are not right for a standard hour plan (incomplete method standardization, difficulty in determining standard, frequent variations in process/materials) or when the workforce is being switched from day rate to incentive, a family of plans called "sharing" plans might be considered.

These plans begin incentive pay at something less than full standard, but then raise the piece rate at less than 1:1 (line A) or in a diminishing curve with increased production (line B) (see Fig. 10-4). The concept is that in return for incentive pay below standard (earnings protection during learning new product), employees share some of their high-end productivity with the company.

Best candidates. Conditions existing in a company that would justify going to a piece rate sharing compensation program are as follows:

- The manufacturing engineering capability is insufficient or the work mixture/situation resists standardization.

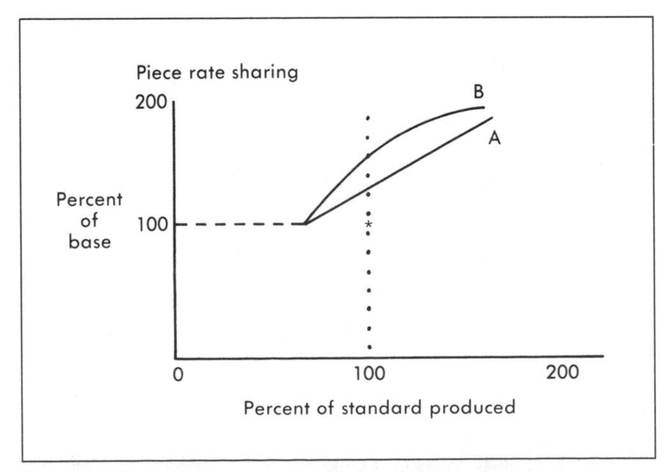

Fig. 10-4 Typical piece rate sharing/production plan.

- Inefficient (50-70% of standard) day workers being introduced to piecework. It gives an early taste of success and plenty of room and reward for improvement. The diminishing high-end returns are unlikely to be encountered often in such an environment.
- The manufacturing engineering capability is insufficient or the work mixture/situation resists standardization.

Drawbacks. Common drawbacks associated with piece rate sharing are as follows:

- Individual high producers would be better off on straight piecework.
- Philosophical arguments are likely to eventually arise about the fairness of getting credit for less than 100% of their high-end output.
- The nonlinear formulas used in piece rate sharing are difficult to communicate and awkward for administration and payroll.

Confident workers and company would tend to move from a sharing plan toward a standard hour or similar plan if standardization and engineering conditions can be met. If the company is characterized by frequent significant changes in products and employees go through a relearning/adjustment process often, they will tend to appreciate the protective feature of sharing.

Small Group Production Incentive Plans

Small group incentive plans make sense when there is an identifiable group of interdependent workers. Group incentives create rewards for sharing know how and for working together to iron out problems in the group's work flow.

Creation of an incentive group offers an opportunity for team building and exertion of some leadership by management. The key management tasks are to structure the groups and the program design to reward the right things and to establish identity between group "good" and company "good." For better or worse, the group has power to influence the work behavior of its members. A group incentive makes it worthwhile for a creative employee to share his or her productivity secrets with the group and for the group to intelligently work out any problems in their work flow or internal division of labor. An incentive group may be more receptive than individuals to new methods and technology.

Almost any of the piecework or standard hour plan formulas could be adapted to group incentives. Determining factors in formula selection would be the nature of the production process and the level of engineering support for standard setting and auditing.

Experience indicates that a small group incentive program will get greater productivity from that group than will plant-wide incentives.

Best candidates. Circumstances in a company that would warrant implementing a small group incentive plan are as follows:

- Small groups of interdependent workers perform sequential operations on the same end product, in a reasonably stable environment.
- A management group needs to improve cooperation and cohesiveness.

Drawbacks. Common drawbacks associated with small group incentive plans are as follows:

- Groups will act in self-interest, sometimes to the detriment of other groups, departments, or the company as a whole. There is no premium on cooperation outside the group.
- To maximize their long-term returns, production groups may restrict production at a bonus level just high enough to avoid triggering a revision of the standard. Key issues are standards, trust, and leadership.
- An individual high producer may feel unrewarded for contributing more than his or her share or may be pressured to restrict output to avoid making co-workers look bad.
- Compensation equity issues can become major problems when incentive groups are mixed in with non-incentive workers.

One group of solutions to these problems attempts to constructively harness employees' ability to manipulate the system by switching to a plan that directly links employee prosperity to either the productivity or profitability of the whole plant or company.

Large Group Production Incentive Plans

Large group production incentive plans can be divided into profit-sharing and risk/reward sharing (gain-sharing) programs. While profit-sharing programs augment employee compensation through the sharing of company profits, gain-sharing programs link additional employee compensation with productivity improvement. Two common types of gain-sharing plans in use today are the Scanlon and Rucker plans. Other customized types of plans are also available through compensation consultants.

Profit-sharing programs. Profit-sharing plans aim at creating a sense of employee participation in an enterprise and of sharing in a common destiny with everyone involved. Currently, there are more than half a million profit-sharing plans in the U.S. about evenly divided between cash plans and deferred benefit plans. Deferred benefit plans tend to function (and be federally regulated) as pension plans rather than motivation devices. As such, they are treated as benefits and not further addressed here.

Cash profit-sharing plans, especially if there are frequent payouts, tend to be more "real" to employees. The payment periods offer greater opportunity for feedback to employees on the state and needs of the business. How effectively management makes use of this opportunity varies widely.

Management. Profit sharing is compatible with any management style from paternalistic to participative. One major goal is to give employees an interest in participating in the success of the business.

Management's willingness to nurture and make use of that interest will largely govern the degree of return on the profit-sharing investment. The ideal outcome is the growth of a performance-oriented culture where every employee is continuously concerned with maximizing company results for the sake of his/her share of the bonus.

Reward systems. Distribution of the rewards of a profit-sharing system is usually governed by a formula based on factors like salary and length of service. A key economic fact is that profit sharing is operated only when management decides there are sufficient profits to share; in contrast, base salaries are relatively fixed expenses.

Best candidates. Conditions existing in a company that would warrant implementing a profit-sharing plan are as follows:

- A desire to add "rewards for membership" in the organization. (Highly compatible with paternalistic management style.)
- The need to emphasize/build individual/group cooperation even more than individual performance.

Poor candidates. Profit-sharing plans should be avoided if the following conditions exist:

- The real goal is performance incentive. The link between performance and result is too tenuous to give much leverage on individual performance.
- Profitable years are too rare to meet employee expectations and make the plan "real."

Attempts to focus and fine tune the desirable effects of profit sharing to create more inspired employee performance lead to development of measurements of performance and formulas for reward. The presence of these turn profit sharing into a risk/reward (gain-sharing) sharing program.

Scanlon plan. The Scanlon plan has been in existence since the mid-1930s. Basically, it is a company-wide productivity improvement plan that usually covers all employees, including managers and supervisors.[3] The basic Scanlon formula is:

$$\text{Standard \%} = \frac{\text{Historical payroll costs included}}{\text{Historical value of sales \& inventory change}} \quad (1)$$

The standard percent is applied to the actual value of sales and inventory change to produce the "allowable labor" cost for the current period.

$$\text{Allowable labor cost} = \begin{array}{l}\text{Standard \%} \times \text{actual value of} \\ \text{sales and inventory change}\end{array} \quad (2)$$

If the actual labor costs for the period are less than the allowable labor cost, the dollar difference is the productivity gain that becomes the bonus pool.

$$\text{Bonus pool} = \text{Allowable labor cost–actual labor cost} \quad (3)$$

Results are calculated monthly, and typically the reserve amount is set at 25% of the monthly gain, with 75% of the balance going to the employees and 25% to the company.

The most important feature of this plan is its philosophy of involvement. To encourage employee participation and cooperation in raising productivity and reducing costs, a committee system made up of departmental committees and an overall screening or steering committee is set up to solicit, examine, and act on suggestions for improvement. In most installations, departmental committees are chaired by supervisors, and committee members are elected by department employees.

Because the Scanlon plan is company-wide, such committees will exist in both office and plant departments. Departmental committees meet at regular intervals and review all suggestions for improvements submitted by employees. The committees usually have budgets and are allowed, where necessary, to spend money on worthwhile suggestions. Suggestions that affect quality, product design, or other departments, or which require capital outlays in excess of the committee's budget, are referred to the screening committee. Suggestions that are approved at the departmental level are implemented immediately, whenever practical, because rapid action on suggestions will help in maintaining employee interest. The committee can also deal with other matters of importance to the department.

Best candidates. Scanlon plans have the best chance of success if the following conditions exist in a company:

- Product lines are relatively stable.
- Management style is "Theory Y" (participative) or determined to become so.
- There is a reasonably good employee/management climate.
- Intent is for whole-company involvement.

In addition, formula-keying on sales makes this plan readily adaptable for service industries.

Poor candidates. The Scanlon plan should not be implemented if the following conditions exist in the company:

- Management style is "Theory X" (autocratic).
- Management lacks confidence in its first-line supervisors.
- Price competition is likely to mask results of internal productivity gains.
- Accurate and timely inventory and labor cost information is not available.
- Layoffs are foreseeable. (They could be unfairly blamed on suggestions arising from employee participation in the program.)

Drawbacks. Some of the drawbacks associated with the Scanlon plan are as follows:

- The uncomplicated formula is easy to communicate, but reacts sharply to changes in product pricing.
- Employee bonuses are not sheltered when the company encounters hard times. This is realistic and cost effective, but also discouraging if effective efforts continually go unrewarded.
- Wage increases require redetermination/renegotiation of the standard.

Rucker plan. The Rucker plan carries less ideological baggage than the Scanlon plan. It uses a formula with more cost accounting to focus employee improvement efforts on materials costs as well as on labor productivity.

The productivity measurement used is based on the "value added" by manufacturing for every dollar of payroll cost, where value added by manufacturing is defined as the difference between the sales dollars received for goods produced and the cost of the materials, supplies, and outside services consumed in the production and delivery of those goods.[4] Payroll costs are defined as all employment costs paid to or on behalf of the employee group covered by the plan.

A powerful feature of the Rucker plan is that there are many areas where improvements can affect the plan results. Important ones are use of materials, use of supplies, output per worker hour, machine utilization, energy consumption, product design improvements, and reduction of downtime and set-up time. However, it is difficult for the employees to see readily what impact they can have on the plan results, and it is necessary, as in the Scanlon plan, to set up a structured program to administer the plan, promote interest, collect ideas, solve problems, provide feedback to the participants, and maintain momentum. Through this program, employees are encouraged to submit ideas for cost reduction and to identify problem areas that have a detrimental impact on costs and productivity. These are given to investigators for resolution, and a program coordinator ensures that all submissions are dealt with quickly, effectively, and fairly. Departmental and plant-wide committees are also formed to review progress, discuss problems and issues, resolve conflicts, monitor plan results, and provide the necessary feedback.

Plan results are usually calculated monthly or quarterly, and any bonus earned is distributed on the basis of hours worked by those included in the plan. Usually a 10-30% reserve is withheld to offset negative months. While this plan appears more difficult than others to install and administer, it has been used successfully by many organizations, and it certainly warrants close examination by managers who are concerned about productivity levels in their manufacturing plants.

Best candidates. Indications are that Rucker plans work best for:

- Plant-level orientation, where end use/sale is remote from production.
- Industries where materials/waste costs are high and employee attention to quality can substantially cut costs.
- Service industry or manufacturing companies supported by good cost accounting systems. (Cost-conscious service businesses should find the cost reduction orientation of this system highly attractive.)

Poor candidates. The Rucker plan should be avoided if:

- Analysis over a 5-7 year period does not show a clear and stable relationship between labor and production value added.
- Adequate and timely measures of cost and inventory are not available.
- Foreseeable layoffs are likely. (They could be unfairly blamed on suggestions arising from employee participation in the program.)

Drawbacks. Some of the drawbacks associated with the Rucker plan are as follows:

- Demands accurate and timely data on detailed production/material costs and inventory composition/value.
- Fluctuations in product price and material costs can overwhelm the effect of internal improvements. Extended loss of bonus can create discouragement.
- Performance ratio will be affected by wage increases. Because the standard is historical, it is not reset by each increase. Only price increase or productivity gain will prevent bonus reduction.

The newest major gain-sharing program (developed by Mitchell Fein in 1974) comes at the productivity problem from the perspective of manufacturing engineering. IMPROSHARE (IMproved PRoductivity through SHARing) has been growing rapidly in recent years, partly because its mechanics do not involve commitment to changing a company's management style.

A major difference between IMPROSHARE and earlier gain-sharing plans is that it measures productivity in hours rather than in dollars. It relies on comparison of actual work hours to engineered standards for each product line produced. This also insulates employee productivity increase bonuses from changes in product line pricing, material costs, or production methods.

IMPROSHARE seems to be marketed with an eye to appealing to traditional and Theory X organizations. Much is made of the fact that management maintains all rights and control. The same plan mechanics, however, could be easily embellished by highly participative programs if so desired.

The current state of employee/management relations is not critical for implementation of this plan. As a unilateral management offer of a benefit, it could even be installed in the middle of a contract period, with or without union cooperation. In the long run, such a program should help even troubled relationships by giving management and labor a common goal with a sizable incentive potential.

IMPROSHARE can be installed over, around, or in place of existing incentive systems. With any system, there must be some channel of communication to the employees on where labor reductions will increase the bonus, and from employees as to what management can do to capture the potential savings. IMPROSHARE requires at least a conventional suggestion system. Anything beyond that is optional.

The IMPROSHARE standard is the historical number (and direct/indirect mix) of labor hours needed to produce a unit of a product. Standards are developed for each product line and combined as required by the current production mix. Once established, IMPROSHARE standards are frozen. The infrequent changes arise only from major capital investment or when long-term productivity above the bonus ceiling triggers a buy-back of the standard.

Labor gains are calculated in hours, across a running 4-week average, and are usually shared weekly. The split is 50/50 employee/company. Employee bonuses can go as high as 30% of base wages (reflecting a 60% increase in production). Weekly excess above the bonus ceiling is banked against future weeks.

The company contributes minor investments and improvements without adjusting the standard. The company gets 50% of the labor gain plus the benefit of any material/cost savings that may arise in any nonlabor area.

The best candidates for IMPROSHARE would be the following:

- Companies that have a strong manufacturing engineering orientation and capability.
- Company's strongest interest is in reducing the labor content of products.
- Company is not currently profitable and needs to reward labor productivity with a plan that will pay under such conditions.

The program should not be implemented under the following conditions:

- Company cannot capture present or historical labor hour data identified by product line.
- Company lacks a strong manufacturing engineering capability.

The drawbacks of IMPROSHARE are as follows:

- There is little element of risk sharing in this plan. Productivity bonuses are earned in good times or bad; however, each bonus dollar paid also reflects an additional dollar saved for the company.
- Introduction of each new product line requires additional engineering attention to the plan formula database.

Skill-Based Pay for Production Workers

Skill-based pay (SBP) is a tool for creation of a flexible, efficient workforce that can be shifted to cover operational needs without job description questions or craft jurisdiction getting in the way. In a plant setting, SBP might accomplish this by paying employees top wages for being able to run six machines, sweep, pull materials, inspect, and ship product. Employees should have no objection to any day's assignment because if they are qualified to do the job, their pay already reflects it.

The employee/management relations situation should be such that this tradeoff of pay for flexibility will be acceptable. To date, almost all skill-based pay implementations have been in start-up situations, rather than replacing a previous arrangement.

Reward structure. Typically, a worker is trained to proficiency in a single position and, as opportunities can be made available, is cross-trained into additional positions. Attainment of proficiency in each new position is rewarded with an increase in base rate. Some Japanese companies also add pay whenever an employee completes a degree or similar milestone. In return, the worker should be willing and able to be assigned within the range of positions for which he is certified proficient.

Management. Because individuals are paid for their full capabilities (used or unused), management must make sure that the workforce is assigned/utilized to the fullest advantage.

Cross-training opportunities are at a premium and require management control to balance productivity requirements against employee ambition. In practice, limitations are sometimes imposed on the numbers of people in the highest classification if there is no operational need for more. A "holding rate" is also sometimes used if employees have no opportunity to actually use the highest of the skills for which they are certified.

Skill-based pay is generally believed to drive up overall labor costs, so management must capture the potential efficiency savings or the program becomes self-defeating. While a participative management style is not a mechanical requirement, more personnel flexibility could be attained much more cheaply by training and paying a small percentage of the workforce as "utility" workers.

Reaping the fullest possible benefit from skill-based pay requires management willingness to encourage and make use of the suggestions resulting from the increased perspective and process knowledge that extensive cross-training gives employees. A recent American Productivity Association study indicates that firms using SBP also report nearly twice the usual rate of employee participation in quality circles, QWL, and other employee involvement programs.

Many firms happy with SBP have also coupled it to a cash profit-sharing plan or gain-sharing plan of some sort. Knowledge-based pay affects the base rate, so it is mechanically and philosophically compatible with group incentives.

Indications are that SBP, as part of an overall style of participative management, will more than pay its own way.

Best candidates. A skill-based pay system can be successful in a company when the following conditions exist:

- New plant start-up in an environment where labor costs are a low percentage of product cost and where there will be a variable demand for a high density of skills.
- Organizations use a slow or crisis period to make a wide variety of changes and management improvements.

Drawbacks. A skill-based pay system would not be practical for high-pressure organizations that could not afford the learning curves of large numbers of employees, or where having a cross-trained workforce is of no major value.

With constant rotation and cross-training going on, the workforce may not be as smoothly proficient as one made up of workers with years of experience in a single function. This could result in either quality or volume problems.

When installing an SBP system, short-term productivity will suffer because cross-training will result in having a large percentage of the workforce in the process of learning new jobs.

COMPENSATION

Nonmonetary Incentive Plans

To be fully effective, an incentive program must provide compensation or other benefits that the employee considers valuable and can see on a short-term basis. The further away the payout, the less likely the incentive will produce a full-incentive work pace.

For example, a home ventilating manufacturer uses time off of the job as a very effective motivator. For example, individual employees and groups who have made the full standard for that period are allowed to leave their job and walk to the lunchroom or their break area 5 minutes early. Employees and groups who have cumulatively made standard by the morning break, lunch time, the afternoon break, and the end of the shift are allowed to leave their job 5 minutes early in each case. Employees are expected to continue to produce even though things are going well and the rate may have been exceeded, except for the 20 minutes of free time that can be earned. With this approach, management has been able to achieve close to an incentive work pace in a number of areas without having to put in the manufacturing engineering and accounting effort that policing a piecework system demands.

SALARY PROGRAMS

There are several programs that a company may use to positively supplement the income of its salaried employees. Two of the more common methods are bonus programs and knowledge-based pay programs.

Bonus Programs

Bonuses for managers and professionals that are linked with profitability are in widespread use throughout American industry. Recipients of these bonuses may include first-level supervisors or may be restricted to higher levels of management.

In general, a percentage of profit above a certain minimum return for shareholders is set aside in a bonus pool. Payments are often made after audited financial results for the year are available. However, plans with quarterly or semiannual payouts have more of a motivational impact. Annual salary reviews at times other than the bonus delivery time can also provide for some of the benefits of more frequent bonus distributions.

As the tightly focused reward arm of a performance-oriented management system, bonuses can be effective in focusing individual effort on organizational objectives when more than one individual or department is responsible for a desired result. Parallel objectives can be effective in departmental and individual cooperation. The management feat is in defining and tracking objectives both vertically and horizontally within the organization so that people will be rewarded for moving the company toward its goals.

Studies have shown that for a bonus program to have the proper motivation, the incentive (bonus) should be at least 15% of the individual's base pay. The reward for top performers should be at least twice the reward received by just acceptable performers. Marginal performers should receive no bonus, which is a difficult task for some evaluators, but necessary to obtain the full motivational impact.

Too often, bonus pools are distributed on a subjective basis. For a bonus program to provide the proper motivation, it should be based on specific, agreed-on objectives between the bonus participant and his or her supervisor. These objectives should include the meeting of performance criteria for the participant's job responsibility as well as achieving specific results. Some examples of these results are as follows:

- 10% improvement in assembly labor productivity through a program of methods analysis.
- 5% reduction in maintenance expenditures through the institution of a preventive maintenance program.
- 5% reduction in purchase material cost through joint cost analysis with vendors, negotiating long-term supply contracts, and resourcing.
- 5% reduction in product cost for a specific group of models through value analysis and product redesign.
- Improvement of customer service through improved scheduling, which will reduce stockouts by 25%.
- Design of a new generation product meeting specified cost, quality, and performance criteria by a specified time.
- Reduction of the number of workers in a nonproductive department by 15% through changes in personnel assignments.

The setting of objectives should be a formalized process. One common approach is for top management to outline the company's long-range (5 years or more) objectives and provide a 1 or 2-year business plan. Each level of management is then asked to provide a half dozen or so specific objectives (written action plans) that will meet or exceed the company's objectives. After discussion between the individuals and their supervisors, the written action plans are solidified. Succeeding levels of management coordinate these action plans, with adjustments taking place in an interactive mode until a cohesive plan is in place. These action plans should be reviewed monthly or quarterly to monitor progress; as action plans are completed, new ones should be added. When conditions change, the various action plans may be changed, added, or deleted. To be effective, the setting of performance objectives must be a living process and capable of changing with the dynamics of the company and business situation.

In the 1960s and early 1970s, a somewhat similar approach, referred to as management by objectives (MBO), was popular. In its extreme form, MBO attempted to allocate all of a company's resources with interlocking objectives and action plans. Many companies found that this approach created unwieldy and burdensome reporting. In many cases, there was inadequate provision for change.

Knowledge-Based Pay for Professionals

An ideal research and development organization would be staffed by a core of high-level designers/researchers and technical leaders, followed by a much larger group of junior professionals who would implement the design and learn in the implementation process.

In practice, there are often extended periods in the development cycle or the business cycle where not much high-level work is necessary. Senior people then spend their time performing tasks that do not require their full skills and nominal titles.

The time is approaching when the great bulk of the technical workforce will consist of mid-career professionals who have design and leadership capabilities that cannot be continuously utilized. Between now and the year 2000, demographic changes will make entry-level and junior professionals increasingly scarce (and, therefore, relatively more expensive). The "technical pyramid" will be shaped more like a watermelon.

A form of knowledge-based pay (KBP), plus a program of group and individual incentives, offers a possible way to provide career path and recognition to individuals while making the carrying of a large pool of senior-level talent more affordable to the organization. Careers would then progress by meeting an objectively defined set of experience, education, and assignment milestones.

Reward structure. The economic key in a KBP program is an affordably low base pay policy for senior titles. This would be supplemented by variable compensation such as group incentives based on project milestones, individual incentives when serving as a leader or top-level designer, and perhaps a cash profit-sharing program involving the larger organization.

The very highest technical steps would probably be limited by economics and reserved for a tiny group of key individuals used almost continuously in high-level roles.

Management. Technical managements in companies that have downsized or had hiring freezes are already in a position very similar to that which would be created by KBP, only without the redeeming cost containment features. There is already a difference between title and use, which knowledge-based pay would legitimize.

The perennial concern for protecting key people from competitors would be well addressed by the individual incentives resulting from the nearly continuous selection of those people for lead or key contributor roles. Lesser lights would be carried at a reasonable base rate, supplemented by project incentives and profit sharing.

The wider incentives are necessary not only to spread variable compensation widely, but to preserve some unity between technical and nontechnical employees and between most-favored contributors and the technical rank and file.

Drawbacks. The result of a technical career ladder, independent of organizational structure, could ultimately be a top-heavy technical group, but demographics are likely to make that situation happen anyway.

In a matrix organization where professionals time-share between different projects and roles, the incentives associated with KBP could become quite complex. Many organizations would have to greatly improve their objective setting and tracking ability to be able to operate such an incentive program.

SUMMARY

The preceding discussion covers most of the popular compensation programs in use today. Many combinations, twists, and variations of these programs exist. Some of these programs have been developed by consultants and are marketed under names such as IMPROSHARE (IMprove PROductivity through SHARing), Pay for Performance, and Gainsharing. Companies that are interested in developing an incentive compensation program may want to work with one of the many compensation consultants who are active in this field.

As one looks at compensation practices in America, a startling number of negotiated or forced wage freezes, wage reductions, and work rule changes are seen. In most instances, these changes have resulted from complacent managements that have succumbed to aggressive unions. Whatever a company's current situation may be, it is essential to get full value for each hour of labor worked. When all else has been done (if that can ever be said), it may become necessary to face up to competitive reality and work with the workforce to achieve realistic wage structures and work rules. Wage rates and work rules cannot be considered the ''territory'' or prerogative of the labor relations or personnel staffs. Manufacturing management must do everything possible to ensure that the competitive wars are not lost because the element of compensation was left to others.

BENEFITS AND SERVICES

No clear-cut distinction exists between benefits and services. For the purpose of this discussion, benefits are considered part of compensation that results in financial gain or security. The employee may or may not contribute to the payment for the benefits received. Services are considered those extras that do not consist of a financial gain to employees, but do fulfill social, recreational, or cultural needs of the employees and at the same time make life easier because of their convenience.

BENEFITS

Employee benefits, often referred to as ''fringe benefits,'' are an indirect form of compensation and are provided by employers for several of the following primary reasons:

- Protect employees from the financial consequences of retirement, disability, or illness.
- Encourage an orderly replacement of older workers.
- Create a positive corporate image among employees.
- Take advantage of risk pooling and economies of scale in purchasing insurance coverage.

While the employer often pays for these benefits, this form of indirect compensation is tax effective to both the employer and employees.

The type of benefits provided by a company should reflect its philosophy and objectives as well as the needs of its employees. In general, most benefit programs improve the quality of work life for employees. This improved work life usually helps to increase cooperation and productivity. The value of an employee benefit program as a motivational tool also should not be underestimated.

Benefit programs have grown significantly over the years, with the average benefit program representing as much as 40% of an organization's payroll cost. Because of these costs, companies should periodically conduct cost analyses of expenditures in each benefit area. To serve their intended purposes, benefit programs should also reflect the changes that are occurring in the society. One way that these changes can be accommodated is through flexible benefit programs.

Flexible benefit plans, or so-called ''cafeteria programs,'' allow employees to pick and choose their benefits to better meet their individual needs. Flexible plans have become increasingly popular because of the growth of two-worker households. These programs allow employees to eliminate unnecessary, duplicated benefits, such as medical insurance provided to two workers, and give the employees the opportunity to instead select needed or desired benefits of an equal value. Flexible benefit programs appear to be greatly appreciated by employees and have acted as a means of retaining employees. Taking a creative, nontraditional approach to benefits can act as a positive mental factor for the employee at a relatively small cost to the employer. Flexible programs have allowed employees to secure benefits that probably were not previously available, such as dependent life insurance and employee assistance programs.

Employee benefits can be classified in a number of different ways. For discussion purposes, this chapter will divide benefits into categories of mandatory, pay for not working, insurance, pensions, and miscellaneous. In addition, health awareness programs and employee assistance programs will also be discussed.

Mandatory Benefits

Mandatory benefits are those benefits that an organization is required by law to provide for its employees. These benefits include Social Security insurance, worker's compensation insurance, and unemployment insurance. Mandatory benefits account for approximately 10% of an organization's payroll cost.

Social Security. The Social Security Act was passed in 1935, providing an insurance plan designed to indemnify individuals against loss of earnings from various causes.[5] This loss of earnings may result from retirement, unemployment, disability, or in the case of dependents, from the death of the person supporting them. Thus, as in the case of any type of casualty insurance, Social Security does not pay off except in the case where a loss of income through loss of employment is incurred.

Worker's compensation. Worker's compensation provides payments to workers or dependents in the event of job-incurred injuries, illnesses, or death.[6] Medical and rehabilitation services are also provided. These benefit programs are administered by the state. The total cost is borne by employers, although the method of insuring may vary from state to state and may involve a state fund, approved insurance companies, or special funds set up by employers. Additional information on worker's compensation can be found in Chapter 12, "Management Concerns for Occupational Safety and Health," of this volume.

Unemployment compensation. Unemployment compensation is administered under a dual system of state and federal laws.[7] The purpose of these laws is to provide emergency income to people when they are unemployed and to encourage employers to provide stable employment. The Federal Unemployment Tax Act of 1935 as amended requires employers of four or more employees performing types of work covered by the act to pay an unemployment insurance tax to the federal government. The federal government then makes refunds to the various states. Typically, unemployment compensation benefits are available from state governments for a period of up to 26 weeks. Individual states can authorize or prohibit the payment of unemployment compensation to workers who are on strike.

Pay for Not Working

This category of benefits includes paid vacations, paid holidays, paid sick leave, paid rest and cleanup periods, paid military and jury duty, paid funeral leave, and paid time off for other personal reasons. The cost of these benefits is approximately 10-12% of an organization's payroll cost.

Vacations. The length of paid vacations offered to employees usually varies with the length of service and nature of the position. In general, companies provide their employees with 1 week of paid vacation for up to 5 years of service; 2 weeks paid vacation for 5-15 years; and 3 weeks paid vacation for service more than 15 years. The amount of money an employee receives for this time off is usually a straight-time rate or an average rate if the employee is paid based on some type of production incentive plan.

Holidays. Ten paid holidays a year is a fairly common norm in industry. Some companies also provide their employees time off between Christmas and the New Year. If it becomes necessary for an employee to work on a paid holiday, he or she may receive double or triple pay, depending on the union agreement or company policy.

Sick leave. There are several ways in which employees may be compensated during periods when they are unable to work because of illness or injury. Most companies provide their employees with a certain number of days each year to cover absences for physical reasons. These reasons may be due to personal illness or injury or to sickness or injury of an immediate family member.

When sick leave is provided, the unused leave can generally be accumulated to certain limits to cover prolonged absences. In addition, many companies are providing employees with sickness and accident insurance as well as long-term disability insurance. Loss of income during absences resulting from job-incurred injuries can be reimbursed, at least partially, by means of worker's compensation insurance.

Rest and cleanup periods. Rest periods usually allow employees time to smoke or procure food and drink. Cleanup periods allow workers to clean up their work area, put tools and equipment back in place, and wash their hands and possibly change their clothes on company time.

The rest period is usually divided into two parts for the two halves of a shift; each period is customarily 10 minutes. In industries where the work is difficult (heavy lifting and such) and the environment hot, rest periods are essential and may total several hours in a day. As an extension of the rest period, some companies allow for lunch on company time.

Other paid leave. This group of paid leave takes into consideration those absences resulting from military duty, jury duty, the death of a family member as well as other personal reasons not covered under the previously mentioned absences.

Most companies provide pay for absence from work to fulfill military training or duty commitment. While the majority of companies allow 12 days per year for this type of absence, some companies allow time off as needed.

Paid time off for jury duty is usually provided by most companies as needed. A common practice is to make up the difference between the employee's pay and the court's allowance.

In the event that a death occurs in an employee's immediate family, some companies provide their employees with paid leave to attend the funeral. In many instances, the time allowed for an out-of-town funeral is greater than that allowed for a local funeral.

Insurance

The three main types of insurance provided to employees are life, health, and dental. In addition, companies may offer long-term disability insurance. Payment of the premiums for these insurances may be by the company, employees, or both parties. When paid for fully by the company, employee insurances account for about 10% of an organization's payroll cost.

Company-sponsored insurance plans are typically group plans that provide benefits at lower cost than if coverage were purchased for individuals. Because all or most employees are included, the employer can negotiate favorable rates with the insurance company or health-care provider.

Life insurance. Group life insurance is one of the oldest and most popular employee benefits. This type of insurance provides death benefits to beneficiaries and may provide for accidental death and dismemberment benefits.

As a rule, the amount of insurance provided for an individual employee is based on his or her annual salary. The most common method of tying life insurance protection to annual salary is to multiply the annual salary by a factor of 1 or 2 and round the product to the next $1000.

Health insurance. The benefit that often receives the greatest amount of attention from employers today because of sharply rising costs is that of health care. In the past, these benefits covered only hospital and surgical expenses. Today, employers are under additional pressure to include prescription drugs and dental, vision, and mental health care.

One of the fastest growing forms of medical benefit coverage is through a health maintenance organization (HMO).[8] For a monthly fee, these organizations provide total health care, including physical examinations, laboratory tests, surgical and hospital services, consulting-nurse and emergency-room services, and prescriptions filled at an HMO-approved pharmacy.

Dental insurance. Dental insurance has grown rapidly in the past decade.[9] Most dental plans provide their coverage through a separate plan, but some companies include it as part of their major medical coverage. Nearly all dental plans cover a wide range of services, including examinations, X-rays, and restorative procedures such as fillings, periodontal care, and inlays.

Long-term disability insurance. Long-term disability (LTD) insurance continues the income of employees during extended periods of disability. Generally, LTD begins after sick leave when sickness and accident insurance are exhausted and continues as long as the employee remains disabled, or until the employee reaches retirement age.

Long-term disability benefits are usually 50-60% of monthly pay. Most plans also have a maximum payment limitation—commonly $1500-$5000 a month.[10]

Pensions[11]

In many cases, Social Security payments provide only a minimum income for retirement and must be supplemented with other funds. Therefore, many organizations offer pension benefits beyond those required by the Social Security Act, representing a significant cost in the typical benefit program.

There are many kinds of pension plans. Under *contributory plans*, both employees and employers are required to contribute to the pension fund. Under *noncontributory plans*, pension funding is the sole responsibility of the employer. The actual payments received during retirement may be specified on the basis of an employee's salary level and length of service; in noncontributory plans where contributions are based on company profits (*deferred profit-sharing plans*) accumulated funds are usually allocated on the basis of salary (and the size of the pension fund depends on profit size and its investment results).

All private pensions are now regulated under the Employee Retirement Income Security Act (ERISA) of 1974, which was passed to help ensure that pension plans are adequately funded and protected against failure. Although the act does not *require* a company to establish a pension plan, it does benefit workers in a number of significant ways. First, the law creates a tax incentive for employers to set aside certain funds for pension plans and establishes an insurance program to protect those funds. Second, the law broadens employee participation in pension plans by prohibiting overly strict eligibility rules. A pension plan must now include all employees who are at least 25 years of age and who have worked in the organization for at least 1 year. Third, the law establishes minimum *vesting* standards, under which an employee who has worked a certain length of time or reached a certain age is entitled to receive the employer's contributions to the pension fund even if the employee leaves the job before retirement. *Portability*, however, or an employee's ability to transfer vested benefits to another employer (to maintain all retirement moneys in one account) or to an individual retirement account, requires the agreement of the employer.

A new pension benefit program adopted by many organizations in recent years is the *salary reduction plan*, also known as the "401(k) plan" (or known as a "403(b) plan" for nonprofit organizations). Under Section 401(k) of the Internal Revenue Code, an employer may establish a retirement savings plan to which employees can contribute through payroll deductions. The employee's contributions reduce the employee's taxable income, and no income taxes are paid until the employee starts drawing from the fund after retirement. Frequently, companies match employees' contributions, and employees can choose which of several investments they wish to participate in, such as money market funds or company stock purchase plans.

Employees can also save for retirement and pay reduced taxes by participating in an *employee stock ownership plan* (ESOP). Under these plans, employees may purchase company stock through payroll deduction or installment plans, usually below market price. Generally, the employee is eligible to sell the stock or withdraw dividends only on retirement or termination of employment, when he or she must pay taxes on those assets.

Miscellaneous Benefits

Many other benefits are found in manufacturing organizations. Again, it is necessary to emphasize the importance of providing benefits based on employee needs. Some of the different types of benefits offered are as follows:[12]

- Free or subsidized parking.
- Full or partial payment of educational expenses.
- Supplemental unemployment benefits.
- Subsidized commuting.
- Prepaid legal services.
- Child care.
- Company-sponsored reimbursement accounts.

Health Awareness Programs

People are like futuristic machines that can fix themselves when they break down, identify a developing problem and correct it, and perform preventive maintenance to minimize the effects of wear, tear, and old age. Health awareness or wellness programs at the workplace focus on the prevention of disease and the promotion of healthy employees (and in many cases, family members as well). Employers are becoming interested in the personal and private issues of someone's health for the following three main reasons:

1. *Increasing health care costs.* It has become a tradition for employers to offer health insurance as a benefit of employment. Although inflation has been held in check through the mid-1980s, health care costs have increased annually at almost double-digit rates. Employers are bearing the brunt of this increase by paying higher health

insurance premiums, while everyone has to pay higher taxes to support indigent medical care. Improved technology and new knowledge about illnesses save and extend many lives, but the cost of modern equipment and research is great.

2. *Increasing cost of replacing employees temporarily or permanently*. As the workplace relies more and more on specialists, the cost of replacing someone can exceed 20-50% of his or her annual compensation and benefits.
3. *Research has shown that some illnesses can be prevented or controlled*. Polio, smallpox, and yellow fever have all but been obliterated by modern medical science. Diseases or illnesses can have many different etiologies: genetic, bacterial or viral, the normal aging process, nutritional, accidental, social or environmental conditions, emotional, or a combination of these. Some of these causes can be controlled. Employers are realizing that if employees are kept up to date about the latest information and make informed choices each day in areas that promote health, there will be a hard dollar return in decreased absenteeism, decreased health care costs, fewer accidents, and increased productivity.

Health awareness program components. The human machine is like a vehicle that can travel under or on water, on land, in the air, and in outer space. It needs constant input and assessment of the external and internal conditions that affect our situation. Health awareness programs provide data to make "in-flight" corrections or add new maintenance programs to improve or maintain peak performance. These programs focus on areas involving personal decisions to help our internal operations work better. A typical health awareness or wellness program consists of the following two main components.

1. *Educational services*. This is perhaps the best-known component of health awareness programs. Lunchtime or after-work sessions with health educators, posters, newsletters, paycheck messages, desktop accessories, T-shirts, and other gadgets and media are some of the ways information is shared.
2. *Early detection and screening*. Many fatal illnesses have silent, but inexpensively discovered, warning signs that indicate when something is wrong. Simple blood pressure and cholesterol tests can indicate the development of possible circulatory problems such as a heart attack or stroke. Both of these problems respond to lifestyle changes in diet and increasing exercise, even before medication is indicated. Blood tests can detect the onset of diabetes, permitting changes in diet to minimize the effects of this progressive disease, and delay the need for medication or other medical treatments. The first line of defense for breast and prostate cancer is examination by a physician, which can provide early interventions to arrest the spread. In addition, research on the genetic predetermination of certain ailments helps identify who is at greater risk for certain diseases, increasing the efforts at preventive measures and early detection.

Health awareness program goals. There are generally four main goals in a successful health awareness program, as follows:

1. *Improve nutrition habits*. Each person has a unique metabolism and environment. Although many people eat a lot, they do not always eat what they need to maintain a healthy body and frame of mind. The keys to a healthy dietary plan are choosing a balanced selection from the major food groups on a daily basis and avoiding excesses, especially sugar and salt, that could unbalance the metabolism. For those with genetic predisposition to certain illnesses or with unusually high test results, specialized nutritional counseling and planning is extremely helpful.
2. *Increase physical activity*. In today's environment, people use their bodies less and less for work or for personal care. Not only do muscles atrophy, but the body does not produce enzymes and hormones that keep one healthy and feeling good. Stress, tension, and weight build up, causing a strain on the muscle and bone structures and leading to physical ailments. Simple stretching exercises and regular cardiovascular stimulation are mandatory in today's sedentary environment.
3. *Decrease addictive behavior*. There is no definitive answer to why certain people can ingest a substance and take it or leave it, while others become uncontrollably addicted. Research is following three major directions looking for answers: genetic, psychosocial learning, and biochemical. Cigarettes are closely tied to terminal lung disease and heart conditions and are a major target for health awareness programs. Other addictive substances—alcohol, street drugs, and prescription drugs—will be discussed further in the section that follows later on employee assistance programs.
4. *Learn to manage stress*. Stress is unavoidable. Thousands of years ago, stressors often meant physical danger from animals or natural catastrophes. The body would secrete adrenaline and other biochemicals to allow for the "fight or flight" response. Today, much stress is environmental or relationship based. The natural response to run or fight is counterproductive to the more modern response of negotiating.

There are numerous techniques that assist the body and mind to return to a balanced chemical and emotional state. It is important that each person find a formula to manage stress so that the cumulative effects do not result in physical or relationship ailments.

Program evaluation. Like any other project, health awareness programs must be rigorously evaluated for return on investment. The first step is to determine target problems, such as absenteeism, turnover, health costs, and accidents. Then apply the standard model of evaluation: gather baseline data on previous experience, establish goals for improvement, monitor the program, and measure the outcomes.

A successful program also requires follow up with each individual who participates. Health is not something one gets and keeps like a piece of art or jewelry. It is more like a voyage where many decisions are constantly being made. People need support from professional contacts, incentives, and self-help groups, to name a few techniques, if they are going to change the way they live.

Starting a health awareness program. When starting a health awareness program in a company, it is important to recognize that this will take a concerted effort from all departments. This cannot be delegated to the human resources department and ignored by everyone else. Most programs have advisory committees to make sure the program is meaningful to all people in the company. It is also beneficial to talk with colleagues in other companies who currently have a program. They can provide valuable input as to what programs have and have not worked.

Professional associations or trade groups, such as the Association for Fitness in Business or the local chamber of commerce, might have local groups or committees to help develop health awareness programs. County health departments, colleges or universities, or local hospitals might offer free or fee-for-service programs. Private firms offer program development consultation and service products.

Employee Assistance Programs

In reality, most people cannot live up to the expectation that they leave their personal lives at the door of their workplace. A chronically ill parent, a call from the child care center, and last night's argument over how much one is drinking are all brought to the workplace in one form or another. These life crises are unpredictable as to when they will afflict a person, and no one is immune. Some of the events are joyous, such as a wedding or graduation, but the joy is often part of a double-edged sword that brings financial issues and old family hurts to the fore.

Employers began to recognize that helping employees with personal problems was good for business. The roots of employee assistance programs (EAPs) goes back to the late 1800s, while the major momentum began in the early 1940s, when alcoholics began returning to their employers claiming they had a new way of staying sober. By attending Alcoholics Anonymous (AA) meetings, each alcoholic can give and receive support in the battle against this chronic, progressive disease. Several large corporations such as Du Pont, Kodak, and Metropolitan Life set up occupational alcoholism programs and formalized a role for recovering people to help fellow employees.

In the mid-1960s, federal funds became available to promote mental health (instead of just treating mental illness) by detecting emotional or relationship problems early and providing information to prevent or minimize problems from getting worse. EAPs are the natural marriage of the occupational alcoholism initiative and the attempt to promote mental health.

What is an EAP? An EAP is designed to assist employees and family members who have personal problems. The goal is to prevent these problems from interfering with job performance. The EAP is set forth as a clearly defined set of policies and procedures adapted to each company's needs and culture. The EAP does not interfere with existing personnel or operational policies such as hiring, firing, and so on.

Counseling services. The heart of an EAP is the contact between the employee or family member (the client) and the counselor. For the initial point of contact, a 24-hour crisis line is the ideal arrangement, so that clients can contact the counselor as soon as a problem emerges. Personal problems can take years to grow and develop, but the window when people are hurting enough to seek help is often open only briefly.

Once the initial contact is made, the counselor assesses the problem to determine the severity, chronicity, and possible underlying causes. Counselors must bring to bear all the experience and training they have to be certain that the assessment is complete and accurate. Sometimes, extensive diagnostic workups are required to pinpoint the problem before moving to the plan of action.

The plan of action must be mutually agreeable if it is to be of any assistance to the client. In addition to the universal benefit of having someone be committed to a goal, a major element of addictive diseases is denial. Especially when confronted for the first time, an addict must accept the assessment and course of treatment, or much time and money could be wasted and future rehabilitation made more difficult. The plan of action could include referral of the client to a hospital or outpatient clinic, an educational program, a mental health professional, or a self-help group.

Once the plan of action is implemented, the counselor follows up with the client to make sure the problem gets resolved. Over time, new problems may develop, or the old ones may return. The EAP is there for as long as it is needed. Even if the staff changes, employees always know there is a place to turn if they are in need of help.

A critical element of any EAP is strict adherence to confidentiality. Although a company and the counselors can promise that the EAP is strictly confidential, only state or federal law can truly grant this promise. Confidentiality must be taken into account when developing an EAP for one's company or plant.

Promotional campaign. People in crisis find their thoughts, feelings, and behaviors out of whack. It is unreasonable to expect someone who has just been informed of the death of a loved one, or has been served divorce papers because of a drinking problem, or is experiencing excessive job stress, to remember about the EAP. It is imperative that an ongoing promotional campaign be developed to constantly remind people about the EAP and how they can use it. Existing communication patterns and media should be utilized whenever possible, minimizing disruption to the workplace for training. Lunchtime or after-work programs on topics such as selecting day care, dealing with aging parents, talking with your children about drugs, AIDS, and so on provide a valuable vehicle for reminding employees about the EAP and informing them about the types of problems the EAP can assist with.

Evaluating the EAP. Many EAPs were started because it seemed to be a good, inexpensive benefit. If these programs are to be more than another human resources fad, there must be proof that they provide a financial return on investment. Each company should review its human resources plan to see what areas need improvement. Studies have indicated that EAPs save money through decreased absenteeism and tardiness, reduced accidents, and decreased health care costs for mental health and substance abuse services and, over time, for stress and addiction-related illnesses.

EAP program models. There are several broad decisions and many smaller programmatic ones that go into developing an EAP. There is no research that indicates which model is best for any particular workplace. The best approach is to take the time to gather ideas and plan the program that most meets the needs of the employees and matches the culture of the company.

Larger companies and some large international or local unions either hire EAP staff or assign existing personnel on an add-on or new-assignment basis to provide EAP services. There are national, regional, and local companies that provide comprehensive EAP services on a contractual basis. Fee for service and capitation are the two most prevalent methods of charging. A number of clinics and hospitals have established EAP services. They may charge in a fashion similar to the contractual service or offer a reduced or no-fee service in exchange for referrals to the clinical service.

A fourth model combines the in-house and contractual models. In some cases, there is a corporate coordinator or team that oversees contracts with vendors on a local basis. Some companies use their own staff for case-finding and program promotion and refer clients to outside professionals for the counseling component. Yet another variation is a counselor-coordinator at headquarters working with a contractor or contractors to provide services to dispersed locations.

The Association of Labor-Management Administrators and Consultants on Alcoholism (ALMACA) is the largest national group of EAP counselors. In addition to establishing the first credentialing process for EAP professionals, ALMACA created the EAP Clearinghouse as a source of current data on EAPs. Current information on EAPs can be obtained from the following address: ALMACA EAP Clearinghouse, 1800 North Kent Street, Suite 907, Arlington, VA 22209, (703) 522-6272.

SERVICES[13]

All of the items just discussed as benefits were included under that heading because they provide some element of financial return to employees. Services, on the other hand, are provided by the company to fill some particular need of some or all of the employee group. Some services may be more necessary than others in any particular situation; some are simply conveniences and may not be "necessary" at all. Companies provide services to instill a feeling of group solidarity and general loyalty to the company. Pleasant "off-job" relationships probably make for more efficient "on-job" relationships. The purpose of this section is not to cover all possible services that a company might offer. Instead, some of the more common and more important ones will be discussed, with emphasis being placed on how they fit into the total industrial relations picture.

Communications

The general problem of communications in a manufacturing concern is much too complex to attempt to cover here, but one of the aspects of this problem is getting information to employees. Two common methods of accomplishing this are the bulletin board and the company magazine or newspaper. Both of these usually come under the jurisdiction of the personnel department, which accounts for their inclusion under employee services.

The bulletin board is usually reserved for official notices on company or union affairs. Some companies provide a true employees' bulletin board where anyone can post a notice signifying the desire to sell a car, buy a boat, give away a kitten, and so forth. The company newspaper or magazine may contain some official information, but it is more apt to include material of general interest to employees. General news items concerning fellow employees, standings of the company softball team, and a feature article concerning highlights from an employee's vacation trip may be of interest.

Social Programs

Social programs vary in extent all the way from the annual company picnic to the employees' social club offering all varieties of social activities throughout the year. Parties, dances, and general get-togethers are designed in such a manner that every employee will find some feature enjoyable. Games and contests are arranged for all ages, with appropriate prizes being awarded to the winners. Families become an important part of these affairs, as well as the employees themselves.

An overall employee social and recreational club may oversee the entire social program, or there may be small clubs appealing to people with specialized interests. Garden clubs, stamp clubs, rifle clubs, and drama clubs are only a few of the possibilities. Such clubs would be separately organized, each carrying on its own program in its own field of interest. The nature of the club to which any one employee belongs would usually be determined by his or her hobby interests.

Musical activities are receiving increasing attention as part of social programs. Choral groups and orchestra groups provide enjoyment not only for the actual participants, but also for other employees who enjoy listening.

Athletic Programs

For those who desire active physical exertion, athletic programs are offered by many companies. Bowling and softball are the two of the more popular sports because they do not require the specialized skill of basketball or football. Competition is maintained between departments within a plant and also between plants. Industrial leagues for both bowling and softball are quite active in many communities.

Although the major sports may receive most of the attention and publicity, there is still a lot of room in the industrial recreation program for minor sports. Some, such as tennis, horseshoes, table tennis, and badminton, have a special appeal, for often they can be enjoyed during the noon hour and do not require after-hours arrangements. Other sports, such as swimming, hockey, and golf, have their appeal to those interested, but they require special facilities and are definitely after-hours or weekend activities.

While the athletic programs mentioned will provide physical exercise for the participants, it is fairly common today for companies to have exercise programs and even provide the necessary facilities. These programs exist not only for recreation but for health reasons.

Eating Facilities

Companies situated in isolated areas or operating around the clock find it almost mandatory to offer restaurant or cafeteria services to their employees. To expect everyone to carry their lunch is too much in these days of emphasis on hot meals and good nutrition. Even without any special set of circumstances, companies provide eating facilities simply as a convenience. Less time is consumed in obtaining and eating the meal, and oftentimes the prices are less than in commercial establishments because the cafeteria is operated on a nonprofit basis. There is never any compulsion to buy food in the company cafeteria, although all are invited to eat there even though they bring their lunches from home.

A variation of the company cafeteria is the lunch wagon that circulates throughout the plant as a convenience to those employees who must eat on the job. Such wagons also provide the coffee and doughnuts for the daily rest period. This service eliminates the necessity of employees rushing to the neighborhood coffee shop and back.

The vending machine has had a phenomenal growth as a factor in food service and may replace both the lunch wagon and the cafeteria. Vending machines are now able to provide both hot and cold foods, and to the extent that their operators can provide good food at reasonable prices and keep the machines from breaking down, the vending machine may be the answer to the in-plant eating facility problem.

Professional Services

Medical insurance discussed previously is a professional service, but there are a few others commonly provided. Legal advice is one of these and can be a real service because employees seldom know what to do when faced with legal difficulties. Company legal departments do not carry out court actions, but simply make recommendations and perhaps help in selecting competent legal counsel to carry out the court action.

Another professional service sometimes offered is personal counseling. The counselor might be anyone from a white-haired grandfather to a trained psychologist or ordained minister. The point is that someone is available to hear the employee's story and maybe offer a little advice. Experience has shown that a good listener can often be of more value to an employee who has a personal problem than an expert who gives advice freely.

WORK SCHEDULING

The scheduling of work in any factory is a fundamental factor affecting productivity. As volumes and throughputs increase, effective scheduling is necessary just to maintain operations.

Chapters in this handbook are devoted to MRP II and just-in-time systems that provide programs to schedule the factory. In addition, newer approaches, such as OPT and Q systems, are reported as being favorable by a number of factory managers. And, of course, there is the ancient and much maligned order launch and expedite approach that helped U.S. manufacturers set the seemingly impossible production records of World War II.

Manual ordering and scheduling approaches are still widely used among smaller companies today. Where manual systems work well, caution should be exercised in converting to a computerized system. While there are many advantages to computerization, there are many pitfalls as well. The more complex the proposed system, the more important is careful study by knowledgeable factory people. Even apparently simple and straightforward computer systems can produce disastrous surprises if proper planning and training are not completed prior to start-up.

The following material provides an overview of computerized planning, scheduling, and control systems in general use today.

COMPUTERIZED FACTORY SYSTEMS

Chapter 20 discusses in depth MRP, MRP II, and the JIT approach to controlling manufacturing operations and the scheduling of work in the plant. There is a sound basis for believing that a marriage of MRP II (which provides for material receipt and management information for overall control) with JIT (which provides a philosophy of lead time and waste reduction) will provide an excellent approach for many or most CIM-oriented companies

There are, however, other software programs and approaches to controlling and scheduling factory operations that may warrant consideration, particularly in complex operations with multiple bottlenecks, like aircraft engine plants.

OPT (optimized production technology) has had considerable acceptance in such environments. It is also described as a system of "synchronized manufacturing" that uses proprietary software to, among other things, optimize schedules in complex factories with multiple bottlenecks. The OPT program has reportedly been simplified and made applicable to a broader range of companies. Its advocates say a major contribution is the development of "thoughtware" that aids in the understanding of complex environments. As with MRP II, JIT, and other such programs, OPT tends to be touted as "a new way of factory life" that will lead to world competitiveness and all kinds of human benefits. More can be learned about OPT by reading *The Goal* by Eliyahu M. Goldratt.

If a plant does not have a computerized material receipt and management information system and/or has not taken a look at a JIT-type set-up, an inventory and waste reduction program, or is not making satisfactory progress with present programs, a study of available systems is in order. APICS (American Production and Inventory Control Society) offers a variety of materials, seminars, and meetings on MRP II. The organization has been biased, however, toward the more rigid classical approach to MRP II to at least the partial exclusion of some of the JIT and OPT concepts. MAI Basic Four has some good MRP packages for small and medium-sized companies, as do other organizations. The Society of Manufacturing Engineers publishes numerous books and papers and conducts extensive seminars and conferences on CIM and other new technologies. SME is far and away the best source of unbiased information of this kind.

To pick the best program, or combination of programs, for a plant will require sufficient study to understand what is being done elsewhere and what is available. Most important is a complete understanding of the present operation. It may well turn out that by applying the logic learned and making relatively minor changes to a present system and plant, all or most of the improvements can be achieved that are promised by the purveyors of complicated, expensive packaged programs. Also, programs may take a year or years to implement and involve trauma as well as expense.

If a particular program appears to have merit, visits to several users are recommended. Talks with management and a quick walk through the factory may not be sufficient. To get more than a partial picture of a proposed system, it is important to talk to schedulers, supervisors, and people on the factory floor. Top management and materials management may present an unreal picture of how well "their baby" is performing. A net-change MRP II system may make for a great talk to visitors, but is it producing more paper than the plant can handle? Would a less sophisticated batch system get the job done? Insightful questions are required at this and every stage of the investigation. The real world is often not exactly the way salespeople and some users represent it to be.

Computerization has made possible many improvements in factory operations. This trend will continue as computers become more user friendly and control devices are made more adaptable to inclusion in an overall CIM network. Even the smallest factory operation is using, or will make some use of, computers. It is indeed a challenge to choose the right computerized factory system for a company at its present state of growth and the present state of the art. Delaying too long can have a very negative impact on a company's competitiveness. Picking an inappropriate system, particularly one that is too complex for the employees to use, can have disastrous results. It must be emphasized that thorough study and considerable caution are needed to assure an acceptable decision.

WORKFORCE HOURS

Most factories represent sizable capital investments. Generally speaking, the profitability and productivity of a factory will be at a maximum when it is scheduled for three-shift operations of capital-intensive equipment. For full optimization, time must be provided either on a rotating basis or on weekends for proper preventive maintenance of equipment. Production schedules must provide for a normal amount of downtime on key equipment. Some additional slack must be provided on bottleneck operations to allow for handling changes in scheduling and abnormal breakdowns. Generally speaking, scheduling well-maintained equipment 20 hours per day for 5 days is about the maximum that can be realistically planned on a day-in, day-out basis. Weekends are available for other than daily maintenance and to provide productive capacity to meet surges in demand. When there is exceptionally heavy seasonal demand, such as in the air conditioning industry, full Saturday schedules with Sunday maintenance for several peak months can optimize factory output.

From time to time, companies have attempted to operate discrete parts manufacturing facilities on a 7-day basis. Experience has been that after a relatively short time operators will work Saturday and Sunday for overtime pay, but will take time off during the week to recuperate and attend to personal business. Even when overtime is paid only after 40 hours, high absenteeism can be expected. When a full 7-day output is needed on key equipment, a better approach is to rotate personnel with other operations or to schedule crews on a rotating 5-day basis, as is often done in the process industries and utilities.

In departments, operations, or full plants where three shifts are required, the usual practice is to schedule each shift for 8 hours and to allow a 20-minute paid lunch break. When it is necessary from the process standpoint to keep equipment operating continuously, as in the case of a paint system, it is customary to provide relief operators to cover breaks and lunches. These may be specifically assigned relief operators or they may be paint mix or paint equipment people, group leaders, or others whose job functions must be performed, but do not have to be carried out continuously during the shift.

When maximum capacity is needed out of bottleneck equipment that is not fully automated, such as machining centers and punch presses, relief operators may be provided by taking people from noncritical operations and shifting their break times and lunches so that they may operate the essential equipment while the regular operator is on relief or lunch break.

When operations are seasonal in nature, as in the air conditioning industry, maximum output may only be needed for 3-6 months of the year. It may make sense to operate for 24 hours, 5 or 6 days of the week and assemble for two full shifts.

It is usually relatively easy to add people to mostly manual assembly operations to avoid a third shift. This is generally a less expensive approach and one that makes communications and quality maintenance less complicated.

Obtaining trained people for peak season can be a serious problem. One approach is to maintain a cadre of trained personnel that would include repair people, group leaders, and certain other key operators during the slow season on the day shift at their full pay rate. This makes them immediately available when the time comes to substantially increase or double production. A variation of this approach is to maintain a smaller cadre and to build up the second shift more gradually perhaps over 2 to 4 weeks.

The amount of training required for people to be effective in a repetitive manufacturing operation has generally been overstated. The American factory worker, including young people right out of high school (not necessarily graduates), is generally capable of mastering simple manual operations in a few hours. A good orientation program is important for new employees. Clear job instructions and appropriate followup by the supervisor or a designated trainer is necessary. If this approach is followed, operations can be increased quickly without sacrificing quality or productivity.

Occasionally one hears, usually from people in small operations of 50-100 employees, about the difficulties of two-shift production. Often it is said that, because of quality and perhaps other operating concerns, it just isn't practical in that particular business to institute a second shift. When an operation is small or in a start-up situation, there is an outside possibility that one shift-only production makes sense. As soon as additional space is needed or a second piece of a particular type of equipment is required, economics may show that a second shift is better than making an additional capital investment.

If there should be resistance to establishing a second shift, a third shift, or continuous operations, an economic review of additional hours vs. the cost of additional capital equipment should quickly resolve the matter. For example, if a $250,000 machine must be duplicated or a second or third shift added to produce according to customer need, it is easy to see that one-alternative has the cost of the depreciation on $250,000, the cost of the space to install it, and additional maintenance and supply costs. The other alternative only costs the hourly shift premium, lighting the area, and perhaps some ventilation. The question of supervision may arise. In modern industrial areas, it is usually possible to find trustworthy personnel who will work alone or in small groups without supervision. However, proper instructions and points of contact in case of trouble must be provided. Daily review of results is also important.

When an economic evaluation indicates that a partial second shift should be added and only one or two operators are involved, a question of safety arises. This can usually be resolved by installing a security personnel route box that must be activated at periodic intervals and then tying this to a monitored security system. If the time clock is not punched within a prescribed interval of time, an investigation will be made. In the case of remote operations by a single employee, the employee can be required to phone periodically to a more fully staffed part of the operation. Some say that this cannot be counted on to take place on a regular basis. This may be true when management has abandoned its prerogative to manage. In a well-disciplined plant, a phone call log that is periodically inspected by supervision will suffice.

SHIFT SCHEDULING

If one were to ask why a factory starts each of its shifts at a particular time, the answer will usually be either "we've always started at this time" or "everybody else in the area starts at this time." This is fine until traffic congestion becomes a significant problem. Factory managers in a number of areas have eliminated major traffic problems by offsetting some shift starting times by as little as 15 minutes. Some will say that this makes it inconvenient for people who like to ride together. When there are enough people to cause traffic congestion, it is usually not too difficult for people to work out alternate ride-sharing programs.

The question of whether or not to butt the first and second shifts or allow perhaps a half-hour interval has been debated extensively. Leaving an interval may be a necessity when parking space is limited. It also avoids the problem of people from the second shift coming in and socializing with people at work to the detriment of production. This should not be a problem in a well-disciplined shop, where people can be directed to the lunchroom or areas away from operations. When discipline is a problem, some companies keep people outside the plant until 5-10 minutes before the shift starting time. This is hardly practical in inclement weather and can cause ill feeling.

Butting shifts makes it easier for supervision, group leaders, and key operators to discuss any problems that may be occurring and to provide instructions and cautions as well as to verbally clarify written instructions left for the oncoming shift. In the writer's mind, the advantages of this personal contact outweigh any disadvantages that may occur from butting shifts.

Clear and concise work schedules and written instructions are important for the second and third shifts; otherwise, required work may not be performed and people may consider themselves to be left out of much that is going on. Most well-managed second and third shifts develop an esprit de corps and pride in the fact that they handle all manner of situations and emergencies without the support and perhaps overmanagement of the day-shift people. This pride often produces second shifts with higher productivity than the day shift. It can also produce some surprising and not necessarily ideal results when clear instructions are not provided. Good nightly written reports from the second shift are important to keep lines of communications open and help the day shift understand the surprising conditions they sometimes find. Written second-shift reports are always a good practice, especially when there is no third shift available to bridge the communications gap.

Unusual Schedules

During the 1974 oil embargo, the federal administration asked industry to consider 4-day work weeks to conserve energy. Sullaire Corp. in Denver and a number of other companies worked four 10-hour days and reported that their employees prefer this arrangement. Some have even gone on to three 12-hour days or other variations that cram the workweek into even shorter periods of time. Broan Manufacturing Co. in Wisconsin experimented with the 4-day workweek for a 3-month period. Broan's experience and reports from others provide the basis for the following advantages and limitations.

Advantages. The advantages of four 10-hour days are as follows:

- Younger members of the workforce enjoy having a 3-day weekend. This can lead to a high esprit de corps, high quality, and high productivity.
- When extra output is needed, working the fifth day can provide up to a 25% increase in weekly output.

Limitations. The limitations of such a workweek are as follows:

1. Older members of the workforce and mothers with young children find 10-hour days objectionable on a regular basis.
2. The younger people can be resistant to scheduling overtime on Fridays once they become acclimated to having the long weekend off (and this acclimatization took less than the 3-month trial period in Broan's case).
   - Office service and shipping operations working on the fifth day can be handicapped by not having factory personnel available.
   - Overtime premiums may be required.

The conclusion is that unusual programs of the type just described have their place, but it is a very small niche indeed.

Scheduling of Overtime

Each workforce tends to have its own personality or characteristics. It will usually include a number of "overtime hogs" who will literally work around the clock unless told to go home. Unless the situation is urgent, it is difficult to get full effort out of people for more than 10 hours. Beyond that, overtime should be restricted to emergencies or essential activities that cannot be otherwise handled. It would be difficult to fully evaluate the effectiveness of a maintenance worker who goes 14 hours night after night, but one would suspect that there are extended periods of limited or negative effectiveness. Control of overtime is an important part of controlling cost and improving overall productivity. Periodically requesting detailed accounting reports of all overtime activity can be useful. Requiring specific prior approval for all overtime in slow periods is often appropriate.

Some people may have real problems with working overtime, including those who share rides, have children at home, and so on. Some may have medical problems that make working overtime inappropriate. Consideration must be given to all these situations if a true rapport and team spirit is to be built with the workforce. Advance notification of overtime where possible (and it usually is), resolves a lot of problems. When due consideration has been given to the needs of the people and the very few special circumstances that prevent them from working overtime, it is the factory manager's job to insist that people come to work as scheduled. Strong counseling and appropriate discipline may be necessary to ensure that a full work team is available as required. Most people can be made to understand that customers can buy elsewhere if their company does not produce to the customer's demand and will accept the necessity of overtime. The extra pay from overtime greatly helps to assuage any negative feelings.

There are cases when concern for human beings and common sense requires the acceptance of a person's request to be excused; the couple who have planned a Saturday wedding are a case in point. Certain religious beliefs prohibit work on Saturday. In such cases, people need to be assigned to areas where Saturday overtime is infrequently required or dispersed so that operations can continue on Saturday without their presence.

There are numerous stories of the effect of opening day of deer season on factory operations in the Midwest. In many plants, permission to exercise this "inalienable right" was refused. The result was impossible excuses, screwed-up operations, and disciplinary procedures that only created turmoil. The answer to this legendary problem was simply to ask ahead of time who was going to be off and schedule the plant accordingly. There is a need for forcefulness in the factory manager's makeup, but there is no need to fight "unwinnable" battles. Even before this era of teamwork and human relations, smart managers developed two-way communications and took care of their people's reasonable needs—imagined or otherwise.

MOTIVATION AND COMMUNICATION

This section of the chapter is broken into three parts. The first part discusses some of the psychological factors that affect employee motivation. It also outlines psychologically based motivation approaches to the individual and groups. The second part provides an overview of some of the common motivational programs being used today. While it does not discuss every conceivable program, this part should stimulate a manager with ideas on how a particular program can be implemented in an organization. Finally, the third part emphasizes the importance of communication to the workforce, thus ensuring motivation.

PSYCHOLOGICAL FACTORS AND APPROACHES

People who make a living by working in manufacturing are often overwhelmed by the complexity. At times, they even wonder how anything gets done at all. Delivering even what appears to be a simple product to the marketplace requires integrating materials, machines, people, and information at all stages of the manufacturing process. All of this must be well managed and coordinated to deliver the highest quality goods to customers.

It is amazing to look back over the past several decades and see the tremendous advances that have taken place in technology and manufacturing methods. Entirely new conceptual models have been developed that could not have been dreamed of 10 or 20 years ago. Advances in computer applications, integrated manufacturing, flexible manufacturing, robotics, and many other areas are testimony to amazing progress, but also to society's ingenuity to overcome technical problems. It is a great credit to human beings that any type of technical problem can be overcome. For example, in July 1969 the U.S. launched the world's most powerful engine and landed men on the moon. The goal was set by President John F. Kennedy in 1961 and was achieved in 9 short years. This achievement, like many others throughout history, are monuments to the human capacity to achieve great things by overcoming technical hurdles.

Unfortunately, this same ingenuity used to solve complex technical problems falls short of the mark when it comes to solving people problems. Somehow, even with the accumulative knowledge of thousands of years, the entity that is least understood is ourselves. Management struggles to comprehend why people behave and do the things they do. Given a choice between solving a technical problem related to machinery, materials, or information and solving a "people problem," managers and engineers choose the technical problem.

If there are problems in organizations that people think persistently defy a solution, then they are problems of human motivation. The important question is how does one get people to do the things that need to be done? Of course, the even greater question is how to get people to do the things that need to be done and, at the same time, feel satisfied.

This is a practical section on understanding why people do the things they do and what can be done about it. It will help a manager determine whether the behavior problems he or she has with people are either motivation problems or training problems. Personal goals that motivate people into action and how these can be integrated with work goals will be explored. Finally, this section will take a look at how to design, or engineer, jobs that have high motivating potential scores.

This section does not present another motivational program; enough of these already exist. Over the past 20 years, managers have been bombarded with leadership programs, one-minute-management techniques, employee involvement, quality circles, management by objectives, incentive systems, and a host of other things. Every program promises the solution to all of motivation. Somehow, the results achieved with these programs fall short of one's expectations and the promises made.

All of these programs are, in fact, sound. They should work. Why don't they? From this author's perspective, they fail for at least two reasons. First, managers do not implement human management programs well. Things are done in implementing these programs that would never be done when implementing technical solutions to problems. The second cause of failure is that the people who use these programs, and possibly at times the people who sell them, do not understand why they should work. They do not understand the theory behind the practice. Therefore, they modify the program, or take shortcuts, that actually make the process a failure. It is like the old saying of "throwing the baby out with the bath water."

Causes of Behavior Problems

Motivation problems really do not exist in organizations; the problems are behavior problems. It is the things that people do, or don't do, that give management problems. When a behavior problem is encountered, such as a person doing something incorrectly, or failing to do something, the first step is to sort out the cause. Determine if the behavior problem is caused by a lack of ability or if the cause is motivational. A classic behavioral science formula that shows the relationship of ability and motivation in determining behavior is as follows:

$$\text{Behavior} = \text{Ability} \times \text{motivation} \quad (4)$$

The model shows that behavior is a combination of two factors: ability and motivation. In one instance, a person in an organization may have all the ability needed to do the job. Let's say this is a factor of 1, or 100%. Then, if this person has no interest in doing the job, the motivation value would be 0. One hundred percent of ability times zero motivation equals no behavior. Here the cause of the behavior problem is motivation. The solution to a motivation problem is not training. Many people try to solve motivation problems with training and they fail. The solution to solving motivation problems lies in managing better and engineering jobs that are motivating.

There are times when people have zero ability but 100% motivation. The outcome is not 0 as might be expected from the formula. Remember, this is a behavioral science formula. The outcome of zero ability times 100% motivation is trouble. There is nothing worse in organizations than when people have no idea what they are doing and are highly motivated. They cause a great deal of trouble. In this instance, the cause of the behavior problem is a deficiency in ability. One might say that the goal is to "demotivate" these people until they are trained. This would be correct. Given that a behavior problem has an ability cause, the solution is training. Better management and performance systems may not help all that much.

Diagnosing Ability Problems

In its simplest form, ability is a combination of a person's aptitude, skills, and experience. If a person is not performing or achieving the results desired, the goal is to find out if the problem can be solved by training. The best way to find this out is by asking a series of diagnostic questions, which are as follows:

1. Has the person ever performed correctly?

   If the answer is yes, then the cause is not a lack of ability. Training would not help. Why? Because if the individual has performed correctly at one time and the job has not changed, then he or she must have the skills. Spending the money on training would be a bad investment.

   If a situation exists where a person has performed well for awhile and then his or her performance deteriorates, there are two possible solutions. A deterioration of performance over time may mean that the person is not getting timely feedback. He or she may think they are doing well, but in fact they are not. They need information on their performance. This type of problem can be solved by: (1) developing a system they can use to check their own performance, (2) helping them get information on their performance from people who use their products or services, and/or (3) by developing short-interval performance reviews with their supervisor.

   In another instance where a person's performance gets worse after they have initially performed well, the best solution may be more practice. This often occurs when the skill is not used or a job is done infrequently. For example, a worker may do a particular type of machine set-up every several weeks; a presentation is given only once a year; a quality audit is done on a part that is made once a month. In situations similar to these, periodic practice between times when the skill is used prevents a deterioration in performance. This is a common solution to prevent many everyday performance problems. Children practice fire drills in school. Workers practice safety routinely even though the adverse event may never occur. Police officers practice at the target range although they may never shoot their pistol in the line of duty. Football players practice all week for a 2-hour game at the end of the week. Practice makes perfect, especially in those activities that are done infrequently.

   If the answer to this first question is no—the person has never performed correctly—then a potential ability problem exists, and feedback and simple practice may not work. It is then necessary to go on to the second question.
2. Does the person have the aptitude to learn the skill if trained?

   If the answer is no—the person does not have the aptitude, or capacity, to learn even if trained—then it should be obvious that training is not a correct solution. If a manager finds himself or herself in this situation, three options are available: (1) transfer the person to a job that he or she can do; (2) redesign the job so that the certain skills are not needed, thereby eliminating the need for training; or (3) if solutions (1) and (2) cannot be followed, the manager must either tolerate poor performance or terminate the person's employment.

   If the answer is yes—the person can learn the skill if trained—then training is the correct solution. The investment in training has a payoff for the company as well as for the employee. Training can be done in many ways. It can be done using a show-and-tell style where an experienced employee demonstrates the procedure and the person then tries it. The person can also be sent to workshops or schools, or he or she can work with an experienced person for awhile. New interactive videotapes make it possible to learn complex skills at home, in the office, or on the shop floor. Additional information on employee training can be found in Chapter 9, "Workforce Development," of this volume.

Motivation

The relationship between ability and motivation indicates that managers must first determine if poor performance is caused by a deficiency in ability. If performance cannot be improved by practice, feedback on results, redesigning the technical aspects of the job, or training, then a motivation problem exists. Training is not the solution.

Motivation is the more complex factor in the equation than ability. It is differences in motivation and not ability that account for the great variability in performance between people. Generally, it has been found that people within the same job classification do not vary significantly in their skills and knowledge. They are much the same. What does vary significantly is their motivation.

Understanding motivation means to understand what moves a person into action, what directs the person's behavior, and finally, what stops the behavior. Motivation is generally defined as behavior that is under voluntary control and is purposeful. Motivation can best be defined as goal-directed behavior. A person begins to behave when he or she has a goal in mind. His or her actions are directed toward attaining the goal. And finally, a person stops when the goal is achieved.

Motivation problems occur when there is a discrepancy between the personal goals of employees and the organization's goals. Organizations by their very nature are formal, goal-oriented, social systems. The objective of an organization is to achieve its goals more efficiently than other organizations. Doing this means that they are more competitive. If organizations are more competitive, they not only survive, but they grow and prosper.

Sometimes in the competitive pursuit of the organization's goals, the personal goals of people are neglected. To manage an organization as if it were only a financial institution, and that people only want money, is to eventually cause its demise. In the long run, managers must ensure the fit between personal goals and organizational goals to maintain competitiveness, prosperity, and innovation.

This idea of fit between personal goals and organizational goals can be represented using the models in Fig. 10-5. The model in view *a* shows a large discrepancy between personal goals and organizational goals, which means the person is less motivated to perform to achieve the organization's goals. The model in view *b* shows less of a difference between personal and organizational goals. In this case, the person is more motivated. Why? Because as he or she works to achieve organizational goals, his or her personal goals are being satisfied at the same time. There is more of a fit.

Let's explore this a little further. It has been said that Thomas Edison, Henry Ford, and others like them were highly motivated—they were always working. It is said that Steven Jobs, founder of Apple Computer, and Thomas Monaghan of Domi-

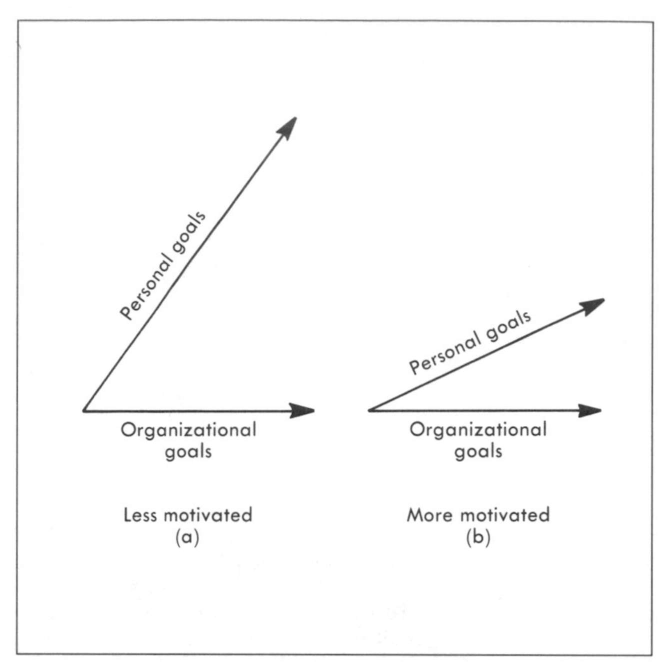

Fig. 10-5 Personal/organizational goal models.

no's Pizza are also highly motivated. How would one draw a model for these men? Would there be a difference between personal goals and organizational goals? No! Personal goals and organizational goals would be on the same line. Therefore, they are always working and always feeling satisfied. They are highly motivated individuals.

Many years ago it was said, "People live to work!" Today it is said, "People work to live!" What's the difference? The first saying implies that many years ago work had much more meaning in a person's life. Personal goals were embodied in the achievement of organizational goals. Doing a good job was fundamentally important, more important than just earning wages. The second saying implies that people in today's modern workforce tolerate work so that they can satisfy their personal goals on their time off. This is why shorter workweeks, flex-time, and longer vacations are so desirable. Today, there is a bigger discrepancy between personal and organizational goals. Therefore, managers encounter less motivation for achieving organizational goals.

Finally, let's look at public schools. These are organizations that have formal goals to achieve. What percentage of students go to school daily to achieve the organization's goals? Most people say a small percentage. The usual personal/organizational goal model for students in school is shown in Fig. 10-6.

All of this helps managers to understand motivation as goal-directed behavior. The task for managers is to create a work environment where the person satisfies his or her personal goals while achieving organizational goals. Now, let's identify and examine the personal goals that people bring to work with them.

Personal Goals for Motivation

The three psychological goals that form the basis of human motivation are affiliation, power, and achievement.[14] Affiliation is the goal of being liked or accepted by others. It is the social goal. Power is the goal of being in control of one's own actions, the actions of others, and/or in control of things—driving a car

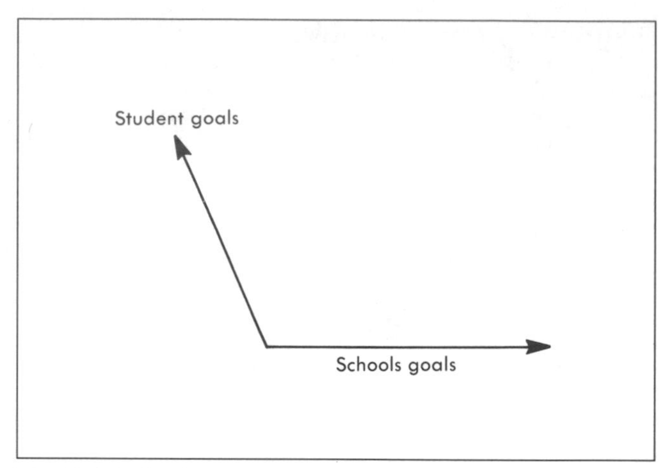

Fig. 10-6 Student/school goal models.

gives some people a feeling of great power. Achievement is the goal of doing the best or being the best.

No matter where they live, where they work, or how old they are, the behavior of people at work, at play, at school, and at leisure can be traced back to these three personal goals. For example, some people may submit a suggestion for productivity improvement because they want approval from others. Their goal is affiliation. Others may submit similar suggestions, and their underlying motive is power. They want to enhance their status—to be seen as important. And other people may submit productivity improvement suggestions because they want to personally improve their performance. Their motivational goal would be achievement oriented. In each of these cases, the behavior of submitting a suggestion is the same, but the goal, or motive, is different.

The goals of affiliation, power, and achievement are an integral part of everyone's motivational patterns. When people come to work in the morning, they have with them their goals of affiliation, power, and achievement. Some people are more achievement oriented, others are more power oriented, and others are more affiliation oriented. But regardless of their dominant motive, every person has some degree of power, achievement, or affiliation goals. Even people with very high needs for achievement still have goals for power and affiliation. So, the things that one may do specifically for a person with high affiliation goals are also appropriate for people with high goals of achievement or goals of power.

Affiliation goals. People whose behavior is motivated primarily by affiliation, or social goals, find that maintaining close, friendly relationships are important to them. They are more motivated to do things if they can see that the behavior, or the performance, leads to approval and being liked by others. Many people obey rules, are outstanding at their work, and do great things for the approval of others. On the positive side, the affiliation motive leads to behavior that is associated with teamwork, cooperation, dedication, loyalty, and cohesiveness.

The same goal of affiliation, however, can lead to undesirable behavior. People engage in inappropriate behavior and do unacceptable things to get approval. This is easy to see in teenagers who have very strong goals for affiliation. They often commit acts of delinquency to get approval from their friends.

One should not think that the negative behavior associated with the affiliation goal is only active in young people. People at work restrict the rate of their production, come back late from lunch, break rules, and fail to perform certain tasks just to get approval from peers.

People with high goals of affiliation make many friends. They become distressed if they think anyone thinks badly of them. They keep in touch by writing letters and making phone calls. They put people before the job. People with high goals of affiliation tend to sympathize or console others who are having difficulties. They communicate in terms of feelings rather than logic. They perform better in teamwork settings. They see work groups as a means for social interaction. People with high goals of affiliation take negative feedback on their performance as personal. They have a hard time being objective when they get negative feedback on their performance.

If employees behave in the ways just described, then it is likely that their dominant motive is affiliation. If this is the case, the following guidelines may be helpful:

1. Include them in group meetings for planning.
2. Rely on positive reinforcement practices and praise for incremental performance improvement. They will work for your approval.
3. Try not to use criticism or punishment.
4. Assign these people to jobs where they can be of help to others.
5. Assign them to jobs that require teamwork.
6. Allow them opportunities to interact socially on the job.
7. Give them jobs as coordinators.

Doing these little things satisfies the personal goal of affiliation while the company's organizational goals are being met. This works best if getting and engaging in these affiliated-related things are made contingent on one's expectations of high performance.

Power goals. Power is often called the great motivator. To be in control of people, events, and things has been known to "turn people on." On the positive side, the power goal in people motivates such behavior as risk taking, entrepreneurism, and leadership. People with high goals for power are stimulated to make changes and to make things happen.

On the negative side, the power goal, like the goal of affiliation, can lead to inappropriate behavior. People who become consumed by power begin to use their influence or positions of authority to satisfy their own personal needs rather than to benefit others. For these people, the goal of power becomes the retention of power. This phenomenon parallels the notion that the more power you get, the more power you want.

People with high power goals are mainly concerned with their reputation and their position. They can be observed seeking ways to impress others and becoming the center of attention. People who have high power orientations are preoccupied with seeking positions of leadership—being in charge. They place themselves in jobs where they can influence others to perform tasks. They may even find themselves in training or teaching roles. They tend to take either very high risks or very low risks. People with high goals of power give unsolicited advice. They collect high-prestige objects. They also seek, withhold, and use information to control others.

If it is determined that the power goal is the dominant motive of certain employees, then the following guidelines can help satisfy the power goal:

1. Invite them to participate in establishing goals.
2. Give them clear decision-making authority when assigning projects or jobs.
3. When they perform well, give them additional visibility through symbols of power and prestige such as plaques, press releases, and privileges.
4. Allow them to be involved in leadership positions outside of the work, such as being an officer of professional organizations and associations.
5. Put them on special projects that have high value and high visibility.

Getting a person involved in these types of activities goes a long way to meeting their power goals while producing high motivation to perform for the organization.

Achievement. People with high achievement goals want excellence as an end in itself. They think about ways to perform better, they find new ways of doing things, and they enjoy long-range planning. They set realistic goals and take moderate risks. High achievers do not get drawn into situations where excellence is a matter of luck. They need frequent and specific feedback on their performance. They tend to choose co-workers who are competent rather than friendly. High achievers have a strong career orientation. They talk often about their next step on the career ladder. People with high achievement goals want jobs that give them personal responsibility. They seek situations that allow for self-control and are not usually interested in controlling others. They like challenging tasks and are very persistent.

On the positive side, it is obvious that high personal goals for achievement leads to results for the organization; however, people with very high goals of achievement are very self-centered. They work best alone. They take little satisfaction in team efforts.

If certain employees have dominant needs for achievement compared to their needs for affiliation or power, the following are constructive things to do:

1. Set performance goals together that are measurable, time specific, realistic, and of moderate risk.
2. Delegate tasks in which they have complete responsibility and accountability.
3. Give frequent, specific, and descriptive feedback on performance.
4. Develop clear job definitions.
5. Give them access to experts, information, equipment, and resources.
6. Give them opportunities for personal growth, advanced education, and career development.
7. Give them projects that require skill.

Motivating Job Design

It is now time to put motivational theory into practice and focus on how to design motivating jobs. Too often, jobs are "designed" to meet the needs of the machinery, without regard for the humans who must do the work. In fact, the quality of the human/machine interface, both physical and psychological, can make a critical difference in the productivity of a plant. The motivational aspects of a job are important. Jobs can be designed, or engineered, to be more motivating.

Jobs with high motivation potential scores incorporate aspects of affiliation, power, and achievement along with the experienced meaningfulness that people get from their work.

MOTIVATION AND COMMUNICATION

A formula for job design is as follows:

$$\text{Job design} = \text{Interdependencies} + (\text{task identity} + \text{skill variety} + \text{task significance})/3 + \text{autonomy} + \text{feedback}$$

What follows is a survey that is included to help managers better understand the idea of jobs designed with high motivation potential scores (see Fig. 10-7). It is suggested that the survey be taken using one's own job as a frame of reference. Then read through the following sections to understand the factors in the equation and to calculate the motivation score of the job.

Interdependencies. Jobs are more motivating if they require interaction with others. People are motivated when they know how what they do fits in with what other people do. Most people want to spend at least part of their work time in interactions with others about work-related matters. Some jobs cannot be completed without the help of others. This is what is meant by the factor of *interdependencies*. Those jobs that require people to work in isolation with little social interaction are less motivating than those that require a sense of interdependence.

Interdependence in job-design technology is related to the affiliation goal in motivation theory. Building interdependencies and interaction into the job satisfies the affiliation goal.

Skill variety, task identity and task significance. These three variables increase the person's experienced meaningfulness of work by providing a sense of a whole piece of work being done. Jobs are more motivating if a person has responsibility for doing a larger portion of the work. This is what is meant by "job enlargement."

The first variable of *skill variety* refers to the degree to which the job includes a number of different activities that require the individual to use a range of the talents and abilities. When a job includes skill variety, the person is likely to experience it as challenging and, thus, personally meaningful. It is important, however, that the skills required are ones that the employee values. Simply giving someone more boring, repetitive tasks to do will not necessarily be experienced as an increase in skill variety or personal meaningfulness.

The second variable included within the concept of a whole piece of work is *task identity*. This refers to the employee doing something that is an identifiable piece of work. A job that requires an employee to create a product from start to finish would be a job with high task identity. The extent to which employees feel that they have been able to complete a task with a visible outcome constitutes the amount of perceived task identity in that job.

The third variable is *task significance*, which refers to the degree that employees perceive the job as having an important effect on their lives or work or on other people. When people feel that their product or service is important to others, they tend to be more motivated to do the best they can.

In the formula, skill variety, task identity, and task significance are divided by 3 to give an average score for the experienced meaningfulness of work. The more meaningful people perceive their work, the more they are motivated.

Autonomy. The dimension of *autonomy* is similar to the personal goal of power. Autonomy is the extent to which a person has authority to make decisions about the work. Building autonomy into the job is what is meant by job enrichment, where the person's decision-making authority is increased. As autonomy increases, so does motivation.

Feedback. Jobs that give people immediate *feedback* on results are more motivating than jobs that do not supply such information. The feedback dimension in the job-design formula is related to the personal goal of achievement. Feedback on performance can come from others and from the job itself.

To calculate the motivating potential score, add the scores for each factor. This gives the motivating potential score for a particular job. It shows how the personal goals of affiliation (interdependency), power (autonomy), and achievement (feedback) can be incorporated into the design of jobs. Scores in the 20s are good. In practice, if scores are too low, ways should be developed to improve the factor(s) that is deficient.

MOTIVATIONAL PROGRAMS

For many years, conventional personnel management theory held that if people were trained or otherwise qualified to perform a job, supervised appropriately, and paid well, they could be expected to perform at satisfactory or good levels. Good fringe benefits and recognition, particularly through advancement, were the icing on the cake that got superior results. Superior performing companies from the beginning of time took a broader view, recognizing that good leadership, personal involvement, and recognition by management were keys to increasing performance. It has taken the Japanese with their quality circles and worker participation techniques to get across to American management how important the involvement of individual workers is to achieving results that will let a manufacturing enterprise survive in today's highly competitive environment.

More than two decades ago, Professor Herzberg classified pay, benefits, and working conditions as "maintenance" items. These do not of themselves inspire superior performance. If they are perceived to be inadequate by the employees, they can be demotivators resulting in less than adequate performance. True motivators include incentive pay; individual involvement, such as the opportunity to be part of a dynamic group, have their input considered, and to be recognized for the part that they play in the team or operation; the opportunity for advancement or learning additional skills; and very importantly, good leadership, which recognizes and encourages individuals and recognizes their contributions appropriately. Today, there are a wide variety of motivational programs used in industry. Some of these are developed in-house. Others use programs developed by industry groups and consultants. The following is a description of some of the more popular programs.

Quality Circles

Quality circles, more or less in the Japanese style, exploded across the country in the early 1980s. The approach was "if the Japanese are doing so much better a quality job, we had better do the things they are doing." In many companies, quality circles have made a real contribution to improving quality and increasing employee involvement to the benefit of the business. In many other companies, after a brief infatuation with a "new toy," management has gone on to other things, and the quality circles have either languished, making a minimal contribution to the company, or have disappeared altogether.

Much has been written about quality circles. There are many consultants and trade associations that put on programs and will assist in developing a quality circle. Generally speaking, the approach is to appoint or hire a facilitator or coordinator who will develop the quality circles, keep them on track, and assist in presenting their proposals to management. Small groups of

Job design questionnaire

Instructions

Please mark your answers on the accompanying questionnaire. This questionnaire asks you to describe your job as objectively as you can. Make it as accurate a description of your job as you are able. A sample question follows.

Sample:

A. To what extent does your job require you to make telephone calls?

1	2	3	4	5	6	7
Very little; the job requires almost no use of the telephone			Moderately			Very much; the job requires almost constant use of the telephone

You are to circle the number which most accurately describes your job. If, for example, your job requires you to use the phone a good deal of the time, but also requires some paperwork, you might circle the number 5, as was done in the example above.

A. To what extent does your job require you to work closely with other people (either clients or people in related jobs in your own organization)?

1	2	3	4	5	6	7
Very little; not necessary in doing job			Moderately; some dealing with others is necessary			Very much; it is crucial to job

B. How much autonomy is there in your job? That is, to what extent does your job permit you to decide on your own how to go about doing the work?

1	2	3	4	5	6	7
Very little; I have almost no personal "say"			Moderate autonomy; many things are not under my control, but I make some decisions about my work			Very much; I have almost complete responsibility

C. To what extent does your job involve doing a "whole" and identifiable piece of work? That is, is the job a complete piece of work that has an obvious b eginning and end? Or is it only a small part of the overall piece of work, which is finished by other people or by automatic machines?

1	2	3	4	5	6	7
My job is only a tiny part of overall piece of work; results of my work can't be seen in final product			My job is a moderate-sized "chunk" of the overall piece of work; my contribution can be seen in final outcome			My job involves doing whole piece of work from start to finish and my work is easily seen in final product

D. How much variety is there in your job? That is, to what extent does the job require you to do many different things at work, using a variety of your skills and talents?

1	2	3	4	5	6	7
Very little; I do the same routine over and over			Moderate variety			Very much; I do many different things and use many skills

(continued)

Fig. 10-7 Job design questionnaire (*based on the work of J. Richard Hackman, Yale University; Edward Lawler III, University of Michigan; and Greg R. Oldham, University of Illinois*).

E. In general, how significant or important is your job? That is, are the results of your work likely to significantly affect the lives or well-being of other people?

1	2	3	4	5	6	7
Not very significant; outcomes of my work are not likely to have important affects on people			Moderately significant			Highly significant; outcomes of my work can affect people in important ways

F. To what extent do managers or co-workers let you know how well you are doing on your job?

1	2	3	4	5	6	7
Very little; people almost never let me know how well I'm doing			Moderately; sometimes people may give me feedback; other times they may not			Very much; managers or co-workers give almost constant feedback

G. To what extent does doing the job itself provide you with information about your work performance? That is, does the actual work itself provide clues about how well you are doing—aside from any feedback co-workers or supervisors may provide?

1	2	3	4	5	6	7
Very little; the job itself is set up so I could work forever without finding out how well I'm doing			Moderately; sometimes doing the job provides feedback to me; sometimes it does not			Very much; the job is set up so I get almost constant feedback as I work about how well I'm doing

Motivating potential score

$$\text{Interdependencies} + \frac{\text{Task Identity} + \text{Skill Variety} + \text{Task Significance}}{3}$$

+ Autonomy + Feedback

$$\text{Job Design Motivation Potential Score} = A + \frac{C + D + E}{3} + B + \frac{F + G}{2}$$

$$\text{Now Score} = \_\_\_\_ + \frac{\_\_ + \_\_ + \_\_}{3} + \_\_ + \frac{\_\_ + \_\_}{2} = \_\_\_\_\_\_$$

Fig. 10-7—*Continued*

perhaps 10-12 people are organized into a quality circle to work on quality problems in their work area. Problems are analyzed and solutions proposed. One of the responsibilities of the facilitator is to train or obtain training in the area of problem analysis and solution for the group. Meetings, usually weekly, are held usually on company time and at company expense.

A key to the success to this or any similar program is a responsive management that will assure that action is taken where it is required outside of the work group involved.

The predominant problem with quality circles has probably been lack of management followup on valid proposals. If supervision is not thoroughly integrated into the program, it may view the quality circle as an usurption of its authority and have a chilling effect on the whole process. Chapter 23, ''Achieving Quality,'' contains more specific procedural information on quality circles and related programs.

Suggestion Award Programs

These programs originated decades ago and are still widely used in American industry. In 1985, suggestions saved American companies at least $1.25 billion. Companies awarded their employees more than $128 million for those suggestions. American companies often receive as few as 25 suggestions per 100 employees per year. The average adoption rate in American companies is 25%.

Some Japanese companies claim as many as 100 suggestions per year per employee with 95% acceptance rates. These suggestions would include moving a wastepaper basket so it is easier to throw the paper in. Toyota and Hitachi are both supposedly in this range. Before Toyota started its quality circle program in 1961, it was receiving an average of 1 suggestion per employee per year and was adopting 38% of the suggestions.

The typical suggestion program works as follows:

1. A suggestion committee of senior or middle management is appointed as a steering committee for the suggestion program.
2. A full-time administrator/coordinator is appointed to organize the program according to the policies set down by the steering committee. In smaller companies, this may be a part-time responsibility.
3. An administrative procedure is developed and approved by the suggestion committee. A suggestion form is developed. This provides a place for the employee to explain his or her idea and to list the benefits or cost savings to result from it. There is usually a place for the suggestor's supervisor to make comments and recommendations on where the suggestion can be implemented at the department level. It may be passed along, marked approved, and implemented.
4. Suggestions are date stamped in the suggestion department and routed to appropriate departments for comments, recommendations, and/or approvals. Manufacturing or product engineers may be asked to calculate the value of the savings that will accrue. The accounting department is usually asked to confirm these numbers.
5. Approved suggestions are usually paid a percentage of the first-year savings after making provision for depreciation of required equipment or other implementation costs in cases where significant investments are required to produce major savings.

Suggestion award programs can significantly increase the amount of employee involvement. The degree to which this occurs depends to a great extent on the atmosphere/morale in the plant. If there is an open, cooperative management style that encourages employee participation, good results can be expected. Suggestion campaigns can increase the amount of participation. In an authoritarian setting, 100% participation may be achieved because the supervisor almost literally writes the suggestions for half or more of his or her people.

When cash awards are paid based on the value of the suggestion, a significant amount of manufacturing engineering and accounting time is required to calculate and verify savings. Because large sums of money may be involved, depending on the cap or maximum award value permitted, a reluctance to approve suggestions may occur on the part of some people or some departments. There can be a lot of wrangling and seemingly inordinate delays in processing suggestions. To minimize or prevent such problems, a number of variations of suggestion plans substitute recognition for cash awards.

Suggestion Program Variations

Joy Manufacturing Co. was an early pioneer in developing a suggestion program that substituted recognition for cash. Such programs represent only a small minority of American firms, according to the National Association of Suggestion Systems. They do, however, offer significant advantages.

Broan Manufacturing Co.'s ''fresh idea program'' (FIP) is an example of this approach. The Broan program, established in 1982, has the standard suggestion steering committee, coordinator, and processing approach. Suggestions, which may be made by individual suggestors or groups, receive chances for a prize drawing. Suggestions of indeterminate value or below $200 in value receive one chance, those valued between $200 and $1000 receive two chances, and those valued at $1000 or more receive five chances.

The first chance received by each individual is his or her ticket to attend the annual FIP banquet and prize drawing. Because the company gives away only chances for a drawing, conflicts between two suggestors as to who was first can easily be resolved by giving each suggestor a chance or chances for the drawing. When the suggestors argue about the value of the suggestion, the company can lean over backward to value the suggestion at the highest possible level without giving away the company store. To give proper recognition to those making major contributions, a special drawing is held for those people who contribute cost savings above $1000. In this drawing, everyone is assured of a prize, which can range from a television to a week at Disney World for the whole family.

The first suggestor making a suggestion with a value approaching $100,000 used the company's open-door policy to pressure everyone from the company president down for a cash award. The company held its ground, despite considerable uneasiness in the personnel department. The suggestor won the trip to Disney World and the concerns that he would be a demotivated employee proved to be groundless. This approach has resulted in a give-and-take, ''let's have fun while we're getting on with the job'' approach that resulted in 1262 ideas from 338 employees (80% of those eligible) and national awards in 1985 and 1986. Savings approach half a million dollars on an annualized basis.

At Broan's first annual FIP award banquet, the company provided prizes costing more than 10% of the value of the suggestions. In addition, top management at the banquet and on a continuing basis during the course of the year gave personal recognition to those who made significant contributions to the program, and on a more limited basis to all. This winning combination quickly brought the FIP at Broan to its current status as one of the most successful employee involvement and motivation programs in the country.

In addition to the FIP program, Broan has instituted a number of quality circles that play an important role in expanding employee involvement and participation.

Job Enlargement and Job Enrichment

Job enlargement and job enrichment are terms that were very popular during the 1960s and early 1970s. Job enlargement was primarily an effort to give the assembly line worker a broader scope than just shooting 14 screws in 30 seconds. By designing the job with a greater variety of work over a longer cycle, it was hoped to relieve some of the job boredom.

Job enrichment encompasses the concept of building some autonomy into the job. This might be accomplished by assigning the complete building and testing of a subassembly to an individual worker. Starting with Volvo in the late 1960s, the concept of work teams that build a major subassembly or complete unit began to reemerge in the literature. The practice must certainly go back to Roman and Egyptian times and has been carried forward by many firms through the "dark ages" of American manufacturing, as it is popular now to describe the 1950s and 1960s. All of these approaches that make work more interesting and provide more control of the work to the individual employee can have positive motivational benefits. They should be considered by manufacturing engineers and manufacturing management when laying out manufacturing processes. A caution is in order, however. There are many people who believe factory work is drudgery that must be accepted to provide better things in life for themselves and their families. This group of people can be very resistant to change because it requires them to think or take a more active part in the dynamics of the factory operation. A factory manager must understand the thinking and temperament of his or her workforce. Particularly when there is a workforce that has been operating in a very structured environment for an extended period of time, these motivational changes can be threatening.

To the degree possible, people need to be involved ahead of time in the planning of such changes or at least be made aware of them. A lot of acclimatization or training may be necessary for such programs to be fully successful. To paraphrase an old axiom, "You must teach an old employee new tricks very carefully." The selection of the employees for pilot programs in job enlargement, enrichment, or production teams is very important. Preference should be given to those who may be searching for more responsibility and opportunity.

"Off-Beat" Motivational Tools

A manager of a sizable Spanish-speaking workforce in southern California taught his employees excellent housekeeping by occasionally throwing pieces of paper on the floor that said in Spanish, "Bring this paper to the office for a $5 bill; this is number 1 of 20 such pieces of paper."

A Midwestern manufacturer in a nonunion shop has achieved close to incentive work pace by allowing 5 minutes off the job before breaks, lunch, and at the end of the shift to all employees and groups who are accumulatively beating standard. If an employee is not meeting standard at the morning break, he or she may still leave for lunch 5 minutes early providing that performance is brought up to standard by noon. The work ethic in the area is such that employees will run over standard where conditions permit rather than pacing production, so as to just gain the 20 minutes available social time. Seeing their peers walk by is a significant motivational factor for those who are not working up to standard.

When there are multiple assembly lines or groups doing similar work, a quality or productivity contest may be in order. A safety contest between all departments can also be productive. One company offered an additional holiday if the entire plant could go 1 year without a lost-time accident. After a 2-year period during which a series of bad breaks had prevented achieving the goal, the program lost its motivational impact. It was then reintroduced as a half day off after 6 months. This redesigned approach created a lot of thought and positive action among the workforce.

Companies with multiple plants performing similar functions have long compared every aspect of operations. These comparisons are used to motivate the management team. In a participative environment, they can also provide motivation to the hourly workforce. Productivity, scrap, and other elements of performance that are not normally discussed with the hourly people could be included. Americans love to compete. In almost any area where there is proper statistical data, two or more roughly comparable groups can be the basis for competition. Scrap, productivity, elements of cost, and many other areas of information that have not normally been shared with the hourly workforce can provide the basis for setting up a competition. This type of promotion or contest might be run for relatively short periods, with the more successful ones repeated in the future.

Summary

An attempt has been made to outline some of the standard programs and tricks of the trade that can be used to motivate a factory workforce. The negative motivators like discipline and separation that must be used, albeit sparingly and with serious consideration in even the most enlightened factories, have not been discussed. Not everyone is cut out for factory work. Even the best screening programs occasionally let slip through a fine human being who should be in a different career. Then there are people who for a variety of reasons, which have nothing to do with the factory environment, become recalcitrant and slip into poor work habits. If these are not dealt with promptly and firmly, performance of the whole workforce can degenerate.

On occasion, it is the supervisor or factory manager's right and duty to show genuine wrath and indignation. There are things that cannot and should not be tolerated in a department or factory. Properly expressed (and one has to know and understand his or her audience), this can have positive motivational effect on the vast majority of employees who are trying to do a good job. However, anyone who reads the literature knows that better performance can be achieved by using positive motivational tools than by using fear. This is true in the vast majority of situations, but utopia is still far enough over the horizon so that it is too early to totally discard the use of fear from a factory manager's set of tools.

To summarize, a number of positive and one negative motivators have been discussed. The following are essential to optimizing the use of these singly or in combination:

1. Management must understand and respect the workforce as individuals and human beings. Without respect and understanding on the management side, one cannot

expect to receive respect and understanding from the workforce, let alone all the positive motivational benefits previously described.

2. There must be complete honesty on management's part. What it tells the people about the competitive situation must be factual. In all programs, management must do what it says it is going to do. If for some unexpected reason it can't carry through, management must promptly explain why not.
3. Motivational programs must be maintained for an extended period of time. Factories make terrible demands on the time of management. Too many quality circles and other programs have withered away because management did not organize and program the time to provide the attention necessary to make the program a success on a continuing basis.
4. The manager in a smaller plant needs to spend time on the factory floor understanding the people and what is going on in the factory. The people on the floor need the leadership and reassurance that somebody is tending the store. In a very large factory, this role may be assumed by a operation manager or superintendent. It has, however, been standard operating procedure in the most competitive manufacturing organizations that you cannot run the factory sitting at your desk in your office.

Hopefully, these ideas have provided some thought-starters that will lead managers to some original motivational programs or combinations of existing programs. The better the manager and his or her management team understands the employees, the easier it will be to figure out "what will turn them on." Experimentation, including an occasional failure, are a part of the process.

COMMUNICATIONS

The section on employee benefits and services discusses communications with employees through the use of bulletin boards, magazines, and/or newspapers. It also discusses social programs that bring employees together away from work. Properly used, all of these can reinforce the concept that the company is a team competing with a number of similar teams for customers who will ultimately decide the company's degree of success or lack thereof. Today, with competition coming from many foreign companies as well as from long-established domestic competitors, every company is or will soon be facing intense competitive pressures. Good communications with the factory workforce so that it understands the realities of the business can provide motivation and a team effort. All of the programs previously mentioned provide opportunities for communicating on this and all other subjects that are of interest and importance to the team.

Periodic meetings with supervision, perhaps down to the group leader level, and with all employees to report on the state of the business can be good communication tools. They may be expensive. It may be necessary to hire an auditorium. One may want to hold a night meeting for supervision and provide dinner and cocktails. It may be appropriate to cover the entire workforce by having a series of meetings with groups of a size that the in-house lunchroom or other local facility can accommodate. Budgeting for two or three such informational meetings a year is necessary to the job of building a participative team. And, of course, there is no better communication than the boss talking to an individual or small group as he or she tours the factory or department.

This section did not discuss the information communication systems that make the factory function. Elements of these systems are discussed in other sections of this Handbook. It is essential, however, that the objectives, plans, and marching orders be clearly communicated to those who need to know on a regular basis, including daily and hourly when appropriate. One of the primary complaints in most factories is "nobody tells me." With the participative team environment, the manager will have plenty of feedback on the quality of communications within the organization. It will only be a matter of sorting the through the messages to determine what, if anything, needs to be done to improve communications.

PERFORMANCE STANDARDS AND THEIR USE IN CONTROLLING FACTORY OPERATIONS

For a factory to operate efficiently and for products to be costed properly, there must be accurate knowledge of the cost of materials and labor, both direct and indirect. The process of developing standards begins with cost targets and preliminary cost estimates during the development phase of a product. During this phase, product engineering, manufacturing engineering, and marketing should work closely together. There will be give and take on design approaches to make it possible to select the most cost-effective manufacturing process. Marketing must maintain saleable product specifications and features as determined by its marketing analyses. These, ideally, have been comprehensive enough to give good information as to what customers will desire and purchase now and in the future and at what prices. A thoroughly knowledgeable marketing group will be expert enough to consider cost-benefit tradeoffs in specifications and features that will optimize the value and quality for the customer. Manufacturing engineering and product engineering will be evaluating manufacturing processes and tooling in an interactive process. This interactive process has been going on in the better manufacturing organizations over the years. It is currently being rediscovered by many manufacturing organizations and termed, among other things, "simultaneous engineering." This process may involve large staffs in big organizations, or it may occur largely in one person's brain in a small, entrepreneurial shop. In the latter case, customers, an outside salesperson, and/or personal visits to the field can provide knowledge of the marketplace. Manufacturing knowhow may exist in the mind of the entrepreneur or outside vendors, or consultants and other sources may be contacted. The point is that the process must take place.

CHAPTER 10

PERFORMANCE STANDARDS

When the design reaches the stage where facility and tool procurement should begin, more refined estimates are normally available for review and approval by management. Before production begins, it is normal to have accurate bills of materials and engineered labor estimates as a basis for material procurement, capacity planning, labor staffing, and shop scheduling. Material and plant scheduling are discussed in other sections of this volume. This section will deal with how labor standards are developed and used to manage the workforce. The estimates that become preliminary standards will generally be based on a combination of historical data, standard data, information from vendors on process and machine capabilities, and extrapolation of this data into new areas. In a mass production or flow shop, there should be detailed labor routings with time estimates for each operation. In a job shop or smaller operation, these estimates may be less detailed. The degree of accuracy and coverage can have a significant impact on the productivity of the factory. In a mass production situation where thousands of units are produced per day, it is the responsibility of manufacturing engineers to review the estimates and assign engineered standards to the labor routings promptly. The next section briefly covers methods for development of labor standards.

SETTING PERFORMANCE STANDARDS

The method and time standards developed by standards engineering affect many aspects of the manufacturing operation. In addition to the traditional use of standards to control direct labor costs and to implement incentive plans, standards are increasingly being applied to indirect labor operations. Improvements in product design and tool and fixture design may also be affected by the proper use of engineering standards.

Hourly Employees

In the absence of formal standards, there will still be a work standard developed by tacit agreement between the worker and the first-line supervision concerning what constitutes an acceptable production level on any task. The quality of supervision and the influence of the worker, or union, will often have a considerable effect on the level at which this acceptable standard is set.

Several methods of determining labor time standards exist, each of which have certain advantages and disadvantages. The methods discussed in this section are based on historical data, estimated standards, time study, work sampling, and standard data.

Historical data method. Many companies use historical data, taken from informal records kept perhaps by the accounting department, to establish time standards for manufacturing tasks.

The principal advantage of using historical data is its availability, because most companies keep some record of the time spent on various tasks. The major disadvantage is the difficulty in setting higher goals than those achieved in the past. In addition, there is seldom any documentation of the operations or equipment that have been used in the production processes. The lack of this information makes it extremely difficult to adequately update standards that have been developed from historical records.

Estimated standards method. The next step in the evolution of work standards is from the use of historical data to the use of estimated standards, which are essentially projections from historical data. From either formal or informal records, estimates are developed to indicate permissible times for new jobs.

Here again is the advantage that records usually exist indicating the time required previously on similar jobs, thus permitting the development of standards without a formalized work measurement program. The disadvantages are similar to those for the historical standards in that estimates based on historical data tend to perpetuate the inefficiencies or poor resource utilization of the past. Also, standards tend to become tied to individual operators or machines that may or may not reflect a consistent performance over a variety of jobs.

Time study method. Time study, first used as a work measurement technique in the early 1900s, is still used in many work measurement systems. In time study, as it is generally practiced, a job is broken down into short elements of work that are timed individually by an observer using a stopwatch. Any manually controlled operation is performance rated, or leveled, to eliminate differences in the skill and effort applied during the observation of the task.

The time study technique has the advantage that the procedures and equipment are known, making it easier to maintain work standards when time revisions are required because of changes in procedures or equipment. When time study is used to set standards on individual jobs, inconsistencies may occur because of the different procedures used in similar jobs and also because of the subjective judgment involved in rating the operator's performance. One disadvantage of this method of developing standards is the difference of opinions regarding the operator performance evaluation and the accompanying arguments, or bargaining, over the validity of the time standards developed.

Micromotion analysis. One time study technique used in determining the times required to perform given operations is micromotion analysis. This technique, which produces extremely accurate job times, involves making a motion picture of the operation, either at normal exposure speed or with a high-speed camera, and counting the frames to make a detailed analysis of the motion pattern. Because micromotion analysis is expensive and time consuming, it normally can be justified only for the development of basic motion times or on extremely short cycle and highly repetitive operations.

Controlled experimentation. Controlled experimentation can also be applied to highly repetitive operations. The exact manufacturing operation and tooling are duplicated in a laboratory situation, with the operation being performed by specially trained people. This technique can be used in connection with proposed operation changes or layouts to determine which one is the most economical prior to making the operation change in the shop.

Predetermined time systems. Predetermined time systems, originally developed to overcome the problem of rating operator performance, have gained wide acceptance in industry. The essential principles of predetermined time systems are the identification of each basic body motion and the assignment of a predetermined time to that motion. Predetermined elemental times have been developed by micromotion analysis over a large number of plant operations and are generally accepted as being a consistent source for the building of time standards. The analysis is made by identifying the motions used either by observation or by synthesis of the elements of a proposed procedure and then by applying the predetermined times to the recorded motions. The major predetermined time systems are methods time measurement (MTM), work factor, and motion time analysis (MTA). In addition to these, there are numerous proprietary systems that have been developed by companies for their own use.

The primary advantage of a predetermined time system is the elimination of the necessity for performance rating and the use of a stopwatch to determine manual times. The stopwatch, of course, is still required to record machine-controlled or process times, but there is understandably less worker resentment to recording a predetermined time analysis than there is to making a stopwatch study of the same activity. The use of a predetermined time system also forces a comprehensive and detailed look at the methods being used, which frequently results in improvements. Predetermined time systems have the disadvantage that to determine a single standard may require more time than by the use of time study.

Work-sampling method. Work sampling is a technique used primarily in determining delay allowances or in checking the activity in a particular situation. It is based on the statistical principle that the recording of a series of random observations of any activity will reflect the overall percentage of that activity.

The use of work sampling has a distinct advantage over time study in that a greater number of meaningful observations can be obtained in a given period of time. This is because, in part, time studies may include machine downtime, operator break time, and so on, during the period of observation. Work sampling is particularly useful when there is a large group of people or operators to be observed, when the operations being used are varied, or when the sequence of motions varies. Studies that would be too expensive if done by any other means are possible using this technique.

One drawback of work sampling is that the methods recording has a minimal amount of information, and therefore, the maintenance and audit of subsequently developed time standards is more difficult.

Standard data method. As was previously stated, the use of either time study or predetermined times to determine individual standards is time consuming and may lead to inconsistencies. The use of standard data is a method that has been developed to overcome these difficulties. While the use of standard data is not actually a measurement technique, such data is used by most plants to reduce the industrial engineering effort required to obtain coverage of operations with work standards. In the development of standard data method, the elemental times, derived from either time study or predetermined times, are arranged and combined in such a way that they cover the complete range of operations or parts that may occur on a particular type of equipment or product. Setting an individual standard then becomes a matter of listing those elements that are required to accomplish the job being studied and adding up the previously assigned time values for those elements. The use of standard data in addition to predetermined time data permits the establishment of operations standards prior to the initiation of a process in the shop.

In a broad sense, all predetermined time systems are a type of standard data because given elements of work (in this case, basic motions), have been assigned predetermined times. The use of predetermined times as elemental data, however, is usually not justified, except in those cases where there are extremely large runs of the same parts across the same operation. One method that has been developed to overcome this barrier is the grouping of the basic motions into somewhat larger building blocks, for example MTM General Purpose Data (GPD), which requires approximately one-third of the industrial engineering time to apply as MTM.

The theoretical basis for much of the available work measurement data has been substantiated by the extensive laboratory research of the MTM Association at the University of Michigan, by the statistical analysis performed by the Society for the Advancement of Management in the development of their performance rating films, and by the many studies that have compared different predetermined time systems with each other and with time study standards. All of this literature provides a sound, rational basis for developing engineered standards that will yield predictable results.

Short-interval scheduling. Short-interval scheduling was a buzz-word many years ago. The origin of this approach goes back to the work done by W.B. Taylor at the turn of the century. It essentially is the scheduling of work for assembly lines, groups, or individuals on an hourly basis and recording worker performance each hour. Any factory that schedules and records worker output on an hourly or bihourly basis is following the practice of short-interval scheduling. Variations of this technique are applicable to groups of indirect and clerical workers as well.

A number of consulting firms have short-interval scheduling programs that they will recommend to factories or adapt for clerical or other functions. The basic approach used in short-interval scheduling is sound, yet implementation can range from thorough to simplistic. Work is often done by nontechnical people using crude observation techniques and tools. An alternative is to employ professionals in-house or go to engineering firms for the manufacturing or industrial engineering talent needed to develop a standards system. Careful consideration should be made before employing one of the so-called short-interval scheduling firms.

Salaried Employees

Measurable performance standards for salaried employees fall into several categories. The annual variable budget is considered by some to be a contract with each department supervisor and a higher level of management. Failure to achieve budget has even been considered by some to be equivalent to stealing. Unless conditions change beyond the ability of a variable budget to be met, the budget is a primary tool for measuring salary performance. Its value is enhanced if the supervisor or manager has an opportunity to participate in the give-and-take goal setting that should be one element of the budgeting process. In today's competitive world, budgets should anticipate improvement. Managers are more likely to meet or surpass budgets if they are given some opportunity to participate in planning the ''stretch'' elements.

A second important measurement tool is the setting of specific objectives or action plans for individuals, highlighting a few areas that require a focused effort. These may be part of a structured bonus plan or management by objective program. These should be developed during interactive planning sessions. While not essential to the process, action plans will be enhanced to the degree of personal commitment involved. Of course, it is likely that management will have its areas of emphasis. Senior managers will pick these up in their own action plan/objectives and solicit supporting action plans from subordinates. In an ideal situation, direction will be provided by the factory manager. This will be supported by a broad base of action plans at the supervisor level. There will be only a half dozen or so action plans for individuals at the first level, with the number increasing at each higher level. These will highlight specific areas where special effort can make a difference. They will not simply list activities that must be performed in any case, such as ''meet expense budget.'' Rather, they will focus on specifics like ''reduce cost of sanding belts 5% by working with purchasing to qualify a new, lower cost supplier.''

Project performance should be formally reviewed at least monthly at the supervisor/engineer level. When reports are available on a daily or weekly basis, it can be profitable to hold more frequent reviews at the departmental level. Most good supervisors look at items like scrap on a daily or weekly basis. Action plans, at least at the lower levels, should be reviewed monthly. Top-level reviews are often made bimonthly or quarterly.

When action plan programs are introduced, there may be concern among some managers, particularly at the lower levels. The senior manager should spend time with each subordinate working out the opportunities for improvement in his or her area that can be translated into action plans. As information flow improves and organizations are ''delayered,'' management spans of control will increase from a half dozen or so subordinates to as many as 15 or 20. This will put a premium on departmental managers managing their own areas and will make less frequent reviews appropriate. If realistic plans are agreed on and assigned, enthusiasm will build as the year, and succeeding years, progress. The program can then become a very positive motivator. This is particularly true when bonus payouts are tied to the program.

An action plan program is active and ongoing, and it does not start and stop with the annual budgeting process. Completed action plans are reported on and dropped from the list (except for the final report at bonus review time). Plans are added as opportunities arise or as necessary to maintain a reasonable list for each individual. If circumstances change and plans are dropped, then no individual should be penalized at bonus review time. There are a number of performance measurement tools for design engineers, manufacturing engineers, and other knowledge workers. Objectives and action plans can be developed for a number of elements like conformance to design schedule milestones, meeting design cost objectives, and so on. Manufacturing engineers can set cost reduction targets in specific areas. In a number of companies, manufacturing engineers working directly with the shop floor are expected to develop cost reduction programs saving 3-5 times their annual salary or more. Areas of cost reduction potential can be worked into action plans by design engineers as well. Of course, there must be allowance in any measurement plan to give high marks to brilliant product or process designs. These may be difficult to quantify, but deserve appropriate credit. They are the lifeblood of a manufacturing operation.

CONTROLLING FACTORY OPERATIONS

Preceding chapters have covered the planning process in a manufacturing company. Good practice generally includes the following:

1. A strategic plan looking out 5 years or more provides the direction and basis for shorter term planning.
2. Annual, biannual, or longer capital investment plans provide the basis for medium-term planning and the funding for new facilities, processes, equipment, and tools.
3. The annual budget provides a basis for tracking operations down to the departmental and subdepartmental levels. Most budgeting in high-production operations is done on a variable basis. Costs other than direct material and direct labor are specified as either fixed or variable. Depreciation and the minimum management required to keep the factory in operation are examples of fixed costs. Costs that can be controlled upward and downward with fluctuations in volume are treated as variable. All or almost all indirect labor and supply materials are examples of costs that are normally treated as variable.
4. A monthly forecast of sales extending out to or beyond the lead times for critical materials is updated monthly and used as the basis for material bring-in and personnel planning. When there are short-term fluctuations in volume or seasonal conditions, management has the information necessary to plan an inventory build-up to level employment or to cover demands for seasonal product. Conversely, a basis is provided for using overtime or temporary employees to handle short-term peak requirements. A monthly meeting is customary to review and finalize the factory schedule and personnel and inventory plans.
5. An MRP II or other system develops master and departmental schedules and provides for material receipt.
6. In high-volume shops that build inventory for off-the-shelf sales to customers, a daily schedule or ''build sequence'' is essential. This schedule may extend for several days or weeks, but must be updated daily or as required by the needs of the business. Inputs come from factory or material problems, perpetual inventories of finished goods (if available), daily sales reports, calls from the sales department, or urgent requests from customers. All of these are coordinated in a morning meeting.

The preceding discussion is intended to generally outline the flow of information that is used to manage the factory and the workforce. In job shops or in companies producing long lead time capital equipment, information may be formatted differently and flow at a different pace. The end result in any case will be a flow of information that specifies what should happen on the factory floor during any hour, day, or given time period.

Direct Labor Control

In a mass production or flow shop, the daily schedule will provide for production of perhaps a thousand or more finished products per day and a much larger number of parts spread over two or three shifts. It is often desirable to report production at least to the departmental superintendent and perhaps the factory manager on an hourly basis. An excellent approach to controlling operations in this environment is to provide each supervisor with this information so that it can be formatted into an hourly ''count sheet.'' This sheet provides the information as to the expected hourly output of an individual or of a group of individuals. It clearly shows what must be done to equal or exceed 100% efficiency based on the standard or estimated time. The supervisor is then expected to check the production of his or her people hourly or bihourly. This might sound to some like overcontrol in an era of worker participation, motivation, and team concepts, but it achieves an important purpose. By communicating with all of his or her people on a regular basis throughout the day, the supervisor can give a smile and a pat on the back, when everything is going well, or if there is a problem, can lend a helping hand or take whatever measures are necessary to solve the problem or bring in additional support. In the present, not-so-perfect world, this is one way to ensure that productivity is maintained and schedules are met. A lot of people like to solve their own problems without calling for help. If one part of a complex production process is in trouble and overlooked for any length of time, it can have a devastating ripple effect throughout the rest of the process.

Other excellent approaches for keeping production on schedule have come out of the just-in-time concept. Red and amber lights along an assembly line, which allow workers to signal when there is a problem (amber light) or to indicate where a line stoppage is occurring (red light), direct attention to problem areas. The "pull" system of triggering production can quickly highlight trouble areas. With these, there must also be a means of reporting production and worker performance. This can be done electronically in many cases, but in the writer's opinion, personal contact on a regular basis should never be eliminated.

A daily production meeting is extremely important in a high-production shop. The factory manager or his or her designate should chair this meeting. It will include representatives from all of the factory production departments, material control, quality assurance, manufacturing engineering, maintenance, tools, and the warehouse or shipping department. It may even be worthwhile to invite sales or customer service staff to keep them up to date and to include them in the team effort. One approach is to have each superintendent or area manager report the productivity of his or her department and to discuss, and attempt to resolve, any problems that may have occurred since the last meeting. Many data processing systems provide daily production and efficiency reports. It may be imposing on each supervisor to calculate the efficiency and productivity of his or her crew personally, yet asking them to do so ensures that they will stay on top of what is going on in their area and reminds them of management's concern about productivity. After all, there is a no more perishable commodity than an hour of human labor.

The material control or scheduling department can report on any schedule changes that may be required due to material availability, production problems, or customer demand. If any such changes are required, all of the people needed to ensure team implementation are present.

Quality assurance can report quality problems that may have occurred or been reported from the field. They can also report on the results of the quality audit program if there is one in place.

The daily production meeting is intended to be informational. If there are matters that require detailed attention, it is usually better to handle these in a separate meeting called specifically for this purpose. This meeting should not be a long, drawn-out affair. Ten minutes is usual at a plant with a schedule of more than 10,000 consumer products per day.

Several approaches are used to calculate direct labor performance. Three approaches are summarized as follows:

1. *Efficiency*. Efficiency is defined as the percentage of standard time produced by an individual or group. If an assembly line of 10 people working a full 8 hours each and producing 84 standard hours in a shift, they have operated at 105% efficiency.
2. *Utilization*. Many companies report the percentage of time that direct workers are assigned to direct labor tasks as utilization. If a punch press operator works 6 hours producing parts and 2 hours on a set-up for which there is no standard, his or her utilization would be 75%.
3. *Productivity*. The total standard hours produced divided by the total actual hours of the work group or individual is productivity. This is the product of efficiency and utilization. The author believes that this is the only real measure to use in controlling the direct labor force. If management stresses efficiency, it is easy for supervision to shift people to nonproductive jobs. If utilization is today's push, it is easy to account for more people on direct labor. Productivity cannot be easily toyed with. It provides a sound basis for evaluating improvement and raises a flag when something has gone wrong. To ensure that productivity is correctly accounted for, only transfers of people (hours) from the department to projects that have prior management approval should be permitted. An example of this would be building pilot models of a new product; efforts in this and similar areas should not be charged to departmental productivity.

Departmental supervisors should normally graph or otherwise track productivity on a daily basis. The factory manager should keep tabular and/or graphic records of productivity by major areas to spot continuing problems. General management may want weekly or even daily reports.

Some budget programs permit calculating productivity based both on current labor standards and on the original standard at the time the product was brought into production. This original standard is the frozen or "price study" standard on which product costing was based. The frozen standard is normally made to equal the current work standard at the time the annual budget is set. Measuring both current work standard productivity and frozen standard productivity provides an excellent measure of improvements made throughout the year. It clearly indicates which supervisors have been most aggressive in implementing improvements and cost reductions. A supervisor who maintains 95% productivity throughout the year (100% is unlikely unless all labor in a department is on standard) may look good when compared to a peer who is at 90% productivity after having implemented improvements equal to 15% of his or her direct labor. Calculating the frozen standard productivity shows the latter is the real hero at 105%.

Much of the preceding discussion is primarily applicable to mass production or flow factories. Factory managers should strive for as much current reporting of direct labor per activity as is feasible in their environment. In a job shop, with a system that permits comparisons only after a job is complete and where the jobs are long cycle, effort should be made to calculate interim performance. With short-cycle jobs, labor performance should be recorded promptly after the job is completed and trends noted and used as an incentive to improve performance. Control of job shop operations is covered in more detail in Chapter 4, "Cost Estimating and Control," of this volume.

Indirect Labor

The historical approach to measuring and controlling indirect labor was to look at the factory operating requirements in the context of the annual budget or perhaps to look further into the future at the business trend. Staffing was based on historical performance, with additions as necessary to cover planned programs or volume increases. Reductions were based on planned labor saving programs or "stretch factors" agreed on by the factory team or mandated by top management. A determination was made as to which indirect employees would be considered fixed or variable. The former are necessary regardless of the operating volume, and the numbers and expense of the latter would be expected to fluctuate in direct relationship to the standard hours or other budget-generating variables.

A complementary approach to controlling indirect labor is the head-count report. This is particularly valuable when there are no indirect labor standards or only very rudimentary ones. Often the manager's only gage of indirect labor performance is personal observation supplemented by work-sampling techniques. This approach has its drawbacks, however, which can be summed up by the law that states, "Work will expand to fill the available hours of the workforce." Because everybody is happily employed does not necessarily mean that useful work is in progress. Many seasoned factory managers consider the head-count report as the most valuable single tool available for controlling indirect labor.

For a considerable period of time, much has been done toward developing standards for indirect labor. As far back as the 1960s and before, programs were developed for the dispatch of maintenance and tool-support personnel. Times were recorded for individual jobs. Time studies were made of repetitive jobs. Maintenance supervision and manufacturing personnel worked on methodizing these jobs and providing estimates of the times repetitive jobs should take. This database was then used to measure performance of maintenance personnel. Routine jobs, such as lubrication, were time studied and put on standard.

Much work has been done in the area of warehousing and shipping, particularly where there is reasonable uniformity in size and weight of packages. Historical records can be used to determine the number of packages warehoused per person per shift, the number of line items picked, the average line items per order, and the number of orders shipped. The line items per order can usually be pulled out of the records with reasonable precision. This information can be used to set historical standards and develop plans for more efficient operations. It is generally useful to have a manufacturing or industrial engineer review procedures and evaluate the current performance of personnel. If there is good consistency in the flow of materials, then direct standards can often be developed. In warehouses that have not had the benefit of standards, an engineered approach to arranging the work can easily produce a 20-40% improvement in labor performance.

For a long time, firms with extensive clerical functions and a number of consultant companies have been developing clerical standards. These cover the gamut of office functions from pages typed to orders processed to claims handled. Factory office personnel probably have enough variety in their assignments to make standard-setting difficult. If there is a large pool in an operation, manufacturing or industrial engineers should take a look at it. They may have the expertise to provide standards, or they can suggest an outside consultant.

As the development of CIM has progressed, MRP II and just-in-time systems have provided information flows that have permitted some companies to greatly reduce both salary and hourly indirect labor in their factory operations. General Electric calls the process "delayering the organization" and is pointing toward operating with fewer levels of management between the division general manager and the factory worker. This is in contrast to 8 or more levels in some classical organization structures. The GE approach involves spans of control of up to 20 people because "if the boss has 20 people reporting to him, he won't have time to nose into all the details and 'micro manage.' " To some more advanced companies, this is not a new approach. CIM and the team concepts that are now popular make this approach and all the benefits that go with it a more important prerequisite to fully competitive manufacturing. A word of caution: A number of companies have installed complex MRP II systems based on the forecast of significant savings in inventory and personnel and with improved customer service. With the more complex systems that have not had enough practical factory experience incorporated into them, results have sometimes gone in the opposite direction. The factory manager of today is going to have to carefully watch the kinds of programs that are installed in his or her operation to be sure that they do truly lead in the direction of better communications and fewer indirect people if he or she is going to manage the factory of the future.

Decades of intense effort have squeezed direct labor down to 5% or so of cost in many factories. Indirect labor now offers correspondingly greater opportunities to improve competitiveness. It is becoming an area no factory manager can ignore.

IMPROVING PROBLEM SOLVING AND PRODUCTIVITY*

How can manufacturers get more of the results they want with fewer resources? This is the central question of productivity improvement, and it is one that has been asked countless times during the past 5 years both in the United States and abroad.

The usual approach to increasing productivity is to concentrate on the externals: better tools and equipment, lower scrap and rework rates, and more efficient work procedures and operating methods. All of this makes sense, but it is not enough.

In fact, there is a more direct way to improve productivity with greater payoffs and longer lasting results. There is no mystery or gee-whiz gimmicks to this approach. It involves sharpening the thinking skills of managers and workers. Improving the quality and precision of problem solving in an organization releases a productive force far greater than that of equipment, machinery, and tools, however vital these are to productivity improvement.

Most approaches to problem solving exhort managers and workers to "get the facts," define the problem, test conclusions, and go ahead and make the fix. This is all sound advice, but how should one go about these tasks?

IMPROVING PROBLEM SOLVING

One only has to recall the last problem-solving meeting he or she attended. Something went wrong. Maybe sales were off, or the reject rate went up, or a staff unit just was not performing. If it was a typical problem-solving meeting, discussion wandered. Some of those attending wanted to learn more about the problem, others proposed taking action, and still others focused on how to avoid similar situations in the future. Probably there was some finger pointing and the usual parry and thrust of charges and countercharges.

These meetings are usually deadly. One reason for the disarray is the lack of a clear problem-solving process or track. Discussion becomes tangled as participants butterfly between trying to solve the problem, taking action to correct it, and preventing future trouble.

To begin, it is necessary to define what a problem is. A problem exists when reality or the "actual" deviates from the norm or "should", when the cause for the deviation is unknown, and when the situation requires that the cause be discovered. When these elements are present, a problem exists. Based on this definition, a problem can be diagrammed as shown in Fig. 10-8.

Contrast the problem situation with those situations requiring a choice or decision among alternatives. In this case, finding the "cause" is not the issue; it is making the best choice. Or contrast problem solving with a situation where one is looking into the future to avoid potential trouble. Here, looking at the past to find "cause" is not a concern. Rather, it is to anticipate and protect oneself against future problems.

Each of these transactions—problem solving, decision making, and planning—requires a different track or process that must not be confused. The Kepner Tregoe problem-solving approach is one of the most widely used formal approaches to problem solving. It has been used in 44 countries and in several thousand organizations, in both the public and private sectors. This approach is a logical, practical, step-by-step *process* for answering the questions: Why did that problem occur, what's the cause? The five basic problem-solving steps in this approach are: (1) problem recognition, (2) problem definition or specification, (3) developing causes, (4) testing for most probable cause, and (5) verification.

PROBLEM RECOGNITION

Many people do not recognize a problem until it really begins to hurt. The recognition step is accomplished when someone perceives that actual circumstances deviate from expected performance norms (the "should") and that the cause of the deviation must be found.

Problem recognition involves being clear about what the problem is. To help clarify the problem, it is important to write out a concise statement that identifies the deviation to be solved. The deviation statement should be brief, indicating the object, person, process, or area and the nature of the deviation. Some examples of deviation statements are as follows:

- Soda bottle caps are twisted.
- Reports on sales performance are late.
- Customers are complaining about premature product deterioration.

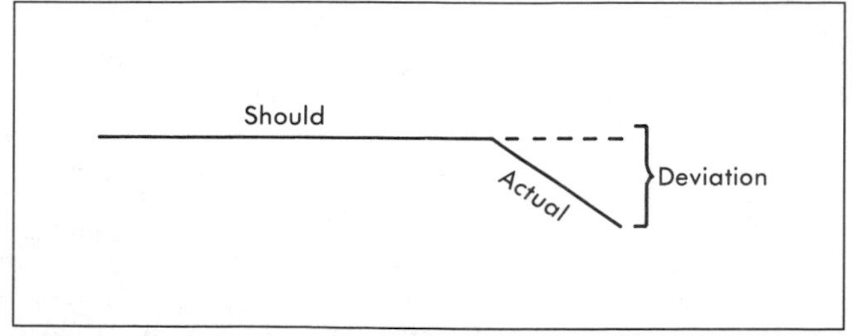

Fig. 10-8 Diagram of a typical problem.

PROBLEM DEFINITION OR SPECIFICATION

Every problem has a certain personality. That personality is fully captured when one knows the identity, location, timing, and magnitude of the problem in two dimensions: (1) where the problem is actually occurring and (2) where it could have been expected to occur but, in fact, did not.

A list of questions that can be used to develop a precise problem specification is as follows:

- *Identification, or what?* What unit, thing, person, condition is involved? What is the difficulty?
- *Location, or where?* Where in terms of geography are the objects observed? Where does the deviation occur on the object itself?
- *Time or when?* When in clock or calendar time were the objects with the deviation first observed? When have the defected objects occurred since then? When, in terms of the lifecycle of the object, did the defect first occur?
- *Magnitude, or extent?* How many units, parts, people, and departments are involved? What is the size, shape, and degree of the defect? What is the trend?

This list of questions must be adapted to meet a manufacturer's specific situation. However they are phrased, each question is essential for pulling out essential information about the problem.

Specifying a problem would be incomplete without knowing where that problem ends. Finding the limits of the problem can best be done by identifying what "is not" the problem. A good general question to ask to pull out the "is not" information is as follows: What normally would be expected to be part of the problem, but is okay this time? For example, consider the deviation statement "soda bottle caps are twisted." A possible specification for this problem is shown in Fig. 10-9.

DEVELOPING CAUSES

Hit-or-miss problem solving is just too costly. The trick is to "home in" quickly on the high-probability possible causes of the problem. How can this be done?

To begin, a problem solver must probe for *distinctions*. A distinction is a quality, feature, or characteristic that makes the problem—the "is" area in the specification—different from that which "is not" the problem.

The true cause must be related to some distinctive feature of the "is." Otherwise, the "is not" area would be affected as well. For each "is" and "is not" pair of the specification, the problem solver should ask what qualities, features, or characteristics are unique to the "is" and different from the "is not." Thus, in the example shown in Fig. 10-9, distinctions between the first and second shifts might be differences in supervision, operators, and throughput. Distinctions narrow the source of the cause.

A second critical step in developing possible causes involves the search for *changes*. At one time, the "should" and the "actual" were together. Now they diverge. Something happened or changed to cause the deviation.

Deviation statement: Soda bottle caps are twisted

IS	IS NOT
What: Soda bottle caps Twisted	Juice bottle caps Scratched, off-size
Where: Discovered at QC Twisted at rim	Found at customers Twisted at center
When: January 9 and continuing First shift	Before January 9 Second shift
Extent: 15% of caps on January 9; 18% on January 10	Less than 15%, more than 18%
Twisting is less than 1 mm off increasing	More than 1 mm Stable

Fig. 10-9 Specification statement for twisted soda bottle caps.

Hundreds of changes occur inside and outside the workplace every day. That is why merely making a laundry list of changes can be defeating. Rather, it is necessary to ask what has changed in, on, or around each distinction. Thus, in the example of the soda caps being twisted, a probe for distinctions reveals that a change in one distinction—different operations on the first shift—involved the hiring of two new operators on that shift during the first week of January. This change in personnel is one candidate for a possible cause of the problem, along with other changes that come from the other distinctions. These changes should be converted into hypotheses concerning the cause of the problem, which are then tested.

TESTING FOR MOST PROBABLE CAUSE

At this point in the process, the problem solver is in a position to identify the one cause that can explain all the "is/is not" dimensions of the problem. This is not a matter of luck or haphazard guesswork. It is a matter of making full use of information to test each possible cause against the specification.

For each possible cause revealed by the analysis one should ask, if this is the true cause, how does this cause account for each "is" dimension of the specification compared with its "is not" counterpart?

To illustrate, consider the "new operator" possible cause. One theory might be that the two new operators are not operating the line speed consistently, causing the bottle capper to turn out faulty caps. This theory is now tested against each of the "is/is not" pairs in the specification to see if it fits the facts with fewer assumptions than any other possible cause.

VERIFICATION

The development and testing of possible causes are based on the best current information that the questioning uncovers. However, to know for sure if a possible cause is the real culprit, more hard facts are needed. This is the time for empirical validation of the "most probable cause," the one that best passes the "is/is not" testing. Thus, in the example, the line supervisor may closely watch the two new operators to see how they are handling line speed. Or, physical testing may be conducted to see how varying line speeds can contribute to increased defects.

With the validation completed, the problem solver can now prepare to make the fix with far greater confidence and probability of success.

CONCLUSION

Problem solving is a discipline that can be improved if the problem solver is conscious of the problem-solving process. Once a problem solver knows how to approach a problem, from identifying that problem through to verification, individual and team problem-solving skills can be improved. This is the best way to boost performance, improve morale, and take a giant step forward in meeting the productivity challenge.

PERFORMANCE IMPROVEMENT AND COST REDUCTION PROGRAMS

Tapping the creativity of the workforce, both salaried and hourly, is an important part of the manufacturing manager's job. A well-organized approach will produce results that play an important part in keeping the enterprise competitive and profitable. The more successful this year's program is, the more likely it is that next year's program will also produce outstanding results. The only factory that does not provide continuing opportunity for cost reduction and improvement is one that has been mothballed. In today's dynamic environment, managers who believe they have milked the greatest part of the possible savings from their operations are a boon to their competition and poor prospects as long-term breadwinners for their own families.

Almost every company of any size has or has had one or more productivity improvement and cost reduction programs over the years. Some companies prefer titles like "performance improvement," which may appear less greedy and controversial, particularly to a workforce with a strong union heritage. One company uses the title "value analysis program" to cover all of their performance improvement and cost reduction activities. Whatever the title, effective programs have elements in common, as described in the following paragraphs:

1. *Everyone in the organization is encouraged to participate.* Supervisors at every level are given an opportunity to report improvements in their operations and changes in organization that reduce cost and/or improve quality, customer service, or other elements of their operations. Design engineers are provided with the means to report savings and improvements from new designs and from cost reductions in redesigns of existing products. Manufacturing engineers report savings resulting from new processes and equipment as well as from their day-to-day activities in improving production operations and taking cost out of the product.

   Purchasing personnel report savings through vendor negotiations and sourcing changes. Vendors are encouraged to submit proposals for reducing cost through changes that take advantage of their expertise and processing at no reduction in product quality and serviceability. Often unnecessarily tight specifications have created unnecessary cost. Marketing people report changes in specifications and features that can reduce cost or increase sales through improved customer acceptance.

   Data processing, finance, and human relations personnel and all others not specifically mentioned are included and provided a means to report their contribution to improving the profit and performance of the organization. Hourly rate and clerical personnel are also given opportunities to participate through suggestion or quality circle type programs.
2. *A mechanism is provided to be sure that ideas are properly evaluated and not rejected out of hand for political, territorial, or other reasons.* The process may provide for review by a high-level steering committee or an individual designated by the manager who becomes personally involved in questionable rejections.
3. *Contributors to the program receive individual recognition.* A pat on the back by the boss is an essential for every submission. Accepted contributions may receive a letter from the general manager and their contributors may be recognized by an invitation to an annual meeting or banquet. Significant contributions may receive monetary rewards. Monetary awards programs can, however, require considerable administrative, engineering, and management time over and above that which is essential to a successful program. Outstanding contributions can be further recognized in company magazines and newsletters and, of course, at merit increase and promotion times.

   Some programs simply provide a mechanism to encourage, process, and record employee ideas for improving profitability. Other programs set goals for the year that may or may not be included in financial forecasts. Reasonable exclusions of the present year's profit-improvement program results may include the following:

- Engineering design changes that are in progress or will be studied over the next year. These would include proposals from marketing and manufacturing.
- Equipment and process changes are considered potentials for future study.
- Savings targets for individual manufacturing engineers depending on the environment. It may be reasonable to include a multiple of the engineer's salary as savings that can be anticipated as a result of his or her regular factory support or other assignment.
- Purchasing savings either planned or anticipated as a result of vendor negotiations, resourcing, and so on. Some companies give recognition to purchasing negotiations that result in price increases below industry average for the particular commodity involved.
- Forecast savings from hourly rate suggestions programs based on history and trends.

Setting objectives for the year's profit improvement program can help to obtain outstanding results. Departmental detail is desirable. With specific objectives, management can compare each month's results to its objective and to prior years' programs. Management effort can be directed to removing roadblocks and encouraging individual creativity and participation.

Sometimes projects with excellent potential are held up because of lack of resources or because of territorial conflict. Perhaps a promotion is needed now to remind the hourly workforce of the reward that will be given out at the next annual banquet for their suggestions. Perhaps encouragement is needed for the quality circles or groups that are working on improvement in certain areas. Most people like to play on a winning team. If they know the objective and are given encouragement if they fall behind, surprising results can often be obtained.

If a standard cost system, which measures materials cost variances, is available, overall objectives can be set for the purchasing department, and these then can be broken down into goals for individual buyers. One company lists the 10 major material categories for its products with the overall dollar percentage of each used and an 11th category for all others. These are listed down the left-hand side of a 12-month spreadsheet. A forecast is made of expected price changes by commodity by date. These are entered into the spreadsheet and the effect of each change is calculated by multiplying the percentage increase or decrease by the percent usage of that commodity. Changes for each month are totaled and accumulated starting with a zero base at the beginning of the year. Actual percentage changes for each commodity group are entered on a monthly basis and compared to the forecast.

Percentage changes in the index of industrial commodities are calculated monthly. The cumulative percentage change in the index for the year is evaluated to obtain a measure of purchasing performance against the profit-improvement goal established at the beginning of the year. The forecast is updated quarterly and used as a basis for price commitments to customers. This approach provides a valuable tool for forecasting and controlling materials costs. It helps the purchasing department see areas of need and opportunity and to work with vendors to achieve economies through specification and other changes that will help avoid price increases. It is an important part of a company's performance improvement program.

One of the most effective elements of a profit-improvement program is a sound value analysis program. Value analysis is discussed briefly in Chapter 13 of this volume. Essentially, it involves having people with design, engineering, purchasing, and marketing experience take the time to do a thorough review of the function and cost effectiveness of each part of the product. Products of the firm's competitors should be available for disassembly and comparison. A brainstorming approach should be taken in the beginning. Negative reactions to any idea should not be permitted. Each part and function of the product should be looked at. The first question should be whether the function or part contributes value to the product. If it contributes value, could this value be achieved by combining functions, parts, or making other changes? Does a competitor do this in a lower cost? Or with a better method? Can we learn from the competitor?

Once all of the ideas have been discussed for a given component, function, or assembly, a professional opinion as to their practicality can be rendered based on the team's review. Some of the unconventional ideas from the beginning of the process may trigger practical approaches or may, on further evaluation, prove to have merit of their own.

The value analysis approach should be used during the design phase of a new product, particularly where the change is to be evolutionary and ideas from competitors can be included. While patents should not be violated, there are all kinds of mechanical and electrical approaches in the public domain. If someone has put together a combination of these, it is there to use and improve on. The company that does not take a hard look at what competitors are doing can easily be left behind in the competitive race.

Even if an excellent design and value analysis job is done up-front and cost reductions are pursued vigorously, a good value analysis review after a year or two will almost surely develope further opportunities to reduce cost and/or improve the quality or function of the product.

PERFORMANCE APPRAISAL AND REVIEW

The annual performance appraisal is a ritual avoided or delayed as long as possible by many supervisors. Done hastily or without considered thought, it can be negative and demotivating. Conversely, a ''Mr. Nice Guy'' supervisor can help institutionalize poor work habits and low productivity. A well-conducted performance appraisal with appropriate follow-up can be very helpful to employees—particularly those who have ambition and potential for growth. It can also be used to point out areas where performance should be improved and then provide the necessary direction for improvement.

One of the benefits of the performance review is that it provides an opportunity for an employee and his or her supervisor to sit down in a quiet, uninterrupted atmosphere and discuss the employee's thoughts about the company as well as the supervisor's thoughts about the employee's progress. In the hectic factory environment, these opportunities do not occur with great frequency. The performance appraisal and review process guarantees at least one such opportunity each year for every employee.

The following outline covers the policy for nonexempt performance reviews at a Midwest manufacturing company. It does a good job of covering the objectives and do's and don'ts of hourly rate performance reviews.

The performance of nonexempt personnel will be formally reviewed and evaluated by their supervisors at periodic intervals. The results of such reviews will be discussed with the individual as an aid to improvement and advancement on the job.

1. The formal performance review system is designed to:
   - Maintain or improve each employee's performance, job satisfaction, and morale by letting him or her know that their supervisor is interested in their job progress and personal development.
   - Serve as a systematic guide for supervisors in planning each employee's further training.
   - Assure considered opinion of an employee's performance rather than snap judgment.
   - Assist in planning personnel moves and placements that will best utilize each employee's capabilities.
   - Assist in determining and recording special talents, skills, and capabilities that might otherwise not be noticed or recognized.
   - Provide an opportunity for each employee to discuss job problems and interests with their supervisor.
   - Assemble substantiating data for use as a guide, although not necessarily the sole governing factor, for such purposes as wage adjustments, promotions, disciplinary action, and re-assignments.
2. Frequency.

   Performance reviews for new nonexempt employees will be completed immediately prior to the completion of the first 60 days of employment, at 6 months of employment, and on the first anniversary of employment. Thereafter reviews will be conducted in conjunction with the employee's anniversary date. A review may also be held on other occasions whenever desirable in order to achieve one or more of the above objectives.
3. Salaried supervisors shall conduct performance reviews.

   If appropriate, input from the employee's immediate group leader will also be solicited.
4. Maintaining objectivity in rating.

   Ratings should be supported by citing specific examples whenever possible. The following tendencies should be recognized and avoided in order to keep ratings as objective as possible.

   a. ''Halo effect''—The ''halo effect'' is created by therater's tendency to rate an employee good in every factor if their general performance is satisfactory, thus more or less endowing them with a halo—or vice versa, finding nothing good about an individual because their general performance is poor. It's wise to remember in this connection, that no one is all good nor is anyone all bad.

   b. Bias—This is the emotional tendency to reward friends by giving them good ratings and, conversely, punish with a poor rating the ones that are personally disliked.

c. Undue credit for length of service—This is the tendency to overrate long-service employees and to underrate short-service employees based on the assumption that the longer an employee performs a job, the more knowledge he or she acquires and the more proficient he or she becomes. This assumption is not necessarily true! Length of service in itself is not a measure of performance. In a performance review program, *how an employee performs their job* is the object of the evaluation, not length of service on the job.

d. Personal projection and self-identification—Refers to the tendency of raters to instinctively, but unconsciously feel that those people who resemble them in any way, whether it be in personality, physical appearance or work habits, are somewhat superior to those who are different. The rater should avoid this chip-off-the-old-block attitude and search in their ratings for good performance rather than for a carbon copy of themselves.

e. Loose rating—This results when a rater, like a cheerful giver, bestows on everyone, ratings that are higher than are realistic or accurate in an easy-going desire not to hurt anyone's feelings. Such inexact ratings are harmful to both the rater and the employee. Areas of weakness in the employee's performance remain uncorrected if not mentioned in a performance review. These ratings can also backfire on the rater at a later date when the employee who has been loosely rated is considered for a promotion and is found to lack the knowledge, skills, and/or attitudes their rating indicated, or when the employee is suddenly subject to disciplinary action for instances of poor performance which have occurred over a period of time, but which have never been reported in a performance review.

f. Tight rating—This results when the rater feels that everyone he or she rates falls short of standards. No one but a genius in their estimation should ever receive a good rating. It rarely is true that a rater would have such a poor assortment of subordinates so as not to produce one good man.

g. Personal competitiveness—This represents a rater's unconscious identity of him or herself as a member of the group they are rating, or their feeling that the individual rating represents a direct challenge to their own job. In such instances, the rater must keep in mind that they are merely an observer and detached from the work situation they are evaluating.

5. Performance review interview.

   a. Importance.

   The most important part of the rating program is the discussion of the review form with the employee—in private. If this discussion is conducted wisely, sincerely and sympathetically, it should:

   1. Build a better understanding between the employee and the supervisor.
   2. Clarify the mutual objectives of the employee, the supervisor, and the company.
   3. Give the employee a feeling of satisfaction about the phases of their work which have been effective, an understanding of how their performance can be improved, and a determination to improve.

   b. Points to remember.

   Remember through the discussion that it is about a person's performance, not themself. Be humble, too. It may be that some shortcoming of theirs is directly due to the employer's failure to guide, instruct, and encourage them. Be ready to admit this.

   c. How to discuss the report with an employee.

   1. Give the employee a "self-rated" performance review form in advance of the scheduled interview. This completed form will help in starting a meaningful two-way conversation.
   2. Plan the interview—Decide on a time and place. Get all the facts. Plan the approach to suit the individual.
   3. Put the employee at ease—Talk first about their outside interests or about the general idea of the progress interview.
   4. Explain the purpose—Each time be certain to point out one or two ways in which it benefits them.
   5. Talk about good points first—Then cover each point in detail. Avoid starting out on a weak point.
   6. Summarize—Summarize strong and weak points. Develope plans for improvement.

The material in Fig. 10-10 is also generally applicable to salary performance appraisals and reviews. To make the performance review meaningful for salaried personnel, particularly at higher levels of management, requires considerably more preparatory work by the supervisor. Jobs tend to be more complex and egos more apparent. An effective time to do a salaried employee's performance review is at the time of a salary change. This is true whether policy is to make salary changes at the time of the person's employment anniversary date or at one specific date each year. Supervisors are very likely to have the employee's undivided attention when money is discussed. Even if the salary increase is a large one, which will be highly acceptable, a supervisor can discuss with the employee what must be done to earn an as big or bigger merit increase during the next year. If there is no increase or a very small one, it is an excellent time to make the employee understand where his or her performance is substandard and agree on a program for bringing the performance to or above standard.

If the company has a management by objective program for salaried employees, it may be desirable to ask the employee to report on the standing of his or her action plans during the early part of the review. It may even be desirable to conduct an action plan review a few days before a scheduled performance review. This will give the supervisor an opportunity to organize his or her thoughts as to the personal development programs for the coming year that might be discussed during the performance review.

While considered of relatively low value by many manufacturing supervisors (outside of the personnel department), the performance appraisal and review process can be an effective communication and motivation tool. It takes relatively little additional effort to make it so.

Name ______________________ Job Classification ______________________

Department ______________________ Supervisor ______________________

Date of last review ______________________ Date of current review ______________________

Review Reason________ 60 day________ 6 month________ Annual________ Other________

FACTORS OF PERFORMANCE	PERFORMANCE RATING				
	UNACCEPTABLE	ACCEPTABLE	SATISFACTORY	COMMENDABLE	EXCELLENT
JOB KNOWLEDGE					
QUANTITY OF WORK					
QUALITY OF WORK					
JUDGEMENT					
VERSATILITY					
ATTITUDE					
SAFETY					

Overall Performance Rating

Unacceptable	Acceptable	Satisfactory	Commendable	Excellent

Attendance:______ Days absent in past______ Times Tardy______ Attendance is improving______ Deteriorating______

Comments on overall performance: What are this person's strong points? How has the employee performed against the generally accepted standards of this position? (Use additional sheets, if necessary.)

When performance is less than satisfactory, what plan of action has been developed and agreed upon to correct this situation?

Rated by ______________________ Employee ______________________

Reviewed by ______________________ Reviewed by ______________________

Fig. 10-10 Employee performance appraisal sheet.

COMPANY POLICIES

A company's policies governing management of the workforce are usually found in employee handbooks and/or company manuals. In unionized shops, union contracts spell out much of the relationship between management and the workforce. These generally cover benefits, services, compensation, and other material, which are discussed in previous sections of this chapter. This section deals with policies relating to the conduct expected of employees and the management approaches to ensure that such conduct does in fact occur. A typical employee manual might include sections on personal conduct, employee relations, housekeeping, shop rules, discipline, management rights, substance abuse, and grievances.

PERSONAL CONDUCT

It may be a company's policy to place as few restraints and restrictions on its workers' personal conduct as is possible. However, for the protection of its property, business interests, and other employees, the company establishes reasonable standards of conduct for its employees. The supervisor will tell a worker about the regulations that apply to him or her, and they will be informed of any changes as they occur. The company depends on its workers' good judgment and sense of responsibility to conduct themselves with dignity and good behavior.

EMPLOYEE RELATIONS POLICY

A company realizes that its fundamental strength and growth depend directly on the contribution made by each person within the organization. Better productivity and efficiency result from real job satisfaction and from the opportunities each person receives as part of his or her self-development.

The company should believe in good employment practices; fair compensation, challenging work, recognition for contributions, pleasant working conditions, opportunity for advancement, and job security. These essentials are the foundation on which a company wants to build its forward program of employee relations.

A personnel policy may, therefore, be designed to do the following:

1. Place a worker, insofar as possible, in a position that best suits his or her natural aptitude and skills.
2. Offer a worker the opportunity for self-development and advancement through training and education assistance.
3. Recognize the importance of a worker's output to the overall company goal.

An employee relations policy of the company should be an open-door policy under which each employee has the right to deal directly with supervisory administration in reference to all working conditions.

HOUSEKEEPING

In business, "housekeeping" means more than the word implies elsewhere. It is a way of working. It means keeping workplaces neat, clean, and free of articles not being used. It means keeping equipment clean and in proper places, disposing of waste in proper containers, and storing materials and equipment in an orderly manner and in designated places only. By practicing good housekeeping, employees are contributing to the safety program. The kind of housekeeping required for most operations cannot be accomplished by periodic cleanups. Like safety, the maintenance of good housekeeping is the daily personal responsibility of each employee.

SHOP RULES

Every factory, regardless of size, should have shop rules that clearly spell out behavior that will not be tolerated. Well-run factories will require infrequent use of disciplinary procedures. But without clear and intelligently enforced rules, good operations cannot be expected to continue for any length of time.

A typical set of shop rules for which employees may be subject to disciplinary action are as follows:

1. Dishonesty, deliberate deception, fraud, deliberate damage to company property, using company machinery, or materials for commercial use or personal gain.
2. Removal of or possession of property without permission from the company.
3. Willfully holding back, hindering, or limiting production of the company.
4. Giving misleading or false information, circulating false or malicious rumors.
5. Altering or changing a time card in any way, punching the time card of another employee. Where this is arranged between employees, both will be subject to immediate dismissal.
6. Disorderly conduct, use of abusive language, fighting, threatening bodily injury, engaging in horseplay or immoral conduct, behavior which is offensive or disruptive to co-workers or the workplace.
7. Coming to work under the influence of alcohol or any controlled drugs or bringing alcoholic beverages, controlled drugs, explosives, or firearms onto company premises.
8. Intentionally giving false or misleading information in applying for employment.
9. Refusal to comply with instruction within the scope of the job from management representatives, including supervisors, or refusal to adhere to company rules or policies.
10. Inattention to duties, idling on the job or elsewhere during working hours.
11. Failure to remain in one's assigned work area and/or regular rest areas during working hours.
12. Frequent tardiness or absence from work without permission; failure to give proper notice of absence; overstaying a leave of absence or vacation without notification; stopping work before break periods, lunch period, or quitting time.
13. Negligence resulting in equipment damage, excessive scrap, or inferior work, or wasting materials or supplies.
14. Deliberate abuse of the company's tools, equipment, or wasting of its materials.

15. Working for another employer that in any way lessens attendance, efficiency, or reputation of the company.
16. Soliciting of memberships, contributions, pledges, or subscriptions, distribution of literature, or conducting anything other than company business during an employee's working time are in violation of company rules.
17. Violation of fire, safety, or health rules.
18. Failure to report an accident, or removing, locking out, or making inoperative any safety device.
19. Disobedience and/or insubordination.
20. Violation of these rules may even result in an employee being terminated.

DISCIPLINE

One definition of discipline is "training that develops self-control, character, orderliness, and efficiency." To be fully competitive, a plant needs to have a good level of workforce motivation and participation. There needs to be a high level of teamwork where people are able to count on one another in a disciplined atmosphere. Unfortunately, a second definition of discipline is "to punish." There must be a procedure for taking corrective action when rules are not followed. For many years, the most widely accepted approach was "corrective discipline." This involved increasing penalties based on the seriousness of the offense and previous violations of any of the shop rules. Corrective discipline would normally proceed from a written reprimand through a series of disciplinary suspensions without pay up to discharge. Discharge was normally reserved for extremely serious offenses or repeated violations of even minor offenses. This approach can be effective when administered evenhandedly. It can work in even the toughest union shops as long as management actions are well supported by the facts.

POSITIVE DISCIPLINE

As manufacturing management has increasingly sought to increase worker participation, other approaches to discipline have developed. These take into account that the vast majority of the workforce will respond as good citizens, particularly when they are treated with individual respect and dignity. Such treatment is a keystone to building a participative workforce.

A company usually has certain rules and regulations for the common good relating to discipline. The traditional approach to discipline, i.e., oral warnings followed by written warnings which lead to suspension without pay and eventual termination, may be replaced by a discipline procedure without initial punishment. One Midwest manufacturing company uses a five-step approach: oral reminder, written reminder, written warning, decision-making leave, and termination. A discussion of each of these steps follows.

Oral Reminder

When casual conversations fail to improve poor workmanship, absenteeism, or violation of company rules, the supervisor issues an oral reminder. This is a friendly but serious discussion spelling out of why the company needs to have its employees meet certain standards. This short, private conversation closes with the supervisor's making sure that the employee knows what is expected from then on.

Written Reminder

If the problem continues, the supervisor holds a second, private conversation still in the same serious but friendly vein. Immediately afterward, the supervisor sends the employee a written reminder. This memo confirms the conversation about the problem in a matter-of-fact, nonthreatening tone and states the need for immediate improvement.

Written Warning

If the problem still continues the supervisor will hold another private conversation followed by a strongly written memo that outlines the problem and expected improvement. This step *may* be repeated depending on the seriousness of the rule infraction, at management's discretion.

Decision-Making Leave

Discharge does not really solve anyone's problem. The company loses a potentially valuable employee, and the individual loses a job and source of income. However, sometimes discharge is the only solution when performance cannot be improved or the rule violation is so severe.

If the first three steps fail, the supervisor again talks privately with the employee. This time the employee is told to leave the building immediately and spend the balance of the work shift at home deciding whether the employee can solve the problem and follow all the rules or not. The employee is told that he or she will get full pay for the balance of the shift as a demonstration of the company's hope that he or she will decide to correct the problem and become a good performer. The employee is told to return at the beginning of the next shift to give their supervisor a decision. The employee is also told that any future violations will result in discharge.

Termination

If another disciplinary problem arises, the employee is discharged. From time to time, circumstances may occur whereby the company is not able to follow the above disciplinary procedures, and the company reserves the right to assess discipline up to and including termination of employment without engaging in the above steps of discipline for any violations of its rules and regulations.

MANAGEMENT RIGHTS

Most every union contract and employee handbook has a management rights or management responsibilities clause.

Management's responsibilities include the management of a plant and the direction of the working forces, including the types of production; the location and number of plants; methods, processes, and means of production; size of the workforce; assignment of work; hours of employment; the right to hire, suspend, or discharge for cause, promote, or transfer; and the right to relieve employees from duty because of lack of work or other legitimate reasons.

These responsibilities should be presented as a matter of information only and not be intended to create nor be construed to constitute a contract between the company and any or all of its employees.

In too many union and nonunion shops, manufacturing management has relinquished some or many management rights by not enforcing them. It takes time and extra effort to reestablish rules and procedures that have fallen into disuse as a result of lax management.

ABSENTEEISM

Few things are more disruptive to a factory's operation than people who show up late or not at all. There will always be the occasional car breakdown, death in the family, or case of the 24-hour flu that prevent the best of workers from being on time. Most people in these circumstances will call in either before the shift or as soon as is practical thereafter. Harassing workers with legitimate excuses will only cost goodwill and damage relationships that may have been built over long periods of time. If a campaign is to be undertaken against absenteeism, care must be taken that harassment does not occur in some overzealous supervisor's area.

If unexpected absences exceed 1 or 2%, particularly on Mondays or Fridays, then it can be assumed that there is an absentee problem. There are a number of positive approaches that can be taken. One company credits each employee with a $100 attendance bonus at the beginning of each year. The account is reduced $20 for each absence and $5 for each tardiness. At year end the balance in the account is delivered to the employee. This has provided a good talking point for management and supervision on the need for and value of good attendance and has provided a positive motivation for most of the workforce.

In shops where there is good morale and team spirit, simply talking to people about how important attendance is can do a great deal. In one case some years ago, a large number of employees were hired from the inner city. After the first few paychecks, attendance plummetted. The supervisor took these young people with little or no prior work experience aside and asked "Do you play basketball?" After a resounding affirmative he asked, "Do you play with four people?" When they answered "no," he asked, "How do you expect us to build our products with four-man teams?" The point was made, and attendance improved significantly.

A key to limiting absenteeism is good recordkeeping. The manufacturing administration office, supervisor, or personnel office should have a record of each employee's attendance every day of the year. If these records are reviewed each time there is tardiness or unexpected absence, it will be easy to determine whether or not there may be a problem. Even if there is no apparent problem, a conversation stressing the company's concern for the worker's attendance should take place. Where other measures fail, positive discipline, including a discharge in chronic cases, may be the only solution.

SUBSTANCE ABUSE

Alcoholism and drug addiction can be very serious problems. Accidents, poor quality, and low productivity are likely to result if people are allowed to work under the influence of alcohol or drugs. Anyone reporting for work or observed at work in this condition should be sent home or assisted home at once. If this is not done, the employee, other employees, equipment, and the work in process are endangered. People regularly reporting for work with some indication of substance abuse need to be carefully observed and talked to with appropriate follow up if behavior is not improved. Absentee patterns can also give indications of drug or alcohol abuse problems.

Employee assistance programs are mentioned earlier in this chapter. They can produce excellent results with some people. An alternate approach is to apply a positive or corrective disciplinary procedure until the behavior is corrected or the employee is terminated. Drug and alcohol testing may be appropriate in some situations though there are widespread restrictions or prohibitions on their use.

COMPLAINTS AND GRIEVANCES

Two-way communications are essential in managing a world-class factory. Ideally, the immediate supervisor will act as the communications channel for his or her people. Their misunderstandings and complaints will either be handled by the supervisor or directed to the appropriate person for an answer or resolution. Suggestion programs, quality circles, and the like provide outlets for expression by employees on many subjects and serve as excellent communication channels. Management should take steps to train supervision and establish a procedure so that questions and complaints will be listened to and answered or acted on as appropriate. A typical complaint procedure for a nonunion shop follows.

"Differences of opinion occasionally arise between individuals in any organization regarding their job and treatment of the individual. A procedure, therefore, is required whereby differences of opinion and/or feelings, and the position of the employee and the company, may be impartially and effectively evaluated, discussed, and hopefully resolved.

A complaint procedure that satisfactorily and expeditiously resolves these differences make for a better place to work. The steps are as follows:

1. Discuss the problem with the immediate supervisor. Should the problem, however, be of a personal nature or a counseling problem which the employee feels may be embarrassing to discuss with the supervisor, he or she is invited to discuss it in strict confidence with members of the personnel department.
2. In the event the employee fails to secure a satisfactory answer from his or her supervisor, a meeting may be requested with the supervisor, department head, or next layer of supervision.
3. If the problem is still unresolved, a meeting may be requested with the department manager, who will discuss the matter with the plant manager and manufacturing officer. (Step 3 for all nonmanufacturing areas will involve a meeting with the respective company officer.)

Employees are encouraged to request assistance from the personnel department in any of these steps."

In a union shop, employees additionally have the grievance procedure available to them. Better supervisors, however, handle most situations without grievances. This definitely does not mean giving away the store to avoid a grievance. If the right relationships have been built up, most people, particularly in today's environment, will be inclined to work with a supervisor who treats them with respect and fairness and acts on their complaints when there is a legitimate problem.

MANAGING IN UNION AND NONUNION ENVIRONMENTS

The preceding sections of this chapter have discussed the many elements of managing the workforce. Except where constrained by law, union contract, or the dynamics of a particular union situation, the approaches presented are equally applicable to both union and nonunion environments.

The following section cover some of the specifics that must be considered in managing nonunion and union shops. Chapter 11 details the legalities that managers must observe with an emphasis on the union environment.

Except in greenfield start-up and potential decertification situations, manufacturing managers must accept the status in their particular shop. Should one move from a nonunion to union plant or vice versa, it should be comforting to know that good management skills developed in either situation are almost fully transferable. There will be a need to quickly study the union contract in one case and the applicable company policies, procedures, and rules in both cases. With this intensive study that is essential to a successful move, cultural shock will be at a minimum.

MANAGING IN A NONUNION ENVIRONMENT

Most managers, given a choice, would rather manage in a nonunion environment. Managing employees in a nonunion environment, however, requires a commitment on the part of management to maintaining a nonunion management program. A successful nonunion management program involves establishing company policies and procedures, providing supervisor training, and developing channels of employee communication.

Advantages

Managing in a nonunion environment has many advantages for first-line supervisors, middle-level managers, and top management. The main advantage of operating in a nonunion environment is the fact that the company, through its supervisors and managers, is able to direct the workforce without any interference from a third party. In a unionized operation, by contrast, the union attempts to limit management's discretion with respect to hiring, firing, discipline, work assignment, wage reviews, transfers and promotions, and complaint resolution.

Hiring. Many union collective bargaining agreements require that a company first seek to obtain new employees from the union hiring hall before it hires new employees from any other source. In a nonunion operation, the company is free to hire anyone it wishes, providing they do not violate applicable state and federal laws with respect to discrimination in employment.

Firing. Most collective bargaining agreements limit the employer's right to terminate employees. Many collective bargaining agreements specify certain procedural steps that must be followed before an employee can be terminated. Managers operating in a nonunion environment are free to terminate employees without these artificial restrictions.

It is essential, however, that all discipline and particularly discharge is administered in a fair and evenhanded manner. It is also important that the basis for firing and severe applications of discipline are made known to other employees.

Discipline. Many collective bargaining agreements limit the employer with respect to disciplining employees. In a nonunion operation, managers are free to discipline employees provided that they are reasonable in doing so.

Work assignment. Most union contracts attempt to limit management's ability to assign work. Under the terms of many agreements, employees assigned to a particular job classification are not permitted to perform the work of employees working in other job classifications. In a nonunion plant, management has the option of cross-training employees and of reassigning work depending on the production needs of the company. This ability to assign and reassign work freely in a nonunion operation often results in substantial labor savings for the company.

Wage reviews. In a nonunion operation, the management of the company is responsible for reviewing employee wages. The company is free to adjust or modify wages based on an employee's demonstrated ability to perform the job. By contrast, in a unionized operation, all employees are compensated according to the wage scale set forth in the collective bargaining agreement.

Transfer and promotion. In a unionized operation, management's ability to transfer or promote employees to different jobs is generally limited by the seniority provisions of the contract. The manager in a nonunion operation has the advantage of being able to select the best qualified employee for a position that becomes available.

Complaint resolution. In a unionized operation, employee complaints are resolved by means of a formal grievance procedure. The union, through its business agents or shop stewards, participates in the presentation and resolution of employee complaints. In a nonunion operation, the responsibility for resolving employee complaints and problems falls squarely on the shoulders of management. There should be a procedure for appealing to higher levels of management.

Additional Responsibilities

With the advantage of nonunion status comes additional responsibilities for supervisors and managers. In a unionized plant, the employees often tend to blame the union for the employees' problems and discontent. By contrast, in a nonunion operation, the management of the company is held responsible for employee problems and complaints.

Because the first-line supervisors and managers are responsible for maintaining employee morale, their actions with respect to directing and controlling the workforce have a significant impact on whether or not the company remains nonunion. Accordingly, supervisors and managers in a nonunion operation must be sensitive to employee concerns.

In addition, the successful maintenance of nonunion status requires a commitment on the part of the company to establish and maintain personnel practices and policies that are responsive to employees' needs.

Required Elements

The maintenance of a nonunion operation requires the commitment and hard work of all levels of management. Management, of course, benefits from the establishment of a sound employee relations program. Sound employee relations results in greater productivity, better quality control, reduced absenteeism, and a spirit of cooperation that makes it easier for a plant to achieve its operational objectives.

Employees who work in a stable employment environment enjoy job security, regular improvement in wages and fringe benefits, and generally better working conditions. In addition, they come to appreciate the part their efforts play in the success and growth of the company. The following elements are important to the establishment and maintenance of a successful nonunion operation.

Supervisory and managerial personnel identification. The identification of supervisory and managerial personnel is absolutely critical to the operation of a successful nonunion plant. Management must know who its supervisory and managerial employees are because these employees at all levels have the authority to bind the company with respect to employment and labor relations matters.

The National Labor Relations Act defines a supervisor as follows:

> Any individual having authority, in the interest of the employer, to hire, transfer, suspend, lay off, recall, promote, discharge, assign, reward or discipline other employees, or responsibly to direct them or to adjust their grievances or effectively to recommend such action, if in connection with the foregoing the exercise of such authority is not of a merely routine or clerical nature, but requires the use of independent judgment.

The definition of a supervisor is written in the disjunctive. If an individual exercises any one or more of the criteria of supervisory authority listed in the definition, that individual is a member of management for purposes of implementing employee relations matters.

Company policy. As part of a comprehensive program for maintaining nonunion status, a company should issue a written policy discussing the company's philosophy with respect to unions. The company may take a strong position that a union would not be in the best interests of the employees. Employees should be aware of the company's position on unionization. It should be included in the employee handbook and discussed during the employee orientation. It is equally important that the company's nonunion policy be understood and accepted by all levels of management. Accordingly, the policy should be conveyed to supervisors and managers as part of their management training program and should be included in the managers' operational manuals.

Supervisory training program. Another important component of a nonunion employee relations program is the establishment of a formal program of supervisory training. Such a program should be a comprehensive and ongoing program integrated into the company's personnel and labor relations programs. Supervisory and management training should concentrate on the following three areas:

1. Training in the various federal and state laws that affect employer relations, such as NLRA, EEOC, OSHA, and wage-hour legislation.
2. Training in the techniques and strategies of dealing with a union organizing drive.
3. Training in personnel skills in such areas as employee communications, administration of discipline, methods of counseling employees, and resolution of employee problems and complaints.

Supervisory manual. To assist supervisors and managers in implementing the company's nonunion program, the company should prepare a supervisor's guide that clearly sets forth the company's policies and procedures with respect to employees, principles of good employee relations, and an outline of the supervisor's rights and responsibilities under various laws affecting employee relations.

Policy and procedure monitoring. On a periodic basis, top management should review the company's policies with respect to equal pay; treatment of women, minorities, and handicapped employees; hiring and promotional policies; employee evaluation procedures; disciplinary practices; employee communications programs; employee complaint resolution procedures; exit interview procedures and techniques; and other personnel policies such as overtime, vacation time, leave time, transfer procedures, and maternity leave. Employees judge their employer based on the company's published policies and the manner in which they are applied. Accordingly, the manner in which a company implements and applies its personnel policies and procedures affects the way in which the employees look on their employment relationship.

Employee handbook. An employee handbook is perhaps the most valuable written form of communication between a company and its employees. A well-drafted employee handbook should set forth the company's philosophy on personnel relations and outline what the company expects from its employees with respect to performance and general conduct.

It should also outline what the company has to offer its employees. The handbook, for example, should set forth the policies and procedures that the company has established to ensure fair and uniform treatment of all employees. Procedures for resolving employee complaints or making employee suggestions should be covered in the handbook.

The handbook should also set forth the opportunities for career development and promotions. A clear and concise explanation of employee benefits is an essential part of a well-written employee handbook. The handbook should be an easily accessible reference for employees when they have questions about company policies, procedures, or benefits. In addition to the many practical aspects of an employee handbook, such as a means of communicating benefits, policies, and procedures, an employee handbook can serve as a kind of written contract between management and its employees.

Communications program. One of the keys to successfully managing in a nonunion environment is the establishment of a sound and effective employee communications program. Every conceivable opportunity to talk with employees should be used by management and supervisors, such as during employee orientations, assignment of job duties, corrective disciplinary interviews, and discussions concerning wages and benefits.

Likewise, every conceivable vehicle for communicating with employees should be utilized by managers and supervisors, such as one-on-one conversations, bulletin board notices, letters to the home, and employee suggestion groups. Managers and supervisors should attempt to provide employees with as many opportunities as possible to speak with them. By making efforts to improve upward communications, management will learn from employees what problems and concerns are most important to them.

Having an open-door policy is often not enough. To successfully maintain a viable communications program, managers have to take the initiative and seek out employees rather than wait for employees to come through their office door.

Employee orientation program. A thorough employee orientation is another key element of an effective employee relations program. If a new employee is oriented properly, his or her performance and success on the job will be enhanced. An employee orientation program should be comprehensive, covering all aspects of the employee's relations with the company.

Employee evaluation techniques. Employees want to know, and have a right to know, how they are doing with respect to their job performance. When performance reviews are conducted in an objective, constructive manner, employee satisfaction and morale is increased.

Disciplinary procedures. Every company needs a disciplinary policy for those few individuals who either cannot or will not comply with plant rules. However, to be effective, a disciplinary procedure must be clearly understood and exercised in a consistent and nondiscriminatory manner. It must be positive and constructive in nature, rather than punitive. It is widely recognized that there are two essential aims of a disciplinary policy: (1) to provide the framework within which self-discipline can develop and (2) to take prompt and fair action against the minority of the workforce who are transgressors.

Key elements of a progressive, constructive disciplinary procedure are as follows:

- A clear, lawful, and reasonable list of rules of conduct.
- Instructions to all employees as to what is expected of them, in terms of both observance of rules and established standards of job performance.
- A procedure for telling employees how well they are meeting job standards and plant rules.
- Careful investigation of the background and circumstances of each case before taking disciplinary action, when apparent breaches of conduct or poor performance do occur.
- Prompt, consistent application of disciplinary measures.
- A system of progressive discipline when appropriate.

Exit interview procedure. As a means of determining employee dissatisfaction, it is sometimes useful to explore the various reasons why employees voluntarily leave a company. It is important to inquire into areas such as possible dissatisfaction with working conditions, benefits, pay, supervisor, shift assignments, fellow employees, and scheduling. The exit interview is a means to ascertain this vital information.

Complaint problem-solving procedure. Employees must be given an opportunity to have their questions, problems, and complaints aired and resolved, within the plant, so that the need to seek the assistance of outside third parties is minimized. The procedure must be workable and designed in part to provide redress beyond the individual employee's direct supervision. There are certain prerequisites to the effectiveness of such a procedure, as follows:

- Employee belief in the viability of the system.
- Managers and supervisors committed to the fair and consistent application of the procedure.

In addition, there are certain technical requirements of the procedure that should be noted, as follows:

- The employee should have an opportunity to bring a complaint to his or her immediate supervisor without fear of retaliation.
- The employee should be given a prompt response by the supervisor.
- If the complaint is not settled by the immediate supervisor to the satisfaction of the employee, the employee should have the option of presenting the complaint to the next level of management.
- If still not resolved, the employee should have access to the top level of management or ownership for final resolution of his or her complaint.

Wage and benefit survey. Management should study its present salary and benefit structure against those of similar operations in the area. Such a study is necessary if the company wishes to remain competitive.

Although a company does not have to be an industry leader in wages and benefits, a compensation package should be competitive with comparable union and nonunion employers in the area.

The key to managing successfully in a nonunion environment is to create a workplace environment where employees recognize that their interests and those of their employer are closely aligned. This objective is reached through a solid, structured labor relations program.

Responding to a Union Organizing Drive

Successfully responding to a union organizing campaign involves three elements: communications, communications, and communications. Union representatives, in attempting to organize employees, act in many respects like door-to-door salespeople. They tell employees whatever they want to hear to get their support. Often, union organizers rely on half-truths and misconceptions about unions to win employee support.

It is the job of the managers and supervisors to give employees the facts about what unionization of the plant will mean to the company and to the employees. Employees should be given the opportunity to weigh the good and bad aspects about unionization and then decide whether they want to remain union-free or elect to be represented by a union. Most employees, once they understand the overwhelming disadvantages of unionization, will elect to remain union-free. Most companies that lose union elections lose because the management of the company did not communicate to employees the disadvantages of unionization.

In a union organizing situation, the facts about unionization that should be communicated to employees are as follows:

1. Unionization is not inevitable.
2. Financial obligations of union membership.
3. Other obligations of union membership.
4. Union negotiations.
5. Possibility of strikes.

Unionization is not inevitable. Many employees assume when a union begins organizing that it is inevitable that the union will succeed. The union supporters attempt to foster this impression so that undecided employees will support the union under the misimpression that everyone in the plant is supporting the union. Management's first task is to tell employees that the company intends to defeat the union organizing drive. Furthermore, management should advise employees that unionization is not inevitable, because unions lose more union elections than they win. The latest statistics compiled by the U.S. Department of Labor show that in the United States unions lose about 55% of the elections in which they are involved.

Financial obligations. Employees sometimes believe that the union is some kind of charitable institution and that it will represent them without charge. Nothing could be further from the truth. Unions are a business, and they expect to be paid for representing employees in the form of union dues, fees, fines, and assessments.

The first charge that an employee will face as a union member is an *initiation fee* charged to new union members. The fee varies among different unions, but can range from a low of $25 to a high of several hundred dollars.

Union members are also required to pay union dues as a condition of employment. Most union collective bargaining agreements, in non-right-to-work states, contain a "union shop" clause in the agreement. This clause requires all employees after a 30-day grace period to become union members and pay union dues as a condition of continued employment. If an employee fails to pay his or her monthly union dues, the employer is required under the agreement to terminate the employee. Monthly union dues range from $10 to $50; over the course of a 3-year union contract an employee is obligated to pay several hundred dollars to the union to keep his job.

In addition to union fees and dues, unions can also charge members *monetary assessments*. Under most union constitutions or bylaws, unions have the right to levy monetary assessments against members whenever they are in need of additional funds to pay for such things as strike funds, legal fees, or building expenses.

Not only can unions charge members for services in the form of union fees, dues, and assessments, but unions can also levy *fines* against members for violating the union's constitution and bylaws. Union members are fined for such things as crossing a picket line, working during a strike, disturbing the peace and harmony of a union meeting, criticizing a union official, or anything else that the union doesn't like. The amount of these fines is unlimited and, unfortunately, union fines are enforceable in court. It is not uncommon for a union to fine a member several hundred or several thousand dollars for violating one of its rules.

Once employees begin to realize the full costs of union membership and compare them with the dubious advantages of union representation, they are less eager to become involved with a union.

Other obligations. In addition to the financial obligations of union membership, there are other obligations of union membership of which employees should be aware. As union members, employees are obligated to follow all the rules in the union's constitution and bylaws. They are also required to obey the directions of union officials, including the plant shop steward. Employees can be directed by the union to participate in union-sponsored demonstrations or picket line activities during the employee's nonwork time. Employees can be directed by the union not to report to work during a union-called strike or work stoppage. Obeying the union's directives can result in a significant loss of income and loss of personal freedom for the employee. Moreover, because the union's directives are enforced by means of union fines, union members have little free choice as to whether they will obey such directives.

Union negotiations. Union organizers lead employees to believe that in union negotiations their wages and benefits will be improved. Unfortunately, this is not always the case. In union negotiations, one of three things can happen. Employee wages and benefits can go up, they can go down, or they can remain the same. Because union negotiations involve a great deal of horse-trading on the part of management and the union, the results of negotiations cannot be guaranteed. Negotiations in the airline industry and in basic mining and steel manufacturing demonstrate that, in some cases, employees wages and benefits can decrease as a result of union negotiations. Regardless of the results of negotiation, union members are required to pay union dues, fees, and assessments. When employees realize that improvements in their wages and benefits are not guaranteed by union representation, the costs of union representation in terms of loss of money and personal freedom become less acceptable.

Possibility of strikes. With union representation comes the possibility of becoming involved in a union-called strike. One rarely sees employees engaged in strike activity without a union being somehow involved in the situation. During a strike, employees do not receive wages or benefits because an employer is not obligated to pay employees who withhold their services. Not only does an employee's paycheck stop, but health insurance coverage may also stop unless the employee elects to pay the premium.

In a strike situation, the employees may receive some strike payments from the union, but these payments are often quite small and do not approach the amount that the employee would receive from the employer while working. Because an employee's regular bills continue during the strike, employees who are involved in a lengthy strike often find themselves faced with financial hardship. Once the consequences of union membership are put into perspective for employees, employees find the option of union representation less appealing.

METHODS FOR COMMUNICATING FACTS ABOUT UNIONIZATION

Managers and supervisors have the right to communicate with employees concerning unionization, provided their communication does not contain a promise of benefit or threat of reprisal. Members of management can communicate lawfully with employees in union organizing situations if they adhere to the following rules:

1. No *threats* of reprisals or other unfavorable treatment because of support for a union.
2. No *interrogation* of an employee concerning his or her union activities or sentiments, or the union activities or sentiments of others.
3. No *promises* to employees to bribe them into abandoning their support for the union.
4. No *surveillance* of employees' concerted union activities.

Managers and supervisors can remember the four things to avoid when engaging in communication about union organizing by remembering the acronym "TIPS." Provided the communication does not violate the above restrictions, any form of communications is lawful. Members of management can communicate with employees in the following ways:

- One-on-one conversations with employees.
- Group discussions with a small group of 3 or 4 employees.
- Group talks to an assembled group of 20 or more employees.

- Written communications to employees in the form of handouts or cartoons.
- Written communications to employees' homes in the form of letters to the employee and his or her family.
- Visual displays, such as a display showing the amount of groceries that can be purchased with an amount of money equivalent to 1-year's union dues.

Communications programs during a union organizing campaign should be similar to communications programs under a well-structured labor relations program. If managers and supervisors in a nonunion company take the time to communicate with employees on a regular basis before any signs of union organizing emerge, implementing an effective communications program during a union organizing campaign will be much easier.

MANAGING IN A UNION ENVIRONMENT

Chapter 11 discusses in depth the current law as it applies to labor relations. Management's responsibilities with regard to unions and a union environment are detailed. Unionized facilities are covered by contracts or are in the process of negotiating same. Because these contracts vary widely, it is not appropriate to discuss contract specifics. There is usually a labor relations department or outside consultants to advise on legal and contractual requirements. They may represent management in grievance proceedings and contract negotiations. If expert advice is not available, it is recommended that it be obtained.

All of the preceding material on managing in a nonunion environment and on good management practice is applicable except as specifically prohibited by the union contract.

Management should make an effort to have reasonably cordial relationships with union stewards, committee members, business agents, and other union officials. This will often make it possible to handle without grievances those matters that could be subject to varying interpretations in the contract or might be desirable in spite of contract provisions. To further such relationships, management may decide to promptly comply with reasonable requests that do not adversely affect operations, discipline, or set precedents that will subvert management prerogatives in the future.

Stewards are human beings. Management can make their work easy. They can also make life extremely difficult by giving grudgingly only what the strictest interpretation of the contract demands and by applying a strict interpretation of the shop rules to the individual's conduct. In a large operation, it may be possible to enhance the standing of cooperative stewards by granting their reasonable requests while refusing or delaying similar requests from those who do not choose to cooperate. The employees may conclude that the uncooperative steward is ineffective and vote him or her out of office. These techniques are frequently used to completely change the tenor of labor relations. For example, the methods were very effective in one plant of 3000 people where attitudes were so hostile that it was said, only half in jest, that it was unsafe for management to walk alone in the punch press department. To solve the problem, all levels of plant management spent time on the floor explaining the need for cooperation and quality. The economics of the company's position vs. competition were discussed with individuals and small groups many times a day. Cooperative committee members were encouraged to work closely with first-line supervision. Their reasonable requests and complaints were promptly and favorably handled whenever feasible. Uncooperative committee members were closely supervised and disciplined as appropriate. The result was a slowly accelerating transformation. Within something over a year, productivity had reached good levels, operations were profitable, and quality and delivery were acceptable and improving.

Granting seemingly reasonable requests may establish precedents that reduce management's rights to manage. This is a potential danger that must be considered. While it is desirable to be cooperative much of the time, the overall climate of labor relations in the particular operation must be considered. Well-managed plants have good communications between first-line supervision, upper management, and the labor relations function or consultant. In this environment, the union relationship is managed—not left to chance.

If a group of people within a bargaining unit decide that remaining in the union is not in their best interest, they may petition the National Labor Relations Board (NLRB) to conduct a decertification election. While not a frequent event, a number of decertifications do take place. When a decertification petition has been granted, other unions may solicit the employees, and management may campaign for a union-free shop. Management is severely constrained by law in the area of encouraging a decertification movement. Should the climate for decertification develop, management should seek expert labor relations counsel.

SUMMARY

All of the good management and human relations practices covered in this handbook apply to both union and nonunion operations. There may be a few exceptions because of legalities, contractual requirements, or the attitudes or actions of some employees in a union shop.

A number of studies have shown that productivity tends to be higher in union-free shops; the difference can amount to 50% or more. Even in the most cooperative union environment, some management time is required to handle grievances, deal with union matters, and periodically negotiate contracts. Operations can get out of control in any factory, but it can happen more quickly in a union operation. This is particularly true if management is not alert to the risk and gets in the habit of making small, seemingly unimportant concessions, which establish precedents that undermine productivity. There are many well-managed union shops. To achieve and maintain this status requires the constant attention of a capable, alert, and aggressive management—attention that might produce improved productivity if it were invested elsewhere.

The importance of strong and expert labor relations counsel cannot be overemphasized. In smaller shops, this may be provided by occasional reference to outside labor relations experts. Larger firms will probably have an in-house staff. In either case, manufacturing management must make sure that the counsel given is as supportive of their objectives as the contract and the law allow. It cannot be theoretical or based solely on the divergent objectives of a human relations department.

SUMMARY AND CONCLUSIONS

In this chapter, most of the elements involved in managing the workforce—hourly, salaried, and management—have been discussed. To be effective, any organization or team must have structure. Sometimes this can be low key and quite informal as, perhaps, in a small research laboratory. In a factory, particularly when large numbers of people and high rates of production are involved, there must be sufficient structure for people to understand their roles and make their contributions in the myriad of activities that must take place, often with very precise timing. Factory management is not a precise science. Though some consultants will say otherwise, there are no empirical formulas that will assure success. There are often a number of approaches that can be applied to each of the many elements of workforce management. Provided here are a variety of materials and insights—some unbiased, some not—which hopefully will stimulate thinking to develop the organization, procedures, and programs—the structure—that encourages the teamwork that is a fundamental part of a world-class manufacturing operation.

Choosing the best organizational structures and programs will be influenced by the nature, background, and personality of the community, the workforce, and the management. Strongly engrained feelings and beliefs take time to change, if change is appropriate. People in a community with a strong work ethic and a well-rated education system will respond quickly to quality programs and good leadership.

On the other hand, people from an inner city or highly unionized or suspicious backgrounds may require different programs and more patience and determination. There are many cases on record of strong leaders developing top manufacturing operations in all environments. It can be done.

Is any of this new? Companies like Lincoln Electric, and others before it, have had and still have programs and organizational structures that involve their people and get the best out of them. Tom Peters pointed out many companies practicing excellent manufacturing management in his book *In Search of Excellence*. It is really a tragedy that so many organizations have forgotten the things that made them great. The Japanese are not teaching anything new—just lessons from the American manufacturing past with a few new twists and improvements.

As a supervisor looks at an area in the factory, there will always be an opportunity to improve the use of human resources as well as the technology. Whether work is being done to improve a factory department or subgroup, a manufacturing engineering department, or the entire factory, good programs alone won't do the job alone. Most factories are not all that gentle even in the best of times, so a little hard-nosed determination may be one of the essential ingredients in a factory manager's character. Every engineer, supervisor, and manager who is intelligently involved in the world-class manufacturing organization must strive to try new approaches, suggest improvements, and plan and carry out change.

References

1. Harold T. Amrine, John A. Ritchey, and Colin L. Moodie, *Manufacturing Organization and Management*, 5th ed. (Englewood Cliffs, NJ:Prentice-Hall, Inc., 1987), p. 423.
2. *Ibid.*, pp. 423-426.
3. Carl Heyel and H.W. Nance, *The Foreman/Supervisor's Handbook*, 5th ed. (New York: Van Nostrand Reinhold Co., Inc., 1984), p. 307.
4. *Ibid.*, pp. 308-310.
5. Herbert J. Chruden and Arthur W. Sherman, Jr., *Managing Human Resources*, 7th ed. (Cincinnati: South-Western Publishing Co., 1984), p. 473.
6. Wendell L. French, *Human Resources Management* (Boston: Houghton Mifflin Co., 1986), p. 472.
7. *Ibid.*
8. *Ibid.*, p. 474.
9. Chruden and Sherman, *op. cit.*, p. 472.
10. Craig T. Norback, ed., *The Human Resources Yearbook*, 1987 ed. (Englewood Cliffs, NJ: Prentice-Hall, Inc., 1987), p. 2.81.
11. French, *op. cit.*, pp. 472-473.
12. Norback, *op. cit.*, p. 2.73.
13. Amrine, Ritchey, and Moodie, *op. cit.*, pp. 387-390

Bibliography

Drucker, P. F. "The Coming of The New Organization." *Harvard Business Review* (January-February 1988).

Goldratt, E. M. *The Goal*. New York: North River Press, 1986.

Herzberg, F. "One More Time: How Do You Motivate Employees?" *Harvard Business Review* (September-October 1987).

__________. *Work and the Nature of Man*. Cleveland: World Publishing Co., 1966.

Kepner, C. H., and Tregoe, B. B. *The Rational Manager: A Systematic Approach to Problem Solving and Decision Making*. New York: McGraw-Hill Book Co., 1965.

"Manufacturing Management's Contribution to Strategic Planning: Optimizing Your Role." 1984 Manufacturing Management Update Conference. Dearborn, MI: Society of Manufacturing Engineers, 1984.

"Manufacturing Strategy as a Competitive Weapon." 1985 Manufacturing Management Executive Conference. Dearborn, MI: Society of Manufacturing Engineers, 1985.

Peters, T. J., and Waterman, R. H., Jr. *In Search of Excellence*. New York: Harper & Row, 1982.

Schonberger, R. J. *Japanese Manufacturing Techniques: Nine Hidden Lessons in Simplicity*. New York: The Free Press, 1982.

__________. *World Class Manufacturing*. New York: The Free Press, 1987.

Tregoe, B.B., and Zimmerman, J.W. *Top Management Strategy: What It Is and How To Make It Work*. New York: Simon and Schuster, 1980.

Wrich, J. T. "Beyond Testing: Coping with Drugs at Work." *Harvard Business Review* (January-February 1988).

__________. *The Employee Assistance Program, Updated for the 1980s*. Minneapolis: Hazelden, 1980.

® IMPROSHARE—Mitchell Fein, Inc.
™ Gainsharing—Hay Management Consultants

LEGAL ENVIRONMENT FOR LABOR RELATIONS

No one can deny that the Industrial Revolution has permanently altered the way the modern world thinks and works. In the United States, the full flow of commerce and full production of the economy is deemed essential to the country's national interest. To prevent or minimize interference with the normal flow of commerce and the full production of articles and commodities for commerce, Congress enacted the National Labor Relations Act (Wagner Act) in 1935, which was subsequently amended in 1947 by the Taft-Hartley Act, also known as the Labor Management Relations Act (LMRA). In 1957 the law was further amended in part by the Landrum-Griffin Act, also known as the Labor-Management Reporting and Disclosure Act. For the purpose of this chapter, these assorted acts and amendments will be referred to as the Labor Management Relations Act, the Labor Act, or simply the Act.

These laws were enacted to define the legitimate rights of both employers and employees in their relations affecting commerce, to encourage collective bargaining, and to eliminate certain practices by management and labor that are inimical to the general welfare. While the Labor Act primarily focuses on union-management relations, it does have an effect on employee-employer relations whether unionized or not, blue collar or white collar, as well as professional employees who may remain nonunion.

Because of the limited discussion of these topics, the services of a competent, professional person should be sought before actions covered by this discussion are undertaken.

CHAPTER CONTENTS:

RIGHTS OF EMPLOYEES

The rights specifically provided in the Labor Management Relations Act, as amended, do not apply to certain categories of workers such as independent contractors, supervisors, and public employees. An individual is an employee if under the direct control of the employer, while an independent contractor, although physically working for the employer, is called on to produce a result without the employer controlling or directing the means or methods used to accomplish it. A worker is deemed a supervisor if he or she has the authority to recommend and/or cause another employee to be hired, promoted, discharged, rewarded, or disciplined. Additional characteristics of supervisory status include the authority to direct the job duties of rank-and-file employees, authorize overtime, authorize use of sick leave, and generally schedule work.

SECTION 7 RIGHTS

The basic rights of employees under the Labor Management Relations Act are defined in Section 7 of the Act:

> "Employees shall have the right to self-organization to form, join, or assist labor organizations, to bargain collectively through representatives of their own choosing and to engage in other concerted activities for the purpose of collective bargaining or other mutual aid or protection; and shall also have the right to refrain from any or all such activities except to the extent that such right may be affected by an agreement requiring membership in a labor organization as a condition of employment as authorized in Section 8(a)(3)."

Under Section 7, employees cannot be prevented from: (1) forming or attempting to form a union among employees of a company, (2) assisting a union in organizing, (3) striking to secure better working conditions, and (4) refraining from organizing should they so choose. It is important to recognize that the rights specified in Section 7 are rights granted to *individuals* who choose to act collectively.

THE UNION SHOP

Although Section 7 of the Labor Management Relations Act grants to employees the right to refrain from activity in behalf of a union, this right is limited and subject to the provisions of Section 8(a)(3) of the Act.

***Contributor of this chapter is: Louis A. DeGennaro, J.D.**, Associate Professor of Management, Lawrence Institute of Technology.*

***Reviewers of sections of this chapter are: Paul F. Geene**, Corporate Personnel Manager, U.S. Manufacturing Corp.; **Frank Guido**, General Counsel, Police Officers Association of Michigan; **Douglass V. Koch**, Associate Professor of Management, Lawrence Institute of Technology; **J.N. Moore**, Director of Administrative Services, Smith, Hinchman, & Grylls Associates, Inc.; **Carol M. Mumford**, Supervisor—Personnel Services, Detroit Edison; **Joe Richardson**, Project Leader, Manufacturing Engineering Sciences, Garrett Turbine Engine Co.; **Dr. Herman A. Theeke**, Assistant Professor, School of Management, The University of Michigan-Dearborn; **Hayward Thomas**, Consultant, Retired—President, Jensen Industries; **James E. Vanderbrink**, Director of Industrial Relations, DeVlieg Machine Co.*

RIGHTS OF EMPLOYEES

Section 8(a)(3) permits, under certain conditions, an employer and union to make an agreement (union security agreement) that would require all employees to join the union to retain their jobs.

A union security agreement cannot require union membership as a condition of hire, but can only require that employees become union members after a certain period. This grace period cannot be less than 30 days, except in the building and construction industry where employees may be required to join the union after 7 full days.

THE RIGHT TO STRIKE

As noted earlier, Section 7 of the Labor Management Relations Act permits employees to engage in concerted activities for the purpose of collective bargaining. Section 13 of the Act ensures that, except in a few circumstances such as health care institutions, an employee's right to strike may not be impeded or diminished.

An employee's right to strike assumes the strike is lawful, which will depend on the purpose of the strike, its timing, or on the conduct of the strikers. A strike that has a lawful purpose, such as higher wages or better working conditions, may become unlawful because of the conduct of the strikers, such as blocking the entrance or exit of a plant, threatening violence against nonstriking employees, or attacking management representatives. Employees who participate in an unlawful strike or unlawful strike activities may be discharged and are not entitled to reinstatement when the strike ends.

Employees striking for economic concessions such as higher wages or improved working conditions are economic strikers. Although they cannot be discharged, they can be replaced during the strike. They are not entitled to reinstatement if the employer has hired permanent replacements. They may be entitled to recall when an opening occurs. The exact terms will usually be the subject of the collective bargaining between the employer and the union.

The Labor Management Relations Act defines certain practices by an employer as unfair labor practices. Employees who strike to protest a company's unfair labor practice cannot be discharged or permanently replaced. When the strike ends, the strikers are entitled to immediate reinstatement even if replacements hired during the strike have to be discharged.

EMPLOYEE REPRESENTATION

Before examining the keystone of the LMRA—the collective bargaining process—two primary questions must be answered: (1) Who shall be represented? and (2) Who shall be the representative?

Section 9(a) of the Act permits a majority of employees in an "appropriate" bargaining unit to designate the representative (union) for the purpose of collective bargaining. A bargaining unit consists of two or more employees who share common employment interests; that is, those who have the same or similar interests concerning wages, hours, and working conditions. Excluded from the bargaining unit by law are agricultural laborers, independent contractors, supervisors, and managerial employees. The National Labor Relations Board (NLRB) determines the appropriate bargaining unit.

The representative (union) who will be negotiating on behalf of the bargaining unit must be the choice of a majority of the employees. Although no particular procedure for choosing a representative is required by law, a basic method of selecting the bargaining representative is by a secret ballot election conducted under the auspices of the NLRB upon the filing of an appropriate petition. Also upon proper petition, the Board can be called on to conduct a decertification election. If successful, such a decertification permits employees to select a different union or to work in a union-free setting.

Where there exists a valid collective bargaining contract, the National Labor Relations Board has adopted a hands-off policy with regard to elections. Under NLRB rules, a valid contract for a fixed period of 3 years or less will bar an election for the period of the agreement. This contract-bar rule does not always apply. A contract not signed or in writing but not ratified would not prevent an election. Likewise, where the bargaining unit is inappropriate or the collective bargaining agreement is racially discriminatory, an election would not be barred.

The NLRB has also promulgated a rule that when a union has been certified by the Board, the certification will be binding for at least 1 year. Petitions for another election filed before the end of the certification year will be dismissed.

Elections are usually held within 30 days after they are directed. A different election date may be set in cases of seasonal drops in employment. An employee on the payroll immediately prior to the election date is eligible to vote. Likewise, employees who are ill, on vacation, or temporarily laid off are eligible to vote. Where employment is typically irregular such as in the radio, television, recording, and movie industries, employees would have their eligibility to vote based on the number of days worked during the year preceding the election date rather than whether they were on the payroll at the time of election.

Regional directors of the NLRB are given the authority to conduct the election and determine such details as time, place, and notices.

An election may be set aside by the Board if there is sufficient evidence to establish that the atmosphere of the election or conduct of the election was such as to interfere or tend to interfere with the employees' free choice. Such interference may exist when an employer threatens employees with loss of jobs or benefits; where management or labor materially misrepresents essential facts during the election campaign; or where management or labor threatens or uses physical force against employees.

Management is certainly permitted to campaign to convince employees that they should remain nonunion. This management campaign may be aggressive provided that each individual's right to decide is not threatened, that the material presented is factual, and that individual supervisors and managers express their opinions only when asked. Management can be successful in such campaigns provided employees are treated with respect, open communication can be established, and a climate of trust is recognized by the employees.

THE COLLECTIVE BARGAINING PROCESS

Congress believed that if workers had the right to associate and organize to negotiate the terms and conditions of their employment, industrial strife would be minimized.

The process of collective bargaining as mandated by law requires an employer and the employee's representative to meet at reasonable times to confer in good faith with respect to wages, hours, and other terms or conditions of employment. Under the Labor Management Relations Act, neither management nor labor may refuse to bargain collectively with the other. It should be noted, however, that the obligation to bargain collectively does not require either party to concede or agree to a proposal by the other.

Where there exists a valid enforceable collective bargaining agreement, certain steps must be taken before the contract can be terminated or modified. First, a party seeking modification or termination must provide the other party written notification 60 days prior to the expiration of the agreement of a proposal for termination or modification; second, the party must offer to meet and confer with the other party to negotiate a new collective bargaining contract; third, after notice to the other party, the Federal Mediation and Conciliation Service must be notified of the existence of a dispute; and fourth, neither party can resort to a strike or lockout until 60 days after notice to the other party.

UNFAIR LABOR PRACTICES OF MANAGEMENT

Congress, in enacting the Taft-Hartley amendments to the Act, believed that certain practices by employers or unions would be inimical to the free flow of commerce and accordingly were declared illegal. The following represent some practices of employers that are prohibited by law.

INTERFERENCE WITH EMPLOYEES' RIGHT TO ORGANIZE

Employers may not "interfere with, restrain, or coerce" employees in the exercise of their rights to organize to form or join a union or bargain collectively. Under this prohibition, the following practices by management would be unlawful:

1. Threatening employees with loss of jobs or benefits if they join a union.
2. Threatening to close down a plant if a union is organized.
3. Spying on union gatherings.
4. Granting wage increases designed to discourage and deter union organization.

ILLEGAL ASSISTANCE AND SUPPORT OF UNIONS

The Labor Management Relations Act makes it illegal for an employer "to dominate or interfere with the formation or administration of any labor organization or contribute financial or other support to it." Prohibited are "company unions," that is, unions created by management or so influenced by management that employees have not been given a free choice in selecting their bargaining representative (union). Thus an employer may not: (1) take an active part in organizing a union; (2) compel employees to join a particular union, except where there exists a lawful union security agreement; or (3) grant special privileges to one of competing unions to the detriment of other unions.

DISCRIMINATION AGAINST EMPLOYEES

The Labor Management Relations Act makes it unlawful for an employer to discriminate against an employee because of the employee's union activity. Discrimination would include such actions as refusing to hire, discharging, demoting, assigning to a less desirable shift or job, or withholding benefits. Additional examples of illegal discrimination would include the following:

1. Refusing to rehire employees because they took part in a union's lawful strike.
2. Discontinuing a plant operation and its employees and relocating with new nonunion employees.

FAILURE TO BARGAIN IN GOOD FAITH

An employer is under a duty to bargain in good faith. This duty covers all mandatory subjects of bargaining, which include all matters related to wages, hours, or other terms and conditions of employment. Examples of mandatory subjects of bargaining include: rates of pay; wages; hours of employment; pensions; bonuses; group insurance; grievance procedures; safety practices; seniority; procedures for discharge, layoff, recall, or discipline; and the union shop.

Good faith bargaining requires an employer to meet at reasonable times and negotiate with the union's designated representative. An employer's good faith obligation includes the duty to supply on request "relevant and necessary" information essential to the union's representative to bargain effectively as to wages and hours. It does not require the employer to provide financial details, unless the employer states that the company is unable to grant wage or benefit increases because of its financial state.

For example, an employer who refuses to supply the union with data concerning a group insurance plan would be violating its duty to bargain in good faith. Likewise, an employer breaches its duty to bargain in good faith when it subcontracts work to another employer without giving the union an opportunity to bargain over such subcontracting.

UNFAIR LABOR PRACTICES

UNFAIR LABOR PRACTICES OF UNIONS

Although early legislation was directed only at management by declaring certain practices unlawful, subsequent amendments likewise prohibited certain practices of unions.

COERCION OF EMPLOYEES

Labor organizations are prohibited from engaging in conduct calculated to restrain or coerce employees in their Section 7 rights. Unlawful coercion would include such acts as physical assaults or threats of violence against employees. Additional prohibited acts would include the following:

1. Threats of violence to nonstriking employees.
2. Use of mass picketing to prevent nonstriking employees from entering the plant.
3. Expelling union members for filing an unfair labor practice against the union.
4. Failing or refusing to process a grievance in retaliation against a union member who criticized or challenged union authority.
5. Charge excessive or discriminatory fee as a condition of becoming a union member.
6. Attempt to scare an employee to pay for services that are not performed or not to be performed (featherbedding).

CAUSING AN EMPLOYER TO DISCRIMINATE

The Labor Management Relations Act makes it unlawful for a union to cause an employer to discriminate against an employee in regard to wages, hours, and other conditions of employment. Where no valid union shop agreement exists, it would be unlawful for a union to demand or conspire with management to discriminate against employees because they are not union members. Additionally, it would be illegal for a union to cause an employer to fire an employee who has made anti-union comments.

REFUSAL TO BARGAIN IN GOOD FAITH

The duty to bargain in good faith applies to labor as well as to management. As previously discussed, the duty to bargain requires that the employee's representative meet at reasonable times with the employer and confer on matters pertaining to wages, hours, or other conditions of employment. As to non-mandatory subjects of bargaining, a union may seek voluntary bargaining but cannot insist on bargaining with management about such subjects.

The good faith obligation imposed on a union applies also to its relationship with the employees. A union must fairly represent its members. This duty of fair representation would be violated, for example, if the union arbitrarily refused to process a union grievance or negotiated a contract with racially motivated provisions.

ENGAGING IN ILLEGAL STRIKES OR BOYCOTTS

The Labor Management Relations Act makes illegal certain kinds of strikes and boycotts and prohibits a union from inducing others to strike or boycott. For example, it would be illegal for a union to picket a construction site to force the removal of a nonunion subcontractor.

A secondary boycott occurs when a union has a labor dispute with a primary employer and brings pressure on that employer by causing a second employer to stop dealing with the primary employer. It is important that the second employer be neutral with regard to the dispute and not be an ally of the primary employer. The second employer may be considered an ally of the first employer if the businesses were jointly owned or operated. The secondary boycott prohibition in the LMRA would not apply where such ally relationship exists. Likewise, the secondary boycott prohibition would not apply where the primary employer and secondary employer share a common site, provided the picketing is directed solely against the primary employer. Nor would the Act prohibit a union from informing the public, such as with handbills, that a product is produced by an employer with whom the union has a labor dispute.

THE ENFORCEMENT OF THE LABOR MANAGEMENT RELATIONS ACT

In passing the Labor Management Relations Act, Congress created the National Labor Relations Board to ensure that the rights granted and contained therein may be exercised.

The National Labor Relations Board (NLRB) is composed of five members appointed by the President with the advice and consent of the Senate for 5-year terms. The Board also includes a general counsel who has final authority on behalf of the Board with regard to the investigation of charges and the filing of complaints. The Board has regional offices under the supervision of the general counsel located in major cities.

The primary function of the NLRB is to conduct and certify representation elections and prevent union and management from committing unfair labor practices. The Board's jurisdiction extends to those enterprises that are engaged in interstate commerce or substantially affect interstate commerce. The Board may choose to decline jurisdiction unless certain "dollar" standards are met. For example, retail businesses must have at least $500,000 gross annual volume of business; newspapers, $200,000; and manufacturing, $50,000.

PROCEDURE BEFORE THE BOARD

A union seeking certification under the Labor Management Relations Act does so by filing a petition with the NLRB regional office in the area where the unit of employees is located. If appropriate, there will be an election by secret ballot with the results certified by the NLRB. When a union receives a majority of votes (a tie vote is a win for the employer), it is entitled to be certified. If a union does not receive a majority of votes, it is barred from seeking an election for one year. A union must be recognized by the employer as the employees' exclusive bargaining agent after it is certified. An employer commits an unfair labor practice by not recognizing the certified union.

A person, whether an employee, an employer, a union, or other labor organization, alleging an unfair labor practice may file a "charge" with the regional NLRB office where the unfair labor practice was committed. The regional office will investigate the charge and issue a complaint if the charge has merit. A formal hearing is conducted before an administrative law judge appointed by the Board. The judge will make findings and recommendations to the Board.

AUTHORITY AND POWER OF THE NATIONAL LABOR RELATIONS BOARD

The Board has broad remedial powers under the Labor Management Relations Act. If the Board determines that an unfair labor practice has been committed, it can do the following:

1. Reinstate employees with or without back pay.
2. Dissolve an employer-dominated union.
3. Refund illegally collected union dues or fees.
4. Seek relief in a federal court to enjoin prohibited picketing and secondary boycotts. Additionally, the Board may require the employer or union to post a notice conspicuous to the employees that the unfair labor practices will cease.

The Board is empowered under the LMRA to seek enforcement of its order in the U.S. Court of Appeals. The order may be modified, set aside, or enforced by the appellate court. Failing to comply with the decision of the appellate court constitutes contempt, punishable by fine or imprisonment. Where the dispute has significant consequences, the Supreme Court of the United States may permit further review.

ARBITRATION

The collective bargaining contract will often provide for arbitration of a labor dispute or union grievance.

By arbitrating their grievance, the parties have chosen to resolve their dispute without government or court intervention. They usually agree that the decision will be binding. The grievance is brought before a mutually agreed arbitrator who will make a final decision. The decision of the arbitrator is called an award and is generally accompanied by an opinion of the arbitrator setting forth the reasons for the award. Arbitration has been proven to be less costly and less time consuming than litigation. The grievance is presented in an informal hearing without a strict adherence to rules of evidence.

WAGE AND HOUR LEGISLATION

The Fair Labor Standards Act (FLSA) is the primary federal legislation regulating wages and hours of businesses. This Act has undergone major overhaul since its passage in 1928, with the 1987 minimum hourly rate at $3.35 per hour.

Under the Fair Labor Standards Act, 40 hours in a week is the maximum number of hours a nonexempt employee can work without being paid overtime. Overtime payment under the Act is one and one-half the regular hourly rate. For employees receiving a shift premium, the overtime rate would include the employees' regular hourly rate plus the shift premium.

What constitutes time worked for purposes of computing compensation has been subject to much confusion and interpretation under the FLSA. Interpretative rulings have said that an employee who voluntarily starts early or stays late with the employer's knowledge and acquiescence is entitled to have work time computed from beginning to end. Likewise, rulings have said that an employee who is on call and required to remain on the premises is working for purposes of computing wages and hours.

Additionally, short rest periods, such as coffee breaks, must be counted as time worked. However, bona fide meal periods do not constitute work time, providing an employee is completely relieved of his or her work duty during this period.

The Fair Labor Standards Act specifically excludes executive, administrative, professional, or outside sales jobs from its coverage. The meaning and definition of the exempt coverages has been subject to numerous administrative interpretations and court decisions, not infrequently with the court disagreeing with the administration of the FLSA.

FAIR EMPLOYMENT PRACTICES

Title VII of the 1964 Civil Rights Act prohibits discrimination in employment based on race, color, religion, sex, or national origin. Subsequently added to the broad range of employment discrimination by separate legislation and amendment was age and handicapped discrimination. Additionally, all but a few states have chosen to supplement Title VII with their own legislation. Along with state legislation and subsequent amendments to Title VII, antidiscrimination laws extend to virtually every commercial enterprise, including labor organizations and employment agencies.

Title VII specifically exempts from its broad coverage those instances where race, color, religion, sex, or national origin is a bona fide occupational qualification reasonably necessary to the normal operation of that particular business or enterprise.

An aggrieved person under Title VII need not be an employee, as coverage under the Act extends to applicants who have been denied employment for discriminatory reasons. Often the employment application provides considerable evidence of discriminatory motivation; questions on an application regarding marital status, religion, citizenship, or arrest record would certainly provide proof of discriminatory hiring practices.

Questions or hiring policies that are ostensibly neutral may have a discriminatory effect and therefore are unlawful. For example, requiring a high school diploma as a condition of hire, although neutral, would disqualify black applicants at a higher rate than white applicants. Thus an employer must establish that the diploma is job related to avoid a successful challenge under Title VII. Requiring a college degree and 500 hours of flight time for the position of airline pilot would be job related because a high degree of skill is necessary to minimize the human risk.

Employers using ability tests must be ready upon challenge to show that such ability tests are likewise job related even if there is no intent to screen out minority groups for hiring and promotion. Employers must establish that such tests can be statistically validated and evidence a high degree of utility, and that alternate hiring or promoting procedures are not available. An employer may validate a test by establishing a statistical relationship between test performance and job performance or by establishing that the test consists of significant or directly related job skills, such as a typing test. Although very difficult and frequently expensive to do, an employer can establish that applicants must have identifiable characteristics deemed essential to successful job performance. A statistically validated test may still be challenged if it produces a disparate effect in hiring minorities.

EQUAL EMPLOYMENT OPPORTUNITY COMMISSION

Five members appointed by the President with Senate approval comprise the Equal Employment Opportunity Commission. As with the National Labor Relations Board, there is also a general counsel who serves the Commission. Originally, Title VII gave the Commission only the power to investigate and conciliate charges of unfair employment practices. By subsequent legislation, the Commission now has enforcement power; that is, the power to bring suit in the federal courts.

Typically, an administrative proceeding begins by the filing of a charge by a person alleging employment discrimination. To encourage reconciliation and expeditious resolution, the parties at this stage are anonymous in that their identities are not publicly disclosed.

Because most states have enacted similar legislation and have created similar agencies, the Commission has adopted a deferral policy providing the states the first opportunity to resolve the grievance.

After a charge is filed with the Commission and notice of the charge is served on the employer, the Commission must investigate to determine if there is reasonable cause to support the charge. If no reasonable cause exists, then the charge will be dismissed. Finding of reasonable cause requires the Commission to attempt conciliation. If conciliation fails, the Commission may bring suit in federal court or permit the charging party to bring suit individually. Where the aggrieved individual is one of many similarly treated, a class-action suit may be brought, provided that the class is sufficiently large with common questions of law and fact and that the class is fairly and adequately represented.

PROCEDURE AND REMEDIES

As with most litigation, the burden of proof is on the complaining party. After sufficient evidence is introduced to establish a prima facie case of discrimination, the burden shifts to the employer to negate or rebut the presumption of discrimination. This may be done by the employer showing that the reasons for the decision were nondiscriminatory. The applicant or employee is given the last chance of showing that the reasons given by the employer in its defense were merely pretext or a coverup for discriminatory action.

Title VII provides a number of remedies for victims of employment discrimination. A court can enjoin the prohibitive behavior and order such affirmative action as the reinstatement or hiring of employees with or without back pay. In addition, Title VII permits the courts to award reasonable attorney fees to the prevailing party.

PROHIBITED PRACTICES

Under Title VII, it is unlawful for an employer or union to discriminate against a person because of his or her race, color, religion, sex, or national origin with respect to compensation, terms, conditions, or privileges of employment.

Race Discrimination

Evidence of racial discrimination may be found not only in the employment application and educational standards and testing procedures, but in certain employment practices such as word-of-mouth recruiting or preferential hiring of relatives of white workers. Additionally, segregated locker rooms or lunchrooms would be the basis of unlawful discrimination, as well as the creation of all-white or all-black work crews.

It is possible that a seniority system would perpetuate the consequences of racial discrimination. Notwithstanding, Title VII, as interpreted by the Supreme Court of the United States, specifically exempts bona fide seniority systems from discriminatory attacks.

FAIR EMPLOYMENT PRACTICES

Religious Discrimination

An employer has an obligation to make reasonable accommodations for an employee's religious needs, where such accommodation does not produce an undue hardship. The duty to "make reasonable accommodations" does not require an employer to violate a bona fide seniority system in a collective bargaining contract that requires the most junior employees to work on a religious observance day.

Sex Discrimination

A sampling of court decisions and commission rulings suggest the following with regard to employment practices that relate to treatment between the genders:

1. Unless sex is a bona fide occupational qualification, segregated "help wanted" advertisements are unlawful.
2. It is no defense to a sex discrimination charge to argue that state laws impose restrictions with regard to working conditions to protect women. Such state legislation has been preempted by Title VII.
3. It is illegal to base payment of employment benefits, such as pension and retirement plans, on gender.
4. Customer preference generally cannot be the basis of a bona fide occupational qualification.
5. Women cannot be forced to take maternity leave at a specified time in the pregnancy, provided they continue to be able to adequately perform their regular duties and that the work environment does not pose a health hazard.

It is now established that sexual harassment constitutes sex discrimination under Title VII. The Commission's guidelines state that unwelcome sexual advances, requests for sexual favors, and other verbal or physical conduct of a sexual nature constitute sexual harassment when submission to such conduct is explicitly or implicitly made a term or condition of an individual's work performance; when it creates an intimidating, equal pay discussion; or when it creates an offensive or hostile working environment. A sexual harassment situation may also exist when an uninvolved party is denied opportunity that is given to another person who is bestowing sexual favors.

The sex discrimination prohibitive contained in Title VII, as well as in the Equal Pay Act of 1963, makes it unlawful for an employer to establish wage differentials based on sex. Men and women performing equal work requiring equal skill, effort, and responsibility under similar working conditions are entitled to equal pay. The equal pay standard applies where the work performed is "substantially" equal. Skill is based on the training, education, and ability required in doing the job. Effort measures the mental and physical exertion necessary to do the work. Responsibility pertains to the degree an individual is accountable for the particular job. The equal pay obligation would not apply if the wage differential was based on a seniority system, merit system, or a system that measures pay by quantity or quality of output.

National Origin Discrimination

Certain job requirements, such as height and weight restrictions, must be shown to be job related because such requirements can discriminate against members of some ethnic groups. For example, height and weight restrictions may exclude a larger number of Spanish-surnamed American men. Likewise, an employer must maintain a working environment free from ethnic harassment.

Age Discrimination

The Age Discrimination Employment Act of 1967 and its subsequent amendment was passed to ensure that those over the age of 40 not be subjected to discharge or demotion at the whim of an employer. Arbitrary or capricious action by employers has been found to produce serious adverse economic hardship to older workers. The prohibition against age discrimination does not apply where there exists a bona fide occupational qualification reasonably necessary to the normal operation of the business or where there exists a bona fide seniority system, which in fact tends to favor more experienced employees of any age.

Handicapped Worker Discrimination

A handicapped individual is a person who has a physical or mental impairment that substantially limits one or more of such person's major life activities. "Life activities" would include ambulation, communication, self-care, and so on. "Substantially limits" refers to the degree that the impairment affects employability. Additionally, an individual who is regarded as being impaired or who has a history of impairment even though not "impaired" is deemed a handicapped individual. The Vocational Rehabilitation Act of 1973 requires employers who deal with the federal government to take affirmative steps to hire the handicapped. The obligation to hire the handicapped is limited to those that are deemed "qualified handicapped workers." A qualified handicapped worker is an individual who may have recovered from a physical or mental impairment or who may not have been impaired but, because of the employer's attitude, will have difficulty in obtaining or retaining employment.

EMPLOYEE BENEFITS

The social insurance provisions of the federal Social Security Act are those that govern the employer-employee relationship. These provisions consist of old age, survivor's, and disability insurance; hospital insurance; and unemployment insurance. In addition, an employer or an employee injured while on the job would be entitled to work-loss benefits.

What is necessary before these benefits apply is that there exists an employer-employee relationship, and an employee must be distinguished from an independent contractor. Although employees and independent contractors must pay Social Security tax, the employer's obligation to withhold tax extends only to the employee and not the independent contractor. An employer is personally liable for the payment of the tax that must be withheld from employees' wages.

The obligation to provide unemployment and worker's compensation benefits would likewise extend only to the employee and not independent contractors. Worker's compensation laws provide that an injured worker receive a monetary amount determined in advance by a state agency. The injured employee's remedy is solely under the worker's compensation law, and once an award is received, the injured worker is not entitled to further compensation and cannot bring suit against the employer. The amount of disability benefits will depend on whether the injury is partial or total, permanent or temporary. Depending on the laws of the state where the employer does business, the employer may be required to obtain insurance protection from a private carrier or to be established as a self-insurer. Worker's compensation laws have abolished several defenses which employers had used; therefore an injured employee may gain an award even though the injury results from his or her own carelessness or intentional disregard for safety. Of course, benefits will only be paid if the injury arises in the course and scope of employment. Worker's compensation is discussed in greater depth in Chapter 12, "Management Concerns for Occupational Safety and Health," of this volume.

A worker who becomes unemployed suffers a financial injury in a different way. The loss of a paycheck has very real and severe consequences, and federal and state unemployment compensation laws were passed to reduce the adverse impact of a sudden job loss. An unemployment insurance fund is created by requiring employers to pay a tax on the payroll of its employees. To receive unemployment benefits, a person must have been employed, be unemployed, be available for work, and have filed a claim.

Bibliography

Avideeson, Howard. *Primer of Labor Relations*. Washington, DC: Bureau of National Affairs, 1975.

Davey, Harold W. *Contemporary Collective Bargaining*, 4th ed. Englewood Cliffs, NJ: Prentice-Hall, 1982.

Estey, Martin. *The Unions*, 2nd ed. Orlando, FL: Harcourt Brace Jovanovich, Inc., 1976.

Hagburg, Eugene C. *Labor Relations*. St. Paul: West Publishing, 1978.

Kochan, Thomas A. *Collective Bargaining of Industrial Relations*. Homewood, IL: Richard D. Irwin, Inc., 1980.

Yoder, Dale, and Henemen, Herbert, eds. *Employee and Labor Relations*, Vol. III. Washington, DC: Bureau of National Affairs, 1976.

MANAGEMENT CONCERNS FOR OCCUPATIONAL SAFETY AND HEALTH

HISTORICAL DEVELOPMENT OF INDUSTRIAL SAFETY IN THE U.S.

No record was kept of injuries sustained by industrial workers in the United States before the 19th century. Factories were growing at an unprecedented rate with expansions in product lines late in the 19th century. These factories, however, often were inferior in proper consideration of human values such as health and safety. Injuries and deaths frequently were accepted as a part of industrial progress.

ORGANIZED LABOR

The change in safety attitudes within manufacturing is due in large part to organized labor. Labor fought for the compensation of accident victims and their families and for the correction of serious hazards in the workplace. Changes often were achieved only after long and hard struggles with employers or through the enactment of state and national laws. Because safety in part became identified with the overall struggle of labor unions, many employers ignored the demand for a safe environment along with the demands for higher wages and shorter working hours. Working conditions were at times deplorable, reflecting the little care industrial management had for workers.

The law at that time favored the employer. The *fellow servant rule*, a doctrine of common law, removed the employer's liability for injuries that resulted from the negligence of fellow employees. Also, the employer was not liable for *contributory negligence*—injuries due to the employee's own negligence. The employer was not liable because the employee was aware of the job risks and hazards prior to employment.

Later in the Industrial Revolution, the extent of poor working conditions and industrial accidents became more obvious to the public. This resulted in changes in governing policies. A great change in mining safety came about in 1869 when the Pennsylvania legislature passed a mining safety law requiring at least two exits from each mine. Massachusetts drafted a law in 1877 requiring the safeguarding of hazardous machinery, and the Employer's Liability Law was passed, making employers liable for worker injuries. In 1907 alone, more than 3200 miners had been killed in accidents, so the Bureau of Mines was created by the U.S. Department of the Interior to investigate accidents, study health hazards, and identify appropriate corrections. In 1911, Wisconsin passed the first effective worker's compensation law; similar laws were thereafter passed by other states.

SAFETY ORGANIZATION

The Association of Iron and Steel Electrical Engineers was organized in the first decade of the 20th century and devoted much attention to safety problems in its industry. In 1911, a request came from this association to call a national industrial safety conference. The first Cooperative Safety Congress met in Milwaukee in 1912. A year later, at a meeting in New York City, the National Council of Industrial Safety was formed. It began operation in a small office in Chicago. At its meeting in 1915, the organization's name was changed to the National Safety Council, and its function was broadened to include safety in all areas of life in addition to industrial accident prevention. The safety movement as we know it was initiated because of concern for safety and the dedication of the people involved in those early meetings.

Today the National Safety Council (NSC) has members among manufacturing firms, construction companies, utility companies, agricultural organizations, labor unions, insurance companies, educational institutions, and government agencies. All of these groups share an interest in the safety and well-being of people in the workforce. The NSC council has safety training programs for people on the beginner level and for professionals alike. The courses cover industrial subjects, agricultural safety, hearing conservation, industrial

CHAPTER CONTENTS:

Contributor of this chapter is: ***Steven A. Cousins**, Development Engineer, DuPont Engineering Development Laboratory, E.I. DuPont de Nemours & Co., Inc.*

***Reviewers of sections of this chapter are:* John D. Brazil**, *Corporate Employee Safety Specialist, Chrysler Motors;* ***Carrol E. Burtner**, Director—Office of Mechanical Engineering Safety Standards, Occupational Safety and Health Administration, U.S. Department of Labor;* ***Jack H. Dobson, Jr.**, Director of Technical Services, American Society of Safety Engineers;* ***Thomas H. Gibbons**, Safety and Health Manager, The Trane Company;* ***Arleen Hayes**, Manager—Occupational Health Services, Wausau Insurance Co.;* ***Frederick W. Lang**, Assistant Director, Occupational Safety and Ergonomics, General Motors Corp.;* ***Julian Olishifski**, Retired—Manager, Occupational Safety and Health, Alliance of American Insurers;* ***Hayward Thomas**, Consultant, Retired—President, Jensen Industries;* ***Theodore M. Wire**, Manager—Occupational Safety, Deere & Company.*

HISTORICAL BACKGROUND

hygiene, and traffic safety. The NSC compiles statistics on causes of accidents through its membership organizations, federal agencies, and its special studies, and it disseminates this information along with accident prevention techniques. Two of the NSC publications are *Accident Facts*, printed yearly with information on accidents, and *Safety and Health*, a monthly magazine with current information in accident prevention and health promotion.

DEVELOPMENT OF ACCIDENT PREVENTION

As the drive for a safer work environment continued to influence industry, the "Three E's of Safety" were developed: Engineering, Education, and Enforcement. Among the discoveries realized by industry were the following:

- Improved engineering could prevent accidents.
- Employees were willing to learn and accept safety rules.
- Safety rules could be established and enforced.
- Financial savings from safety improvements could be reaped by savings in compensation and medical bills.

Between the two World Wars, industrial safety practices received wide acceptance. Also, the federal government encouraged its contractors to maintain a priority for safety. The connection between quality and safety was discovered. Also discovered was the seriousness of the effect off-the-job accidents had on manufacturing. This was especially realized during the labor shortage brought on by World War II.

The 1960s saw passage of significant occupational safety legislation, including the Service Contract Act of 1965, the National Foundation of Arts and Humanities Act, the Federal Metal and Nonmetallic Mine Safety Act, the Federal Coal Mine Health and Safety Act, and the Contract Workers and Safety Standards Act. However, these laws affected a limited number of employers, namely those who had obtained federal contracts or were of a specific industry. More significant federal legislation affecting the safety of workers in all areas of commerce would soon be passed (see the discussion of the Occupational Safety and Health Act of 1970 in the following section of this chapter) because:

- Generally, the states legislated safety requirements only in specific industries, had inadequate safety and health standards, and had inadequate budgets for enforcement.
- The injury and death toll due to industrial mishaps was still considered to be too high. In the late 1960s, more than 14,000 employees were killed annually in connection with their jobs.
- Work injury rates were taking an upward swing.

Today, employers are making a commitment toward off-the-job safety programs because of recently passed state laws on compulsory insurance that make the employers financially responsible for injuries that originate off the job. Companies have also discovered that operating costs and production schedules are adversely affected when workers are injured on or off the job. Companies with sound safety policies have found that off-the-job safety programs naturally complement on-the-job safety programs.

American workers today are better off than their counterparts of the last century. A worker's chance of being killed in an accident today is less than half that likelihood 60 years ago. Indeed, there is no doubt that the trend of declining accident rates is because of the continued safety movement in the manufacturing sector.

SAFETY LEGISLATION

The Occupational Safety and Health Act of 1970 and the Federal Mine Safety Act of 1977 are the two most significant pieces of federal legislation on occupational safety affecting the manufacturing community. The OSHA legislation will be given attention here.

THE OCCUPATIONAL SAFETY AND HEALTH ACT

The Occupational Safety and Health Act (OSHAct) was enacted to assure safe and healthy working conditions for every working man and woman in the nation. The Act applies to every employer in the U.S. or in U.S. possessions who has any number of employees and who engages in a business affecting commerce. Federal employees are covered by this Act, but employees of state and local government are excluded. Operators of mines covered by the Federal Mine Safety and Health Act of 1977 are excluded from the OSHAct. Also excluded are operations where a federal agency (other than the Department of Labor) has authority to dictate or enforce occupational safety and health regulations or standards.

Employers covered by the Act have the obligation of complying with the safety and health standards connected with the Act. Also, the OSHAct has a "general-duty clause" that obligates employers to provide employees with a workplace free from recognized hazards that are likely to cause death or serious physical harm to the employees. This legislation makes on-the-job safety and health a management responsibility. The law places on every employee the duty to comply with safety standards; however, final responsibility for compliance remains with the employer.

OCCUPATIONAL SAFETY AND HEALTH ADMINISTRATION

The administration and enforcement of the Occupational Safety and Health Act are vested with the Secretary of Labor and the Occupational Safety and Health Review Commission. Investigation and prosecution are performed by the Secretary of Labor, and the Review Commission makes decisions in contested cases during the enforcement procedure. The National Institute for Occupational Safety and Health (NIOSH), established within the Department of Health and Human Services (DHHS), carries out safety and health research and related functions. Statistical data relating to injuries and illnesses are compiled by the Bureau of Labor Statistics within the Department of Labor.

The Occupational Safety and Health Administration (OSHA) has the authority to institute and revoke safety and health standards, conduct inspections and investigations, issue citations and penalties, place requirements for recordkeeping of safety and health data, and to petition the courts to act against employers with dangerously hazardous work environments. OSHA also has the authority to provide employer and employee training, implement voluntary protection programs including injury prevention consultation, grant funds to states for the operation of safety and health programs, grant funds to private organizations to develop safety and health training programs, and develop and maintain occupational safety and health statistics programs.

To carry out the responsibilities of OSHA, 10 regional offices have been established (see Fig. 12-1). Each Regional Administrator supervises, coordinates, evaluates, and executes all OSHA programs in the region. There are area offices within each region, headed by an Area Director, which carry out the compliance program of OSHA in each area. These area offices also function as local information resources for safety and health.

The Occupational Safety and Health Review Commission is a semijudicial three-member board that is appointed by the President and confirmed by the Senate. The function of the Commission is to make decisions in employer-contested cases resulting from an OSHA enforcement action against that employer. OSHA notifies the Commission of contested cases when they arise, and the Commission hears all appeals on action taken by OSHA concerning citations and proposed penalties and then determines whether the employer has violated the Act or any of its standards. The employer has a stay of any penalties assessed by OSHA during the commission review period. Any person adversely affected by a final order of the review commission has 60 days to appeal to the U.S. Court of Appeals; the courts will allow an appeal only of issues considered by the review commission.

NATIONAL INSTITUTE FOR OCCUPATIONAL SAFETY AND HEALTH

The National Institute for Occupational Safety and Health (NIOSH) is part of the Department of Health and Human Services' Centers for Disease Control. NIOSH is engaged in research, education, and training related to occupational health and safety. NIOSH develops criteria and recommendations for health and safety standards, conducts safety research experiments and demonstrations, and conducts educational programs to train personnel to carry out the OSHAct. The OSHAct also

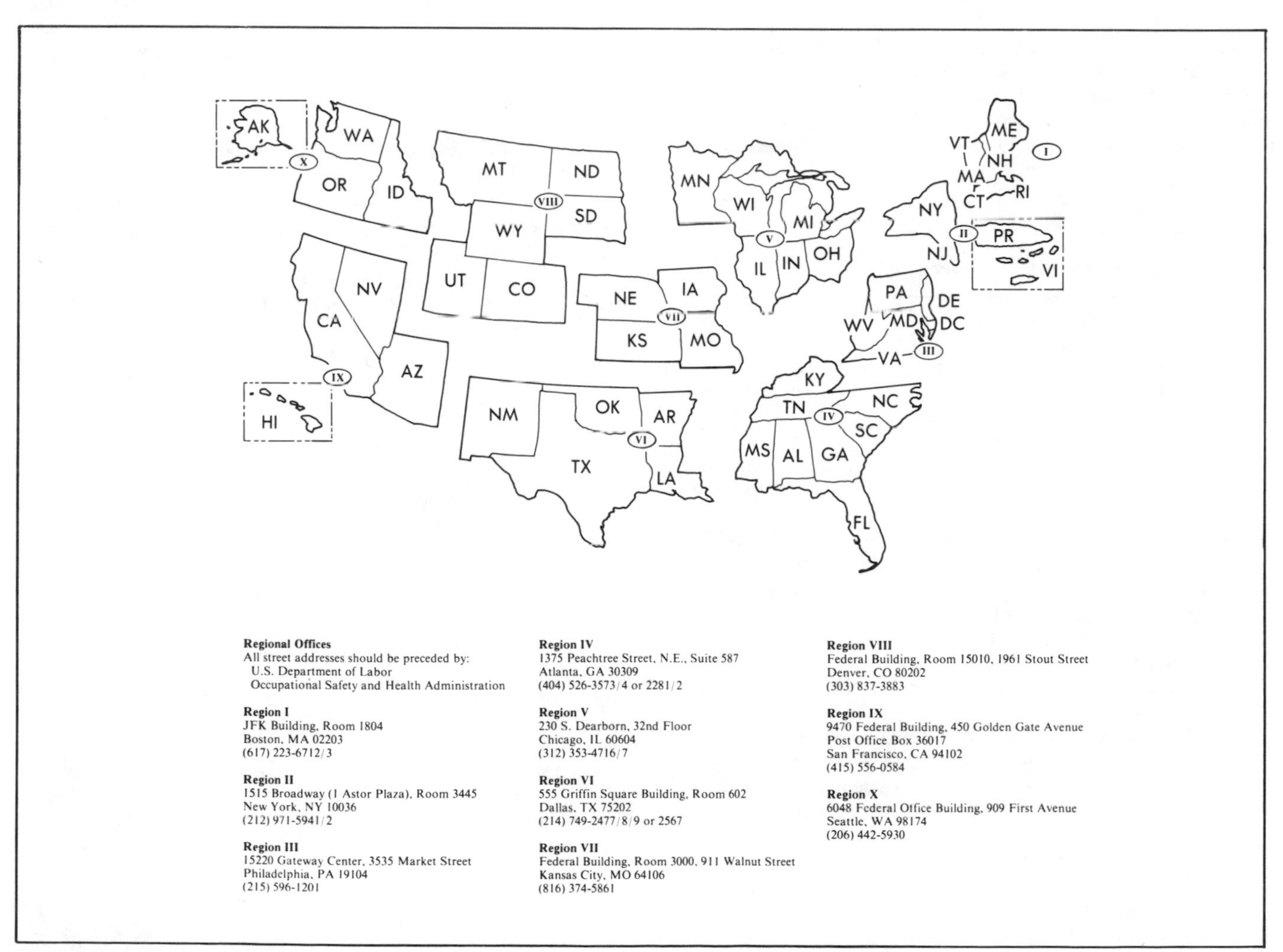

Fig. 12-1 Breakdown of 10 OSHA regions and listing of regional offices. (*Copied from* Health and Safety Guide for Textile Machinery Manufacturers, *NIOSH, U.S. Department of Health, Education, and Welfare (January 1978), pp. 107-108***)**

requires NIOSH to annually publish a listing of all known toxic substances and their toxic thresholds.

NIOSH offers the following technical services to employers through its Division of Technical Services in Cincinnati:

1. On-site hazard evaluation of potentially toxic substances.
2. Technical information concerning health and safety issues in the workplace.
3. Technical assistance in reducing on-the-job injuries by evaluating problem areas and offering corrective techniques.
4. Technical assistance in industrial hygiene and recommendations for engineering controls.
5. Assistance in solving workplace problems in occupational medical and nursing areas.

EMPLOYER AND EMPLOYEE DUTIES UNDER THE OSHAct

The employer's rights and responsibilities covered by the OSHAct are listed in Table 12-1. Employee rights are detailed in Table 12-2. Figure 12-2 shows the OSHA poster that informs employees of their rights under the OSHAct. A full-sized 10 x 16" (250 x 400 mm) version of this poster must be displayed in the workplace where employee notices are normally placed. The OSHA poster is also available in Spanish.

It is the employer's duty to abide by the applicable safety and health standards promulgated by OSHA. The *Code of Federal Regulations* (CFR), published in paperback volumes every year, contains the rules and regulations that have been released. The CFR is divided into 50 different titles, representing broad subject areas of federal regulations, and each title is further divided into parts and subparts covering specific regulatory areas. Title 29, for example, pertains to all regulations promulgated by the Department of Labor. Standards contained in Part 1910 of 29 CFR apply to general industry; Part 1926 relates to construction; Parts 1915 through 1918 are applicable to ship repairing, shipbuilding, shipbreaking, and longshoring; and Part 1928 pertains to agriculture.

OSHA can promulgate emergency temporary standards if employees are exposed to grave hazards. These emergency standards take effect immediately on publication in the Federal Register and serve as a proposed rule for a final standard, which is required to be published within six months. Any person has the right to challenge a standard issued by OSHA by petitioning the U.S. Court of Appeals within 60 days after its release.

It is possible, on some occasions, for employers to be unable to meet the standards that are in place. The OSHAct empowers OSHA to grant variances from the standards when the purposes of the Act are fulfilled. A variance may be temporary or permanent. The variance is granted if the employer demonstrates with actual evidence that alternative measures will provide a place of employment that is as safe and healthful as that provided by the standard. Affected employees must be given notice of the variance application, as well as an opportunity to participate in a hearing.

TABLE 12-1
Employer Rights Under OSHA Law

1. Seek off-site consultation, as needed, through contact with the nearest OSHA office.
2. Receive free on-site consultation service to help identify hazardous conditions and determine corrective measures.
3. Request and receive proper identification of the OSHA compliance officer before an inspection takes place.
4. Realize that OSHA's right to inspect is not unlimited. Request and receive advice from the OSHA compliance officer for the reason for the inspection.
5. Receive an opening and closing session with the OSHA compliance officer, if an inspection takes place.
6. Receive protection of proprietary trade secrets observed by an OSHA compliance officer.
7. Realize that they are not necessarily in violation of the law when a citation is issued. The owner can first request a conference with the Area Director. A Notice of Contest can be filed within 15 working days from receipt of a citation and proposed penalty to challenge the citation.
8. If unable to comply with a standard within the required time, application to OSHA can be made to request extension of abatement period for a short time. This is called a petition for modification of abatement (PMA).
9. Assist in developing safety and health standards through participation with OSHA Standards Advisory Committees, through standards writing organizations, and through public comment and public hearings.
10. Use Small Business Administration loans to help bring establishment into compliance, if applicable.

OSHA HAZARD COMMUNICATION STANDARD

On November 25, 1985, the OSHA hazard communication standard (29 CFR 1910.1200) took effect, requiring chemical manufacturers and importers to assess the hazards of chemicals they sell, affix warning labels to containers they ship, and provide material safety data sheets (MSDS) to their clients in the manufacturing sector. On May 25, 1986, almost all companies had to have written hazardous material handling procedures in place, readily available MSDSs, chemical hazard training and education programs in place, and certain in-plant containers appropriately labeled. Although the initial standard had a limited scope, OSHA was directed by the courts to expand this scope. At the time of this writing, OSHA is preparing additional rulemaking to extend the standard to all industrial classifications in which employees are exposed to chemical hazards. OSHA has established a minimum of 2300 chemical substances as hazardous, but it is possible that many more chemical products also fall under the standard's mandate.

The manufacturing manager who wishes to know how best to comply with the standard should read the entire standard—the preamble, the summary, and the guidelines prepared for compliance officers—or consult professionals in the field. The manager can find OSHA's definitions to key terms within the standard.

By the standard, manufacturers must compile a list of hazardous materials in the workplace and cross reference them to their MSDS by chemical and common names. The written material handling procedures must outline methods to inform workers about the hazards of tasks not performed often and the hazards associated with chemicals in piping systems. Manufacturers must also inform contractors and employees about the hazardous materials in the workplace and their proper handling procedures. This written information must be available to employees and OSHA inspectors on request.

Manufacturers must label bags, boxes, drums, storage tanks, and the like that are used with hazardous chemicals. The label must contain the identity of the hazardous chemical and appropriate warnings. The label should be a link to the more comprehensive material safety data sheet. Manufacturers are also responsible for how well the workplace hazards have been communicated to the employees. This calls for a suitable training and educational program. Training of individual employees should be based on actual and potential exposure within their work areas.

OSHA has compiled official interpretations of the hazard communication standard. This document reprints letters requesting interpretations or clarifications of key provisions of the standard, along with the responses by OSHA. The order number for this document is PB86-220456, and it can be obtained from:

National Technical Information Service
5285 Port Royal Road
Springfield, VA 22161

OSHA RECORDKEEPING REQUIREMENTS

The OSHAct requires most employers to maintain in each establishment records consisting of a log of occupational injuries and illnesses, a supplementary record of each occupational illness and injury, and an annual summary of occupational injuries and illnesses. To relieve small businesses of recordkeeping requirements, OSHA has ruled that an employer who had

TABLE 12-2
Worker Rights Under OSHA

1. An employer cannot punish or discriminate against a worker (by firing, demotion, harassment, or reassignment) for job safety or health activities, such as complaining to the union or OSHA or participating in union or OSHA inspections or conferences.
2. Employees can privately confer and answer questions from an OSHA compliance officer in connection with a workplace inspection.
3. During an OSHA inspection, an authorized workers' representative may be given an opportunity to accompany the compliance officer to aid the inspection. An authorized worker has the right to participate in the opening and closing inspection conferences with pay.
4. If an employer fails to correct a hazard causing a dangerous situation, the worker can contact OSHA to make an inspection. OSHA will not tell the employer who requested an inspection if this is the worker's preference.
5. An employee can notify OSHA or a compliance officer in writing of a potential violation before or during a workplace inspection.
6. A worker can give OSHA information which could affect proposed penalties by OSHA against that employer.
7. If OSHA denies the inspection request of a worker, then the worker must be informed of the reasons in writing by OSHA. The worker may request an OSHA hearing should he object to the OSHA decision.
8. If an OSHA compliance officer fails to cite an employer concerning an alleged violation submitted to him in writing by a worker, then OSHA must furnish the worker with a written statement of the reason for the disposition.
9. Workers have the right to review an OSHA citation against their employer. The employer must post a copy of the citation at the location where the violation took place.
10. Employees can appear to view, or be a witness in, a contested enforcement matter before the Occupational Safety and Health Review Commission.
11. If OSHA fails to take action to rectify a dangerous hazard, and an employee is injured, then that employee has the right to bring action against OSHA to seek appropriate relief.
12. If a worker disagrees with the amount of time OSHA gives his employer to correct a hazard, then the worker can ask for review by the Occupational Safety and Health Review Commission within 15 days of when the citation was issued.
13. Employees can request OSHA to adopt a new standard or to modify or revoke a current standard.
14. Employees may take action for or against any proposed federal standards and may appeal any final OSHA decisions.
15. An employer must inform his employees when he applies for a variance of an OSHA standard.
16. Employees must be given the opportunity to view or take part in a variance hearing and have a right to appeal OSHA's final decision.
17. Workers have the right to all information available in the workplace pertaining to employee protections and obligations under the Act and standards and regulations.
18. Employees involved in hazardous operations have a right to information from the employer regarding toxicity, conditions of exposure, and precautions for safe use of all hazardous materials in the establishment.
19. The employer must inform the worker should the worker be overexposed to any harmful materials, and the worker must be told of the corrective action taking place.
20. If an OSHA compliance officer determines that an alleged imminent danger exists, he must tell the affected workers of the danger and further inform them that court action will be taken if the employer fails to eliminate the danger.
21. If an employee makes a request for access to records covering his exposure to toxic materials or harmful physical agents which require monitoring, then the employer must meet his request.
22. Workers must be given the opportunity to observe the monitoring or measuring of hazardous materials or harmful physical agents, if OSHA standards require monitoring.
23. An employee can make a request in writing to NIOSH for a determination of whether or not a substance used in his workplace is harmful.
24. If a worker requests to review the Log and Summary of Occupational Injuries (OSHA No. 200), then the employer must post or provide it for his perusal.
25. An employee is entitled to a copy of the Notice of Contest and to participate in hearings on contested hearings.

JOB SAFETY & HEALTH PROTECTION

The Occupational Safety and Health Act of 1970 provides job safety and health protection for workers by promoting safe and healthful working conditions throughout the Nation. Requirements of the Act include the following:

Employers

All employers must furnish to employees employment and a place of employment free from recognized hazards that are causing or are likely to cause death or serious harm to employees. Employers must comply with occupational safety and health standards issued under the Act.

Employees

Employees must comply with all occupational safety and health standards, rules, regulations and orders issued under the Act that apply to their own actions and conduct on the job.

The Occupational Safety and Health Administration (OSHA) of the U.S. Department of Labor has the primary responsibility for administering the Act. OSHA issues occupational safety and health standards, and its Compliance Safety and Health Officers conduct jobsite inspections to help ensure compliance with the Act.

Inspection

The Act requires that a representative of the employer and a representative authorized by the employees be given an opportunity to accompany the OSHA inspector for the purpose of aiding the inspection.

Where there is no authorized employee representative, the OSHA Compliance Officer must consult with a reasonable number of employees concerning safety and health conditions in the workplace.

Complaint

Employees or their representatives have the right to file a complaint with the nearest OSHA office requesting an inspection if they believe unsafe or unhealthful conditions exist in their workplace. OSHA will withhold, on request, names of employees complaining.

The Act provides that employees may not be discharged or discriminated against in any way for filing safety and health complaints or for otherwise exercising their rights under the Act.

Employees who believe they have been discriminated against may file a complaint with their nearest OSHA office within 30 days of the alleged discrimination.

Citation

If upon inspection OSHA believes an employer has violated the Act, a citation alleging such violations will be issued to the employer. Each citation will specify a time period within which the alleged violation must be corrected.

The OSHA citation must be prominently displayed at or near the place of alleged violation for three days, or until it is corrected, whichever is later, to warn employees of dangers that may exist there.

Proposed Penalty

The Act provides for mandatory penalties against employers of up to $1,000 for each serious violation and for optional penalties of up to $1,000 for each nonserious violation. Penalties of up to $1,000 per day may be proposed for failure to correct violations within the proposed time period. Also, any employer who willfully or repeatedly violates the Act may be assessed penalties of up to $10,000 for each such violation.

Criminal penalties are also provided for in the Act. Any willful violation resulting in death of an employee, upon conviction, is punishable by a fine of not more than $10,000, or by imprisonment for not more than six months, or by both. Conviction of an employer after a first conviction doubles these maximum penalties.

Voluntary Activity

While providing penalties for violations, the Act also encourages efforts by labor and management, before an OSHA inspection, to reduce workplace hazards voluntarily and to develop and improve safety and health programs in all workplaces and industries. OSHA's Voluntary Protection Programs recognize outstanding efforts of this nature.

Such voluntary action should initially focus on the identification and elimination of hazards that could cause death, injury, or illness to employees and supervisors. There are many public and private organizations that can provide information and assistance in this effort, if requested. Also, your local OSHA office can provide considerable help and advice on solving safety and health problems or can refer you to other sources for help such as training.

Consultation

Free consultative assistance, without citation or penalty, is available to employers, on request, through OSHA supported programs in most State departments of labor or health.

More Information

Additional information and copies of the Act, specific OSHA safety and health standards, and other applicable regulations may be obtained from your employer or from the nearest OSHA Regional Office in the following locations:

Atlanta, Georgia
Boston, Massachusetts
Chicago, Illinois
Dallas, Texas
Denver, Colorado
Kansas City, Missouri
New York, New York
Philadelphia, Pennsylvania
San Francisco, California
Seattle, Washington

Telephone numbers for these offices, and additional area office locations, are listed in the telephone directory under the United States Department of Labor in the United States Government listing.

Washington, D.C.
1985
OSHA 2203

William E. Brock, Secretary of Labor

U.S. Department of Labor
Occupational Safety and Health Administration

Under provisions of Title 29, Code of Federal Regulations, Part 1903.2(a)(1) employers must post this notice (or a facsimile) in a conspicuous place where notices to employees are customarily posted.

Fig. 12-2 A copy or facsimile of the OSHA poster must be posted in place of employment.

10 or less employees at any time during the calendar year need not follow the recordkeeping requirements.

No employer, however, is relieved of the obligation to report any fatalities or multiple hospitalization accidents to OSHA. Within 48 hours after the occurrence of an accident (that is fatal to any employee(s) and/or requires hospital treatment of at least 5 employees due to one single incident), the employer must report the occurrence orally or in writing to the nearest OSHA Area Director. The report must relate the circumstances of the accident and the extent of fatalities and injuries.

The required injury records to be kept are described below:

- *Log and summary of occupational injuries and illnesses.* A log of all recordable occupational injuries and illnesses must be maintained. OSHA Form 200 or any private equivalent may be used (see Fig. 12-3). The employer must enter each injury and illness no later than 6 working days after receiving information that a recordable case has occurred. Logs are to follow a calendar year. A log must be kept in an establishment for 5 years after its calendar year.
- *Supplementary record.* Each employer must have available for inspection a supplementary record for each injury or illness for that establishment. The details of OSHA Form 101 must be completed. Other reports are acceptable alternatives if they contain the information required by OSHA Form 101 (see Fig. 12-4).
- *Annual summary.* An annual summary of occupational injuries and illnesses based on OSHA Form 200 must be compiled for each establishment. The annual summary must be posted in each establishment by February 1 of the following year and remain posted until March 1. Failing to post this summary may result in the issuance of citations and assessment of penalties.

These records must be available for inspection and copying by OSHA compliance officers during any inspection, investigation, or for statistical computations.

WORKPLACE INSPECTION

OSHA inspections are almost always conducted without prior notice. An employer representative and employee representative can accompany the compliance officer during an inspection. These representatives can also participate in the opening and closing conferences. The OSHA priorities for investigations are as follows:

1. *Imminent danger investigation.* An imminent danger is a condition or practice within a place of employment that could reasonably be expected to cause death or serious physical harm before the danger can be removed through the enforcement procedures of the OSHAct. Except in extreme situations, a health hazard is not normally an imminent danger. Imminent danger allegations will ordinarily provoke an inspection within 24 hours of notification.
2. *Catastrophic or fatal accident investigation.* An accident will be investigated if it is of the type that requires OSHA reporting, or draws significant publicity, or is of the type that calls for investigation under an OSHA special program.
3. *Employee complaint investigation.* Investigating priority is given first to those complaints of imminent danger, then to serious situations. The inspection by the compliance officer is normally restricted to the area of complaint. However, the officer is required to cite any other violations observed during the inspection.
4. *High hazard industry inspections.* Industry classifications with high death, injury, and illness incidence rates receive priority for inspection.
5. *Reinspections.* Employers who have been cited for alleged serious violations are normally reinspected to determine if the hazard has been controlled.

The compliance safety and health officer (CSHO) will begin the inspection with an opening conference with top management and an employee representative. The CSHO will state the purpose of the investigation, provide copies of the OSHAct as necessary, outline the scope of the inspection, and answer questions. The officer will then normally inspect the entire operations and will have the appropriate instrumentation for monitoring. Any apparent violation of standards would be verbally indicated and recorded by the CSHO. During the closing conference, the CSHO would advise of all conditions and practices that may be a safety or health violation and indicate the appropriate standards violated. The CSHO will normally advise that citations may be issued for alleged violations, with penalties for each violation. The CSHO would attempt to obtain from the employer a time estimate for correcting the alleged violations. This may affect the time for abatement for the citation.

VIOLATIONS

There are four types of violations:

- *Imminent danger.* The OSHAct defines imminent danger as "any condition or practice in any phase of employment which is such that a danger exists which could reasonably be expected to cause death or serious physical harm immediately." An employer will be seen as having abated an imminent danger if employees are removed from the danger area until the hazard is eliminated or if the hazardous operation is removed altogether. When the employer voluntarily eliminates the danger, an imminent danger procedure is not instituted and no Notice of Imminent Danger is issued. If the employer does not abate the alleged imminent danger, the CSHO will inform the employer and employees that a civil action will be recommended to shut down the operation.
- *Serious violation.* A serious violation is one where there is a strong probability of death or serious harm from a hazard and that the employer knew or should have known about it.
- *Nonserious violation.* A nonserious violation is one where a condition exists that is likely to cause an injury but not a death or serious physical harm, or that the employer did not know of the hazard.
- *DeMinimis violation.* A violation of a standard without having a direct relationship to safety or health. An example is compliance with a more recent issue of a voluntary consensus standard that is referenced in an OSHA standard, resulting in equal or greater protection to the employee.

CITATIONS AND PENALTIES

An employer may receive a written citation from the OSHA Area Director if an inspection reveals a condition that is alleged to be a violation of the standards. The citation would state the standard allegedly violated, specifically how the condition violates that standard, and the time allowed for correction. A penalty may not be proposed when a violation is not a serious one, but proposed penalties always accompany a serious violation.

A nonserious violation may have a penalty of up to $1000 for each violation. A penalty assessment of up to $1000 per violation must be brought against the employer who has received a serious violation. For willful or repeated violations, a penalty of up to $10,000 may be proposed. In cases of imminent danger, penalties may be proposed even if the employer immediately corrects the situation. An employer may receive criminal penalties if a standard is willfully violated and this violation causes death to an employee. Any employer who fails to correct an uncontested citation within the abatement period may receive a proposed penalty of a maximum of $1000 per day for each day that the violation continues beyond the abatement period.

EMPLOYER PREPARATION FOR A CONTESTED CASE

If an employer feels that an OSHA response is unjustified, then the action can be contested, whether it be a proposed penalty, a notice of failure to correct a violation, the correction time allowed, or a combination of any of these. Before contesting a citation, the employer should seek professional counsel

Bureau of Labor Statistics
Log and Summary of Occupational
Injuries and Illnesses

NOTE: **This form is required by Public Law 91-596 and must be kept in the establishment for *5 years*. Failure to maintain and post can result in the issuance of citations and assessment of penalties.** *(See posting requirements on the other side of form.)*	**RECORDABLE CASES:** You are required to record information about every occupational **death**; every nonfatal occupational **illness**; and those nonfatal occupational **injuries** which involve one or more of the following: loss of consciousness, restriction of work or motion, transfer to another job, or medical treatment (other than first aid). *(See definitions on the other side of form.)*

Case or File Number	**Date of Injury or Onset of Illness**	**Employee's Name**	**Occupation**	**Department**	**Description of Injury or Illness**
Enter a nonduplicating number which will facilitate comparisons with supplementary records.	Enter Mo./day.	Enter first name or initial, middle initial, last name.	Enter regular job title, not activity employee was performing when injured or at onset of illness. In the absence of a formal title, enter a brief description of the employee's duties.	Enter department in which the employee is regularly employed or a description of normal workplace to which employee is assigned, even though temporarily working in another department at the time of injury or illness.	Enter a brief description of the injury or illness and indicate the part or parts of body affected. Typical entries for this column might be: Amputation of 1st joint right forefinger; Strain of lower back; Contact dermatitis on both hands; Electrocution--body.
(A)	(B)	(C)	(D)	(E)	(F)
					PREVIOUS PAGE TOTALS →
					TOTALS (Instructions on other side of form.) →

OSHA No. 200

Fig. 12-3 An incomplete OSHA Form 200. Instructions on how to complete this form are given on the reverse of the actual form.

and/or attempt to resolve the matter informally with the OSHA Area Director or Assistant Regional Director. An informal conference may be requested within 15 working days after the receipt of a citation.

Should an employer decide to contest a proposed penalty or an alleged serious violation, the site or organization safety officer is in many cases the key to successful preparation of an OSHA case. Many cases are based on a question of feasibility—of how the workplace should be made safer, how much safer it should be made, at what cost, and by what timetable. Usually, the safety professional has the training and experience to identify the relevant facts, determine their importance in the workplace under question, and evaluate the possible alternatives. Employees familiar with the operation involved may offer significant insight into the details of the process as well as possible solutions.

The first step in preparing a case is to gather the necessary factual data. This includes photographs of the establishment, test data, and employee statements. This information gathering should begin soon after the inspection, should the OSHA compliance officer indicate that a citation is proposed for alleged violations. Do not wait for an official citation to arrive by mail. The employer has 15 working days from receipt to contest the citation, penalty assessment, or abatement period.

Once the facts have been collected, a conference should be held between management, the safety officer, and the attorney

U.S. Department of Labor

For Calendar Year 19 ______ Page ___ of ___

Company Name

Establishment Name

Establishment Address

Form Approved
O.M.B. No. 44R 1453

Extent of and Outcome of INJURY						Type, Extent of, and Outcome of ILLNESS												
Fatalities	Nonfatal Injuries					Type of Illness							Fatalities	Nonfatal Illnesses				
Injury Related	Injuries With Lost Workdays				Injuries Without Lost Workdays	CHECK Only One Column for Each Illness *(See other side of form for terminations or permanent transfers.)*							Illness Related	Illnesses With Lost Workdays				Illnesses Without Lost Workdays
Enter **DATE** of death. Mo./day/yr.	Enter a **CHECK** if injury involves days away from work, or days of restricted work activity or both.	Enter a **CHECK** if injury involves days away from work.	Enter number of **DAYS** *away from work.*	Enter number of **DAYS** of *restricted work activity.*	Enter a **CHECK** if no entry was made in columns 1 or 2 but the injury is recordable as defined above.	Occupational skin diseases or disorders	Dust diseases of the lungs	Respiratory conditions due to toxic agents	Poisoning (systemic effects of toxic materials)	Disorders due to physical agents	Disorders associated with repeated trauma	All other occupational illnesses	Enter **DATE** of death. Mo./day/yr.	Enter a **CHECK** if illness involves days away from work, or days of restricted work activity, or both.	Enter a **CHECK** if illness involves days away from work.	Enter number of **DAYS** *away from work.*	Enter number of **DAYS** of *restricted work activity.*	Enter a **CHECK** if no entry was made in columns 8 or 9.
(1)	(2)	(3)	(4)	(5)	(6)	(7) (a)	(b)	(c)	(d)	(e)	(f)	(g)	(8)	(9)	(10)	(11)	(12)	(13)

INJURIES ILLNESSES

Certification of Annual Summary Totals By ______ Title ______ Date ______

OSHA No. 200 POST ONLY THIS PORTION OF THE LAST PAGE NO LATER THAN FEBRUARY 1.

SAFETY LEGISLATION

OSHA No. 101 Form approved
Case or File No. ____________ OMB No. 44R 1453

Supplementary Record of Occupational Injuries and Illnesses

EMPLOYER

1. Name ______________________________
2. Mail address ______________________________
   (No. and street) (City or town) (State)
3. Location, if different from mail address ______________________________

INJURED OR ILL EMPLOYEE

4. Name ______________________________ Social Security No. ______________
   (First name) (Middle name) (Last name)
5. Home address ______________________________
   (No. and street) (City or town) (State)
6. Age ____________ 7. Sex: Male__________ Female__________ (Check one)
8. Occupation ______________________________
   (Enter regular job title, *not* the specific activity he was performing at time of injury.)
9. Department ______________________________
   (Enter name of department or division in which the injured person is regularly employed, even though he may have been temporarily working in another department at the time of injury.)

THE ACCIDENT OR EXPOSURE TO OCCUPATIONAL ILLNESS

10. Place of accident or exposure ______________________________
    (No. and street) (City or town) (State)
    If accident or exposure occurred on employer's premises, give address of plant or establishment in which it occurred. Do not indicate department or division within the plant or establishment. If accident occurred outside employer's premises at an identifiable address, give that address. If it occurred on a public highway or at any other place which cannot be identified by number and street, please provide place references locating the place of injury as accurately as possible.
11. Was place of accident or exposure on employer's premises? ____________ (Yes or No)
12. What was the employee doing when injured? ______________________________
    (Be specific. If he was using tools or equipment or handling material,
    ____
    name them and tell what he was doing with them.)
    ____
13. How did the accident occur? ______________________________
    (Describe fully the events which resulted in the injury or occupational illness. Tell what
    ____
    happened and how it happened. Name any objects or substances involved and tell how they were involved. Give
    ____
    full details on all factors which led or contributed to the accident. Use separate sheet for additional space.)

OCCUPATIONAL INJURY OR OCCUPATIONAL ILLNESS

14. Describe the injury or illness in detail and indicate the part of body affected. ______________
    (e.g.: amputation of right index finger
    ____
    at second joint; fracture of ribs; lead poisoning; dermatitis of left hand, etc.)
15. Name the object or substance which directly injured the employee. (For example, the machine or thing he struck against or which struck him; the vapor or poison he inhaled or swallowed; the chemical or radiation which irritated his skin; or in cases of strains, hernias, etc., the thing he was lifting, pulling, etc.)
    ____
    ____
16. Date of injury or initial diagnosis of occupational illness ______________________________
    (Date)
17. Did employee die? __________ (Yes or No)

OTHER

18. Name and address of physician ______________________________
19. If hospitalized, name and address of hospital ______________________________
    ____

Date of report ______________ Prepared by ______________________________
Official position ______________________

Fig. 12-4 An incomplete OSHA Form 101. Instructions on how to complete this form are given on the reverse of the actual form.

to discuss possible lines of defense based on fact. All opinions must be based on accurate and thorough knowledge and documented facts. Opinions can also be formulated in part based on any additional outside investigations.

Employees or their authorized representatives are entitled to receive a copy of the Notice of Contest and to participate in the hearing. Their participation in the preparation of the hearing presentation is also encouraged.

Because OSHA hearings are usually scheduled promptly, each party working on behalf of the employer's case should work as a team. The safety professional can obtain pertinent statistical and industrial data. Any hired consultants can visit other facilities for comparison. The attorney can obtain pertinent information as available from OSHA and other sources. This careful investigation can sometimes reveal solutions that may be acceptable to all parties.

After opinions are finalized, the hearing presentation can be prepared. A conference should be held to discuss the presentation of the facts and opinions. Visual aids are important in a hearing. Professional photographs of work areas and equipment, mechanical drawings, illustrations, and graphs can make difficult technical evidence easy to understand. An OSHA case should be prepared as thoroughly and professionally as any other litigation. A poorly prepared case could prevent the full and fair consideration of all the factors affecting the situation and create a bad precedent for an entire industry.

MANAGEMENT POLICY TOWARD SAFETY

Every organization, no matter what the size, should make a written safety policy statement and follow it. It should publicize the policy to everyone in the organization, from all supervisory levels to employees. All top-level managers must agree with the policy, and each level of management must understand who is responsible for safety. The safety policy statement should be succinct and reflect the attitude of management. The effectiveness of any safety policy depends directly on the active support of management. The policy comes from the president of the company or the person with final responsibility, such as the plant manager. The policy will lay out a plan for making a safe workplace. It should require the keeping of accident investigation records and direct that a training program will be in place for employees and supervisors. Specific goals should be set forth, and personnel responsible for safety matters must have access to company and government standards. An example of a corporation's safety policy is summarized in Fig. 12-5.

The OSHAct spells out minimum standards for safety and health and does not give details of safe treatment of all workplace hazards. Hazardous conditions or practices that are not covered by specific OSHA standards are covered by the general-duty clause of the Act. This clause says: "Each employer shall furnish to each of his employees employment and a place of employment which are free from recognized hazards that are causing or are likely to cause death or serious physical harm to his employees." In all cases, recognized hazards need to be eliminated in the workplace. If the intent of the law is being met, a variance from the applicable standard can be requested from OSHA. Active support from management means enforcing the policy, recognizing good safety performance, continuously reviewing safety performance, and setting a good example. A safety program must always start at the top management level, and each level must accept their responsibility for their role in the company's safety objectives. Employees will usually commit to safety at the minimum level displayed by their supervisors.

ECONOMIC REASONS FOR SAFETY PROGRAM

The employer should give the health and safety program equal importance with other business functions such as manufacturing, sales, marketing, cost control, and customer service. If the health and safety program within the organization is effective in the prevention of accidents and illnesses, it can save money for the company.

Accidents cost an organization money in insured and uninsured costs. The *insured costs* (also called "direct" costs) are covered by the employer's liability insurance premium and include medical treatment costs, cost of lost work time for the injured person, and cost of payment to the injured person. The *uninsured* (or *hidden*) costs (also called "indirect" costs) of accidents can be outlined as follows:

1. Cost of an accident investigation.
2. Cost of replacement labor.
3. Cost of business interruption.
4. Cost of damage repair.
5. Cost of damaged product.
6. Miscellaneous cost of photographs, transport, administrative overhead, and so on.

An effective health and safety program also helps the organization save money in addition to reducing pain and suffering.

MANAGEMENT STRATEGIES FOR SAFETY

Regardless of the size of a manufacturing organization, the integration of the safety ethic throughout the business is essential if the safety function is to be effective. If safety is a separate function, its effectiveness is doomed. Safety is not a concern for only the manufacturing personnel; it is a necessity for the engineering, marketing, accounting, shipping, industrial relations, and customer service groups. Participation within the safety function must be obtained across plant political or organizational lines.

Management within a moderately sized organization or larger must admit the necessity for a safety staff function. Safety and health programs are becoming increasingly complicated. Safety and health standards are becoming voluminous and difficult to interpret. No organization can afford to rely on existing personnel within the site and trained in other disciplines for competent and expert safety advice. A safety staff group should be present in the large company, and a staff position should be present in the small company. In a small company, a competent individual may be assigned the responsibility for safety in addition to other duties. Even with the presence of staff safety professionals, the individual who dele-

MANAGEMENT POLICY

The Du Pont Safety Philosphy:
"We will not make, handle, use, sell, transport or dispose of a product unless we can do so safely and in an environmentally sound manner."

Ten Principles of Safety:

1. All injuries and occupational illnesses can be prevented. At Du Pont, we believe that this is a realistic goal and not just a theoretical objective. Our safety performance proves that this goal is achievable, as we have plants with over 1000 employees that have operated for over ten years without a lost-time injury.

2. Management is directly responsible for preventing injuries and illnesses, with each level accountable to the one above and responsible for the level below. This includes all levels, from the chairman, who is also the chief safety officer, through to the first line supervisor.

3. Safety is a condition of employment; each employee must assume responsibility for working safely. In Du Pont, safety is as important as production, quality, and cost control.

4. Training is an essential element for safe workplaces. Safety awareness does not come naturally—management must teach, motivate and sustain employee safety knowledge to eliminate injuries. This includes establishing procedures and safety performance standards for each job or function.

5. Safety audits must be conducted. Management must audit performance in the workplace to assess the effectiveness of facilities and programs, and to detect areas for improvement.

6. All deficiencies must be corrected promptly, either through modifying facilities, changing procedures, bettering employee training or disciplining constructively and consistently. Follow-up audits are used to verify effectiveness.

7. It is essential to investigate all unsafe practices and incidents with injury potential, as well as all injuries.

8. Safety off the job is just as important as safety on the job. Du Pont was a pioneer in tracking the off-the-job safety of employees beginning in 1953. As a result, the Company initiated programs to dramatically improve off-the-job performance.

9. It's good business to prevent illnesses and injuries. Serious illnesses and injuries involve tremendous costs—direct and indirect. The highest cost is human suffering.

10. People are the most critical element in the success of a safety and health program. Management responsibility must be complemented by employees' suggestions and their active involvement in keeping workplaces safe.

Fig. 12-5 Example of corporate directive stating the safety policies and management commitment to employee safety. (*E.I. DuPont de Nemours & Co., Inc.*)

gates responsibility is still accountable for the results. Although safety functions are performed by the site safety group, the ultimate authority and responsibility for safety must rest with their immediate supervisors.

A prominent function of the safety staff is to identify safety and health hazards and to recommend countermeasures. The identification of hazards with respect to existing regulations is to be uncovered by staff through plant inspections. The identification of hazards unique to system idiosyncrasies is the responsibility of the line manager. The staff personnel cannot be as intimately familiar with the various operations as the line manager. Staff safety personnel can implement special countermeasures beyond the normal operational controls for a hazard, once they are made aware of them.

RESPONSIBILITIES OF MANAGEMENT

When a line supervisor is made responsible for safety, the end result can only be for the good. If the supervisor has knowingly contributed to an accident, he or she should share in responsibility for it. If a supervisor is to be held accountable for safety, he or she must have the authority to make procedural or physical changes to improve safety. If upper management does not back up the supervisor, employees will blame management for accidents.

There should be an unbroken chain of accountability for safety, from the line supervisor to the president or owner of a company. A manager at each level should be able to evaluate the safety efforts of each subordinate manager. A good system can operate in accordance with the following assignment of safety responsibility:

- The *supervisor* is accountable for the safety of all subordinates and for the safe condition of the work area under his or her responsibility.
- The *department head* is responsible for the establishment of good housekeeping practices in the department and for the safety training and development of each supervisor.
- The *superintendent's* safety commission is to see that the corporation's safety program is administered at the plant level. He or she is responsible for all safety activities in his or her administration.
- The *plant manager*, who is generally accountable to a vice president at the corporate level, is vested with the responsibility for safety performance at the plant. The plant manager should see that members of the staff complete safety training, which will continue to develop safety awareness. The plant manager must establish the corporate safety program within the plant and, if necessary, clearly spell out the specific safety duties of subordinates.

HEALTH AND SAFETY PROGRAM FUNDAMENTALS

In addition to humanitarian concern and a natural desire to obey the law, there are purely economic incentives for organizing and maintaining safety and health programs in any organization. Escalating worker's compensation costs and potential civil litigation are two significant reasons. As previously stated, the safety program must have the backing and commitment of management at all levels. The boss's attitude toward job safety and health will be reflected by the employees.

Two main features of any safety program are accident and illness prevention and controlling potential losses. The keys to minimizing accidents are to have engineered hazard-eliminating features on plant machinery and to have employee training programs with instruction of general work practices. Scheduled periodic inspections can identify and lead to correction of unsafe conditions and work practices. Communicating safety and health information is effectively done through joint labor-management safety and health committees. Should an accident occur, an accident response procedure must be in place to discover the cause, take corrective action to prevent future injuries, and break the cascading nature of accident-related costs. Maintaining good relations with injured workers is essential to preventing increased losses.

THE SAFETY PROFESSIONAL

Administration of the health and safety program at the plant level is usually done by a safety professional, personnel manager, or line supervisor. If the plant or company is of substantial size, top management will place the safety program administration in the hands of a safety coordinator/manager. Since the passage of the OSHAct, the trend is to employ a full-time safety professional because of the high degree of employee involvement in developing safety policies and because of the constant changes in plant machinery and processes. The safety manager is normally accountable to the plant manager or other appointed representative and is responsible for the following:

- Executing the details of a program for safety and industrial hygiene within the plant.
- Directing the safety inspections within the plant.
- Developing the details of and directing safety training programs.

The safety professional should be effective in reducing the frequency and severity of accidents and costs stemming from accidents, and in planning and implementing an efficient safety program on the plant site. Table 12-3 lists the typical duties of a safety professional. This professional provides advice and guidance to plant management about safety and industrial hygiene appropriate to plant installations, processes, and procedures. He or she provides guidance and training for employees about preventing accidents in specific jobs and work areas and also maintains relationships with professional safety and organizational groups.

SAFETY AND HEALTH COMMITTEES

With the support of management and the involvement of people across the organization, safety and health committees can greatly aid accident prevention. A safety and health committee is a group of regular employees that advises plant employees on safety matters by site educating, investigating, evaluating, and possibly monitoring. A part of the basic charter of every safety committee is also to create and maintain interest in safety and health, thereby reducing accidents. Committees that represent various constituencies within the establishment (or joint committees) indicate by their very nature that safety commitment is a shared responsibility. Every employee can potentially make positive contributions to the company's safety program. The committee serves as a forum for discussing incentive ideas, changing regulations, and identifying possible site hazards. Employees can communicate problems and concerns directly to management. The knowledge and experience of many individuals can be combined to translate ideas into action and to accomplish the objectives of maintaining a hazard-free workplace.

Likewise, all site employees can motivate each other to avoid accidents when away from work by their active participation in off-the-job safety programs. Off-the-job safety does not have to be a separate plant safety program, but rather can be an extension of a company's on-the-job safety program. Companies have a moral responsibility to try to prevent injuries away from the job, because all injuries are a waste of a valuable resource—people. Maintaining level operating costs is another reason for off-the-job considerations in the overall safety program. Production schedules can be upset severely if employees are injured away from work. Absences may result in lost sales, late deliveries, and loss of customers. Employers generally bear the cost of

TABLE 12-3
Duties and Responsibilities of the Safety Professional

1. Implementing and administering the company's safety policy statement in the plant.
2. Promulgating safety and industrial hygiene procedures and standards and enforcing their use in the plant.
3. Constantly auditing the site facilities and equipment to catch unsafe conditions or acts which could cause damage or injury.
4. Developing and presenting safety training programs to benefit supervisors in the charge of their safety responsibilities.
5. Seeing that unsafe acts and conditions are corrected by follow-up with appropriate supervisory personnel.
6. Conducting regular (monthly) meetings of the executive safety committee and attending or giving regular (monthly) site safety meetings to assure a high level of safety awareness.
7. Keeping statistical records of accidents, near accidents, frequency of injuries, and costs of accidents and injuries.
8. Writing and editing already written safety bulletins and notices and posting or distributing these documents.
9. Obtaining appropriate safety publications for plant personnel and presenting audiovisual material to help minimize accidents.
10. Testing or outlining test programs for safety devices in the workplace.
11. Maintaining a program for dispensing safety shoes and safety glasses and other protective equipment such as respirators.

off-the-job accidents by paying medical costs and/or wages to absent workers.

The employer has to depend more on education and behavior modification or motivation to keep an employee safe at home. Because new and fresh approaches to the safety campaign are important to keep employees at a high level of safety consciousness, a committee is needed to brainstorm and generate different approaches to accomplish its goal. The safety committee's approach in motivating employees should be positive, not negative. Any activity can be performed safely if it is thought through, if safe procedures are followed, and if the proper equipment is used.

Safety committees can take advantage of national programs. Seasonal activities or topics like Fire Prevention Week, Poison Prevention Week, and Safe Boating Week can be discussed. The committee can also urge workers to enroll in local swimming and lifesaving classes, safe hunting classes, first aid courses, driver improvement courses, and other such programs in the community. The safety committee can interest employees in the off-the-job program by education through on-the-job programs.

SAFETY TRAINING

It is management's duty to see that safety training programs are developed for all employees—from the director level to the shop-floor worker. There are numerous types and aspects of employee training programs. Employees should:

- Understand the need for constant attention to their work environment.
- Know how hazardous substances encountered affect the body.
- Be instructed in proper procedures for handling, storing, mixing, and disposing of hazardous substances.
- Be trained in the proper use of protective equipment.
- Be aware of plant emergency and evacuation procedures.
- Understand the importance of good housekeeping practices. (Good housekeeping can reduce accidents and fire hazards.)
- Be trained in the use of plant equipment.
- Be instructed in the use of portable fire extinguishers.
- Be instructed in the safe technique for lifting.
- Be thoroughly aware of general plant safety policies.

NEW EMPLOYEE TRAINING

An effective new employee orientation program will include safety training of the general and overall safety policies. Before the employee actually begins work, while he or she is forming initial attitudes about the company, a good safety indoctrination will lay a foundation for initial and continued safe performance. To have a good start in safety training, management needs to stress the following:

- Management is very interested in keeping every employee in the workplace safe.
- It is possible to prevent accidents.
- Unsafe conditions should be immediately reported.
- If unsafe procedures or equipment are recognized, then management is willing to make immediate corrections.
- No employee should perform a job without the proper training and a clear understanding of any potential hazard.
- All injuries, no matter how small, should be reported at once.

The plant personnel department often gives the preliminary instruction to individuals or groups. Sometimes the safety professional or management executive gives instruction. Presentation by management adds force and interest. The initial instruction should be prepared and presented carefully to have the maximum impact on employees. Verbal instruction should be given in an earnest and helpful fashion. The important role that the supervisor plays in the safety program should be emphasized. So that there is no omission or contradiction with the information given in the initial safety orientation, the supervisor should know what has been discussed in the session.

Once the employee gets to his or her own work group, the immediate supervisor should reinforce the safety orientation with additional instruction pertinent to the job. The supervisor should explain safety regulations pertaining to the particular area of the plant and furnish the worker with personal protective equipment for the job. The supervisor can also provide for training in safe work procedures. Here is where scheduling for training courses on equipment operation can be arranged. This type of treatment is the least that a new employee should receive if he or she is to understand the importance of safety on the job and to appreciate the company's attitude about safety.

METHODS OF SAFETY TRAINING

Both formal and informal training methods are needed to establish and maintain a high level of safety awareness in employees. Formal methods are used during orientation and instruction for specialized tasks. Informal training should take place regularly on the job in one of the following forms:

- One-to-one contact with supervisors to go over the fine points of daily tasks relevant to work.
- A group of individuals and the supervisor informally discussing a matter of common interest, such as housekeeping, reporting accidents, protective clothing, the need for guarding, and so on.

Formal training methods have the best results when they consist of the following main steps:

- Indicate objective of the training.
- Brief the trainee on what will be covered.
- Present the training session using visual aids when appropriate.
- Review what was presented.
- Provide trainee follow-up. Follow-up is as important as the initial training and probably should be done by the supervisor to reinforce the training message.

Visual training methods with the verbal instruction can be effective because a person remembers:

- 10% of what is read.
- 20% of what is heard.
- 30% of what is seen.
- 50% of what is seen and heard.
- 70% of what is seen and spoken.
- 90% of what is said as he or she does a thing.

From this description, it is apparent that involvement sessions are of the most benefit. The best training methods in order of effectiveness to persuade employees to follow safe practices are:

1. Role playing.
2. Films.

3. Posters and charts.
4. Verbal scare tactics (not normally an accepted training practice in the safety profession).
5. Discussion.

A safety film can be effective by presenting a carefully planned message in a consistent manner. Its safety message will be told the exact same way to every new employee.

OSHA TRAINING REQUIREMENTS

The requirements of the Occupational Safety and Health Act place great importance on employee training. An index of OSHA training requirements is given in Table 12-4.[1] It lists the major parts of the OSHA regulations from the *Code of Federal Regulations* that are applicable to most industrial manufacturers.

OSHA has published voluntary training guidelines that follow a model consisting of:[2]

1. Determining if training is needed.
2. Identifying training needs.
3. Identifying goals and objectives.
4. Developing learning requirements.
5. Conducting the training.
6. Evaluating program effectiveness.
7. Improving the program.

Copies of these guidelines can be obtained from:

OSHA
Office of Training and Education
1555 Time Drive
Des Plaines, IL 60018

TABLE 12-4
Partial Listing of OSHA Training Requirements

Following is a partial list of the major parts of the OSHA training regulations that suit most industrial manufacturers

Part	Subpart	Section	Hazard
1910.179	N	(m)(3)(ix)	Cranes and derricks
1910.180	N	(h)(3)(xii)	
1926.803	S	(a)(2)	Decompression or compression
1926.803	S	(b)(10)(xii)	
1926.803	S	(e)(1)	
1910.109	H	(g)(3)(iii)(a)	Employee responsibility
1926.609	U	(a)	
1910.217	O	(f)(2)	Equipment operations
1926.20	C	(b)(4)	
1926.53	D	(b)	
1926.54	D	(a)	
1910.252	Q	(c)(6)	
1916.32	D	(e)	Fire protection
1917.32	D	(b)	
1926.150	F	(a)(5)	
1926.155	F	(e)	
1926.351	J	(d)(1) through (5)	
1926.901	U	(c)	
1910.218	O	(a)(2)(i) through (iv)	Forging
1910.109	H	(d)(3)(i) and (iii)	Gases, fuel, toxic materials, explosives
1910.111	H	(b)(13)(ii)	
1910.266	R	(c)(5)(i) through (xi)	
1910.106	H	(b)(5)(vi)(v)(3)	
1916.35	D	(d)(1) through (6)	
1926.21	C	(a) and (b)(2) through (6)	
1926.350	J	(d)(1) through (6)	
1926.21	C	(a)	General
1915.57	F	(d)	Hazardous material
1916.57	F	(d)	
1917.57	F	(d)	
1910.94	G	(d)(9)(i) and (vi)	Medical and first aid
1910.151	K	(a) and (b)	
1915.58	K	(a)	

(continued)

TABLE 12-4—*Continued*

Part	Subpart	Section	Hazard
1917.58	F	(a)	
1926.50	D	(c)	
1910.94	G	(d)(11)(v)	Personal protective equipment
1910.134	I	(a)(3)	
1910.134	I	(b)(1),(2), and (3)	
1910.134	I	(e)(2),(3), and (5)	
1910.134	I	(e)(5)(i)	
1910.161	K	(a)(2)	
1915.82	I	(a)(4)	
1915.82	I	(b)(4)	
1916.57	F	(f)	
1916.58	F	(a)	
1916.82	I	(a)(4)	
1916.82	I	(b)(4)	
1917.57	F	(f)	
1918.102	J	(a)(4)	
1926.21	C	(b)(2) through (6)	
1926.103	E	(c)(1)	
1926.800	S	(e)(xii)	
1910.217	O	(e)(3)	Power press
1916.37	D	(b)	Radioactive material
1910.96	G	(f)(3)(viii)	Signs—danger, warning, instruction
1910.145	J	(c)(1)(ii)	
1910.145	J	(c)(2)(ii)	
1910.145	J	(c)(3)	
1910.264	R	(d)(1)(v)	
1910.252	Q	(b)(1)(iii)	Welding
1910.252	Q	(b)(1)(iii)	
1915.35	D	(d)(1) through (6)	
1915.36	D	(d)(1) through (4)	
1916.35	D	(d)(1) through (6)	
1917.35	D	(d)(1) through (6)	
1917.36	D	(d)(1) through (4)	

HAZARD ANALYSIS AND ACCIDENT PREVENTION PROGRAMS

Accidents are caused by unsafe acts, unsafe conditions, or a combination of both. On occasion, an accident is caused by a series of errors. This should be considered when an accident is being scrutinized for its cause. An employee's mental state should also be considered. Ignorance and improper attitudes frequently contribute to accidents. In addition, management must recognize that internal faults, such as weak leadership, failure to enforce rules, and poor maintenance, may also be contributing factors when the unexpected happens. Because management defines the system in which production takes place, they must bear the major responsibility for accidents.

Unsafe acts, in more cases than unsafe conditions, are the cause of accidents. An "unsafe act" is a human action that deviates from a standard or written job procedure, safety rule or regulation, instruction, or job safety analysis. Most times, unsafe acts are done because the worker did not receive proper instruction, lacked attention to the job, was looking for a shortcut, or was poorly coordinated, distracted, or under stress. An "unsafe condition" is an arrangement that increases the odds of an employee having an accident. The actions of employees themselves are the greatest contributor to unsafe conditions.

"Unsafe acts and conditions" are often referred to as "hazardous acts and conditions," or simply "hazards." There are three steps in the proper management of hazards: (1) recognition, (2) evaluation, and (3) control. Supervision at all levels has a responsibility to eliminate unsafe acts and conditions, or hazards. This means management has a responsibility to recognize, evaluate, and control hazards that could cause injury or damage.

HAZARD ANALYSIS

The first step toward controlling a hazard is to recognize the hazard. Management must work with the site safety professional to determine locations with a high accident potential as well as to identify severe hazards with a low occurrence likelihood. When specific hazards are being evaluated, both the probability of accident occurrence and the severity of injury or property damage are factored in. Once this is done, it becomes possible to rate hazards across a site to determine the priority of corrective action. This is also useful in cost-effectiveness evaluations. As hazards are being evaluated, countermeasures should be considered. Immediate action, unencumbered by procedure, should take place to eliminate hazards when the remedies are known and cost is not a factor.

An overall look at the setting under consideration is sufficient for a general hazard analysis. The purposes of this approach are to discover moderate potential hazards and to identify critical areas in need of detailed analysis. When performing a general area analysis, the determination of corrective priority order must be qualitative and possibly subjective. Numerical analysis is not necessary at this point.

Each hazard should be ranked according to its probability to cause an accident and the severity of that accident. The ranking here is relative. Some hazards identified might be placed into a category for further analysis. A useful practice for early sorting is to place catastrophic severities together, followed by critical, marginal, and nuisance hazards, respectively. Then, with each item, indicate the probability of occurrence-considerable, probable, or unlikely. Rate the correction of these hazards next by cost. This rating will be very useful when conducting a cost-effectiveness evaluation for hazard elimination.

COST EFFECTIVENESS WITH GENERAL HAZARD EVALUATION

The main purpose of cost-effectiveness evaluations is to stimulate rational thinking and decision making on the part of management. There are numerous procedures for performing cost-effectiveness evaluations. One will be discussed here. Whatever the technique used, it is essential that several alternatives for resolving each hazard be suggested and that reasonable constraints be placed on the problem. The selection of alternatives that will provide the greatest degree of safety in the face of recognized hazards within given budget constraints will be given careful consideration. With careful consideration, several countermeasures can be employed and cost effectiveness becomes useful.

The matrix display is a form of evaluating general hazards. The purpose of this tabular presentation of alternatives is to provide cost-effectiveness information in a concise, logical format. The matrix format arranges information so that consideration can be directly given to important problems. A matrix display of alternatives for safeguarding a power press is shown in Fig. 12-6.

With the matrix format, specific hazards are determined by the general analysis and are listed at the top of the page that will contain the matrix. The alternate countermeasures are placed in order of effectiveness vertically downward on the left margin. The cost of each alternative is listed just to the right of the alternative. In the spaces that are created by each combination of the matrix, a symbol can represent the effectiveness of the countermeasure. The symbols listed below might be used:

— Elimination of the hazard.
R Reduction of the hazard.
X No change in the hazard.
I Increase in the hazard.

The purpose of this arrangement is to set up a simple way to find those controls that are not cost effective. A control at the top of the matrix with a low cost would be a preferable alternative. An expensive control at the bottom of the matrix would be an inferior selection. Because the matrix display shows several controls that can be simultaneously viewed, it helps the reasoning process of the decision maker.

WORKPLACE HAZARD ANALYSIS

To identify occupational health hazards with a specific task, a job health hazard analysis can be made. A job health hazard analysis survey, like the one shown in Fig. 12-7, can be used. The survey should list the substances used, the number of employees at risk, entry routes, hazardous byproducts, and methods of controlling exposure. Activities, such as maintenance or service operations, should be examined when the survey is done to determine other health hazard potential. After completing the survey, evaluate all exposure substances for hazard potential. Determine if the existing controls are ade-

		Specific hazards at working area			
Countermeasures	Cost	Flying objects	Pinch points	Escaping lube sprays	Electrostatic discharges
1. Lexan guard	$30	—	—	—	—
2. Solid metal guard	$10	—	—	—	R
3. Expanded metal guard	$20	R	—	R	R
4. Capacitance field	$500	X	—	R	I
5. Light curtain	$150	X	—	X	X

—Elimination of hazard
R Reduction of hazard
X No change in hazard
I Increase in hazard

Fig. 12-6 Power press cost effectiveness matrix display.

quate and, if not, then any required controls should be provided.

DETAILED HAZARD ANALYSIS

There are formal analytical methods for the evaluation of hazards. Generally, these methods are only employed where the hazard(s) is so severe that a fatality or multiple injuries are possible. There may be value in using these methodologies for less hazardous situations, however.

Job Hazard Analysis Survey for Spray Painting Operation

Operation:____________ Page:______ Date:______

Number of employees	Job title	Exposure substance	Form[1]	Route of entry[2]	Control[3]
4	Spray painter	No. 4 red primer	M	I	LV
			V	I	LV
			L	S	G (rubber)
				S	O (apron)
		Xylene	L	S	G (rubber)
					O (apron)
			V	I	LV
			M	I	LV

1. Form: D = dust, L = liquid, V = vapor, G = gas, F = fume, M = mist.
2. Route of entry: S = skin, I = inhalation.
3. Control: LV = local ventilation, GV = general ventilation, R = respirator (type), G = gloves (type), F = face protection, O = other protection (type).

Fig. 12-7 Job hazard analysis survey for spray painting operation, (*Copied from* Health and Safety Guide for Textile Machinery Manufacturers. *NIOSH, U.S. Department of Health, Education, and Welfare (January 1978), p. 11*)

One inductive method of consideration is the failure mode and effect analysis (FMEA). In this form of analysis, the malfunction of each component in the system is considered, one at a time, including the mode of failure. The effect of each failure is traced through the system for the overall effect on the system. From these effects, critical failures can be detected and how component parts contribute to the failure determined. Further safeguards can be appropriately placed within the system to minimize the likelihood of undesirable occurrences. A failure and effect analysis can be performed by completing a form similar to the one shown in Fig. 12-8.

One deductive method that can indicate how failures can occur is fault tree analysis (FTA). With the FTA, an undesired event is selected, all of the happenings that contribute to this event determined, and a diagram depicting the event is made. This diagram is in the form of a tree. On this tree diagram, probabilities for the occurrence of individual events are placed at the symbols representing the event. The fault tree is a graphic model of the various sequential and parallel combinations of system faults that could result in an undesired failure of the system, causing harm. This form of analysis is rigorous but thorough. An example of a fault tree is shown in Fig. 12-9.

Hazard analysis, whether general or detailed, should be jointly conducted by personnel representing various viewpoints to be fully effective and reliable. Management should see to it that this is accomplished. The insights provided by employees of differing areas of expertise need to be recorded. This input may be extremely valuable when selecting alternatives to minimize dangers.

CONTROL OF HAZARDS

A continued program of hazard control is a necessary part of the management process. Management commitment toward the control of hazards is expressed through evaluations, procedures, and audits. A hazard control program, with predefined specifics, assures that sound design and operating procedures, operator training, test programs, inspections, and hazard communications are addressed. The hazard control program must emphasize shared responsibilities among departments of a facility. This is possible only if supervisors cooperate.

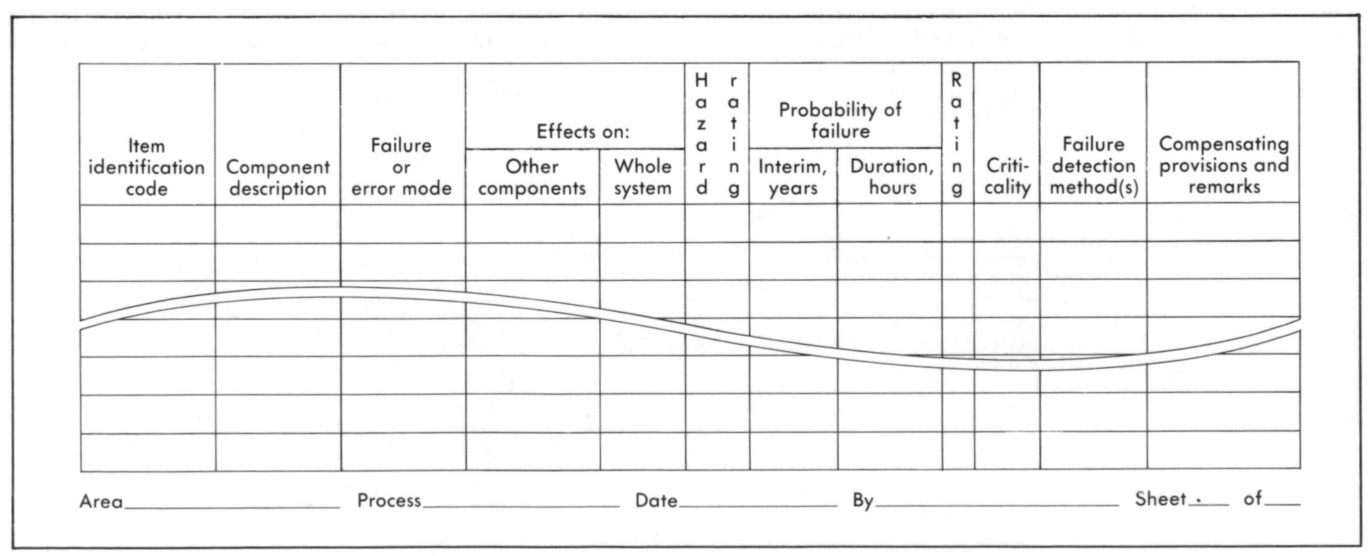

Item identification code	Component description	Failure or error mode	Effects on: Other components	Effects on: Whole system	Hazard rating	Probability of failure: Interim, years	Probability of failure: Duration, hours	Rating	Criticality	Failure detection method(s)	Compensating provisions and remarks

Area__________ Process__________ Date__________ By__________ Sheet____ of____

Fig. 12-8 Sample copy of a failure mode and effect tabulation sheet.

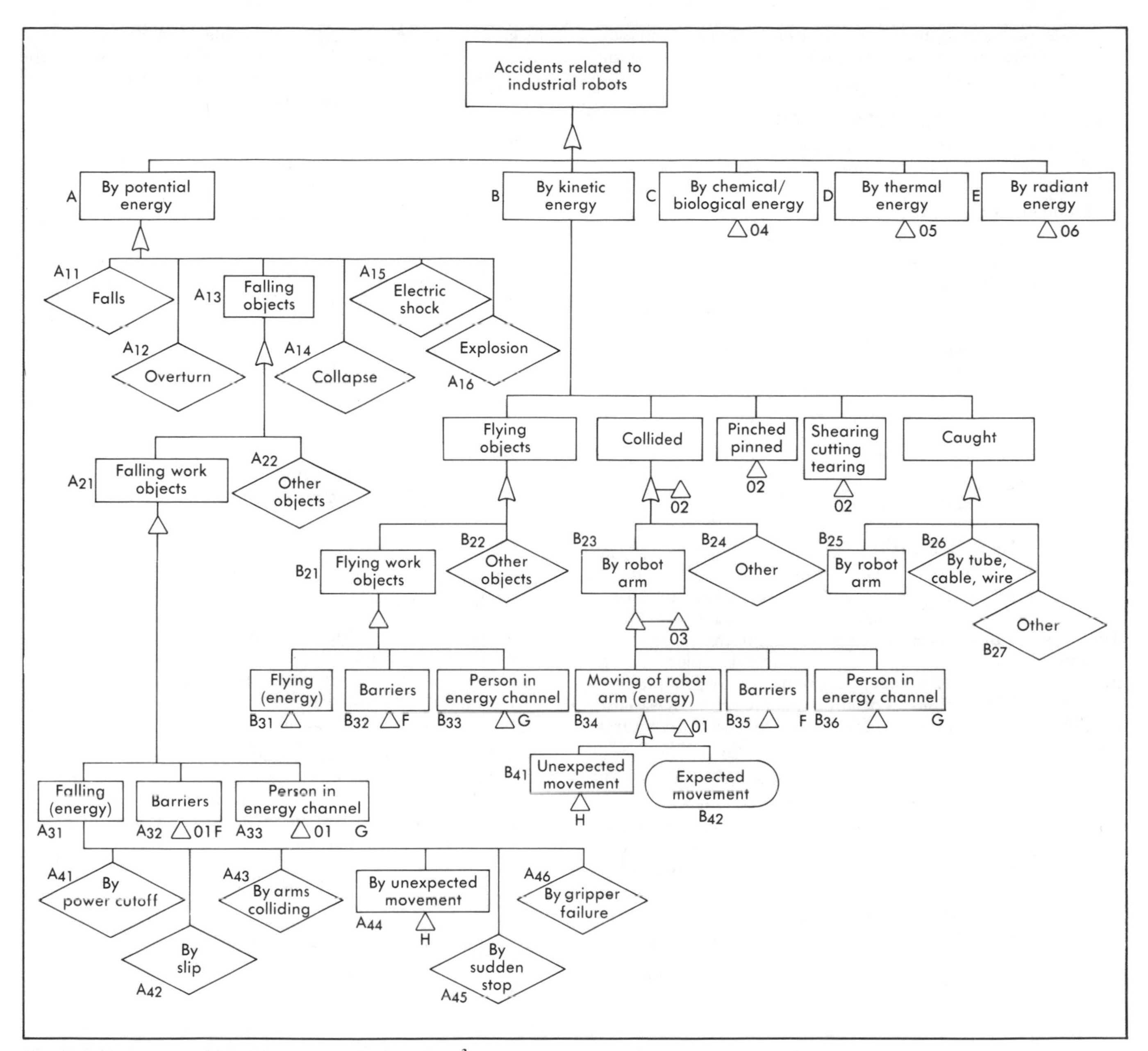

Fig. 12-9 Fault tree analysis of hazards created by robots.[3]

Accident and hazard control requires a team effort by employees and management. Groups within the same facility can work together to successfully accomplish this end.

The following points should be considered:

- The manufacturing engineering department should design facilities to control hazards. It should aid other departments with hazard identification and analysis. Machinery and equipment specifications should include safe control circuitry, safeguards, noise control, and ergonomic and maintenance considerations. It is easier and less expensive to design for safety than to add it on at a later date.
- Through process changes and job hazard analysis and control, the manufacturing department can reduce hazards.
- The quality control department can work to improve product safety and quality, as well as the safety of employees, by conducting studies to determine if an alternate design, material, or manufacturing method would result in improvements.
- The maintenance department can perform planned preventive maintenance on systems and machinery to prevent health and safety hazards.
- The purchasing department can make sure that incoming process constituents and equipment meet established safety and health standards.
- The human resources department can administer health

and safety programs by assisting engineering and manufacturing process personnel with safety and health standards and regulations.

HAZARD CONTROL METHODS

Various means can be used to reduce or prevent employee exposure to hazards. Some of these methods are listed below, in order of preference. These methods can be used singly or in combination:

1. Eliminate the source of the hazards.
2. Substitute a less hazardous equivalent.
3. Reduce the hazards at the source.
4. Remove the employee from the hazards (i.e., automating the process).
5. Isolate the hazards (i.e., by enclosure).
6. Dilute the hazards (i.e., ventilation).
7. Reduce employee's exposure to hazard by administrative control such as employee rotation.
8. Use personal protective equipment.
9. Train employees in the proper methods used for hazard avoidance.
10. Practice good housekeeping.

SAFETY INSPECTIONS

Inspection, a vital management tool, is an essential part of hazard control. Inspection is the organization's monitoring activity to locate and report hazards that may cause accidents. The purposes of safety inspections are: (1) to detect potential hazards for correction before an accident takes place, (2) to increase operational efficiency, and (3) to improve profitability.

For safety inspections to be effective, corrective steps must take place when hazards are located. The following are several types of safety inspections:

- *Ongoing inspection* is conducted as part of the job responsibilities of supervisors and maintenance personnel. Also, continuous inspection by all employees of personal protective equipment is important. Continuous inspection is a feature of successful health and safety programs. Supervisors can do well in spotting hazards by having another supervisor audit the area. For example, inspections at the start of the shift and immediately after lunch are particularly effective.
- *Planned periodic inspection* is an inspection that is conducted at scheduled intervals, like the monthly inspection. In facilities where the accident potential is high, formal inspections should be more frequent. The advantage of planned periodic inspection is that it covers a preselected area and is scheduled regularly to correct hazards either before or shortly after they occur.
- *Intermittent inspections* are made at irregular intervals. These unscheduled safety inspections are done as the need arises. They can be made when a new piece of equipment is installed, when a process change is instituted, or when a health hazard is suspected.
- A *general inspection* is a planned inspection of an area that is not inspected regularly, like overhead areas, fencing, and the parking lot.

For management to prepare an effective safety inspection plan, there must be:

1. A thorough knowledge of the site.
2. Knowledge of applicable codes, regulations, and standards.
3. A list of potential hazards to look for.
4. Reporting procedure for findings.

The following five elements should be considered and known before a scheduled safety inspection plan is formulated:

1. What to inspect.
2. Aspects that need examining.
3. Conditions to be inspected.
4. How often to inspect.
5. Who will inspect.

The fourth item of this list—how often to inspect—is determined by the following:

- The probability for injury to employees.
- The likely severity of potential injury or loss.
- The estimated time between inspections in which the area can become unsafe.
- The past history of the area.

Also, OSHA regulations dictate the safety inspection frequency for some equipment items.

The fifth item of this list—who will inspect—depends a great deal on the operation. Any inspector on a safety audit should possess the following qualities:

- Awareness of past site accident history.
- Knowledge of the standards that apply.
- Familiarity with potential hazards that could occur.
- Ability to suggest corrective actions.
- Knowledge of the organization and its personnel.
- Diplomacy.

It cannot be overstressed that safety inspections are part of the duties of management. Management should actively participate in inspections, not just delegate the task. When employees know that management is coming to inspect, the area stays ship-shape.

ACCIDENT INVESTIGATIONS AND REPORTS

If and when an accident takes place, its circumstances must be investigated so that steps can be taken to keep history from repeating itself. Investigation of an accident is a means to uncover the direct and indirect cause, document the occurrence and follow-up action, provide cost information, and stress safety. Management should include accident investigating as a part of the organization's approach to hazard control.

The primary reason for investigating an accident is not to identify a scapegoat, but to determine the cause of the accident. The investigation concentrates on gathering factual information about the details that led to the incident. If investigations are conducted properly, there is the added benefit of uncovering problems that did not directly lead to the accident. This information benefits the ongoing effort of reducing the likelihood of accidents. As problems are revealed during the investigation, action items and improvements that can prevent similar accidents from happening in the future will be easier to identify than at any other time.

A report, which is a tool of management personnel to implement corrections, is generated after the investigation. It documents the facts involved in an accident that can be useful in compensation matters and litigation. The report produced becomes the permanent record of facts surrounding the accident and allows the reconstruction of an accident situation. It also contains information that is useful in determining the direct and indirect costs of the accident.

Accident investigation indicates management's sense of accountability for accidents and the organization's concern for a safe work environment. When the investigation is a cooperative effort between management and labor, it further demonstrates management's commitment to safety.

WHEN TO INVESTIGATE

Supervisory personnel must take a serious approach to investigating an accident, no matter how trivial it seems. All accidents, no matter how small the consequences, are candidates for thorough scrutiny. Management must be aware that serious accidents arise from the identical hazards as minor accidents. Therefore, an immediate, on-the-scene investigation is best. This will prevent other important nonrelated matters from prolonging the investigation, and the facts generated will be most accurate and useful. The longer the time elapsed prior to examining an accident scene, the greater the possibility of gathering incomplete or erroneous information. The scene of the accident changes, people involved forget, and witnesses change their stories to agree with someone else's interpretation. Besides these reasons for a prompt investigation, the swift investigation expresses management's concern for the safety and well-being of employees.

WHO PARTICIPATES

There is no correct, universal list of personnel who should participate in an accident investigation. The participants are selected, to a great extent, based on the type of accident that occurred. It is sometimes necessary to call on a consultant, a specialist in the technology involved, or someone who is very familiar with the accident site area. It is important, however, that the following three parties participate in the investigation:

- *Management*. Senior-level management and/or the subordinate supervisor and the supervisor of that area of the site should help investigate an accident that resulted in lost work days or significant property damage. When the accident investigation reveals the specific corrective action to be taken, management has the responsibility to implement such action and to follow through to be certain that it was done.
- *Safety professional*. The safety professional may already be aware of hazards prevalent to the area that could be encountered at the accident scene. He or she could advise on the dangers that may remain at the scene of the accident and provide the proper protective equipment. Also, the safety professional is probably more aware of site safety procedures and can indicate if they had an impact on the accident in any way.
- *Physician*. A physician may help establish whether the injured party was physically and mentally capable at the time of the accident. He or she could help indicate if the personal protective equipment was adequate or if it was utilized. He or she could help to evaluate the rescue, first aid, and emergency medical response teams. He or she could also assess the nature and degree of injury sustained.

WHAT TO INVESTIGATE

During an investigation, many questions need to be answered. It is best to have a prepared procedure to help direct an investigation. It is also best to have a supervisor lead the investigating team. If at all possible, sketches should be made and photographs taken during the investigative work. Table 12-5 contains a list of questions that are generally applicable in most accident investigations.

THE ACCIDENT REPORT

The accident report should be impartial and objective and should accurately present the facts in a clear fashion. All of the facts should be presented; nothing should be omitted. It should say which unsafe acts and unsafe conditions specifically caused or contributed to the accident. Ambiguous terminology should not be used because it may mean something different to various people. The accident report is the product of the investigation, and it should be carefully prepared to adequately justify the conclusions and corrective action being suggested. The report should be understandable to anyone who reads it. Also, it must be issued soon after the accident.

The report should receive circulation through the supervisory arm of the organization. Supervisors can then keep workers informed of the findings and preventive measures exe-

TABLE 12-5
Accident Investigation Checklist

The following list of concerns can be used for consideration in accident investigations:

1. What type of work was the injured performing? Precisely what was he doing at the time of the accident?
2. Was the injured person familiar with the task that he was engaged in? Was he authorized to work on the equipment or process?
3. Were there other workers in the vicinity at the time of the accident? If so, what were they doing?
4. Was the task being performed properly? Proper equipment being used? Proper energy sources locked out?
5. Was the injured employee new on the job? Is the process/operation/task new?
6. Was the injured person being supervised? What role did supervisory personnel play? Was this role adequate?
7. Had the injured employee received adequate or proper training?
8. Is it apparent that any site safety rules were being violated?
9. Where did the accident take place? What was the condition of the area at the time of the accident?
10. What could have prevented the accident? Short-term solution? Long-term solution?
11. Has an accident like this occurred before? Was there any corrective action recommended? If so, was it adopted?

cuted. The report may be summarized, mentioning the causes and recommended action, and posted throughout the facility. The actual report should be maintained on file for at least 5 years. This report ought to be used by management as a tool to boost the safety program. Past mistakes can be used to improve future operations.

SUPERVISORY PLANS FOR EMERGENCIES

The management of any organization should assess the potential emergencies that could occur at the establishment and put these into a plan of action. There must be written procedures in connection with fire, first aid, toxic release, explosion, and other possible emergencies. It is sometimes necessary to involve employees in this planning process. The procedure must be well documented and must be communicated to all employees through training.

FIRE

In all premises, employees should be instructed what to do in the event of fire. They should understand fire precautions and have regular fire drills. Written fire emergency instructions should cover the following:

1. What to do if a fire is discovered.
2. Location of fire alarm boxes and how they are triggered.
3. How to call the fire brigade. (Emergency telephone numbers and the conditions under which they should be dialed.)
4. What to do if a fire alarm is heard.
5. The location and use of firefighting equipment. If firefighting equipment is used, hands-on training, with periodic refresher training, must be conducted.
6. Proper escape routes that are clear of obstruction for emergency egress.
7. The need to keep fire doors closed.
8. How to shut down machines and processes that could be dangerous if unattended or otherwise left operating or in service.
9. Isolating appropriate power supplies.

Fire emergency procedures should be prominently displayed at conspicuous points throughout the workplace. Practice fire drills should take place at least once a year, and practice drills simulating a blocked escape route are a good practice. Also, fire emergencies in the light of new equipment installations should be examined during new equipment safety audits. Fire alarms should be tested on a monthly basis, and an audibility test should be taken to be certain that they can be heard when equipment is operating. Also, portable fire extinguishers should be inspected monthly.

FIRST AID[4]

First aid programs in industry are required under OSHAct/70. A physician should help develop and direct the program, including approving the first aid supplies (as required by the OSHA rules), selecting the ambulance services to be used, planning routines for handling the various types of injuries and illnesses that may occur, and overseeing the records that will be needed.

Smaller companies may want to consider part-time medical service to provide not only first aid and injury care, but to bring to the company and its people the broader benefits of an occupational health program.

First Aid Training

First aid is the early treatment that is given until the injured or ill employee can obtain professional medical care. Emphasis in first aid training is placed on getting the individual to medical care. Because the law permits only licensed practitioners to provide definitive medical care, the first aider is limited as to scope of care provided. The first aider should provide immediate, temporary care within the scope of his or her training to relieve pain and suffering and enhance recovery. The first aider should not engage in any continuing treatment.

Acceptable first aid training can be provided by the American Red Cross through one of its local chapters. At least two employees on each shift should be trained, one to be assigned the primary responsibility and the other to assist as needed and to cover when the primary first aider is absent or otherwise unavailable.

Some companies have provided supervisors and a major segment of the employees with training in first aid and cardiopulmonary resuscitation (CPR). This provides additional trained support personnel, and it assures that actions by workers at the scene before the responsible first aider arrives will be helpful and not harmful.

The Consulting Physician

OSHA regulations require "...the ready availability of medical personnel for advice and consultation on matters of plant health." Further required are "First aid supplies approved by the consulting physician..." These two requirements of OSHA are an expression of the logic of involving a physician in the beginning of emergency planning in an occupational health program.

It is necessary to find a physician who not only can assist in devising the emergency plan, but also provide or arrange for the needed care. If the medical resource is a local clinic, a group practice, or a hospital, it is important for management to ascertain which specific physician is to be designated as the "consulting physician." Otherwise, the individual industrial case is likely to become submerged in the flow of clinical cases.

The services of the consulting physician should include the following:

- Recommendations for medical policies and procedures.
- Assistance in developing an emergency plan.
- Medical directives for the first aiders and guidelines for referring cases requiring medical care.
- List of supplies and equipment for first aid facilities.

The consulting physician should become acquainted with the accident and health hazards in the workplace. Familiarity with the physical demands and environmental factors of the various jobs will enable the physician to make return-to-work decisions and specific work placements.

First Aid Facility, Equipment and Supplies

The kind of first aid facility that is appropriate depends on such factors as the size of the company, number of employees, severity of accident and illness hazards, and the distance to a clinic or hospital.

The facility should include a cabinet for first aid supplies, a sink with hot and cold running water, a chair, an adjustable light, a basin, covered waste container, paper towels with dispenser, and paper cups with dispenser.

Maintenance of the first aid facility should be an assigned responsibility of the first aiders on each shift. Additional equipment and supplies should be obtained in accordance with the recommendations of the consulting physician.

If the size of the company warrants it, consideration should be given to setting aside a separate first aid room to accommodate a bed or cot and other equipment that the consulting physician deems appropriate.

The basic first aid supplies would include: a first aid manual, adhesive tape, bandage strips, bandage scissors, gauze bandage rolls, sterile gauze squares, antibacterial soap, elastic bandages, triangular bandages, safety pins, splinter forceps, and ammonia inhalants. Additional first aid kits should be available to field crews, either by assigning a kit to one worker or by placing the kit in the crew vehicle when it remains on the site.

Emergency Transportation

Emergency planning should include the investigation and evaluation of ambulance or medical services in the communities. Many communities have a comprehensive emergency transportation system—the ambulances carry sophisticated equipment and are staffed with emergency medical technicians capable of assessing needs and initiating prompt emergency care in the ambulance, if indicated. Communication with the emergency care physician in the hospital or clinic is readily available for advice and guidance.

An investigation into what, how, when, and where qualified ambulance services can be obtained should be made. The selection of ambulance service should be based on the qualifications of the ambulance driver, training of ambulance attendants, condition of the vehicle, and the emergency care equipment.

Awareness of other community services that are available, such as the fire department, rescue squad, and burn center, should also be considered in the emergency care planning.

Names and telephone numbers of the ambulance service, rescue squad, fire department, hospital, and consulting physician should be posted in appropriate places throughout the plant or building, in addition to being posted at the first aid facilities.

First Aid Records

A record should be kept of all persons receiving first aid. These records may have worker's compensation implications and may be used to verify worker's compensation claims. These records can also be used to identify patterns of injury occurrence. OSHA Form 200 must be kept for cases involving time lost from work, temporary limited work assignments, and cases referred to the physician for continuing care.

The Occupational Health Nurse

Companies with sufficient numbers of employees to warrant it should consider part-time or full-time in-house nursing services. The nurse will assume responsibilities beyond taking care of injuries. The scope of the occupational health nurse's function could include the health assessment of new and present employees, conducting or arranging for medical monitoring procedures, educating workers to the exposures and measures for protecting themselves from injury or illness, and maintaining the medical record system.

The Role of the Insurance Carrier

The responsibility for providing a safe workplace is solely that of the employer. The worker's compensation insurance carriers have traditionally provided assistance to policyholder management in safety matters, and so it has been logical for policyholders to look to their carrier for guidance in the safety aspects. The occupational health aspects are now receiving increasing attention, and the worker's compensation insurance carrier can provide guidance to policyholder management in setting up an adequate occupational health program.

OTHER EMERGENCY PROCEDURES

Management of every organization should be certain that effective plans are in place for all foreseeable emergencies. Possible emergency situations are bomb threat, chemical spill, toxic gas release, natural disaster, loss of power, civil disturbance, plane crash, explosion, or out-of-control production process. The emergency instructions should incorporate procedures to deal with the cause of the emergency, evacuation of nonessential personnel away from the danger, arrangements for the coordination of emergency services, news releases to the public, and restoration of operations.

In facilities where extreme hazards exist, an emergency control center should be predesignated. Information kept at the control center must include the following:

- Telephone numbers of emergency services.
- Telephone numbers of essential personnel.
- Layout drawings of the facilities identifying power lines, water mains, emergency services, and so on.
- Hazardous material information and material safety data sheets (MSDS).
- List of explosive and toxic materials and indication of their locations.

AN OVERVIEW OF WORKER'S COMPENSATION

In historical perspective, worker's compensation laws as they are now are a recent development. Until the Industrial Revolution, there was no need for a formal system to identify the liable party in a workplace injury. The great change in

CHAPTER 12

WORKER'S COMPENSATION

technology brought about by the Industrial Revolution did not create parallel changes in the social justice system. The unsafe, poorly lit, poorly ventilated mines and factories with long hours and child labor created the need for a system of worker's compensation. This developed first in Europe during the 1800s.

Before worker's compensation laws existed in the United States, an employee received no compensation for damages for an industrial accident or occupational disease, even when the following circumstances prevailed:

- The worker was disabled or died from an occupational injury or disease.
- The injury was a natural expectation from the risks and hazards associated with that occupation.
- The injury was caused by worker negligence on the part of the injured party or another worker.

A worker had to file suit in a court of law and prove the employer's negligence to recover damages from that employer. Even an employer who had been negligent could forego recovery if it could be proved that the injury was due to ordinary risk, negligence of a fellow worker, or negligence on the part of the injured party. Employees encountered tremendous difficulties in obtaining compensation; there were long delays in bringing cases to trial, the outcome often went against the employee, and negligence suits were very expensive.

The first worker's compensation law in the United States was passed in 1908 and applied only to federal employees. The first state worker's compensation law was enacted in 1911. In 1916, the U.S. Supreme Court declared worker's compensation laws to be constitutional. All states had some form of worker's compensation by 1948. Today, all 50 states, the District of Columbia, Guam, and Puerto Rico have compensation acts. Also, U.S. government employees are covered by the Federal Employee's Compensation Act, and maritime workers by the Longshoreman's and Harbor Worker's Compensation Act.

MODERN WORKER'S COMPENSATION

Generally, worker's compensation is a no-fault arrangement. The employee's negligence does not affect the determination of liability. A compromise was reached with the worker's compensation system; for the compensation, the employees have given up their right to legally pursue the employer for unlimited damages due to pain and suffering. The aims of worker's compensation benefits are to rehabilitate the worker and minimize his or her losses because of the reduced ability to compete in the labor market.

It should be noted that many states consider each physical or chemical-related disability as a bodily injury by disease. This is commonly referred to as "occupational diseases." Disorders that are the result of an extended, long-term attack on the body may be called cumulative injuries. According to the California Labor Code, Sec. 3208.1, a cumulative injury is defined as "occurring as repetitive mentally or physically traumatic activities extending over a period of time, the combined effect of which causes any disability or need for medical treatment."

A difficulty associated with the cumulative injury claim is that the disability usually manifests itself as a disease rather than a wound. Many disorders due to occupational overexposures involve the joints and may appear similar to symptoms of aging. Therefore, a portion of the solution to the difficulties presented by the cumulative injury claim relies on the safety and health professional. If a worker's compensation claim goes to court, it is vital that a close liaison be established by the safety and health professional with the physician and attorney involved in the case to ensure a just and equitable settlement as well as to rehabilitate or return the disabled employee to gainful employment.

Because a cumulative injury-type claim has a long time in which to develop, it is even more important to identify its origin (usually a poorly designed work operation) and correct it before the worker's health is affected.

OBJECTIVES AND CHARACTERISTICS OF WORKER'S COMPENSATION

The commonly accepted objectives of worker's compensation are the following:

1. *Replacement of income.* This is the first major objective of worker's compensation—to replace the lost wages that a job-related injury or illness causes. This replacement should be prompt and adequate. The program should replace the present and projected lost earnings, minus taxes and commuting costs. Most state statutes mandate a two-thirds income replacement ratio, but the program must treat all workers fairly. All workers should have the same proportion of their wages replaced. The employer's worker's compensation program should provide income replacement as soon as possible after disability. The employee should know in advance what he or she would receive if he or she were to become disabled on the job. These benefits are to continue even if the employer's business were to discontinue.
2. *Restoration of an injured worker.* A second objective is rehabilitation, both medical and vocational, to return the employee to the workforce. Medical care should be provided for the employee at no cost until he or she is restored as fully as possible. The program should attempt to positively motivate the employee to return to work.
3. *Prevention of accidents.* Safety programs should reward good safety practices and discourage dangerous operations. A third objective of worker's compensation is occupational accident prevention and reduction.
4. *Cost allocation.* A fourth objective of worker's compensation is to allocate the costs of worker's compensation programs in accordance with the employers and industries responsible for the losses. Because this allocation tends to shift resources from hazardous employers and industries to safe employers and industries, it motivates safety improvements.

Who is Covered?

Most state worker's compensation laws fail to cover all forms of employment. It is also important to realize that the laws vary greatly between states. An employer can reject worker's compensation coverage in states where laws are elective and can relinquish indemnity to common law defenses. This gives an employee a legal action in negligence against an employer, and the employer may not plead defenses of assumed risk, fellow servant negligence, or contributory negligence. Many laws contain exemptions, such as employment at charitable or religious institutions. Others not protected by worker's compensation are volunteers, unpaid family workers, and the self-employed. Compensation benefits, for those who are covered, are limited to injury caused by conditions arising out of and in the course of employment.

Benefits

Benefits are paid by three methods: commercial insurance policies, state insurance funds, and self-insured employers. Benefits include the following:

- Medical service.
- Cash payments to the worker while disabled.
- Burial allowance should the worker die.
- Allowances to the worker's dependents.
- Allowances for a nursing attendant (some states).
- Special costs for prosthesis (some states).

Income Replacement

Although a large percentage of worker's compensation cases are for temporary total disablement, these cases account for one fourth of cash benefits. Income benefits to workers for permanent partial disabilities account for nearly two thirds of the total dollar amount. Cash benefits are payable as a wage-related benefit.

ADMINISTRATION OF WORKER'S COMPENSATION

Almost all of the states have agencies to handle the administrative responsibilities of worker's compensation. Worker's compensation claims may be contested or uncontested. The administrative agency usually has jurisdiction over contested cases. It is the agency's responsibility to supervise the processing of all cases. Administration by a division within the labor department, by a board, or by a commission can effectively operate to assure compliance with the law and guarantee an injured worker's rights. The state agency sees to it that worker's compensation payments commence promptly. It also sees that the worker gets the full benefit that is due. Some states require signed receipts with every compensation payment, or the filing of final receipts, to permit an audit of individual payments.

POTENTIAL EMPLOYER LIABILITIES

In the case of some possible workplace injuries, the employer may be liable beyond worker's compensation. The following is a listing of potential employer liabilities:

1. *Criminal liability* should the employer fail to train employees about on-the-job hazards, neglect to furnish adequate protective equipment against hazards, or disregard complaints of hazardous working conditions.
2. *Liable for aggravation of injuries (dual capacity)* should the employer negligently aggravate a workplace injury when performing functions other than an employer, as in providing improper medical treatment.
3. *Product liability (dual capacity)* should the employee become injured from the use of a faulty product that the employer manufactured.
4. *Liable for intentional assault* should the worker become injured as a result of physical attack by a manager.
5. *Liable for damages to immediate family* should the employee's injury cause loss of consortium, loss of companionship, and negligent mental suffering to the injured worker's immediate family.
6. *Liable corporate subsidiaries*. A California court (Gigax v. Ralston Purina Co., 136 Cal App 3d 591-1982) ruled that "a host of cases hold an employee of a wholly owned subsidiary, who has obtained worker's compensation benefits from the subsidiary, may obtain an action in tort against the parent corporation and this is so even though the parent and subsidiary are covered by the same worker's compensation policy."

INFORMATION SOURCES

Management personnel can find up-to-date specialized information on the various topics covered in this chapter from several sources. Various information sources are listed in Table 12-6. Professional societies, trade associations, and standards organizations are also sources of helpful information, although this listing is by no means complete.

TABLE 12-6
Organizations and Professional Societies Offering Information on Occupational Safety and Health Concerns

SERVICE ORGANIZATIONS

American Red Cross 17th & D Streets, N.W. Washington, DC 20006 (202) 737-8300	Industrial Health Foundation 34 Penn Circle W. Pittsburgh, PA 15206 (412) 363-6600	National Safety Council 4444 North Michigan Avenue Chicago, IL 60611 (312) 527-4800	National Society to Prevent Blindness 79 Madison Avenue New York, NY 10016 (212) 684-3505

STANDARDS GROUPS

American National Standards Institute 1430 Broadway New York, NY 10018 (212) 354-3300	American Society for Testing and Materials 655 15th Street, N.W. Washington, DC 20005 (202) 639-4025

(continued)

BIBLIOGRAPHY

TABLE 12-6—*Continued*

FIRE PROTECTION ORGANIZATIONS

Factory Mutual Engineering and Research 1151 Boston-Providence Turnpike Norwood, MA 02062 (617) 762-4300	National Fire Protection Association Batterymarch Park Quincy, MA 02269 (617) 770-3000	Underwriters Laboratories 333 Pfingsten Road Northbrook, IL 60062 (312) 272-8800

PROFESSIONAL SOCIETIES WITH SAFETY CONCERNS

American Association of Occupational Health Nurses 3500 Piedmont Road, N.E. Atlanta, GA 30305 (404) 262-1162	American Board of Industrial Hygiene 302 S. Waverly Road Lansing, MI 48917 (517) 321-2638	American Conference of Governmental Industrial Hygienists 6500 Glenway Avenue Cincinnati, OH 45211 (513) 661-7881	American Industrial Hygiene Association 475 Wolf Ledges Parkway Akron, OH 44311 (216) 762-7294
American Occupational Medical Association 2340 S. Arlington Heights Road Arlington Heights, IL 60005 (312) 228-6850	American Public Health Association 1015 15th Street, N.W. Washington, DC 20005 (202) 789-5600	American Society of Safety Engineers 1800 East Oakton Street Des Plaines, IL 60016 (312) 692-4121	Board of Certified Safety Professionals 208 Burwash Avenue Savoy, IL 61874 (217) 359-9263
Board of Hazard Control Management 8009 Carita Ct. Bethesda, MD 20817	Health Physics Society 1340 Old Chain Bridge Road, Suite 300 McLean, VA 22101 (703) 790-1745	Human Factors Society P.O. Box 1369 Santa Monica, CA 90406 (213) 394-1811	International Healthcare Safety Professional Certification Board 5010 A Nicholson Lane Rockville, MD 20852 (301) 984-8969
National Safety Management Society 3871 Piedmont Avenue Oakland, CA 94611 (415) 653-4148	SAFE Association 15723 Vanowen Street, Suite 246 Van Nuys, CA 91406 (818) 994-6495	Society of Fire Protection Engineers 60 Batterymarch Street Boston, MA 02110 (617) 482-0686	System Safety Society 14252 Culver Drive, Suite A-261 Irvine, CA 92714 (714) 551-2463
	Veterans of Safety 203 N. Wabash Avenue, Suite 2206 Chicago, IL 60601 (312) 346-3835		

References

1. *Training Requirements in OSHA Standards*, OSHA 2254 (Washington, DC: U.S. Department of Labor, Occupational Safety and Health Administration, 1979).
2. *Voluntary Training Guidelines*, 49FR30290 (Des Plaines, IL: OSHA, Office of Training and Education, 1984).
3. Noboru Sugimoto and Kunitomo Kawaguchi, "Fault Tree Analysis of Hazards Created by Robots," *13th International Symposium on Industrial Robots and Robots 7 Conference Proceedings*, held April 17-21, 1983, in Chicago, IL (Dearborn, MI: Society of Manufacturing Engineers, 1983), pp. 9-13 to 9-28.
4. *First Aid and Emergency Plans*, Report to Management No. 4 (Schaumburg, IL: The Alliance of American Insurers, 1981).

Bibliography

All About OSHA, OSHA 2056 (To obtain a copy of this publication, contact the nearest OSHA office as indicated in Fig. 12-1.)

Anton, Thomas J. *Occupational Safety and Health Management*. New York: McGraw-Hill Book Co., 1979.

Brown, David B. *Systems Analysis and Design for Safety*. New York: Prentice-Hall, 1976.

Chemical Hazard Communication, OSHA 3084 (To obtain a copy of this publication, contact the nearest OSHA office as indicated in Fig. 12-1.)

Consultation Services for the Employer, OSHA 3047 (To obtain a copy of this publication, contact the nearest OSHA office as indicated in Fig. 12-1.)

Employer Rights Following an OSHA Inspection, OSHA 3000 (To obtain a copy of this publication, contact the nearest OSHA office as indicated in Fig. 12-1.)

Fenton, Jack R. "Worker's Compensation Safety Programs." *Safety Law* (1983), pp. 17-19.

Gartner, Ludwig B. "How to Prepare an OSHA Case—The Employer's Viewpoint." *Safety Law* (1983), pp. 73-74.

Health and Safety Guide for Textile Machinery Manufacturers. Washington, DC: NIOSH, U.S. Department of Health, Education, and Welfare (January 1978).

Job Hazard Analysis, OSHA 3071 (To obtain a copy of this publication, contact the nearest OSHA office as indicated in Fig. 12-1.)

McElroy, Frank E. *Accident Prevention Manual for Industrial Operations—Administration and Programs*, 8th ed. Chicago: National Safety Council, 1981.

OSHA Inspections, OSHA 2098 (To obtain a copy of this publication, contact the nearest OSHA office as indicated in Fig. 12-1.)

Peters, George A. "The Employer's Liability Beyond Worker's Compensation." *Safety Law* (1983), pp. 25-27.

Rader, Robert E. "The Employer's Rights on Inspections Under OSHA Law." *Safety Law* (1983), pp. 63-67.

Reid, Robert. "Manufacturers: Are You Ready for OSHA's Hazard Communication Standard?" *Occupational Hazards* (April 1986), pp. 55-58.

Ridley, John R. *Safety at Work*. London, England: Butterworths.

"Safety and Occupational Health: A Commitment in Action." *National Safety and Health News* (October 1985), pp. 55-59.

Stepkin, Richard L., and Mosely, Ralph E. *Noise Control*. Park Ridge, IL: American Society of Safety Engineers, 1984.

What Every Employer Needs to Know About OSHA Recordkeeping, BLS Publication 421-3 (To obtain a copy of this publication, contact the nearest OSHA office as indicated in Fig. 12-1.)

MANUFACTURING/ ENGINEERING INTERFACE

SECTION

4

DESIGN FOR MANUFACTURE

Before the Industrial Revolution, customer needs and the products and production systems used to meet these needs were simple. By and large, individual craftworkers were able to integrate all aspects of customer needs, product design, and manufacturing. The design process was essentially optimized by a single worker through the combination of simplicity and comprehension of the interdependencies existing between the various facets of the manufacturing system.

By comparison, today's manufacturing systems and the products they produce are by orders of magnitude more informationally dense and complex. They require vast amounts of specialized knowledge, all focused on single-product problems. Superimposed on this is an exponentially increasing amount of new information and technology coupled with ever-shortening product lifecycles and increasing global competition.

Continual and rapid change is one of the leading characteristics of the modern Information Age. New materials, process refinements, and customer requirements are emerging and evolving at ever-increasing rates. Coping with rapid change has become a major problem for industry. In industries driven by technological change, coming to market 9-12 months late can cost a new product half of its potential revenues. Even in calmer businesses, companies that drive from lab to market more swiftly than the competition have the luxury of starting later so they can employ up-to-date manufacturing methods and the latest customer requirements.

Business goals, on the other hand, have not changed. The business goals of most companies are twofold: (1) make as good a product as possible, in as short a time as possible, and for as little cost as possible; and (2) sell as many as possible, as fast as possible, for as much as possible. Accomplishing these goals under today's conditions of rapid change requires change in many of the ways that companies currently make decisions and operate.

Perhaps one of the most difficult and challenging problems created by this required change is the need to design products in a new way. Addressing this problem requires a new awareness of all aspects of an enterprise, one of them being a new awareness of manufacturing systems. Defined as the conversion of raw material into end products, the manufacturing system spans all activities associated with a modern industrial enterprise, including conception, development, and design of the product; production, marketing, sales, and distribution of the product; and ultimately, support of the product in use.

Embedded within the manufacturing system are a large number of distinct processes or stages that, individually and collectively, affect product quality and cost, as well as the time and effort required to produce the product and the time required to introduce a new version of the product. The interactions between the various facets of the manufacturing system are complex, and decisions made concerning one aspect have ramifications that extend to the others (see Fig. 13-1).

In its broadest sense, design for manufacture (DFM) is concerned with comprehending these interactions and using this knowledge to optimize the manufacturing system for effective quality, cost, and delivery. More specifically, DFM is concerned with:

1. Understanding how the process by which a product is designed interacts with the other components of the manufacturing system and using this understanding to design better quality products that can be produced for lower cost and brought to market more quickly.
2. Understanding how the physical design of the product itself interacts with the components of the manufacturing system and using this understanding to define product design alternatives that help facilitate "global" optimization of the manufacturing system as a whole.

Ultimately, the goal of design for manufacture is to facilitate the design of functionally and visually appealing products with mechanical reliability, to manufacture these products effectively, to introduce the products, and to market them in a timely manner. The purpose of this chapter is to explore some of the challenges involved in achieving this goal; develop a basis for "how-to-do design" for effective quality, cost, and delivery; and provide an overview of several DFM methodologies. The chapter begins with a brief review of the design process and other design fundamentals that form the basis of the DFM philosophy.

CHAPTER CONTENTS:

***Contributor of this chapter is: Henry W. Stoll**, Manager, Design for Manufacturing, Industrial Technology Institute.*

***Reviewers of this chapter are: Vaidis Draugelis**, Manager, Xerox Design Institute, Advanced Reprographics and Design Technology, Xerox Corp.; **Louis Gills**, RMA Engineer, Office of the Assistant Secretary of the Navy (S & L) RM & QA; **John P. Hinckley, Jr.**, Manager, Advance Manufacturing Engineering, Project Liberty, Chrysler Motors; **Gerard H. Hock**, Project Manager, Design for Simplicity, Corporate Engineering/Manufacturing Staff, Engineering Consulting, General Electric Co.; **Bernard J. LaVoie**, Director of Manufacturing and Quality, Department of the Air Force, HDQTRS Electronic Systems Div., Hanscom Air Force Base; **James L. McConnell**, Staff Vice President, Advanced Manufacturing Engineering, Research and Engineering, Whirlpool Corp.; **Ronald S. Peterson**, Manager, Space Station Operations, Astronautics Div., Lockheed Missiles & Space Co.; **Louis G. Sportelli**, Director of Quality and Operations, Federal Systems Div., IBM.*

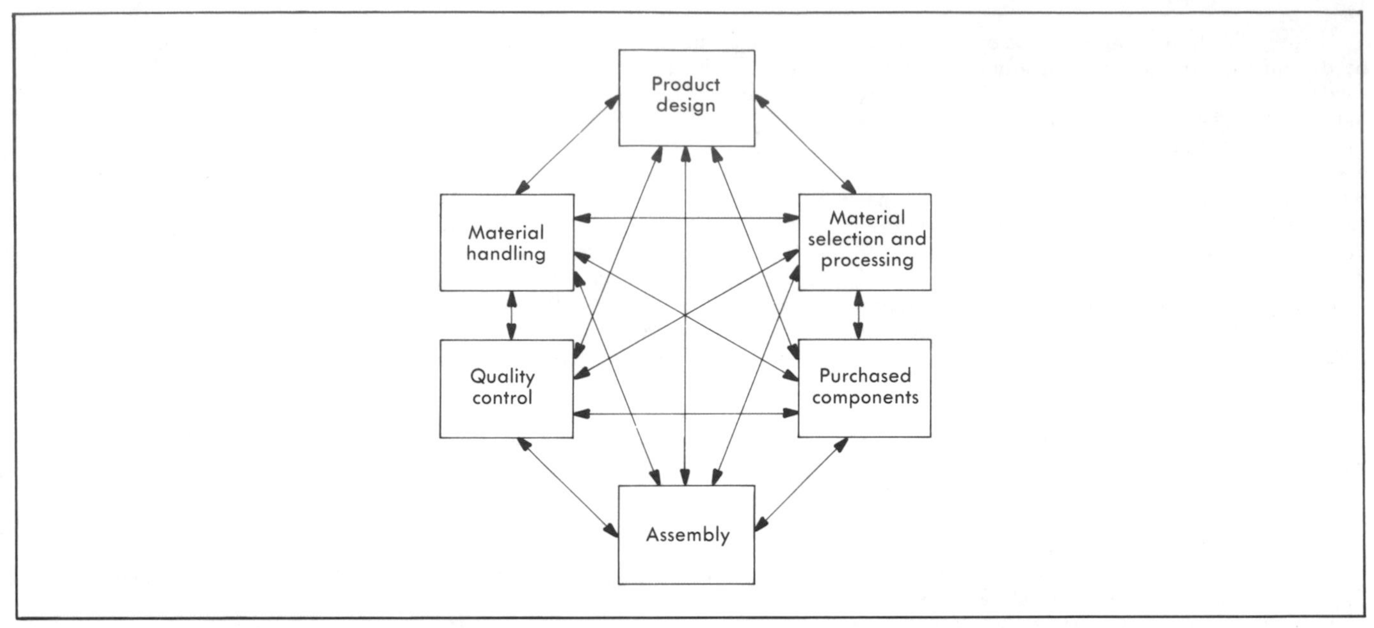

Fig. 13-1 Manufacturing interactions. (*Industrial Technology Institute*)

DESIGN BASICS

A description of the mechanics of the engineering design process has existed for many years. The book *Introduction to Design*, written in 1962 by Morris Asimow, presents one of the first definitive treatments on engineering design.[1] Since then, many textbooks on this subject have appeared, each telling basically the same story. Additional references are listed at the end of the chapter for general review.[2,3]

A study of design basics reveals much about design. It shows why conceptual product design is such a pivotal step in the lifecycle of a product, where current practices have gone wrong, and what steps need to be taken to get back on track. As was mentioned previously, the heart and foundation of DFM is a fundamental understanding of the design process.

THE DESIGN PROCESS

To design is to pull together (synthesize) something new or arrange existing things in a new way to satisfy a recognized need. Design is a method or scheme of action or a way proposed to carry out a system or design. It is also purposeful planning as revealed in or inferred from the relations of parts to a whole. Hence, in a very general sense, design is synonymous with plans and planning. The design engineer and manufacturing engineer produce plans that enable other components of the manufacturing system to produce the product.

Design engineers are decision makers. To carry out this activity, the design engineer uses various structured techniques, scientific principles, and norms learned. Typically, design decisions involve the form, function, and fabrication of parts, components, devices, and systems that satisfy a set of specified or implied requirements. In most engineering design situations, design decisions are constrained by materials, manufacturing techniques, and economics. Economic constraints relate not only to the cost of product manufacture, but also to the limited resources of time, money, and talent available for carrying out the design.

The essential purpose of design is to satisfy need. Need arises in many forms. Products ranging from ball bearings to satellite communication systems, from deep sea oil drilling platforms to manufacturing systems all fulfill human needs. Often the particular needs involved determine the design strategy followed. For example, the needs to be satisfied in the design of a simple welding fixture are far different from those of a mass-produced washing machine or multifunctional interplanetary space probe. Needs also exact various prices. Certain functions require certain materials and various amounts of energy. Safety, ecological, and societal considerations increase complexity. Light weight often implies costly materials as well as extensive analysis, research, and development. Quality implies consistency and adherence to design intent. Efficient design requires an optimal balance between cost, function, appearance, convenience, maintainability, and life. This balance is generally determined by the needs the design is intended to fulfill.

In general, engineering design begins with the recognition of a need and the conception of an idea to meet this need. It proceeds with the definition of the problem; continues through a program of directed research, analysis, and development; and often leads to the construction and evaluation of a prototype. It concludes with the effective manufacture and distribution of a product or system so that the original need may be met wherever it exists. Because many decisions made along the way must be made in the absence of needed information, the engineer seldom arrives at an acceptable solution the first time around. As the final design evolves, many decisions are reexamined because additional information becomes available. This process of reexamination is the iterative nature of design and is recognized as an essential part of the process. Iteration permits inadequate decisions to be improved under increasing knowledge.

Engineering design is accomplished through a process of problem solving. Problem solving in design is a process that follows a logical sequence. The complete process, from start to finish, is often outlined as shown in Fig. 13-2.[4] The process begins with a decision maker—the engineer or manager—recognizing a need and making a decision to do something about it. After many iterations, the process ends with plans for satisfying the need.

In design, a "problem" is often not the clear, incisive, thorough statement of numbers and facts usually associated with physical or mathematical exercises. Rather, a "problem" in this context might better be termed an "unresolved technical need," a "perplexing situation," or "something that must be done to achieve progress/success/survival."

It is also important to note that design problems generally have no unique answer. For example, two alternative approaches for supporting a radar antenna are shown in Fig. 13-3. Both solutions are intended to hold the antenna stationary under the action of external forces such as wind loads. The alternative in view *a* does this by utilizing a very rigid and massive support structure. The alternative in view *b*, on the other hand, provides a stable support using a light and fairly flimsy support structure protected from wind disturbances by an equally light and flimsy dome structure. Both alternatives provide the needed function, but in different ways. Both the nature of these alternative concepts as well as the actual solution selected are determined by the needs that must be satisfied (problem definition).

GOVERNING CHARACTERISTICS

The design process previously discussed can be summarized by the following definition:

Design is an *iterative, decision-making* activity involving the use of scientific and technological information to produce a system, device, or process intended to meet *specified needs*.

The italicized words in this definition represent intrinsic, governing characteristics of the design process. At the outset of a design project, usually very little is known about the design problem and what the eventual design solution will actually consist of. Hence, early design decisions must often be made under a great deal of uncertainty. Two mechanisms are available to offset this uncertainty: (1) accurate problem definition (careful and thorough specification of needs, examination of past experience, and a study of the competition) and (2) iteration.

Problem Definition

Problem definition is the process of going from a primitive statement of need to a clear, exact statement of the problem to be solved expressed in engineering terms. In most cases, the solution to a design problem is directly determined by the problem definition. For this reason, it is essential to understand and define the problem before any solutions are sought. Developing an accurate and thorough definition of the problem before any lines are put on paper avoids much of the uncertainty associated with early design decisions.

Accurate problem definition in appropriate terms leads to the most desirable solution. To illustrate this, consider the problem faced by a certain TV station. The problem was that ice would form on the antenna tower during certain types of weather and subsequently fall off, causing harm to people and damage to automobiles below. Concerned about this, the station manager approached a design consultant with the following problem definition: How to prevent ice from forming on the TV tower.

Although solutions to this problem statement such as installing costly heating elements on the tower structural members could easily be imagined, the consultant, experienced in the need for accurate problem definition, asked the following questions:

- What would happen if ice did form?
- What harm would such formation do?

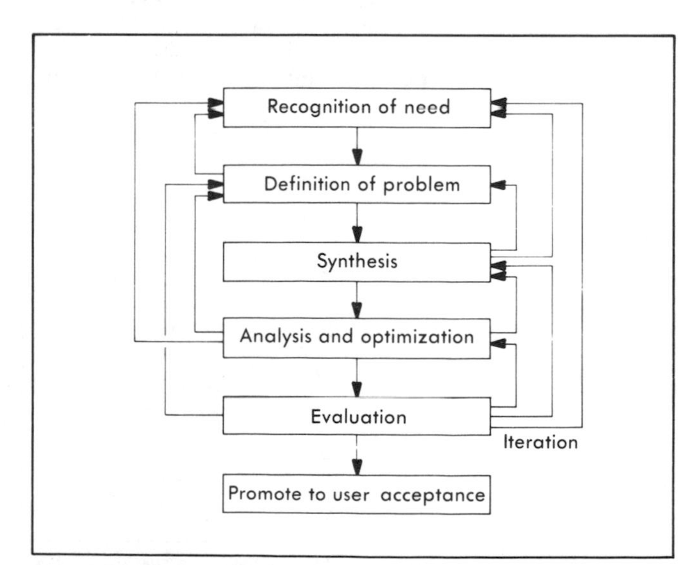

Fig. 13-2 Schematic diagram showing the different phases of design.

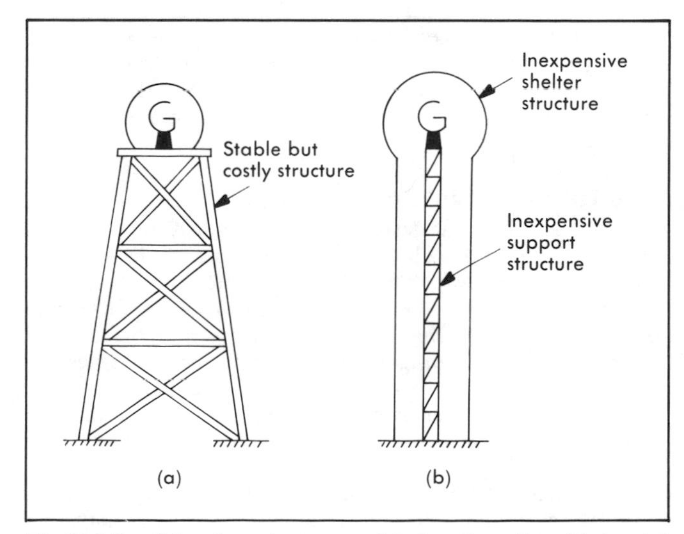

Fig. 13-3 Possible radar antenna mount design alternatives. (*Industrial Technology Institute*)

Following this line of reasoning, the design consultant formulated a much broader problem definition: How to prevent ice that forms on TV tower from doing harm or damage to people and equipment below the tower.

This second problem definition led to an obvious and cost-effective solution that was quickly accepted: build a shed-like structure to protect against falling ice. Not only was this selected solution a clearly superior solution functionally, but it also avoided all the costly and time-consuming design and development effort that would have been required by solutions to the first, less accurate problem definition.

The importance of problem definition can be further emphasized by considering the problem definitions that resulted in the alternative radar antenna support solutions shown in Fig. 13-3. In this case, the need was for a radar antenna that does not move under adverse weather conditions. Two of many possible problem definitions are given as follows:

1. Design a support structure that stabilizes the radar antenna.
2. Design an antenna system that provides support and protects against weather disturbances.

The first problem definition is stated in such a way that it implies a solution. It might be said that, in this case, the problem and its solution are coupled. Such a problem definition leads naturally to the stable but costly structure shown in view *a* of Fig. 13-3. The second problem definition is stated in a way that separates the problem from the solution. The design result is the more cost-effective solution shown in view *b*. In this design, the lightweight dome structure deflects in the wind, but protects the, light antenna support structure from wind and weather disturbances.

These examples illustrate that time and effort spent "up front" on careful and thorough definition of the problem can pay big dividends in achieving a design solution having acceptable quality, cost, and delivery. Problem definition translates into product planning and development in the larger context of a manufacturing system. It is the all-important first step leading to high quality in early design decisions. It improves the quality of early design decisions by providing more complete information on which to base a decision. In the final analysis, the benefit of thorough product planning lies in the reduction in uncertainty that it produces and the guidance that it provides in leading the design team more quickly and accurately to the effective solution. In other words, it is essential to get the customer requirements right and then translate them to engineering terms.

The Iterative Nature of Design

Iteration allows the design to be continuously improved and optimized over time as better and more complete design information becomes available. What constitutes "more complete information" can vary depending on circumstance. For example, more complete information could come in the form of improved understanding gained from a prototype test or revealed through analysis or innovation; it could be a better definition of customer or market needs or a change in those needs; or it could involve a new material or emerging manufacturing technology or some other new discovery.

Accompanying iteration are local design changes made to improve the design. These local changes generally propagate in a "ripple effect" throughout the design because they require that each part of the design affected by the change be reexamined. An unavoidable consequence of design iteration is therefore "engineering change," caused both by the design iteration itself and the ripple effect it produces. In the early phases of a design project, engineering changes are handled fairly easily because the design is fluid, hardware is still remote, few people are involved, and constraints and interactions have not become tight. The cost of engineering changes in the early stages of design is usually just the direct labor costs involved.

In the later stages of the project, engineering changes become much more difficult and costly to handle; many engineers, designers, and drafting personnel are committed, several components of the manufacturing system are usually actively involved, and much has been designed and irrevocably fixed. Because of the ripple effect, an ill-chosen solution for even a relatively minor problem can put the whole project in jeopardy. For these reasons, the range of solutions to a problem discovered late in a project are severely limited, and even minor changes are likely to result in both undesirable deviations from the original design intent and in suboptimal design.

Cost of engineering changes made late in the project are high for a multitude of reasons. Direct cost is high because of time delays and the large amount of personnel involved. In addition, there are indirect costs incurred because of the suboptimal solutions that often must be accepted to minimize or contain direct costs associated with the change. Suboptimal solutions generally lead to higher manufacturing cost as well as a degradation in product quality and manufacturing productivity.

When engineering changes occur after production release or originate as a result of production difficulties, the costs incurred include additional indirect costs such as scrap parts, wasted material, and idle machinery and workforce. After sales, engineering changes usually follow high warranty or service costs, which are also indirect costs that must be taken into account. Invisible costs, such as loss of reputation and declining workforce morale, must also be taken into account. Hence, the "real" cost of engineering changes made late in the design project and especially after release to production include both visible (direct and indirect) and invisible costs.

The cost of engineering changes can be plotted as a function of product lifecycle (see Fig. 13-4). This curve is based on the "law of 10s," which assumes that the cost of an engineering change increases by a factor of 10 with each subsequent stage in the product development cycle following concept decision. Recognizing that iteration is an inherent part of the design

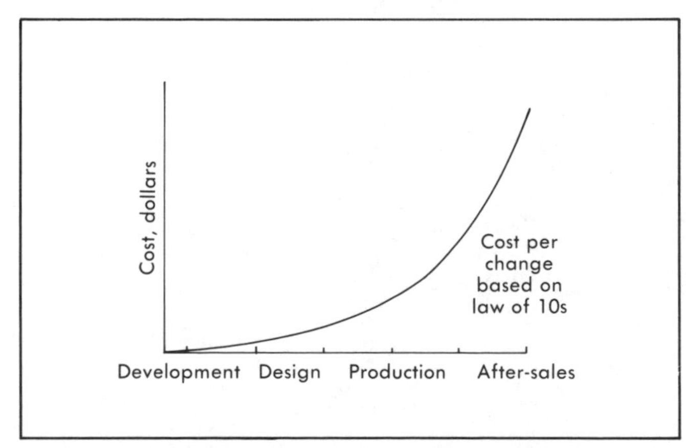

Fig. 13-4 Real cost of engineering changes based on the law of 10s. (*Industrial Technology Institute*)

process, Fig. 13-4 shows that a concerted attempt should be made to converge quickly to the desired solution in the early stages of the design project.

Poor quality design decisions and design iteration result from a lack of complete design information. Time and money spent at the onset on comprehensive product planning, research, and development can generate much of the information needed to reduce the amount of iteration required; move engineering change to the earlier, less costly stages of design; and ultimately improve and enhance the quality, cost, and delivery of early design decisions. Much of the remaining discussion in this chapter is directed at this same goal by exploring additional ways to improve design decisions made at each stage of the product development cycle.

NEED FOR CHANGE AND CHANGING NEEDS

The organizational practices and procedures of groups of people who perform design work, as well as the understanding of what physical characteristics and properties of a product constitute good design, are continually evolving. In many ways, design for manufacture is just one more iteration of this evolution. The need for DFM is largely a response to rapidly changing human needs and the capabilities available for meeting these needs.

Awareness that there is a need for change in the basic way industry accomplishes product delivery began to surface in the late 1970s. It was at this time that many established and successful manufacturers realized that their preeminent position in the global marketplace was being challenged by declining productivity, shortening product lifecycles, global competition, and the emergence of new manufacturing technologies. This precipitated a quest for short-term solutions that quickly focused on fledgling technologies such as robotics, vision and optical processing, and flexible manufacturing systems. Failure to achieve promised productivity improvements using these and other advanced manufacturing technologies in a variety of industrial settings has taught much about product design and its relation to the rest of the manufacturing system.

Implicit in this experience is the realization that there is both a need to change and changing needs; that is, coping with the demands of the present-day market requires that changes be made in the way companies do work. In addition, this same market is imposing new product requirements that also demand a new approach to design. Along with market demands is the availability of advanced manufacturing technologies that promise highly desirable productivity improvements that cannot be fully realized without changes in design practice. In the following sections, some of the needed changes in current practices are discussed along with some of the changing needs that are making it imperative for changes in the design approach.

ORGANIZATIONAL AND PROCEDURAL ISSUES

Perhaps the most important lesson learned from the failures of the late 1970s is that the design, function, and implementation of advanced manufacturing technology is directly related to the product being manufactured. When implementing these technologies, design of the product and design of the equipment processes that produce it can no longer be treated as separate entities. Manufacturing goals and requirements must be included, from the very beginning of the project, as part of the product plan (problem definition). To minimize design iteration and move engineering changes back into the early stages of design, the product and process design must also proceed, from the start, hand in hand as one common, integrated activity.

Recognizing the need to integrate product and process design is only the first step. The pivotal second step—actually doing it within the constraints imposed by the organizational structures and procedural processes of today's large corporations—is the challenge. The difficulties associated with integration of product and process design are due, in large part, to the traditional phased approach (also organizational/geographical separation of design and process engineering functions) to product development and design practiced by many companies. The traditional approach is illustrated by the simplified classical manufacturing model shown in Fig. 13-5. In this model, concept decision, product design, and testing are performed prior to manufacturing system design, process planning, and production. This requires the product design team to make design decisions, many of which are irreversible, on the basis of incomplete information regarding the capabilities of the manufacturing process and the constraints imposed by production requirements on the product design. The resulting counterflow of engineering changes produced by the unnatural separation of design and manufacturing in the traditional approach is

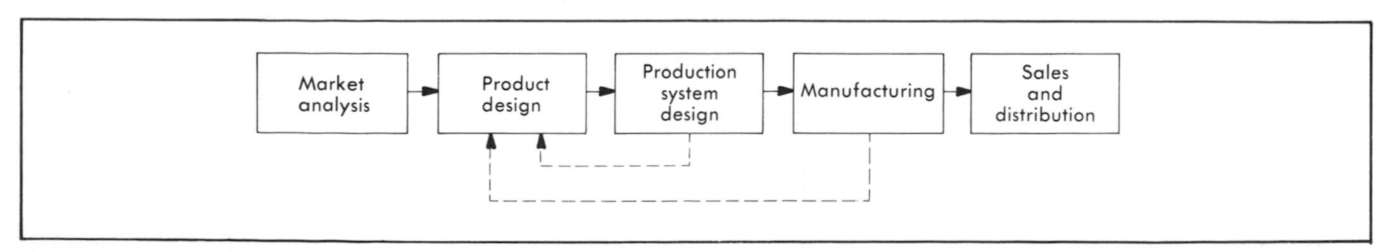

Fig. 13-5 Classical manufacturing model. (*Industrial Technology Institute*)

predicted and explained by the iterative nature of design (see Fig. 13-6).

The serial nature of the traditional approach prevents integrating design of the product and process, even when manufacturability is recognized as being of paramount importance. The result is suboptimal design and operation of the manufacturing system; extra effort, time, and money spent solving avoidable manufacturing problems; and extra manufacturing costs that continue for the life of the product. To correct this problem, it is necessary to understand the factors that make change hard to implement and the interplay between these factors. From a design for manufacture point of view, the principal factors are complexity, specialization, and functional organization.

Complexity is perhaps the main underlying cause of the problem. Prior to the Industrial Revolution, a single craftworker understood and implemented all of the functions of the manufacturing system. With the onset of the Industrial Revolution, complexity of the manufacturing system grew to the point where the work, skills, and knowledge required to go from design to production had to be split up into a multitude of specializations, each organized and compartmentalized within different companies, divisions, and departments. As businesses grew and further specialized, many companies became organized along lines of functional specialization (see Fig. 13-7).

The benefits in dealing with complexity produced by specialization are enormous. Specialization leads to the in-depth knowledge about many aspects of design and manufacture needed to solve complex problems. On the negative side, specialization tends to cause each manufacturing system activity (marketing, design, and manufacturing) to be performed separately, without benefit of interaction and input from other activities and often without vision of the manufacturing system as a whole. Although optimization occurs within each activity, without comprehension of the total system, global optimization is seldom possible. Specialization has also been largely responsible for the phased or sequential nature of the traditional product design and development approach (refer to Fig. 13-5). With the modern Information Age, complexity has now increased to the point where these two shortcomings—local optimization and phasing of specialized activities—are dominating and obscuring the benefits of specialized knowledge and the economies of scale on which it is based. This phenomena is depicted graphically by the solid line in Fig. 13-8.

Because of the complexity of the products that are now required and desired, and because of the complexity of the manufacturing systems that are needed to produce these products, it is unlikely that meaningful reductions in specialization can be made. Also, even though the need exists for restructuring many industrial organizations and operational procedures, the immense cultural and social inertia that must be overcome makes rapid change in this area unlikely. Nevertheless, to maximize the quality of early design decisions and thereby minimize the amount of engineering change, input is needed from as many manufacturing systems activities as possible and as early

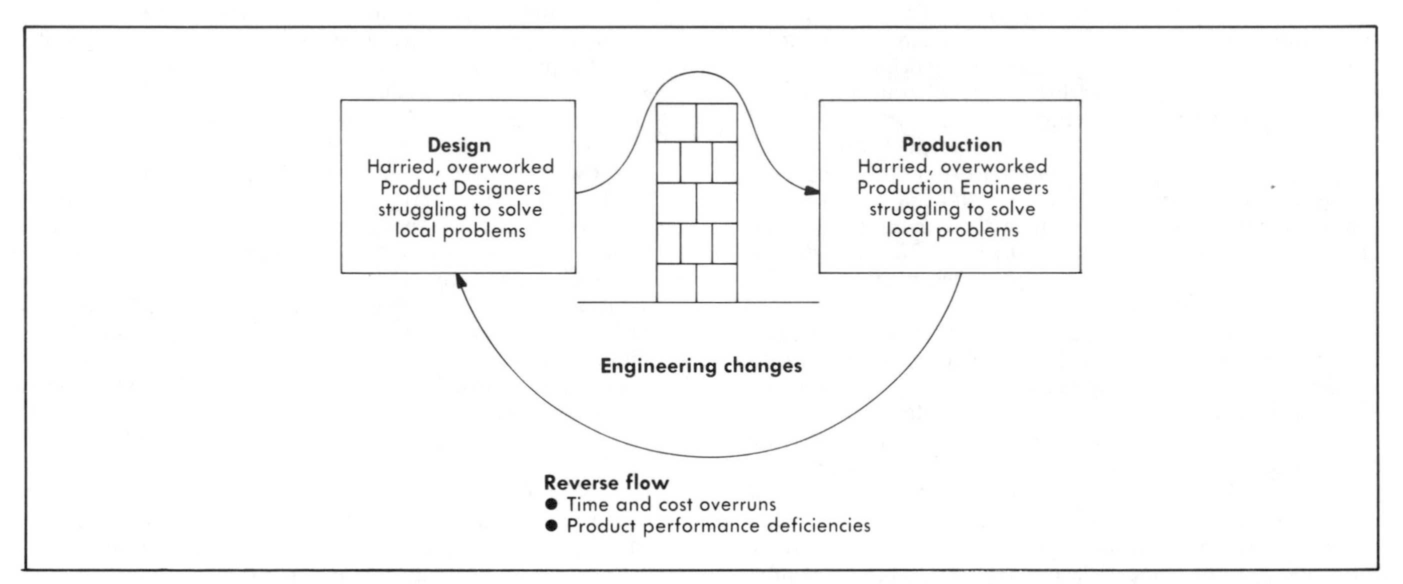

Fig. 13-6 The consequences of the phased product development and design approach results in a counterflow of costly engineering changes made late in the project. (*Industrial Technology Institute*)

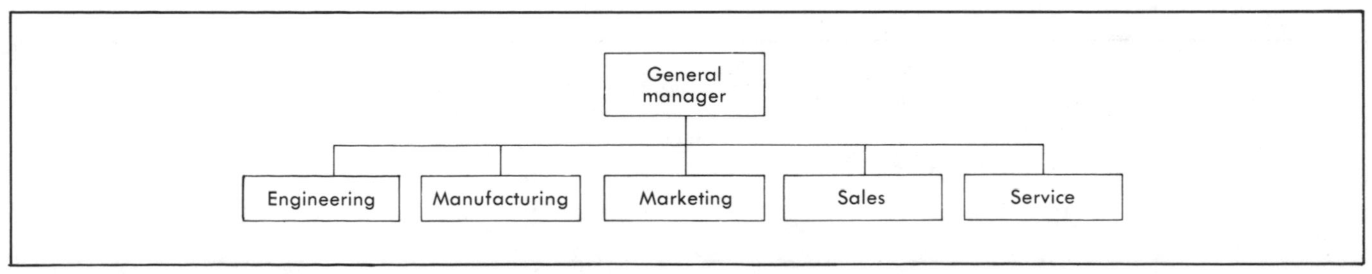

Fig. 13-7 Diagram of a typical hierarchical organization. (*Industrial Technology Institute*)

in the design process as possible. This requires the simultaneous engineering model (see Fig. 13-9). The challenge is to do this in a way that levers the advantages of specialization, while avoiding its pitfalls, to produce the preferred effect depicted by the dashed line in Fig. 13-8. Any new approach must also be compatible with existing organizational structures and business practice. While it is true that existing organizational and cultural structures may need to change, this change usually does not occur overnight.

COST REDUCTION, QUALITY AND PRODUCTIVITY

The importance of manufacturability in product design has been recognized for years. Just how important is illustrated by the well-known fact that up to 80% or more of production decisions are directly determined by the product design. This leaves little freedom of choice for process planning, especially

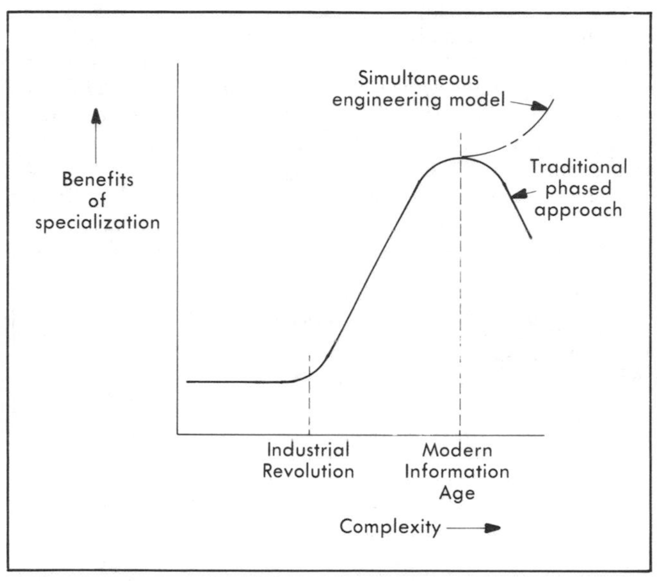

Fig. 13-8 Graph showing the benefit of specialization versus increasing complexity. (*Industrial Technology Institute*)

when process planning is often performed downstream of concept decisions (refer to Fig. 13-5). In spite of this, traditional practice has been to first optimize product function and life. Then, in detail design, component cost is scrutinized to the last detail. Component cost has been, and to a large extent still is, the primary product design consideration involving manufacturability. Another important driver is the time schedule.

Ultimately, however, it is total manufacturing system cost and product lifecycle cost that determine the bottom line (profit). These larger, harder to define costs are seldom analyzed or known until well after many irreversible tooling commitments and processing decisions have been made. When these costs do become available, cost reduction programs involving product redesign, process optimization, and work simplification studies are quickly instigated. Unfortunately, by this time, it is too late, and even obvious improvements to reduce cost are often very costly or impossible to make. In fact, many improvements must be rejected because the cost of implementation exceeds the cost savings that would be realized by the improvement.

What this all-too-frequent scenario shows is that truly effective cost reduction begins when the first inklings of a new product or product enhancement or product redesign begin to emerge into the corporate consciousness. As shown in Fig. 13-10, early design decisions affect lifecycle cost far more than years of manufacturing improvements made subsequent to concept decision and detail product design. The elimination of a part, for example, or a machining direction or separate fasteners, by design can result in the elimination of stations, operations, fixturing, and quality risks, in addition to the direct costs accompanying them, for the life of the product. No amount of optimization of speeds and feeds or work simplification or advanced manufacturing technology can match the overwhelming benefits of a product correctly designed, from the start, for manufacture.

A major reason why early design decisions affect product lifecycle cost so heavily is that they directly affect manufacturing system productivity. Product designs that are based on a design concept selected for its inherent ease of manufacture and composed of components that have been carefully designed to be easy to make, easy to fixture and handle, easy to assemble, and are carefully matched in process and materials to the manu-

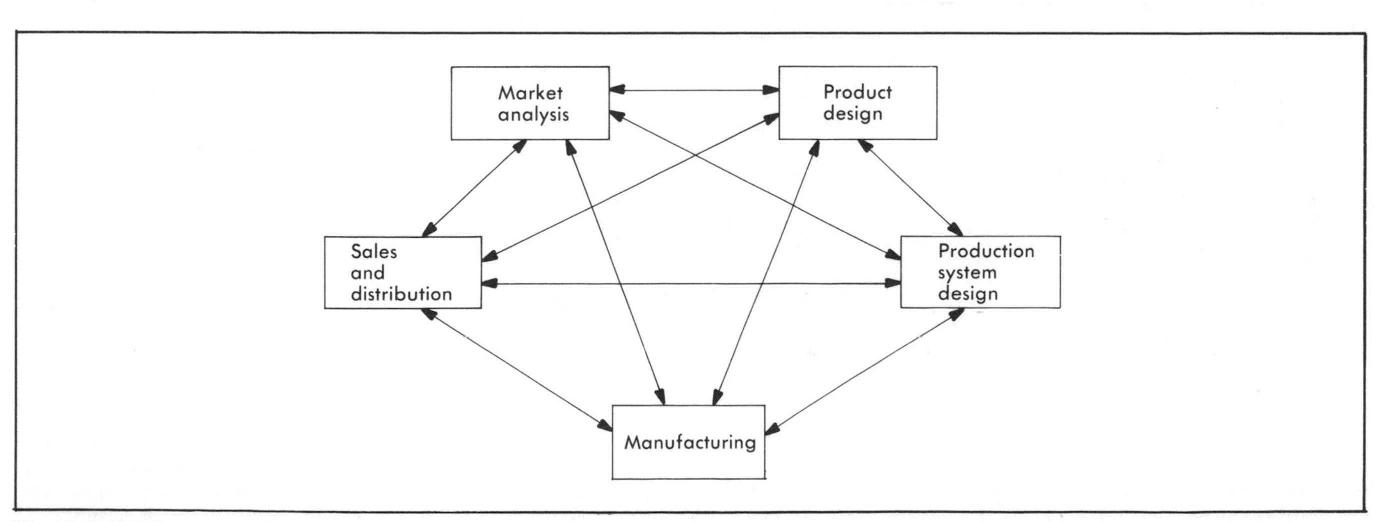

Fig. 13-9 Simultaneous engineering model. (*Industrial Technology Institute*)

CHANGE AND CHANGING NEEDS

facturing processes that produce them naturally result in productivity improvements. Along with lower cost and enhanced productivity comes quality improvements. Quality is the third term in the cost reduction equation. Improved productivity requires error-free manufacture, which in turn requires design for quality and leads to quality improvements, both real and perceived.

The synergism between cost, productivity, and quality is the real key to cost reduction. Taking the time to properly plan and design the product right the first time makes it easier to manufacture. Ease of manufacture reduces both direct and indirect manufacturing costs. Because the product is easy to manufacture, there is less quality risk and deviation from design intent. This results in a better, more desirable product, which in turn leads to increased sales and market share. Also, because the product has more perceived quality, it can be sold for a higher price. Lower total manufacturing cost plus increased sales and sale price translate into bigger profits, the ultimate goal of business.

But the benefits of implementing cost reduction from the start do not end with just lower manufacturing cost and higher quality. By spending time in the beginning to develop more complete design information, less engineering changes are required along the way to production and sales. This means less downstream compromise and chaos and, hence, a product that more closely represents original design intent. The result again is a better, more desirable product that further compounds the cost, quality, and productivity benefits previously described.

Less engineering changes also carry the promise of significant savings in development and design costs. Comparison of the total cost of engineering changes to a design project implemented using the traditional approach (refer to Fig. 13-5) and that of a project where considerable more time was spent defining the problem of design is shown in Fig. 13-11. It should be noted that the law of 10s discussed previously (refer to Fig. 13-4) is used to estimate the total cost of engineering changes shown in the figure.

A final cost reduction comes from the time saved by requiring fewer engineering changes. This saved time can translate into more timely delivery of the product as well as all the other benefits that this can provide in the marketplace.

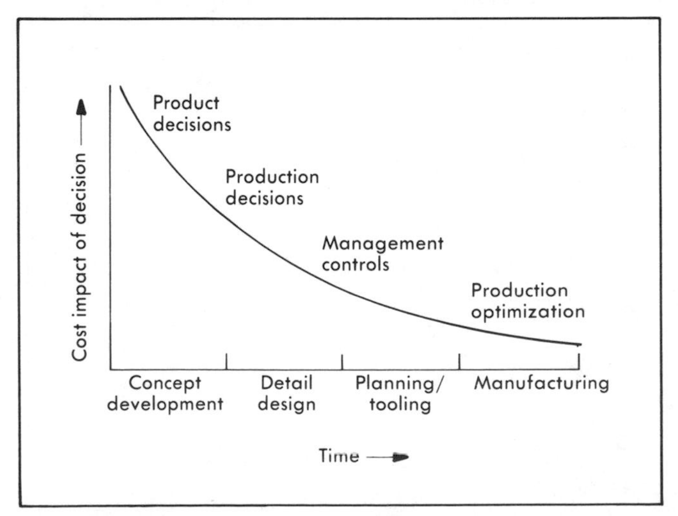

Fig. 13-10 Graph showing cost impact of decisions. (*Industrial Technology Institute*)

FLEXIBILITY

The modern Information Age, together with ever-increasing electronic computing capabilities and applications, has created an entirely new manufacturing system dimension that significantly affects the design process. This new dimension is flexibility, the ability to adapt quickly and easily to changes in product or production conditions and/or requirements. Flexibility is manifested on the manufacturing floor in the form of programmable automation and new approaches to material handling and part fixturing.

A robot is a typical example of flexible automation. Because it is programmable (under software control), the robot can be programmed to perform a variety of different tasks within its envelope of capability, without change or modification to its physical form. Hence, in a new manufacturing application, instead of designing a special piece of equipment, a standard robot, programmed to perform the new task, can be readily placed on line.

Kitting is an example of a flexible approach to material handling used for simplification and automation for assembly. A kit is nothing more than a pallet designed to hold parts at known locations and orientations with respect to an established reference frame on the pallet. A different kit, tailored to hold the parts for each product or model variation, is used to supply parts to the assembly worker or programmable automation. In this way, the assembly station is supplied with the right parts, at

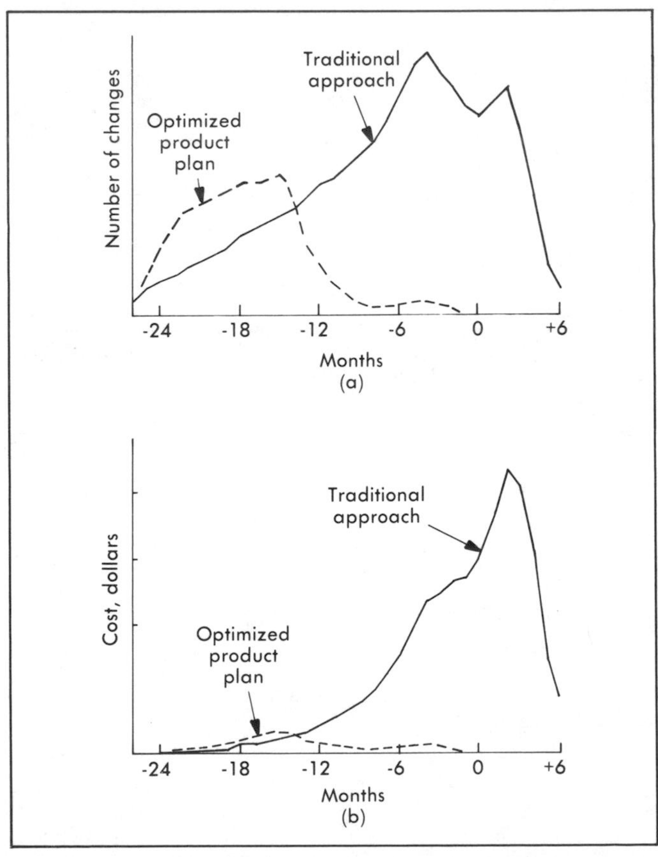

Fig. 13-11 Graphical comparison between the traditional product development and design approach and the DFM approach: (a) number of engineering changes processed and (b) total cost of engineering changes processed. (*Industrial Technology Institute*)

the right time and in the right orientation, to build the specific product or model variation required. In the case of programmable automation, the station needs only read a bar code and/or touch a tooling post on the kit to know what product is to be built, the location and orientation of parts on the kit, and how they should be handled and manipulated for assembly. For manual assembly, the assembly worker often needs only to look at the kit to know what product variation is to be built.

The ability to manufacture to customer order, produce a correct first part, manufacture a variety of different products or product models in any sequence and/or quantity, and rapidly introduce a new product or change an existing design are a few of many important capabilities promised by flexible, computer-based manufacturing. Realization of this promise requires a process-driven product design methodology (model) together with a viable means for electronically transferring design and process planning information between the computer-aided design (CAD) environment in which the product is designed and the computer-integrated manufacturing (CIM) environment in which it is manufactured, assembled, and tested. Additional information on CAD and CIM can be found in Chapter 16 of this volume.

The process-driven design model is needed to ensure product conformance with processing requirements and constraints. For example, unless the parts are designed to be handled by a robot and/or assembled from a kit, these technologies are likely to fall short of their promise or fail altogether. Similarly, the CAD/CIM link is needed to facilitate the rapid exchange of information required to achieve production flexibility. The electronic bridge allows the robot or assembly automation to be quickly reprogrammed to handle different parts or interface with a new or different kit.

MAINTAINING OPTIONS

From the standpoint of design, flexibility implies more than the ability to manufacture to customer order in any sequence and lot sizes of one or more. It also implies the ability to maintain product and process options over time. As discussed previously, change is an inherent and fundamental property of the design process. Customer needs and perceptions change, new product innovations and technology breakthroughs occur regularly, competition is constantly challenging and pushing current products, and new materials and processes are continually emerging. Maintaining options under these conditions means designing the product and process so that either can be easily and quickly changed without major cost or timing consequences.

Designing this kind of flexibility into the manufacturing system is fast becoming a primary design requirement, especially in industries involving very high volume production. In these industries, economies of scale have historically dictated fixed (nonprogrammable) automation, dedicated transfer lines, specialized fixturing and tooling, and continuous motion assembly lines. Any change to the product under these conditions implies costly and time-consuming changes to the production system. Similarly, introduction of new manufacturing technologies or innovations is often hampered by the fact that the product was never designed with the new technology in mind.

Maintaining product and process options requires that the engineer and designer look forward in time to anticipate what changes are likely to occur or be required and then to consciously plan for these changes in the design. If benefits are to be maximized and penalties minimized or avoided, all immediate, evolutionary, and revolutionary product, process, and technology changes, improvements, and breakthroughs must be identified and accounted for by design.

PROCESS-DRIVEN DESIGN

Designing a product that is responsive to the needs of flexible manufacturing as well as making it possible to maintain product and process options requires a shift to process-driven design. In process-driven design, the manufacturing process plan is developed prior to performing the product design. Although having manufacturing-led product design represents a radical departure from current practice, the wisdom in doing so becomes obvious when the impact of early design decisions on lifecycle cost of a product is considered (refer to Fig. 13-8). One of the primary reasons that early design decisions affect lifecycle cost as much as they do is because they determine and often severely limit the manufacturing options available for production of the product. This is especially true if advanced manufacturing technologies such as robotics, machine vision, and flexible automation are to be used. Process-driven design keeps the product design from unnecessarily constraining manufacturing by providing up-front guidance to the design team before the design concept is frozen and allows both to converge in a uniform and controlled fashion.

Process-driven design is implemented by specifying process requirements and the preferred methods of manufacture as design requirements before design of the product begins. The product is then designed so that it can be manufactured in this most desirable way. For example, by telling the engineer (or designer) beforehand that the assembly is to be assembled using SCARA (selective compliance assembly robot arm) type robots, the designer can provide the features to make easy use of this type of assembly equipment. As the product design evolves, it is often necessary to modify the proposed method of manufacture. The result is a refined and carefully matched product design and process plan which, by meeting the requirements of both, achieves the goal of "doing it right the first time."

THE DFM APPROACH

Design for manufacture has evolved out of the need to change the way a company performs design and the changing needs of design. The objectives of the design for manufacture approach are:

- Identify product concepts that are inherently easy to manufacture.
- Focus on component design for ease of manufacture and assembly.
- Integrate manufacturing process design and product design to ensure the best matching of needs and requirements.

Meeting these objectives requires the integration of an immense amount of diverse and complex information. This information not only includes considerations of product form, function, and fabrication, but also the organizational and administrative procedures that underly the design process and the human psychology and cognitive processes that make it possible.

Because of the complexity of the issues involved, it is convenient to divide the subject of DFM into two considerations:

1. The DFM approach or process by which a product can be effectively designed for manufacture.
2. The methodologies and tools that can be used to help enable the DFM approach and help ensure that the physical design meets the DFM objectives.

In this section, the first of these two considerations will be discussed. A discussion of DFM methodologies and tools follows in the next section.

In discussing the DFM approach to product design, it is necessary to develop a structured DFM process to help guide the design team toward its goal of a product design for ease of manufacture. Then, certain imperatives for effective DFM can be discussed relating to designer attitude or orientation and to the way the design project might be approached organizationally and procedurally. The section closes with a discussion of implementation of the imperatives.

A DFM PROCESS

Many different versions of the DFM process can be proposed. Each version is likely to be similar in the issues addressed and the concepts embodied. Differences would likely reflect idiosyncrasies imposed by the organization in which a particular version originated and the type of design problem it was meant to address. With this in mind, one proposed version of the DFM process is shown in Fig. 13-12.[5]

This process begins with a proposed product concept, a proposed process concept, and a set of design goals. All three of these inputs would be generated by a thorough, well thought out product plan developed using the team approach. Design goals would include both manufacturing and product goals. For example, the design goals may include goals for product performance improvements and added conveniences as well as manufacturing goals such as the elimination of a particular number of assembly workers and/or processing operations and the ability to substitute alternative manufacturing technologies.

Four activities, each addressing particular aspects of the DFM approach, are included within the process itself. Optimization of the product/process concept is concerned with integrating the proposed product and process plan to ensure that product and process flexibility goals are maintained, that the best match between product and process requirements is attained, and that the integrated product/process developed ensures inherent ease of manufacture. With the integrated product concept and process plan specified, the next activity focuses on component design for ease of assembly and handling and on the simplification of components to further promote ease of manufacture, improve quality, and reduce manufacturing cost.

The third activity is aimed at ensuring conformance of the design to processing needs. For example, if a particular part is to be a plastic injection molding, this step would seek to make sure the part is properly designed for the particular process involved. This would not only include designing the part for ease of injection molding, but also designing the part to simplify the tooling, fixturing, and material handling required to support the process. The last activity, which in the traditional approach has often been one of the first steps, is to optimize the product function. This activity is saved until last to ensure that all of the design constraints, including assembly, processing, and material handling requirements, are known before the optimization is attempted.

Figure 13-12 shows the four activities arranged in a circular fashion. This is to emphasize the iterative nature of the process. For instance, as a result of performing the first three activities, imagine that an approach to optimizing the product function is

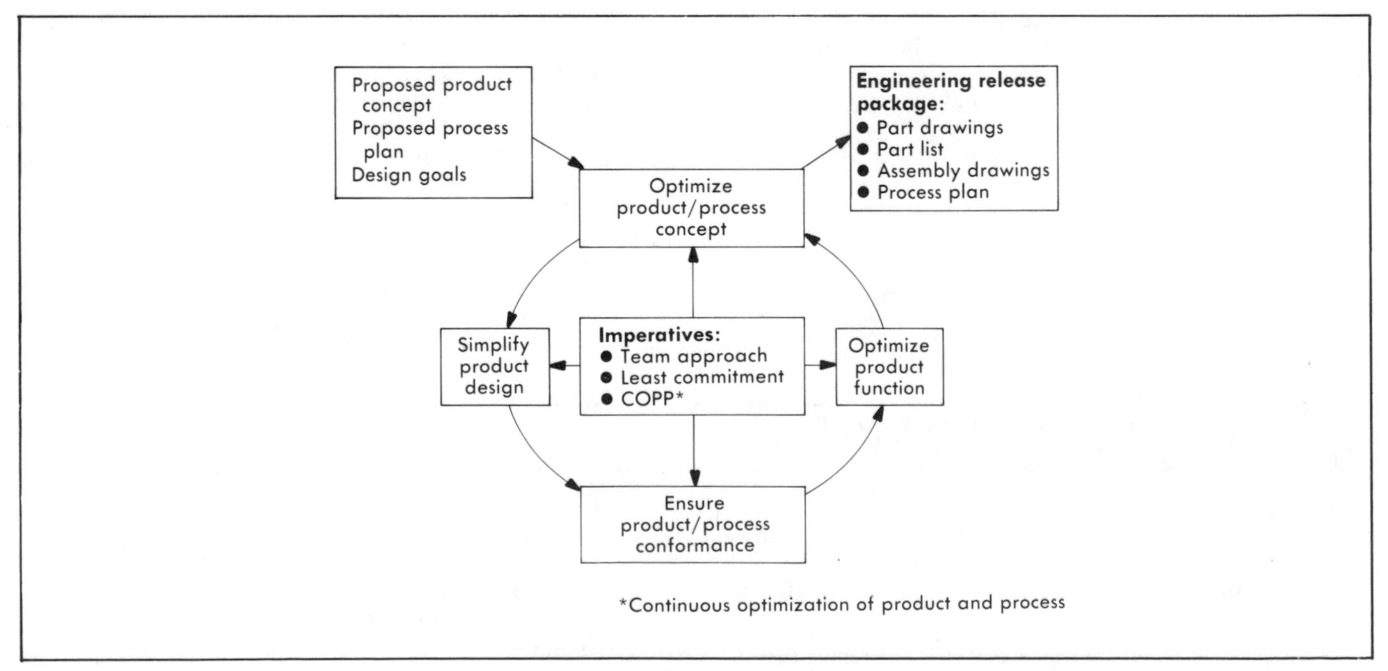

Fig. 13-12 Diagram of one type of DFM process. (*Industrial Technology Institute*)

identified that necessitates changing one or many aspects of the design. Implementing these changes implies a second iteration through the loop. Similarly, as a result of going through the four activities, a means may be found for eliminating camming in a mold, a fastening operation, or a fixture flow. Again, these opportunities would be implemented in the next design iteration. The key is to consider all aspects of the product's design and manufacture in the early stages of the design cycle so that design iteration and accompanying engineering changes are easy and cost effective to make.

One of the major objectives of the DFM process is to ensure that the product (including materials selection) and process are designed together. By beginning with both a proposed process concept and a proposed product concept, and by making sure that all design changes are reflected in both the product and the process, it is possible to include manufacturing recommendations and a process plan as part of the engineering release package. This has great advantages because it leads to few, or no, manufacturing surprises. Instead of being "can-doers" who are handed a set of drawings and told to get on with it, manufacturing personnel have participated in the design from the start, have a clear picture of what must be done, and are likely to be geared up and ready to go into action well in advance of the release. An added advantage of this approach is that both the manufacturing and engineering departments share equally in ownership of and ultimate commitment to the design.

IMPERATIVES FOR EFFECTIVE DFM

Product design is only one of many complex activities that comprise the manufacturing system. From the very beginnings of an idea for a new product or product model to the last step in retiring it from production and use, a great deal of product knowledge, originating within all components and activities of the manufacturing system, is gained. This interdependency between time, design knowledge, and the manufacturing system in which the product is conceived, designed, produced, and eventually brought to the marketplace to be sold and serviced makes it imperative that there be effective communication within the manufacturing system and that there is flexibility to adapt and modify the design during each product stage as new knowledge becomes available. These imperatives, which are essential for good design in the modern Information Age, can be formalized as the following design principles:

1. *Simultaneous engineering*. The design project should, from the beginning, have continued input from all aspects of the manufacturing system because product function, cost, and quality result from the combined efforts of all components of the manufacturing system. To enable this, all relevant manufacturing system components (process planning, facilities planning, manufacturing equipment design, material handling, marketing and sales, distribution, and service) should be designed and/or planned concurrently with the product design. The simultaneous engineering principle is best implemented using a team approach (refer to Fig. 13-9). Pursuit of this principle helps ensure that total product knowledge is as complete as possible at the time each design decision is made.
2. *Principle of least commitment*. The designer or design team should pursue a policy of least commitment because of the iterative nature of design, the ripple effect, and the real cost of engineering change; that is, in progressing from step to step or phase to phase in the realization of a design, *no irreversible decision should be made until it must be made*. This principle permits maximum flexibility in each step, ensures that better alternatives can be implemented if they become available, and when necessary, makes it easier to accommodate engineering changes. Planning for change by designing flexibility into the product and/or process as well as designing in a means for maintaining options are natural extensions of this principle.
3. *Continuous optimization of product and process*. The best design for mass production is seldom arrived at the first time because of incomplete knowledge. Hence, a policy of improvement and simplification of the product and process should be pursued. That is, for mass-produced products, changes that are directionally correct (help optimize the product design together with the other components of the manufacturing system as a whole) should be implemented on a continuing basis. This principle essentially acknowledges that the benefits of global optimization (lower lifecycle cost, better quality, increased market share, and flexibility) always justify, in the long run, the incremental cost required to achieve it. It also implies that incremental changes are usually the most effective. Introducing frequent small improvements based on customer's reactions or improved manufacturability is less risky than taking one great leap forward based on market researcher's prognostications or the need to cut manufacturing costs. To be practical, continuous optimization of product and process (COPP) should be pursued in conjunction with simultaneous engineering and the principle of least commitment.

IMPLEMENTING THE DFM IMPERATIVES

As shown in Fig. 13-12, the design imperatives underly the entire DFM process and are pivotal to its successful implementation. Implementation of these imperatives tends to be hierarchical in nature; that is, the team approach (simultaneous engineering) generally needs to be adopted before the principle of least commitment can be made to work well. Both of these principles need to be operative for continuous optimization of product and process to work.

The traditional approach has some definite advantages from a management point of view (refer to Fig. 13-5). For example, it helps executives control a potentially messy process, it allows functional managers to focus closely on their jobs, and it keeps engineers busy on several different jobs. Use of the team approach, on the other hand, is much more difficult to manage and, in many cases, may even call for some novel management approaches. The team approach may also require costly and complex high-tech tools such as CAE/CAD/CAM and common databases. Hence, just as a successful product requires careful planning, so does the installation of the team approach.

In general, the team should be multifunctional, and efforts should be made to keep the same people on the project from start to finish; in reality, however, this does not always happen. As the design evolves, each team member stays in close communication so that changes and fast-breaking developments are quickly relayed to each component of the manufacturing system as they occur. This allows all activities to progress in parallel, thereby shortening the development cycle and increasing the completeness of information on which design decisions are based. Engineers and designers start work before feasibility testing is complete, and manufacturing and sales begin gearing up well before the design is finished.

One company has resorted to a "60 foot rule"; this rule requires that all disciplines involved in simultaneous engineering/ process-driven design be physically located together. This company has found that people fail to communicate with each other if their desks are more than 60′ apart.

In making up the team, it is important that there be a balance of skills and personalities. Team members should be technically excellent, but broad enough to understand what others have to say, share goals, and exercise good people skills. The team approach also depends strongly on the presence of an experienced, strong team leader (champion).

Hallmarks of the team approach are to cut the volume of communications within the company and to ease the essential flow of information (see Fig. 13-13). Doubling the team size quadruples the communication burden, so the team should be small. Eight to 12 main players is ideal. Also, it should be noted that the team approach saves time and cuts down on late engineering changes by reviewing the design intensely and regularly. But, to do this, the team should be relatively autonomous. Making frequent reports to top management can drive developers to waste weeks polishing their presentations.

Once the team approach is in place, the secret to quick product development is exhaustive product planning, including technology planning and planning for the future. Establishment of checkpoints for a technology's robustness and readiness is essential before letting engineers and designers use it. The key is to make sure that every technological risk has a potential market payoff. Unnecessary risks should not be taken. Often the best course is to reconfigure components already proven in mass production instead of designing everything from the ground up. Also, it is often wise to test new technologies in existing products before putting them all together in a new product. Ultimately, minimizing risk associated with new technology is best done by judicious application of the principle of least commitment.

Knowing what new technologies need to be considered and planning for them in current designs is an important application of the principle of least commitment. Planning several generations of a new product also helps facilitate continuous optimization of product and process through incremental change. Experience has shown that for each increase in technical difficulty, cost and time increase exponentially. Developing technically advanced products by doing an intermediate version, then a final one, can take only one fourth as long as going all the way to the final version in one gut-wrenching effort. This explains why there are so many small improvements at office equipment and electronics shows. The new products are a series of little pops, not big bangs.

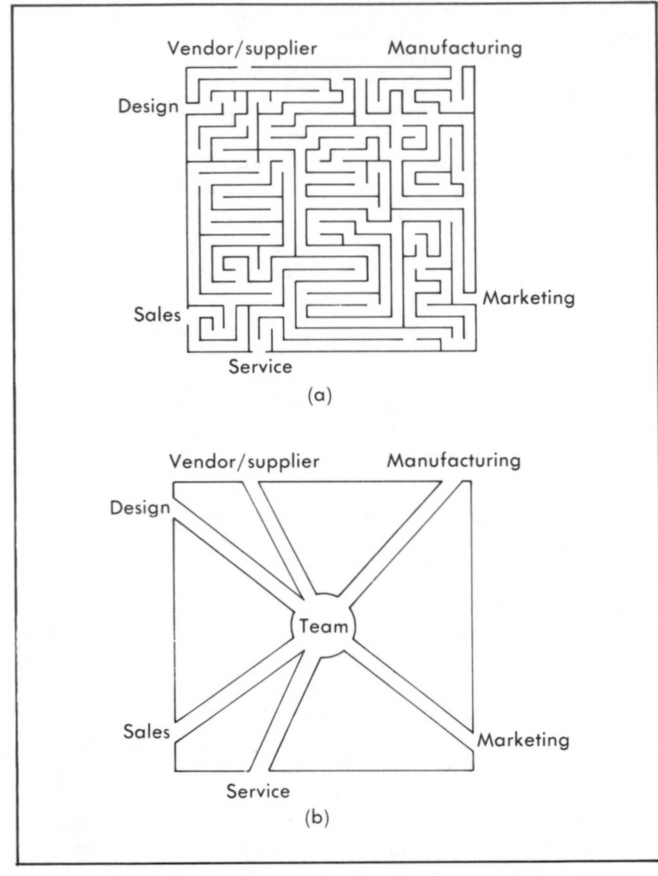

Fig. 13-13 Diagram of communication channels: (a) phased design approach and (b) DFM approach. (*Industrial Technology Institute*)

DFM METHODOLOGIES AND TOOLS

In addition to the DFM process, there are a variety of DFM methodologies and tools that help promote the objectives of DFM by guiding the designer in making better informed design decisions. Figure 13-14 gives a selected list of DFM tools and shows where they might logically fit into the DFM process. The purpose of this section is to review these tools by describing what they are and how they might be used.

DFM PRINCIPLES AND RULES

Over the years, a wealth of product design approaches, techniques, short cuts (rules of thumb), and design tips have evolved out of product design and manufacturing experience that help show the way to good design for manufacture. Knowledge of this information and the ability to correctly apply it has always been one of the hallmarks of the expert design engineer and manufacturing engineer. DFM principles and rules are codifications of this empirical experience into forms that can be directly used to help guide the designer in the early stages of design. The great advantage of using concisely stated design for manufacture principles and guidelines is that they explicitly state what is intuitively obvious to experienced designers and commonly used by good designers. Such a system of principles and guidelines, therefore, provide a firm basis on which design knowledge can be expanded in a systematic way.

Axiomatic Design

An axiomatic approach to design is based on the belief that fundamental principles or axioms of good design exist and that

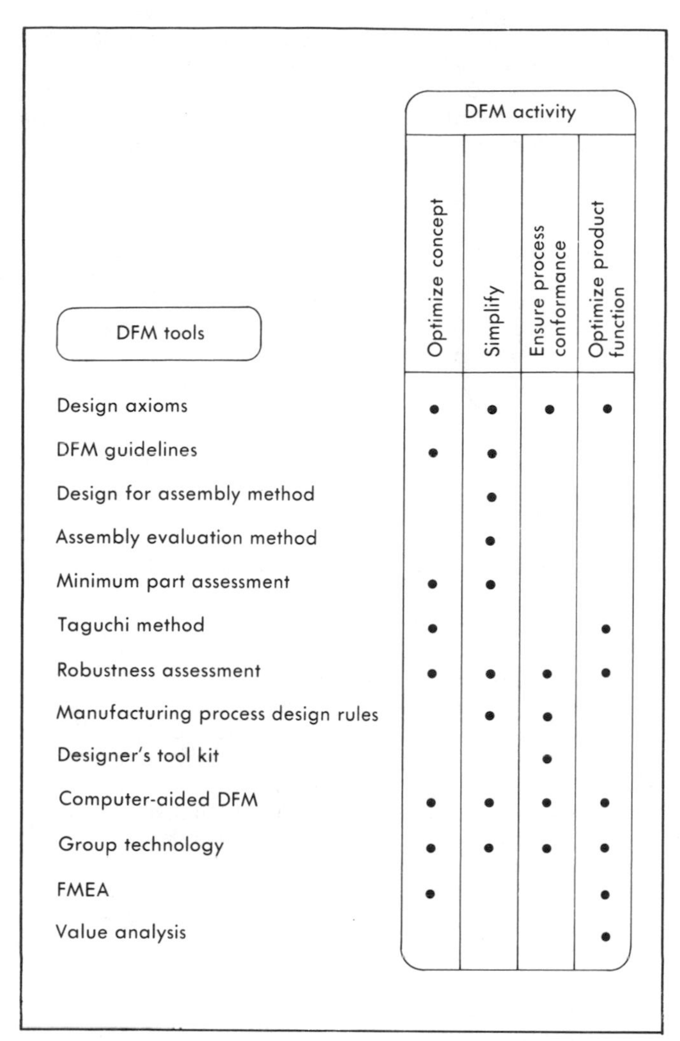

DFM tools	Optimize concept	Simplify	Ensure process conformance	Optimize product function
Design axioms	•	•	•	•
DFM guidelines	•	•		
Design for assembly method		•		
Assembly evaluation method		•		
Minimum part assessment	•	•		
Taguchi method	•			•
Robustness assessment	•	•	•	•
Manufacturing process design rules		•	•	
Designer's tool kit			•	
Computer-aided DFM	•	•	•	•
Group technology	•	•	•	•
FMEA	•			•
Value analysis				•

Fig. 13-14 Design for manufacture tools. (*Industrial Technology Institute*)

use of the axioms to guide and evaluate design decisions leads to good design. By definition, an axiom must be applicable to the full range of design decisions and to all stages, phases, and levels of the design process. Design axioms cannot be proven, but rather must be accepted as general truths because no violation or counterexample has ever been observed.

A study of many successful designs by several individuals in 1977 led them to propose a set of hypothetical axioms for design and manufacturing.[6] Analysis and refinement of the initial axioms has shown that good design embodies two basic concepts. The first of these is that each functional requirement of a product should be satisfied independently by some aspect, feature, or component within the design. The second basic concept is that good designs maximize simplicity; in other words, they provide the required functions with minimal complexity. These two concepts have been formalized as follows:[7]

- *Axiom 1:* In good design, the independence of functional requirements is maintained.
- *Axiom 2:* Among the designs that satisfy Axiom 1, the best design is the one that has the minimum information content.

Use of the design axioms in design is a two-step process. The first step is to identify the functional requirements (FRs) and constraints. Each FR should be specified such that the FRs are neither redundant nor inconsistent. It is also useful in this step to order the FRs in a hierarchical structure, starting with the primary FR and proceeding to the FR of least importance. Once the functional requirements and constraints are specified for a given product or design problem, the second step is to proceed with the design, applying the axioms to each individual design decision.

Beginning with conceptual design and followng through each stage of realization of the design solution, the axioms can be used to provide insight into the design problems at hand, show the way to good design, and form the basis for quality design decisions. Each decision should be guided by the axioms and must not violate them. A product designed by following the axioms will be one that is optimized with respect to all aspects of its lifecycle as well as one that facilitates global optimization of the manufacturing system as a whole.

Application of the design axioms to the analysis and design of products and manufacturing systems is not always easy or straightforward. Because the axioms are quite abstract, their use requires considerable practice as well as extensive on-the-job design and manufacturing experience and judgment. For this reason, the reader is encouraged to study the axioms carefully and to experiment with their use in understanding problems with which he or she is very familiar. As confidence in the correctness and applicability of the axioms grows, the reader is further encouraged to try to use the insight provided by the axioms to redesign or correct a particular situation. Most people who become familiar with the axiomatic approach in this way learn very quickly the power of the method and how to effectively use it to both guide and evaluate design.

To illustrate the use of the design axioms, consider the radar antenna design problem discussed previously (refer to Fig. 13-3). The primary functional requirements of this design can be stated as:

- Provide support for the radar antenna.
- Protect the antenna from weather-induced disturbances such as wind loads.

Examination of the designs in Fig. 13-3 shows that both designs provide the stated functional requirements. In view *a*, however, both functions are satisfied by the same rigid support structure. In terms of the first axiom, it is seen that the functional requirements are coupled in this solution, and hence their independence has not been maintained. View *b*, on the other hand, is an example of "good design" according to the first axiom because the independence of functional requirements in this case is maintained. Hence, using the axiomatic approach, the design in view *b* would be selected in preference to the design in view *a*. The second axiom would then be applied by seeking the simplest possible implementation of the design in view *b*.

One of the important reasons why the axiomatic approach leads to good design is that independence of functional requirements effectively short-circuits the ripple effect. By uncoupling or decoupling functional requirements, changes in a particular requirement or changes made to the design solution affecting that requirement do not affect the other functional requirements satisfied by the design. For example, if the sheltering structure in view *b* of Fig. 13-3 were to be changed in some way, the change should not affect the support structure, provided the change is not a drastic revision of the concept itself.

DFM PRINCIPLES

Similarly, a change made to the support structure should not "telegraph" to the shelter. By contrast, any change made to the design in view *a* would affect both functions.

Use of the second axiom can be illustrated by considering the tradeoffs between hard and flexible automation. Information is contained in both hardware and software. In hard automation, all information is designed into hardware. This accounts for the disadvantage of hard automation, which is its intolerance to major product variations. To offset this, the tendency in flexible automation is to transfer as much information as possible to software. The second design axiom requires that total information content be minimized by judicious distribution of information between hardware and software. In many cases, the information content can best be minimized by clever use of both hardware, software, and perhaps "human ware." For example, designing guiding and self-locating features into the product components and using a simple SCARA-type robot for component insertion involves less information content than using a multi-degree-of-freedom robot combined with tactile sensing and vision. Similarly, if a component or operation is particularly difficult to automate for one reason or another, the low-information solution may be to do it manually. Because hardware includes both product and automation equipment, it is essential that both be considered during design. This is a simple illustration of why the team approach is often essential for successful application of the second axiom.

The design axioms can be used to both explain and provide insight into many well-known product design strategies and tactics. For example, standardization is a way of separating or decoupling product and marketing needs from manufacturing. Standardization also reduces the amount of information required for product manufacture. This can be illustrated by considering modular design. A module is a self-contained component with standardized interfaces to other product modules and to the production equipment and tooling used in the product manufacture. Individual modules can be varied to provide functional and styling diversity. Similar diversity can be provided by using different combinations of standard modules. All of this has no effect on the production line as long as the module/process equipment and tooling interfaces are standardized. Modular design also reduces final assembly information content because there are fewer parts to assemble and each module can be fully checked prior to final assembly.

Multifunctional and multiuse parts are another way of decoupling the product from manufacture and reducing manufacturing information content. For example, the same mounting plate can be designed to mount several different components. By having robot grippers only touch the mounting plate, the particular component being installed is decoupled from the installation process. Information is reduced because the same mounting plate and installation process can be used in a variety of applications.

Reducing or minimizing the number of individual parts in a product is perhaps the most effective way to reduce information content. Put another way, a part that is eliminated costs nothing to design, make, assemble, move, handle, orient, store, purchase, clean, inspect, rework, or service. It never jams or interferes with automation. It never fails, malfunctions, or needs adjustment. It requires no drawing or part number and never needs to be changed.

When viewed in this way, it is seen that many of the rules of thumb and other broadly applicable design for manufacture guidelines are really ways of reducing the manufacturing information content. Reduced information content explains the emphasis on near net shape processes and the need to avoid secondary processing. Often, higher material and/or processing cost can be accepted because the reduced information content leads to lower overall production cost and indirect costs. In automation applications, separate fasteners are difficult to feed, tend to jam, require monitoring for presence and torque, and require costly fixturing, parts feeders, and extra stations. Avoiding separate fasteners eliminates all this extra automation information. The same thing is true for assembly directions and material handling. By designing for *Z*-axis insertion, many needed degrees of freedom and the information that goes with them are eliminated. Similarly, the use of symmetry to assist in achieving proper part orientation and the use of parts magazines, tube feeders, part strips, as well as palletized trays and kitting techniques for preserving orientation greatly reduces the information required to manipulate and handle parts. Finally, use of generous tapers and other guiding, orienting, and locating features in part design greatly simplifies component insertion and assembly because it reduces the amount of information needed to perform these tasks. These concepts aid both automated and manual assembly.

DFM GUIDELINES

DFM guidelines are systematic and codified statements of good design practice that have been empirically derived from years of design and manufacturing experience. Typically, the guidelines are stated as directives that act to both stimulate creativity and show the way to good design for manufacture. If correctly followed, they should result in a product that is inherently easier to manufacture. Various forms of the design guidelines have been stated by different authors,[8,9,10] a sampling of which are provided as follows:

1. Design for a minimum number of parts.
2. Develop a modular design.
3. Minimize part variations.
4. Design parts to be multifunctional.
5. Design parts for multiuse.
6. Design parts for ease of fabrication.
7. Avoid separate fasteners.
8. Minimize assembly directions; design for top-down assembly.
9. Maximize compliance; design for ease of assembly.
10. Minimize handling; design for handling and presentation.
11. Evaluate assembly methods.
12. Eliminate or simplify adjustments.
13. Avoid flexible components.

It should be noted that DFM guideline number 13, "avoid flexible components," is really a subset of guideline number 10. In general, most of the guidelines given can be subdivided into an almost endless list of additional rules that become more and more specific to particular applications and situations. For this reason, many companies have found it advantageous to develop a distinct set of rules that apply more specifically to their particular business.

APPLYING THE GUIDELINES

Application of the DFM guidelines is not always easy or straightforward. They show the way, but do not replace the talent, innovation, and experience of the product development team. They must also be applied in a manner that maintains and, if possible, enhances product performance and marketing goals. Design guidelines should be thought of as "optimal suggestions," which, if successfully followed, will result in a high-quality, low-cost, and manufacture-friendly design. If a product performance or marketing requirement prevents full compliance with a particular guideline, then the next best alternative should be selected. Use of the guidelines in this way helps to both ensure a product design optimized for manufacture and to delineate problem areas requiring special attention. Hints for creative application of the guidelines as well as insights to stimulate innovative design are provided in the following discussion.[11]

Minimize Total Number of Parts

A part is a good candidate for elimination if there is (1) no need for relative motion, (2) no need for subsequent adjustment between parts, (3) no need for service or repairability, and (4) no need for materials to be different. However, part reduction should not exceed the point of diminishing return, where further part elimination adds cost and complexity because the remaining parts are too heavy, too complicated to make and assemble, or are too unmanageable in other ways.

Perhaps the best way to eliminate parts is to identify a design concept that requires few parts. Integral design, or the combining of two or more parts into one, is another approach. Besides the advantages previously given, integral design reduces the amount of interfacing information required, and decreases weight and complexity. One-piece structures have no fasteners, no joints, and fewer points of stress concentration. Conversely, structural continuity leads to high strength and light weight. An example of a single stamping that replaced a two-part assembly is shown in Fig. 13-15.

Plastics are a major key to integral design and are available for making springs, bearings, cam and gears, fasteners, hinges, and optical elements. Powder metallurgy (PM) is a good alternative if plastic parts do not have adequate strength, heat resistance, or cannot be held to the tolerance needed. Brazed, welded, or staked assemblies of stampings and/or machined parts can often be made as one-piece PM parts. Extrusions and precision castings are also good ways to eliminate subassemblies. Although switching to a different manufacturing process may lead to a more costly part, experience with part integration has shown that a more costly part often turns out to be more economical when assembly and overhead costs are considered. Less assembly also often implies less quality risk.

Develop a Modular Design

Designing for modularity requires careful consideration of a variety of needs. To decouple the product from the method of manufacture, it is essential that stable groups be identified and that the interfacing information be specified in a way that facilitates the desired decoupling. Often, standardization of just a few dimensions is all that is needed to decouple product from process. An example of this is the standardization of the base module of the IBM typewriter. Because all the location and automation interface information is concentrated in the typewriter base, changes to other modules do not affect the automation system (provided the module handling interfaces are not changed).

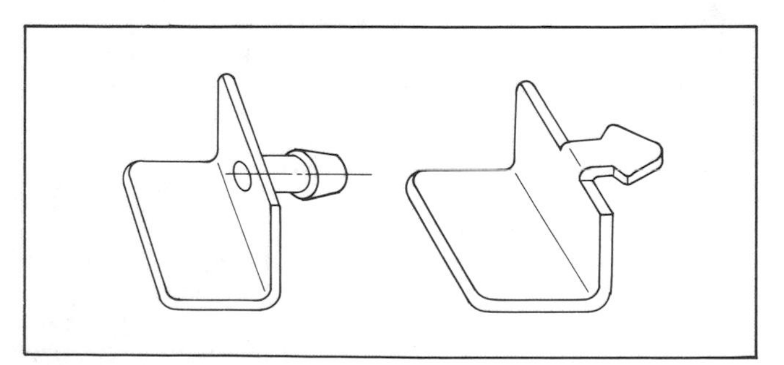

Fig. 13-15 Two-part assembly (left) is replaced by a single stamping.

In seeking to identify the minimum number of stable groups, it is useful to look at the distribution of functional requirements among different product models and lines to see if a particular function could be satisfied everywhere by one module. Looking for common problems within different models and product lines is another approach. Will the same solution work everywhere? Other questions to ask include the following:

- Can customization (diversity) requirements be satisfied using add-on modules?
- How will future product(s) differ from the current design? What commonalities will exist?
- What modular configuration will simplify material handling? Decouple quality from production? Decouple variation in vendor-supplied components from production? Decouple style from production?

Experience has shown that products consisting of 4-8 modules with 4-12 parts per module are most automation friendly. A good design strategy is to keep the product generic for as long as possible during assembly by saving the specialized modules for last. If possible, the modules should be designed to add up to the final product, thereby eliminating the need for a housing or other integrating structure. Also, information content is reduced if all modules (except perhaps the base) are approximately the same size.

Minimize Part Variations

Information content of the product and quality risks are reduced when part variations (such as the types of screws used) are kept to a minimum. It is seldom justifiable, for example, to use several screw sizes or types of metal in one part. Minimizing part variations also simplifies manufacturing by reducing the information content of the production system required to produce the part.

Use of standard (off-the-shelf) components also helps reduce part variations as well as total information content of the manufacturing system. A stock item is always less expensive than a custom-made item. Standard components require little or no lead time and are more reliable because characteristics and weaknesses are well known. They can be ordered in any quantity at any time. They are usually easier to repair, and replacements are easier to find. Use of standardized components enables the supplier to become a part of the design team.

The many advantages of standardization can be further amplified through the concept of standardization and rationalization (S&R). In the S&R approach, standardization is defined as the reduction in number of part numbers used in current and former designs. Rationalization is the identification of the fewest number of parts required for use in future designs. For example, at one company the computer printout of washers available to the designer was 14 pages and contained 448 part numbers.[12] Although the various washers were arranged in size order, it took the designer some time to select the washer because of variations in finish and thickness. A new, rationalized list was developed that contains only 7 washers with bore size ranging from 3 to 16 mm. There is only one material (steel), one finish, and one thickness. This spec has been loaded into the standard library in the computer so the designer can call up the file and select the washer. The washers have also been sized to complement the company's new rationalized standard bolt sizes. S&R is a substantial timesaver to the designer in three ways:

1. Parts are selected from a highly compressed listing that is well organized.
2. The selected part has already been tested, saving a great deal of time. This is particularly true for parts requiring life test, such as switches and bearings.
3. Time does not have to be expended locating a suitable supplier and negotiating a favorable price. Standard parts generally are the volume parts.

Design Parts to be Multifunctional

Combine function wherever possible. For example, design a part to act both as a spring and a structural member or to act both as an electrical conductor and structural member. An electronic chassis can be made to act as an electrical ground, a heat sink, and a structural member. Less obvious combinations of function might involve adding guiding, aligning, and/or self-fixturing features to a part to aid in assembly or providing a reflective surface or recognizable feature to facilitate vision inspection. These latter examples illustrate inclusion of functions that are only needed during manufacture. Such function combinations are often the result of design for manufacture awareness produced by the team approach.

Design Parts for Multiuse

Many parts can be designed for multiuse. For example, the same mounting plate can be designed to mount a variety of components. The same gear can be used for different applications in different products. A spacer can also serve as an axle, lever, or standoff. Multiuse parts reduce manufacturing system information content by reducing the number of different parts or part variations that need to be manufactured. They also produce economies of scale because of increased production volume of fewer parts and economies of scope because the same part is being used in a variety of applications and products.

The key to multiuse part design is the identification of part candidates. Multiuse parts can be created using a group technology database (discussed subsequently) by first sorting all parts (or a statistical sample) manufactured or purchased by the company into two groups consisting of (1) parts that are unique to a particular product or model (such as crankshafts and housings) and (2) "building block" parts that are generally used in all products and/or models (shafts, flanges, bushings, spacers, levers, ball bearings, switches, connectors, and hardware). Each group is then divided into categories of similar parts (part families). Multiuse parts are created by standardization and rationalization of similar parts. This process consists of sequentially seeking to (1) minimize the number of part categories, (2) minimize the number of variations within each category, and (3) minimize the number of design features within each variation. Once developed, the rationalized family of standard parts should be used exclusively in new product designs. Also, manufacturing processes and tooling based on a composite part containing all design features found in a particular part family should be developed. Individual parts can then be obtained by skipping some steps and features during the manufacturing process.

This approach has been used by one company to review 14 commodity product groups that it manufactures.[13] Of the 1919 parts contained in this group of products, 235 were identified as building block parts for use in future designs. Another 250 parts, such as nameplates and external housings, were identified as unique parts. The remaining 1434 parts, or about 75% of the total number of parts examined, were classified as proliferative parts that will be excluded from future designs. Projecting this against the company's current parts database of 50,000 numbers, it is anticipated that this approach will eventually lead to 6000 building block parts, 6500 unique parts, and 37,500 candidates for elimination.

Design Parts for Ease of Fabrication

This guideline requires that individual parts be designed using the least costly material that just satisfies functional requirements (including style and appearance) and such that both material waste and cycle time are minimized. This in turn requires that the most suitable fabrication process available be used to make each part and that the part be properly designed for the chosen process. Use of near net shape processes are preferred whenever possible. Likewise, secondary processing (such as finish machining and painting) should be avoided whenever possible. Secondary processing can be avoided by specifying tolerances and surface finish carefully and then selecting primary processes (such as precision casting, PM) that meet these requirements. Also, material alternatives that avoid painting, plating, or buffing should be considered. This guideline is based on the recognition that higher material and/or unit process cost can be accepted if it leads to lower overall production cost. In other words, adding information content to a particular part is acceptable as long as total information content of the product/process is reduced.

Avoid Separate Fasteners

Separate fasteners involve large amounts of information. Even in manual assembly, the cost of driving a screw can be 6-10 times the cost of the screw. One of the easiest things to do is eliminate fasteners in assembly by using snap fits. If fasteners must be used, cost as well as quality risks can be significantly reduced by minimizing the number, size, and variations used and by using standard fasteners whenever possible. Screws that are too short or too long, separate washers, tapped holes, and round and flat heads (not good for vacuum pickup) should be avoided. Conversely, captured washers should be used for reduced part placement risk and improved blow feeding. Self-tapping/forming/locking fasteners are preferred as are screws with dog or cone (chamfered) points for improved placement success. Also, screw heads designed to reduce "cam-out" problems, bit wear, and fastener damage should be used. For vacuum pickup, screw heads having flat vertical sides should be used (socket head, fillister head, and hex head).

Minimize Assembly Directions

All parts should be assembled from one direction. Extra directions mean wasted time and motion as well as more transfer stations, inspection stations, and fixture nests. This in turn leads to increased cost, increased wear and tear on equipment due to added weight and inertia load, and increased reliability and quality risks. The best possible assembly is when all parts are added in a top-down fashion to create a *Z*-axis stack. Multimotion insertion should be avoided. Ideally, the product should resemble a Z-axis "club sandwich," with all parts positively located as they are added.

Maximize Compliance

Because parts are not always identical and perfectly made, misalignment and tolerance stackup can produce excessive assembly force leading to sporadic automation failures and/or product unreliability. Major factors affecting rigid part mating include part geometry (accuracy, consistency), stiffness of assembly tool, stiffness of jigs and fixtures holding the parts, and friction between parts. To guard against this, compliance should be built into both the product and production process. Methods for providing compliance include highly accurate (consistent) parts, use of "worn-in" production equipment, remote center compliance, selective compliance in assembly tool (SCARA robot), tactile sensing, machine vision systems, designed-in compliance features, and external effects. Although a variety of combinations of these approaches are commonly used, experience has shown that the simplest solution consists of a combination of acceptable (consistent) quality parts, designed-in compliance features, accurate (rigid) base components, and selective compliance in the assembly tool (SCARA robot).

Designed-in compliance features include the use of generous tapers or chamfers for easy insertion, use of leads and other guiding features, and use of generous radii when possible (see Fig. 13-16). A nice approach, if possible, is to design one of the product components, perhaps the largest, to act both as the part base (part to which other parts are added) and as the assembly fixture (avoid need for a special fixture to hold the assembly). A good base design may include features that aid manufacturability, but are not needed for product function. In any case, the part base should be made as stable and rigid as possible to improve insertion accuracy and simplify handling. If fixturing is required, "fixture-friendly" features such as accurate location points, generous tapers, and other guiding features that provide easy compliance between base part and fixture should be provided. Gravity is an extremely useful external effect that assists compliance and costs nothing. In addition to assisting with insertion, gravity is useful for feeding parts and for ejecting finished and defective product.

Minimize Handling

Position is the sum of location (X, Y, Z) and orientation (α, β, γ). Position costs money. Therefore, parts should be designed to make position easy to achieve, and the production process should maintain position once it is achieved. The number of orientations required during production equates with increased equipment expense, greater quality risk, slower feed rates, and slower cycle times. To assist in orientation, parts should be made as symmetrical as possible (see Fig. 13-17). If polarity is important, then an existing asymmetry should be accentuated, or a very obvious asymmetry should be designed in, or a clear identifying mark provided. Orientation can also be assisted by designing in features that help guide and locate parts in the

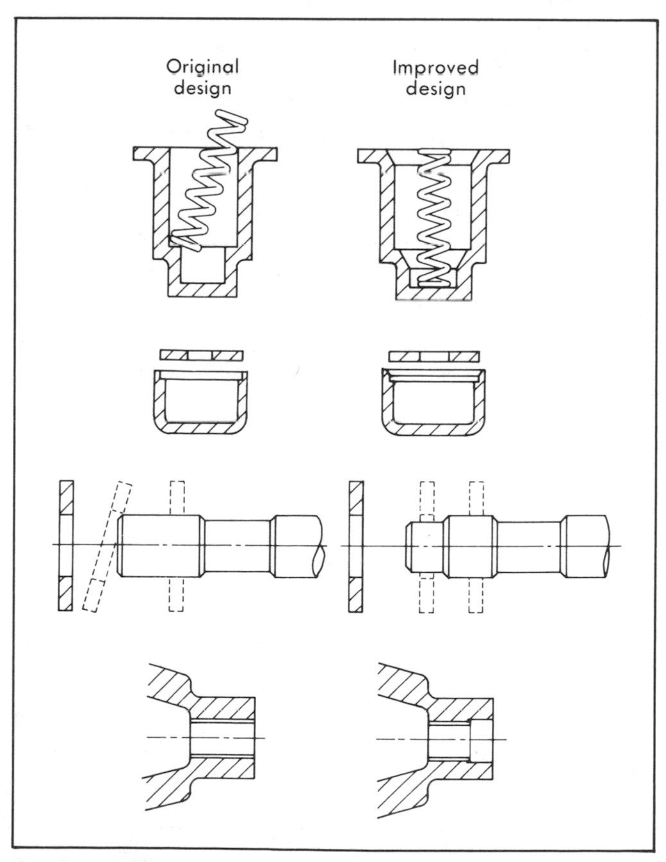

Fig. 13-16 Changes in design that facilitate inserting and mounting of components.

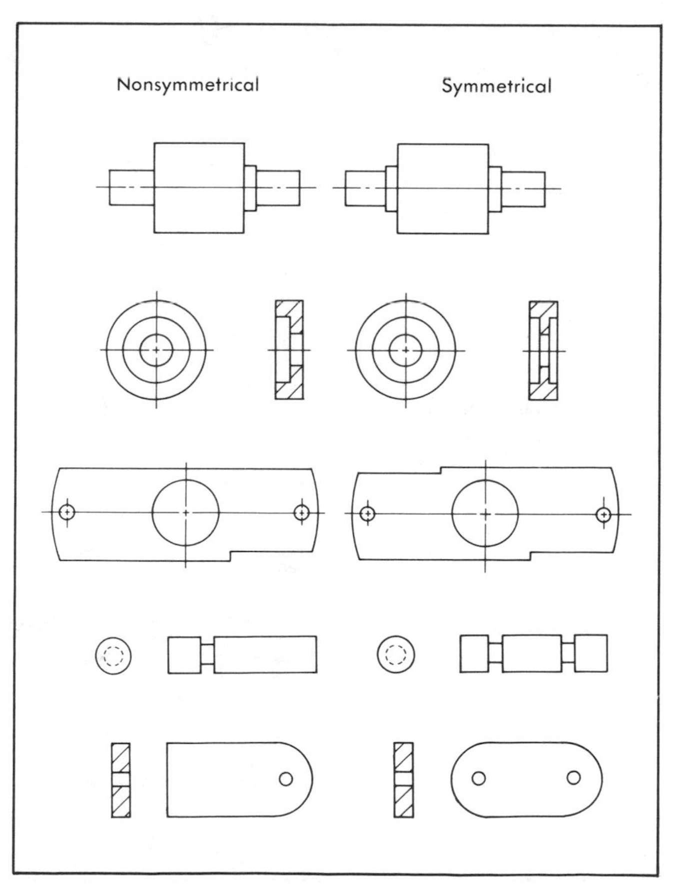

Fig. 13-17 Parts made symmetrical for easier orientation.

proper position (see Fig. 13-18). Parts should also be designed to avoid tangling, nesting, and shingling in vibratory part feeders (see Fig. 13-19).

Robotic part handling can be facilitated by providing a large, flat, smooth top surface for vacuum pickup, an inner hole for spearing, or a cylindrical surface or other feature of sufficient length for gripper pickup. Because parts usually come off the production line properly oriented, this orientation should be preserved by using magazines, tube feeders, or part strips. Palletized trays and kitting are methods for supplying properly oriented parts to the assembly line.

Original design

Addition of external feature to facilitate orientation

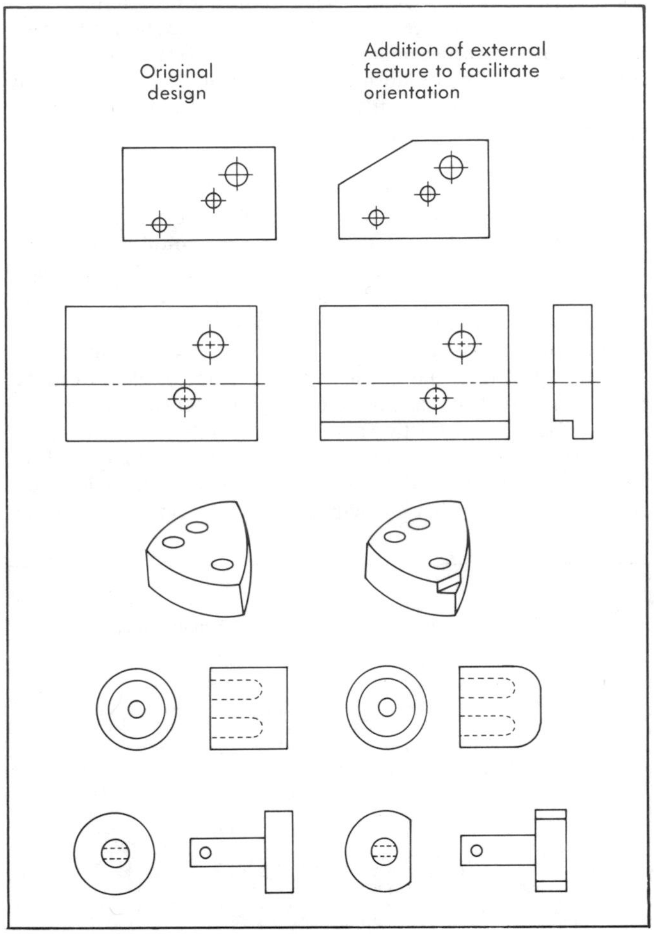

Fig. 13-18 Adding external features such as chamfers, slots, shoulders, radii, or flats can facilitate orientation.

Features should be designed in the product to facilitate packaging. Standard outer package dimensions should be used for machine feeding and storing. Packaging should be designed to adequately protect and ensure quality at all stages of handling and for easy handling. It is important to consider material flows within the production facility including product flow, workspace flow, supply flow, hardware flow, trash or scrap flow, bulk material flow, container flow, and fixture flow. For each flow, the designer should consider how the product, subassembly, component, or part can be designed to simplify or eliminate the flow.

Eliminate or Simplify Adjustments

Manual and automated mechanical adjustments are expensive and a continual source of assembly, reliability, test, and service problems. Also, equipment that goes out of adjustment is one of the biggest causes of customer dissatisfaction. Avoiding adjustments reduces assembly cost, enables automation, and improves service of the product as well as reducing service costs. The need for adjustment can be avoided in a variety of ways by providing natural stopping points, notches, and spring-mounted components, which ensure a preferred location as well as compensate for wear. Often, by understanding the nature of the difficulty caused by a particular adjustment, the designer can find an innovative way to reduce or eliminate the need for the adjustment. Process-driven design combined with the team approach can solve many of these problems.

Avoid Flexible Components

Electrical wires and other flexible components are difficult to assemble because their flexibility makes handling difficult. Connectors in fixed position are good for robotic assembly. Rigid or process-applied gaskets should be used whenever possible. Plugs and connectors can be used to eliminate lead wires. Also, consider using circuit boards in place of cables. Often, locating all connectors at one end of the assembly helps simplify assembly operations. If a cable cannot be avoided, it is often helpful to locate the cable end by having the cable plugged into a dummy connector. A reduction in the use of flexible components can also reduce information content in other ways. For example, a reduction in the number of electrical cables will also

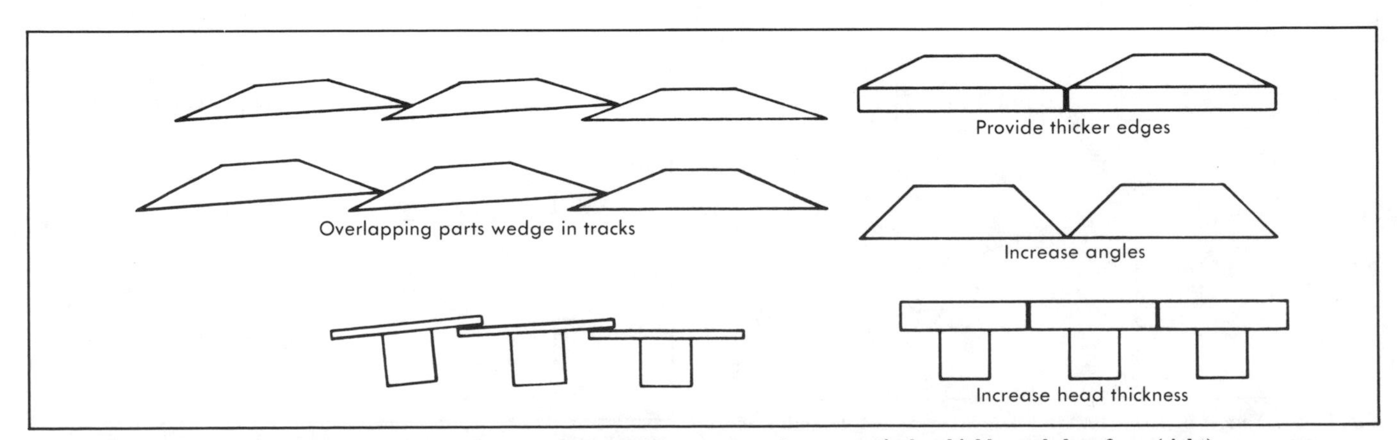

Fig. 13-19 Shingling or overlapping can be avoided by providing thicker contact edges or vertical or highly angled surfaces (right).

reduce cable wiring errors, thereby reducing a major source of assembly test problems.

QUANTITATIVE EVALUATION METHODS

Quantitative evaluation methods form a major part of the arsenal of DFM tools that have been developed in recent years. These methodologies allow the design engineer to rate the manufacturability of the design quantitatively and, in so doing, provide a systematic, step-by-step procedure that helps ensure that the DFM guidelines are being correctly applied. These methods encourage the designer to improve manufacturability of the design by providing insight, stimulating creativity, and rewarding the designer with improved quantitative scores if the DFM guidelines are correctly applied. An important added benefit of these methodologies is that they teach good design practice. Consequently, the need for repetitive use of the methodology diminishes with use.

At present, there are two quantitative evaluation methodologies in widespread use, both of which focus on ease of product assembly. The design for assembly (DFA) method developed by G. Boothroyd and P. Dewhurst is perhaps the most widely used of these methods. The second quantitative methodology, known as the Hitachi assemblability evaluation method (AEM), is less widely used domestically because of its proprietary nature. Available from General Electric Co. through a license from Hitachi, Ltd. (Tokyo), modified versions of this methodology are being used by Hewlett-Packard, General Motors Corp., Caterpillar, and several other companies in the U.S.

Boothroyd-Dewhurst DFA Method

The design for assembly (DFA) method was developed by G. Boothroyd and P. Dewhurst while at the University of Massachusetts. They are presently at the University of Rhode Island. Details of the methodology are presented in the *Design for Assembly Handbook*.[14] The DFA methodology has been adopted by a number of large U.S. manufacturers, among them Ford Motor Co., Xerox Corp., and Whirlpool Corp.

Based largely on industrial engineering time study methods, the DFA method developed by Boothroyd and Dewhurst seeks to minimize cost of assembly within constraints imposed by other design requirements. This is done by first reducing the number of parts and then ensuring that the remaining parts are easy to assemble. Essentially, the method is a systematic, step-by-step implementation of the DFM guideline numbers 1, 7, 8, 9, and 10.

The *DFA Handbook* is divided into three sections dealing with choice of assembly method, design for manual assembly, and design for automatic assembly. The first section provides a procedure for choosing between manual, special-purpose automatic, or programmable automatic assembly. Basic information required includes production volume per shift, number of parts in the assembly, single product or a variety of products, number of additional parts required for different styles of the product, number of major design changes expected during the product life, and the company investment policy regarding labor-saving machinery. This basic information is used to choose the appropriate row and column on a color-coded chart. The recommended assembly method is given at the row-column intersection on the chart. By varying the basic information slightly, different intersections are obtained, thereby giving a feel for what parameters are driving the assembly method choice.

The design for manual assembly procedure consists of comparing an "ideal" assembly time with an estimated "actual" assembly time required for a particular product design. To calculate the ideal assembly time, the theoretical minimum number of parts is first determined by questioning whether each part needs to be separate for good reasons regarding relative motion, whether there is a need for differing materials, and whether there is a need for manufacture and repair. The ideal assembly time is calculated assuming an assembly containing the theoretical minimum number of parts, each of which can be assembled in an ideal time of 3 seconds. This ideal time assumes that each part is easy to handle and insert and that about one third of the parts are secured immediately on insertion with well-designed snap-fit elements.

Assembly for each part is divided into handling and insertion operations. To estimate the actual assembly time for each part, penalties in seconds are assessed for difficulties associated with each operation. The penalties are based on a compilation of standard time study data as well as dedicated time study experiments. This data is tabulated in chart form as a function of part geometry, orientation features, handling features, and method of attachment. Actual assembly time is the sum of handling and insertion times obtained from the charts for each part contained in the actual assembly. The manual assembly efficiency rating is computed as the ratio of ideal assembly time to actual assembly time.

Following evaluation, the assembly is redesigned for ease of assembly by first eliminating and combining parts using the insights gained from the theoretical minimum part count determination. The remaining parts are then designed to provide features that reduce assembly time, again using insights gained from the evaluation. To gage improvements, assemblability efficiencies calculated for the new and old designs can be compared.

The design for automatic assembly analysis consists of four steps: (1) estimate the cost of automated bulk handling and oriented delivery, (2) estimate cost of automatic part insertion, (3) decide whether the part must be separate from all other parts in the assembly, and (4) combine the results of steps 1-3 to estimate the total cost of assembly. Although more computations are involved, the basis for the design efficiency calculation and procedure for product redesign is essentially the same as for manual assembly.

Minimum Part Assessment

Once the basic ideas behind quantitative evaluation methodologies are understood, it is not difficult to begin to imagine other design measures and/or methods for rating a design's manufacturability. The reader is encouraged to consider the following procedure as one way of implementing DFM within his or her organization. The basis for the minimum part assessment is the part count ratio, which is defined as:

$$\text{part count ratio} = \frac{\text{theoretical minimum number of parts}}{\text{actual number of parts}}$$

A procedure for determining the theoretical minimum number of parts, based on the Boothroyd-Dewhurst questions, could be developed as follows:

1. Obtain the best information about the product or assembly. Useful items include:
   - Engineering drawings.
   - Exploded three-dimensional views.
   - An existing version of the product.
   - A prototype.

2. Take the assembly apart (or imagine how this might be done). If the assembly contains subassemblies, treat these, at first, as "parts" and then analyze them later as assemblies.
3. Begin reassembling the product in the reverse order from which it was disassembled. As each part is added to the assembly and regardless of practical or functional limitations, answer each of the following questions:
   a. Does the part move relative to other parts?
   b. Must the part, for good reasons, be made of a different material?
   c. Does the part need to be separate for manufacture or repair?
4. If the answer to any of these questions is yes, then the part under consideration must be a separate part and cannot be eliminated. If the answer to all three questions is no, then the part is a candidate for elimination.
5. The theoretical minimum number of parts for the product is equal to the total number of parts that must be separate as determined by yes answers to the previous critical questions.

Suppose an assembly consisting of seven separate parts was analyzed using the previously described procedure and only three were found to actually need to be separate for theoretical reasons. The part count ratio would then be 3/7 or 0.429. Like the Boothroyd-Dewhurst DFA analysis, opportunities for part count reduction would be identified, and a creative search for a better design would be stimulated. Now suppose, as a result of this, ways were found to eliminate or consolidate three of the four candidates for elimination identified in the analysis. Then the new part count ratio would be 3/4 or 0.75, giving a net improvement in this measure of about 75%.

But this is only the beginning of what could be done with the idea of minimum part count. For example, having determined the theoretical minimum number of parts for the design, the designer could begin searching for ways to eliminate the theoretical reasons why a part should be separate. The designer could also begin to question whether the number of extra parts are actually needed to provide a desired range of product variations and then seek innovative ways to reduce this number. With these possibilities in mind, the guidelines for the minimum part assessment methodology are as follows:[15]

1. In many designs, the product concept or technological approach determines the number of parts. Consider modifying the design or changing the technological approach to make the theoretical minimum number of parts as low as possible.
2. Determine the theoretical minimum number of parts for the design. Seek innovative ways to eliminate the reasons why a part must be separate.
3. Reconsider alternative product concepts. Determine the theoretical minimum number of parts for each concept. Consider adopting favorable aspects of alternative concepts.
4. Modify the design to reduce the number of extra parts needed to provide a desired range of product/model variations.
5. Check all parts for function and modify the design to eliminate redundant parts.
6. Seek ways to modify the design to eliminate those parts that do not need to be separate for theoretical reasons.

Summary

Once a set of design measures such as the Boothroyd-Dewhurst assemblability efficiency or the proposed part count ratio have been agreed on, it might be possible to begin including target values for these measures as part of the design goals of a project (refer to Fig. 13-12). Such an approach could be beneficial both by giving the design team some quantitative design goals to aim at and by providing added basis for judging the acceptability of a particular design. One cautionary note is in order. It is important to remember that the quantitative DFM methodologies are only design tools and not ends in themselves. It is, therefore, very important that such measures are used for what they are—a means for improving the DFM quality of the design—and that they do not become just one more "hoop" the design department must jump through to get a design released. DFM practice must become a natural, not an additional, part of the design process.

ROBUST DESIGN

Robust design implies a product designed to perform its intended function no matter what the circumstances. A rifle that continues to fire reliably even after being dropped in mud might be considered robust with respect to external or environmental factors such as intrusion of dirt and moisture. Similarly, if the rifle continues to fire reliably in spite of years of mistreatment and neglect, it might be considered robust with respect to deterioration over time. If a large number of rifles, each manufactured by the same factory, were to fire reliably initially, and exhibit the same robustness with respect to external factors and deterioration over time, then the rifle might be considered robust with respect to normal manufacturing variation associated with production of the rifle by a given set of manufacturing processes. If rifles manufactured under differing conditions using different methods in different factories all exhibited the same performance in use, then the rifle design specifications might be considered robust with respect to method or place of manufacture. Finally, if changing business or market conditions made it necessary to alter or modify a major characteristic of the rifle, such as its caliber, and the design change was easy to accomplish, both with respect to the product design and the method of production, then the rifle might be considered robust with respect to changing functional or market needs.

As can be inferred from the discussion, robustness is a measure of the sensitivity of a product to change and variation that occurs during manufacture and use of the product over its lifecycle. Every product, no matter what its nature, is likely to be subjected to change and variation over which there is little or no control, but which influence the benefits derived from manufacture and use of the product. Benefits such as performance, quality, reliability, serviceability, and lifecycle cost are determined by design intent. The more closely the product conforms to design intent, the greater the benefits. Robust design seeks to maximize benefits derived from conformance to design intent by seeking to define designs that are insensitive to change and variation.

Although the concept of robust design is well known and its importance is well understood by consumers and manufacturers alike, little has been done to provide the manufacturing or product design engineer with formalized methodologies for doing robust design. Perhaps the greatest contribution in this area to date has been the work of Genichi Taguchi. Better known as the Taguchi methods, the on-line and off-line quality control methodology developed by Taguchi in the early 1950s is, for the most part, a formalized approach to robust design based largely on statistical design of experiment theory.

Much of Taguchi's ideas about robust design directly addresses the problem of off-line quality control.[16] In this work, Taguchi calls variables that disturb the function of a product noise. Noises are classified as follows:

1. *Outer noise.* Outer noise consists of those variables that are operative during product use, such as temperature, humidity, input voltage, dust, external load, time rate of load application, and type of use.
2. *Inner noise, or deteriorating noise.* Inner noise consists of those properties or variables of a product that influence the benefits or function of the product and change during use of the product. Examples include loss of strength due to corrosion or fatigue, wear of mating parts, and deterioration due to operation at elevated temperature.
3. *Variational noise or between product noise.* Variational noise consists of those properties or variables of a product that influence the performance or function of the product and vary from product to product manufactured under the same specifications. Examples include variations in part dimensions, stackup tolerances, and calibration variations.

Functional variation due to deterioration and variation between products are actually, in Taguchi's view, two types of inner noise. The former is referred to as "timewise noise" and the latter as "spacewise noise." For Taguchi,

> Good functional quality means less functional variation due to inner and outer noise, and that the product always functions correctly under a wide range of conditions. Quality is measured by the degree of variation from the target value (design intent)—the nominal value or ideal value identified in the specification.[17]

In performing robust design, the design team seeks to minimize the effects caused by these noise sources. Several strategies for robust design are possible:

- Identify product concepts and process concepts that are inherently insensitive to variation and change (robust).
- For a given product design and process plan, determine the optimum values for the design parameters and process parameters to maximize robustness.
- Minimize the source and cause of variation.
- Provide design and process features that are variation tolerant.

A brief discussion of each of these strategies is given in the following sections.

Identify Inherently Robust Concepts

Perhaps the surest way to achieve a truly robust design is to begin with a design concept that is inherently insensitive to variation and change. Although identifying inherently robust concepts often requires considerable innovation by an experienced product development team, there are some aids that can help. Because "good design" and "robust design" are essentially equivalent, use of the design axioms discussed previously can be an effective approach to identifying inherently robust design concepts. A design concept that maintains the independence of functional requirements is, by definition, inherently robust. In other words, by satisfying functional requirements independently, changes or variations with respect to one requirement will not affect other requirements. Likewise, the DFM guidelines can also be used to identify inherently robust concepts because they are codified statements of good design practice where good design practice implies robust design.

To illustrate how a robust concept could be defined using the design axioms, consider an application where a delicate electrical component having a complex external shape is to be installed robotically in a chassis or frame. Suppose that the component is a purchased part and that it may be desirable to change suppliers from time to time to obtain the best price for the part. Suppose also that the shape and mounting configuration of the part varies from supplier to supplier. One approach to dealing with this situation would be to design special-purpose end-of-arm tooling to handle the component and to design the frame to accept the component currently being purchased. Then, when a new supplier is selected, modify the end-of-arm tooling, the mounting detail on the frame, and reprogram the robot to install the new part. Although feasible, this solution is not particularly robust with respect to change in component supplier because the change is difficult and costly to implement and may have a ripple effect on other aspects of the design.

An alternative approach might consist of using a mounting or adapter plate to interface the component with the frame and end-of-arm tooling. In this case, only the way the new component mounts on the plate need be changed to change supplier. By having the mounting plate interface with the end-of-arm tooling and the frame, no change to these components or to the robot programming would be necessary. In effect, the mounting plate decouples the requirement to change suppliers from robotic assembly and mounting requirements. Other decoupled solutions could also be proposed. Selection of the best decoupled solution would then depend, according to the second axiom, on which solution has the least information content.

Optimize Robustness of a Given Design

If one were to plot the relationship between particular outputs of a product or system and the design variables that affect them, nonlinear relationships would often be discovered (see Fig. 13-20). Optimizing the design by selecting values for the design variables that maximize robustness (minimize sensitivity to variation) is based on this insight. As shown in Fig. 13-20,

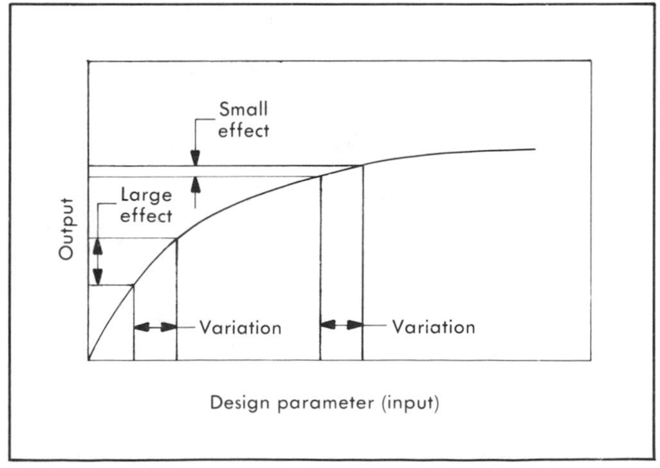

Fig. 13-20 Design to minimize output variation produced by design parameter variation. (*Industrial Technology Institute*)

robustness of the design would be maximized by selecting values for the design variables that lie in regions where variation of the design variables produce little or no variation in the desired output of the system.

Optimizing robustness of a design can be illustrated by considering the needle bearing assembly shown in Fig. 13-21. In this case, an undesirable wobble of the housing results due to tolerance stackup caused by unavoidable variations in shaft, bearing, and housing dimensions. The question of design is to select a value for the length of the needle bearing that will result in assemblies that do not exhibit unacceptable wobble regardless of dimensional variations and tolerance stackup of the assembled parts.

A possible solution to this problem can be provided by developing a "Monte Carlo" simulation of the tolerance stackup that causes the housing wobble. Several PC programs are available for this function. In performing the simulation, a realistic model accurately reflecting the design and based on known part and assembly process capability is created and then used to predict the distribution of wobble that could be expected among different assemblies produced in a production run. By performing similar experiments using different length needle bearings, the curve shown in Fig. 13-21 was generated.

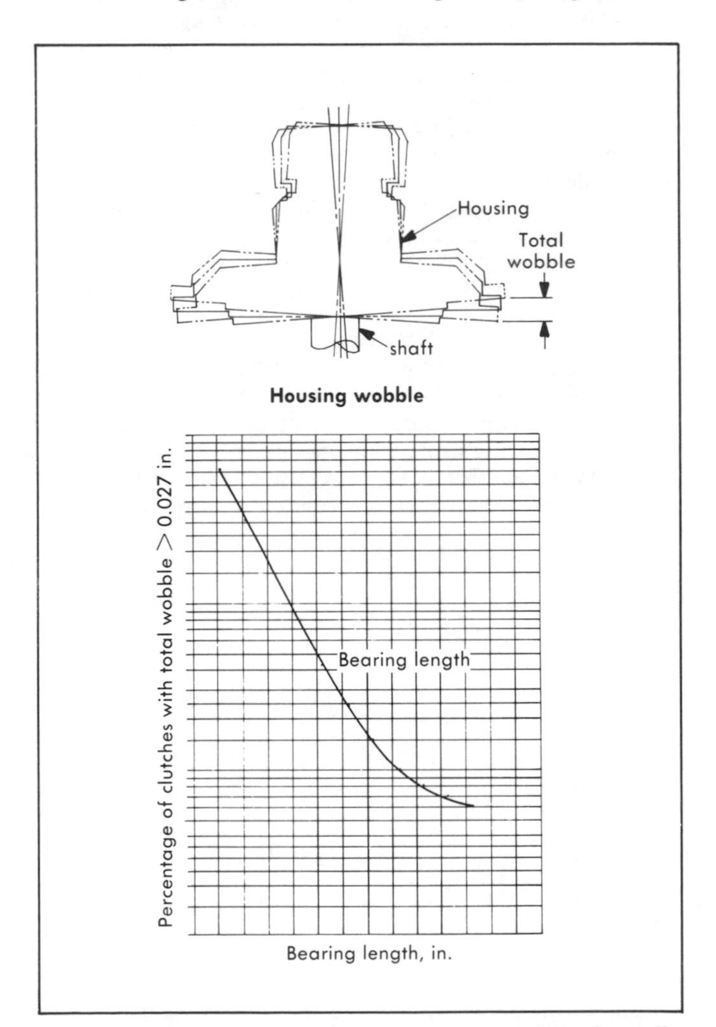

Fig. 13-21 Identifying the robust design for housing wobble of a needle bearing assembly. (*Industrial Technology Institute*)

This curve shows the percentage of assemblies expected to have a total wobble greater than 0.027″ (0.69 mm) as a function of bearing length. As can be seen, sensitivity of housing wobble to tolerance stackup, in this case, is drastically reduced for bearing lengths greater than 1.05″ (26.7 mm). Selection of a needle bearing length greater than 1.05″ could, therefore, be recommended as one way to make the assembly robust against housing wobble caused by tolerance stackup.

Off-Line Quality Control (Taguchi Method)

The previous example illustrates how a robust region for one design variable or parameter might be determined. In most designs or systems, many design parameters are involved and hence the determination of values for each parameter that make the entire system robust can be a challenging task. The task is often made even more difficult by the fact that the robust region for a particular parameter may be quite narrow. Also, the degree of sensitivity of the system output to variation in a particular parameter will differ. In some cases, a clear robust region for a particular parameter cannot be found and means for implementing tight control of variation of the parameter must be found.

The Taguchi method addresses these problems by using statistical design of experiment theory. In particular, the Taguchi method seeks to identify a robust combination of design parameter values by conducting a series of factorial experiments and/or using other statistical methods. Termed parameter design by Taguchi, this step establishes the midvalues for robust regions of the design factors that influence system output. The next step, called tolerance (allowance) design, determines the tolerances or allowable range of variation for each factor. The midvalues and varying ranges of these factors and conditions are considered as noise factors and are arranged in orthogonal tables to determine the magnitude of their influences on the final output characteristics of the system. A narrower allowance will be given to noise factors imparting a large influence on the output.

In establishing the tolerance or allowance range for a particular parameter, Taguchi uses a unique concept defined as a loss function. In this approach, loss is expressed as a cost to either society (the customer) or the company that is produced by deviation of the parameter value from design intent. Depicted graphically in Fig. 13-22, the loss function concept is implicit in all of the Taguchi methodologies. Essentially what the loss function says is that quality equals uniformity around design intent. Any deviation from design intent produces a loss and hence the allowance or permissible deviation should be determined based on the magnitude of the cost associated with this loss. The concept of loss and other Taguchi concepts provide valuable insight into quality and the role design plays in determining the quality of a product or system. Many of Taguchi's original ideas are presented in references 18 and 19 listed at the end of this chapter.

Minimizing the Source of Variation

Minimizing the source or cause of variation or error can be another effective means for implementing robust design. For example, by providing features that help orient parts and facilitate assembly as illustrated in Fig. 13-23, the opportunities for manufacturing error can be greatly reduced. In general, the goal in this approach is to design the product in such a way that it is easier to "do it right" than it is to "do it wrong." This principle is depicted graphically in Fig. 13-24.

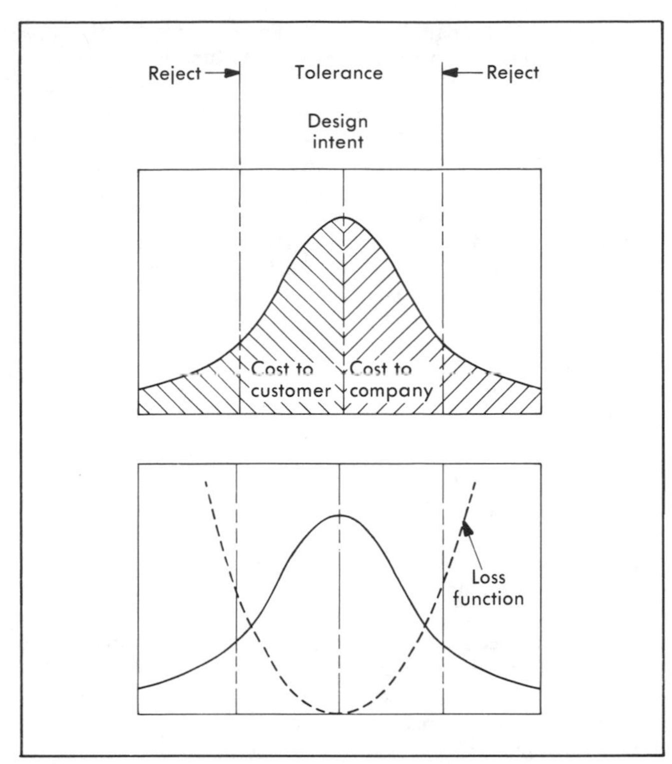

Fig. 13-22 The loss function reflects price paid due to deviation from design intent. (*Industrial Technology Institute*)

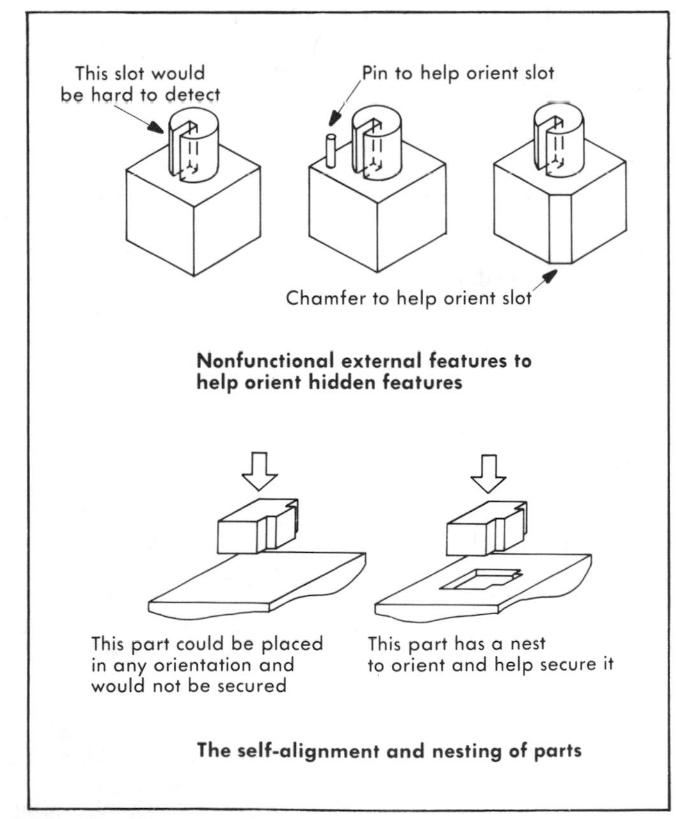

Fig. 13-23 Example designs for reducing variation and error.[20]

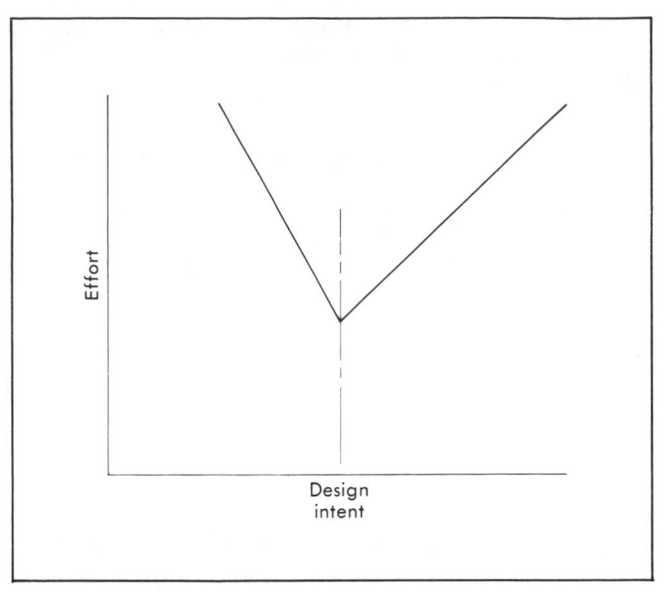

Fig. 13-24 Least-effort design principle. (*Industrial Technology Institute*)

Variation Tolerant Design

Another approach to robust design is to provide features that are variation tolerant. For example, providing a feature that is easily recognized by a machine vision system, regardless of lighting conditions, makes the vision system less sensitive to variations in lighting conditions. Similarly, the use of generous tapers and other guiding features makes part insertion less sensitive to placement accuracy of an assembly robot. Replacement of a mechanical adjustment with a spring-loaded support that automatically compensates for wear not only elim inates the need for adjustment and service, but also helps make the product robust against deterioration over time.

Robustness Assessment

Quality can be defined as consistent conformance to design intent. Robust design, or design that consistently conforms to design intent in spite of external and internal noise, results in a high-quality product. This idea indicates that great opportunity for quality enhancement exists through proper assessment of the robustness of a product concept and process plan in the early stages of design. With this in mind, the following robustness assessment procedure is suggested.[21]

1. Evaluate the robustness of the proposed product concept and process plan with respect to the following principal questions:
   - How might the product change over time? How might market or customer needs change? How would the product design be affected by these changes?
   - How might the process plan or production technology change over time? What effect would these changes have on the product design?
   - What product or model variations are planned? How does the product/ process concept accommodate these variations? What new variations could be introduced in the future? How would these changes affect the product design and process plan?
   - What normal manufacturing variations can be ex-

pected? How will these variations affect product performance and other benefits?
- What extreme or major changes in manufacturing conditions could occur? How would such change affect product performance and other benefits?
- What outer noise will the product or process be subjected to? What will be the effect?
- What inner noise (functional variation due to deterioration and variation between product units) will the product and process be subjected to? What will be the effect?

2. Analyze the results of the evaluation and formulate a strategy for reducing or eliminating the undesirable effects of variation and change identified. Use a combination of the robust design strategies as well as others that might apply.
3. Improve the product design and process plan according to the strategy adopted and reanalyze. Iterate until satisfied.

In using the robustness assessment, the analyst should attempt to answer each question in as much detail as possible, given the product and process definition available. If the product design and process plan are fairly complete, then the assessment will help identify vulnerabilities to variation and change. If the product/process concept is less well developed, then the assessment will help define design direction and goals.

TOOLS FOR PROCESS-DRIVEN DESIGN

Process-driven design seeks to ensure that parts and products are correctly designed to be produced using a particular production process or method. Design requirements for a given process are often stated in the form of design guidelines and rules of thumb. Typically, these guidelines are highly specialized for a given industry or particular process implementation or particular plant or particular equipment installation within a particular plant. Making the designer aware of these process requirements and constraints early in the design process, before concepts are finalized and lines are put irreversibly on paper, is a key goal of design for manufacture. Design tools that help ensure product/process conformance and enable process-driven design can generally be classified either as process specific or facility specific. Each of these categories are discussed in the following sections.

Process-Specific Tools

Process-specific DFM has to do with the design of parts to be manufactured using particular methods or processes such as casting, forging, injection molding, and stamping. Typically, these tools facilitate systematic application of specialized process knowledge in the form of codified statements of design guidelines and rules to the design of parts to be made using a particular manufacturing process or method. Examples include design for casting, design for injection molding, and design for metal stamping.

This area of design for manufacture recognizes that manufacturing variables such as cost and life of dies, time and cost to build tools and fixtures, configuration and complexity of material flows, throughput of the process, process capability, and number of stations and workers required are often established and heavily influenced by detail design of the parts to be produced. It also recognizes that knowledge of the interaction between the product and process is crucial when the designer first puts lines on paper and that many designers, at this stage, must of necessity be more concerned with function than fabrication details. Hence, these tools generally seek to provide the designer with guidance as he or she makes these critical decisions and to help evaluate the suitability of the design once it has been specified.

Development of process specific design tools is just beginning. At present, few well-developed and proven methodologies are available and those that do exist are not generally available because of their proprietary nature. One of the problems is that development of process-specific design tools is often highly dependent on specialized knowledge and experience. In other words, successful processing of parts is more art than science. A further complication results from the fact that much of this knowledge or art is unique to particular employees of certain companies and industries and is, therefore, not readily available to those interested in and capable of developing appropriate DFM tools. For these reasons, much of the development of process-specific DFM tools is currently being done on a proprietary basis within individual companies and industries. For example, some vendors of CAD software or systems are working on the development of such tools. Similarly, large companies with considerable internal facilities and capabilities are developing specific tools that reflect their experience and are intended to meet particular needs associated with the products they manufacture. A more complete discussion of developments in this area is provided in the references listed at the end of this chapter.[22]

Facility-Specific Tools

The second general category of activity involves the development of DFM tools that facilitate correct design of products intended to be manufactured using highly specialized or unique advanced manufacturing facilities. Such tools, which could be aptly described as "designer tool kits," provide design rules, physical examples and models, various design aids, and other specific information about a specialized manufacturing facility in a readily usable form to the designer.

To illustrate a facility-specific application, suppose a company has a flexible assembly system (FAS) that can be programmed to automatically assemble a variety of different subassemblies used in various products manufactured by the company. Such a system would likely consist of several stations designed to perform certain classes of assembly operations. Parts might be handled and inserted by robots using programmable "gripper engines" and fixtured using "flexible fixtures" specially designed for the FAS. For a particular subassembly to be built on a specific FAS facility of this kind, the subassembly and parts making up the subassembly must be correctly designed to be compatible with the assembly operations available, the work envelopes available, and the associated flexible tooling that interfaces the parts and in-process assembly to each station and to the material handling system. The "designer's tool kit" for such a FAS would provide needed process specific information to ensure conformance of the product design to the FAS requirements. It would also facilitate "what-if" optimization and experimentation with alternative design concepts and forms.

Facility-specific DFM tools are often a must if specialized advanced manufacturing systems such as the FAS just described are to deliver the productivity and quality improvements that they promise. Development of manufacturing facility-specific DFM is, at present, in its infancy and is likely to advance very quickly as the relevance of this approach becomes more widely recognized. Typical applications that could benefit greatly from the "designer's tool kit" approach include such diverse situations as flexible assembly and manufacturing system concepts, design of stampings for production on certain classes of triaxis transfer press lines, and design of weldments for production on special flexible welding fixtures or lines.

COMPUTER-AIDED DFM

A major barrier to DFM is usually time. Design and manufacturing engineers are typically operating under very tight schedules and are, therefore, reluctant to spend time learning and using DFM approaches. Computer-aided DFM helps simplify the effort and shortens the time required to implement DFM on a daily basis. Computer-aided DFM also enables the design team to consider a multitude of product/process alternatives easily and quickly. "What-if" optimization allows each alternative to be refined and fine tuned. Together, these capabilities greatly increase the probability of identifying the most desirable solutions during the early stages of design. When properly implemented and applied, computer-aided DFM has the potential to vastly improve quality of early product/process decisions and thereby enhance the design team's ability to do design for effective quality, cost, and delivery.

Another major benefit of computer-aided DFM is the way it fosters team building and the team approach. For example, the use of Monte Carlo simulation to predict the effect of needle bearing length on housing wobble (refer to Fig. 13-21) described in the section discussing robustness optimization could not have been accomplished without a team approach involving participants from design engineering, manufacturing engineering, and the bearing supplier. Input regarding function and permissible wobble had to come from engineering, machining and assembly process capability from manufacturing, and bearing stochastic characteristics from the bearing manufacturer. Beyond this, a team approach was also needed for construction of a realistic simulation model and for correct interpretation of results.

A variety of proprietary computer-aided DFM software packages are currently available. The Monte Carlo simulation uses a commercial software package designed to simplify and facilitate analysis of tolerance stackup in assemblies. In addition, considerable effort is being directed toward the development of new computer-based and/or computer-aided DFM methodologies. This work includes DFM enhancements of CAD systems, research involving conventional interactive computer programming approaches, and research involving artificial intelligence and expert system approaches. A review of current research directions and activities in computer-aided DFM is given in reference 23.

TRADITIONAL DESIGN METHODOLOGIES

In this section, some widely used, but more traditional design tools that can also offer assistance and insight into the problems of design for manufacture will be reviewed. They are group technology (GT), failure mode and effects analysis (FMEA), and value engineering.

Group Technology

Group technology (GT) is an approach to design and manufacturing that seeks to reduce manufacturing system information content by identifying and exploiting the sameness or similarity of parts based on their geometrical shape and/or similarities in their produciton process. GT is implemented by utilizing classification and coding systems to identify and understand part similarities and to establish parameters for action. Manufacturing engineers can decide on more efficient ways to increase system flexibility by streamlining information flow, reducing setup time and floor space requirements, and standardizing procedures for batch-type production. Design engineers can develop an attitude of designing for producibility and help eliminate tooling duplication and redundancy.

As a DFM tool, group technology can be used in a variety of ways to produce significant design efficiency and product performance and quality improvements. One of the most rapidly effective of these is the use of GT to help facilitate significant reductions in design time and effort. Often in design, it is easier to design new parts, tooling, and jigs rather than try to locate a similarly designed part. The ease with which parts can be designed using a CAD system has increased this problem. A GT database helps reverse this tendency by making it possible to easily and quickly retrieve and review existing parts that are similar to the new part being designed. In using a GT system, the design engineer needs only to identify the code that describes the desired part. A search of the GT database reveals whether a similar part already exists. If a similar part is found to exist, and this is most often the case, then the designer can simply modify the existing design to design the new part. In essence, GT enables the designer to literally start the design process with a nearly complete design. For example, the designer may find a gear identical to the one being designed, but of a different thickness. Simply copying the existing design and making minor changes will save substantial design time and effort by helping to prevent "redesigning the wheel."

Group technology can also be effectively used to help control part proliferation and eliminate redundant part designs by facilitating standardization and rationalization (S&R) approaches discussed previously. If not controlled, part proliferation can easily reach epidemic proportions, especially in large companies that manufacture many different products and product models. When these companies implement GT, it is not unusual to discover many versions of the same part. By noting similarities between parts, it is often possible to create standardized parts that can be used interchangeable in a variety of applications and products. For example, a family of gears or shafts or spacers can be created that satisfy the full spectrum of applications within the company as revealed by the GT database. The GT database can also be used to identify the most useful or appropriate dimensions and features. An alternative approach is to create a composite design in the CAD system that contains all the features of a particular part family. A specific design could then be quickly created simply by leaving out unwanted features. S&R pays big dividends because it simultaneously creates both economies of scale by increasing part volume and economies of scope because the same part can be used in a variety of applications.

In addition to facilitating the standardization of part families, GT also facilitates standardization and rationalization of part features and attributes. For example, a study of the GT database may reveal that certain fillet radii or hole diameters are frequently used in a wide variety of parts while others are seldom used. By standardizing on the more frequently and widely used features, total manufacturing system information content can be drastically reduced because of the tooling and processing simplifications produced. The understanding of frequency of occurrence of particular design features provided by

GT can also help supply feedback to the designer regarding increased manufacturing costs associated with using infrequently specified features.

The grouping of related parts into part families is the key to group technology implementation. The family of parts concept not only provides the information necessary to design individual parts in an incremental or modular manner, but also provides information for rationalizing process planning and forming machine groups or cells that process the designated part family. A part family may be defined as a group of related parts that have some specific sameness and similarities. Design-oriented part families have similar design features, such as geometric shape. Manufacturing-oriented part families share similar processing requirements. In principle, part families can be based on any number of different considerations. For example, parts manufactured by the same plant, or parts that serve similar functions such as shafts or gears, or parts all fabricated from the same material could conceivably be grouped into part families.

In implementing GT, the problem that immediately presents itself is how the parts are to be effectively grouped into these families? Three methods are commonly used: (1) visual (ocular) inspection, (2) production flow analysis (PFA), and (3) classification and coding. The first method is obviously very simple, but limited in its effectiveness when dealing with a large number of parts. Production flow analysis is a technique that analyzes the operation sequence and the routing of the part through the machines in the plant. The three steps involved in PFA are factory planning, process planning, and operation planning. Parts with common operations and routes are grouped and identified as a manufacturing part family. Similarly, the machines used to produce the part family can be grouped to form the machine group or cell. For PFA to be successful, it must be assumed that the majority of the parts belong to clearly defined families and the machines to clearly defined groups. One of the advantages of this technique is the ability to form part families without using a classification and coding system. Part families are formed using the data from operation sheets or route cards instead of part drawings. There are also a number of disadvantages associated with the PFA approach. Chief among these are its reliance on existing production data and routing methods and the difficulties involved in manually sorting the production data.

Part classification and coding is perhaps the most effective and widely used method. In this approach, parts are examined abstractly to identify generic features that are captured using an agreed-on classification and coding system. The main advantage of classification and coding is the accuracy it provides in forming part families. The major disadvantage is cost. Classification and coding systems are expensive to develop in-house or to purchase from the outside. In addition, employees must be trained, and the actual coding of the existing parts is labor intensive and time consuming. Additional costs are incurred to achieve data retrieval and analysis. If purchased outside, the system may need to be modified to fit the environment. Finally, there is the cost of associated hardware and its operation. Offsetting this direct cost is the cost reduction potential of S&R, the elimination of proliferative parts, the savings in design time and effort, and the endless list of manufacturing system simplifications that GT makes possible.

Although a number of commercial classification codes are available, none can be considered universally applicable. The best coding system is one that is adapted specifically to the industry or company where it is used. Ideally, a coding system for classifying parts should allow the designer to visualize both the part and its process plan by the code number alone. The level of coding sophistication is one of the first decisions that the user of GT must make. Shorter codes, which are easier to use and less prone to error, are also less flexible. For computerized GT systems, the simplest scheme is often a hard-coded program, in which an interactive routine asks the user a series of questions regarding a particular part. More sophisticated applications use soft-coded programs, such as decision tree coding, where the length of the code depends on which branch of the logic tree is being used.

At present, there are two main methods of coding: attribute-based coding (also referred to as polycodes) and hierarchical-based coding (also referred to as monocodes). In attribute coding, the simpler of the two, code symbols are independent of each other (see Fig. 13-25, view *a*). Hence, codes of fixed length span all parts families, and each position in the code corresponds to the same variable. Because of this, each attribute that is to be coded must be represented by one digit, which can make the code quite long in some cases. One advantage of the attribute-based code is that information can be extracted quickly from the entire parts population. In addition, it is relatively simple to develop. Unfortunately, the price paid for such elegance and simplicity is an extremely long code when many parts families are involved.

A hierarchical code structure is designed so that each digit in the sequence is dependent on the information carried in the digit just preceding it (see Fig. 13-25, view *b*). Generally, the first digit holds the most basic information, and each succeeding digit contains more specific information. This makes it possible to capture a large amount of information in a relatively short code. A disadvantage is that the entire code must be interpreted because no one digit carries any significant information. Monocodes are also more difficult to store in databases that do not use a decision tree approach.

Because both methods have advantages and disadvantages, many modern coding systems rely on a hybrid type of code. One useful compromise is a fixed-length code that is standard for all part families for the first few digits and variable for the rest. A major characteristic of today's GT software programs is the ease with which they can be modified and customized. This is a necessity because each application must be adapted to suit each company's part type.

Codes can be configured to capture a wide range of information. Basic information includes part geometry, dimensions, mechanical characteristics, and manufacturing features. Spe-

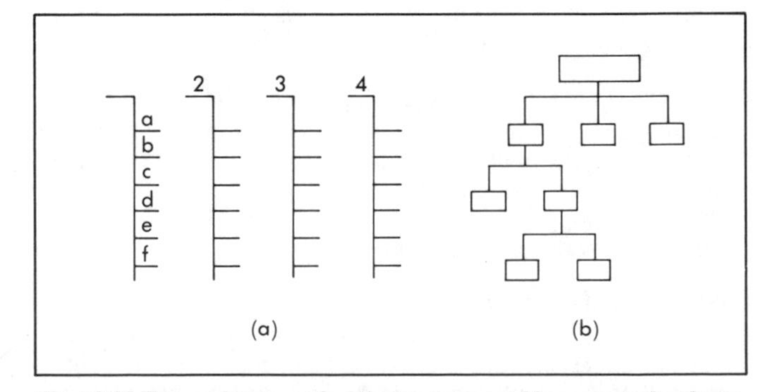

Fig. 13-25 Two main types of coding systems used in group technology: (a) attribute based and (b) hierarchy based.

cific variables may include shape, finish, and size. More sophisticated codes also capture subelements (such as holes, cones, slots), material type, chemistry, finish, and special processes such as heat treatments. Production scheduling can be facilitated by including lot sizes, information on order frequency, relationships between processing steps for particular parts, and assembly or quality control requirements.

In recent years, computer programs using cluster analysis, pattern recognition, and other methods have been developed to enhance the conventional methods of PFA and classification and coding for more effective grouping of part families. Also, while traditional classification and coding systems are still useful and well adapted, new concepts for generative coding and computer-oriented schemes and expert systems are being studied to be more compatible with recent developments for CIM implementation.

The benefits derived from group technology are not limited to DFM and design economy. With increasing emphasis on flexible and integrated manufacturing, GT is proving to be an effective first step in structuring and building an integrated database. Standardized process planning, accurate cost estimation, efficient purchasing, and assessing the impact of material costs are benefits that are also often realized. The competitive edge that group technology can provide when teamed with a company's computer can be significant and should be considered along with other high-technology productivity enhancement equipment such as robotics, CAE/CAD/CAM, and MRP systems. A more complete discussion of GT, especially as it applies to manufacturing, is given in reference 24.

Failure Mode and Effects Analysis

Failure mode and effects analysis (FMEA) is an important design and manufacturing engineering tool intended to help prevent failures and defects from occurring and reaching the customer. It provides the design team with a methodical way of studying the causes and effects of failures before the design is finalized. Similarly, it helps manufacturing engineers identify and correct potential manufacturing and/or process failures. In performing an FMEA, the product and/or production system is examined for all the ways in which failure can occur. For each failure, an estimate is made of its effect on the total system, of its seriousness, and its occurrence frequency. Corrective actions are then identified to prevent failures from occurring or reaching the customer. FMEA asks the following principal questions:

1. How can each part or process conceivably fail?
2. What mechanisms might produce these modes of failure?
3. What would the effects be if the failures did occur?
4. Is the failure in the high-risk direction?
5. How is the failure detected?
6. What inherent provisions are provided in the design to compensate for the failure?
7. What corrective actions should be taken to prevent the failure?

In FMEA, function is defined as the task that a component, subsystem, or product must perform, stated in a way that is concise, exact, and easy to understand for all users. Functions are typically actions such as position, support, seal, retain, and lubricate. Failure is defined as the inability of a component/subsystem/system to perform the intended function (design intent).

Failure modes are the ways in which components/subsystems/system could fail to perform their intended functions. Typical failure modes would be fatigue, fracture, excessive deformation, buckling, leakage, fail to open, fails to close, and requires excessive force. Asking what could happen to cause loss of function is often an effective way to identify failure modes. Such questioning of a simple electrical on-off switch, for instance, leads to failure modes such as fails to switch on, fails to switch off, difficult or impossible to actuate, inadvertent switching, blows fuse, and produces electric shock. Other examples of failure modes could be listed as:

- Component parts: Broken, deformed, worn, or corroded.
- Assembly: Noise level, loose fit, tight fit, or imbalance.
- Process: Oversize, undersize, cracked, omitted, or improper finish.

For each failure mode, there are possible mechanisms and/or causes of failure. Identification of these are an important part of FMEA because it points the way to preventive or corrective action. Examples of design-caused failures include improper tolerancing, improper stress calculations or the use of wrong or unrealistic loads, wrong material specification, vibration, or improper fits. Process-caused failures include broken or worn tools, worn bearings, inadequate gating, insufficient cooling, heat treat shrinkage, improper locating, improper setup, inaccurate gaging, operator error, voltage drop, and pressure drop.

Effects of failure are the outcomes produced by the occurrence of the failure mode on the system or product. In general, there are two types of effects that are considered: local and global. Local effects do not affect other components, whereas global effects can involve other functions and components. For example, a parking light failure or noise from a speedometer cable are minor nuisances and will not disable a vehicle. Such failures produce only local effects. On the other hand, loss of brake fluid or power steering fluid has a local effect of gradual degradation of performance and a global effect that could be loss of function resulting in possible catastrophic failure.

Controls are measures that exist to prevent the causes of failure from occurring or which are intended to detect the causes of failure. Design guidelines, design reviews, drawing checking processes, materials engineering sign off, statistical process control (SPC), visual checks, sampling plans for incoming material, operator training, and torque wrench calibration are typical examples of controls. Least effort design, design for assembly principles, designer's tool kits, and the other DFM tools discussed in this section might also be considered to be controls because their use helps prevent the causes of failure.

The format of an FMEA varies somewhat depending on the objectives, but generally the following sequence of steps is involved:

1. Description of the component, assembly, subsystem, or system whose failure mode is being identified.
2. Description of the function or functions of the component under consideration.
3. Identification and description of the failure modes associated with the component.
4. Determination of the mechanisms or causes of each failure mode.
5. Determination of the effects of occurrence of each failure mode, both local and global.
6. Determination of all current controls that are intended to prevent or detect the causes of failure.
7. Determination of the probability of occurrence for each

failure mode. Statistical data is used where feasible. Otherwise, a qualitative ranking can be used, such as 1 = very low, 2 = low, 3 = moderate, 4 = high, and 5 = very high.

8. Assessment of the seriousness of failure associated with each failure mode. Again, a qualitative ranking can be used, such as 1 = none, 2 = minor, 3 = significant, 4 = high, and 5 = catastrophic.
9. Determination of the likelihood that the defect will reach the customer. Again, a qualitative ranking can be used such as that given in step 7.
10. Determination of a risk index based on the product of factors assigned in steps 7, 8, and 9. Using the qualitative ranking system, the risk index can be between 1 and 125. In general, if the risk index is equal to or greater than 17, some sort of corrective action is recommended. Also, as a matter of general practice, special attention should be given to any failure mode that is assigned a 4 or 5 in any of the steps above.
11. Recommend and describe corrective actions required.

The complete FMEA can be organized and carried out in a methodical way by developing appropriate worksheets. Typical forms used for design and process FMEA are shown in Fig. 13-26 and Fig. 13-27, respectively. In general, to be fully effective, the FMEA should be conducted by an experienced, multidisciplinary team representing all relevant components of the manufacturing system including outside suppliers. The great value of FMEA lies in its systematic approach in analyzing product and/or process performance in critical detail. From a DFM point of view, the interaction between manufacturing and design engineering required while conducting a FMEA can be very effective in helping to ensure product and process conformance at an early stage of design. A more complete discussion of the FMEA approach is provided in references 25 and 26 listed at the end of this chapter.

Value Engineering

Value engineering provides a systematic approach to evaluating design alternatives that is often very useful and may even point the way to innovative new design approaches or ideas. Also called value analysis, value control, or value management, value engineering utilizes a multidisciplinary team to analyze the functions provided by the product and the cost of each function. Based on results of the analysis, creative ways are sought to eliminate unnecessary features and functions and to achieve required functions at the lowest possible cost while optimizing manufacturability, quality, and delivery.

					Effect On				Criticality		Corrective Action Implementation			
Item Part No.	Block Diagram Reference	Function	Failure Modes	Failure Mechanism (cause)	Local	Higher Levels	End Item	Other Effects (interface, etc.)	Severity	Frequency	Detectability	Time Available for Corrective Action	Recommendation	Item Close Sign Off or Rationale for Other Action

Fig. 13-26 A typical form used for design failure mode and effects analysis. (*Copyright American Society for Quality Control, Inc. Reprinted by permission*)

POTENTIAL
FAILURE MODE AND EFFECTS ANALYSIS
(PROCESS FMEA)

Page ______ of ______

Process ____________ Outside suppliers affected ____________ Engineer ____________
Primary process responsibility ____________ Model year/vehicle(s) ____________ Section supervisor ____________
Other div.(s) or PEO(s) involved ____________ Scheduled production release ____________ FMEA date: (orig.) ________ (rev.) ________

						Existing conditions						Resulting					
Part name/ part number	Process function	Potential failure mode	Potential effect(s) of failure	▽	Potential cause(s) of failure	Current controls	Occurrence	Severity	Detection	Risk priority number (R.P.N.)	Recommended action(s) and status	Action(s) taken	Occurrence	Severity	Detection	Risk priority number (R.P.N.)	Responsible activity

Fig. 13-27 A typical form used for process failure mode and effects analysis. (*Ford Motor Co.*)

In value engineering, value is defined as a numerical ratio, the ratio of function, or performance to the cost. Because cost is a measure of effort, value of a product using this definition is seen to be simply the ratio of output (function or performance) to input (cost) commonly used in engineering studies. In a complicated product design or system, every component contributes both to the cost and the performance of the entire system. The ratio of performance to cost of each component indicates the relative value of individual components. Obtaining the maximum performance per unit cost is the basic objective of value analysis.

For any expenditure or cost, two kinds of value are received: use (functional) value and esteem (prestige) value. Use value reflects the properties or qualities of a product or system that accomplish the intended work or service. To achieve maximum use value is to achieve the lowest possible cost in providing the performance function. Esteem value is composed of properties, features, or attractiveness that makes ownership of the product desirable. To achieve maximum esteem value is to achieve the lowest possible cost in providing the necessary appearance, attractiveness, and features that the customer wants. Examples of prestige items include surface finish, streamlining, packaging, decorative trim, ornamentation, attachments, special features, and adjustments.

In addition to the two kinds of value received, additional costs come from unnecessary aspects of the design. Termed waste, these are features or properties of the design that provide neither use value nor prestige value. Typical scales of value for some common, well-known items are given in Fig. 13-28.

A value analysis is generally carried out in two phases, the analytical phase and the creative phase. In the analytical phase, the functional value and prestige value offered by the product is systematically investigated by a team made up of experts representing all relevant components of the manufacturing system. Findings generated in the analytical phase are then used by the team in the creative phase to define innovative design solutions that maintain the desired balance between use and esteem value, maximize these values by providing required functions for the lowest cost, and eliminate identified waste. Some steps in a typical value analysis are as follows:

1. Define the basic and secondary function or functions the product is intended to perform. Basic functions relate to the specific work the product is designed to do. Secondary functions are other functions the device performs that are subordinate to the basic functions. The goal of this step is to develop a clear and precise statement of intended product function so that a design solution that provides exactly what is needed without waste can be defined. In this step, emphasis is placed on understanding the function, not on the hardware that performs the function.
2. Make a functional analysis of all parts and features comprising a proposed product design. This step is usually carried out in tabular form with each component or feature of the design forming a separate row. The percent of total product function contributed by each component or feature is estimated, and its cost as a percent of total product cost is determined. Value is calculated as the ratio of percent performance to percent of total cost. Insight into what aspects of the design constitute waste and should be eliminated and/or where improvements are needed is obtained by comparing the value ratio calculated for each item.
3. Question the need for stated specifications and other design requirements. Compare the cost required to meet specifications with their value.
4. Make a materials analysis to determine the need for material properties used. Consider material substitutions.
5. Make a design analysis to determine alternative ways of performing the functions or eliminating certain functions.
6. Determine the value of high-cost features. Consider modification or elimination of these features.

Tie	5% function	90% esteem	5% waste
Hammer	80% function	15% esteem	5% waste
Tie clasp	20% function	75% esteem	5% waste
Button	90% function	10% esteem	

Fig. 13-28 Estimated scale of value of some common products. (*Industrial Technology Institute*)

An alternative value engineering approach, called "Blast, Create, and Refine" shortens the previous six-step procedure to three steps:

1. Determine the function and state it in the simplest verb-noun form.
2. Find the cheapest system to satisfy the function. This system usually represents fairly decent value because of least cost.
3. Compare more advanced systems to the cheapest system by the ratio of performance to cost.

Stating function in an elemental verb-noun form helps eliminate things that could confuse the real issue. Also, pursuit of the "cheapest system" often leads to a discovery of the real objective of the product and the distinction between necessity and luxury. For example, a ball-point pen is used to write words. A simple, one-piece ball-point pen consisting of an ink tube encapsulated in a plain, unadorned plastic case might be the cheapest solution. Suppose analysis shows that this solution results in a value of 50 inches/cent. For a $15.00 ball-point pen to be as valuable as the cheap pen, it has to be able to write a line 6250′ (1900 m) long. Or the esteem value offered by a more attractive appearance plus other features such as renewable ink cartridge has to justify the additional cost. In evaluating alternative designs, the function and cost for each alternative is tabulated, and the value of every alternative is compared with that of the cheapest design.

Value engineering is very broad in scope. All of the other DFM tools discussed could be considered as partial studies within value analysis. A disadvantage of the value engineering approach is its reliance on cost data that often cannot be accurately estimated until after design decisions have been made. This can detract from its usefulness as a tool for enhancing the quality of early design decisions. On the other hand, in value analysis, function and performance are sought from an abstract and conceptual perspective without implying a particular physical concept or hardware implementation. In other words, the functional analysis step in value analysis forces separation of problem and solution. For this reason, function definition as practiced in value engineering is an effective problem definition technique. Hence, performing a value analysis on an existing product might be a productive place to begin planning the next generation of product or a new version or product model. An in-depth discussion of value engineering principles and procedures can be obtained in the references 27 and 28 listed at the end of this chapter.

COMPARISON OF DFM METHODOLOGIES

A number of different DFM methodologies and tools have been discussed. All of these techniques are effective and, if properly applied, can produce significant improvements in product quality and performance, manufacturing system productivity, and lifecycle cost. Ideally, these methodologies, as well as others that are just beginning to be developed (AI/expert systems), should all be implemented and applied to effectively address DFM needs. The question that arises for many managers is how to begin to do this in as effective a way as possible. Which methodologies can be implemented most easily and quickly? Which are most rapidly effective? What would be a good long-range implementation plan?

To help provide insight into these questions, a comparison of the various DFM methodologies and tools with respect to a variety of different criteria along with specific advantages, disadvantages, and appropriate applications are summarized in Fig. 13-29. Guidelines for using this comparison are as follows:

1. *Implementation cost and effort.* A rating of "better" indicates that the methodology can be effectively implemented simply by creating awareness through seminars and/or brief training and by providing management expectation that it be used. A "worse" rating indicates that implementation may require extensive company-wide commitment, purchase or development of expensive software or hardware, extensive and/or costly training, and possibly extensive preparation for methodology use. An "average" rating indicates that implementation requirements are relatively uncertain and may involve varying degrees of software expense, training expense, and organizational and procedural change.
2. *Training and/or practice.* A rating of "better" indicates that relatively little training or practice is required for effective use. A "worse" rating indicates that extensive training requirements and/or user experience is needed or that effective use is directly dependent on effective training. An "average" rating indicates that a significant commitment to training and/or extensive practice in using the method is required.
3. *Designer effort.* A rating of "better" indicates that little or no additional designer time and/or effort is required to make effective use of the methodology. A "worse" rating indicates that significant additional design time must be allocated for use of the methodology. An "average" rating indicates that the designer must make a commitment to using the methodology, that perseverance may be required, and that some additional design time must be allocated.
4. *Management effort.* A rating of "better" indicates that little or no management effort or expectation is required. A "worse" rating indicates that significant management effort and commitment is required and/or that effective use is directly dependent on management expectation and support. An "average" rating indicates that successful use of the method requires management expectation that the method be used, coupled with good support in using the method.
5. *Product planning/team approach.* A rating of "better" indicates that effective use of the methodology requires good product planning and/or the team approach. A "worse" rating indicates that the methodology neither depends on nor encourates good product planning and/or the team approach. An "average" rating indicates that the methodology *can or may* require and/or foster good planning and the team approach, depending on circumstances.
6. *Rapidly effective.* A rating of "better" indicates that the method is likely to be rapidly effective in producing beneficial results. A "worse" rating indicates that benefits will likely be a long time in coming and use of the methodology therefore requires a long-term view. An "average" rating indicates the methodology *can or may* be rapidly effective, depending on circumstance.
7. *Stimulates creativity.* A rating of "better" indicates that effective use of the methodology tends to require design innovation and creativity. A "worse" rating indicates that use of the methodology in itself is not likely to require or stimulate design creativity. An "average" rating indicates that there is a good potential that design creativity will be stimulated and possibly required to successfully apply the methodology.
8. *Systematic.* A rating of "better" indicates that the methodology involves a systematic, step-by-step procedure that helps to ensure that all relevant issues are considered. A "worse" rating indicates that there is relatively little or no step-by-step procedure involved. An "average" rating indicates that there are aspects of the methodology that involve systematic, step-by-step procedures.
9. *Quantitative.* A rating of "better" indicates that the methodology is primarily quantitative in nature and that one or more quantified design ratings are generated. A "worse" rating indicates that the method is primarily subjective and qualitative in nature and that there are no quantitative ratings generated. An "average" rating indicates that the method has both qualitative and quantitative aspects and that one or more useful quantified design ratings may be generated.
10. *Teaches good practice.* A rating of "better" indicates that use of the methodology teaches good design for manufacture practice and that formal reliance on the method may diminish with use. A "worse" rating indicates that use of the methodology does not in itself teach good practice and that the benefits produced by the methodology depend directly on formal use of the methodology. An "average" rating indicates that the methodology teaches good practice, but it must still be formally applied to achieve intended benefits.

The relative comparisons given in Fig. 13-29 are based on the assumption that no DFM tools are currently being used. This means that the ratings given could change depending on the actual DFM capabilities and level of DFM experience that exists within a particular company. For example, value analysis is rated "worse" with respect to rapid effectiveness. One of the

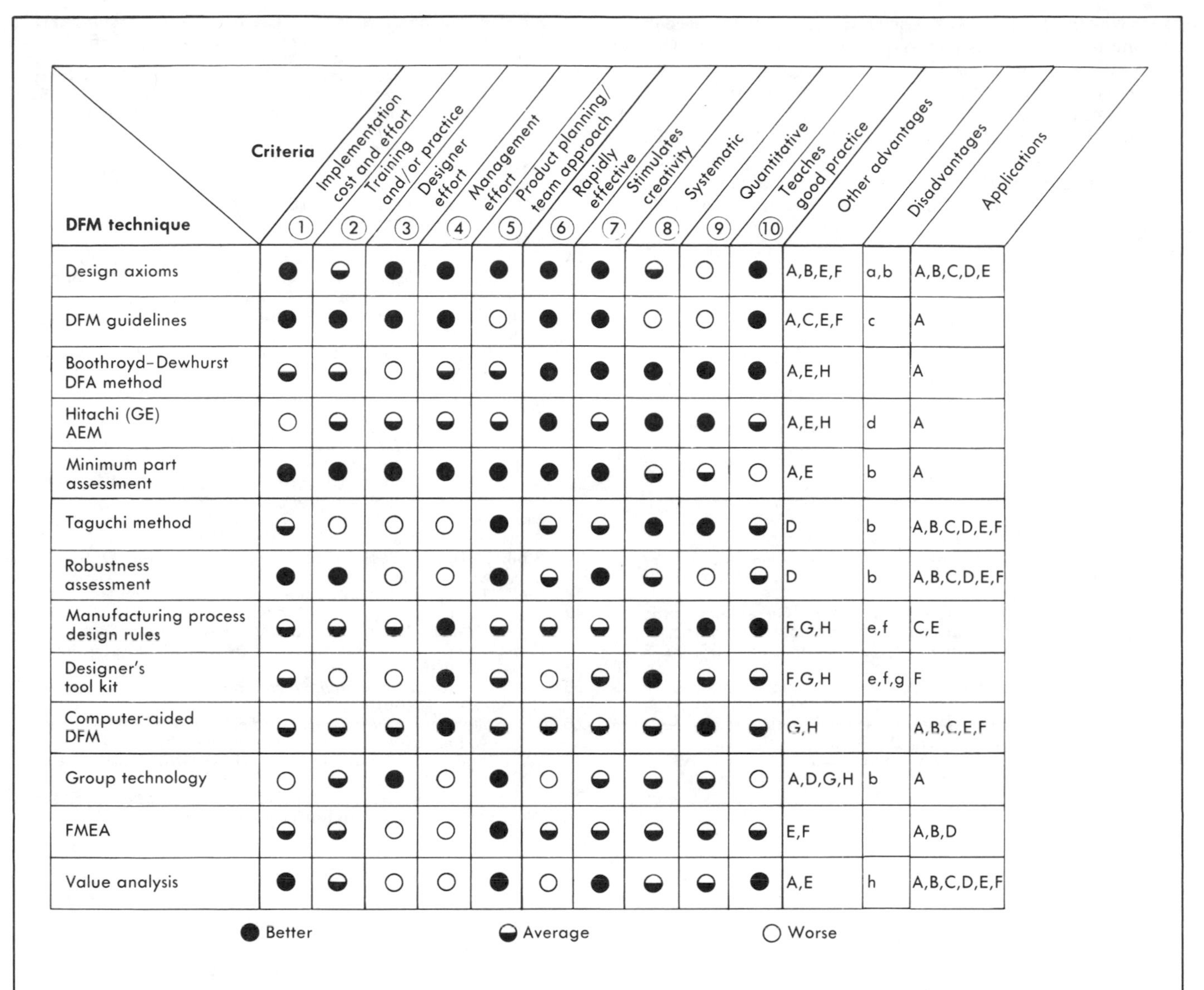

DFM technique / Criteria	① Implementation cost and effort	② Training and/or practice	③ Designer effort	④ Management effort	⑤ Product planning/team approach	⑥ Rapidly effective	⑦ Stimulates creativity	⑧ Systematic	⑨ Quantitative	⑩ Teaches good practice	Other advantages	Disadvantages	Applications
Design axioms	●	◒	●	●	●	●	●	◒	○	●	A,B,E,F	a,b	A,B,C,D,E
DFM guidelines	●	●	●	●	○	●	●	○	○	●	A,C,E,F	c	A
Boothroyd-Dewhurst DFA method	◒	◒	○	◒	◒	●	●	●	●	●	A,E,H		A
Hitachi (GE) AEM	○	◒	◒	◒	◒	●	◒	●	●	◒	A,E,H	d	A
Minimum part assessment	●	●	●	●	●	●	●	◒	◒	○	A,E	b	A
Taguchi method	◒	○	○	○	●	◒	◒	●	●	◒	D	b	A,B,C,D,E,F
Robustness assessment	●	●	○	○	●	◒	●	◒	○	◒	D	b	A,B,C,D,E,F
Manufacturing process design rules	◒	◒	◒	●	◒	◒	◒	●	●	●	F,G,H	e,f	C,E
Designer's tool kit	◒	○	○	●	◒	○	◒	●	◒	◒	F,G,H	e,f,g	F
Computer-aided DFM	◒	◒	◒	●	◒	◒	◒	◒	●	◒	G,H		A,B,C,E,F
Group technology	○	◒	●	○	●	○	◒	◒	◒	○	A,D,G,H	b	A
FMEA	◒	◒	○	○	●	◒	◒	◒	◒	◒	E,F		A,B,D
Value analysis	●	◒	○	○	●	○	●	◒	◒	●	A,E	h	A,B,C,D,E,F

● Better ◒ Average ○ Worse

Key to Advantages
A. Narrows range of possibilities
B. Results in inherent "robustness"
C. Ready reference to "best practice"
D. Emphasizes effects of variation
E. Helps identify and prioritize corrective action
F. Provides both guidance and evaluation
G. Can shorten design/tooling cycle
H. Can reduce tooling and fixturing cost

Key to Applications
A. Mechanical and electromechanical devices and assemblies
B. Electronic devices and systems
C. Manufacturing processes; other processes
D. Software, instrumentation and control, systems integration
E. Material transformation processes
F. Specialized and/or unique manufacturing facilities such as flexible assembly systems

Key to Disadvantage
a. Interpretation not always simple
b. Requires "buy-in" on part of user
c. Exceptions are not indicated
d. Rates only ease of assembly—does not address part handling or other related manufacturing parameters
e. Development requires input from experienced experts familiar with specific process capabilities and needs
f. To be used on a regular basis, implementation must be "designer friendly"
g. Must be developed and/or customized for each specific application
h. Often requires difficult-to-obtain information

Fig. 13-29 Comparison of DFM methodologies. (*Industrial Technology Institute*)

REFERENCES

reasons why this rating was assigned is the difficulty involved in obtaining accurate cost estimates early in a design project. This rating could change dramatically if a group technology database was available for use in estimating cost of new parts based on known cost of existing parts.

Figure 13-29 can be used in a variety of ways. For example, by consulting column 1, it is apparent that efforts to begin using the design axioms, DFM guidelines, minimum part assessment, and robustness assessment can probably be initiated fairly quickly. Also, because design for assembly can be rapidly effective and requires less formal training, it might be selected for implementation before the Taguchi method.

References

1. M. Asimow, *Introduction to Design* (Englewood Cliffs, NJ: Prentice-Hall Book Co., 1962).
2. T.T. Woodson, *Introduction to Engineering Design* (New York: McGraw-Hill Book Co., 1966).
3. W.H. Middendorf, *Design of Devices and Systems* (New York: Marcel Dekker, Inc., 1986).
4. J. Shigley and L. Mitchell, *Mechanical Engineering Design*, 4th ed. (New York: McGraw-Hill Book Co., 1983).
5. H.W. Stoll, "A Design Backwards Approach to Product Optimization," presented at the *SME Simultaneous Engineering Conference*, held June 1, 1987 (Dearborn, MI: Society of Manufacturing Engineers).
6. N.P. Suh, A.C. Bell, and D.C. Gossard, "On an Axiomatic Approach to Manufacturing and Manufacturing Systems," *ASME Journal of Engineering for Industry*, vol. 100, no. 2 (May 1978).
7. M. Yasuhara and N.P. Suh, "A Quantitative Analysis of Design Based on Axiomatic Approach," *Computer Applications in Manufacturing Systems*, ASME Production Engineering Div. Publication, PED-vol. 2 (1980).
8. H.W. Stoll, "Design for Manufacture: An Overview," ASME Applied Mechanics Reviews, vol. 39, no. 9 (September 1986).
9. M.M. Andreasen, S. Kahler, and T. Lund, *Design for Assembly* (Bedford, UK: IFS Publications, Ltd., 1983).
10. S.N. Dwivedi and B.R. Klein, "Design for Manufacturability Makes Dollars and Sense," *CIM Review* (Spring 1986).
11. H.W. Stoll, "Automation: Teacher and Test for Good Product Design," Paper No. MS86-1029, Ultratech-Robots West Conference Proceedings, Long Beach, CA, September 22-25, 1986 (Dearborn, MI: Society of Manufacturing Engineers, 1986).
12. R. Bradyhouse, "The Rush for New Products Versus Quality Designs That Are Producible; Are These Objectives Compatible?", presented at the *SME Simultaneous Engineering Conference*, held June 1, 1987 (Dearborn, MI: Society of Manufacturing Engineers).
13. *Ibid.*
14. G. Boothroyd and P. Dewhurst, *Design for Assembly—A Designers Handbook*, Department of Mechanical Engineering, University of Massachusetts-Amherst (1983).
15. Stoll, "A Design Backwards Approach to Product Optimization," *op. cit.*
16. G. Taguchi and W. Yuin, *Introduction to Off-Line Quality Control* (Nagaya, Japan: Central Japan Quality Control Association, 1979).
17. *Ibid.*
18. *Ibid.*
19. G. Taguchi, *On-Line Quality Control During Production*, (Japanese Standards Association, 1978).
20. J.F. Laszcz, "Product Design for Robotic and Automatic Assembly," *Robots 8 Conference Proceedings*, vol. 1, "Applications for Today," held June 4-7, 1984 (Dearborn, MI: Society of Manufacturing Engineers, 1984).
21. Stoll, "A Design Backwards Approach to Product Optimization," *op. cit.*
22. Stoll, "Design for Manufacture: An Overview," *op. cit.*
23. *Ibid.*
24. Inyong Ham, "Group Technology," *Handbook of Industrial Engineering*, Gavriel Salvendy, ed. (New York: John Wiley & Sons, Inc., 1982), Chapter 7.8.
25. William G. Ireson, *Reliability Handbook* (New York: McGraw-Hill Book Co., 1966).
26. J. J. Hollenback, *Failure Mode and Effects Analysis* (Warrendale, PA: Society of Automotive Engineers, Inc., 1977).
27. W. L. Gage, *Value Analysis* (New York: McGraw-Hill Book Co., 1967).
28. A. E. Mudge, *Value Engineering, A Systematic Approach* (New York: McGraw-Hill Book Co., 1971).

STANDARDS AND CERTIFICATION

The spoken language constitutes a standard, as does the alphabet. Communication would be impossible without agreement as to the meaning of oral sounds. If there is a need to count, measure, weigh, or otherwise describe a symbol, activity, process, component, or product, a standard is necessary. Screws must fit nuts, plugs must fit receptacles, electrical loads must be suitable for connection to a source of power. The mass production of products is possible only with the adoption of standards to describe raw materials, tests, parts, processes, and a myriad of activities essential to the manufacturer.

The definitions of terms commonly used when dealing with the subject of standards and certification are as follows:

certification An offer of proof of conformance to a standard.

self-certification Certification offered by a party attesting to its own conformance to a standard.

third-party certification Certification offered by an independent party with respect to another's conformance to a standard.

consensus Substantial agreement reached by concerned interests according to the judgment of a duly-appointed authority, after a concerted attempt at resolving objections. Consensus implies much more than the concept of a simple majority, but not necessarily unanimity.

standard A standard is usually a written document that aims at achieving the optimum degree of order in a given context. It may define terms and establish definitions, set test methods and requirements, or specify performance and safety.

external standard A standard established by a national or international coordinating organization; technical, trade, or professional group; or government agency.

internal standard A standard established by a company, organization, or agency for its own use.

international standard A standard developed under a process open to relevant bodies from all countries, adopted by an international standardizing/standards organization, and made available to the public.

mandatory standard A standard adopted by an authoritative body that requires compliance.

national standard A standard that is adopted by a national standards body and made available to the public.

voluntary standard A standard for which compliance is obtained by consent of the users.

standardization Establishing provisions for common and repeated use, aimed at the achievement of the optimum degree of order in a given context. Involves the processes of formulating, issuing, and implementing standards.

CHAPTER CONTENTS:

HISTORICAL BACKGROUND

The history of standards is essentially the story of civilization, starting with the agreement as to the meaning of spoken sounds. The development of a written language and the establishment of measuring systems were huge steps. The process of bartering goods with one's neighbors required a rudimentary method of measurement. The development of a money system, linear units, weights, measures, and the ability to make a written record were essential to the development of trade and commerce between peoples.

The problems of measurement start with the choice of a numbering system. The ancient Sumerians devised a system that was partly decimal and partly sexagesimal in nature. They first introduced the notion of 12 subdivisions in measurement by dividing the day into 12 units. This, along with their division of the circle into 360°, is still in common use today.

Most of the ancient systems of linear units were based on the human body. Thus, the cubit is derived from the length of the forearm, and the digit the width of the finger. The foot and palm are obvious measurements, and the fathom was the distance between a person's outstretched arms.

In Roman linear units, 16 digits were equivalent to a Roman foot. However, the foot was also subdivided into 12 parts called *unciae*, which later became inches. The Romans also used the cubit, which they rated at 24 digits. This led to the English yard, which is really a double cubit. For larger distances, the Romans used a unit of 5000 feet, which they called *mille passus* (1000 paces), or a mile.

In the Middle Ages, units of mass kept a close association with monetary units and precious metals. Thus, the pound (or lira in Italy and livre in France) was the standard unit of mass. Charle-

Contributor of this chapter is: ***James Pearse****, Group Vice President of Engineering, Leviton Manufacturing Co.*

Reviewers of this chapter are: ***John T. Benedict****, Standards Engineer—Consultant, Technical Div., Society of Automotive Engineers;* ***Dorothy Hogan****, Vice President Communications, American National Standards Institute, Inc.;* ***Hayward Thomas****, Consultant, Retired—President, Jensen Industries;* ***Bob Toth****, President, R.B. Toth Associates.*

magne established a ratio of 1 pound equal to 20 shillings equal to 240 pence; this standard remained in use by the British until recently. When the pound was used for purposes of weight, it was subdivided in various ways ranging from 6 to 18. The avoirdupois pound with 16 ounces and the pound of Troyes with 12 ounces are still used today. In the fourteenth century, English weights and measures were based on the penny called a sterling, which weighed 32 grains of dry wheat from the middle of the ear; 20 pence made an ounce; 12 ounces made a pound; and 8 pounds made a gallon of wine, with 8 gallons making a bushel of London.

The history of certification parallels that of standards. If a standard is used, users want proof. A manufacturer may attest on its own authority, through advertising and on the product, that the item meets a standard (self-certification), or the manufacturer may submit the product to an outside laboratory for independent certification of compliance with a standard (third-party certification).

In the United States, the National Bureau of Standards (NBS) deals with both standards and certification. NBS was established in 1901 by an act of Congress and was the first national laboratory. NBS develops, maintains, and disseminates hundreds of measurement standards. These include the fundamental standards for temperature, time, mass, volume, length, and electrical measurements. Modern manufacturing could not take place without the means to periodically calibrate gages, scales, instruments, and test equipment to reference instruments "traceable to NBS standards." These standards are an essential part of the certification process.

Underwriters Laboratories Inc. (UL) plays a key role in providing third-party certification of materials and products, and as a developer of safety standards. UL was founded in 1894 as a nonprofit organization to promote public safety. Using more than 500 safety standards, it investigates more than 12,000 product categories for clients who request listing or label service; these services will be discussed subsequently.

The world's largest developer of voluntary consensus standards is the American Society for Testing and Materials (ASTM). ASTM has been in existence since 1898 and was formed for "the development of standards on characteristics and performance of materials, products, systems, and services; and the promotion of related knowledge." In 1986, ASTM had 140 main technical committees with 2034 subcommittees; there were 30,000 active members with 17,900 people serving as technical experts on various committees; and there were 8400 standards comprising 60,000 pages of text.

Many independent testing laboratories also exist for the sole purpose of offering third-party certification to clients in categories that range from industrial hygiene to telephone cable. They perform a valuable service to society.

DEVELOPMENT OF STANDARDS

The development of a standard starts with the identification of a need. For example, a problem may be indicated in interchangeability, quality, or application of a product or process. If the need is urgent, it is brought to the attention of a standards-developing organization that forms or encourages the formation of a committee to address the issue. The committee should be balanced with respect to the interests of producers, users, consumers, and those with general interest, but this is not always the case. For example, the standards-developing organization may be an association of manufacturers, whose committees would not include users, consumers, or those with general interest. It may be a professional organization, with members skilled in some specific branch of science or engineering. Table 14-1 lists the number of standards developed in the U.S. by both government and private-sector organizations. A graphical representation of the number of standards developed by each organization is shown in Fig. 14-1.

Once a committee is formed, work proceeds and a draft document is prepared reflecting the consensus agreement among the committee members. This draft document is then issued for review by others interested, and comments are sought from affected producers, users, consumers, and general-interest parties. These comments are then addressed, appropriate changes are made, and the document then may be published as a standard of the sponsoring organization.

It is generally impossible to obtain unanimous agreement with respect to the final form of the standard. However, unanimous agreement is not required for consensus, which is one of the values of the consensus process for adoption of a standard. To address minority viewpoints, an appeals procedure must be a part of the adoption process. A standard should contain or reference a rationale or reason for its requirements. The rationale may or may not be a part of the standard document itself. The major standards-developing organizations in the U.S. are members of the American National Standards Institute (ANSI), and most of them voluntarily submit standards developed under their auspices to the Institute for approval. The development work is carried out in accordance with ANSI requirements for consensus and due process. When

TABLE 14-1
Current U.S. Standards

Government	
Defense	38,000
Federal	6,000
Other	6,000
60%	50,000
Private Sector	
Scientific and professional	13,300
Trade association	11,800
Standards writing	9,900
40%	35,000
Total	85,000

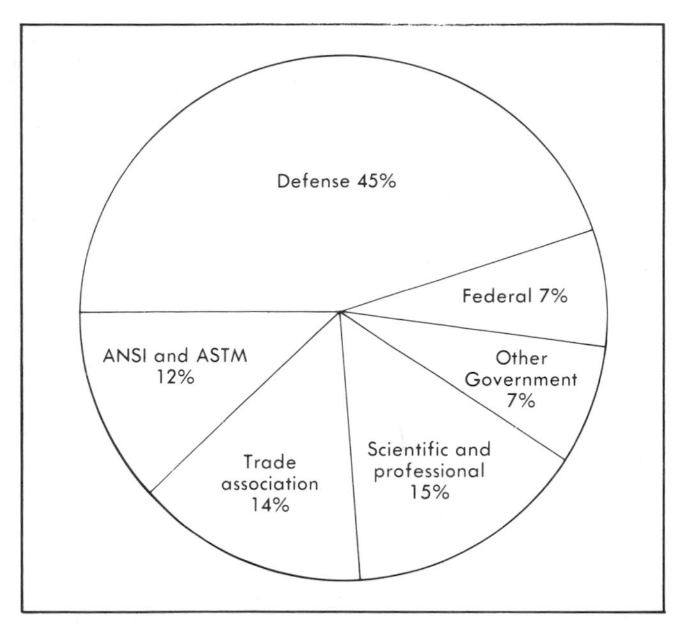

Fig. 14-1 Graphical representation of the percentage of standards developed by U.S. organizations. (*R.B. Toth Associates*)

the Institute verifies that these requirements have been met and approves the document, it is then published by the developer or by ANSI as an American National Standard. In some cases publication may precede submission for approval. Interested parties are now free to adopt the particular standard on a voluntary basis. Assuming that the standard is well written and that the review process has been effective, there should be no reason for failure to adopt it.

Some standards may be of sufficient importance or interest to governmental bodies that they will be adopted as mandatory for certain applications or adopted by code or ordinance into law and made mandatory in that process. An example of this is the National Electrical Code (NEC), produced by the National Electrical Code Committee of the Electrical Section of the National Fire Protection Association and approved by ANSI as an American National Standard. The NEC is offered as a model code suitable for adoption by reference into a law, ordinance, regulation, or similar administrative instrument. Enforcement of the NEC is through inspection agencies for the enforcing authority, such as the building inspection department of a municipality that has adopted the NEC.

As indicated earlier, standards may be submitted for adoption as national standards. In the U.S., the American National Standards Institute performs this function. ANSI serves in the following capacities:

- Coordinator of the U.S. Voluntary Standards System.
- Approval organization for American National Standards.
- U.S. member of the International Organization for Standardization (ISO) and International Electrotechnical Commission (IEC).
- Clearinghouse and information center for national and international standards.

A standard submitted to ANSI for approval triggers a process to ensure compliance with the rules for consensus. This is done through the use of an independent body, the Board of Standards Review, to verify evidence that the rules established by ANSI have been met. To help standards developers establish evidence of consensus, ANSI makes available three separate methods: accredited organization, committee, and canvass. Standards groups apply for authorization to use one or more of them. ANSI does not develop standards; it approves standards submitted to it by standards developers.

The submitter of a standard to ANSI must make an objective attempt to resolve dissenting views and respond to all negative or adverse comments received from its own efforts to develop evidence of consensus. With the submittal of the standard, the sponsoring organization must include a record of voting within the accredited group, the record of public review, unresolved objections and attempts at resolution, and information on steps taken to harmonize domestic proposals with work on parallel international standards activity. The submitter must also certify that ANSI's due process requirements were met, that significant conflicts with other standards have been resolved, and that appeals have been completed.

Finally, if this new standard is of significant interest, the work may be adopted by the ISO or IEC under their accelerated procedures or used as a basis of ISO or IEC standards.

INTERNATIONAL STANDARDS AND CERTIFICATIONS

The two predominant international standards organizations are the International Electrotechnical Commission (IEC) and the International Organization for Standardization (ISO). IEC was formed in 1907 and prepares standards in the electrical and electronic fields. ISO was formed in 1947 and prepares standards in all other fields. Both of them are nongovernmental agencies with headquarters in Geneva, Switzerland. The number of standards developed by ISO and IEC, as well as other organizations, is shown in Table 14-2.

The member bodies of ISO are national standards bodies, predominantly agencies of their respective governments. The member bodies of IEC are national committees with representation from a variety of interests including manufacturers, utilities, testing and certification agencies, government agencies, and others. Their methods of operation and organizational structures differ, primarily because they deal with different

TABLE 14-2
International Standards

International Organization for Standardization	6,400
International Electrotechnical Commission	1,800
22 others	1,500
180 more	1,500
Total	11,200

subjects. Because ISO and IEC become involved in subjects that overlap, coordination between the two takes place.

ANSI is the U.S. member body for ISO and also for IEC through the U.S. National Committee (USNC). The manufacturing manager should be aware of these relationships, but is more likely to be concerned with certification, listing, or approval of equipment to be sold overseas than with the development of the applicable standards.

Some years ago, a plan was established to promote international trade by simplifying the work of testing laboratories in various countries. This was known as the CB Scheme ("certification body" scheme) and was administered by the Commission for Conformity Certification of Electrical Equipment (CEE).

In 1985, CEE was formally integrated into the IEC under the title IEC System for Conformity Testing to Standards for Safety for Electrical Equipment (IECEE). The IECEE system is managed by a management committee responsible to the IEC Council. Under the management committee, a CB Scheme is managed by the Committee of Certification Bodies (CCB), and a Committee of Testing Laboratories (CTL) deals with the uniformity of test procedures and equipment called for in the various standards accepted for use in the system.

Although primarily European in nature, CB Scheme members also include Australia, Japan, Canada, the People's Republic of China, Israel, the Republic of Korea, and the United States. However, with the exception of Japan and Israel, these non-European countries, along with the U.S., do not participate in the CB Scheme. The USNC/IECEE has observer status and is a member of the IECEE management committee.

For countries participating in the CB Scheme, manufacturers submit their products for approval to a national certification body (NCB) in their countries. When specimens have been tested and found to be in compliance with the given standard, the manufacturer is granted a CB test certificate along with a test report. With these, the manufacturer can then submit specimens to those countries where the product will be marketed, to obtain approval to use the national mark or certificate.

The procedure is somewhat different for manufacturers in countries not participating in the CB Scheme and calls for the manufacturer to submit an application to the secretary of the CCB that will require testing in two different countries. If the results of both are favorable, a CB test certificate and report may be issued.

If a company intends to sell products in overseas markets, the products must conform to the needs of those customers. More than that, many countries legally require approval of their national certification agencies before products may be sold there. The particular standards involved must be carefully considered during the design phase of these products, and the manufacturer must be completely familiar with the procedures used to certify compliance with those standards and obtain the required national certificates or use of the national mark.

COMPANY STANDARDS PROGRAM

Very few things can be successful in any company without management's support. Thus, the first step in setting up a standards program is to obtain the necessary management support and a commitment to the essential part that standards and certification play in the success of operations. The need for this standards activity can be developed by means of a list that shows the standards affecting various aspects of the company's business and their degree of importance. A budget should be prepared to indicate the investment that would be made in this activity. This budget reflects an investment for which a return is desired, as is the case for any expenditure. Standards are not an end unto themselves, nor are they the reason why a company is in business, but they are an essential part of the business.

A policy statement relating to standards and certification activity is one way of gaining this support and assuring use of internally developed standards. Nothing can be gained by a program to standardize screws if the next new product calls for new screws instead of using existing screws and slightly modifying the design. Similarly, it does no good to assign an engineer the responsibility to work on an ASTM committee without allowing time to both attend the meetings and process committee work. Management must support the standards program and must be prepared to monitor this program to ensure the competitive edge that this investment will yield.

The nature and activity of the standards program depends on the company size. For a very small company, it is not unusual to find this responsibility as a part-time job split among a few individuals. In very large companies, there may be several specialized standards departments that coordinate activities to ensure that duplicate effort does not take place.

A good standards program has the following five distinct aspects:

1. Internal standards development.
2. (External) voluntary standards participation.
3. Certification.
4. Standardization programs (compliance).
5. Communications.

Prior to proceeding with these different aspects of a standards program, it is necessary to determine the cost and return (justification) of this activity and to ensure that qualified people are in positions of leadership for the program.

JUSTIFICATION

A vigorous standards program affords a competitive advantage to the modern manufacturer. It serves to minimize product cost as well as the overhead activity necessary to manufacture and bring the product to market. The manufacturer who minimizes the number of raw materials and purchased items in a product or process and makes use of standard industry items as well not only benefits from the lowest possible purchase cost, but also from the ability to obtain materials and other purchased items quickly from vendors. The net result is lower product cost and less investment for work-in-process inventory. It is important to keep in mind the distinction between standards and the process of standardization (or adoption of a standard). The savings that come from standardization may be lost by a poor choice of standards.

Internal Standards

Obviously, it is a good thing to standardize on fasteners such as screws, nuts, bolts, rivets, and eyelets. It would also be an obvious step to minimize the different varieties of these fasteners, as well as the materials and finishes. In this computer age, it would also be an obvious step to standardize on the computer hardware and software necessary to carry out business activities. These things can be quantified, costs assigned to the standardization activity, and benefits calculated to determine payback. This is usually true of most company internal standards activity.

For example, consider the steps required to standardize machine screws. Basically, the end use of each type and quantity of screw should be reviewed. If a screw is required, the least expensive one that will do the job should be used. The objective is to minimize the number of screws in inventory and the cost of this inventory. The proper screw standard must be chosen or developed. Product design changes may be required, so the engineering department will be involved. A bill of material for all screws, "where-used" lists, costs, and the ability to project the cost of a change to the screw will be needed. The steps for this process are as follows:

1. Analyze current costs.
   - List the screws by size, length, and strength, with separate lists for different head types, material, surface treatment, and so on.
   - List quantity and cost.
   - Note type of assembly (manual or autofeed).
   - Estimate inventory costs and special factors.
   - Determine what screw thread standards (if any) are now used.
   - Determine if there are any manufacturing problems with the screws.
   - Determine the dollar total on an annual basis.
2. Establish reasonable goals.
   - Review the information in step 1.
   - Many screws will show slight differences in size, length, head type, material, and so on. An estimate of the number that can be combined should be established as a goal.
   - A reduction in the number of screws will permit purchasing or manufacturing savings and a reduction in inventory costs. Set a goal based on the goal for reduction in the number of screws.
   - Will an existing screw standard be adopted or modified or will a new one be developed? Set direction and a goal.
   - If the analysis showed manufacturing problems involving screws, set a goal to eliminate them.
   - Set a goal for annual cost savings. This may be phased in over a period of years.
3. Determine a budget.
   - Six-month payback? One year? What is the policy on any cost-savings programs?
   - Who's going to handle standardization? How long will it take? What is the cost?
   - What engineering time will be involved? What is the cost?
4. Assign the task responsibility.
   - Choose the proper person for the job.
   - Make sure the task requirements are clearly understood, including budget, time fences, goals, authority, and constraints.
   - Arrange for followup reports and final project report.
   - Determine how the program will be maintained.
5. Develop policy.
   - Notify affected managers of the program.
   - Issue a policy to ensure continued compliance with screw standardization.

At step three, the manager is in a position to decide if the effort is cost justified based on the company's criteria for any cost-savings program. However, the benefits of participation in external standards activity may be less obvious and quantifiable.

External Standards

External standards activity in areas that affect a company's business may proceed in a manner that appears satisfactory regardless of whether or not the company participates. Management may reason that good standards activity is not designed to show a competitive advantage to any particular manufacturer. While it is true that the purpose of standards activity is never to limit competition, competitors do not know another company's products and processes as well as that company does and may fail to include material that is important. As a general rule, companies should take a leadership position in the development of industry standards where they are a significant factor in that industry and should monitor standards development where they are an insignificant factor.

Cost justification of external standards programs can rarely be accomplished in the same fashion as internal standards programs. The cost of a program to internally standardize screws can be accurately determined along with the payback and the time frame. However, the payback from participation in an ASTM committee establishing test standards for materials the company manufactures may be difficult to quantify. Failure to participate may result in loss of market share or price erosion if competitors bring material or products conforming to a new standard to the market first. The payback may not occur immediately and will usually involve intangibles, such as customer perception of the company as a "quality house."

External standards activity affords opportunity to participate and influence standards that affect a company's product, operations, and business. It also is an image builder that pays off in long-term benefits. It indicates a "good corporate citizen" who is concerned with quality, technology, and safety—the kind of company customers want to deal with.

ORGANIZATION AND PERSONNEL

In selecting personnel for a standards department, the overriding concern should be knowledge of the company's operations, products, and processes. Unfortunately, standards engineering is not included as a college-level course at technical institutions, with perhaps the exception of one or two. There are no undergraduate or graduate courses that will lead to recognition as a standards engineer for newly graduated technical personnel. Hence, the best source for the recruitment of a good standards engineer from within the company is a person who has the knowledge of the company and products. It is also possible to hire an experienced standards engineer or contract with a consultant to set up a standards program and provide training.

CHAPTER 14

INTERNATIONAL STANDARDS

A good standards engineer must be conscientious, very patient, and detail oriented. If he or she is going to represent the company, he or she must present a good appearance and be able to communicate well. This particularly includes the ability to write clearly.

The question of personnel is best considered in the context of company size and industry, but must include three separate responsibilities: (1) internal standards, (2) external standards, and (3) standardization programs. In a large company, separate departments may exist for each of these functions. In a small company, the work may be carried out as a part-time responsibility by people who "wear several hats." Where external standards are concerned, specialized technical personnel must occasionally "wear two hats" in even the largest companies because of the expertise needed to participate in the development of some highly technical standards.

Internal standards development may be the responsibility of the engineering or manufacturing managers, or each department head may be required to develop standards peculiar to his or her specific area. Because existing personnel may be best suited to these tasks, another "two-hat" situation exists.

Participation in the development of external standards is usually the responsibility of the engineering manager, but may also be assigned to the president or the marketing or manufacturing managers. A full-time standards engineer, if available, is the best choice. It is not unusual for the standards engineer to be accompanied by a technical specialist to meetings where special expertise is required, or for the technical specialist to attend as a company representative or as an individual as a member of a professional organization. In the latter case, consideration must be given to the fact that even if membership in a professional organization is personal, the technical specialist is still an employee. The specialist should represent the company's interests and should be reimbursed for expenses.

Standardization programs are usually the responsibility of the manufacturing manager, although they may be assigned to the engineering manager. Some experienced standards department managers believe the best people for these tasks are those trained as manufacturing engineers because these programs are akin to cost-savings programs.

Management of the standards function in any company must be active. This is somewhat difficult in view of the fact that a complete standards function cannot be segregated into a specific department such as accounting or quality control. This is further complicated by the requirement that a standards program be tailored to fit a specific company by size and industry, or by top management's lack of recognition of the utility and importance of standards while, at the same time, desiring standardization to cut costs.

Management of the standards function consists of the following:

- Setting goals, policy, and budgets.
- Identifying and reviewing standards and standardization programs.
- Identifying qualified personnel to participate in particular programs.

INTERNAL STANDARDS DEVELOPMENT

A manufacturing company may have a number of internal standards and not recognize that fact. There may be an informal agreement (consensus) in the engineering department to indicate angles on shop drawings using only whole numbers (voluntary standard), except that one of the workers uses fractional degrees, so there is no uniformity (standardization). The title block for shop drawings may specify a tolerance of $\pm 0.005''$ unless otherwise indicated (mandatory standard, documented). Some internal standards may be thought of as procedures, specifications, and instructions.

The distinction between a company with a good internal standards program and one with a bad program is quite simple. Questions to ask regarding an internal standards program are as follows:

1. Do the standards respond to the current needs of users?
2. Documentation.
   - Are the internal standards in written form?
   - Are the documents clear, concise, and well prepared?
3. Communication.
   - Are the standards available to the people who should use them?
   - Do designers have a loose-leaf book of standard company practices?
4. Control.
   - Does management control internal standards or are they "developed by custom"?
   - Does the company have a periodic review of internal standards?
   - Are the standards being used? If not, why not?

The existence of internal standards should not be equated with the assumption that the company has been "standardized." For example, a company may have an internal standard that specifies the requirements for machine screws and further mandates that all machine screws must meet this standard. The machine screw standard may be a variation of an external standard or one that was prepared in-house to suit particular requirements. Compliance with this standard is the measure of the company's degree of standardization.

The preparation of internal standards follows the same process given in the section discussing development of standards. There are, however, differences or special considerations. Some recommended steps to follow in this process along with issues to be addressed during each step are listed in Fig. 14-2.

The first step is to find out what has already been adopted by custom, general usage, or informal agreement. This task of gathering and documenting the information will take time and cost money. Because several departments will probably be

1. Identify the need
2. Justification
   - What will it cost to develop?
   - What will it save?
3. Authorization
   - Budget and personnel
   - Instructions, goals, time fences
4. Development
   - Rely on input from potential users
5. Review
   - Acceptance by users
   - Approval from management
6. Compliance
   - Mandated use—standardization
7. Maintenance
   - Who maintains the standard?
   - Assure periodic updating

Fig. 14-2 Recommended steps for the development of internal standards. (*Leviton Manufacturing Co., Inc.*)

affected, individual representatives from those departments should perform the task. Because the information should be gathered in a uniform (standard) manner, these individual representatives should comprise a committee, with a chairperson selected. The committee should receive a clear mandate from management as to its task, including the authority, goal, scope, and completion date. The results should be reported to management along with recommendations.

The committee can now exercise control of existing company practices. This type of exercise always leads to a number of surprises. "I didn't know we were doing this" becomes a common phrase. The examination of existing de facto internal standards of a company in a uniform written format permits management to give additional direction and specific task assignments. Some will be scrapped, others ignored. Most will probably be changed, with deficiencies corrected and omissions filled. There will be obvious voids. A standard may be incomplete or missing entirely. It is important to keep this exercise in perspective with respect to the goals of the company. Standards are not an end unto themselves; they are one of the tools that a company uses to ensure efficient operation.

The format for internal standards can vary to suit the needs of a particular company. If a company is developing internal standards from scratch, it is best to purchase typical external standards, examine the formats, and if possible adapt or adopt one that fits the company's needs. Whatever process is used, it is important to ensure that the standards are followed.

EXTERNAL STANDARDS DEVELOPMENT

As indicated earlier, the choice of whether or not a manufacturer should participate in a voluntary standards program is a long-range issue. Before a decision can be made as to participation and before a budget can be developed, it is necessary to know what standards affect the company's operation. Development of this type of list will usually lead to the identification of an incredible number of industry, trade, and professional organizations that develop standards affecting every facet of the business. To aid in this identification, the reader should review NBS Special Publication 681, "Standards Activities of Organizations in the United States." Some of the developers of industrial standards are listed in Table 14-3.

TABLE 14-3
Developers of Industrial Standards

	No. of Standards
American Society for Testing and Materials	8400
Society of Automotive Engineers	4600
Aerospace Industries Association	2900
American National Standards Institute	1900*
Association of American Railways	1350
Factory Mutual	600
American Society of Mechanical Engineers	590
Electronic Industries Association	550
Institute of Electrical and Electronics Engineers	550
Underwriters' Laboratories	520
American Railway Engineers Association	400
American Petroleum Institute	350
Technical Association of the Pulp and Paper	270
Industry	260
National Fire Protection Association	

* Copyright assigned to the American National Standards Institute.

Determining whether to participate in the development of external standards is straightforward. As in the development of internal standards, the company should make a chart listing its component parts, products, processes, procedures, markets, and means of distribution. For each of the general classifications on the list, a search should be made for a particular standard that might affect that item, along with the organization that sponsors that standard. This information should be listed in a second column on the chart. In a third column, the relative importance of that standard should be entered using a scale of 1 to 10 or other convenient means of priority identification, such as critical, major, minor, or discretionary. A fourth column should be left free for comments. From this, a decision may be made as to the desirability of participating in the identified standards work, and inquiry can be made to the sponsoring organization regarding current activity on that standard. Policy may then be established, and a budget can be developed to reflect the expected costs. Table 14-4 shows the evaluation one manufacturer used to determine its participation in external standards development.

It may be in the interest of the company to subsidize employee dues expense for membership in professional organizations in addition to compensating the employee for time and expense involved in participation in standards produced by the professional organization. Trade and industry association dues may also be considered in this same light. All in all, these expenses reflect the costs of company participation in the development of standards that affect the business.

As a first step in external standards development, a company may want to obtain membership in ANSI, which puts at its disposal the resources of the approval agency for American National Standards. ANSI is the source for 8000 approved American National Standards as well as 11,000 ISO and IEC standards and keeps members informed of standards activity through the *ANSI Reporter* and *Standards Action* publications.

CERTIFICATION

A manufacturer may be asked for proof that a certain product complies with specific standards and that the product is suitable for the customer's intended application. This proof is referred to as certification. Certification may be either self-certification or third-party certification.

With self-certification, the manufacturer sends a letter on company letterhead to the customer stating that the product in question does indeed comply with the requirements of a specific standard(s). Sometimes this letter will have to be notarized and signed by a company officer. It may also be appropriate to

TABLE 14-4
Typical External Standards Evaluation for a Manufacturer of Electrical Switches

No.	Item	Standard	Value*	Comments
1	All switches	UL 20	9	Important for U.S. sale
2	All switches	CSA C22.2#111	10	Required for sale in Canada
3	All switches	Fed. Spec. WS896-E	3	Required for sale to certain customers
4	All switches	NEMA WD-1	4	Required for NEMA compliance
5	Contact material	ASTM B-36	9	Alloy used in device
	(copper alloy)	CDA C-26000	9	Alloy used in device
6	Contact material	ASTM B-617	9	All electrical contacts
7	Molded case	ASTM D-705	2	Purchasing specification
	(urea plastic)	UL QMFZ2	8	Required for UL listing
8	Insulator	ASTM D-709	2	Purchasing specification
	(laminates)	NEMA LI-1	4	Required for NEMA compliance
		UL QMFZ2	8	Required for UL listing
9	Coil springs (music wire)	ASTM A-228	9	Design reference
10	Screws, material	ASTM B-134	9	Alloy used in device
	Screws, material	CDA C-26000	9	Alloy used in device
11	Screws, thread and head	Fed. Std. H-28	9	Alloy used in device
12	Straps	NEMA OS-1	7	Product quality
	(galvanized steel)	ASTM A-526	7	Product quality
13	Solder, wire	ASTM B-32	5	Purchase specification
	Solder, wire	Fed. Spec. QQ-S-571	5	Purchase specification

* Ranked in importance from 1 through 10, with 10 being most important.

include test data showing compliance, references to particular construction requirements, or drawings. Enough information should be included with the self-certification letter, and it should be worded in the proper form, to satisfy the needs of the customer.

Third-party certification brings someone else into the act. This is usually an independent testing laboratory or agency such as Underwriters Laboratories Inc. (UL), the Canadian Standards Association, or even a government agency. For example, UL will accept applications to investigate products; if the investigation and testing shows a product to be in compliance with the applicable standards, UL will authorize the manufacturer to use the UL Listing Mark as evidence (third-party certification) of this fact (see Fig. 14-3). UL has an established factory followup service that the manufacturer is required to agree to as part of its comprehensive service.

In some instances, the customer may want assurance from the product manufacturer that third-party certification will be maintained or that notice will be given if compliance to a particular standard is no longer maintained. This type of situation should be handled by a letter on company letterhead (see Fig. 14-4). The customer may require this letter to be notarized and signed by a company officer. The exact requirements should be negotiated with the customer.

It is important that records of these certifications be kept in a central file supervised by the individual or department responsible for standards engineering. Obviously, an activity such as this should be kept under control. The same type of certification should be furnished, preferably on a form letter, to the same class of customer. It is also important to know what was said in past years in case a customer wants a certification renewal or update. It may be appropriate to adopt a policy with respect to the form of certifications.

Fig. 14-3 UL Listing Marks generally appear in a variety of forms together with the product designation. (*Underwriters Laboratories Inc.*)

Some regulating agencies require third-party certification to sell some products. The electrical industry is typical of this, with Underwriters Laboratories Inc. the leader in performing this service in the United States and the Canadian Standards Association performing this in Canada. A request by a customer for evidence of third-party certification to a particular UL standard can be handled by simply mailing a UL listing card to that customer, along with the notice that evidence of listing of the product is given only by the UL Listing Mark, which must be affixed to the product itself.

Although many independent testing agencies are available for third-party certification, the procedures used by UL are unique in that safety standards are developed, compliance of

Gentlemen:

We certify that our material XYZ meets the requirements of ASTM UVW-1974. Evidence of this compliance in the form of a report by Great American Testing Laboratories dated April 21, 1987, is attached.

We further certify that all material XYZ furnished to you will be in compliance with ASTM UVW-1974 unless or until we furnish you with 30 days notice of noncompliance.

We further certify that material XYZ is suitable to produce your part number BA-19750 in accordance with your drawing dated December 3, 1986.

Yours very truly,

Massive Manufacturing Company

Fig. 14-4 Sample letter to a customer verifying that third-party certification will be maintained on a product. (*Leviton Manufacturing Co., Inc.*)

products is determined, and followup visits are made to the factories. UL is the largest independent testing laboratory in this country; the certification procedures of other independent agencies may vary.

The UL procedure starts with an application and submittal of samples of a product or component to UL for investigation. UL will determine whether or not the product complies with its requirements. In most cases, these requirements will be published in an existing UL standard. UL produces product standards in response to perceived need, generally three or more clients with products in a specific area. When a standard does not exist, a "desk drawer" standard may be put together by UL staff detailing construction requirements and tests that will be necessary to investigate the product for safety and compliance with NEC requirements and other installation standards.

If the product meets the construction requirements and passes the required tests, UL will authorize the client to use the appropriate UL Listing Mark on the product and give the client a copy of the *Report and Procedure*. The Report is a record of the tests performed on the product, and the Procedure is a record showing product construction. The Procedure specifies materials, dimensions, and photographs that are safety related and that are required to enable determination as to whether succeeding products are built the same as the samples submitted. Materials and components used within the product must either have been investigated and listed in UL's Recognized Component Directory or be investigated separately during the product investigation. The Procedure will specify the use only of materials and components that have been so investigated.

UL offers two classes of listing, called Label Service and Re-Examination Service. A "white card" is issued for both classes indicating that the product is complete and ready for installation. For Label Service, labels are purchased from UL or an authorized label maker. Information on the label includes the UL registered mark, the word "Listed," an issue number, and a product identification. Factory visits to inspect the product are scheduled on the basis of the number of labels sold. For Re-Examination Service, the product will have the Listing Mark applied by the manufacturer with essentially the same elements as for Label Service, with the exception of the issue number. UL does not charge for the number of products made using the mark. The number of unannounced factory visits depends on the production volume; in general, there are at least four visits per year.

UL also offers Recognition Service, or "yellow card" listing. This applies to a component or material that will be suitable for installation in another product, but is not suitable for use by itself. Rather than the registered mark consisting of the UL in a circle, a separate mark called the recognition mark consisting of a reversed R with the leg extending into a U is allowed for use on the product.

UL also offers fact-finding investigations for the purpose of developing product or system information or data for use by the applicant in seeking recognition in or amendment of a nationally recognized installation code or standard. Another service offered by UL is Classification Service. In Classification Service, some special use of the product is investigated for compliance to a specific standard.

During UL factory inspections, the UL inspector may find that the product varies from the procedure detailing construction of the product or that it fails to pass a required test. In such cases, a Variation Notice will be given to the manufacturer indicating that the products covered by this notice are not suitable to bear the Listing Mark and that the noted deficiencies must be corrected or the Listing Mark removed prior to sale.

In a manufacturing environment, particularly when it is necessary to maintain third-party certifications to allow the sale of a product, care must be taken to ensure that any product changes are forwarded to UL for investigation, that the factory procedure is up to date, and that product quality is maintained to the level expected by the customers and the third-party certifier. The best way to do this is with personnel assigned to this task as part of the standards engineering function.

COMMUNICATIONS

An essential part of communications with respect to a program of standards and certification is to assure that management has the information needed to authorize and control the activity. "Reporting up" means answering the following questions:

1. What's being done?
   - Internal standards.
   - External standards.
   - Standardization programs.
   - Status of certification.
2. Why is it being done?
   - Justification.
3. Who's involved?
   - List programs and active personnel.
   - Time involvement.
4. How much is it costing?
   - Original budgets.
   - Performance to budget.
5. Where's the return?
   - Cost savings.
   - Market penetration.
   - Corporate image.
   - Effect on operations.

"Reporting down" is equally important. Management must be prepared to perform its traditional duties as follows:

1. Set policy and give clear direction.
2. Set goals, both internal and external.
3. Authorize activities.

BIBLIOGRAPHY

4. Delegate responsibility.
5. Evaluate performance and hold personnel accountable for their work.
6. Offer support and leadership.

Management must continually evaluate the effectiveness of its standards and certification programs and let employees know the progress of their efforts.

In addition to reporting up and reporting down, horizontal communications are essential to a good standards program. The product of a standards and certification program must be useful and used for it to be effective. This not only means passing the information on to those members of the manufacturing community that should use the information, but also passing the information on in a form that will catch their attention and suit their needs. It is not enough to indicate a possible change in a product standard to the design engineer by sending a copy of the proposed revision. The standards engineer should outline the specific revisions for the design engineer and give an analysis of the possible effect on the product or process involved. It is important to follow up on the issued change notice to ensure that the design engineer received and understood the information and that decisions have been made and action taken. There may be a time period involved for implementation of these changes (effective date) that must be met.

Information with respect to standards and certification activity must be available to those who have need of it. This may range from maintaining a library of standards at a central point to arranging for on-line computer information retrieval services that deal with required standards. Individual product managers may work so closely with particular standards that they require their own copies, but the standards engineer should assume the responsibility of making sure that these standards are current. Standards are available in printed form, microfilm, and microfiche. Bibliographic information (citations) is available on standards through computerized on-line information retrieval services. The key is not to centralize the information itself and keep it in one spot, but to centralize control of the information to ensure that it is timely and current. Above all, the standards engineer should "know where to find it" if the need for information with respect to standards should arise.

Bibliography

ANSI's Role in International Standardization. New York: American National Standards Institute, 1985.

Breitenberg, Maureen A., ed. *Directory of International and Regional Organizations Conducting Standards-Related Activities*, NBS Special Publication 649. Washington, DC: U.S. Government Printing Office, 1983.

————, ed. *Private Sector Product Certification Programs in the United States*, NBS Special Publication 703. Washington, DC: U.S. Government Printing Office, 1985.

Burton, William K. *Measuring Systems and Standards Organizations*, SR10. New York: American National Standards Institute.

Directory of Engineering Document Sources. Santa Ana, CA: Global Engineering Documents, 1986.

Fink, Donald G., and Beaty, H. Wayne. "Standards in Electrotechnology," Section 28. *Standard Handbook for Electrical Engineers*, 12th ed. New York: McGraw-Hill Book Co., 1987.

Fleckenstein, Donald C. "International Standardization—Insulator to Semiconductor." *IEEE Electrical Insulation Magazine* (March 1987), pp. 36-38.

Healy, Robert E. *Federal Regulatory Directory 1979-80*. Washington, DC: Congressional Quarterly, Inc., 1979.

Kitzantides, Frank K. "Toward an International Agreement on Testing and Certification of Electrical Equipment." *IEEE Electrical Insulation Magazine* (March 1987), pp. 24-26.

McAdams, William A. "Standards and the United States." *ASTM Standardization News* (December 1986).

National Bureau of Standards: A National Resource for Science and Technology, Special Publication 538. Gaithersburg, MD: National Bureau of Standards, July 1979.

National Electrical Code 1987. Quincy, MA: National Fire Protection Association, 1986.

1987 Progress Report. New York: American National Standards Institute, March 1987.

Standards Forum. Rexdale, Ontario: Canadian Standards Association, 1984.

The United States Government Manual 1981/82. Washington, DC: National Archives and Records Service.

Toth, Robert B., ed. *Federal Government Certification Programs for Products and Services*, NBS Special Publication 714. Washington, DC: U.S. Government Printing Office, 1986.

————, ed. *Standards Activities of Organizations in the United States*, NBS Special Publication 681. Washington, DC: U.S. Government Printing Office, 1984.

————, ed. *The Economics of Standardization*. Minneapolis: Standards Engineering Society, 1984.

Working Together. Northbrook, IL: Underwriters Laboratories Inc.

JUST-IN-TIME MANUFACTURING

American manufacturing in the 1980s and beyond is a picture of struggle. Apparent contradictions between ultimate goals (improved quality, reduced manufacturing throughput and delivery times, and reduced product costs) are a source of management frustration and endless internal conflict. What is not entirely clear about the current situation is that it is also a picture of tremendous opportunity. Not only can the individual goals listed above be reached simultaneously, but they can be addressed as parts of a single, unified management philosophy and program. This approach is called just-in-time (JIT).

Just-in-time is a philosophy that has the elimination of waste as its objective. Waste may appear in the form of rejected parts, excessive inventory levels, interoperation queues, excessive material handling, long set-up and changeover times, and a number of others. Just-in-time highlights the need to match production rate to actual demand and eliminate non-value-adding activities.

A brief review of the success stories created by American applications of this philosophy yields an impressive list of accomplishments. For example, Black and Decker has reported that, with the application of just-in-time concepts and a great deal of hard work, a pilot plant increased volume of throughput by 300%, increased inventory turns from 4 to 40, and reduced manufacturing lead times by more than 50%.[1]

Omark Industries has reported that its version of JIT, called ZIPS (Zero Inventory Production Systems), reduced 6 ½ hour set-up times to 1 minute 40 seconds, cut space requirements by 40%, reduced lead time from 21 days to 3 days, and lowered plant-wide inventories by 50%.[2] All of these accomplishments notwithstanding, Omark also reported 10% productivity improvements, 97% order fill rates, and 35% reductions in manufacturing costs.

General Electric studied and applied just-in-time manufacturing techniques in a factory producing dishwashers, with the following results: 9 miles of conventional conveyor replaced by 2.9 miles of computer-controlled nonsynchronous conveyors and robot handling systems; human material handling reduced from 28 occurrences to 1; internal failure cost reductions of 51%; inventory turns increased from 12 to 34; inventory investment reductions from $9.7 million to $3.7 million; throughput time reductions from 5 or 6 days to 18 hours; plant output increased by 20%. All of this was achieved with a 40% reduction in required floor space.[3]

Harley-Davidson has attributed its very existence to the application of JIT, which it refers to as the MAN (Material As Needed) program. With MAN in place, Harley-Davidson was able to reduce its breakeven point 32% during 1982. Between 1982 and 1984, raw material and work-in-process inventory turns increased from 6 to 17 and by 1985 were around 20. Set-up time reductions of 75% have been realized, and assembly throughput has gone from 3 days to half a day. Direct, indirect, and salaried employee productivity improved 37%. Supplier bases were consolidated by 23%, internal scrap costs were reduced 52%, and rework costs dropped by 80%. Defects per unit are reported to have dropped by 53%, and warranty costs per unit are down 46%.[4]

While the just-in-time philosophy of waste elimination proves helpful in virtually all types of manufacturing (and service) environments, some types of manufacturing offer more opportunity than others. Table 15-1 lists the estimated improvements for different types of manufacturing.[5]

Generally, process industries (paper mills, chemical manufacturers, and food producers) pose a greater challenge in terms of set-up time reduction and process flow improvements because of the "connected" nature of their operations. Repetitive and discrete manufacturers (metal stamping and machining) tend to encompass more opportunity in these areas because of their predisposition toward functional area layouts and extensive/frequent machine set-ups. Conversely, process industries generally have a great deal of opportunity in the areas connected with material planning and procurement. These industries often use relatively large quantities of raw materials and may have significant advantages in leverage to negotiate supplier improvements such as delivery, quality, and cost.

Even within an industry, opportunity levels will vary from company to company because of skill levels, layout styles, organization structures, labor union constraints, management abilities, geographic location, financial resources, and several other factors. The best way to evaluate the opportunities in a specific manufacturing environment is to perform an overall Diagnostic or Opportunity Assessment. This procedure or "phase" of JIT will be discussed later in this chapter. The important point to remember is that opportunity areas vary in size depending on the industry involved and may vary widely within industries depending on the characteristics of the specific company.

CHAPTER CONTENTS:

Contributors of this chapter are: ***Daniel R. Bradley****, Management Consultant, Coopers & Lybrand;* ***William L. Duncan****, Management Consultant, Coopers & Lybrand;* ***Dale H. Zempel****, Partner, Coopers & Lybrand.*

Reviewers of this chapter are: ***James E. Harl****, Senior Planner, Production Planning, John Deere Harvester Works;* ***Charles Kopoulos****, Management Consultant, Coopers & Lybrand;* ***Ken Rydzewski****, Manufacturing Planning and Control Manager, A.O. Smith Automotive Products Co.;* ***George Sutton****, Consulting Engineer;* ***Hayward Thomas****, Consultant, Retired—President, Jensen Industries;* ***Roger G. Willis****, Partner, Management Information Consulting Div., Arthur Andersen & Co.;* ***Dr. George M. Yaworsky****, Consultant.*

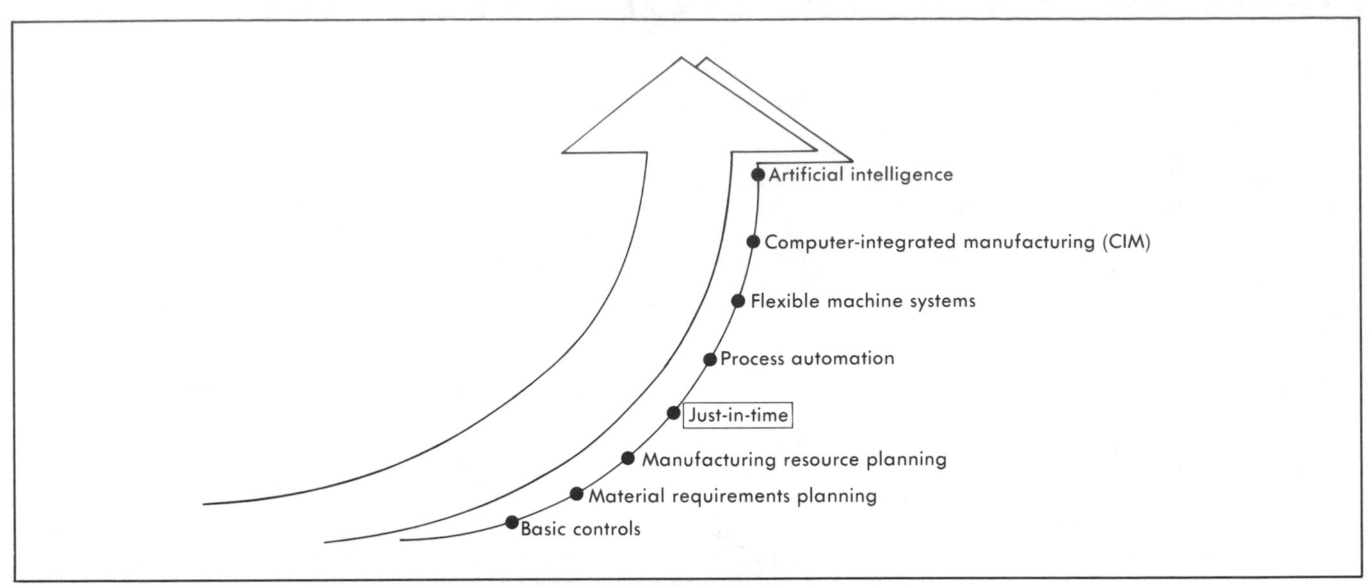

Fig. 15-1 Relationship of JIT with other elements in a manufacturing enterprise. (*Coopers & Lybrand*)

TABLE 15-1
Estimated Percent Improvements for Different Industries as a Result of JIT Implementation

Reductions	Automotive Supplier	Printer	Fashion Goods	Mechanical Equipment	Electric Components	Range
Manufacturing lead time	89	86	92	83	85	83-92
Inventory						
Raw	35	70	70	73	50	35-73
WIP	89	82	85	70	85	70-89
Finished goods	61	71	70	0	100	0-100
Changeover time	75	75	91	75	94	75-94
Labor						
Direct	19	50		5	0	0-50
Indirect	60	50	29	21	38	21-60
Exempt	?	?	22	?	?	?-22
Space	53	N/A	39	?	80(Est.)	39-80
Cost of quality	50	63	61	33	26	26-63
Purchased material (Net)	?	7	11	6	N/A	6-11
Additional capacity	N/A	36	42	N/A	0	0-42

(*Coopers & Lybrand*)

As illustrated in Fig. 15-1, JIT may be considered a prerequisite to automation, flexible machine systems (FMS), and computer-integrated manufacturing (CIM). Simplification before automation is essential to the efficient use of automation dollars and the intelligent planning of automation requirements. Otherwise, non-value-adding operations are automated and good money follows bad. The result is a highly automated but highly inefficient operation.

JIT THEMES AND MODULES

Based on the improvements made through the application of JIT, a sound JIT program should include the following modules:

- Planning and assessing.
- Organization.
- Awareness and education.
- Housekeeping.
- Quality improvement.
- Uniform plant loading.
- Process flow.

- Set-up and changeover reduction.
- Pull system implementation.
- Supplier integration.

Figure 15-2 shows how these 10 modules of a JIT program can be organized in a wheel-like fashion. The wheel illustrates the continuous nature of JIT improvement activities and demonstrates a convenient sequence of implementation. It is important, however, to keep in mind that the time frame and level of involvement of various disciplines and of the total program will be different in each organization.

JIT THEMES

Before discussing the JIT modules in detail, it is advantageous to identify a few underlying JIT themes. These themes are at the center of the JIT wheel illustration shown in Fig. 15-2. The five basic themes are as follows:

1. *JIT is a philosophy and a continuous program.* It is extremely important to understand that just-in-time is *not* a project, because it has no end. Once JIT improvement activities are under way, they should continue indefinitely. The environment should be transformed to one of continuous improvement and of cooperative management/labor endeavor.
2. *The importance of visibility.* JIT involves an aversion to hidden problems. It increases visibility (and encourages elimination) of these problems by gradually reducing work-in-process inventory, queues, and lead times.
3. *The benefits of synchronization, or balance.* This process involves the matching of throughput times from operation to operation during the course of manufacturing and support functions. Production then occurs at a common rate, or "drum beat."
4. *Simplicity—the view that simpler is better.* A continuous effort is made to perform required operations with fewer resources (time, personnel, and equipment) and in a less complicated fashion.
5. *A holistic approach to the program.* It must cross disciplinary lines and deal with the manufacturing process as a whole, rather than with separate parts.

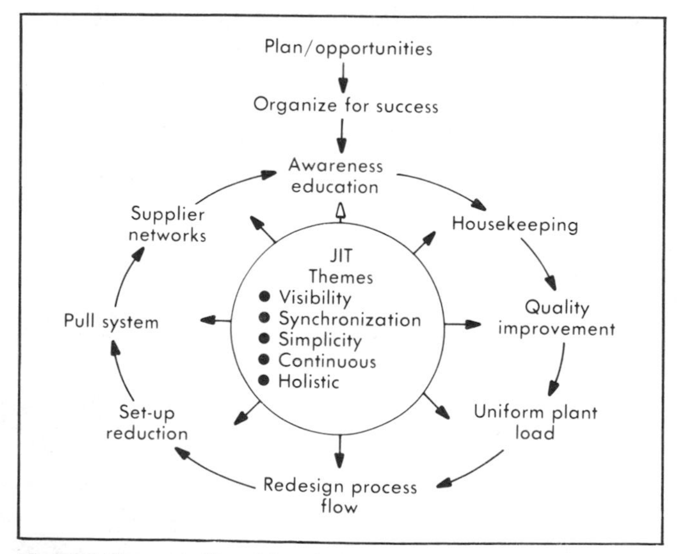

Fig. 15-2 Themes and modules of a JIT program. (*Coopers & Lybrand*)

PLANNING AND ASSESSING

A large percentage of the "failures" associated with some form of just-in-time implementation are attributable (in part or in whole) to inadequate planning and to poor assessment of current operations and opportunities. With regard to the subject of planning, it might do well to consider the following quote from William Penn in 1694:

> "Method goes far to prevent trouble in business; for it makes the task easy, hinders confusion, saves abundances of time, and instructs those who have business depending, what to do and what to hope."

Experience with the implementation and support of JIT programs indicates that knowing what to do, when to do it, and what results can be expected can only come about through thorough, intelligent planning. It is also important to obtain management commitment to the program before starting.

JIT programs generally involve the following progressive phases:

- Diagnostic review.
- Conceptual design.
- Implementation planning.
- Implementation.
- Continuous improvement.

These phases will be discussed in more detail under the section "JIT Program Phases."

Each phase of the program needs to be evaluated in light of the size of the organization, the experience level of the staff and program team, the number of product families involved, the availability of resources, and the expected time frame for implementation.

ORGANIZATION

An effective just-in-time program can be initiated only after an adequate plan encompassing all of the disciplines of the organization has been drawn up and after a group of people has been assembled and committed to the enterprise. From initiation of the program through the implementation phase, a strong leader (or "champion") and a steering committee will be required to direct the JIT team's efforts. Participants on the steering committee are usually senior-level management personnel and could include the general manager, engineering manager, manufacturing manager, MIS manager, and any assistant managers. If a company is unionized, it is also beneficial to have the involvement of upper-level labor leaders, such as a union president, on this committee. This will usually help to resolve the concerns of labor and encourage acceptance during implementation.

Another group of individuals indigenous to the just-in-time environment is the task group, or problem-solving work group. This group is often formed around a nucleus of people devoted to a specific functional area or process. Included with the area-oriented or process-oriented nucleus are representatives from other disciplines that may affect the process being studied by the group, along with a "ringer." A "ringer" is someone who is completely uninvolved or isolated from the process under normal circumstances and can be counted on to ask questions that would be considered or "dumb" by the other group members.

The interactive processes of these groups can be structured in a number of ways. Table 15-2 shows the three groups and their respective responsibilities.

CHAPTER 15

JIT THEMES AND MODULES

AWARENESS AND EDUCATION

The need for awareness, education, and training cannot be overstated. The education and training of all personnel (including direct labor) is essential to the correct application of JIT.

Awareness is comprised of the introduction of JIT concepts, how they have helped other companies, and speculation about how they could prove useful in the current manufacturing environment. Education involves a more detailed explanation of each module of the JIT wheel (refer to Fig. 15-2), with case studies and sample exercises drawn from the actual manufacturing environment being addressed. For example, fictitious data could be utilized to complete sample SPC charts for operations that are recognizable by name and specification type as belonging in the current manufacturing environment. Training will differ with the trainees' needs. Purchasing personnel, for instance, will need to be trained in JIT "co-op" contracting (discussed subsequently) and negotiations, while manufacturing operations personnel will need a heavier emphasis in areas such as set-up reduction. There are some training programs from which everyone can benefit.

Table 15-3 lists the type of education/training required by various workforce levels. Among the general training sessions that frequently prove valuable are the following:

- Value analysis/value engineering.
- Time management.
- Group dynamics.
- Communication skills.
- Problem solving.
- Fundamentals of inventory management.

Among the tailored training sessions that can be utilized are the following:

- Set-up reduction.
- Statistical process control.
- Specific machine maintenance.
- Cost accounting in a JIT environment.
- Make vs. buy analysis.
- "Designing in" manufacturability and standardization.
- Forecasting/order entry in a JIT environment.
- Production planning/scheduling in a JIT environment.
- Purchasing:
  "Co-op" contracting.
  Vendor analysis evaluation.
  Procurement strategy.
  Breakeven analysis.
  Methods of payment.
  Negotiation skills.
  Contractual law.
  Specific commodity training.
- Cost of quality.
- Distribution in a JIT environment.

HOUSEKEEPING

Housekeeping in a just-in-time environment is far more than what has been referred to as a "clean broom award" program. It

TABLE 15-2
Participants in JIT Program and Their Respective Responsibilities

Program Leader	Steering Committee	Task Group
Organize steering committee	Identify opportunities	Identify symptoms
Monitor overall program progress to goals	Prioritize opportunities	Gather relevant data
Report on program progress to upper level management	Define tasks to address opportunities	Prioritize potential solutions
Communicate requests and approvals for large expenditures to/from upper level management	Organize task groups and assign tasks	Perform testing of potential solutions
	Review/approve task group recommendations	Recommend remedies to Steering Committee and obtain approval
	Dissolve task groups	Implement and test rememdies in conjunction with line operations personnel
	Report progress to Program Leader	

(*Coopers & Lybrand*)

TABLE 15-3
Awareness, Education, and Training Stratification by Management Level

	Awareness Sessions	Education Sessions	General Training Sessions	Tailored Training Sessions
Senior management	X			
Middle management	X	X		
Lower level management	X	X	X	
Staff	X	X	X	X
Direct labor	X	X	X	X

(*Coopers & Lybrand*)

is an effort to establish the attitude that each individual "owns" and is responsible for his or her equipment and environs. The worker is accountable for ensuring that necessary tools are in the right places and that those places are effective locations. Tools and materials must be visible and accessible. Lighting, heating and ventilation, and workcenter equipment (racks and trays) must be properly designed and located to minimize reaching, bending, and fatigue. Trash and debris should not be present. Housekeeping attitude can be communicated through the following principles:

- A place for everything.
- Everything in its place.
- Everything visible.
- Everyone involved in cleaning, checking for damage, and anticipating problems.

Exercises dating back to the Hawthorne studies[6] in 1932 indicate that substantial productivity increases can be obtained by focusing on these kinds of workplace-related issues.

QUALITY IMPROVEMENT

To say that quality is an absolutely critical aspect of just-in-time, or of manufacturing in general, is to state the obvious. Successful competition in domestic and international marketplaces requires a total involvement in quality control and continuous quality improvement. In a broad sense, this involvement takes the form of a company-wide commitment to the following three basic concepts:

1. *Fitness for use.* The product must perform, over a period of time, the functions for which it was designed, over a period of time.
2. *Process capability.* Assurance that the processes involved are capable of consistently producing parts that meet or exceed all critical specifications. How to conduct a process capability study is discussed in Chapter 23, "Achieving Quality."
3. *Defect prevention orientation.* A bias toward preventing defects rather than inspecting for them and reworking/scrapping them.

The commitment should be formalized into a corporate mission statement and communicated throughout the organization. How to write a mission statement and other quality management issues are discussed in Chapter 22, "Quality Management and Planning," of this volume. When this statement is communicated and understood, the remaining activities comprising a quality system include the following:

- Setting and maintaining quality standards.
- Monitoring and appraising performance to quality standards.
- Responding to quality performance appraisal results.
- Continuous improvement activities.

TABLE 15-4
Quality Control Management Responsibilities

Group	Responsibilities
Top management	— Allocate adequate resources to fulfill long-range quality needs of the company and the customers — Convey an attitude that quality problems are to be prevented rather than inspected in — Convey an attitude that "low quality is unacceptable" throughout the organization — Overtly encourage the development and maintenance of high quality standards throughout the organization — Hold regular meetings to review quality levels and quality performance
Quality control manager	— Initiate and maintain a defect prevention program — Identify and organize the quality objectives of the various departments within the organization. Ensure that they are consistent and support overall corporate quality goals — Provide adequate inspection, testing, and other quality monitoring equipment and training required by the organization to meet quality objectives — Monitor the performance of the various departments within the organization with regard to their quality objectives and report these findings to top management
Quality control engineers	— Reliability analysis — Develop operator controlled inspection and testing procedures where applicable — Design/specification of test and inspection equipment — Supplier quality evaluation — Reporting of quality performance data — Verification of corrective action — Statistical analysis and defect cause identification — Inspector and operator training
Quality circle members	— Monitor all critical quality specifications — Track frequency and severity of defects via SPC charting, results, and prioritizing occurrences by frequency — Investigate and eliminate defect causes
Other department managers	— Understand, define, and communicate their individual department responsibilities for quality to their subordinates — Regularly monitor the quality performance levels of their staff and take corrective action when defects are identified

(*Coopers & Lybrand*)

Without these controls, competitive capability is lost. Quality costs average from 4 to 30% of sales. A great deal of "fire-fighting" and "inspecting in" of quality occurs. By properly implementing a sound quality control system, however, quality costs can be reduced to an average level of less than 3%, and product costs should then be much more competitive.

The just-in-time philosophy emphasizes the fact that workers are responsible for the capability and quality of their own processes and for inspecting the work of the previous operation. Typical responsibilities for various workforce levels are listed in Table 15-4. In this manner, all defects become visible quickly, and less scrap/rework is produced. Defects uncovered during JIT production operations may result in the stopping of the production line while the source of the problem is discovered and addressed.

To demonstrate how this principle, in conjunction with small lot sizes, can reduce total defect levels, the following example will be used. Layout A in Fig. 15-3 involves three workcenters, a work-in-process (WIP) storage facility, and an inspection station. Assuming that the production of defective material began on workcenter 1, about 500 pieces of defective material (which would involve rework or scrap) would be produced before the defect was discovered. In layout B, the same three workcenters are used, along with reduced lot sizes, operator inspection, and the elimination of WIP. In this scenario, defects produced at station 1 would be identified no later than station 2; this would reduce the amount of scrap or rework required to no more than 10 pieces.

Another aspect of quality control in a JIT production environment is the statistical tracking and tightening of process controls using statistical process control (SPC). A major objective of SPC is to recognize the occurrence of special causes of variance in the presence of the constant system of common-cause variances and to shed light on the nature of special causes, thereby providing a basis for corrective actions. Control charts are one of the tools used to accomplish this objective.

The control chart builds a model that describes the way the process variability pattern is expected to appear when only common causes of variation are at work. Once this model is established through appropriate sampling and statistical data characterization methods, a basis is formed for identifying the occurrence of a variation that does not fit the pattern of common-cause variation.

A variety of control charts are available for monitoring and improving a production operation. The charts are generally classified according to the type of data they are based on. The charts most frequently used are presented in Table 15-5.[7] For additional information on SPC and other statistical methods for quality improvement, refer to Volume IV, *Quality Control and Assembly*, of this Handbook series. Process capability is discussed in Chapter 23, "Achieving Quality," of this volume.

Another important aspect of quality in a JIT production environment is the quality level of purchased goods. This subject will be discussed further under the heading "Purchasing Review" in the section "Diagnostic Review" of this chapter.

The last aspect of quality in a JIT production environment to discuss is fail-safing. Fail-safing is a technique that can be applied in both design and process. Simply stated, a fail-safe design does not allow parts to fit together in any way except the correct way. Wrong parts or the wrong orientation of correct parts do not fit together. In terms of the process, special checking devices and locking devices are utilized (built into the set-up) to ensure that the process produces correctly to specifications. These techniques have been extremely successful in both

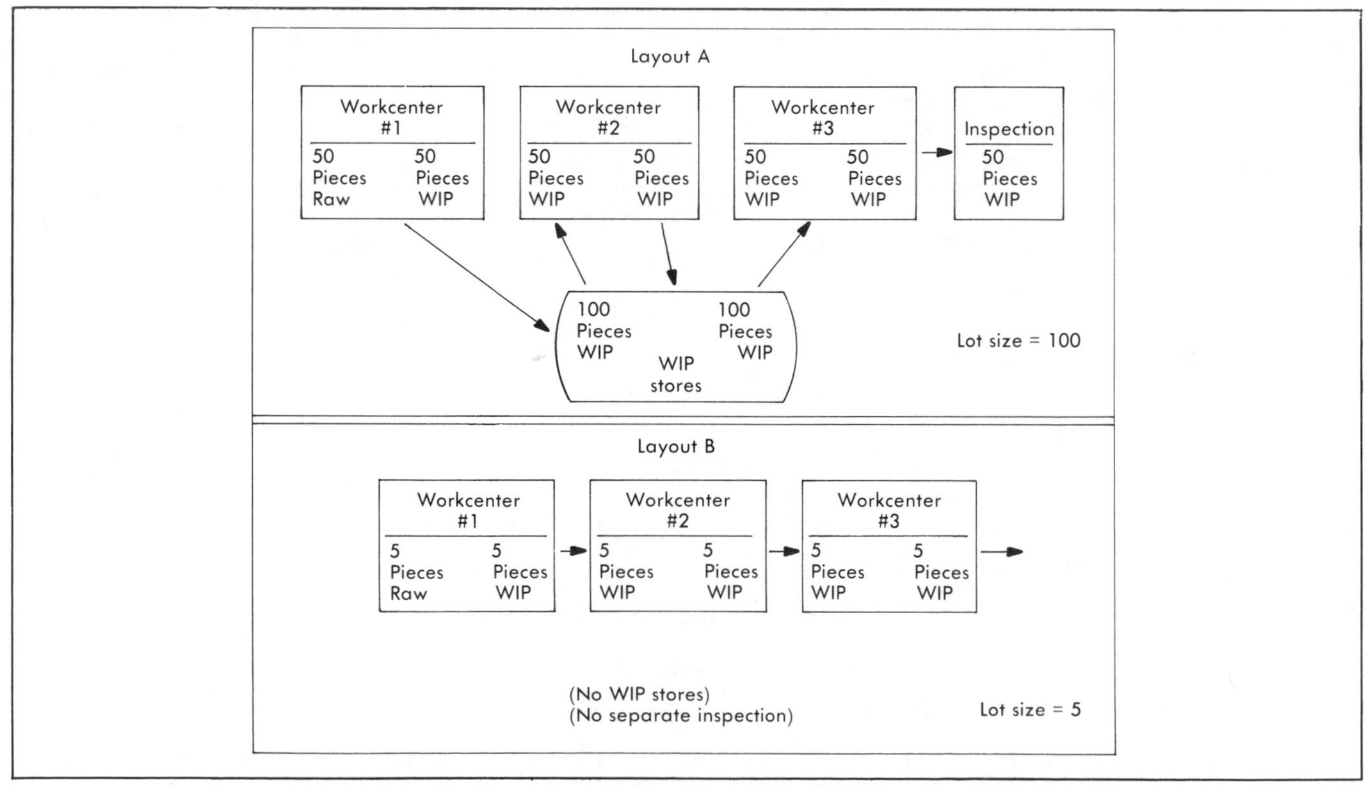

Fig. 15-3 Comparison of work performed in a traditional manufacturing environment to work performed in a JIT environment. (*Coopers & Lybrand*)

TABLE 15-5
Commonly Used Control Charts

Data Type	Chart Name	Value Charted
Variables	X and R chart	Sample averages and ranges
	X and s chart	Sample averages and standard deviations
	X and moving R chart	Individual observations and moving ranges
	Median and R chart	Sample medians and ranges
Attribute	p chart	Proportion or percent of units nonconforming (defective) per sample
	np chart	Number of units nonconforming (defective) per sample
	c chart	Number of nonconformities (defects) per inspection unit
	u chart	Average number of nonconformities (defects) per production unit

Japanese and U.S. just-in-time installations. The subject of designing quality into a product and process is discussed further in Chapter 13, "Design for Manufacture," of this volume.

UNIFORM PLANT LOAD

The concept of uniform plant load (UPL) at its most basic level is simply this: if one sells daily, build daily. Each model that is sold is manufactured on a daily basis in relatively small lots, so build rates match demand rates. An underlying precept of just-in-time manufacturing is that the organization is moving from a "make-to-stock" to a "make-to-demand" environment. Uniform plant load is the cycle time required to match (not exceed) demand. The result of applying UPL is a production rate that is *not* tied to machine rate, current capacity, or some annualized average demand based on forecast or history. It is a picture of demand per day divided by hours of production per day.

The application of uniform plant loading, coupled with a pull system (to be discussed later) should result in a synchronized production volume from workcenter to workcenter from end to beginning of the production sequence. Production occurs as though according to a common "drum beat" that is sounded at the beginning of each cycle. The benefits of this mode of operation are reduction of indirect labor costs (managing and transporting excess inventory between operations) and improved utilization of direct labor in the sense that the number of operators can be varied to match desired production rates. Uniform plant loading needs to vary as demand varies and will then dictate personnel and equipment requirements. Excess personnel in one area are sometimes shifted to other areas or assigned to preventive maintenance in some cases. This often requires cross-training of fewer job classifications and assurances from management that layoffs will not result from short-term order fluctuations.

PROCESS FLOW

The major benefits of UPL can usually be realized best when the process flow is redesigned. This is true because most manufacturing layouts are designed in a functional manner (see Fig. 15-4). The functional approach does not lend itself to rapid throughput times because of the travel distances, material handling, WIP, and storage/retrieval activities involved.

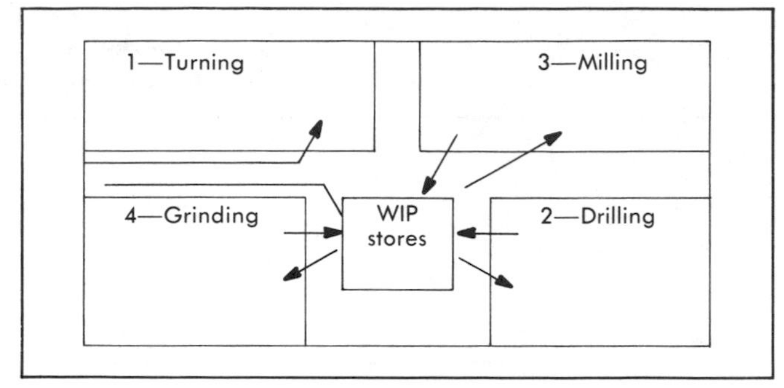

Fig. 15-4 Traditional functional layout of a manufacturing company. *(Coopers & Lybrand)*

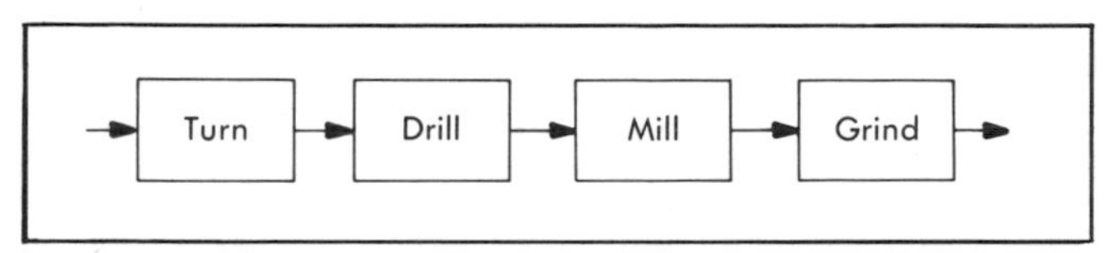

Fig. 15-5 Schematic of typical JIT workcenter layouts. (*Coopers & Lybrand*)

In some respects, JIT layout for repetitive manufacturers is an attempt to turn the production operations into a process. Just-in-time layouts are typified by workcenters that are in close proximity and have little or no inventory between the separate operations (see Fig. 15-5). These smaller process-oriented layouts within the larger production facilities are referred to as cells. One or more cells that comprise all of the product of a given product family may be isolated and run effectively as an autonomous entity from the other factory operations. When this is done, the entity is referred to as a "focus factory" or a "factory within a factory."

The organization of individual cells can be made much more efficient, and provide much greater flexibility in terms of personnel, by designing layouts in U-shape or serpentine configurations (see Fig. 15-6). These designs, and others like them, allow operators to move between individual production operations, performing more or fewer functions as required to meet demand. An important principle to remember here is that the operators must be kept busy, not necessarily the equipment. "Before and after" pictures of plant layouts demonstrate the advantages of redesigning process flow. Travel distances, space requirements, throughput times, and labor expenditures can be reduced dramatically. Paid labor per unit can be stabilized through periods of fluctuation in demand (assuming other uses for direct labor are available) as illustrated in Table 15-6 on p. 15-8.

This kind of approach to process flow design can be extremely beneficial as a precursor to automation because flexible machine systems and robotics are rapidly becoming available to run at different speeds depending on demand and to perform multiple operations within a close physical proximity. Cellular processing requires worker flexibility (cross-training to perform multiple operations) and should include as much of the entire process as possible. It exposes imbalances between individual operation cycle times and work content, thereby encouraging synchronization. Moreover, layouts and personnel levels in cellular manufacturing require periodic tuning to compensate for changes in product design and improvements in technology. Cellular processing is most effectively applied to fami-

TABLE 15-6
Flexible Manning Techniques
(How to Accommodate Changes in Demand)

Sales Per Month	Sales Per Day	Sales Per Hour	Cycle Time, s	Standard Converted to Pieces Per Hour	Crew Size
12,000	600	75	48	15	5*
9,600	480	60	60	15	4
14,400	720	90	40	15	6
16,800	840	105	34.3	15	7

(*Coopers & Lybrand*)

* The crew size is predicated on sales per hour divided by the standard (converted to pieces per hour) or 75 ÷ 15 = 5.

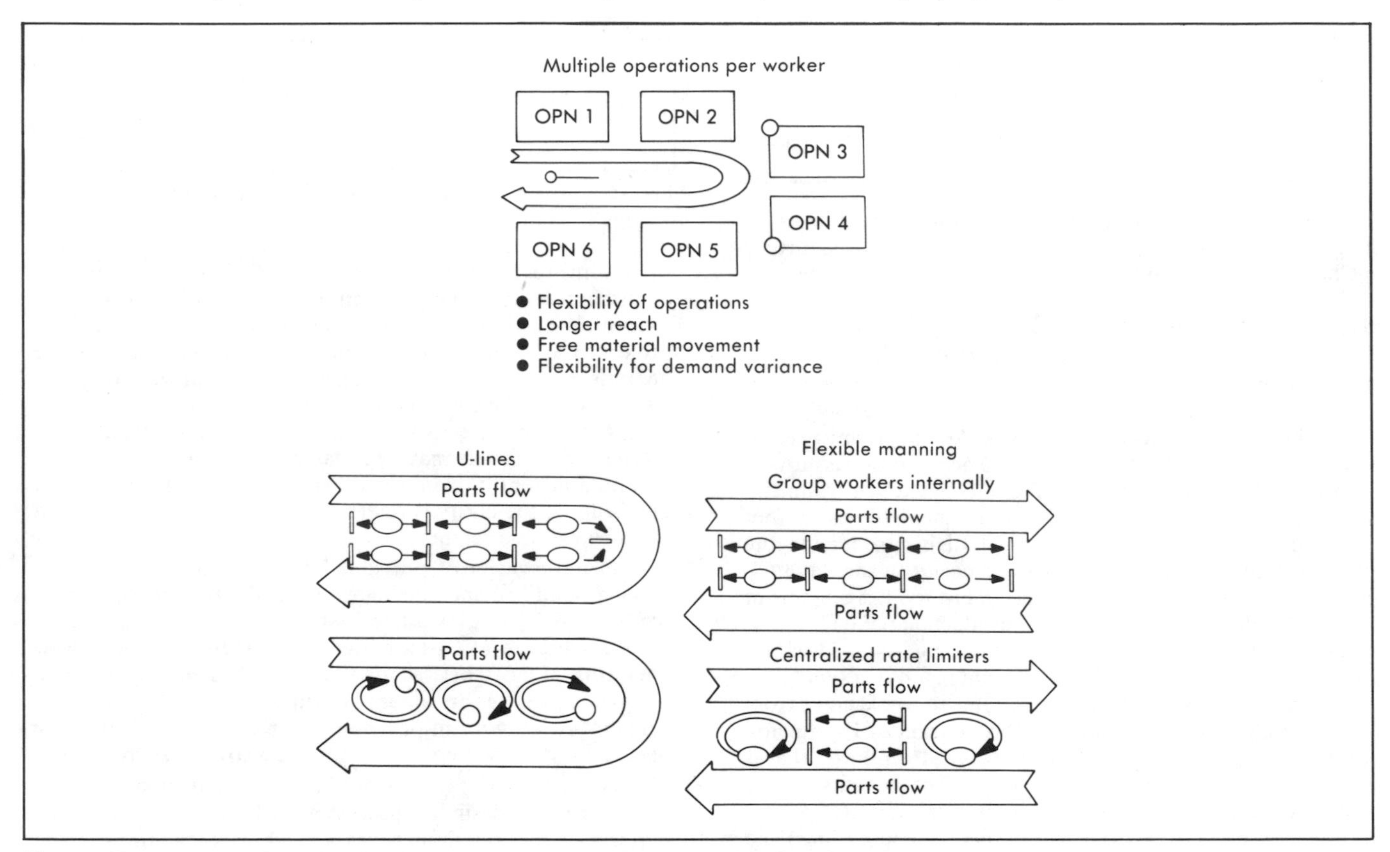

Fig. 15-6 Workcenter configurations in a just-in-time environment. (*Coopers & Lybrand*)

lies of parts because the production of entire families increases flexibility of equipment usage. Leveling plant load generally becomes easier because rate and personnel variability are controlled and planned.

SET-UP AND CHANGEOVER REDUCTION

Before discussing the techniques involved in set-up reduction, it is important to mention that entire set-ups and groups of set-ups may be completely eliminated by dedicating equipment to product groups. There are drawbacks to this approach, and the most serious of these is that the dedication of equipment may lead to a requirement for more equipment; however, as the dedication process occurs, machine capabilities should be reviewed. It is frequently possible to replace extremely expensive and very flexible equipment with less expensive, less flexible, but entirely adequate, equipment. This type of equipment must be capable of producing quality parts at the designated demand rate.

In those cases where multiple components must share the same manufacturing resources, JIT's orientation to small lot sizes and frequent runs will make it very important to the reduction of set-up times. Reduced set-up times will also yield benefits in uniform plant loading, workcenter organization, and pull system installation. Set-up time means the time from when the last good piece is produced on the old set-up to the time when the first good piece is produced on the new set-up. It includes teardown, installation of the new job, and any required first-piece inspection. Reduction refers to reduced time, but not necessarily to reduced cost. Cost is usually, but not always, reduced.

Set-up reduction in the JIT environment is not an engineering project. It should be performed by a set-up reduction team comprised of the following individuals:

- The machine operators.
- The technical people as required from process engineering, tool design, tool maintenance, and general maintenance.
- A "ringer."

The team should have complete implementation responsibilities, including a small budget for equipment purchases and revisions. One of the set-up reduction group members will need to function as a group leader, keeping the activities on track and communicating constantly with the area supervisor. The area supervisor will need to be directly involved or kept closely informed of group activities to facilitate "buy-in" and ensure successful use of the improvements over time.

The actual process of set-up reduction occurs in the following manner:

1. The existing set-up process is documented and studied. This should involve a videotape of the process with a digital clock running in the corner of the picture and a classical work study performed that breaks the process down into work elements. The work elements should be discrete to the extent that they may be categorized and analyzed as to value and time expended. A Pareto diagram can be used to determine which work element should be worked on first.
2. Identify all elements of the set-up that can be made external (done outside the confines of the machine down period). At this point, all of the set-up elements are converted into steps that can be performed while the machine is producing parts. More time is spent presetting tools so that actual machine inactivity is minimized.
3. Identify and reduce or eliminate adjustments. Adjustments may include centering, setting spacing levels, and squaring. The point of this exercise is to maximize the extent to which the machine and any fixtures are self-positioning. When machine settings are analyzed, there are usually a few positions that represent the entire spectrum of current production. In these cases, mechanical stops can frequently be used to increment machine settings in less time and with greater accuracy. Slots, notches, and pins are examples of different types of stops that may be employed. Utilizing these devices in conjunction with clamping mechanisms allows the set-up process in many cases to be reduced to three steps: (1) loosen clamps, (2) move to new stop, and (3) tighten clamps.
4. Replace nuts and bolts with clamping devices wherever possible. Minimize the force required to do the clamping and simplify die or fixture positioning to minimize the number of clamps required. Attempt to minimize the use of tools and motion.
5. Identify and resolve problems in the set-up. Problems are defined as anything that stands in the way of trouble-free and undelayed set-up. The key is to keep asking "why?" until the root problem is unearthed and then eliminate (rather than compensate for) the problem.

The proper application of these techniques has often resulted in set-up reductions of more than 75%.[8]

THE PULL SYSTEM

The pull system is the next logical step in a JIT program when uniform plant loading and process flow revisions have been installed. The pull system has dramatic effects on inventory levels because it does not provide for production of any inventory until it is needed. Pull systems do not allow parts to be produced until "authorization (pull signal)" is received from the subsequent operation. To visualize this situation, and the contrast between traditional "push" and just-in-time "pull" systems, refer to Fig. 15-7.

If product A was produced in a traditional push production environment, the following sequence of events could occur:

1. A problem arises with part B1.
2. Unaware that B1 production is experiencing difficulty, production continues on C1, C2, B2, C3, and C4 in other areas of the factory.
3. Unable to produce part A, the assembly department switches to another product. Expediting of other components occurs, and surplus inventory of C1, C2, B2, C3, and C4 must be moved to a storage area.
4. When B1's are able to be produced again, priorities will need to be realigned, and work will need to be rescheduled on all workcenters involved to compensate for WIP level imbalance and timing discrepancies.

If product A was produced in a just-in-time pull system environment, the following sequence of events could occur:

1. A problem arises with part B1.
2. Because B1 is unavailable, production of A stops.
3. Because production of A stops, demand for (and therefore production of) B2, C1, C2, C3, and C4 stops. Operators converge from these workcenters on B1 to assist in resolving the problem.
4. When B1's are able to be produced again, priorities are correctly aligned and no surplus WIP inventory has been produced.

Pull systems generally take one of two possible forms: overlapped or linked (see Fig. 15-8). Overlapped pull systems utilize empty space as the pull signal or communication device between production operations. This technique is best applied when operations are in close physical proximity. One example is a simple square marked off with tape or painted lines that, when empty, indicates that the following operation is ready for additional material. No material is produced until the square is emptied for replenishment by the subsequent operation.

Linked pull systems are typically utilized when parts compete for the same resource and cannot be made on a one-for-one basis with end-item demand, or when they have to travel signifi-

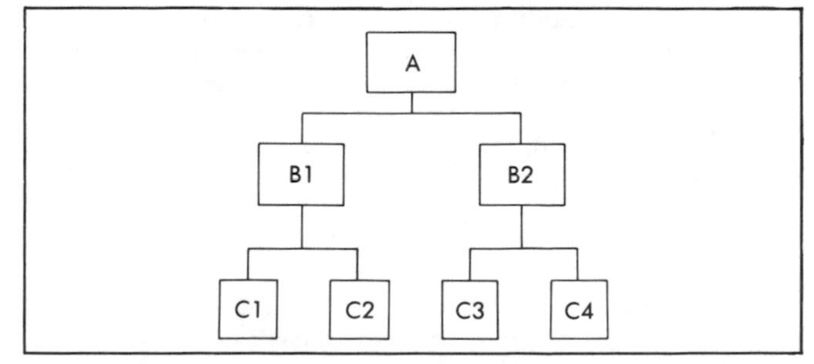

Fig. 15-7 Diagram for comparison between "push" and "pull" production system. (*Coopers & Lybrand*)

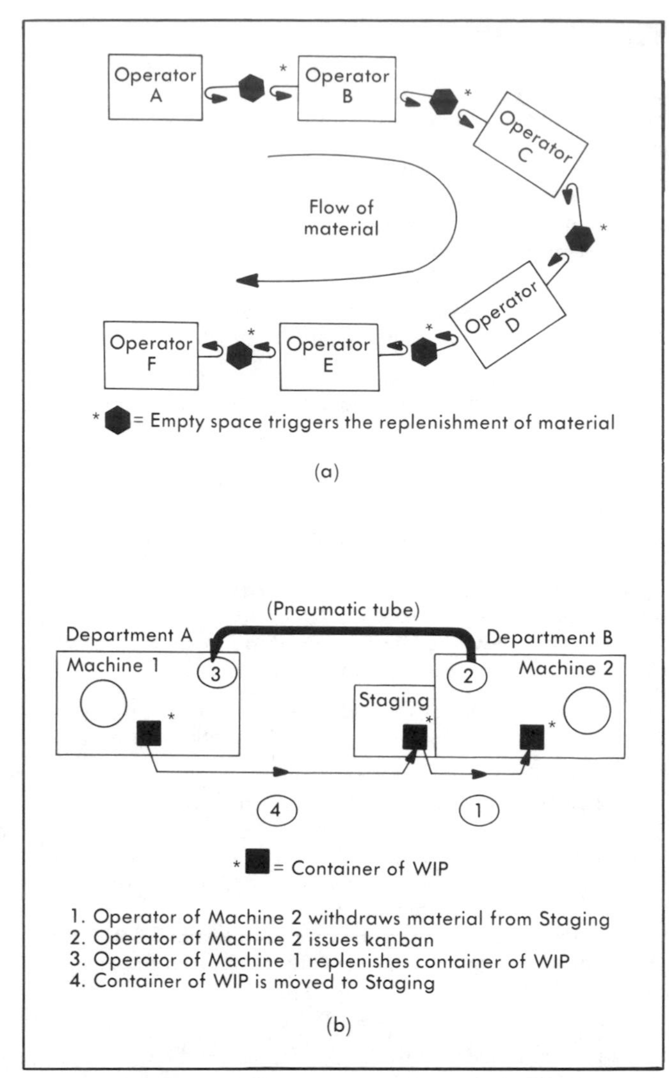

Fig. 15-8 Production system in a JIT environment is usually by the pull system, which can be of the (a) overlapped or (b) linked form. (*Coopers & Lybrand*)

cant distances between operations in a lot (or batch) mode. In these situations, it is typical to utilize a pull signal (or kanban) to trigger the production of components from operation to previous operation. The card or kanban is issued by an operator as a container of material is picked up. The material must be used in a "first-in, first-out" (FIFO) sequence. The kanban is the only authorization for additional material to be produced at the previous operation. When no kanban is issued, no additional components are made. Kanbans may be cards, colored golf balls, or even empty containers. They may be moved by hand, slotted slide, or pneumatic tube. An important principle with regard to the use of a kanban is that it is employed in a FIFO manner.

Regardless of whether the linked or overlapped system is used, the pull system becomes the vehicle for communicating short-range parts requirements and scheduling close-in production. Manufacturing resource planning (MRP II) is still required to perform the longer range planning of material requirements, capacity requirements, and personnel requirements. In short, MRP II is the planning side of production, and just-in-time's pull system is the execution side.

SUPPLIER NETWORK INTEGRATION

The involvement of suppliers in a just-in-time program is the last item on the JIT wheel (refer to Fig. 15-2). Many times, companies who embark on the JIT journey try to make supplier involvement one of the first steps in their program. This approach can be problematic and is not recommended. Just-in-time can most effectively be sold to the supplier community when its benefits can be demonstrated in the manufacturer's own setting. Also, it is easier to provide a clear picture of what is needed in terms of demand rate, delivery performance, quality levels, and point-of-use consignment to the supplier when the manufacturer's requirements are clearly visible. By the time the suppliers are contacted, demand fluctuations should have been smoothed to some extent and points of use should be known.

There is probably no more cogent description of the procurement/manufacturing/distribution continuum than this:

> Henry Ford said, in the book *Today and Tomorrow* (1926), "The time element in manufacturing stretches from the moment the raw material is separated from the earth to the moment when the finished part is delivered to the ultimate consumer."

The essence of just-in-time philosophy applied to the user/supplier relationship is a continuum or process within which procurement, manufacturing, and distribution are merely operations. Like any other just-in-time production process, the objectives include tying the operations together as closely as possible, balancing demand between them, eliminating any queues between them, and shortening the total throughput time. All of these things translate into a broader objective of being flexible and responsive to ultimate customer demand.

Achieving a close relationship between the manufacturer and supplier is not an easy task. When it comes time to approach this aspect of just-in-time, it is important for the manufacturer to have a clear understanding of the following items:

- What is wanted. Typically, this can be broken down into quality levels, delivery performance, responsiveness, packaging, and price.
- What devices will be used to measure whether what is wanted is actually received. Performance measures need to be defined for quality, delivery, response, packaging, and price.
- What will be provided to assist the suppliers in becoming just-in-time manufacturers so that they in turn can support the manufacturer's goals. For example, manufacturers with vendor certification and zero inspection goals should be willing to provide suppliers with SPC training.

With these factors clearly defined, purchased commodities and volumes can be identified. A commodity is defined as a group or family of parts that requires the same resources for their production and/or the same raw material processing. The things that are wanted during this identification phase are those commodities currently purchased, their dollar and quantity volumes, the supplier base currently providing each commodity, lead times, and quality levels, by commodity and by supplier. Using this information, it is possible in many cases to allocate the purchases of entire commodities to single suppliers, "leveraging" volume to achieve an economy of scale. This volume supports better supplier relations because it represents a larger percentage of supplier capacity.

It should be noted that this approach does not preclude more than one supplier per commodity. To act as a hedge against strikes or other unforeseen circumstances, alternative suppliers are usually given smaller, regular volume to preserve the previously stated benefits.

Buying supplier capacity in lieu of contracting for actual part numbers and volumes is a just-in-time objective. It is, therefore, important to reach the point in JIT supplier relationships where that portion of the supplier's capacity that is attributable to one's products may be regarded as an extension of one's own manufacturing facility. This objective can be achieved through co-op contracting. Co-op contracting rests on the following nine principles:

1. The supplier knows the product best.
2. Forecasts are always wrong.
3. A theoretical goal is zero lead time.
4. Single-source resource management.
5. Elimination of non-value-adding activities.
6. A theoretical goal of lot size is 1.
7. Rejects are intolerable.
8. Mutual (supplier and manufacturer) habit of continuous improvement.
9. Open-door policy between supplier and manufacturer.

In the co-op contracting environment, the user in effect controls the manufacturing schedule of the supplier. The supplier is assured of a long-term number of jobs and, given this commitment, will invest in operation improvements. Whenever possible, the user shares long-term schedules with its supplier.

Just as JIT implementations are best handled through a pilot line, co-op contracting is best implemented on a pilot commodity. The commodity selected should not be the toughest commodity. It should represent a moderate amount of purchased dollar volumes and provide a high degree of visibility within the manufacturing organization when it is successful. Components of the co-op contract include the following:

- Quality goals.
- Lead-time reduction goals.
- Cost reduction goals.
- Annual contract renewal (automatic, if goals are reached).
- Indemnification details.
- Joint manufacturer/supplier improvement group participation.
- Open-door policy.
- Variable volumes of individual products, but a fixed percentage of total supplier capacity utilized.
- Commitment to design support on the part of both manufacturer and supplier.
- Alternate sourcing contingencies.

Aside from the co-op contract, there are other aspects of just-in-time that may be applied in the purchasing environment. These include the following:

- *Supplier certification.* A systematic reduction in the sampling of incoming material from suppliers as their quality levels improve, until no inspection is required at all. At this point, the supplier is "certified."
- *Supplier performance evaluation.* The establishment of specific goals and milestones against which suppliers will be measured as to their performance in the areas of quality, delivery, lead time, and price.
- *Revised ordering mechanisms.* An evaluation of how close-in requirements can most effectively be communicated to suppliers (a kind of linked pull system).
- *Buyer training.* A program involving the identification of buyer training needs and compensating for those needs through the development of individual training modules.
- *Restructured buyer performance evaluation.* A review and revision of buyer performance evaluation based on JIT concepts such as vendor delivery performance, lead-time reductions, inventory reduction, and material quality levels.
- *Supplier training programs.* A group of training programs developed to assist suppliers in the adoption of JIT in their own organizations. If this is not feasible, consideration should be given to bringing suppliers into JIT training sessions for the manufacturer.
- *Review and revision of the purchasing organization.* An analysis of how purchasing functions and roles will change and how the purchasing organization can best be structured to take advantage of those changes.

JIT PROGRAM PHASES

As mentioned previously, a JIT program involves a series of progressive phases: diagnostic review, conceptual design, implementation planning, implementation, and continuous improvement. The subsequent sections discuss these phases in general and also provide manufacturers with specific guidelines to follow when performing the particular phase.

DIAGNOSTIC REVIEW

The diagnostic review is that portion of the just-in-time program that allows a manufacturer to identify, assess, and prioritize improvement opportunities. The opportunities can be found in current organizations, current operations, and current layouts. The three fundamental purposes for the diagnostic phase are as follows:

1. Establish a baseline of data (a "benchmark") against which future progress can be measured.
2. Identify gross opportunities for analysis and prioritization.
3. Establish goals and objectives for use during the conceptual design phase.

The diagnostic review should include the development of a project team and an analysis of all of the functional areas.

Project Team Selection

The objective of the project team is to assume full responsibility for the success of the project. A steering committee oversees the progress of the project and intercedes (as appropriate) to resolve any conflicts that may hinder project success.

DIAGNOSTIC REVIEW

The first task in project team selection is to assign a project manager. Table 15-7 lists the recommended qualifications and responsibilities of the project manager during the diagnostic review. Next, a group of individuals is assigned to be part of the team. Responsibilities of the team are listed in Table 15-2.

Once a project manager and team members have been assigned, a time-phased project plan for the JIT program can be developed with due dates and responsibilities. Specific task details should be completed through at least the diagnostic phase when the program is launched. Resource requirements must be estimated, and resources must be obtained.

Organization Assessment

The objective of the organization assessment is to determine the readiness of company personnel to convert to a just-in-time environment and to develop recommendations that will support desired levels of success. The following steps are usually required for this assessment:

1. Compile a list of candidates to be interviewed. This list should include top management as well as middle and lower management personnel and some direct labor employees. Several interviews are usually required to get an accurate picture of most organizations.
2. Schedule and perform the interviews. Questions should include the areas of:
   - Assignment clarity/definition.
   - Performance standards/measurements.
   - Autonomy levels.
   - Perception of the company.
   - Policies/procedures, accuracy, adequacy, and use.
   - Perceived need to improve.
   - Current problem-solving techniques.
   - Consistency of objectives.
   - Reward/recognition levels.
   - Levels of trust.
   - Levels of participation in decision making.
   - Pride in work.
   - Perceptions of management.
   - Levels/frequency of conflict.
   - Ability to exchange ideas.
   - Receptivity to change.
3. Summarize the findings from the interviews.
4. Develop a list of recommendations based on summarized findings.
5. Report recommendations to project manager and senior management.

TABLE 15-7
Recommended Qualifications and Responsibilities of the Project Manager During Diagnostic Review

Qualifications
• Practical knowledge of JIT principles
• Ability to commit to a full-time effort
• Practical knowledge of all functional areas of the facility
• Ability to make decisions
• Receptiveness to change
• Materials management background (preferably)
• Respected by his or her peers
• Dedicated to the progress of the project
• Good communication skills
Responsibilities
• Coordinate and control the activities of the project team, outside personnel, and all other personnel involved in various aspects of the project
• Verify that the manufacturing environment is prepared and ready for the JIT diagnostic review
• Plan and schedule the diagnostic review in an orderly, logical, and timely fashion
• Schedule the education and training efforts
• Verify credibility of all data collection
• Assure schedules are being met on the overall plan
• Submit periodic progress and status reports to the steering committee

(*Coopers & Lybrand*)

Design Engineering Review

The objective of the design engineering review is to understand the current role of the design engineer, especially as it relates to throughput times, bottlenecks, market response, interaction with manufacturing and purchasing, use of standardization, and levels of manufacturability. The following four steps are usually required for this review:

1. Chart the flow of the current design process from conception to adoption and implementation in the factory.
2. Summarize the findings in step 1, identifying "long-lead" activities, major costs, external department involvements, and levels of standardization and manufacturability. Identify all activities that do not add value to the process.
3. Develop recommendations based on the summarized findings, oriented toward reducing product design throughput times/costs, queues, and non-value-added activities.
4. Establish a list of preferred manufacturing and purchasing practices for later use in conceptual design and implementation phases.

Forecasting Review

The objective of the forecasting review is to identify current forecasting throughput times, delays, costs, close-in change levels, and reasons for close-in changes. Another objective is to determine the adequacy and/or reliability of the forecast as related to actual demand and to identify resources/methods currently used. As in the design review, the forecasting review requires the following four steps:

1. Chart the flow of current forecasting processes, identifying area functions, non-value-adding activities, amounts of time expended, and what leading indicators are used.
2. Summarize the results from step 1. Calculate total throughput time, identify any bottlenecks, develop costs associated with the processes, and identify accuracy levels of the leading indicators used (compare actual vs. forecasted schedules). Identify non-value-adding steps involved in the forecasting processes.
3. Develop recommendations regarding what steps might be improved, reduced, or eliminated, how bottlenecks might be resolved, and how close-in changes can be reduced or eliminated.
4. Summarize the recommendations for later use during the conceptual design and implementation phases.

Order Entry Review

The objective of the order entry review is to gain an understanding of the current order entry processes, including throughput times, queues, and costs. It will be important to identify the appropriateness of orders entered as they relate to actual demand. Any orders that are lost or inadvertently omitted should also be identified. The four steps in the review are as follows:

1. Chart the flow of current order entry processes, identifying functions by area, non-value-adding operations, and times/costs associated with each operation. Identify total throughput times/costs, bottlenecks, and any cushions built into the orders for shrink, scrap, and processing delays. Review orders generated against actual historical demand levels to determine their appropriateness.
2. Summarize the findings from step 1.
3. Based on the summarized findings, develop a list of recommendations for reducing throughput times, eliminating cushions, dissolving bottlenecks, and improving ordering accuracy.
4. Summarize the recommendations for use during the conceptual design and implementation phases.

Production Planning Review

The objective of the production planning review is to assess the production planning processes, including their throughput times, delays, and costs. It also determines the adequacy and appropriateness of the data, the methods used, and the value of the output. In summary, the production planning review determines the impact of lot sizing techniques, set-up times, production resource allocation, and process planning in the manufacturing environment.

The production planning process in the current manufacturing environment is responsible for the timely and accurate performance of some or all of the following discrete functions: order launching (their respective quantities and due dates), order/data integrity, inventory adjustments, capacity requirements planning, and the provision of accurate purchasing requirements. In a just-in-time environment, all of these functions will take place, but it is important to note that they may or may not be performed by the production planning department. For example, order launching will undoubtedly be performed, but instead of utilizing the current production planning process of "pushing" orders out into the shop floor, the order launching process may be simply an empty container specifying a part and its quantity returned to the department responsible for its manufacture for replenishment.

The diagnostic review of the production planning department is performed primarily to develop an understanding of all functions being performed. This will be of use during the conceptual design and implementation phases when recommendations are made that may dramatically affect current production planning responsibilities.

The five steps required for this review are as follows:

1. Identify all functions currently performed by the production planning department, such as master scheduling, capacity requirements planning, order launching, economic order quantity (EOQ) and lot sizing techniques, and inventory adjustments.
2. Chart the flow of all of the production planning processes, identifying throughput times, functions and responsibilities, non-value-adding activities, and costs associated with each operation. Identify any bottlenecks or queues.
3. Summarize the findings from step 2.
4. Based on the summarized findings, develop recommendations regarding reductions in throughput times, queues, and costs. Also, consider which functions will need to change in a just-in-time environment, and how. For example, will order launching still be done by production planners? How can the demand most quickly and effectively be translated into shop floor production?
5. Summarize the recommendations for later use during conceptual design and implementation phases.

Production Scheduling Review

The objective of the production scheduling review is to assess the processes of production scheduling and their throughput times, delays, and costs; to determine the adequacy, appropriateness, and accuracy of requirements communication systems and their outputs; and to determine if the current requirements communication system consists of the traditional push system whereby requirements are generated through hot lists, production schedules, and/or supervisor job assignments. If the system is a pull system, requirements are signaled by the use of kanbans (empty containers or any other form of a replacement signal). In addition, the production scheduling review determines the impact of unbalanced capacities and inconsistent priorities from operation to operation within the production route.

Traditional production scheduling processes in the current manufacturing environment include some or all of the following functions: order sequencing and prioritization, tool and die availability verification, operator utilization (expediting orders regardless of due dates to keep the operator busy), maintaining and updating hot lists, attending hot-list meetings, future personnel requirements analysis, and departmental inventory verification. In a just-in-time environment, many, if not all, of these functions will be eliminated.

The basis for the diagnostic review of this area is to develop an understanding of all functions performed in the current scheduling environment. This study will be of use during the conceptual design and implementation phases of the project when recommendations may be made that dramatically change traditional production scheduling functions.

The four steps in this review are as follows:

1. Identify and chart the flow of all functions currently performed by the production scheduling department. Identify functions by area, time expended at each activity, total throughput times, and bottlenecks. Also identify non-value-adding activities and costs associated with the process. Identify imbalances between work content levels from activity to activity in the scheduling processes.
2. Summarize the findings from step 1.
3. Develop and summarize general recommendations from these findings concerning how/where the process can more efficiently be performed employing JIT principles. Maintain this list of recommendations for use during the conceptual design and implementation sections of the program.
4. Collect data on all shortages or hot jobs that occur over a two-week period. Identify the root causes for each shortage. Summarize the list by reason for shortage using a Pareto diagram. Maintain this listing for later use with the summarized findings described in step 3. The listing should include a summary of general recommendations about how the root causes identified should be addressed.

Purchasing Review

The objective of the purchasing review is to assess the current purchasing processes, their throughput times, delays, and costs; to develop an understanding of the supplier base, their locations, volumes, delivery performance, prices, quality levels, multiplicity of sources, and willingness to adopt just-in-time techniques; and finally to develop an overall understanding of the appropriateness and adequacy of the current purchasing environment.

An effective just-in-time purchasing program can be one of the most important elements of the entire project. A great deal of analysis must be performed to develop a real understanding of the entire purchasing process.

The diagnostic for this section is broken down into the following tasks:

- Buyer time distribution study.
- Commodity analysis.
- "ABC" analysis.
- Vendor analysis.
- Buyer interviews (concerns/issues).

These studies will be of use during the conceptual design and implementation phases of the project when gross opportunities have been identified and recommendations may be made that dramatically change traditional purchasing functions. The 10 steps in this review are as follows:

1. Perform a buyer time distribution study identifying where buyer time is expended (see Fig. 15-9). Also perform interviews with buyers directed toward analyzing buyer training needs.
2. Develop flowcharts for all purchasing-related activities, with responsibilities and associated time/costs. Also identify all interface activities that affect purchasing performance, such as forecast accuracy and design engineering responsiveness.
3. Identify process bottlenecks, time expenditure percentages, and non-value-adding steps and list time expenditures by category in descending order.
4. Summarize findings and develop general recommendations for reduced time expenditure and cost through elimination of non-value-added activities.

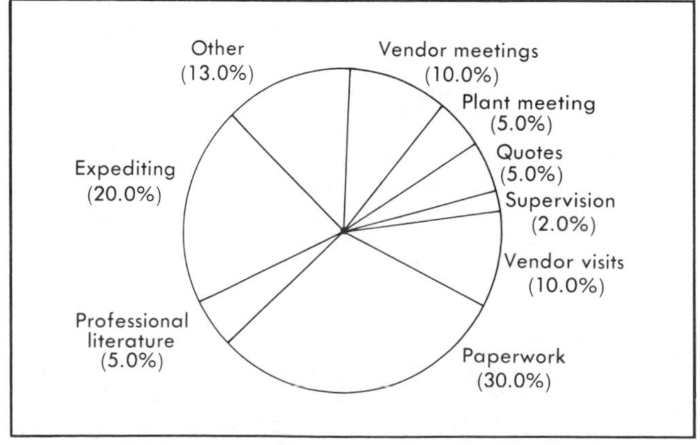

Fig. 15-9 Typical time breakdown in a traditional purchasing department. (*Coopers & Lybrand*)

5. Identify those commodities that comprise 80% of all purchases. Commodities are defined as those items that require the same equipment and/or raw materials to produce. Examples of commodities include investment castings, stampings, die castings, springs, and molded plastic parts.
6. Identify total annual dollar value of purchased goods.
7. Complete the commodity matrix (see Fig. 15-10) and summarize the findings (see Fig. 15-11).
8. Develop a dollar-descending ("ABC") listing of all purchased commodities (see Fig. 15-12). This data should help to identify those commodities that contain the greatest improvement opportunity levels based on cost.
9. Develop a chart depicting all suppliers currently providing purchased material and/or components. On the chart, identify the following data by supplier:
   - Commodity.
   - Yearly purchases by vendor (highest to lowest).
   - Number of miles from facility.
   - Delivery performance (percentage of deliveries late and percentage of deliveries early).
   - Quality levels (percentage of rejected parts).
   - Average lead times.
   - General knowledge of JIT techniques.
   - Receptiveness to JIT concepts.
   - Percentage of vendors' business dedicated to production of the materials/components.
10. Develop a listing of general recommendations for quality improvement, lead-time reduction, supplier base reduction, vendor delivery performance improvement, cost reduction, and vendor qualification and certification programs based on the summarized findings developed in steps 1 through 9. Retain this information for use during conceptual design and implementation phases.

Manufacturing Review

The objective of the manufacturing review is to identify potential opportunity levels with regard to changeover time reduction, quality improvement, layout improvements and capacity requirements, and housekeeping improvements.

Changeover time review. The objective of the changeover time review is to identify existing set-up/changeover times and assess their impacts on workcenter layouts, capacity, personnel requirements, and queues. The four steps in this review are as follows:

1. Obtain a listing of all machine centers and associated set-up/changeover times by product. Summarize the findings by product, longest to shortest.
2. Review how order quantities are calculated based on time allocated to set-ups/changeovers. Review how capacity planning is performed based on set-ups/changeovers. Identify amount of people required to accommodate set-ups/changeovers.
3. Recalculate order quantities, capacity requirements, and personnel level requirements based on set-up/changeover time reductions of 75%.
4. Summarize the resulting opportunity levels for use in the later conceptual design and implementation planning phases.

Quality level review. The objective of the quality level review is to assess current reject rates and their causes, assess current disposition procedures for rejected parts, and calculate current costs associated with maintaining quality levels. The five steps involved in this review are as follows:

1. Obtain a listing of reject rates over the last 6 months and indicate failure causes in each instance. Develop a chart summarizing findings, categorized by cause and listed in descending order by frequency within cause.
2. Obtain a listing of defective parts over the last 6 months and indicate dispositions in each instance. Calculate costs associated with the dispositions.
3. Identify all costs associated with the maintenance of quality over the previous year. These costs include prevention, appraisal, internal failure, and external failure. Typical activities performed by various departments that affect these costs are listed in Tables 15-8, 15-9, 15-10, and 15-11. Additional information on quality costs can be found in Chapter 24, "Quality Cost and Improvement," of this volume.
4. Obtain the cost-of-production figures for the previous year. Obtain gross sales figures for the previous year. Calculate the quality cost as a percentage of the cost of production and as a percentage of the cost of sales. Compare these figures with prior years, if available. Develop a chart summarizing these figures.
5. Identify methods used and parties responsible for measuring/plotting defects. Identify course of action taken on defects (where defect notices go, what is done with them, by whom, and how quickly). Summarize gross opportunities for use in conceptual design and implementation phases of the program.

TABLE 15-8
Typical Activities Performed by Various Departments That Affect Prevention Costs in a Quality System

Marketing	Product Design and Development	Purchasing	Operations	Quality Administration
Marketing research Customer perception surveys Contract review	Quality progress reviews Support activities Qualification test Field trials	Supplier reviews Supplier rating Purchase order technical data reviews Supplier quality planning	Process validation Operations quality planning Operations support quality planning Operator quality education Operator SPC/process control Design and development Quality related equipment	Administrative salaries Administrative expenses Quality program planning Quality performance reporting Quality education Quality improvement Quality audits

TABLE 15-9
Typical Activities Performed by Various Departments That Affect Appraisal Costs in a Quality System

Purchasing	Operations	External
Receiving or incoming inspection and tests Inspection equipment for purchased material Supplier product qualification Source inspection and control program	Performance of planned inspections, tests and audits • labor operations • quality audits • inspection and test materials Set-up and first-piece inspection Special tests (manufacturing) Process control measurements Laboratory support Inspection and test equipment • depreciation allowances • measurement equipment expenses • maintenance and calibration labor Endorsement and certification by outside agencies	Field performance evaluation Special product evaluations Field stock and spare parts evaluation Test and inspection data review

CHAPTER 15

DIAGNOSTIC REVIEW

Commodity ______________________

Part Number	Price per Part	Annual Requests, pieces	Annual Requests, dollars	Quantity on Hand, pieces	Quantity on Hand, dollars	Weeks on Hand	Turns		Vendor Name	MAJOR	MINOR	PRIME	Performance $\pm n$ Days, percent
							Current	Goal					
1.													
2.													
3.													
4.													
5.													
6.													
7.													
8.													
9.													
10.													
11.													
12.													
13.													
14.													
15.													
16.													
17.													
18.													
19.													
20.													
21.													
22.													
23.													
24.													
25. Total													
26.													

Fig. 15-10 Sample commodity analysis worksheet to use when conducting purchasing review. (*Coopers & Lybrand*)

Commodity	Number of Parts	Annual Dollars Purchased	Dollar Value of On-Hand Inventory	Weeks of Inventory on Hand	Number of Turns	Number of Major Vendors	Number of Minor Vendors	Number of Prime Vendors	Average Delivery Performance
1.									
2.									
3.									
4.									
5.									
6.									
7.									
8.									
9.									
10.									
11.									
12.									
13.									
14.									
15.									
16.									
17.									
18.									
19.									
20.									
21.									
22.									
23.									
24.									
25. Total									
26.									

Fig. 15-11 Sample summary worksheet for commodity analysis. (*Coopers & Lybrand*)

Quantity Variance, dollars	Percentage Rejected	Reason for Rejection	Item Master Lead Time (System)	Total Lead Time**	Percent of Lead Time That is Queue	Comments	
							1.
							2.
							3.
							4.
							5.
							6.
							7.
							8.
							9.
							10.
							11.
							12.
							13.
							14.
							15.
							16.
							17.
							18.
							19.
							20.
							21.
							22.
							23.
							24.
							25.
							26.

Fig. 15-10—*continued*

Average Quantity Variance	Average Percentage Rejected	Main Response(s) for Rejection	Average Lead Time	Percent of Average Lead Time That is Queue	Percent of Lead Time Reduction Opportunity	Additional Comments	
							1.
							2.
							3.
							4.
							5.
							6.
							7.
							8.
							9.
							10.
							11.
							12.
							13.
							14.
							15.
							16.
							17.
							18.
							19.
							20.
							21.
							22.
							23.
							24.
							25.
							26.

Fig. 15-11—*continued*

Total Purchases ______________

	Commodity	Number of Parts	Total Annual Purchases	Percentage of Total	Cumulative Percentage
1.					
2.					
3.					
4.					
5.					
6.					
7.					
8.					
9.					
10.					
11.					
12.					
13.					
14.					
15.					
16.					
17.					
18.					
19.					
20.					
21.					
22.					
23.					
24.					
25. Total					
26.					

Fig. 15-12 Sample ABC analysis worksheet. (*Coopers & Lybrand*)

TABLE 15-10
Typical Activities Performed by Various Departments That Affect Internal Failure Costs in a Quality System

Product Design	Purchasing	Operations
Corrective action	Purchased material reject disposition costs	Disposition
Rework due to design changes	Purchased material replacement costs	Troubleshooting or failure analysis
Scrap due to design changes	Supplier corrective action	Investigation support
Production liaison costs	Rework of supplier rejects	Corrective action
	Uncontrolled material losses	Rework
		Repair
		Reinspection/retest
		Extra operations
		Scrap
		Downgraded end product or service
		Internal failure labor losses

Process flow review. The objectives of the process flow review are to develop a clear understanding of the operations required to produce a specific product; document distances of product travel in the current layout; determine the levels of value added to the product at each stage of the process; identify personnel, work-in-process, and capacity levels at each stage of the process; identify queue levels and causes in the process; identify the lead-time distribution by product within the process; and to evaluate work content at each stage of the process.

The seven steps in this process are as follows:

1. Follow each product through its processes, documenting all major steps. Develop a straight-line flowchart of the process, highlighting all major operations. An example of a straight-line flowchart is shown in Fig. 15-13.
2. Obtain a facilities layout (scale drawing or blueprint) of the facility in which all production processes occur. Identify the location of each process on the layout and show routes/distances traveled between each step (see Fig. 15-14). Transfer data to a separate sheet.
3. Document everything that happens to a given part before, during, and after each major operation in the

TABLE 15-11
Typical Elements That Affect External Failure Costs in a Quality System

Customer complaint investigation and resolution
Returned goods evaluation and repair or replace
Retrofit costs
Recall costs
Warranty claims
Liability costs
Penalties
Customer goodwill
Lost sales

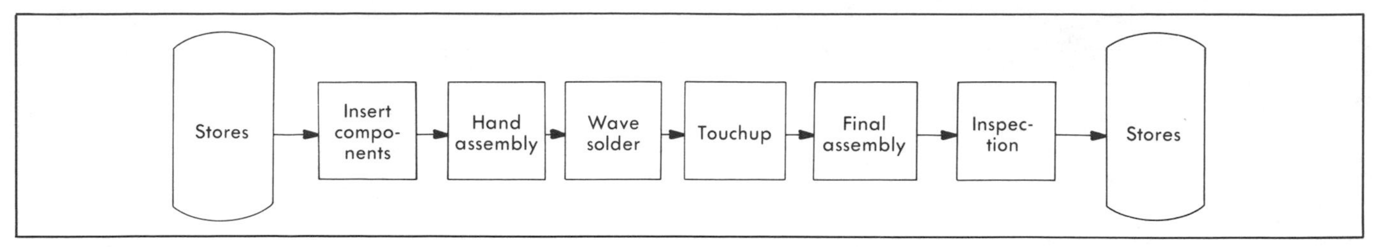

Fig. 15-13 Straight-line flowchart example of a production process. (*Coopers & Lybrand*)

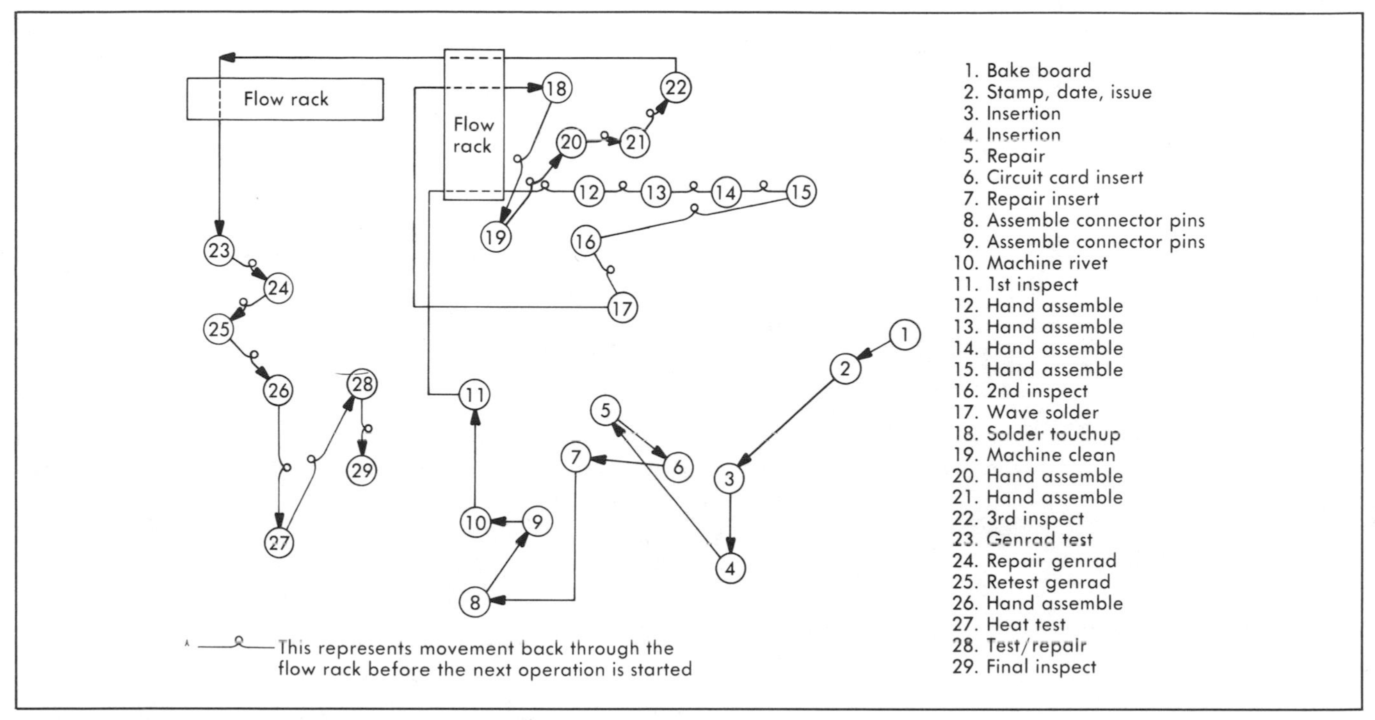

Fig. 15-14 Actual flowchart of a production process. (*Coopers & Lybrand*)

process. Indicate whether each step in the process adds value to the product (see Fig. 15-15).

4. Select a part that is representative of its entire product line in terms of its processes and complete the information designated in Fig 15-16. From this chart, the following information can be obtained:
   - A comparison of the days-on-hand figure with the throughput figure. The days-on-hand figure represents how long it will take to get a part completely through the process (lead time) after utilizing the WIP currently on the floor. The throughput figure represents how long it will take if lot sizes of 1 are utilized and the part moves immediately from one operation to the next without any interruptions. If larger lot sizes will be used, multiply the throughput figure by the new lot size number to get a new throughput time.
   - The identification of bottlenecks, which are those areas with the greatest number of days-on-hand inventories.
   - A comparison of the machines required with the current number of machines available. This comparison is a quick way to identify underutilized (machines that could be used elsewhere) and overutilized (bottleneck) machine applications that can be addressed immediately.
   - The fundamental data required for the calculation of the uniform plant load.

5. Develop a listing of WIP inventory levels between and at each workcenter, listing quantities and reasons for the queues. Call out the equivalent days or hours of production represented by each queue (see Fig. 15-17). Identify potential methods of reducing the levels of queue.

6. Utilizing an operation-by-operation listing of manufacturing throughput time per piece, graphically depict the lead time in the production process. Use X's to represent appropriate units of time such as $X = 1$ hour or $X = 1$ day (see Fig. 15-18). Use this review to identify operations that appear to be logical targets for lead-time reduction.

7. As the part moves through its production process, observe direct and indirect labor activities. Review the data gathered and accumulate times by activity. Then accumulate times attributed to activities that add value to the product. Calculate the percentage of labor time (by operation and in total) that is attributable to value-adding activity.

DIAGNOSTIC REVIEW

Product ______________

Activity	Value Added (Yes or No)	Distance Traveled
1. Unload parts from truck	N	60
2. Move to cage	N	110
3. Store boxes on shelf	N	100
4. Pull boxes from shelf	N	20
5. Reload boxes with 72 pieces each	N	0
6. Deliver skid to rough sanding	N	382
7. Move boxes from skid to table	N	30
8. Rough sand edge	Y	0
9. Move boxes to next workcenter	N	4
10. Rough sand butt	Y	0
11. Move boxes to skid	N	30
12. Move skid to notcher	N	30
13. Move 10 boxes to table	N	10
14. Notch part	Y	0
15. Move boxes to skid	N	10
16. Move skid to staging area	N	82
17. Move skid to grinding area	N	362
18. Move 1 box to workcenter	N	120
19. Grind part and drain	Y	0
20. Return drained parts to box	N	0
21. Move box to inspection cart	N	120
22. Lot sample inspection	N	0
23. Log production	N	0
24. Move box to outgoing area	N	8
25. Move skid to holding area	N	60
26. Move skid to sanding hold area	N	382

Fig. 15-15 Sample worksheet summary for recording distance between production steps and whether step adds value to product. (*Coopers & Lybrand*)

Daily demand __________ Hourly demand __________
Hours per shift __________ Cycle time ______________
Item ____________________ Number of shifts ________

Operation	Standard (hours per 100 pieces)	Rate Per Hour (RPH) (100/standard)	Throughput (TP) (60/RPH), min	Number of Machines	Machines Required (hourly demand/RPH)	Machine Status, Current Required	Personnel Required Per Shift (hourly demand/RPH)	WIP Inventory	Inventory, dollars on hand	Inventory, days on hand
1.										
2.										
3.										
4.										
5.										
6.										
7.										
8.										
9.										
10.										
11.										
12.										
13.										
14.										
15.										
16.										
17.										
18.										
19.										
20.										
21.										
22.										

Fig. 15-16 Sample capacity requirements spreadsheet. (*Coopers & Lybrand*)

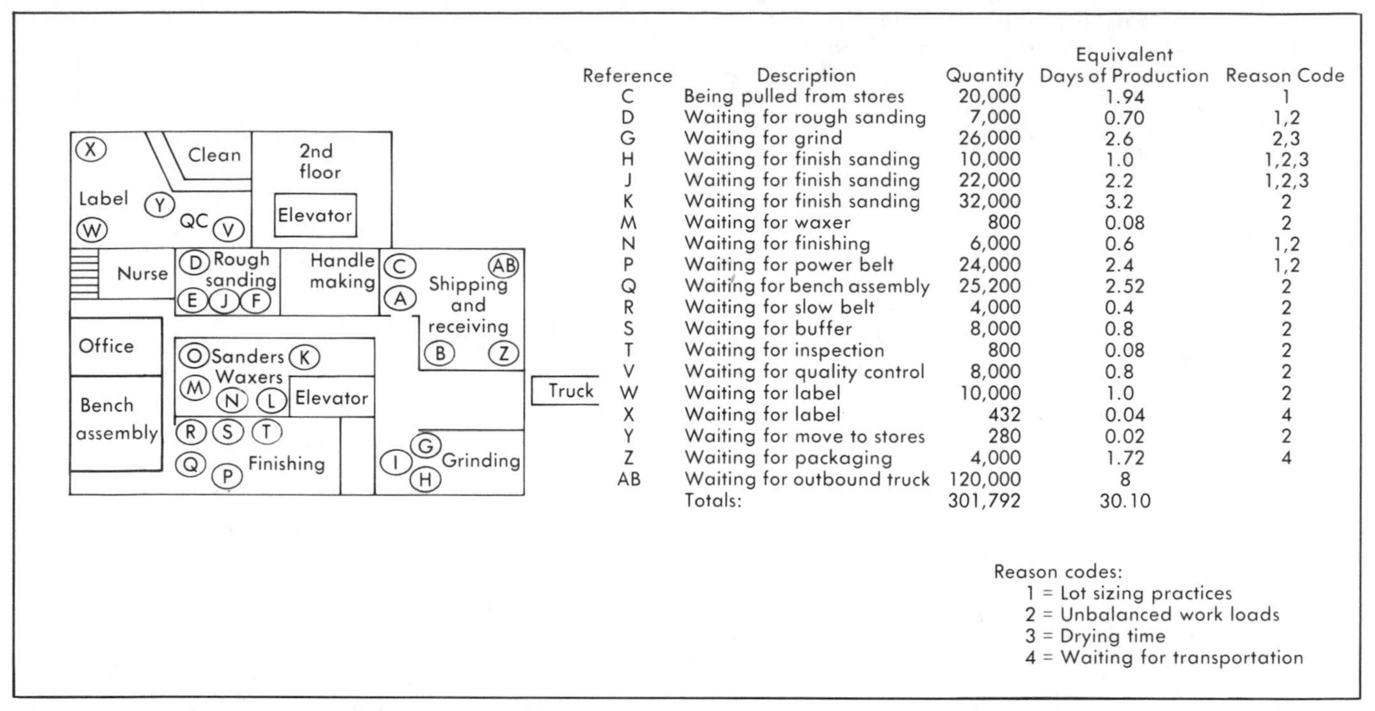

Reference	Description	Quantity	Equivalent Days of Production	Reason Code
C	Being pulled from stores	20,000	1.94	1
D	Waiting for rough sanding	7,000	0.70	1,2
G	Waiting for grind	26,000	2.6	2,3
H	Waiting for finish sanding	10,000	1.0	1,2,3
J	Waiting for finish sanding	22,000	2.2	1,2,3
K	Waiting for finish sanding	32,000	3.2	2
M	Waiting for waxer	800	0.08	2
N	Waiting for finishing	6,000	0.6	1,2
P	Waiting for power belt	24,000	2.4	1,2
Q	Waiting for bench assembly	25,200	2.52	2
R	Waiting for slow belt	4,000	0.4	2
S	Waiting for buffer	8,000	0.8	2
T	Waiting for inspection	800	0.08	2
V	Waiting for quality control	8,000	0.8	2
W	Waiting for label	10,000	1.0	2
X	Waiting for label	432	0.04	4
Y	Waiting for move to stores	280	0.02	2
Z	Waiting for packaging	4,000	1.72	4
AB	Waiting for outbound truck	120,000	8	
	Totals:	301,792	30.10	

Reason codes:
1 = Lot sizing practices
2 = Unbalanced work loads
3 = Drying time
4 = Waiting for transportation

Fig. 15-17 Sample queue causality listing. (*Coopers & Lybrand*)

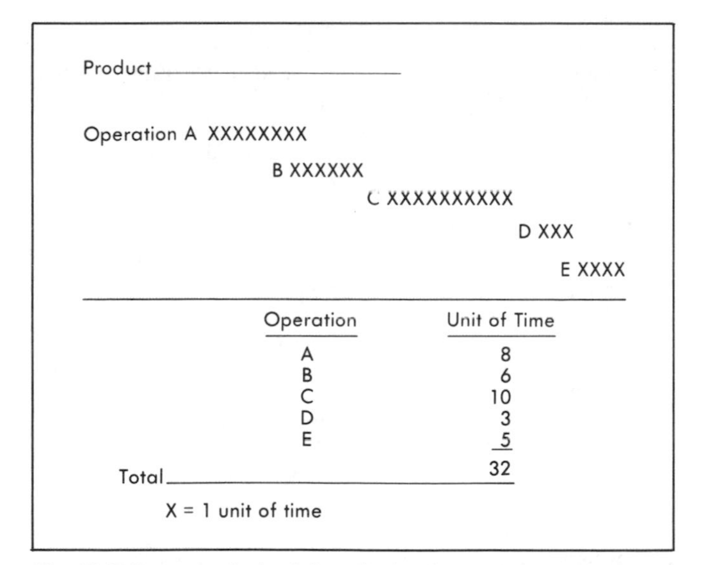

Operation	Unit of Time
A	8
B	6
C	10
D	3
E	5
Total	32

X = 1 unit of time

Fig. 15-18 Example of a lead-time distribution depiction. (*Coopers & Lybrand*)

Retain the data from steps 1 through 7 for use in the conceptual design and implementation phases.

Housekeeping review. The objectives of the housekeeping review are the following:

- Identify current tooling/fixturing requirements.
- Identify current WIP storage requirements and how WIP is identified.
- Determine whether all tooling and fixtures have designated locations and if the equipment is kept in those locations.
- Assess current housekeeping practices in regard to general cleanliness and upkeep of the equipment and facility.

The following two steps are recommended when performing a housekeeping review:

1. Develop a chart that lists all workcenters under review. Break the chart down into the following categories:
   - Dies and fixtures required by machine workcenter.
   - Tools required to perform all set-ups on the specific machines.
2. Tour the production area and look for trash on the floor, personal items on or near workstations, WIP inventory lying around, and poor upkeep of machinery, such as fluid leaks and squeaks/rattles. Develop a list of recommendations to be used during the implementation phase of the project. Typical recommendations include the following:
   - Make operators responsible for the general cleanliness of their work area, including picking up trash, placing personal items in lockers, not allowing inventory in their area until it is required, and cleaning machines during idle times.
   - Awarding a prize for the employee(s) keeping their area the cleanest.

Retain this data for later use during the conceptual design and implementation phases of the project.

Distribution Review

The objective of the distribution review is to review current distribution practices and procedures and assess their appropriateness in light of just-in-time manufacturing/distribution concepts. It is also used to identify gross opportunity levels that may be achieved through the application of JIT techniques. The following nine steps can be performed when conducting a distribution review:

1. Develop flowcharts depicting current distribution processes and functions. At least one chart should describe material flow, and another should reflect paperwork processing activities. In the charts, or on a separate listing attached to the charts, identify the following:
   - Areas performing each function.
   - Number of reviews/approvals in each flow.
   - Amount of time spent in each activity. Queue time (and inventory levels, as appropriate) between each activity should be noted when available. More detailed analysis will follow in the implementation planning phase.
   - Distances traveled.
   - Number of stores. Areas and/or distribution facilities involved and their actions.
2. Summarize the data gathered and review it with all affected personnel to ensure its accuracy. Include estimates of the following:
   - Total throughput times.
   - Percent of throughput times that is queue.
   - Bottlenecks in the flows.
   - Rough cost data associated with the flow.
   - Percent/cost of functions within the flow that do not add value to the documents or material involved and/or are not critical.
3. Assess the gross opportunity levels in these processes for general efficiency/productivity improvements through the reduction of non-value-adding activities and bottlenecks.
4. Document the estimated opportunity levels (potential for improvement) and retain the data for use during conceptual design and implementation phases later in the program.
5. Conduct interviews with distribution management to determine their perception of:
   - Current customer base.
   - Current service level requirements.
   - Nature/use of the product.
   - Special constraints such as packaging, handling, and contracts.
   - Profit contribution by product.
   - Geographic and distribution channel contribution levels.
   - Areas of needed improvement.
   - Appropriate/desirable carriers and distribution channels.
6. Review historical and current data to identify actual customer base and service levels. Compare the results with the data gathered previously.
7. Conduct a cursory review of distributed end item contribution levels and perform the following tasks:
   - Determine those parts making the greatest contribution to sales.
   - List those parts in descending order (if reasonable).
   - On a chart, plot the parts against the cumulative percentage of sales (see Fig. 15-19). The intent of this chart is to highlight (if applicable) the number of parts that comprise the greatest percentage of sales. In the example shown in Fig. 15-19, less than 50% of the part number population represents 95% of total sales.
8. Considering any special constraints and product characteristics, estimate the improvement potential of JIT concepts (queue elimination, travel distance reduction, and simplification) applied to the existing distribution environment. Quantify the improvement potentials to the extent possible and practical.
9. Summarize and document the findings and retain this data for use in the conceptual design and implementation phases later in the program.

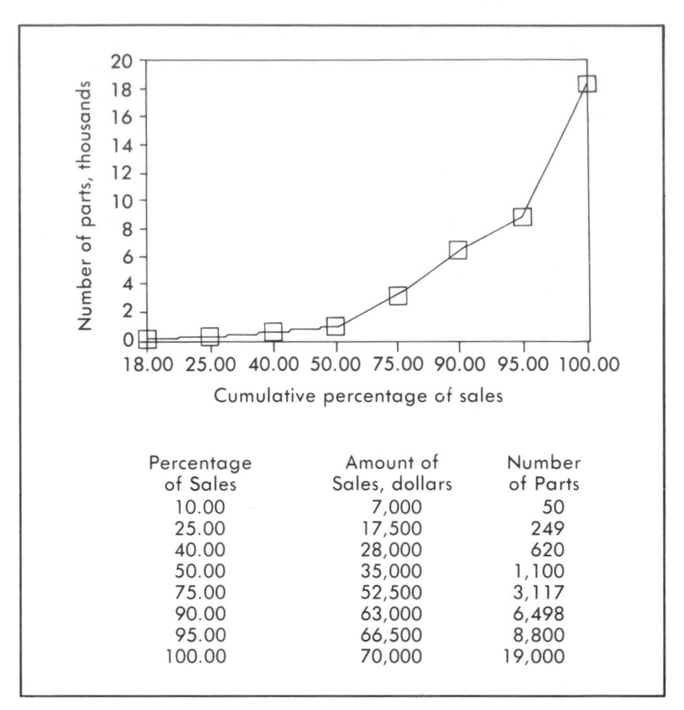

Percentage of Sales	Amount of Sales, dollars	Number of Parts
10.00	7,000	50
25.00	17,500	249
40.00	28,000	620
50.00	35,000	1,100
75.00	52,500	3,117
90.00	63,000	6,498
95.00	66,500	8,800
100.00	70,000	19,000

Fig. 15-19 Graph showing relationship of cumulative sales percentage to number of parts. (*Coopers & Lybrand*)

Accounting Review

The objective of the accounting review is to review current accounting practices and procedures and assess their applicability to JIT manufacturing operations. Also, it identifies opportunities for general productivity improvement in the existing accounting activities that may be addressed through the application of JIT methodology.

An accounting review is generally broken up into two parts: (1) paper/process improvement and (2) philosophical assessment.

Paperwork/process improvement. Four steps can be followed when performing an accounting review in the paperwork and process area:

1. Develop a flowchart that depicts current accounts payable activities, accounts receivable activities, and cost accounting activities. In the charts, or on a separate listing attached to the charts, identify:
   - Area performing each function.
   - Number of review/approvals in the flow.
   - Amount of time spent at each function; queue time between each function should be noted when readily discernible. More detailed analyses will follow in the implementation plan phase.
   - Distance traveled, number of stops/people involved.
2. Summarize the data obtained in step 1. Document the following:

- Total throughput times.
- Estimate of percent of throughput times that is queue.
- Bottlenecks in the process flow.
- Rough cost data associated with the flow.
- Percent of functions in the processes that do not add value to the document and/or are not critical. (Related cost data may be generated if appropriate.)

3. Review the data gathered with all affected personnel to ensure its accuracy.
4. Assess the gross opportunity levels in these processes for improvement through the reduction of non-value-adding functions, queues, and bottlenecks. Document the estimated opportunity levels and retain the data for use during the conceptual design and implementation phases.

Philosophical assessment. Seven steps can be followed when performing a philosophical assessment during an accounting review:

1. Identify major commodities and services currently included and planned for inclusion in the accounts payable system.
2. Estimate the practicality of the elimination of purchase orders, invoices, and receiving transactions by considering the following:
   - Commodities involved and whether a precedent exists for the elimination of these devices in the purchasing and receipt of these kinds of commodities.
   - Any ongoing programs involving electronic data transfer to/from suppliers and/or backflushing bills of material.
   - General openness of management to these concepts.
   - General openness of suppliers to these concepts (see purchasing section).
3. Document gross potential estimates calling out time reductions. Document travel distance reductions, queue reductions, and probably ultimate costs.
4. Summarize the data for use during conceptual design and implementation phases later in the program.
5. Identify major process steps and responsibilities in the cost accounting function. Include the following:
   - What units are costed.
   - Special procedures for stores and/or material handling to facilitate product costing.
   - The frequency and importance of stock pick and issue transactions.
   - A description of the cost "role-up" process.
   - A description of rework and scrap costing.
   - A description of capital equipment justification processes.
   - A description of any piecework or incentive systems in existence or on the drawing board.
   - Historical data on downtime and approximate costs.
   - Methods for distributing fixed and variable overhead costs.
   - Processes for calculation of tooling costs.
6. Summarize the data gathered and estimate the gross opportunities in these areas with the advent of the following:
   - Elimination of WIP costing.
   - Elimination of WIP stores.
   - Formation of workcells.
   - Flattened bill of material structures (fewer levels of assembly within the bill).
   - Reductions in rework levels.
   - Significant reductions in maintenance-related and quality-related downtime.
7. Review the data with all affected personnel to ensure its accuracy. Retain the summarized data for use during conceptual design and implementation phases later in the program.

CONCEPTUAL DESIGN ACTIVITIES

As mentioned earlier, a thorough diagnostic assessment provides excellent groundwork for conceptual design activities. By knowing what the existing conditions, capabilities, and opportunities are, it is easier to intelligently estimate what the business can look like in 1 to 5 years. Of course, a knowledge of what JIT programs typically yield and the magnitude of the improvements is also needed. Some education will be required at this point for those individuals involved in conceptual design activities.

To obtain a sound conceptual picture of the future in a manufacturing organization incorporating JIT, a number of important factors need to be addressed. Table 15-12 lists the factors that need to be addressed in various departments, along with the desired goal. For each factor, the question to be asked is, "How will each factor be measured?"

The question must be answered in both a 5-year and 1-year time frame. The resulting "snapshot" of future operations is comprised of important milestones for the balance of the JIT program. Each milestone must be measurable. For example, improved employee morale is not measurable and, therefore, does not qualify. Milestones are an important indicator of whether planned improvements relate correctly to overall corporate goals and objectives and provide a good infrastructure for divisional and departmental goals and objectives. Four basic steps to follow during the conceptual design phase are:

1. Review the diagnostic data summaries developed for each area during the course of the program. The information to review during this phase is listed in Table 15-13.
2. Provide the conceptual design participants with a review of the advantages of such JIT-related concepts as focus factory, pilot line approaches to implementation, and group incentives, bonuses, and profit sharing.
3. Using the JIT wheel (refer to Fig. 15-2) as a structure, go through each module (housekeeping, quality, and so on) and define 1-year (short range) and 5-year (long-range) goals. The goals should be as specific and measurable as possible.
4. Considering these overall goals, review each functional area involved in the diagnostic phase of the program. In each area, develop specific objectives and measurement criteria for the objectives that support the JIT goals. Again, the objectives should be as measurable as possible.

The JIT goals and the functional area objectives supporting those goals comprise the bulk of the conceptual design. The design should provide enough detail to create a view of future operations without unnecessarily constraining the development of specific layouts, methods, and devices.

Initial implementation of the conceptual design is typically performed in a "pilot line" setting. This approach allows flexibility for revision of layouts, methods, and devices with minimal disruption to overall production operations. In fact, implementation is comprised of two major phases: (1) pilot line

TABLE 15-12
Factors to be Addressed During Conceptual Design Phase

Department	Desired Goal	Factors
Design engineering	Reduced design throughput time Accurate designs that support zero defects and manufacturability	Design change throughput time Levels of manufacturability Levels of standardization
Forecasting	Reduced forecast generation throughput time	Level of forecast accuracy
Order entry	Reduced order entry throughput time Defect-free order entry	Order entry accuracy levels
Production planning	Reduced production planning throughput time	Average lot size
Production scheduling	Reduced production scheduling throughput time	Requirements communication efficiency and effectiveness levels
Purchasing	Reduced supplier base Zero defects in purchased goods Reduced purchase costs Point-of-use delivery Frequent deliveries Willingness to do electronic data interchange (EDI)	Geographic location Multiplicity by commodity Delivery performance Appropriateness of price Quality levels Willingness to adopt JIT techniques
Manufacturing	Reduced setup and changeover times Zero defects Zero inventory Reduced throughput times Reduced cost of goods sold	Quality levels Quality standards Product flow Pull system mechanisms and procedures WIP levels Manufacturing throughput time Production reporting methods, frequencies, and locations Layout attributes Level of work center commonality and utilization Efficiency levels Designation level for materials and tooling Level of trash, personal effects, and debris at work centers
Distribution	Reduced distribution throughput time Zero	Finished goods levels Transportation costs, frequencies, and volumes
Accounting	Reduced accounts payable throughput time Zero errors in payments	Philosophical and operational changes for accounts payable Accounts receivable throughput times Cost accounting throughput times Philosophical and operational changes for cost accounting

(*Coopers & Lybrand*)

implementation and revision and (2) remaining factory conversion.

In addition to developing a preliminary layout of the entire manufacturing facility as it should appear in a JIT environment, a pilot product should be identified and a location specified for pilot line operations. Finally, a broad time frame target should be established for the completion of pilot line implementation and remaining factory conversion.

The conceptual design should, then, be comprised of a functional area breakdown of the following:

- Objectives in narrative form that detail throughput times, philosophies, responsibilities, and organization structures to provide a snapshot of projected future operations.
- Process flows (paperwork and material) as they are projected to be in a 1-year and 5-year context, incorporating the objectives identified.
- A high-level schedule, assigning dates and responsibilities, for pilot line identification, pilot line implementation, and remaining factory conversion.

The conceptual design typically requires several iterations, based on revisions stemming from the experience gained in pilot line operation.

IMPLEMENTATION PLANNING

The diagnostic review establishes where the organization is today, thus establishing benchmarks against which future progress can be measured. To provide the company with a target of where it should be in 1 to 5 years, a conceptual design of the entire organization is performed. The implementation plan, along with a lot of hard work, provides the means for the organization to move from point A to point B.

Within the implementation plan, conceptual design objectives (described in the previous section) for each area are broken

TABLE 15-13
Recommended Reports/Summaries to Review During Conceptual Design Phase

Report on organizational readiness for JIT
Design engineering recommendation listing
Listing of preferred manufacturing/purchasing practices
Forecasting recommendation summary
Order entry recommendation summary
Production planning recommendation summary
Production scheduling recommendation summary
Shortage analysis recommendation summary
Buyer time recommendation summary
Commodity analysis gross opportunity listing
Purchased commodity ABC listing
Vendor analysis chart(s)
Set-up reduction opportunity summary
Quality level opportunity summary
Straight-line process flowchart
Actual process flow diagram
Value added/distance traveled summary
Layout/capacity gross opportunity summary
Queue reduction recommendation listing
Recorded lead time analysis
Work content analysis
Tool/fixture recommendation summary
Housekeeping recommendation listing
Accounting paperwork/process improvement opportunity listing
Distribution paperwork/process opportunity listing
Distribution philosophy opportunity listing

(*Coopers & Lybrand*)

down into required tasks and subtasks. Each subtask, task, and ultimate goal is then listed with completion due dates, responsibility assignments, and deliverables in an implementation plan format. Table 15-14 shows a section of an implementation plan devoted to the accounts payable process.

To obtain a proper perspective on the various tasks in the implementation plan, and the sequence in which they should be performed, a Gantt chart should be developed. Figure 15-20 shows a Gantt chart of one manufacturer's approach to the implementation plan. Although this particular chart covers a period of 94 weeks, the time frame will vary in other organizations based on the size and the scope of the JIT program.

The implementation plan should be developed with the input of all personnel involved to maintain user support and commitment. Some general rules to guide those involved in implementation planning are as follows:

1. The tasks defined in the plan must be specific and measurable. At the due date prescribed, activities should be clearly finished or unfinished.
2. The tasks should be reasonable, given the resources and time available to perform them.
3. When due dates are missed, as with any good project management exercise, the subsequent activity due dates must be adjusted or additional resources and expediting must be applied.

The following sections provide a generic plan for the implementation of JIT in a manufacturing facility. The approach for this plan is that all aspects of JIT are to be included. It may be applied first to a pilot product line and later "leveraged" into the other product lines at the facility until the entire factory (including all support functions) is converted to JIT. Some of the items in the plan may already be completed or under way in the pilot area and may be disregarded until the next product line is addressed. Another factor to be considered is the current posture with regard to focus factory operations. If any of the manufacturing operations will be handled as a focus factory (factory within a factory), with dedicated support staffs, the focus factory portions of the plan should be utilized.

It is important to note that while the following plan addresses the implementation of JIT in a fairly concise and incremental fashion, it is not possible to include every detail of each step in a work plan short of publishing a book on the subject. Therefore, it is recommended that specialists in each area be utilized to direct the specific improvement activities in those areas, with oversight of the entire implementation project handled by an individual with JIT implementation experience and project management skills. Because of the interdisciplinary nature of many JIT activities, an integrated approach is very important.

Implementation Organization

In addition to a program leader, a JIT implementation organization should be comprised of a steering committee and an implementation team. The responsibilities of these members are listed in Table 15-2. Participants on the implementation team are typically middle to lower level management personnel and may include members of the direct workforce as appropriate. Some of the skills required of the team members include a general understanding of JIT, prior JIT experience, product-specific knowledge, familiarity with overall company policy/procedure, technical leadership, intellectual leadership, ability to identify opportunity, and the ability to incorporate change.

TABLE 15-14
Partial Implementation Plan for Accounts Payable Process

Activity	Responsibility	Due Date	Deliverables
Identify and document current accounts payable activities flow and responsibilites	John Smith	12/01/87	Completed flowchart with times and responsibilities
Identify all queues, duplication of effort, and/or data and unnecessary (non-value-added) activity in the accounts payable process	John Smith	1/15/88	Listing of all non-value-added activities

(*Coopers & Lybrand*)

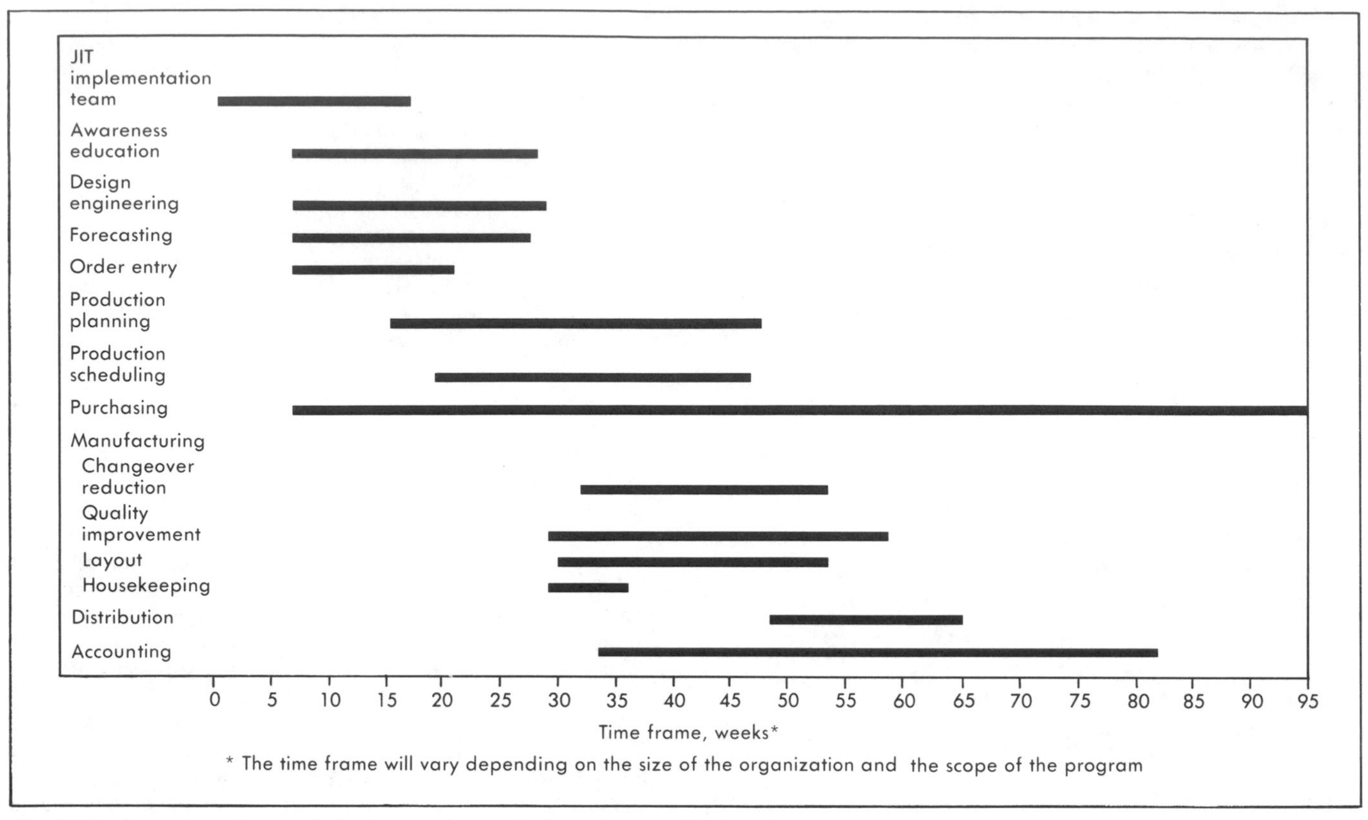

Fig. 15-20 Gantt chart of a generic implementation workplan. (*Coopers & Lybrand*)

Once the implementation team members have been selected and their responsibilities assigned, education and awareness sessions should be conducted.

Awareness, Education and Training

Because the JIT concepts affect a variety of people in the organization, it is necessary to identify who those people are that will be affected and then determine the type of education/training required. Depending on the individual's involvement in the JIT program, the information he or she may require ranges from a general awareness of the program to an in-depth understanding. Table 15-3 lists the type of education/training required by various workforce levels.

Awareness sessions should cover broad JIT philosophical views, general areas of JIT activity, brief case history data from other JIT implementations, and a general overview of the implementation plan at the company. In addition, awareness sessions should indicate who is heading up the program (program leader), the time frame, and the expected benefits.

Education sessions should cover the more detailed JIT philosophies, an outline of the JIT methodologies, and specifics about the implementation plan. Specifics about the implementation plan include oversight responsibilities for all phases of JIT implementation activities, specific delineations of affected and unaffected areas and/or functions, time frames for each phase, and the basic methodologies for each area.

Topics to be included in training sessions vary based on the person's involvement in the implementation plan. General and tailored training sessions that have been useful in different companies are listed in the first section of this chapter under "Awareness and Education." Training should be designed to evolve into an ongoing program for compensation of personnel changes.

Design Engineering

Just-in-time activities in the design engineering area fall into process and product categories. Process improvements are generally addressed by a program called design process improvements. Product improvements are generally addressed in three modules referred to as early manufacturing involvement (EMI), current product review (CPR), and value engineering (VE).

Design process improvements. The objective of the design process improvement module is to speed up and improve the efficiency of the design/implementation process itself. Based on the flows developed during the diagnostic review and any other additional detail, the following steps can be used as a guide when implementing design process improvement plans:

1. Identify all queues, duplication of effort and/or data, and any non-value-added activity. This information should largely be available from activities performed during the diagnostic phase.
2. Select the most efficient/effective processing alternatives.
3. Document improvement potentials and obtain approval to make changes.
4. Write procedures and perform training.
5. Implement changes and then monitor and report design throughput times.

Product-related improvements. The objective of EMI, CPR, and VE programs is to improve the effectiveness of the design by designing the product with an understanding of how it will be manufactured. Additional information on how to design a product for manufacture is found in Chapter 13, "Design for Manufacture," of this volume.

Early manufacturing involvement is a program that involves the review of new designs and design revisions by a group of representatives from product engineering, purchasing, manufacturing, quality control, and industrial (or manufacturing) engineering disciplines. The review should be done on a regular basis, varying meeting locations between the manufacturing and engineering facilities as appropriate. The objectives of this review are to increase standardization, improve manufacturability, improve quality levels, improve interdisciplinary communication, and reduce product cost.

Current product review is a program that allows manufacturers to rethink how current products are designed. The program (like an EMI program) is intended to promote the simplification/standardization; to improve the manufacturability, quality levels, and interdisciplinary communication; and to reduce the product cost of current parts. This program is most effective with the participation of individuals from purchasing, manufacturing, and marketing, as well as specific vendors.

Value engineering provides a systematic approach to evaluating design alternatives that is often very useful and may even point the way to innovative new design approaches or ideas. Also called value analysis, value control, or value management, value engineering utilizes a multidiscipline team to analyze the functions provided by the product and the cost of each function. Based on results of the analysis, creative ways are sought to eliminate unnecessary features and functions and to achieve required functions at the lowest possible cost while optimizing manufacturability, quality, and delivery.

The following steps can be used as a guide when implementing any of the product-related improvement plans:

1. Identify and document the objectives of the program, considering current levels of standardization and manufacturability defined in the diagnostic review and the objectives defined during conceptual design.
2. Identify and document the participants and participant responsibilities in the program.
3. Establish and document procedures for the ongoing operation of the program.
4. Establish a schedule for regular program activities.
5. Implement the program and then monitor and report progress based on the goals established during conceptual design.

Forecasting

Just-in-time activities in the forecasting area fall into two categories: (1) paperwork process flow and (2) forecast process components.

Paperwork process flow. The objective of the paperwork process flow module is to speed up and improve the efficiency of paperwork processing. Based on the identification of non-value-adding activities, queues, and duplication efforts from the diagnostic review, the following steps can be used as a guide when implementing this part of the forecasting plan:

1. Identify, analyze, and select most appropriate alternatives for eliminating non-value-adding activities, queues, and redundancy in the process.
2. Develop a proposal for improving the forecasting process.
3. Verify that the proposal is workable by checking with all affected parties.
4. Present the proposal to management for approval.
5. Assign implementation responsibilities and due dates.
6. Monitor/report progress, through achievement of objectives established during conceptual design.

Forecasting process components. The objective of the forecasting process components module is to select the most efficient forecasting model(s). Based on the data gathered about forecast accuracy during the diagnostic review, the following steps can be used as a guide when implementing this part of the forecasting plan:

1. Identify the units most conducive to forecast.
2. Identify potential leading indicators, analyze historical data to determine accuracy levels of all potential indicators, and select the most accurate indicator or indicators.
3. Identify potential forecasting models, analyze historical data to determine accuracy levels of all potential forecasting models, and select most accurate forecasting model or models.
4. Establish and document forecasting methodology incorporating the most accurate leading indicators and forecasting models.
5. Identify data gathering and forecasting responsibilities and write forecasting procedures.
6. Train forecasting personnel.
7. Implement new forecasting program and then monitor and report forecast accuracy.

Order Entry

The objective of applying JIT techniques to the order entry area is to accelerate and improve the efficiency of the order entry process, minimizing non-value-adding activities. In essence, this means moving the order from the customer to the shop floor with the least amount of time and effort expended. Based on the data obtained during the diagnostic review, the following steps can be used as a guide when implementing the order entry plan:

1. Identify all queues, duplication of effort and/or data, and non-value-adding activity.
2. Identify and select the most efficient and effective processing alternatives to reduce or eliminate all non-value-adding activities.
3. Obtain approval, write procedures, and perform training.
4. Implement changes and then monitor and report order entry throughput times.

Production Planning

The purpose of applying JIT techniques in the production planning area is to establish an accurate production plan for factory operations that can be efficiently constructed and modified and that reflects consolidated user requirements in an understandable and timely fashion. For the purpose of this discussion, production planning is divided into three categories: (1) master scheduling, (2) production planning, and (3) capacity requirements planning (CRP).

Master schedule process. Based on the information obtained during the diagnostic review, the following steps can be used as a guide when implementing the master schedule plan:

1. Identify all queues, duplication of effort and/or data, and unnecessary activity. This data should have been obtained

during the diagnostic review.
2. Identify and select most efficient and effective processing alternatives to reduce or eliminate non-value-adding activities.
3. Write procedures and perform training.
4. Implement changes and then monitor and report master schedule throughput times.

Production planning process. Based on the concepts of work orders and lot sizes in a JIT environment and the information obtained during the diagnostic review, the following steps can be used as a guide when implementing the production plan:

1. Determine the role of production planners in a JIT environment, considering carefully the objectives established during conceptual design.
2. Identify any required changes in production order identification logic, order launching logic, and lot sizing logic.
3. Determine resources and time required to make appropriate systems changes and obtain necessary approvals.
4. Implement systems changes on a test system.
5. Perform user tests and obtain approval on systems changes in test system.
6. Write procedures and conduct user training.
7. Turn on the revised system and then verify output.
8. Analyze system change effects on production planning personnel levels and make necessary adjustments.
9. Monitor progress.

Capacity requirements planning. Based on the concept of CRP in the company's current environment and in a JIT environment, as well as the information obtained in the diagnostic review, the following steps can be used as a guide when implementing the CRP plan:

1. Define any required alternative methods for the identification of capacity requirements in a JIT environment at the company.
2. Select the most efficient and effective method of capacity requirements planning in a JIT environment at the company facility.
3. Determine input and output requirements and system change requirements for capacity requirements planning in a JIT environment at the company.
4. Determine resources and time required to perform appropriate systems changes and obtain necessary approvals.
5. Write procedures and conduct user training.
6. Implement systems changes on a test system.
7. Perform user testing and obtain approval on systems changes in test system.
8. Turn on the new system and then verify live system output.
9. Monitor and report progress.

Production Scheduling

JIT typically produces major changes in the way production priorities are developed (scheduled) and communicated. Scheduling requirements and associated changes are addressed in the module on production scheduling process. Communicating the production priorities through kanban or some other replenishment signal will be described in the section "Requirements Communication" later on.

Production scheduling process. Based on the concepts of production scheduling, order identity, and material movement as well as information obtained in the diagnostic review, the following steps can be used as a guide when implementing the production scheduling process plan:

1. Determine those activities that will need to be performed and who will be responsible for those activities as they pertain to production scheduling processes in a JIT environment.
2. Obtain management approval.
3. Write procedures and perform training.
4. Implement changes and then monitor and report progress.

Requirements communication. Based on the process flow information obtained during the diagnostic review, the following steps can be used as a guide when implementing the requirements communication plan:

1. Identify alternative mechanisms for accurate, timely communication of finished goods requirements to the appropriate subsequent assembly point. Select the most efficient alternative.
2. Identify maximum allowable queues between each operation and distances between successive operation workcenters. Items to consider are physical room available, amount of queue required to balance work load (UPL), and desired rate of improvement regarding WIP reduction.
3. Coordinate the design and procurement of containers that will facilitate flexible queues for level machine loading and that can be used to communicate requirements to preceding operations. When it is not possible to use empty containers to trigger preceding operation production, identify other more appropriate triggering mechanisms.
4. Obtain management approval.
5. Repeat previous steps for raw materials and/or purchased parts requirements.
6. Procure required containers and/or trigger mechanisms (kanbans).
7. Write procedures and train users.
8. Implement requirements communication devices and then monitor and report progress.

Purchasing

The purpose of a purchasing implementation plan is to establish an ongoing purchasing system that reduces throughput times, improves quality and reliability, and is generally consistent with JIT operating philosophies. Specific categories of activity included in this plan are buyer training, buyer and interface activity performance evaluation, purchasing organization, focus factory issues, purchased part/raw material requirements and ordering, procurement plan, vendor qualification, vendor certification, and vendor training.

Buyer training. Based on information obtained in the diagnostic review, the following steps can be used as a guide when implementing the buyer training section of the purchasing plan:

1. Identify current buyer competence levels with regard to general purchasing principles, purchasing policies procedures, departmental functions/responsibilities in those areas which have JIT purchasing interfaces, general production and inventory control (P&IC) principles, JIT techniques, value engineering, focus factory operations,

commodity knowledge, and vendor evaluation techniques (discussed subsequently).
2. Design specific training modules to address all appropriate areas.
3. Structure modules into training programs for individual buyers.
4. Assign responsibility for buyer training.
5. Prioritize buyer training requirements by criticality.
6. Implement training program.

Buyer and interface activity performance evaluation. Measurements are necessary to give the respective buyers a baseline from which to measure their progress. Because many of the buyers' actions are predicated on activities and decisions made by departments outside of their realm of responsibility, it is important to evaluate the information provided them from these interface departments. The following steps can be used as a guide during this evaluation:

1. Identify criteria to be used for buyer performance evaluation considering data gathered during diagnostic review and objectives established during conceptual design. Possible criteria include vendor delivery performance, lead-time reduction, cost reduction, early manufacturing involvement, inventory level reductions, part number reductions, vendor quality performance, supplier base reductions, supplier profit improvement, supplier visits, and professional education efforts.
2. Identify interface activities that require performance measurement. Activities that should be considered include forecast accuracy, lead-time violation from production control, design engineering responsiveness, accounts payable turnaround time and flexibility, and quality control responsiveness.
3. Identify mechanisms that will be used to monitor performance, watching for consistency with conceptual design objectives.
4. Identify frequency and format of monitors and reviews.
5. Obtain management approval.
6. Write performance evaluation procedures.
7. Implement buyer performance measures.
8. Monitor and report progress through achievement of conceptual design goals.

Purchasing organization. Because a JIT program will typically simplify/streamline a department, it is important to review how the purchasing organization will look in the future. Approval procedures will also typically change in this new environment. The following steps will address these issues:

1. Identify structure and reporting relationships of focus factories (if they are part of the conceptual design).
2. Determine reporting relationships of buyers in corporate, general factory, and focus factory environments.
3. Define and document approval processes and expenditure limits at each level within the reporting structures.
4. Assess the need for quality assurance and engineering personnel to report to purchasing.
5. Define the roles/responsibilities of any quality assurance and engineering personnel reporting to purchasing. The functions that should be considered are listed in Table 15-15.

Focus factory issues. If during the layout portion of conceptual design it is determined that a focus factory approach will be taken, the issues in Table 15-16 will need to be addressed with regard to the purchasing organization and its function.

Purchased part raw material requirements and ordering. The ordering process itself, once purged of duplicate/non-value-adding activities, should evolve into one of the most efficient means of communicating a manufacturer's needs to the respective vendor. The following steps review the information, then help determine the most appropriate/expedient means of passing the information on:

TABLE 15-15
Responsibilities of Quality Assurance and Engineering Personnel During the Purchasing Implementation Plan

Quality Assurance
• Implement and oversee the operations of the vendor certification program
• Coordinate new product launch to ensure manufacturing process capability at supplier
• Coordinate engineering changes to ensure manufacturing process capability at supplier
• Support ongoing supplier quality improvement program
• Identify and prioritize rejection data for summary review
• Coordinate supplier TQC training
• Conduct supplier visits with buyers to assist in qualification and selection
Engineering
• Coordinate new product launch
• Coordinate engineering changes
• Clarify engineering specs
• Clarify drawings
• Design simplification
• Act as liaison to supplier to ensure least-cost manufacturing
• Develop part number reduction program
• Ensure selected JIT suppliers are specified by engineering
• Act as liaison between engineering and suppliers
• Early manufacturing involvement participation

(Coopers & Lybrand)

TABLE 15-16
Issues to be Addressed in Purchasing Organization When Focus Factory Approach is Considered

Early Manufacturing Involvement (EMI) Program
• Define buyer EMI roles, responsibilities
• Perform any required buyer EMI training
• Write EMI procedures
• Initiate EMI activities
Purchasing organizational support
• Define roles of corporate buyers, centralized factory buyers, and focus factory (liaison) buyers
• Determine personnel requirements
• Obtain management approval
• Fill personnel requirements
• Assign responsibilities
• Write procedures
• Initiate activities
• Monitor and report progress as consistent with related conceptual design objectives

(Coopers & Lybrand)

1. Identify available forecast and order data. The types of data to include are minimum required forecast, hard requirements, and actual order horizons. It is also important to include current order entry to purchase order release cycle times and process components.
2. Assess current process to identify queues that could be reduced/eliminated.
3. Conceptualize alternative ordering methods and estimate attendant cycle times and costs.
4. Review order entry to purchase order release cycle time goals as established in conceptual design. Determine most effective and efficient alternatives, methodology, and specific processes.
5. Implement order processing and release improvements, including resource procurement, installation and interface construction, education and training, procedure writing, initial data load and testing, and cutover.
6. Monitor and report progress through achievement of objectives established during conceptual design.

Procurement plan. To develop an effective procurement plan within an organization, it is imperative to review the information gathered during the conceptual design. From this information, the manufacturer should gain insight on how parts can be grouped together and classified as specific commodities. Once the commodities are established, it is important to develop a plan regarding how they can be procured in a fashion that is beneficial to both the manufacturer and the respective vendor(s). The following steps can be used as a guide when implementing this program:

1. Identify existing commodities.
2. Review relevant data for each commodity gathered during the diagnostic review. Types of data required are shown in Figs. 15-10 through 15-12.
3. Document measurement methods for tracking data.
4. Assign commodity improvement responsibilities to buyers that are consistent with conceptual design objectives.
5. Establish commodity improvement goals that are consistent with the conceptual design.
6. Obtain management approval.
7. Initiate improvement group activities.
8. Establish progress measurement and reporting formats, time frames, and organizational details.
9. Monitor procurement plan and report progress.

Vendor qualification program. A vendor qualification program helps determine those vendors that will be considered as a supplier of JIT materials. The following steps can be used as a guide when implementing this program:

1. Identify/develop qualification requirements. A list of recommended vendor qualifications is given in Table 15-17.
2. Develop qualification checklist.
3. Establish qualification responsibilities.
4. Obtain management approval.
5. Write qualification procedures.
6. Implement procedures with any new vendors.
7. Prioritize existing vendors by importance.
8. Implement procedures with existing vendors in order of importance, in a manner consistent with any related objectives established during conceptual design.
9. Monitor and report progress.

TABLE 15-17
Recommended Qualifications for Vendors Being Considered as JIT Materials Supplier

- SPC/quality record
- Length of relationship
- Internal manufacturing controls
- Geographic location
- Delivery reliability record
- Financial stability
- Willingness to share data
- Willingness to make long-term commitments
- Ongoing quality improvement programs
- Ongoing VA/VE programs
- Delivery frequency/timing flexibility
- Vendor size with respect to proportion of vendor business
- Warranty and cooperation on rejected goods
- Capacity
- Willingness to aggregate into sets/kits
- Willingness to do special labeling
- Ability/willingness to do blanket purchase orders
- Ability/willingness to do paperless purchase orders (electronic data transfer)
- Willingness to do special containering
- Willingness to work on lead time reduction
- Current lead times
- Preventive maintenance practices
- Willingness to tie prices to independent indexes
- Willingness to accept consumption-driven receipt/payment
- Break-even point suitability
- Legal history/pending lawsuits

(*Coopers & Lybrand*)

Vendor certification. A vendor certification program helps determine those vendors who have achieved preferred vendor status. The following steps can be used as a guide when implementing this program:

1. Identify current (baseline) quality levels on purchased commodities from diagnostic review data.
2. Identify current (baseline) delivery reliability levels.
3. Establish ongoing quality and delivery reliability measurement devices.
4. Establish quality and delivery reliability goals consistent with the conceptual design.
5. Obtain management approval.
6. Write vendor certification (preferred vendor status) procedures.
7. Prioritize existing vendors by importance.
8. Implement procedures with existing vendors in order of importance.
9. Recognize preferred vendors. Possible criteria to use in this selection process are quality, responsiveness, technical competence, geographic location, size, current price, delivery, and packaging.
10. Monitor and report progress.

Vendor training. A vendor training program is used to help vendors achieve preferred vendor status and implement JIT in their own facilities. The following steps can be used as a guide when implementing this program:

1. Identify current vendor competence in the areas of policies and operations, JIT techniques, value engineering

techniques, and rough-cut capacity planning. Determine vendor training requirements.
2. Prioritize vendor training candidates by importance.
3. Design vendor training programs.
4. Obtain management approval.
5. Assign training responsibilities.
6. Implement training program.
7. Monitor and report progress.

Manufacturing

The purpose of applying JIT techniques to the manufacturing process is to match production rates to end demand. The objective is to produce products at the least possible cost, highest possible quality level, and at rates that are tied directly to demand. The specific categories of activities in the manufacturing area are the changeover time reduction process, changeover time reduction techniques, a quality improvement program, and process flow layout.

Changeover time reduction process. The following steps will help a manufacturer develop an effective ongoing program designed to continually reduce set-ups by identifying those machines, workcenters, and/or processes most critical to the success of a JIT implementation:

1. Obtain a listing of all current set-up and teardown times using data gathered during diagnostic review.
2. Sort the list by length of time, longest to shortest.
3. Analyze workcenter configuration to determine whether any parts of the set-up or teardown processes can be handled off-line without stopping the machine.
4. Analyze on-line set-up and teardown processes for potential simplification and potential utilization of quick-change fixturing.
5. Analyze set-up, teardown, and maintenance tooling locations for proximity to the machine or workcenter.
6. Determine the potential benefits of duplicate tooling or special (preadjusted or machine-specific) tooling.
7. Select the most efficient and effective set-up and teardown alternatives, utilizing off-line activities, simplification, quick-change fixturing, improved tooling locations, and duplicate or specialized tooling to reduce overall changeover times.
8. Identify required resources and personnel to perform the designated changes, and obtain approval.
9. Train the users, and correct production standards and costs where appropriate.
10. Implement the changes and then monitor and report progress.

Changeover time reductions through tooling and fixturing standardization and preventive maintenance techniques. To continually strive to simplify/standardize the changeover process, the following steps may be employed to establish a baseline for further improvement:

1. Identify all required fixtures and tooling for each operation performed at a given workcenter.
2. Analyze the fixtures and tooling for potential reductions through standardization or machine center design alteration. Enhance commonality of fixtures and tooling from operation to operation and part to part within workcenter wherever possible.
3. Identify potential applications of flexible machines and automation to retain multiple set-ups at the workcenter. Analyze cost and benefit potential of any flexible manufacturing systems (FMS).
4. Select the most efficient and effective standardization alternative for each workcenter's spectrum of production.
5. Develop a preventive maintenance program that promotes machine and tooling credibility. Some elements of this program could be:
   - Operator responsibility for general appearance of his or her machine.
   - Operator responsibility for guaranteeing that all required dies and bits are properly maintained prior to use.
   - Dedicated maintenance personnel to specific machines and focus factories.
6. Identify required resources and personnel to perform the designated changes and obtain approval.
7. Train the users and correct standards and costs.
8. Implement the changes and then monitor and report progress.

Quality improvement program. The purpose of a quality improvement program is to develop a plan to identify, eliminate, and prevent defects; zero defects is the ultimate goal. The following steps can be used as a guide when implementing this program:

1. Define required operator knowledge base, utilizing diagnostic review data and any other required information such as SPC, machine or process-specific training (including preventive maintenance), blueprint training, and general total quality control (TQC) principles.
2. Identify training personnel (trainers and training candidates).
3. Develop training package.
4. Obtain management approval.
5. Implement training package.
6. Monitor and report progress.
7. Develop quality "brainstorming" sessions with operators, quality and product engineers, and line supervision to address quality problems at their source.
8. Provide a vehicle to recognize individuals responsible for plotting defect data, identifying the defect, and providing the suggestion used to eliminate the problem.
9. Develop a program to coordinate the efforts of the direct labor quality data compilation. Such activities may include verifying the accuracy of the data, providing assistance for more detailed defect analysis, providing current techniques to resolve quality problems, and initiating an immediate feedback mechanism to the shop floor and/or vendor when specific defects or trends are appearing.
10. Establish a measurement mechanism to monitor the decrease or increase and severity of quality problems. All defect resolutions must be implementable without adding lead time to the current process, such as adding an inspection station or test machine.
11. Develop a package to continually cross-train all operators on different types of manufacturing skills. Keep a matrix chart listing all operators and their applicable skills in a highly visible spot on the shop floor.
12. Establish a communication link with all personnel responsible for the research and development of the product with the intent of establishing recommended

manufacturing practices. Typical elements could include:

- The utilization of as many standard off-the-shelf components as possible.
- The discouragement of unique parts except for technological advantages.
- The involvement of supervisors and/or operators in the design change (or implementation) process.
- A list of preferred manufacturing practices, such as use threaded fasteners only when absolutely necessary, avoid labels whenever possible, and develop operator time expenditures on preventive maintenance.

Redesign of process flow (manufacturing layout). The following JIT philosophies should be adhered to (when possible) prior to the development of the layout itself: minimum space allocation; inclusion of only those operations absolutely necessary to manufacture the product; linked and/or overlapped processes, operations, and/or machines; shortest possible throughput; operator and machine flexibility; cellular (serpentine) design; if appropriate, high visibility and effective operator communication. The following steps will help to accomplish the aforementioned goals:

1. Identify those operations that are absolutely necessary to the manufacture of the end product:
   - List all possible operations and processes for all parts which are to be manufactured on the line.
   - Collectively (manufacturing, engineering, and so on) review each operation and process. Then make a determination as to whether it adds value and must be retained, does not add value but must be retained (final testing), or does not add value and can be eliminated.
   - Collectively analyze how those operations or processes that do not add value can be eliminated.
   - Relist those operations and/or processes that have been identified as necessary for the successful manufacture of the end item.
2. Design the layout following steps outlined in Table 15-18.
3. Identify any new equipment that will be required to accommodate the new layout.
4. Based on the approved changes, develop a Gantt chart of the milestones and dates of all changes required for the new layout. Some of the things to include on the chart are the following:
   - What machines will be moved, when they will be moved, and who will be responsible for moving them.
   - What workstations will be moved, or if they are new workcenters, who will order them and when will they be delivered.
5. Determine how the schedule will be affected because of the move. Based on the Gantt chart of the facilities move, determine what machines will be inoperable and for how long.
6. Determine how to compensate for adverse impacts on the schedule.
7. Identify improvement potential from new layout. Typical improvements would include the reductions in space requirements, throughput times, WIP, and personnel. Improvements in meeting daily schedules will also occur.

TABLE 15-18
Recommended Steps for a JIT Plant Layout

1. With the schedules and rates known for all of the processes identified as "necessary," the actual layout planning can begin by determining what machines are required to produce the end item, how many of each machine is required to produce the schedule, and how many workstations are required for hand assemblies
2. Link (on paper) all appropriate workstations together in a straight line in a sequential manner from beginning to end. This identifies all physical processes that will be included in the JIT layout
3. Identify all subassemblies that are included in the manufacture of the end item
4. Identify all processes included in the subassembly and sequentially list them from beginning to end
5. Determine whether the subassembly can be moved from its current physical manufacturing location to the end item manufacturing location
6. Link the completion of the subassembly directly to its point of use on the final assembly line
7. Identify what major components (bulk items) will be purchased or delivered from other areas of the factory
8. Determine how many hours or days of stock will be physically contained in the process
9. Calculate how much space will be required to accommodate the components
10. Determine whether a single major drop area (storeroom) is needed for the layout or if several smaller drop areas will be used for specific components, or a combination of the two
11. Design a layout that allows for easy access to other workstations and provides an outlet for worker communication
12. Locate component part drop areas as close to their point of use as is functionally possible
13. Where possible, design the layout with the machines and workstations placed together in a cell design
14. Where possible, link the operations together to allow the operator of the subsequent operation to reach the part without moving from his or her workstation
15. Design the layout to keep the travel distance between various workstations at a minimum, keeping operators within the cell
16. Design the layout to allow for easy access to components
17. Design the workstations and (if possible) the machinery to be functionally mobile. Design individual workstations which can be added to or taken from the assembly process commensurate with the daily and/or weekly schedule

(*Coopers & Lybrand*)

8. Execute the Gantt chart activities, monitoring and reporting progress.
9. Identify all workcenters and attending operators.
10. Identify all required tooling and equipment for each workcenter.
11. Develop a housekeeping training program for operators and perform training.
12. Review workcenter listing of necessary items with operator and revise as required.
13. Implement operator-controlled workcenter housekeeping program.
14. Monitor housekeeping and recognize operators with outstanding housekeeping performance.

Distribution

The purpose of the distribution plan is to distribute finished goods to the customer as quickly and as efficiently as possible, incorporating JIT techniques to eliminate queues, duplication, and unnecessary effort at each step in the distribution of goods. The distribution plan is broken into philosophy and process.

The distribution philosophy. The following steps will allow a manufacturer to develop a clear understanding of a customer's base and how one's products are distributed to them. This information should enable a company to develop a program that allows it to satisfy the customer's requirements more efficiently and effectively.

1. Identify all current carriers.
2. Perform distribution analysis on all current customers' products and producers (see Table 15-19).
3. Identify contribution data by product, product group, customer type order size, geographic area, and distribution channel.
4. Rank contribution list in dollar-descending order.
5. Determine percent of contribution of entire product line for each item.
6. Identify high and low-contribution producers.
7. Determine most efficient and effective carriers, product distribution mix, width of distribution, length of distribution channel, location(s) for distribution facilities, and methods of distribution by applying JIT principles to the data accumulated.
8. Design improved distribution system philosophy and basic operating methodology.
9. Identify and document improvement potentials.
10. Obtain approval for changes.
11. Implement changes.
12. Monitor and report progress.

TABLE 15-19
Items to Consider During Distribution Analysis

Customer Considerations
• Quantity, type of customer
• Location of obvious territories
• Service requirements
• Average order sizes
Product Evaluation
• Definition of product/price
• Nature of product (perishability, and so on)
• Use of the product
Producer Review
• Location of producer
• Trade agreements, contractual requirements, and so on
• Organization's technical capabilities
• Relationship to competitors

(*Coopers & Lybrand*)

The distribution process. To streamline/standardize the distribution process itself, it is important to purge all unnecessary (non-value-adding) activities from the current day-to-day activities and concentrate on only those activities that are deemed necessary to perform this function. The following steps can be used as a guide when implementing this program:

1. Identify and document current distribution activities, flow, and responsibilities.
2. Identify all queues, duplication of effort and/or data, and unnecessary activity in the distribution process.
3. Select the most efficient and effective alternatives.
4. Document improvement potentials and obtain any required approvals.
5. Write procedures and perform training.
6. Implement changes and then monitor and report distribution throughput times.

Accounting

To properly deal with accounting issues in a JIT environment, it will be necessary to review a number of traditional accounting perspectives in areas such as WIP inventory valuation, incremental labor activity costing, capital equipment justification, and so on. In addition, the regular accounts payable, accounts receivable, and cost accounting processes will need to be reviewed and purged of wasted time and effort.

Accounts payable philosophy. Based on the information obtained during the accounting review, the following steps can be used as a guide when implementing the accounts payable philosophy part of the accounting plan:

1. Analyze each commodity in the current accounts payable system to determine which ones would be most conducive to the elimination of purchase orders, invoices, and receiving transactions. (Identify transaction volumes and points of receipt and use to support this analysis as required.)
2. Identify alternative accounting methods for the remittance of payment regarding purchased goods.
3. Select the most efficient and effective alternative, facilitating the elimination of purchase orders, invoices, and receiving transactions wherever possible.
4. Document improvement potentials, and obtain any required approvals. Include any potential head-count reductions, interface changes, and systems support requirements.
5. Write procedures and train users.
6. Implement changes.
7. Verify accuracy of system data following any required manufacturing information system (MIS) changes.
8. Monitor progress.

Accounts payable process. The following steps can be used as a guide when implementing the accounts payable process part of the accounting plan:

1. Identify and document current accounts payable activities, flow, and responsibilities.
2. Identify all queues, duplication of effort and/or data, and unnecessary activity in the accounts payable process.
3. Identify alternatives to the steps in the current process that will potentially reduce and eliminate queues, duplication, and unnecessary activity.
4. Select the most efficient and effective alternatives.
5. Document improvement potentials and obtain any required approvals.
6. Write procedures and perform training.
7. Implement changes.
8. Monitor and report accounts payable throughput times.

Cost accounting philosophy. The following steps can be used as a guide when implementing the cost accounting process part of the accounting plan.

1. Educate the cost accounting contingent of the JIT implementation team about the cost accounting issues associated with just-in-time. Some of the issues to include are the:
   - Elimination or modification of stock issue transactions on the shop floor.
   - Costing by flow or process as opposed to discrete production stages and/or workcenters.
   - Drastic reductions of WIP inventory valuation.
   - General "flattening" of structured bills of material.
   - Substantial reduction in product rework levels.
   - JIT impact on capital investment decisions.
2. Establish an interdisciplinary team, led by cost accounting, to address each of these issues in the context of the newly defined JIT environment. Among the questions to be considered by the team are the following:
   - What units of the process will be costed, and to what extent will they be aggregated in terms of labor and WIP inventory?
   - Are storage areas required, and if so, how will they be arranged to facilitate costing (visibility anu fixed quantities)?
   - Is it possible to eliminate stock issue transactions by the elimination of storage areas, fixed stores quantities, or some other method? If so, how will inventory accountability be verified and reported, and what systems changes would be required?
   - Will a traditional "role-up" of costs be required, and if so, how will the process change with alterations in incremental measurements and flattening of bill of material structures?
   - How should rework costs be applied when they are rare and are readily traceable to a specific source?
   - How should capital equipment decision making change as equipment investments begin to affect and include material transport systems, robots, and computer hardware and software?
   - How, where, and when should pay mechanisms change from piecework to hourly rates and/or possible group incentives?
   - What will be the impact be of substantially reduced downtime costs and machine set-up costs?
3. Identify and document improvement potentials, taking into account needed systems changes and all other resource requirements. Delineate costs and benefits. Assess impact on interface activities.
4. Obtain approval for changes.
5. Implement changes.
6. Verify the accuracy of cost system data following any required MIS changes.
7. Monitor progress.

Accounts receivable and cost accounting processes. The following steps can be used as a guide when implementing these parts of the accounting plan:

1. Identify and document the current accounts receivable and cost accounting activities, frequency, flow, and responsibilities.
2. Identify all queues, duplication of effort and/or data, and unnecessary activity in the processes. Include any unnecessary reports.
3. Identify alternatives to the steps in the current processes that will potentially reduce and eliminate queues, duplication, and unnecessary activity.
4. Select the most efficient and effective alternatives.
5. Document improvement potentials and obtain any required approvals.
6. Write procedures and perform training.
7. Implement changes.
8. Monitor and report throughput times.

IMPLEMENTATION

It is useful to perceive actual implementation activities in two steps: (1) pilot line implementation and (2) remaining factory conversion.

The pilot line approach to just-in-time implementation focuses initial activities in a relatively easily controlled environment. It minimizes disruption to total factory operations while providing a highly visible incubator for "hatching" and developing improvements and solving problems.

Pilot lines are most successful when they encompass the entire manufacturing process of at least one product family. It is best to keep pilot line production volumes at or under 20% of total production because start-up is usually slow and many problems will surface.

Because the pilot line is a microcosm (in most cases) of total factory operations, the problems that arise and are resolved are typical of what can be expected during remaining factory conversion. Therefore, it is also reasonable to expect that the implementation and primary revisions of the pilot line will usually require more time than subsequent line start-ups. This time differential will vary with product and process similarity.

The pilot line implementation will involve most of the major layout decisions, initial designs of pull mechanisms, quality controls, and housekeeping practices that will eventually be adopted in some form throughout the factory. It also provides an incentive for management to begin wrestling with such concepts as cost accounting, engineering, and purchasing in a JIT environment.

The remaining factory conversion aspects of implementation involve leveraging the technology developed during pilot line operations into more or less similar lines throughout the rest of the factory. Like any other systems implementation, there is a "cutover" period during and beyond which dual support systems (such as cost accounting) may need to operate in parallel. This will almost certainly be the case when JIT pilot lines and conventional production areas are operating simultaneously. Frequently, the remaining factory conversion process will encompass most, but not all product lines. There are often a few products, representing very little volume, that are very different in production process from the rest of the products. These products can be isolated from the JIT environment and produced as they always were until decisions are made as to what action is appropriate. They may eventually be produced through JIT methods or even dropped entirely.

CONTINUOUS IMPROVEMENT

When conceptual design and implementation objectives have been reached, it becomes all too easy to rest on the successes of the program and allow JIT activities to taper off. As noted at the beginning of this chapter, just-in-time is a program with no end. It is a continuous process, *not* a project with a completion date. However, as objectives are reached, new targets for improvement must be established. In many cases, the efforts are refocused, and organizations within the company must evolve to deal with these changes.

At the end of the remaining factory conversion, the entire organization from forecasting through distribution should be more streamlined and efficient. In successful JIT programs, one of the challenges at this stage of the process is dealing with the diversity of activity levels and directions. Some groups and individuals still in the throes of success will display nearly wild enthusiasm and require firm direction and control. Others will simply be tired of the continual strain of waste elimination and productivity improvement efforts. One of the best ways to deal with this situation is to allow the organization controlling JIT efforts to evolve, redistributing people to new groups that are more closely oriented to the new environment. The new group structures will generally mix wildly enthusiastic members with other personnel and allow groups to be more closely aligned with the issues of the daily work environment.

For example, when the organization is comprised of several focus factories, a structure incorporating the concept of focus factory improvement groups (FFIG) may be utilized (see Fig. 15-21). The purposes of the FFIG are to identify problems in the focus factory, prioritize the problems, assemble task groups to deal with the problems, monitor task group activities through problem resolution, and dissolve the task groups. The FFIG is, in effect, a problem-resolution steering committee. It is typically chaired by a representative of manufacturing from the focus factory involved and contains representatives from purchasing, quality assurance, engineering, and other disciplines.

In terms of manufacturing problem resolution, workcell employees may be organized into quality circles to address quality issues or potential improvements in cost reduction and throughput time. By organizing these groups around workcenter locations, meeting times can be more readily determined that are conducive to all members of the group. Also, common problems are more quickly identified and resolved.

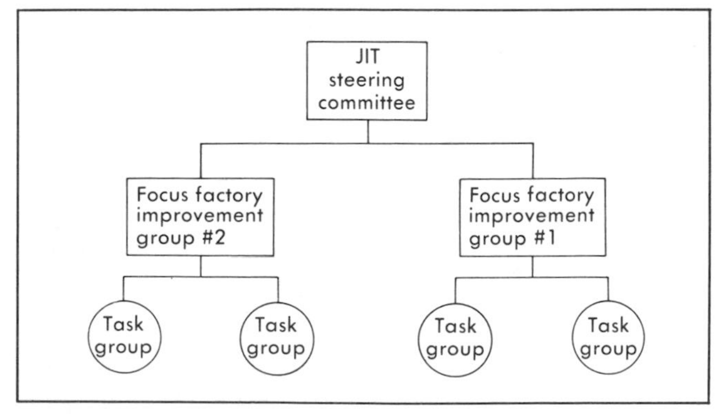

Fig. 15-21 Sample organizational diagram for focus factory improvement groups. (*Coopers & Lybrand*)

Beyond the evolution of the JIT organization and administration, continuous improvement activities revolve around the following:

- Problem identification and resolution.
- Constant revision and pursuit of goals and objectives related to quality-level improvement, queue reduction, and throughput time reduction.

These activities may be structured under separate programs, such as a preventive maintenance program, an SPC program, a part number reduction program, or they may be folded into the overall JIT program as subelements. Regardless of the approach, it will be important to maintain an oversight entity such as the JIT steering committee to coordinate all of the activities and provide status reports to senior management.

The underlying point to the continuous improvement phase is simply that the improvement activities are to be pursued in a continuous and organized fashion. The organization must always strive to perform its functions, from forecasting through distribution, with less and less waste.

References

1. Albert G. Wordsworth, "Survival," *Strategic and Tactical Issues in Just-in-Time Manufacturing*, 1985 Conference Proceedings (Wheeling, IL: Association for Manufacturing Excellence, Inc.).
2. Michael J. Rowney, "Omark Industries' Zero Inventories Production System (ZIPS)," *Strategic and Tactical Issues in Just-in-Time Manufacturing*, 1985 Conference Proceedings (Wheeling, IL: Association for Manufacturing Excellence, Inc.).
3. Raymond L. Rissler, "The General Electric Dishwasher Plant at Louisville, Kentucky," *Strategic and Tactical Issues in Just-in-Time Manufacturing*, 1985 Conference Proceedings (Wheeling, IL: Association for Manufacturing Excellence, Inc.).
4. Thomas A. Gelb, "The Material as Needed Program at Harley-Davidson," *Strategic and Tactical Issues in Just-in-Time Manufacturing*, 1985 Conference Proceedings (Wheeling, IL: Association for Manufacturing Excellence, Inc.).
5. William A. Wheeler III, *Straight Talk on Just-in-Time* (New York: Coopers & Lybrand).
6. Christine Ammer and Dean S. Ammer, eds., "Hawthorne Studies," *Dictionary of Business and Economics* (New York: The Free Press, 1984).
7. Jacob Frimenko, *Statistical Process Control: Fundamental Concepts*, SME Technical Paper MF84-472 (Dearborn, MI: Society of Manufacturing Engineers, 1984).
8. Edward J. Hay, "Reduce Any Setup by 75%," APICS Seminar Proceedings.

COMPUTER-INTEGRATED MANUFACTURING

INTRODUCTION

CHAPTER CONTENTS:

The importance of computer-integrated manufacturing (CIM) to the future of U.S. manufacturing cannot be overstated. It is a key ingredient in improving the productivity, efficiency, and profitability of the U.S. industrial base and in regaining a competitive position in the world marketplace.

Computer-integrated manufacturing is the view of manufacturing that recognizes that the different steps in the development of a manufactured product are interrelated and can be accomplished more effectively and efficiently with computers. These relationships are based not simply on the physical part or product being produced but also on the data that define and direct each step in the process. Controlling, organizing, and integrating the data that drive the manufacturing process through the application of modern computer technology effectively integrates all the steps in the manufacturing process into one coherent entity. Such integration should yield efficiencies not possible from a more segmented approach to manufacturing.

Although CIM implies integrating *all* steps in the manufacturing process, in practice many companies have achieved significant benefits from implementing only partial CIM systems (those in which only some manufacturing steps are integrated). In fact, it is likely that no company has achieved full integration to date. Both situations—those in which all or only some manufacturing sequences are integrated—are accurately called CIM systems.

One of the pioneers in CIM was Joseph Harrington, Jr., who presented his concept of CIM in the book entitled *Computer Integrated Manufacturing*.[1] Dr. Harrington saw an increased understanding of the science of manufacturing, which the CIM concept supports, as key to the industrial health and competitiveness of the nation. Now, private industry, educational institutions, and governments worldwide have come to recognize the value of CIM. Some of the more progressive companies are already enjoying the benefits of successful partial CIM implementation. Many others are struggling to manage the technological and business implications of a manufacturing philosophy that defines manufacturing as a continuum encompassing the entire gamut of activities from product concept to maintenance of past products in the field.

This chapter will discuss the importance of CIM, its current state of implementation, and its component technologies and their interaction in the CIM environment.

DEFINITION OF CIM

To translate CIM from a manufacturing theory to a manufacturing tool, a comprehensive definition is required. The interrelationships that exist in any manufacturing environment are many. Technical disciplines such as design engineering and manufacturing engineering interact; departmental issues such as marketing schedules and production schedules interact; and business interests such as financial concerns and strategic concerns interact. All these relationships and many more must be part of a workable definition of CIM.

The National Research Council defines CIM as follows:

> CIM includes all activities from the perception of a need for a product; through the conception, design, and development of the product; and on through production, marketing, and support of the product in use. Every action involved in these activities uses data, whether textual, graphic, or numeric. The computer, today's prime tool for manipulating data, offers the real possibility of integrating the now often fragmented operations of manufacturing into a single, smoothly operating system. This approach is generally termed computer-integrated manufacturing.

Because of the complexity of CIM, some organizations that promote the understanding and

***Contributors of sections of this chapter are: Ted J. Egan**, Senior Consultant, Div. of American Financial Consulting Co., Greenwich Technologies; **Alice Greene**, Senior Consultant, Arthur D. Little, Inc.; **Dr. Charles M. Savage**, CIM Consultant, Digital Equipment Corp.*

***Reviewers of this chapter are: Paul Brauninger**, MIS Manager, Cone Drive Operations Div., Textron; **Donald B. Ewaldz**, Director, Ingersoll Engineers, Inc.; **Tom Gunn**, National Director of Manufacturing Consulting Group, Partner—Arthur Young & Co.; **Fred W. Jones**, Computer Aided Manufacturing Program Manager, Development Div., Martin Marietta Energy Systems, Inc.; **Steve Miller**, Assistant Professor, Graduate School of Industrial Management, Carnegie-Mellon University; **George P. Sutton**, Associate Division Leader, Material Fabrication Div., Lawrence Livermore National Laboratory; **Hayward Thomas**, Consultant, Retired—President, Jensen Industries; **Joseph Tulkoff**, Director of Manufacturing Technology, Lockheed-Georgia Co., A Div. of Lockheed Corp.; **Roger G. Willis**, Partner—Management Information Consulting Div., Arthur Andersen & Co.; **Dr. George M. Yaworsky**, Consultant.*

CHAPTER 16

DEFINITION OF CIM

implementation of CIM have found it useful to define CIM graphically by diagramming how technologies and disciplines work together as a unified whole. One of the most inclusive representations is the CIM Wheel developed by the Computer and Automated Systems Association of the Society of Manufacturing Engineers (CASA/SME), as in Fig. 16-1.

The CIM Wheel is composed of five fundamental dimensions:[2] (1) general business management, (2) product and process definition, (3) manufacturing planning and control, (4) factory automation, and (5) information resource management. Each of these five dimensions is a composite of other more specific manufacturing processes that have shown a natural affinity to each other. As a result, each dimension is seen to be a family of automated CIM processes that has emerged naturally because of an affinity among what used to be independent, stand-alone "islands of automation."

The general business management family of processes is arrayed around the periphery of the Wheel. Although seen as an integral part of the manufacturing enterprise, this family of processes was viewed as being the primary link between the rest of the enterprise and the outside world. The general business management family of applications includes a wide range of automated processes from general and cost accounting to marketing, sales, order entry, human relations, decision support, program scheduling, cost status reporting, and labor collection.

The second, third, and fourth families have been arrayed as thirds of the inner circle of the CIM Wheel. The processes represented in these families, however, do not occur in series, moving clockwise from the product and process definition dimension. In reality, all these activities are happening at the same time. It is important to recognize that even though the individual processes in each family of manufacturing processes have a natural affinity for one another, they also are closely tied to processes in other families represented in the Wheel. Most of the interfamily integration occurs through information resource management, the Wheel Hub. Because of the way the technology has grown in most manufacturing environments, *inter*family integration is much more difficult than *intra*family integration.

In many ways, the hub is the most difficult part of the Wheel to comprehend. It has to do with a set of functions called information resource management (IRM).

In the 1980s, information has emerged to be a major management issue in the manufacturing enterprise. In previous decades, information management was thought to be a problem that had to be solved by each organization for itself. In CIM, information is viewed as a management opportunity for the enterprise as a whole. A tremendous amount of time and attention is being devoted to understand exactly what manufacturing information is and to developing new and better techniques and tools for managing it.

At the heart of CIM information resource management is the concept of data management. Manufacturing productivity is believed to be linked to the notion of shared or common data, especially between engineering and manufacturing. One of the objectives of CIM information resource management is to break down the walls that have traditionally existed between those two organizations, rendering them, in effect, into one organization through a common database. There are basically two aspects to CIM information resource management. One aspect is intangible and the other is tangible. The intangible (or logical) aspect is the information itself. The tangible (or physical) aspect includes the tools of information such as the computers themselves and also the disk storage devices, printers, plotters, tapes, communication devices, operating systems, and terminals that are used to store, exchange, and process the information.

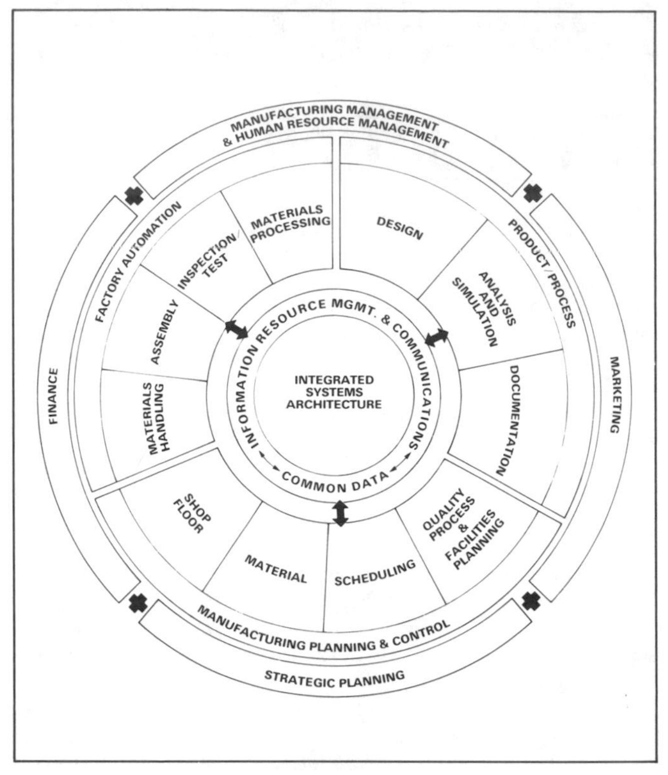

Fig. 16-1 The CIM Wheel emphasizes that CIM encompasses the total manufacturing enterprise. (*CASA/SME*)

This graphical description of CIM illustrates how the manufacturing process is a single entity with many segments and subsegments that do not and cannot operate without some impact from or to other segments. This is true whether or not a computer is part of the manufacturing environment.

Even in a nonautomated manufacturing environment, these same segments exist and are related to each other. No product or part has ever been physically made without its design having preceded it. That design could have existed in the mind of the individual making the object, as a sketch on a sheet of paper, or as a geometric model in the database of a computer-aided design (CAD) system. The generic functions of design and manufacture are interrelated. This interrelation is further discussed in Chapter 13, "Design for Manufacture," of this Handbook volume.

Similarly, material movement and material management are interrelated whether they are accomplished by someone carrying a piece of metal or a box of bolts to a workbench where the two will be combined or by a material requirements planning (MRP) program generating a pick list and sending it electronically to an automated storage and retrieval system (AS/RS) that, in turn, directs the physical movement of the selected parts to the appropriate workstation.

Individual segments of the manufacturing process are interdependent. In the CIM concept, computer technology serves as the integrating mechanism that connects these segments so that their inherent data-driven interactions can be exploited and streamlined to the benefit of the aggregate manufacturing enterprise. Definitions of CIM, both graphic and verbal (textual), are an attempt to describe the interrelationships that exist in manufacturing and how computer technology can combine all aspects of the manufacturing process into a smoothly running whole. Graphic definitions also provide a visual framework against which new developments can be understood.

DRIVING FORCES FOR CIM

For many years following World War II, the United States enjoyed the position of being the world's leading manufacturing country. Mass production techniques had revolutionized manufacturing, enabling fast and consistent production of complex products greater than had been possible before. Abundant human resources and raw materials put the U.S. comfortably ahead of its worldwide competitors. Some would say that the U.S. manufacturing community allowed itself to feel too comfortable for too long.

The changes in manufacturing in the 1950s and 60s were many; they were market related, technology related, and managerial. Markets became global. Countries smaller than the U.S. realized that an important key to becoming a world power was a strong economy. A strong economy meant a strong industrial base, one that relied on a healthy export level. Several countries, Japan being the most notable, focused their research, engineering, and manufacturing and substantial government resources to producing high-quality products that could compete in world markets.

Technology also changed the nature of manufacturing. Computers moved out of the accounting office to the manufacturing floor. Numerical control was introduced, although it was expensive and clumsy at first. Airframe manufacturers experimented with computer-aided design. Production and inventory control systems began to tackle material management. Advances in computer technology came at a pace too fast for all but the most sophisticated manufacturers to keep up with.

Traditional management techniques struggled to cope with increasing competition from all over the world and with the rapid pace of technological improvements. Many companies floundered and fell victim to the reality of a growing worldwide, technology-based manufacturing environment. Some entire industries moved offshore. As U.S. manufacturing entered the 1970s and 80s, it faced some serious handicaps. Higher labor rates made it difficult to compete with lower cost producers. A tough regulatory environment demanded investments for safety and pollution control, while inflation and the high cost of money made modernization capital harder to find.

The most sophisticated manufacturers were already addressing these handicaps. Greater productivity and improved quality was necessary to match foreign producers, and a greater and faster responsiveness to market needs was necessary to regain a leadership position. These are the driving forces that caused the CIM pioneers to rethink their business strategies, to adopt a long-term business perspective with improved manufacturing efficiency as a priority, and to use technology as a tool for change.

Essential to the drive to revitalize manufacturing was and is the need for more timely, accurate data: market data, design data, manufacturing data; data on material availability; data on work in process; data on production machinery availability; data on shop floor problems; and data to enable management at all levels to make informed decisions. Recognizing the data-driven nature of manufacturing and the power of computer technology to capture, analyze, manipulate, transmit, and communicate this data made computer-integrated manufacturing a logical avenue for foresighted organizations to pursue.

INDUSTRIES IN THE FOREFRONT OF CIM IMPLEMENTATION

The early adopters of CIM technology charted new waters in manufacturing. Some of them succeeded, while many others failed. The successful companies have been repaid with increases in product quality and productivity, reductions in inventory and product lead time, and many, many more benefits that are qualitative as well as quantitative. Perhaps the principal benefit cited by many of these successful pioneers is a business that is able to succeed despite increased competition.

The first industries to take advantage of the potential benefits of partial CIM implementation were aerospace, off-highway equipment, electronics, and automotive, as well as several government facilities. Although the relative size of the companies within these industries varied, as did the number and type of products produced, they all had pressures to be competitive, productive, and responsive. The early participants were companies with the knowledge and the resources to invest in a new methodology. They believed in CIM and planned a systematic approach to its implementation. In addition, the more successful companies had specific improvement areas that were targeted. To this day, larger companies have more readily accepted CIM, although the benefits of CIM apply equally to smaller manufacturers.

Although no company has successfully implemented CIM in its entirety, many companies have made great strides toward that goal. In fact, partial CIM systems have proven very cost effective for many manufacturers. Partial CIM systems may be: those in which a number of key functions are supported by stand-alone computer systems or are still performed manually; those in which there is only some shared data and it is not widely shared throughout the organization; or those in which the integration is limited to linking only some of the functions of engineering, design, manufacturing, marketing, purchasing, and distribution. Essentially all CIM systems today fall into the latter category. For many companies this is a useful and viable solution that is usually easier to implement and less expensive than full CIM implementation.

A 1984 study by the Committee on the CAD/CAM Interface, formed by the Manufacturing Studies Board of the Commission on Engineering and Technical Systems of the National Research Council, described the experience of leading companies that have made significant progress toward integration.[3] In this study, the CIM efforts of five companies were examined. The benefits realized by these companies during the integration process are shown in Table 16-1. These results are preliminary, representative of 10 to 20-year efforts at integration. Further benefits are expected to accrue as companies approach full integration.

It should be noted that not all the improvements described in Table 16-1 are entirely due to CIM. Some would have occurred over time as a result of improved manufacturing practice, without the application of computer technology.

In contrast to the benefits achieved by successful CIM implementation are the examples in which CIM implementation has failed. Most CIM failures have resulted from insufficient planning, commitment, preparation, and training. Some guidelines for successful implementation are presented later in this chapter.

TABLE 16-1
The Benefits of CIM Implementation

Reduction in engineering design cost	15-30%	Increased productivity of production operations (complete assemblies)	40-70%
Reduction in overall lead time	30-60%	Increased productivity (operating time) of capital equipment	2-3 times
Increased product quality as measured by yield of acceptable product	2-5 times previous level	Reduction of work in process	30-60%
Increased capability of engineers as measured by extent and depth of analysis in same or less time than previously	3-35 times	Reduction of personnel costs	5-20%

CIM TECHNOLOGIES

The technologies or building blocks that comprise CIM are discussed in this section in four groups: (1) those that relate to the beginning of the product cycle, (2) those that relate to the physical manufacture of the product, (3) those that plan and control the manufacturing process, and (4) those technologies that tie all the others together. It is important to remember that although each technology is discussed as an entity, its operation in a CIM environment enhances the benefits the technology has to offer a manufacturer.

While this section will discuss all of the major CIM technologies, the discussions on just in time (JIT) and manufacturing resource planning (MRP II) will be limited to the role of these technologies in a CIM environment and their importance to the overall CIM concept. Additional information on JIT can be found in Chapters 15, 20, and 21. MRP is discussed more fully in Chapter 20.

BEGINNING THE PRODUCT CYCLE

Computer-aided design (CAD) is the principal technology that begins a product design cycle. Computer-aided design actually comprises a number of technologies involved in the creation and analysis of a design—whether that design is a three-dimensional part to be machined, a printed circuit board, or a plant layout. In the CIM environment, the design data generated on the CAD system is used by other CIM technologies in the manufacture of the product.

Once largely used as a drafting tool, CAD has become a pivotal technology for those who use it properly. The design data generated on the CAD system interacts with numerous other automation systems in a modern manufacturing company. And the part geometry captured in the system's engineering database can be used to design the physical work environment in which that part will be produced, including fixturing, tooling, and transportation equipment. Some CAD systems can, as well, simulate the operation of the workcell that will produce the part and generate the required machine tool and robotic programming.

Computer-Aided Design

Computer-aided design systems are often referred to as computer-aided design/computer-aided manufacturing (CAD/CAM) systems. However, because this technology performs primarily a design function, the systems are more appropriately called CAD systems. In fact, the majority of installed systems are used primarily for the computer-aided drafting application and have little interaction with manufacturing.

Computer-aided design equipment allows a designer to create images of parts, integrated circuits, assemblies, and models of virtually anything else—from molecules to manufacturing facilities—at a graphics workstation connected to a computer. These images become the definition of a new design, or the modification of an existing one, and are assigned geometric, mass, kinematic, material, and other properties simply by the user interacting with the computer.

The image on the screen replaces the paper on the drawing board. The data that describes and defines the geometry displayed on the screen is calculated by the computer and stored in the computer's database. CAD allows the designer to be more creative by making it easier to experiment with many different designs and to be more efficient by using existing design data already stored in the same computer database.

In simpler applications, CAD reduces the amount of paperwork for drafting and design. Complete engineering drawings can be created without ever lifting a pen. The savings in drafting time and expense alone have often paid for a CAD system in a very short time. More important than the reduction in the direct cost of drafting is the reduction in the possibility for error that exists when different versions of a drawing are circulating throughout an organization.

A number of types of CAD systems are available. They range from small, two-dimensional systems that simply automate the drafting function to three-dimensional systems that create geometry from lines, surfaces, and solid shapes. Generally, CAD is composed of four distinct functional fields: (1) geometric (product) modeling, (2) analysis, (3) testing, and (4) drafting.

Computer-aided design systems have evolved over the years. Ten years ago, a prospective CAD user had to either buy a turnkey solution (typically minicomputer-based hardware and software) or to construct a system using a mainframe computer, a limited spectrum of licensed software, and internally developed software.

Today, personal computer-based systems offer most of the simpler mechanical capabilities required by small manufacturing companies. Workstation-based systems put the analytical power of a minicomputer (32-bit) at the disposal of an individual engineer. Turnkey solutions are still available, although turnkey offerings have been augmented with both personal computer and workstation features. The mainframe has been recognized as the necessary database manager for the engineering data generated on any CAD system, and almost all systems can provide a communications path to a mainframe computer.

And a variety of third-party software is available to run on almost any imaginable configuration.

The broadest breakdown of applications for CAD includes mechanical design; electronics design; architecture, engineering, and construction (the AEC industry); and mapping. The mechanical design area has been the largest single application since CAD was introduced. It will continue to represent the largest single market share for many years. Figures 16-2 and 16-3 represent the breakdown of CAD applications.

Mechanical CAD systems. Mechanical design software allows a user to translate ideas on a product design into a geometric model of that design through the use of interactive graphics. The model can be created in two, two and one half, or three dimensions, depending on the capability of the CAD system and the type of part to be modeled. While two dimensions are usually adequate for flat parts, two and one half give more information for parts that don't need the side walls detailed; the most sophisticated modeling is done in three dimensions. Because most mechanical parts exist in three dimensions, three-dimensional models are required to adequately describe their features. Three-dimensional capability is also necessary for multiaxis machining, and it gives better accuracy for subsequent NC programming.

Geometric information is stored in a database that can then be used for other manufacturing functions by other software. The geometric database is critical to the implementation of the CIM concept through proper database and data flow management. In the ideal CIM environment, the same data that describes one part, for example, can be used to create the sequence of operations that will make that part; the data can be used to design the fixtures that will hold the part; and the data can input other programs to aid in scheduling shop floor operations or in ordering raw materials. In other words, the data can be used not only by design engineering, but by manufacturing, administration, purchasing, and maintenance personnel. And because that data is entered once and is invariant, the opportunity for error throughout an organization is dramatically decreased.

CAD systems may even change the way a designer approaches the design process. Traditionally, a designer with pen and paper would draw an image of the design in multiple (three) views. By looking at these views and synthesizing the information they provided about the design, the designer could visualize the part that would result from that design. Using the CAD system simplifies this process by presenting the designer with a pictorial view of the part as it is created. It also expedites the design process by performing the synthesis directly. As a result, the designer not only saves time, but has a much better understanding of the part. If that part is a complex casting or forging, the designer can have much greater certainty that the design incorporates all the required features. Moreover, the design concept can be more quickly and easily communicated to other functions that need to use the design.

Computerized systems greatly reduce the time to create a new design. They allow the user to view the design from a number of perspectives, machine it on the screen, create an NC program to produce the part, and test that program without developing a prototype part. The availability of multicolor graphics displays allows a user to distinguish different areas of the design, to distinguish the tool path from the part, or to differentiate mating parts of an assembly. CAD can also be used to design the tooling and fixtures that hold a part in place on a machine tool bed.

Solids modeling is a relatively new feature of CAD systems. This capability goes well beyond the visualization of a solid shape on the screen. Solids modeling provides the most complete geometric and mathematical description of a part geometry possible to date. This data is particularly important if the model is to be used as the basis for computer-aided engineering (CAE), for generating mass properties of the part, or for generating the NC data to machine the part.

Electrical/electronics CAD systems. The electrical/electronics design area includes the printed circuit board (PCB), integrated circuit (IC), large scale integrated circuit (LSI), and very large scale integrated circuit (VLSI) industries. As recently as 5 years ago, this area was served mostly by some of the two-dimensional CAD systems available. These systems largely concentrated on

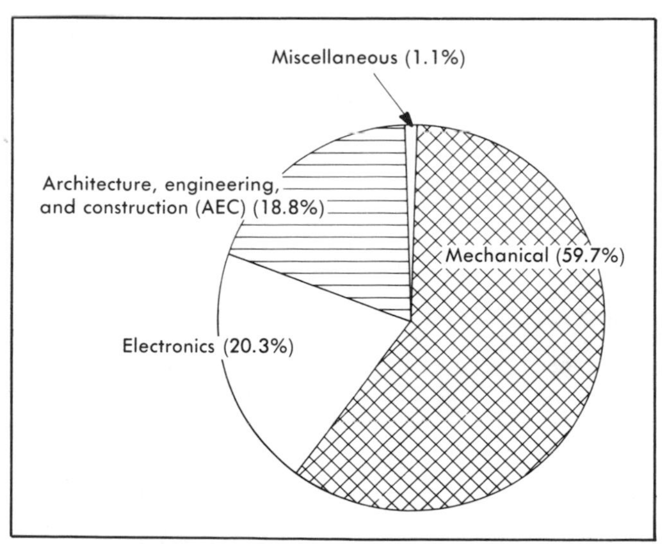

Fig. 16-2 CAD/CAM/CAE market segments. (*From CAD/CAM/CAE:* Survey Review and Buyers' Guide, *1986; available from Daratech, Inc., Cambridge, MA.*)

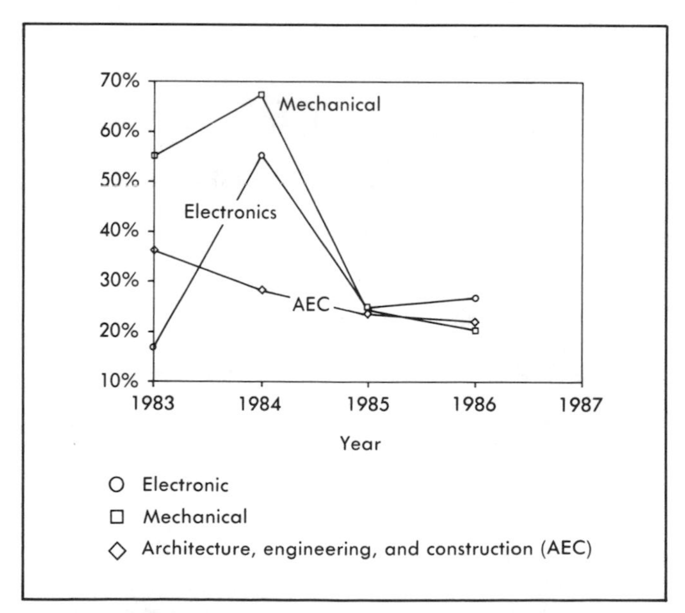

Fig. 16-3 CAD/CAM/CAE segment growth. (*From CAD/CAM/CAE:* Survey Review and Buyers' Guide, *1986; available from Daratech, Inc., Cambridge, MA.*)

the layout of these electronics devices. Since that time, the electronics CAE industry has emerged to address the more complex tasks associated with electronic design.

Electronics design may involve such activities as circuit analysis, optimum placement of components on a substrate or circuit board, layout of the conductive interconnections between those components, testing of those interconnections against design rules, and generation of the artwork that is used to create the actual circuit patterns on a PCB or IC. The artwork layers become the production tools that are used as masks to transfer the pattern onto the board through electroplating, photoresistant coating, or etching in the case of printed circuit boards. Similar steps are used in the case of ICs or VLSIs, but on a microscopic level. The value of CAD systems applied to electronics design lies in the time saved and in the quality of the finished products.

Traditional CAD systems for electronics design handled the schematic capture, component placement and routing, and the generation of the artwork as well as the necessary documentation. Circuit analysis and simulation was done on a mainframe computer or by hand. Today's electronics CAE systems can encompass all the tasks involved in electronics design, from circuit design to analysis, design verification, and simulation. The ability to handle the complete design cycle, plus the flexibility of analyzing different designs, determining the cost differential associated with substituting different components in a design, is driving the market from CAD to CAE.

Architecture, engineering and construction CAD systems. CAD systems applied to architectural applications provide the tools to model a design concept, produce the architectural production drawings, perform space and facilities planning, produce the contract drawings, and to produce schedules and bills of materials that are required for the electrical, plumbing, and structural work in a facility.

Technical publications. Technical publications is a growing area of importance for CAD systems. Almost every manufacturer is faced with the need to provide technical documentation on products, write proposals for sales or other business activities, and to publish product literature as well as annual reports.

Specialized software allows a CAD system to take a design (a product design or an illustrated parts breakdown) that resides in the CAD engineering database, merge that design with text on the CAD screen, and through this merger, develop the exact format of a document that can then be printed out on a laser printer or sent to a phototypesetter for subsequent publishing. CAD system software exists that also allows business graphics to be generated and halftone photographs to be merged with text.

CAD Standards

Over the past 10 years, the use of CAD systems has increased dramatically. At the time many of these systems were installed, they were selected to solve specific problems; integration or communication between systems was not an issue. The consequence of these diverse acquisitions was that several varieties of heterogeneous CAD systems were employed, within each industrial enterprise and within the industry as a whole.

As a result, CAD systems used by the engineering department of a company to design and define products could not electronically communicate this information to CAM systems used in manufacturing or other downstream functions. Rather, engineering was required to produce transitional paper drawings so that design data could be transmitted to manufacturing.

As organizations began to realize the value of integrating different business functions, the need to communicate between the computer-based systems that controlled those functions became obvious. Several efforts were initiated in the early 1980s to address this problem. One such effort resulted in today's IGES organization and standard. IGES, or Initial Graphics Exchange Standard, was intended as a standard for exchanging graphical and textual data between heterogeneous CAD systems. After diligent work by some 70 companies and hundreds of individuals, the first versions of the IGES standard were released under the auspices of the National Bureau of Standards. IGES has been adopted as an ANSI standard.

At this point, Version 3.0 is being specified in a number of DOD, DOE, and NASA procurement contracts and is consequently the subject of widespread user and vendor attention. IGES Version 3.0 specifies a standard format for exchanging graphic entities such as lines, curves, circles, 3-D wire frames, and surfaces as well as textual data. It is used principally for transmitting engineering drawings, technical orders, schematics, and similar graphic documentation between CAD systems.

As useful as IGES is, it has some drawbacks. The more sophisticated manufacturers have seen a need for another standard that will allow exchange not just of graphic entities, but of complete product definitions. Consequently, a new effort has been launched, called PDES or Product Definition Exchange Specification. The PDES program was begun in 1985 as a research initiative within the IGES community. PDES concerns not just graphic and textual entities, but the entire array of product definition information relating to the complete lifecycle of the product. PDES information is to be based on solid geometric model representations of detail parts logically associated with a wide range of related information to directly support design, analysis, manufacturing, and logistics applications.

It is essential that IGES and PDES are individually important but essentially different products. They are based on different technical approaches and are presently in widely separated stages of development.

Computer-Aided Engineering

Computer-aided engineering (CAE) is the technology that analyzes a design and simulates its operation to determine its adherence to design rules and its performance characteristics. Today, CAE is almost two separate technologies: one is CAE applied to mechanical design and one is CAE applied to electronics design. Common elements, however, do exist. Both of these subdivisions involve subjecting a design to extensive analyses, simulating its performance, and verifying the design against physical laws and/or accepted industry standards. But the differences that lie in performing these steps on a mechanical design and on an electronics design are quite significant. Until recently, different companies supplied CAD software and systems for mechanical CAE and electronics CAE. In fact, the growing demand for electronics CAE spawned a new industry with new participants and new hardware/software concepts.

Mechanical CAE. As was mentioned previously, mechanical CAE takes the mechanical part design and subjects it to a variety of engineering analyses. While analyses have been performed by computer for many years (CAE is actually the oldest aspect of CAD), the preprocessing required to put design data in an acceptable form for a computer was a lengthy procedure. With computer assistance, these techniques are easier and faster to execute. Color graphics also simplifies interpretation.

Mechanical CAE programs include finite element analysis (FEA) to evaluate the structural characteristics of a part and advanced kinematics programs to study complex motions of linkages and mechanisms. Other programs might evaluate thermal stress or fluid mechanics. The CAE analysis data may be displayed in graphic form.

The ability to design a part and test it before ever cutting a piece of metal has obvious economies. Less tangible are the benefits in design freedom, the benefits of allowing a designer the flexibility to look at several alternate designs, and the ability to use existing designs of similar parts stored in the same computer database. The more critical the structural integrity of the product, the more important is the use of mechanical CAE technology.

As the manufacturing industry becomes more aware of CAE's role in ensuring quality and as the cost of CAE tools decreases, the use of CAD and CAE will become routine.

Industry still needs the technology to transfer a geometric image from CAD to CAE (not all of today's systems have this capability) to CAI (computer-aided inspection), NC programming, and CAT (computer-aided testing). More CAM functionality is also needed in commercial CAD systems.

Electronics CAE. If CAE is an option in mechanical design, CAE is a necessity in electronics design. It would be impossible to design a complex integrated circuit without the benefit of a computer. And more and more, complex printed circuit board designs are being captured and simulated on CAE systems.

Electrical/electronics CAE capabilities include design capture, design verification, functional simulation, circuit simulation, timing verification, and netlist generation. Each one of these technical analyses are required to make sure that a printed circuit board or a VLSI operate properly. These functions had traditionally been performed by computer, but until recently on a mainframe in a batch mode. CAE systems provide the engineer with the ability to interact with the design and see the effects in a graphical form.

The current trend is the interaction between CAD and CAE systems to provide a more unified design system with more unified design management. The next few years will bring a number of new software systems on the market to further automate and integrate the electronic design process.

Computer-Aided Process Planning

One of the bigger hurdles that CIM must overcome is closing the gap between CAD and CAM. Technologies exist to minimize this hurdle, two of the most important being computer-aided process planning (CAPP) and group technology (GT).

Computer-aided process planning systems are, in effect, expert systems that capture the knowledge of a specific manufacturing environment plus generic manufacturing engineering principles. This knowledge is then applied to a new part design to create the plan for the physical manufacture of the part. This plan specifies the actual machinery employed in the part production, the sequence of operations to be performed, the tooling, speed and feeds, and any other data that is required to transform the design to a finished product. To use CAPP most effectively in a CIM environment, the design should originate on a CAD system and be electronically transferred to the CAPP system from the database.

CAPP draws on the geometric model of the part to be produced, generated in the CAD system, and matches the characteristics and components of that part to the production machinery on the factory floor. This technology will develop the process sheets or routings needed to manufacture a part.

Because a computer-aided process planning system contains information on all parts being produced at any given time, it is able to contribute to more efficient machinery utilization. Because CAPP determines how a part will be made, it is a factor in determining the cost of manufacturing the product.

Process planning is not a new technology. Successful process planning has been performed in many manufacturing companies by individuals with a great deal of experience in manufacturing and in the manufacturing resources and practices of the individual company. However, such a manual system is time consuming and totally dependent on the knowledge and experience of individuals who will eventually retire and leave the organization. CAPP systems capture the knowledge and experience of these individuals. They also allow a planner to generate a process plan for a new part by modifying an existing plan for a similar part.

A CAPP system will provide a set of instructions on how to make a part, in what sequence the process steps shall be executed, and what machines, tools/fixtures, workcenters, and labor skills will be required. A process planning system, manual or automatic, must be constantly updated to reflect a company's total manufacturing capability. Process plans must change as, for example, new machinery is procured and new quality control procedures or robots are introduced.

Computer-aided process planning systems are of two basic types: variant and generative. The variant method of process planning is the most commonly used method today. This method develops a process plan by modifying an existing plan that is selected using group technology principles, namely coding and classification. Some form of parts classification and coding is essential to expedite the location of previously developed plans, which are to be adopted as is or modified and adopted, and to specify the processing plan for a new part.

Generative process planning systems incorporate into their database a body of manufacturing logic, the capacities of existing machinery, standards, specifications, and the like. Based on the part description (geometry and material) and finished specifications, the computer then selects from this stored knowledge the optimum method of producing the part and automatically generates the process plan.

Automated process planning is useful to small and medium-sized companies as well as those with large manufacturing facilities. The more complex the planning function and the more factors to be considered, the better suited a computer is to the job. Some aircraft manufacturers, for example, may generate many hundreds of plans in a week for parts that will have limited production runs. In this kind of situation, automated process planning speeds the planning process by as much as 75% and makes the plans consistent. It also saves labor, scrap, and rework. Therefore, CAPP substantially lowers the cost of getting the design to the manufacturing floor and allows user companies to be more flexible in their parts production.

CAPP systems are actually expert systems that capture the knowledge of process planning and allow that knowledge to be applied to new situations. CAPP systems are probably the major application of artificial intelligence principles in manufacturing.

Group Technology

This discussion on group technology (GT) is limited to its role in the CIM environment. Group technology will be discussed in more detail in Chapter 13, "Design for Manufacture," and Chapter 19, "Equipment Planning."

Group technology, like CAPP, uses previous experiences in engineering and manufacturing to streamline current practices. Group technology is more accurately defined as a methodology rather than a technology. It is a method of organizing part designs and their manufacturing processes so that they can be grouped into families that have similar characteristics. GT is also a method of organizing a manufacturing facility by cells according to the types of parts manufactured and the types of equipment used to manufacture them. It is a software-oriented rather than a hardware-oriented technology.

Group technology is based on the fact that most manufacturers produce a lot of similar products and products that have a lot of similar components. If these products and their component parts are classified, their similarities can be used to avoid, if not eliminate, redundant steps in the design and manufacturing processes. Once parts are classified, parts standardization becomes easier to implement.

In the discussion on variant CAPP systems, it was pointed out that a new process plan can be derived from a previous plan for a similar product if the previous plan can be retrieved. Classifying each plan, each product, and each component in those products simplifies this retrieval process. This simplification applies equally to the design function and the manufacturing function. It is impractical and costly to design a new product from scratch when the design already exists for a similar product. Likewise, manufacturing operations should be grouped together to handle families of parts.

Classification systems should accommodate both the design attributes and the manufacturing attributes of a product. The design-related characteristics such as size, shape, volume, tolerances, and materials can be used to minimize the time and effort required to design a similar part; the manufacturing characteristics such as machining, heat treating, and welding can be used to develop a process plan for a part with similar manufacturing characteristics. Because the parts, the processes, and the machinery represent the kernel of the manufacturing operation, the structure that contains and organizes these elements becomes the cohesive force in integrating factory operations. To this extent, every other technology in a CIM environment is affected by GT principles.

The benefits attributed to GT are many. It can be used to reduce the number of parts in a company's database, new part introduction costs, product design lead time, scrap, and overall design cost, and to increase effective capacity utilization. Other reported group technology benefits include:[4]

- 52% reduction in part design.
- 10% reduction in numbers of new shop drawings through standardization.
- 60% reduction in industrial engineering time.
- 40% reduction in raw material stocks.
- 62% reduction in work-in-process inventory.

MANUFACTURING THE PRODUCT

The physical manufacture of a product involves a number of interrelated technologies. By using CAD and CAE to create and analyze the design and by using CAPP and GT to organize, plan, and control the individual manufacturing steps, the manufacturing enterprise must now control the processing of the physical materials that will become a product or a part.

The production process is complex. Raw materials, fixtures, tools, or components must be delivered to the specified production machinery in a timely fashion, dictated by a production control system. Materials and/or components must be accurately loaded and fixtured, if necessary, onto a machine bed or other work surface. The production machinery itself must be properly tooled for the part to be made, and it must have directions on how to perform the required operation. After machining, assembly, or whatever other operation is performed, the finished or semifinished part must be moved to the next logical step in its production, be that another machine station, a test stand, a packaging operation, a loading dock for shipment, or a storage location.

In an automated environment, all these operations must be monitored and controlled. Progress, as well as discrepancies, through the production process must be reported, at least in summary fashion, to manufacturing management.

Unlike the design technologies, which are predominantly software related, the physical manufacturing or production operations are a combination of hardware and software. The hardware has been around for as long as manufacturing has existed. Refinements have advanced conventional machine tools into complex machining centers or flexible manufacturing systems (FMSs), lift trucks into automated guided vehicles (AGVs), and shop floor data collection devices from clipboards into data entry terminals and bar code readers.

While automation is available for almost every manufacturing function, many such functions do not need and cannot justify the use of automation. A lot of traditional machinery is still being used effectively. Planning for CIM involves recognizing when automation should be applied for the most benefit and not just for the wholesale replacement of traditional equipment.

Because these changes are not as obvious as the changes brought about by software-only technologies, the efforts to integrate these islands of automation have not been consistently as strong as it has been with the design functions. Computer-aided design users and vendors have been at work for many years to develop a communications standard, IGES (Initial Graphics Exchange Specification), so that graphic data could be exchanged from system to system. Yet only after General Motors Corp. demanded a communications standard for factory floor equipment did efforts for a Manufacturing Automation Protocol (MAP) standard begin; MAP will be discussed subsequently.

Manufacturing Machinery

Manufacturing machinery encompasses machine tools, manufacturing cells, flexible manufacturing systems, automated assembly equipment, transfer lines, and inspection equipment.

Machine tools include machinery for metalcutting operations such as drilling, milling, and boring and for metalforming machinery such as presses, forges, and extrusion machinery. These machines have been the backbone of manufacturing for decades. Through the 1950s, most machine tools were operated by skilled machinists who learned their trade through a long apprenticeship. Product quality associated with manual machining operations could be excellent depending on the skill of the operator, but consistency and repeatability from one part to another could pose a problem. Numerical control (NC) was developed in the early 50s and began to be applied in the aerospace industry in the mid-50s. NC solved the problem of repeatability for the part volumes typical of aerospace.

MANUFACTURING MACHINERY

Manufacturing cells and flexible manufacturing systems are so closely related that it is difficult to discuss them separately. Computer technology was first applied to very high volume and very low volume production operations. High-volume repetitive manufacturing implemented "hard automation," typified by automotive transfer lines. Low-volume aerospace-type industries were pioneers of numerically controlled machine tools. The middle ground of manufacturing, with volumes too high for stand-alone machine tools and too low for dedicated automation systems such as transfer lines, went for a long time before adapting existing automation technologies to its particular production requirements. This is somewhat surprising because mid-volume manufacturing represents quite a significant portion of the total manufacturing universe.

The increasingly competitive environment that demanded higher manufacturing efficiency, higher quality products at lower costs, and faster response to market demands triggered the application of automation to this mid-volume area of manufacturing. Manufacturers were faced with the necessity of increasing the flexibility as well as the efficiency of their production operations. The greater efficiency that automation could provide in production operations would improve quality and cost of finished products. Increased flexibility would also allow a manufacturer to turn out a greater variety of similar products in the same time frame. Mid-volume manufacturers also needed a solution that would combine the speed, consistency, and production efficiency of dedicated automation with the flexibility of the stand-alone NC machine. Manufacturing cells and flexible manufacturing systems provided that solution for many companies.

Historically, machining cells are small groups of machine tools linked by material handling equipment. All elements of the cell may be computer controlled, and the individual controls may be able to communicate with each other as necessary to coordinate the operation of the cell. It is possible to have a cell of co-located machines with manual material transfer and without common computer control, although current usage of the term *machining cell* most often refers to more automated configurations. A manufacturing cell may or may not have a central control, depending on its complexity. In most cases, cells are configured to machine/process a variety of parts in a batch mode. Lot sizes are typically small to medium.

Manufacturing cells are considered by many to be the most important manufacturing advance of this decade, particularly for batch manufacturing. The cell is, actually, a basic concept of group technology.

The definition of a manufacturing cell and its use in the CIM environment has been affected by trends in the related technology of flexible manufacturing systems. When FMS technology was first introduced, systems were large and complex, often containing special customized machinery. More recently, manufacturing companies, looking for the benefits of FMS without an enormous initial investment (approximately $10 million-$20 million), have opted to begin an investment in FMS with a small, cell-like system using standard machinery and then grow it into a larger FMS configuration. This development blurs the distinction between cell and system technology into one of semantics. Whatever formal definitions are used, manufacturing cells and manufacturing systems have become the cornerstone of modern manufacturing technology.

Flexible manufacturing systems, like manufacturing cells, are groups of computer-controlled machine tools linked by an automated material handling system, controlled by a common supervisory computer control, and capable of processing parts in random order. Variations in the parts handled are accommodated by the FMS machinery directed by the supervisory control. Figure 16-4 illustrates a typical FMS installation that combines machining centers with head-indexing machines.

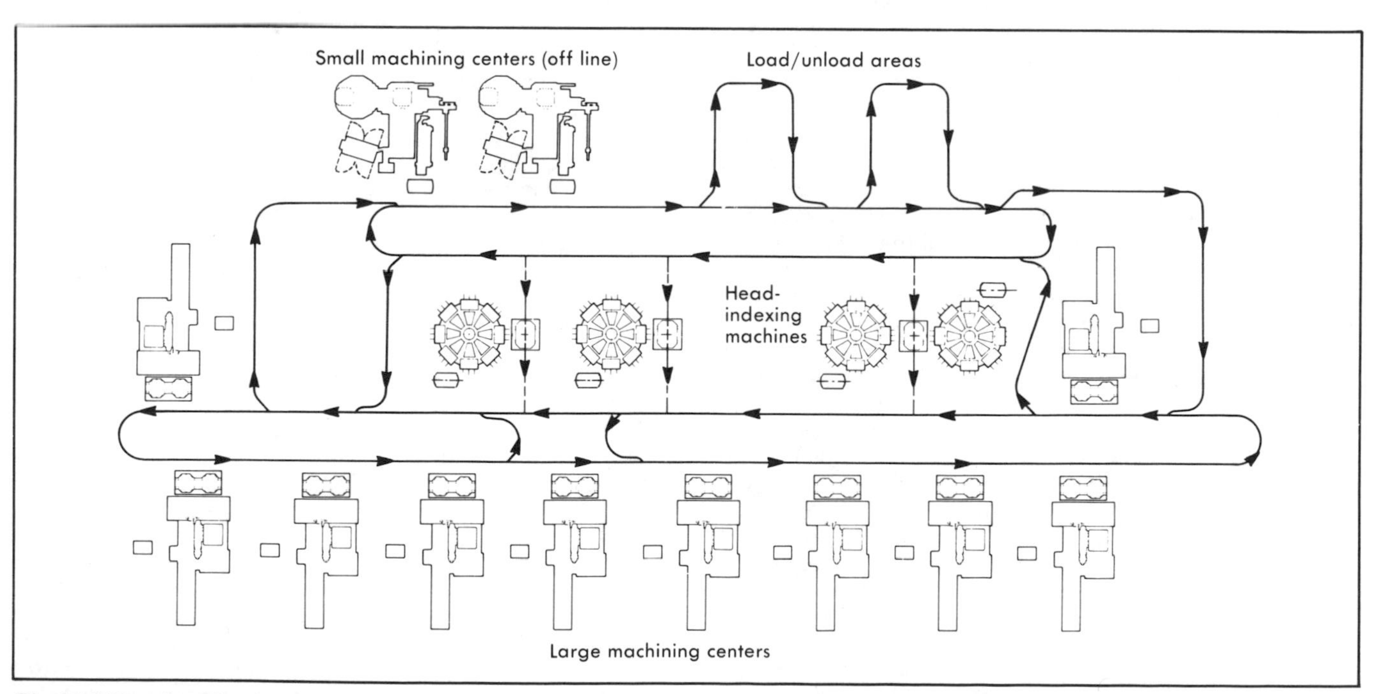

Fig. 16-4 Plan view of typical flexible manufacturing system (FMS). Large machining centers and head-indexing machines are linked through an automated fixture cart transportation system under computer control. NC part programs are downloaded to CNC machining centers from memory of the central computer. (*Kearney & Trecker Corp.*)

MANUFACTURING MACHINERY

The successful implementation of the concept of flexible manufacturing involves proficiency not only in the level of integrating the physical material or part processing (machining and transporting), but in the level of integrating the information flow that determines the physical operation of the system. For optimum efficiency, the operation of an FMS is integrated into a broader automation environment. In this environment, which is not yet a reality, parts to be produced are designed on a CAD system that produces both the bill of materials to allow a manufacturing management and control system to schedule production and the NC part programs to accomplish that production. Also part of that environment is feedback from the FMS to management on the quality and productivity of the system's operation.

In this sense, it typifies CIM. Like CIM, computer system data drives physical machinery which, in turn, drives other computer systems. The goal of CIM is that hardware and software systems work in synergy to support the business goals of the organization: competitive products at competitive prices.

The first industries in the U.S. to take advantage of the benefits of FMS were the truck, construction, and farm equipment industries; aerospace followed soon after. The first systems were really developed jointly by the user industries and the machine tool builders. Experience gained by both users and suppliers of these early systems has influenced the way modern systems have developed.

Most of the first systems were very large and expensive. They were also highly structured, being designed and engineered for the specific application. The cost of such systems was prohibitive for smaller companies and even for larger companies during periods of poor economic growth. The knowledge gained in the early systems allowed the machine tool builders to devise smaller and more modular FMSs using standard machine and control components to offer the advantages of FMS to large and small manufacturers at a more reasonable cost. These so-called "starter systems" represent a major current trend in FMS technology.

Flexible systems technology was initially applied to the production of prismatic parts, which still represent close to 80% of the parts produced on FMSs in the U.S. However, FMS principles have also been applied to turning operations, metalforming operations, assembly equipment, and welding and riveting.

Automated assembly machines and transfer lines are interconnected groups of machinery dedicated to the high-volume assembly or production of parts (see Fig. 16-5). This type of equipment has been the mainstay of the automotive industry. These systems had been characterized by fixed tooling and hard-wired controls. They provided the advantages of maximum machine utilization and minimum direct labor. They were, however, limited in the numbers of different products handled. It was not uncommon for transfer lines to produce a few different parts, but the changeover from one to another was generally planned well ahead, and the changeover itself was largely a manual and time-consuming operation.

Today's more competitive environment has changed the way even the high-volume industries address the market. More flexibility in manufacturing and assembly is required to provide the product diversification and differentiation that both industrial and consumer customers demand. Technological changes, also, such as more and better control technologies, automatic tool-changers, and robots, now make it possible to increase the flexibility of both transfer lines and assembly machines without compromising their traditional benefits of speed and machine utilization. Assembly machines and transfer lines are moving in the direction of flexibility. Of course, flexibility in transfer lines can only be achieved economically if the changes in product design or quantities can be anticipated; planning is a must.

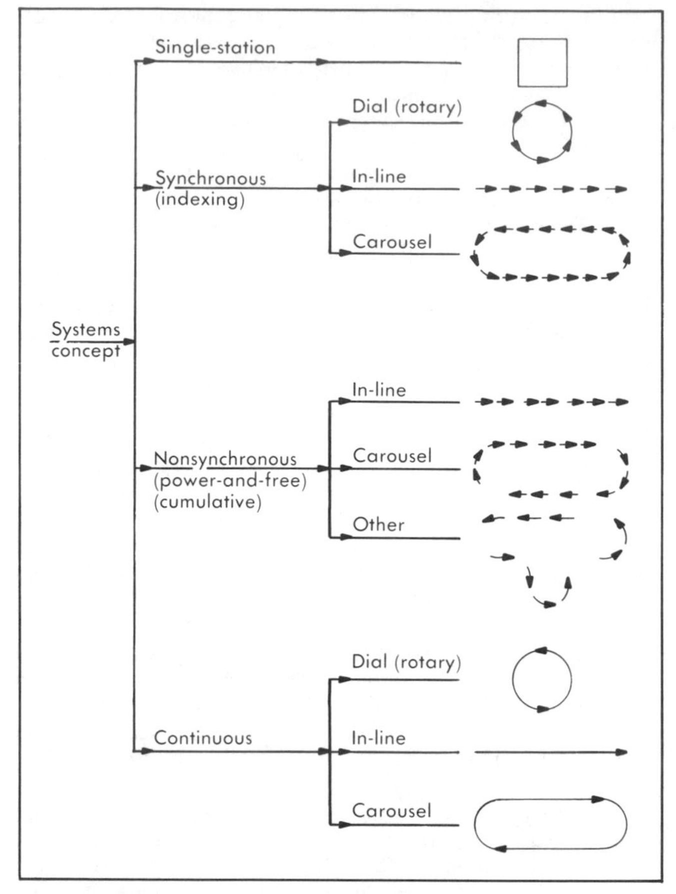

Fig. 16-5 Basic concepts for automated assembly systems.

With the flexibility of newer transfer line technology, which includes numerically controlled stations and automated quick-change tooling, it is now possible to make the changeover while the line is running. It is not uncommon now for an installation designed to make diesel engines for small trucks to run six-cylinder models at the end of the line while four-cylinder models enter the line. The cost of such lines is still high for many passenger car manufacturers, but costs are dropping and their flexibility makes them attractive to many medium to high-volume manufacturers.

Assembly machinery is also headed toward more flexibility. In this technology, many of the advancements come from the application of robotics. A widely publicized robot-assisted assembly system, the Robogate system, uses a number of robots to weld a car unibody (see Fig. 16-6). The robots are free-standing, unlike the rigid welding systems they replace. Reprogramming allows the robots to switch from one car model to another. The sequencing of operations on this line is controlled by programmable controllers. Each robot has its own control to direct its positioning/welding. Programs are downloaded from a supervisory computer that directs the overall operation.

As in FMS, a modular approach can often provide cost advantages to the user. Modern automatic assembly machinery

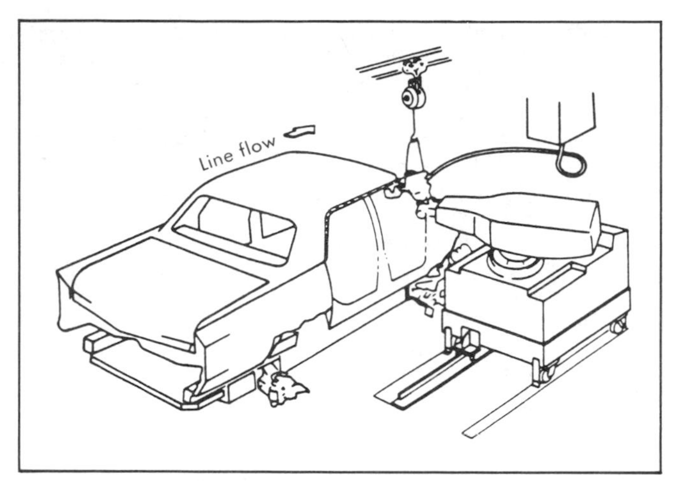

Fig. 16-6 Spot welding guns are positioned accurately and automatically by means of robots.

is often modular. Standard machinery modules include the machine base, the drives, lubrication systems, and wiring harnesses. This type of machine design satisfies the need for flexibility and at the same time shortens installation time. As tooling and toolchanging advances to allow for handling a larger variety of parts, automatic assembly equipment will have even broader usage.

In assembly machines, as in transfer lines and even standalone machine tools, the aim of automation is to allow one machine or one line to accommodate more types of products. Flexible automation devices, whether they be NC controls on machine tools or microprocessors or programmable controllers on assembly lines, enable changeover between models to be largely a matter of reprogramming. Machinery utilization increases, and the useful life of special machinery is also increased dramatically.

Auxiliary Manufacturing Machinery

Auxiliary manufacturing machinery is the machinery that enhances the efficiency of machine tools and assembly equipment by coordinating material movement and machine loading and unloading so that the overall production flow is smooth. These technologies include automated material handling equipment and robotics. Both of these technologies are discussed in more detail in Chapters 19 and 21. Therefore, this discussion will limit its coverage of these technologies to their role in CIM.

Automated material handling is the technology that is responsible for the physical integration of the production process. It includes the computer-based systems that store materials and parts before, after, and sometimes during processing and the systems that transport materials, parts, and assemblies from the time they enter the plant, through the various production stations, and until they leave the plant as finished assemblies or products. As such, this technology plays a very important role in a CIM environment, a role that is very often underestimated and undervalued.

The principal system that automates material storage in a manufacturing environment is the automated storage and retrieval system (AS/RS). These types of systems are the modern counterparts of storage racks and storage bins. Adding computer control to the storage function permits faster retrieval, more efficient use of storage space, buffer storage to feed machine tools, and, most importantly, better control.

On the transportation side of automated material handling, there are many technologies such as automated guided vehicles (AGVs), conveyors, towlines, and cart systems. Automated guided vehicles are not constrained in their movement by rigid travel paths (chains or tracks in the floor) and therefore have the potential to contribute more to a modern manufacturing environment where flexibility is important.

Automated storage and retrieval systems. Automated storage and retrieval systems fall into two categories based on their size and the size of the goods they store. The larger high-rise AS/RS is often referred to as a unit loader, while smaller versions that handle binnable materials are called miniloaders. Storage carousels are similar in concept to AS/RS. They are often used for tool, fixture, and fastener storage.

An AS/RS is really a rack into which parts are loaded and from which parts are unloaded automatically. It typically consists of input/output staging areas, the storage/retrieval (S/R) machines which shuttle loads in and out of storage, the control system, and the storage racks themselves. Some system designs require humans to perform the picking function (physically removing material from a rack), but the state of the art is the unmanned system. Newer systems make use of bar code readers and optical character readers to identify the correct load to be picked. Stored parts may be palletized or placed in bins or pans.

The devices that service the input/output staging areas may be conveyors, lift trucks, automated guided vehicles, or other transport devices. It is important that these devices mesh well with the operation of the AS/RS to get the maximum efficiency from the system. The size and configuration of the AS/RS and of the individual storage compartments in it are customized to the specific installation.

Control of an AS/RS can vary widely. A manual input station controls storage and retrieval through prepunched cards or pushbuttons. In a more sophisticated system, AS/RS controls can be a combination of computers, microprocessor controls, and programmable controllers working together under a supervisory control, which is normally a computer. This type of system not only controls the moving of loads in and out of storage, but can provide a host of data to other computers for managing inventory, for coordinating other plant activities, or for providing management information systems.

Many companies are exchanging data between their manufacturing resource planning systems and AS/RS. The MRP II system generates a pick list that is electronically transmitted to the AS/RS supervisory computer to enable the automatic retrieval of the necessary parts and materials for the current production cycle. The AS/RS computer tells the MRP II system when the picking has been successfully completed so that inventory records may be updated.

The advent of just-in-time (JIT) systems seems to be at odds with a large AS/RS. Although companies want to reduce and keep inventories at a minimum, they still need the control that such automated systems provides. Storage systems need not connote excess inventory. One major food processor operates a huge AS/RS in which goods are kept a maximum of three days.

In general, the benefits of an AS/RS can be further enhanced if the system is located near the operations it serves. If the manufacturing facility is planned so that the AS/RS is next to or in the center of the manufacturing stations, it can serve to reduce in-process inventory as well as perform the storage function. With planning, access stations can be placed at convenient locations along the sides of the AS/RS to serve specific lines or operations for even more efficiency.

For those who cannot plan manufacturing near the AS/RS, the current trend toward smaller AS/RS (one or two aisles) allows companies to bring the storage to the manufacturing site.

Included in the list of advantages of a well-designed AS/RS are security of stored material, utilization of cube storage space, increased productivity (more picks per day), more efficient use of prime manufacturing space, real-time inventory control, and more accurate stock balance and location information. In addition, a well-designed AS/RS with a well-planned computer interface reduces production interruptions, lowers shop cost, and improves customer service.

Automated guided vehicles. Automated guided vehicles (AGVs) are driverless trucks that operate under computer control. Most systems in use today follow a path defined either by a wire embedded in the plant floor or by a reflective stripe of paint or tape. Because of the computer control, they are able to interface automatically with a variety of other factory floor equipment, including machine tools, production machinery, the AS/RS, and robots. Their principal advantage is that AGVs are able to move over "flexible" routes to deliver materials.

Guided vehicles can perform a variety of tasks in a factory environment. They can be used to tow trailers that hold rolling loads up to 50,000 lb (25 tons) or they can be used as unit load transporters and carry loads up to 12,000 lb (6 tons). Lift tables can be built into some AGVs to increase their flexibility in delivering loads; some vehicles can place loads in elevated storage locations, up to 30′ (9 m) high. Vehicles with integral forklifts can transport pallets. Vehicle controls can trigger other activities such as opening doors or signaling elevators, if desired. AGVs can be either unidirectional or bidirectional.

A few years ago, the automobile industry began to use AGVs as assembly platforms that transport semifinished workpieces to a variety of workstations where some operation is performed on the workpiece while it remains on the AGV. This application has become very widespread in the auto industry, and its practice is moving to other industries.

The motion of AGVs is consistently smooth, as is braking, so the truck components are subjected to minimal wear, and the cargo they carry has less chance of falling off or being damaged. And safety features are built in to protect plant personnel and equipment.

Controls for automated guided vehicle systems vary. They can be mounted on board the vehicle and directed by programming stations located throughout the plant or at one central location. On-board controls may consist of switches or pushbuttons that correspond to plant locations or workstations to which the vehicle is being directed. Safety features, such as collision avoidance, can be built into the control systems.

A remote programming station may by used to assemble loads and then transfer them to the AGV with the program that governs the vehicle's travel route and destination. A central control location will probably keep track of all the vehicles in the system and correlate their activity with other activities in the plant, such as an AS/RS or a material tracking system. Graphic displays at this central location may give an overview of the plant and all the material transport systems in use. Vehicle alarm conditions can be monitored at the central location.

New guidance systems are being developed that will further enhance AGV usefulness; these include radio frequency and laser-based systems and self-navigational systems.

Robotics. Robotics technology is one of the most versatile of the CIM technologies. Robots can function as material handling devices to load and unload machinery; they can function as manufacturing machines to perform machining, painting, or welding operations; equipped with vision capabilities, they are performing assembly operations for small parts manufacturing, particularly in the electronics industry, and in inspection tasks for a variety of industries.

Robots do not function alone; they must be integrated into plant operations. To be most successful, their operation must be integrated with other computer-based systems in the plant. Integration, even the limited systems development required to load and unload a machine tool, takes planning to be most effective. That simple operation must fit into an overall manufacturing strategy that governs production flow. Robots are very flexible, but a company must be ready to use that flexibility properly.

Within the concept of CIM, robots serve many functions. They are both manufacturing machinery and auxiliary manufacturing machinery. They range in complexity from very simple pick-and-place types to complex, sensor-based systems. They are an integral part of computer-aided manufacturing and of automated material handling. Their operation in a manufacturing cell or FMS can be simulated on some CAD systems, and some robotic programming can be performed on some CAD systems. Newer developments in programming languages will allow the robot and the workcell in which it operates to be programmed at the same time. Clearly this has the potential to simplify workcell programming and improve the cell performance. In this environment, the robot control, like a machine tool control, can be linked with other controls in a DNC-type arrangement. Table 16-2 shows the future trends in robotic applications.[5]

Material handling is and will continue to be a major application for robots. In this function, robots interface with an AS/RS and AGVs. In this environment, the robot operation will be coordinated with the overall system operation, and the robot control will be interfaced to and directed by the system central control. A robot arm is sometimes attached to a guided vehicle to facilitate loading and unloading. AGVs are themselves robotic vehicles.

Robotic controls vary with the type of robot. Simpler programmable logic controllers with some motion control are ade-

TABLE 16-2
Future Trends in Robotic Applications

	Distribution of Robot Sales in U.S., %		
Application	1985	1990	1995
Machine tending	16	15	15
Material transfer	16	15	15
Spot welding	26	15	10
Arc welding	10	10	9
Spray painting	10	10	7
Processing	5	7	7
Electronics assembly	6	12	14
Other assembly	5	8	12
Inspection	5	7	10
Other	1	1	1

quate for applications where movement is limited to one axis at a time. Complex CNC-type controls are more widely used to coordinate the multiaxis movement of the robot arm and end effector. Robotics continues to offer significant advantages for manufacturers, including a 98% uptime compared to 75% for an average worker. But the key to successful implementation is to consider the robot as an integral part of the factory automation scheme.

Controls for Manufacturing Machinery

Computer control enables manufacturing machinery to communicate and coordinate its activities with other computer-based systems in the CIM environment. A variety of control types exist; all rely on the power of the microprocessor. Microprocessors allow sophisticated controls to be small enough to be embedded in machinery if required, to operate on the factory floor, and to communicate the status and performance of the equipment they control to other machine controls on the floor and to supervisory-level control systems.

Computer numerical control. Computer numerical controls (CNC) are outgrowths of the hard-wired numerical control systems first developed in the 1950s. Today the terms *NC* and *CNC* are used interchangeably.

Although the addition of a computer added considerably to the capability of NC control, the basic function of the systems is the same: to control the operation of a machine tool through a series of coded instructions that represent the tool path to be followed, the depth of cut, coolant flow start, tool change, the machine speeds and feeds associated with the operation, and emergency stop. NC and CNC took some of the art out of machining and put it in the machine control so that repeatability and consistency were achievable from part to part.

Modern CNC technology can go beyond controlling a single machining sequence on a particular workpiece to encompass multimachining operations. It can direct an automatic tool-changer to replace a cutting tool in preparation for the next operation. Instructions are also included that trigger the operation of a robot to load and unload the workpiece and then signal to the cell controller the successful completion of the operation, the condition of the cutting tool, the time elapsed, and a host of other data that can be used to refine the operation of the cell. Computer control has changed manufacturing technology more than any other single development. It introduced the concept of automated machine control. In so doing, it broke the ground that enabled manufacturing cells, FMS, robots, coordinate measuring machines, and programmable logic controllers to be developed.

NC and its successor CNC did more than just control machinery. It captured the machining expertise of individuals; it raised machining to the level of a science—controllable, repeatable, and measurable. This fact became more important as years went by and fewer young people chose to enter a long and arduous machining apprenticeship.

As it developed, CNC was seen to have other advantages: it improved quality because machined parts were consistent; inspection time was reduced; it improved machine uptime because the operator didn't have to stop to think about how to machine a part; it reduced set-up time on machine tools because tooling and fixturing could be specified ahead of time; and most of all, CNC enabled the machine tool to be linked with the rest of the factory environment through its control system.

Numerical control was first applied in the aerospace industry, where precision in machining critical engine parts could mean the difference between survival and disaster. Once proven in the aerospace environment, other industries began to investigate this technology.

In the beginning, numerically controlled machine tools received their part programming instructions from punched cards, punched tape, or magnetic tape. But cards and paper tape are easily damaged, and any storage media that resides outside the computer runs the risk of being either physically damaged or technically obsolete. Most manufactured parts are changed from time to time to reflect improvements in design, in manufacturability, and in maintainability. The most satisfactory method of managing the numerous revision levels that characterize a machined part is to store the latest revision electronically and to load its associated part program into the machine tool each time it is required. This requirement was one of the principal reasons behind the development of distributed numerical control.

Distributed numerical control. Distributed numerical control (DNC) is the outgrowth of direct numerical control, which is the concept of linking a computer containing the part programs and associated information to the NC control attached to a machine tool. In this concept, the computer would download programs as needed. Unfortunately, computer technology was not as reliable in the early days, and if the DNC computer went down, so did the machine tool.

Distributed numerical control is a much more practical concept because it connects several CNC machine tools to a higher level DNC computer. A DNC system allows all the part programs for one facility to reside in one central location and be downloaded as required to individual CNC machine tools. This arrangement facilitates management of part programs and part program revisions. At the same time, a DNC installation can allow data from the machine operation to be automatically passed to a higher level computer for management information and interaction. Such data can include production piece counts, machine downtime, quality control information, or part program enhancements. Figure 16-7 depicts a typical DNC hierarchy.

Initiating as it did the idea of centralizing the control of several machine tools, DNC paved the way for the development

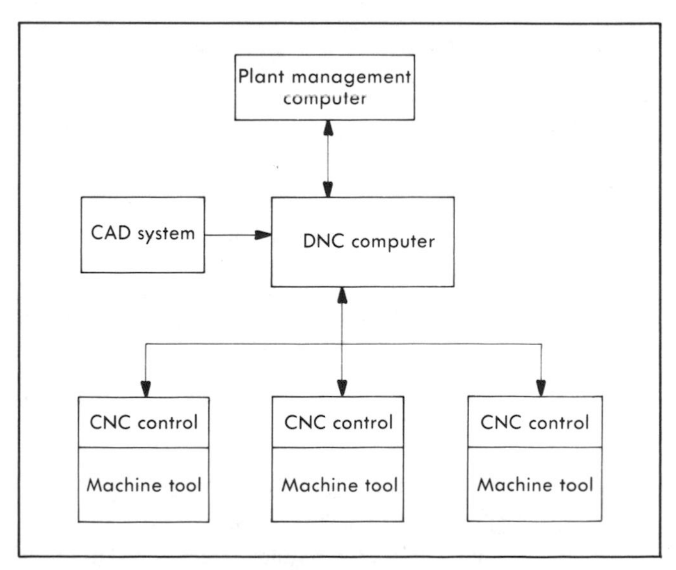

Fig. 16-7 Typical distributed numerical control layout.

of manufacturing cells and flexible manufacturing systems. Despite the proven efficiencies that CNC and DNC have brought to metalworking manufacturing, the technologies have been slow in being applied. CNC machine tools still represent a minority of the U.S. installed machine tool base. Indications are that as U.S. companies automate to remain competitive, many of the aging population of U.S. machine tools will be replaced with CNC tools. Some simpler machine tools are not suited to CNC control, such as some punch presses and thread rollers. For this type of equipment, another category of machine control exists, the programmable logic controller (PLC).

Programmable logic controllers. Programmable logic controllers are a very important control element in a factory automation environment. PLCs are computers specifically designed to be rugged and to withstand the heat, dirt, and electrical noise that is endemic to the factory floor. PLCs originally replaced relay panels in the automotive industry and were primarily used as sequencing controls. Developed at the request of General Motors Corp., they were designed to be programmed in relay ladder logic so that they could be programmed and maintained by electricians. Built-in diagnostic capabilities further simplified maintenance.

The acceptance of programmable logic controllers was so strong that the devices were enhanced by more and more features, such as arithmetic functions, timing, data storage and processing capabilities, and subroutine capabilities. All these features made PLCs more computer-like. Newer higher level languages have replaced the ladder diagrams in some applications, and sophisticated computer graphics provide a window into the operation being controlled.

Following the trends in computer technology, PLCs have evolved into a hierarchy of controls. Some models of PLCs are quite simple; handheld models with the control equivalent of five relays are commonplace. At the same time, some have incorporated sophisticated, and even redundant, data processing and control capabilities and are serving as supervisory controls in processing lines and manufacturing cells. Many PLCs are of modular design so features can be added as needed. Communications capability enhances the ability of the PLC to interact with other PLCs and with other factory control systems. The PLC manufacturers were among the first to accept the manufacturing automation protocol (MAP) and to begin to incorporate MAP features in their products.

The flexibility of PLCs allows them to be used in a number of different applications. These can be as simple as controlling valves to limit the flow of material in a processing plant or embedded in a packaging machine. They can be as complex as controlling a robotic welding line. PLCs are widely used in transfer lines, assembly machines, and material handling systems, as well as in the simpler machine tools mentioned previously.

Because of their flexibility, their ease of use, and their built-in communications capability, PLCs are widely implemented in manufacturing today. These same qualities will increase their implementation even more in the next several years. As simple stand-alone controls, they serve at the lowest level of a factory control hierarchy and in their more sophisticated configurations they extend upward into the cell control level.

PLANNING AND CONTROLLING THE MANUFACTURING PROCESS

No matter how efficient the physical production operation appears to be with computer-aided machinery cutting and shaping metal, assembling components, and moving material and parts around, they may not add up to real efficiency unless all these operations are planned, scheduled, and coordinated in view of the overall business objectives. The CIM technologies that perform these manufacturing management functions are manufacturing resource planning (MRP II) and, more recently, just-in-time (JIT) systems. Because these technologies are the subjects of separate chapters (see Chapter 15, "Just in Time," and Chapter 20, "Production Planning and Control"), discussions in this chapter will be limited to the role they play in the CIM environment.

Manufacturing Resource Planning Systems

Manufacturing resource planning systems have been called the central nervous system of the manufacturing enterprise. Contained within these systems are the software modules that plan and schedule manufacturing operations, allow production and material schedules to be explored for better alternatives, monitor operations against the plan, allow operating results to be projected, and tie financial reporting to operating figures.

These manufacturing systems allow management to control an entire organization and give management the timely information it needs to control and direct the organization in accordance with its established business plan. In a CIM environment, modern MRP II systems have the ability to tie real-time information from virtually the entire business enterprise together—design engineering, manufacturing, accounting and finance, marketing, distribution—so that this data may be used to support the overall business strategy. Relatively few of the installed systems take advantage of the full capabilities of the MRP II software.

The importance of such systems to successful CIM implementation is obvious; through the data they generate, collect, and manage, MRP II systems establish and maintain links between the shop floor, the engineering department, and the front office. At this point in time, very few commercial MRP II systems have established direct links with CAD systems, NC programming, or CAPP. However, this integration is currently being discussed by most of the leading MRP II vendors.

Just-in-Time Systems

Just-in-time (JIT) production, a technology related to MRP II, has caused many U.S. companies to rethink their approach to material management; many have experienced significant benefits after adopting this philosophy.

One of the basic tenets of JIT is to produce only what is needed when it is needed, thereby reducing inventory, particularly work-in-process (WIP) inventory, and inventory costs. Purchased parts or raw materials are delivered directly to the production line, several times a day if necessary. Internally, one production area makes only as many components or subassemblies as the next portion of the production operation can use that day or that shift. The philosophy turns inventory into products as quickly as possible. It's a 180° change from the philosophy of keeping a full supply of spare parts in storage just in case they were needed.

To be successful, however, JIT requires close working arrangements with suppliers; quality must be assured because poor quality parts or materials result in manufacturing problems, and JIT allows no time for checking incoming parts. When properly applied, JIT can mean reductions of inventory of more than 75% and equivalent improvements in product quality.

To date, JIT has been implemented primarily in repetitive manufacturing environments because they offer the most stable plant load. The popularity of this technique, however, is attracting the attention of many other industries.

CONNECTING THE ISLANDS OF AUTOMATION

The previous sections of this chapter have attempted to describe the CIM concept and how its component technologies fit into that concept. This section will discuss the technologies and the technological trends that are enabling integration to be realized. These technologies center around computer technology and telecommunications standards; trends include the drive toward integration of all business activities.

Computing Technology

Computing technology is the technology that integrates all of the other CIM technologies. Computing technology includes the range of hardware configurations as they are used in the CIM environment and the software-related functions that effect integration, namely, database management systems, linkages between technologies, and telecommunications. Figure 16-8 illustrates the control hierarchy in a manufacturing environment.

The automated factory environment is composed of and controlled by a number of types of computers from very simple microprocessors to highly complex mainframes and even supercomputers. In CIM, these different products often operate in hierarchical fashion with the lowest level receiving instructions from higher levels and, in turn, transferring data upward. There is also communications within levels in addition to upward/downward communications.

The lowest level in the control hierarchy is the machine control level. This level consists of microprocessor-based products that directly control machinery. These products can be PLCs or other microprocessor-based controls.

The next level is the cell level. At this level, several machines act together, and although they may each have their own control, their operation is coordinated by a central computer. At this level, the control computer may be a sophisticated PLC or a microcomputer. A small manufacturing cell is typical of this level of complexity.

Moving up the control hierarchy is the area control. This computer will monitor the operation of an area of a plant such as an assembly line or a robotic welding line. These control products are typically minicomputers or superminicomputers.

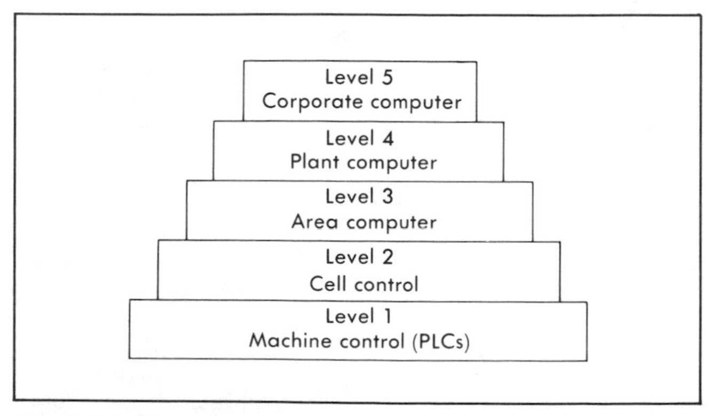

Fig. 16-8 Computer control hierarchy. (*Arthur D. Little, Inc.*)

The VAX line of products from Digital Equipment Corp. is very strong at this level of control. The area computer can direct the operation of a number of controls. In turn, it can collect data from a number of operations and synthesize this data for transfer to a plant-wide computer, the next level of control.

The plant-level computer serves more of a management function. Although planning must be done at several levels, the plant level has responsibility to plan and schedule shop floor operations. The plant-level computers collect operations data, including machinery and personnel performance data, and direct and manage maintenance operations. The process planning system resides at this level also; however, it interacts with several other levels.

At the top of the control hierarchy is the corporate computer wherein resides the corporate database and the financial and administrative programs that run the company. One of the most important functions of the corporate computer is to organize the corporate database so that data can easily be stored, retrieved, and manipulated.

Database management is critical to CIM; improved availability, accessibility, and accuracy of data can support all other operations. Because manufacturing is a data-driven process, integrating and controlling the data associated with manufacturing goes a long way toward integrating and controlling the process. The importance of database management can not be overstated. The strength of certain companies in this area has been responsible for their success in other areas of a manufacturing enterprise. IBM's prowess as a data manager, for example, has made it the de facto standard for corporate data management to the extent that every other computer company adapts its products to communicate with the IBM database. Recent developments in database technology envision a central knowledge base of engineering, manufacturing, and other data that is relational and can be accessed by a variety of systems at various hierarchical levels.

Communications between systems is vital in a modern manufacturing environment. A hierarchy of computers that communicate with each other implies at least commonality in communications protocols. One of the most significant trends in computer and control communications is the recent initiative by General Motors Corp. to develop the Manufacturing Automation Protocol (MAP). This subject is so important that it is discussed in a separate section in this chapter.

Computer-Integrated Business

Computer-integrated business is the logical extension of the trend toward communications and integration in a manufacturing enterprise. Considering that all the operations of a facility depend to some degree on the same data, it is shortsighted to limit CIM-type activities to the design and manufacturing technologies. If Dr. Joseph Harrington's definition of computer-integrated manufacturing is accepted, then all business activities, direct or indirect, that are involved in the transition of a product from concept to after-sales service are part of CIM. They are linked by the data that defines the business.

MAP/TOP

Computer-integrated manufacturing is the art of integrating computers into the manufacturing process. How well this integration is accomplished is directly related to the quality of the communications network design. The Manufacturing Automation Protocol and Technical and Office Protocol (MAP/TOP) networking standards provide a framework that can be used to integrate computers, and the software which runs on those computers, into the manufacturing process. While MAP specifies functional network protocols for the factory floor, TOP specifies them for information processing in technical and business environments. Protocols are rules governing the interaction between communicating entities. This section discusses MAP/TOP from three perspectives: (1) an internal view of the Open Systems Interconnect (OSI) seven-layer model and the protocols selected by MAP and TOP; (2) a view of MAP and TOP networks from the perspective of an application program, which will cover some of the services available through the application layer; and (3) an implementation perspective that will illustrate how MAP/TOP networks can be designed using the available standard "building blocks."

Internal View of MAP and TOP

MAP and TOP are the first network standards that conform to the OSI seven-layer reference model introduced in 1977. This model presents a structure that can be used to organize the services provided by a network into seven layers or modules. The reason behind organizing the services into modules is that a module can be replaced as long as the functionality of the module is not compromised. For example, a network that conforms to the OSI model could use fiber-optics as the transmission medium in place of broadband without affecting the ability to detect and correct transmission errors.

To be as useful as possible, network protocols must provide a variety of services. These services are as follows:

- Allow the transmission of data between application programs or processes on the network.
- Provide for control mechanisms between software and hardware on the network.
- Insulate programmers from the intricacies of communicating over the network.
- Be modular so that choices between alternate protocols can be made with minimal impact.
- Allow for communication with other networks.

One of the first tasks faced by the International Organization for Standardization (ISO) was to produce a framework for the establishment of these standards. It agreed to produce a layered model, with each layer representing a set of services so that the implementation of services at one layer could be changed without affecting the other layers. The seven-layer model for Open Systems Interconnect, or OSI reference model, was proposed and eventually adopted by the ISO.

As can be seen in Fig. 16-9, the layers are organized in a very specific way. The services provided by the model can be thought of as user services at the top layers: application, presentation, and session. The network services are provided by the bottom four layers: transport, network, data link, and physical.

User services are those services that are concerned with providing a uniform method for communicating processes to use the network, a method of converting the way data is represented on one device to the way data is represented on another, and some means to establish and control the dialog between communicating entities.

Network services are those services that are concerned with the details of the network: control the flow of data between end systems, route messages, prioritize messages, access to the network, represent data electronically on the network, and physically attach devices to the network medium.

Services Provided by the Seven-Layer Model

It is important to note that the seven-layer model is still functional even if the services provided by one or more layers are left undefined. The idea behind the use of a seven-layer model was that it could provide a path for migration toward OSI compatibility for all networking systems then in existence, even though the layered network architectures that were available at the time did not use all seven layers.

For example, if a network does not need the services of the presentation layer, the presentation layer could be left empty, or null, and the network will still be OSI compatible. If a system was later added that required the services of the presentation layer, those services could be added without affecting the other six layers.

Physical layer protocols. If devices are to communicate with one another through a communication medium, some standard means must exist for connecting those devices to the medium and for representing data electronically on that medium. The protocols governing attachment to the medium and the modulation techniques are found in the standards for the physical layer.

MAP and TOP have chosen different protocols for the physical layer. The designers of the MAP protocols chose broadband coaxial cable as the preferred physical medium because of the superior immunity to electromagnetic interference and multichannel capability found in this medium. Broadband can best be thought of as a way to guide radio waves along a piece of cable. Devices attached to a broadband system transmit all messages at low frequencies and receive all messages at high frequencies. The conversion between low and high-frequency transmissions takes place at a headend or translation device. Broadband is the system used in cable television systems (CATV).

The designers of the TOP protocols took a different approach than that chosen by the designers of the MAP protocols. Because very few factories had networks in place (and those who did usually had broadband closed-circuit TV systems), the MAP Task Force was free to specify any medium it felt was appropriate. The designers of the TOP protocols, on the other hand, were faced with a large installed base of baseband coaxial cable. The use of baseband simplifies the process of installing a TOP network in those offices that are contemplating replacing their existing network.

Data link layer. In a networking method that uses a single communications medium shared between the devices on the network, a method must be specified for controlling access to the physical medium and for identifying stations on the network. Control mechanisms have to be found to ensure that a station does not transmit more data than the receiver can digest at one time, and to detect and recover from errors caused by problems on the physical medium. In the OSI seven-layer reference model, the standards for these areas are found in the standards for the data link layer.

The data link layer is organized into two sublayers: media access control (MAC) and logical link control (LLC). The MAC sublayer is concerned with controlling access to the medium; the LLC sublayer is concerned with the size of frames, or chunks of data transmitted on the network, and the priority of transmissions.

Media access control. The standards for controlling access to the medium and for identifying stations on the network are functions of the MAC sublayer. There is a heated debate about whether this sublayer should be a part of the data link layer or

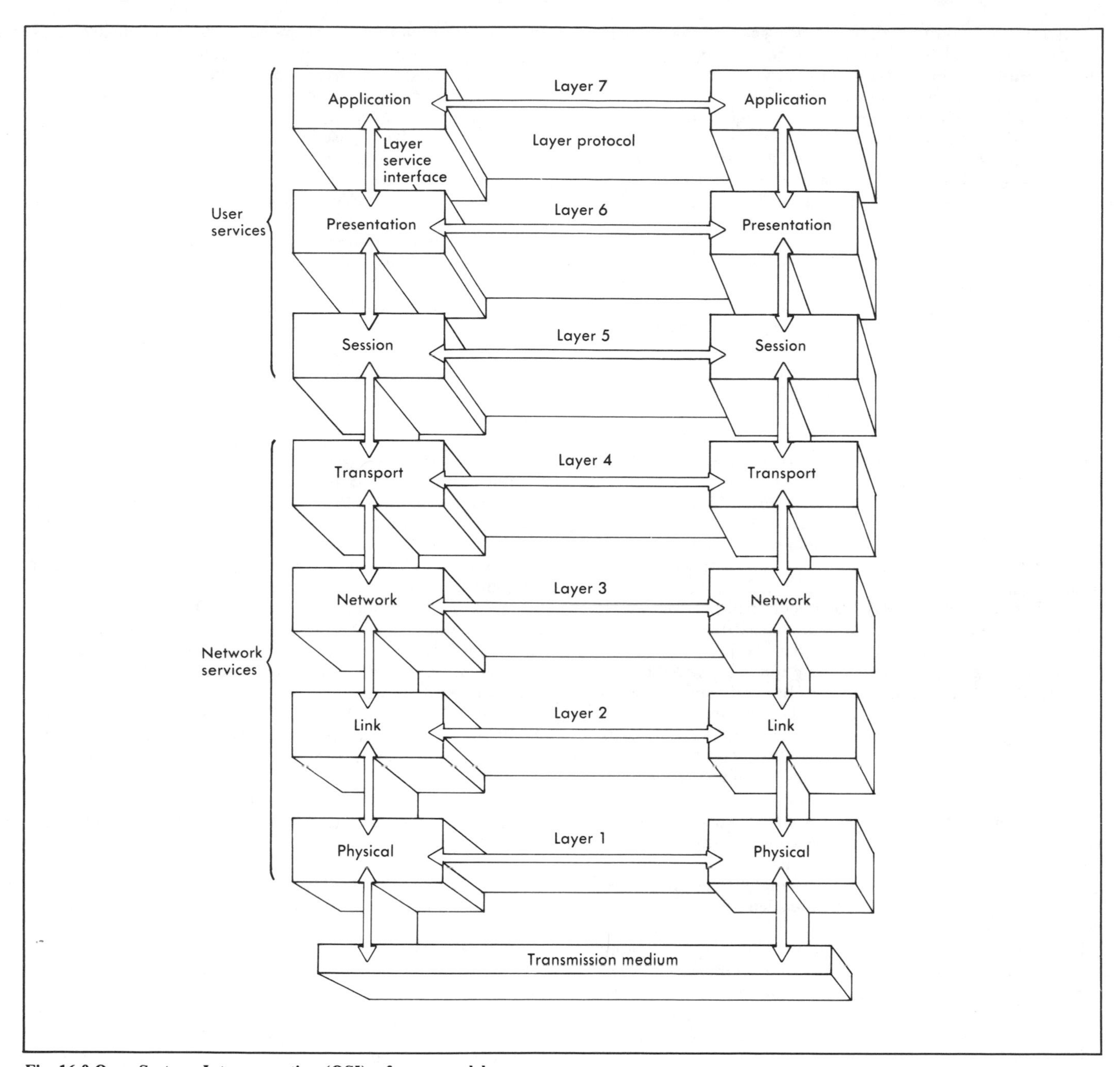

Fig. 16-9 Open Systems Interconnection (OSI) reference model.

the physical layer. The debate is really meaningless from an implementation point of view because the sequence of layers and sublayers is unaffected by the placement of the MAC sublayer at the top of the physical layer or the bottom of the data link layer.

MAP and TOP use different protocols for controlling access to the medium. MAP uses a method called Token Passing on a Bus (specified in IEEE 802.4), while TOP uses Carrier Sense Multiple Access with Collision Detection (CSMA/CD, specified in IEEE 802.3). Once again, the reasons for the difference in choices has to do with the different environments where the networks will be used.

Under the Token Passing on a Bus protocols, a special control frame called a "token" is passed from station to station on the network. When a station has the token, it can transmit messages on the network for a preset time. Because access to the network can be determined by calculating the length of time it takes the token to be passed around the network, token-passing networks are considered deterministic.

The TOP standard, CSMA/CD, allows any device that wishes to transmit a message to do so at any time, providing the medium is not busy. Because it is possible for two or more stations to perceive silence (nonuse) on the medium at the same time, each station must listen to its own transmission to determine if another station began transmitting at the same time. If another station also began transmitting, both stations detect the collision, cease transmitting, and wait a random amount of time before trying to transmit a second time.

While CSMA/CD works very well on lightly loaded networks, it can result in very long delays when the network is busy, because there will be fewer periods of silence, and the probability of collisions during nonsilence will increase. As a result, the network is nondeterministic; there is no way to predict how long it will take to transmit a message because access to the network is essentially random.

CSMA/CD does have several advantages. If the network is lightly loaded, access can be instantaneous. Because the protocols that govern the access method are very easy to implement, the cost of connecting a station to the network is considerably lower than the cost of implementing other access schemes. The designers of the TOP protocols felt that an access delay of a few seconds under heavy network loads is offset by the lower cost of CSMA/CD.

Logical link control. The second sublayer of the data link layer is the logical link control (LLC) sublayer, which is always found in the data link layer. This sublayer is concerned primarily with frame size and access class, or priority. MAP and TOP both use the Class 1, Type 1 LLC protocols specified in IEEE 802.2. Class 1, Type 1 allows the exchange of messages between stations without the need to first establish a connection and without flow control.

The designers of the MAP protocols realized that some factory networking applications would need faster response times than could be achieved with the overhead of the full seven-layer stack. They designed an option to the MAP protocols that allows a station with a short, time-critical message to bypass the upper layers of the stack and interface directly to the data link layer. This high-performance option is known as enhanced performance architecture (EPA). The EPA option specifies Class 3, Type 3 LLC protocols that use single, frame-acknowledged services to achieve faster response times.

Network layer. It is sometimes desirable for a device on one network to communicate with a device on another network. This can happen in a large network that is divided into smaller subnetworks or when the communicating devices are on different local area networks (LAN) connected by a wide area network (WAN).

Because the addresses used at the data link layer (or more correctly, the MAC sublayer) are not sufficient for this purpose, there must be standards for accomplishing this internetwork routing over public data networks. The protocols that standardize the methods used for this global routing are found at the network layer.

TOP and MAP both use the protocols specified in the ISO internet standard 8473 for the network layer. This specification establishes formats for network addresses that are unique on a global scale and provides a base for resolving protocol differences between different OSI seven-layer networks as well as providing an internetwork routing capability.

Transport layer. There are often several routes a message can take to arrive at its destination. It is not unusual in these situations for a portion of a message to arrive at its destination before earlier portions that took different, and longer, routes and thus were delayed in arriving at their destination. In real life, this happens when a passenger's luggage takes a different airline route than does the passenger. As a result, the receive buffer on a device attached to the network must be divided into frames, with each frame corresponding to an anticipated message packet. Some means must be provided to ensure that packets are not lost in transit or reassembled in the wrong order. The standards for these transport problems are addressed in the protocols for the transport layer.

MAP and TOP both use Class 4 transport protocols. Class 4 is the largest and most complex class of transport protocols; it allows for connection establishment, data transfer, and connection disestablishment. Class 4 handles flow control, detection of protocol data units that are missing or received out of sequence, and the transfer of a limited amount of expedited data. The specific options to the protocols are negotiated between stations during connection establishment.

Session layer. Two communicating application programs or processes must have some means to create and control the connection between themselves. This control involves establishing, maintaining, and terminating the connection between the application programs. The standards for these services are covered in the protocols for the session layer.

Both MAP and TOP specify the use of the session kernel protocols with the full duplex option at the session layer. The session kernel contains the minimal subset of session layer functionality; it allows the establishment of a connection, the transmission of data over the connection, and the release, either orderly (normal) or through an "abort" service.

Presentation layer. Different computer manufacturers use different sequences of binary digits to represent characters and numbers. If an IBM computer using EBCDIC wants to talk to a DEC computer using ASCII, some translation must occur so the data will be meaningful to the computer receiving the message. Translation services such as this are covered in the standards for the presentation layer.

At the present time (Version 2.2 of MAP and Version 1.0 of TOP), the presentation layer is null; translation services must be handled by the application programs that are involved in the dialog.

Application layer. Finally, standards need to be established for specifying the interface to the seven-layer stack and for establishing the context, or meaning of the data, of an information exchange. The protocols for the application layer standardize this interface and allow communicating application processes to specify the context for the communication session.

An Application View of MAP and TOP

When two application processes communicate with one another, they have an association in OSI terminology. The services available at the application layer are standardized for some of the more common types of associations. There is a great deal of work still going on to further define the context of application associations.

TOP Version 1.0 and MAP Version 2.2 specify the protocols for a small subset of the possible application associations. When Version 3.0 is released in 1988 for both network standards, there will be a great deal of functionality added to the application layer.

Currently, both MAP and TOP specify a limited subset of file transfer access method (FTAM), which specifies the protocols for transferring files between systems. MAP Version 2.2 and TOP Version 1.0 allow only bulk file transfers using FTAM. When Version 3.0 is available, FTAM will be upgraded to allow record-level access as well as bulk file transfer.

Early MAP versions also specified manufacturing message format service (MMFS) for control of factory devices at the application layer. In mid-1987, manufacturing message standard (MMS) (EIA RS 511) became available as an option. While MMFS was a standard created by General Motors Corp., the MMS standard is an international standard. Because TOP networks are designed for use in the office, there is no companion standard for TOP.

By 1988, both the MAP and TOP specifications will include virtual terminal, which will support process to terminal communication by specifying a standard for interfacing terminals to the network. Using a single set of control characters for terminals holds the translation between the control characters used by terminals from different manufacturers to a minimum. Where the number of translations required for N different terminals on a network would be N $(N - 1)$ without virtual terminal, there will be only N translation algorithms required after virtual terminal is included in the specification.

TOP Version 3.0 will also include a standard for electronic mail, the electronic equivalent of the postal service. The standard is based on the message handling system (MHS) application profile that is based on the CCITT (Comite Consultatif Internationale) X.400 recommendation. The MHS standards define two services: (1) basic message transfer service, which supports the general application independent services, and (2) interpersonal messaging, which uses the basic message transfer service to support electronic mail. The TOP specification addresses only interpersonal messaging.

TOP Version 3.0 will also include the product definition exchange standard (PDES). This standard will specify the format for storing information about a product during its entire lifecycle, from design through customer support. The actual transfer of product information is expected to use such existing protocol standards as FTAM and MHS.

TOP Version 3.0 will include a standard for document architecture, which addresses the requirement for interchanging electronic compound documents composed of multiple content types such as character text, raster graphics, and bit graphics. The TOP office document architecture recommends the eight-part ISO 8613 office document architecture and interchange format (ODIF). Currently there are two specified document classes, memo/letter and report. Two others, free-form and technical specification, are under study.

Building Blocks and Future Options

Both the TOP and MAP Version 3.0 specifications will include optional media and access methods. TOP will have an option for a token passing bus on broadband; the MAP specification will include token passing on a bus with carrierband as the medium. There is some pressure on the MAP Users Group to include CSMA/CD as an optional access method in the factory. Both groups are considering fiber-optics, which will probably use the token ring access method. The token ring access method is very similar to token passing on a bus.

The set of application layer protocols available on TOP networks will allow network designers to select "building blocks" of application layer options (see Fig. 16-10). These building blocks will be described by their name and three attributes: function, specification reference, and binding rules. The shaded areas in Fig. 16-10 represent the building blocks, superimposed over the OSI model layers and elements of the TOP specification. There are two sets of these building blocks: (1) user building blocks at the upper layers and (2) network building blocks at the lower four layers. A complete TOP end system will include one or more network building blocks from the bottom four layers and at least one user building block covering the upper three layers.

Selections from the user network building blocks are optional. If two end systems that use different selections at the bottom layers want to intercommunicate, they can do so through a device called a router as long as the protocols used above the data link layer are compatible. A router is a device that translates between different protocols at the physical and data link layers as long as the protocols used at the network layer and above are compatible.

The network designer is also free to choose from the building blocks available at the upper layers, as long as two communicating application processes have at least one building block in common. These sets of standardized building blocks are what give MAP and TOP networks their functionality. A network designer can select building blocks that complement the functionality of application processes without relying on a single vendor to supply all data processing and equipment needs—the true promise of international standardization is freedom from dependence on vendors.

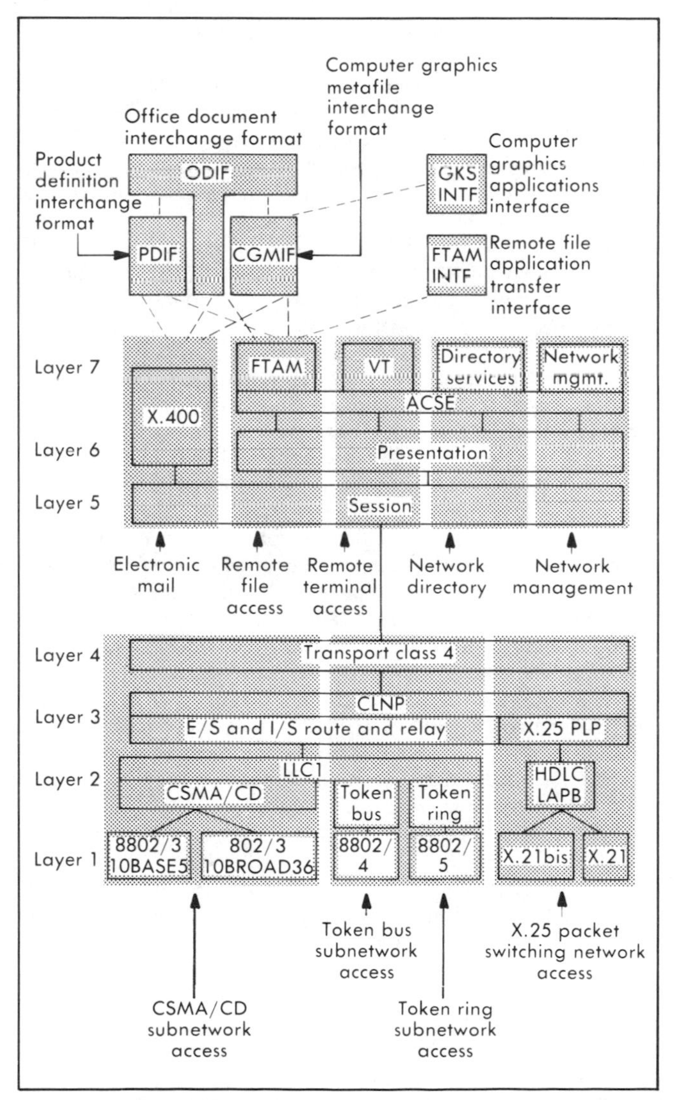

Fig. 16-10 Building blocks available on TOP 3.0.

IMPLEMENTATION OF CIM

Computer-integrated manufacturing is the way of the future for U.S. manufacturing. It has the promise to restore the competitive advantage the U.S. enjoyed for so many years. But CIM is not an easy cure. It requires careful planning; it requires an enlightened management to weigh the investment in CIM in terms of its value as a strategic tool, not just another capital investment. For those companies who choose CIM, the benefits of CIM are real; it can mean the difference between success and failure.

PLANNING FOR CIM

CIM is a manufacturing philosophy adopted by organizations to remain competitive or increase market share. As such, CIM is more of a direction or a management program that may involve the planning for and implementation of a great many technologies in a time-phased manner. Planning for CIM involves planning for its component technologies with CIM as the goal or reference.

The objective of CIM planning is to focus individual projects toward a well-defined set of goals usually established at the highest management level. Ultimately, all systems are expected to work synergistically, supported by information systems, i.e., computer hardware, system and application software, and telecommunications technology and common databases.

Because CIM is so broad in its definition and because CIM implies the implementation of several diverse technologies according to each company's needs, it is probable that no two companies' CIM plans will be alike. Adding to the individuality of the planning process is the fact that few companies can afford the human and capital resources to simultaneously progress on all fronts of CIM. Each company will assign priorities according to its own needs and cost constraints. One thing is certain, the CIM planning process will never be completed. As technologies progress there will always be new technologies to add, older technologies to fine tune, and new strategies to address.

Effective planning requires involvement and cooperation from many different functional parts of an organization, usually in a team or taskforce approach. By drawing from various departments and disciplines, it is possible to assemble a team that understands the many diverse CIM technologies, how they relate to one another, and how they can work together. A team or taskforce approach also has the advantage of being able to better identify benefits of CIM that will occur throughout the enterprise; this is a great help in the justification process.

A CIM checklist has been developed that may be useful for companies contemplating CIM implementation.[6] The list that follows reflects the types of issues companies must address:

- Clear statement of business strategy and goals.
- List of critical success factors (CSF) to support the business strategy. What key things must go right to achieve this strategy?
- Clear statement of CIM strategy that supports business strategy and CSF.
- CEO/GM charter for CIM strategy and implementation program and top support from all functions.
- Identification of the company's starting point:
    1. Level I—Integration, policy and data architecture.
    2. Level II—Interfacing, standards.
    3. Level III—Isolated islands.
- Strong commitment to overall integration, not just excellence in functional islands or convenience in interfacing activities.
- Survey existing systems to determine levels of compatibility. How are many software and hardware systems capable of communicating together?
- Agreement to manage data as a corporate asset.
- Support from the MIS department and good cooperation with engineering and manufacturing.
- Commitment to move information between all functions in digital form.
- Agreement on standards and protocols.
- Use of integration methodologies like those that have grown out of the USAF's ICAM Program.
- Development of an architecture for a common logical database.
- Development of phased process to support the transition to CIM:
    1. Goals/plans.
    2. Conceptual design.
    3. Detailed design.
    4. Implementation.
    5. Benefits tracking and finetuning.
- Ability to learn from and receive support from the hardware and software vendors.
- Ability to learn from experiences of other companies.
- Willingness to streamline or simplify existing operations. (Reduce the complexity index.)
- Reviewing use of just in time and group technology.
- Adjust departmental charters and work assignments to support a networked organization.
- Ability to move from an individual application-bound approach to a data-driven approach, where the same data can be used by different functions for their own unique needs.
- Use of outside resources (universities, professional associations, and consultants).
- Identify potential financial benefits.
- Develop new ways of working with and exchanging information with vendors and customers/clients.
- Make use of public-domain information from the various federal projects, such as NASA's IPAD, USAF's ICAM, and the Advanced Manufacturing Program of the National Bureau of Standards.

JUSTIFYING CIM

Justifying the multimillion dollar expenditure that is involved in CIM is a very difficult task. Although most leading manufacturing companies recognize the need for a better system than the strict financial methods that have been used historically, very few have devised a justification methodology that adequately considers the strategic and nonquantifiable benefits of CIM.

Too often, it remains the job of the CIM taskforce to plan the CIM investments in a time-phased fashion, so that financial paybacks can be achieved and so that qualitative benefits recognized during the course of one phase can be used to simplify the justification of subsequent phases.

In too many cases, company management looks at CIM investments as decisions with only short-term implications. They are, in fact, strategic decisions that will affect the future, perhaps the existence, of the organization. Some of the problems associated with using traditional justification methods include failure to consider the changing business environment and failure to recognize the intangible benefits of CIM.

In the first case, capital expenditure requests are oftentimes evaluated against a "base case" of no new investment. Moreover, the base case typically assumes a continuation of the status quo regarding the external competitive environment. In reality, the external environment is constantly changing, and many of the more important changes are quite predictable. Some of these changes are as follows:

- *Adoption of new technology*. Once new process technology becomes available, it will likely be adopted by leading domestic and international manufacturers. In particular, new process technology tends to be rapidly introduced into the manufacturing operations of companies participating within the international marketplace in the so-called "early adopter" industries of automotive, electronics, and aerospace. These companies are likely to reap benefits in terms of quality, flexibility, responsiveness to customer requests, and the ability to attract and retain engineering talent.
- *Market share changes*. Companies adopting new process technology are likely to reap benefits in terms of quality, flexibility, and responsiveness to customer requests. These companies will enjoy, at least temporarily, comparative advantages over their competitors.
- *Prices*. History has taught that, in other than depressed markets, prices tend to increase over long periods of time. Ignoring the virtual certainty of price increases leads to underestimates of revenue and more difficulty in justifying the new investment.

A valid cost/benefit analysis taken over a reasonable length of time, say 10 or so years, must incorporate "best judgment" estimates of changes in the external environment. The assumption of continued cash flows for the base case is not realistic. A more valid assumption is that the base case of no new investment will result in declining cash flows as the company falls behind its competitors.

It is well accepted that CIM technology yields benefits that are difficult to quantify, that may be indirect, and that are oftentimes characterized as "intangible" benefits. The oft-used "intangible" label is unfortunate because many of the typically cited benefits such as improved quality and reduced inventories are quite tangible.

These benefits can be difficult to quantify accurately and incorporate into a discounted cash flow analysis. Some—such as quality, manifested by known scrap rates, rework activity, and warranty expenses—may indeed have quantitative data available from which to formulate informed estimates of benefits. Others—such as managerial control, responsiveness, or quality—are much more subjective and so are virtually impossible to assign numerical values.

Perhaps a better approach to justifying CIM investments lies in developing a system model of the business, using ICAM Definition Method (IDEF) or other available simulation modeling methodologies. Any such model should include the various functions and transactions that typically take place in a manufacturing enterprise. IDEF is discussed in greater detail in Chapter 8, "Management of Technology," of this volume.

A completed systems model can then be used with a simulation tool (discrete event simulator or spreadsheet software package) to examine the effects of introducing new technologies into the manufacturing operations. Pro forma financial statements for both "As Is" and "To Be" situations can be developed, as can credible cost/benefit analyses.

As noted earlier, the development of such a comprehensive model will generally fall beyond the scope of any one department. The creation of a credible business model, encompassing multiple interrelated functional activities and driven by accurate data and realistic market scenarios, requires strong top-down management commitment and the dedication of a working-level group. Additional information on equipment justification can be found in Chapter 3 of this volume.

BENEFITS OF CIM

Although the qualitative benefits of CIM are not often factored into a cost justification equation, it is well accepted that CIM technology yields many unquantifiable benefits. Among the most important benefits of CIM are improved productivity, faster introduction of new and/or modified products, and improved traceability of specific jobs and processes. Some of the more important strategic benefits of CIM are contained in Table 16-3.

TABLE 16-3
Strategic Benefits of CIM

Benefit	Description
Flexibility	Ability to respond more quickly to changing mix/volume requirements
Quality	Resulting from automated inspection and greater consistency in manufacturing
Lead times	Substantial reductions resulting from information integration efficiencies
Inventories	Reduced WIP and finished goods inventories due to lead time reductions and timely access to accurate information
Managerial control	Reduced control resulting from information access and computer implementation of production decisions
Floor space	Reductions resulting from more efficient layouts and integration of operations
Options	Prevents potential pre-emption by competitors by maintaining option to exploit new technology

CIM AND FIFTH-GENERATION MANAGEMENT*

It makes little sense to install third, fourth, and fifth-generation computer technology in second-generation organizations. Yet this is precisely what many manufacturing companies are doing. Why? Because they have always installed new technology in their existing organizations. It has not occurred to many that the logic of a computer-networked technology fundamentally changes the way organizations work.

This chapter has introduced the key notions of computer-integrated manufacturing (CIM), its definition, and its related technologies, key components, and the networking approaches such as MAP and TOP that will bring it about. The impression is often given that CIM is 80% technical and perhaps 20% organizational. Yet CIM practitioners are quick to realize that the implementation of any CIM system cuts across many different turfs. It is not long before there is more politics than technology. Soon it becomes clear that CIM is 80% organizational and 20% technical. Why?

CIM cuts across all functional and departmental boundaries. It involves everyone in new ways. Most are not ready for this. In fact, it goes against the grain of the ''common logic'' of the Industrial Era. The idea has always been to take a large process and divide it up into its smaller steps, like Adam Smith suggested with the pin-making factory in the *Wealth of Nations* (1776). At first this applied to production workers, but as organizations grew, the notion of the division and subdivision of work was applied to the managerial ranks as well. Alfred Chandler has chronicled this shift in his classic study, *The Visible Hand, The Managerial Revolution in American Business*.Chandler documents the rise of a hierarchy of middle and top-salaried managers who monitor and coordinate the work units under their control. The railroads, telegraph, telephone, and large manufacturing concerns ushered in these managers with well-defined job responsibilities and clear departmental charters. It is not surprising that there should be resistance to a technology that blurs traditional functional distinctions, as CIM and networking does. And it is not surprising that managers should resent the invasion of their turf and prerogatives that they had so carefully established over the years. Modern *luddites* are not the workers, but middle managers who feel threatened by all the changes. The irony is that they do not need to fear the changes if they can grasp the logic of the new technology.

If the changes can be dealt with, the organization is likely to be a much more humanly challenging and exciting place to work. It will rely on a great deal more human creativity and initiative to take advantage of the built-in flexibility in the organization. These changes need to be put into perspective, and it becomes clearer as to what the benefits will be. This leads to an examination of the nature of Fifth-Generation Management.

Fifth-generation management (FGM) assumes a well-developed and flexible infrastructure of networked functions, together with their computers and applications, capable of referencing a common data architecture. Each of the functions becomes a node, or decision point, on the network.[7] These nodes become reference points or knowledge centers, capable of teaming with other nodes in the support of the enterprises business strategy. Rather than being held together by a ''command and control structure,'' FGM is coordinated by its shared vision, values, and culture. This arrangement frees up the organization from being locked into the management of ''sequential handoffs'' and provides a way of coordinating the ''iterative dialogue'' that must go on between functions, such as ''design for assembly.'' In short, FGM makes it possible to effectively leverage the information infrastructure for competitive advantage.

FGM calls for a new leadership style and expects more of all employees in dealing with business variety, so windows of business and technical opportunity can be more effectively met. Moreover, it recognizes human interaction as an important quality issue, as is product quality, if competitive advantage is to be achieved. Finally, it taps into the power, wisdom, and insight of its managers, professionals, and employees who have, as a team in a nodal network, learned to use computer-based resources to enhance their decision-making capabilities.

These and related notions will be explored in this section. First, the discussion of fifth-generation management needs to be put in perspective. Therefore, the next section defines the generations of computers and generations of management.

FIVE GENERATIONS

Thanks to the Japanese initiative, there is much talk about the fifth-generation computer. The five generations of computer technology are generally defined as shown in Fig. 16-11.

Essentially, the first four generations of the computer must pass all information through a single central processing unit (CPU). This single CPU has been described as the ''von Neumann bottleneck.'' It is named after John von Neumann, computer pioneer and mathematician, who essentially designed the architecture that became embodied in the first through fourth-generation computers.

The key to the fifth-generation computer, as has been pointed out, is in the networking of multiple processing units (PU). This linking provides a new task, that of dividing the problem so that these PUs can work on portions of the same problem concurrently and in parallel, then piece together the whole solution.

There is not a well-defined list of the five generations of manufacturing enterprises, but a possible list of these is shown in Fig. 16-12.

In the first four generations of enterprise management, raw materials and information are passed *serially* from one department to the next, as in Adam Smith's pin-making factory with its division and subdivision of labor. These concepts are reinforced by Frederick Winslow Taylor in *Scientific Management*. Moreover, the hierarchical mode of organization predominates, even in third and fourth-generation management. Matrix management does not fundamentally challenge the basis of power in second-generation organizations. Fourth-generation management does help to build the necessary infrastructure for a new way of doing

* This section on ''CIM and Fifth Generation Management'' was written by Dr. Charles M. Savage of Digital Equipment Corp. (Marlboro, MA).

CIM AND FIFTH-GENERATION MANAGEMENT

●First:	Electronic vacuum tube
●Second:	Transistor
●Third:	Integrated circuit
●Fourth:	Very large scale integration
	----------------------------------
	von Neumann bottleneck

●Fifth:	Parallel networked process units and symbolic processing

Fig. 16-11 Generations of computer technology.

●First:	Small/entrepreneurial
●Second:	Hierarchical/functional/divisional
●Third:	Matrix
●Fourth:	CIM I—Computer-interfaced manufacturing

	Smith/Taylor bottleneck

●Fifth:	CIM II—Computer-integrative management of the manufacturing enterprise

Fig. 16-12 Generations of enterprise management.

business, although it does not overcome the habits and practices of second-generation management.

Fourth-generation management, digital interfacing, is what should be called *Computer-Interfaced Manufacturing* (CIM I). So much of what is called CIM is, in reality, *Computer-Interfaced Manufacturing*.

Fifth-generation management, nodal networking, assumes the *Computer-Integrative Management of the manufacturing enterprise* (CIM II). First, its focus is not just on the manufacturing function, but the entire enterprise. Second, it is integrative, not integrated, because manufacturers are involved in a continually evolving integrative process. No one will awaken one day to find everything ''integrated.'' Third, each of the departmental functions and subfunctions become nodes, or decision points, in a network capable of bringing their accumulated knowledge to bear in an interactive mode as the functions work in *parallel*.

Interfacing is not the same thing as *integrating*, and yet the confusion between the two is leading to deep frustrations and, in some cases, a rejection of the idea of CIM itself. CIM I will fail to solve the fundamental problems of the hierarchical organization because it, like matrix management, is an overlay on second-generation management.

Manufacturing Automation Protocol (MAP), Technical Office Protocol (TOP), and the Initial Graphic Exchange Specification (IGES) are helping second-generation organizations move toward fourth-generation management. They will make possible the digital interfacing of information between functional departments or between organizations. MAP and other networking approaches will play a large role on the shop floor; TOP will help tie engineering and manufacturing. IGES provides a neutral file format for passing information between dissimilar CAD systems. But by themselves they will not lead to fifth-generation management because they remain technologies and do not begin to address the values and cultural aspects of integration. This leads to a brief consideration of the clash in logic between the first four generations of management and fifth-generation management.

CLASH IN LOGICS: CIM I AND CIM II

Goaded on by promises about computer-integrated manufacturing, companies are focusing on the technology and forgetting the logic of management. The new computer-based technology is being put into traditional, Industrial Era companies.

The logic of Industrial Era enterprises (1776-1976) has been to divide and subdivide the production processes to manage the sequential flow of raw material through work in process (WIP) to finished product, as in Adam Smith's pin-making factory and Taylor's *Scientific Management*. This makes it easier to assign responsibility and maintain accountability. This organization is usually pictured as a hierarchical arrangement with lines and boxes. It is the source, unfortunately, of many political turf battles, even as the informal organization gets the work out the door. Too often, the written company policy is superseded by a series of accommodations (unwritten IOUs) between functional departments.

But this has led, in many instances, to what George Hess, vice president of systems and planning for Ingersoll Milling Machine Co., has called ''human-disintegrated manufacturing,'' because humans have had to learn to conform to the idiosyncrasies of an environment that has been slow to fully appreciate and use the full range of their talents. Moreover, the hierarchical organization has made meaningful communication between departments cumbersome at best.

If CIM is confused with CIM I, then problems may arise. Many companies are indeed stumbling along with their CIM efforts. As long as the emphasis is primarily on the COMPUTER and MANUFACTURING with a bent for *interfacing*(CIM I), this trouble will persist. The difficulties are caused because each functional department has its own dialect that is difficult to understand by the other functions without adequate translation.

In the traditional manual manufacturing approach, human translation takes place at each step of the way. As information is passed from one function to the next, it is often changed and adapted. For example, manufacturing engineering takes engineering drawings and red-pencils them, knowing the product can never be produced as drawn. The experience and collective wisdom of each functional group, usually undocumented, is an invisible yet extremely valuable company resource. Computer-interfaced manufacturing (CIM I) bypasses this reservoir of valuable knowledge.

Part of the problem is that each functional department has its own set of meanings for key terms. It is not uncommon to find companies with 4 different parts lists and 9 bills of material. Key terms such as part, project, subassembly, and tolerance are understood differently in different parts of the company. When files with these and other terms are expected to be used directly in other departments, there will be problems because of the following four conditions:

1. The same words are used, but they have different meanings.
2. Different words are used, yet they may have the same meaning.
3. And the same words have differing shades of meaning.
4. The same words take on different meaning, depending on the context in which they are used.

The traditional human translators—seasoned employees—can pick up these nuances and compensate for them as in Fig. 16-13. This is a very valuable company resource, although it is not usually recognized.

The International Organization for Standardization (ISO) has developed the seven-layer Open Systems Interconnect (OSI) model that serves as a reference model for the interfacing of various applications (see Fig. 16-9).

Direct digital interfacing short-circuits the valuable service of the human translation, as illustrated in Fig. 16-13. So when digital files, sent from one department to the next, are expected to be used directly, there can be both major and minor misunderstandings concerning the definitions of key terms or data fields (see Fig. 16-14). This is because the different applications have been developed by different people at different times, usually without reference to a common reference model.

It is not enough to simply interface these nodes or decision points. They need to be related to a common reference context: strategic vision, values, common reference data architecture, group technology, and so forth. This common reference context (the management context thread that will be discussed subsequently) has three time elements: the strategic business vision (the future orientation), the flexible infrastructure of networked computers and professionals (present orientation), and stored knowledge and common and agreed-on definitions of key terms (past orientation). These three elements will be discussed in greater detail in a subsequent section.

Now consider an eighth layer composed of three sublayers. The adding of this layer represents the transition to CIM II, which supports fifth-generation management. The new logic assumes a networked infrastructure and adds a way of coordinating the managers, professionals, and employees in a dynamic and flexible manner, allowing for reconfiguration of the business.

The Industrial Era organization was held together by a "command and control" structure where everyone's job was clearly defined. The new Knowledge Era requires a new approach to control and coordination. As the functions use the computer-networked infrastructure to communicate in an iterative manner, the management challenge becomes one of giving focus and direction. Rather than being separate boxes on an organizational chart, the functional decision points (nodes) overlap in their responsibilities in a flatter and more participative environment. There is potential ambiguity in this type of structure. The management challenge is to manage this ambiguity and variety and "thrive on chaos," as Tom Peters has stated in his book *Thriving on Chaos*. A clearly articulated strategic business vision becomes the glue of organization. An understanding of the enterprise's "critical success factors" helps to orient diverse activities. Fifth-generation management takes place at the eighth level of the model (see Fig. 16-15).

When attention truly shifts to the process of integrating the functions in a more dynamic whole, then the stage will be set for more rapid progress. Certainly CIM I, the interfacing of the key functions, is a necessary precondition for CIM II. It is part of the process of building the necessary communications infrastructure. But the physical linking of functions will not lead to true integration, hence the distinction between CIM I and CIM II. The infrastructure of CIM II is not only concerned with "connectivity," the physical linking of computer nodes, but it is also "attitudinal" and "referential." Attitudes of inclusion between functions, openness, trust, growth, and intellectual honesty are essential starters. The referential context (as articulated in the strategic business vision) the critical success factors, and common reference data architecture are also key factors.

In short, the distinction between CIM I and CIM II is as important, if not more so, than the shift from material requirements planning (MRP), which dealt with just production and inventory control, to manufacturing resource planning (MRP II), which includes most all aspects of the manufacturing planning and control.

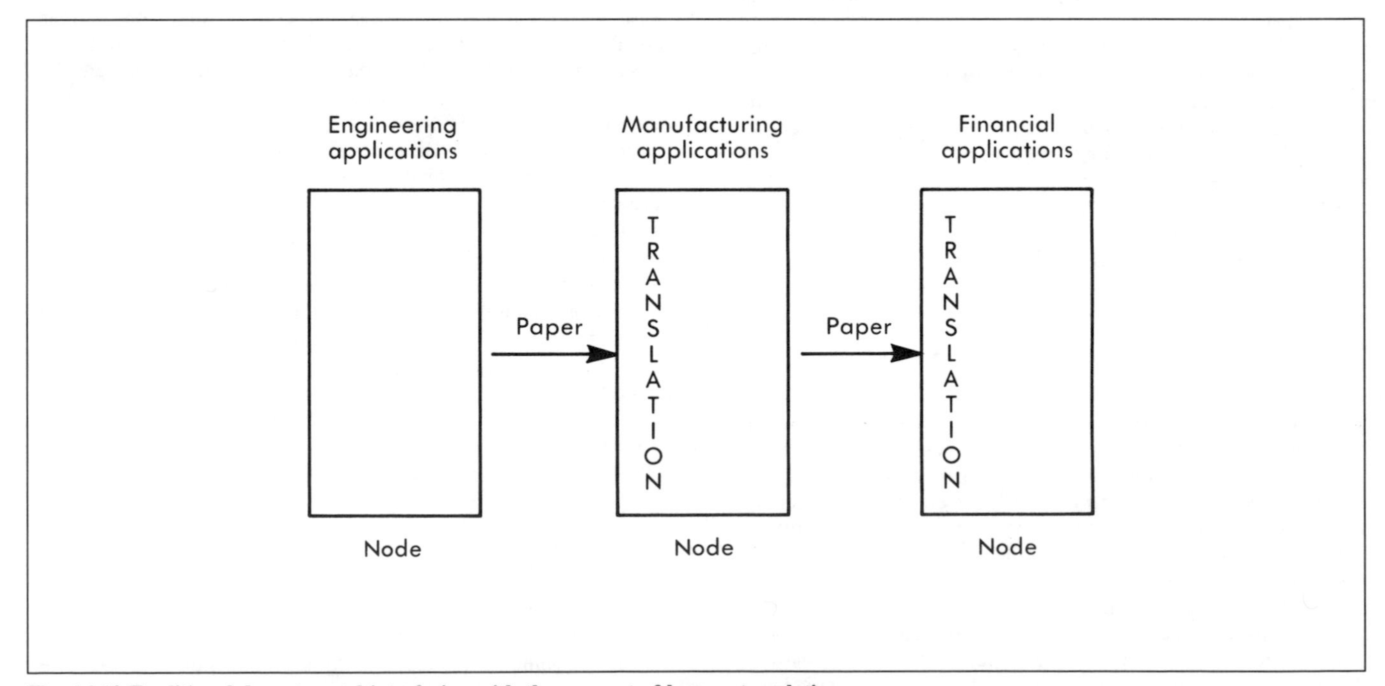

Fig. 16-13 Traditional departmental interfacing with the support of human translation.

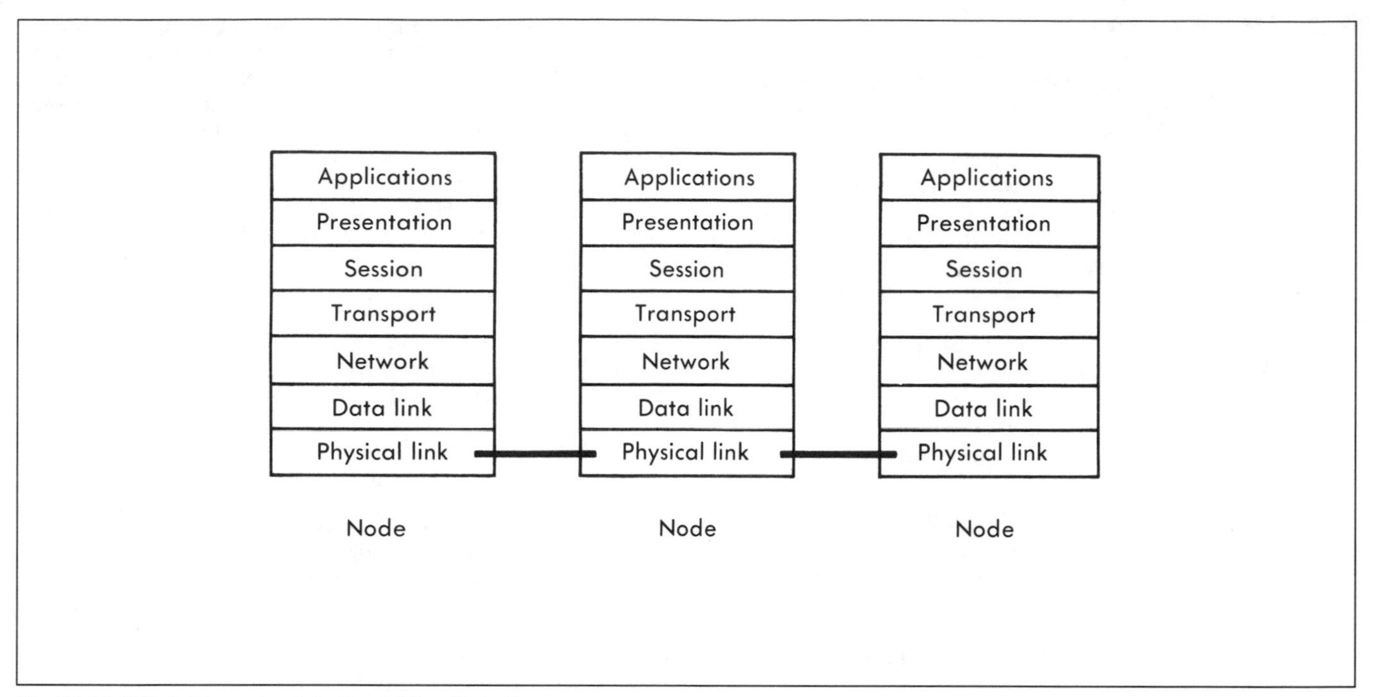

Fig. 16-14 CIM I: Computer Interfaced Manufacturing.

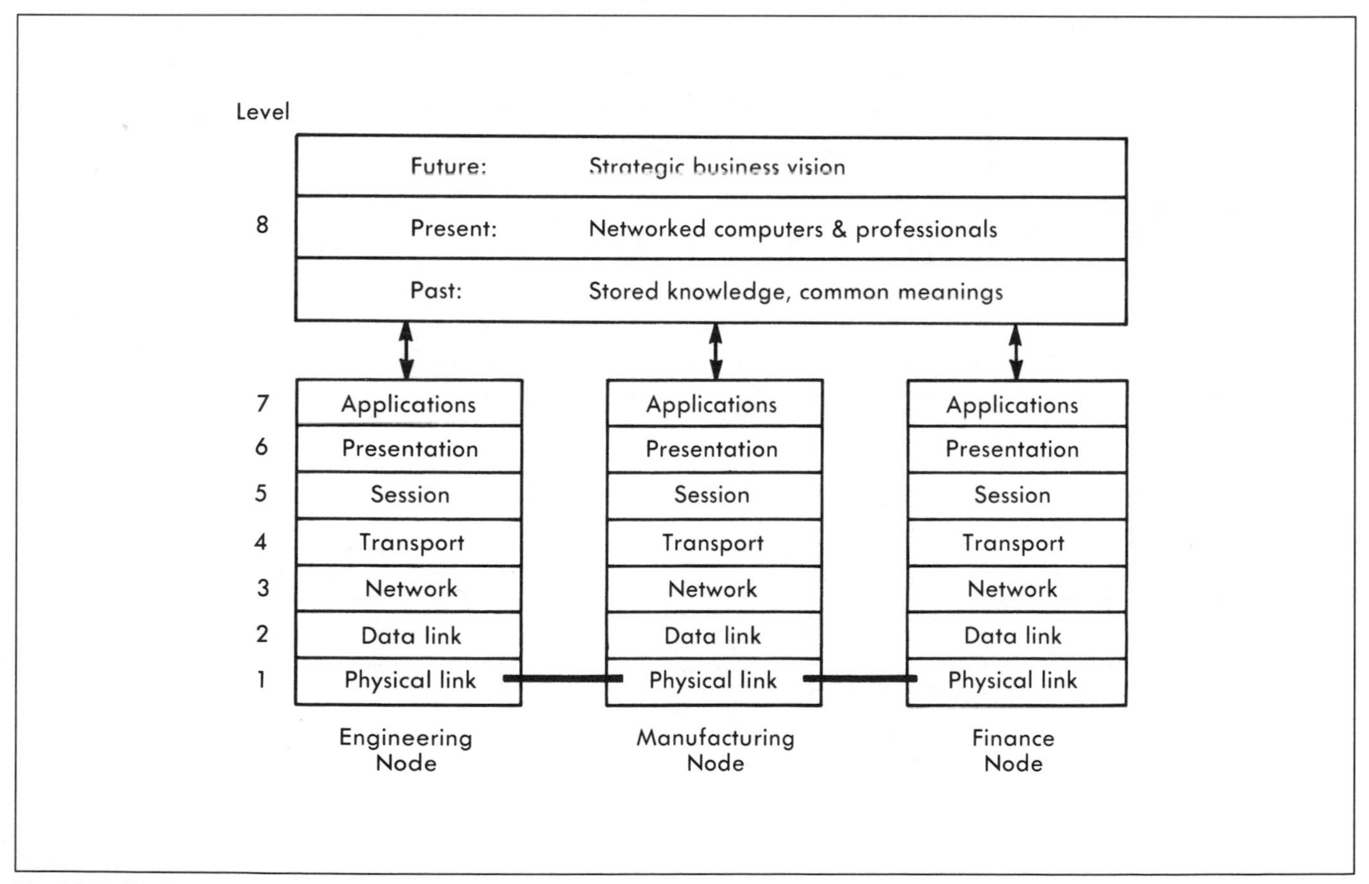

Fig. 16-15 CIM II: Computer Integrative Management Eighth Level Model.

CHAPTER 16

CIM AND FIFTH-GENERATION MANAGEMENT

TROUGH OF CONFUSION

Are manufacturers facing a contradictory situation? The promises of CIM led manufacturers to expect a more wonderful today, yet the reality is different. Some executives are remarking: "The technology is not what was sold to us, and our organization is confused by the new logic implicit in it." As in any era of major transition, as people leave the tried-and-true way of working there is a period of confusion and uncertainty.

Certainly, there have been benefits to CIM I. The computer has had a significant impact on the operations of individual functional departments. Yet from a historical perspective, CIM I will be seen as a bottleneck, just as the von Newmann architecture for computers is a bottleneck that must be broken through if the fifth-generation computer is to be made a reality. Breaking the bottleneck has its price. CIM I brings with it the "trough of confusion" as new relations get worked out, new modes of operating are discovered, and new norms are established. Figure 16-16 illustrates this point.

THE LONG TRANSITION PROCESS

As has been pointed out, the first four generations of computer technology have been based on the notion that everything must pass through a single central processing unit that operates in a *serial* fashion, step by step. The fifth generation assumes the networking of multiple processing units (PU) in *parallel* to allow new memory organizations, new programming languages, and new operations for handling symbols and not just for handling numbers.

The Japanese have set out on a 10-year transition process during which they are hoping to develop their fifth-generation computer. In a similar vein, the shift from CIM I to CIM II and ultimately to FGM will be just as time consuming in the transition. In fact, it may well take longer, although any manufacturing nation cannot afford to wait that long for more integrative management of the manufacturing enterprise. Hopefully, the process can be set in motion to near completion in a 5-year time period, as is suggested by Fig. 16-17.

The transition to FGM will certainly not occur overnight. There is so much more that needs to be understood. At this point in time, there is only a very sketchy vision of FGM. The transition to integrative management that is possible under FGM will be no easy matter. The weight of past practices (legacy) is strong and will certainly carry on well beyond 1992. This is why Fig. 16-17 shows a continuing presence of the legacy of the early generations of management that continues to flow along with the present.

The next section briefly sketches out some of the changes taking place in the manufacturing environment.

MANUFACTURING: A CHANGING CONTEXT

Traditionally, manufacturing enterprises have been thought of as hierarchies managing activities through a process of *serialized* handoffs of product and information from one function to the next. As mentioned, Smith's pin-making factory, with its division and subdivision of labor, and Taylor's *Scientific Management* have served as convenient paradigms for this process, as seen in Fig. 16-18.

However, there are challenges to this model. First, companies are producing more with less direct labor (3-15% of costs, depending on the industry). Second, they are finding they need

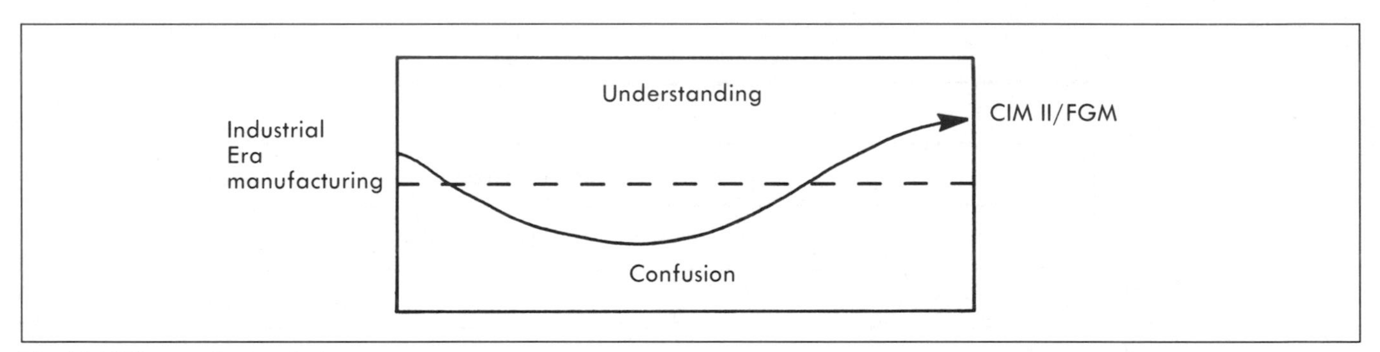

Fig. 16-16 The trough of confusion.

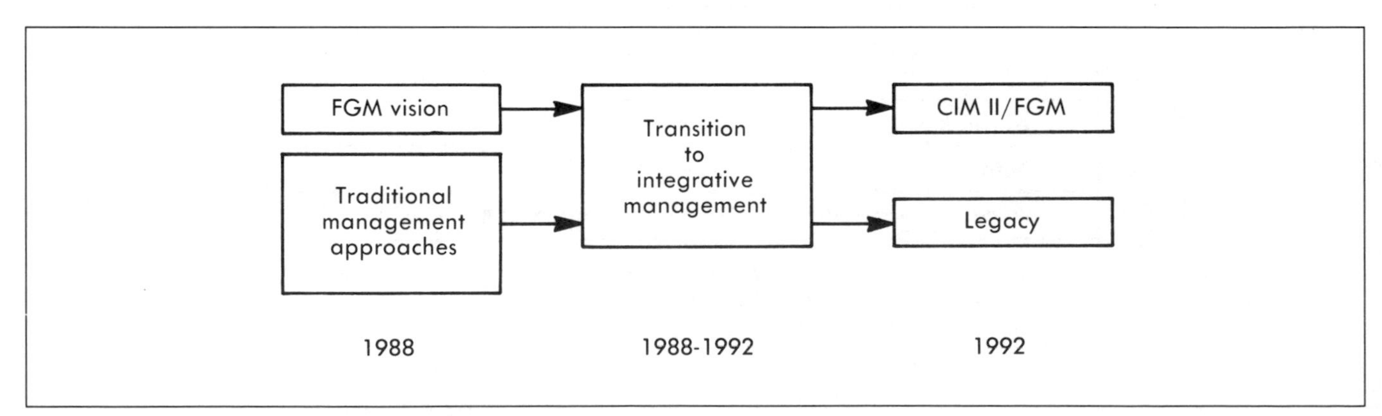

Fig. 16-17 The transition process to CIM II/FGM.

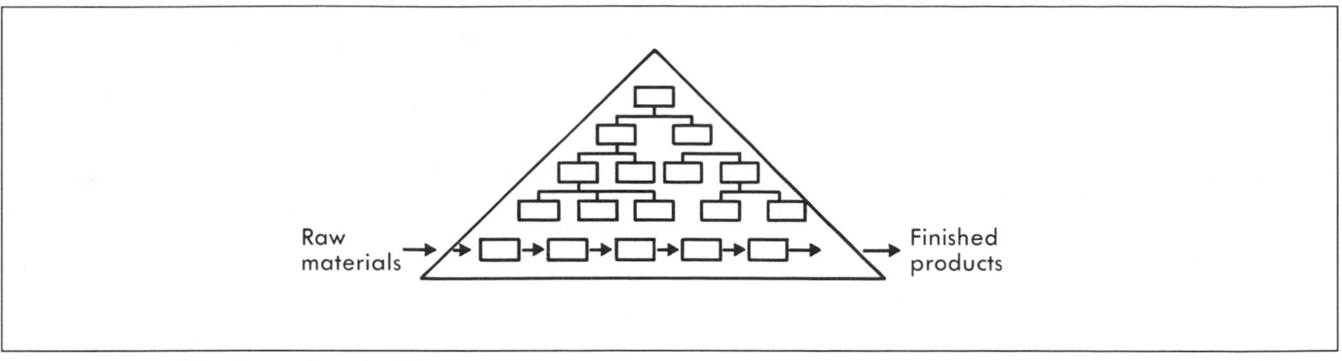

Fig. 16-18 The traditional manufacturing hierarchy.

fewer levels of management (down from 8-12 levels to 4-6 levels). Third, the transformation of data and information into useful knowledge is now as important, if not more so, than the transformation of raw material into finished goods. Only a few touch the product, but almost everyone touches the information about it.

How does the organization begin to look when the direct labor base shrinks and the levels of management are thinned out? Figure 16-19 shows the changes.

As these changes take place, a new organizational shape, the spade, begins to emerge. It is flatter in shape, and the direct labor base has shrunk. The question arises: How is this new-shaped organization to be managed?

Manufacturing in the 1990s will demand a qualitatively different approach to management from what has been known in the past, just as the shift from a von Neumann to a non von Neumann architecture (fifth-generation computer) requires radical new approaches to computer science. What type of management will this be? This question is at the heart of the quest for fifth-generation management.

Joseph Harrington, the father of CIM, set everyone on a new course with the integration of manufacturing into a seamless whole. Unfortunately, as has been indicated, many people mistake computer-interfaced manufacturing (CIM I) with computer-integrative management of the manufacturing enterprise (CIM II). CIM I leaves the traditional structure in place and simply wires together (networks) the various functions through digital communication. It still operates in a *sequential* fashion. CIM II is based on the various functions working together in *parallel*.

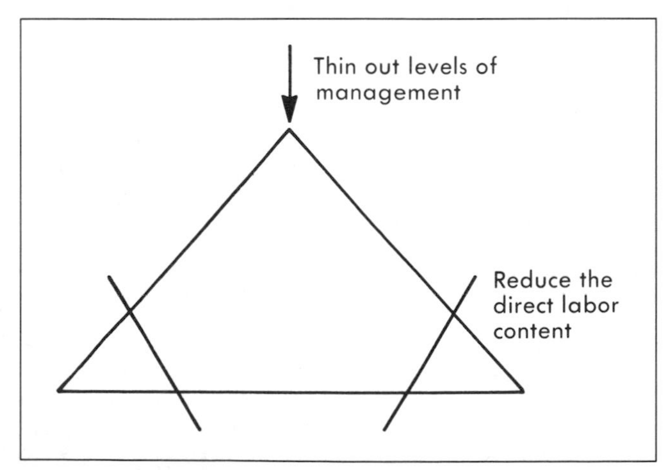

Fig. 16-19 Enterprise management in transition.

To repeat, the logic of traditional manufacturing management, even with CIM I, is almost 180° opposite from the logic of CIM II. If this goes unrecognized, then new technology will continue to be "stuffed" into old organizational skins.

The logic of CIM II rests on a much deeper understanding of organizational integration. This leads to the search for understanding FGM, which needs its own perspective.

In FGM, what has to be integrated? How is this integrative process to be carried out? To simplify a complex process, FGM will require the interweaving of five threads.

FIVE THREADS

There are five major areas in a manufacturing enterprise that need to be dynamically interrelated and integrated. Traditionally, each area lives in its own world, with its own professional societies and training. Each has its own educational feeder systems. The task is to weave a tapestry that leverages the strengths of the entire garment.

The five threads of CIM II include the management context thread, the business thread, the technical thread, the information architecture thread, and the production systems thread. Refer to Chapter 6, "Organization," for a detailed list. The notion of the weaving of four threads (all except the management context thread) comes from work done under the U.S. Air Force ICAM program as well as work at LTV Corp. (Dallas, TX) and Advanced Machining Systems—General Dynamics (Fort Worth, TX). The points of intersection of the tapestry should be thought of as nodes where decisions are made.

Figure 16-20 illustrates the interrelationship of these five threads of CIM II. Many of the elements under each thread may be represented by an applications program. Manufacturing companies usually have a large portfolio of different software applications to assist the various functions. It is the weaving of these applications together within the management context that is the challenge of fifth-generation management.

Companies are and will be experimenting with how best to weave the thread and their constituent applications together. To date, most of the emphasis has been on the interfacing of the business, technical, and production threads with their applications. For example, a few companies are pulling out their bills of materials directly from their digital CAD drawings. They are also driving these directly into their NC systems.

Traditional management information systems (MIS) have been primarily concerned with aspects of the business thread

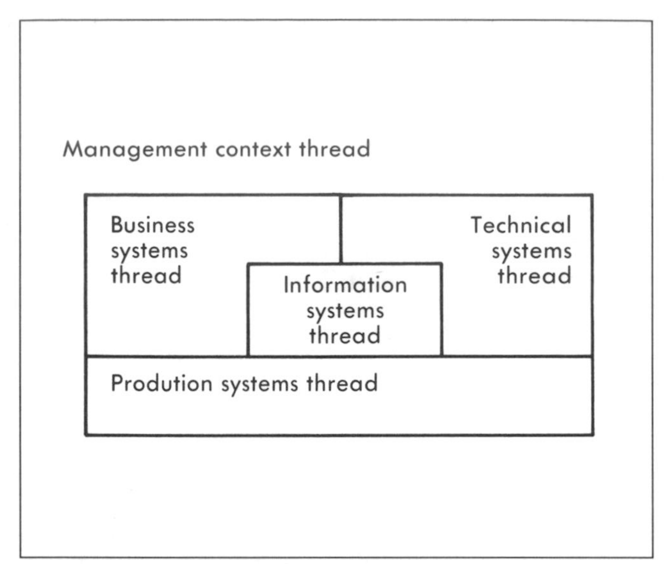

Fig. 16-20 The five threads of CIM II/FGM.

such as the financial, payroll, and MRP applications. Successful CIM implementations show that the MIS is deeply involved with supporting engineering and manufacturing. This team spirit of MIS needs to be expanded to all companies, not just the few companies who are industry leaders.

The problem is that most large manufacturing companies have from 50 to 1000 different application packages that can be tied together only with extreme difficulty, if at all. Each application has its own captive database. Although there is redundant data in most of these databases, it cannot easily be shared because the definitions of the data elements are slightly different. This is primarily because there has been no universal architecture for information systems.

This architecture has to be redone as new systems tools become available, such as computer-aided software engineering (CASE) and computer-aided data engineering (CADE), which can leverage the power and flexibility of relational databases, artificial intelligence, and beyond. These are the cornerstone developments of CIM II and will make true integration possible. Systems will have to be rewritten for this to be accomplished, so they can be structured rather than just mashed together.

For it to be possible for these systems to be integrated, particular attention will have to be paid to the management context thread. It is not possible in this short study to discuss the other threads in any detail, except to explain the naming of the information architecture thread.

The information architecture thread has in its name the word ''architecture'' instead of ''systems'' because information requires more than the hard-wiring of diverse systems. It requires the building of an explicitly designed and structured infrastructure. This infrastructure is composed of four elements: the hardware, applications software, communications capabilities networks, and data architecture (with its supporting glossary of definitions and relationships as well as the DBMS). When this is done, the necessary connectivity (physical network) will be established together with an adequate referential infrastructure defining the key information elements needed to run the business, allowing the applications to be run against this common data architecture.

The next section will briefly review the highlights of the management context thread because it sets the ''context'' for the entire enterprise. The management context thread has to be carefully designed, grown, and nurtured within the manufacturing enterprise.

THE MANAGEMENT CONTEXT THREAD

There are two new key concepts that lie at the heart of the management context thread. They deal with ''nodal management'' and a rediscovery of the nature of ''human and organizational time.'' Once these two concepts are understood, then the other responsibilities of the management context thread will gain a new perspective.

Nodal Management

Traditional Industrial Era hierarchies assure that the functions mesh together like ''cogs in a gear'' as they hand off work sequentially from one function to the next. Everyone has their assigned place. If properly ''trained,'' the whole will function well. Unfortunately, this approach does not give the enterprise the flexibility and resiliency it needs to stay competitive in a changing economic environment. This ''mechanical'' view of organizations is no longer adequate in the Knowledge Era. Another model is needed, especially when there is less direct labor and fewer levels of management.

In the Knowledge Era, every person in the enterprise plays a more significant role. Much of the routine can be off-loaded to automated devices. Each person becomes a decision point—a node—within a larger network of persons who can work in parallel in an iterative manner. This change affects everyone from the worker on the shop floor to the executive at the desk in the front office.

H. Chandler Stevens has perceptively captured the significance of this shift from ''cogs'' to ''nodes.'' In a brief poem entitled ''The Networker's Creed,'' he writes:

I'd rather be a node in a network
 than a cog in the gear of a machine.
A node is involved with things to resolve,
 while a cog must mesh with cogs in between.

The task of managing a network of nodes or decision points is qualitatively different from managing a standardized set of functions, as is done in second-generation organizations. The glue that holds the enterprise together is different. In hierarchical/functional organizations, the responsibilities of everyone should be predefined. Work is done at one function, then handed off in a serial manner.

FGM requires nodes (people as decision makers) to interact on an ongoing basis as new products and processes are developed. Marketing perceives a new market opportunity, so it has engineering sketch out a possible design. Manufacturing engineering is asked to simultaneously sketch out the production process, and finance assists in costing it out. These functions work back and forth to refine their concepts and determine the probability of market success. Each of the functions serves as a node and ''is involved with things to resolve.''

In a nodal network, it is a give-and-take or back-and-forth environment, rather than a sequential handoff process, as is true in a traditional manufacturing enterprise. This difference means there is a need for continually focusing the efforts of the nodes. This is where the need for a good, clear strategic business vision comes in. As has been mentioned, this strategic business vision becomes the new glue to hold together and focus the decisions being made by the various nodes.

In all companies, a strategic business vision is not an off-the-shelf commodity. It must be crafted and grown through the interactions of many key persons in the organization. It must be engineered, in the creative sense of the word, by a process that may be called "visioneering." Often professionals and middle managers have innovative solutions to marketing's perceived needs, and these individuals can help develop this strategic vision in dialogue with top executives. This is the strategic vision process that supports the development of the enterprise's business strategy and planning efforts.

In the traditional organizational chart, responsibilities are clearly delineated by the boxes on the chart. This responsibility is usually narrowly defined. Each function guards its own turf.

In a nodal organization, however, there is a great deal of overlapping of responsibilities. The horizon of concern extends beyond the narrowly confined borders of individual functional responsibilities. For example, engineering has to worry about the way a product will be serviced, a new concern for many. The black-and-whiteness of second-generation management is replaced by gray and fuzzy borders in FGM. The regimentation of the Miami Dolphins is replaced by the fluid maneuvering of the Boston Celtics.

The gray and fuzzy borders of overlapping nodes suggest that there are problems to be faced in developing marketable products and that processes cannot be easily compartmentalized. Most business challenges require the insights and experience of a multitude of resources, which need to work together in both temporary and permanent teams to get the job done. This is why the discipline of good project management needs to become a more integral part of a manufacturing enterprise. The role of project management will be discussed further subsequently.

Rather than being thought of as conventional functions, as in second-generation management, the nodes in FGM are really knowledge centers. They unite around the challenge to bring the insights of their disciplines to its solution. This means that the nodal knowledge centers do not wait for responsibility to be assigned to it. Instead, managers and professionals assume responsibility for projects after the proper internal negotiations.

The quality of human interaction between professionals and managers becomes of paramount importance in this context. In fact, the "quality of human interaction" needs as much attention and nurturing as "product quality." Both pay handsome dividends. This quality interaction is helped by the type of management and leadership style present in the enterprise.

Management and leadership style. Integration requires more than just technology. FGM and CIM II cannot be achieved without strong executive leadership. Success demands the true integration of digital technology with the realignment of departmental charters, reward systems, accounting practices, organizational design, career paths, and management style.

The CEO or general manager, together with the vice presidents of marketing, engineering, manufacturing, finance, human resources, and information systems, faces a critical and creative challenge. FGM requires the integration of most of the aspects of the enterprise, so the rewards systems, motivation strategies, and accounting practices will complement and support the growing digital infrastructure being developed. Moreover, it will demand a new and more imaginative style of leadership at the top.

The contrasts between leadership styles of second-generation management and fifth-generation management are striking. Second-generation management has often taken the notion of division and subdivision of labor to such an extent that the fragmented functions become divisive, looking out for their own interests. This leads to a sense of disengagement because the fragmented functions do not have a clear sense of the whole. Management is usually by declaration, and commands come from the top down.

In contrast, FGM nurtures an integrative atmosphere where the knowledge centers are expected to work together in an iterative fashion. In this context, management can more effectively lead interrogatively—that is, by well-placed questions. In fact, a well-chosen question can be a much more powerful motivator for a manager or a professional. It allows them room to "show what they know." Well-placed questions can also be extremely effective in keeping the various working teams focused on the business objectives of the organization.

Top executives can also use questions to bring the various functions into closer working relationships. Too often, CIM is an undertaking of the vice president of manufacturing, or the vice presidents of manufacturing and engineering. But the litmus test of true CIM II effort will be based on the involvement not only of manufacturing and engineering, but also marketing, finance, human resources, and information systems.

For example, if a company has a "CIM program," does it begin to address questions such as the following:

- Is finance redesigning the accounting system to get beyond the simplistic burdening of direct labor, especially because this base is continually shrinking?
- Is human resources redefining jobs to build in more career growth? Is it redefining compensation systems to reward the sharing of knowledge across functional boundaries?
- Is marketing working with engineering and manufacturing to ensure "designing for marketability" and not just "designing for producibility"?
- Is management redefining responsibilities in an overlapping or nodal, rather than functional, manner with the use of effective project management to maintain focus and momentum?

These are a few of the questions that might be asked of the top management team as it works together with middle management and other professionals to redefine basic relationships. Their meaning will become clearer in subsequent sections.

Politics, cultures and values. Needless to say, these changes can be threatening and can upset the culture and delicate balance of political accommodations between functional groups. Surely, many people in the organization will not want their turf violated. Change brings more uncertainty than most people are ready to cope with, thus hampering the transition to FGM. Moreover, what will happen if the "strong rudder" of the command-and-control management style is somehow changed?

Unfortunately, too many companies take their organizational chart too seriously. Each functional department feels the box to which it has been assigned is like a piece of fee-simple real estate. It is a piece of turf that is private and exclusive. Temporary alliances might be made between various functions to get the job done, but the scorecard of accommodations is tallied and remembered so future favors can be requested.

Traditional second-generation management cultures have assumed that the key to success is in managing the "routine." This has often led to an underutilization of the talents of managers, professionals, and workers because they are expected to "fit in" like cogs.

CIM I puts additional stresses and strains in organization, but it is not likely to change the traditional balance of power. CIM II, on the other hand, will bring a new mode of operation. Preliminary indications suggest that without openness and trust, the organization will falter in its attempt to reach a more integrative environment. This calls for the articulation and formulation of new values. They, together with the strategic visioneering, will play an increasingly important role in holding the organization together.

Well-defined bureaucratic structures hold traditional second-generation manufacturing enterprises together. Yet these companies lack the quickness of mind and nimbleness of foot to adapt to changing competitive environments. Automation is seen as a way to further eliminate people and "routinize" the processes even more. The tasks of second, third, and fourth-generation management have been to master the routine, while FGM is focused on managing "variety." Figure 16-21 illustrates the shifting differences.

FGM, on the other hand, is a creative response to the need to manage variety, ambiguity, uncertainty, and even chaos. Therefore, managers, professionals, and workers need to learn to work as overlapping nodes and decision points in a larger network.

If, in FGM, the key managers and professionals need to work together as teams of nodes with issues to resolve, then narrowly defined job responsibilities and restrictive departmental charters will be a hindrance. Instead, a climate of trust, openness, and information sharing is essential, made possible by changes in rewards systems, job definitions, and departmental charters. CIM II provides the technical resources to enhance human capabilities to manage the variety of constantly changing competitive conditions. They are supported by a broad range of decision support systems.

Key operating norms. In addition to the shift to a more nodal form of management, the management context thread refers to the values and operating philosophy established by executive management. In manufacturing companies over the last 5 years, acceptance of a whole new set of operating values has begun. This includes just-in-time (JIT), total quality, and design for assembly. JIT has evolved in some companies into a "Total War on Waste" (TWOW). That is, activities that do not add value to the process or product should be eliminated. This is what James Lardner of Deere and Co. refers to as reducing the "complexity index." It makes little sense to computerize operating inconsistencies, confusions, and poorly defined procedures.

Companies are realizing that the cost of poor quality is much higher than conventional wisdom has assumed. The commitment to quality throughout the manufacturing community is rapidly increasing. JIT and total quality are now becoming more widely accepted. In fact, they can often serve as excellent lead-ins to CIM II because they expose the real problems and opportunities within the manufacturing enterprise.

The manufacturing community is just waking up to the need to radically revamp its outdated accounting approaches. Burdening direct labor may have made sense when it was 50% of the product cost, but it is hardly realistic when it is 5%.

Traditional reward systems have tended to encourage people to keep their information close to their chest. There will be a need to reward managers and professionals for a more open sharing of information across functional boundaries. Companies are experimenting with approaches to support closer functional and team cooperation, based on trust and openness.

One of the most exciting ideas of FGM is the shift from an organizational architecture of narrowly defined functional responsibilities to one of overlapping nodes. Figure 16-22 illustrates this shift.

The traditional architecture of second-generation management assumes a fairly static environment where responsibilities can be fairly well defined. The best metaphor for this organization has been the "organization chart" drawn on a piece of paper. This serves as a "spatial" representation of the key responsibilities in the organization. The spatial representation of an organization is too easily translated into the notion that a manager has responsibility for a given piece of real estate or functional area. But an organization exists more in time than space. Time—especially human and organizational time—is a more difficult metaphor to describe and picture.

New tools are needed to focus the nodes on the business and technical challenges of the company. A shared vision needs to emerge to help sharpen this focus. It will come not only from the top executives, but also from within the enterprise. It will get translated into concrete results through the use of "project management" techniques.

Project management. The construction industry has long used project management to pull together diverse disciplines around a common project. This kind of capability is vitally needed by FGM to dynamically network key resources and

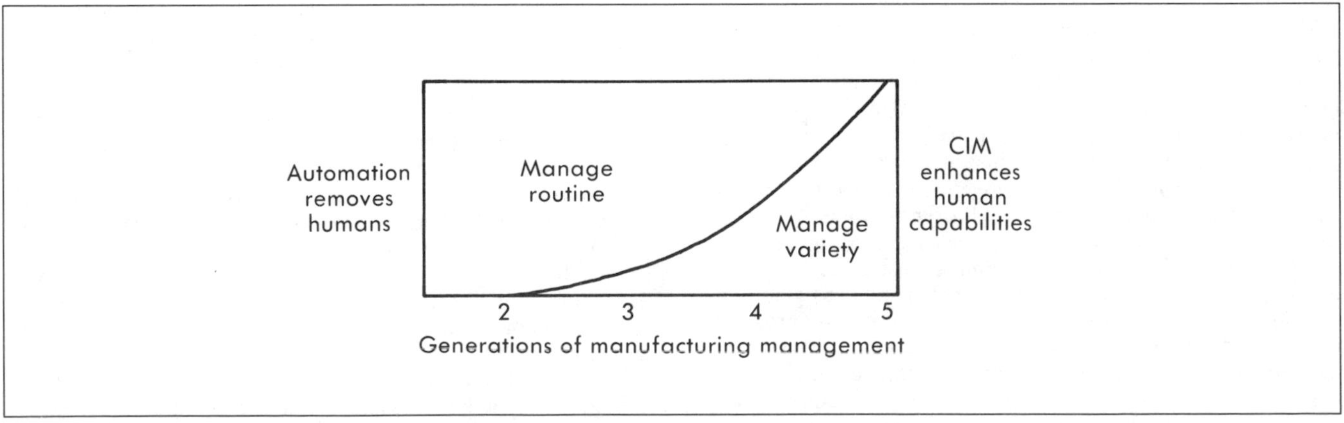

Fig. 16-21 The shifting task of management.

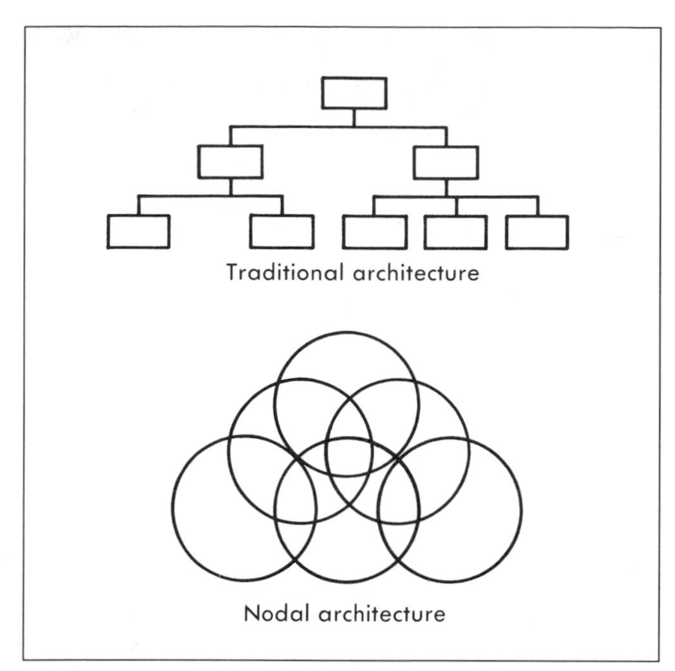

Fig. 16-22 Shifting management architectures.

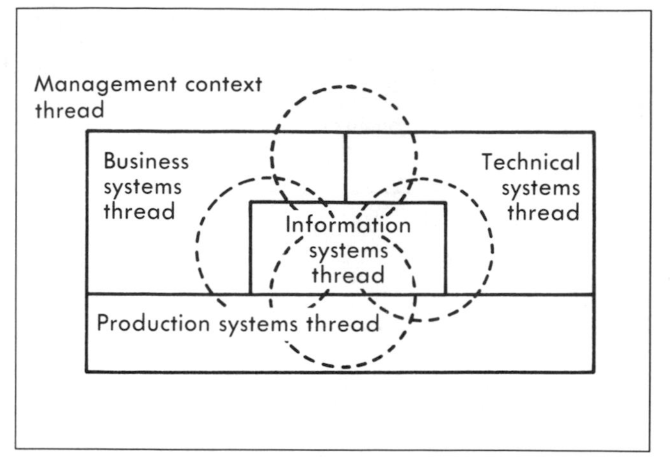

Fig. 16-23 Nodal project team linking the enterprise resources.

harness the entrepreneurial energies of the project. Nodal networking companies need these tools and the discipline of project management to concurrently design new processes and products in parallel. The emerging concept of simultaneous engineering captures the trend toward concurrent design.

Project management helps overcome "foxhole management," where the functions are more intent on politically protecting their turf than listening to and sharing with the other functions. In a project management context, the focus of loyalty shifts to the project at hand, and issues get resolved.

Envision the diagram illustrating the five threads, but this time see the nodes superimposed on it as in Fig. 12-23. This may be a project team that has been assembled to develop a new product and process concurrently.

A few manufacturing companies that are using a project management approach are achieving exciting results. This is primarily because they are using the talents of their professional resources more effectively.

In addition to the development of new product and processes, nodal project management teams can be used for a variety of other undertakings, such as the following:

- Realignment of resources for JIT.
- Building of total quality approach.
- Development and coordination of a vendor strategy.
- Development of a design for marketability approach.

Certainly, the use of teams is not a new phenomenon in manufacturing. Groups are often mobilized, on an ad hoc basis, to tackle problems not normally within their traditional functional roles. The structure, however, remains hierarchical and function oriented. FGM tips the scales in a new direction. The general disciplines remain, but within the context of a nodal organizational architecture. The nodal project management teams are supported by an evolving digital information infrastructure, together with well thought out standards and protocols so that it is easy to pass meaningful information between nodes.

This nodal project management approach does not need narrowly defined job slots and exclusive departmental charters. Instead, it requires open-ended careers and overlapping departmental responsibilities. There must be room for human organizational growth that then creates a higher level of engagement and commitment on the part of all employees. Moreover, motivation is, in part, self-generating as the quality of human interaction improves.

Human and Organizational Time

Conventional thinking about time assumes that the past leaves us behind. Conversely, the future is out there yet to come. It is as though there is physical space between past and present, and physical space between present and future.

What is past has been left behind. What is future is not yet here. This is a spatialized view of time (clock time). A linear approach has been crucial in coordinating the various activities in the sequential handoffs of Industrial Era enterprises.

However, there is another way of thinking about time. Just as individuals will grow in their abilities, so the enterprise will grow. In fact, decisions made today to build "good habits" into the organization, such as group technology, engineering and manufacturing standards, and quality consciousness become an integral part of the enterprise. They "flow with" the company, just as an individual's knowledge flows through the years.

As humans and as organizations, our past flows with us, either as a resource to nurture us or as a drag to haunt us. If we are able to capture and organize our information in a meaningful and retrievable manner, then it is possible to leverage this information as an asset.

The past flows with an organization for good or ill, as is shown in Fig. 16-24. If the organization has sorted out its engineering drawings through group technology, then they are readily accessible. If not, usually the work gets redone and the past is a poor resource. It is like a hallway closet that is scrambled full of junk.

The future vision informs present actions today. Our consciousness of the future strongly influences present actions.

Our present actions may be made in the vacuum of the present, or they may grow on the resources developed in the past, the resources that flow with us. Moreover, present actions need to be tempered by a realistic vision of the future.

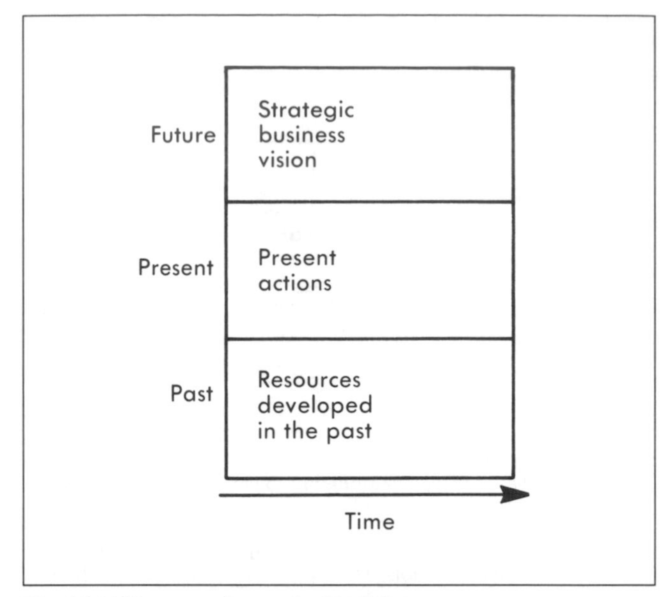

Fig. 16-24 Human and organizational time.

This model suggests a new way of thinking of networking. Usually, networking is thought of as a spatial activity, linking geographically diverse persons and organizations. However, it may have a deeper and richer meaning, the networking of human and organizational time. The ability of an individual and an organization to draw on the resources developed in the past and have a vision of the future is a powerful determinant of the "quality" of management's actions.

In fact, true leadership is born in the intersection of the past, present, and future, all which flow with us. The leader has an ability to "see" the significant patterns and act decisively. He or she has the ability to tap the visions of the other nodes of the system, as they too bring together past experiences, future visions, and present capabilities.

In short, if persons are nodes in a larger network, they are not only linking themselves spatially, but they are linked in terms of human and organizational time. It is very important to understand this distinction between spatial relationships and human time because it is likely to be the key in unlocking the power of fifth-generation management.

FGM networks active and creative teams. Therefore, the leadership demands are qualitatively different from first through fourth-generation management.

It is time to begin thinking of the traditional departments as "knowledge centers" (nodes in the network). Knowledge centers become pools of resources from which to draw the nodal project management teams. The vice presidents or managers of the traditional functional departments will evolve their responsibility for these knowledge centers, and they need to become skilled at identifying and growing talent as well as establishing focused norms of operation.

The project management approach provides the "focusing discipline" so essential for a growing organization. It introduces the notion of the "management of events" rather than the "management of timc," as Warren Shrensker of General Electric has pointed out. Just because half the time has gone by doesn't mean that half the project is completed. Instead, the nodal project management team sets up target gates through which it must pass to accomplish the task. The gates are the "critical events" that must be managed.

The nodal project management teams—and there are multiple teams working on difficult projects—require managers and professionals to be innovative in their problem-solving abilities. Their management style must be able to thrive in a context where variety and ambiguity abound.

Additional information on project management can be found in Chapter 17 of this volume.

TENTATIVE CONCLUSIONS

Taken together, these shifts suggest several important contrasts between traditional management (second through fourth generations) and FGM (see Fig. 16-25).

Functional departments will begin to give way to knowledge centers built around a more explicit understanding of human and organizational time. They will actively develop GT, JIT, engineering standards, and the like so that these resources can flow with the organization over time and be there to draw on as a resource. It is likely in the future to have vice presidents of major knowledge centers much as vice presidents of functional departments exist today.

These vice presidents will function differently than today's executives. Rather than giving orders, they will be sharing their vision. Moreover, they will spend a lot of time identifying and developing the talent they need for their centers.

This leads into the second point; people will more explicitly seek "careers" rather than "jobs." A job is a slot to fill; careers, on the other hand, presuppose intellectual involvement in the tasks at hand and expect the person and teams of persons to be "decision nodes."

This is also why education, rather than training, is key. In the "functions-and-slots" model of industrial organization, people had to be trained to fit into the organization. In FGM, people will be expected to "grow in understanding" of their professional responsibilities.

In the first four generations of management, the system has established "bogeys" and then managed against variances to these norms. The problem is that they are often out of date even when they are first established. They also build in a false sense of confidence that things are under control.

FGM, on the other hand, assumes that teams need to be formed to focus on changing opportunities. These teams are pulled from the knowledge centers on a temporary basis. They act as nodes in a larger network. Good project management helps to keep them focused and disciplined.

The problem with members of first through fourth-generation management is that they have been poor learners. They have not taken the time to sort out their visions of the future and their

Second to fourth-generation management	Fifth-generation management
Functional departments	Knowledge centers
Jobs	Careers
Training	Education
Management by variance	Nodal project management
Informational amnesia	Informational memory
Disposable data	Data as an asset

Fig. 16-25 Contrasting characteristics of FGM.

experience, lessons, and knowledge of the past. Their weakness is that they have imploded into the present, creating a black hole of understanding.

This leads to informational amnesia. Without strong roots in the past and without the broadening vision of the future, which are both part of the present, amnesia sets in. It is little wonder that the manufacturing organization should be anemic. There is little spirit and inspiration in the organization.

Finally, data (that is, information in context) is assumed to be disposable. To be sure, records are kept, but retrieval is cumbersome at best. At this point, no controller will stamp an asset number on a piece of data. Yet this is a major opportunity because an organization needs to have ready access to one of its most important assets. The 1990s will be the decade of data. Just as the past flows over time, this data also flows over time. It may be in a garbled mess, or it may be in an easily accessible form, as is offered by relational database technology.

Fifth-generation management can offer a whole new way of organizing and managing manufacturing enterprises. Rather than the divisive fragmentation so common in companies, a dymanic interaction is in the offing that will challenge everyone to be more creative as businesses build and grow.

References

1. Joseph Harrington, Jr., *Computer Integrated Manufacturing* (Malabar, FL: Robert E. Krieger Publishing Co., 1979).
2. *Introducing the New CIM Enterprise Wheel* (Dearborn, MI: The Computer and Automated Systems Association of the Society of Manufacturing Engineers, 1985).
3. *Computer Integration of Engineering Design and Production: A National Opportunity* (Washington, DC: Manufacturing Studies Board, National Research Council, 1984).
4. Marvin F. DeVries, Susan M. Harvey, and Vijay A. Tipnis, *Group Technology, An Overview and Bibliography*, Manufacturing Systems Report MDC 76-601 (Cincinnati: Metcut Research Associates, Inc., 1976).
5. Donald N. Smith and Peter Heytler, Jr., *Industrial Robots Forecast and Trends*, Delphi Study Second Edition (Dearborn, MI: The Society of Manufacturing Engineers and Ann Arbor, MI: The University of Michigan, 1985), p. 15.
6. Dr. Charles M. Savage, ed., *A Program Guide for CIM Implementation* (Dearborn, MI: The Computer and Automated Systems Association of the Society of Manufacturing Engineers, 1985).

Bibliography

AUTOFACT '86 Conference Proceedings. Dearborn, MI: The Computer and Automated Systems Association of the Society of Manufacturing Engineers, 1986.

Bartee, Thomas C., ed. *Digital Communications*. Indianapolis: Howard W. Sams & Co., 1986.

Beeby, W., and Collier, P. *New Directions Through CAD/CAM*. Dearborn, MI: Society of Manufacturing Engineers, 1986.

Behringer, Catherine A. "Steering a Course With MAP." *Manufacturing Engineering* (September 1986), pp. 49-53.

Chiantella, Nathan A., ed. *Management Guide for CIM*. Dearborn, MI: The Computer and Automated Systems Association of the Society of Manufacturing Engineers, 1986.

CIMTECH '86 Conference Proceedings. Dearborn, MI: The Computer and Automated Systems Association of the Society of Manufacturing Engineers, 1986.

CIMTECH '87 Conference Proceedings. Dearborn, MI: The Computer and Automated Systems Association of the Society of Manufacturing Engineers, 1987.

Cooper, Edward. *Broadband Network Technology*. Mountain View, CA: Sytek Press, 1984.

Groover, Mikell P. *Automation, Production Systems, and Computer-Aided Manufacturing*. Englewood Cliffs, NJ: Prentice-Hall, Inc., 1980.

Hyer, Nancy Lea, ed. *Capabilities of Group Technology*. Dearborn, MI: Society of Manufacturing Engineers, 1987.

Pickholtz, Raymond L., ed. *Local Area and Multiple Access Networks*. Rockville, MD: Computer Science Press, 1986.

Ranky, Paul G. *Computer Integrated Manufacturing*. Englewood Cliffs, NJ: Prentice-Hall, Inc., 1986.

Rembold, Ulrich; Blume, Christian; and Dillman, Ruediger. *Computer-Integrated Manufacturing Technology and Systems*. New York: Marcel Dekker, Inc., 1985.

Scott, David C. "Making MAP a Reality—A Users View." *MAP/TOP Interface* (Summer 1987).

Sherman, Kenneth. *Data Communications, A Users Guide*. Reston, VA: Reston Publishing Co., Inc., 1981.

Stuck, B.W., and Arthurs, E. *A Computer and Communications Network Performance Analysis Primer*. Englewood Cliffs, NJ: Prentice-Hall, Inc., 1985.

Teicholz, Eric, and Orr, Jeel N., eds. *Computer Integrated Manufacturing Handbook*. New York: McGraw-Hill Book Co., 1987.

Thacker, Bharat. "TOP 3.0 Update." *MAP/TOP Interface* (Spring 1987).

Yeomens, R.; Choudry, A.; and ten Hagen, P. *Design Rules for a CIM System*. New York: Elsevier Science Publishing Co., Inc., 1985.

PROJECT MANAGEMENT

The arena of project management is fast becoming recognized as a complex and demanding field of practice. There are professional societies and journals that focus on the needs of project managers. Many companies and professional organizations offer training in how to become an effective project manager. In some organizations the role of project manager has been given special status, equivalent to middle-level line management, in recognition of the importance of project management and the scarcity of organizational talent who are truly good at it.

Most project managers have grown into their roles from engineering backgrounds. Increasingly, however, more are coming from business administration backgrounds, and some with combined engineering/MBA degrees. Whatever the background, the project manager must be well rounded. Most important is that project managers have a good understanding of the product and the environment in which it is to be developed. Knowledge of the technologies involved and of financial and contractual matters is also very important. Little can be accomplished, however, without strong human relations and communications skills. The project manager must be able to lead, to inspire, and to introduce discipline into the project team so that project success is achievable and everyone can share in that success.

This chapter will provide managers with an overview of the concepts and practices associated with project management. It is not meant to be a comprehensive treatment of the subject. Several publications that discuss the various aspects of project management in greater detail are listed at the end of this chapter.

The chapter begins by clarifying what is meant by the term *project management* and then discusses the task of project management, project planning, and project execution and control. To help clarify some of the abstract concepts presented, a fictitious example is given on how a project is executed. The chapter concludes by discussing the legal issues of which a project manager should be aware.

CHAPTER CONTENTS:

TASK OF THE PROJECT MANAGER

Although the terms *project* and *program* are used by the general public interchangeably, there is a clear distinction between the two that should be made before discussing the particular issues of project management. A *project* refers to all activities associated with the achievement of a set of specific objectives that are regarded as important and worthy of financial support. By its very nature, a project has a finite life, requires specific resources, and has a clear definition of when the job is complete (closure). Moreover, a project's goals require accomplishing something that, in one sense or another, has not been done before; a project has uniqueness. A *program*, on the other hand, is regarded here as an ensemble of related projects, that may be conducted together or sequentially. Programs tend to be large, open ended, self-sustaining, and enduring over periods of time much longer than project timetables.

To take this distinction a step further, program management is essentially the ongoing general management over a certain business area. Project management, on the other hand, has certain special characteristics owing to its well-defined set of end goals and its finite lifetime. The responsibility of the project manager is to manage the project to completion, not to build and sustain a program.

RESPONSIBILITIES

The appropriate person to head the project, once chartered, is a matter of major importance. Except in unusual circumstances, the project manager is a member of the business or governmental organization whose business is performing services to the client paying for the project. There can be many reasons for making the selection of the project manager. Certainly experience with the technology and the organizations involved are high on the list. In addition, successful project managers are persons who have distinguished themselves as good managers of time, assets, and people and who have highly developed communications skills. The intangible, but all-important, quality of leadership is also vital to the project management function. In most cases, the project manager is one who has been either responsible for or close to the formulation and marketing of the proposal that led to project funding.

Successful project management requires careful attention to meeting the technical objectives, meeting the project time requirements, and meeting the budget. These three factors characterize any project and compete with one another. To be successful, the project manager needs to understand this inherent conflict and how to manage effectively within a competitive environment.

***Contributor of this chapter is: Dr. Philip H. Francis**, Director, Advanced Manufacturing Technology, General Systems Group, Motorola, Inc.*

***Reviewers of sections of this chapter are: John T. Benedict**, Standards Engineer—Consultant, Technical Div., Society of Automotive Engineers; **Donald F. Condit**, Assistant Professor of Management, Lawrence Institute of Technology; **Edwin E. Lindahl**, Engineering and Management Consultant, Edwin E. Lindahl Associates; **Larry Strom**, Chair, Applied Science Div., Yavapai College; **Joseph Tulkoff**, Director of Manufacturing Technology, Manufacturing Technology—Dept. 15-01, Lockheed-Georgia Co.*

CHAPTER 17

TASK OF MANAGER

The project team can always do a better or more thorough job in meeting the technical objectives if the time and budget constraints are relaxed. Conversely, the project can be completed less expensively, and perhaps more quickly, if the technical objectives are pared back. Some projects do not permit any flexibility in the technical, time, and cost objectives initially set forth. In other cases, there are opportunities to modify these objectives in light of developing knowledge about the project and because of changes in business and political environments that surround the project. These are just some of the interesting challenges that confront the project manager.

In most cases, there are four constituencies to whom the project manager is responsible: (1) line management, (2) the client, (3) the project team, and (4) the outside suppliers and contractors.

Responsibilities to Line Management

The project manager is responsible to his or her immediate management in the parent organization for the economic utilization of resources and the coordination of effort with other project managers within and for the benefit of the organization. The project manager must accept direction from management and keep management informed of the project's progress. As an element of the entire organizational system, the project manager controls assets (people, space, facilities) that are loaned or allocated to the project and often shared with other projects. Accordingly, the project manager must coordinate both across functional lines, which supply resources, and project lines, which share these resources. Additional information on this topic is discussed in the section "Project Organization."

Responsibilities to the Client

The client or project sponsor is the person(s) authorizing and paying for the project. The client may be the firm itself or a business unit thereof, or it may be an outside organization or agency. In either case, it is important that the project manager has a clear understanding of just what is expected. This would include those requirements that are written in the project work statement as well as those that are desired in the client's unwritten (political) agenda. There are times when the client wishes to influence the process, the outcome, or the presentation of the project results for reasons quite apart from the project's stated purposes. The project manager needs to determine at the outset if there is an "unwritten agenda," and if there is one, what it is, how it would influence the conduct of the project, and how it should be reflected in the statement of work. This calls for sensitivity, candor, and tact on the part of the project manager.

The client may or may not wish to become directly involved in the project. In most cases, it is useful to involve the client in the project because it enhances communication and provides a mechanism for the client to "buy into" the project and thereby make its success more likely. The client can become involved in the project by serving as a direct technical contributor, as a consultant, in an oversight capacity, or by participating in project or design reviews. Whatever the client's role in the project is to be, it should be thoroughly discussed and agreed to by both client and project manager; it should not be so strong or controlling as to usurp the authorities and responsibilities of the project manager.

The project manager must also establish a means to keep the client informed of significant progress, difficulties, and accomplishments during the life of the project. It is never good practice to withhold bad news from the client, who may be helpful in correcting the problem or at least will be more tolerant and understanding should the problem magnify later on. Also, if the project team overcomes a difficult and unexpected problem of which the client had been apprised, the success is all the more sublime. Keeping the client informed of the project's progress usually involves periodic (weekly, monthly, or quarterly) written progress reports as well as oral presentations and informal conversations and meetings.

In most cases, the client, the person who authorizes and pays for the project, is the same as the customer to be served. However, there are cases where the client and the customer are different parties, as when the company or agency funds a project that is intended to serve the needs of other groups. In such cases, the project manager must attend to the interests of both client and customer. The customer may be one or many; if many, there may be conflict among the needs expressed by the various customer constituencies, and these may not be sympathetic to the interests of the client. Such situations demand great tact on the part of the project manager.

Responsibilities to the Project Team

The project manager has a variety of responsibilities to the project team so that team members can accomplish their work with full support and understanding of what is expected. In the case of large projects involving many persons and perhaps geographic separation of project team members, it may not be possible for the manager to have personal and frequent interplay with each team participant. Then an organizational chain of command must be created to link, directly or indirectly, each project contributor with the manager. Rapid and faithful communication is essential to a successful project, and the project manager must be easily accessible to and in touch with each participant, directly or indirectly.

Managing the project, of course, means much more than communication. The project manager must bring the appropriate people to the project when needed and provide them with the resources they need on a timely basis. Resources can take the form of data, supplies, personnel assistance, and travel. In projects where task managers are assigned and carry the responsibilities of a project manager over a limited portion of the project, such task managers must have timely and accurate financial management information in addition to staff and other resources. They, together with the project manager, must have access to current cost accounting data, including cost encumbrances, to track the project's financial performance. Project staffing, discussed subsequently in this chapter, requires that the right people are firmly committed (assigned) to the project when needed. Typically, project team members are not under the administrative control of the project manager, so the project manager must negotiate with appropriate line managers to ensure that the needed commitments are made and honored.

Beyond these management responsibilities, the successful project manager must have a repertoire of leadership skills. At various times the manager must be a motivator, counselor, idea-generator, and referee, in addition to being the decision-maker for the project team. Leadership is the art of management and can make the difference between success and adversity in a difficult project.

Responsibilities to Outside Suppliers and Contractors

In addition to the needs of the immediate project team, the project manager must attend to the needs of outside vendors, contractors, consultants, and others who are responsible to him/her and on whom the project relies. Here again, planning, communication, and leadership are the key elements in meeting these responsibilities. Two questions that must be posed during planning are: (1) What are the deliverables, the timing, the costs, and the quality control/assurance measures? (2) What are the contingency arrangements in the case of work stoppages, price or delivery changes outside the contractor's control, or unforeseeable and serious technical problems?

Communication is important because outside contractors need to understand their role in the overall project requirements. Their contributions must fit together smoothly and on time with other project developments.

Finally, leadership is just as important in dealing with outside contractors as it is with the central project team. Contractors must be treated fairly and be given every reasonable opportunity themselves to succeed, for the benefit of the project. In essence, they should be made to feel a part of the project team.

COMMON DIFFICULTIES

Even the most experienced project manager is apt to encounter difficulties in managing and controlling a project. Problems arise when the issues of limited authority, moving targets, client satisfaction, and inadequate management information systems are not properly addressed.

Limited Authority

Typically, the project manager must negotiate with various line managers to secure commitments for needed project personnel. This is the essence of matrix management, which is discussed subsequently. Although matrix management permits more flexibility in the assignment of personnel, it suffers in that the project manager does not have continuing line control over project team members. Team members, on the other hand, must serve two masters, and the lines of authority can easily become blurred. For this reason, project managers should have rapport with and be favorably regarded by other line managers in the firm to command the influence needed to assemble and direct the project team.

Moving Targets

Managing a project is a much more dynamic process than managing a product line. Because the project is structured to close itself out, the environment is constantly and rapidly evolving. Projects usually involve some form of innovation, technology development, or transfer. Such processes do not respect careful and deliberate planning and organization; surprises can and do arise, lessons are learned, directions that were planned must be altered in light of the experience gained from problems encountered. Moreover, expected performance from suppliers and vendors can deteriorate for reasons beyond the project manager's control. Although such difficulties cannot always be anticipated, the project team can minimize their impact by being alert to their early signs and by developing alternate strategies for coping with potential problems.

Client Satisfaction

The relationship between the project manager and the project officer charged with looking after the client's interest is extremely important because it affects the outcome of the project. For example, the client may expect results different from those obtained or may expect a "breakthrough" rather than a mere adaptation of state-of-the-art technology. As mentioned earlier, this unwritten agenda is often veiled and may not be apparent to the project manager.

Some projects come under the purview of a committee that is chosen to represent the client's interests. This approach is common in projects sponsored by federal agencies and professional associations. A committee is often more difficult to please than a single project officer because of the different perspectives, expectations, influences, and fields of expertise among individual committee members. Projects overseen by committees demand even more attention to communication needs than those overseen by a single client contact.

Inadequate Management Information System

Prompt and accurate management information systems (MIS) must be available to support project managers. This often presents a problem, especially in those firms whose financial reporting systems are not developed to serve the needs of an occasional project. Project MIS needs are different from those which support general management requirements. The project MIS (PMIS) needs to display information in a different form than standard financial reporting systems may be able to provide. Projects have a need not only to track costs, purchases, and encumbrances, but to relate these to time and milestone performances that are part of the project. Without reliable PMIS, good performance of even the best-managed project is risky. For additional discussion on project MIS, refer to the section "Project Execution and Control."

PROJECT PLANNING

The importance of careful, deliberate, and detailed planning in advance of the initial execution work cannot be overemphasized. Although this may appear obvious, planning is often treated casually in the rush to begin work and save time. The fact is, however, that time invested in the front end can pay real dividends in reducing overall cost and time. Planning provides a mechanism for coordination and mutual understanding of project requirements. It helps to minimize risk by identifying likely problem areas and alternative strategies for dealing with them. Planning also permits the project to be broken down into tasks and milestones to be achieved; these tasks and milestones can be phased in parallel (for independent efforts) and in series (for dependent efforts) so as to minimize the overall performance period. In this section, the more important elements and tools of project planning are discussed.

STATEMENT OF OBJECTIVES

As mentioned earlier, any project embodies three often conflicting factors: (1) technical performance, (2) time, and (3) cost. Because of these competing requirements, the project's charter or statement of objectives must address these factors. The statement of objectives is the foundation on which all project planning and execution is built. These objectives must be agreed to mutually by the client and the project manager. They should be drawn so as to define the desired outcome of the project, such as the result of a competitive product assessment or the design and prototype development of a pump with improved thermodynamic efficiency. The objectives themselves may or may not be quantifiable, but they should be supported by measurable goals.

Goals

Goals are specific statements that are intended to quantify the project objectives. Every objective should articulate one or more goals, and these goals should be as precise as possible. Platitudes such as "The prototype shall perform with high accuracy" should be restated as "The prototype shall exhibit control accuracy of 1% within ± 0.2 mm of nominal as measured by Standard XXX."

Feasibility

The objectives should be written with due understanding and appreciation for their feasibility. Unless this is done, the objectives may prove to be unachievable within the resources available to the project.

When project feasibility is being determined, three things must be taken into consideration. First is *technical feasibility*. The questions to be asked are whether or not the requisite technology is available to enable the objectives to be met. If not, can it be adapted from another field or application and at what risk? Projects that rely on the development of technology are better regarded as exploratory research initiatives, not technical projects.

The second consideration is *operational feasibility*. Here the issue is whether the resulting system or process is likely to operate as expected. What are the most likely problems and approaches to their solution? Finally, project objectives should address the issue of *economic feasibility*. Economic feasibility determines whether project resources are adequate for successful project completion. It may also mean whether the project's result (prototype system or process) will itself prove to be an economic success by whatever financial analysis is used to measure such success.

Timing

Timing, one of the three basic factors involved in any project, should be addressed in the statement of objectives. Timing involves both the overall period of performance, as well as the major milestones that the project comprises. As with all objectives, timing should be supported by measurable goals. These, in turn, may be developed through analytical planning models such as network techniques (PERT, CPM) described later in this chapter.

TASK PLANNING

The statement of objectives, and the goals developed to support them, are used to break the project down into a set of tasks. A *task* is a specific and contributing element of the project that has an identifiable beginning and end. The ensemble of tasks should be necessary and sufficient to satisfy all project objectives. Tasks may be arranged both in series and in parallel, according to the mutual dependencies of the tasks.

Tasks should be linked directly to *milestones*. Milestones are events of significant accomplishment—the start or completion of tasks and jobs, the achievement of objectives, the completion of customer reviews and approvals, and the demonstration of prototype performance. Milestones are convenient points at which to report status or to measure and evaluate progress.

Many approaches are used in task planning. At the one extreme, projects that are small and of short duration can be broken down easily and obviously into tasks; no sophisticated tools are needed. At the other extreme, large complex projects can profit from systematic planning using modeling techniques. Here again, several approaches are used. The most common, however, are *network* approaches to task planning, especially as exemplified by critical path method (CPM) techniques.

The development of a work breakdown structure (WBS) can be of great help in task planning. The WBS represents the project functionally, much like an organization chart describes the functions of a business unit, and gives the hierarchical relationships between work elements (design, hardware, software, and services). The project comprises major systems and subsystems. Each subsystem is decomposed into lower-level subsystems as an aid to planning, implementation, and control. The smallest subsystems, known as work packages, define specific tasks that contribute to clearly defined and specific accomplishments required to meet system objectives. An example of a project WBS is shown in Fig. 17-1.

Network techniques are methods of defining and arranging project tasks so that the project may proceed as efficiently as possible. Efficiency may mean minimizing the overall period of performance, smoothing and minimizing the human resources

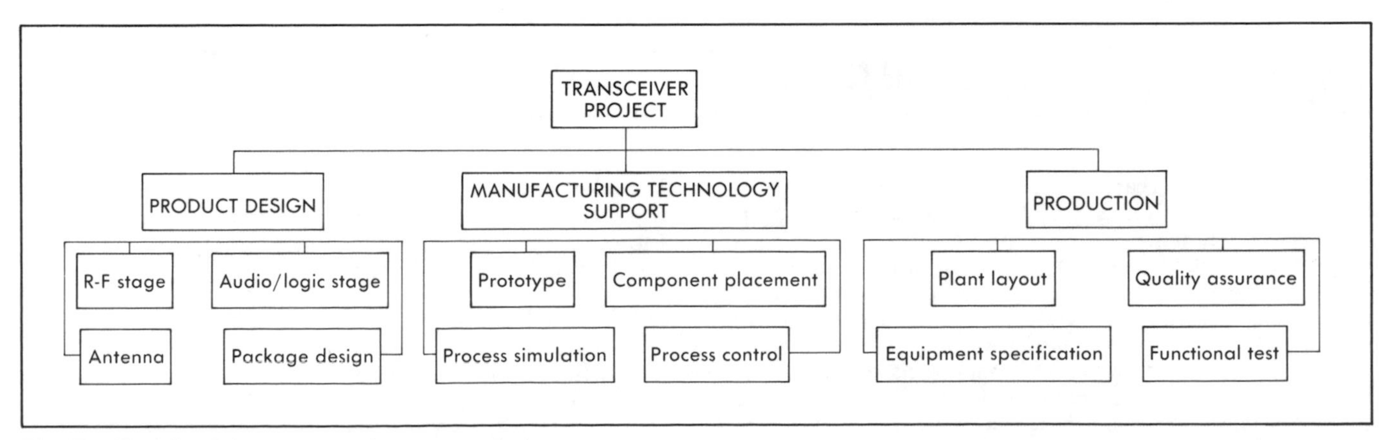

Fig. 17-1 Work breakdown structure for a new product.

requirements, or some combination of both. CPM techniques are used to define the tasks required to accomplish a project, the sequence of these tasks, and the time needed to complete the project. Foremost among the CPM techniques is the program evaluation and review technique (PERT). The Du Pont Co. used CPM as early as 1956 for construction projects, and PERT was first applied in 1958 to manage the Navy's Polaris submarine program.

PERT is well suited for both project planning and project control. The heart of PERT is the construction of a network that comprises all tasks and milestones of the project. These tasks and milestones are then linked together in a flow network represented by circles called *events*; successive events are connected by arrows indicating the direction and content of the activity. Each activity has a beginning and end point. These activities, or tasks, may be related in series or in parallel relationships, as required of the project plan.

PERT requires that three time estimates be associated with the completion of each activity. The unit for time is generally in weeks, but days and months are also used. Based on these time estimates, the expected time to complete an activity between two successive events can be calculated using the equation:

$$t_e = (t_o + 4t_m + t_p)/6 \qquad (1)$$

where:

t_e = expected time
t_o = the most optimistic (shortest) time
t_m = the most likely time
t_p = the most pessimistic (longest) time

These estimates are often difficult to determine. In cases where such work has not specifically been done before, the estimates may be mere "guesses." Also, unforeseen problems may intervene to discredit even the pessimistic estimates of t_p. Nonetheless, estimates must be made with the best experience available; such estimates can be modified as the project progresses.

This network plan can then be analyzed by linear programming computer methods to determine the most time-consuming path that may connect project initiation and completion. This information serves as a management flag to suggest reallocation of resources to improve overall project performance.

Figure 17-2 illustrates a PERT network for a project consisting of nine events. The boldface arrows describe the critical path

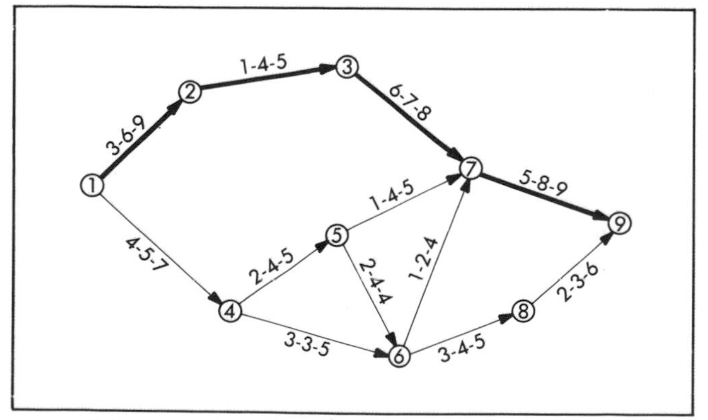

Fig. 17-2 PERT network for a project containing nine events. The three time estimates included on the linkages between the events are given in weeks. The bold line denotes the critical path.

over which the total estimated time for project completion is 24.3 weeks. Other paths, as can be shown by calculation, require less time; these are termed *semicritical paths* or *slack paths*, depending on how close they are to the critical path in terms of total time required. For example, the estimated time for path 1-4-5-6-7-9 is 22.5 weeks; this path may be regarded as semicritical. Path 1-4-6-8-9, by contrast, requires an estimated 15.8 weeks to complete and is called the slack path. All other possible paths lie between the critical and slack paths. By reallocating some of the human resources from the slack to the critical and semicritical paths, it may be possible to reduce the overall project completion time and project costs. Unless such adjustments are made, the duration of the project will be governed by the critical path.

The CPM approach has adaptations that are quite similar to PERT. In one such procedure, only two time estimates are used to characterize the activity linking successive events. The first is the "all normal" estimate (the same as the t_m estimate in PERT), and the second is the "all crash" estimate for the activity completion time when no cost is spared to reduce the time to a minimum. In some circles, CPM is understood as a specific technique in which only a single number, the expected time to completion, is associated with each activity arrow. In this sense, CPM is merely a special case of PERT, with $t_o = t_m = t_p = t_e$.

CPM and PERT methods are best suited to planning and controlling large and complex projects. One practical use of these methods is that they permit management by exception. The project manager need pay little attention to activities that are on schedule; the technique identifies those tasks that are not being accomplished within the time estimates. Management by emergency is thereby eliminated because emergencies are highlighted and thus can be dealt with.

RESOURCE REQUIREMENTS

The project manager and the project team must be wise architects of the project during the planning phase. The two main items that must be taken into consideration at this phase are human resource requirements and capital investment. In addition, the project manager should decide whether any other supporting services would be required during the life of the project.

Human Resources

The project manager and the project planning team should develop detailed, realistic plans for project staffing. Typically, the staff required are not all under the line control of the project manager, and therefore human resources must be drawn out from other units of the firm to staff the project. Moreover, there may be a need for outside subcontractors or consultants. All these anticipated resources need to be defined in the project planning phase.

The planning function begins with identifying the particular skills needed to accomplish the project's task. Depending on the size, duration, and kinds of specialties needed, the skills requirements may be defined by position descriptions, or specific individuals may be tentatively identified. For each task, the mix of people and skills required should be planned along the entire timeline of the task. Personnel needs will rarely be constant throughout the task. The staffing plan should be realistic and should be as conservative as possible. Because of unexpected problems that inevitably arise, most projects usually underestimate personnel requirements. Obviously, personnel requirements have a direct and strong influence on project financial budgeting.

PROJECT PLANNING

After the planning function has been completed, the project manager should negotiate internally for the human resources required by the plan. When it is possible, written commitments or intentions should be obtained from appropriate managers to minimize the risk of untoward surprises later on. These written commitments should also state any special conditions or arrangements such as training, location, protracted overtime, and extensive travel under which project personnel are to work. For individuals assigned full time to a project for an extended period of time (say, a year or more), agreements should be made between the project and line manager regarding issues such as performance reviews that will affect the project member. As insurance against future events that may abrogate such commitments or intentions, the project manager ought to make contingency provisions, especially when key individuals are being counted on to perform a specific task.

Human resource planning for projects should be taken very seriously. Nevertheless, with the best of plans there always will be problems in matching planned and actual personnel allocations. Personnel commitments may be retracted; the project may be redirected, necessitating different technical skills than were planned; and the timing of project plans may also slip. Human resources planning should anticipate and allow for unexpected bottlenecks and scheduling problems.

Capital Investment and Expenses

If the project requires either the acquisition or delivery of equipment, plans must be made to accommodate such needs. The first decision to be made is whether to make or buy certain equipment. This decision is usually dictated by the nature of the project, the availability of suitable equipment from a supplier, and cost. In any event, a detailed set of performance specifications should be drawn up when planning the project.These specifications should be rigorous enough so that any conforming equipment will perform adequately. On the other hand, overly tight specifications may discourage consideration of available equipment that may be adequate for the project. Another important factor to evaluate is the setting in which the equipment is to operate. Consideration should be given to space and to special power needs, to effluents and other contaminants that may have an environmental impact, and to other support needs such as specially trained or dedicated operators and software development. These kind of considerations are easily overlooked or underestimated unless provided for in detail. Plans also should be made for the disposition of the equipment at the completion of the project. The question of ownership, warranties, and personnel training have to be thoroughly considered before the project begins.

In some cases, it may be advantageous to rent or lease available equipment rather than to purchase it. There are many benefits to such arrangements, including the avoidance of maintenance, servicing, and depreciation costs (if treated as a capital asset). On the other hand, if plans call for building equipment, sufficiently detailed plans must be developed so that expected performance falls within the budgeted resources.

Supporting Services

A project will often require specialized services from sources outside the performing firm. The project planning phase should anticipate and provide for such requirements. Most common is the engagement of suppliers of materials and equipment, and of contractors for the delivery of specialized skills and services. Other occasions may call for retaining private consultants from universities or elsewhere. Such persons can provide expertise on specific problem areas at cost-effective rates. In still other instances, partnerships may be formed with other organizations to conduct the project as a joint venture where a single enterprise may not be able to undertake the project by itself.

PROJECT BUDGETING

The next step in project planning is to develop the detailed budget plan. The budget is based on the task descriptions and the resource requirements previously discussed. It should reflect the most realistic estimate of all expected costs without being overly conservative (high) or optimistic (low). Those costs that can be predicted accurately should be budgeted as such. Elements of the project whose costs cannot be reliably predicted should be budgeted at expected levels, with contingency costs posted in the budget to provide for costs that may exceed expected values. Such an approach to project budgeting builds integrity into the project and makes it defensible to management who will approve the project. It also provides a means to manage excessive costs through effective project control measures.

Budgeting

A variety of approaches exists for constructing the project budget. Conventions may differ among firms or because of industry practices. For example, direct/indirect labor and overhead mean different things to different people. Organizations have different rules for establishing what is to be capitalized and expensed. Internal transfer costs are likewise treated in different ways. Despite these and other seeming inconsistencies, all project budgets must account for the same elements of cost, one way or another. The five main elements are labor, expense, capital, overhead, and profit.

Labor costs comprise both direct and indirect categories. Direct labor may consist of both exempt and nonexempt employees. Exempt personnel would include managers and technical and other professional employees who are exempt from overtime compensation unless specifically requested and approved before the fact per the Fair Labor Standards Act (FLSA). Nonexempt employees, such as clerical and skilled trades personnel, must be compensated for any overtime that they work in accordance with the FLSA. Direct labor costs should be estimated for both of these categories and for each task of the project.

Indirect labor includes administrative and general (A&G) labor costs. These costs are common to many or all activities of the organization and therefore cannot be directly allocated to individual projects. Typical of these are costs of senior management, finance, legal, personnel, and health care. Indirect labor costs are often computed as a percentage of direct labor. It should be cautioned that in some manufacturing organizations the term *direct* is used synonymously with nonexempt and *indirect* with exempt. These distinctions are of little consequence, however, as long as all costs are appropriately included.

Expense costs refer to all nonlabor costs that are created for accounting purposes as expenses (not capitalized). Expenses may include travel, expendable supplies, project materials, and consultant fees. Also, equipment costs up to the threshold set by the company as requiring capitalization (often $5000) are expensed. However, equipment costs that are below the capitalization threshold are treated as capital if such equipment is a permanent and integral part of a capital resource. Capital costs, then, refer to fixed assets owned by the company that can be depreciated over their useful lives as determined by standard accounting practices.

Overhead costs include all nonsalary indirect costs, such as fringe benefits, depreciation, taxes and insurance, and heating and lighting. Overhead rates are set by cost accounting practices and are usually figured as a percentage of direct labor.

Project budgeting should include provision for a profit or fee where appropriate. For example, when the project is being funded by a separate company or agency, a profit or fee is appropriate. Such a fee depends on the type of contract involved (cost-type or fixed-price), what the market may bear, and industry practices.

The project budgeting process just described represents the "bottom-up" approach to budgeting. The project is broken down into a fine structure of tasks, each task is costed out according to all the cost elements involved, and the project cost is the sum of all the cost components. The project manager then must evaluate this cost estimate: Does it appear realistic? Does it "feel" too high or too low? Often, "top-down" budget pressures will suggest revisions be made to bring the cost in line with available resources or with what is regarded as fair and reasonable. The challenge, of course, in such budget tuning is not to cut the cost estimate without reducing the cost elements commensurately, as by adjusting the task descriptions.

Value Analysis

Projects that involve major capital purchases or production, or services, can profit by the application of the well-known approach of value analysis or value engineering. Value analysis is a team approach that is based on the system concept and carries through from design to production and field service. Its purpose is to ensure not only that the least overall investment is made consistent with system (project) requirements, but in a larger sense that as much value and capability as possible be built into the system as designed and developed. It addresses the cost/value problem functionally so as to build as much advantage into the project as possible.

The results of value analysis are often difficult to quantify, although the benefits may be undeniable. The process proceeds in steps, beginning with the product or system concept. It is important that all parties having a business stake in the project such as engineering, manufacturing, and purchasing contribute to the value analysis. This team is responsible for gathering information, identifying costs, and evaluating functional requirements. Meetings, focused on concept development and critical design reviews, should be held to establish and refine the system concept. Alternative approaches to meet functional specifications at lower cost are developed and ranked. This formal approach can produce cost savings of 25% or more, often while simultaneously enhancing system performance.

MANAGEMENT PLAN

The management plan is an essential component of the project planning activity. It sets out all the responsibilities and authorities, the control systems, and the communications necessary for the project to meet the expressed goals successfully. It also gives assurance to those funding the project that the management functions have been well conceived and developed, thereby further lessening risk.

Project Organization

Ultimate responsibility for the success of the project resides with a single person, the project manager. The project manager serves as the chief executive officer of the project and should have full authority from the highest authorities within the performing organizations to match the responsibility. All the functions associated with line management are carried out by the project manager: planning, organizing, directing, and controlling. The project manager normally is the person who has been involved in developing, marketing, and planning the project. He or she, and others involved in managing the project, work cooperatively toward the goal of project completion and thus toward project obsolescence.

As compared with line management, project management brings a high degree of flexibility to bear on the problems being solved. Projects and team members are transient, and the project manager has certain latitude in who is employed and when; resources are tailored to the specific needs of the project. It also provides a means for persons not yet having line authority to develop leadership and management skills. On the other hand, project management has its special difficulties. Principal among these is that the project manager must assemble and direct the project team by negotiation and persuasion rather than by command. Also, in practice, the project manager may not have complete authority over the project as should be ideally. Line management may usurp some authorities of the project manager.

Staffing

Before the project officially begins, the project manager should secure appropriate commitments for staff from within existing organizational units. This is easier said than done and often commitments made are later recanted because of other pressures regarded as having higher priority. This is where the art of negotiating for project staff comes in. The project manager must cultivate alliances with line managers so that needed staff or replacements can be secured when needed. Also, if the project plan calls for outside consulting assistance, or for subcontracting with other organizations, appropriate agreements must be drawn to cover these requirements.

Project management usually takes place within the framework of a matrix management environment. The term *matrix management* refers to the two dimensions required in staffing a project. One dimension is the pool of employees assigned to line departments, and the second is the group of projects that must draw on these human resources temporarily for support. The matrix management concept is illustrated in Fig. 17-3. The line organizational units needed to support the various projects going on are arranged horizontally, and the projects themselves are arranged vertically. The intersecting points, or nodes, designate where line resources are needed to support particular projects. These nodes must specify specific labor efforts, by category or by individuals, that are to be loaned to the project for particular periods of time. Adding the labor allocations horizontally across a set of nodes reveals the total labor commitments made to the project by the organization over the project lifetime. On summing the allocations along a vertical set of nodes, the total commitment of personnel over time made by a line department to all projects is determined. Balance should be designed into the matrix system to avoid extreme surges in personnel to be absorbed by line units; these units usually have other work of their own to conduct as well.

The key to making matrix management work is effective communications and a clear understanding by all parties of the importance of the project to the client (customer). Information must flow continuously back and forth from a node to the

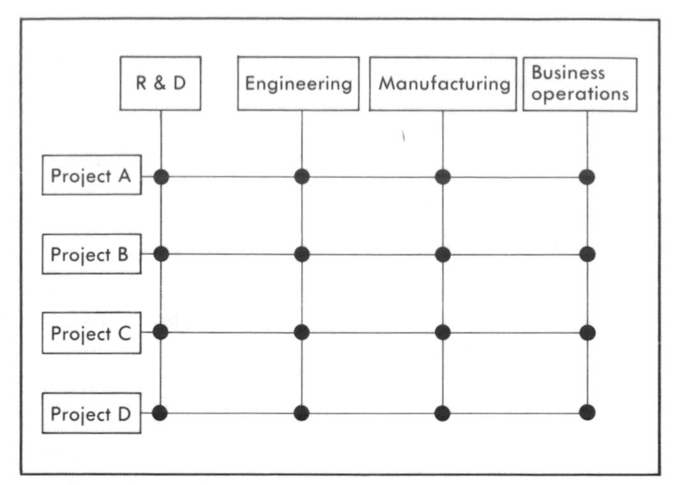

Fig. 17-3 Matrix management structure for four projects drawing personnel from four departments.

project manager and the line manager. Only in this way can line managers anticipate and update their personnel allocations across all project requirements so that other work can proceed on schedule. Likewise, the project managers must have timely information from line managers to schedule project activities and to deal effectively with unplanned deviations from personnel commitments already made.

Matrix management is, in fact, characterized by dual relationships, most particularly dual authority and balance of power. Dual authority refers to project team members having to report both to line and project managers; balance of power indicates the equilibrium between the authorities of the line and project manager groups. These features can easily become unbalanced and at best may result in a fuzzy organizational structure in which some people have trouble working. Several issues often arise. For example, performance reviews, merit increases and promotions (or terminations), and vacation leaves are normally the responsibility of the line manager. Under the dual authority and balance of power in matrix management, such events should be approached cooperatively and with the employee knowing at the outset just how these events will be handled. Another problem is planning for re-entry of an employee to his or her line unit (or reassignment to another project) after a long assignment to a project. Some employees simply do not want to accept assignments to projects because it represents a disruption in work content, possible physical relocation, and can be perceived as diverting one's pursuit of a chosen career path.

Project Reviews

One of the most common and effective tools for controlling the technical progress of a project is to hold project reviews. A project review is simply a meeting of a carefully selected group of persons who will provide a constructive critique of a project as presented by the project team. Notice of this meeting should be given well in advance. The project review (sometimes known as a design review) has the following three essential purposes:

- To enhance confidence in the technical development of the project.
- To build greater reliability into the products of the project.
- To lead to cost reductions (value engineering).

Many projects require more than one project review. For example, a review may be required at concept stage, preliminary design, general or intermediate review, and final review. The particular nature of the project (period of performance, complexity, or risk) will indicate how many reviews should be held and how they should be organized. These details should be spelled out in the project's management plan and incorporated into the statement of work. A more thorough discussion of project reviews can be found in the section "Project Execution and Control."

Communications and Reporting

The management plan should specifically define the lines of communication and the kinds of reporting to be made during the project. It should indicate which project team members are authorized to communicate both internally and externally on behalf of the project. The management plan should also specify the nature and frequency of the oral and written reports to be made. Reports may include periodic (monthly) financial management/milestone and technical progress reports, quarterly reports, semiannual interim reports, and a final report. Appropriate provisions for review and approval of project reports should be set forth in the management plan.

PROJECT EXECUTION AND CONTROL

Careful project planning, as has been mentioned, can greatly enhance the likelihood of a project's success. Ultimately, however, whether or not a project will meet all its stated objectives will depend entirely on the resources allocated to the project and how well the project is executed and controlled. Execution is the coordinated effort of all participants according to the project plan and focuses on identifying and solving problems along the way. Control is the process of comparing project status (technical, financial, schedule) with the plan and making suitable adjustments when the two differ. The project plan itself must be viewed as a dynamic plan and updated as appropriate in light of changing project requirements. Without control, a project is an open-loop process whose successful outcome depends more on chance than good management.

PROJECT CONTROL TOOLS

Many tools are available to assist the project manager in controlling the project. Some are generic, as will be discussed here; others are specifically designed by the firm as part of its overall management information systems (MIS). The purpose of any control tool is to alert the manager to deviations from the plan and to identify problem areas as early as possible. It is up to the manager to take appropriate actions to resolve the problem by bringing the project back on track; there is no automatic pilot for managing a project.

Gantt Charts

One of the simplest, and probably the most commonly used, of all project control tools is the Gantt chart, named after Henry L. Gantt, who first proposed it. The Gantt chart is a type of bar chart; each project task is represented by a horizontal bar, the length of which represents the total time expected for task completion. At project initiation, all bars (tasks) are represented by open or unshaded bars. As time proceeds and work is accomplished on some tasks, the corresponding bars are shaded along the bar in proportion to how much of the task has been completed. The project manager then can tell at a glance the status of all project tasks. Figure 17-4 is an example of a simple project represented by a Gantt chart.

Gantt charts are useful because they are simple to construct and to maintain/update. They show at a glance the relative staging of all tasks and how each measures up to progress expected by the current date. For example, suppose in Fig. 17-4 that the project is into its sixth month (the unit of time on the horizontal axis). It is immediately clear that Task 1 should be complete, but the chart shows that it is only two-thirds complete. Similarly, Tasks 2, 3, 5, and 6 are also behind schedule. Only Task 4 is not behind; indeed it is a month ahead of schedule. Based on the current status, the project manager would need to evaluate the situation and then take the appropriate action. In this example, some of the actions that the project manager could take are to reallocate resources from Task 4 to other tasks that are lagging, recruit additional resources, or revise the project plan if it is now regarded as unrealistic.

There are disadvantages to using Gantt charts; the disadvantages stem precisely from their advantage, that is, their simplicity. For example, Gantt charts only provide gross information about project status. Much detail is obscured or lost, such as reasons for schedule deviation, technical and financial performance, and what is likely to happen in the near future. It is also often difficult to determine how the degree of shading really relates to task performance. Does the length of shading indicate the proportion of task resources already expended or the degree of technical accomplishment or some subjective measurement of both? Another drawback is that the Gantt chart does not reveal mutual interactions or dependencies among project tasks. Rather, the chart treats tasks as mutually independent activities.

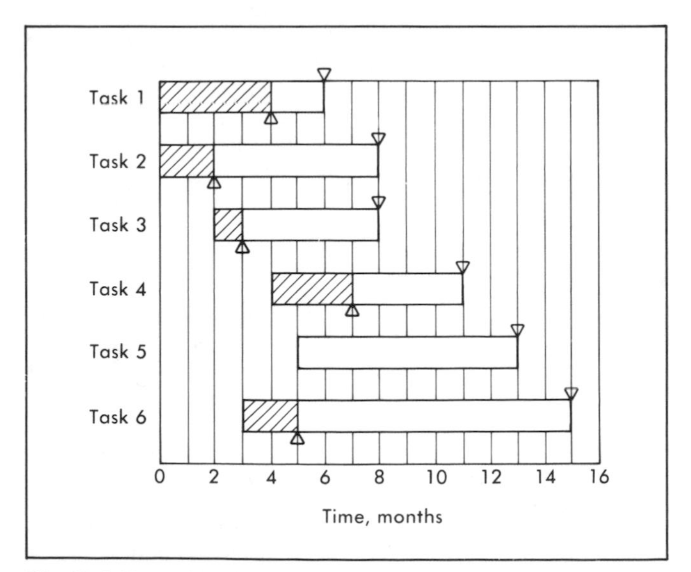

Fig. 17-4 Gantt chart for a project with six tasks to be performed.

Nevertheless, the Gantt chart is probably the most fundamental and useful tool for project control. Most federal contracts governing project work require periodic management reports that present actual-to-planned performance in the form of Gantt charts.

PERT Charts

PERT (program evaluation and review technique) charts were described earlier in this chapter. PERT is a network technique that involves the development of temporal interconnections of project events or tasks, which are represented as arrows. Linkages between nodes (arrows) correspond to activities to be completed to accomplish the tasks. Time estimates (pessimistic, expected, and optimistic) are assigned to the linkages, thus permitting a quantitative estimation of the critical path and overall project time requirements to be made.

Unlike Gantt charts, PERT charts do not lend themselves to easy visual identification of project status. On the other hand, they are richer in information content and demand a more detailed analysis of the task's status for periodic updates of the network. PERT networks are primarily used for major projects comprising many tasks that are mutually interconnected. Small projects comprising up to, say, 10 tasks are more usefully represented by Gantt charts, especially when there is a high degree of independence among the tasks.

CPM Charts

Critical path methods (CPM) are network diagrams that are very similar to PERT charts except that only the expected time needed to complete a task is represented on the arrow connecting successive events, rather than the three statistical estimates required of PERT. As in PERT, CPM enables the critical path (longest time) and slack paths to be calculated. CPM networks are used in the same way as PERT charts.

Project MIS

The planning and control tools previously described are very effective in assisting the project manager to keep the project "on track." These tools, however, should be a part of, and coordinated with, a more general business system within the firm, generically known as the management information system (MIS). A true MIS is a computer-based system that collects, updates, and manages information coming in from all elements of the firm, such as market forecasting, sales, manufacturing/production, and accounting/finance. This information is packaged and made available, either on-line or in periodic report form, to aid executives and managers in their decision making.

The current view of MIS programs is even somewhat broader, in that it includes decision-making tools. Many large organizations today have decision support groups as part of the MIS. Their role is to develop, use, and transfer quantitative tools for decision making. Examples of these tools are statistical inference, dynamic forecasting, and other specialized methods of representing data to enable trends to be examined.

The project manager must have an MIS available to provide timely and reliable data for project control purposes. The data requirements for managing a project are different from those of other corporate operations. Project reporting formats must present information on progress to date, time spent by project team members (the project team itself may change with each reporting period), and project expenditures in a convenient form useful to the project manager. Many firms, whose activities involve conducting formal projects, have developed specialized MISs for project management. In this discussion, this is referred to as a project management information system (PMIS). PMIS software is also available commercially for use on personal computers. To a large extent, the firm's business environment dictates the form of information that the project manager requires, and therefore, generic PMIS software templates may or may not suit a particular purpose.

Typically, the project manager needs the kind of information summarized in Table 17-1. This information may be provided periodically, say twice monthly, or ideally it may be continuously revised and updated for access by the project manager on-line. Whatever the system, the PMIS must be timely and accurate. It must also be structured so as to be easily read. Experience has shown that technically oriented project managers will often ignore PMIS reports that are bulky or difficult to decipher, thus defeating the whole purpose of the PMIS.

TABLE 17-1
Data Requirements for a Project Management Information System

Headline Data	
Project number	Period of performance
Project title	Project budget
Project manager	Brief statement of project objectives
Project customer	Distribution of this report
Cost Data*	
Labor: Hours and dollars expended, by person	
Overhead: Fringe, burden, and overhead costs	
Supplies: Expenses, referenced to purchase order numbers	
Travel: Expenses, referenced to purchase order numbers	
Capital: Expenses, referenced to purchase order numbers	
Encumbrances: Supplies, travel, and capital committed to but not yet purchased	
Performance Status	
Milestone achievement	
Report requirements	
Summary Data	
Budget-to-actual expenses; highlights of areas of significant discrepancy	
Percent cost accumulation/percent time expended	
Delinquent reports; late milestones	

* Costs to be broken out both cumulatively and for the period.

PMIS Software

Personal computers (PCs) are well suited as PMIS tools because of their speed and power. Already more than a score of software tools are commercially available that run on PCs, and the number of such products may be expected to expand rapidly in the near future. Current products range in price from about $200 to $6000, with most in the $500 to $1000 range.

The selection of PMIS software requires a careful evaluation of features and needs. Among the technical features to be considered are operating systems, random access memory (RAM) and read only memory (ROM) requirements, and communications links with peripherals and local area networks. Other considerations that affect the user more directly are of equal or greater importance, however. Questions that should be asked about the software system from the user's perspective are as follows:

- Is the program easy to learn and use (is it "user friendly")?
- How difficult is it to enter data and get from one portion of the program to another?
- Can the software reside in a file server for use on a multiuser local area network?
- Does the software permit cross project leveling for matrix management?
- What type of support services does the software supplier provide?
- Is there a telephone hotline that the user can call to have questions answered?

The software program should be more than a planning tool; it should allow project tracking and control as well. The PMIS software must also be compatible with the centralized corporate MIS mainframe computers to permit dynamic updates on finances and human resources. Table 17-2 lists some of the kinds of features currently available in PMIS software. Some of these features are available only in high-end software, and therefore a careful evaluation of needs should be made prior to committing to any software.

DIRECTING THE PROJECT

The art of project management comes into play in directing the project. All the project planning and control tools, and all the PMIS, amount to nothing unless the project manager uses them to move the project ahead in timely and effective pursuit of the objectives. Directing is a part of any manager's repertoire. It also includes planning, organizing, and controlling. Directing has been defined as "coordinating employees' activities and, through effective leadership, ensuring proper execution of group assignments." The word *leadership* is the key to directing.

Identifying Problems

PMIS data can be very helpful to the project manager in identifying problems such as excessive costs and missed milestones. Such systems cannot, however, reveal other kinds of equally serious problems. For example, excessive costs may be due to poor utilization of human resources or they may be due to unforeseen technical problems. The PMIS will only highlight the symptom; the project manager must diagnose the cause and then act to implement a cure.

The project manager should be equally alert to potential as well as real problems. Project financial reports may hint at a problem of growing proportions, such as untimely contractor

TABLE 17-2
Features That are Currently Available in PMIS Software

• Gantt charts	• Variable scheduling
• CPM/PERT charts	• Resource leveling
• Highlights critical path	• Calendar modifications (for holidays, etc.)
• Number of tasks/project	• Spreadsheet integration
• Kinds of resources/project	• Resource loading report
• Multiple billing rates	• Project status report

supplies or slipped project schedules. The effective project manager will read such signs carefully and coordinate well with project team members to anticipate problems, not merely react to them.

Solving Problems

Ideally, the project manager has full authority over all project resources. In practice, however, there are usually limits on such authority. For example, project team members are borrowed from other line or staff operations and are subject in part to their home-based management control. Solving problems requires creative insight to do what will work best, and the courage and style to do it swiftly and effectively. Identifying the solution may involve adjustments in personnel assignments, or changes in the terms of the contract or project plan with respect to time or money, or it may involve changes in technical objectives due to problems or breakthroughs experienced. In any case, the solution to the problem will usually affect persons outside the project team. The project manager should do all that is possible to inform and work with such persons so as to minimize suspicion or ill will generated by the solution itself.

Customer Coordination

One of the maxims of corporate etiquette is never to surprise your boss with bad news. In the context of a project, the client is the boss, whether the client is an external sponsor or an internal customer. The project manager should inform the customer not only of the project progress, but also of developing problems and the steps to be taken to overcome them as early as possible. The implementation of this principle depends on the relationship between the project manager and the customer. In some cases, the customer wants an arms-length relationship to the project; such a customer approves the project, provides the resources, and then stays away from the project until its completion. However, it is usually best for the project manager to advise this customer of potential problems that may affect the project's outcome. If the problem either is corrected effectively or fails to materialize, the project manager and team emerge in a favorable light. On the other hand, if the problem proves real and affects resulting project strategy, the customer has been offered the opportunity to be party to the solution, and any blame ascribed to the project manager and team is minimized.

At the opposite extreme is the customer who insists on being closely involved (sometimes to the point of interference) in the day-to-day project activities. In such cases, the customer will readily know of emerging difficulties, and the project manager may have a rough assignment in limiting the customer's involvement and influence in correcting the problem.

PROJECT REVIEWS

One of the most effective tools for overall project management is the use of project reviews. The purpose of a project review is a critique of the project at an early stage by interested parties outside the project team. It is especially useful in projects where a complex system is to be designed and prototyped. The project review process provides a forum for the project team to present the design and defend it before an audience of various persons, including experts and specialists and, of course, the customer. New points of view are brought out; assumptions and approaches are challenged. The project's status is presented in considerable detail, and discussions should be probing and constructive. By so doing it is usually possible to identify certain improvements in the design before it is "frozen," thus resulting in a better project outcome.

When conducting a design review, the following proven guidelines should be used:

Scheduling—The project plan itself should identify the point within the project when the project review is to be held and whether more than one is to be scheduled. Typical milestones to be considered are design requirements, initial design concepts, and final design.

Planning—The project review should be carefully planned well ahead of time by developing the agenda, time, place, presenters, and participants. Presenters, particularly, should be given ample time for preparation of information.

Invitations—A list of those individuals who are expected to attend should be compiled. Those individuals who are coming (or designated alternates) should give written commitments to be present for the entire project review. The participants should represent technologists (engineering and manufacturing), the customer, internal operations (marketing and finance), and perhaps respected outside independent authorities. The program should be limited to as small a number as needed to provide the balanced representation and to facilitate communications. The invitations should indicate the purpose, agenda, time, location, participants, and ground rules for participation/interaction.

Location—The ideal location for a project review is off-site, to create a psychological and physical distance from routine pressures such as ringing telephones, competing meetings, and other business distractions. It should be in a congenial setting and with adequate audiovisual and materials reproduction facilities. A typical project review may run from 1 to 3 days in length.

Atmosphere—The meeting coordinator (the project manager or designee) should establish an atmosphere that encourages spontaneous interaction between presenter and participants, and one of constructive criticism. Excessive cheerleading should not be condoned nor should character assassination be tolerated.

Transcript—A person should be assigned to make a tape and written transcript of the meeting, with decisions and action items highlighted. At the very least, a written summary of meeting highlights and action items should be made. The written transcripts or summaries should be promptly distributed to all attendees.

A properly planned and run project review meeting can be extremely beneficial to a design project. Much of the benefit comes from the fact that assumptions are challenged, requirements are sharpened, outside perspectives are brought in, and creative thinking is induced. The project review also serves as an important public relations tool by informing interested constituencies and getting the participants to become a part of the project's success.

REPORTING AND COMMUNICATIONS

The project statement of work (SOW) should specify the nature and frequency of the project report requirements. Technical progress reports are often required on a frequent (monthly) basis. These are brief summaries of achievements, problems, and near-term plans for the report period. More extensive quarterly, semiannual, or annual reports are often required in the case of longer term projects, and a final comprehensive report is essential for any project. Depending on the sponsor's policy, additional financial reporting may be required. In the case of federal contracts, a number of other kinds of reports may be needed. Project reports are significant, and their content and deadlines should be given high importance.

Other areas of communication that should not be overlooked are the informal reporting channels. These channels consist of scheduled meetings, informal conversations, telephone calls, and telecommunications messages that are needed to maintain contact, information, and confidence in the project's progress. The project manager should pay equal attention to these channels of informal communication and inform project team members of their importance.

CLOSING OUT THE PROJECT

The closing out of a project is frequently given scant attention in both the planning and the termination of the project. All efforts are focused on achieving the project's objectives, and when they are met, attention is diverted to new or other project activities. Consequently, the importance of proper closing out procedures is usually overlooked. Experience shows that approximately 5% of the project effort should be given to closeout. These requirements address several important needs. First, there is the need for document storage. All project materials should be stored in a manner and for a time consistent with the firm's records management policies. There is no single period legally governing statutes of limitations for project records storage (the *Federal Register* of the National Archives of the United States daily reports on federal regulations governing the retention of business records). Civil suits can be filed at any time, reaching back for years, for cases involving patent infringement and product liability. On the other hand, records storage is expensive, underscoring the importance of developing a balanced corporate policy for records management. Microfiche can be a cost-effective medium for records storage, and new developments in laser-based computer records storage promise further cost reductions.

Another important element in project closeout is transitioning the project team to other assignments within the firm. This activity requires obvious advanced planning. Another factor is the delivery or handoff of the final products of the project to the customer. This may require training, troubleshooting, help in maintenance, or other downstream assistance in hardware or software systems developed during the project. Finally, attention must be given to the intellectual property developed during the course of the project. Patent, license, trademark, copyright, and trade secret protection requirements should be identified and properly handled through corporate counsel.

EXAMPLE OF PROJECT MANAGEMENT IMPLEMENTATION

The following example is instructive in describing some of the primary concerns a project manager faces in executing a project of moderate complexity. This example, although fictitious, bears likeness to several projects currently under way in the U.S. to improve productivity and competitive position by implementing new manufacturing technology. The discussion is necessarily brief and is intended to highlight those issues important to project management.

PROJECT DESCRIPTION

An industrial corporation, a leading producer of specialized hardware systems marketed to both private and governmental sectors, has made a top-level decision to develop a flexible manufacturing system (FMS) for use in its future manufacturing operations. The FMS will be the company's first such system and is intended to replace and consolidate several old, dedicated production lines that are labor intensive and scattered geographically across several states. This project, code named Project Mentor, carries a major financial commitment and will be highly visible within the corporation. Management is aware of the risk involved in implementing what is for the company a new technology, while using internal technical personnel who may be marginally prepared to meet the challenges. If the project proves successful and the desired productivity measures achieved, Project Mentor would spawn other flexible manufacturing operations throughout the corporation and change the entire complexion of corporate operations.

Project Mentor is to be a flexible and integrated manufacturing system created by modernizing an existing facility. Management has established a goal of 30 months from project authorization through to process verification by making the product in a preproduction mode in the new Project Mentor factory. A preliminary budget has been established, and management has identified a set of performance goals that are to be met during the course of Project Mentor. Some of these goals refer to market-driven and financial production targets: product mix, product quantities, inventory reductions, and (improved) quality measures. Beyond these, the following two additional objectives were established:

1. Inasmuch as the new factory will be a modernization of an existing productive facility, the project is to be planned to minimize the time the facility is down. Target downtime is 9 months.
2. As much of the planning, engineering, and factory setup work as possible (ideally all) is to be done with existing company personnel to maximize the experience and knowledge acquired in the course of Project Mentor.

EXAMPLE PROJECT

The factory modernization program involves a thorough assessment of existing space and capital facilities to determine what is to be retained for use in the new integrated factory and what is to be disposed of. Such determination, of course, must be coordinated with the developing plans for the new plant's factory architecture.

PROJECT ORGANIZATION

The project begins (and the 30-month clock starts ticking) with management's formal announcement of the project and the appointment of the project manager. The project manager is chosen on the basis of his/her broad knowledge of corporate operations, especially in manufacturing, and for specific knowledge of advanced flexible manufacturing technology. This person should have proven management and communications skills. Perhaps above all the project manager should be the kind of person who, if approached concerning this opportunity, would seize on it with enthusiasm, optimism, and total dedication. Persons having this repertoire of experience and talent are rare in the corporation, and management must demonstrate its own commitment to Project Mentor by finding the best person and doing whatever is necessary to relieve him/her of all other responsibilities. After all, this project may be the key to the company's future.

Once appointed, the project manager's first task is to organize the project planning team. This team, working with and under the direction of the project manager, will be responsible for all project planning and will deliver to management a detailed proposal of the work to be accomplished, costs, and timelines. The project planning team should be as few in number as necessary to provide adequate representation by all functional areas concerned: product design, manufacturing and manufacturing technical support, marketing, finance, and human resources. Ideally, all members of the project planning team should be fully dedicated to this phase of Project Mentor, even to the point of being physically located in a common set of offices free from nonproject distractions. This type of a "skunk works" environment is often difficult to establish, however, and compromises usually must be made. In the case of Project Mentor, there are no other projects going on within the company that would compete for these team members' time; in other words, matrix management is not an issue. Because of this, the project manager's job is simpler than it otherwise would be. The project manager thus moves swiftly to commission the project planning team using the authorities given by top management to borrow the needed talent and move most of the team into common quarters. Exceptions occur in the marketing and human resources areas, where such persons are not available to be dedicated to a project. The project manager accordingly negotiates to have one experienced person from each of these areas available on a 50% regularly scheduled basis to support the planning team.

The project manager meets with the project team and outlines the project's objectives, goals, timelines, and each team member's specific contributions to the planning phase of Project Mentor. Some members of the planning team will be kept on to participate in the project execution phase; others will not. Final decisions are to be deferred until the completion of the project planning phase draws near.

The planning team begins its work by developing a work breakdown structure (WBS), which defines the project organizational structure by functional responsibilities. The WBS for Project Mentor is shown in Fig. 17-5. Note that the WBS is not just for the project planning activity, but should represent the organizational needs throughout the entire project. It is a preliminary or initial version to the extent that it may be difficult to identify the names of task leaders not yet chosen for tasks to be undertaken later in the project lifecycle.

Next, the planning team develops an initial PERT chart for Project Mentor (see Fig. 17-6). This chart will be subject to future revisions as the project develops. However, it is important to produce the initial chart now, at the very beginning, for the following reasons:

1. The project planning team must organize the entire project into a set of phases, goals, and milestones.
2. The chart integrates the team and highlights mutual dependencies among project elements.
3. The chart provides the first realistic evaluation of management's 30-month project lifecycle. Is it realistic? Can it be beaten?
4. The initial chart orients team members to this planning tool and imposes discipline on the team.

The initial PERT chart in Fig. 17-6 indicates that the critical path is 134.8 weeks, or about 31.4 months. The slack path is 113.2 weeks, or about 26.3 months. To these timelines must be

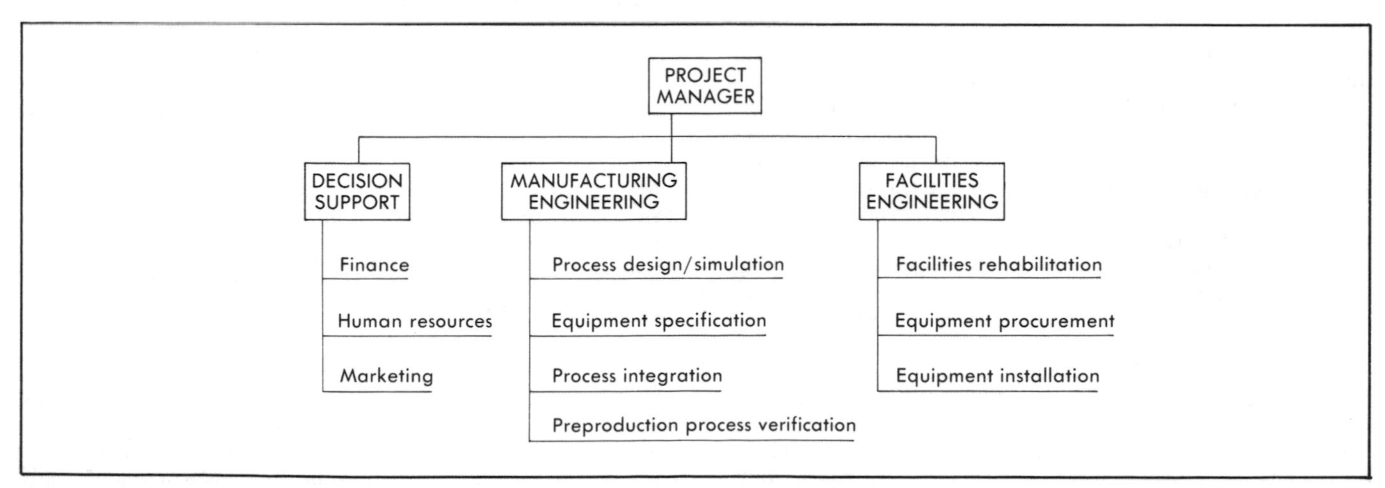

Fig. 17-5 Work breakdown structure for Project Mentor.

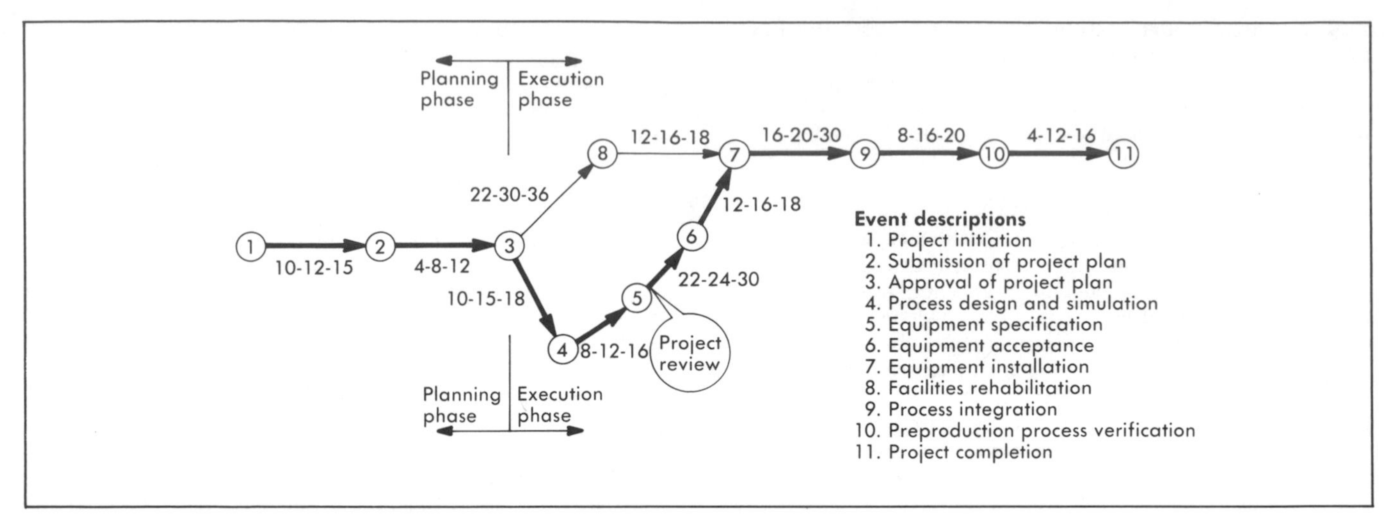

Fig. 17-6 Initial PERT chart from Project Mentor. The bold line denotes the critical path.

added the approximately 2.6 months expected from Preproduction Process Verification to Project Completion. These estimates are made on the basis of the project planning team's best collective early judgment and suggests that management's target of 30 months may be achievable.

While developing the initial PERT chart, the planning team struggles with the issue of how and where to continue to make the product while the factory is down for modernization. According to the plan, facilities rehabilitation is to begin about 18 months into the project and will continue for about 16 months through to project completion. The team, working through corporate manufacturing channels, develops a plan to maintain an uninterrupted supply of product during these 16 months by permitting a temporary surplus of product inventory and temporarily making the product at other production facilities.

The project planning team now sets out to do the initial design of the manufacturing process. This entails detailed considerations of inventory and work-in-process storage and retrieval, material handling, cutting and forming operations, assembly, process control, and inspection methodologies. Other considerations are the level of human production support (both maintenance of line and touch labor), buffer capacity, and process redundancy to bring the production consequences of equipment failure down to acceptable levels. The project manager decides to use an available software package to perform dynamic process simulations and to identify bottlenecks that can occur under a variety of product mix and equipment downtime scenarios, thereby enabling the process design to be modified to accommodate such circumstances. Because of time limitations, the project manager may seek outside consulting assistance in developing the process plan.

As the project planning team makes progress in developing the process design, the project manager begins to realize that additional engineering strength experienced in setting up computer-integrated flexible production systems may be necessary for the success of the project. Aware of the difficulty in recruiting such people, a plan is worked out with the human resources department to search out the needed talent and to have them on board as soon after approval of the project plan as possible.

As the process plan develops, it becomes more evident what existing capital equipment can be retained and incorporated into the new FMS and what is to be outplaced. Aided by staff finance support, decisions are made to keep a few major and relatively new CNC machining facilities. Because of the substantially reduced inventory requirements of the new system, the total floor space needed for the new factory is less than that occupied by the existing factory, and plans are made accordingly to downsize the facility.

Meanwhile, the decision support activities for the planning stage of Project Mentor swing into action. Marketing assists by providing long-range forecasts of product mix and production load levels. Finance provides equipment justification guidance through forecasts of new overhead allocations, internal rate of return on new capital assets, and expected direct labor costs. The human resources department plays an important role in evaluating how to handle the expected reductions in direct labor needs and future increased needs for experienced automation maintenance technicians and software and manufacturing engineers.

The project manager arranges for all these diverse inputs to be merged into a clear and comprehensive plan to present to management. The plan is issued in the form of a proposal to management, written by the project planning team. The proposal should be formatted so that it can be quickly scanned by those needing only an overview of Project Mentor's objectives, supporting plans, and costs, or so that it can be inspected in depth by those critiquing the methodology and details underlying the proposal. The project manager and the three key project planning team members (refer to Fig. 17-5) arrange to make a formal presentation of the proposal to senior management. Management responds by giving tentative approval subject to suggested refinements to the plan. These modifications are made to the proposal, the proposal is accepted, and the project planning phase is completed.

PROJECT MANAGEMENT TOOLS

As Project Mentor transitions into the execution phase, the project manager assembles the procedures and tools to help coordinate and manage the project. One aspect of this has already been dealt with: the development of the work breakdown statement (refer to Fig. 17-5) and the PERT chart (refer to Fig. 17-6). These diagrams must be regarded as dynamic tools and are to be updated regularly as better information becomes available. This is especially true of the PERT chart. By comparing the correct performance to the original chart, a person can quickly evaluate whether the project is gaining or losing ground overall and what the sources of these changes are. It is decided to review and update the PERT chart for Project Mentor monthly.

The project manager next establishes plans for formal communications. This plan comprises three essential elements. First, weekly staff meetings will be held with key project team members. The purpose of these meetings is to share current information on task status, current and anticipated problems, and solutions to problems. These "standing meetings" are regarded as required attendance, and attendees are expected to give them top priority when arranging their individual schedules. Although no written agenda is distributed beforehand, a one or two-page summary of decisions, concerns, and action items is distributed within two days to all participants. Second, the project manager develops plans to hold one full project review immediately following the equipment specification event (Event 5 in Fig. 17-6). This is to be a two-day off-site meeting of key team members of Project Mentor, management, and representatives from other selected functional areas of the corporation. It is decided to contract for an independent consultant also to participate in the project review and to submit his/her observations and recommendations directly to the project manager.

The third element in the communications plan is to issue periodic written reports to management. These reports, to be distributed quarterly, are intended to summarize project status and performance and to justify any significant changes required in either schedule or resources.

In addition to these communications mechanisms, the project manager next selects the project control tools that will be used. One of these, the Gantt chart, is to be updated monthly and distributed to all team members. Figure 17-7 illustrates the format of such a Gantt chart for a certain period in the project. The tasks are taken from the PERT chart (refer to Fig. 17-6), as are the planned timelines.

The project manager decides also to implement a PC-based project management information system (PMIS) software package. The company conducts projects comparable to Project Mentor only rarely and accordingly does not have financial information protocols set up to aid the project manager. Several software packages are commercially available, and after consideration, the project manager selects one of them. The project manager then works with the financial staff so that they can provide the kind of information needed to monitor and control the project. In this example, the information was provided on a weekly basis. Figure 17-8 gives the format of the output from the PMIS package. In addition, the package permits the project manager to produce graphics of plans-to-actuals and expenses over time as needed for reports and presentations.

PROJECT EXECUTION

During the project execution phase, it is up to the project manager to use the communications and project control systems already in place, together with his/her own management abilities, to keep Project Mentor on track. There are many difficult technical obstacles to be overcome. Some are unforeseeable in the early stages of the project, and still others likely will prove more difficult than first expected.

The Gantt chart for Project Mentor (refer to Fig. 17-7) sets aside some 42 weeks, or about 10 months, for process integration. This is the process of linking the manufacturing equipment (material handling systems, machine tools, and robots) together electronically so that they can communicate among each other. In the case of Project Mentor, the process integration activity goes beyond establishing working communications among manufacturing equipment. The plan calls for linking the manufacturing system to the new computer-aided design (CAD) facility to enable downloading of design data into a design database, which in turn can instruct the manufacturing system on what it is to make and the priorities to set. This is an ambitious undertaking and involves such technical issues as control architecture, network technology (broad-band versus

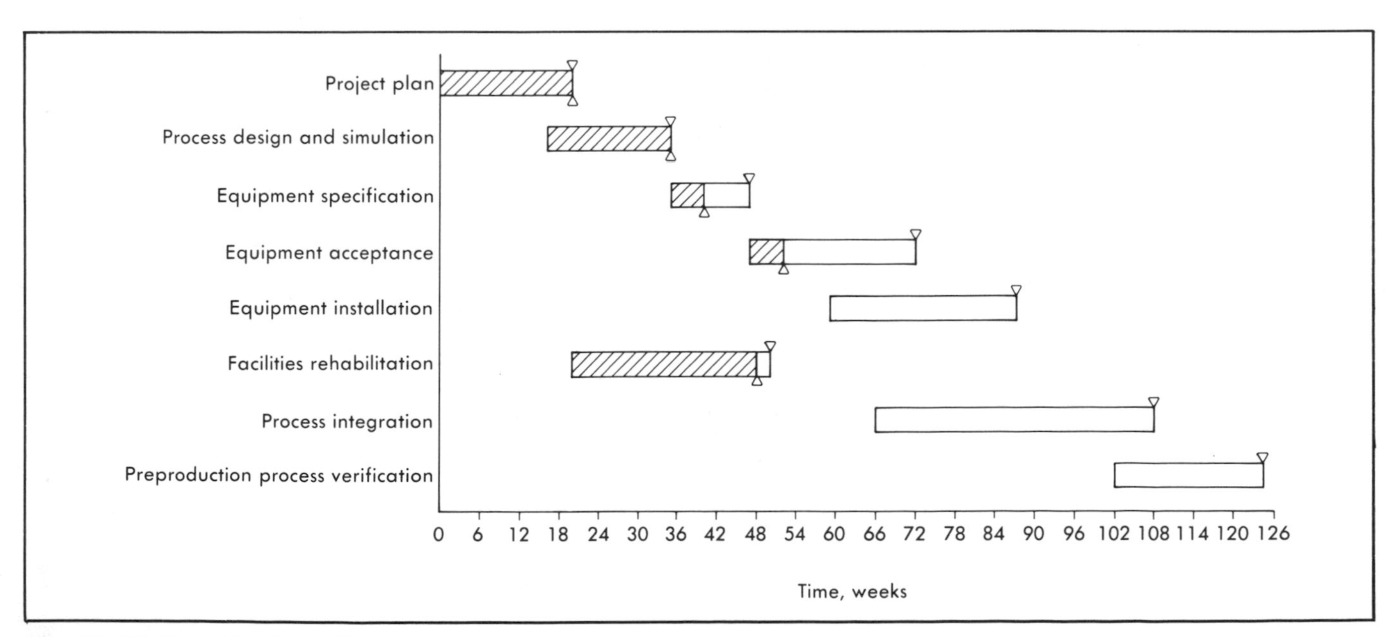

Fig. 17-7 Gantt chart for Project Mentor.

base-band communications), building a design database, establishing a group technology program to minimize parts count, selection of sensors, process control methodology, and a statistical quality assurance program. It is almost impossible to overestimate the time and effort required to work through these problems of process integration. Fortunately, the project manager in this case has wisely chosen to purchase the software and consulting expertise that forms the backbone of the integration task, rather than attempt to develop it in-house. Even so, the project manager will probably find the 10 months allotted to this effort to be too little and may have to consider changes in the planning schedule and budget to keep Project Mentor on track overall.

Another problem that is apt to be overlooked is the task of selecting and training the skilled personnel required to run and maintain the new factory. Project Mentor is built around the principle of a "pull" system rather than the conventional "push" system. In the pull system, the line operator passes work to the next station downstream only when it is asked for. This represents a fundamentally different approach to manufacturing in that it requires closer cooperation among operators, higher technical skills, and acceptance of the fact that an idle worker is not necessarily an unproductive worker. The point to be made is that unless the project manager has gone through this experience before, or has had some good professional advice, training and acculturation needs may not be planned for adequately. If not, problems likely will show up when the new system is evaluated by making a mix of products in a preproduction mode.

CONCLUSION

The example of Project Mentor has highlighted the sweep of responsibilities and tasks required of the project manager. The project manager's role is analogous to that of the head of a company: he or she is responsible for the planning and execution of a process that requires objectives to be met at specified milestones. It requires leadership, the talent to see "the big picture" and how action affects other aspects of the project, the

PROJECT MANAGEMENT INFORMATION SUMMARY

Project name: Mentor	Project security status: Confidential
Project manager: R.A. Parsons	Project number: M86.01
Project initiation: 06JAN86	Total project budget: $10,000,000
Project completion: 01JLY88	Report date: 09SEPT86

Labor Utilization

	Period			Cumulative		
Employee	Budget Hours	Actual Hours	Actual Cost	Budget Hours	Actual Hours	Actual Cost
—	—	—	—	—	—	—
—	—	—	—	—	—	—
—	—	—	—	—	—	—

Expenses

Category	Purchase Order No.	Cost (Period)	Cost (Cumulative)
Travel	—	—	—
Supplies	—	—	—
Other	—	—	—

Capital

	Period			Cumulative	
Item	Purchase Order No.	$ Spent	$ Encumbered	$ Spent	$ Encumbered
—	—	—	—	—	—
—	—	—	—	—	—
—	—	—	—	—	—

Accounting summary

Category	Period	Cumulative	Budget
Labor	—	—	—
Expenses	—	—	—
Capital	—	—	—
	Sum =	Sum =	Sum =

Fig. 17-8 PMIS report format for Project Mentor.

ability to make decisions in an uncertain environment, and total commitment to success. As is the case with companies, all projects are different and operate within their own unique systems of pressures, politics, and constraints. It is, therefore, not possible to give a prescription for success that will be universally applicable.

LEGAL ENVIRONMENT

Project managers need to be conscious of a number of legal matters that bear on the project's performance and that continue even well beyond project completion. The remainder of this chapter presents a broad introductory treatment of some of these legal issues. Although several of them, especially contracts, are of significance mainly to projects sponsored by outside clients, others such as liability and intellectual property are of concern to all technical projects. Obviously, because of the limited treatment given to these complex matters, appropriate counsel should be sought before making decisions.

CONTRACTS

One of the legal issues that a project manager must be aware of is the matter of contractual requirements. He or she must be familiar with the different types of contracts that exist and when a certain type of contract should be used. The project manager must also know how to modify an existing contract based on changes that occur in the project.

Common Types of Contracts

Project managers should be familiar with cost-reimbursement contracts and fixed-price contracts, as well as certain hybrid contracts that exhibit features of both. Each of these occurs in several variations, as discussed below.

Cost-reimbursement contracts. The cost-reimbursement contract is an agreement whereby the buyer promises to reimburse the contractor for all reasonable and allowable costs, up to a maximum amount stated in the contract, plus an agreed-on fee or profit. The contractor, in turn, pledges to do its best in working efficiently and productively toward the project's objectives, but does not guarantee any specific results other than meeting reporting requirements. The cost-reimbursement type of contract is best suited to R&D projects, where the uncertainties and risk are too high for the contractor to develop, say, a working prototype system meeting certain specifications. Because there is no strong incentive to perform efficiently, other than moral considerations and reputation, this type of contract is usually written when the nature of the required effort precludes a fixed-price or other contract where the profit incentive is stronger.

Many variations of cost-reimbursement contracts exist. Of the several types of cost-reimbursement contracts normally used, perhaps the most common is the *cost-plus-fixed-fee* (CPFF) contract.* In this, the contractor is reimbursed for a best-efforts performance for reasonable and allowable costs incurred and is given a fee or profit, computed as a percentage of cost. The CPFF contract is used primarily for research studies and other limited R&D requirements when objectives can be stated only in rather broad terms. Closely related to the CPFF contract is the *cost-without-fee* contract. The only difference between the two is that in the latter no fee is paid. It is otherwise applied in the same situations as the CPFF contract, but is used in connection with educational and nonprofit R&D foundations as well as captive facilities wholly owned and funded by the government.

Various types of cost-reimbursement contracts have been devised in an effort to overcome the primary drawback of the CPFF contract, namely, the lack of incentive to control contractor costs. One type is the *cost-plus-award fee* (CPAF) contract. The contractor's fee in this contract is composed of a fixed amount plus an amount based on the buyer's subjective evaluation of the quality of performance as compared with contract specifications. The *cost-plus-incentive fee* (CPIF) contract provides another, similar means of building in cost-control incentives. In this contract, the buyer pays for costs incurred up to the contract limit, and the fee is computed from a formula that compares the contractor's final cost to the budgeted cost. The fee can result in a profit to the contractor if costs are controlled so as not to exceed the estimates, or a loss if costs are excessive. Like the CPAF contract, this method of fee allocation is appropriate only if the project goals and approaches for goal attainment are set forth with sufficient clarity to enable an objective assessment of performance.

Another type of cost-reimbursement contract that is sometimes used is the *cost-sharing* contract. Cost sharing, or "buying in," indicates the contractor's offer to receive no fee and to assume some portion of the expected costs in conducting the program, that is, to perform the required development at a net loss. A contractor might offer a cost-sharing arrangement if it believes this will buy the first contract in an area that will open up new horizons in the future. The cost-sharing contract is often used when the contractor in question is not among the leaders in a particular technology and finds competition otherwise too strong to be competitive. Cost sharing is also used as a loss leader that will help guarantee the award of a contract for a system design with the broader objective of winning a lucrative follow-on production contract.

The practice of cost-sharing contracts is controlled by the Federal Acquisition Regulations (FAR),** which generally discourage their use on grounds that the long-term effects may be to diminish competition and/or result in poor contract performance. When there is reason to believe that buying-in has occurred, contracting officers must give assurance that costs excluded from an initial contract are not recovered in follow-on contract work. The FAR also prohibit a contract award to a

*Excerpted by permission of the publisher from *Principles of R&D Management*, by Philip H. Francis, pp. 189-213. Copyright 1977 AMACOM, a division of American Management Association, New York, NY. All rights reserved.

**The Federal Acquisition Regulations replaced the Defense Acquisition Regulations, which in 1982 replaced the Armed Services Procurement Regulations. The FAR policies regulate procurement in all U.S. government agencies; supplements to FAR exist that prescribe additional regulations specifically to individual agencies.

bidder if the sole reason for its being selected is its willingness to share costs. Cost sharing, however, is sometimes specifically required of the contractor, especially in cases where the contract effort is in the mutual interest of the government and the contractor (as in a government-owned, contractor-operated R&D facility). In these cases, the FAR place limits on cost participation.

Fixed-price contracts. Compared to cost-reimbursement contracts, fixed-price contracts shift the burden of risk over to the contractor. For this reason, both federal and commercial R&D procurement agencies prefer a fixed-price contract whenever appropriate, even though it may ultimately mean somewhat higher R&D costs associated with higher contractor risk. In contrast to cost-reimbursement contracts, a fixed-price contract *guarantees* contract performance, regardless of whether or not the contractor makes a profit. This guarantee not only provides the buyer with protection against risk, but simplifies the budget process.

Under the provisions of a fixed-price contract the buyer pays a set price in return for the contractor's guarantee of performance. In contrast to the cost-reimbursement type of contract where the fee is set a priori, the contractor's fee under a fixed-price contract is directly related to its efficiency and ability to control costs. Profit is simply the difference between the contract cost and the incurred cost. This profit can range from handsome to a loss, depending on performance.

The most common type of fixed-price contract is the *firm-fixed-price* (FFP) contract. In this, the contractor assumes all risks and is paid a set price in return for assured performance. The FFP contract is applicable when the work scope and objectives are clearly stated and when sufficient historical cost data are available on which to base the cost estimate. It is used, for example, in military production contracts when detailed design specifications are available and in commercial procurements when the specifications are clear and the technology is well established. In preparing the cost estimate for a fixed-price proposal, the contractor estimates costs and then adds on a projected fee derived from an assessment of the risk in fulfilling contract obligations, the confidence in projected cost rates, an evaluation of the competition, and other factors. Several variants of the FFP contract are available when uncertainties exist regarding costs or when it is desired to reward superior contract performance.

There are many instances in basic and exploratory R&D where a fixed-price procurement is desired for work of a study nature. The *fixed-price-level-of-effort* contract provides for a firm fixed price for a specified level of effort over a stated period of time. Here, the contractor is paid for a best-efforts performance rather than for results achieved; therefore, as in the case of the CPFF contract, there is little incentive for effective cost control. This type of contract is especially suited for R&D efforts whose end results cannot be clearly defined and may even be unknown.

Other contracts. There are some contractual situations that are suited neither to cost-reimbursement nor fixed-price agreements. One of these is the *basic ordering agreement*, sometimes called a *task-type* contract. This contract is used when a buyer wishes to have a certain contractor available for quick responses over a period of time to perform various tasks that cannot be specifically foreseen. The contract is written to permit authorization of specific tasks, when needed, by letter, thereby avoiding time-consuming formal procurement processes each time. The basic ordering agreement contains price and time limits and usually specifies the broad technical area under which the tasks are to fall.

Another commonly used contract is the *time-and-materials* (T&M) contract. This contract is a hybrid in that it is neither a cost-reimbursement nor a fixed-price contract, yet has elements of both. It provides for costs (including fees) of labor rates and materials used for work of a continuing or repetitive nature over a specified length of time. Because the total scope and quantity of work is unknown, it is neither a fixed-price contract nor a cost-reimbursement contract. It is used frequently in service agreements, for equipment repair, overhaul, maintenance, and similar work, and also for certain types of R&D where flexibility in work scope is desired. There is little direct incentive for efficiency because all costs are covered on any given job; for that reason, the T&M contract is often used when no other contract type is applicable. To protect the buyer against runaway costs, a ceiling cost is usually established, but this is subject to increase should the quantity of work to be done exceed initial estimates.

Contract Modifications

It is not unusual for contracts to be modified in course because of unforeseeable circumstances that arise during the contract's performance. There are four aspects to this problem that occur quite frequently: (1) changes in scope, (2) time extensions, (3) terminations, and (4) cost overruns.

The overriding message when confronting any of these issues, especially when the contract is with an outside agency, is to involve the contracts or legal officer with the problem at the very earliest date. In too many cases, the project manager has neglected to address the contractual commitments, either ignoring the problem or coordinating needed changes to the scope of work only with the customer's technical contract officer.

Changes in scope often occur. For example, the system operating specifications may be changed, usually at the customer's request to tighten them up for improved performance, such as increasing the number or range of tests in an experimental project or expanding the range of alternative technologies to be studied before committing to a design. In most cases, the requested change can be accomplished, and the project manager may be so inclined to please the customer that the change may be agreed to informally. From a contractual viewpoint, however, any such change involves the risk that the added tasks will be more difficult to achieve than anticipated and that costs will increase (or decrease) according to the nature of the change. Accordingly, good practice indicates that any change of scope be coordinated officially between the two authorized signatories to the contract.

Similarly, when the project will require more time to complete than agreed to, a formal request for a no-cost time extension should be initiated from the contractor's legal officer. If this step is ignored and the project overruns in time, not only can ill will result, but the contractor may face legal actions for default of contract. The project manager should initiate a contract modification for a time extension at the earliest possible time.

The project manager should also be aware of the implications of contract terminations. Terminations "for cause" arise from disputes over project performance. Terminations "for the convenience of the government" occur when funding disappears. Although the contractor has little recourse in this latter case, terminations "for cause" often result in litigation.

Cost overruns are an especially important and sensitive area of contract modification. If the project manager has been diligent in informing the customer of progress periodically during the course of the project, the element of surprise will be minimized. Nevertheless, the customer may be unable or unwilling (or both) to commit additional funds to complete the agreed-on scope of work. Unless a mutually acceptable arrangement can be made, the contractor may be forced into an overrun position, which means an operating loss to the project. This risk is especially high in the case of fixed-price contracts.

INTELLECTUAL PROPERTY PROTECTION

Intellectual property is protected through the use of patents, licenses, copyrights, trademarks, and trade secrets.

Patents

From the earliest time, governments have attempted to stimulate creativity and inventiveness by granting patents and making inventions visible to the public to induce further inventiveness. A patent is a grant of specified rights to an individual, individuals, or an organization by the government of a particular nation. In the United States, a patent may be issued only to an individual or individuals; in some foreign countries a patent may also be issued to a company. Patent rights generally consist of the exclusive right, for a limited time, to manufacture, use, and sell the patented invention and to exclude others from so doing. U.S. patent law has been defined as that which "confers on the patentee a limited monopoly, the right or power to exclude all others from manufacturing, using, or selling the invention. The extent of that right is limited by the identification of the invention, as its boundaries are marked by the specifications and claims of the patent."

The patentability of an invention depends on three essential criteria: (1) newness or novelty, (2) usefulness or utility, and (3) nonobviousness. Of these criteria, the last is the most subjective. The U.S. patent code provides that a patent may not be obtained "if the differences between the subject matter sought to be patented and the prior art are such that the subject matter as a whole would have been obvious at the time the invention was made to a person having ordinary skill in the art to which said subject matter pertains." It is intended to deny patent protection to trivial modifications of existing products or processes.

In the United States, a patent is granted for a term of 17 years from the date of issue and will be granted only to the inventor or inventors. In many countries (but not in the United States), a periodic payment of fees is required subsequent to the filing of a patent to maintain the patent in force. This system is intended to ensure that patents in which the owner is no longer interested will be dropped from consideration as a subsisting monopoly.

The infringement of a patent consists of the unauthorized exercise of any rights granted to the inventor, such as unauthorized manufacture, use, or sale. When these exclusive rights have been infringed, the inventor may file suit in court to recover damages as well as to obtain an injunction prohibiting future infringement. In deciding an infringement suit, the court may not only adjudicate the infringement question, but may also reconsider the validity of the patent itself.

Federal policy generally intends to protect the public interest by granting the government title to inventions wherein the nature of the work or the government's past investments in the field favor full public access to resulting inventions. In this policy, however, is the recognition that the public is better served by giving exclusive patent rights to a contractor in cases where it is likely that the contractor having exclusive rights will develop the technology further than would otherwise happen if the invention becomes freely available to all.

In the case of the federal government, policy vests in the government entire right and title to and interest in all inventions made and developed by any government employee during working hours, with government equipment, funds, or information and which bear a direct relation to the inventor's prescribed duties. If, however, the government's interest in the inventions is insufficient to warrant its ownership of the patent, the agency concerned may release to the inventor all rights and title to and interest in the invention subject to a reservation by the government of an irrevocable, nonexclusive, and royalty-free license to practice and have practiced the invention for government purposes throughout the world.

Licenses

In many cases, an organization that owns all rights and title to and interest in a patent as a result of an assignment or a purchase agreement may not be in a position to capitalize on the patent and derive the fullest potential benefit from it. The organization may not be in a position to manufacture and sell the invention because of limited manufacturing or marketing capabilities or because the invention may not be compatible with existing product lines. If this is the case, several options may be available to the organization to realize profit potential from the invention. Two of these are the following:

- Manufacture the invention and license other organizations to market it.
- License other organizations to manufacture, use, and market the invention.

Either of these courses may be chosen with the hope of realizing immediate gain, depending on the licensee's manufacturing and marketing capabilities.

If an organization holds a patent on an invention and decides to manufacture and/or market it through a licensing agreement or series of agreements, it will recover part of the profit potential in royalties. These are cash returns resulting from authorizing an outside organization, by license, to manufacture, use, or sell the patented product.

Copyrights

A *copyright* is a right granted by the federal government to an author, inventor, composer, artist, and so forth, whereby he or she may control the work or realize benefit from it. The copyright is designed to stimulate the creation and dissemination of creative work by providing recognition and reward to the author. Present U.S. code defines a copyright as "original works of authorship fixed in any tangible medium of expression, now known or later developed, from which they can be perceived, reproduced, or otherwise communicated, either directly or with the aid of a machine or device." Protection extends to written, pictorial, audio, and theatrical works, but not to ideas, processes, principles, and the like. The owner of a copyright has exclusive rights to do (or authorize to do) reproduction, preparation of derivative works, or distribution of the copyrighted work for sale, lease, rental, or transfer of ownership.

Whereas a patent can be granted only after meeting the criteria of novelty, utility, and inventiveness, a copyright can be granted for any original and expressive work as long as it was done independently, whether or not similar works exist. Even though the criteria for granting a copyright are less stringent than for granting a patent, a copyright enjoys substantial protection and is in force for the author's entire life and 50 years thereafter.

Trademark

A *trademark* is a distinctive mark used to distinguish the products of one producer from those of other producers. As defined in the Trademark Act of 1946, a trademark "includes any word, name, symbol, or device, or any combination thereof adopted and used by a manufacturer or merchant to identify his goods and distinguish them from those manufactured or sold by others." Its basic function is to identify the origin of the product to which it is affixed. Herein lies the primary distinction between a patent and a trademark: the purpose of the patent is not to indicate origin, but to protect the investors' rights in a new product or process, irrespective of origin.

As with patents and copyrights, trademarks enjoy legal protection. A trademark offers a monopoly of sorts, because it provides the owner with the exclusive right to use the mark. Its value is derived from its function as a visual assurance of the source or manufacture of the product bearing the mark, thereby both creating and maintaining a demand for the product.

Trade Secrets

Patents are the most common form of protection for new products or processes of technology. In addition, however, the device known as the *trade secret* offers a legitimate and effective means of market protection. A trade secret is simply a means used in the business world for restricting information on formulas, designs, systems, or compiled information, thus giving the organization an opportunity to obtain an advantage over competitors who do not know or use the secret. Courts have held that a bona fide trade secret must involve information that exhibits "a quantum of novelty and originality, is generally unpublished, and provides a 'competitive advantage' over competitors who do not use it."

Notwithstanding the legal aspects of trade secret law, trade secrets offer an attractive practical alternative to patents and copyrights as a means of protecting proprietary information. In many cases, an organization would rather avoid the time, expense, and risk involved in the more common patent procedures; they choose therefore to enshroud the project with secrecy and capitalize on it immediately. After all, once a patent has been issued, the entire file becomes public information, making it easy for competitors to analyze and imitate the product or process, thereby weakening the advantages the patent is supposed to provide.

A trade secret generally is considered to be a property right that is afforded protection under law from fraudulent access by outside parties. However, it is quite legitimate to uncover a trade secret through the practice of *reverse engineering*, that is, starting with a finished product or process and working backward in logical fashion to discover the underlying new technology. This practice is common within the automotive and electronics industries. For example, manufacturers regularly purchase and tear down their competitors' vehicles to keep abreast of (and frequently adopt) the others' new technology.

LIABILITY

Of mounting concern to producing industries and the public is liability incurred by both producers and consumers through the use or misuse of products. This aspect of civil law has recently undergone radical change. Project managers need to understand how these changes in product liability might affect their responsibilities. While much attention is paid to consumer products, product liability questions arise often in connection with materials, systems, and other products of advanced technology.

Product liability generally arises from a defective design, manufacture, instruction, label, service, installation, or application that should have been foreseen by the designer, manufacturer, or seller. To have a valid legal claim, the injured party must demonstrate that the physical or mental injury or property damage resulted from such negligence or defective design.

Courts presently are moving in the direction of holding a product defective in design if it is unfit for all uses and abuses that can reasonably be foreseen. In practice, one frequently finds this attitude goes hand in hand with the so-called "deep-pockets" theory of tort liability: the party with the deepest pockets (the most money) carries the burden of liability. However, the current trend in law is to hold that all damages suffered by the consumer or user are the responsibility of the organization that placed the product into channels of commerce. That is, the one who profits is responsible.

The best way to limit litigation related to product liability claims is by using design and manufacturing procedures that will result in fault-free products and services. While this is an ideal, certain methods can be employed to approach the fault-free standard.

The first area to address in limiting product liability is that of product or systems design. The overriding design objective in product liability should be to anticipate problems that might arise through reasonable use or misuse. Design should be based on accepted principles using verified performance data that can be recalled and supported at a future date. This places a responsibility on the designer not only to generate data using accepted procedures, but also to document permanently any data that may have been acquired indirectly and that could be held to be of doubtful validity or applicability.

In addition to the design area, product liability also is affected by the kinds of measures taken in product manufacture. The approach here is to develop and regulate a good quality assurance (QA) program. The QA program involves monitoring all stages of the manufacturing process in which value is added to the product. This requires that both the product and the equipment used in the various stages of production perform in accordance with established standards. Thus, all instrumentation and calibration devices used to check tolerance or performance should be traceable to the specifications of the National Bureau of Standards or other accepted authority. When nondestructive testing (NDT) methods are used, such as ultrasonic, radiography, or magnetic particle testing, they should be applied under strict adherence to standards of good practice. Whenever the QA program discloses a problem, efforts should be made to determine the cause of the malfunction and rectify it.

The legal basis for liability suits lies in the concept that a professional should be held accountable for incompetence and/or negligence. Of the two faults, *incompetence* (the lack of adequate qualifications or abilities to perform to accepted standards) is the easier to address. Approval of design work or advice by an engineer who is duly registered and licensed to practice by appropriate state or national authority virtually eliminates liability through incompetence in most cases of practical interest. To an increasing extent, professional registration or certification is being recognized as a definition of competency.

Negligence is a more difficult charge to defend against. It is often taken to mean anything short of the most exhaustive and exacting conformance to standards, whereas practical realities of time and cost nearly always demand that some sacrifice in testing and analysis be accepted in favor of sound technical judgment. No specific contingency or potential anomaly in materials, assembly, geological characteristics, environment, and other parameters can be accounted for with 100% assurance. Nevertheless, it is a fact that, in the majority of legal suits brought against engineers, some negligence has been proven to the satisfaction of the courts.

In some instances, the engineer may be subject to prosecution under conditions where *strict liability* applies. Strict liability embraces the concept that a professional should be held liable if his or her work is deficient, without the need to prove negligence or intent. The courts thus far have rejected strict liability in most cases involving engineering designers. However, current trends in medical malpractice suits suggest that the door may soon be opened to strict liability litigation in connection with technical design work.

The engineer can exercise self-protection by practicing defensively in two general directions. The first of these has to do with contract language. More attention is being paid to current interpretations of traditional contract wording. Construction engineers who are contractually required to "design and supervise" the projects and who, for reasons of cost, are limited in the amount of time they can spend in on-site supervision, have been held liable for construction deficiencies that are normally considered to be the fault of the contractor. Thus the trend has been toward replacing the wording "design and supervise" with "provide advice" to reduce the engineer's responsibility for such errors.

Another area of contract language that is undergoing change concerns the guarantee of performance. Here again, the engineer's responsibility for backing the guarantee can be cushioned by giving the client assurance that the construction "to the best of the design professional's knowledge, information, and belief" will meet specifications. In this way the client is called on to share the burden or risk for the project with the contractor. Obviously, engineers remain responsible for cost-cutting actions that could be held unwise or dangerous if not justified by sound technical judgment. One of the best safeguards against legal allegations of negligence is through the strict adherence to accepted test procedures or design practices.

The second way in which the engineer can practice defensively is to maintain thorough and orderly records. This caveat applies not only to chronologies of technical data and calculations, but also to memoranda and correspondence that can help to reconstruct the evolution of a project if the need to do so should arise. Bound notebooks of technical results should be kept if possible and should be signed, witnessed, and dated as information is entered. Instruments and test equipment should be calibrated by certified inspectors at regular intervals using instruments that are in turn traceable back to approved standards.

Bibliography

Cleland and King. *Project Management Handbook*. New York: Van Nostrand Reinhold Co., Inc., 1983.

Davis, Edward W., Jr., ed. *Project Management: Techniques, Applications & Managerial Issues*, 2nd ed. Norcross, GA: Institute of Industrial Engineers, 1983.

Dean, B.V. *Project Management: Methods and Studies*. New York: North-Holland/Elsevier Science Publishing Co., Inc., 1985.

Fallon, William K., ed. *AMA Management Handbook*, 2nd ed. New York: American Management Association, 1983.

Francis, P.H. *Principles of R&D Management*. New York: American Management Association, 1977.

Gido, Jack. *Project Management Software Directory*. New York: Industrial Press, Inc., 1985.

Kerzner, Harold, and Thamhain, Hans J., eds. *Project Management Operating Guidelines*. New York: Van Nostrand Reinhold Co., Inc., 1986.

Levine, Harvey S. *Project Management Using Microcomputers*. Berkeley, CA: Osborne-McGraw, 1986.

Lock, Dennis. *Project Management*. New York: St. Martin Press, Inc., 1984.

Meredeth, J., and Mantel, S.J. *Project Management: A Managerial Approach*. New York: John Wiley & Sons, Inc., 1985.

Rosenau, Milton D., Jr. *Project Management for Engineers*. New York: Van Nostrand Reinhold Co., Inc., 1984.

Stuckenbruck, Linn C. *Implementation of Project Management: The Professionals Handbook*. Drexel Hill, PA: Project Management Institute, 1981.

For further information, the interested reader is referred to The Project Management Institute, Drexel Hill, PA. PMI is a professional society promoting improvements in the practice of project management and offers professional certification in the field. PMI publishes, five times a year, the *Project Management Journal*, which contains technical articles, current awareness information, and advertises new books that are of interest to project managers.

Butterworth Scientific, Ltd., in the U.K. publishes the *International Journal of Project Management*, a quarterly technical publication. This journal also sponsors the publication of the annual *International Project Management Yearbook*. The yearbook contains a directory of products and services and discusses project management techniques and analysis trends. Butterworth also publishes a series of handbooks on special topics.

RESOURCE UTILIZATION

SECTION

5

FACILITIES PLANNING

In planning for and assembling an efficient manufacturing process, the people responsible for results frequently overlook the importance of the facility housing the operation. This is understandable because machines, tools, and a workforce—and how to make them interact most efficiently—are uppermost in the mind of most production managers.

The building, for its part, is thought of literally and figuratively as an "overhead" item, passively protecting its contents from the elements while more or less controlling temperature extremes sufficiently to satisfy the occupants and the manufacturing process. And, of course, the building should be available at the least possible cost. Preoccupation with manufacturing procedures makes it easy to overlook that the facility must do more than provide inert shelter to the people, equipment, and process. The building can and should contribute actively to an operation's success. It is particularly important that the plant operating staff contribute whenever possible during the planning stage of a new plant or an older facility to be upgraded. The improved space, after all, is meant for their benefit, and no one can better delineate the operational requirements.

"Cast in concrete" is a frequently heard expression. It implies permanence, and the term came about for good reason. Before the concrete is cast in the new manufacturing facility, it behooves those affected to take an interest in where the concrete is placed because they may have to live with their oversights for a long time.

Chapter 18 examines seven areas of concern to the planner in establishing and operating the facility itself. They are:

- Site selection.
- Plant layout.
- Housekeeping.
- Disaster control.
- Security.
- Energy conservation.
- Pollution abatement and environmental protection.

Site selection, plant layout, and housekeeping are planning items that render a continuing benefit—or inflict an ongoing penalty—on the manufacturing operation. Disaster control and security are overhead items until needed, at which time their role may become vital. Energy conservation is always of concern even when the preferred energy source is abundantly available; attention to minimizing energy loss may pay handsome dividends on a continuing basis. Pollution abatement and environmental protection are necessary plant design considerations for compliance with government regulations; careful planning may provide economic benefits as well.

CHAPTER CONTENTS:

SITE SELECTION

Making a business location decision is one of the most critical moves a firm can make in determining the long-term success of an operation. This section discusses most of the major considerations encountered by manufacturers when faced with a location decision. It is not designed to be an all-encompassing guidebook. Although selecting a business location is not a task easily summarized in this brief format, this section covers some of the key items to consider before expanding into a new area or relocating an existing operation. This discussion is primarily concerned with manufacturing establishments, but many of the factors presented can also be applicable for distribution, research, or commercial facilities. Because of all the facts required in determining a plant location, it may be advantageous to hire a consultant.

Although this section deals with the factors that a firm would consider during a nationwide, or even worldwide, search, the discussion centers on a North American evaluation only. If the desired search area has already been narrowed to a few communities or a portion of a state, for instance, then some of the macroeconomic criteria may be less pertinent.

As with most business decisions, selecting an operating location must be approached in an orderly, sequential manner to produce the desired results. The chronology of the necessary evaluation steps is as follows:

***Contributors of this chapter are: James R. Bingham**, Manager, Facilities Location, The Austin Co.; **A. Leonard Schade**, Associate Director of Planning, Automation and Productivity, The Austin Co.; **Robert L. Shumaker**, Project Engineer, The Austin Co.; **William Taylor**, Senior Research Associate, The Austin Co.; **Robert A. Will**, Manager, Technical Services Div., The Austin Co.; **Steven S. Zagor**, Associate Director of Planning, EDP Facilities, The Austin Co.*

***Reviewers of this chapter are: Joseph F. Auclair**, Associate Director of Planning, EDP Facilities, The Austin Co.; **G. F. Bryant, Jr.**, President, Bryant Security Systems; **Norman R. Harper**, Consultant, Retired—Figgie International; **Mark R. Murphy**, Manager, Lease Review, Real Estate Development Dept., CSX Transportation; **Joe Richardson**, Project Leader, Garrett Turbine Engine Co.; **Larry Strom**, Chair, Applied Science Div., Yavapai College; **George P. Sutton**, Associate Division Leader, Materials Fabrication Div., Lawrence Livermore National Laboratory; **Dr. H. E. Trucks**, Consultant.*

CHAPTER 18

SITE SELECTION

1. Planning, project justification, and criteria development.
2. Macroeconomic analysis, which identifies a country, state, or province (or several) as most practical for a proposed operation.
3. Specific labor market, community, and site evaluation.

PLANNING/PROJECT JUSTIFICATION

Certain preliminary tasks must be addressed prior to initiating a location evaluation. Although this may sound simplistic, a firm must conduct sufficient market research to identify and verify current customers as well as to establish projections for future growth into either new products or geographic areas. Also, necessary budgeting and capital allocation must be addressed before any detailed analysis is performed.

Other major items that must be addressed in a project's initial stages are the development of facility requirements and location characteristics. Facility requirements include specific future building characteristics, utility needs, and desired site features. Location considerations usually include such items as a community's profile; access to certain industrial services; proximity to a major freeway, airport, or port facility; and numerous other potential attributes. Once these "criteria identifications" have been performed, a location evaluation can begin.

MACROECONOMIC ANALYSIS

The following paragraphs discuss the methods whereby a firm identifies a particular state or other geographic region to consider. This phase of the evaluation is performed by comparing operating costs and conditions at selected potentially favorable locations. For example, a search of the eastern one-half of the U.S. or an entire quadrant of the country would entail this macroeconomic approach.

Transportation Considerations

Most firms should give considerable emphasis to freight costs in the location process. Access to raw material supplies should be emphasized, even if the costs of transporting that product are not directly incurred by the subject firm. Access to markets is also critical. When looking at raw material suppliers for a proposed operation, it is important to investigate alternate suppliers from those currently utilized for an existing operation. The results of a nationwide or regional freight cost comparison are often skewed because of the incorrect assumption that no alternate sources of raw materials exist.

No company can be successful if proper attention is not paid to customer location during the location evaluation process. It must be remembered that, above all else, customers must be satisfied with service and price. A firm that locates a manufacturing operation without regard for customer proximity cannot, except in unusual circumstances, provide a competitively priced product. Table 18-1 shows the sum of inbound and outbound freight costs for various geographic locations based on a particular firm's requirements. These costs pertain to a metalworking manufacturing operation that obtained raw materials (inbound) from the Midwest and served a nationwide market. As can be seen from the table, centralized locations such as Chicago, Cincinnati, and St. Louis were most economical for this particular operation. Figure 18-1 graphically illustrates those costs.

The transportation industry has experienced some significant changes since 1980, particularly because of deregulation. With reduced Interstate Commerce Commission (ICC) control, rail, trucking, air, and barge firms have been freer to establish rates based on the pure economics of providing a given service. Also, with the frequent mergers and acquisitions in the transportation industry, "megafirms" and huge "intermodal" firms have emerged. It is important to determine the impact of these significant industry changes when performing a macroeconomic location analysis.

Labor Considerations

Labor-related factors have long been among the most important criteria in location evaluation. Labor and fringe benefit costs vary considerably by geographic region throughout North America.

TABLE 18-1
Estimated Total Annual Freight Costs

Trial Location	Inbound Freight	Outbound Freight	Total Cost	Penalty Over Base	Index (Base = 100)
Baltimore, MD	$45,100	$132,700	$177,800	$21,800	114
Baton Rouge, LA	39,800	144,000	183,800	27,800	118
Birmingham, AL	37,300	127,900	165,200	9,200	106
Charlotte, NC	41,100	132,000	173,100	17,100	111
Chattanooga, TN	36,200	127,100	163,300	7,300	105
Chicago, IL	27,400	128,600	156,000	0	100
Cincinnati, OH	31,600	128,400	160,000	4,000	103
Dallas, TX	35,000	140,800	175,800	19,800	113
Denver, CO	39,300	147,000	186,300	30,300	119
Little Rock, AR	35,800	133,900	169,700	13,700	109
Louisville, KY	32,000	127,800	159,800	3,800	102
Omaha, NE	35,100	140,300	175,400	19,400	112
Richmond, VA	46,600	135,100	181,700	25,700	116
Sacramento, CA	38,100	172,300	210,400	54,400	135
St. Louis, MO	30,600	127,800	158,400	2,400	102
Tulsa, OK	33,700	136,900	170,600	14,600	109

(*The Austin Co.*)

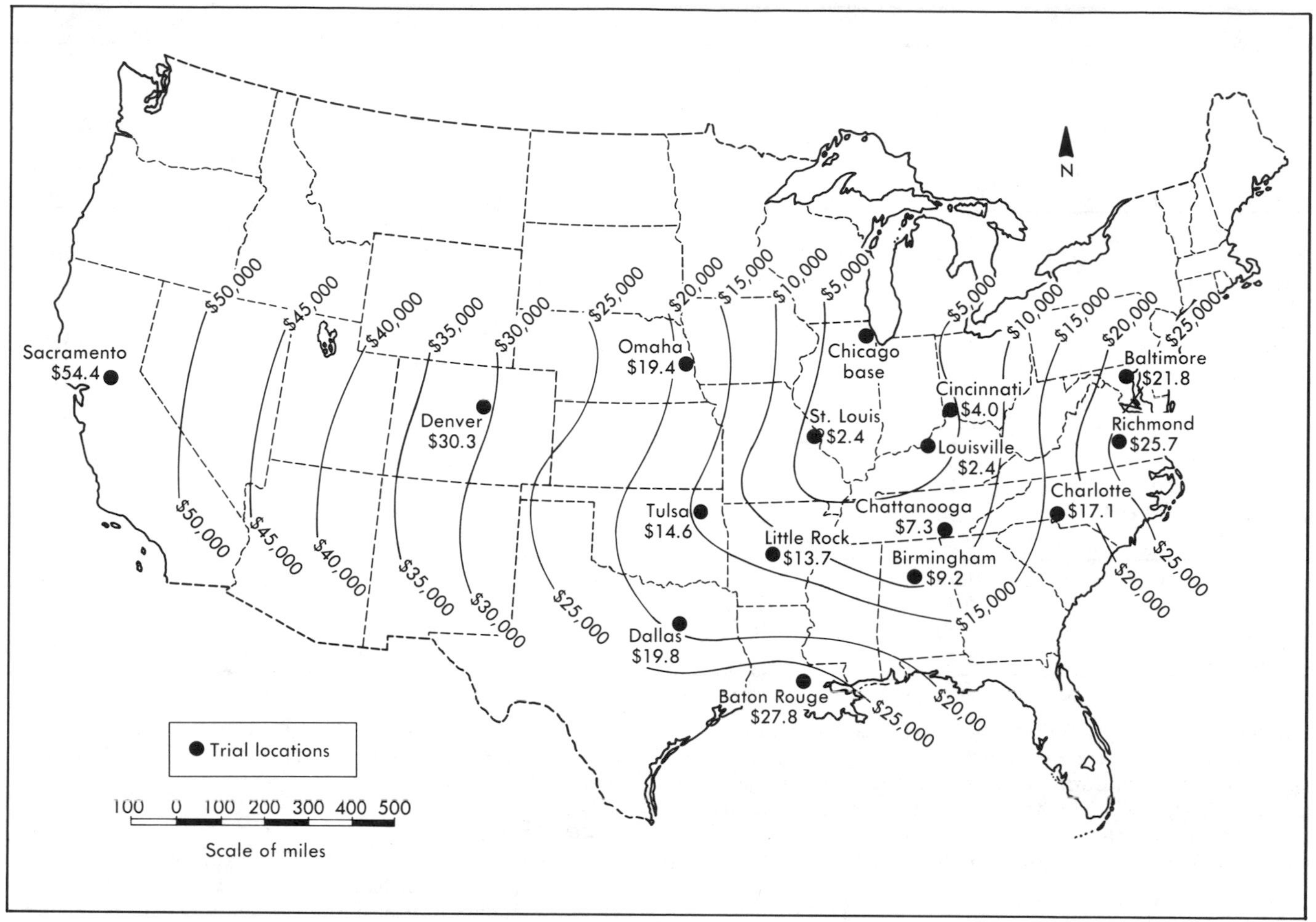

Fig. 18-1 Graphical representation of total annual freight cost penalties over the base location in Chicago. The freight costs at different locations are in thousands of dollars; the isolines are at intervals of $5000. (*The Austin Co.*)

Although the significance of these costs has been reduced in recent years because of increased factory automation and the resulting decrease in numerical labor requirements, experience indicates that labor and fringe benefits often comprise more than 50% of geographically variable costs. A recent analysis of nationwide labor and fringe benefit cost trends by a consulting firm indicates that the Southeast and Mid-South regions have maintained a cost advantage in these important categories. Table 18-2 shows a sample labor cost comparison for selected metropolitan areas. The wages and salaries are based on information provided by the U.S. Department of Labor and the consulting firm's experience in the study area; the rates reflect November 1986 levels. It is important to recognize that local communities within the metropolitan area can vary significantly from the average.

The incidence of unionization is also an important factor to many firms when evaluating different locations for plant investment. This subject is difficult to evaluate on a large scale because worker attitudes and historical union preferences can vary on a city-to-city basis. Some of the most unionized states in the Mid-Atlantic and Great Lakes regions also have numerous nonunion-oriented communities. Conversely, certain less-unionized southeastern states have various mining and manufacturing centers with a long history of union activity.

Several other labor-related issues can be significant during the macroevaluation stage, namely worker attitude, workers compensation rates, unemployment compensation requirements, and industrial skill training assistance.

Workers compensation rates and the administration of related policies vary considerably in the U.S. The costs to an employer depend on the state under consideration and the specific industry involved. Historical industry injury and severity statistics are, of course, used in establishing the industry-variable rates. The geographic variation in rates and administration is on a state-by-state basis. Several states, particularly in the central and southeastern U.S., are perceived by many as more pro-management and less confrontational regarding the awarding of workers compensation benefits. Additional information on workers compensation can be found in Chapter 11, "Legal Environment for Labor Relations," and Chapter 12, "Occupational Safety and Health," of this volume.

Unemployment compensation requirements of employers also vary considerably among the states in the U.S. Unemployment compensation is also discussed in Chapter 11.

If skill requirements are substantial, many firms have utilized state-subsidized industrial education programs that are available in many states. These programs are most often administered through a vocational or technical school system

TABLE 18-2
Estimated Labor Costs *

	Metropolitan Area					
Trial Location	Average Skilled Hourly Wage	Average Semi-Skilled/Unskilled Hourly Wage	Average Clerical Weekly Salary	Total Annual Cost †	Penalty Over Base	Index (Base = 100)
Baltimore, MD	$11.00	$10.25	$360	$2,761,700	$609,400	128
Baton Rouge, LA	12.30	11.25	380	3,034,600	882,300	141
Birmingham, AL	9.45	8.75	340	2,377,500	225,200	110
Charlotte, NC	9.35	8.00	350	2,229,600	77,300	104
Chattanooga, TN	9.00	7.75	330	2,152,300	0	100
Chicago, IL	11.65	10.40	400	2,844,600	692,300	132
Cincinnati, OH	11.25	10.50	380	2,834,000	681,700	132
Dallas, TX	10.25	9.40	380	2,566,700	414,400	119
Denver, CO	11.05	10.00	390	2,730,400	578,100	127
Little Rock, AR	9.45	8.25	370	2,294,300	142,000	107
Louisville, KY	11.50	10.25	375	2,794,500	642,200	130
Omaha, NE	10.25	9.50	330	2,560,500	408,200	119
Richmond, VA	11.70	10.50	355	2,843,500	691,200	132
Sacramento, CA	9.27	9.27	390	2,497,400	345,100	116
St. Louis, MO	11.70	10.60	370	2,871,100	718,800	133
Tulsa, OK	11.15	10.25	380	2,779,600	627,300	129

that has facilities located throughout the state. Local access is thereby available to state-supplied instructors and instruction space, if necessary, during plant construction. In the case of manufacturing skills, which are specialized to a particular firm, funds may be available from the state to reimburse the firm if its own employees must be used to instruct others.

Utilities

The availability, reliability, and cost of utilities are often overlooked during the macroevaluation phase. However, it is critical that certain utility-related issues be given close attention, even during this initial stage of analysis.

Electric utilities must be researched as to past, current, and anticipated future generation sources (percent of coal, nuclear, hydroelectric, and other sources); historical trends in rate increases; rate requests pending with a state utility commission; and other comparative factors. It is also important to ensure that all candidate electric utilities be of sufficient size and have the experience to meet the specific power requirements.

Various energy sources such as coal, electricity, natural gas, propane, and fuel oils may be necessary for either space heating or process purposes. The availability and cost of these materials must, of course, also be addressed. As has been seen in recent years, these energy sources may be in abundance or extreme scarcity, depending on economic and political realities. Therefore, many firms have provided for access to multiple sources of these fuels if they are crucial to their operation.

Although water and sanitary sewer capabilities should be evaluated on a municipal basis, firms with very large requirements—more than 250,000 gallons per day—or firms with rigid water quality needs or restrictive sewer effluent characteristics should look at these factors on a macrogeographic basis as well. The availability of water has been a serious problem in portions of the Southwest, the Southeast, the Plains states, and the Northeast. For firms that dispose of process wastewater into municipal sewers, it is important to realize that certain states have wastewater disposal regulations that are more stringent than federal standards.

Taxes/Finance

Although the new tax law of 1986 is expected to increase the nonresidential tax burden, taxes as a location criteria will remain secondary to transportation, labor, customer location, and utility considerations for most firms. Income and sales taxes are the two state corporate taxes of most significance. Rates and schedules vary by state and should be evaluated in tandem with a firm's financial structure and decision-making experience.

Property taxes vary on a site-to-site basis. However, firms with a considerable planned investment in a new operation should estimate a local property tax in all states being considered. This can be accomplished by obtaining all necessary tax rates on a particular parcel with the characteristics preferred by the subject firm and applying the estimated land, building, machinery, and inventory values against those rates. Local assessment practices must also be taken into account. Many states exempt manufacturers' and distributors' machinery and/or inventories from local property taxes; this consideration is of particular interest to capital-intensive firms.

Many states and individual communities offer certain tax or other financial incentives to qualifying firms. Frequently, these incentives are available on a case-by-case basis and may not be routinely offered to any interested party. These incentives include temporary moratoriums on a portion of local property taxes, low-interest financing, grants for site clearance, lower tax rates on pollution control equipment, and many other items.

Industrial revenue bonds have been a popular source of financing for new operations for many years. Because of their popularity and because of abuses of eligibility through various nonconforming uses, additional restrictions on their availabil-

Nonmetropolitan area **					
Average Skilled Hourly Wage	Average Semi-Skilled/ Unskilled Hourly Wage	Average Clerical Weekly Salary	Total Annual Cost †	Penalty Over Base	Index (Base = 100)
$9.60	$8.90	$320	$2,404,300	$419,100	121
9.75	9.00	290	2,415,900	430,700	122
7.85	7.30	290	1,985,200	0	100
8.70	7.50	325	2,085,300	100,100	105
8.60	7.40	315	2,055,400	70,200	104
10.50	9.40	360	2,568,800	583,600	129
9.00	8.40	300	2,265,100	279,900	114
8.95	8.20	335	2,241,300	256,100	113
10.40	9.40	360	2,563,800	578,600	129
8.50	7.35	335	2,050,900	65,700	103
8.90	8.00	290	2,175,900	190,700	110
9.35	8.65	300	2,332,000	346,800	117
8.70	7.90	280	2,140,900	155,700	108
9.27	9.27	390	2,497,400	512,200	126
8.90	8.30	300	2,240,400	255,200	113
8.70	8.00	300	2,171,100	185,900	109

(*The Austin Co.*)

* Wages and salaries are based on information provided by the U.S. Department of Labor and Austin experience in the study area updated to reflect November 1986 levels.

** Nonmetropolitan figures reflect the lower labor costs that often exist in smaller communities outside larger metropolitan areas.

† Based on 4 skilled maintenance workers and 20 skilled certified welders at 2080 hours annually; 15 semiskilled metalworkers and 80 unskilled assemblers and shippers at 2080 hours annually; and 10 clerical employees working 52 weeks annually.

ity have surfaced. However, industrial revenue bonds will continue to be an important source of funding for certain major projects.

Enterprise zones are fairly recent entrants in the economic development incentive scene. Specifically, zones are designated by state or local authorities as "economically depressed," usually on the basis of unemployment rates, income levels, housing conditions, and several other related factors. Firms locating within these zones may be eligible for myriad tax breaks and exemptions. However, experience has shown that locating in this type of environment has certain "hidden" costs, which are not easily identifiable during the location evaluation study.

Environmental Considerations

The environmental impact of a proposed operation emerged as a locational factor in the early 1970s. For example, federal guidelines on air emissions were established in 1970 with the Clean Air Act. Although later amended, the guidelines control the emission of five specific pollutants by manufacturers. This legislation was aimed primarily at major point sources of particulate matter, hydrocarbons, sulfur dioxides, and selected other pollutants less pertinent to manufacturing. Counties and selected subareas throughout the U.S. were tested and designated as either attainment or nonattainment areas for each of the five pollutants in comparison with the standards. Those areas identified as nonattainment areas have greater restrictions on the subject pollutants for new operations than do attainment areas. Because large areas and even entire states are nonattainment areas for certain pollutants, this can be a locational factor for foundries, refineries, and other manufacturers with significant pollutant levels.

Other environmental considerations during this macroevaluation phase include the permissibility of process effluent discharge into area streams, the availability of disposal and/or treatment facilities for toxic or hazardous waste materials, and any requirements for environmental impact assessment studies to be performed. The requirements often vary on a state-to-state basis because several have enacted regulations more stringent than those required by the federal government.

Macroeconomic Analysis Summary

After evaluating and comparing geographic and any other variables relevant to a particular operation, a determination can be made of the state or region that is most economical and practical for a proposed operation. The task is most easily performed by setting up a cost matrix in a spreadsheet-type format for ease of comparison. Table 18-3 presents a typical total cost matrix. A weighting factor is frequently attached to certain noneconomic issues that can also have an impact on the success of an operation. These issues might include climatic conditions, perceived living conditions, or the level of cooperation between business and governmental agencies, to name a

TABLE 18-3
Estimated Annual Variable Costs for Metropolitan Locations

Trial Location	Inbound Freight	Outbound Freight	Labor	Fringe Benefits	Natural Gas	Electric Power
Birmingham, AL	$1,293,700	$2,260,800	$2,826,700	$467,100	$ 977,400	$1,542,400
Charlotte, NC	1,025,800	2,157,200	2,481,200	444,000	995,700	1,453,200
Cincinnati, OH	903,300	2,337,300	3,187,100	504,600	920,000	1,502,000
Dallas, TX	1,837,100	2,323,500	2,885,200	476,800	786,000	1,708,400
Los Angeles, CA	2,769,100	3,413,300	2,932,000	541,500	1,395,400	2,707,200
Memphis, TN	1,339,800	2,417,100	2,789,300	473,400	853,000	1,528,900
New Orleans, LA	1,675,000	2,416,900	2,950,000	501,300	710,000	1,482,000
Norfolk, VA	1,043,100	2,134,000	2,728,400	488,900	1,208,400	1,500,000
Omaha, NE	1,867,200	2,168,600	3,006,900	486,000	800,000	1,347,800
St. Louis, MO	1,156,300	2,301,400	3,301,000	487,800	938,000	1,440,300
Savannah, GA	1,129,200	2,197,800	3,212,800	441,600	902,000	1,424,800
Seattle, WA	2,769,500	3,890,000	3,622,300	641,200	1,016,000	747,400
Syracuse, NY	732,400	2,170,400	3,324,400	571,000	918,800	1,522,200
York, PA	756,000	2,105,600	2,724,500	488,100	1,240,000	2,214,000

few. The result of this macroevaluation is called a "favorable area", which should now be subjected to a detailed field analysis.

COMMUNITY AND SITE ANALYSIS

All potentially suitable communities (as many as 50 or even 100, depending on size of area) within the favorable area should be screened in-house prior to any field investigation. This screening should emphasize the following issues:

1. Community access to required services and facilities.
2. General community population characteristics.
3. Compatibility of proposed operation to existing firms in candidate communities.
4. Transportation facilities including rail services, interstate access, and trucking capabilities.
5. General utility capabilities of the community.
6. General labor statistics of interest, such as unemployment rates and average area manufacturing wages.

For certain operations, other items are pertinent and should be checked prior to the field investigation.

Labor

Often the most important factor in community evaluation is the labor environment. As was stated previously, labor costs generally comprise 50% of variable costs. But in labor-intensive operations, labor costs may comprise up to 80% of total geographically variable costs (freight, labor, utilities, and taxes). Although the increasing automation of today's manufacturing operations is reducing the impact of labor costs on many balance sheets, labor will remain critical to most manufacturing location decisions. Local manufacturers are generally the best sources of wage information, although certain local development agencies may provide access to community wage surveys.

Equally important in the labor realm is the availability of necessary skills to satisfactorily operate a proposed facility. Firms with unique technical or mechanical skill requirements should determine if a history of development of those particular skills exists in a community. This skill development may have occurred through the efforts of local technical schools, the presence of other firms in the area, or, more likely, a combination of these two factors. Again, local manufacturers, as well as state employment offices, are crucial sources of this information.

Each firm has different expectations regarding the possible unionization of a proposed operation. It is important that local conditions be thoroughly checked prior to making a location decision. Local economic development representatives can provide general data that can be substantiated in discussions with local firms.

Services/Amenities

The services/amenities category is difficult to summarize because requirements vary greatly by specific project. Almost always, however, the following service/amenity factors should be checked:

1. Demographic and socioeconomic characteristics and trends.
2. Availability of required industrial services such as equipment maintenance and industrial supplies.
3. Adequacy of city services including water, sewer, police, and fire.
4. Availability and quality of educational and recreational facilities.
5. Availability and cost of housing.
6. Access to desired cultural amenities.

Of course, many other factors that are not mentioned in the preceding list could be important to a particular operation. Table 18-4 shows several typical community considerations presented in a matrix format.

Although difficult to quantify, the relationship or level of cooperation between government and business in a community

Total	Penalty Over Base	Index	State Taxes	Property Taxes	Total With Taxes	Penalty Over Base	Index
$ 9,368,100	$ 811,000	109	$559,800	$ 835,500	$10,763,400	$ 655,400	106
8,557,100	Base	100	593,300	1,237,400	10,387,800	279,800	103
9,354,300	797,200	109	777,900	1,308,800	11,441,000	1,333,000	113
10,017,000	1,459,900	117	479,800	1,434,500	11,931,300	1,823,300	118
13,758,500	5,201,400	161	975,300	722,600	15,456,400	5,348,400	156
9,401,500	844,400	110	683,200	1,334,900	11,419,600	1,311,600	113
9,735,200	1,178,100	114	521,000	1,046,800	11,303,000	1,195,000	112
9,102,800	545,700	106	513,700	491,500	10,108,000	Base	100
9,676,500	1,119,400	113	621,000	1,611,500	11,909,000	1,801,000	118
9,624,800	1,067,700	112	283,000	1,458,700	11,366,500	1,258,500	112
9,308,200	751,100	109	561,400	960,000	10,829,600	721,600	107
12,686,400	4,129,300	148	833,200	528,800	14,048,400	3,940,400	139
9,239,200	682,100	108	602,700	771,500	10,613,400	505,400	105
9,528,200	971,100	111	845,300	576,900	10,950,400	842,400	108

(*The Austin Co.*)

(or county or state) must be checked during a community evaluation. This can be determined by conversations with regulatory authorities as well as with existing businesses in a community who have had an opportunity to interact with the public sector for some time. For obvious reasons, it can be difficult to operate in a confrontational or uncooperative regulatory environment.

Taxes/Finance

Once specific communities and properties are identified, tax rates, ratios of assessment, and local assessment practices can be identified to more accurately estimate local property tax burdens. Communities that participate in any local option exemptions on inventories, machinery, or other categories can be identified and factored into the cost comparison at this time. Items that should not be overlooked include local sales and income taxes.

Other financial incentives can be discussed as the choice of communities narrows. Depending on the economic and employment impact of a proposed operation, a community may offer various incentives such as low-cost financing, temporary reductions on property taxes, grants for property grading, and many other possibilities. Many incentives are offered on a case-by-case basis and are not necessarily part of a formal economic development program. A firm's leverage in negotiating the availability of incentives depends, of course, on the impact of the project on the community. It is important not to place excessive priority on incentive availability because that may minimize the emphasis given to more important items such as labor availability, market access, or reliability of electric power.

Transportation

The transportation characteristics of a community and specific property area can be critical to the ultimate success of an operation. In today's deregulated air, trucking, and rail environment, estimated rates should be obtained for all communities under consideration, even if they are only a few miles apart. Access to major highways can have obvious benefits for freight-intensive operations. Firms requiring rail service should investigate, in detail, the financial stability of the serving railroad, determine any plans for the lines serving the property, and ascertain the capability and cost of rail service to the property. Other transportation-related considerations such as air freight services or access to ports may warrant investigation for particular operations with those requirements.

Site Characteristics

The suitability of the property must be given priority over all other factors in a local evaluation. Physical features such as access soil characteristics, acreage, configuration, easements, topography, and drainage are among the items to be checked.

Contact should be made with authorities from all appropriate utilities (electric, natural gas, water, and sewer). Specific facility needs should be thoroughly reviewed with these authorities to verify adequate supplies, quality, and treatment capabilities. Rates can also be substantiated at this time. Letters of commitment for any necessary line extensions or upgrades should be obtained, and any underground trunklines or interceptors should be identified.

Municipal zoning ordinances vary significantly by community. A manufacturing use permitted on a parcel zoned "light industrial" in one community may not be permitted on an identically zoned parcel in another town. Therefore, the specifics of a zoning ordinance must be thoroughly investigated. If a variance is required, it is important to determine whether or not it can be obtained before taking an option on any property. If variances are needed, delays are almost certain to occur. The community's master plan, if one exists, should also be reviewed to see if problems could arise in the future.

Finally, close attention should be given to the environment near any site being considered because the image and value of a

TABLE 18-4
Comparative Community Data

Community County	Population 1980	Population Percent Change 1970-80	Percent Minority	Transportation Truck Lines/ Terminals	Nearest Interstate Highway	Nearest Major Port	Nearest Commercial Airport
Cullman, AL Cullman County	13.069 61,642	+ 3.6 +15.0	0.2 1.2	12/1	I-65 (Local)	Decatur (31 miles)	Birmingham (53 miles)
Hopkinsville, KY Christian County	27,318 66,878	+22.7 +16.0	25.7 28.9	20/5	I-24 (13 miles)	Nashville (75 miles)	Nashville (79 miles)
Monroe, NC Union County	12,639 70,380	+12.0 +28.6	36.6 16.6	30/1	I-85 (20 miles)	Wilmington (175 miles)	Charlotte (30 miles)
Shelby, NC Cleveland County	15,310 83,435	- 6.2 +13.1	36.3 20.9	38/5	I-85 (12 miles)	Wilmington (245 miles)	Charlotte (40 miles)
Smithfield, NC Johnston County	7,288 70,599	+ 8.4 +12.6	31.6 19.7	11/0	I-95 (Local)	Morehead City (115 miles)	Raleigh- Durham (45 miles)
Gaffney, SC Cherokee County	13,453 40,983	+ 1.5 +11.4	32.4 19.5	45/1	I-85 (Local)	Charleston (225 miles)	Charlotte (45 miles)
Union, SC Union County	10,523 30,764	- 2.3 + 5.2	32.9 29.5	37/0	I-26 (26 miles)	Charleston (190 miles)	Greenville- Spartanburg (50 miles)
Covington, TN Tipton County	6,300 35,500	+ 8.1 +21.0	36.0 27.2	20/2	I-40 (18 miles)	Memphis (38 miles)	Memphis (38 miles)
Dyersburg, TN Dyer County	15,856 34,663	+ 8.5 +12.3	19.0 12.0	23/5	I-155 (Local)	Memphis (80 miles)	Jackson (45 miles)
Martinsville, VA** Henry County	18,149 57,654	- 7.7 +11.6	31.1 23.4	39/6	I-85 (45 miles)	Norfolk (215 miles)	Greensboro- High Point (45 miles)

* Combined police and fire protection = 62.
** Martinsville, VA, is an independent city, surrounded by Henry County; population figures in Henry County do not include Martinsville.
NA = Not available.

manufacturing operation can be affected by the compatibility of its neighbors. Furthermore, a manufacturing operation can be hindered by excess traffic from densely developed commercial areas or be faced with public relations problems if residential areas are located too near a site.

Once a site is selected, an option should be taken on the property so as to conduct a title search, arrange for soil analysis, and make any final checks of local conditions. Other considerations are pertinent if there is an investigation of existing buildings, but they are beyond the scope of this section.

PLANT LAYOUT

Plant layout/design is one of the most interesting and important phases of industrial engineering. It has a direct bearing on quality and profitability because it deals with the arrangement of the physical facilities and the manpower required to operate it profitably and still produce a quality product. The objective in plant layout is to plan the arrangement of facilities and personnel to be the most cost effective by minimizing the movement of both materials and personnel during the conversion process.

Planning a layout or arrangement necessitates keeping an open mind. Some of the common goals to keep in mind when planning a layout are the following:

- Integration—integrating the many factors being considered that affect the layout.
- Utilization—effective use of process flow, personnel, machinery, site, and space available.
- Expansion—the process envelope must be easy to expand for future business growth.
- Flexibility—easy to modify to suit changes in business trends; adaptable to changes in product design and process improvements.
- Symmetry—a developed condition of regular or straight division of areas and somewhat even sizes of areas, especially when separated by walls, floor, and main aisles.
- Transportation—minimize distances for movement of all materials.
- Safety—safe working conditions for all employees.
- Support functions—to be convenient to process flow and to complement the site where applicable.

ECONOMIC IMPACT

Materials movement within the facility is a fundamental part of initial planning. Manufacturing cost is directly proportional

Education			Police		Fire				
Public Elementary/ Junior/Senior (Private)	Pupil:Teacher Ratio in Public Schools	Hospitals/ Beds	Full-Time	Squad Cars	Full-Time/ Volunteer	Pieces of Equipment	Fire Insurance Rating	Average Daily Water Usage/ Capacity of Water System, million gal per day	Average Daily Sewer Usage/ Capacity of Sewer System, million gal per day
2/1/1 (4)	NA	2/269	32	5	30/2	6	6	6.75/10.00	3.0/6.0
13/2/2 (5)	19:1	1/216	55	8	59/0	7	5	4.5/7.0	3.0/6.0
3/1/1 (2)	18:1	1/124	*	NA	*	5	6	7.0/12.0	4.85/7.00
4/1/1 (3)	18:1	1/300	45	NA	34/20	6	4	6.6/10.0	3.0/6.0
2/1/1 (0)	18:1	1/180	29	NA	2/31	7	6	3.0/6.0	3.7/4.0
12/2/2 (1)	16:1	1/164	29	8	20/NA	7	5	6.75/12.00	4.00/8.15
3/2/1 (1)	19:1	1/143	23	7	8/23	5	5	8.0/8.0	3.1/7.3
1/4/2 (2)	20:1	1/100	21	7	12/18	5	5	1.5/3.0	2.5/4.5
2/1/1 (0)	18:1	1/225	32	11	52/0	9	5	3.7/10.9	3.5/6.3
4/1/1 (1)	19:1	1/264	48	16	26/35	9	5	5.0/7.0	4.0/6.0

(*The Austin Co.*)

to the distance material must travel and the time required to process the product through the manufacturing cycle.

Materials movement (material handling, material transfer, and material flow) must be considered as the most important of all the necessary planning functions when considering a new facility arrangement or revising an existing one.

Material flow is the principal starting point in the planning function. An optimized continuous flow is ideal; however, the process may require synchronous or asynchronous functions during completion of a component or a product. This could introduce the need to modify the continuous flow process.

Many industries do not consider material handling, material transfer, or material flow a major cost, and therefore, these costs are inaccessible in their standard cost system. This can cause a significant adverse impact on profits.

To minimize the impact of this expense on a business, it is necessary to apply the fundamentals of engineering economy, manufacturing engineering, and business management to optimize profitability. Good manufacturing practices can add several percentage points to the profitability of a concern by optimizing material handling, material transfer, and material flow. The reverse is also possible; that is, reduced profits by poor application engineering in these areas.

FACILITY ARRANGEMENTS

Three distinct or major types of facility arrangements are commonly used in industry: (1) function or process, (2) product process or production line, and (3) fixed position or fixed material components. Each arrangement has advantages as well as limitations that make it suitable for certain applications. The two primary factors that influence facility arrangement are the physical size and quantities of the products to be manufactured. In general, most manufacturing facilities combine all three types of arrangements or a variation of these basic types.

Most individuals involved in facility planning and design will probably use some form of computer-aided design to develop the arrangement of the facility. Although the software programs perform many functions, the facility planner must be knowledgeable in product process design requirements before attempting the decision process.

Function or Process Arrangement

Layout by function or process is the most predominant and frequently used facility arrangement. In this arrangement, all operations for the same function are grouped together (see Fig. 18-2). For example, turning, drilling, and milling are in their respective areas; painting is generally in an area specifically designed for this function, as are the heat treatment process components. In other product fabrication such as shoe or shirt manufacturing, the cutting function is separated from the molding and stitching or sewing functions.

This type of arrangement requires less capital equipment to purchase and install; increases equipment utilization; is easier to automate, using either robotic or other hard automation and cells; is adaptable to group technology and schedule changes; and reduces material handling and material transfer costs.

PLANT LAYOUT

Product Process Arrangement

In the product process arrangement or layout, one product or one family of products is produced in one area where group technology is implemented. This type of arrangement is utilized as greater volumes and velocity of product or components are required to meet the schedule and consumer demand.

The material moves from operation to operation for this arrangement (see Fig. 18-3), with one operation immediately adjacent to the next. The equipment required to manufacture (fabricate) the product, regardless of the process, is arranged to follow the process flow diagram or the sequence of operations.

This type of arrangement provides the lowest possible material handling and transfer costs within the confines of the process. With state-of-the-art technology, there is flexibility when using group technology principles. Using the just-in-time inventory philosophy, little or no in-process material is required. This reduces the amount of money invested in inventory. The continuous operation is the easiest to automate with robotics or other hard automation. Controls and communications provide some flexibility and opportunities to reduce inventories.

Local area networks (LAN) provide measures to monitor the process and communicate to the management information system (MIS). This provides an opportunity to implement a paperless facility.

The product process arrangement also provides greater labor efficiencies and cost reductions. It requires the least amount of floor space to perform a function, and less supervision is necessary.

Fixed Position Arrangements

In the fixed position or fixed material arrangement, the material, components, or assembly remain at a fixed location or station. This is referred to by some as station assembly or station fabrication. This arrangement is utilized by those manufacturers/fabricators whose production volume is low or assembled unit is large.

Major components may be fabricated elsewhere and transported to the assembly station where assemblers custom-fit and fabricate parts to make the complete assembly. Examples of the types of components manufactured using this arrangement are low-volume machine tools and airplanes.

This type of arrangement permits the skilled workers to complete their tasks at one location, thus saving labor. It is flexible because modifications in design and/or procedures can be accomplished readily. It is also adaptable to a varied group of products. Finally, the fixed position arrangement does not require much planning or engineering for the production process within the facility.

LAYOUT FUNDAMENTALS

Plant layout, using the broad definition, is the master plan that shows the integration of all systems and subsystems into a homogeneous, cohesive, dynamic, and functional operation. Functions included in this master plan are the following:

- Site or grounds.
- Utilities.
- Departments.
- Process equipment and machinery.
- Process methodology.
- Material handling, transfer, and storage.
- Service facilities.
- Production flow.

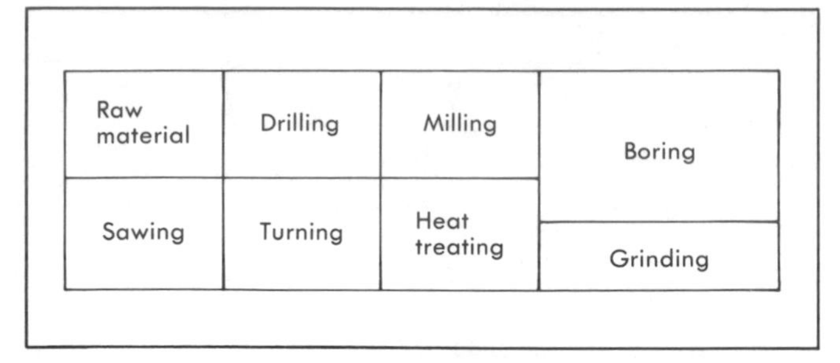

Fig. 18-2 Schematic of a plant arranged by function or process.

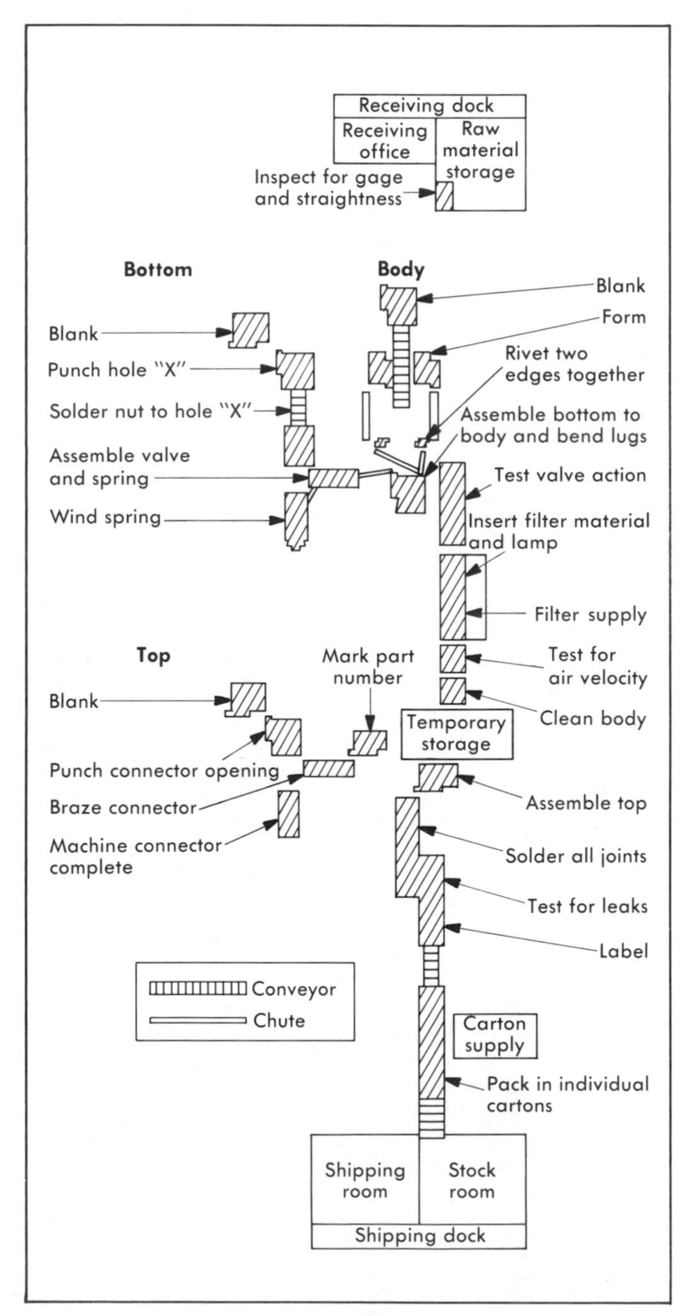

Fig. 18-3 Schematic of a plant arranged by product process.

- Labor utilization.
- Receiving/shipping.
- Distribution of finished product.
- Employee amenities.

Before starting a new plant layout or modifying an existing layout, the planner must recognize certain factors. Each of these factors affects the planning of an efficient layout to a significant degree. Some of the factors that must be taken into consideration are the following:

- Ergonomics within the facility and workplace.
- Products produced—lot quantities or continuous process.
- Processes—type, space requirements, and material flow.
- Layout—flexibility, location of services, support functions, and special requirements of departments.
- Materials—material handling, material transfer, material staging, and material storage.
- Miscellaneous—supervision offices, production offices, cafeteria services, restrooms, and aisle width (main and service).

The flow of materials is a major influence in the determination of the functional association. However, compromise due to undesirable functions and support functions is always necessary.

Adherence to the flow of materials required to produce product reduces material handling and transfer distances and, therefore, presents the opportunity to improve productivity and efficiency of the operations. In addition, material handling and transfer costs are reduced.

COMPUTERIZED FACILITIES PLANNING

A prerequisite in designing and planning a facility is flexibility. The design must be cost effective for a range of changes in products, designs, and production techniques. Computers and related software serve as tools in flexible facility design and enable the user to develop the optimum arrangement of the process.

Facilities planning using computerized integrated technologies is available in the form of software programs for use on both mainframe and personal computers. These software programs range from simple to complex and require expertise in the use of the computer terminal and peripheral equipment. In addition, a knowledge of the methodologies of manual structuring of facilities arrangements is also required to develop the database used with these programs.

The greatest benefit of using computerized facilities planning is the ability to compare alternative arrangements. This enables the CAD operator to interchange selected functions and input data to develop the most cost-effective facility. Further, the database, together with other inputs, can be manipulated to perform simulation of pertinent production, material handling, and material transfer systems, including automation and robotics, within the facility for additional cost reductions.

Computer software programs for planning arrangements or layouts have been in existence for more than a quarter of a century. Initially the output from these programs was not all that useful because it only assisted in a small portion of the planning work. The programs dealt with major changes to the arrangement or layout and were of little use for minor rearrangement. Because minor projects outnumbered major projects by approximately 10 to 1, most rearrangement projects were best serviced manually.

During the past decade, technology has played catch-up with the needs of industry. Development of advanced computer software, displays, and interactive operation, together with the increased requirement for efficient low-cost facilities planning, have produced myriad cost-effective, practical tools with a wide range of capability and complexity. There are so many that it is often difficult to determine which program is most appropriate for the application and planning needs. The titles of some of the software programs currently available are ALDEP, CRAFT, CORELAP, COFAD, PLANET, and CALFAS. Details pertaining to the use of these programs can be obtained from the Institute of Industrial Engineers.

Three major technologies that make up computerized facility planning are: (1) the decision database system, (2) computer-aided design, and (3) the management information system. The decision database system uses statistics, modeling, algorithms, and calculations. The results are displayed as graphics (charts, graphs, block diagrams) that are then interpreted. These interpretations are applied to a project's early phases for assigning locations, sizing, block layouts, and feasibility studies. The compiling of the input data into the project's database requires a great amount of manual effort, but it is necessary and must be accomplished prior to using the decision database system.

Computer-aided design (CAD) is used during a project's intermediate phases to produce plans, alternatives, dimensioned layouts, drawings, and visualizations. Visualizations may include static views or two and three-dimensional dynamic simulations. If CAD is used to design structures and mechanical systems, it may interface with engineering routines for various load, stress, and sizing calculations. Nearly all such systems provide ancillary takeoffs and area measurements.

A management information system (MIS) includes the routine storage and retrieval of data with and without graphic displays. During a project's final phases, the MIS is used for scheduling, controlling, and communicating. The MIS usefulness continues after a facility is functional to assist in cost reduction audits and space and equipment tracking. It also provides communication and controls to management for primary functions. In most cases, data in the MIS is used to provide input data to create the database necessary to implement the decision support systems.

APPROACH TO FACILITY LAYOUT PLANNING

The maxim that "Simple is Better" should be utilized when planning changes in an existing or a new arrangement. The reason for any change will be stated in a great many ways, but the objective will always be to improve safety, quality, and profitability.

Layout planning is not a one-person task. A facilities design/layout team should be formed utilizing the in-house expertise available. A consultant who is proficient in the field of facilities design and layout should also be engaged to take advantage of the leadership and exposure to the state-of-the-art technology.

Layout planning can be simplified by compiling a thorough and complete database that will be utilized for all decisions made for the project. This database consists of, but is not limited to, the following:

- Business plan for the company.
- Manufacturing strategy.
- Growth projections, preferably a five-year forecast.

- Status or percent of market being enjoyed at present and what is anticipated/desired.
- New products in design to be introduced to the marketplace.
- Obsolete products to be deleted from production.
- Present utilization of existing capacity and proposed improvements.
- Inventory-carrying philosophy and customer service level to be maintained. Is just-in-time a consideration?
- Standard costs and cycle times of components and assemblies affected.
- Equipment lists.
- Group technology and cell concept information for flexible manufacturing approach.
- Simulations.
- Annual inventory turn data.
- Warehouse systems and inventory information, including receiving and shipping systems. If the warehousing systems are subject to change and improvement, all material movement and throughput must be compiled together with physical inventory data and the cubic size requirements of line items.
- Inventory stock status reports.

This basic data becomes the database for the entire project and should be placed in its proper category, cataloged, indexed, and placed in a binder for easy, quick reference. It is to be used extensively as the analysis progresses for reference and decision making.

If the arrangement is more than a modification to plan space for an additional piece of equipment, then it is an opportunity to investigate state-of-the-art systems that are on the leading edge of technology and evaluate them for possible implementation. This would include all communications and controls that could improve any systems, from the order entry system through the conversion process and including the inventory and shipping systems. Included also would be the computer networking systems for closed-loop communications and controls.

Now that the database is complete and available for reference, the following steps are to be completed, either manually or with the computer-aided design graphics terminal:

1. Product flow diagrams—use the ladder-type flow diagram format to minimize confusion. Include all functions required for every product line.
2. From the product flow diagrams, develop a facility flow diagram showing all functions identified in the product flow diagrams.
3. Develop a relationship chart for the facility, using priority evaluation.
4. Do the capacity planning function using information from the business plan and standard data to convert into generated hours and space requirements.
5. Create a block diagram and alternates using items 3 and 4. This can be accomplished manually or with the use of computer-aided design software. When this step is completed, the planner will have the basic configuration of the facility and the size necessary to maintain the flexibility level desired. It is important to remember that in this world marketplace flexibility is required to remain competitive.
6. Prepare the plant layout using the equipment required to perform each function. Create a cell library using the computer-aided design software. The cell library can be two or three-dimensional; however, two-dimensional is usually adequate unless there is a clear-height problem. Then, using the flow diagrams and relationship charts, withdraw the equipment templates from the cell library and place them in flow patterns dictated by the process requirements. When this is completed, print this arrangement on an electrostatic plotter at convenient scale, and then begin to create the alternatives. When this has been done, select the optimum arrangement by using the appropriate software. Finally, plot out the arrangement at a working scale, usually ¼″ = 1′.
7. Simulation of facility designs is necessary in today's age of high-technology production facilities. The high cost associated with the design and installation of a high-tech facility can be prohibitive. It is therefore advisable to simulate all of the systems involved using the computer and appropriate software to perform these functions. The success of a new high-technology facility most often hinges on the successful design of material handling and material transfer systems to integrate the various automation cells. The layout is then modified based on simulation results.

HOUSEKEEPING

Since the industrial revolution, the United States has been one of the world leaders in the development of manufacturing and industrial processes. During more recent years, however, foreign companies have expanded their products and markets to blanket the United States and the world. To counter this trend, many U.S. industries are developing new ideas, upgrading their technology, and attempting to expand their economic base. During this process, many industries have realized that it is necessary to build new facilities or upgrade their existing manufacturing facilities so that they are efficient, functional, and attractive. An important part of achieving these objectives is to design building facilities that can be efficiently maintained. The selection of materials and systems for the facility should promote housekeeping.

TYPES OF FACILITIES

Industry uses a wide range of facilities to make its products. The largest category, at least in size, is heavy industry, including steel mills, foundries, and forge shops. These industries use heavy equipment to move large volumes of heavy materials. Inside the buildings are furnaces, pits, rolling mills, casting machines, and presses. Housekeeping equipment should be heavy and rugged to maintain a clean environment.

The next category is medium-to-light industries. A variety of products are produced in these industries, ranging from heavy equipment, machinery, and automobiles to fractional horsepower motors, furniture, and appliances. Facilities that house these industries have diversified equipment for fabrication, welding, machining, assembly, painting, and heat treating.

The last category is high-tech industries. In these industries, the environment and facilities are required to be kept clean, such as in the pharmaceutical and microchip industries. In recent years, high-tech industries have been an expanded market and have placed high demands on maintenance and housekeeping.

From heavy through high-tech industries, the selected materials that make up the buildings have an impact on maintenance and housekeeping.

Exterior Walls

The requirements for exterior walls vary with the industry. For heavy industry, the walls should be durable; by the nature of the industry, they will receive little maintenance. In most cases there is little requirement for insulation to prevent heat loss or heat gain. The material most commonly used for the exterior walls of a heavy industry facility is corrugated metal panels with or without insulation. Masonry walls are also used for many buildings. Other materials used are corrugated panels of cement-bound mineral fiber; this offers less protection than corrugated metal panels and requires some repair and maintenance. Because it is unlikely that these walls will be cleaned periodically, consideration should be given to the need to repaint them in the future. Repair of damaged panels should also be considered.

For medium and light industry, the outside walls are formed sandwich—an exterior metal face panel that is ribbed, flat, or corrugated, with insulation between panels for reduced heat loss in the winter and heat gain in the summer. This type of exterior wall is durable, cleanable, and cost effective. Corrugated panels of cement-bound mineral fiber may provide less protection from impact damage and are more difficult to maintain and replace. Concrete block with brick veneer should be considered in light industry where durability and esthetic qualities are required. This type of construction requires a minimum of maintenance.

The two types of materials commonly used in exterior walls for the high-tech industry are customized metal panels covered with powdered coated epoxy paint and concrete block with brick veneer and epoxy paint for interior finish. These materials require little maintenance and promote a clean appearance for this type of industry.

Roofs

The metal roof system is frequently used in the heavy industry because it has a long life under extreme conditions and is virtually maintenance-free. It should be noted that these roofs are not always watertight and airtight. This does not create problems for some applications.

The roof commonly used for medium-to-light industries and high-tech industries is EPDM (ethylene propylene diene monomer) roof. The single rubber elastomeric membrane is applied in large sheets to reduce field seams, thus minimizing future maintenance. Long-term elasticity permits the material to withstand normal building movement without cracking or tearing. These products have a fairly good track record in minimizing leaks. It is important to have good flashing and to pay attention to details. When reroofing or refacing, the membrane often may be applied right over the existing roof with no costly tear-off required.

The third type of roof that has been used extensively over the years is the three or four-ply built-up roof. It is popular because of the low initial cost. However, the higher maintenance cost is a consideration when selecting this type of roof.

Floors

Heavy industry has narrowed the choice of floors down to monolithic concrete slabs with metallic reinforcing and various metallic toppings. Variables include slab thickness and type and size of steel reinforcing. These variables are a function of the weight that the floor must support. The type and thickness of the topping is based on the type of floor traffic.

In the medium or light industries, conventional concrete slabs should be considered. The concrete mixture and type of topping promotes dense concrete and minimizes dust. A metallic shake topping may be placed on the surface of freshly poured concrete to give a harder surface and improved resistance to abrasion/wear and to aid in maintenance. Solid-end grain wood blocks and ceramic tile floors have been popular in medium-sized industry throughout the years. Although the impact resistance is superior, the initial cost, daily maintenance, and repair requirements make these products less attractive.

Generally, high-tech industry requires two types of floors: access floors and seamless floors. Access flooring panels, with a high-pressure, plastic laminate finish, are commonly used for clean room applications and computer rooms. These floors require vacuum cleaning and light dust mopping. Seamless floors are of a resinous matrix and should be considered when acid, chemicals, and solvents are used. The maintenance department must have the appropriate supplies available to repair and clean this material.

Interior Walls

Concrete masonry offers designers a vast array of sizes and shapes suitable for virtually any application. In addition, this type of wall provides many advantages such as low maintenance, noise resistance, fire resistance, elimination of formwork, and structural integrity. One disadvantage is the inflexible nature of these walls. Because flexibility, expandability, and cost are important considerations in the contemporary office or plant, gypsum wallboard systems are becoming increasingly popular. In recent years, demountable drywall systems have been used. These systems are more flexible and cost effective than fixed-partition gypsum wallboard.

Interior walls for the high-tech industry, such as for a clean room, are generally customized steel panels covered with powdered epoxy paint or gypsum wallboard covered with epoxy paint. When a high degree of cleanliness or sanitary conditions are required, steel paneling is recommended because of the economy of cleaning.

Ceilings

Except for the high-tech industry, manufacturing space in general does not have a need for ceilings. However, this is changing rapidly because of the wide use of sophisticated equipment, computer-controlled equipment requirements, and the need to produce quality products. Exposed deck and structural systems facilitate flexibility in most manufacturing processes. Under normal circumstances, these beams, girders, bar joists, and decks are not cleaned or do not require any maintenance during the life of the building.

In the high-tech industry, ceilings often are required, and they are either smooth or perforated sheet metal, acoustical tile, or gypsum board covered with epoxy paint. Like walls, they require routine maintenance and cleaning.

CHAPTER 18

HOUSEKEEPING

Grounds and Roadways

The grounds and the roadways of a facility are the areas seen by the public, customers, and clients. How well they look determines to a large extent the general impression that the community and visitors have of the facility and the company.

Ground cover is divided into three categories: (1) natural setting, (2) lawns, and (3) special cover. The natural setting type of ground cover requires a minimal amount of maintenance. Well-kept lawns are attractive, but require cutting, weeding, fertilizing, and watering. The special type of ground cover could be myrtle, pachysandra, or other ground cover, all of which require special attention.

It is important that the walkways, driveways, and parking areas consist of the best material within the company's budget. The least expensive capital cost is gravel, but it also has the highest maintenance cost. The next least expensive from a maintenance cost standpoint is asphalt paving. Concrete is the least expensive to maintain.

DESIGN CONSIDERATIONS FOR HOUSEKEEPING

During the design of most facilities, considerations for keeping the facility properly maintained and clean begin too late. With a little forethought and careful planning by the design team, the owner and the maintenance managers of every building can become as efficient as is technically possible. This section discusses some of the things that should be taken into consideration in the design phase of a new or remodeled facility to reduce maintenance cost and promote an efficient/profitable operation.

The entrance to the plant or office area should be designed to prevent dirt, mud, sand, snow, water, or salt from being tracked into the facility. Recessed or surface walk-off mats should be considered. Prefabricated mat systems are becoming increasingly popular. These consist of a grating with rectangular bars placed over a catch pan that will remove the gross soil from people's feet. The value of additional walk-off mats during inclement/wintry weather should not be underestimated. Collecting snow, water, and mud at the entry will increase the life of carpeting and other floor materials in adjacent areas as well as reduce the amount of cleaning required.

The type of flooring installed in a facility should be selected based on the work being performed. Embossed tile and light-colored floors should be used selectively because they may require special cleaning equipment and materials. Floors should also be kept flush and at one level for safety as well as for ease of cleaning. Concrete floors should have a troweled finish and be properly cured. Vinyl or rubber baseboards should be used to provide a scuff-free surface; cove bases facilitate cleaning.

Washrooms should be laid out to permit a circular flow of traffic, with wash basins located near doors. They should also be away from busy areas to equalize the traffic load on floors and be well ventilated. A floor drain should be installed in the washroom floor in case of emergency spills and to permit spray cleaning. Fixtures and partitions should be suspended from the walls and ceiling to leave the entire floor area accessible for quick, effective care. To prevent cigarette butts from being thrown into toilet fixtures, durable ashtrays should be installed between them. Lockers should be placed on concrete or ceramic bases and have slanted tops to facilitate cleaning. Washroom walls should be painted with enamel paints or should be made of ceramic tiles. Paper dispensers should hold an ample supply, and waste containers should be large enough to collect all the daily waste products. Double toilet tissue dispensers hold a reserve supply of paper, thereby minimizing complaints and reducing the amount of time for service. Soap dispensers should be placed over sinks to prevent the soap from dripping on the floor. Liquid-type soaps should be considered.

The materials used for the construction of interior walls and ceilings should be capable of being dusted as well as wet cleaned. Soft-blown mineral acoustic ceilings, for example, are very difficult to clean. Walls should be painted with durable and washable paint or covered with other durable materials such as vinyl wall coverings. If concrete block is used, the surface should be treated with a block filler coated with an epoxy resin. Structural glazed tiles and ceramic tiles should be used on the walls of custodial closets, abuse-prone hallways, or stairways.

The window frames should be made from aluminum rather than wood or painted steel, except in applications where the metal is exposed to a corrosive environment. In areas of high humidity, thermal-break metal frames minimize condensation. Operable windows should be installed on multistory buildings so that both surfaces can be washed from the inside; washing can even be performed at night or during inclement weather. It is also less expensive to wash windows from the inside than from the outside.

Metal, plastic, or composition furnishings are more easily maintained than wooden furnishings because the material can be spray or steam cleaned. Waste receptacles should be placed in the areas where the waste is created and should be accessible to some type of mechanized cart so that they can be dumped without removing them from the area. They should be sized appropriately and furnished with covers when odor-producing materials, edibles, or dusty materials are discarded. Plastic liners minimize odors and receptacle cleaning.

Drinking fountains, cigarette urns, waste receptacles, telephone booths, and coat trees should be hung from the wall rather than placed on the floor. This facilitates cleaning and helps to prevent damage to fixtures and floors. Vending machines and other equipment should not be placed immediately against the wall to allow room for cleaning. It is also important to include sufficient electrical outlets throughout the building to facilitate floor vacuuming and buffing.

KEEPING A FACILITY CLEAN

The type and number of people involved in housekeeping determine to a large extent how well a facility is maintained and cleaned. It is important to remember that most people cannot routinely go about daily maintenance and cleaning duties without allowing boredom and haphazardness to occur. To minimize the effects of boredom, good employee selection procedures, proper supervision, and good equipment are necessary. It is also important to provide the maintenance and housekeeping personnel with the proper training to bring together the personnel, materials, and machinery for effective results. Management should also consider contracting the housekeeping work with an outside firm. This may be cost effective and minimizes personnel problems.

Several methods are used to determine the number of people necessary for keeping a facility clean. One method determines the number of hours required to perform all the tasks. From this statistical information averaged over a number of plants, the number of workers is calculated. In the arbitrary staffing method, a fixed number of workers is determined based on past experience, budgetary limitations, and preconceived opinions. Another method determines the number of workers required by trial and error. This approach is impractical because it tests various staffings until one is found that seems to work all right. The final and recommended method of staffing is called creative analysis: the entire custodial function is evolved from a creative standpoint and an optimum method of staffing is arranged. In this approach, a study is made to determine the cleaning needs, collect data, set the objectives, make the analysis, set frequencies of cleaning, and develop standards based on conditions and the most productive techniques. These methods and standards should be recorded for the various operations and explained to the housekeeping staff.

Equipment

The maintenance persons, janitors, or custodians, or whatever title they are given, must be equipped with the proper tools of good quality, along with proper control and care for repeated use. The following discussion describes typical cleaning equipment needed by the housekeeping staff, starting with the basic needs.

Custodial carts allow the popular cleaning equipment and supplies to be kitted and moved from one area to another, saving walking and searching time.

The fountain-type window washer unit, with a telescoping handle, permits quick washing of up to four levels above the ground while standing on the ground. This type of washer eliminates staging and reduces the hazards of high-level working.

Drum vacuums consist of a vacuum head on top of an open-head steel drum on a dolly or coaster. They are useful for picking up water from a leaking roof, sprinkler leakage, or draining problem.

Automatic scrubbing machines offer a means of rapid cleaning in open areas where preservation of surface is not of great importance or where the surface will be buffed.

Scoop vacuums are used for wholesale water removal areas where automatic scrubber use is not desirable.

Carpet equipment is necessary because most facilities have some areas that are carpeted. Available are powerful surface vacuums, medium and heavy-duty pile-lifter vacuums, dry cleaning equipment, and shampooers.

A floor machine with a single disk is still the workhorse for moderately congested floor areas. These machines remove wax without using detergent.

Steam cleaning units are used in cleaning vinyl upholstered furniture and floor machine pads, removing wax buildup from other equipment, degreasing vents, descaling shower walls, and for many other applications.

Litter collectors, powered with a lawn mower engine, are good for picking up leaves, paper, and litter inside the facility as well as outside. In some areas, litter collectors have replaced power sweepers, but their use is limited because of noise and fumes.

Ultrasonic cleaning equipment removes soil from surfaces by introducing high-frequency waves into a cleaning solution. This type of equipment has been successful with such items as egg crate louvers and filters. Ultrasonic cleaning is useful for cleaning a large quantity of parts that can be placed in a container with ultrasonic capabilities.

Wall-washing machines can be used effectively on uninterrupted wall space where a good painted surface exists.

Spray buffing, probably the most significant floor care technique developed in recent years, has prompted the manufacture of various spray devices. They can either be pistol-grip hand sprayers or mounted on floor machines. This can be referred to as high-pressure cleaning.

Power sweepers are available in many sizes, both walking and riding types. Some models are convertible to automatic scrubbing use. This is a must for medium-sized plants.

Spray buffing and power sweepers are essential in the maintenance of a modern manufacturing facility. Therefore, the facility must be designed and laid out with this in mind.

Waste disposal equipment performs functions of shredding, pulping, incineration, packaging, blowing, bagging, and compacting

Chemicals

Cleaning chemicals, like cleaning equipment, are tools for the job. Whenever chemicals are used, it is necessary to have some type of decontamination spray or shower available to the users. Chemicals must be safe, from the standpoint of avoiding both a hazard to the workers (such as a slippery floor) as well as a hazard to the user (damaging to the skin or dangerous when breathed). Chemicals should be easy to use, not require excessive stirring, and be conveniently packaged so excessive time is not required for decanting and measuring.

Although chemicals amount to just a few percent of the total cleaning cost, the proper dilutions should be used to save money. The distribution system for these materials and supplies should be well thought out, planned, and controlled. The housekeeping and purchasing departments should establish good communication, be alert for new products, and use quantity discounts. Setting specifications on chemicals is advisable, but does not always ensure the selection of the best products. It is possible to evaluate rather accurately by sight and touch whether a detergent is doing an effective job in cleaning. To determine whether a germicide is killing germs demands carefully controlled samplings and testing procedures. Laws require that such germicides comply with definite minimum standards of effectiveness and be accurately labeled.

Waxes. The new polymeric waxes consist of blends of acrylics, polyethylene, polystyrene, and resinous materials and perform better than the old carnauba and ouricuri waxes. Polymers have an immediate gloss upon drying, resist scuffing, and demand less care than do waxes. Floors that require stripping four or more times each year may now be maintained for three or four years between stripping. This provides economics in labor and materials.

Detergents. Cleaning was once accomplished through the use of natural soap. Now specialized detergents and blends of synthetic detergents with natural soap are used. The type of detergent used will depend on the soils, surfaces, and methods of cleaning.

Disinfectants. Disinfectants, germicides, or bactericides are products that kill harmful germs and inactivate some harmful viruses. The three most widely used classes of disinfectants are the phenolics, the quaternary ammonium chlorides known as "quats," and the iodophors. Time can be conserved in housekeeping operations by using a combination cleaner disinfectant or germicidal detergent; one operation will both clean and disinfect an area simultaneously. Both phenolics and quats can be formulated to provide effective cleaning and effective germ control.

Sealers. Sealers are available for wood as well as materials containing cement and stone. Of interest are the oil-modified and moisture-cured polyurethanes as well as the epoxy seals.

Strippers. Strippers are commonly used in housekeeping for wax and finish stripping. Most are formulated with ammonium hydroxide or organic amines to give pH levels as high as 12.5. It is important to select the proper stripper to avoid damage to the floor material.

Specialized chemicals. Other specialized chemicals are available and should be part of the housekeeping supplies. Some of these special-purpose materials are for suds control, stain removal, carpet shampooing, glass washing, metal polishing, scouring, descaling, scum removal, mop treatment, deodorizing, pest control, hand cleaning, and antisoiling.

DISASTER CONTROL

A disaster can be identified as any occurrence that negatively has an impact on the economic or profit posture of a manufacturing facility. The assessed dollar value of loss determines whether it is classified as a disaster or incident, and this will vary with different industries.

TYPES OF DISASTERS

Disasters can be categorized into two groups, those that originate from the influence of humans (sometimes animals) and others that are caused by acts of nature. Incidence of disaster can be kept to a minimum by implementing various preventive measures and developing control features that reduce the duration of a disaster.

Disasters of Human Origin

In most instances of fire disaster, the cause can be traced back to poor or improper housekeeping. For a fire to occur, it is necessary to have three ingredients: fuel (something that will burn), an ignition source (something to ignite the fuel), and oxygen (to maintain combustion). Good housekeeping and maintenance will significantly reduce the risk of a fire occurring.

Explosions represent another form of disaster. Though short in duration, they can be very devastating to both life and property. Most explosions are caused by a single volatile product that when ignited results in an explosion.

A nuclear contamination disaster is probably the least likely disaster that would be confronted. If a plant location falls within a 40 mile (65 km) radius of a nuclear reactor or beyond and downwind in the direction of the prevailing wind, measures should be taken to develop an evacuation and monitoring plan for the duration of the emergency. Likewise, it is also important to investigate and become familiar with the process of decontaminating a facility once it has been exposed to radiation.

A labor dispute that causes a strike and, in turn, a work stoppage of significant duration can be classified as a disaster. This can be compounded further if employees and/or management are divided in their opinion and physical violence erupts, resulting in human injury and/or damage to property.

Disasters Caused by Nature

Extreme changes in weather can be a strong contributor to major disasters developing in the manufacturing circle. The best countermeasure against this is to keep abreast of the weather forecast. Early warning allows the manufacturer to prepare.

Tornadoes and hurricanes are enough alike in behavioral character that the resultant effects of the two create a disaster that cannot be distinctly differentiated. They both produce a high-velocity wind force that victimizes anything that is not properly anchored, breaks glass-like material, and destroys any poorly assembled structures. To thwart the possibility of any of these results from occurring, the owner has two options. The first approach would be to cover those items that are breakable and to tie down loose elements. This approach, however, is only a temporary measure and must be repeated for each occurrence. The second approach would be to upgrade existing building components through a construction effort so they can repeatedly withstand the high-velocity winds; this is a one-time, permanent solution.

Storm and flood disasters vary directly with regard to the geography of the site location. If an existing building is located in a valley versus a hilly site, the possibility of a problem exists. In selecting sites for new structures, it is very important to determine the vulnerability to flooding, as well as the downstream proximity to a dam should there be a structural failure. The integrity of a dam or levee should be considered, whether it is earthen or concrete.

Earth movement or seismic hazard is an inherent subsurface soil stability condition that is defined and related to geographic areas of the country. Maps defining fault areas can determine how vulnerable an existing building or anticipated site would be to seismic impact. High-rise structures are more vulnerable to damage than low-rise buildings. Different construction techniques and materials survive earthquakes in varying degrees. These elements should be evaluated from an engineering standpoint before setting out on a new construction plan or even major remodeling. Most seismic disasters can be avoided through an evaluation of conditions and commensurate design of facility as well as site selection.

Vessels or piping that transport liquids and are subject to freezing can contribute to a disaster in a number of ways. Control of freezing liquid is very basic and can be addressed in either of two methods. If the function of the liquid is to circulate and cool operating machinery, the remedy is to dilute the water or replace it entirely with an antifreeze solution. On the other hand, if the liquid is the manufactured product or used in process development, the remedy is to insulate the carriers of liquid or house them in a heated environment. A third approach may be considered in limited application: to install the piping with a heat trace such as an electrical element heating tape.

PREVENTIVE PLANNING

Risk analysis is the process of evaluating the possibility and severity of an economic impact occurring. Risk evaluation determines the cost effectiveness of either experiencing the disaster or taking partial or complete precautionary measures that would intercept and control the problem.

Fire Prevention

Fire prevention has many facets that are applicable to the various types of potential fire sources that occur in industry. The first approach to fire prevention is avoidance. There are two principal areas of effort that can contribute measurably in preventing fires from occurring. The first is to comply with current building codes in design and construction. The second is to operate with an effective policy toward maintenance and housekeeping.

The best efforts of fire prevention sometimes fall short and fires do start. When this occurs, it is important that early detection devices respond to the incident in one of the four stages of fire. The four progressive stages of fire are as follows:

1. *Incipient stage.* There is no visible smoke, and chemical decomposition takes place with tiny particles being emitted that rise to the ceiling. If fire is detected at this stage, very little damage will occur.
2. *Smoldering stage.* If the fire continues beyond the incipient stage, the quantity of combustion particles increases and the mass becomes visible smoke.
3. *Flame stage.* With further progression of the fire, ignition occurs and flames appear.
4. *Heat stage.* This occurs just moments after flames appear and is now a fully developed fire. At this point, smoke, flame, heat, and gases are all present.

Detection devices are grouped into four categories that correspond approximately to the four stages of fire: (1) ionization, (2) photoelectric, (3) flame, and (4) heat (see Fig. 18-4).

Extinguishants come in a number of forms, from water (still the most common) to powders, chemicals, and inert gases. The carriers that are used to deliver the extinguishants also are quite varied. The most basic is the handheld extinguisher, which is dedicated to the small fire. Even at that scale, extinguishers are available with one or a combination of ingredients to effectively combat the three classes of fires. Figure 18-5 defines the classification of fires as to their makeup and the appropriate extinguishant to be considered.

Another simple method of combating fires is with a hose and water. This arrangement can be a permanent installation or a portable hose that would be connected to fire hydrants located conveniently on the exterior and/or interior of the building. Requirements for the type and quantity of extinguishers, fire hoses, and hydrants are derived from local and statewide codes.

Automatic sprinklers have been in use since before the turn of the century, first as a wet-pipe system, later as a dry-pipe system. The wet-pipe system utilizes a piping network fully charged with water. When a sprinkler head is opened, there is an immediate and direct discharge of water from the respective head. In the dry-pipe system, the sprinkler piping system is air filled under slight pressure. When the sprinkler head opens, the pressure in the system is reduced and this triggers the dry valve to open, allowing the water to flow through the system. The dry-pipe system was developed for areas where freezing may occur.

The preaction system is utilized in areas where the potential exists for a water-charged system to develop a leak and thus damage the building contents with water. The preaction system consists of fusible-link fitted sprinkler heads and connecting piping system. The system is activated by a detection device that opens a valve to permit water to flow into the piping system and discharge from any open sprinkler.

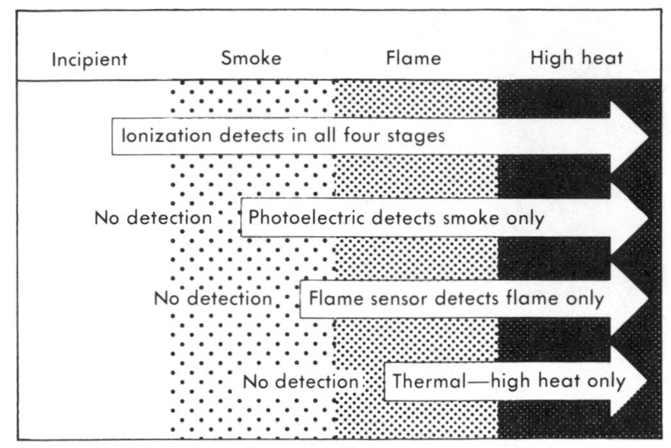

Fig. 18-4 Comparison of four main types of fire detection devices.

In areas where large quantities of highly flammable solids and liquids are stored, the deluge system is most appropriate to combat a fire. In principle, the system functions much like the preaction system and consists of the same components except the fusible link in the sprinkler heads. With open sprinkler heads and water charged into the system, the result is an overall deluge.

Other types of sprinkler systems also exist. In principle, they are variations of those systems previously described and designed for specific performance. For specific standards relative to fire detection and suppressions, the National Fire Protection Association (NFPA) offers complete information in its "Standard for the Installation of Sprinkler Systems." This standard offers guidelines for such things as pipe sizes, sprinkler head spacing, types of devices, and equipment, all in accordance with occupancy hazards.

Another effective suppression system is a network of discharge nozzles, all connected through piping and supplied by storage tanks containing gas, that are designed to flood a room or chamber. The two types of gases that are used predominantly are halon and carbon dioxide. These gases are used in rooms that house expensive and delicate electrical equipment that otherwise would suffer major damage if doused with water. A typical application would be in an electronic data processing (EDP) facility of a manufacturing plant. For more information regarding the specific application of gas extinguishing systems, refer to NFPA volumes.

Halogenated agents are commonly referred to as halons and represent a family of hydrocarbons. All halogenated agents are nontoxic below specific concentrations and over certain time periods. For example, at 7% concentration or less and for an unlimited time period, halon 1301 is not harmful to humans, but is still effective in its role of flame suppression. Halons are also used in portable or hand-carried fire extinguishers; halon 1211 is typically used in unconfined spaces.

Carbon dioxide (CO_2) gas is odorless, noncombustible, and effective in extinguishing fires by reducing the oxygen content in the area of discharge. Because the gas is nonconductive and contributes practically no damage to contacting surfaces, it is highly recommended for extinguishing electrical fires and flammable liquids. There is, however, a concern that should be recognized in the application of carbon dioxide. The discharge of large amounts of carbon dioxide can create hazards to personnel such as oxygen deficiency and reduced visibility.

CHAPTER 18

DISASTER CONTROL

Classification of Fires	Pressurized Water	Multipurpose Dry Chemical	Carbon Dioxide	Halon 1211
Class A fires Paper, wood, cloth, and other ordinary combustible materials where quenching by water or insulating by dry chemical is effective	Yes Excellent	Yes Excellent Forms smothering film, prevents reflash	Small surface fires only	10, 15, and 20 lb sizes only
Class B fires Flammable liquids and gases (paint, oil, cooking fat, gasoline, and natural gas) where smothering action is required	No Water will spread fire	Yes Excellent Smothers fire, prevents reflash	Yes Carbon dioxide has no residual effects on food or equipment	Yes Excellent CO_2 is a non-conductor, leaves no residue
Class C fires Fires involving live electrical equipment (motors, switches, and appliances) where the nonconductivity of the extinguishing media is vital	No Water is a conductor of electricity	Yes Excellent Nonconducting, smothering film screens operator from heat	Yes Excellent Evaporates rapidly to blanket fire, chemically interferes with chain reactions	Yes Excellent Halon 1211 is a non-conductor, leaves no residue

Fig. 18-5 Extinguishants that are applicable for different fire classifications.

A drawback of the gas suppression systems is that a deep-seated fire may give the appearance of being extinguished yet, after the gas has dissipated, a smoldering fire could flare up again.

A significant fact should be made clear when deciding on a fire suppression system. An automatic water sprinkler system should not be substituted for a gas system and vice versa. Each is suited to its own application. Gases are intended to protect valuable electronic equipment, and sprinklers are intended to safeguard material content as well as minimize building damage.

The intent of codes is to guide and influence property owners to meet the minimum requirements for their facilities. For example, codes establish where "fire-rated" construction shall be utilized, with the rating for various hazards being identified by a resistance factor to fire in hours. The hourly fire resistance rate can be found on an "Underwriters Label" (usually referred to as U.L.) on various construction products and materials.

The "National Fire Codes" are a collection of eight volumes that deal with various aspects of fire regulation. The complete set contains the codes, standards, recommended practices, manuals, and guides developed by technical committees of the National Fire Protection Association and can be found in most technical libraries.

Storage of Combustibles and Explosives

It is important to properly identify combustibles and explosives and to understand their characteristics of flammability and/or explosion. The product's vapors may be as volatile as the product itself. Most combustible products are found as a gas, liquid, or powder.

Storage vessels should be designed for structural integrity and located in areas void of external influences of ignition. The vessels should also allow for expansion or contraction of the product due to temperature change. This can be accomplished by properly venting product vapors.

Combustible products can be stored below ground, above grade within fire-rated enclosures, or remote from other structures. Liquid storage vessels located above grade should have an earthen or masonry dike constructed around the perimeter that is capable of containing the liquid should there be a leak or spill.

It is important to understand the behavior and characteristics of explosives to design the storage container and determine the safest location. One source of information for determining the requirements of storing explosives is the National Fire Codes as compiled by the NFPA. Because a variety of products classed as explosive exist and because each product is different in its triggering behavior, reference to the NFPA series of volumes or the explosive manufacturer is recommended. These codes provide information for determining the explosive characteristics and appropriate measures that are necessary for storage in a safe atmosphere and should be adhered to without exception. Permits are also required when purchasing and/or storing explosives. A permit signifies that the persons handling explosives are knowledgeable about explosives and will exercise the necessary care and caution to ensure the safety of life and property.

A critical time in handling combustibles and explosives is when they are being transferred from one container to another. This situation warrants monitored activity in the immediate vicinity of transfer with a controlled environment designed to prevent a disaster. It is also important to maintain an up-to-date record and inventory of combustibles and explosives to be aware of any shortage or misplacement.

Evacuation Because of Hazard

Regardless of what causes a hazardous condition, once it has been declared all employees should exit the facilities in an orderly manner following designated routes and exits. Prior to the actual emergency situation, there should be practice drills on exiting and performing any other necessary tasks. Employees should be assigned to evacuate the building through designated halls and aisles leading to the nearest exit relative to their workplace. Plans of evacuation should be posted throughout the facilities at key visible points to assist visitors or new employees.

When a hazardous condition occurs, it may be necessary to secure portions of the facility to protect items of value. Security tasks should be preassigned to specific employees and backup personnel in the event the primary person is absent. Management may elect to assign personnel to operate fire extinguishers and fire hoses that are located throughout the facility. This fire brigade would handle small problems and then rely on the local fire department for assistance, or call on the fire department exclusively for anything major.

If the facility has a guard force to maintain security control, they can be assigned the role of monitoring the evacuation of personnel from the premises. The guard force can also be utilized to maintain security over product or material of significant value as well as guard or secure irreplaceable documents.

Flood Plain

If an existing facility or anticipated new structure is located within a flood plain, there are two options that can be considered as a solution to the problem. The simplest approach would be to use water pumps that are activated when the water reaches a preset level. During operation, the pumps would be capable of maintaining that level until the water subsides. A second method is to construct a dike or wall around the facility that would prevent water from entering the area.

Emergency Power

Emergency power is an alternate source of power, other than the utility power, and has the capability of offering power for a short period of time. Total power outages of any duration can create a disastrous condition if they interrupt the production process.

It may also be necessary to establish a clean and/or uninterrupted power supply for maintaining continuous activity despite power variation or outages. This can be achieved through multiple storage batteries capable of automatically sending power to the system for various durations. Batteries provide switch-over time for backup generators to supply power without interruption.

It is also necessary to store the fuel to power the generator. Fuel storage should be located so it does not create a hazardous condition (see previous discussion on the storage of flammable products).

Power outages could be a result of an event such as an explosion or fire. Regardless of the cause, it may become necessary to evacuate the facility. To minimize the hazards and panic that could result from exiting the facility in the dark, emergency light should be located along the avenues of escape with emergency-powered exit signs located enroute and at the exits.

Insurance

Insurance premiums have a positive influence on reducing or eliminating possibilities for disaster. Because the insuring company is not interested in paying out claims for incidents of violation or neglect, they make conditional demands on the insured to eliminate or minimize situations that potentially would result in an insurance payout. These demands can be in the form of voiding coverage if safety precautions are not maintained or increasing the premium cost if incidents of claim become excessive. No business manager wishes to lose insurance coverage or have the premium rate escalate beyond reason. Insurance rates, therefore, contribute to favorable conditions of operation, which reduce the hazards precipitating a disaster.

Published Procedures

Experience has proven that actions in an emergency are rarely effective unless they have been planned in advance. A plan that has been prepared carefully, documented in writing, properly implemented, and modified to meet current conditions can be a very important part of any plant's disaster prevention program. An ideally prepared plan that is ignored could be worse than no plan at all because it creates a false sense of comfort. In the event of an emergency, an unprepared staff would tend to compound the confusion rather than alleviate it. This would add to the disaster, having a further impact on life, property, or interruption of production.

Procedures to be followed prior to and in the course of an emergency need to be presented to all employees in written form with assignments noted. These procedures should also list backup support for absenteeism and other critical areas requiring alternate teams for providing relief due to physical exhaustion.

All documented procedures should be posted in strategic locations for the benefit of visitors and employees. On an unannounced basis, at some interval of time, an emergency drill should be exercised on behalf of the employees to maintain a degree of alertness and an understanding of what is expected of everyone.

From time to time, changes within facilities occur that warrant revising the emergency plan. A member of management should be assigned the responsibility of making periodic evaluations of the plan and implementing the necessary modifications to keep it current. This is another method for management to express its concern to prevent and/or control the probability of disaster.

Some of the suggested directives that could be published and posted are as follows:

- Method of announcing an emergency to the staff.
- Evacuation plan relative to type of emergency.
- Emergency escape routes and procedures.
- Procedure for plant operation shutdown.
- Procedure to preserve vital documents and material.
- Method of accounting for personnel that have vacated the facility.
- Rescue and first aid duties for appropriate persons.
- Role of the guard force.

SECURITY

All security programs should begin with a philosophy expressed in writing by management indicating what, where, and how security measures shall be instituted in the facility. It is recommended that a member of senior management be

SECURITY

appointed as sponsor or team leader of the security program. The person in this role acts as liaison in conveying management ideas to employees and conveying outside input.

Management should define the purpose and scope of security in general terms. This is accomplished by establishing what assets are to be secured, such as the building site, the building itself, its contents, or any combination of the three.

Next, management should select the degree of security they wish to exercise. This is accomplished through the selection of various types of equipment, devices, and procedures. The scale ranges from state-of-the-art devices, depicting a high degree of security, to simple locks, bars, and bells.

Finally, management should select security equipment devices that establish the objective degree of security and also satisfy the scope of security; these devices must also be cost effective. In other words, an evaluation is necessary to attain that delicate balance where security implementation equates economically to the security objective. More or costly security devices do not necessarily result in the best security system. More often than not, security overkill can be a hindrance to plant production or operation even though still meeting security criteria. To achieve the cost-effective plateau, it may be necessary to seek outside aid, such as from unbiased vendors or technical security consultants. As is the case in the medical profession, when in doubt, seek a second opinion.

PLANNING

The planning process determines where security measures are required. The greatest asset the so-called management "security sponsor" can have in planning a security program is to have a security consultant available for participation. The selection of a security consultant is extremely important from the standpoint that, like other professionals, confidence, capability, and experience are necessary requisites in the array of candidates available.

The title of security consultant is often self-appointed and is usually based on years of experience in the field of security. Many times, training in the area of criminal behavior or a long-time background in law enforcement and the military also provide strong qualifications in nontechnical areas. One of the more prominent organizations that promotes security information distribution is the American Society for Industrial Security. ASIS, as it is commonly referred to, has an international membership consisting of vendors, certified protection professionals, and security consultants. These individuals are brought together by the common interest in research, education, and promotion of good security practice.

Security Levels

The levels of security are established by determining the importance of various activities within definable bounds or limits that take place on the site or within the plant. The basic levels of security, going in the order from minimum to maximum requirement, are the building site boundary, the building exterior walls and roof, and various internal levels in the building that are determined by specific activity.

It is easy to understand that the area requiring the greatest degree of security should be located as central as possible within the building structure. This places the elements requiring maximum security into a position of greatest distance and density from the exterior walls. In some circles this is referred to as "layering" or "insulating," which is merely creating a design with many obstacles to deter penetration from the exterior. At this point, it becomes evident that there is a correlation of equality between a level of security and the related security devices that results in a cost-effective installation.

Limited Access Zones

A limited access zone is a defined area of activity, on the interior or exterior of the facility, to which entry is controlled by only allowing access to authorized persons. An access zone can be made up of single or multiple operational activities that are established as having a common level of security. In new construction, it is recommended that access zones incorporate as many activities of the same level of security as well as those areas of activity that interface with each other. The principle is that the fewer limited access zones there are, the fewer security devices are required for access control. From the standpoint of security and economics, ingress and egress points should be kept to a minimum. If code requires it, there can be more means of exit from a zone than there are means for entry. Exits would allow passage only in one direction and could be equipped with physical and electronic security hardware to monitor traffic.

Priority

When planning a security program in a facility, it sometimes becomes necessary to make concessions in either the area of production or security. When a conflict between the two occurs, it becomes necessary to prioritize the objectives so a harmonious plan of activity can evolve.

Security procedures or devices should not be instituted in a manner where production or costs are affected, unless all other options have been explored. Conversely, it may be necessary to make adjustments in production on a one-time basis to create an effective long-term security program that will satisfy all requisites. The key to success of any group effort is teamwork; the role of team leader is executed by the "security sponsor."

Building Integrity

Building integrity—its resistance to uninvited penetration—can be best incorporated into the facility prior to or in the process of construction. Existing structures can be modified or added to in the effort of introducing features of security, but costs may escalate to a premium because of the need to meet and accept fixed physical conditions.

The following list represents target areas on a building exterior that are vulnerable to intrusion:

- *Doors.* Doors, frames, and hardware should be of substantial makeup to resist forceful entry.
- *Windows.* Windows can be glazed with break-resistant material or be fitted with metal grille to prevent entry.
- *Openings.* Mechanical and/or electrical openings should incorporate design characteristics that discourage penetration.
- *Roof.* Roof hatches or skylights are also vulnerable to entry. Hatches should be substantial in their makeup and fitted with a medium to high-security lock. Skylights should have physical deterrents of entry or be eliminated.

One of the strongest assets for building integrity is to minimize the number of openings and glass areas on the exterior, thereby reducing the points of penetration. The type of material that is utilized on the exterior skin of the building should be substantial in its makeup. Materials such as sheet metal, gypsum board, or other like products serve as poor protective barriers on exterior surfaces where secured areas occur internally. When considering the makeup of the exterior of a building, it is prudent to be mindful of the fact that an intruder or thief is going to target entry at the path of least resistance.

IMPLEMENTATION

It is through implementation that the "how" of the "what, where, and how" scenario is addressed and accomplished. A good working security system is made up of two major elements. One element can be identified as the security devices or hardware that have been referred to as the "bells and whistles" portion of security. The other is recognized as the human input, which consists of properly trained personnel exhibiting diligent security awareness.

Security Devices

Security devices can be broken down into three categories that are defined by the function they perform: (1) detection, (2) surveillance, and (3) access control. It is not the intent in this text to specifically define the technology and makeup of the referenced security devices, but rather to offer a simple overview of the devices available and their general application.

A detection device is used on both exterior and interior areas of a facility; the objective is to know when and where an intruder has entered a protected area. Contact and magnetic switches are commonly used on doors and operate on the principle that a break in the electrical contacts or magnets will create a signal.

Another group of devices that monitor intrusions have the common characteristic of "sensing" an intruder. The ultrasonic sensor operates on the principle of sending out inaudible sound waves. When an intruder alters the wave pattern, an alarm is triggered. The active infrared sensor directs an invisible light beam at a receiver, and when an intruder interrupts the beam, the result is an alarm. A microwave sensor transmits a radio frequency field into an area that is to be protected. The motion of the intruder within the field activates the alarm. Vibration sensors employ sensitive elements that will sound an alarm when mechanically stressed, as in the act of a break-in. Other variations of the sensing principle are used in intrusion detection, but the objective remains the same for all, that is to establish the ability to apprehend a person who is found to be in an unauthorized area.

The primary purpose of closed-circuit television (CCTV) in the security field is to maintain remote surveillance of multiple locations from a central source. Subject conditions vary, such as interior and exterior, day and night, dimly lighted areas, and outdoor weather conditions. All of these conditions place special demands on the camera. A variety of lenses are required to monitor the image with good resolution (sharpness of image).

Some of the other options available are *pan*, which is the allowance for horizontal movement, and *tilt*, which allows vertical movement. *Zoom*, which performs much like a telephoto lens with a variable focal length, is a feature of the lens and controls image size. Some of the accessories available for outdoor camera installation are an all-weather housing, heaters, fans, and electric lens wipers.

Probably the most important and useful companion to the TV monitor is the time-lapse videotape recorder. The videotape is most useful as a historic record, offering the opportunity to review any strange activity that may contribute to determining the source of theft or sabotage. Videotape, providing that it also records date and time, is acceptable evidence in a court of law to press charges against a perpetrator whose act has been recorded on tape.

CCTV systems can be augmented with external motion-sensing devices or purchased with built-in motion sensitivity that provide the capability of focusing the attention of the observer and video tape recorder on activities that have been programmed as being abnormal.

Access control encompasses a variety of procedures and equipment capable of monitoring and recording traffic through access portals in either direction. The most common application for access control is to allow authorized personnel to enter a controlled access zone and keep unauthorized personnel out.

The simple principle of controlling traffic through a door opening is to control the locking mechanism. The operation of the lock is based on having the "bolt" or the "strike" powered electrically to an open or closed position. Various devices are used for verification of an individual and to authorize a lock to open and allow entry or exit.

The most basic access control is a card reader. A person is issued a card with a code that is unique to that person's identity, and the code is incorporated into the card. When the card is inserted into the reader and the code is verified as acceptable by the connected control processor, a signal is sent back to the lock to release the mechanism, thereby unlocking the door. Unfortunately, if this card is lost, the finder, an unauthorized individual, could gain entry to a restricted area.

A simple method of overcoming this problem is to add a key pad arranged with numbers much like that of a pushbutton telephone. Using the card reader and PIN (personal identification number) in tandem, it now becomes necessary for the entrant to present a card and number that are related. Because the personal identification number is issued secretly to the recipient, this format has now elevated the level of security.

The key to an enhanced security is for the access controller to verify that the person is who he or she claims to be. Cards can be lost, codes can be broken, and PIN numbers can be intercepted. As a result, a medium known as "biometrics" has emerged. Biometric technology records a biological characteristic of a person, such as fingerprint or handprint, voice, handwriting sample, or retinal pattern of the eye, into a computer data storage bank. An individual now seeking entry not only needs an identification card, but also must submit the unique physical trait required for comparison and verification. Thus far, biometric devices coupled with keypads and/or coded cards represent the highest level of security available in access control.

Despite the sophistication of access control devices and any combination they are installed in, there is another weakness that needs to be overcome and that is "tailgating" or "piggybacking." This is the practice of additional persons other than the authorized employee passing through a controlled access portal. This problem can be overcome by introducing a "mantrap" to the entry door. A mantrap configuration is the creation of a small room or chamber with a capacity of one person maximum, and interlock capability on the two pass-through doors, where only one door can be unlocked at a time. Currently, there are turnstiles and revolving doors manufactured and marketed that can also serve as mantraps. They have an advantage of taking less floor area than the field fabricated trap, but on the other hand their cost would be greater.

Another dimension of access control deals with vehicles. These range in makeup from a simple lift-arm gate to a motorized swing or horizontal sliding gate. Tire puncturing spikes can also be installed in the pavement. The spikes are normally in a raised position until entry is authorized, at which time they can be lowered to allow nondamaging entry. To resist the passage of large truck-type vehicles attempting to crash through an entry, there are large steel barriers installed in the pavement that will resist an impact force load of up to 5 tons at 50 mph (80 km) when in a raised, defensive position. This barrier is powered by a hydraulic power system for raising and lowering, with options of manual or automatic control.

Guard Force

A guard force is a vital tool in the overall makeup of a security system because it is the only element that possesses a "reasoning" capability that can make a judgment decision. A guard force is either proprietary or contract in its makeup. The former represents a force that has been selected by and is in the employ of company management. The contract force is an outside organization with an available guard force that is hired by contract to perform a specific service.

There is no issue or debate as to which service is better. Each has its advantages that are applicable for respective types of security, with company management making the decision of choice. Whether the guard force is proprietary or contract, areas of activity that must be established are as follows:

- Required training and qualification.
- Specific task in a normal and emergency mode.
- Interface with local police and fire departments.
- Interface with immediate community area.
- Whether the guard force will be armed.
- Whether guard dogs and vehicles will be part of the force.
- What decisions they will be permitted to make.

Security Control

Security control is the marriage of security forces and hardware that is usually housed in a single room and referred to as the security control center (SCC). This is the central monitoring location for all aspects of security hardware that have been referred to previously in this section. The type of hardware used determines how it fits into the SCC activity.

Intrusion detection. The sensing devices and door contacts in their remote locations report back to the SCC when activated. This is accomplished in one of two ways. When an activity is recorded, it can appear on a screen or a printer identifying its location. The activity can also be visually viewed from a map display panel. This is a panel with a layout to graphically simulate the site and/or the building plan with all sensing devices noted by a light. When a particular location is activated, the corresponding light is illuminated and the problem source is located.

Closed-circuit television. All CCTV cameras provide a picture of their respective viewing areas on a bank of television monitors in the SCC. Because there are usually more cameras than there are monitors, a switcher offers the capability of selecting a specific camera or group of cameras for viewing.

Access control. All activity of access control devices indicating successful entry, attempted entry, and the option of exiting can be monitored on a CRT screen and recorded on a companion printer. The printer offers an opportunity of belated search in determining the occupancy of a specific room or the travel an access portal, with related time and date noted. This record is referred to as a computerized "audit trail."

Vendor and Contractor

The security vendor and contractor play an important role in a new or existing security program. The security equipment vendor is committed to sell and promote the product of the manufacturer he or she is bound to represent. When the need and the product submitted are a match, there is no problem. On the other hand, when there is a mismatch, first the problem has not been solved, second the client is stuck with a security device that is not needed, and third the client has paid out money for a device that is not contributing to the security of the facility. It is therefore important to be certain that the analysis of the problem warrants the need of the product or that the recommendation of the product application comes from a sound and capable source such as a security consultant. Security equipment vendors' recommendations are to be respected because many of them are quite knowledgeable in their area of activity. Therefore, the burden of responsibility in that situation falls on the vendor to satisfy the requirement specified by the client.

The selected contractor installs the security equipment that has been recommended or specified. It is necessary to ensure that the contractor is capable of properly installing and maintaining this equipment. It is also a good practice to prequalify contractors even before they are allowed to bid on the project. This helps to avoid the situation where the low-bid contractor turns out to be incapable of performing the required installation. Contractors should possess the following attributes before they are considered qualified:

- Have the proper tools that are necessary for the type of work to be engaged in.
- Possess knowledge of the security equipment to be installed and how it is to operate.
- Be proficient in installing equipment appropriate to the client's needs.

Two points should be considered when preparing the bid documents for vendors and contractors. First, the vendor should be required to factory test the equipment to ensure that it is operational before delivery; a written certification is also important. Second, the vendor should guarantee when the equipment will be delivered and the contractor should guarantee when the installation will be completed, in accordance with the certification.

Maintenance

The security program and related equipment is only as good as the maintenance staff that services it. If the security equipment is not operational, then the integrity of the security effort has been compromised. Maintenance personnel should take advantage of security equipment modification or installation by a contractor as a learning experience by observing the work. Also, from time to time, maintenance personnel should update their knowledge through video tapes or classroom activity, which can take place on-site or at the manufacturer's facility.

In areas of sophisticated security equipment maintenance, it may be beneficial to exercise a service contract with the manufacturer or contractor if it is offered. This provides around-the-clock service and eliminates the need for full-time specialty service technicians. Conventional day-to-day maintenance would still be handled by in-house staff. Although production equipment is essential, security equipment should be given a slight edge in priority for maintenance service over production equipment for the simple consideration that it is no good to manufacture a product without security to protect it.

ENERGY MANAGEMENT

Energy, in its various forms, is a commodity used for building systems, processing, manufacturing, and transportation. Facilities planning includes requirements for energy distribution systems in a plant.

Energy is a commodity with a market price and is available from more than one source. Two common sources are electric power and natural gas, both of which are subject to government regulations. Oil and coal are also major energy sources.

Energy costs are manufacturing costs. Energy audits of thermodynamic cycles and efficiency of processing equipment often reveal opportunities for reducing energy use of processing equipment, plus the heating, ventilating, air conditioning, and industrial support systems. Facilities planning can reduce energy use and costs by providing facilities that use less energy and by reclaiming energy from manufacturing processes.

ENERGY CODES AND STANDARDS

The fuel crisis caused by OPEC and the natural gas shortages that accompanied cold weather in the Midwest in the 1970s prompted federal and state governments to adopt energy codes. Code compliance is mandatory and must be considered in facilities planning.

Every state has adopted energy codes for new buildings that set standards for building construction, lighting, heating, ventilating, air conditioning, and service hot water components and systems. These codes were adopted in part because the federal government withheld money grants to those states that did not comply. Although state energy codes vary, the basic document is the ANSI/ASHRAE/IES Standard 90A Energy Conservation in New Building Design. This standard sets prescribed standards for various components based on present technology. Work is continuing on this standard with the purpose of allowing alternate paths for compliance.

At present, processing and manufacturing operations are not covered by energy codes. However, reclaimed heat from exhaust stacks and waste streams can be cost effective and can reduce energy costs.

The United States Department of Energy (DOE) sponsors research on energy use and conservation. These studies include building energy performance standards (BEPS), resource utilization factors (RUF), resource impact factors (RIF), energy use computer programs (DOE-2.1B), and research on alternate energy sources including solar, wind, coal gasification, tar sands, geothermal, and other potential sources.

Building energy performance standards are based on the best available technology for the components used for building envelope, lighting, heating, ventilating, air conditioning, and service water heating systems. Resource utilization factors (RUF) attempt to account for the energy used for mining, cleaning, processing, and transporting plus energy used to manufacture the equipment used to provide these services. In effect, RUFs are an energy burden factor. Resource impact factors attempt to evaluate loss of depletable sources, waste disposal and pollution effects on the environment, and dependence on foreign or unproven reserves. Because of the difficulty of setting fixed numerical values for RUFs and RIFs they are not included in codes at this time.

The DOE-2.1.B computer program, with the commercial name BLAST, is used on government and private work to calculate building energy demand loads and energy use and to simulate operation of building systems. Computer programs for energy demand, energy use, and equipment and system simulation are also published by major equipment manufacturers, engineering consultants, and computer software suppliers. Most, but not all, of these programs use methods and algorithms developed by American Society of Heating, Refrigeration and Air Conditioning Engineers (ASHRAE) for calculation of energy demand and use.

ENERGY AUDITS

An energy audit is a review of energy use to determine sources of energy supply, where and how much energy is used, where energy use can be reduced by more efficient machinery and thermodynamic cycles, where energy may be reclaimed from stacks and waste streams, and where nondepletable sources (solar, wind, hydro) can be substituted for depletable sources.

Some jurisdictions require an energy analysis on residential and commercial buildings before issuing a building permit on new and renovated buildings. These codes designate energy use limits for various purposes. Model energy audit program guidelines are published by the DOE. An energy audit is recommended for facility planning to establish demand, energy use, and sources.

Many new buildings include energy monitoring of lighting, heating, ventilating, and air conditioning systems. Central utility plants, serving manufacturing plants, maintain records on energy distributed as steam, hot water, and compressed air. These records are an excellent historical record for an energy audit.

Manufacturing and processing companies keep records of utility bills and delivered fuel costs. Some keep an index of cost in dollars per million Btu's or cents per kilowatt hour to allow selection of the lowest market price. Some companies assign an energy use factor to products manufactured in the plant. This energy use factor may be used for an energy audit and to forecast energy use. Energy use factors may have been assigned on an estimated basis (not metered). A change in product mix and quantity or manufacturing processes can introduce errors. Energy use factors should be verified during the energy audit.

ENERGY USE FORECASTS

Energy use estimates and forecasts should be part of a facilities plan. However, there are limitations to the accuracy of all energy use forecasts because of the interaction of all the variables. For example, production cycles vary with market conditions. Daily, seasonal, and annual cycles of production and product mix affect energy use and internal loads of heating, ventilating, and air conditioning systems. Weather (temperature, humidity, solar effects, rain, snow) varies from day to day and year to year. The weather data used for energy use estimates and forecasts are based on a statistical average.

Heat storage of a building and its contents reduces transient loads, but is difficult to measure and calculate. There is interaction between systems; shutting off lights may increase heat supplied by the heating systems. Tempering the outside air required to replace room air for pollutant control can be a major part of the building heating load. An operating paint spray booth increases energy supplied by the heating plant.

Although energy use forecasts do not predict next month's or next year's utility bill, these estimates are essential to allow evaluation of alternates for facility planning. Because of the complexity of these estimates and tedious work required for manual calculations, computers are used for this work. Energy estimating methods are discussed in detail in Chapter 28 of the *1985 ASHRAE Handbook*.

More than one method exists to estimate energy use. The degree-day method, with many variations, is often used as a convenient, easy-to-use method. This method must be used with caution because it might not be accurate. Degree-day methods are not recommended for manufacturing plants, except as a first approximation.

Better estimates can be obtained using bin methods. Weather data is presented in number of hours at set temperatures in 5° F (3° C) increments for three shifts per day for each month, with yearly totals. Basic weather data for the bin method has been compiled in *Engineering Weather Data* compiled by the Departments of the Air Force, the Army, and the Navy. Compiled weather tapes for PC microcomputers are also available from ASHRAE.

Modifications are used to reduce calculations by combining bins. Computer programs based on procedures developed by ASHRAE committees provide quick, reasonably accurate energy use estimates.

Annual hour-by-hour computer programs are considered the best method available for energy use estimates. However, variations of up to 25% can occur between programs and by variations of input data selected by different engineers. Averaged weather tapes are available for these calculations from ASHRAE, the National Climatic Data Center, and private meteorology consultants. There is considerable amount of calculation time required for 8760 hours in a year, and a computer run on a complex facility takes time and can be expensive. Input should be carefully checked and systems carefully selected before running these programs. If heat pumps, heat reclaim, or heat storage are used in heating, ventilating, and air conditioning systems, it may be necessary to check short-term effects of extreme weather in other-than-average years.

Energy use estimates that use a series of design days or graphs of major loads by month are used in some cases. Under the right conditions, these methods will produce satisfactory estimates with minimum effort. Designers who use repetitive designs for buildings in similar climate areas have had good results with simplified methods.

None of the methods are guaranteed to accurately predict next month's or even next year's utility bills. However, an energy use forecast is essential for determining lifecycle costs and comparing heating, ventilating, and air conditioning.

Energy management monitoring and control systems are recommended for manufacturing facilities. With the microprocessor hardware and software now available, records may be kept and trends plotted to monitor energy use on all building systems and to allow better planning of new facilities. Data from energy audits of existing plants may be used to check and verify if energy use forecasts are reasonable.

ENERGY SOURCES AND USE

Manufacturing plants use energy to provide light, heat, and mechanical work for processing, industrial support, and building systems. The energy is delivered to the plant as high-voltage electric power, natural gas, fuel oil, coal, wood and other fuels, plant wastes converted to fuels, and by solar irradiation. Windmills, hydroturbines, and hot springs are sometimes used. Energy is distributed within the plant as low and medium-voltage electric power, natural gas, fuel oil from storage tanks, steam, hot water, heat transfer fluids, hydraulic power systems, and compressed air. There may still be some processing lines that use line shafts.

Energy is used to provide mechanical work using electric motors, steam turbines, hydraulic and pneumatic operators, and tools. Energy also provides heat for melting, boiling, chemical reactions, heat treating, cleaning and sterilizing, and heating outside air for ventilation. Light is provided by electric power, supplemented by daylight from windows and skylights.

Before selecting an energy source, the availability, equipment requirements, pollution abatement, and energy cost of the source should be considered. Availability is determined by selecting a source that can supply current and future energy requirements. Equipment modifications may be required for alternate energy sources and distribution systems. This can be a problem if existing equipment will be used in a new facility.

Regulations of the federal Environmental Protection Agency (EPA) and state environmental protection departments for pollution abatement affect the choice of the energy source. Initial and operating costs of pollution abatement devices can be a significant cost burden on the energy source.

Energy costs should allow for the following:

- Initial cost of equipment for receiving and storage, cleaning and preparation, distribution, ash removal, pollution abatement, and heat reclaim.
- Capital recovery and finance costs as well as taxes and insurance for electric power and heat generating systems and energy distribution systems.
- Operating and maintenance costs including operating labor and supplies, plus costs to provide preventive maintenance, overhaul, repair, and replacement of equipment.
- Electric power purchased from utility.
- Fuel prices paid to suppliers.

If all costs are not considered, the most economical energy source might not be selected.

ENERGY CONSERVATION

Facilities planning reduces energy use by providing facilities that use less energy and by coordinating energy conservation with manufacturing. Historically, the low cost of energy meant that there was a cost advantage to increase power input rather than provide more efficient machinery and systems. For example, machines, fans, and pumps were designed for lowest cost, not lowest power input. Heat exchangers were designed with high flow rates, low temperature rise, and forced convection. This type of design reduced the heating surface and construction cost, but required more energy to operate the fans and pump. Ductwork and piping systems were also designed to have high flow velocities and high pressure drops. Although this design reduced the size and cost of the system, the power input to fans and pumps increased. This thinking is still reflected in equipment design and selection. By carefully selecting the machines and systems in a plant, energy use can be reduced without excessive cost penalties.

The energy used in compressed air systems can be reduced by operating at lower pressures and repairing air leaks in the system. For example, a compressor operating at 80-100 psi

(552-690 kPa) requires less power than a 125 psi (862 kPa) system. Air receivers (accumulators) should be used near loads to maintain minimum pressure rather than maintaining high pressures at the central plant. Compressed air blow guns should also operate at 80 psi (552 kPa) for safety and energy conservation.

Another way to reduce energy use is to improve the thermodynamic cycles of equipment. Boilers should be fired to match loads and shut down when the heat demand is low. Insulation should also be kept in good repair and added where necessary to improve efficiency. It is also good practice to return uncontaminated condensate back to the boiler rather than to the drain. Leaking steam traps and line leaks should be repaired immediately.

Absorption refrigeration cycles use low-temperature heat sources, but generally are not efficient users of energy when compared to refrigerant compressors or outside-air heat exchangers. However, absorption chillers may be justified if low-temperature heat is reclaimed from a waste stream.

Energy can be saved in the heating, ventilating, and air conditioning systems by reducing the amount of makeup air used. This reduction can be achieved by removing pollutants with local exhaust systems rather than a general ventilation system. General ventilation requirements can also be reduced by enclosing hot and dirty operations in a separate area. Interlocks between the exhaust system and supply air reduce the amount of makeup air required when the exhaust systems are not running. Ventilation systems, which remove excess heat in summer months, should not be operated during cold weather. Exhaust air streams can also be cleaned and then recirculated in the plant.

Energy use may be reduced by considering energy conservation in building design. Items to consider are as follows:

- Provide more insulation and maintain envelope integrity on buildings and hot and cold piping systems. Envelope integrity increases in importance when insulation is increased.
- Limit windows and skylights. Provide designs that permit the efficient use of solar energy.
- Provide vestibules for doors and shipping and receiving docks to reduce outside air infiltration.
- Follow published lighting guidelines and use low lighting levels whenever possible to reduce power used. Consider daylight options.
- Use efficient heating, ventilating, and cooling systems. In manufacturing plants, supply air delivered to workspace provides more comfort than air delivered to truss space above lights.
- Discharge waste streams at low temperatures. Heat can be reclaimed from hot gas exhaust stacks, coolant, and condenser water systems. This heat can then be used for preheat of makeup air and fluids or for other process use.
- Use variable volume by step or proportioning control to reduce power delivered to distribution systems to meet reduced demand.
- Minimize wall and roof area.
- Orient building so that openings face away from prevailing winds.
- Consider cogeneration, discussed subsequently.

Current energy costs are higher than in the past; however, costs are not so high that every heat reclaim device or system is cost effective. Heat recovery systems must provide heat for a useful purpose, at the time needed, and at a reasonable cost.

Heat recovery methods that have been successfully used include the following:

- Cleanup and recirculate ventilation air and fluid wastes. High-efficiency collectors, filters, and other devices reduce emissions from the plant and save some energy.
- Blend outside air and recirculated air to heat makeup air in cold weather and reduce cooling in mild weather.
- Install a heat exchanger in the exhaust stream and the makeup stream so that it can be used with exhaust gas stacks and liquid wastes.
- Install heat exchangers on steam turbines and refrigeration machines to transfer heat from exhaust gases and liquid waste streams to air, water, or a heat transfer fluid. Some of the common types include revolving wheels, plate heat exchangers, studded or finned tube waste heat exchangers, and double-bundle condensers.
- Increase temperature of heat transfer fluids for processes that require higher temperatures than can be generated by exhaust streams with heat pump cycles.

COGENERATION

With the increasing cost of electric power, as well as federal and state government incentives, cogeneration may be considered for facilities with a high electric power rate, a low fuel cost, and a need for processing heat. Cogeneration is defined as simultaneous generation of electric power and useful heat or mechanical work. Most systems are interconnected with a local electric utility, but this is not necessary. Standby plants used to limit demand (peak shaving) with heat recovery are grouped with cogeneration plants. To decide the feasibility of cogeneration, careful evaluation is required.

The only justification for cogeneration is useful heat generation. Small electric generating plants cannot match the efficiency of large utility electric generating plants. However, large utility electric generating plants throw away considerable amounts of heat. A cogeneration plant is more efficient because this heat is used to serve a useful purpose.

Heat from a cogeneration plant can be used to provide low and medium-pressure steam for process heat and to run mechanical drive turbines and hot water for process heat and building heating. Low-pressure steam and hot water can be used to generate chilled water for air conditioning and process cooling with absorption cycle chilled water machines.

Power Demands

The engineer must realistically evaluate electric power and heat demands on a plant. Some of the questions that must be answered are as follows: What are the heat and electric power demands the plant must supply? What temperatures are required for the process? How do heat demands balance with electric load? Are times of demand in phase? Do process and electric loads vary? By season? By hour of the day? By production rates?

Maximum heating and air conditioning and process loads may occur only a few hours each year. The plant must run efficiently at a range of demands. Load balance and temperatures required for the process dictate the choice of equipment and the thermodynamic cycles used.

With process loads, all sorts of combined cycles are possible. The engineer must select a cycle that makes good thermodynamic and economic sense over the range of loads the plant demands. Heat storage systems are limited by economics to very short terms. Perhaps the best way to balance loads is through an export-import control. This type of control allows sale of power to the utility company during times of high heat demand and the purchase of power from utility company at periods of low heat demand.

Although there is considerable cost for synchronizing gear, the advantages of cross connection for balancing loads and ensuring reliability of operation make a cross connection almost mandatory.

Selling Cogenerated Power

To sell power, the plant should be qualified under the Federal Energy Regulatory Commission (FERC) regulations and state regulations to receive tax benefits and be protected by the Public Utility Regulatory Policies Act (PURPA). This act, which has withstood court tests, makes it mandatory for public utilities to accept power from qualified cogeneration plants at a nondiscriminatory price and relieves cogeneration plants from being regulated as public utilities.

Some state regulations encourage cogeneration. California's regulations allow a cogenerator to buy gas at a reduced price and sell power at a profit. However, it would be prudent to check the buy-back price with local utilities when making economic feasibility studies. The nondiscriminatory buy-back price is generally lower than selling prices shown on rate sheets to the user. There also may be a standby demand charge.

It is also important not to forget EPA and various state environmental regulatory authorities. For example, burners and engines require emission controls. Water injection or catalytic converters, the best available technology for nitrous oxide (N_2O) control on engines, increases fuel consumption and reduces exhaust gas temperature.

Plant Operation

Economic benefits depend on long-term operation at high efficiency rates. Maintaining a reliable power plant requires skilled operators and maintenance personnel. Automatic controls help, but do not replace a good well-trained operator.

Standby Power

If standby power is required, economic studies may be based on the cost of upgrading a standby plant to a continuous-duty plant if the client agrees.

Fuels Used and Associated Costs

Natural gas and light fuel oil are used in small plants. Coal, heavy oil, and waste products may be used in large industrial plants.

Fuel costs must include fuel preparation and storage, furnace maintenance, ash removal, and air pollution abatement. These costs can be substantial. Some "zero" cost wastes are high-cost fuels. If a waste fuel has a salvage value, include the salvage value in the cost of the fuel. Waste fuel may also have a negative cost if there is a charge for removal of waste. If electric power cost is high and fuel is available at bargain rates, cogeneration can be a good choice.

Equipment

Manufacturing plants can use a gas turbine or diesel engine generator to generate electricity with natural gas or number two fuel oil in a combined cycle plant. Gas turbine or diesel engine exhaust is used to generate superheated, high-pressure steam in a waste heat boiler. Steam is used for processing or for driving a steam turbine. The exhausted low-pressure steam or hot water from a condenser can be used for process heat or for generating chilled water in an absorption refrigeration cycle. Heat is also recovered by heating hot water with engine coolant on diesel engines. High efficiency can be maintained by balancing loads.

For large industrial plants, coal, heavy oil, and waste fuels are used to generate superheated high-pressure steam in boilers. Electricity is generated with steam turbine generators. Low and medium-pressure steam is used for process heat and for driving mechanical drive turbines. The output of these plants includes more heat per kilowatt of power produced than combined cycle plants.

The use of cogeneration will be limited as long as power rates remain low. A cogeneration plant requires a substantial investment, but this investment is low when compared to nuclear plants now under construction.

POLLUTION ABATEMENT AND ENVIRONMENTAL PROTECTION

Pollution abatement and concern for the environment should be considered in facilities planning to provide a safe workspace and reduce the impact on environment. The concern of all levels of government for health, safety, and general welfare has resulted in many laws and regulations that apply to the manufacturing plant and that must be complied with. When toxic or carcinogenic substances are involved, it is not enough to follow rules and regulations. Workers, the general public, and the environment must be protected from harm.

CODES AND STANDARDS

The federal Occupational Safety and Health Administration (OSHA) oversees safety in the workspace and indoor environment. This agency has adopted various industrial standards and enforces safety in the workspace by inspections and hearings. Although operations are the responsibility of manufacturing, facilities planning must consider OSHA standards in design and construction. Occupational health and safety is discussed in Chapter 12 of this volume.

The federal Environmental Protection Agency (EPA) is responsible for the cleanup, monitoring, and regulating of environmental pollution. Major pollution sources must have operating permits from the EPA. A facility that emits more than 100 tons of air pollution per year is considered a major pollution source.

State environmental protection departments interpret and enforce federal regulations on pollution emitters of less than federal limits and enforce state laws and regulations. State laws and regulations may be more restrictive than federal regulations. Operators must obtain construction permits and operating permits for new facilities and equipment.

State and local authorities enforce laws, regulations, and codes concerning environment and public health. Many of these authorities have overlapping jurisdiction. An abbreviated list of these authorities and their jurisdiction is given in Table 18-5.

All facilities that emit pollution must provide the governmental authorities with information about the type and amount of pollutants emitted. Criminal charges can be filed against company executives for filing false information or withholding pertinent information. Because of this, a code review should be a part of facilities planning.

Engineering and industry societies can provide design criteria and data on the best available technology. Some of these societies are the American Society of Sanitary Engineers, American Conference of Governmental Industrial Hygienists, American Industrial Hygiene Association, and Institute of Environmental Sciences.

PERMITS AND REPORTS

Construction permits are required from the EPA and other governmental authorities before installing or starting construction of pollution abatement systems. An application for a construction permit, along with engineering calculations and estimated quantities of emissions, effluents, and wastes, should be filed early in the planning cycle to allow for review and filing of additional information requested by authorities. It is also important to check proposed regulations during the planning cycle to be aware of any new regulations that may be implemented in the future.

Work should not be started on pollution abatement systems until construction permits are obtained. If new, unproven technology is used for pollution abatement systems or processes producing pollutants, a delay may be encountered for engineering evaluations and tests to ensure that the system can meet performance requirements.

If a construction permit is granted, it does not guarantee that an operating permit will be granted. Tests may be required on completed installations to prove performance. If regulations are changed during the construction period, pollution abatement equipment might have to be modified or replaced to meet new regulations.

Operators of the systems may be required to submit reports on incidents and accidents that result in the release of pollutants. Monitoring systems and safety devices should be included in process systems.

TABLE 18-5
Jurisdiction of the Various State and Local Authorities that Enforce Environment and Public Health Laws, Regulations, and Codes

Authority	Jurisdiction
Building departments	Building codes, life safety codes, plumbing codes, and electrical codes
Fire marshal	Fire and life safety codes and storage tanks for toxic and flammable materials
Protection agencies	Fish, game, wildlife, and endangered species protection
Sewer districts	Storm, sanitary sewers, and industrial waste sewers and sewage treatment
U.S. Army Corps of Engineers	Issue permits for discharging cooling water and plant wastes into navigable streams and harbors
Public health departments	Public health
Federal Occupational Safety and Health Administration	A safe workplace
State labor costs	Health and safety in the workplace and labor relations
Environmental Protection Agency and State environmental protection departments	Protection of outdoor environment

AIR POLLUTION

Air quality inside a plant is regulated by OSHA and state codes to protect the health of workers. In general, toxic emissions must be limited to safe, threshold limits. The limits for known carcinogens, inert particles, gases and vapors, and toxic fumes and dust are published in OSHA regulations and in data sheets provided by the suppliers of these materials. Any pollution emission exceeding these threshold values must be collected rather than emitted in the plant atmosphere.

Outside air is regulated by the EPA. The EPA has classified geographic areas as noncompliance areas, compliance areas, and unclassified areas using monitoring stations and computer models on air pollution. In noncompliance areas, pollution effects of new equipment must improve conditions. Existing pollution emissions must be eliminated or reduced in an amount equal to or greater than the capacity of a new facility. In compliance areas, pollution emissions are allowed within limits as published in the EPA regulations. These limits are often based on the best available technology.

In unclassified areas, an environmental impact study must be prepared in accordance with EPA guidelines before construction is permitted. Hearings may be held and concerned citizen and intervenor groups may challenge the proposed process and use of the property. A public relations group from the plant should participate in these hearings. If a construction permit is granted, it is normally at the same conditions as compliance areas. However, the permit may also require specific requirements for the particular site. Early resolution of site requirements is recommended for facilities planning.

Industrial Ventilation, a manual of recommended practice published by the American Conference of Governmental Industrial Hygienists, provides engineering data and specific applications for control of pollutants in industrial applications. Another publication, entitled *Air Pollution Engineering Manual*, provides engineering data and specific applications for a variety of processes. Although this publication is prepared by the Air Pollution Control District in Los Angeles County, which has conditions not matched in other areas, it is an excellent source.

WATER POLLUTION

To protect rivers, lakes and streams, and groundwater supplies from water-carried contamination, water-carried waste must be treated to reduce pollution. Various governmental and private companies operate plants for biological treatment of sanitary wastes. These plants must have waste disposal permits from the EPA and other governing authorities that place limits on types and quantities of waste treated.

Manufacturing and processing plants generate wastes that may or may not be acceptable to these plants. Separate treatment or pretreatment will be required if process wastes exceed sewage treatment plant limits for total flow, dissolved and suspended solids, biological oxygen demand (BOD), chemical oxygen demand (COD), or toxic and hazardous wastes. Treatment for removal of heavy metals, chromates, phosphates, ammonia, soluble organic materials, spent acid and alkaline wastes, oily and solvent wastes, cyanide, and other hazardous wastes is required for processes that use these materials.

All treatment and pretreatment plants, in-plant or off-site, require construction and operating permits issued by the EPA and may treat only waste covered by the permit. Industrial plants can be held liable for damage if hazardous wastes are detected in storm or sanitary drainage systems. Plants that have pretreatment systems are also required to have monitoring systems to ensure that untreated hazardous wastes are not discharged to sewers.

Groundwater contamination from leaking underground storage tanks is a widespread and complex problem. The EPA requires underground tanks to be registered to determine the extent of this problem. Facilities planning that includes storage of gasoline, fuel oil, and other hazardous wastes in underground tanks should include protection to prevent spills and leaks during the life of the tank as well as equipment to detect leaks that may occur. Removal and replacement of existing tanks should be included in facilities planning. If tanks have been leaking, contaminated soil must be removed and replaced with clean backfill.

Water runoff from coal piles, salt storage piles, parking lots, and vehicle wash facilities can pollute storm drainage systems. Interceptors should be provided to remove wastes.

When water from rivers, lakes, or streams is used for process cooling in a plant, the temperature of the water returning to the source is subject to thermal pollution standards. Water must circulate through evaporative coolers or cooling towers to remove excess heat before returning to the original source. Dry, air-cooled heat exchangers may also be used.

HAZARDOUS WASTES

Hazardous wastes or product spills, including solids, liquids, and toxic and noxious gases must be disposed of in a safe manner. These wastes may be treated in-plant or shipped off-site for treatment and disposal; some wastes can be treated and recycled.

Facilities producing hazardous wastes can be held liable for improper transportation and treatment. It is, therefore, important to select contractors using approved methods for the transportation, treatment, and disposal of wastes. Contractors should have current EPA operating permits.

If hazardous wastes are transported off-site, the Department of Transportation requires invoices and manifests showing types of wastes and their origin. Small-quantity generators (between 10 and 1000 kg/month) are no longer exempt. Since August 1985, small-quantity generators have been limited in the amount of hazardous wastes that they can store on-site, and they must complete a manifest for all materials shipped off-site. Plants that have been storing these materials on-site will have to make arrangements for removal and treatment of these wastes. Landfill operators will probably refuse to accept tanks, drums, and other containers containing unknown materials because of the penalties for accepting hazardous wastes.

Nuclear power plant and radioactive wastes are regulated by the Nuclear Regulatory Commission (NRC). When radioactive materials are used, access to those areas should be restricted. The materials and workers should also be shielded and the amount of exposure monitored. Disposal of radioactive waste requires special knowledge provided by consultants in this field.

EXISTING CONDITIONS

When planning to build a new facility or remodel an existing one, it is necessary to consider what effect pollution cleanup will have on the project. New construction permits will be denied until any existing pollution violations are corrected.

When purchasing property on which a new plant will be built, a pollution violation may exist because radioactive wastes or gas-producing wastes have been used for landfill. Radioactive wastes must be removed, disposed of in a safe location, and then new landfill added.

When purchasing an existing building, it is important to determine if any materials have been stored on-site, if underground tanks had been used, and if asbestos was used in the building construction. Materials stored on-site, such as maintenance supplies, must be identified and reclaimed or disposed of in a safe and acceptable manner. Underground tanks that are no longer in use must be removed. If it has been determined that the tanks were leaking, the soil must be removed, decontaminated, and replaced. Asbestos in existing buildings must be removed or encapsulated.

NOISE ABATEMENT

In the machine shop, noise is a complex combination of sounds from many sources. Typically, there is sound from tooling, machining processes, material handling, ventilating fans, air compressors, and hydraulic pumps. Although it is difficult to separate one source from another, major noise sources can be identified, characterized, and ranked in order through systematic analysis with sophisticated electronic equipment.

In 1971, OSHA promulgated regulations covering employee daily noise exposures. These regulations state that protection against the effects of noise must be provided when exposure to the sound levels exceeds the allowable duration. When employees are subjected to sound levels exceeding those listed in Table 18-6, feasible engineering or administrative controls must be utilized. According to OSHA, engineering controls pertain to

TABLE 18-6
OSHA Noise Exposure Limits

Hours of Exposure	Sound Level, dB (A)
8	90
6	92
4	95
3	97
2	100
1 1/2	102
1	105
1/2	110
1/4 or less	115

BIBLIOGRAPHY

the facilities, design, and operation of suitably quiet equipment and systems and to changes in the work environment. For example, mufflers and noise abatement barriers could be installed on process equipment to reduce noise levels. Administrative controls pertain to the time limitation of worker exposure to or "dose" of OSHA-specified noise levels. If such controls fail to reduce sound levels within the levels specified, personal protective equipment must be provided to reduce sound levels in conformance with the stipulated requirements.

The EPA and local authorities also set limits on noise levels at the property line. These limits are usually set to protect nearby residential areas, hospitals, or similar facilities. To reduce the noise level, setbacks, wooded strips, or other sound absorbers are incorporated into the facility plan.

Bibliography

Hales, H. Lee. *Computerized Facilities Planning: Selected Readings*. Norcross, GA: Institute of Industrial Engineers, 1985.

Lewis, B.T., and Marron, J.P. *Facilities & Plant Engineering Handbook*, New York: McGraw-Hill, Inc., 1973.

Muther, Richard. *Systematic Layout Planning*, 2nd ed. Van Nostrand Reinhold Co., 1973.

National Fire Codes, vol. 1. NFPA 13. Quincy, MA: National Fire Protection Association, 1983.

Tompkins, James A., and Moore, James M. *Computer Aided Layout: A User's Guide*. Norcross, GA: Institute of Industrial Engineers, 1977.

Tompkins, James A., and White, John A. *Facilities Planning*. New York: John Wiley & Sons, 1984.

The following materials are sources of additional information on topics discussed in this chapter.

Energy Management

ASHRAE Handbook—Applications, Ch. 58—"Computer Applications." Atlanta: American Society of Heating, Refrigeration and Air Conditioning Engineers, 1982.

ASHRAE Handbook—Fundamentals, Ch. 28—"Energy Estimating Methods." Atlanta: American Society of Heating, Refrigeration and Air Conditioning Engineers, 1985.

"Energy Conservation in New Building Design," ANSI/ASHRAE/IES Standard 90A, ASHRAE/IES Standard 90B, and ASHRAE Standard 90C.

"Energy Efficient Design of New Buildings Except Low-Rise Residential Buildings," Proposed ASHRAE Standard 90.1P—Second Public Review Draft.

Engineering Weather Data. Manual NAVFAC-P-89, U.S. Department of the Navy; Publication AFM-8, Ch. 6, U.S. Department of the Air Force; and Technical Manual TM-5-785, U.S. Department of the Army. Available from Superintendent of Documents, U.S. Government Printing Office, Washington, DC.

DOE-2 BDL Summary—Users Guide. Lawrence Berkeley Laboratory, prepared for the U.S. Department of Energy, Washington, DC.

DOE-2 Reference Manual. Los Alamos National Laboratory, prepared for the U.S. Department of Energy, Washington, DC.

Model Energy Audit Program Guidelines. DOE-DE-FG4-80R510197. U.S. Department of Energy, Washington, DC.

U.S. Department of Commerce, Washington, DC.

U.S. Department of Energy, Washington, DC.

User Manual—TRACE. The Trans Co.

User Manuals—HCC III. APEC—Automated Procedures for Engineering Consultants.

Pollution Abatement and Environmental Protection

Air Pollution Engineering Manual. Air Pollution Control District, County of Los Angeles. Published by the Environmental Protection Agency, Washington, DC.

Environmental Protection Agency publications. Offices of Air and Water Programs, Air Quality Planning and Standards, and Noise Abatement and Control, Washington, DC.

Industrial Ventilation. American Conference of Governmental Industrial Hygienists, Committee on Industrial Ventilation.

Noise Pollution Control Regulations. Illinois Environmental Protection Agency.

Occupational safety and health standards. Occupational Safety and Health Administration, U.S. Department of Labor, Washington, DC. Published in the Federal Register.

Professional Trade Publications

ASHRAE Journal. Atlanta: American Society of Heating, Refrigeration and Air Conditioning Engineers.

Business Facilities. Belmar, NJ: Bus Fac Publishing Co., Inc.

Building Operating Management. Milwaukee: Trade Press Publishing Co.

Industrial Engineering. Norcross, GA: Institute of Industrial Engineers.

Industrial Management. Norcross, GA: Institute of Industrial Engineers.

Plant Engineering. Barrington, IL: Technical Publishing.

Plants, Sites, and Parks. Coral Springs, FL: Plants, Sites, and Parks Inc.

EQUIPMENT PLANNING

"The only sure way to build a long term competitive edge in manufacturing is by developing a strong infrastructure that reinforces a strong structural base, consisting of facilities, technology and suppliers."[1]

This chapter addresses the embodiment of technology—equipment—and is composed of six sections. As part of the material in Volume V of this Handbook, it deals with the management aspects of manufacturing; as part of the material in Section V of this volume, it focuses on resource utilization. This chapter provides insight and guidelines for the planning, acquisition, and use of manufacturing and support equipment; to address the need for and approaches to integration; and to present the changing role of maintenance in support of the factory.

The six sections address the key concerns that a manufacturing manager has to face as new equipment is considered and brought into the operation. The topic of each of the sections and its purpose are as follows:

1. *Manufacturing in the company strategy.* Orient the reader to the differences in manufacturing types, to the role that manufacturing plays in support of the company strategy, and to the role that equipment plays in manufacturing strategy.
2. *Equipment selection and sequence.* Look at the considerations affecting equipment selection and the process behind making equipment selections.
3. *Material handling.* Investigate the means of physically integrating a manufacturing process and the process used to make good material handling equipment decisions.
4. *Systems integration.* Investigate the meaning of integration and the issues that have to be faced while integrating a manufacturing operation. Guidelines are offered to make this process as smooth as possible.
5. *Equipment installation.* Address the relationship between the facility and the equipment in the facility. Key facility/equipment interface concerns are addressed.
6. *Maintenance.* Explain the new role of maintenance and the basics of preventive and predictive maintenance.

It is important to be concerned about these topics because a good understanding of these topics helps the manager to understand and improve the operation performance measurements that determine profit for the company. These measurements, which will eventually be used as criteria for new equipment selections, generally include the following:

- Total product.
- Quality.
- Product flexibility.
- Volume flexibility.

The preceding criteria are not listed in the order of importance.

CHAPTER CONTENTS:

MANUFACTURING IN THE COMPANY STRATEGY

Orientation is the necessary first step in deciding how to get someplace. In other words, one has to know where they're at and going before deciding how to get there. Although it sounds trivial, it isn't. One only has to consider the many "horror stories" about the misapplication of equipment and technology to realize how important this step is. In the majority of the cases, the problem was not in the functionality of the equipment, but in the misapplication of that functionality. Why does that happen? It often happens because some managers don't know enough about the relationships among their own operations, the needs of their industry, and the technology that they employ.

Part of the needed orientation is found in the answer to the question "How would you describe the operation?" In fact, there are a couple of ways that make sense.

Hall presents two ways of looking at manufacturing,[2] from the point of view of the product and from the point of view of the variety in the process. Manufacturers may be classified as "continuous or discrete" manufacturers. Discrete manufacturing deals with the fabrication and assembly processes

***Contributors of this chapter are:* Robert Lasecki**, *Vice President, Austin Consulting;* **William R. Welter**, *Vice President, Austin Consulting.*

***Reviewers of this chapter are:* Kerry A. Baker**, *Manufacturing Engineering Supervisor, Plough, Inc.;* **Donald M. Chipman**, *Manager of Design Engineering, Plough, Inc.;* **Arthur P. Kalemkarian**, *Group Industrial Engineer, Boeing Helicopter Co.;* **Sunder Kekre**, *Assistant Professor, Graduate School of Industrial Administration, Carnegie-Mellon University;* **Edward M. Stiles**, *President, Ed Stiles & Partners;* **Larry Strom**, *Chair, Applied Science Div., Yavapai College;* **Hayward Thomas**, *Consultant, Retired—President, Jensen Industries;* **Robert A. Walk**, *Managing Consultant, Cresap, McCormick and Paget.*

and, consequently, with the production of pieces. Continuous manufacturing, on the other hand, deals with a continuous process, as with an oil refinery.

Hall has further supplied rather succinct definitions of the concepts of repetitive manufacturing, job shops, and projects. Repetitive manufacturing is the production of discrete units planned and executed by a schedule. Material moves in a flow. (If the product is a gas, liquid, or powder, this is usually considered continuous production.) Job shop manufacturing is the production of discrete units planned and executed in irregular-sized lots. Material moves in integral lots through production. Project manufacturing is producing units of such size and complexity that each one generally requires planning and control by its own project organization.[3]

Hayes and Wheelwright differentiate producers by describing the pattern of the workflow, combining both the type of product and the variety of the operation.[4] They classify these workflow patterns as the following:

- Project.
- Job shop.
- Batch (decoupled) line flow.
- Assembly line.
- Continuous flow.

Project workflow is associated with customized products, and according to Hayes and Wheelwright, the key to success is the management of the jobs. Job shops have to contend with small batches and considerable variety with respect to the routing of the product(s). This environment requires good control over work in process. A batch facility is more of a standardized job shop and includes such products as specialty chemicals, heavy equipment, electronic devices, and metal castings. Assembly lines contend with higher volumes but less variety. The workstations are laid out in the needed sequence, and the whole operation is geared toward efficiency for the production of a few different items in large volumes. In general, this is a very inflexible arrangement. Finally, continuous flow deals with the processing of nondiscrete products, such as with oil refineries and food processing plants.

Table 19-1 compares typical performance measures of different types or categories of manufacturing. Many of these relationships are as old as manufacturing itself and have long been accepted as "true." Many are changing or at least in need of change. For example, the traditional intersection of the "assembly" column and "flexibility" results in the traditional problem of being "equipment-bound." The present-day need is for manufacturing managers to understand the capabilities of programmable (computer-based) equipment, which will certainly expand the bounds. The focus, however, is more on management of technology than on the technology itself. Consider the following.

A 1984 study of flexible manufacturing systems compared 35 systems in the United States and 60 in Japan.[5] The average types of parts produced per system (product flexibility) ranged

TABLE 19-1
Comparison of Different Performance Measures for Different Types of Manufacturing

Typical Performance Measures	Manufacturing Types				
	Project	Job Shop	Batch Line	Assembly Line	Continuous Flow
Cost per unit	High labor content Variable capital	High labor content Low-mid capital	Variable labor Variable capital	Low labor Mid-high capital	Very low labor, High capital
Quality	Requires good project management and reporting	Based on craftsmanship	Requires timely information and ability to stop process and correct		Requires process control-feedback and feed forward
Efficiency	Based on ability to grow/shrink workforce	Based on steady flow of work	Based on ability to maintain control of quality and bottlenecks		Based on equipment and quality of raw material
Delivery reliability	Dependent on good project management and workforce	Dependent on ability to manage backlog and estimate content of new jobs	Dependent on good priority planning and control	Dependent on capacity planning and control and good maintenance practices	
Product and volume flexibility	Very flexible	Flexible within bounds of general equipment	Flexible within product family	Inflexible	Inflexible
	Need large labor pool	Labor intensive—flexible within ability to add overtime and second shift		Equipment bound—not very flexible (changing?)	
Lead time	Variable—depends on workforce, project management	Variable—depends on backlog	Variable but predictable within scheduling capabilities	Predictable within capacity limits	

(*Austin Consulting*)

from 10 in the U.S. to 93 in Japan. The U.S. equipment may be more management-bound than technology-bound.

THE BUSINESS AND MANUFACTURING STRATEGY

Understanding the actual relationship between the strategy that a company uses to compete and the role that manufacturing plays in that strategy requires an assessment of the proactive versus reactive nature of manufacturing-related decisions. At one extreme is the position that manufacturing has to respond to the needs of marketing and finance, and therefore, it is placed in the position of "don't mess up." The other extreme is the position of trying to market the strengths of the company's manufacturing capabilities—to use these capabilities as a competitive weapon.

In the course of ongoing competition a company states, possibly in words but ultimately by its actions, how it intends to compete in the marketplace. This statement (words or actions) is the business strategy that it pursues. However, a set of functional strategies is needed to support the goals of the higher level business strategy. For example, the business strategy of price competition requires a manufacturing and distribution strategy of low-cost production and distribution. Another example would be the business strategy of product variety that is supported by the manufacturing strategy of flexibility.

At the operational level, a manufacturing strategy can often be defined by the combination of decisions that a company makes with respect to capability and capacity. This strategy is most often seen through the decisions made regarding equipment, facilities, vendor relationships, systems, and the workforce. Capability refers to what the operation can accomplish and how well it performs; capacity refers to the amount of acceptable product per unit time. (Capacity is a rate and therefore time dependent.) All of these decision categories interrelate and, therefore, determine the competitive strength of the manufacturing operation. Refer to Fig. 19-1 for examples of decisions within these categories.

Consider the addition of a new machine tool to a small fabrication and assembly operation. By adding this machine, management has made decisions regarding equipment capability and capacity. It has decided to make, rather than buy, certain parts; determined the de facto relationship to the existing scheduling and reporting systems; and added a tool that may, or may not, be in the working capabilities of the present workforce. Designing manufacturing operations as an integrated system presents the benefits of smoother operations and responsibility to realize that there are no more "stand-alone" decisions.

GROUP TECHNOLOGY AS A MANUFACTURING STRATEGY

Group technology (GT) is simply the philosophy of looking at the similarities of things—parts or processes needed to make the parts. This approach is also discussed in Chapter 13, "Design for Manufacture," and Chapter 16, "Computer-Integrated Manufacturing," of this volume.

The use of GT in manufacturing is based on the concept of "families of parts"; these families can be oriented toward the product or the manufacturing capabilities needed to make the product. In the case of product orientation, the similarities include shape, material, or function; manufacturing similarities relate to the processes or equipment used to produce a part family. In this case, the parts produced are not necessarily similar in appearance.

The point to be made here is that group technology (and the associated concepts of "cellular manufacturing" and "flexible manufacturing systems") is, in itself, a strategy that can provide manufacturing benefits. In fact, it is a grouping of many of the decisions that determine the manufacturing strategy of a company.

For example, moving from a functional layout (all mills in one area, all grinders in another) to a cell or FMS layout (group the machines needed to make a part family) means that manufacturing management is embarking on a change in manufacturing strategy. It will have to address facility and material handling equipment changes, the probable need for fewer labor classifications, and production and inventory system changes (possible bill of material changes). Again, this is not a stand-alone decision.

The grouping of people and equipment into cells forces equipment, facility, and workforce decisions and simultaneously sets a level of capability (the cell is designed for a particular part family or part families) and capacity (the number of equipment items is set and balanced). In one respect, it allows great flexibility to run any part within a specified family; however, the cell is designed for a particular family and may result in lower total efficiency for the "oddball" parts. Individual machines may be used to make the oddball parts, but they may tie up the cell while doing so.

There are three methods of grouping parts into families and, consequently, manufacturing capabilities into cells. These methods are shop experience, flow analysis, and classification and coding.

Shop experience is the simplest way to establish manufacturing cells—knowledgeable people review the parts and routings and group the parts by common routings. Flow analysis is a bit more formal (and tedious). To perform this analysis, a matrix of parts and required machines is produced. The matrix is then sorted by machine and machine groups to determine the

	Capability	Capacity
Equipment	A	B
Facilities	C	D
Vendor relationships	E	F
Systems	G	H
Workforce	I	J

A. HP, spindle speeds, stock size, programmable
B. Quantity, hours of uptime
C. Clear heights, floor loads, column spacing
D. Area, capability of expansion
E. Degree of vertical integration
F. Vendor capacity, number of vendors, materials management strengths
G. Functions performed by the system
H. Network transactions capacity, input/output capacity
I. Cross training
J. Size of workforce, overtime policy

Fig. 19-1 Interrelationship of typical manufacturing strategies with capability and capacity. (*Austin Consulting*)

part families. Classification and coding is much more formal and is usually used for large-scale applications. This method, using developed software packages, requires that parts be coded to reflect their relationship to a series of interested classifications. These classifications could relate to such things as tolerances, threads, heat treatment requirements, shape, size, and material.

Flexible manufacturing systems go even further toward deciding on a group of decisions; equipment, facilities, vendor relationships, systems, and workforce decisions are all made in the process of defining an FMS. Automated production and material handling equipment are integrated with a scheduling system. NC programs are written and, ideally, made part of a distributed numerical control (DNC) system. Tradeoffs are made between skilled machine operators and semiskilled machine tenders. Facilities are altered to provide the infrastructure for automatic guided vehicles or automated monorail systems.

ISSUES TO BE ADDRESSED

The technical aspects of equipment decisions are generally rather well defined and easy to understand. The managerial issues, however, are where one has to be most careful and, naturally, are the hardest to make. Nothing is clear-cut when it comes to tradeoffs and organizational issues.

Some of the relationships between equipment decisions and the rest of the decision categories (facilities, vendor relations, systems, and the workforce) are shown in Fig. 19-2. Each of these relationships influence capability or capacity, and all have to be addressed.

Figure 19-2 is a simple matrix that can be used to stimulate thought about the relationships that arise in an integrated manufacturing enterprise. Block number 10, for example, represents the intersection of systems and workforce lines of thought and may represent the planning that many managers overlooked in the early 1970s. It was then that American manufacturing management became aware of the MRP panacea (material requirements planning or manufacturing requirements planning or manufacturing resource planning). For the most part, the focus was on the software and associated computers. The impact that the *use* of this technique would have on the supervisors who had grown up with paper-based schedules and "hot lists" was totally ignored. This simple matrix can be used to avoid repeating history—for the most part, companies did *not* experience the promised inventory reduction and just-in-time manufacturing of needed subassemblies.

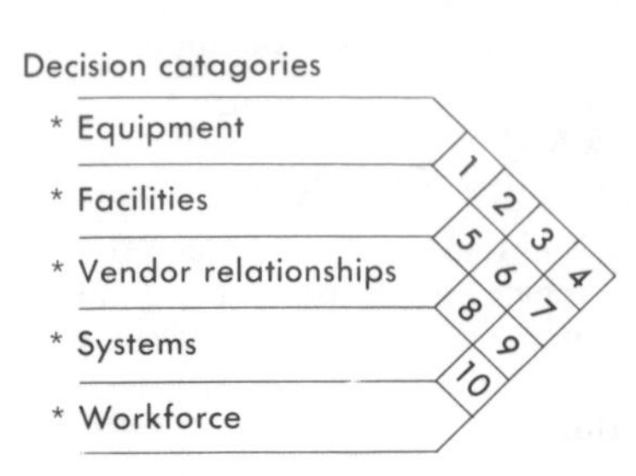

Relationships

1. Layout, safety, utility requirements, pollution requirements, installation
2. Capabilities, capacity, cost, inventory
3. Level of automation and integration, networks, databases, monitoring, statistical process control
4. Training, level of expertise, safety, flexibility
5. Scope, capabilities, location, cost
6. Networks, maintenance
7. Working environment, quality of work life
8. Vendor network, order processing, scheduling, receiving, cost
9. Size, capability, outsourcing, cost, union relationships
10. Computer literacy and acceptance, understanding of procedural impact

Fig. 19-2 Interrelationship matrix for manufacturing strategy issues. (*Austin Consulting*)

EQUIPMENT SELECTION AND SEQUENCE

How does one know what equipment is needed for a particular operation? What are the variables with which one has to contend? Why and how should one use consultants and vendors? What is the selection process for equipping or modernizing a manufacturing facility, and finally, how can one plan for the evolution of the equipment in one's facility? This section will provide some insight into these questions and, through the use of an example, will address some of the related issues, both tangible and intangible.

AN EXAMPLE

Let's assume that one has been assigned the responsibility to plan for the equipment purchases of a new precision gear manufacturing facility, a "greenfield" facility. Where does one start?

First of all, it is necessary to put this problem in the proper perspective. Some very fundamental manufacturing strategy decisions involving capacity will be dealt with, and therefore, it is necessary to understand the basic issues with which the planner will have to cope. Looking at past decisions, both good and bad, the following three fundamental issues can be proposed:

1. A facility's capacity is mix dependent and will change as the product mix moves through its elemental lifecycles. Therefore, the present and forecasted workload has to be understood from a lifecycle point of view.

2. A facility's total capacity depends on the interactions of all the strategic elements because capacity has to be based on "acceptable output" out the shipping door. Let's assume that the systems are in place and will be supported for a high-quality, rapid response (near just-in-time) operation.
3. Capacity can be "stored" as finished goods and work in process. Therefore, the capacity decision is intertwined with inventory decisions. Because this operation is to attempt JIT, work in process is to be minimal.

Mix dependency means that the planner needs to understand the materials to be machined and the gear features and dimensions that the facility will have to produce. This will determine the basic type of needed equipment. Materials can range from standard steels to exotic alloys. Features can range from simple profiles to unique tooth forms. Some of the probable features for this example, and the associated production equipment, are shown in Table 19-2. A word of caution, however, is in order. Process planning is not a trivial task and, for the most part, requires extensive experience because there will often be more than one way to produce a desired part feature, resulting in a varying mix of equipment that can be used.

For the sake of this example, assume that the equipment for this greenfield facility must be capable of producing about 100 different gear-like parts. The process engineering task is substantial, and there is not enough time to process all 100 parts in detail before equipment selection begins. However, using the fundamentals of group technology, it is possible to group these parts into several distinct families. The planner is then able to process the most complex part and unique features from other parts in the family. Understanding this constraint enables the planner to develop the preliminary list of needed equipment in a much shorter period of time.

If a person had to contend with a very complicated gear, he or she might have need for all of these equipment items. The layout would then be determined by the process routing for this particular gear. However, what if one had to do this for 100 different gears incorporating some or all of the potential gear features? Furthermore, consideration would have to be given to the mix and volume many years into the future; it would be very short-sighted to assume that the product mix and volume would remain constant.

The problem has now become compounded. There is a need to consider some of the fundamental decisions that, as mentioned in the previous section of this chapter, will determine the manufacturing strategy. This, then, is the second concern: how will the equipment selections affect, and be affected by, the other areas of manufacturing strategy? What are the issues, and how do they lead to the development of equipment selection criteria?

The issue that the manufacturing community is just beginning to face is that, in spite of all the talk about "flexible" manufacturing, problems exist in going to a *lesser* degree of flexibility. Since the beginnings of the Industrial Revolution, operations have been integrated by the ultimate flexible resource, the human operator. Manufacturers have, however, tacitly agreed to the tradeoffs of cost and variable quality. Therefore, before looking at the criteria associated with manufacturing strategy, it is necessary to briefly review the issues.

TABLE 19-2
Potential Equipment for Producing Certain Gear Features

Gear Features	Potential Equipment
Spiral bevel teeth	Spiral bevel generator Spiral bevel grinder
Parallel axis teeth	Hobber Shaper Broach Grinder
Splines	Hobber Shaper Grinder
Profiles	Turning center Vertical machining center Horizontal machining center
Ground surfaces	ID grinder OD grinder Hone Gear teeth grinder
Threads	OD grinder IO grinder Turning center
Case hardening	Carburizing furnace Carbonitriding equipment Induction hardening equipment

(*Austin Consulting*)

CRITERIA DEVELOPMENT

How does one know if they are making a good or bad equipment decision? Simply, what criteria is used to evaluate the decision process?

In general, all criteria can be expressed in terms of the following four general criteria (not necessarily listed in order of importance):

- Total product.
- Quality.
- Product flexibility.
- Process (volume) flexibility.

Safety is actually a fifth and absolute criteria.

Cost has to be thought of in terms of total cost, which means that it is not enough to simply consider the cost of production. The cost of rework and warranty work have to be included, as well as the cost of all overhead, including inventory carrying cost. Therefore, the criteria of cost has to consider the total impact of new equipment. Figure 19-2 shows that many of the relationship-based issues include cost. Note further the cost implications of the remainder of the issues.

Quality used to be the usual tradeoff to cost; you could either get a good part or you could get it for less cost. This tradeoff no longer exists. Quality is no longer a "nice-to-have" feature; a manufacturer either has quality or loses business.

A question that may be asked is, What is quality and how is it affected by equipment-related decisions? One simple definition is to make it the way it was designed. In other words, if the part, subassembly, or assembly was designed with certain features and dimensions, then it is a quality operation that can regularly produce parts with those features and dimensions. This means a few things as far as equipment is concerned. First, numerically controlled equipment is usually capable of consistent performance, and this performance can be traced back to the actual part design through the use of integrated computer-aided design and computer-aided manufacturing equipment. Second, the equipment will need the capability to correct its own operating performance through the use of sensors and adaptive controls tied into a statistical process control system.

Product flexibility is becoming more of a way of life for many manufacturers. Product lifecycles are shortening, and innovation is the norm for many companies. Consequently, the criteria of product flexibility is moving more and more into the forefront of equipment decisions.

Lifecycle refers to the sequence of product introduction, acceptance, maturity, and decline. For example, consider the need for manufacturing capabilities for V-8, diesel, four-cylinder, and V-6 automobile engines and the shift from carburetion to fuel injection.

Just what is product flexibility? It's the ability to make or produce something different than what is currently being made in a time frame that is acceptable to the customer. "Making something different" may mean customized products or even brand new products; and, depending on the product, the acceptable time frame, which is determined by one's customers and competitors, could be anywhere from weeks to years. It is important to keep in mind that this flexibility is itself bounded by the basic process technologies in place. For example, it would be reasonable to expect a company like Deere and Co. to make golf carts in the same facility as riding lawn mowers; it would not be reasonable to expect that same facility to blend custom lubricating oils. The basic technologies in place are just too different from those that are needed.

Volume flexibility is the basic tenet behind the just-in-time movement —make what you need, no more and no less. The real difficulty here is handling all of the perceived and real tradeoffs that are associated with good volume flexibility. U.S. companies have thousands upon thousands of manufacturing managers who were trained by the "economic lot size" theory. This theory simply states that the most economical unit cost for a manufactured item is found at that quantity where the cost of setting up the piece of equipment is balanced against the generally unknown but assumed cost of carrying inventory. This mentality, and the basic lack of attention placed on reducing set-up times, leads down the road to the "economies of scale" scenario that was so prevalent with the feelings about the inevitability of mass production in the 1950s and 1960s. If U.S. manufacturers learn anything from their Japanese counterparts, it should be that set-up is not something to be avoided (and ignored), but rather that it is something to be attacked and made as efficient as possible.

Additional criteria are often used for ease of understanding and because they apply more directly to the problem at hand. Ease of cleaning and maintaining equipment will be reflected in the cost criteria. Set-up relates to both flexibility and cost. The repeatable accuracy of equipment will affect the level of attainable quality. Durability is another of the cost criteria. Other examples are found in each and every company.

CONSULTANTS AND VENDORS

Technology transfer is the conscious (or unconscious) adoption of technology from sources outside a manufacturer's normal sphere of awareness. Simple examples of this range from the use of personal computers in the home (transfer from the business environment to the home environment) to the use of composites in the trucking industry (transfer of composites knowledge from the aerospace industry to the trucking industry) to the transfer of just-in-time operating techniques from Japanese companies to U.S. companies. This process is depicted in Fig. 19-3.

The problem that often occurs with this process is that too many manufacturing executives go straight from the first step, Awareness, to the third step, Decision to Adopt. The result is the misapplication of technology.

Consider the robot. A few years ago this wonderful piece of programmable automation was on the cover of *Time* magazine and was proclaimed to be the savior of world manufacturing. There is no doubt that many manufacturing executives decided that "that's what we need" before giving much thought to how the technology would be applied. The evaluation step was shortchanged. Unfortunately, many executives neglected to ask for the specifics of how the robots would be used, why they were better than the previous method, and how the robots would pay for themselves. Too many decisions were made on the "faith" in new (not understood) technology.

One of the keys to proper technology transfer is the determination of what new technology is applicable to one's business. What is right and works for one company or even industry does not necessarily work for another, even with modifications. Consultants and vendors, because of their exposure to a number of companies and industries, are a way to apply technology transfer to a company's equipment needs. The key, of course, is to select the consultant or vendor that can help with the application of the technology in one's manufacturing environment. Relevant experience in technology application should be a major factor when selecting a consultant/vendor.

THE SELECTION PROCESS

The equipment selection process shown in Fig. 19-4 has both managerial and technical aspects. Therefore, it is rarely carried through by just one person.

The first two steps are fundamental to the process engineering function and answer the questions "What does this part have to look like?" (What features are needed?) and "With what material do I have to work?" To keep this whole methodology in a broad context, the features should be thought of in a wide range of terms from actual physical features such as profiles and splines to such features as location when planning for material handling equipment. Likewise, the material properties could range from the metalworking properties associated with 9310 steel to such things as "transport in small tote" for material handling.

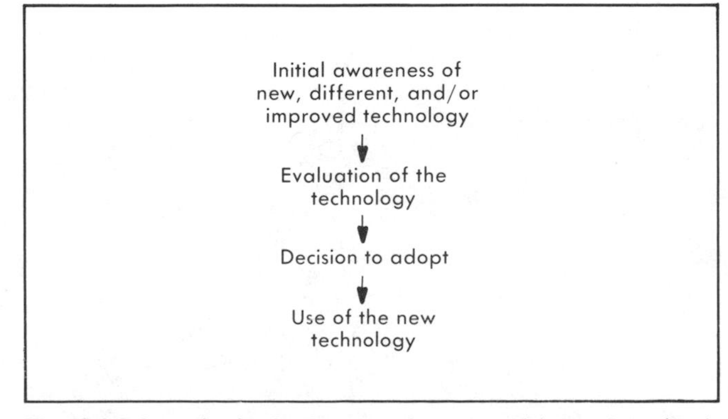

Fig. 19-3 Schematic of technology transfer process. (*Austin Consulting*)

The next step, determining the type of equipment, is essentially a research step and, as such, one in which there is likely to be considerable involvement with equipment vendors. Another good source of information is trade publications. During this step, the evaluation needs to be as open as possible to nontraditional techniques and technologies. For example, have the advantages and limitations of electrical discharge machining (EDM) been considered? If not, why not?

Evaluation criteria are then applied to the candidate list of equipment. The critical point here is that management should either have defined the criteria or should be aware of them and support them. Remember, these criteria should be the same that will be used to evaluate the ultimate success of the selected equipment. Criteria examples include installation cost, commonality with existing equipment, ability to interface with a computer control system, "American made," and so on. Naturally, there will be multiple criteria and not all equipment will meet all criteria uniformly. Therefore, it may be necessary to weigh the various criteria so that the candidate pieces of equipment can be ranked as objectively as possible. Even so, this process generally calls for a heavy dose of judgment and is not as clear-cut as many engineers would like it to be. This process should at least result in a "short list" of candidates.

The process of selecting the make and model of the needed equipment now has to consider the future. For example, an NC lathe with an X-inch (mm) swing may serve today's needs, but with a little market research, it may be determined that an $X + 2$ swing is what will be needed in the near future. After all, most equipment has a useful life measured in decades, not months. It always pays to spend some time to assess the future impact of present decisions. Financial aspects of equipment decisions are discussed in Chapter 3, "Planning and Analysis of Manufacturing Investments," of this volume.

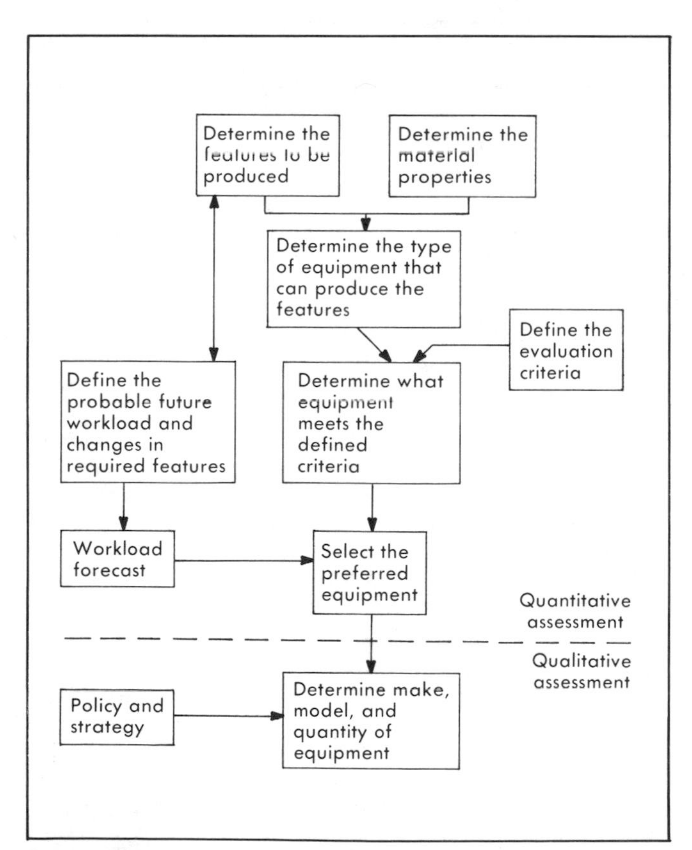

Fig. 19-4 Diagram of steps in equipment selection process. (*Austin Consulting*)

Determining the required quantity of equipment is another prime area for technical and managerial assessment. Equipment utilization projections can be made based on standard industrial engineering techniques. These techniques do not, however, take management policy into account. For example, management may not allow single quantities of unique/special equipment as insurance against breakdown. Likewise, the adoption of a cellular manufacturing philosophy may require multiple pieces of underutilized equipment to fully develop needed cells. The key here is understanding and appreciating the fallacy of high equipment utilization as the primary criteria. This has, in fact, been one of the biggest problems with adoption of the JIT manufacturing philosophy. Many conventional managers just can't see the advantage of letting equipment sit idle so as to *not* build inventory.

The final big step before signing the purchase order is to assess how this particular solution will fit with policies and strategies of the company and, therefore, move the whole process from the realm of quantitative assessment to the managerial realm of qualitative assessment.

EQUIPMENT PROFILE EVOLUTION

Many manufacturing managers are or will be faced with the problem of shorter and shorter product lifecycles and, consequently, the need for flexible manufacturing capabilities. This need has led, in turn, to the call for more programmable automation and less fixed automation. At the same time there has been a general call to move away from "islands of automation" and to address the need for integrated operations. Naturally, these influencing forces go hand in hand. The term *islands of automation* generally applies to automated equipment that is not physically or electronically tied into the manufacturing system.

The response is to address the needs for manufacturing within an "envelope" of capabilities. As mentioned previously, manufacturing managers have to look at the probable future requirements when addressing the specific needs of today. In light of this, some leading manufacturers have defined their manufacturing capabilities in terms of such statements as "we can manufacture prismatic parts that fall within a dimensional cube of X x Y x Z inches. This sets the limits on what they are prepared to fabricate and forces the need to react rapidly within those limits. In fact, fast reaction time is rapidly becoming a manufacturing requirement.

What does "rapid response" mean in the manufacturing world? It might mean schedules based on the latest inventory status and part production. It might mean having the ability to quickly change equipment set-ups. It might mean passing the latest NC data to the shop floor. No matter what it means, it requires that the equipment profile evolve into one that is tied together and operated as a system.

Rapid response is inherent in the capabilities of a networked manufacturing environment. General Motors saw this in its historic move to force the development of devices adhering to a specific manufacturing protocol on the vendor community. General Motors had tens of thousands of computer-based manufacturing devices and found that, for the most part, they could not communicate with each other. Rather than wait for an industry standard, which might never appear, GM forced the issue with its buying power "either adhere to the standard or we won't buy from you", and forced the development of MAP, Manufacturing Automation Protocol. As of this writing, the development is still under way and gaining momentum.

MANUFACTURING AUTOMATION PROTOCOL

Integrated manufacturing, specifically computer-integrated manufacturing, requires communication capabilities if the devices and the humans using the devices are to share information. Doing this in a factory setting requires the development and use of a local area network. Simplistically, this is just the data "circuit" that the computer-based tools use to transfer and share data.

In the early 1980s, General Motors found itself with about 40,000 programmable tools, controls, and systems installed at its facilities. The major problem with this situation was that only 15% of these devices could communicate with one another. The basis of this problem was that noncompatible methods of communication existed among the various vendors, and no one was going to change. IBM could communicate with IBM, DEC could communicate with DEC, HP with HP, and on and on. General Motors was faced with the desire to increase the amount of programmable equipment by 400-500% during the next 5 years; the communications problem would quickly get out of hand.

Assessing the situation led to the following three alternatives:

1. Continue to buy equipment from different vendors and develop custom solutions.
2. Buy all equipment from a single supplier.
3. Develop a standardized approach to plant floor communication.

GM chose the third alternative and set up a taskforce in 1980 to investigate the use of a model developed by the International Organization for Standardization (ISO) as the basis for the third alternative. GM representatives met with various computer manufacturers in 1981 and began to explore solving the problem through the use of local area networks based on the model; this model was to describe a set of communications protocols that would become known as the Manufacturing Automation Protocol. A protocol is a set of rules that govern the format and timing of messages that are exchanged between two communicating devices. In 1982 the GM taskforce developed an initial model composed of existing and newly defined layers for the protocol. (Without going into the architecture of MAP, just consider these levels as modules for the protocol.) The first MAP users meeting was held in 1984, and the development continues today.

As MAP becomes more of a reality, GM and the other users of MAP will have a standard set of protocols with which to access the networks that tie their machines into flexible manufacturing systems; the basic building block of CIM. The end result is that GM is slowly changing its equipment profile to an integrated manufacturing capability. Additional information on MAP as well as TOP (Technical and Office Protocol) can be found in Chapter 16 of this volume.

MATERIAL HANDLING

In the development of manufacturing facilities, the material handling operations must be considered as a critical factor in the design process. Material delivery and product movement philosophies have a dramatic effect on quality, facility requirements, equipment layout, and the physical integration of the operation. Manufacturing operations that are developed without consideration to material handling most often result in compromised performance well below the potential that could have otherwise been achieved. The implementation of complex systems that result in world-class operations is infrequent because of the lack of planning needed for the required coordination and interaction between production, material handling, and facility departments within a company.

The key to developing a successful integrated system is to rigorously follow a structured system development methodology. One such approach that has proven quite successful is shown in the flow chart in Fig. 19-5. This methodology works equally well independent of the operations performed within a facility. The seven steps begin by segmenting the facility into logical operations and conclude by restructuring the facility with definitive subsystems that can be implemented in a time-phased manner.

The progressive definition of the material handling systems and operations are described in detail in the following sections. To help illustrate the approach, a gear manufacturing operation will be used as an example. For initial considerations, the manufacturing operations have been grouped into cells, a small lot size is desired, and the facility is to be automated as much as possible.

SEGMENTATION

The first critical step is to segment the facility into logical operations that require or interface with material handling functions. These may include, but are not limited to: receiving storage, process workstations, packaging, and kitting. Each operation must then be defined in terms of internal function, requirements, and interrelationships with adjacent areas. Essentially, a minisystem functional specification is prepared for each segment or subsystem. The development of these accurate operational performance specifications is a critical element most often overlooked or ignored. Attention to detail at this point generally results in significant reduction in effort required in the later stages of facility development. The most significant benefit in segmentation of the facility into logical operations is that the design tasks are divided into manageable sections. In addition, the engineer is not faced with a monumental database that could result in losing sight of the original objective.

FLOW DEVELOPMENT

An accurate representation of material flows cannot be developed without a clear understanding of the functional operation and requirements of the subsystem. This understanding is the earliest benefit realized from the preparation of performance specifications. Steps 2 through 4 are performed independently for each subsystem.

In these steps, the material flows and the data flows are independently developed, although the data flow definition generally is preceded by the material flow development. All load movements are first defined in a "From-To" matrix for each subsystem as well as the complete facility. These movements are then plotted on a layout of the facility if for a retrofit situation, or as a functional block diagram (see Fig. 19-6) for a new facility development. Consideration must be given to volumes transported, frequencies of moves, and the envelope and characteristics of the items being moved. Just-in-time (JIT) movement of parts and supplies to an assembly station reduces work in process and line stock utilization of floor space at the expense of the complexity of the material handling system, which must deliver smaller volumes very frequently.

Implementation tradeoffs can be evaluated, and significant benefits can be attained if process and material handling operation approaches remain somewhat flexible. Once the material and product movement flows have been determined, including peak and cyclical factors, the required data transactions and flows can be established to effectively support the material and product flows. The data flows will account for inventory monitoring, material dispatch, and tracking.

IDENTIFY IMPLEMENTATION CANDIDATES

For each subsystem, a range of implementation techniques can be defined as alternatives for evaluation. Various candidate technologies may be applicable for any given implementation. All potential candidates should be evaluated for cost, flexibility, performance, and ease of implementation tradeoffs. Alternative implementation candidates can be manual, semiautomated, or fully automated. Examples of candidate technologies in each of these classifications are shown in Table 19-3.

Multiple factors must be considered when selecting implementation candidates for evaluation. Local configuration, orientation, and delivery frequencies can restrict traffic because

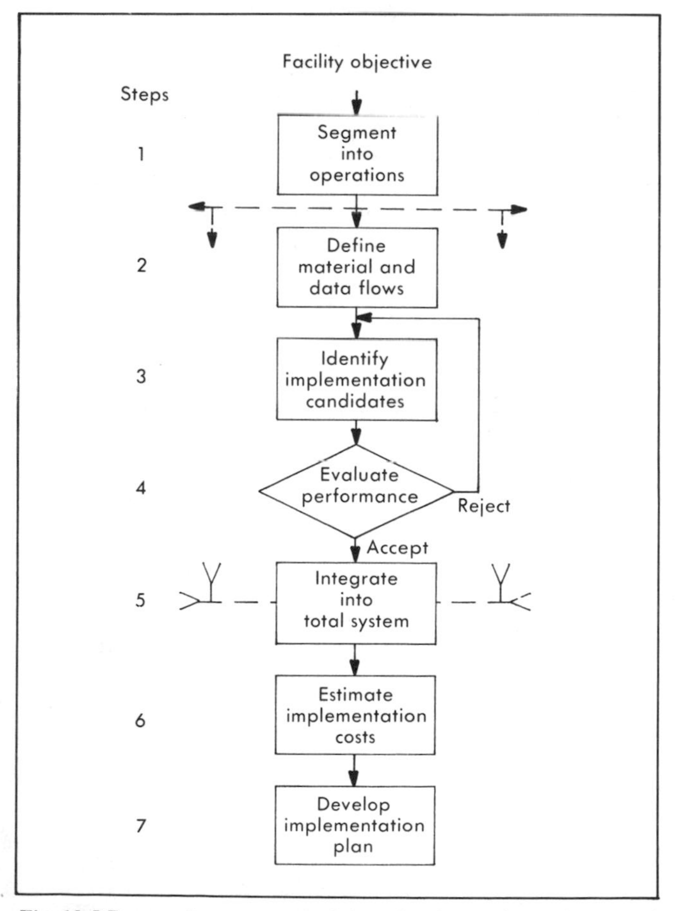

Fig. 19-5 Proposed system methodology for developing an integrated material handling system. (*Austin Consulting*)

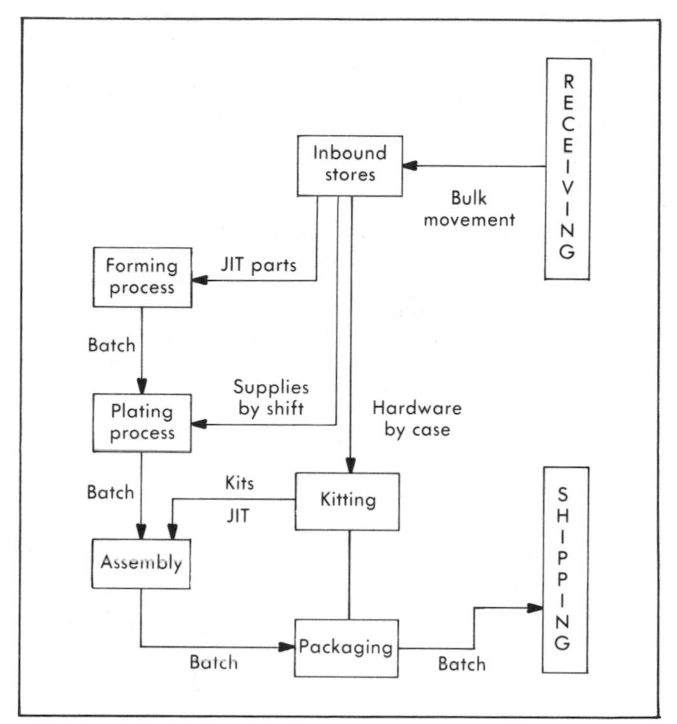

Fig. 19-6 Functional block diagram of material flow in a manufacturing facility. (*Austin Consulting*)

TABLE 19-3
Material Handling Movement Implementation Candidates

Manual	Semiautomated	Fully Automated
Pallet jack	Automatic guided vehicle (AGV) pallet jack	Towline
Fork truck	AGV tugger	Conveyor
Fork truck with data terminal	AGV stock selector	Power-and-free monorail
		Automated monorail
		AGV tugger with automatic trailers
		Unit-load AGV
		Multiple-load AGV
		AGV fork truck
		Specialty AGV

(*Austin Consulting*)

of throughput limitations of load and unload stations. Process requirements at the manufacturing workstations, such as precise load positioning, can also affect material handling choices. The key consideration is to develop a series of candidate configurations that will provide the appropriate balance between delivery frequencies and work in process on the production floor. Facility configuration may also significantly affect the range of candidates available for evaluation. A facility that has low internal ceilings or one that cannot support structures from overhead would probably not be able to accept power-and-free monorail or automated monorail systems.

Various techniques are available for technology selection. One effective method that results in an unbiased selection of practical alternatives is to use a weighted matrix and perform a numerical analysis. With this method of selection, evaluation criteria must be established and weighted values assigned. Numeric rankings are assigned to each candidate against each criteria. The analysis is a mathematical evaluation that depicts which technology best satisfies the composite criteria. Criteria can be either quantitative or qualitative. Examples of quantitative criteria include throughput, cost, and speeds. Examples of qualitative criteria include product quality, safety, expandability, and reliability. Caution must be exercised to ensure that professional input is used to determine appropriate values for the rating matrix. This is most important because the value used must be periodically adjusted to reflect introduction of new products as well as advancements in technologies. Also, the facility factors reflect the specific characteristics of the application area and must be assigned based on experience.

An example of a technology selection criteria matrix is shown in Table 19-4. Once the matrix has been established, the numeric analysis is then performed. The analysis compares the variance of each candidate's weighted value from the desired criteria value. The technology having the least variance is most preferable. Multiple high-ranking candidates are then selectively evaluated based on engineering application experience.

EVALUATE PERFORMANCE

Once candidate technologies are chosen, a preliminary implementation layout is formulated. Performance is evaluated against the performance specifications developed when the subsystem segmentation was performed. Operational performance is best characterized through the use of computer simulation. Simple operations can be modeled using mathematical networks that only yield statistical results. More complex subsystem implementations require analysis of detailed flows in areas such as critical intersections and subsystem interfaces. This detailed analysis cannot readily be determined from statistical data. Animated graphics are then implemented to

TABLE 19-4
Technology Selection Criteria Matrix

Selection Criteria	Input Value
Safety	9
Travel distance	5
Throughput	4
Routing	4
Compatibility	5

	Evaluation Criteria				
	Compatibility	Safety	Alternate Routing	Travel Distance	Throughput
Weight factor	6	10	8	10	10
Manual pallet jack	2	4	9	3	1
Manual forklift	9	5	9	4	2
Maximum tractor w/trailer	7	5	9	7	6
Maximum forklift with complete dispatch	9	5	9	4	2
In-floor towline	1	8	1	3	2
Power-and-free monorail	7	7	2	5	6
Automated conveyor	5	8	1	3	9
Automated monorail	8	7	6	6	7
AGV pallet jack	2	8	8	9	2
AGV tractor	7	9	8	9	8
AGV with automatic trailers	8	9	8	9	8
AGV forklift	9	3	8	9	4
AGV platform	7	8	8	9	6
AGV with unit roll deck	8	9	8	9	6
AGV with dual roll deck	8	9	8	9	7
Inertially guided AGV	9	1	9	9	7

(*Austin Consulting*)

visually depict the operation so that flows may be observed and conditions leading to bottleneck situations can be identified. The benefit of using simulation as a design verification tool is the ability to test and evaluate conditional performance before committing to expensive hardware.

Having established conformance with operational criteria, an assessment can be made of cost effectiveness of the final implementation. If the performance, either in flows or cost, is deemed unsatisfactory, another implementation is chosen and the process repeated.

This process is performed for all subsystems individually. The estimated performance of each subsystem can be used as a benchmark at the time of installation test at the subsystem level.

INTEGRATION OF SUBSYSTEMS

Having completed the definition of all subsystems, the process of system integration begins. This process begins with the interface evaluation when any two subsystems are interconnected. The interfaces that must be evaluated are both informational and physical. Care must be taken to ensure that appropriate and timely information transfer is provided and that data handling means are not overloaded. In terms of physical aspects, load transfer mechanisms, orientation, positioning, and transfer times that could restrict flows must be considered. Individual subsystems are then added to the initial integrated pair one at a time. System performance should be reviewed after each subsystem integration definition is completed.

ESTIMATE COSTS

Overall costs for the system can now be considered. The costs for implementation should include not only the purchase price of the hardware and software, but project management, installation, and various administrative costs such as purchasing the installation supervision. The ongoing cost of personnel and maintenance should also be considered. These estimates can be evaluated against improvements in production and displacement of personnel to ascertain the payback rate and cost effectiveness of the facility. If the cost effectiveness is determined to be acceptable, then implementation funding can be pursued. Equipment justification is discussed in detail in Chapter 3 of this volume.

PLAN IMPLEMENTATION

The implementation plan should be developed to provide installation and testing in a phased manner. This will permit a smooth transition into full system operation, minimize risk, and provide early payback on portions of the installed system resulting from the benefits of early usage.

Material handling system development is still an art as much as it is a science. This is because of the applied creativity of the people who develop the systems. There are as many options and tradeoffs as there are opportunities. Virtually every installation is unique and, as such, will have unique solutions that depend as much on the operating business issues as on the facility. System development and evaluation should be performed or at least supervised by experienced professionals.

SYSTEMS INTEGRATION

Consider how the dictionary defines the terms *integrate* and *integration*:

Integrate—to make into a whole by bringing all the parts together.
Integration—the act of integrating.

The definitions sound so simple. Why then is this concept so difficult to comprehend and perform when it is put into the phrase "computer-integrated manufacturing" or "systems integration"?

Before getting all wrapped up in the technology of computer-integrated manufacturing (CIM), a manager should think about any manufacturing operation in which he or she has worked. It was integrated—human integrated for the most part. People wrote process plans, set up machine tools, withdrew and issued inventory, posted schedules, calculated payroll and incentives, and on and on. These people responded to interruptions by changing the machine tool set-ups, or expediting the missing raw material, or getting a replacement for one of the workers who went home sick. Tasks were prioritized by looking at the backlog, finding the customer orders on the shop floor, and putting "HOT" tickets on the work in process that was needed that day. This scenario was repeated day in and day out and all it cost the company was some inventory, some lead time, and occasionally, some parts made to the wrong engineering release. But what was right in the past is wrong for the future. Like it or not, the rules have changed, and consequently, computing technology in the factory is here to stay.

What then is systems integration? One concept is that it's using a computer and some wires to connect the various parts of the manufacturing operation together so that the manufacturer can respond to and interrupt and prioritize tasks very fast. A more accurate description is that it is the gathering of dissimilar computer-based products or technologies of multiple vendors into an operating entity. Easy? Only if the person can understand a new language that includes terms like protocol, token ring, twisted pair, and controllers. These terms define elements of the needed local area networks.

The purpose of this section, then, is to lend a bit of understanding to this new language. To do this, a very small shop will be created to work with and integrate. Let's assume that the shop has two machines, a front office, a tool crib, and a stockroom. A schematic of this operation and a listing of the kinds of things that go on in this arrangement is shown in Fig. 19-7.

If this operation reflected the reality of years ago, the problems associated with the need for shorter lifecycles, shorter lead times, and smaller lot sizes would not be present. However, that reality is no longer true. Everything is changing and the biggest problem that manufacturers have to face is another aspect of integration—the problem of getting the timing right. A new drawing release won't be of much use if the operator is

CHAPTER 19

SYSTEMS INTEGRATION

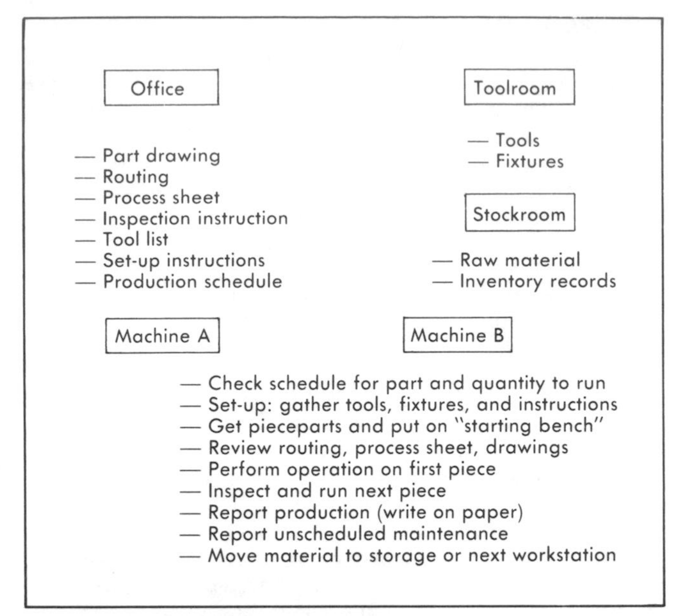

Fig. 19-7 Schematic of a small machine shop operation. (*Austin Consulting*)

using an old "red-lined" drawing on the shop floor. Likewise, breaking a set-up to expedite a hot job might not be necessary if one knows that the tooling isn't ready.

This little machine shop will work all right as long as there are only a few parts to contend with and a good job is done of keeping all of the paperwork up to date and matched among the shop floor, office, and stockroom. However, as the business grows, the number of parts and tools and the amount of paper will grow at a phenomenal rate. It now becomes harder to keep things up to date—it becomes harder to run the operation in an integrated fashion. What then are some of the key facets of integration and how might they be accomplished?

INTEGRATION

One integration need is to make sure that the shop floor makes the products to the latest specification. Face it, it's a waste of time, material, and labor to make "revision A" when the office has just released "revision B." Under the manual method of integration, drawings are hand drawn, filed, passed to the manufacturing engineer for the development of instructions, and finally passed to the shop floor for the operator to use. Problems arise when the design engineers start to respond to the marketplace with changes or when the machine operator realizes that the part can't be made the way it's shown on the drawing and changes the actual part without informing the design engineers.

If the drawings are created on a computer-aided design (CAD) system, run off on the plotter, and then passed to the manufacturing engineer and the shop floor, the only thing that has been accomplished is a slight reduction in drafting time. However, what if the paper drawings could be eliminated and one electronic master maintained for use by all functions? This may sound very simplistic, but that is the essence of integrating CAD with numerical control tools—use the computer to "file" the matching set of drawing and NC instructions and a network to send the data to the appropriate machine tool. This is a good start, but it may not be enough.

Integrating CAD with NC machine tools allows the manufacturer to make the correct parts, but it doesn't necessarily help to make the parts that are needed or make them when they are needed. This involves another kind of integration, the integration of schedule requirements with the status of inventory and work in process in a timely manner. To do this, a manufacturer might look at the applicability of software packages designed to do manufacturing resource planning (often referred to as MRP II).

The use of MRP II is an attempt to integrate the current production requirements with the status of the factory and the company vendors. The essential objective is to have accurate data to ensure the accomplishment of the production schedule and to determine the ability to change the schedule.

EXAMPLE

Probably the major cause of integration failure is that the concept is so easily understood and is so very agreeable—"of *course* we need to integrate this operation, let's hook up the machines and go for it!" While the concept is easily understood, the implementation is rarely grasped by manufacturing management because so much of systems integration is really a major data processing project. The example in this section is based on the earlier scenario of a two-machine tool operation, the difference is that there is a distributed numerical control (DNC) environment. The phase or steps in this process are product introduction, downloading, operator functions, and machine monitoring. Different software modules are also used to support the DNC operation.

Figure 19-8 shows the conceptual design for a DNC system for the small operation in this example. Part designs are created and maintained on a CAD system, NC tape images are maintained in the office, tool and inventory information is maintained in the production database, and the machine tools have been fitted with machine controllers capable of accepting NC directions.

Product Introduction

After the design of the part has been finalized by the design engineering department and released to manufacturing engineering, a process plan is prepared, tooling fixtures are designed, NC tool path is generated, and cutting tools are selected. The parts programmer specifies the type of operation, the tool path, and the cutter. Verification of the tool path is usually visual, and changes are made interactively. Once the parts programmer is satisfied that the tool path is correct, a request will be made to convert the tool path to automatically programmed tool (APT) source language. The APT statements are processed through the language processor and subsequently on to the NC machine-specific post processors. The tool path is then reverified. The output from the post processor and coded machine instructions are saved in the Tape Image Database.

Downloading to the Shop Floor Workstation

When a work order for a particular part is issued by production control, the manufacturing engineering department will transmit the tape image file and the other elements of the NC package to the DNC file server. The status of the NC tape image stored on the DNC file server will be easily identified as either production (proven) or unreleased. Unproven programs will not be loaded to the operator's workstation. Transmitted data is checked for integrity and is retransmitted in the event of an error.

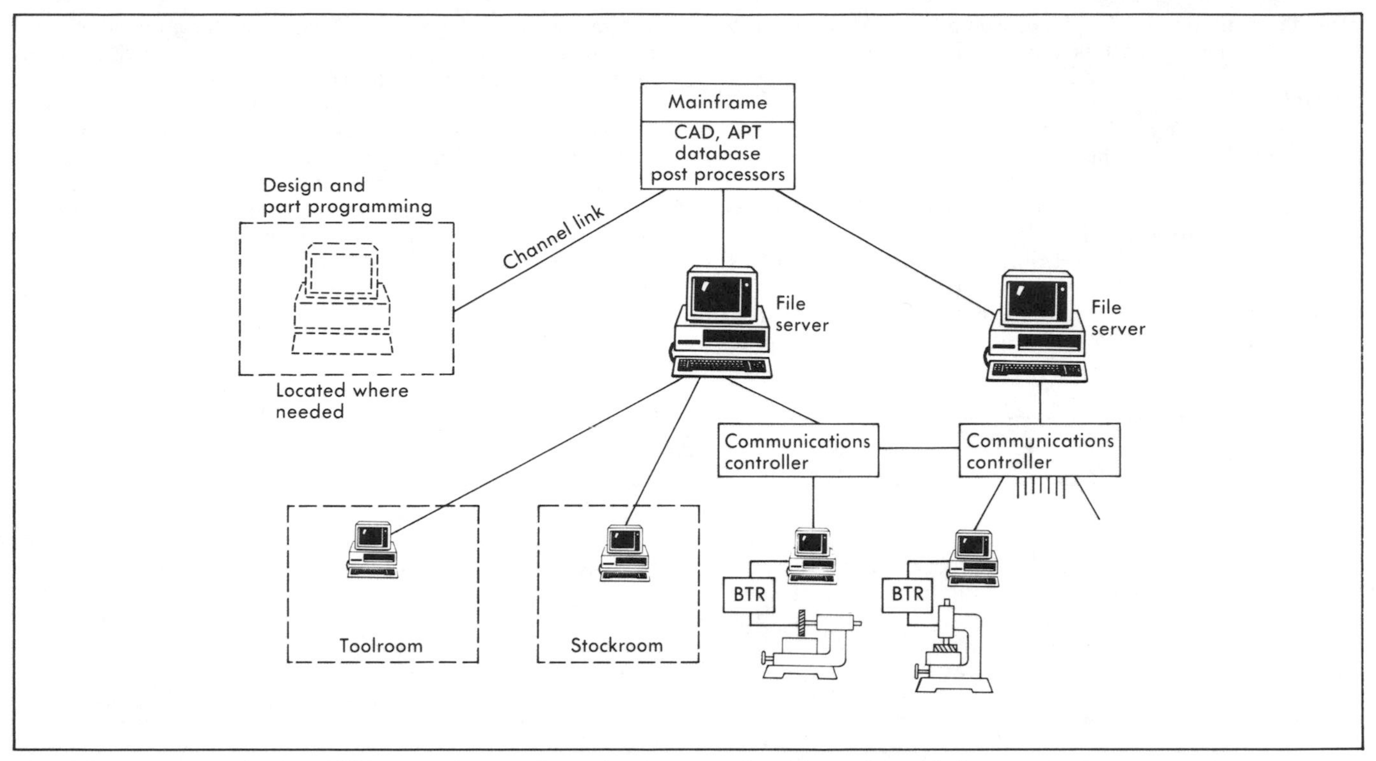

Fig. 19-8 Conceptual design for a DNC system for a small machine shop operation. (*Austin Consulting*)

Operator Functions

The machine operator may query the shop floor workstation to list those jobs ready for production. Once a particular job is selected, the operator may display part geometry with dimensions, part number, program image, program length, cutting tool descriptions, fixture numbers, and set-up instructions.

The operator is allowed to load NC program images designated for that specific machine authorized by NC programming. The downloaded NC program package is stored at the shop floor workstation until the operator is ready to transmit it to the machine controller. Once the operator begins loading the program image into the machine controller, he or she continues the machine tool operations as usual.

Software Modules

A variety of functions are performed by software in support of the DNC operation. The shop traveler software builds the relationship between tool lists, graphic images, machine tool number, job history, and tape number. All of the associated documents can be referenced by the part number and downloaded to the shop floor workstation.

Machine maintenance software automates repair requests and assigns priority levels. It also schedules preventive maintenance for each machine tool. Maintenance management software should also be capable of keeping historical records by machine of repairs and modifications made as well as performing the scheduling function. These historical records are essential to the establishment of an efficient predictive maintenance program.

Automated tool presetting software ties toolroom operations with the DNC host computer. As a tool is measured, the program updates the tool number and tool description in the cutting tool database. When the tool presetter enters the tool number, the program updates the database with the measurement of the actual tool dimensions.

Operational Machine Monitoring

Operational monitoring of NC machines is accomplished by collecting data from the machine control, capturing data from sensors, or from manual data entry by the operator. The electronic signals being monitored are sampled and stored as raw data values. Both analog and digital information is recorded. At periodic intervals, the shop floor workstation polls the data and transfers it to the DNC file server where it is converted into a more usable format. This converted data is used to produce management reports.

Manual data entry by the operator is initiated through a menu display on the shop floor workstation. The operator specifies machine tool status information by selecting an option from a list such as the following:

- Job set-up.
- Tape tryout.
- Fixture problem.
- Time out.
- No work.
- Work in production.
- Part inspection.
- Tool inspection.
- No tools, material, operator, or tape.
- Maintenance problem.

Each time the operator selects one of the menu items, the system clock records the "start time" of that event. The event is stopped whenever the operator selects the next event. The elapsed time is computed and recorded by event and is used to produce a variety of machine tool utilization reports.

RESULTS

The integration of the factory has implications beyond simply eliminating the need for punched tape or manual keyboard/console entry. One only has to think of the implications of the manufacturing engineer being able to answer fundamental questions, such as "If we make this design change, what will be affected?" By accessing the material requirements database (containing the item master, "where used," and inventory data) the engineer will be able to assess the impact that the change will have over the expected life and volumes of the part. By accessing the CAD database, the engineer will be able to determine whether or not this change will require changes to other parts (such as mating parts). By accessing the purchasing database, the engineer will be able to assess the impact that the change might have on purchase agreements and blanket orders. And the process goes on and on.

In the final analysis, the integration of the factory will result in the ability to exchange information and share resources. Doing this means being able to leverage manufacturing capabilities and makes the shop more responsive to changes that are made in the product. With CAD/CAM files and integrated systems, changes may be integrated into the product itself more rapidly.

EQUIPMENT INSTALLATION

The purpose of this section is to look at the relationship between equipment and the facility. Although some of this can be considered from the plant engineering point of view, it is more appropriate here to consider it from the plant management point of view. This is best accomplished by answering a series of broad questions about the equipment under consideration. Consider the following questions:

- What does this equipment need?
- What does this equipment produce?
- Where should this equipment fit in the manufacturing flow?
- What is needed if it breaks?

NEEDS

Consider the list of equipment needs and realize that someone from the organization will have to supply those needs. These needs include the following:

- A floor strong enough to support the weight of the equipment in operation.
- Enough clear height to allow access to the workpiece.
- Enough operator space to allow maintenance and operation of the equipment.
- Power, water, and air.
- Floor drains for cleaning purposes.
- Elevators strong enough to move the equipment into place.
- Doorways with enough height and width to allow movement.
- Access to automated material handling equipment.
- A level floor for the required material handling and storage equipment.
- Isolation from vibration caused by other equipment.
- Vent hoods and drains.

PRODUCTION

Another aspect of the equipment installation process is answered by addressing what the equipment produces. Consider the following topics and questions:

- Product.
  How will the product be moved from the machine?
  Where will the product be stored?
- Chips and waste cutting oils.
  Will underfloor chip conveyor tote pans be used?
  How would chip conveyors help or hinder equipment movement flexibility?
- Heat.
  What heat load will this equipment place on the air conditioning system?
- Fumes.
  Are hoods needed?
  Are air scrubbers needed?
- Waste water and tramp oils.
  Can the waste water and tramp oils enter the municipal or company sewer system or will it have to be treated first?
- Noise.
  Are baffles needed?
  Will ear protection and operator screening be required?
- Vibration.
  Is this a high-tolerance machine? Will this machine need to be isolated from the vibration of nearby machines?

MANUFACTURING FLOW

Answering the question of fit in the manufacturing flow is not as easy or as straightforward as the earlier questions. Nevertheless, it is important and often overlooked. For example, most companies find the open space when new equipment is considered. The impact of the new equipment on the overall layout is usually not considered. Likewise, a company rarely updates its layout as the products change over time. What might have been an ideal layout for the old product line may be inappropriate for the present or future products. Whether one looks at the operation from a product or process point of view, it is more easily understood when considered in light of "From-To" considerations and the application of group technology principles. Consider the following questions:

- Where do the workpieces originate?
- Where do the workpieces go?
- How is the tooling delivered?
- Does work in process stay at the machine or does it go to inventory?

MAINTENANCE PLANNING

Deciding what to do when the equipment breaks is a form of contingency planning. It raises the following issues:

- Who will repair the equipment? Does the organization have the needed skills or will people have to be trained? For example, are the electricians in need of electronics knowledge?
- What repair inventory is needed? How long are the lead times for the critical repair items? What are the unique items needed for repair and should they be on hand to prevent a shutdown of the production line?
- How does this equipment fit in the scheme of the preventive maintenance program? What aspects of the equipment have to be considered for a predictive maintenance program?

MAINTENANCE

One of the requirements of the JIT approach to manufacturing is to have the equipment available when needed. Equipment may be listed as an asset on the accounting records, but it is a liability if it is down and in need of repair. Also, as factories become more and more integrated, a single piece of broken equipment could cripple the entire process. (There is no inventory to act as a buffer.)

Past manufacturing practices have taken the position of "if it's not broken, then don't fix it." A more enlightened approach, however, is to recognize the relationship between maintenance costs and the cost of production losses caused by equipment failures. This has led managers away from the older assumption of equating maintenance with emergency repairs to the acceptance of the principles of preventive and predictive maintenance.

It is beyond the scope of this chapter to investigate the details of maintenance management. However, because of the relationship between productivity and good maintenance management practices, a quick overview is in order.

MAINTENANCE PRINCIPLES

The maintenance function in a factory has four basic resources with which it must be concerned: (1) maintenance labor, (2) plant equipment, (3) maintenance inventory, and (4) maintenance information. Considering the role of these resources in the management of a manufacturing enterprise leads to four fundamental principles.

Principle One

Maintenance labor can only be used in an efficient and effective manner if it is considered in light of a scheduled maintenance program. A policy of "repair maintenance only" leaves this resource in a reactive mode. The three types of maintenance efforts are repair, preventive, and predictive maintenance.

Repair maintenance is the unscheduled servicing of equipment after a problem is discovered. Preventive maintenance (PM) is the application of scheduled work to reduce maintenance costs through a reduction in repair needs. Predictive maintenance is the application of analytical techniques to reduce both maintenance costs and production downtime through identification of upcoming equipment failures. This is accomplished through measuring devices and the use of statistical techniques. In the future, this approach may be fertile ground for the application of artificial intelligence techniques.

Principle Two

The dictates of increased productivity require maintenance workers to know and understand the way in which a task is to be accomplished and how long it should take to accomplish it. This requires a work order system that organizes and maintains the following information:

- Description of the work to be accomplished and where it's to be done.
- Time estimates developed through experience, historical data, or standard data.
- Material and equipment requirements.
- Priority designation.
- Authorization.

Principle Three

Like manufacturing, the organization of the maintenance function has to respond to the needs of advanced technology by supplying the skills that are needed. The maintenance organization must provide those skills based on the existing structure and job descriptions as well as from the standpoint of determining the best qualified person or group from a trained skill viewpoint. Consider the "traditional" crafts associated with a maintenance organization: electricians, millwrights, pipefitters, carpenters, painters, sheet metal workers, and general laborers. The question today is who will be called when a fiber-optic cable has been severed or when a solid-state machine controller is malfunctioning. The maintenance department will also need its share of computer-literate (and trained) workers. Determining who does what may be a function of who is properly trained for the task and is not assigned another equal-priority task.

Principle Four

Principle Four is the requirement for properly documenting any and all actions taken with respect to maintenance or modification of a piece of equipment. This documentation assists other maintenance personnel in troubleshooting similar pieces of equipment as well as providing the statistical basis for a sound predictive maintenance program and for justification of replacement equipment in the future. Such documentation can also assist in personnel planning and scheduling of maintenance personnel and actions.

BENEFITS

Many studies of maintenance expenditures have shown that the total cost of maintenance is subject to improvement through the proper application of maintenance efforts. As shown in Fig. 19-9, by reducing costs due to equipment being out of service and repair costs, an effective predictive and preventive maintenance (PPM) program reduces total controllable maintenance costs. The savings are accrued through extended equipment life, lower repair frequency, reduced lost production time, and reduced employee exposure to malfunctioning equipment.

BIBLIOGRAPHY

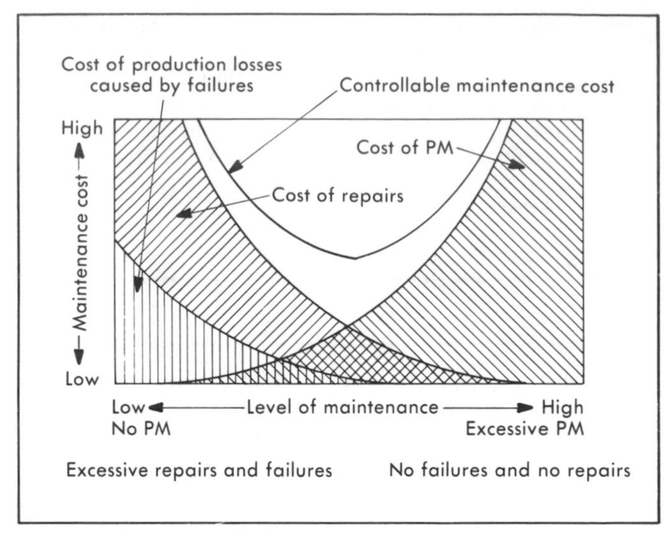

Fig. 19-9 Cost tradeoffs of a maintenance management program. (*Austin Consulting*)

DEVELOPMENT AND IMPLEMENTATION

The development and implementation of a productive and cost effective PM function is a two-step process. The first step involves the determination of the facility's actual equipment PM requirements and the methods and procedures needed to ensure proper results. Step two involves the implementation of a formal PM program that meets the equipment needs for ensuring maximum reliability and minimum emergency breakdowns. The steps require an understanding of the cost of equipment downtime and the ability to forecast the resources required to avoid an unacceptable level of downtime. Although it would be prohibitively expensive to avoid all emergency repairs, a properly designed PM program will return its annual cost many times over.

References

1. Robert H. Hayes and Steven C. Wheelwright, *Restoring Our Competitive Edge—Competing Through Manufacturing* (New York: John Wiley & Sons, Inc., 1984).
2. Robert W. Hall, *Zero Inventories* (Homewood, IL: Dow Jones-Irwin, 1983).
3. *Ibid.*
4. Hayes and Wheelwright, *loc. cit.*
5. Ramchandran Jaikumar, "Postindustrial Manufacturing," *Harvard Business Review* (November-December 1986).

Bibliography

Chiantella, Nathan A., ed. *Management Guide for CIM*. Dearborn, MI: Computer and Automated Systems Association of the Society of Manufacturing Engineers, 1986.

Harrington, Joseph. *Understanding the Manufacturing Process*. New York: Marcel Dekker, 1984.

Miller, Jule A. *From Idea to Profit—Managing Advanced Manufacturing Technology*. New York: Van Nostrand Reinhold Co., Inc., 1986.

Moss Kanter, Rosabeth. *The Change Masters*. New York: Simon and Schuster, 1983.

Skinner, Wickham. *Manufacturing: The Formidable Competitive Weapon*. New York: John Wiley & Sons, Inc., 1985.

PRODUCTION PLANNING AND CONTROL

One of the essential manufacturing support activities is that of production planning and control. Very often, the principal group of personnel involved in these processes is in a group called production and inventory control. They may be a part of an overall materials management department or they may report directly to the organization's manufacturing manager.

The required interfaces for this group include essentially every other department in the organization. Its coordination of planning and control activities requires knowledge and skills related to demand forecasting, aggregate production and inventory planning, material requirements, and personnel/machine capacity planning for shop floor functions.

The tools used by this support group may or may not include the use of material requirements planning (MRP) or manufacturing resource planning (MRP II) systems. If an organization is implementing or using just-in-time manufacturing techniques, its focus is reduced relative to shop floor control activities, but its planning and communication roles remain essential.

The main topics in this chapter discuss the various techniques that constitute key tools used in performance of the coordination activities that support the performance of the operations group.

CHAPTER CONTENTS:

FORECASTING

The production and inventory planning process begins with forecasting. All techniques of production and inventory control require some calculation of quantities, which represent future demand. The specific needs of each application are determined by the lead times inherent in the manufacturing processes being supported. Short lead time processes, including material procurement, may be supported quite well utilizing current open orders and only a few weeks' worth of estimated demand. Very long process requirements (12 months or longer) are most likely supported by a contractual order process, and all material planning and workcenter planning will be based on values calculated to support the more or less "known" requirements. Most manufacturing applications fall somewhere in between. To their customers, they present one of two faces:

- Make to stock: Orders have short turnaround—either hours, days, or weeks.
- Make to order: Orders will be scheduled, manufactured, and shipped per customer order within *X* weeks.

Both types of business operations, including combinations of the two, require some technique of forecasting for business planning and personnel decisions. The anticipated demands, and the forecast error rates experienced, will determine the basis for the systems, inventory policies, purchasing practices, and shop scheduling techniques used to support production requirements. Make-to-stock operations may also use forecasts to determine quantities stocked. Independent and dependent demands are discussed in Chapter 21, "Materials Management," of this Handbook volume.

Marketing and sales organizations are the logical choice for the placement of the responsibility for demand forecasting. They are also responsible for the reliability of the forecasts. Because forecasts are always subject to some level of error, it is essential that the best possible combination of techniques be used to minimize the negative impact on production flow and inventory investment.

THEORY

The basic theories behind forecasting depend on some ability to use historical data and arrive at a guess (an estimate) or a calculation that will be

***Contributors of this chapter are: W. David Lasater**, Managing Consultant, K.W. Tunnell Company, Inc.; **Kenneth W. Tunnell**, President, K.W. Tunnell Company, Inc.; **Richard W. White**, Senior Consultant, K.W. Tunnell Company, Inc.*

***Reviewers of this chapter are: Dan Ciampa**, President, Rath & Strong, Inc.; **William F. Cunniff**, Vice President, Management Consulting Div., GAINS Systems Group, Div. of Benton Schneider & Associates, Inc.; **Ronald M. Hutchinson**, Managing Consultant, K.W. Tunnell Company, Inc.; **Dennis E. Kaufman**, President, Vacuform Corp.; **Leo Roth Klein**, President, Manufacturing Control Systems; **Bernard J. LaVoie**, Director of Manufacturing and Quality, Department of the Air Force, HDQTRS Electronic Systems Div., Hanscom Air Force Base; **Hayward Thomas**, Consultant, Retired —President, Jensen Industries; **Harold Zeschmann**, Director of Manufacturing, Circle Seal Controls.*

FORECASTING

useful in the planning activities that are necessary to the competitive operation of the business. Three possible groupings of forecasting techniques are expert opinion, intrinsic methods, and extrinsic methods.

The theoretical basis of the expert opinion technique lies solely in the past practical demonstrations of the ability of involved persons to assimilate a combination of experience and a wide variety of inputs from the marketplace. This is the oldest method known and can be quite reliable for aggregate forecasting, such as quarterly forecasts of total company dollar volume.

Intrinsic methods depend on statistical interpretations from historical demand data from within the organization. These methods have been used to support both purchasing managers and production and inventory control managers for many decades. The only statistic required, in the simplest form, is an average period quantity.

Extrinsic methods include marketplace factors from outside the business as key variables in the forecasting process. Prime examples include housing starts, automobile production, and gross national product. The use of extrinsic methods is both more complex and more costly and, therefore, is used mostly in the domain of the largest organizations. As with any technique of forecasting, the cost of the method selected can only be justified through a reduction in the forecast error experienced.

The basis of all forecasts may be described as an extrapolation of a demand pattern over some future period. Regardless of the method or combination of methods chosen, a typical forecast can be depicted as shown in Fig. 20-1. In the sections that follow, linear forecast models will be discussed, with or without trends and with or without seasonal influences. The final forecast developed will often be the result of a combination of methods; the most likely is one based on statistical interpretation of intrinsic data modified by expert opinion.

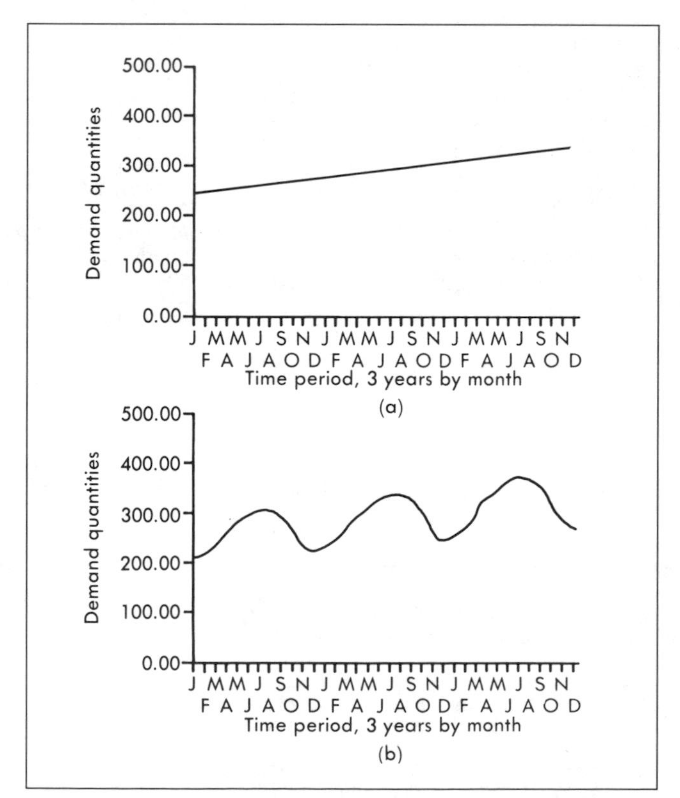

Fig. 20-1 Demand patterns: (a) trend pattern and (b) trend-seasonal pattern. (*K.W. Tunnell Company, Inc.*)

Several general principles can be described regarding forecasts, regardless of their origin. It is important that both marketing and manufacturing staffs understand the importance of these principles and the impact they should have on routine decision processes. Four key concepts and principles describing forecasts are as follows:

1. Accuracy of the forecast is indirectly proportional to the length of time to the forecasted period; the shorter the forecast period, the more accurate the forecast.
2. Accuracy of the forecast is directly proportional to the number of items in the forecast group. The total company forecast can be expected to be more accurate than the corresponding forecast for a given product line, which, in turn, will be more accurate than the corresponding forecast for a single part number in that product line.
3. Forecast error is always present and should be estimated and measured on all forecasts.
4. No single forecast method is best; alternate methods should be tested periodically to determine if another method would result in a smaller forecast error.

For those companies using computerized forecasts, some software packages will vary the model for an item using calculations based on the forecast error experience on the item.

Table 20-1 gives an example of the time and group concept expressed in terms of forecast error. The percentages shown are representative expectations resulting from the principles just discussed.

SEASONALITY

A study of seasonality for a forecast group, whether it is an individual part, product line, or company total, requires a sufficient amount of historical data to support the calculation. The result of the calculation is referred to as a deseasonalized demand pattern. Very simply, the removal of the seasonal component of the demand pattern is accomplished by using the average of the historical period percentages (divided by 1/12) as corresponding divisors to calculate the theoretical demand pattern that would be present if no seasonality influence existed. It is not possible to test the randomness of variation without first removing the impact of the seasonal variable, if one is present. An example of such a calculation is shown in Table 20-2. The calculated values show the underlying change in the overall demand pattern; these values had previously been masked by the swings in demand caused by a demonstrated seasonal pattern (see Fig. 20-2).

TABLE 20-1
Time and Group Concept Expressed in Terms of Forecast Error, Percent

Forecast Group	Month 1	Month 2	Month 6
Company total	+/- 1	+/- 4	+/- 8
Product line 01	+/- 3	+/- 9	+/- 15
Part number 1010101	+/- 10	+/- 20	+/- 40

(*K.W. Tunnell Company, Inc.*)

TABLE 20-2
Seasonality Analysis—Deseasonalized Demand History

Month	Year X-2	Year X-1	Year X	Three-Year Total	Monthly Average, %	Deseasonal Factor	Deseasonalized Demand for Year X
JAN	100	110	120	330	5.0	0.595	202
FEB	110	120	140	370	5.6	0.668	210
MAR	110	130	140	380	5.7	0.686	204
APR	160	180	200	540	8.1	0.974	205
MAY	180	210	230	620	9.3	1.119	206
JUN	210	240	260	710	10.7	1.281	203
JUL	220	230	260	710	10.7	1.281	203
AUG	210	240	270	720	10.8	1.299	208
SEP	190	220	240	650	9.8	1.173	205
OCT	180	180	200	560	8.4	1.011	198
NOV	170	170	200	540	8.1	0.974	205
DEC	160	170	190	520	7.8	0.938	202
Total	2000	2200	2450	6650	100.0	12.000	2450
Average	167	183	204	185			204

(*K.W. Tunnell Company, Inc.*)

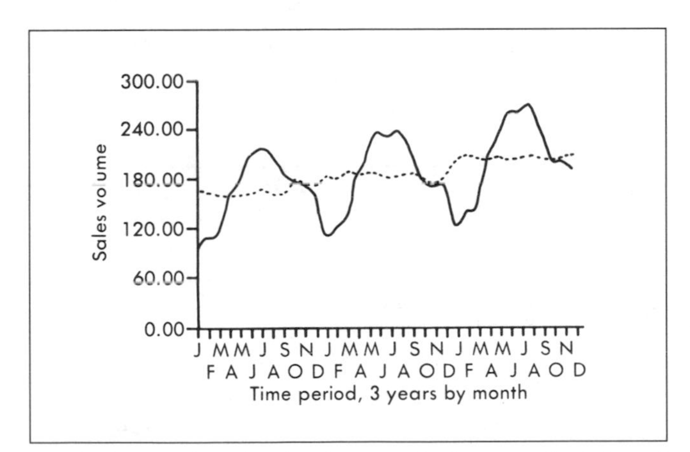

Fig. 20-2 Sales data showing deseasonalized curve. (*K.W. Tunnell Company, Inc.*)

STATISTICS

Statistical approaches can also range from quite simple to extremely complex. The basic premise of statistical forecasting lies in the assumption that the behavior of individual part demand variations will be random. This is expected whether or not the total forecast is stable or is subject to trend and seasonal factors. Such factors must be treated separately to expose the underlying randomness of individual part demand variations. The basic statistical techniques used in intrinsic forecasts will be examined in the paragraphs that follow. The principles involved also apply to extrinsic forecasting, but the statistical treatments appropriate to multivariate analysis are beyond the scope of this discussion. All of the components of a demand model are shown graphically in Fig. 20-3.

Figure 20-3 shows the various components of monthly historical demand. After statistically treating this demand data to remove the component of variation that can be recognized (classified) as ''seasonal,'' as was done in the last section, the remaining variation can be examined to determine if there is some recognizable trend or pattern of growth or decline. Finally, because all forecasts are subject to some error, the residual forecast errors can be examined and a judgment can be made on the validity of the forecast model.

Trend Patterns

The deseasonalized demand pattern can likewise be calculated for the prior year's data by month in preparation for an analysis of any trend pattern that may exist. This requires the selection of a method or calculation technique for the linear curve-fitting problem. The 12-month moving average can provide the data points for this calculation. Simple techniques of determining the slope of a line include the use of graphical methods and solving for an equation based on two points on the plotted trend line (see Fig. 20-4).

Another method of calculating the trend line for use in a forecasting model is to use the least square regression technique on the deseasonalized monthly data in the previous example. If the number of months is plotted as the X variable and the monthly quantities as the Y data, the graph in Fig. 20-5 results. To solve for the slope of the line using linear regression, the following calculation can be made:

$$Y = 159.0 + (1.389 \times X)$$

Graphically, the apparent linear relationship representing growth of volume over time is demonstrated by also plotting the equation on the same graph. By assigning the values 1 to 42 to the X variable, the trend line can be extended 6 months beyond the 36-month history as a trend forecast of deseasonalized demand.

When the equation for this model is used to project future months' deseasonalized demand and is multiplied by the appropriate monthly season factors, a forecast quantity can be determined. Table 20-3 shows results of the calculations extended for a 6-month period.

Residual Variation—Randomness

The example set of data in Table 20-3 can also be used to look at residual variation not explained by the model. This data can

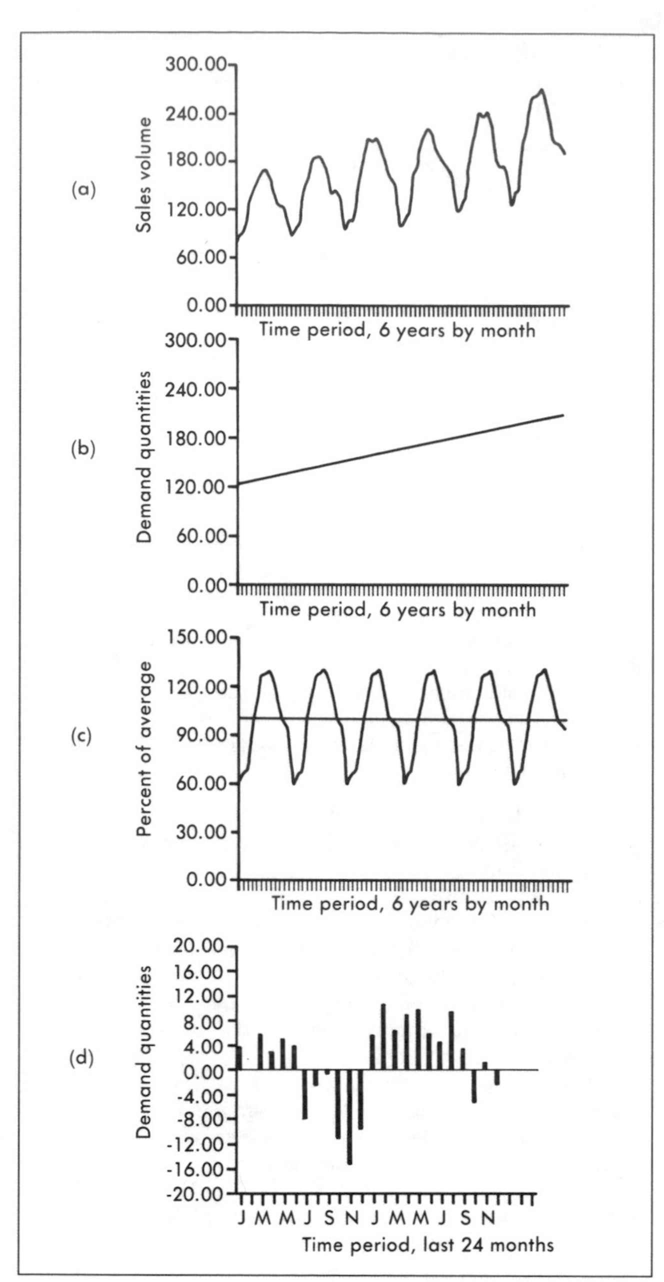

Fig. 20-3 Components of demand: (a) sales history, (b) trend line of demand, (c) seasonal factors, and (d) residual error. *(K.W. Tunnell Company, Inc.)*

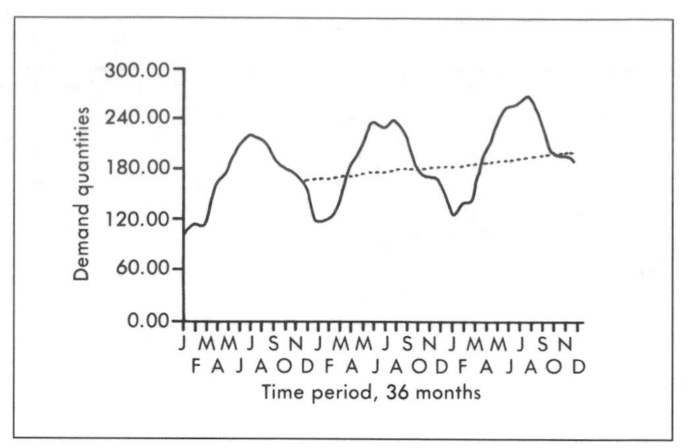

Fig. 20-4 Trend line graph from 12-month moving average. *(K.W. Tunnell Company, Inc.)*

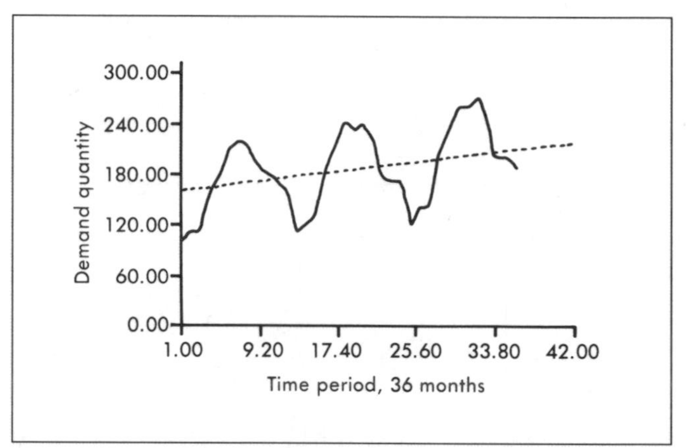

Fig. 20-5 Trend line forecast using the least square regression technique. *(K.W. Tunnell Company, Inc.)*

TABLE 20-3
Product Quantity Forecast Extended for Six Months

Future Month	Deseasonalized Demand Forecast	Seasonal Factor	Resulting Forecast
37—JAN	210	0.595	125
38—FEB	212	0.668	142
39—MAR	213	0.686	146
40—APR	215	0.974	209
41—MAY	216	1.119	242
42—JUN	217	1.281	278

(K.W. Tunnell Company, Inc.)

then be studied through a simple histogram to determine if the variation appears to be random. If so, the model depicts the statistical definition of the demand pattern. If not, it is necessary to look for other causes that are contributing to the demand pattern. The goal is to establish a tool or technique that can be expected to yield good results until new variables enter the statistical population that makes up the demand universe. This testing of the model, or any set of techniques used to prepare a forecast, must be reviewed routinely to determine if further improvement can be made. This will be measured by the forecast error rate actually experienced (see Table 20-4).

When the trend-seasonal model is compared to the actual results given in the example, the deseasonalized values experienced in the past 36 months show residual variation from the trend forecast (see Fig. 20-6). This distribution of errors is certainly not a perfect example of a normal distribution because of the slightly low frequency on the bar in the range of -1 to +2. There is a prevalent indication of the desired bell shape exhibited in the histogram. It is also known that small groupings of actual data from any normal distribution will usually not yield a perfect

TABLE 20-4
Tabulation of Results From Trend-Seasonal Forecast Model

Past Months	Actual Demand	Deseasonal Actual	Seasonal Factor	Deseasonalized Trend Forecast	Forecast Error
1—JAN	100	168	0.595	160	-8
2—FEB	110	165	0.668	162	-3
3—MAR	110	160	0.686	163	+3
4—APR	160	164	0.974	165	+1
5—MAY	180	161	1.119	166	+5
6—JUN	210	164	1.281	167	+3
7—JUL	220	172	1.281	169	-3
8—AUG	210	162	1.299	170	+8
9—SEP	190	162	1.173	172	+10
10—OCT	180	178	1.011	173	-5
11—NOV	170	174	0.974	174	0
12—DEC	160	171	0.938	176	+5
13—JAN	110	185	0.595	177	-8
14—FEB	120	180	0.668	178	-2
15—MAR	130	190	0.686	180	-10
16—APR	180	185	0.974	181	-4
17—MAY	210	188	1.119	183	-5
18—JUN	240	187	1.281	184	-3
19—JUL	230	180	1.281	185	+5
20—AUG	240	185	1.299	187	+2
21—SEP	220	188	1.173	188	0
22—OCT	180	178	1.011	190	+12
23—NOV	170	174	0.974	191	+17
25—DEC	170	181	0.938	192	+11
25—JAN	120	202	0.595	194	-8
26—FEB	140	210	0.668	195	-15
27—MAR	140	204	0.686	197	-7
28—APR	200	205	0.974	198	-7
29—MAY	230	206	1.119	199	-7
30—JUN	260	203	1.281	201	-2
31—JUL	260	203	1.281	202	-1
32—AUG	270	208	1.299	203	-5
33—SEP	240	205	1.173	205	0
34—OCT	200	198	1.011	206	+7
35—NOV	200	205	0.974	208	+3
36—DEC	190	202	0.938	209	+6

(K. W. Tunnell Company, Inc.)

Range	Frequency	Histogram
-14 to -17	1	*
-10 to -13	1	*
-6 to -9	6	******
-2 to -5	9	*********
-1 to +2	6	******
+3 to +6	7	********
+7 to +10	3	***
+11 to +14	2	**
+15 to +18	1	*

Fig. 20-6 Distribution of errors from the trend forecast. *(K.W. Tunnell Company, Inc.)*

picture. This simple analysis technique is, however, sufficient for practical testing of assumptions. As long as there is no convincing evidence that the dispersion is not bell shaped, the premise that the data appears to be from an essentially normal distribution can be accepted.

DEMAND/ORDERS

It is important to reconcile the forecast generated with the likely impact it may have on the company using the forecast. The most straightforward usage results from using order history data in preparing a forecast of order demand. This gives the best measure of the marketplace demand that can be expected to be presented to the manufacturer as new orders.

If the order history data is based on shipments and the shipping performance is poor, or if there is a significant offset between order receipt and order shipment, a similar offset must be used to arrive at an appropriate timing for the predicted order volume. The best measures of true demand will also include input that corresponds to lost sales volumes.

It is always important to ''consume'' a forecast with incoming orders to calculate a balance expected in the projectionperiod. This allows for early detection of forecast errors and appropriate reaction. It is equally important to measure error percentages as a part of the learning experience that is essential to an ongoing improvement of prognostication ability. The only way that forecasts ever improve is by forecasters using their forecast errors as signals to guide the improvement of the methods applied.

SPARES/SERVICE PARTS

To most manufacturers, the most troublesome independent demands are those that partially overlap with the core of dependent demand items. Spare parts requirements that do not overlap with the calculated requirements for components are often prone to be small in quantity and infrequent in occurrence. For low-volume, infrequent-demand items, an expert opinion forecast manually loaded into the system by period is usually the best approach. Demand patterns of similar items or groups are also useful in some cases.

For service parts or spare parts that have a history of repetitive demand, the same techniques of forecasting used for the primary products are appropriate. Some form of lifecycle analysis is also useful in understanding the demise of service parts requirements. It is important that the historical data representing this form of demand be maintained separately from the usage history on the same parts that is associated withdependent demand patterns that have been exploded (calculated) through a bill of materials (see Fig. 20-7).

OTHER TECHNIQUES

Numerous other forecasting techniques and variations of techniques are in use today by manufacturers and distributors of products. The advent of the computer in business applications in the later 1950s made many forms of analysis practical that were not particularly useful for more than half of this century.

One of the more interesting applications of the calculating power of computers dedicated to forecasting problems is described in *Focused Forecasting* by Bernard Smith.[1] The principle of his method is to arrive at as many models as possible to forecast the demand of any item, calculate a short-term forecast using each model, and then measure the actual results for a few past periods against the forecast calculated by each model. The comparison of results to forecast is then used to ''select'' the forecast model that will be used in the next period for the official forecast. The technique is highly adaptive, always focusing on minimizing forecast errors, but, as might be expected, it can consume large amounts of computer power.

Independent demand	Dependent demand
FORECAST = 450	USAGE = 2000
(Spares orders)	(Bill of materials explosion)

Part number 10101
TOTAL DEMAND equals - - - - - - 2450

Fig. 20-7 Example of a bill of materials for 1 month's demand. (*K.W. Tunnell Company, Inc.*)

Among the more popular techniques of forecasting that is associated with the advent of commercial computer application to forecasting problems, and often connected with material requirements planning (MRP) systems implementation, is the technique called exponential smoothing. Based on the mathematics associated with moving averages, various forms of the technique may be used with deseasonalized data to arrive at another form of the trend-seasonal model.

In exponential smoothing, a set of techniques is used to develop part forecasts using only a small amount of historical data.[2] An estimate of the moving average demand is developed each period based on a weighting of the current demand and the previous forecast. It is a lagging-indicator form of forecast, but the amount of responsiveness to shifts in demand is easily controlled through the selection of the factor alpha, a. In its simplest form, called *first-order smoothing*, the formula is:

$$NF_1 = OF + a\,(CA - OF) \qquad (1)$$

where:

NF_1 = new forecast (first-order smoothing)
OF = old forecast
a = alpha
CA = current actual forecast

Continuing the use of the data from the previous example in this chapter, the fourth-period forecast can be calculated if a value for alpha is chosen and the forecast for the third period is assumed to be 112 as determined by the curve-fitting trend-seasonal model. With 110 given as the third-period actual demand, then the calculation is as follows:

$$NF = 112 + 0.5\,(110 - 112) = 112 - 1.0 = 111$$

Because the actual demand for period 3 was 110, this small variation caused only a small change in the forecast, from an old forecast of 112 to the new forecast for period 4 of 111.

The alpha factor is the key variable in determining the sensitivity that the model has to the actual demands occurring in the most recent periods. The alpha factor choice can be equated to an approximation of the number of periods to be included in the ''weighted average'' estimate to be calculated. The formula for this relationship to the number of periods is:

$$\text{alpha} = 2 / (\text{periods} + 1) \qquad (2)$$

Based on Eq. (2), the value of the alpha factor for various periods is given in Table 20-5. The more stable the demand pattern, the smaller the value of the chosen alpha. An alpha factor of 0.5, as chosen in the example calculation, gives heavy weighting to the current actual demand as it attempts to estimate the moving average for only the past three periods. The smaller the alpha factor, the greater the ''lag'' in responsiveness to changes experienced in actual demand.

Second-order smoothing introduces a factor to adjust the forecast for a trend and to eliminate part of the lagging associated with the technique. The forecast using second-order smoothing techniques can be calculated with the following equation:

$$NF_2 = 2\,NF_1 - NB \qquad (3)$$

where:

NF_2 = new forecast (second-order smoothing)

NF_1 = new forecast (first-order smoothing)

NB = new "B" correction

The value for NB can be calculated as follows:

$$NB = OB + a\,(NF_1 - OB) \qquad (4)$$

where:

OB = old "B" correction

First-order and second-order calculations on the deseasonalized data calculated for a 36-month example, and extended for an additional 3 months, are summarized in Table 20-6 on p. 20-8.

TABLE 20-5
Estimate of Weighted Averages

Number of Periods to be Included	alpha Factor
2	0.67
3	0.5
4	0.4
5	0.33
6	0.29
7	0.25
8	0.22
9	0.2
.	.
.	.
19	0.1

(*K. W. Tunnell Company, Inc.*)

The curves for the forecast track very well to the actuals experienced using a short-cycle influence on the moving average estimate with an alpha factor of 0.5 (see Fig. 20-8). In this example, smaller values of alpha increase the forecast error because of the significant trend component that is present.

There are several ways that the accuracy of a forecast can be measured. The simplest is the *maximum deviation* occurring when the fit of the model is tested to the actual historical data. Other methods include calculation of the *average error* or the *standard deviation* of the estimate. For the example data, Table 20-7 on p. 20-8 lists the errors resulting from the use of exponential smoothing with several alpha factors. The analysis of deviations is helpful in determining the alpha factor to be chosen, which will represent the best model of actual demand.

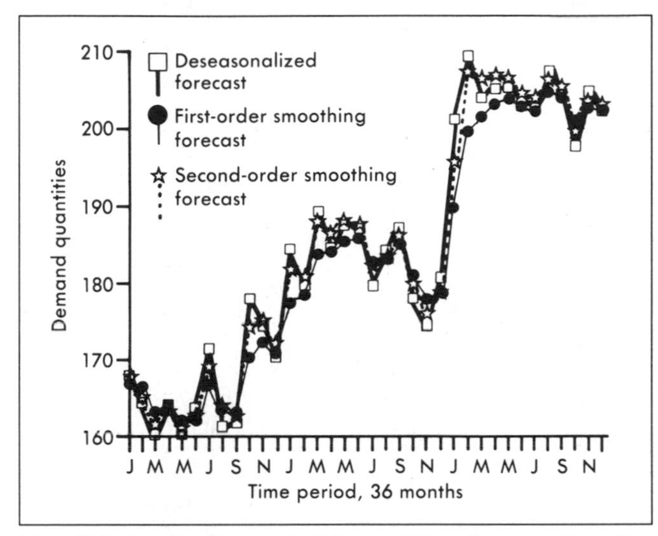

Fig. 20-8 Graph of exponential smoothing forecast data. The second-order smoothing graph lags the actual graph less than the first-order smoothing graph. (*K.W. Tunnell Company, Inc.*)

AGGREGATE PLANNING AND MASTER SCHEDULING

Top-level planning in a manufacturing organization is essential to the smoothness of both its floor and support operations. The performance of actual shop order completions vs. the detail plan as well as the material flow generated through the combination of the production/inventory control and purchasing functions can only be as successful as the top management direction that guides those detail processes.

Business requirements planning, or production, sales, and inventory (PSI) planning, is the activity that feeds management priority data into the master scheduling segment or top level of a material requirements planning approach (see Fig. 20-9 on p. 20-9). These are the decisions that determine how finished goods are going to be controlled. The material requirements planning (MRP) system or any other detailed parts planning system can then be used to control all levels governing the part requirements support for finished goods production.

SALES PLAN

A myth has long existed that the master schedule is the business plan. The source of the master schedule must begin at an executive group level that operates with product group data only and must be finalized by a top management review process. This responsibility may be vested in a single executive or it may be a formal mechanism for negotiation tied to the sales forecasting processes.

If it is a formal process that involves sales and marketing negotiations with production management, ideally with finance providing the link to the overall business plan, it utilizes the chief operating executive as arbitrator for the natural conflicts arising from the specific objectives of each function. The master schedule itself is not the business plan, but should be formally developed to meet the objectives set forth in the plan proposed by top management.

TABLE 20-6
Deseasonalized Demand Forecasts—Exponential Smoothing Models

Past Months	Actual Demand	Seasonal Factor	Deseasonalized Demand	First-Order Forecast	Second-Order Forecast
1—JAN	100	0.595	168	168	168
2—FEB	110	0.668	165	166	166
3—MAR	110	0.686	160	163	162
4—APR	160	0.974	164	164	163
5—MAY	180	1.119	161	162	161
6—JUN	210	1.281	164	163	163
7—JUL	220	0.595	172	167	169
8—AUG	210	0.668	162	165	164
9—SEP	190	0.686	162	163	162
10—OCT	180	0.974	178	171	174
11—NOV	170	1.119	174	173	175
12—DEC	160	1.281	171	172	172
13—JAN	110	0.595	185	178	182
14—FEB	120	0.668	180	179	181
15—MAR	130	0.686	190	184	188
16—APR	180	0.974	185	184	186
17—MAY	210	1.119	188	186	188
18—JUN	240	1.281	187	187	188
19—JUL	230	0.595	180	183	182
20—AUG	240	0.668	185	184	184
21—SEP	220	0.686	188	186	187
22—OCT	180	0.974	178	182	180
23—NOV	170	1.119	174	178	176
24—DEC	170	1.281	181	180	179
25—JAN	120	0.595	202	191	196
26—FEB	140	0.668	210	200	208
27—MAR	140	0.686	204	202	207
28—APR	200	0.974	205	204	207
29—MAY	230	1.119	206	205	207
30—JUN	260	1.281	203	204	204
31—JUL	260	0.595	203	203	203
32—AUG	270	0.668	208	206	207
33—SEP	240	0.686	205	205	205
34—OCT	200	0.974	198	202	200
35—NOV	200	1.119	205	203	203
36—DEC	190	1.281	202	203	203
37—JAN		0.595		203	203
38—FEB		0.668		203	203
39—MAR		0.686		203	203

(*K. W. Tunnell Company, Inc.*)

TABLE 20-7
Forecast vs Actual Deviations Analysis

	First-Order Smoothing			Second-Order Smoothing		
alpha	Maximum	Average	Standard	Maximum	Average	Standard
0.5	22	5.1	7.2	22	5.2	7.0
0.4	22	5.3	7.5	23	5.1	7.1
0.3	23	5.9	8.0	22	5.1	7.3
0.2	25	7.1	9.1	21	5.4	7.6
0.1	29	9.9	12.1	23	6.6	8.6

(*K. W. Tunnell Company, Inc.*)

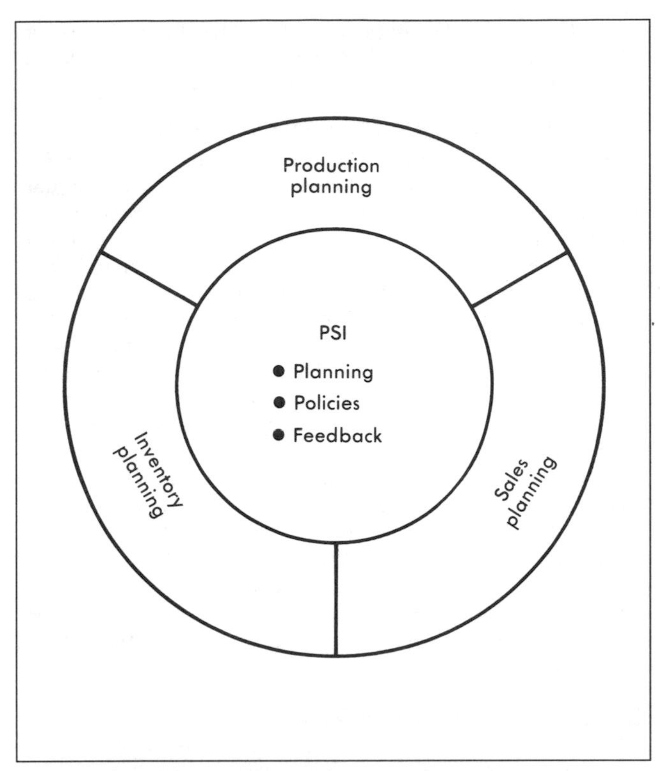

Fig. 20-9 Production, sales, and inventory (PSI) planning process schematic. (*K.W. Tunnell Company, Inc.*)

There have been many instances of companies attempting to operate an MRP system by driving a sales forecast directly into the master schedule. The one thing certain about any forecast is that it is subject to error. Therefore, when it comes to trying to run a plant schedule that is made solely from a forecast exploded against lead times, then directly translated into ordering actions, the schedules will need to be changed constantly, and expediting will be used to keep the plant running.

The sales plan begins with a shipments forecast, but should end with a process that recognizes the need for tradeoffs withproduction management and matching of flexibility requirements with true factory capability. This is necessary to achieve a reliable plan that supports the customer service requirements, objectives, and delivery commitment communication needs. The most important sales advantage of reliable planning is the consistent achievement of deliveries made on the promised ship dates. Customers, usually seen as another company's purchasing department, judge the reliability of a suppliers' promises as a critical factor in determining the choice of a supplier.

Techniques for developing demand forecasts were covered in the previous section. It is important to remember the principle of forecasting related to groups vs. detail. Management can usually accurately estimate the product group demand for the short-term planning horizon. This level of forecasting is appropriately used as an internal measure of management's control of its operations. Using an assessment and negotiation process, the critical resources of the business can be dealt with in a proactive manner, and the true capability of the manufacturing unit can be matched to the flexibility requirements of the marketplace to the maximum degree possible. When management uses some form of this negotiation process in preparing for the authorization of a master production schedule, it forces itself to deal with foreseeable problems before they occur.

INVENTORY PLAN

The inventory plan element of the top-level planning process is a critical tool in a management negotiation process. By considering the variations in requirements of the marketplace, inventory policies and overall inventory levels can be set that will allow the operations unit to satisfactorily meet those requirements. Further, this allows the investment in inventory to be managed, not just happen as a collective result of detail-level policies and procedures.

The level of inventory investment should be determined by the amount of mismatch between current manufacturing capability and current market flexibility requirements. If the manufacturing operation can quickly change to meet the needs demonstrated by orders entered, little finished or semifinished product inventory is required. If there is a large seasonal component in the demand function, it may be necessary to build up significant inventories to level the production requirement. This is particularly true if there is a high degree of skilled labor involved in the production process or if the in-process lead times are long.

This planning focus on inventory provides a basis for calculating tradeoff costs between (1) inventory-carrying authorization and (2) changes in the plant or in the workforce intended to shorten its response time capability. It can certainly be argued that as a plant approaches the ability to shift its detail plan in a matter of hours, or maybe even days, almost no finished product inventory is required. Decisions to carry fixed levels of long lead time materials or categories of semifinished product often result from the recognition of the associated investment tradeoffs.

PRODUCTION PLAN

The production plan is the final result of the top-level planning process. Once the changes in anticipated demand are recognized through analysis of the product group forecasts and once the major inventory planning decisions are determined, the production plan is essentially decided. In truth, this is the final test of the decisions already made. If the production plan, so derived, is still not feasible, other changes will need to be negotiated. If shortfalls to certain forecasted demands are still predicted in the final plan, sales and marketing may need to make direct contacts with customers in search of opportunities to delay some of the problem demands without harm. This method of reconciliation of demand to capacity is far superior to order delays caused by missed shipping dates.

Given the stability of a firm plan that has been equated to a realistic assessment of production capability, there is a responsibility to measure actual results against the plan, just as one would measure forecast error for each of the product groups. An excellent indication that good management ability is available, demonstrating skills in both planning and execution, is if both measures attain an accuracy record of 90-95% or better on all product groups (when actual results are measured against the last month's plan). Until such reliability is achieved in the short-term planning processes, the deviations from plan can be utilized as indexes of the weaknesses in predictive capacity of those that control the processes. This form of discipline in top-level planning leads to prioritization of need resolution regarding opportunities for stabilizing the production environment.

Production plans so developed will have the benefit of having already been communicated through the top management ranks as part of the process. Likewise, the strengths and weaknesses of the plan are equally known, as is the probability of satisfactory achievement of the support functions such as purchasing. The production and inventory planning and control function is thus positioned to develop achievable detail-level plans.

AUTHORIZATION OF MASTER SCHEDULE

The key link between a management team's negotiated plan and the detail processes is the *master production schedule* (MPS). The purpose of the MPS is to plan the replenishments for the lower level components and assemblies to meet the production plan. The latter plan is usually expected to be updated on a weekly, but sometimes monthly, basis. The master schedule must translate the product group-level plan into one that operates at a part number level. This plan will consider open orders and will use part number level independent demand forecasts that may or may not be derived using the same techniques as those used for the product group forecasts.

The master scheduler, or person responsible for this detail-level process, must also summarize the detail plans at the product group level to ascertain compliance with the production plan set forth by the management team. If deviations to the management plan beyond an agreed tolerance level are indicated, the master scheduler must obtain approval from management. It is appropriate to include the master scheduler in meetings leading to the production plan so that he or she may be completely aware of the reasons behind the decisions contributing to a given plan.

ROUGH-CUT CAPACITY PLANNING

The detail-level tests of the feasibility of the master schedule must include an evaluation of the production resource consumption by day or week. This is referred to as rough-cut capacity planning. It may be conducted using product group data if workcenter loading factors have been developed accordingly. This testing of capacity gives an indication of the loading levels, both for personnel and for machine groups involved. The concept of underscheduling may be introduced at this point, allowing a small portion of capacity to be allocated to emergency or other orders.

Many software packages sold to handle the MRP system contain a module that addresses rough-cut capacity planning from a part number level. The difference between the rough-cut requirements planning and detail capacity planning functions lies in the definition data used. Rough-cut planning uses a simple bill of labor (and/or machine) hours instead of the finite definition or operations specified in the process routings data. Only critical resources are generally planned to simplify the calculation process. Further, rough-cut planning does not consider the netting of available work-in-process inventories as a reduction of the load, as is done in post-MRP runs where inventory netting has occurred at all levels of the bill of materials.

FUNCTION OF THE MASTER SCHEDULE

The primary documentation of a factory plan of operation for a period of time equal to its longest lead times is the master schedule (see Fig. 20-10). The master production schedule, sometimes abbreviated MPS, is the actual top level of input into an MRP system. This technique uncouples factory demands from the sales forecasts and order entry elements of the planning information chain. Before this process can be effective, the planning exercise at the management level that determines in broad terms what will go into the master schedule must be completed. This framework, or production, sales, and inventory plan, is the basis for the master scheduler's actions in allocating specific orders and stock units to time slots for production.

The master schedule is not a production budget. Once a year it may be tied to the production budget to provide calculations related to the production plan, but from then on the plan is also considered to be a forecast. Without the top-level planning approach and without some discipline and philosophy in this aspect of demand planning, results from the MRP system will be very nervous in terms of action messages, thereby harming both the shop floor and outside suppliers.

PLANNING TIME FENCES

A final definition is required to understand the elements of the factory production planning process. "Time fences" are merely designations of time periods that allow different rules to be related to production planning and scheduling. The first time fence, or end date of a period beginning with the current day, defines the duration of the period for which shop orders have or can be released. This implies that whatever documentation is required on the shop floor to initiate and control the performance of work is already published to cover this period.

The second time fence defines the end of the period that begins immediately after the released period and includes the time in which all orders are firmed. The orders in both the released and firmed period are expected to be completed on time and without the interference of schedule changes. Both periods are usually only a few weeks in duration. Purchasing support must provide material for both "released" and "firm" orders.

The final time fence is the same as the end of the planning horizon. This planned order period begins immediately after the "firmed" period. Figure 20-11 pictorially clarifies the relationship of the time periods defined by time fences.

Time fences provide for policy-level statements regarding the period usually referred to as the "frozen schedule" period. This

	Periods..........												
Part Number	1	2	3	4	5	6	7	8	9	10	11	12	13
A1	35	40	45	55	45	35	30	25	30	30	30	30	30
A2	50	45	35	30	35	45	45	50	50	50	50	50	50
Totals	85	85	80	85	80	80	75	75	80	80	80	80	80
	Released		.... Firm				.... Planned						

Fig. 20-10 Example of a master production schedule. (*K.W. Tunnell Company, Inc.*)

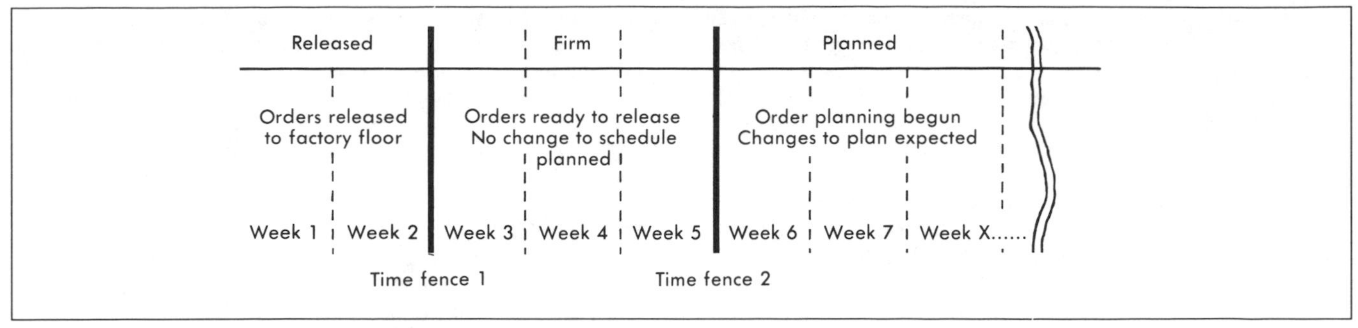

Fig. 20-11 Schematic of production planning time fences. (*K.W. Tunnell Company, Inc.*)

policy recognizes that there are limitations inherent in the factory environment that require some period of stability in the scheduling. Little or no flexibility exists for change in the first period. The firm period has only the small amount of flexibility that has been created by underscheduling to allow for emergency order processing from standard or common materials. In the final period, it is anticipated that almost any reasonable level of demand pattern change can be accommodated.

Ignoring the current ''true capability'' of a plant is an extremely poor management practice. Improving that capability to minimize the stable schedule period requirement is a reasonable management priority. This allows a faster response to the changing needs of customers without the costs associated with disruptive expediting.

PULL METHOD DIFFERENCES

The discussions of master scheduling and time fences assume that an MRP system is used as the tool to develop part schedules. Another way to schedule the part replenishment release authorization is with a pull system. Pull systems focus on the production of only those parts the next period or to replace those that have just been taken for another area. If the pull methods are used to schedule the part replenishment release authorization for parts supply, master scheduling will be used to drive only the demand planning functions related to capacity planning and supplier communications regarding materials forecasts. It is not necessary to implement full-blown MRP systems to gain the benefits of aggregate planning (PSI planning), master scheduling, and rough-cut capacity planning.

REQUIREMENTS AND CAPACITY PLANNING

The detail planning elements associated with production and material control are essentially those efforts that contribute to the balancing of limited resources with requirements that are generated through the sales activities. Timing is always the critical variable. Costs associated with any set of alternatives are a function of the time elements. Materials can be expedited at additional costs. Labor resource levels can usually be fluctuated significantly in short periods of time, but this adds costs related to hiring, training, and the loss of quality and productivity during training periods.

In this section, the various means and techniques for planning related to personnel and machinery used to support production activities will be examined. Capacity planning provides input into the management process that begins with forecasts of sales activity and is combined with productivity-related projections from manufacturing engineering. Significant process changes, such as the introduction of automation into an area formerly based on benchwork, can create a major shift in personnel requirements and often in overall manufacturing lead time. Changes in internal lead time requirements can have dramatic effects on inventory investment and storage requirements.

DETAIL PRODUCTION PLANNING

The process of production planning involves tradeoffs between changes in production rates and inventory investment. Seasonal demands require that a choice be made from the alternatives available. Some of the alternatives are as follows:

- Hold a level production rate and build up an inventory sufficient to cover the period of peak demand.
- Hold a near-level inventory investment and fluctuate the labor supply to meet the sales rates.
- A combination of the previous two, which allows some use of overtime instead of hiring for a portion of the labor fluctuation.

Production plans must consider the economics and feasibility of the alternatives as they relate to inventory investment, storage capacity, purchased component availability, and personnel availability. The longer the total manufacturing process time and corresponding lead time, the more complex the problem. Likewise, the problem is more pronounced if the labor requirement is skilled rather than unskilled. Production scheduling alternatives are always limited to some degree by the lead times required to obtain parts and materials from suppliers. In some cases, the final restriction comes from the possibility of exceeding the capacity of a key material supplier.

PERSONNEL PLANNING

Rough-cut capacity planning gives the first look at personnel requirements occurring during the period defined by the planning horizon. Budget planning may require a separate but similar analysis taken from a business plan forecast of product group demands. It is always wise to develop several key experience factor tables for use in personnel planning decisions. It is not sufficient to depend solely on the detail outputs of the capacity planning module of a manufacturing resource planning system.

Because personnel requirements vary with the inherent productivity of the resources and the processes used, shifts in mix can be equally as important as major changes in the processes. The top-level index of personnel required may be as simple as a calculation of the number of direct (or touch) labor resource hours required to produce a standard level of output—for example, 1000 units, $100,000 in sales, or 1000 "standard" process hours. A separate index for the amount of indirect support labor may be defined using the same factors or using a secondary factor such as the number of indirect hours required per 1000 direct hours or number of indirect "heads" per direct "head." An extremely useful refinement of these gross measures may be developed by using a similar set of factors for each product group contributing to the total. This allows for an early indication of the implication of product line mix changes in a manner that is easily handled in the development of management input into the routine planning process. Table 20-8 is an example of the group-level capacity planning tool. Such planning is generally done using the rough-cut capacity planning tools associated with the master production schedule.

The material content may vary significantly in the product cost buildup, and even minor changes in the dollar volume planned for a manufacturer can cause problematic swings in personnel requirements. The next period planned for in the previous example shows the significance of such a change in the forecasted demand (see Table 20-9). If the production scheduler waits until all of the inputs have been processed through some computer system to begin to take action, valuable time may be lost and that may ensure that the change in plan is not feasible. Using group-level data in the examples provided in Tables 20-8 and 20-9, it is possible to estimate that a volume increase of 3.0% with a shift in mix will require a 7.1% increase in direct labor hours supported by a 4.1% increase in indirect labor hours.

MACHINE LOADING

Loading of machine groups may be treated in a similar fashion for early warning of potential short-term capacity problems, which might be easily overcome by the use of outside resources if sufficient time is available to make such arrangements. If the manufacturing environment is one that combines make-to-stock on its higher volume items with its basic make-to-order operations, additional flexibility is available at the expense of additional inventory. Table 20-10 is a simple analysis table used to describe anticipated loads on key workcenters resulting from the demands presented in Table 20-8 after an allocation of machine hours to the appropriate groups.

The next period described in the personnel plan example called for an output of $515,000 (refer to Table 20-9). Although the overall change was only 3.0%, Table 20-11 shows an increase of 4.6% in the machine group loads. If the situation was further complicated, as usually happens in reality, by identifying the fact that the capacity of machine group L01 was almost completely overloaded with a capacity of only 2304 machine hours, an overloading situation exists in this area by 10.9%. Unless the processes are changed on some of the work or some of the operations are subcontracted, some of the shop orders planned for the period will be late. Such is the usual dilemma of the production scheduler. If this example were identified as

TABLE 20-8
Group-Level Capacity Planning Tool Example

Product Group	Personnel Hours per $1000 Output	Indirect Ratio	Current Plan	Direct Required	Indirect Required
01	2.5	1:5.0	$200,000	500	100
02	20.0	1:4.0	$100,000	2000	500
11	10.0	1:2.5	$120,000	1200	480
14	7.5	1:8.0	$ 80,000	600	75
		Totals	$500,000	4300	1155

(K. W. Tunnell Company, Inc.)

TABLE 20-9
Revised Personnel Requirements Plans

Product Group	Personnel Hours per $1000 Output	Indirect Ratio	New Plan	Direct Required	Indirect Required
01	2.5	1:5.0	$194,000	485	97
02	20.0	1:4.0	$115,000	2300	575
11	10.0	1:2.5	$110,000	1100	440
14	7.5	1:8.0	$ 96,000	720	90
		Totals	$515,000	4605	1202

(K. W. Tunnell Company, Inc.)

TABLE 20-10
Analysis of Anticipated Loads on Key Workcenters

Machine Group	Machine Hours per $1000 Output	Current Plan	Machine Hours Required
A01	12.0	$130,000	1560
D01	9.0	$180,000	1620
L01	16.0	$ 95,000	1520
L02	24.0	$ 95,000	2280
Totals		$500,000	6980

(*K.W. Tunnell Company, Inc.*)

TABLE 20-11
Revised Machine Load Analysis

Machine Group	Machine Hours per $1000 Output	New Plan	Machine Hours Required
A01	12.0	$194,000	1494
D01	9.0	$115,000	1638
L01	16.0	$110,000	1668
L02	24.0	$ 96,000	2502
Totals		$515,000	7302

(*K.W. Tunnell Company, Inc.*)

being part of a just-in-time manufacturing operation, the schedule change is likely not feasible because there are no buffer inventories for protection. This demonstrates the need to have a stable schedule, particularly in a JIT shop, that is at least equal to the real capability to respond to desired changes.

If a plant is not cellularized and operates as functional departments preparing manufactured components for use in subsequent manufacturing processes, the part demands at lower levels will be lumpy and infrequent. This adds to the variation of the normal demand deviation and is more difficult to estimate through simple methods. The nature of this form of capacity planning requirement is the one that can benefit from the detail capacity planning module that follows an MRP run. However, it is usually late enough in the planning process that it should be considered to be very much a reaction-oriented solution process; hence, expediting reigns supreme in the traditional manufacturing environment.

The ultimate solution to the production planning process as it relates to machine loading may well be in the development of cellular flows that have a primary focus on minimizing set-up time and overall lead times. The concept of the unitary lot size is as well founded in reducing the coordination and planning costs associated with functional alternatives as it is with the floor control simplicity. The fundamental difference in the JIT manufacturing philosophy from that of batch processing oriented MRP approaches is related to the manufacturing logistic problems that are never solved when lead times and lot sizes are allowed to exist at greater than the lowest possible levels.

BOTTLENECK RECOGNITION

Processes that control the output of a manufacturing entity are true bottlenecks. Often found by piles of in-process inventory sitting in the queue, these islands are always a trouble spot for the planner, the scheduler, and the operator. Eliyahu Goldratt has summarized the situation of bottlenecks in *The Goal* by indicating that an hour saved in a bottleneck operation is equivalent to an hour's worth of total output of a factory.[3] This perspective is as appropriate to the planner or scheduler as it is to the operations chief. Every possible technique is appropriate if it causes a true bottleneck operation to be "running" in lieu of "in set-up," "waiting for material," or "waiting for an operator." From the capacity planning viewpoint, totals of standard machine hours, including set-up, would be expected to be near or in excess of 100%.

Solutions to bottlenecks are usually costly. At the top of the curve, duplication of the bottleneck machine or machine group facilities would be expected. At approximately the same total cost, it might be possible to acquire a more capable replacement of the equipment. Much lower cost alternatives might be accomplished through shifting some of the demands from "make" requirements to "buy." Often some of the least costly, but perhaps partial solutions, come in concentrated efforts to reduce set-up time, to give special attention to the bottleneck's scheduling and coordination requirements, and to stagger breaks and lunch periods.

CAPACITY ALTERNATIVES

The long-term view of capacity considerations requires decisions that are strategic in nature. In a world economy that has numerous examples of overcapacity situations, more is not necessarily better. Short-term capacity alternatives must consider outside vendors as potential solutions. The possibility also exists for improvement of the productivity of current facilities through efforts involving all employees using techniques for set-up reduction. Many firms have found that previously mothballed equipment is an inexpensive way of providing equipment for manufacturing cells. Such usage typically places less demand on individual pieces of equipment than on the traditional or functional layout approach, which depends on larger lot sizes and longer runs.

New products and new product lines often offer the basis for reasoning that justifies the addition of plant capacity. The decision processes utilized weigh the associated risks as with any speculative venture. In industries where overcapacity situations already exist, this alternative for expansion involves higher risks. As in other production planning processes, a combination of approaches is often used to minimize the potential impact of the decision without giving up all of the perceived benefits.

Other forms of capacity restrictions include warehousing, floor space, and critical labor shortages. Each form requires its own approach.

PURCHASED PARTS PLANNING

For as many decades as industrial production has existed, a special focus has been required for the coordination of purchased materials and component part inputs to an individual manufacturer's process. An early written mention of a "materials man" dates from the 1820s. Glass and pottery factories were highly adept at acquiring required materials, both in the U.S. and abroad, during the 19th century. Early in the 20th century, Henry Ford introduced not only the assembly line process, but a high degree of vertical integration as well. The conversion processes took raw material input as iron ore and prepared parts such as automobile frame components that were then used on the assembly line only a few days later.

In the 1950s, many purchased parts planning techniques that had not been feasible because of the clerical and/or analytical time required became possible for many manufacturers because of the decreasing cost of the computer. Time-phased order-point techniques were an outgrowth of this new business capability. From this beginning, the techniques called material requirements planning (MRP) and eventually manufacturing resource planning (MRP II) evolved.

In fact, MRP can be thought of as a simulation tool that describes the interactions and dependencies involved in the material planning and material flow process. The technique of chaining the elements of a product's structure together as relationships between components and their ''parent'' part numbers led to the achievement of workable computer-based, time-phased requirements planning and manufacturing control systems. The ability to offset demand variables with lead time gave rise to a realistic representation, in a printed description, of the material flow.

MRP AND JIT

The planning and control systems techniques that are based on computer-aided manipulation of data in time-phased ''buckets'' are now about 25 years old. It has become common practice to refer to such systems as some form of an MRP system. The introduction of just-in-time manufacturing principles into U.S. manufacturing facilities during the past few years has created a different set of needs for such tools than those concentrated on before 1980.

MATERIAL REQUIREMENTS PLANNING

This section describes the elements, evolution, and application of the popular time-phased manufacturing control systems, usually referred to as material requirements planning (MRP).

Typical MRP Systems

The *material requirements plan* is the source of the name of the techniques based on time-phased planning, and it originates from the components requirements planning needs of a manufacturing organization (see Fig. 20-12). While this is one of the major outputs of a manufacturing control system, MRP does not create a good manufacturing plan. It is a tool that can be used to provide the means to assist good planning and communication procedures.

It is important to recognize that MRP is not for all businesses. There are single-product firms with simple bills of materials that do not require MRP for manufacturing management activities related to planning and scheduling.

MRP approaches can be mixed with other techniques. Most software houses with MRP products are working on new ''back-end'' reporting techniques that are more appropriate to factories implementing Japanese-like manufacturing approaches. The traditional shop floor control approach has probably been the most difficult module of an MRP system to manage. It has also been the cause of failure in many implementation efforts. It does *not* work well with just-in-time methods of shop floor execution and must be replaced by other techniques.

MRP is not just for replenishing inventories. The past decade has seen the focus shift from simpler material requirements planning systems to total integrated manufacturing control systems. Most current references are to ''MRP II,'' implying *manufacturing resource planning*. Most of the implementation efforts that are in process today are utilizing MRP II software resources. The capacity planning functions and the activities related to personnel planning are critical to the real needs of a manufacturing organization. To avoid further confusion with the acronym, MRP will be used as a generic notation covering both types of systems—material requirements planning and manufacturing resource planning.

MRP is an exception-message, action-oriented system (see Fig. 20-13). It is not a clerical-burdened system unless it is improperly implemented. The discipline required in the management of this process becomes one of managing the variables' data associated with the control process. MRP techniques provide a method for routine calculation of:

- The ''when'' (timing of an order release).
- The ''how much'' (quantity required for order release).
- ''Priorities'' updated with the results of all purchasing and shop activities in a traditional or functional environment.

Properly utilized, this systems approach provides shop loading data to help plan capacity and to make judgments about the alternative uses of available capacity. It allows for the varying of demands and/or rules and/or lead times as a means of simulating results by asking ''what if'' questions.

MRP is a set of procedures, a set of decision rules and policies that govern many of the routine decisions required in setting the manufacturing schedule. As such, it provides a highly disciplined approach for arranging lower level factory schedules. Its exception-action orientation is not a clerical system in nature, although when out of control, an MRP system can become a tremendous clerical burden. One of the critical aspects of the definition is that it is a highly disciplined management process; MRP depends on shop events happening just as they were simulated by the computer system based on the plans entered as the MPS and the policies and operations data loaded in their databases.

MRP system logic. The end-item replenishments shown in the master production schedule (MPS) are exploded using the *bill of materials* (BOM), which contains records describing the parent/component and material relationships (see Fig. 20-14 on p. 20-16). The *process routings* describe the operations to be performed and labor/machine hour content. The *workcenter* definitions for those operations complete the basis for shop loading information.

MRP elements. Given the rules and procedures that have been implemented for a given company, MRP determines the time and the quantity for order releases at lower levels and part manufacture requirements supporting the finished product schedules. MRP is not for controlling finished goods. All MRP systems are driven by a master scheduling approach, which is a finished product scheduling technique that must be developed in some manner to create the top-level demands that will be supported by MRP's subsequent arithmetic calculations.

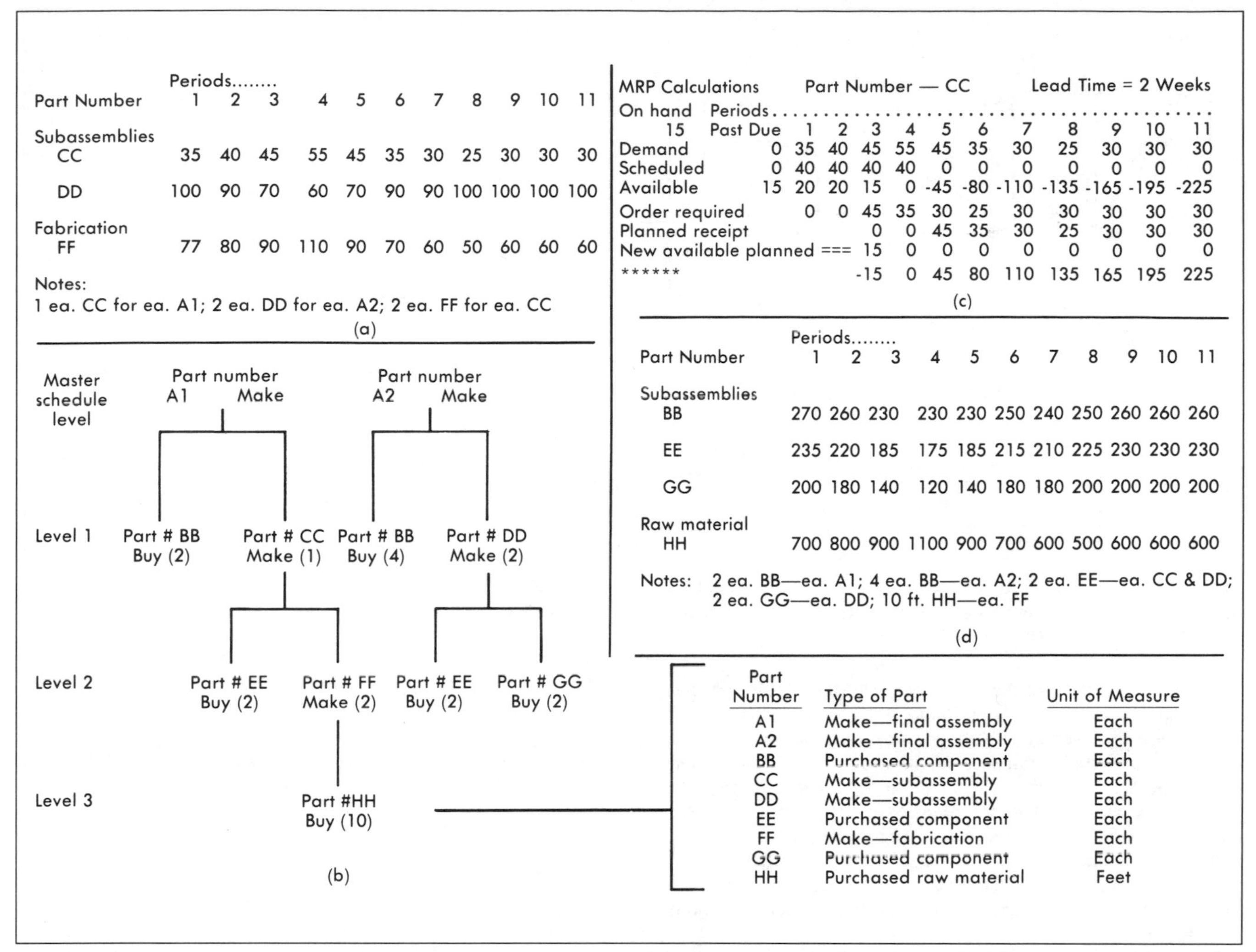

Part Number	1	2	3	4	5	6	7	8	9	10	11
	Periods........										
Subassemblies											
CC	35	40	45	55	45	35	30	25	30	30	30
DD	100	90	70	60	70	90	90	100	100	100	100
Fabrication											
FF	77	80	90	110	90	70	60	50	60	60	60

Notes:
1 ea. CC for ea. A1; 2 ea. DD for ea. A2; 2 ea. FF for ea. CC

(a)

(b)

MRP Calculations Part Number — CC Lead Time = 2 Weeks

On hand 15	Periods........ Past Due	1	2	3	4	5	6	7	8	9	10	11
Demand	0	35	40	45	55	45	35	30	25	30	30	30
Scheduled	0	40	40	40	40	0	0	0	0	0	0	0
Available	15	20	20	15	0	-45	-80	-110	-135	-165	-195	-225
Order required		0	0	45	35	30	25	30	30	30	30	30
Planned receipt				0	0	45	35	30	25	30	30	30
New available planned ===				15	0	0	0	0	0	0	0	0
******				-15	0	45	80	110	135	165	195	225

(c)

Part Number	1	2	3	4	5	6	7	8	9	10	11
	Periods........										
Subassemblies											
BB	270	260	230	230	230	250	240	250	260	260	260
EE	235	220	185	175	185	215	210	225	230	230	230
GG	200	180	140	120	140	180	180	200	200	200	200
Raw material											
HH	700	800	900	1100	900	700	600	500	600	600	600

Notes: 2 ea. BB—ea. A1; 4 ea. BB—ea. A2; 2 ea. EE—ea. CC & DD; 2 ea. GG—ea. DD; 10 ft. HH—ea. FF

(d)

Part Number	Type of Part	Unit of Measure
A1	Make—final assembly	Each
A2	Make—final assembly	Each
BB	Purchased component	Each
CC	Make—subassembly	Each
DD	Make—subassembly	Each
EE	Purchased component	Each
FF	Make—fabrication	Each
GG	Purchased component	Each
HH	Purchased raw material	Feet

Fig. 20-12 Material requirements planning example: (a) MRP demand calculations for parts to be made, (b) bill of materials, (c) requirements for part number CC, and (d) MRP demand calculations for purchased materials. (*K.W. Tunnell Company, Inc.*)

Part #	Period	Action Required	Quantity
..	..	..	..
..	..	..	..
..	..	..	..
CC	Week 3	Place order for delivery in Week 5	45
CC	Week 4	Place order for delivery in Week 6	35
CC	Week 5	Place order for delivery in Week 7	30
CC	Week 6	Place order for delivery in Week 8	25
CC	Week 7	Place order for delivery in Week 9	30
CC	Week 8	Place order for delivery in Week 10	30
DD	De-expedite	Reschedule order due in Week 2 to Week 3	125
DD	De-expedite	Reschedule order due in Week 3 to Week 4	125
DD	De-expedite	Reschedule order due in Week 4 to Week 6	125
DD	Expedite	Reschedule order due in Week 10 to Week 8	150
DD	Week 7	Place order for delivery in Week 9	100
DD	Week 8	Place order for delivery in Week 10	100
..	..	..	..
..	..	..	..
..	..	..	..

Fig. 20-13 Examples of MRP exception messages. (*K.W. Tunnell Company, Inc.*)

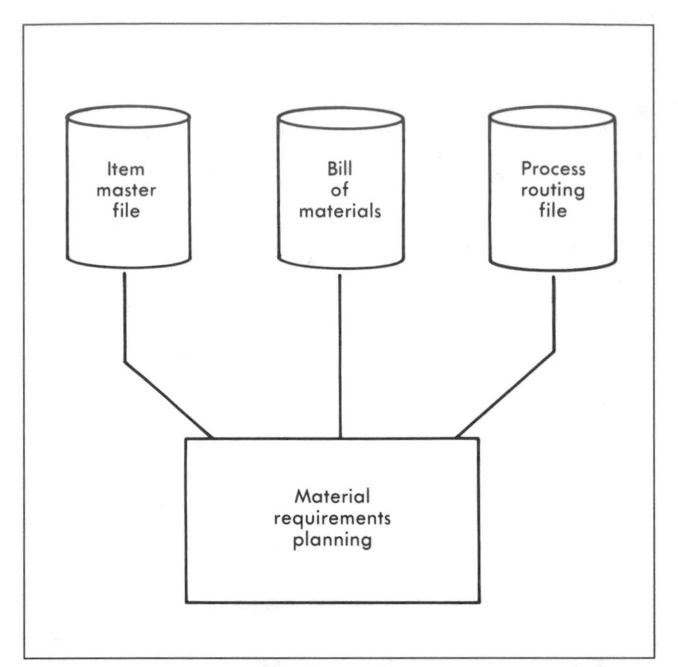

Fig. 20-14 Diagram of item master, bill of materials, and routing for an MRP system. (*K.W. Tunnell Company, Inc.*)

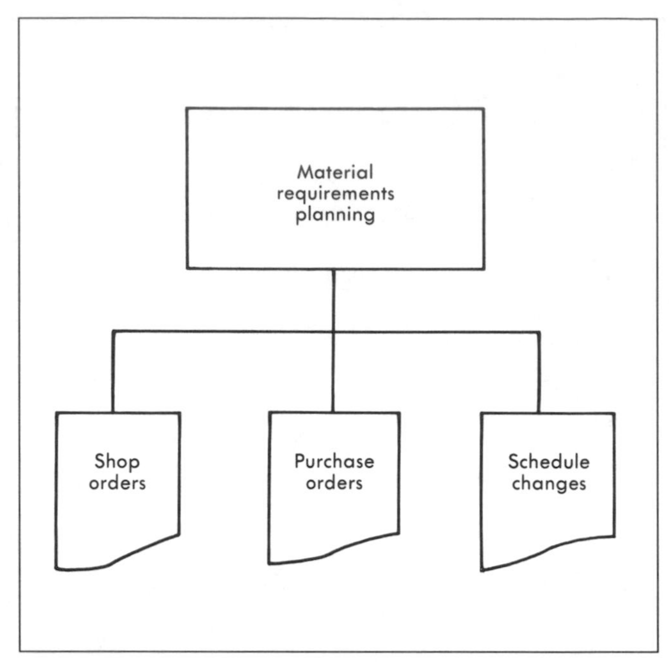

Fig. 20-15 Output of an MRP system. (*K.W. Tunnell Company, Inc.*)

The first calculations that must be performed are those that explode the demands into requirements by time period for all lower level subassemblies and component requirements. All demands are netted against available orders and committed replenishments (both shop orders and purchase orders), to calculate the actions that are required to support the master schedule. Two types of messages with associated quantity calculations result: (1) new order requirements and (2) changes required to existing orders. Order change messages may indicate data change needs only, quantity change requirements, or both (see Fig. 20-15).

Material requirements planning is not highly complex in the basic logic. Essentially, it works with arithmetic statements of demand and replenishment on a time-series basis.

Some MRP systems are also used for both capacity planning and shop execution feedback. These systems are capable of maintaining priorities on work that is active within the shop through a rapid feedback process, either on-line data reporting or frequent batch updating. This allows the system to have a view of the current status of all orders that are in process and, if done without fault, report back to its users a reliable picture of work in process and current priorities (see Fig. 20-16).

The data used to plan capacity utilization results from the shop loading on orders that are already released and netted against all inventories supporting the requirements of those orders (see Fig. 20-17). Most systems have an ability to answer some ''what if'' questions (a simple simulation technique) by allowing variation of the inputs regarding policies, rules, and lead times.

Disciplines required. MRP cannot create a good manufacturing plan. An achievable plan requires the combination of good top-level business planning with its requirement for good master scheduling rules to translate the requirements to the system. MRP can support the execution of a good plan, but only if its requirements for near-perfect data reporting disciplines are met.

Workcenter: 01 Fabrication Shop Date: 87200

Shop Order	Due Date	Quantity Ordered	Balance Due	Set-up Req.	Run Time Req.	Next Workcenter
W123	87202	75	30	0.0	15.0	02
W129	87209	75	75	0.2	37.5	02
W139	87216	75	75	0.2	37.5	02
W147	87223	75	75	0.2	37.5	02
		300	255	0.6	127.5	

Fig. 20-16 Example of a shop order priority list by workcenter. (*K.W. Tunnell Company, Inc.*)

The basis for improving operations with the use of an MRP system requires that the functions related to scheduling be integrated and that it be driven by a valid master production schedule. The people who operate and manage the variables of the MRP system must also be qualified—that means educated and trained in not only the techniques of a given system, but also in terms of the principles that they are dealing with. Very often, a different organization from that which may have existed prior to an MRP implementation is needed. Disciplines, again, can be the singular downfall of any MRP implementation. If record-keeping is not accurate and if actions and exceptions that result from an MRP system are not carried out, the system will fail.

Users of MRP systems cannot maintain bills of materials and routings in a haphazard manner without having a dramatic effect on the manufacturing operation. The disciplines related to the introduction of part numbers for new products, prototypes, and engineering change data must be accurately input into the system

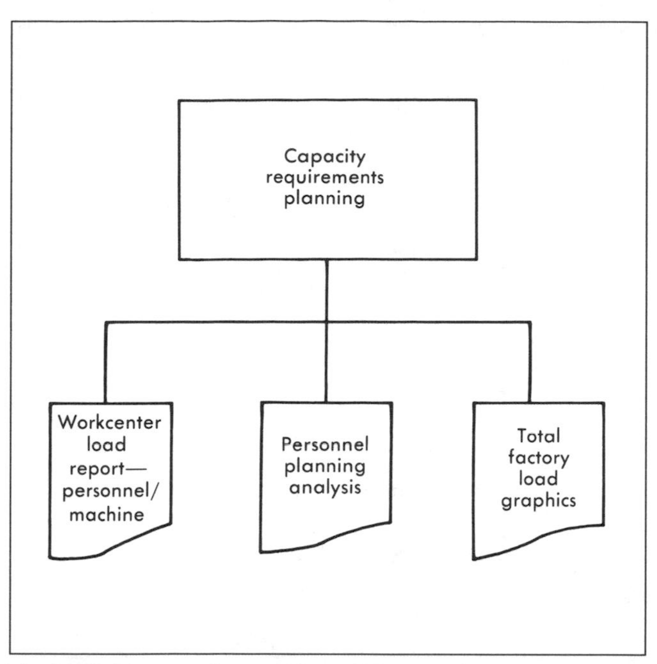

Fig. 20-17 **Diagram of outputs in an MRP capacity planning system.** (*K.W. Tunnell Company, Inc.*)

if the shop loading or material planning is to be calculated properly. The development of the engineering database elements for new products must be handled as manual override inputs to the system until all data is ready to be structured into the appropriate databases.

Past-due orders must not be allowed to overload the master schedule. Allowing past-dues to accumulate under the guise of giving good visibility (if they are past due, we will focus on getting solutions) gives disastrous results. Intentional overstuffing of the master schedule to create an extra load in the first month of the planned operation (equal to some percentage above the capability of production) automatically guarantees the carrying of excessive inventories.

Procedures and controls for data accuracy are primarily operating discipline issues. The single largest failing in most MRP systems is the lack of discipline in the day-to-day activities that maintain data integrity within the system. A system that attempts to emulate the total production environment within the computer depends on accurate data in the following elements:

- Inventory balances.
- Bills of materials.
- Process routings.
- Shop order status by operation.
- Purchase order status by item and date.

It is easily seen in the shop activity whether a continuous effort is applied to eliminate errors (see Fig. 20-18). An effective procedure is to track errors that are introduced in the system, determine the cause of that error, and institute a reaction that is designed to prevent that particular error from occurring again. This focus on what the causes of errors are, followed by a determination of methods to prevent the same errors from recurring will, in time, reduce the overall error rate.

Some users of MRP systems tend to leave shop orders open to collect actual costs for some period of time after the

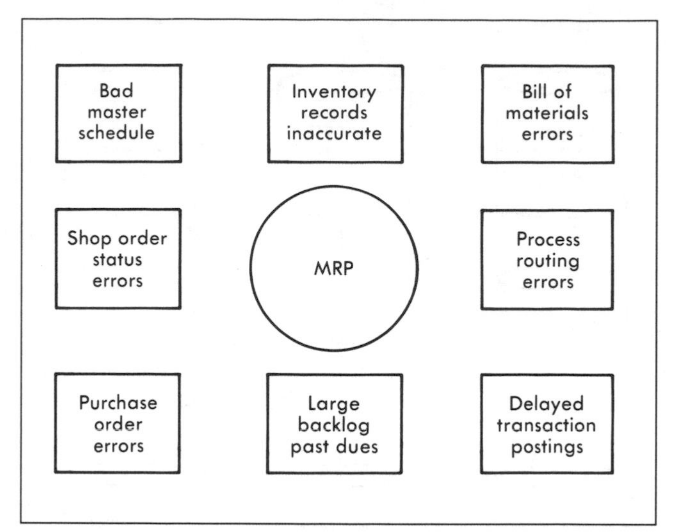

Fig. 20-18 Schematic showing types of MRP system errors. (*K.W. Tunnell Company, Inc.*)

production activity has ceased. There may be residual cost collection activity that can be improved slightly with this approach, but the confusion factor added to the overall environment requires a better technique. Production counts must be accurate and timely, and scrap must be accounted for at the time an order is truly active. Inventory data integrity is the fundamental rationale for locked stockrooms. This has long been used to display a management attitude toward good disciplines associated with the control of inventory counts throughout the operation and can be used to enhance the probability of achievement of near 100% data accuracy.

Past-due orders must be rescheduled. An MRP system that is "lied to" will respond in kind and chaos will result. Past-due orders that are not rescheduled cause inventory buildups and unreliable loading information for workcenters. Lack of credibility in an MRP system implementation will lead to nonuse.

Failing to eliminate errors, letting little errors slide by, letting errors go until the next physical inventory, or working around a problem on the shop floor for a specific purpose spell disaster for the user of an MRP approach. It is a myth that "reasonable" production counts are acceptable. In all cases with an MRP system, the objective must be 100% accuracy in data recording. Zero defects is a more appropriate term to define the goals for the accuracy of data of all types in an MRP system.

MRP does not, in and of itself, reduce inventory or cause a reduction in inventory, nor can it improve customer service or productivity or reduce costs. As a tool, however, it can provide the means for management to gain those benefits.

Database Considerations

MRP databases require integration. Not only do they have to be more accurate than other systems, the magnitude of the functions affected also requires that efforts be made to allow a single reporting of an event to cover all needs for that data. Often, inventories and production activities depend on this tool to control millions of dollars. In a multiplant, multilocation environment, the management of the MRP database is most often controlled by data processing specialists. This administrative function controls the integrity of the system by managing the database elements (see Fig. 20-19).

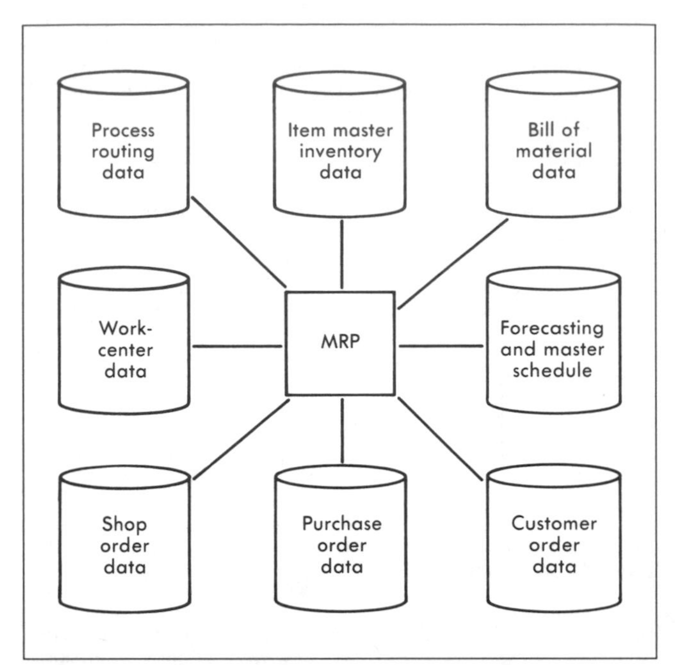

Fig. 20-19 Diagram of elements in an MRP database. (*K.W. Tunnell Company, Inc.*)

Control database
- Item master file (inventory data)
- Workcenter master file

Planning database
- Bill of materials file (product structure)
- Process routing file

Functional modules
- Master production scheduling
- Rough-cut capacity planning
- Material requirements planning
- Shop order release
- Purchase order control
- Production activity control
- Standard costing
- Job cost reporting

Potential integrated modules
- Accounts payable
- Labor reporting
- Payroll
- Customer order processing
- Sales activity reporting
- Accounts receivable

Fig. 20-20 Modules of a manufacturing resource planning system. (*K.W. Tunnell Company, Inc.*)

The database administration tasks, not database updating, are appropriately done on a centralized basis; it is the kind of technical activity that benefits from centralization. This helps to keep the system simple by using a single definition for the entire company. This need is best enforced if a company has a cohesive plan to execute of its manufacturing philosophy and strategy.

A company's database for MRP ultimately needs to be integrated with other systems requirements. The lack of integration usually results in many software interfaces being written and implemented. The continuing nature of the necessity of reacting with systems and database changes every time there is a change in another area of the business will diminish the reliability of the manufacturing resource planning system.

Lack of accuracy in the database will cause failure in any MRP system. Accuracy must be present not only in count data, but in structures and other relationships, such as operation routings. A single company bill of materials, not multiple versions of parts lists and/or cost structures, should form the basis for the overall database plan. Systems providers have found that it is difficult to get many companies to operate with a single database structure in the initial implementation efforts, just as it is difficult to bring a company into a team planning focus operating from a single company forecast.

Modular Implementation

MRP systems are adaptable to modular implementation (see Fig. 20-20). The planning and control database elements and their associated maintenance capabilities form the first implementation focus. It is wise to orchestrate a project plan based on system modules in an order that does two things: (1) minimizes the number of key users involved in a start-up activity at one time and (2) allows the opportunity to begin collecting some benefits of the system capabilities at a relatively early date.

As was previously stated, MRP is not a system that can be said to be appropriate for all businesses. Sometimes it is not appropriate because of size, and sometimes because the business is founded on a single product with a simple bill of materials. In such cases, MRP is not required as a technique for controlling the scheduling activity. MRP is not a total information system, but is very adaptable to being integrated during its implementation into a total information system (see Fig. 20-21). The database requirements of a total system are quite compatible with an MRP operating environment.

The total systems requirements design approach, which includes MRP modules, is quite suitable in developing an overall systems strategy, particularly if the business is either converting from essentially all manual systems, or if it is replacing an array

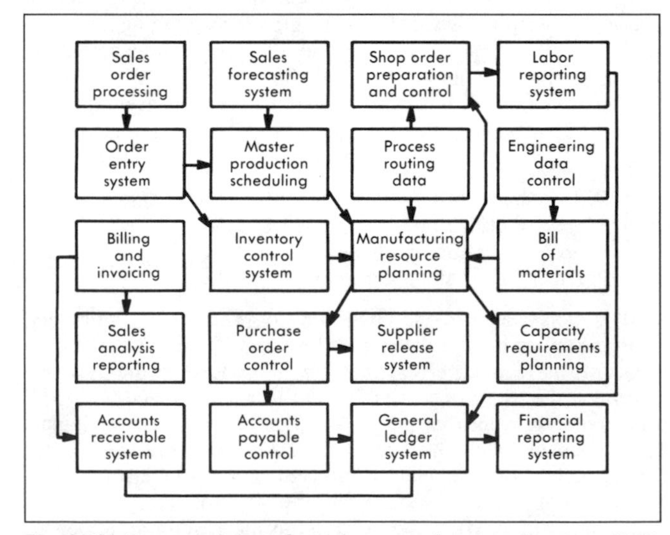

Fig. 20-21 Integrated manufacturing control system diagram. (*K.W. Tunnell Company, Inc.*)

of system pieces built over a long period of time that no longer match the business requirements.

Historically, MRP systems have been the subject of many failed implementation efforts, whether designed for updating to occur on an on-line basis or on a fast response batch update cycle. This is not usually a function of the software. The failures of implementation efforts seem to be mainly from lack of management commitment and internal discipline. Occasional failures stem from detail design not fitting the need, insufficient education and training, or problems in the organizational backing of the endeavor.

Implementation projects tend to be quite long in duration because many organizational boundaries must be crossed with an integrated approach. Some smaller companies have totally replaced their internal systems in 6 months or so, but many large firms have been involved in implementation projects exceeding a year, often even two or more.

Computer hardware planning and selection relative to an MRP system is a process that should follow the software decision, not precede it. Hardware purchase decisions should-support the software that best satisfies the requirements defined. Many systems failures can be attributed to approaches that were in the wrong order. Most users now buy software and customize some of its reporting. A technical error contributing to systems implementation disasters comes from modifying the core logic of software packages. Such a plan is rarely successful in the long run because, among other reasons, the software house is no longer liable for upgrades or maintenance after user changes have been made to portions of the main logic of the programs.

JUST-IN-TIME MANUFACTURING

The focus on the manufacturing philosophies and techniques that have, in the decade of the 80s, come to be generically called *just-in-time* (JIT) methods are founded on the study of techniques that have been demonstrated so successfully in Japan. As a set of tools for creating environments of manufacturing excellence, even their experience is based mostly from the late 1960s to the present. In a world marketplace, this tiny country, with severely limited natural resources, has become a dominant power in many products.

Based largely on the teachings of Dr. W. Edwards Deming, the management techniques employed by Japanese manufacturing managers attacked a number of problems inherent in the "typical" manufacturing environment:

- Raw material quality problems.
- High scrap rates.
- Frequent rework requirements.
- Long operation set-up times.
- Large areas of space occupied with inventory.
- Poor visibility of priorities at the manufacturing floor.
- Manufacturing driven by expediting efforts.
- Poor labor productivity.
- Poor employee morale.

Using statistical process control (SPC) techniques applied by all employees, Deming led the change in manufacturing management toward that of a continuous improvement process.

The techniques and philosophies implemented into a rapidly growing number of manufacturing enterprises have sought to achieve improved results in three categories:

1. Maximize the productivity of assets, particularly inventories.
2. Improve labor productivity simultaneously with the continuous improvement of product quality.
3. Minimize waste costs (inventory is considered to be a liability, and nonessential movement of materials is a waste).

Because of the importance of inventory in describing the differences in these Japanese methods, other names appeared as banners in attempting to attract the efforts of American managers responsible for improving their own operations, many of which were/are suffering directly from the success of these techniques in the Far East. Early explanations of the process described inventory as a pool hiding the real problems (see Fig. 20-22).

Improvements in manufacturing performance are achieved through implementation of specific techniques and operating methods including statistical process control, simplified systems, manufacturing cells, flexible automation, and flow control modifications consistent with JIT manufacturing practices. It is essential that some portion of future production requirements are forecasted to be repetitive individually or in product families to encourage the introduction of these manufacturing concepts.

There are significant organizational effects associated with the implementation of a just-in-time manufacturing strategy. It is very much an operator-centered approach, and traditional designs that have created separate functional entities, both on the factory floor and in the support organizations, will require significant modification to achieve the desired results. In the early stages of implementation, there is a pivotal need for an increase in the availability of manufacturing engineers to support the operator teams in accomplishing the waste cost elimination efforts. There is an equal need for additions in purchasing staff to launch vendor training programs targeting quality results, which provide the required reliability of flow from the supplier base. On the other side of the ledger, there are usually significant reductions, sometimes total elimination, of quality control inspectors. Material handlers, as a separate work group, are reduced drastically. Staff support related to production and inventory control is reduced as material flow control is converted to visual control methods on the shop floor. Additional information on JIT can be found in Chapter 15, "Just-In-Time Manufacturing," of this volume.

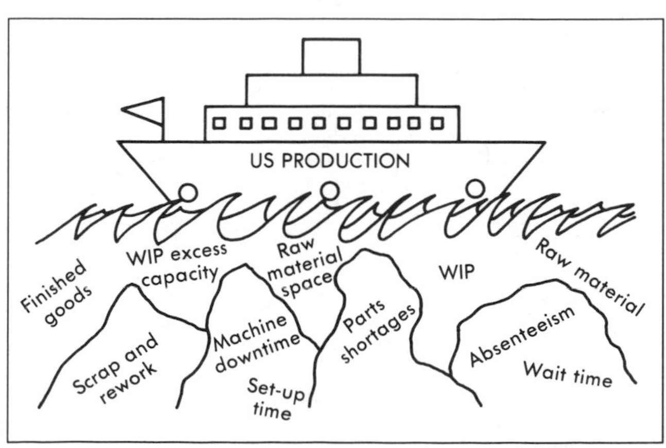

Fig. 20-22 Problems hidden by inventory but brought out when applying JIT techniques. (*K.W. Tunnell Company, Inc.*)

CHAPTER 20

JIT

Planning Requirements

A focus at the top management levels on production, sales, and inventory planning is important to establish a stable pattern of production to successfully utilize JIT approaches. The results in month-to-month sales, inventory, and production showing the conformance to the first month's plan may be measured as an indication of the effectiveness of the management team. This instills a view that planning is as important as execution. Management expectations must be matched to real capabilities of production and true flexibility demanded by the customer. In utilizing just-in-time techniques, managers must be able to judge which problems and opportunities are truly "floor" related vs. those that are "management" related to effectively drive out waste costs.

The lowest level of top-level planning is the functional equivalent of the master production schedule by product or by capacity group. This plan, which may be called a final assembly schedule, must be strictly followed to achieve the required matching of flexibility with existing capabilities. Such a schedule provides for the time elements required to establish the reliability of flow through the manufacturing organization by allowing for the current true lead time elements, both internal and external. This schedule can contain a small percentage of "shortened" lead time response capability using the concept of underscheduling, but only to the degree that material requirements can be met as arranged through the forecasting efforts, as opposed to scheduled order follow-through.

Signal Methods

One of the most predominant elements of the Japanese methods, which created excitement in both production control groups and purchasing groups throughout the United States in the early 1980s, was the talk of new methods of communicating needs for parts and materials between departments and even to outside suppliers. The first technique to be explored by American businesses was the Toyota technique called kanban (see Fig. 20-23). This is a card system that uses two pull signals: (1) move authorization and (2) production authorization.

Other techniques of implementation require the use of a preset number of empty containers (see Fig. 20-24). The empty containers specify the source and movement of materials required. These containers always use some form of visual counting technique to describe the quantity required. This technique can be very effective in communicating release-for-shipment messages to suppliers. Many unique part-carrying devices have been formed from plywood and other inexpensive materials by innovative companies.

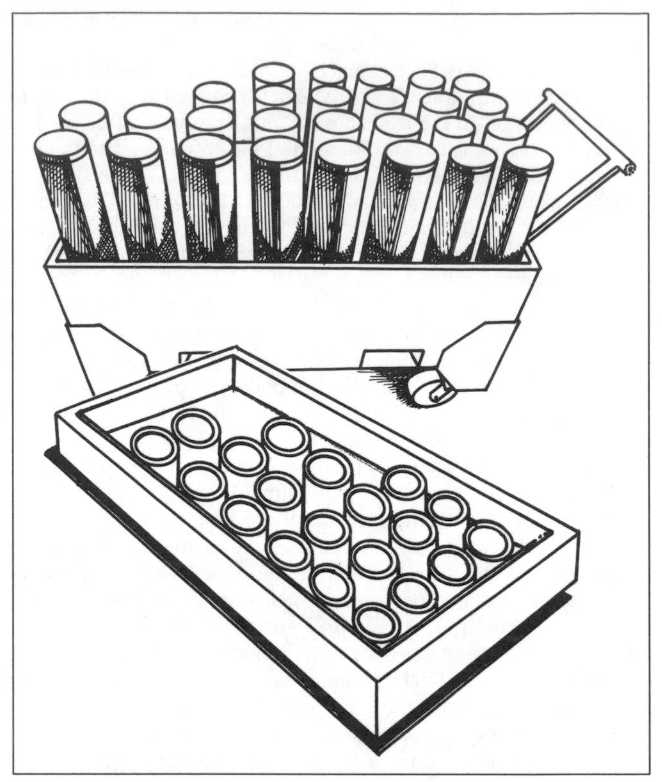

Fig. 20-24 Sketches of special containers used to signal the source and movement of required materials. (*K.W. Tunnell Company, Inc.*)

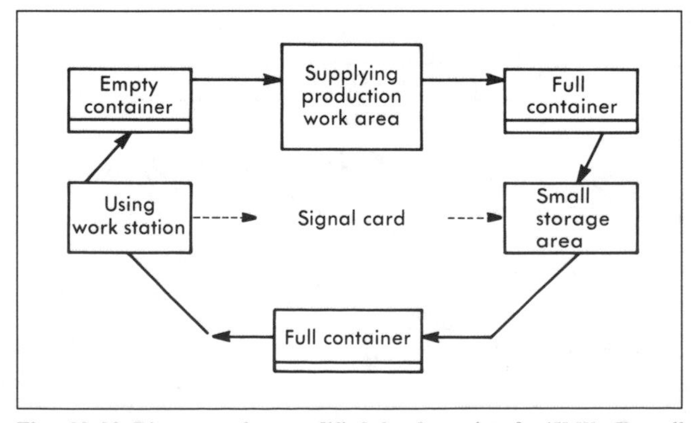

Fig. 20-23 Diagram of a modified kanban signal. (*K.W. Tunnell Company, Inc.*)

Both of the techniques just described, as well as other variations, determine an absolute maximum for inventory that can be contained within the total production system.

Pull methods are often implemented simultaneously with manufacturing cells. As cells are set up, the methods for causing the movement of material are converted to visual or signal processes, which minimizes the paperwork required and restricts material flow to the true production requirements. If possible, this signal system is carried all the way back to the suppliers of raw material. Material handling and conveyance methods usually change during this process.

Control Implementation

Four elements are required for successful JIT manufacturing control. The first requirement is a stable, short-term schedule equal in length to the facility's true manufacturing and procurement lead times (see Fig. 20-25). Both elements of lead time are the first focus of reduction efforts. The shorter the lead time, the more flexible the operation. Without the protection of safety stocks, flow of materials must be extremely reliable and cannot tolerate a nervous or unsure schedule during the "inflexible" period. The master production schedule, and the consumption of that schedule with specific orders as received, is, therefore, a critical ingredient of control. In make-to-stock environments, the consumption is done with internal allocation to individual finished product production authorizations.

The second element of control is the communication of requirements to the various workcenters. This communication

Part Number	Periods.......... 1	2	3	4	5	6	7	8	9	10	11	12	13
A1	44	44	44	44	34	34	34	34	29	29	29	29	29
A2	40	40	40	40	45	45	45	45	50	50	50	50	50
Totals	84	84	84	84	79	79	79	79	79	79	79	79	79

| Released || .... Firm .... || .... Planned .... |

Fig. 20-25 Example of an even flow schedule in a JIT manufacturing environment. (*K.W. Tunnell Company, Inc.*)

may take the form of daily or weekly production expectations by product or part, or it may be the flow and queuing of production signals provided by a kanban-like card or empty containers authorizing allocation of the manufacturing resources. No production in excess of that which is absolutely required is allowed in any workcenter. The key principle employed is that the control techniques make it very easy to see the requirements and counts without the use of computer reports and paperwork.

The third element relies on visual control techniques, such as spaces, pins, and holes built into the material handling and conveyance aids utilized in the manufacturing and material flow combination. All workers involved are able to "see" the work requirements and utilize team efforts to accomplish that which is truly necessary. When one part of an operation gets "in trouble," resources are applied from other parts of that workcenter to the area in trouble to solve the flow problem.

This type of control is most easily implemented in manufacturing cell environments (see Fig. 20-26). Employee teams in these cells usually work through set-up reduction projects and process improvement projects to improve the flexibility and reliability of their individual cells. Such cells may have been created for either individual high-volume products or for "families" of similar products.

The fourth method of control deals with reliability of flow of production and vendor supplies. These elements usually require significant amounts of concentration. Most often the solution to the problems that have contributed to the lack of reliability are attacked by employee teams. In some companies these are still referred to as quality circles, but many have changed to focus groups, which involve both management and labor employees in team approaches. These teams attack problems related to quality, productivity, or any form of waste cost inherent in the processes that they operate together.

The specific problems chosen by team members are consistent with the management priorities. Equipment moves may be required. Material handling methods usually change. Inventory policy changes and raw material supplier actions require staff support. Changes in the process usually require support from manufacturing engineering staffs. This is usually referred to as elements of a "continuous improvement process" environment (see Fig. 20-27).

Set-Up Reduction

Just-in-time applications in manufacturing may start with set-up reductions in a pilot workcenter. As set-up is reduced using operator team building methodologies, the need for large order quantities disappears. Experience shows that set-up times of 8-10 hours may be reduced to unit minutes. When the order quantity can be reduced to near 1, several things occur.

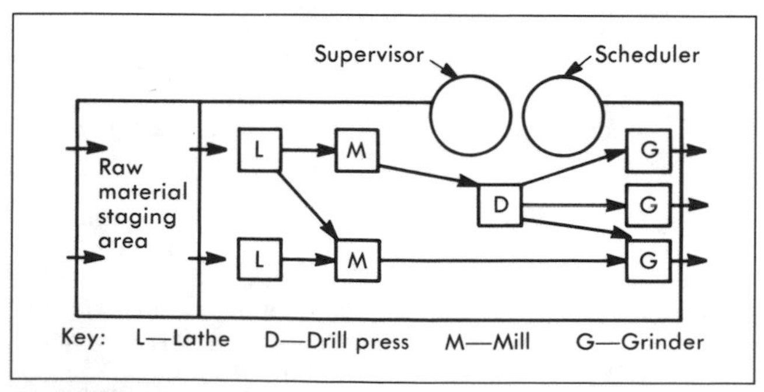

Fig. 20-26 Schematic of a manufacturing cell. (*K.W. Tunnell Company, Inc.*)

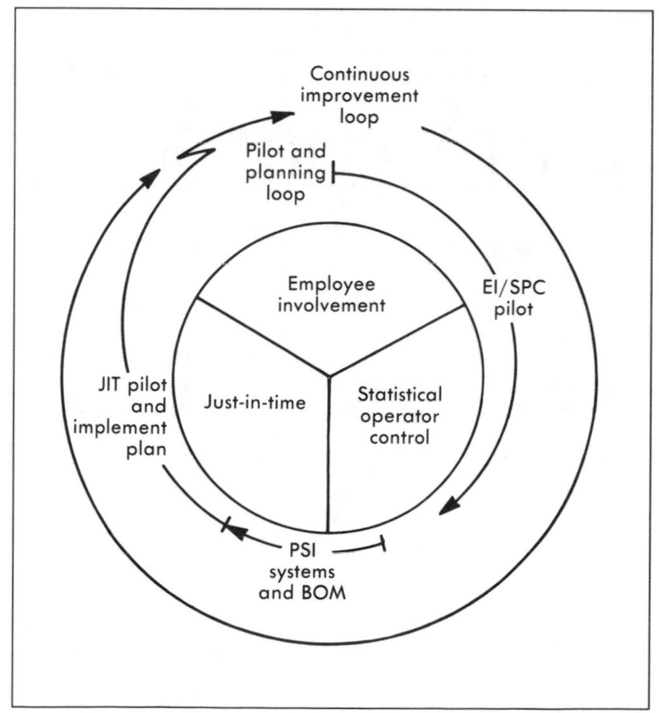

Fig. 20-27 Continuous improvement process diagram. (*K.W. Tunnell Company, Inc.*)

First of all, products can be made as needed because there is minimal set-up required. Departmentalization of functions gives way to the creation of machine cells organized in group technology oriented families. Moving machines closer together in a cell permits the following:

- One or more persons may perform multiple operations, eliminating the usual queue and move times between operations.
- The elimination of most of the material handling required in moving parts from one operation to another.
- Greater utilization and productivity of direct labor.

Machine cells also help to significantly improve quality because a bad part made one at a time can be caught immediately, not weeks later when a whole lot is withdrawn from the stockroom.

Total Quality Management

Just-in-time manufacturing requires a quality strategy that is designed not only to produce high-quality products for its customers, but also to satisfy its internal requirement for reliability of parts flow. This further implies that the user of just-in-time manufacturing concepts must also help its suppliers to attain the same reliability of flow with regard to the products and materials it purchases. Thus, the strategy and the associated action plans are divided into the following three focus areas:

1. Vendor-supplied items.
2. Work-in-process flow items.
3. Finished product feedback.

The only proven way of providing the tools for accomplishing the required quality to achieve the essential reliability of parts and products flow is through the use of statistical quality control techniques. These techniques require training of all employees to provide a common language and understanding of the processes utilized. The focus of a continuous improvement process is the foundation of the success of the Japanese approach.

Priorities may be focused on either product or functional relationships and needs; they must address both internal and external improvement opportunities. The principles and practices implemented are operator centered. Management's role lies in its commitment to communicate priorities and true support for the operators. Participants are taught to use SPC as a tool for understanding what is going on in the processes operated. Through this common understanding, a front line of employee involvement in problem solving is achieved; this is essential to low-cost producer operations created through a continuous improvement environment.

Supplier Requirements

The purchasing department is responsible for supporting the manufacturing operations in developing programs for suppliers. Part of the technique usually employed is the reduction of the number of vendors utilized. Suppliers who cannot or will not support its cost and quality objectives are eliminated. All suppliers that do participate in the development of reliable just-in-time flow of materials are provided with support and assistance. This type of program is necessary to ensure that quality products are always shipped to meet a precise schedule, without the added costs associated with receiving inspection.

Programs developed between suppliers and their customers always include a focus on specification requirements and on communication methods. Further, it identifies the measures of capability to be demonstrated by the supplier and the definition of a method of ascribing quality performance. Methods of working together toward the elimination of costs and quality problems are integral to the nonadversarial relationships sought.

SCHEDULING AND PRODUCTION ACTIVITY CONTROL

Scheduling and production activity control needs also change drastically with the introduction of JIT manufacturing principles into an organization. The basic elements of these tools and techniques are discussed in the balance of this chapter. It is important to understand the language and characteristics of these planning and control techniques to be able to assess and select the best methods to support an individual manufacturer.

PRIORITIES

The first order of business for the production scheduler and the production and inventory control department is the establishment of the "right" sequence of jobs to be run in the shop.

Determination of Priorities

Many manufacturing organizations rely on a formal shop scheduling and feedback reporting system. Before discussing the various ways in which priorities can be determined in such a system, it is important to understand what a good shop priority system should accomplish. The requirements of a good priority system are as follows:

1. A priority system should specify which jobs should be done first, second, third, and so on.
2. A priority system should allow for easy and fast updating of priorities, inasmuch as priorities will change after a very short period of time because actual conditions change.
3. A priority system should be objective. If jobs are overstated, an "informal system" will be developed to determine which jobs are really needed.
4. A priority system should be as simple as possible; visually controlled "pulls" work for JIT.

A major objective of any computer-supported shop scheduling system is to develop the relative priorities of every shop order in a workcenter. Some companies use sophisticated schemes, while others do not. Examples of the most commonly used priority schemes are first in/first out, start date, due date, critical ratio rule, slack time ratio, and queue ratio.

Each of the priority schemes has strengths and weaknesses. The priority calculations are somewhat limited in scope because they cannot consider other factors, such as machine capacity, availability of needed work skill, whether the job is for stock or customer orders, and the importance of attaining the monthly shipping budget. Most companies, after experimenting with several of the priority rules, will realize these shortcomings and provide as much information to their production staff as possible so that production goal decisions can be made to work.

First in/first out method. The first in/first out (FIFO) method is the simplest priority rule. It assumes that the first shop order to enter a workcenter is the first shop order to be worked on. The major advantage of this rule is that it does not require a computer or other sophisticated system to determine priorities. The major disadvantage is that it assumes that all jobs have the same relative priority. It normally does not allow for the redistribution of priorities, nor does it permit an order that was released late to be moved ahead of other orders in the schedule.

Start date method. The start date priority rule is really a subset of the FIFO rule insofar as the shop order with the earliest start date is the first job to be worked on, and so forth. Naturally, this scheme assumes that all shop orders are released on time. The start date can be calculated from a backward rather than a forward scheduling technique. Forward scheduling is "a scheduling technique in which the scheduler proceeds from a known start date and computes the completion date for an order usually proceeding from the first operation to the last."[4] Backward scheduling, on the other hand, is "a scheduling technique in which the schedule is computed starting with the due date for the order and working backward to determine the required start date. This can generate negative times, thereby identifying where time must be made up."

Due date method. The due date is the date when the material is needed to be available. The due date priority rule is the most popular priority technique that is used in manufacturing industry, particularly with the advent of MRP-type systems. When properly used and kept up to date, the due date rule can be very effective. It is, of course, possible to overstate the master production schedule in an MRP system and destroy the credibility of the due date priority. If the master schedule is well maintained and kept up to date with actual conditions from the shop floor as feedback into the planning and scheduling system, the due date technique is a straightforward tool to determine shop priorities.

Critical ratio rule. The critical ratio priority considers the total standard lead time remaining to complete the job relative to the total time remaining to the due date of the order. The control ratio is computed as follows:

$$\text{Critical ratio} = \frac{\text{Due date - today's date}}{\text{Lead time remaining}} \quad (5)$$

Lead time is defined as the sum of the processing time, set-up time, move time, and queue time. Any order with a critical ratio of less than 1.0 is behind schedule, while an order with a critical ratio of more than 1.0 is ahead of schedule. An order with a critical ratio of 1.0 is right on schedule. Using this technique, shop orders with the lowest ratio have the highest priority. Conversely, the orders with the higher ratios have the lowest priority.

Slack time ratio method. The purpose of the slack time ratio is to assign priorities to jobs in queues at various workcenters. The slack time ratio is computed as follows:

$$\text{Slack time} = \frac{\text{Due date - today's date - processing time}}{\text{Time remaining}} \quad (6)$$

If two or more orders have the same slack time ratio, the computations may be modified to consider the number of operations remaining as follows:

$$\text{Slack time} = \frac{\text{Due date - today's date - processing time left}}{\text{Number of operations remaining}} \quad (7)$$

This alternative encourages the completion of orders having the most operations remaining.

Queue ratio method. The same queue ratio calculates the relationship between its slack time remaining and the queue line originally scheduled between the start of the operations being considered at the scheduled due date. The ratio decreases as the shop order becomes late. The queue ratio is calculated as follows:

$$\text{Queue ratio} = \frac{\text{Slack time remaining}}{\text{Original queue time}} \quad (8)$$

where:

Slack time = Due date - today's date process time remaining

and

Original queue time = Due date - scheduled date - standard process time remaining

Work Authorization

The method used to authorize work will often depend on both the type of production (whether it is job shop, process, or repetitive manufacturing) and on the degree of sophistication in the manufacturing planning and control system. There are generally three ways that work is authorized in the shop: (1) verbal, (2) shop order, and (3) dispatch list.

Verbal. Verbal authorization is the oldest method of assigning work. Using this method, the person responsible for the work assignment simply tells the appropriate supervisor which jobs to work on next. The method is straightforward, but is often based on limited information. It is also subject to loss of control and generally depends on expediting to get critical jobs done.

Shop order. Many companies use shop orders to authorize and track the production of parts, subassemblies, and end products through the plant. In many systems, every workcenter is specified in the writing of a shop order. At the time the order is released, the quantity to be produced is shown, as well as the time the part is expected to arrive at a given workcenter. These dates depend on the start time, process times, and move times associated with the particular job. These work orders are generally used for planning personnel, set-up, and machine loading levels. Specific operation due dates can be used in calculations relative to the shop order due date and expediting that is communicated to the individual workcenters.

Dispatch list. Manufacturing companies frequently issue "dispatch lists" or "shop schedules" at regular time intervals, particularly if a computerized system is being used. The dispatch list shows all shop orders to be worked on by the various workcenters in a priority sequence. Normally, companies that issue dispatch lists will not show any start date or due date on their shop orders when they are released to the individual workcenters because all copies of the work order would have to be collected and the dates modified as the dispatch list dates were reviewed and revised.

The dispatch list document always indicates the part number, description, shop order number, quantity, number of stated hours required, and priority rank of the order. The dispatch list must be continually updated and is usually issued daily. Priorities in the dispatch list can be recalculated periodically based on the latest information from the plant floor regarding each order. The priority rank shown in the dispatch list can be based on any of the priority rank calculations previously discussed, or it may be based simply on the start and due dates that are assigned to each operation.

Order Launching and Expediting

This ''launching'' or release of the shop orders involves the actual dispatch of the shop order to the workcenters that will be involved in production activities. Workcenters receive copies of each shop order when the order is released to the workcenter for the first operation specified in the routing. Only orders with a high probability of being completed on time should be released. Typically, shop orders are not released to the first operation until the following conditions are met:

1. All of the material required to produce the part is available.
2. Sufficient capacity for the job is available.
3. Tooling at the first operation is available.

The releasing of the order authorizes the start of production and/or the distribution of the routing to the shop floor. Components and materials can now be moved to the work areas.

Companies without a formalized priority system tend to use some type of manual or visual method of communicating problems with job flow. Informal systems such as ''hot lists,'' ''shortage lists,'' and various expedite tags are evidence of some of the methods that are used to determine priorities and expedite shop orders through a plant. Although such approaches to priority planning are widely used, they are quite ineffective as methods of determining priorities. Once there are large numbers of orders released to the shop floor, the many dates on many open jobs on the plant floor soon become suspect. When everything tends to become suspect, then nothing is perceived as ''right,'' and scheduling chaos results.

Operation Scheduling and Tracking

To accomplish production scheduling, the production control department must have methods to know what to make, when to start it, where to make it, how to make it, what to make it from, how much time is necessary to make it, and when the order is due. The scheduling methodology can be expected to differ by the basic type of manufacturing processes that are involved.

The frequency of reporting production feedback and the corresponding updating of schedules must satisfy all of the stated objectives. Schedules must be updated often enough to ensure that the outputs of the scheduling system are meaningful for the operating personnel. If updating is not done frequently, people will soon work around the system instead of use it. Eventually, this will lead to a deterioration in data accuracy in the formal scheduling system as shop people revert to their informal manual workaround notes and methods to get product through the shop.

Continuous-process manufacturing. In a continuous-process manufacturing company, work flows through the manufacturing process. The rate of flow can usually be varied, although there are processes in which the rate can be varied little if at all. The production rate in continuous-process production can be varied in several ways:

- The line equipment may be operated for more or fewer hours per day.
- The line rates may be varied, faster or slower.
- The production line may be operated intermittently.
- The line may be operated with more or fewer personnel.
- The line may be operated continuously or only in stages by accumulating in-process inventory within the process.

Generally, the scheduling of a continuous manufacturing process is simpler than scheduling an intermittent production shop because production can be monitored more accurately by utilizing automatic means to continuously measure the flow in terms of weight and length.

Job shop production. A plant with intermittent production is defined as a job shop (see Fig. 20-28). Intermittent production is defined as ''a production system in which jobs pass through a functional department in lots.''[5] There are a number of problems that must be faced in scheduling and controlling a job shop operation that are different from a continuous flow manufacturer. Examples of the information to be collected manually include:

- When will the order be finished?
- Is it behind schedule?
- Where is it in the process?
- How long will the next operation take?
- Is the machine down for repair?
- How are the dies or tools replaced?

Other basic problems are how does one decide what to schedule and how does one determine the best way to schedule the choices. Should products be scheduled? Or should machines be scheduled? Perhaps a combination of both methods? These decisions determine how data must be grouped from the forecasts and orders to allow the production activity scheduling group to visualize the potential load on existing capacity. Should all products or machines be scheduled or just a portion of them? Should the control focus of scheduling be departments, workcenters, or individual machines? What are the scheduling needs, and how should they be addressed? The solutions may be almost as varied as the manufacturing enterprises themselves.

Repetitive manufacturing. Repetitive manufacturing involves the production of discrete units that are planned and executed according to a schedule and involve manufacturing at a relatively high volume and speed. Material moves in a sequential flow. Production feedback for control is based on the number of

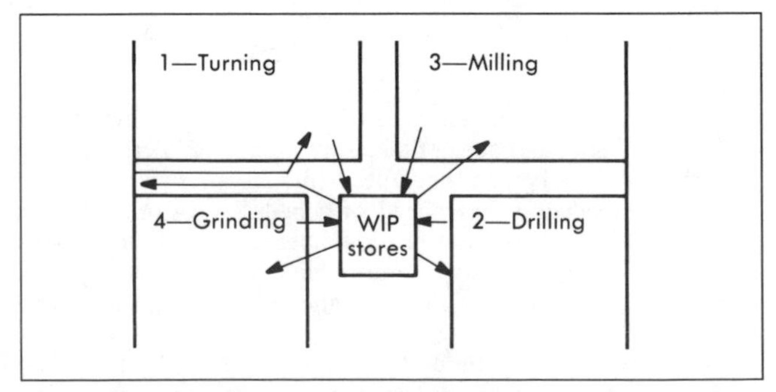

Fig. 20-28 Diagram of a plant with intermittent production. *(K.W. Tunnell Company, Inc.)*

items that pass a predetermined control point. Regardless of the type of industry process that is being scheduled, there are several basic rules that should always be followed in a repetitive manufacturing environment:

1. Process similar orders in an appropriate sequence to minimize set-up time and reduce changeover times.
2. Schedule the input (released shop orders) to meet the planned production rates. If an individual workcenter is consistently not producing to meet its plan, the amount of work released to that center should not exceed the actual "experienced" capacity.
3. Keep released backlogs off the plant floor. On-floor work order backlogs are more difficult to control, make engineering changes more difficult to implement, generate more expediting, and create space problems.
4. Sequence the orders based on the latest requirements, not the requirements data estimated when the shop order was first written.
5. Schedule to the shortest cycle possible, daily if possible, to obtain the latest and most accurate feedback and requirements data on the orders released.

A reliable production scheduling system should provide efficient, smooth, cost-effective production runs with minimal raw material and in-process inventories. It should deliver products on time with minimum lead times and react quickly to changes created by normal production interruptions, such as machine breakdowns, late arrival of raw materials, and maybe even a few rush orders from customers. At a minimum, the system should provide accurate and timely feedback of results obtained on the production floor.

The schedules that are issued to the plant need to be frequently updated so that management can assess how closely actual production is to the plan, determine the status of any given shop order, and reschedule the plant based on actual conditions. Production feedback from the plant floor is used not only to update schedules and the status of all open shop orders that have been worked on, but the data may be used in the payroll, efficiency, or labor reporting systems and cost-tracking systems.

Performance Measurements

The prerequisite for the evaluation and measurement of performance to a predetermined expectation includes clearly defined objectives, clearly defined time elements for measuring performance, reasonable delegation of authority adequate to the performance of the tasks to be measured, and mutual agreement that the measure can be attained between the person being measured and the person doing the measuring. It is the responsibility of management to measure the performance of its production resources. Ideal measurements are quantitative, but should extend beyond simple output quantity measures. The three basic tasks of performance measurement are as follows:

1. An evaluation of individual performance in a manner resulting in the motivation of the workforce.
2. An evaluation of the general assumptions used as the basis for performance measures (assumptions that are really an estimate of overall capabilities).
3. An evaluation of the system as a whole, recognizing that performance measures are based on an estimate and that the overall performance of the systems may require a restating of the original goals.

One of the most critical measurement activities in a manufacturing company is related to its production operations. Typically, the production department will be measured on some of the following objectives:

- Shop orders completed on time.
- Output hours vs. plan.
- Direct labor productivity.
- Machine utilization.
- Actual costs vs. budget.
- Customer service vs. stated plan.
- Quality level vs. plan.
- Production rates vs. capacity plan.
- Personnel vs. plan.
- Input/output backlog vs. plan.

It should be recognized that performance measurement, particularly individual productivity measurements, may become so important internally as to sacrifice the objectives established for overall company inventory and/or customer service objectives. In fact, the measurement of production productivity can in some cases contribute directly to the failure of time-phased material planning systems, usually referred to as material requirements planning (MRP) or manufacturing resource planning (MRP II) systems. Scheduling for operations using just-in-time (JIT) manufacturing philosophies requires that measurements be made at the team level to maintain the focus of employee involvement in problem-solving processes. As the concepts of JIT, sometimes called zero inventory, are embraced by increasing numbers in the manufacturing community, it is becoming necessary to continually rethink the traditional forms of performance measurement. The emphasis on good manufacturing control is moving away from the measurement of worker performance to one of creating a smooth flow of small quantities of first-time quality work through the production workplace in a predictably reliable manner. The more important measures in the long run are judged not to be the short-term production out the back door, but rather the long-term measures associated with continuous improvement and low-cost, defect-free production.

In the future, the emphasis on compliance with individual time standards must change to team targets of reduction of set-up and lead times, to flexibility of the workforce, and to the ability of the production functions to manufacture a high-quality product reliably to a scheduled completion date. Traditional performance measurement will continue to change to meet these changing objectives.

WORK FLOW PATTERNS

The physical organization of the manufacturing facility has a dramatic impact on the expected costs of production. In recent years, a variety of changes affecting the direct labor resource have been implemented. Layout and flow can also contribute greatly to the reduction of indirect labor costs associated with material movement and on-floor work-in-process materials.

Traditional/Functional Layouts

The traditional layout of a factory is based on groupings of operations according to manufacturing function, such as drills in one department and lathes in another. This form of plant layout results in a back-and-forth type of physical flow of material throughout the factory. It is not a flow-oriented design. Operations may be spaced as far apart as from one end of the facility to the other. Functional layouts tend to demonstrate the poorest flow patterns after many years of operation (see Fig. 20-29). Growth patterns seldom follow in the manner of the initial plant layout logic. This can cause parts or products to travel great distances and accumulate significant amounts of non-value-added activities, such as material handling and queue.

Production is scheduled in batches that move from one operation to another. There is generally no overlap in operations, inasmuch as a succeeding operation will not begin production of its first piece until the entire batch or lot has been processed at the previous operation and the material physically transported to the next operation.

Cellular/Focused Layouts

Cellular manufacturing is based on the linking of operations according to part families or according to similarities of the manufacturing processes employed in production of a group of parts. Some companies have created cellular manufacturing units to form a complete production operation, from raw material to finished part, subassembly, or final product. The linking of manufacturing operations reduces work-in-process inventory between operations and permits the overlapping of production between operations, although the production may still be scheduled in batches or lots. Cellular flow distances are usually quite short when compared to the functional process routings replaced.

Cellular manufacturing is based on techniques usually described as group technology. This is basically a set of methods whereby functionally oriented production plants are studied for rearrangement into process layouts. Manufacturing cells are those equipment arrangements generally established to produce one or more families of parts. Manufacturing cell layouts permit the parts to be manufactured from start to finish in one location rather than being transported around the factory from workcenter to workcenter or department to department (see Fig. 20-30). Cellular manufacturing can be accomplished by taking the continuous layout principles of long standing and applying them at the component part level. In such a layout, machines are grouped according to the sequence of the required operations. Additional information on group technology can be found in Chapter 13, ''Design for Manufacture,'' and Chapter 19, ''Equipment Planning,'' of this volume.

Despite the attractiveness of cellular manufacturing, it has not yet been widely accepted in the United States. American companies have been reluctant to adopt cellular manufacturing because of the effort, time, and/or cost requirements of changing from a functional or traditional layout in which a department or workcenter performs a single function and the product moves from area to area before it is completed. Equipment within cells must be dedicated to a limited number of components (often a single part or part family), often requiring an increased number of machines. Until the competitive position of a company falls to the point where it has surplus equipment, it has been difficult to show justification for such a change in the manufacturing environment. This is largely the fault of current accounting practices that do not capture many elements of cost.

Cellular manufacturing can be an element of the overall solutions sought in reducing set-up time and obtaining produc-

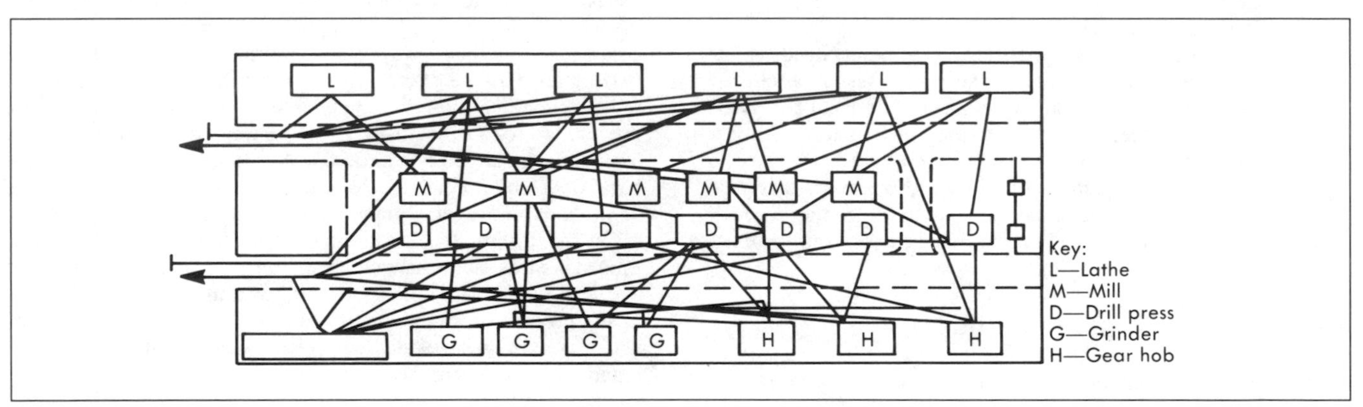

Fig. 20-29 Example of a functional/traditional plant layout. (*K.W. Tunnell Company, Inc.*)

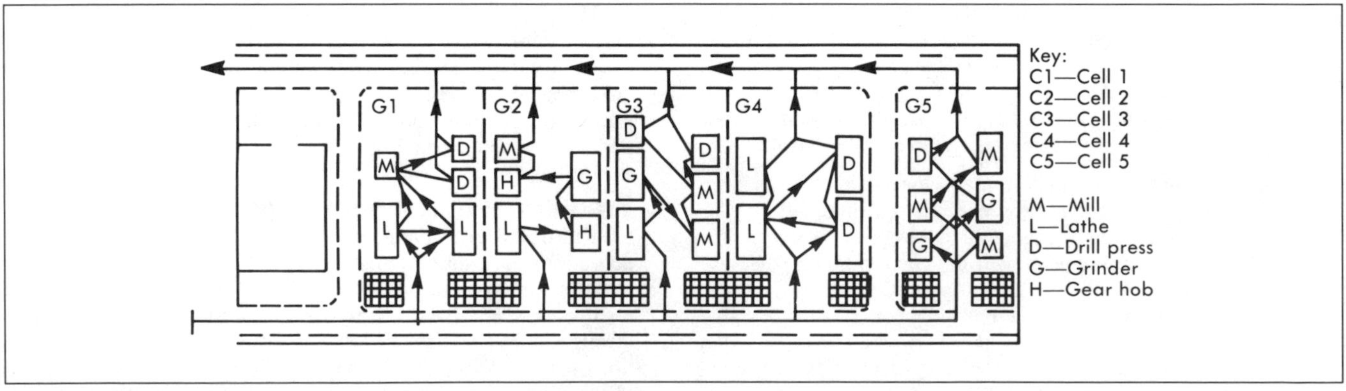

Fig. 20-30 Layout of plant in Fig. 20-29 after it has been changed to a cellular manufacturing layout. (*K.W. Tunnell Company, Inc.*)

tion efficiencies. It certainly is the most logical layout to take advantage of employee involvement problem-solving teams on the manufacturing floor. Most American manufacturers process parts on a serial basis, completing each manufacturing operation for an entire lot quantity before beginning the next. The minimum lead time under this method consists of the summation of the time required in the two operations. Cellular technology, however, utilizes overlapped or concurrent operations to achieve a high degree of throughput and shortened lead times.

In-Process Queues

One of the reasons that manufacturing companies in the United States have large work-in-process inventories is their emphasis on producing "economic lot sizes," which seek to balance inventory costs against the set-up costs created by changing from one part to another. In contrast, the Japanese believe that inventory is, by definition, a liability and a creator of waste costs. Therefore, they strive to avoid the rationale of large batch production by directing their attention and ingenuity to reduction of set-up costs, thereby permitting the economic production of smaller lots. As set-up and ordering costs approach zero, the calculated economic lot size also approaches zero or, for practical purposes, a lot size of 1.

A secondary reason for large in-process inventories in American companies is the method of scheduling operators somewhat independently, which results in batches of in-process work sitting in front of machines, particularly bottleneck operations, waiting to be worked on. This is commonly referred to as a push system. In contrast, pull systems focus on the production of only those parts that are required for use in the next period or to replace those that have just been taken for use in another area.

There are two essential differences between a push and pull system of material control. In a push system, the parts are moved to the next location immediately after completion. In a pull system, the workcenter making the part takes its cue or signal from the user of the part and not from some central planning source. A workcenter does not receive a schedule as in a push system for a repetitive manufacturer or a dispatch list as in job shop environments. In a pull system, the parts that are made sit at the location where they were made until needed rather than automatically being moved to the next operation. A pull system eliminates queues of work before workcenters because the parts are not called for until needed.

The final assembly or production output schedule is a key element in making a pull system work. When possible, the final assembly or final production station signals the flow of all components throughout all the manufacturing operations. A stable master schedule that is translated to a final assembly schedule or production output schedule is critical to the pacing of the overall operation. The objective of the pull system is to have an even-paced flow of material through all operations coming into final operations only as required. If parts are not needed, production resources are not applied to the false bargain of inventory building.

SIMULATION

As a planning tool, simulation of operations activities aids in the testing of alternatives. MRP systems software usually offers some level of capability to assist in tactical planning evaluation through simulation. This is most often accomplished by allowing the system user to execute functional portions of the system using temporary work files and/or "no update" options. Software for small systems is more likely to require that unique sets of files be established for use in "alternative testing," or simulation runs, of the production programs. The use of simulation for planning and analyzing manufacturing investments is discussed in Chapter 3, "Planning and Analysis of Manufacturing Investment," of this volume.

Demand

If demand at the MPS level is varied between a minimum and maximum level around the live production planning forecast, the shop loading and material requirements changes required to respond to such a change in demand can be reviewed. This can be extremely important in developing specific plans for coping with an opportunity to handle unusually high demand patterns that might result from a competitor's strike or closure. To effectively respond in the shortest possible time period without diminishing service to existing customers, specific planning must be done to support decisions regarding the acquisition of additional labor and material resources.

Using simulation techniques, it is possible to load a new forecast into the system and measure the amount of change in processing requirements by workcenter. The amount of material that needs to be expedited and the additional material ordering activity required by that forecasted demand can be calculated. If the output gives strong indication that the level of increase being considered cannot be supported, a second execution of the simulation calculations with different timing and volumes in the trial forecast can be tried. The goal of any simulation activity is to give visibility of the likely results of a set of decisions without committing financial resources to the decision.

Planning for new product-line introduction also gives rise to the need to estimate the impact on an ongoing operation. Considered as an increase in forecast demand, the same iteration of calculations is appropriate in preparing detail planning for the manufacturing operations introduction timing. Handled at a pre-master scheduling level, product group data can be used if a corresponding set of definition data is prepared as family bills of materials and bills of labor.

Production

Given any level of forecasted demands, another tactical planning support activity is required in anticipation of major changes in the production environment. The addition of new capacity through the installation of some new level of automation in a process requires an assessment of the work-in-process flow that will be expected in the new environment. Likewise, variations in worker-hours and machine-hours for the new operation can be calculated using modified process routing data that must, in any case, be prepared before implementation.

If product group data is to be used for these "what if" exercises, data must be prepared that corresponds to the average load profiles represented by each group. If product-level forecasted demands are to be used, the calculations will be based on product structures and process routings that are loaded into the manufacturing control system database.

Another common use of simulation techniques is in the assessment of the likely impact on a manufacturing facility that would result from a major shift in product mix. Changes in material flow, in-process queues, and existing workcenter loading can be predicted through such calculations. This can be valuable input to the marketing organization in determining the risks associated with a proposed change in their strategies and tactics related to the promotion of certain product groups.

The two primary results from "what if" calculations related to the production environment are:

1. Changes required in labor and equipment to achieve the new demand and processing alternative.
2. Feasibility of the timing of the changes proposed in the "what if" statement being tested.

Testing of alternatives with simulation techniques should be considered as a paper evaluation of the realities that might be expected from a proposed significant change to the existing norms. If evaluated through calculations instead of actual attempts to implement the change, expensive errors and premature investments in inventories and other resources can be avoided.

Standard Costs/Standard Hours

Because most MRP application packages include modular support for the standard costing requirements of a manufacturer, changes in structures, routings, and material costs can be used to derive forecasts of financial results related to the changes in any of these database elements. This may be as simple as the calculation of a new cost for an individual product caused by a change proposed in any of those elements. It could also be a prediction of new product group margins that should be expected from the set of alternatives being evaluated. Purchasing departments often need such information to justify a potential change in a material alternative when the change being considered affects a large array of products. Standard costs changes also affect inventory investment levels in raw material, work in process, and finished goods. These too can be estimated through the simulation processes.

Standard hours are used as the primary manufacturing measure reviewed in the calculated results of a simulation run. This measure helps to answer the tactical questions of feasibility and increase/decrease requirements resulting from a proposed change. The organizational unit addressed in the potential consequence study being performed is the workcenter. Standard hours calculations are based either on the process routing data at a part level or a load profile prepared to represent the average requirements at a product group level. A review of capacity planning approaches is helpful in deciding what potential reports might be useful as outputs from a simulation run.

Modeling

More complex approaches to simulation include the use of various modeling techniques. In a manufacturing environment, these include the study of moves and queues. Specialized computer software exists, allowing an engineer or analyst to describe the elements of a process for simulation tests. These software packages use statistical approaches based on random number generation and probability distribution patterns to study a set of production demand and flow alternatives.

Additional specialized software can be used to determine the optimal or improved facility layout alternatives for a given pattern of in-process move requirements. These models attempt to minimize the total distances required for material and work-in-process moves by rearranging the rectilinear representations of workcenter space.

References

1. Bernard T. Smith, *Focus Forecasting* (Boston: CBI Publishing Co., 1978).
2. Robert G. Brown, *Statistical Forecasting for Inventory Control* (New York: McGraw-Hill Book Co., 1959).
3. Eliyahu M. Goldratt and Jeff Cox, *The Goal* (Croton-on-Hudson, NY: North River Press, Inc., 1984).
4. *APICS Dictionary*, 5th ed. (Falls Church, VA: American Production and Inventory Control Society, 1984).
5. *Ibid.*

Bibliography

Buffa, Elwood S. *Modern Production/Operations Management*, 6th ed. New York: John Wiley & Sons, Inc., 1980.

Burnham, John. *Japanese Productivity: A Study Mission Report*. Falls Church, VA: American Production and Inventory Control Society (APICS), 1983.

Deming, W. Edwards. *Quality, Productivity, and Competitive Position*. Cambridge, MA: Massachusetts Institute of Technology, 1982.

Greene, James H. *Production and Inventory Control Handbook*, 2nd ed. APICS. New York: McGraw-Hill Book Co., 1987.

Jansen, Robert L. *Production Control Desk Book*. Englewood Cliffs, NJ: Prentice-Hall, Inc., 1975.

Magee, John F. *Industrial Logistics*. New York: McGraw-Hill Book Co., 1968.

Mayer, Raymond F. *Production and Operations Management*, 3rd ed. New York: McGraw-Hill Book Co., 1975.

Plossl, G.W., and Wight, O.W. *Production and Inventory Control*. Englewood Cliffs, NJ: Prentice-Hall, Inc., 1975.

Albert Ramond and Associates. *Controlling Production and Inventory Costs*. New York: McGraw-Hill Book Co., 1967.

Readings in Zero Inventory. APICS 27th Annual International Conference Proceedings. Falls Church, VA: APICS, 1984.

Sepehri, Mehran. *Just-in-Time, Not Just in Japan*. Falls Church, VA: APICS, 1986.

Teichroew, Daniel. *Introduction to Management Science: Deterministic Models*. New York: John Wiley & Sons, Inc., 1964.

Wilson, Frank C. *Production Planning and Control Handbook*. Englewood Cliffs, NJ: Prentice-Hall, Inc., 1980.

MATERIALS MANAGEMENT

The production planning and control function and some of its key tools were explored in Chapter 20. Materials management usually incorporates this function with the other manufacturing support services that address the total product logistics support issues. These issues include all of the material and product replenishment decisions and may include all activities related to materials movement.

The inventory management responsibilities include the policies and rules that govern the aggregate inventory management decisions. Other manufacturing support roles usually included under the overall materials management umbrella are purchasing (procurement), receiving, warehousing (storage), and material handling. The environment and organizations of companies utilizing just-in-time manufacturing principles differ from those used in support of traditional manufacturing approaches. This chapter focuses on most of the key support responsibilities.

CHAPTER CONTENTS:

INVENTORY MANAGEMENT

Of all the disciplines represented by the field of materials management, none is more prominent than inventory management. But inventory management is, itself, a varied collection of many disciplines. In its simplest forms, the inventory management responsibilities may include the establishment of policies and procedures as well as the maintenance of manually posted card records. Thousands of materials managers depend heavily on computerized forecasting and manufacturing resource planning (MRP II) systems to control the flow of inventories. In the most competitive and innovative manufacturing companies today, newer methods utilizing the pull concepts of just-in-time (JIT) and frequent supplier communications orchestrate the management of inventories that may be turning over 20, 40, or more times per year. The following sections cover a number of key techniques that relate to the management of the inventory asset.

USE OF FORECASTING

Inventory is one of the most important financial assets present in manufacturing companies. Stocks of raw materials, work-in-process inventory, and finished goods constitute the focus of control for the time they are held before being converted into sales dollars. The shorter the period that inventory is held, the more productive the asset. Inventory affects the financial health of a company in the following two ways:

1. As an asset representing stored value that, when sold, will produce income and, hopefully, a profit.
2. As a major investment that is financed by equity or debt.

Therefore, inventory levels directly affect a company's rate of return on its total assets. The cost of financing the inventory may be quite visible if interest is being paid on debt or it may be quite difficult to quantify if it is solely the loss of other investment opportunities, such as increased automation that could improve profitability. Hence, forecasting of product demand that is as accurate as possible is one of the key tools used in the management of the inventory asset.

Inventory may also exist as a result of lot sizing rules. For example, it may be more economical to produce more of a product or work-in-process item than is immediately required. Another cause for inventory results from discrepancies between supply and demand. Types of inventories usually found include the following:

- Work-in-process and in-transit inventories.
- Raw material.
- Finished goods or semifinished products, manufactured to cover anticipated demand and prone to significant forecast error.
- Inventory buildup in anticipation of a new product introduction or special promotion.
- Purchase of a stockpile inventory in anticipation of a supply interruption such as an impending strike, or in anticipation of a substantial price increase.
- Manufactured products to cover seasonal demands that exceed near-level production requirements.

***Contributors of this chapter are:* W. David Lasater**, *Managing Consultant, K.W. Tunnell Company, Inc.;* **Kenneth W. Tunnell**, *President, K.W. Tunnell Company, Inc.*

***Reviewers of this chapter are:* Adnan Aswad**, *Professor and Associate Dean, School of Engineering, The University of Michigan-Dearborn;* **Paul Chapman**, *Associate, Transportation and Logistics Group, Temple, Barker, & Sloane, Inc.;* **Ronald M. Hutchinson**, *Managing Consultant, K.W. Tunnell Company, Inc.;* **Dennis E. Kaufman**, *President, Vacuform Corp.;* **Leo Roth Klein**, *President, Manufacturing Control Systems;* **Ken Stork**, *Corporate Director of Materials and Purchasing, Motorola, Inc.;* **Hayward Thomas**, *Consultant, Retired—President, Jensen Industries;* **Roger G. Willis**, *Partner, Management Information Consulting Div., Arthur Andersen & Co.*

CHAPTER 21

INVENTORY MANAGEMENT

Therefore, forecast accuracy is a key determinant of the size of the inventory asset. Demand for products and parts inventories present themselves in a variety of ways. Some inventory items have only one source of demand. Other items receive demands from multiple sources. The following paragraphs discuss the nature and importance of the demand functions on the management and control of the inventory resource.

Independent Demand

Demand for an item is considered to be ''independent'' when that demand is unrelated to the demand for other items. Demands for finished products arriving through orders are the principal element of this type of demand. This form of demand is usually associated with the primary revenue source for the manufacturer and is the subject of the bulk of the forecasting effort. Demand for items that will be consumed in destructive testing and service parts requirements are likewise independent demands.

Dependent Demand

Demand for parts or raw materials are considered to be ''dependent'' demands when they are derived directly from the demands for other items. The usual source of these requirements is the output of a bill of material ''explosion'' (see Fig. 21-1). These demands are then accumulated as component and material requirements by time period. Such demands are, therefore, calculated and should not be forecasted independently. Some items are subject to both independent and dependent demands.

Demand Variability

The demand for products, parts, components, and materials is almost never stable over time. Forecasting addresses the trend and seasonal causes of variation and anticipates additional random variation in actual demands. Demands may vary from plan as a result of forecast error and lead time variation. Forecast error is always anticipated. Response to this demand variability may be dealt with either by inventory buffers or by improvement in the capability and flexibility of the manufacturing processes. Lead time variation may occur in both the input and output sides of the manufacturing processes. In both cases, the best efforts toward good management of the inventory investment are those that are applied to minimizing demand variation. Forecast error variation is difficult to control, but through meaningful feedback and involvement of people contributing to the variation, such error can be minimized.

Lead time variation is automatically minimized as lead times are shortened. It is possible to theorize that lead time and the lack of reliability in the flow of quality products are the only real causes of problems with variation. As lead time approaches zero, flexibility exists to a degree that the anticipated levels of forecast error variation can be tolerated (see Fig. 21-2).

Replenishment Policies

Inventory is always replenished according to some set of rules, either formal or informal. In small companies, there is a tendency to use variations of the simple two-bin system. When one bin is emptied, the second is pulled into the picking position and a signal is given to begin the production or procurement cycle to replenish the first. Bins or storage containers in a pure two-bin system must be able to hold enough inventory to cover all demands that are expected during replenishment lead time.

The objective of inventory policies must always be to balance the cost of carrying inventory with the service level required. The principal measure related to this activity is called inventory ''turns.'' Inventory policies and procedures determine what turns can be achieved. The turns ratios experienced by companies using conventional manufacturing philosophies, functional layouts, and large lot sizes will be far less than those experienced by companies using JIT manufacturing methodologies. The difference in ratios might be expected to fall in the range of 2 to 10 for the conventional manufacturer vs. ratios of 10 to 50 or more for the latter. Calculating inventory turns is simply a measure of annual usage at cost divided by the average inventory usage, also at cost. The equation for this calculation is as follows:

$$\text{Inventory turns} = \frac{\text{Annual inventory usage \$ at cost}}{\text{Average inventory \$ at cost}} \tag{1}$$

Turns may be calculated separately for various categories of inventory such as raw material, purchased parts, work in process, or finished goods. It is not possible to add or average any set of individual inventory groupings to arrive at an overall turns ratio because there is only one usage number that is truly meaningful. That is the total standard cost of goods sold—the same dollar measure that is used to value inventory. Individual managers may be responsible for achieving a specified turns rate on a segment of inventory, but the overall inventory policies governing the entire logistics process will determine the total ratio for the company.

Safety Stocks

Minimum inventory balances may be specified as protection against stockouts when the cost of such an event is quite high. Protection may be purchased with additional inventory investment against an anticipation of an unusually large upward swing

Level	Part Number	Description	Quantity per Unit
1	A1	Make—final assembly	1 Each
2	CC	Make—subassembly	1 Each
3	EE	Purchased component	2 Each
3	FF	Make—fabrication	2 Each
4	HH	Purchased raw material	10 Feet
2	BB	Purchased component	2 Each
1	A2	Make—final assembly	1 Each
2	DD	Make—subassembly	2 Each
3	EE	Purchased component	2 Each
3	GG	Make—fabrication	2 Each
2	BB	Purchased component	4 Each

Fig. 21-1 Example of an indented bill of material (explosion). ***(K. W. Tunnell Company, Inc.)***

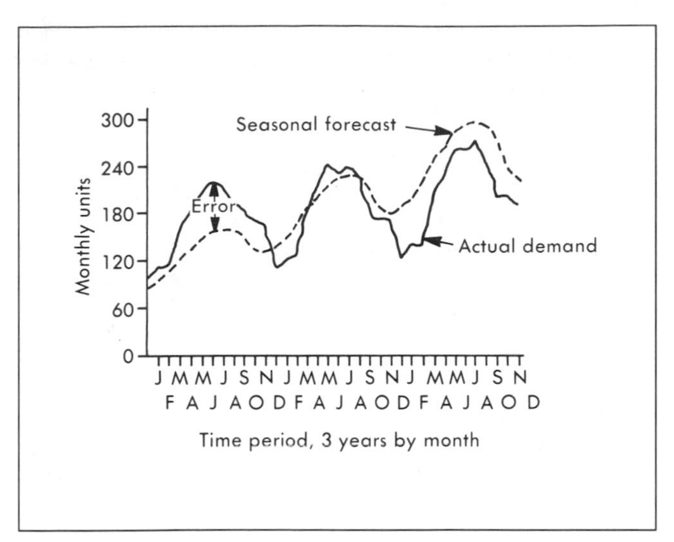

Fig. 21-2 Graph comparing seasonal forecast demand with actual demand for a three-year period. The error between the two is indicated. (*K.W. Tunnell Company, Inc.*)

in demand or lead time. If safety stocks are specified by the inventory replenishment policy, such stocks should be given separate stocking locations.

The theoretical basis for the calculation of appropriate safety stock levels to reduce the probability of a stockout to near zero requires the tracking of demand or usage deviations from plan caused jointly by forecast and lead time errors or deviations. Once the mean or average is known and an estimate of the standard deviation measuring variability of the forecast error is calculated, a quantity can be determined that will protect against stockouts (see Table 21-1). Table 21-1 assumes that the variability of the demand error is random and may be described as a normal distribution. A simple test of this assumption may be made by plotting a histogram for the demand variation experienced or by plotting the distribution on normal probability paper.

Safety stock should always be considered an ''added-cost'' investment, and methods of reducing the risk of stockouts, such as improving the associated lead time, should be considered first. The use of safety stocks can always be expected to lower the inventory turns ratios while increasing the probability of higher customer service measures. The technique should only be considered to be equivalent to an insurance policy, not a cure for the underlying lead time related problems that are inherent in the manufacturing environment being protected.

TABLE 21-1
Safety Stock Determination

Standard Deviations of Demand Variations	Probability of No Stockout, %
1.0	84.1
1.5	95.3
2.0	97.7
2.5	99.4
3.0	99.8
3.5	99.9

(*K.W. Tunnell Company, Inc.*)

ORDER POINTS/ORDER QUANTITIES

Of the key quantitative tools used in production and inventory control procedures designed to carry out the execution of stated inventory policies, various forms of reorder point (ROP)/reorder quantity (ROQ) logic are used. Many processes can be described as variations of generic ROP/ROQ logic.

The simple two-bin system may be said to have an order point of ''one bin full (remaining)'' and an order quantity equivalent to the volume of one bin or container. The old manual inventory card methods gave the names for the logic and used a handwritten or typed pair of numbers: order point and order quantity. Material requirements planning (MRP) logic uses a set of rules that is capable of describing the quantity of zero or any time-dependent quantity like ''two weeks'' supply as ROPs and many varieties of fixed and variable quantity statements as ROQs. Further, MRP extends the arithmetic calculations out into future periods so as to give a projection of all the times that an ordering action must occur during the planning horizon, often as much as 12 months in advance. Even kanban can be described as having a reorder point of one or more ''card's'' worth and a reorder quantity of one ''card's'' amount. When a card's worth of material is consumed, the move authorization causes a production authorization to be given to the first operation for exactly that many more.

Empirical Assignments

Both order points and order quantities are subject to the application of empirical quantity or time period assignments. In manual and computerized systems alike, all methods of inventory replenishment require the establishment of rules (policies and procedures) governing the signal to begin replenishment (when to reorder) and rules determining the quantity for the next addition to inventory (how much to order).

Empirical rules may be simple quantity measures or they may be time and demand data dependent. Quantitative rules can be expressed in a manner similar to the following statements: ''when the inventory balance drops below 100, order 200 more'' or ''when the last container is pulled from the warehouse, notify X to order 3 more containers.'' This type of rule is just as easily implemented in a small manufacturing operation with all manual systems as it is in those with computerized support.

The more complex arbitrary rule is the one stated as a function of time, such as ''when the inventory balance drops below the demand forecasted for lead time weeks plus one week for safety, order two forecasted weeks' supply.'' Many MRP systems that are used for looking at demand and replenishment data on a time series basis utilize this form of quasi-empirical rule assignment. Of course, it is also possible to combine the previously described rules with a procedure dictating: ''order 200 when the current inventory balance drops below the forecasted or calculated usage over the stated lead time period.''

The kanban technique, discussed previously as an ordering technique associated with just-in-time manufacturing, is a variation of the empirical order point/order quantity rule assignment, which signals the need for replenishment by visually requiring that ''when this container of 12 parts is emptied, return the empty container (or signal card) to the source location to trigger the production of 12 more parts.''

Theoretical Basis

The basis for inventory management theory and replenishment calculations is the basic sawtooth model describing inventory consumption and replenishment (see Fig. 21-3). Order points are shown as signal points on the straight-line curve showing the diminishing total of inventory on hand. The amount of the replenishment is the order quantity. All methods and systems attempt to start the replenishment process at a point in time where there is sufficient inventory to cover the demands that will occur during the replenishment lead time.

The order quantity used determines the height of the inventory curve upon stock receipt. The effect of halving order quantity or cutting lead time is easily visualized in this model. In the statistical model shown in Fig. 21-4, the inventory requirements are reduced toward 1 when lead time and set-up and ordering costs approach zero.

Statistical Calculations

It is helpful for the practitioner to explore several practical methods related to the calculation of the ROP/ROQ quantities. These methods are based on the statistics that are suitable to the sawtooth models described in Figs. 21-3 and 21-4.

Order point. The statistical basis for order point is the Poisson distribution, which has had its primary application in studies involving random arrivals. In the inventory control application, both the average number of orders received during the lead time period (N) and the average number of units on each order (D) can be tracked. Then, using Table 21-2, a service factor variable (F) can be extracted and a value to be used as the order point (OP) can be calculated using the following equation:

$$OP = D\,(N \times F \times \sqrt{N}) \qquad (2)$$

where:

OP = order point
D = average demand
N = average number of orders in lead time
F = service factor (from Table 21-2)

The Poisson distribution can only be used to estimate the number of orders that will arrive during lead time, not the total demand that will occur during that period. Further, Eq. (2) is extremely sensitive to the average order quantity experienced because it is the primary multiplier. Finally, the shorter the lead time, the lower the inventory required due to the reduction in the average number of orders occurring during that period.

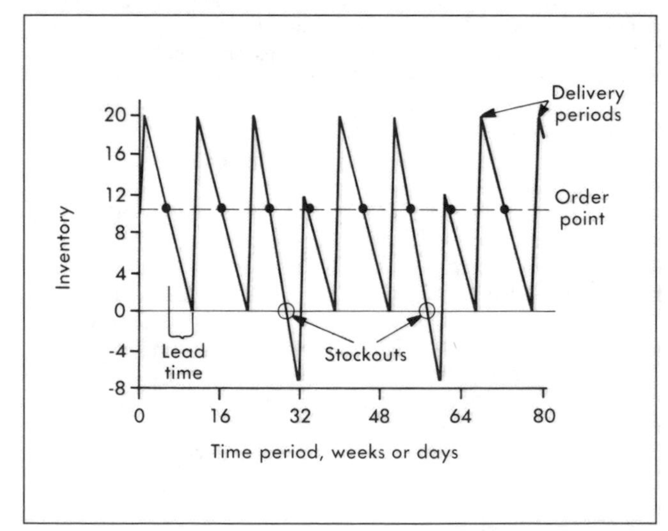

Fig. 21-3 Example of a sawtooth inventory curve having an order quantity of 22, an order point of 10, and a lead time of 6. (*K.W. Tunnell Company, Inc.*)

Order quantity. The most common use of a statistical calculation in inventory control has, for several decades, been that of the standard economic order quantity (EOQ) formula, which is based on the model shown in Fig. 21-5. This formula attempts to balance inventory carrying costs, with the ordering costs. Annual usage in pieces is required as the first estimate in the calculation. The approximation of ordering costs must include set-up costs if the part is a manufactured item rather than a purchased one. Inventory carrying costs result from the multiplication of the cost of one item by the management policy variable. This last factor describes the interest rate percentage believed to be appropriate as a forecast of appropriate costs, including the cost of money, the cost of storage, the cost of handling, the cost of storage loss, and the costs associated with inventory obsolescence.

Although the EOQ formula serves as a near-perfect model of the theoretical inventory management problem, it has grown to be accurately criticized as the square root of an estimate times an approximation divided by the product of an average unit cost and

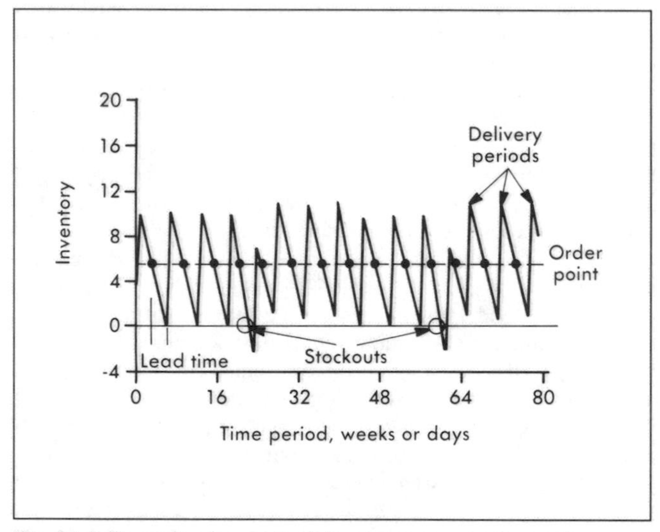

Fig. 21-4 Example of a sawtooth inventory curve when order quantity is reduced to 12, order point is reduced to 5, and lead time is reduced to 3. (*K. W. Tunnell Company, Inc.*)

TABLE 21-2
Poisson Service Factor Table

Backorder Acceptable, %	Service Factor	Customer Service Expected, %
20	0.8	80
15	1.0	85
10	1.3	90
5	1.7	95
2	2.1	98
1	2.3	99
0.1	3.1	99.9

(*K.W. Tunnell Company, Inc.*)

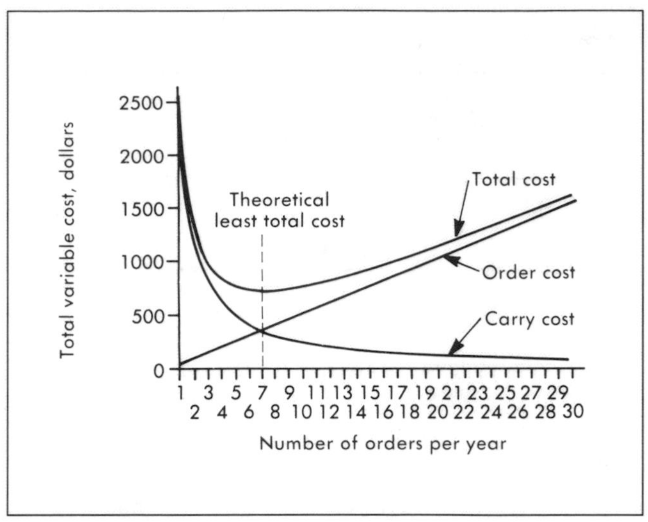

Fig. 21-5 Order cost vs. inventory carry cost curve. (*K.W. Tunnell Company, Inc.*)

a management variable (a guess). It is included in this discussion to help understand inventory cost variables and because it is useful in certain preliminary analyses. It is not recommended as a practical shortcut to order quantity calculation because itdepends on estimated values. Yet it is calculated in a manner that gives rise to the delusion of scientific "accuracy." The formula for EOQ is as follows:

$$EOQ = \sqrt{(2AS)/(ic)} \quad (3)$$

where:

EOQ = economic order quantity
A = annual usage
S = order and set-up costs per order
i = interest and storage costs percentage
c = unit cost of one part

Example calculation results are shown in Table 21-3.

Stockouts/Shortages

Stockouts on finished products and component or material shortages are symptoms of failures of the production and inventory control process. In materials management, this is the usual complaint. Why this occurs is often not well understood by members of the management team. There are a variety of reasons why the occurrence of stockouts is so common.

One of the potential causes is simply human error; when hundreds or thousands of items are being scheduled by one or more persons, errors and oversights are likely occurrences on some percentage of the total. In order of magnitude, it is more likely that frequent errors are caused by poor top-level planning, bad inventory policies, inadequate systems, or a combination of factors. It is impossible to avoid parts shortages without excessive inventory investment if the forecasting, management planning, and master scheduling efforts are not a focus of continuous improvement. In times of inventory reduction pressures, management teams often attempt inventory policies that are unrealistic when taken arbitrarily and without corrective action on the real causes for the inventory that exists. Master scheduling must be timely, operate within the umbrella of the management planning focus, and be accurately maintained with good data disciplines to minimize detail part-level shortages. Refer to Chapter 20, "Production Planning and Control," of this volume for a detailed discussion on master scheduling.

Inventory management systems are also possible shortcomings in material management's focus on the balancing of service rates and inventory investment. If the organizational approach and the system's tools do not match the current operating environment of the manufacturer, rampant occurrence of stockouts and shortages will be evident. This is often caused by a gradual change in the basic business over a period of years with too little change in the methodologies applied to managing the production and inventory requirements. Growing businesses outgrow their old organizations and systems; both can become totally inadequate.

Expediting is a precursor symptom to parts and component shortages or stockouts. Often individual heroics in the production and inventory control function become standard practice, always trying to orchestrate little miracles to keep the factory operating and the customers out of trouble. This is a very costly solution. Costs are always added when parts are acquired through expediting, whether sourced in the shop or at vendor locations. But it must be remembered that expediting is only a symptom of the underlying problem; only at its worst does it get added to the list of causes—outdated systems, inappropriate policies and techniques, and people, including management.

If a manufacturer attempts to implement JIT principles without solving the planning and people problems, the resulting shortages will cause production interruptions because there will be little or no buffer stocks for protection. Stockouts and parts shortages are intolerable in the JIT environment. Expediting is even less of a solution in this environment. JIT manufacturing requires the reliability of flow that is associated with statistical control of processes (and the ownership of the quality and flow responsibility) by the persons who actually control the activities (the operators and buyers).

TABLE 21-3
Economic Order Quantity Calculation Results for a $20 Item

Annual Usage	Order and Set-Up Costs	Carry Cost Rate Estimate, %	Economic Order Quantity
20,000	$50.00	12	913
20,000	$50.00	20	707
20,000	$50.00	30	577
20,000	$200.00	20	1414
20,000	$500.00	20	2236
50,000	$50.00	20	1581
100,000	$50.00	20	2236

(*K.W. Tunnell Company, Inc.*)

ABC Analysis

A popular technique that lends itself to good management of the inventory asset dollars is the classical "ABC" analysis, which results in the coding of items by categories called A, B, and C (see Table 21-4). This technique requires the sorting of all items by the amount of dollar demand (at cost) recorded over some past period or from the output of an MRP system projected over some future period. It is based on the principles set forth by the Italian economist Vilfredo Pareto (1848-1923). The policies

TABLE 21-4
ABC Analysis of a $1,000,000 Inventory Usage at Cost

Number of Parts	Annual Usage Cost	Annual Usage Percent	Cumulative Usage Cost	Cumulative Usage Percent	Inventory Category
6	$681,000	68	$ 681,000	68	A
15	$182,000	18	$ 863,000	82	B
79	$137,000	14	$1,000,000	100	C

(K.W. Tunnell Company, Inc.)

associated with the use of the ABC analysis technique focus on maximizing the human resource attention on the "vital few" described by Pareto. It is usually observed that only about 20% of the items in any inventory will be involved in 80% of the usage measured by dollars. If this top 20% is managed carefully, the lower dollar items can be handled less often with little effect on the total dollar investment. Therefore, it is appropriate to base inventory policy statements on the basis of ABC analyses as a method of establishing an inventory plan. Basing inventory policies on the ABC analysis results in items being given replenishment rules like the following:

1. Review A items weekly, and order 1 week's supply when less than a lead time plus 1 week's supply remains.
2. Review B items biweekly, and order 4 week's supply when less than a lead time plus 2 week's supply remains.
3. Review C items monthly and order 12 week's supply when less than a lead time plus 3 week's supply remains.

The result of such policies and procedures is that the high dollar volume items get the most attention. In this example, the A items will be individually reviewed 4 times as frequently as the B items and 12 times as frequently as the C items.

Warnings

There are many ways in which the materials management function can fail to meet its objectives of service to its operations organization. It may be helpful to review a few of the potential causes and focus on appropriate actions.

A typical forecast will always contain some level of error. Materials management resources should be able to educate, train, and guide forecasters into a program of continued improvement of their prognostications. Likewise, it is important that the operations side of the house be encouraged to continually focus on improving the true flexibility of the manufacturing facility. The more rapidly operations can respond routinely (the shorter the manufacturing lead time), the less problematic the variation in demand. Inventory, or "stored capital," is a very expensive way to provide for the reliable flow of parts and materials into the manufacturing processes.

Inventory policies must reflect the current state of the manufacturing enterprise as to its achievement of overall flexibility. However, it should be recognized that reorder points, the ROP portion of the routine calculations, are a function of inherent lead times and that lead time is essentially the root of all inventory. As lead times approach zero, inventory requirements do likewise. The reorder quantity portion of the calculations, ROQ, is a function of set-up and ordering costs. A wise manufacturer will invest more heavily in set-up reduction and less heavily in inventory.

Just-in-time manufacturing (JIT) is growing in popularity as more and more management teams become aware of its benefits in a worldwide marketplace. But it cannot be achieved without some very basic changes in the establishment of the reliability of flow through aggressive and continuous improvements in process capability and first-time quality. There is no tolerance for rework or replacement of bad lots; cycle times need to be short. The benefits far outweigh the investment, but JIT is not an approach that can be implemented overnight. Many experiences exist that prove that it takes two or more years to dramatically change the manufacturing culture in a single organization and its unique group of suppliers.

There has been a tendency over the past few decades to develop an overdependency on computer systems. Management teams have been repeatedly deluded into believing that the solution to their problem can be purchased and "plugged in." This has led to complexity in approaches far beyond that which is actually required. Computers should be used only as tools to aid in the planning and communications processes. Shop floor execution is usually better with much simpler and more visual methods for controlling the flow of work in process because computer systems must assume a preciseness in execution that is not likely to be achieved. The small number of "Class A" MRP system users as a percentage of all users of time-phased manufacturing resource planning systems is overwhelming testimony that the disciplines built into the model are seldom duplicated in reality.

JUST-IN-TIME (JIT) INVENTORY

Since 1980, the increasing popularity of techniques associated with the management of materials and other manufacturing resources has led to a large increase in the publication of books, articles, and speeches on the subject generally known as just-in-time manufacturing. The popular beliefs about the methods, the conditions of its success, and the potential in American manufacturing remain diverse. The most popular misconception is probably the one that says "...it is a method based on someone else holding inventory for you until you need it."

Definition

Just-in-time (JIT) inventory is usually referred to as Zero Inventory (ZI) by the American Production and Inventory Control Society in its education and promotional programs. Different mnemonics have been used to describe the implementation efforts of different manufacturers: ZIPS for Zero Inventory Production System by General Electric, CFM for Continuous Flow Manufacturing by IBM, MAN for Material As Needed by Harley-Davidson, and many others.

The basic premise is based on the concepts, procedures, and techniques for achieving high enough in-process quality to operate with near-zero inventory between processing steps within a factory. Workers and line management in Japan, particularly those at Toyota, Toshiba, and Hitachi, chose to focus on the pursuit of quality perfection in their manufacturing processes. In fact, they did achieve quality levels that allowed them to begin to measure parts defects per million instead of percent defective. This was accomplished through a continuous improvement process that focused on making all elements of the manufacturing processes reliable and predictable.

This focus found its way into the United States through a few firms such as General Electric in 1980. Its ZIPS approach sought to flow material through the plants with a minimum of intervening queues. Small-lot production began to become a new way of factory life. With no batch or lot queues, and essentially no defects to cause flow interruption, near-zero inventory levels become attainable. Without the prerequisite reliability of parts flow, production stoppage is the predictable result.

Manufactured Parts

In manufacturing, the attainment of near-perfect quality requires small lots to allow early detection of problems and prompt corrective actions. The focus of quality assurance efforts becomes supportive defect prevention instead of defect detection. JIT production and the achievement of near-zero in-process inventories is a never-ending continuous improvement process (see Fig. 21-6). Responsibility for quality must lie directly with the operator. With this responsibility, the operator must also have the authority to shut a process down that is producing unpredictable results.

Low-cost approaches for detecting defects during the process followed by immediate root-cause analysis are critical elements of this approach. The achievement of the targeted near-zero inventory depends totally on reliable and predictable processes. This requires an ever-present mode of operation that takes every possible action to prevent defects. When defects do occur, they must be caught almost immediately and used to help isolate the causes so that a cure can be implemented to prevent a recurrence. In any manufacturing environment, it is soon learned that inventory exists for only the following two reasons:

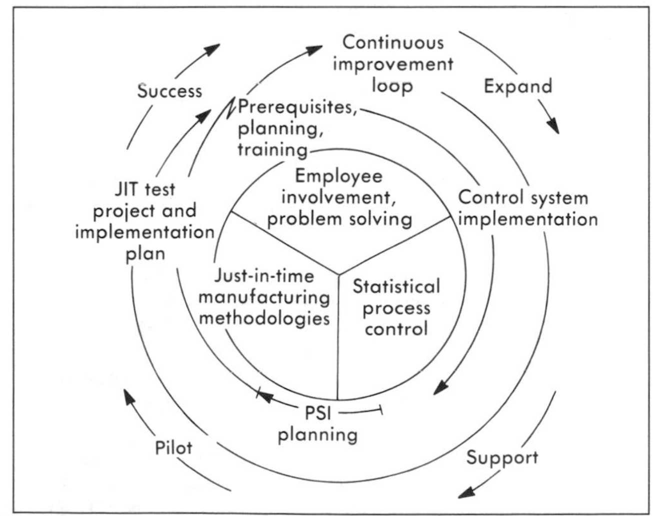

Fig. 21-6 The JIT continuous improvement process diagram. (*K.W. Tunnell Company, Inc.*)

1. To cover the uncertainties associated with material flow related to lead times.
2. To cover the risks associated with the failure of prior processes to deliver quality materials.

Therefore, to eliminate the inventories, one must eliminate the reasons for their existence.

The first cause is approached through a continuous reduction in the lot size requirements. Because of the impact of "economic" job sizing, the factors contributing to order costs must be eliminated. Even in the old standard economic order quantity (EOQ) formula, it can be seen that as set-up or order costs approach zero, the calculated economic lot size approaches the practical lowest limit, 1. This leads to set-up reduction efforts as part of the continuous improvement process. If there is a willingness to reconsider past ways of thinking about equipment utilization, or have surplus equipment that can be easily dedicated to manufacturing cells or dedicated processes, set-up times may, indeed, drop to zero.

The second cause is appropriately addressed by implementing methods that allow the manufacturer to gain control of a process and make its output statistically predictable as "good" parts. This requires the use of statistical process control (SPC) techniques in the manufacturing workplace. It further requires the involvement of all employees striving for a common goal—production that is free of all defects and free of all waste costs. This necessitates a significant change in management practices and the entire manufacturing culture. Additional information on SPC techniques can be found in Volume IV, *Quality Control and Assembly*, of this Handbook series.

Purchased Parts

To achieve near-zero inventories on purchased materials, suppliers must use the same statistical quality management techniques in their operations as does the manufacturer. Defects, or missed delivery dates, on vendor-supplied parts and materials can very quickly shut down a production activity that is not buffered by inventory.

Supplier programs are likely to take even longer to implement than the internal programs because support is made more difficult by the typical remoteness of the supplying facilities. There are several steps that can improve the opportunity for near-zero inventory on purchased materials:

1. A reduction in the overall quantity of suppliers.
2. Long-term partnership programs designed to make both vendor and customer more profitable.
3. Devoting human resources in the purchasing department to long-term cost and quality gains, not adversarial negotiating and expediting.
4. Concentration of the supplier base near the manufacturing facility.

It is likely that a company initiating such a program through its purchasing organization will be required to provide some level of training and coordination assistance to its suppliers. There are corresponding benefits caused by the successes of the program in the receiving inspection area and perhaps in rework or other internal waste cost sources. Once again, small lots, short lead times, and no defects remove the need for inventory buildups in the supply chain.

Reliability of Supply

The elimination of causes of defects includes the control of processes used to generate the materials manufactured. Statistical process control (SPC) is the most likely candidate as a technique to support this endeavor, but many management processes come into play. Prevention of defects, not after-the-fact detection, is the focus. The rules for success as stated by Dr. W. Edwards Deming are summarized in Fig. 21-7. These rules were adopted in Japan at least 15 years before a few U.S. companies began to aggressively pursue them in defense of their manufacturing *raison d'etre* in a world marketplace. It is important that both internal and external sources of production understand that a JIT or zero inventory program is a continuous pursuit of perfection, that some of the rules that have been taught for years must be rethought, and that defects can only be eliminated by removing the causes at the source and by prevention augmented by detection at the time of occurrence.

Visual Control

The increased responsibility for quality and production placed in the hands of the operator requires that the control mechanisms become increasingly simple. Visual control methods are the most effective. If a count is needed, a container should be created that makes the count instantly visible; this can be accomplished in a variety of ways. Some of the containers use pins, or holes, each designed to hold one part. By lining up rows of 2 by 5, 4 by 5, or 5 by 10, it is easy to arrive at total counts per "deck" at a glance. This also provides a method of communicating production and move authorizations as well, an improvement over kanban cards.

1. Create constancy of purpose for improvement of product and service.
2. Adopt the new philosophy (toward quality).
3. Cease dependence on inspection to achieve quality.
4. End the practice of awarding business on the basis of the price tag.
5. Constantly and forever improve the system of production and service.
6. Institute training on the job.
7. Improve supervision (focus only on problem solving).
8. Drive out fear (help people to know and understand what is really needed)
9. Break down barriers between departments.
10. Eliminate slogans, exhortations, pictures and posters for the work force.
11. Eliminate numerical quotas.
12. Remove barriers that stand between the hourly worker and his pride of workmanship.
13. Institute a vigorous program of education and retraining.
14. Create a structure in top management that will push every day on the above 13 points.

Fig. 21-7 Deming's points for management.

PROCUREMENT

For many manufacturers, the most significant portion of their total costs lies in the material cost. Often in today's environment, the material component of the cost of goods sold will range from 80 to 90% of the total (see Fig. 21-8). This may amount to a total of 30-70% of every sales dollar. Purchasing management is in the process of change. For decades, the tradition has been an adversarial relationship between supplier and customer, each trying to get "a little extra" for their side. Negotiation and leverage were the only tools considered to be important in the professional buyer's toolbag. The pressures of world competition have made manufacturers realize that partnerships with suppliers can be far more productive than the price-only posture of the past.

VENDOR SELECTION

The process of procurement begins with the selection of a supplier. The tools used include the "request for quotation," often referred to as the RFQ, but the use is changing the nature of the tool. The vendor interview, the plant visits, and the information requested on the quotation are beginning to be oriented to a view that there are many elements of total cost that are difficult to measure. The focus of most enlightened manufacturers today is to employ a smaller and smaller total number of vendors, but to work with them on a continuous basis toward mutual advantages in profitability and competitiveness.

Quality

If the first concern is quality instead of price, how does one begin the process? The real need is to establish whether or not the potential supplier has processes that are capable of supplying the desired quality and reliability of flow. For vendors to be considered as potential candidates, they must be able to do the following:

- Prove the capability of controlling processes with SPC.
- Show that the operators are responsible for the quality of their own output.
- Demonstrate that the manufacturing environment is one that uses SPC to drive out any waste costs in the processes.
- Communicate an attitude of continuous improvement.

A supplier that meets these criteria can contribute annually to decreased costs as a long-term partner with the customer. In the recent start-up of a new foreign-owned plant, some of the parts were sourced in the United States. The primary reason for selecting particular American suppliers was the perception of the "right" attitude on the part of the potential supplier's plant or general manager.

During the past 15 years, it has been proven by Japanese manufacturers that it does not cost more to have fewer defects, and it is not possible to inspect quality into a product. Since 1980, a growing number of American manufacturers are proving to themselves that these truths are universal. Techniques involving all employees are the "secrets" that lead to the process of continuous improvement in manufacturing. It is most important in the acquisition of new suppliers for long-term relationships to determine if these new ideas are prevailing attitudes, instead of the philosophy that "managers think and workers work."

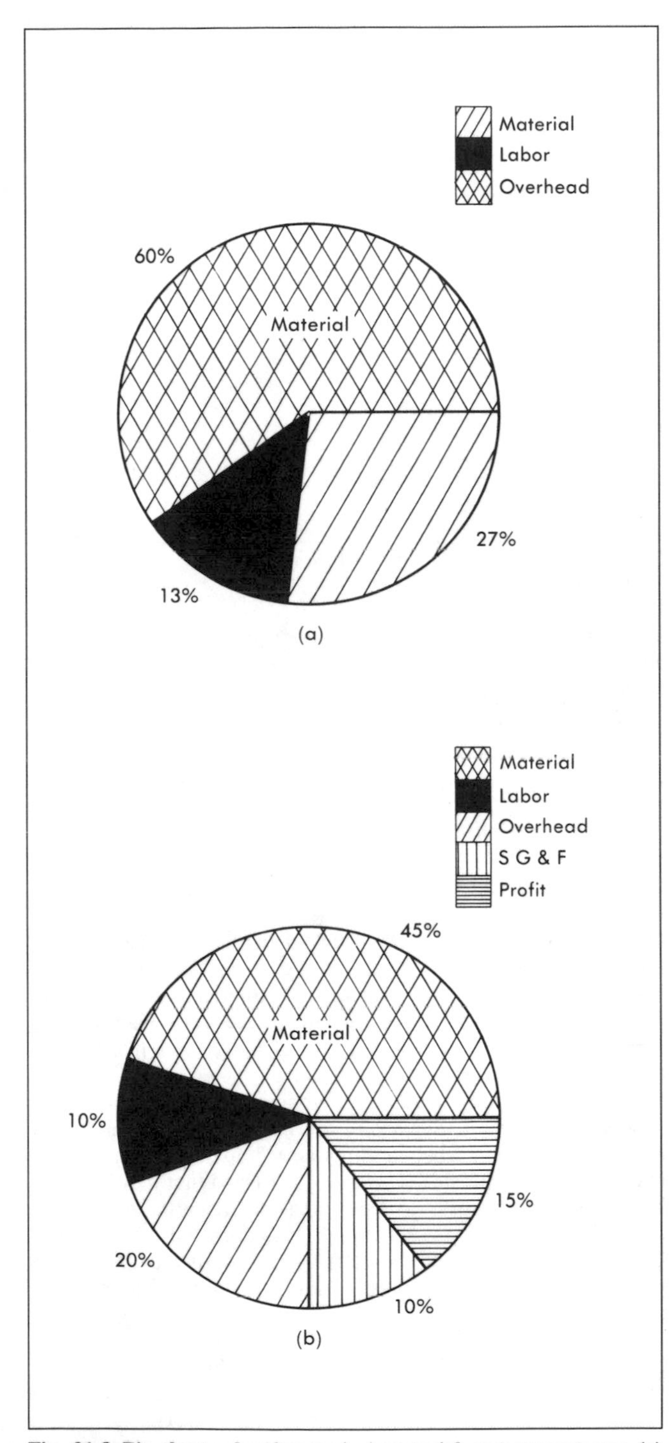

Fig. 21-8 Pie charts showing typical material cost percentages: (a) manufacturing costs and (b) total revenue. (*K.W. Tunnell Company, Inc.*)

Delivery

If minimum inventory and least total manufacturing costs are the objectives, reliability of flow is a key element in the sourcing of any part or material. If the manufacturer seeking a vendor is in the process of implementing just-in-time manufacturing philosophies, it is absolutely essential. The flow of parts and materials cannot be reliable at low costs unless the same JIT techniques of achieving quality are used. If the alternative approach of carrying lots of just-in-case inventory is chosen, then the lowest cost production objective must be sacrificed. Quality described by a supplier in terms of parts per million instead of percent defective is evidence of the capability to deliver on time, without rejects, without rework, even without receiving inspection, and with a minimum delivery lead time.

Without reliability in the material flow from one element of the logistics process to another, the receiving manufacturing plant will be forced into a series of shutdowns and start-ups. Without the protection of banks of inventories, the root problems that were previously hidden must be solved through the continuous efforts of all involved.

Price

Price is important, not just this year's price, but the price that is expected in the years to come. If the supplier chosen adopts the partnership mode of operation, value engineering approaches will become a standard element of the relationship. Often, "partnership" suppliers bring opportunities for standardization, simplification, and minor design changes into the process of cost reduction year after year. Only a few years ago it was common among American manufacturers to speak of the expected average price increase for the current year over the past. Now it is possible to interview many purchasing managers who tell of receiving successive price decreases on those same parts three or more years in a row.

An example of the elements of a request for quotation designed to meet the needs of a JIT manufacturer reveals the amount of open communication that is common to the initiation of this partnership in effective quality and cost purchasing (see Fig. 21-9).

VENDOR RELATIONSHIPS

To achieve new standards of partnership with one's suppliers, new tools and practices must be introduced into the process. As more manufacturers attempt this change in their mode of purchasing parts and materials, the potential of nonadversarial procurement becomes known in the marketplace.

Communications

Supplier communications are an important part of the satisfactory long-term relationship. Contracts and purchase orders will provide the umbrella for the material acquisition communication, but the daily or weekly routine of communicating releases or those "infrequent" expedites is another matter of concern. Many companies have begun the process of improving their ability to communicate rapidly and frequently with their suppliers. Another advantage of fewer total suppliers is the reduced cost for this activity. Computer-to-computer communications began between large companies and their larger suppliers in the 1970s. Now all sizes of companies communicate their requirements from one personal computer to another. Telex and facsimile transmission are used by thousands more.

Request for Quotation Detail
Manufacturing Corporation

Supplier ________ Part number ________
Part description ________

Date ________	Total material cost	________	**
	Total labor cost	________	
Per unit	Total package cost	________	**
	Total freight cost	________	
Enter .00 here	Total tooling cost	________	
if billed in thirds	Overhead & margin	________	
	Price quotation	________	

Price authorized by ________ Good until ________
*Details below — If bracket pricing is used, please provide separate schedule.

	Description	Weight per Unit	Material Costs
1.			
2.			
3.			
4.			
5.			
6.			

Raw material lead time ________ Factory lead time ________

Packaging Considerations Expendable ____ Returnable ____
Package size ________ Cost per package ________
Package type ________ Pieces per package ________
(Returnable 1st ________ ONLY) Cost per unit ________
Package standards are printed on reverse of this sheet.
Circle quantity per package: 1 2 4 5 10 20 25 50 100
or 100 multiple ____ (Max. 10) or 1500 or 1000 multiple ____

Selected mode of transportation: Truck load LTL Other ____

Fig. 21-9 Example of a request for quotation form used for JIT purchasing.

A few years ago the automotive industry began a focus on standardization in several areas with its Automotive Industry Action Group (AIAG). One of the great benefits to all manufacturers of this effort came in the area of supplier communication. The formats for communicating weekly release information against blanket orders or contracts were standardized to allow ease of use and integration with existing computer systems (see Fig. 21-10). A number of software developers have provided packaged software for use on a variety of computers, from personal computer to mainframe. This form of data transmission is critical to the communication of manufacturing planning data to all involved suppliers on a rapid and frequent basis. To do so in a standard data format allows cost-effective assimilation into the supplier's computer to be used in the manufacturing planning and control system. The order entry, acknowledgment, shipping advice, invoicing, and payment transactions are all part of the capabilities planned for this level of organization standardization. More effective communication does not have to result in more paperwork.

Long-Term Considerations

The change in the procurement focus to long-term relationships with a smaller number of total suppliers is, perhaps, the most important change in the psychology of purchasing today. As more and more companies seek profitable partnerships with a small portion of their supplier base, competition increases and the ineffective vendor is driven out of the marketplace. It is no longer just the Far East that can boast of quality and price advantages. A growing number of American suppliers are positioning themselves to regain some of the world market share that they lost to the earlier followers of individuals like Dr. W. Edwards Deming (refer to Fig. 21-7).

The move toward a focus on long-term relationships with vendors also began to surface in the United States around 1980 as a few companies began to pursue implementation of JIT manufacturing principles. It was soon learned that it is impossible to staff a purchasing department sufficiently to improve communications and work through problems and opportunities with the large number of suppliers typically utilized. More time was required with each of these vendors. Joint training programs that focused on establishing a long-term cost-reducing environment took even more resources. Many such departments have been a part of providing the training in SPC or total quality management (TQM), which is essential to a smooth operation that no longer depends on large banks of inventory.

The biggest concern expressed by most procurement managers, as the material supply environment changes again, is that of assurance that there is enough competitive pressure in this new

Purchasing Release Schedule

Vendor: 87572 Goodyear tire   Ship to: 01 Atlanta GA
Release No. 87-5   Date: 2/12/87

Part Number Description	Order No:	Cum Shipped	Last Ship Date	Behind Schedule	Cumulative Shipping Requirements Through: Week of 2/14/87	Week of 2/21/87	Week of 2/28/87	Week of 3/7/87	Week of 3/14/87	Week of 3/21/87	Week of 3/28/87	Week of 4/4/87
24680 Rev B Tire 20.5 x 4	430872	18253	1/30/87	0	17250	19750	22250	24750	27250	29750	32250	34750

	Cumulative Authorizations Fabrication	Material	Cumulative Forecast Requirements Through: Month of 4/87	Month of 5/87	Month of 6/87	Month of 7/87	Month of 8/87	Month of 9/87
Weeks	7	11						
Quantity	32250	42250	44250	54750	65750	76750	87250	97750

Fig. 21-10 Example of an AIAG purchase release form.

mode of operation to ensure that the "best" price is being obtained. It is true that there is a great deal more "trust" involved in this new customer-supplier relationship, but it is also true that price was never a total measure of costs associated with parts and materials obtained from the supplier base.

The old style of "micropurchasing" can be described according to essentially a "buy low" philosophy. The new style of "macropurchasing" focuses much more on the total cost (price, cost of supplier-caused scrap and rework, transportation, handling, and the effects of long lead times in the supplier's operation).

Performance Measurement

Once communication and long-term considerations are worked out, it is important to consider performance measurements of the vendor. Some of these measures will be directly related to the vendor's specific performance on parts or materials supplied. However, it is just as important to conduct some internal measures that reflect the overall success of the quality-oriented procurement environment.

The overall success of the procurement activity is best judged as it relates to the competitive position of the company served. It is appropriate to note that some of the measures, such as the buyer ratios, may get worse before they start to improve. To make a significant transition in the culture of a manufacturing organization, it is often necessary to increase purchasing and manufacturing engineering resources during the implementation period. These increases are usually temporary in nature and are always offset with decreases in other areas, such as quality control. The focus must be changed from defect detection to defect prevention.

Supplier measures. For each part or material purchased, records of certain statistics need to be maintained for each supplier. A few of those that can be related to specific parts or materials are as follows:

- Number of shipments received.
- Number of shipments received on time.
- Quantity of items received.
- Quantity of items received on time.
- Defects incurred in subsequent operations.
- Subsequent costs incurred due to defects.
- Cost-reducing suggestions received/implemented.
- Process improvement or design changes suggested/implemented.
- Manufacturing interruptions caused, if any.
- Periodic process capability measurements.
- Tracking of actual prices billed.

The revelation of hidden costs is shown in Table 21-5.

If the information for Table 21-5 is collected from routine systems, a track record can be established for the nonprice benefits of the individual relationships. Equally important, however, are the collective measurements that result from the overall procurement program, including average material cost per sales unit.

Internal measures. The new approach to purchasing becomes a mixture of those elements that have been in use and those that may have been previously thought of as belonging to either quality assurance, accounting, or some other staff function. If the buyer is held responsible for the quality of the parts purchased, just as the operator is held responsible for his or her output, there is less and less dependence on other overhead functions. The following list suggests that the potential impact of the buyer is, indeed, significant to the total operation:

- Total dollars of sales revenue per buyer.
- Total number of suppliers.

TABLE 21-5
Comparison of Real Costs of Procurement

Part 101010 Cost Element	Vendor A	Vendor B
Price	$1.10	$1.11
Freight	$ .02	$ .01
Scrap	$ .02	$ .00
Rework	$ .05	$ .00
Other	$ .03	$ .00
Totals	$1.22	$1.12

(K.W. Tunnell Company, Inc.)

- Total dollars of purchases per buyer.
- Total cost of scrap and rework on vendor material.
- Total number of receiving inspection personnel.
- Material cost percentage of sales revenue.
- Inventory turns on raw materials and purchased parts.
- Number of parts eliminated through standardization.

CONTRACTS

The traditional purchasing tool used in conjunction with the repetitive acquisition of materials or parts from a given supplier is the purchasing contract. Contracts may be used to define the result of negotiations covering a given quantity, a year's procurement plan, or a longer term agreement. Contracts are used as an attempt to establish a definition for all of the obligations of both the supplier and the customer for a stated procurement situation. The legal implications of formal contracts in purchasing include not only the expectations regarding price and delivery, but also the penalties to be suffered by each party in the event of any default to the conditions of the agreement. Simple forms of contracts used for the procurement of goods and services include those discussed in the following paragraphs.

Blanket Orders

One simple form of a purchasing contract is the document referred to as a blanket order. It is sufficient in its usual content to define both the details of a specific procurement requirement and the pricing and delivery schedules or rules covering a period of weeks or months.

The nature of a blanket order is very much like a standard purchase order in content. The key difference is that delivery schedules are relegated to a separate purchasing release process or they are spelled out in a manner that indicates that revisions may be made to the initial quantities and dates through subsequent change orders. Quite often, blanket orders are issued once a year for the purpose of documenting overall planning volumes and price agreements covering that period. Automotive release standards published by the Automotive Industry Action Group (AIAG) provide one of the accepted methods of communicating release changes on a weekly basis.

Umbrella Agreements

Umbrella agreements represent another form of purchase contract documentation. Again, the actual documentation will most likely appear quite similar to a standard purchase order. The function is very much like that of the blanket order, but this form of agreement covers not just a single part or material, but rather an entire commodity or a group of similar parts. The reason for selecting an umbrella agreement is usually one of providing standardized documentation for the results of negotiated prices and terms. Very often, the umbrella agreement is a sufficient contract to support the agreement on terms and conditions for procurement of a commodity or a family of parts.

The actual quantities and dates for shipment or delivery schedules on individual items included in the agreement must be handled by some form of purchase release activity. In the case of the umbrella agreement, the release must also select which part number or material identification within the group of items covered is being released for shipment.

The identification of the part or material to be shipped was not absolutely necessary with the release against a standard blanket order because each order number specifically designated a single material specification. Therefore, the release information could have just as easily specified the blanket order number on the release instead of the part or material number. The umbrella agreement is, however, equally suitable to contractually cover items purchased using AIAG release standards.

Purchase Orders/Releases

Any purchase order can constitute a valid procurement contract. It is typical for simple purchase order documents to specify a single delivery and the current price purchase order document for infrequently ordered production materials not eligible for coverage under a blanket or an umbrella agreement. Standard purchase order documents are also used frequently for services, tools, lubricants, machinery, and equipment, and for various ''maintenance, repair, and other (MRO)'' items that are required by a manufacturer.

Purchase releases may also be communicated using the standard purchase order form, though it is preferable to use other techniques for this communication. The AIAG standard for release transmission is based on cumulative requirements and simplifies the reconciliation of in-transit material. Other forms of computer-aided or facsimile transmission supported communication are preferred methods. Simple computer printouts of the time-phased requirements regenerated for the next few weeks or months and mailed weekly can suffice for the communication need, if agreeable to both parties.

Supplier Scheduling

The critical need for each manufacturer regarding the scheduling of suppliers is related to the need to minimize inventory investment without loss of the ability to respond to marketplace requirements. The use of various forms of contracts in purchasing increases the flexibility of the parties regarding the subsequent manufacture and delivery requirements support without causing undue risk for either party.

Consideration must also be given to the scheduling impact on transportation and handling costs. In a JIT environment particularly, it is important to schedule deliveries in a way that is both economical to ship and consistent with a near-zero inventory philosophy of operation. It is better to have two parts shipped together in a truckload once a week than it is to schedule a truckload of each to ship in alternate weeks.

Further, packaging of parts and materials affect the supplier scheduling and possibly even the negotiation of price. Smaller packages of components that can be handled internally without the aid of forklifts, for example, are supportive of the elimination of waste costs typically present where material handling is a major internal service requirement. Small boxes can be designed to palletize easily for the ease of loading and unloading trucks without the necessity of pallet load moves and storage in work areas. Many JIT suppliers have learned to use an overpack/underpack arrangement to ship different part numbers on a single pallet as a part of daily or weekly deliveries to support a customer's component part scheduling (see Fig. 21-11).

Reliability of flow of materials becomes increasingly important as buffer inventories are squeezed toward zero. The purchasing contract, which provides the guidelines for the constant communication of the ''best'' knowledge available as to requirements, is critical to this process. As contractual arrangements cover longer and longer periods of time, the supplier is increasingly willing to make specific investments that will help to drive out unnecessary costs in the material being supplied. The added benefit of close communication in the supplier scheduling process is in the improvement in the management of manufacturing flexibility, which results in reduced overtime and production interruptions. The nature of this type of customer-supplier relationship is capable of generating more profitable operations for both companies than the adversarial low-bid, single-order competition that has no long-term implication. The costs associated with expediting at both ends of the supply chain should be avoided completely; both quality and close communications are essential to that accomplishment.

QUALITY AND CERTIFICATION

Part of the vendor qualification effort must deal with the true capabilities that can be demonstrated. This requires specific analysis and a thorough understanding of a potential supplier's processes and procedures.

Process Capability Studies

As quality requirements are redefined from the old standard acceptable quality level (AQL) definition toward zero defects, purchasing agreements and specifications are beginning to aim toward defects targeted in terms of a new measure, parts per million (PPM). A goal of 100 PPM is 100 times better than an acceptance specification of 1% AQL. Purchasing programs striving to achieve quality levels of this order of magnitude require significant changes in approach. One of the specific techniques used is the process capability study.

Explicit part or material design and specification requirements documentation is absolutely critical to this technique. The processes (including personnel, machine, material, and methods) that are targeted for production of the parts or material to be procured must be tested using statistical management techniques to determine whether or not those processes are capable of meeting the specifications essentially 100% of the time. The supplier must use SPC techniques to hold the processes in control all of the time to ensure the predictability of the process. Thus, the operators of such processes must be given the tools and training required to produce "perfect parts" all the time. Gages, test equipment, and sampling procedures must be in place to maintain a process that is well centered near the nominal values that were targeted.

Process capability studies are used to check the amount of variability in the critical measurements on production output by comparing the percentage of common-cause variation vs. the engineering tolerance. If a process is well centered and the percent of the engineering tolerance used by the process variation is small, the capability is judged to be good to excellent (see Fig. 21-12). Only under these operating conditions can quality measured in defect PPM be achieved on a consistent basis.

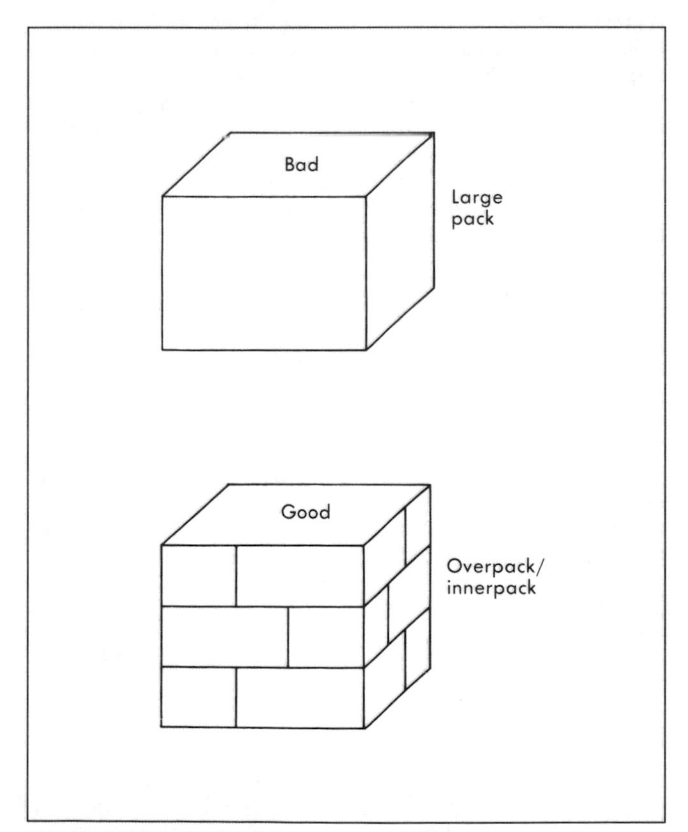

Fig. 21-11 Example of different packaging alternatives. (*K.W. Tunnell Company, Inc.*

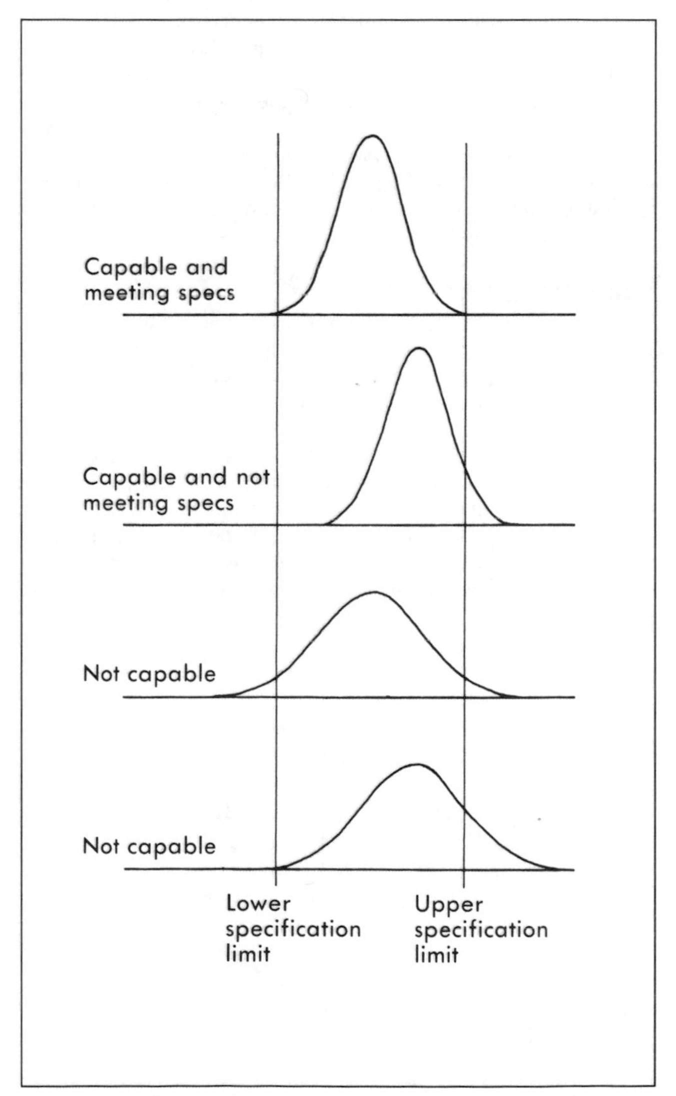

Fig. 21-12 Bell curves of processes that are capable and not capable of meeting specifications. (*K.W. Tunnell Company, Inc.*)

Additional information on process capability studies can be found in Chapter 23, "Achieving Quality," of this volume.

Supplier Quality Assurance

For the manufacturer, the buyer must be held responsible for the quality of procured materials, just as the operator in both the manufacturer's and the supplier's plant must be held responsible for the quality of production. The method by which the supplier ensures quality is its continuing program of process control. It is no longer appropriate to consider that incoming inspection at the receiving location is a suitable method of providing vendor quality assurance. U.S. manufacturing's failure to remain competitive has proven that "inspecting quality in" is a costly and ineffective technique.

Sound quality purchasing practices are consistent with the requirements of any manufacturer's vendor quality assurance programs. They are absolutely essential to companies that are attempting to implement just-in-time manufacturing techniques. Vendors that are willing and able to embark on joint programs designed to achieve this increased level of assurance will be invaluable in making certain that both parties understand the specifications and the criteria of good parts or materials.

A different set of management attitudes is required in both the supplier and customer companies if the desired level of quality assurance is to be achieved. Supplemental (outside) resources may be required to set up process control-based quality assurance, but fewer resources (therefore, lower costs) are required to maintain good quality after the training and methods implementation are completed.

Documentation

With process-controlled quality assurance methods, the documentation is different from the old inspection sampling records. It is more important to understand how well the manufacturing processes are being controlled and the capability weaknesses that need to be worked on than it is to know the statistics of inspection sampling. Therefore, the meaningful documentation sought is provided by copies of SPC charts that were prepared by the operators while the parts were being produced and by the problem-solving documentation that supports the continuous improvement efforts of the employee teams involved in those operator-controlled processes. An example of a process that is in control is shown in Fig. 21-13.

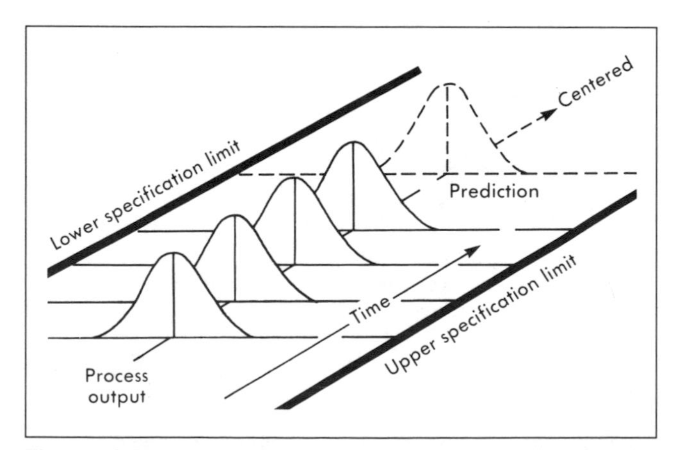

Fig. 21-13 Process control chart that is centered and predictable. (*K.W. Tunnell Company, Inc.*)

The essential element of ongoing quality achievement requires that the focus shift from the detection of defects to the prevention of defects. The interaction of the customer with its supplier regarding the control charts and other documentation will be related to an understanding of the causes of excess variation, shifts in the process, a review of the special causes of variation, and the actions taken to eliminate such variation.

It may be necessary for the supplier or the buyer to pursue their analysis and conclusions from the process documentation into design or specification change proposals. The availability and willingness of design and manufacturing engineering resources to participate in this striving for perfect quality are important elements of the ongoing process.

OFFSHORE PROCUREMENT

From the 1960s until the recent new beginnings of a "Buy American" mentality, importation of parts and materials by manufacturers increased dramatically. The ability to buy quality products at substantial savings in cost influenced many companies to engage in considerable foreign sourcing activity.

Direct Contact

Large manufacturing companies around the world, particularly in the Far East, established an effective sales presence within the United States to assist in the handling of the direct contact demand. Thus, direct contact in many commodities was made quite easy for the purchasing departments of many manufacturing enterprises. In many cases, inventories were brought into the U.S. as stock for foreign-owned distribution centers. This further increased the attractiveness of the international sourcing because the lead time concern related to ocean shipment delays had been removed.

As companies began to be more aware of the methods being used in Japan during the late 1970s, a large number of purchasing executives made routine trips to the countries involved, thereby facilitating the rapid growth of foreign sourcing. This direct contact created immense enthusiasm in procurement for the following two reasons:

1. Most manufacturers identify purchased components and materials as the largest element of their cost of goods sold.
2. Purchasing department executives and managers were (are) most frequently measured by favorable "purchase price variance" or material cost as a percentage of the total.

As a result of this period of focused search and buying from international sources for quality and cost advantages, a large number of manufactured commodities shifted from a position of American dominance to foreign supplier control. In the mid-1980s purchasing environment, some components and materials have become extremely difficult to source competitively anywhere on the North American continent. The U.S. manufacture of still other commodities has begun again only since the value of the dollar began to weaken in 1986.

Trading Companies

Smaller manufacturers on both sides of the ocean found that trading companies have been extremely useful in expanding the supply base to those foreign companies that could not afford a resident representation in the United States. Operating as an agent for a number of foreign plants, trading company organizations made local contact and representation conveniently available to U.S. purchasing professionals.

Trading companies have also established warehousing and distribution stateside, but not to the same degree as the larger foreign manufacturing enterprises. Trading companies are more likely to use a U.S. warehousing site as a means of consolidating ocean freight to a break-bulk type of operation than they are to provide "off-the-shelf" stock for domestic consumption.

Offshore Buying Offices

Large U.S. corporations, particularly those that operate as multinational manufacturers and sellers, have found it beneficial to establish foreign buying offices in support of their purchasing contract negotiation and administration activities. In a recent survey, it was estimated that at least half of the current Fortune 500 industrial companies have an offshore purchasing organization. These organizations are usually small in terms of staffing, but significant in terms of dollars of annual procurement. In most cases, the offshore purchasing unit is considered as an extension of, and is managed by, the U.S. company's centralized purchasing organization. They often serve as one arm of a corporation's contract sourcing to supply many plants by consolidating like demands to increase purchasing leverage in the world market.

RECEIVING, INSPECTION AND STORAGE

Physical control of parts and components is responsible for one of the largest groups of indirect manufacturing labor usage and frequently occupies as much space as the direct manufacturing activity. Receiving and storage typically constitute the closest contact services to operations of all the materials management functions. This results in these functions being viewed internally as the principal materials services rendered. Inspection usually rests with a quality control or quality assurance department that polices the quality of incoming materials through inspection sampling.

DOCUMENTATION ACCURACY

From a materials management standpoint, physical control activities represent the front lines of count control. Not only must the records related to receipts be completely accurate to control the associated accounts payable function, the usual purchasing activity depends on a very rapid response from its affiliated receiving group.

The accuracy of recordkeeping in the receiving function must include all of the inputs from the inspection activity as part of its support to the manufacturing operation. Its records documenting quantities received must be totally accurate as to the part numbers or material identification numbers, as well as to the purchase order data, if that information is to be depended on by the planning and procurement arms of the overall materials support organization.

Inventory accuracy, or the lack thereof, has been one of the most important reasons for the success or failure of the materials systems used. Material requirements planning (MRP) systems are extremely intolerant of any errors in the documentation of inventory and purchase order activities. The inventory control function may use techniques such as cycle counting to verify inventory record accuracy and to make corrections as required. In short, the success of the total materials management function depends on the accuracy of inventory transaction documentation. This responsibility is greatly simplified when self-counting devices are used to aid visually controlled JIT materials replenishment by "pull" signal methods.

CONTROL DISCIPLINES

Disciplines related to the control of inventory data always cross departmental boundaries. Although much of the physical control activity may be organizationally tied to the materials management group, the end use is usually in support of the operations function that provides input to the recordkeeping as to its usage of components and materials and to its production of work-in-process, semifinished, or finished stock output.

Control disciplines include the maintenance of data related to a wide variety of documentation. These may be principally seen as computer records accessible through terminal screens, or they may be a collection of manually posted records. Included in the controls are the following:

- Purchase requisitions.
- Receiving reports.
- Production and assembly orders.
- Material withdrawal requisitions.
- Material movement transactions.
- Stock production or put-away records.
- Customer order shipping records.

The standard measure for sufficiently accurate inventory data useful in the production and inventory planning and control activities is 95% or greater. Of course, there are other staff-supported records that play a critical role in the overall result. These include the part identification records, the bill of material data (or parts lists), and the process routings involved. Note that several levels in a bill of material, even a 95% level of accuracy, accumulate to a total level of "inaccuracy" that will make the support of an assembly process highly unreliable.

Often, added security has been required to ensure that the accuracy of records is maintained. This is usually seen as locked receiving areas, locked storerooms, and locked special "cages" for identified high-risk items. With a just-in-time manufacturing philosophy, the control techniques that have been traditional are not appropriate or manageable. The near-zero inventory levels force an enterprise to be extremely accurate, but only with critical, easy-to-count, small supply stocks that are in-house at a certain time.

Physical control disciplines also include those related to the placement and location of parts and materials so there is little chance of "lost" components in the inventories stored in the stockrooms. Location recordkeeping systems are critical to the larger functional department manufacturing company.

INSPECTION SAMPLING

Inspection activities are increasingly moving back toward the operations involved in the manufacture of parts and materials, but many firms still rely heavily on receiving or incoming material sampling. In these environments, a set of rules governing the acceptance of lots of materials is used. These rules are usually related to some acceptance sampling rationale. Although it is less costly to accept materials by inspecting (if required) only limited sample quantities, there are always risks of accepting materials that do not meet specifications. Even 100% inspection cannot ensure quality of all parts because of boredom and human error factors involved in the process. A variety of standard sources for sampling plans exist, including those related to military specifications (MIL SPEC) and tables used for acceptable quality level (AQL) (see Table 21-6).

The critical need for inspection may come from a variety of reasons, the chief one being the procurement from sources that cannot or do not choose to use statistical quality management techniques in controlling their processes. The highest risks to the manufacturer using such materials exist when:

- The manufacturer of the supplied materials is known to experience low process yields.
- Materials have been supplied by a supplier known to include significant quantities of parts in shipments that do not meet specifications.
- Operations are about to be performed in the manufacturing process where relatively high value is added.
- Operations are about to be performed that make component or material defects difficult to find or repair.
- New designs or new sources are being introduced into the manufacturing processes at the supplying or using location.

In these cases, some receiving inspection and the added cost of quality control inspection is usually more effective than reliance on the performance of the supplier. As much care as possible should be taken to ensure that the materials being sampled are from homogeneous lots, including size, producing machine, operator, and tooling.

MATERIAL HANDLING

The function defined as material handling deals with the movement of materials from the receiving areas, through all manufacturing processes, and finally to the shipment of the finished products. The activities involved in the physical handling of materials, like the other materials management functions, add no value to the products. Therefore, it is essential that all such support activities be done as efficiently as possible. This precipitates decision alternatives regarding both people and equipment and the methods in which they will operate.

The shorter the cycle times, the less opportunity exists for material handling and storage to become an added element of cost. It has already been described how just-in-time manufacturing approaches recognize these effects and minimize the costs of inventory storage and movement. In traditional manufacturing environments, there is often a significant investment in material handling equipment. Forklifts, cranes, and specialized conveyors represent expensive alternatives that are related to conventional manufacturing. Material handlers are usually critical elements of the move-and-wait job queues that are always present in the traditional job shop and some repetitive manufacturing environments.

The shorter the distances, the lower the overall costs of material handling alternatives. As cycles and distances are decreased in approaching a just-in-time manufacturing environment, it is often possible to change the unit quantities and weights of the material receipt to allow specialized carts to become the primary one-move conveyance into the manufacturing area. These known savings opportunities are also being included in the redesign of traditional manufacturing support functions. The simplicity of methods and procedures related to material handling are also critical to the material control system interfaces. Material handlers and support personnel often bear the primary responsibility for carrying out the disciplines required for accurate recordkeeping, which may ultimately determine the successfulness of the support function as a whole.

Finally, order picking, accumulation, and containerization for shipment to customers represent the last opportunity a manufacturer has to ensure the quality of its products. This requires skills in packaging, the choice of packing materials, and the tactics required for safe loading of the transportation vehicle. Material handling cannot contribute to the value of the product, but, in more ways than one, it can contribute to the cost and to the delivered condition.

TABLE 21-6
Example of an Acceptable Quality Level Table Based on MIL-STD-105D

Lot Size	Level II CODE	Sample Size	1% Reject Number	4% Reject Number
2 to 8	A	2	1	1
9 to 15	B	3	1	1
16 to 25	C	5	1	1
26 to 50	D	8	1	1
51 to 90	E	13	1	2
91 to 150	F	20	1	3
151 to 280	G	32	1	4
281 to 500	H	50	2	6
501 to 1200	J	80	3	8
1201 to 3200	K	125	4	11
3201 to 10,000	L	200	6	15

(K.W. Tunnell Company, Inc.)

WAREHOUSING FACILITIES

Storage facilities and equipment are required to receive, store, and ship raw material, work in process, and finished goods in support of the total logistics responsibilities of the materials management function. Factors in designing the warehouse facility include the nature of the materials handled and the volume of transactions. Aisles must be wide enough to accommodate the material handling equipment utilized. It is desirable that warehousing space have the tallest clear ceilings of any space available to the manufacturer. This allows the maximum storage of product for the floor space consumed.

Storage space should be planned for the highest volume of activity to occur with the minimum distances traveled. This can be achieved through a study of the anticipated transaction volumes on all materials and components. If an ABC analysis is available, any A materials that will travel through the warehouse should be stored near the "front" of the warehouse space (nearest the manufacturing space). In contrast, the slower moving C items should be allocated space in the remotest sections of the floor plan. Final layout of storage equipment will depend to some degree on the placement of any columns that are present in the building.

Adequate attention must be paid to the design and location of docks to be used by both shipping and receiving. Depending on the type of outside space available, decisions are required in selecting the minimum cost combination of indoor docks, sawtooth docks, open docks, and/or flush docks. Dock doors and levelers are accessory decisions that depend largely on the capital available, the weather protection needed, and the loading/unloading requirements. Finally, lighting equipment is desirable to facilitate loading and unloading and is essential if any night operations must be supported.

Storage equipment should be adequate to handle the weights and sizes expected and should be designed and placed to minimize the effort involved in put-away and picking. There are many types of pallet racks that can be utilized in the warehousing application. These may have drive-under sections to facilitate forklift movement. Multiple heights of shelving required are achieved through bolt pattern designs of the rack equipment. Not all sections of the warehouse need to be configured the same.

Specialized racks include gravity flow systems and feed-through racks loaded from one side and unloaded from the opposite. Bins and shelving are also utilized to handle smaller and lighter materials.

High-volume storage and retrieval activities may justify automatic storage and retrieval systems (ASRS). However, with the focus of JIT implementation, many previously justified installations are now sitting vacant. Carousels offer still another semiautomated storage and picking alternative.

Handling equipment may vary from overhead cranes to small pushcarts. There are many different types of forklift equipment for handling loading and unloading of pallet loads of material from both transportation equipment and storage racks. Standard forklift equipment may be customized for specific nonpallet handling assignments requiring large clamps (barrel or drum handlers, paper roll handlers, and large case clamps). Fixed-base varieties including narrow-aisle sideloaders are available if tall reach requirements call for equipment to be dedicated to one or more warehouse aisles.

Because it is not desirable to have forklift equipment moving through manufacturing space, the movement of materials required outside of the warehouse may be handled by wheeled carts, dollies, or small hand-operated hydraulic lifts. For the JIT environment, specialized carts may travel to and from the vendor location on company-owned or contracted trucks and essentially bypass the warehousing operation. These carts must be easily handled, provide adequate protection for the parts being moved, and provide for easy visual counting; they may even signal the replenishment requirement in a paperless system.

WAREHOUSING CONTROLS

Paperwork control for the warehousing activities may include all transactions related to receiving, internal movement, and shipment of finished product. Not only are there requirements for the physical or custodial controls, but the warehousing organization must have adequate administrative controls that are accomplished through careful execution of its recordkeeping responsibilities.

Physical control should include accurate identification and location of the materials stored. As material withdrawal occurs, the most satisfactory approach is to adhere to a physical first-in-first-out (FIFO) discipline. The custodial responsibility also includes the appropriate care of all materials free from any form of handling damage. The receiving duties associated with most warehousing activities require that the warehouse personnel ensure that no damage to packages and materials has occurred when the material is unloaded from the carrier and then that no damage occurs thereafter.

The administrative controls will likely include various transactions that drive the material control and accounting support functions associated with both replenishment of stock used and with the shipment of product. Specific tasks may include verification of the quantities received to orders and to packing lists. This is usually followed by an updating of inventory records. Both receipt and put-away transactions require the updating of location records. Finally, all accounting inputs for payable activities on receipts and billing on shipments must be judiciously managed.

REPETITIVE MANUFACTURING

Job shop operations represent approximately 50% of the manufacturing in the United States. Little differentiation has been identified in this chapter for special classes of manufacturing activity. There are other specific types of production that, although similar to job shop approaches, warrant further discussion. This section will only discuss the differences in repetitive manufacturing.

History indicates that Henry Ford implemented the first repetitive manufacturing line in the early part of this century, although it is likely that there were earlier unsung innovators prior to Ford. Since then, repetitive practices have been applied to the manufacture of many products, including automotive, electronics, consumer goods, and many other high-volume items. Many of the practices used in repetitive shops vary significantly from those of traditional job shops, although in practice, most repetitive companies apply job shop techniques to a portion of their activities, usually components or subassemblies. Repetitive manufacturing also employs many of the concepts of the just-in-time philosophy discussed in Chapters 15 and 20 of this volume.

The distinguishing characteristics of repetitive manufacturing are dedicated facilities/fixed routings, production planning/master scheduling, line supply, labor reporting, and purchasing.

DEDICATED FACILITIES/FIXED ROUTINGS

Products are assembled in workcenters that are dedicated to the manufacture of a single product or product family. Production runs are lengthy or often continuous. Personnel are assigned to workstations, and the work content is distributed evenly among the number of stations according to a technique known as "line balance." The line balance determines the production rate per shift. It is difficult to develop alternative balances; therefore, changes in planned production rates require careful preparation and planning. Component material is structured in the appropriate bills of material to the operation at which it will be consumed.

PRODUCTION PLANNING/ MASTER SCHEDULING

Production planning and master scheduling is usually expressed in rates per day for each product or product line. The rate is generally stable over a reasonable period of time and coincides with the rate established in the line balance plan.

Master schedules for specific models and options are generated by exploding product family schedules using "planning bills of material." Planning bills relate optional components (for example, manual and automatic transmissions) to the family according to a forecast percentage.

Because this concept can provide only approximate requirements for use in supplying components and assemblies, a separate daily assembly schedule is then prepared to determine the detail manufacturing plan. When products are made to specific order, as in the case with automobiles, the daily schedule consists of a listing of all orders scheduled for assembly each day. The horizon for the daily schedule is determined by the nature of the business; usually it ranges from just a few days to two weeks in length.

LINE SUPPLY

A major difference between job shop and repetitive techniques is the manner in which components are issued from stores to the production area. In job shop environments, components are picked and issued to a manufacturing order, covering a discrete number of finished items. Repetitive shops do not use manufacturing orders that require excessive effort to maintain accurately. Instead, they pick and issue components to the shop floor on a continuous basis.

Computers are used extensively to support this process. Stock issues are transferred from stores to work-in-process accounts for each component. Reductions in work in process (WIP) occur by exploding the bill of material according to actual production recorded at key workstations, or passed into finished goods, through a process known as "backflushing." Care is taken to accurately record and relieve all scrap and material used in rework operations. The result is a perpetual WIP inventory balance that, when matched against projected requirements, generates new replenishment orders for the stockroom.

When operating in a JIT environment, key components and assemblies bypass the stockroom and are delivered directly to the production line. Replenishment needs are either computer-generated, as with stockroom supplied items, or manually transmitted using turnaround documents, called kanbans in the Japanese style. The key objective is to synchronize the supplying center with the short-term production requirements and reduce the amount of work-in-between-processes to a minimum.

LABOR REPORTING

In an environment with no work orders, time cannot be charged to a manufacturing order, as with conventional job shop practices. In many cases, labor cannot even be reported against the end product because the final configuration may not be known as the work is performed. Therefore, in a repetitive system, production is recorded at the workcenter level; this data is then used for productivity and other production analysis reporting.

Cost accounting is typically performed at standard; labor variances, calculated from efficiency deviations at the workcenter, can be allocated to product costs on a proportional basis. It should be noted that the issue of cost allocation is in a process of change. Major accounting firms as well as accounting associations are not sure how standard and actual costs should be measured. Much of the contemporary writing attempts to explain this issue.

PURCHASING

Repetitive purchasing typically involves placement of blanket orders or contracts for major items covering requirements for a calendar or model year. Release schedules are issued frequently to communicate requirements to suppliers; often these schedules are telecommunicated on a computer-to-computer basis. This technique serves to involve the supplier in a close working relationship, as well as to reduce the cost and speed the flow of information between companies.

This method of communication is facilitated when vendors transmit advance shipping notifications (ASN) as product is shipped. This practice permits reconciliation of quantities in transit. Protocols for the establishment of computer-to-computer communications have been established by the Automotive Industry Action Group (AIAG) and adopted by the American National Standards Institute (ANSI).

Bibliography

APICS Dictionary, 5th ed. Falls Church, VA: American Production and Inventory Control Society (APICS), Inc., 1984.

Buffa, Elwood S. *Modern Production/Operations Management*, 6th ed. New York: John Wiley & Sons, Inc., 1980.

Burnham, John. *Japanese Productivity: A Study Mission Report.* Falls Church, VA: APICS, 1983.

Deming, W. Edwards. *Quality, Productivity, and Competitive Position*. Cambridge, MA: Massachusetts Institute of Technology, 1982.

Greene, James H. *Production and Inventory Control Handbook*, 2nd ed. APICS. New York: McGraw-Hill Book Co., 1987.

Heinritz, Stuart F., and Farrell, Paul V. *Purchasing: Principles and Applications*. Englewood Cliffs, NJ: Prentice-Hall, Inc., 1971.

Jansen, Robert L. *Production Control Desk Book*. Englewood Cliffs, NJ: Prentice-Hall, Inc., 1975.

Johnson, Ross H., and Weber, Richard T. *Buying Quality*. Danbury, CT: Franklin Watts, Inc., 1985.

Magee, John F. *Industrial Logistics*. New York: McGraw-Hill Book Co., 1968.

Mayer, Raymond F. *Production and Operations Management*, 3rd ed. New York: McGraw-Hill Book Co., 1975.

Albert Ramond and Associates. *Controlling Production and Inventory Costs*. New York: McGraw-Hill Book Co., 1967.

Readings in Zero Inventory. APICS 27th Annual International Conference Proceedings. Falls Church, VA: APICS, 1984.

Rosaler, Robert C., and O'Neill Rice, James. *Standard Handbook of Plant Engineering*. New York: McGraw-Hill Book Co., 1983.

Sepehri, Mehran. *Just-in-Time, Not Just in Japan*. Falls Church, VA: APICS, 1986.

Smith, Bernard T. *Focus Forecasting*. Boston, MA: CBI Publishing Co., 1978.

Teichroew, Daniel. *Introduction to Management Science: Deterministic Models*. New York: John Wiley & Sons, Inc., 1964.

Wilson, Frank C. *Production Planning and Control Handbook*. Englewood Cliffs, NJ: Prentice-Hall, Inc., 1980.

QUALITY

SECTION

6

QUALITY MANAGEMENT AND PLANNING

Out back in many plants, where the final inspection area meets with the shipping dock, an event still takes place that might be labeled the American "Industrial Dance" syndrome.[1] The American "Industrial Dance" syndrome is characterized by several individuals moving about the final inspection area viewing discrepancy reports that cast doubt on the desirability of packaging these products for shipment. Feelings of the players involved are evident from the use of excited, loud voices and exaggerated physical gyrations.

The titles of the cast begin with the quality control (QC) or quality assurance (QA) manager, and usually include the production control manager, the operations superintendent, a customer sales representative, the project engineer, and possibly a company executive. The sides will not be evenly drawn, with the QA or QC managers often on the smaller and most likely defeated team. Unfortunately, the final decision to ship will be based on the need to meet monthly shipment quotas even though the product does not meet all of the required quality specifications.

If a thorough analysis of this scenario was made, glaring problems would be evident in the planning for and management of product quality. When a company contracts to deliver the right product, on time, and with the proper quality, but fails to do so, something has gone wrong with the quality planning process during one or more of the planning time frames. The time frames and questions that should be asked within the time frames are as follows:

1. Short Term (recent days)
   - Were the inspection instructions adequate?
   - Did the manufacturing plans provide proper inspection and test steps?
   - Did the manufacturing and inspection personnel have proper tools and adequate training?
2. Intermediate Term (recent weeks)
   - Was adequate notice given to acquire necessary inspection gages and fixtures?
   - Were the customer's requirements properly conveyed to the manufacturing and QA departments?
3. Long Term (recent months or years)
   - Were trained and educated personnel available, having capabilities sufficient to meet the needs of the manufacturing and inspection tasks?
   - Did upper management continually convey a clear message that all QC steps, for all departments, would be properly performed and not bypassed?

The answers to these questions necessary to carry out the execution of quality management and planning will be covered in this chapter.

At the onset, it is necessary to understand where one is going with the quality function and how it will integrate with the overall company goals. Why should planning and management of the quality function be of interest to any company in the first place? The prevailing reason is best provided in Dr. W. Edwards Deming's first point of his 14 Points of Management Obligations:[2]

> Create consistency of purpose for improvement of product and service with a plan to become competitive and to stay in business.

CHAPTER CONTENTS:

QUALITY PLANNING HIERARCHY

Central to managing the company's quality function is the planning effort. Commitment to quality planning and its implementation must occur at every level of the company. From the corporate directors to the operator performing work, everyone must do their part to ensure that products meet the customer's expectations.

Quality planning reflects both time-phased considerations and changes in substance from conceptual ideas to specific directives, as addressed

***Contributor of this chapter is:* Richard B. Stump**, *Manager, Quality, National Computer Systems, Inc.*

***Reviewers of this chapter are:* Major Leslie Anderson**, *Director of Quality, Department of the Air Force, Hanscom Air Force Base;* **Dr. William L. Berry**, *Associate Dean, Management Sciences Department, College of Business Administration, The University of Iowa;* **Richard W. Bigg**, *Consultant, Quality Practice, Coopers & Lybrand;* **Dr. C.L. Carter**, *Vice President, Quality Assurance and Human Resources, C.L. Carter, Jr. & Associates, Inc.;* **Nathan D. Hollander**, *Manager, Management Consulting Services, Coopers & Lybrand;* **Theodore J. Janssen**, *President, Theodore Janssen & Associates, Inc.;* **Bernard J. LaVoie**, *Director of Manufacturing and Quality, Department of the Air Force, HDQTRS Electronic Systems Div., Hanscom Air Force Base;* **John M. Liittschwager**, *Professor, Industrial and Management Engineering, College of Engineering, The University of Iowa;* **Robert W. Peach**, *Principal, Robert Peach and Associates, Inc.;* **Hayward Thomas**, *Consultant, Retired—President, Jensen Industries;* **Arthur R. Thompson**, *Professor (retired) and Consultant, Advanced Manufacturing Center, Cleveland State University;* **Joseph Tulkoff**, *Program Manager of Factory Modernization, Lockheed-Georgia Co., A Div. of Lockheed Corp.*

CHAPTER 22

QUALITY PLANNING HIERARCHY

by the various company work levels. Time-phased considerations of quality planning tie very closely with the position level in the organization under consideration. The more conceptual the plan is in nature, the more conceptual it is in content; and the more long term the quality planning, the higher up in the organizational structure (or position level) the planning must begin. Moving downward through the organization, each position level provides quality planning that causes the time phases to have shorter horizons, the objectives targeted being more specific and quantitative in content.

An adaptation of Burrill and Ellsworth's description[3] of work output (in this instance, quality planning) and the hierarchical structure to which it belongs, provides the organizational setting, and aids in understanding the time/content considerations (see Fig. 22-1).

Considerations for and emphasis on quality can and should enter into every major planning output of the organization. At each position level of the organizational structure, the individual must focus an appropriate portion of his or her planning output toward the effect that quality has on the desired results. Key managers in the chain of command for the quality organization (preferably starting with a vice president at the executive management level, but in practice typically starting at the middle management level) must provide the starting point of the thought process for product and service quality, influencing thought upward; directing thought downward to meet the needs generated by policies, strategies, and goals; and receiving upward thrusts of ideas through participative management practices.

When the relationships just described exist in a company's culture, management of the quality function operates in the best of all worlds. Executive management and the board of directors set the arena in which decisions relating to product and service quality will be focused and provide a charter for quality man-

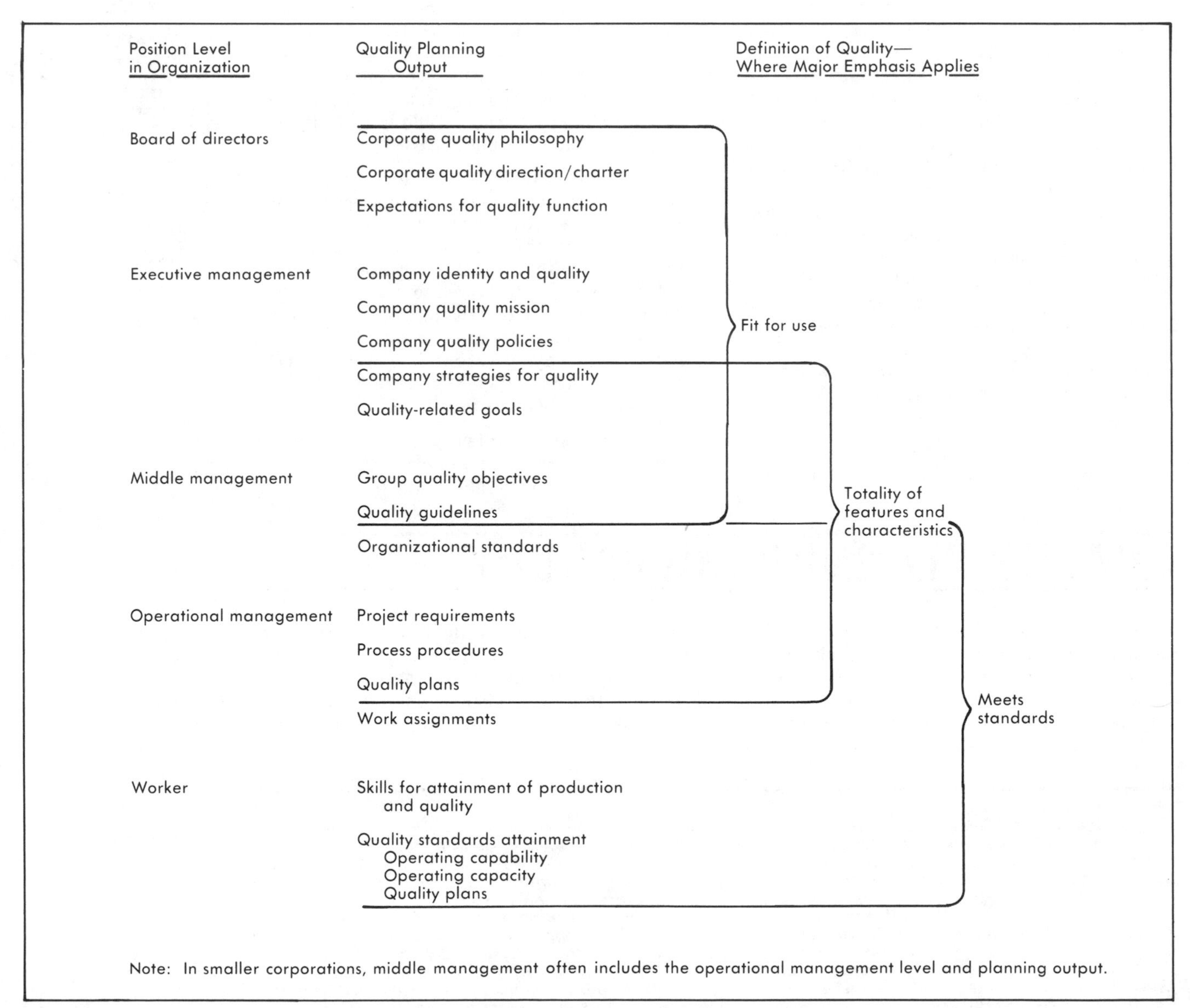

Fig. 22-1 Hierarchical structure of quality planning.

agement. Quality management works with peer-level middle managers to jointly describe the programs and structure needed to meet group objectives and provide guidelines to fulfill the executive charter. This sequence of events sets into motion and helps to define each succeeding position levels' efforts to carry out the specific assignments and tasks that actualize the guidance provided.

Definitions of quality and its applications run parallel to the hierarchy of planning for quality and help in establishing the specific plans at each organizational level. It may seem unusual to consider the definitions of quality in the plural sense, but it is appropriate as one moves downward in the hierarchical structure for the quality planning sequence of events.

"Fitness for use" is the most wide-sweeping definition of quality and is also the most conceptual in nature. It emphasizes the customer's needs and focuses on how the product will be used and how service will be provided; however, "fitness" can be misunderstood and needs to evolve to precise terms that can be measured. What does marketing research indicate is fit for or desired by the consumer? What product regulations affect the design of the product? What consideration should be given to ergonomics? What is the overall focus on fitness? And so on...

Appropriately, the planning output from the board of directors and executive management levels must address the two-edged considerations of quality leadership. Starting with the corporate philosophy, down through and including the company goals (refer to Fig. 22-1), those individuals involved at each level of planning for quality must take great care to provide pragmatic definitions as to what "fitness" and "use" mean as applied to the products and services provided by the corporation. How will the product or service be designed and consequently produced to achieve "fitness" considerations? How will the customer be viewed in the "user" sense? How does the company want the quality of the product or service to be perceived by the user?

The American Society for Quality Control (ASQC) provides a definition for quality that guides the activities of the next lower grouping of company position levels. Note that this definition is less conceptual and more detailed in nature:

> Quality is the totality of features and characteristics of a product or service that bear on its ability to satisfy a given need.[4]

The challenge from this definition is to produce quality planning that guides the applicable organizational position levels to provide specific product or service features and meet detailed needs. Starting with company strategies for quality planning (refer to Fig. 22-1) on to and including quality plans, this definition of quality creates thought processes that consider the product and service itself in more detail (features and characteristics) as part of the quality planning process.

Last, but in no way least, is the definition of quality in terms of "meets standards." This portion of the quality planning process, beginning with the planning output of organizational standards (refer to Fig. 22-1), continuing through all of the quality planning output addressed by the individual worker, encompasses this definition of quality. "Meets standards" is the most common definition used by the organizational position levels closest to the physical producing of goods and rendering of services. A significant portion of current printed materials, relating to the quality function, covers the definition of quality as it relates to an attribute of the product or process activities, not as a factor in design or a product of planning above the middle management position level in the organization.

Each definition of quality is appropriate and adequate to use for the range of quality planning output indicated. Each definition of quality requires those individuals at the applicable organizational position level to think in terms that match their charter and typical areas of responsibility in the company.

The key issue is that each position level understands what must be integrated into its planning process so that their version of quality planning output contains a well thought out relationship to the quality issues and goals for the company and each separate division. A variety of standards exists that can provide guidance for the quality planning activities associated with each position level (see Table 22-1). Also, incumbent on each position level, starting with the board of directors, is the need to challenge each successive level downward to make the same thorough evaluation of how quality fits into its planning. With this challenge goes the responsibility to be responsive to feedback aimed at improvement of that level of quality planning.

These understandings—the relationships between the position levels in the organization, the quality planning output at each level, and the definition of quality at each level in the organization—set the stage for the remainder of the quality management and planning topics to be considered in this chapter. All of the following topics can be placed on this framework.

In conclusion, top executives should not be preoccupied with providing detailed instruction to ensure what specific requirements the product must meet. On the other hand, production workers must ensure that specific quality characteristics have been met, not pondering on a day-to-day basis whether the final product is "fit for use" by the end user. All position levels of the organization may be included in participative management approaches to gather input from a cross section of the organization's employees. But once these ideas become product specifications, "quality" is defined as "meets specifications." And this is the predominant goal at the production worker position level.

MISSIONS, POLICIES AND PLANS

During the past five years, many U.S. companies have come forth with published statements expressing their approach to product and service quality. These statements have appeared in internal specifications, documents, publications, and newsletters; magazine and newspaper advertisements; TV commercials; and customer-directed documents. These declarations have been called quality mission statements, quality policies, quality charters, commitments to quality, and quality statements.

The significance of all this to quality professionals is that corporations have given in-depth thought to product and service quality. The variety of titles used for the corporate quality declarations also indicate that a variety of styles exists to label and embody the output of this thought process.

CHAPTER 22

MISSIONS, POLICIES AND PLANS

TABLE 22-1
Standards for Quality Control

Standard Number	Standard Title
ANSI/ASQC A1	Definitions, Symbols, Formulas and Tables for Control Charts
ANSI/ASQC A2	Terms, Symbols and Definitions for Acceptance Sampling
ANSI/ASQC A3	Quality Systems Terminology
ANSI/ASQC C1	Specifications of General Requirements for a Quality Program
ANSI/ASQC E2	Guide to Inspection Planning
ANSI/ASQC Z1.15	Generic Guidelines for Quality Systems
ANSI/IEEE 730	Software Quality Assurance Plans
ANSI/IEEE 830	Guide to Software Requirements Specifications
MIL-Q-9858A	Quality Program Requirements
MIL-I-45208A	Inspection System Requirements
MIL-STD-45662	Calibration System Requirements
ISO/8402	Quality Vocabulary
ISO/9000 (ANSI/ASQC Q90)	Quality Management and Quality Assurance Standards
ISO/9001 (ANSI/ASQC Q91)	Quality Systems: Model for Assurance in Design/Development, Production, Installation and Servicing
ISO/9002 (ANSI/ASQC Q92)	Quality Systems: Model for Quality Assurance in Production and Installation
ISO/9003 (ANSI/ASQC Q93)	Quality Systems: Model for Quality Assurance in Final Inspection and Test
ISO/9004 (ANSI/ASQC Q94)	Quality Management and Quality System Elements—Guidelines

This section will attempt to provide one way to approach a company's desire to define (or redefine, if new thoughts are needed to indicate advancements since the initial definitions were presented) the official quality declarations, positions, or statements. The methodology to be applied will build on the concepts presented within the quality planning hierarchy, the target output being precise, individualistic statements, beginning with a quality mission statement, of product and service quality.

Three major points to keep in view, whether this recommended process is followed or not, include the following:

1. An effort should be made to produce a statement on quality to serve as a guiding force for all succeeding efforts relating to product or service quality. It is important that the statements *can* and *will* be followed.
2. Top executives and key individuals at the middle management level should be included in the preparation of the company quality statement. This will cause the "shakers and movers" of the company to take an in-depth look at the quality function.
3. All of those involved in the authorship of the quality mission statement must have a personal commitment to it (buy in) and know exactly what it represents when asked to champion the causes resulting from its publication.

Starting with the company planning efforts and cascading on to those who actually carry out the supporting detailed actions, a thorough understanding of the key terms is needed. The impact of the entire quality planning process can be lost if those involved, at each position level, lose direction of their own objectives.

QUALITY MISSION

The statements of purpose needed to formulate the quality mission can be determined once the corporation answers two initial questions:[5]

- What business are we in?
- Why are we in business?

Ideally, the quality mission will include or reflect on portions of the overall company mission or charter statements, adapting them to the description of the specific quality positions taken.

Once the overall statements of purpose are provided, another set of questions must be asked to cause focus on the objectives of the company quality mission:

- What particular quality needs of the "business we are in" affect how we address quality issues?
- How do we view the quality function as it intertwines in the various people-oriented beliefs and actions?
- What part does quality play in support of our competitive edge in the marketplace?

When the quality mission statements are being prepared, product end uses should be kept in mind and used to shape some of the statements. Once completed, typical applications of the published quality mission include the following:

- Providing guidance for each individual, at each company position level, as they in turn provide specific quality planning output or actually perform to the planning dictates.
- Giving vendors and suppliers information that directly

converts to company expectations about purchased goods and services.

- Telling customers something about the company's attitude and practices that affect the product quality they will receive.
- Telling customers something about the way company people will deal with them on issues relating to the product and service quality they will receive.
- Presenting a favorable marketing image based on the perceived product quality, enhancing the potential to attract new customers.
- Providing concepts that are benchmarks for the next quality planning output—quality policy.

The company quality mission should be a "living document" reflecting changes in the company culture, markets, and environment and the need to change stated positions over time. Changes to the quality mission should be expected and be considered normal in the long term. The top executives should include a review of the quality mission whenever the overall corporate mission is revised, when strategic plans are created or altered, and when company restructuring takes place, significantly altering reporting among groups.

An example of a current quality mission is provided in Fig. 22-2. Many companies have provided quality policies that address the questions posed and end uses just described.

QUALITY POLICY

Evans and Evans indicate the word *policy*, when properly used, should "refer not to permanent principles but to courses of action. Primarily, it means a definite course of action adopted as expedient or from other considerations (our policy is to give the customer what he wants)..."[6] This is an excellent starting point. For no matter how a company chooses to describe and apply policy (and this can be varied depending on preference) the "course of action" considerations should be the primary focus. The courses of action should establish the long-term, overall intents of the company while focusing on the quality function (as contrasted to the short-term, specific day-to-day decisions). It must describe how people at all position levels will interact both within and without of the company in the interests of product and service quality.[7] Several examples of these are provided in Figs. 22-3, 22-4, 22-5, 22-6, and 22-7.

Again, quality policy exists in a varied number of forms as different companies see their needs and assert personal preference. This will be described later in this section. No matter which style is chosen, it is important to make sure that the company produces a quality policy that is a result of a wide participation of key individuals, high-level managers, and executives. Finally, a preferred course of action will be presented in this section to guide decisions relating to the quality planning function.

Quality Policy/Mission Statement

Many quality policies are described in terms similar to the description of the quality mission previously provided. This probably occurs because of the expediency needed at the time the company decides to produce such a statement. The time available to research the "mission-policy-procedure" chain of relationships often prohibits a more detailed understanding to emerge for consideration. The impetus to provide a quality policy should come from upper management rather than from marketing. When the idea to provide a quality policy comes from management, it usually is because such a statement was deemed essential to explain current emphasis on the quality function and its role as an integral part of the company plans and operations. When it comes from marketing, the reason is

Information Services Quality Mission

Information Services is in the business of providing products and services for the collection, processing, distribution, and reporting of information. These products and services are important to individuals, education institutions, government agencies, and commercial/industrial businesses. We believe that the best way to meet the needs of our customers is to be the quality leader in our industry. We will accept nothing less.

Information Services' business mission is planned, steady, profitable growth—we believe that a key way to get new business is to do well the business that we have.

Quality requires the commitment to a job well done by every employee.

Quality thrives in an environment of positive attitudes, good communications, cooperative effort, and an open, participative management process.

Quality does not just happen, nor can it be inspected into our products and services. Quality is built by all employees doing their individual jobs right the first time.

Quality is too important to be assigned to just one department. Each department is responsible for the quality of its own work. Quality results from a total, division-wide effort to deliver to our customers the same high quality of products and services that we ourselves insist upon when we buy in the marketplace.

Fig. 22-2 A quality mission of a data processing bureau. (*National Computer Systems*)

We shall strive for excellence in all endeavors.

We shall set our goals to achieve total customer satisfaction, and to deliver error-free, competitive products on time, with service second to none.

Fig. 22-3 A quality policy in the annual report of a major electronics-based information systems corporation. (*Burroughs Corp., now a part of UNISYS*)

Quality Policy

We have established a quality policy for the corporation—which gives direction to quality efforts. The policy makes five basic points:

1. Quality is our cornerstone.
2. The objective is to provide products and services which are defect-free.
3. Each of us must learn to do things right the first time.
4. Each job—each stage of the process—must be defect-free.
5. Quality is truly everyone's job.

Fig. 22-4 The quality policy of a major computer manufacturer. (*IBM*)

Data Processing Quality Policy

1. We at X Company deliver defect-free products and services to our customers, on time. Because quality to us is conformance to requirements, we create clearly stated specifications. Each one of us does the job right the first time, in accordance with those specifications, or cause them to be changed by mutual agreement.
2. Quality is the fulfillment of a commitment to produce a product or provide a service that meets the customer's expectation as expressed in specific or functional specifications with measurable values.
3. Each of us in information systems has an obligation to provide quality products and services.
4. Our products and services must be free of errors, exemplify pride in workmanship, and must be offered at a reasonable cost and delivered on time. Finally, it must be supported with reliable and reputable service.
5. The information systems department is committed to providing quality products and services, and recognizes that quality is an individual as well as an organizational responsibility.

Fig. 22-5 A generic quality policy for a company in the data processing industry. (*Quality Assurance Institute*)

Quality Policy

1. A key factor in consumer value is quality, and the consumers are the judges of quality.
2. The goal of the company is to deliver to the consumer products of consistent, uniform quality meeting specifications based on consumer values and produced at affordable costs.
3. The determination of the quality attributes which result in superior consumer value (satisfaction) will be determined by consumer surveys which will be updated every three years.
4. Written quality specifications will be prepared which include the permissible levels and tolerances for each quality attribute.
5. The permissible tolerance for each quality attribute will be based on data reflecting system capabilities and consumer preferences.
6. Quality standards (specifications) will be reviewed at least once each year.

Fig. 22-6 A quality policy for a major food-producing corporation. (*United Brands Co.*)

primarily a customer attention-getter or a counteraction to a competitor's release of quality policy/mission statement.

Quality Policy—Company-Wide

Having produced a company quality mission statement, a company-wide quality policy is the next link in the chain of events needed to convey high-level, conceptual company issues to the specific, point-by-point quality activities in terms that can be readily acted on by certain individuals.

Policy
Quality Management Function

It is the policy of our company that the function of quality management shall exist in each manufacturing and service operation to the degree necessary to ensure that:

1. The acceptance and performance *standards* of our products and services *are met.*
2. The *cost* of quality *goals* for each operation are achieved.

The company general manager is responsible for obtaining agreement with the corporate director of quality on the proper degree of quality function to be established in each operation. The general manager shall issue a quality policy for the operation, quoting this document, and shall take affirmative steps to ensure that the employees understand that the quality policy of the company is to perform exactly like the requirement or cause the requirement to be officially changed to what we and our customer really need.

To ensure its effectiveness, the quality function must be exercised in an objective and unbiased manner. As such, the head of the quality function in each unit shall report directly to the general manager and be on the same organizational level as those functions whose performance is being measured. The head of the quality function shall represent the company in the quality councils.

The quality function shall be staffed with professionally qualified personnel, and their responsibilities shall include:

- Product acceptance at all levels
- Supplier quality
- Quality engineering
  - Data analysis and status reporting
  - Corrective action
  - Planning
  - Qualification approval of products, processes, and procedures
  - Audit
- Quality education
- Consumer affairs
- Product safety

The company shall produce *cost of quality reports* in accordance with the comptroller's procedure and will create regular quality status reports for presentation to all management personnel. Standard practices to support the details of all activities mentioned in this policy have been prepared.

Fig. 22-7 A recommendation for a quality policy. (*Reproduced with permission from McGraw-Hill Book Co. from* Quality is Free, *by P.B. Crosby*)

When following this approach to prepare a company-wide quality policy, these thoughts should be kept foremost in mind:

- Policy is a course of action, not a permanent principle.
- Policy provides the summary-level issues required for the overall quality program.

The company quality mission focuses first on considerations of the following questions:

1. *Who* are we and *how* do we relate to product and service quality?
2. How does the issue of quality fit in our approach to doing business and to the marketplace?

Then the company-wide quality policy focuses on the question: Given our quality mission positions, *how* will we go about our jobs to fulfill the obligations described so our self-described image is attained?

Quality Policy—Specific Areas

Another version of the company-wide quality policy also can be found in a quality manual or similar functional document. Quite often an individual policy is provided as a preface for each section of the company's quality manual. In this fashion, each "preface" quality policy establishes a frame of reference for all the quality assurance and quality control procedures assigned to that section of the QA manual.

A few additional considerations to keep in mind when working on the company-wide quality policy and on those specific areas of application include the following:

- Company management must state its standards concerning the subject of quality, couched in words typically used within the company...[8]
- Management must clearly state the organizational responsibility and reporting relationship of the quality department. The quality department, in turn, must issue an operating policy concerning all activities that take place there.[9]
- Once a policy is determined, goals will become self-evident. These goals must be expressed concretely, in quantitative figures whenever possible.[10]
- Policies and goals must be put in writing and widely distributed. The more thoroughly that policies and goals reach the position levels of the employees in the organizational chart, the more information must be provided, with emphasis on explicitness and concreteness. At the same time, all policy and goal statements must be mutually consistent.[11]

Another possibility for having a quality policy for a specific area outside of the quality assurance department is the case where other company functions formulate their own policy relating to quality. Most frequently observed is a quality policy prepared for purchased goods and services. Quite often the verbiage is provided by the purchasing or materials management department itself. The quality of purchased goods and services ultimately reflects on the company's quality posture. This fact places specific quality-oriented demands on the vendor-vendee relationship. This situation creates the same needs for the procuring function to insist on high quality, just as the internal quality function provides an impetus for overall quality (including purchased goods and services).

An example of a quality policy for purchased goods and services is shown in Fig. 22-8. This quality policy provides an extension of the internal quality policies, requiring suppliers to perform within the quality framework provided for all activities.

It is important to use and display the quality policy, whether issued by the quality function, purchasing function, or whatever the source of origin. When it is a stand-alone document (and appropriate), the quality policy should be typeset, printed on textured paper, and displayed in a frame. If the policy is used as part of a "Policies and Procedures Manual," it is important to make sure that the QA procedures support and reference specific policy statements. It is also necessary to explain how it will be carried out. Because much time and energy has been spent to provide a viable policy, it is important to advertise it and use it.

Quality of Purchased Items

It is the policy of the division to seek an optimum price and delivery for items that conform to their defined requirements. It is not the policy of the division to trade-off quality, price, and delivery.

The primary objective of the division's purchased item quality system is progressively to reduce the proportion of purchased items that are defective (i.e., that are nonconformant to the defined requirements).

The division will purchase items only from approved suppliers.

The division will establish a relationship of cooperation with its approved suppliers, with the purpose of helping them to supply products conforming to the defined requirements.

The division will approve only suppliers who are honest. It will disapprove suppliers who show that they are not honest. The emphasis of the division's purchased items quality program is on cooperating with honest suppliers, and not on policing dishonest suppliers.

In turn the division will treat its suppliers in an ethical, honest manner.

The division's policy of establishing cooperative rather than adversarial relationships with suppliers should minimize the probability of contractual disputes with suppliers. Nonetheless the division will ensure that contractual requirements are clearly defined.

It is the policy of the division to comply with contractual requirements from its customers, which in turn affect its suppliers. It will work with its customers and suppliers to improve the effectiveness of quality systems affecting both.

Fig. 22-8 Statements from a model purchasing quality policy. (*TRW*)

QUALITY PLANS

Quality plans provide the "line of demarcation" where all of the preceding mission statements and policies become actualized. Quality plans embody that point where the concepts used to describe the quality function translate to the specific directions needed to accomplish them. As such, quality plans are at the last stop in the hierarchical structure of the overall quality planning process described in Fig. 22-1.

Quality plans can appear in many forms. Most commonly they include the following:

- The quality assurance manual, which sets the directions for the total quality program.
- Quality plans for special programs.
- Quality planning for control of processing.
- Quality objectives in personal appraisal planning.

The Quality Assurance Manual

The quality assurance manual is the cornerstone of planning for the quality function that is under primary control of the QA management. Contents cover the major areas of concern for quality-related activities, whether performed by personnel in or out of the quality-titled departments.

A well thought out, well-maintained QA manual provides the following:

- Descriptions of how each organizational department or function views and acts on its quality-related assignments and responds to the needs of its interface groups.

- Specific detail for inspection, test, and audit activities, including tools and equipment used; sampling plans; product characteristics to consider; sequence of events; and documentation needed for all employees, not just those in the QA/QC departments.
- Quality reports to be provided.
- Various coverage provided by QA, QC, inspection, and audit personnel.
- Means for attracting prospective customers and portions of proposals for those actively pursued.
- A constant reminder of quality-related activities to all reviewers of draft proposals and recipients of final copies.
- A thought-provoking document, requiring various levels of employees to think carefully about current and potential applications for quality activities.
- A training tool for new or transferred personnel.
- A description of the commitment to customer quality requirements and how the company will strive to fulfill them.

Typically, the QA manual contains two types of documents: policies and procedures. Quality policies explain how the company views the quality function as it applies to the area of interest and what will be done to provide a quality product or service. Quality procedures support a specific policy and explain how the company will carry out the policy.

The QA manual typically approaches the needs established by one of two major product and service interests: military/government or commercial. The military and government-regulated industries have, over many years of evolution, required contractors to meet the needs of strictly worded, precisely defined quality program requirements, perhaps best described by the table of contents shown in Fig. 22-9.[12] In contrast, a typical table of contents for a QA manual directed at production of commercial goods and services is shown in Fig. 22-10.[13]

The major issues for the quality department to tackle when planning the production of a QA manual include the following:

1. Who has the requisite experience to write such a manual?
2. Who will be able to provide the written documentation from a time-demand standpoint?
3. Who will provide the maintenance activities after initial distribution is made?
4. How can it be designed so that necessary changes can be made without a major rewriting of the manual?

Quality Plans for Special Programs

A QA manual that emphasizes the standard operating procedure (SOP) of the company may require the preparation of a supplemental document to address special program needs. Such coverage is initiated for contracts requiring atypical customer requirements, pilot programs, or special development studies. A good way to "layer in" these requirements, yet allow for an oversight capability by the QA department, is to provide a special quality plan that emphasizes how departures from SOP norms will be handled.

The quality plans for special programs often spawn a series of additional requirement sheets or specifications that are stand-alone documents in the areas or departments where applied or are attached to an SOP document, traveling with it, and being acted on by appropriate personnel. The main control and responsibility, however, resides with the QA department.

Part 1 GENERAL PURPOSE QUALITY ASSURANCE MANUAL
- Title Page
- Table of Contents
- 1.0 Introduction
- 2.0 Administration
  - 2.1 Quality Procedure and Bulletins
  - 2.2 Stamp and Signature Administration
  - 2.3 Audit and Surveillance Administration
  - 2.4 Training and Certification Administration
- 3.0 Quality Program Management
  - 3.1 Organization and Functions
  - 3.2 Quality Planning
  - 3.3 Work Instructions
  - 3.4 Records and Reports
  - 3.5 Corrective Action
  - 3.6 Costs Related to Quality
- 4.0 Facilities and Standards
  - 4.1 Drawings, Documentation and Changes
  - 4.2 Measuring and Test Equipment
  - 4.3 Production Tooling Used as a Media of Inspection
  - 4.4 Use of Contractor's Inspection Equipment
- 5.0 Control of Purchases
  - 5.1 Candidate Supplier/Subcontractor Quality Evaluation and Selection
  - 5.2 Active Supplier/Subcontractor Quality Compliance
- 6.0 Manufacturing Control
  - 6.1 Materials and Material Control
  - 6.2 Production Processing and Fabrication
  - 6.3 Completed Item Inspection and Testing
  - 6.4 Handling, Storage and Delivery
  - 6.5 Nonconforming Material
  - 6.6 Statistical Quality Control and Analysis
  - 6.7 Indication of Inspection Status
- 7.0 Coordinated Government (Customer)/Contractor Actions
  - 7.1 Government (Customer) Inspection at Subcontractor or Vendor Facilities
  - 7.2 Government (Customer) Property
- Appendix 1 Quality Bulletins

Part 2 QUALITY ASSURANCE FORMS
- List of Quality Assurance forms
- Quality Assurance Forms (in form-number sequence)

Part 3 INDICES
- Numerical Index of Procedures and Forms
- Alphabetical Index of Procedures and Forms

Fig. 22-9 Table of contents for a QA manual of companies involved in military or government-related industries. (*Reproduced with permission from Prentice-Hall, Inc. from the* Manual of Quality Assurance Procedures and Forms, *by R.D. Carlsen, J.A. Gerber, and J.F. McHugh*)

Quality Planning for Control of Processing

This type of quality planning usually is provided in support of systemic programs initiated by departments that interface with the quality assurance department. Often the systems involved include the manufacturing-sequence quality control program for internal activities and purchase order quality requirements for purchased goods and services.

The quality SOPs provided by the QA manual and supplemented by quality plans for special programs is translated from the more general descriptions and applied to the specific part, process, or service. As the manufacturing sequence-of-events is provided by manufacturing SOPs, so are the inspection needs for each operator at each step, plus specific inspection station, patrol inspection, testing, and audit requirements.

Figure 22-11 shows a typical quality planning form that is suitable for use at any production step and includes the inspection activities as part of the manufacturing flow.

Annual Plan for Quality

Some companies in the U.S. have begun to make planning for the quality function an annual event, as previously has been done for other functions such as marketing, finance, and human resources.[14] The IMPRO Conference on Quality Improvement

1.0 General Quality Policy
2.0 Organization Chart and Division of Responsibilities to Implement This Policy
- 2.1 Sales Division
- 2.2 Product Engineering Division
- 2.3 Technical Division
- 2.4 Purchasing Department
- 2.5 Manufacturing Division
- 2.6 Manufacturing Engineering Division
- 2.7 Industrial Engineering Department
- 2.8 Finance Division
- 2.9 Industrial and Community Relations Division
- 2.10 Data Processing Department

3.0 Quality Control Inspection/Audit Functions
- 3.1 Incoming Inspection
- 3.2 In-Process Inspection/Audit
- 3.3 Process Capability Analyses
- 3.4 Pre-Control Audits
- 3.5 Assembly Inspection/Audit
- 3.6 Final Product Inspection
- 3.7 Indication of Inspection Status
- 3.8 Shipping Audit

4.0 Quality Assurance Management Functions
- 4.1 Product Quality Records
- 4.2 Product Quality Reporting
- 4.3 Motivational Aspects
- 4.4 Quality Cost Reporting

5.0 Corrective Action
6.0 Pre-Production Sample Inspection and Approval
7.0 First Production Shipment Approval
8.0 Drawing Change Control
9.0 Measuring Equipment Control Procedure
- 9.1 Technical Division Measuring Equipment
- 9.2 Manufacturing Division and Quality Assurance Division Measuring Equipment
- 9.3 Production Tooling Used as Inspection Media

10.0 Segregation of Nonconforming Material
11.0 Engineering Configuration Control
12.0 Customer Returns
13.0 Audits of Quality Assurance System
14.0 Appendix
- 14.1 Sampling Plans
- 14.2 Copies of Forms

Fig. 22-10 Typical table of contents for a QA manual of companies producing commercial goods and services. (*Reproduced with permission from Rath & Strong, Inc. from the* Quality Assurance Handbook, *by D. Shainin, et al.*)

SUBJECT: QUALITY PLANNING	Effective Date	Number
	Page of	Revision

QUALITY PLAN			Plan No.____ Page __ of ____	
Program or Project			Contract No.	
Part or Assy. Name			No.	
Planned By Date		Approved By Date		
Mfg. Oper. No.	Inspect./ Test Point No.	Attributes to be Verified and Method	Inspection/ Test Equipment	Estimated Hours per Point

Fig. 22-11 An example of a quality planning form. (*Reproduced with permission from Prentice-Hall, Inc. from the* Manual of Quality Assurance Procedures and Forms, *by R.D. Carlsen, J.A. Gerber, and J.F. McHugh*)

lead by the Juran Institute each year emphasizes this concept.[15]

The quality function, like other key business functions, must be given specific attention and guidance throughout the management process. Quality plans must be made for the year to come and integrated into the overall business plans. The cascading effect of the overall plans for the quality function provides challenges for departments and individuals.

The Japanese have led the way in establishing the "habit of improvement" as each year builds on the advances and successes from the prior years to further improve product and service quality. Upper management leadership there has provided the impetus for the organization to establish a "pursue the last grain of rice" mentality.

Some U.S. companies are adopting the Juran model and are having considerable success at improvement of the product quality learning curve.[16] Other companies have had success with programs prescribed by Phil Crosby and A.V. Feigenbaum. The key issue here is to find or establish a well-defined quality program that is compatible with the company culture, then, as Tom Peters recommends, "Follow some rigorous system religiously."

Quality Objectives in Personal Objective Planning

Another very important concern is the involvement of each individual in carrying out the quality plans and obligations of the company. This involvement includes not only the direct quality function personnel but all of those responsible for activities that combine to provide the quality products and services. In recent years, more stories have been told where certain departments outside of quality have been challenged with specific performance objectives in the quality area, along with their other achievements to be measured throughout the year.

Figure 22-12 shows an application of this planning style as used by a major hospital supply company. The individual quality objectives were a product of this kickoff session and several others, as each level of marketing personnel understood specifically how he or she would contribute to the organization's overall goals that year.

While the emphasis on personal objective planning is placed last in the sequence of plans for the quality function, application should be made at each position level in the organization. The "hands-on" worker strives for zero defects at each process step. Each level can thus establish objectives for its contribution to improvement of product, process, or service quality.

MARKETING QUALITY OBJECTIVES PROGRAM
Agenda

I. Introduction of Program

II. Quality as a Marketing Strategy

III. Relative Quality Program
Introduction to Presentations of Divisional Objectives

IV. Marketing Quality Objectives Presentations by Divisions
A. Product #1
B. Product #2
C. Product #3
D. Product #4
E. Product #5
F. Product #6

V. Next Steps, Followup

VI. Comments

Fig. 22-12 A planning program example used to establish objectives for the marketing function. (*Travenol Laboratories*)

ORGANIZATION

Early craftworkers controlled all of the components of the quality function within their own operations. All of the issues involved with the quality of the product were handled by the craftworker. Quality was designed into the product, the processes were planned to provide quality output at each operation step, the final product was inspected and tested, and even customer feedback when products did not meet expectations was handled by the craftworker. The craftworker was the total quality organization all in one.

As manufacturing moved toward mass production, the processing steps became more distant from one another, divided among several individuals and workstations. The concern for product quality evolved into the inspection function as the Industrial Revolution dawned.

INSPECTION

Well into the 1960s in the U.S., the sum of the inspection and test activities were tantamount to the quality function for many manufacturing facilities. Support for the inspection-and-test mentality came in the form of emphasis on sampling plans and, to some degree, statistical process control. The inspection and test function sorted the bad from the good, plus was the enforcer in a game sometimes called "cops and robbers," where the production personnel ("robbers") followed the manufacturing-prescribed methods only as they saw fit to do so. Inspection ferretted out bad product and admonished the offenders. Often poor quality product was shipped, because the general atmosphere on the manufacturing floors either produced more problems than inspection could identify or, even when properly identified, defective products were given an "OK to ship" over any objections.

In the general area of manufacturing, upper management often looked at the inspection and test function as a necessary evil, something to provide a modicum of conscience toward product quality. Inspection hours to production hours expended were contrasted in ratios, and the dictate of the ratio balance was met. As shown in Table 22-2, other rules of thumb were provided for how many inspectors were needed to monitor and control quality in a manufacturing operation.[17]

As far as the inspection and test functions were concerned, everything was predictable, controlled, and terribly inefficient. Finding and putting out fires were the order of the day. Preven-

TABLE 22-2
Guidelines That Used to be Used When Determining the Number of Inspectors for a Manufacturing Operation

Job Shops	Mass-Production Shops
One inspector per 10 direct workers	One inspector per 30 direct workers

tion of quality problems was virtually unheard of and therefore seldom attempted.

CHANGES IN ORGANIZATIONAL STRUCTURE

Many events have occurred since the early 1970s that have affected the way manufacturing companies now look at the quality function. The impact of change has improved the way that the quality function is organized. Perhaps the best conceptual description of how the quality function can be organized for maximum effectiveness is found in three related functions that together form an extended, uniform network, though each has its own entity. This is referred to as the Juran Trilogy. Each of these organizational processes of the quality function are described as follows:[18]

1. *Quality planning function.* The group responsible for preparations to meet quality goals. The end result is a process that can meet quality goals under operating conditions. The quality planning activities emphasize efforts that:
   - Identify the customers, external and internal.
   - Develop product features that respond to customer needs. Products include both goods and services.
   - Establish quality goals that meet the needs of customers and suppliers alike and do so at lowest combined costs (combined in the sense that quite often the total cost is not just the purchase price, but includes maintenance, repairs, and downtime).
   - Develop a process that can produce the needed product features.
   - Prove that the process is capable, that it can meet the quality goals during operation.
2. *Quality control function.* The group responsible for documenting or certifying attainment of quality goals during manufacturing operations. The result is conduct of manufacturing operations in accordance with the quality plan. The quality control activities emphasize efforts that:
   - Choose control subjects (what to control).
   - Choose units of measure.
   - Establish measurement methods.
   - Establish standards of performance.
   - Measure real performance.
   - Interpret the difference, real vs. standard.
   - Act on the difference.
3. *Quality improvement function.* The group responsible for breaking through to unprecedented performance levels. The results provide operations at levels of quality distinctly superior to planned performance of previous periods. The quality improvement activities emphasize efforts that:
   - Identify specific projects for improvement.
   - Organize to guide the projects.
   - Organize for diagnosis, for discovery of causes.
   - Diagnose to find causes.
   - Provide remedies in the forms of corrective and preventive actions.
   - Prove that the remedies are effective in operation.
   - Provide for control to hold the gains.

An up-to-date company will be structured to provide groups responsible for each of the three functions—planning, control, and improvement. With the addition of a group responsible for vendor quality, the coverage is complete. The vendor quality group ensures that the vendors have replicated the three quality functions (planning, control, and improvement) within their own organizations.

Major changes have occurred that have challenged and allowed companies to rethink the organization of the quality function. There are no given rules of thumb to follow in setting the number of inspection or auditing personnel. As the focus on inspection of product quality has evolved into a function of operator control, fewer people are required in the quality control function known as "inspection and test." Comparatively, more people are required in the quality planning function (working toward integration of customer needs and setting up total quality systems) and in the quality improvement function working in such areas as producibility and quality engineering.

Monitoring the costs of quality (the categories of prevention, appraisal, and failure) in a formal program provides a management tool that can help to understand how the company's costs relate to (1) both improvement and maintenance of quality, (2) avoidable quality costs, and (3) waste due to poor quality. The quality costs program also shows how the categories interrelate, so that the optimum combination can be studied and attained.

The first action is directed at understanding the recent years' quality costs for the purpose of:

- Identifying the absolute cost of quality. The objective then is to reduce this amount as it relates to such categories as total revenue and labor hours expended.
- Determining the proportions of prevention costs, appraisal costs, and failure (internal and external) costs as they combine to equal 100% of the total cost of quality. The objective then would be to reduce the proportion of failure costs through increases in prevention costs and adjusting appraisal as needed, usually up at first to identify the magnitude of problems, then down as prevention activities become effective.
- Laying the groundwork to estimate the optimum quality costs from the absolute and proportional cost analyses.

Total quality cost can be determined using the following equation:[19]

$$T(q) = f(q) + p(q) \tag{1}$$

where:

$T(q)$ = total quality cost
$f(q)$ = total failure costs (internal and external)
$p(q)$ = total appraisal costs and prevention costs
q = variable quality level (0-100% good product)

ORGANIZATION

Conceptually, total quality cost is minimized when the first derivative of Eq. (1) is set equal to 0, such that $dp/dq = -df/dq$. At this point, an additional dollar invested in prevention and appraisal produces exactly one dollar's worth of reduced failure costs.

A graph using Dr. J.M. Juran's model for optimum quality costs is shown in Fig. 22-13. It is the company's decision where to establish the optimization point. In recent years, companies have moved the optimization point closer to 100% good product as reject rates have been lowered to parts per million for some products. This analysis, even in rough form, can help a company determine how much will be spent both on quality maintenance costs and improvement efforts to attack the avoidable costs of poor quality.

Recently, it has been recognized that emphasis on prevention during product design and process development has resulted in drastic reduction (in some cases elimination) of defectives, thereby driving appraisal costs toward zero and approaching "perfection" at a bounded cost, contrary to the graph in Fig. 22-13. Therefore, Juran's model of optimum quality costs can be misleading because it indicates that appraisal and prevention costs are infinite at 100% quality and thus a program of zero defects cannot succeed. Juran's model, established in earlier years of quality assurance theory, assumes that no feedback to product design or process controls occurs to eliminate errors at their source, thus requiring appraisal methods to screen out defectives.

As Schneiderman points out, these assumptions may not be valid in many industries today, and therefore a minimum (if not optimum) in total quality costs can occur at 100% quality levels as shown in Fig. 22-14.[20]

Dr. Juran himself has commented that "In some areas we have undergone dramatic changes. Quality improvements have eliminated failure costs at their source and have thereby eliminated the need for appraisal...and perfection becomes available at finite costs."[21]

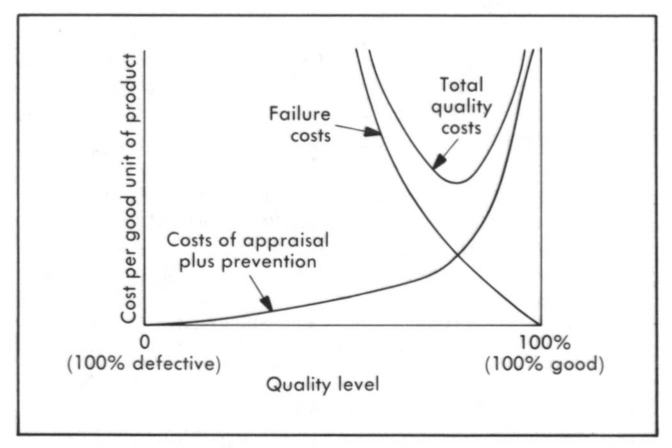

Fig. 22-13 Juran's model for optimum quality costs. (*Reproduced with permission from McGraw-Hill Book Co. from* Quality Control Handbook, *3rd ed., by J.M. Juran*)

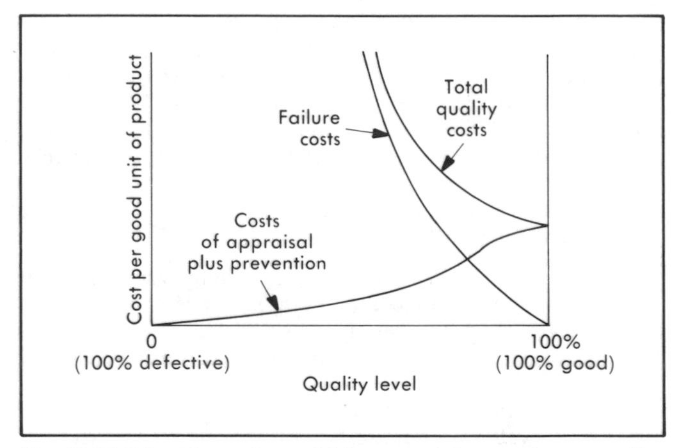

Fig. 22-14 Schneiderman's model for zero defects optimum quality level. (*Copyright American Society for Quality Control, Inc. Reprinted by permission*)

STRATEGIC PLANNING

Strategic planning is presented apart from the planning process detailed earlier. This is done because strategic planning is significantly different from the earlier described, more smoothly flowing planning process, where the higher level mission and policies set the stage for the supporting list of planning practices. Strategic plans disrupt this interrelated process. Perhaps this is why strategic planning is only done in earnest every three to five years by many company executives and managers. Additional information on strategic planning can be found in Chapter 1, "Planning," of this volume.

Tregoe and Zimmerman define strategy as "the framework which guides those choices that determine the *nature* and *direction* of an organization."[22] Included in this definition is the issue of future self-definition, setting direction to *what* the organization wants to be in the future (to realize its future vision). Once this direction is determined, the stage can be set for the relationship of the quality function to strategic planning. It is important to understand that strategy deals with dramatic change in the quality function and existing quality programs. It does not deal with projections or regression-line-based data as the essence of this planning effort. Nor does it deal with the incremental goals and steps leading to the vision.

The quality function has existed in just this atmosphere in recent years for many companies. Quality has been included in the company's strategic planning in one of two ways: (1) integrated as part of a major company thrust in another function or program or (2) implemented as a major company thrust itself.

As such the quality function has been included as a strategic thrust to achieve longer term objectives. More and more, strategic planning for quality is included as a unifying plan, tying all parts of the company together; a comprehensive thrust, covering all major aspects of company; and an integrating effort compatible with each part of the company.[23]

AS PART OF ANOTHER PROGRAM

Only recently has the quality function been given consideration on a widespread basis when practitioners of production and operations management techniques planned programs and set them in motion. Recent books on production management, inventory, processing control, and manufacturing operations now include a full chapter or more on the importance of the quality function as it abets and integrates with the overall effort. The realization is that strong support from the quality function contributes to the successful attainment of the overall company goals. Conversely, less than strong quality support creates downside risk, if not doom, for the targeted improvements.

Just-In-Time Production

Just-in-time (JIT) production systems have fascinated many U.S. manufacturing companies. JIT emphasizes a three-pronged attack to: (1) eliminate all waste in the product lifecycle (waste is defined as anything that does not contribute to the production of or add value to the final product), (2) simplify all operations and methods, and (3) promote relentless problem solving in all company functions. Additional information on JIT can be found in Chapters 15, 20, and 21 of this volume.

There is a straightforward involvement of a total quality control (TQC) program, emphasizing attainment of excellence in each step, in a successful JIT program. One leading authority on manufacturing has stated that "No TQC = No JIT."[24] This means that lack of strong emphasis on the quality function can prevent a company from having a strong or viable JIT program.

An effective quality control system signals the alarm when problems occur, describes the parameters of the problem, and leads the attack to resolve the situation, restoring the status quo of the producing system. Often scrap and rework data indicate those targets for the quality control problem solvers to attack.

Because of JIT, special challenges present themselves for attention by the materials management function. Supplier-provided goods and services must arrive as planned, meeting all quality requirements. These needs demand that the supplier's quality system assure product and service quality equal to the purchaser's expressed needs. It is essential that the purchasing function obtain support from the quality function to work with vendors, assisting them to attain adequate quality system capabilities, and later certify those capabilities.

A total quality control system is considered so essential to the success of JIT that the two concepts are often described as the two sides of a coin, inseparable and both needed to complete the whole.

Manufacturing Resource Planning

The concept of manufacturing resource planning (MRP II) as well as the earlier version, material requirements planning (MRP), assumes that the benefits of a successful quality function program are in place. Quality problems left to themselves, not attacked by the JIT/TQC two-pronged approach, can negatively affect the planning efforts of MRP II to the point where the output predictions are not reliable. When this occurs, the users develop a low opinion of the value of MRP II that is not actually a fault of that system. Additional information on MRP II can be found in Chapter 20 of this volume.

Cost Containment

Cost containment focuses internally and emphasizes the old cliche, "Do it right the first time!" Doing it right begins as far back in the product lifecycle as understanding the customer's needs. Companies that have implemented strong quality programs or recently strengthened existing programs have found out that substantial cost savings occur as a major benefit because less material, labor, and usage of processes and machinery are needed, plus delays are eliminated or lessened dramatically.

A diversified electric and electronic products and systems producer anchors its "Best Cost Producer Strategy" on a dedication to quality. The best cost producer is defined as having "the best product performance and the highest quality."[25]

QUALITY AS A MAJOR THRUST

Annual reports of major corporations announcing that leadership in product quality has been designated as one major corporate strategy have blossomed in the 1980s.

One of the Big Three U.S. automobile manufacturers' strategies for the future is "to produce high-quality products that provide the customer good value, ...meet customer requirements and are competitive in cost and quality with any in the world."[26] A worldwide producer of components and systems that control motion states that it will "remain the quality leader in the broadest line of motion-control components and systems."[27] The guiding principles of another producer of power and motion-control products states that "...the combination of absolute quality, cost-effectiveness, service and marketing...gives us the edge in our business."[28]

Why do these major U.S. corporations emphasize the quality function so strongly in their strategic plans? The Strategic Planning Institute (SPI) provides some answers by showing that the higher quality producers also are typically higher achievers in:[29]

- Return on investment.
- Obtaining market share.
- Obtaining premium prices for products.

Statistical Process Control

The application of statistical process control (SPC) quite possibly is the most widely used statistical quality control technique in the United States today. Two major thrusts have caused this to happen. The widespread management rediscovery of the works of Dr. W. Edwards Deming in the early 1980s and his work that emphasizes the use of control charting is a primary first cause. The second major push for SPC came about through applications by the automobile industry.

Manufacturing companies who previously had little knowledge of how to measure process quality now espouse the SPC program in their shops for several significant reasons. First, SPC returns control of the operation to the person most capable of understanding why changes in output occurred and what to do to make corrections (the operator). Additionally, the burden on the inspection department is relieved in terms of less need for checking product, thereby eliminating the "cops and robbers" mentality. Third, SPC improves profits for both the supplier and purchaser, by cutting the costs of scrap and rework as well as reducing the possibility of shipping defective products to the purchaser.

A major concern over the emergence of SPC is that it is often viewed as "be-all and end-all." Sometimes little concern is given to understanding the issues of (1) realistic tolerances of products; (2) capability studies of the equipment or process used, in the activities prior to the SPC applications; or (3) the problem-solving activities needed after SPC signals to restore control of process or product parameters. Additional information on SPC can be found in Volume IV of this Handbook series. Capability studies are also discussed in Chapter 23 of this volume.

Total Quality Control

In his book, *Total Quality Control*, Armand V. Feigenbaum moved the conceptual image of the quality function ahead in a quantum leap (including specific application of techniques for those companies moving from conceptual to actual practices). Total quality control (TQC) is defined by Dr. Feigenbaum as:

> "...an effective system for integrating the quality-development, quality-maintenance, and quality-improvement efforts of the various groups in an organization so as to enable marketing, engineering, production, and service at the most economical levels which allow for full customer satisfaction."[30]

Beyond this initial work, the Japanese have claimed to redefine total quality control, and Dr. Richard J. Schonberger has the expanded application of TQC horizons in his book, *World Class Manufacturing*.[31]

Through Feigenbaum's concept of TQC and the many varieties of this theme, companies have learned how to improve the quality of their products and services through improvement of production efficiency and effectiveness. An additional benefit gained through achieving a TQC program has been a reduction in the costs to manufacture products in the form of lowered avoidable costs due to poor quality, such as lowered scrap, less rework, and lowered returns.[32]

MONITORING THE OUTCOMES

While still in the management planning process for quality, provisions must be made for monitoring the impact of plans and policies in terms of actual quality results. It is important to plan ahead and anticipate the results, so that they will be quantitatively measured either in absolute measures, ratios, or trends. Employees at all levels want to know "How well am I doing?" Higher up in the organization, the desire to see end results in monetary and financial terms is greatest. At lower organizational levels, interests are in terms of counts of good vs. bad, percent yield from the amount processed, and so on.

AUDITS

Much like the accounting function, a primary way for the quality function to determine whether policies and plans are actually being carried out as planned is to audit the company's ongoing practices. The more specific and detailed the quality plan, the more detailed the quality audit can be. Specific product or process audits typically represent this kind of activity. A major issue is "Are we meeting specifications and the quality plans and procedures provided to assure them?" These quality audits are typically performed by experienced managers, quality engineers, quality specialists, and quality auditors.

Quality audits can also focus on higher level quality plans or policies. These audits require an understanding of the quality function that is less involved with the specific details and more alert to the indicators or nuances that indicate compliance to the higher level plans and policies. These audits are performed by higher level managers, quality specialists, and consultants from inside or outside the company.

A discussion on the various types of quality audits and techniques for conducting the audit can be found in Chapter 23 of this volume.

QUALITY PERFORMANCE DATA

A variety of absolute, ratio, and trend data can be used to measure the impact of quality policies and plans once the company has begun the new approaches to product or service quality or the quality function itself.

Primary on any list of organized approaches used by a company to measure quality and the quality function is the "quality costs" technique. Quality costs demand several systems of quality data collection from many areas of the company. When properly implemented, these quality costs support systems will provide measurements that are valuable in identifying and then solving problems, thus eliminating waste at various key areas. It is important to be sure that the need for accurate data is thoroughly understood. Several of these reporting systems, essential in support of quality costs tabulation include:

- Scrap and rework in relation to production.
- Repair costs required to bring nonconforming products back up to standards.
- Excessive costs due to defective vendor-supplied goods.
- Warranty costs.
- Costs related to customer return of goods.
- Costs of salaries dedicated to inspection, test, and audit.

Once the dollar amounts attributable to the four categories of quality costs (preventive, appraisal, internal failure, and external failure) are obtained, the ability to manage the quality function is enhanced. Each category can be analyzed to attack those avoidable costs resulting from poor quality products and practices. And the long-term effort can begin to zero in on the optimum balance of the four categories as they relate to one another. Guidelines for this effort commonly follow this sequence:

1. Overall, reduce the absolute cost of quality. Be sure to target a realistic and, in the long run, a beneficial optimum, because driving the quality costs of prevention and appraisal too low as part of the overall effort can create adverse reactions.
2. Initial targets should include the largest dollar losses due to waste in the cost of failure (internal or external), establishing specific projects, expenditures, time frames, and responsibilities for each category.
3. When external cost of failure is shrinking (and especially if it never was too large in relationship to other cate-

gories) the attack should focus on internal waste resulting from not "doing it right the first time."

4. When the amount of defective material is reduced, coupled with implementation of strong systems and controls to prevent further generation of defective products, the appraisal effort should be scrutinized for (1) reduction in amount of effort expended and (2) conversion to auditing.
5. Coincident with step 4, the concept of operator control must be established.
6. Throughout all of the steps, considerations must be given to investments in prevention costs. Encouragement must also be given to acquire those systems, people, equipment, and consultants that will focus on prevention of defects. Often a "pump priming" is required in the prevention cost category to provide the impetus to overcome inertia in the company. This pump priming is accomplished by providing visible proof that management supports the desire to change.

Monitoring the outcomes of the quality cost data requires more than just an interest in minimizing them. Measurement of quality costs includes a need to understand the contributors to avoidable costs vs. those costs required to provide the desired or targeted product or service quality. Ratios of quality costs to other key business indicators provide additional insight to the trends and successes of the overall program. Additional information on quality costs can be found in Chapter 24 of this volume.

CONCLUSION

A vigorous approach to product and service quality has become mandatory for companies to successfully compete in many industries. Intelligent management of the quality function demands an emphasis on strong planning.

The sequence of events for thoroughly integrating quality into the company's mainstream is as follows:

- Define quality for the company's products and services as applied to the various position levels.
- Provide a quality mission to support the overall business mission and summarize how the business relates to product and service quality.
- Link a strong quality policy to the mission, describing how to carry out the activities necessary to fulfill the mission statements.
- Strategically fit the quality function to long-term plans.
- Actualize the conceptual ideals and ideas of the quality mission and policy through establishing quality goals, plans, and procedures.

Obviously this chain of events results from people at each company level who have the requisite education and experience to provide a total quality function within the company structure. When this occurs, the impact of the quality planning function, the quality control function, and the quality improvement function will unite to ensure that the quality challenges of the past have been met and that progress is made to anticipate and prepare for opportunities of the future.

Measurement of the current status of the quality function requires audits and performance data to cover both subjective and objective evaluations, respectively, of "how well we are doing." The use of quality costs has evolved as an effective way to measure the financial impact of the quality function, plus provide accountants another way to monitor progress and plan for expenditures.

The issues of product and service quality, combined with the impact of the quality function on the economic success of the company, have risen in the list of concerns of company leaders since the 1960s. Quality management and planning must be thoroughly understood and implemented by those companies where a criterion of competition focuses on product and service quality. The list is growing.

References

1. M.B. Spangler, "Syndromes of Risk and Environmental Protection: The Conflict of Individual and Societal Values," *The Environmental Professional*, vol. 2 (1980), p. 284.
2. W. Edwards Deming, *Quality Productivity and Competitive Position* (Cambridge, MA: Massachusetts Institute of Technology, Center for Advanced Engineering Study, 1981), p. 16.
3. C.W. Burrill and L.W. Ellsworth, *Quality Data Processing* (Tenafly, NJ: Burrill-Ellsworth Associates, Inc., 1982), p. 88.
4. *Glossary and Tables for Statistical Quality Control* (Milwaukee: American Society for Quality Control, 1973), p. 5.
5. W.F. Glueck and L.R. Jauch, *Business Policy and Strategic Management*, 4th ed. (New York: McGraw-Hill Book Co., 1984), p. 51.
6. B. Evans and C. Evans, *A Dictionary of Contemporary American Usage* (New York: Random House, 1957), p. 377.
7. ———, "What is Quality?" *Quality* (March 1979).
8. P.B. Crosby, "Management and Policy," *Quality Management Handbook* (Milwaukee: Marcel Dekker, Inc., 1986), p. 6.
9. *Ibid.*
10. K. Ishikawa, *What is Total Quality Control? The Japanese Way* (Englewood Cliffs, NJ: Prentice-Hall, Inc., 1985), pp. 60-61.
11. J.M. Groocock, *The Chain of Quality* (New York: John Wiley & Sons, 1986), p. 181.
12. R.D. Carlsen, J.A. Gerber, and J.F. McHugh, *Manual of Quality Assurance Procedures and Forms* (Englewood Cliffs, NJ: Prentice-Hall, Inc., 1981).
13. D. Shainin, et al., *Quality Assurance Handbook* (Boston: Rath & Strong, Inc., 1972).
14. J.M. Juran, "Product Quality—A Prescription for the West," *Management Review* (June/July 1981).
15. "The Juran Report, Number Two" (New York: Juran Institute, Inc., 1983).
16. *Ibid.*
17. Shainin, *op.cit.*, pp. 3-6
18. "New Course for Quality: The Glory Road?" *Production Engineering* (October 1986).
19. A.M. Schneiderman, "Optimum Quality Costs and Zero Defects: Are They Contradictory Concepts?" *Quality Progress* (November 1986).

BIBLIOGRAPHY

20. *Ibid.*
21. "J.M. Juran Comments on the Quality Cost Optimum Model," Letters column, *Quality Progress* (April 1987), pp. 7, 9.
22. B.B. Tregoe and J.W. Zimmerman, *Top Management Strategy* (New York: Simon and Schuster, 1980).
23. Glueck and Jauch, *op. cit.*
24. R.W. Hall, APICS Seminar "Just in Time" (Cedar Rapids, IA: March 1985).
25. *1986 Annual Report*, Emerson Electric Co.
26. *Annual Report 1985*, Ford Motor Co.
27. *Annual Report 1986*, Parker Hannifin Corp.
28. *1986 Annual Report*, Trinova Corp.
29. R.D. Buzzell, *Product Quality* (Cambridge, MA: The Strategic Planning Institute, 1978).
30. A.V. Feigenbaum, *Total Quality Control*, 3rd ed. (New York: McGraw-Hill Book Co., 1983), p. 6.
31. R.J. Schonberger, *World Class Manufacturing* (New York: The Free Press, 1986).
32. H.W. Kenworthy, "Total Quality Concept: A Proven Path to Success," *Quality Progress* (July 1986).

Bibliography

Beels, Gregory J., ed. "SPC, A Prerequisite for World Quality." *Quality* (August 1985), pp. Q1-Q30.

Buzzell, Robert D. "Product Quality." *The Pimsletter on Business Strategy*. Number 4. Cambridge, MA: The Strategic Planning Institute, 1978.

Crosby, Philip B. *Quality is Free*. New York: McGraw-Hill Book Co., 1979.

Gale, B.T., and Klavens, R. "Formulating a Quality Improvement Strategy." *The Pimsletter on Business Strategy*. Number 31. Cambridge, MA: The Strategic Planning Institute, 1984.

Guide for Reducing Quality Costs. Milwaukee: American Society for Quality Control, 1977.

Hall, Robert W., with the American Production and Inventory Control Society. *Zero Inventories*. Homewood, IL: Dow Jones-Irwin, 1983.

Juran, J.M., and Gryna, F.M., Jr. *Quality Planning and Analysis*, 2nd ed. New York: McGraw-Hill Book Co., 1980.

Peters, Tom. *Quality!* Palo Alto, CA: The Tom Peters Group, 1986.

Principles of Quality Cost. Milwaukee: American Society for Quality Control, 1986.

"Quality—A Management Gambit." *Quality* (August 1980), pp. Q1-Q24.

Quality as a Strategic Weapon. Cambridge, MA: The Strategic Planning Institute, 1984.

Upper Management and Quality, 4th ed. New York: Juran Institute, Inc., 1982.

ACHIEVING QUALITY

CHAPTER CONTENTS:

Why should manufacturers strive to achieve quality? Just trying to define the term can be an exercise in frustration. However, most people have an idea of what quality is and know that in the era of intense international competition, the absence of quality in a product or service spells doom.

Quality professionals use a variety of definitions for quality. One school of thought equates quality with satisfying the customer's needs. Another adopts a more modest goal and states that quality is conformance to specifications. J.M. Juran has popularized the concept known as "fitness for use" as the primary interpretation. W. Edwards Deming says, "Good quality means a predictable degree of uniformity and dependability, with quality suited to the market."

Whatever definition is used, there are several factors that must be present to achieve quality. Careful and realistic planning, well in advance of the delivery date, is an obvious must. When new products are developed, engineering and marketing must consider a host of factors which, if ignored, will prevent quality from occurring, or at least will drive up the cost of achieving it to a prohibitive level. A thorough understanding of all these interrelated factors is obtained only by close coordination in the new-product planning phase.

The largest single section of this chapter focuses on the process capability study. While the other sections provide useful overviews, the section on process capability provides detailed instructions so that the manufacturing engineer can become prepared to perform this key element of achieving quality. The process capability study depends on an understanding of basic statistical methodology and control charts, which are both discussed in Volume IV, *Quality Control and Assembly*, of this Handbook series. By focusing heavily on the process capability study, the importance of developing and validating one's process thoroughly, before calling upon it to perform routine production is reemphasized.

While the process capability study is a largely technical procedure dealing with mechanical operations and data analysis, it is not an end in itself. Manufacturers must recognize that in planning, in capability studies, in troubleshooting, and in routine production, people can make or break quality every time. Thus, the human factor must be addressed. There are many different types of quality programs that can help train, motivate, and integrate a company's workforce into a successful quality effort. Workforce development and management are also discussed in more detail in Chapters 9 and 10 of this volume.

Quality also depends on standards. The era of the individual craftworker who took complete responsibility for his or her product from beginning to end is part of the nostalgic past. Today, specialists meet regularly to reach consensus on standards of performance, testing, and operations. These standards allow manufacturers to work confidently within established systems and ensure that their results will merge harmoniously with the work of others. The world grows smaller as standard methodologies are agreed on and translated into normal manufacturing operations. The consumer is the ultimate beneficiary from this work.

Finally, quality audits are recognized as a step toward achieving quality products. The concept of auditing reflects manufacturing's understanding that as systems of production change and grow more complex, it is necessary to periodically verify that earlier decisions are in fact being followed. Auditing supports the system of process improvement and standardization. It provides a balanced compromise between complete trust that "everything will turn out all right" and complete rechecking or redoing of all operations.

Achieving quality requires manufacturers to put one foot in the technical camp of specifications and processes and the other foot in the managerial camp of plans and people. The manufacturing engineer must at times work as a quality engineer, and at other times call on a specialized quality engineer, to achieve the desired end result, which is a quality product delivered at the right time and at a reasonable price.

***Contributors of this chapter are:* Roger W. Berger**, *Professor, Department of Industrial Engineering, Iowa State University;* **Richard Copp**, *Supervisor, Statistical Methods Staff Quality Assurance, Holley Automotive Div..*

***Reviewers of this chapter are:* Alex M. Burgess, III**, *Manager, Statistical Programs, Gleason Corp.;* **Richard Bigg**, *Consultant, Coopers & Lybrand;* **Dr. C.L. Carter**, *Vice President, Quality Assurance and Human Resources, C.L. Carter, Jr. & Associates, Inc.;* **O.P. Gupta**, *Vice President, Quality and Productivity, Gleason Corp.;* **Nathan D. Hollander**, *Manager, Coopers & Lybrand;* **C.J. Marty**, *Director, Productivity, Energy and Advanced Technology Group, Westinghouse Electric Corp.;* **Paul Ockerman**, *Consultant to Westinghouse;* **Robert W. Peach**, *Principal, Robert Peach & Associates, Inc.;* **Harrison M. Wadsworth, Jr.**, *Professor, School of Industrial and Systems Engineering, Georgia Institute of Technology;* **Steve R. Wall**, *Manager, Quality Systems, Gleason Corp.;* **William O. Winchell**, *Associate Professor of Industrial Engineering, School of Engineering, Alfred University;* **Fred W. Woods**, *Manager, Supplier Quality, Gleason Corp.*

NEW PRODUCT DEVELOPMENT

The old adage, "You can't inspect quality into a product; you have to design it in," is true. Therefore, it is important to consider a product's quality from the design stage right through manufacture and use. Increased use of automation forces quality control "upstream" into design. At the design stage, quality really has a double meaning as described by the following:

1. The quality of design of the potential product, assuming that all the processing and marketing assumptions are valid. This can be thought of as theoretical or idealistic quality.
2. The quality of conformance to the theoretical design. This is the more pragmatic view of quality, which actually gets out of the design realm and into the realm of managerial control. This represents all of the quality control work that must be done after the designer has proven a design concept and before production begins.

When both design quality and conformance quality are achieved, the term "fitness for use" is sometimes used.

ASSURANCE OF QUALITY IN NEW DESIGNS

Many inputs are required to develop new designs and specifications. The product definition from the marketing area is a key input. Other considerations required are the quality policy, goals, and objectives of the company. The ability of the company to produce similar products should be fully explored, as well as the ability to produce those products to meet the customer's requirements. Past experience and customer input should be fully utilized to prevent previous problems from reoccurring. A detailed design and associated validation plan, with specific approval points, should be prepared to ensure that the necessary steps are taken in a timely manner.

New-design quality can be defined as the establishment and specification of the necessary cost quality, performance quality, safety quality, and reliability quality for the product required for the intended customer's satisfaction, including the elimination or location of possible sources of quality troubles before the start of formal production.[1] To accomplish these tasks requires a number of tools and techniques. The first is an organizational tool.

Quality Control Team for New Design

To channel the creativity of design into a productive end result, careful organization of effort is needed. For example, a design engineer labors long and hard to develop the prototype of a new product. After successful testing of this prototype, carefully crafted in a laboratory environment, drawings are prepared and sent to manufacturing. Soon, sparks begin to fly back and forth between engineering and manufacturing. According to engineering, the factory is incompetent because it cannot produce the design that has already been proven feasible. But manufacturing claims that engineering has developed another design that can only be produced under the most ideal conditions, which do not exist in most shops, and at prohibitive cost.

This age-old dispute between engineering and manufacturing can be resolved by forming a quality control team, adding one person who can speak to the issues of design intent and manufacturability from an unbiased viewpoint. That person is, of course, the quality control engineer. Along with the new organizational device comes a more explicit way of resolving the inherent difficulties of converting a design to production.

New-Product Cycle

The new-product design-production cycle can be broken into the following 16 phases:[2]

1. A new market opportunity to serve customers is identified, and a new design is contemplated.
2. Technical, production, customer-use, and marketing analysis are made of the marketplace and the design. Cost targets, production volumes, and price levels are established.
3. General specifications are written. They may be:
   - Sales propositions in the case of job-lot production.
   - Rough functional specifications for products that will be manufactured in mass quantity.
   - Broad identification of the coverage of the quality program for the product.
   - Overall outline of product service and maintenance objectives, quality performance requirements, product lifecycle targets, and other related product goals.
4. Preliminary design is made.
5. First prototypes are made. An extensive program of testing the characteristics of this design is carried out, including the components and subassemblies to be used. For products with electronic computing modules, the software will be evaluated and testing begun.
6. Preliminary design review takes place. Preliminary classification of characteristics of the design process (including components and subassemblies) is done, test procedures are evaluated, manufacturing and assembly capability are assessed, cost targets are reviewed, quality levels are identified, design changes are defined and reviewed, process and manufacturing considerations are identified, and a failure mode and effects analysis (FMEA) is conducted on critical items.
7. Intermediate design is made, including production drawings, and prototypes are built from these drawings.
8. Tests are made on this intermediate design, and design review takes place. Action continues on classification of characteristics and on manufacturing, assembly, and test requirements. Marketing and pricing estimates are reviewed. Design changes are defined and reviewed.
9. Final design is completed along with final specifications, standards, tolerances, guarantees, quality planning, and production drawings. Life and performance tests are culminated before final design completion. Component, subassembly, and assembly specifications are completed; assembly inspection plans are developed; tool design and procurement are completed; and costing is finalized.
10. Sample production units are built from the production drawings. This is best done in a simulated or actual manufacturing environment.
11. Shipping and service procedures are defined.
12. Capability studies are made of new and current machines, equipment, and processes.

13. Supervisors and production employees are trained. Pilot runs are made using samples composed of production units. The results of the tests of these samples are incorporated into the design and manufacturing specifications.
14. Final design is reviewed. Test results for product, software (when appropriate), equipment, processes, facilities, and development are analyzed by those functions that need to become familiar with the plans and that can make constructive inputs. The basic product cost targets and lifecycle cost objectives are reviewed to ensure the goal of "design to cost." Product qualification tests are satisfactorily completed. Release for manufacture of production tools and facilities is given, consistent with final design review approval and completion.
15. Marketing announcements are confirmed; product information manuals, service publications, and training aids are completed, all with thorough attention to quality considerations.
16. The unit is released for active production.

These 16 steps provide a generic approach to the development of any new product. The degree of application of each step in the sequence will depend on the particular product, its state of sophistication, and its degree of departure from its predecessors. The one essential element for the sequence is active cooperation among design, manufacturing, and quality personnel.

THE QUALITY PLAN

For projects relating to new products, services, or processes, management should prepare, as appropriate, written quality plans consistent with all other requirements of a company's quality management system.[3]

Quality plans should define the following elements:

- The quality objectives to be attained.
- The specific allocation of responsibilities and authority during the different phases of the project.
- The specific procedures, methods, and work instructions to be applied.
- Suitable testing, inspection, examination, and audit programs at appropriate stages (such as design and development).
- A method for changes and modifications in a quality plan as projects proceed.
- Other measures necessary to meet objectives.

A quality plan that defines these elements will yield a desirable product. Additional information on developing quality plans can be found in Chapter 22, "Quality Management and Planning," of this volume.

FAILURE MODE AND EFFECTS ANALYSIS

Failure mode and effects analysis (FMEA) is an important design and manufacturing engineering tool intended to help prevent failures and defects from occurring and reaching the customer. It provides the design team with a methodical way of studying the causes and effects of failures before the design is finalized. Similarly, it helps manufacturing engineers identify and correct potential manufacturing and/or process failures. In performing an FMEA, the product and/or production system is examined for all the ways in which failure can occur. For each failure, an estimate is made of its effect on the total system, of its seriousness, and its occurrence frequency. Typical failure modes include temperature, humidity, vibration, shipping conditions, and negligence. Effects include damage to other property, loss of life, increased operating cost, repair, loss of product, and increased maintenance. Corrective actions are then identified to prevent failures from occurring or reaching the customer. Information on how to conduct an FMEA is given in Chapter 13, "Design for Manufacture," of this volume.

DEALING WITH SUPPLIERS

Ford Motor Co. uses the following scheme for new-product quality in dealing with suppliers:

1. Use a team approach.
2. Use analytical techniques:
   - Design of experiments.
   - Failure mode and effects analysis.
   - Cause and effects studies.
   - Statistical process control.
3. Develop a consensus schedule (four phases):
   - Feasibility analysis.
   - Control plan development.
   - Process capability.
   - Manufacturing sign-off.

The feasibility analysis has three parts:

A. Design information:
   1. Engineering drawing.
   2. Engineering specification.
   3. Design FMEA.

B. Manufacturing information:
   1. Process flow diagram.
   2. Process FMEA.
   3. Floor plan.
   4. New equipment list.

C. Historical warranty and quality information:
   1. Reports from field experience.
   2. In-house reports.
   3. Market studies.

The control plan is the implementation of a quality system. A quality system checklist may be required. Additional information on how to deal with suppliers to ensure quality in purchased products can be found in Chapter 21, "Materials Management," of this volume.

DESIGN REVIEW

As indicated previously, high achieved quality is the result of solid quality plans. One powerful planning technique is the design review.

Ideally, the design review forces careful study of a design before it is released, but also avoids adversely affecting the creativity of the designer. This is a delicate balancing act.

The design review team consists of a chairperson and a group of specialists. One suggestion of the individuals on this team and their functions is shown in Table 23-1.[4]

A schedule of design reviews should be included in the new product development plan. Depending on the complexity of the product, the number will range from one to five formal reviews. Although product and process design reviews may be done separately, at some preselected points in the new product development cycle the reviews must be conducted jointly.

Process Design Review

A process must be designed, just like a product. However, the design requirement changes more and is usually not as stringent.

TABLE 23-1
Suggested Members of a Design Review Team

Member	Responsibility
Group chairperson	Calls group meetings and ensures that reports are issued promptly
Reliability engineer	Evaluates design for optimum reliability consistent with goals
Quality control engineer	Ensures that the functions of inspection, control, and test can be efficiently carried out
Manufacturing engineer	Ensures that design is producible at minimum cost and schedule
Field engineer	Ensures that installation, maintenance, and user conditions are included in the design
Procurement representative	Ensures that acceptable parts and materials are available to meet cost and delivery schedules
Materials engineer	Ensures that materials selected will perform as required
Tooling engineer	Evaluates design in terms of tooling cost required to satisfy tolerance and functional requirements
Packaging and shipping engineer	Ensures that the product can be handled without damage in shipping
Marketing representative	Ensures that customers' requirements are realistic and understood by all parties
Design engineer(s)	Constructively reviews adequacy of design to meet specified requirements. These individuals are not associated with the design of the item being reviewed
Consultants	Evaluates design from a designated specialist viewpoint. These individuals are called when a special viewpoint is required
Customer representative	Expresses customer viewpoint and requests additional study where appropriate. This individual may be considered optional

This is because a process is under the continuous control of engineers and users, while a product must be shipped off to a customer where it operates in a relatively unknown environment. Thus, there is more of an opportunity for process failure as well as greater flexibility for adaptation. Engineers expect to make minor process adjustments. Careful design review minimizes the number and extent of these adjustments.

Process Failure Mode and Effects Analysis

The FMEA technique is a key element of prevention and should be incorporated into every design review, whether for product or process. Basically, the product or process is examined to determine all of the possible ways that failure can occur, and the likely outcome of each of these ways (modes) of failure. Refer to Chapter 13 for additional information on this subject.

FAULT TREE ANALYSIS

The primary focus of the design review is on how the product will perform. But consideration must also be given to what happens if the product does not function properly. In addition, consideration must be given to the kinds of consequences that may occur and how likely they are to occur. A fault tree analysis (FTA) addresses these issues.

A fault tree analysis first identifies all of the undesired events that might occur. It then works backward to determine the cause of each event. This gives a trouble-and-remedy diagram that shows the probable causes of each symptom of system failure. The FTA then provides a checklist for managers to seek corrective action or develop improvement programs.

Fault tree analysis is used primarily with products demanding high reliability, having significant safety considerations, and having high technological content. Besides helping to design a superior product, FTA can significantly reduce the probability of product liability.

The fault tree has branches that are connected together. The two types of connections are (1) "AND" gates (all inputs must occur for the output to occur) and (2) "OR" gates (occurrence of any one or more of the inputs will cause the output to occur). Like a cause-and-effect diagram, the fault tree diagram provides an effective framework for creative thought and interaction between people. It requires people to think more deeply and to systematically record their results. Several iterations, along with brainstorming sessions and pilot tests, might be involved in producing the final fault tree. This would then be used by design or manufacturing engineers in rethinking their plans.

The fault tree analysis shown in Fig. 23-1 looks for causes of implantable lead failures in heart pacemakers.[5] This failure can occur for either of two reasons: insufficient current reaches the electrode (event A) or the lead fails to conduct current to the heart tissue (event B). Because either event can cause the unwanted output event, "Lead does not pace as required," they are connected to it through an "OR" gate. They are also outputs for other events. For the lead to fail to conduct, two events must happen simultaneously: (1) the lead is displaced from the heart tissue and (2) there is a lack of conduction fluid. These two events must connect to event B through an "AND" gate.

The five stated events are all interconnected as shown in Fig. 23-1. Note the differing shapes of the "AND" and "OR" gates. These symbols are used in all kinds of logic diagrams. The actual diagram presented in the cited reference is much more complex than in Fig. 23-1. For additional information on fault tree diagrams, refer to the papers by Hammer and Eisner.[6,7]

QUALITY FUNCTION DEPLOYMENT

Quality function deployment (QFD), a relatively new planning technique, is one of the most all-encompassing and is especially suited to large-scale consumer items such as automobiles and appliances. These products have heavy tooling and design costs as well as many optional features.

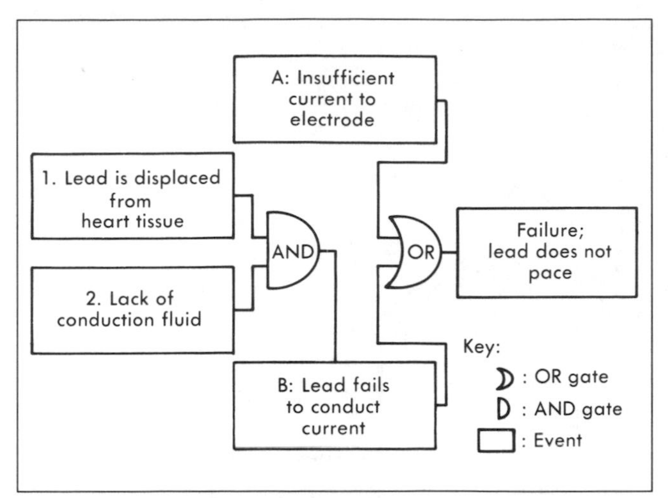

Fig. 23-1 Simplified fault tree analysis used for determining implantable lead failures in heart pacemakers.

QFD begins with market research to obtain consumer feedback on product features, performance, and service.[8] This feedback, referred to as "the voice of the customer," is then put into part specifications and manufacturing parameters upon which an engineer can act. Thus, QFD ensures that engineering activities focus on meeting customers' likes rather than on the dictates of upper management.

QFD is sometimes called "The House of Quality" because of the distinctive matrix diagram that appears as the roof of a house (see Fig. 23-2). The House of Quality lists what the customer wants and matches that against how a company will meet those wants. It also encompasses competitive assessments, such as best-in-class comparisons. This allows complex relationships to be displayed in a way that people with different disciplines and in different areas of a company can understand them and keep their focus on what the customer wants. This minimizes "opinioneering."

Each House of Quality is referred to as a case study, and as a company builds up a library of case studies, it is also building up a knowledge base that can be used in future product developments or to train new engineers.

MATERIAL REVIEW BOARD

In early production, there are frequently adjustments that must be made in the use of material. Often, specifications are assigned in haste, without sufficient preliminary data. As a result, material outside specifications is not necessarily defective; it must be reviewed for suitability. If the specifications are wrong and the item will suit the purpose, then it should not be rejected. However, the specification should be reviewed and then changed. These tasks are the function of the material review board (MRB).

A major pitfall with material review boards is the tendency to concentrate on the function of dispositioning material and to neglect the corrective action function. Corrective action should be a corollary of every single MRB decision. There is also an unfortunate tendency to bring the same deviation request to the MRB over and over again. One partial solution to this problem is a rule stating that the third time the same deviation is requested, the deviation automatically becomes the specification. Such a rule helps focus attention on the consequences of the deviations and forces the different parties to resolve their differences. It is also helpful to assign corrective action to a specific individual, with a specific, published completion date.

RELIABILITY PLAN

Product reliability is, fundamentally, quality extended through time or over a number of cycles of operation. Although most physical products will eventually fail, the length of time to failure and the mode of failure are critical questions that must be answered before mass production begins. A reliability plan is established at the design stage to lead to these answers. This plan should include a specific reliability objective stated in mean time between failures (MTBF), or other appropriate parameters, and the method that would demonstrate that this objective has been met. For additional information on reliability planning, refer to Chapter 8 in reference 9.

PROCESS CAPABILITY

Providing high-quality goods and services requires uniformity around well-defined target values. Product planners determine customer wants and needs and translate them into specifications, commonly expressed as a target value with some range (tolerance) of acceptable deviation from that target value.

When planning the manufacturing process, adequate processes must be developed for critical product characteristics. Given that the designer has correctly determined the best target value, those processes most consistently hitting the target value will produce the best product at the lowest cost.[10] Efforts at continuous quality improvement are synonymous with understanding and reducing variation from the target value.

CONCEPTUAL DEFINITION

A process capability study quantifies the variation in the output from a process and is the first step in understanding variability. No process produces all identical items. Each item varies, perhaps immeasurably or perhaps a great deal. However, a stable process will produce a stream of product, the individual units of which will fall within some predictable range (see Fig. 23-3). That range is the process capability. J.M. Juran defines process capability as "the measured, inherent reproducibility of the product turned out by a process."[11]

Numerous reasons exist for quantifying process capability. Major reasons include comparing process output to specification requirements, planning for process control, determining the suitability of machines for producing specific products, making purchasing decisions on new machinery, and for various contractual purposes between a customer and supplier.

Just as there are numerous reasons for quantifying capability, many methods for estimating process capability exist and fall under the general term "capability studies." Capability studies

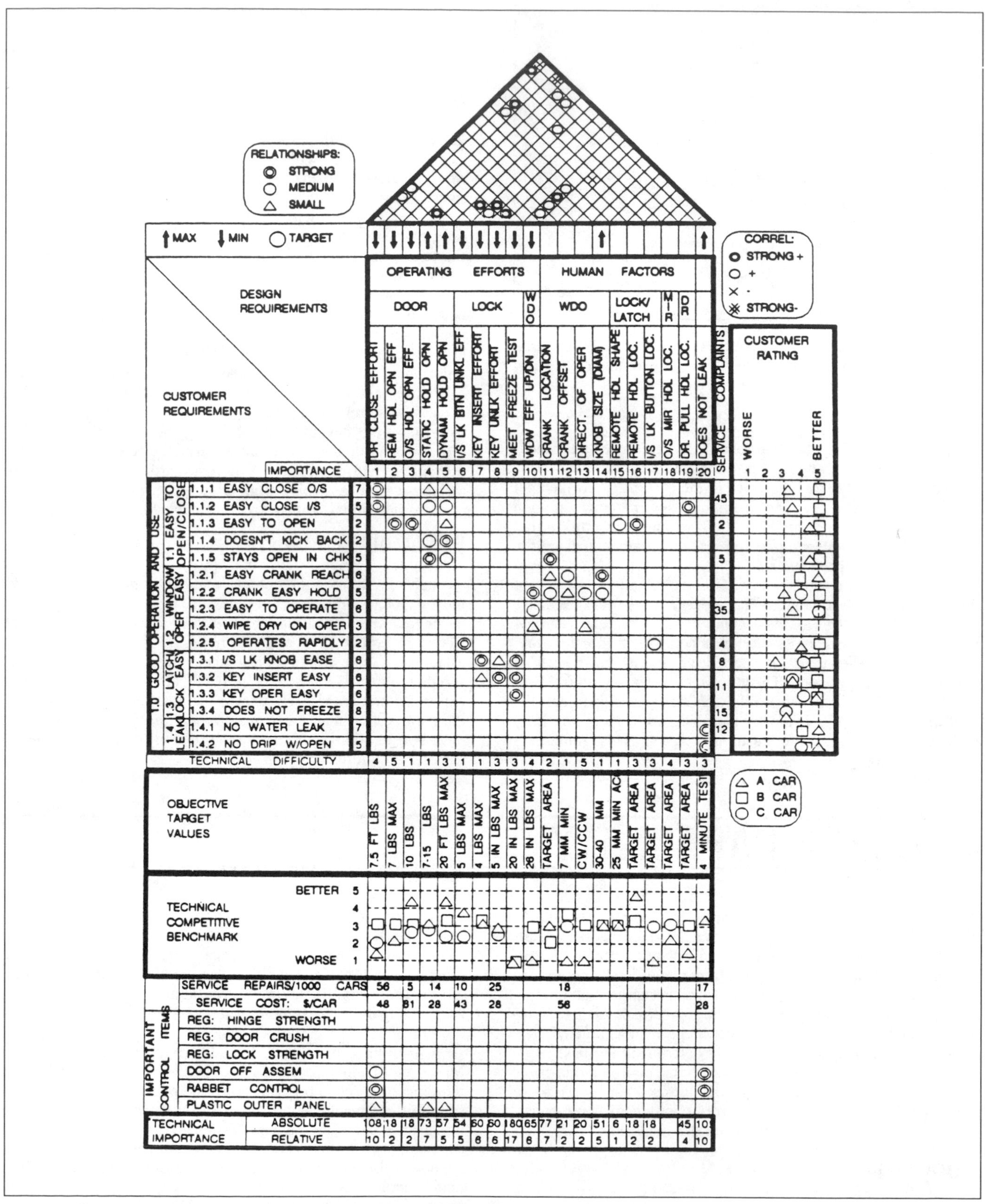

Fig. 23-2 Example of a House of Quality diagram used in QFD. (*American Suppliers Institute*)

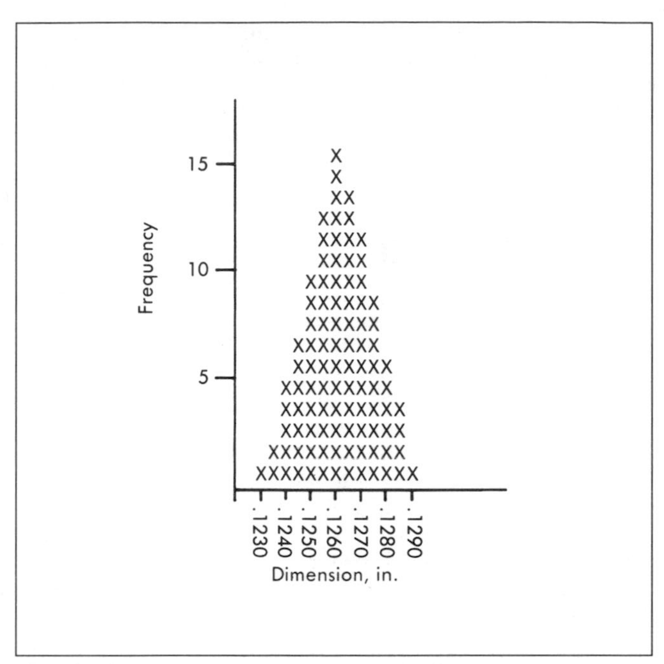

Fig. 23-3 Frequency tally of 100 parts coming from a process with a target of 0.1265".

may be short-term efforts for gathering information on new products, machinery, and processes. Or they may be long-term studies of ongoing production to determine process stability. A truly meaningful statement about process capability can only be made about a process that is stable and whose "range of repeatability" remains consistently predictable over time. Control charts are generally used to monitor process variations. The types of control charts used and how they are constructed and interpreted is discussed in Volume IV, *Quality Control and Assembly*, of this Handbook series.

ANALYTICAL DEFINITION

To find out how the process is behaving in terms of an output quality characteristic, samples can be drawn from the process, and the information can be used to estimate process behavior as a whole. The whole unique process, if it exists, is often referred to in statistical terms as a population. It may be described by a set of measures called population parameters. A process (population) is said to exist if all the elements of the population are subject to a fixed set of common causes of variation. Usually a population can be adequately characterized by a few simple measures such as the mean and standard deviation and a probability distribution. These measures are defined by population parameters that are seldom known and hence are estimated by corresponding measures calculated from sample data. Because sample measures, called statistics, are based on only part of the population, they are uncertain estimates of the population parameters. Consequently, each kind of statistic follows a sampling distribution of its own, which is different from the parent population of individual measurements.

A measure of central tendency of a distribution is a numerical value that describes the central position or location of the data. The most commonly used measure is the arithmetic average or mean value of all the data in the sample or population.

The true mean of a population (a parameter) can be calculated by the equation:

$$\mu = \sum_{i=1}^{N} X_i/N \tag{1}$$

where:

μ = true mean of the population
Σ = the sum of
N = the total number of observations in the population
X_i = the i-th value of the individual observation

In practice, however, the true mean of the population is seldom calculated because it is rarely possible and unnecessary to measure or count all items in the population. In most instances, the true mean is estimated by the sample mean or average (a statistic). The sample mean can be calculated by the equation:

$$\overline{X} = \sum_{i=1}^{n} X_i/n \tag{2}$$

where:

$\overline{X}$ = the sample mean (pronounced X-bar)
n = the total number of observations in the sample

The variance is an important measure of the variability in data. It is the average of the sum of the squared deviations of the data from their mean. The true variance, σ_X^2, can be calculated by the equation:

$$\sigma_X^2 = \sum_{i=1}^{N} (X_i - \mu)^2/N \tag{3}$$

The square root of the variance called the standard deviation, σ_X, is a more commonly used measure of variation.

In practice, both the true mean and true variance of the population are seldom known and are therefore estimated from sample data. The sample variance, s^2, can be calculated by the equation:

$$s_X^2 = \sum_{i=1}^{n} (X_i - \overline{X})^2/(n - 1) \tag{4}$$

Like the true variance, the square root of the sample variance, called the sample standard deviation, s, is a more commonly used measure of variation.

Most electronic calculators are capable of calculating the sample mean and sample standard deviation. Also many software programs exist for statistical process control for use with personal computers.

Another important measurement of variability is the range, which is the difference between the largest value and the smallest value of the data within a sample. The range is calculated by the following equation:

PROCESS CAPABILITY

$$R = X_l - X_s \quad (5)$$

where:

R = the range
X_l = largest value in the sample
X_s = smallest value in the sample

The range is often used to obtain an estimate of the population standard deviation (σ_X). Mathematically, an exact relationship between R and σ_X has been established by the equation:

$$\sigma_X = R/d_2 \quad (6)$$

The value of d_2 varies depending on the size n of the sample/subgroup. This relationship also depends on the assumption that the observations of X come from the normal distribution. Moderate departures from the assumption, however, do not markedly erode the effectiveness of this relationship.

As was mentioned earlier, the standard deviation is a measure of the variability of the data. It is especially useful in conjunction with the normal distribution. The normal distribution has many useful mathematical properties. The most important property affecting capability analysis is the fact that the normal distribution is unimodal (has one large hump) and symmetrical (has tails of equal length) as shown in Fig. 23-4. Also, because of the symmetry of the normal distribution, all three measures of location—the mean, median, and mode—have the same value. In practice, no collection of data is perfectly normal. However, from a practical standpoint, if the data set has a single mode and is fairly symmetrical, straightforward use of the normal distribution as a model for capability analysis is justified. Methods for dealing with non-normal data will be briefly discussed later.

A formula containing the average and standard deviation provides the analytical (mathematical) definition of process capability. Process capability is expressed as the range defined by the following:

$$UNL = \bar{X} + 3s \quad (7)$$

$$LNL = \bar{X} - 3s \quad (8)$$

where:

UNL = upper natural limit
LNL = lower natural limit
$\bar{X}$ = average or sample mean
s = sample standard deviation

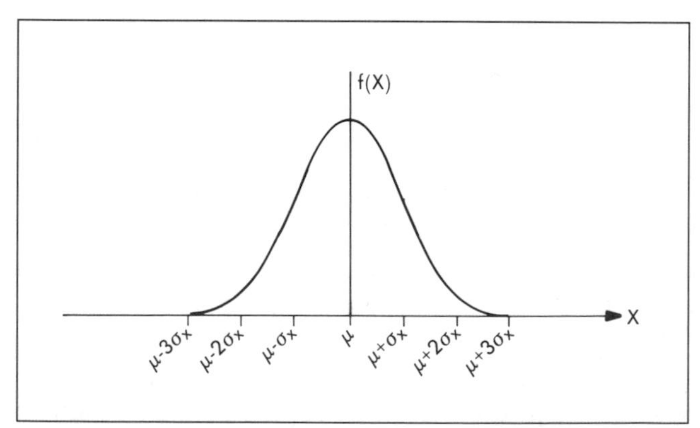

Fig. 23-4 Normal distribution curve; μ is the true mean and σ is the standard deviation.

Note that this formula contains information on both the location (average or $\bar{X}$) and dispersion (standard deviation or s) of the process. In some cases, process capability is expressed simply as the $6s$ range independent of the average.[12] The $6s$ range is often called the "natural tolerance" (NT) of the process, where:

$$NT = UNL - LNL = 6s \quad (9)$$

For processes in which the average level is readily and economically adjustable, expressing the capability simply as the $6s$ range is acceptable. In contrast, for processes in which adjusting the average presents technical or economic difficulty, the capability should be expressed using both the average and standard deviation because the information defining the location of the process output is essential.

One might ask why the natural tolerance is defined as the $6s$ range. Why not $8s$ or $10s$? Or why not $4s$? In any distribution, the area under the curve corresponds to the quantity of output. In the normal distribution, 99.73%—almost 100%—of the process output falls within $\pm 3\sigma$ of the average (refer to Fig. 23-4). Increasing the multiplier above 3 would have no practical value. Thus, the accepted standard for process capability defines capability in terms of 3 standard deviations ($3s$) taken either side of the average, or a total of 6 standard deviations ($6s$).[13]

Determining process capability only estimates the range within which the process will operate. Some practitioners discuss capability with great precision, as though they can predict with great certainty that 3 out of 1000 pieces will be outside the natural limits.[14] Such interpretations of capability estimates attempt to make theoretical mathematical abstractions applicable in actual situations. For all practical purposes, one should consider that nearly 100% of the process output falls within the $6s$ natural tolerance range.

To make a meaningful and valid capability study, one must accurately estimate the standard deviation and quantify any important factors affecting the average. Several methods are currently in use for conducting capability studies and performing capability analysis. Different approaches are appropriate for different processes and situations. In all cases, the process must be shown to be in statistical control to perform a valid study.

PROCESS CAPABILITY AND SPECIFICATIONS

Ultimately, the process capability and/or the natural tolerance must be compared to some standard or requirement. In general, if the limits of the process capability, $\bar{X}$-$3s$ and $\bar{X}+3s$, fit within the specification limits, the process is said to be "capable." Many situations require a more severe standard of capability: $\bar{X}$-$4s$ and $\bar{X}+4s$ must fit within the specification limits. This provides an extra margin of safety in critical situations. If the range of the process capability does not fit within the specification limits, then the process is termed "incapable." The reason for not being capable may be due to one of three possible conditions: (1) the dispersion (measured by the standard deviation) may be too great, (2) the average may be off target, or (3) a combination of excessive dispersion and the off-target condition (see Fig. 23-5).

Up until this point, all the discussion has assumed that the characteristics being considered can be measured with variable data. Many cases exist, however, in which only attribute data is available, as when GO/NOT-GO gaging is used. Unlike variable data where the process has a natural tolerance inherent to itself and independent of any specification, capability for attribute data has meaning only when related to the specification. The capa-

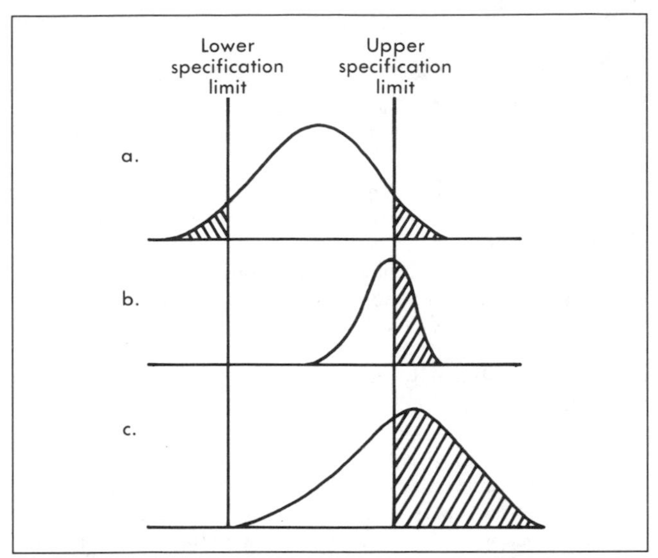

Fig. 23-5 Failure to produce output within the specification limits may be due to: (a) excessive dispersion of the output, (b) an off-target condition, or (c) a combination of excessive dispersion and off-target condition.

bility is simply the average level or percentage of conformance to the specification over time. Attribute data studies require very large sample sizes; the best data would come from an attribute statistical process control chart (*p, np, c,* or *u* chart). So, for a process that produces nonconformities at an average level of 2%, the capability is 98%. In general, one should convert attribute data to variable data. The remainder of the discussion will deal with variable data.

The Capability Index

Several indexes are used to quickly summarize capability with respect to the specification tolerance. In general, the indexes are a ratio of the natural tolerance and the specification tolerance. One common index is C_p and is estimated using the following equation:

$$C_p = (USL - LSL)/6s \tag{10}$$

where:

C_p = capability index
USL = upper specification limit
LSL = lower specification limit
s = standard deviation

As mentioned previously, the $6s$ range is independent of the location of the process and is only appropriate in situations where the average level of output is easily and economically adjusted. While C_p is used by many manufacturers, its reciprocal, referred to as the capability ratio, is also commonly used.[15] Some manufacturers also express capability as the percentage of tolerance used by the $6s$ range.[16]

Another index, C_{pk}, which was originally developed for one-sided specifications such as minimum weld strength or maximum surface finish, is now recognized as a useful index for bilateral specifications. A convenient feature of C_{pk} is that it takes into account the location (average) of the process as well as its natural tolerance. C_{pk} is the preferred index in the automotive industry. It is estimated as follows:

$$C_{pk} = \text{minimum of } (USL - \overline{X})/3s \text{ or } (\overline{X} - LSL)/3s \tag{11}$$

where:

C_{pk} = capability index
USL = upper specification limit
LSL = lower specification limit
$\overline{X}$ = sample average
s = sample standard deviation

When applying the C_{pk} calculation to a one-sided tolerance, Eq. (11) can be used along with the appropriate specification limit (*USL* for a maximum or *LSL* for a minimum). Figure 23-6 illustrates C_p and C_{pk}.

Interpreting the values calculated for C_p and C_{pk} is straightforward. If C_p (C_{pk}) is greater than 1, then the process natural tolerance is smaller than the specification tolerance range and the process is capable. If C_p (C_{pk}) is less than 1, then the process natural tolerance is larger than the specification range and the process is not capable. Many authorities consider a process to be "barely capable" if $1 < C_{pk} \leq 1.33$ and "capable" if $C_{pk} > 1.33$.[17]

C_{pk} offers a better indication of process capability than C_p because it indicates the success of maximizing the proportion of output within specification. Figure 23-6 shows the output from two processes, A and B. Process A is a capable process that does not produce any nonconformities, yet it also produces virtually no output at the desired target. On the other hand, process B has a natural tolerance that exceeds the specification tolerance, but it is centered on the target, thus producing a great deal of output at the desired target value, although it may produce a small number of nonconformities. Although process A has a C_p much less than process B, it has a C_{pk} about equal to process B. Process B is the better process because it is producing more "on-target" product. However, if the location of process A can be easily adjusted to improve its C_{pk}, it would become the better process.

When confronted with the need to improve a process, one often must make the choice between reducing the natural tolerance of the process or getting the process to run consistently on target. Using C_{pk} to analyze the options will show that, in

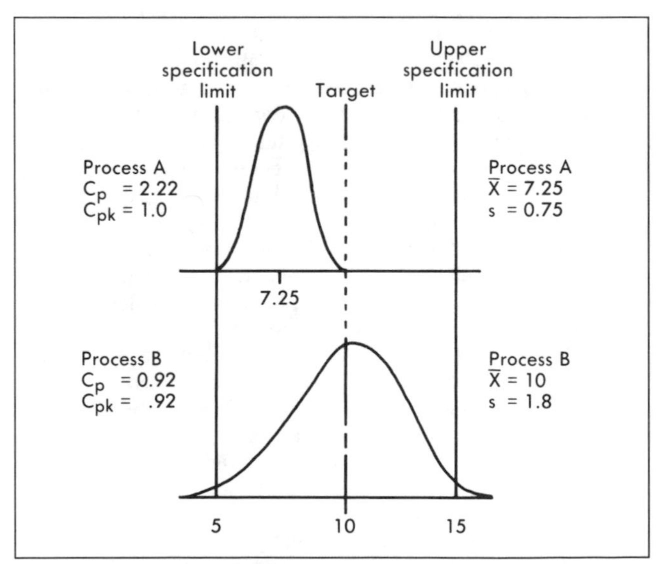

Fig. 23-6 The C_{pk} index indicates that despite the much better C_p for process A, because process B is centered on target, its C_{pk} is virtually equal to process B's capability.

most cases, getting and staying on target does much more to improve C_{pk} than reducing the natural tolerance. Further, reducing the natural tolerance is often an expensive option requiring the use of more precise machinery. Conversely, manufacturing engineers often take great pains to develop a highly precise process (one with a small natural tolerance), but overlook the need to control the process with respect to location. Thus, in spite of great precision, the actual capability (as expressed by C_{pk}) is marginal.

Process Capability Studies and Analysis

In most cases, estimating process capability will require the collection and analysis of new data. In some cases, capability information from a very similar product or process will allow a reasonable estimate of capability. There are many ways to approach the capability study. However, underlying principles exist that apply in any study. The purpose of the capability study must always be kept in mind. The capability study is the initial attempt to determine the natural tolerance of the process and begin to understand the sources of variation that degrade it.

The output from a process will frequently result from the collective effect of numerous sequential operations on different pieces of machinery, or multiple streams of product coming off of the same machine. The first step in performing a capability study is to break the process down into the subprocesses and operations requiring study. For the sake of explanation, the discussion of capability studies will begin with the simplest case, a single operation with a single stream of product, and then move to some special cases.

Capability studies fall into two broad categories—short-term studies and long-term studies. Short-term studies are usually conducted on new products and processes where no opportunity exists to collect data during actual production. Consequently, the short-term study uses relatively small sample sizes (30-100 pieces) to keep costs down. The short-term study is sometimes called a ''machine capability study'' because all variables are held constant in an attempt to isolate the inherent variability of a particular machine. Long-term studies, on the other hand, consist of analyzing data collected from a current production process. Sample sizes are generally larger than those from a short-term study. The analysis must take into account those factors that change with time and may affect the process output.

In the automotive industry, many companies make a distinction between the short-term and long-term capability study. The short-term study is called a process potential study because it predicts the most optimistic view of long-term capability. The long-term study is called, simply, a capability study.

Long-term capability study. In the ideal case, data for a long-term study is plotted on a control chart, usually the average and range chart. A prerequisite to applying the formulas used for capability analysis is that the process must be in a state of statistical control. If a chart shows a lack of statistical control, the process is unstable, and no valid predictions about capability can be made in the presence of instability. The instability causes must be removed before the analysis can be completed.

The analysis of a long-term study is straightforward. Assuming that the data comes from an average and range chart, one need only estimate the standard deviation, then use Eqs. (7) and (8) to calculate the upper natural limit (*UNL*) and lower natural limit (*LNL*). Comparison to the specifications and calculations of the capability index, C_p or C_{pk}, complete the analysis. The sample standard deviation can be estimated using the equation:[18]

$$s \approx \bar{R}/d_2 \tag{12}$$

where:

s = standard deviation
$\bar{R}$ = average range
d_2 = a constant

The value of d_2 varies depending on the size of the sample/subgroup. The numerical value of d_2 for different subgroup sizes can be found in Table 23-2.

An important consideration in the long-term study is what period of time the study should cover. From a purely statistical viewpoint, the rule of thumb offered by most experts is that a minimum of 25 subgroups or plots on the control chart is required for a valid study. Such a rule is general and arbitrary. Actually, the process will dictate what constitutes a sufficient period of data collection to make a valid study. The study should continue until all process variables that may affect the process have changed at least once. Thus, the study should include changes in shifts, operators, raw material, perishable tooling, and warmup periods. About 25 days of operation is desirable. If the process is stable throughout those changes, then the capability study is valid.

Short-term capability studies. Many situations exist in which it is impractical or impossible to perform the long-term capability study using data from a control chart. Often only a ''snapshot'' look at the process is available/needed to make a

TABLE 23-2
Values for Constant d_2 Based on Subgroup Size

Subgroup Size, n	d_2
2	1.128
3	1.693
4	2.059
5	2.326
6	2.534
7	2.704
8	2.847
9	2.970
10	3.078
11	3.173
12	3.258
13	3.336
14	3.407
15	3.472
16	3.532
17	3.588
18	3.640
19	3.689
20	3.735
21	3.778
22	3.819
23	3.858
24	3.895
25	3.931

decision. Examples of such situations include qualifying new machinery prior to final acceptance, estimating capability to determine the appropriate processing for a new product, or investigating a problem in production. The short-term study provides that quick look at the process.

Typically, all process variables are held constant, and a minimum of 30 consecutive pieces are run and inspected. Depending on the cost of inspection, the sample size may go as high as 100 pieces. Another sampling method for the short-term study involves pulling a random sample (30 minimum) from a larger production run. The average and standard deviation are calculated, and then the natural limits and capability indexes are calculated.

During the study, no intentional changes should be made to the process. The same operator, materials, tools, and so forth should be used. The pieces run should be identified with respect to order of production. A log should be kept to record any changes or adjustments to the process. The data should be plotted with respect to time to observe any changes associated with time or events recorded in the study log.

Multiple-stream processes. The previous explanations represent very simple cases. In many cases, however, product comes from multiple-stream processes. In such cases, special procedures for data collection and analysis are required. The help of a quality engineer or statistician is highly recommended.

A four-station chucker will serve as an example of how one approaches the capability study on the multiple-stream process. The purpose of the study is to isolate the ''between-station'' variation from the ''within-station'' variation. Each station must have its location and dispersion quantified, and the collective capability of the machine must be estimated. A statistical procedure called the analysis of variance (ANOVA) is one technique for studying such situations. [19] The goal of those engineering the process is to have all stations homogeneous for the location and dispersion. Figure 23-7 shows frequency tallies from two four-station machines, one that is not homogeneous, the other that is.

Machine 1

	Spindle 1	Spindle 2	Spindle 3	Spindle 4
20		XX		
21		XXXXX	X	
22	X	XXX	XXXX	
23	XX	XX	XXXXX	
24	XXXXX		XX	XX
25	XXX			XXXX
26	X			XXXX
27				XX

Machine 2

	Spindle 1	Spindle 2	Spindle 3	Spindle 4
20				
21				
22	X			
23	XX	XX	X	XX
24	XXXXX	XXXXX	XXXX	XXX
25	XXX	XXX	XXXXXX	XXXXXX
26	X	XX	XX	X
27				

Fig. 23-7 Frequency tallies off of two four-spindle machines. Machine 1 has spindle-to-spindle differences contributing to most of the variation. Machine 2 has very little spindle-to-spindle variation.

Dominance. Juran introduces the concept of dominant variables in a process.[20] A dominant variable is that variable that will cause the greatest change in the process. Juran lists four types of dominance: (1) machine, (2) set-up, (3) operator, and (4) material. In some cases, the dominance is shared by more than one variable. The short-term capability study, if properly planned, can systematically introduce changes in the dominant variables to quantify their effect on the process. The study actually becomes a designed experiment with the dominant variables acting as factors. As with the multiple-stream process study, the analysis of variance (ANOVA) is one technique for analyzing such data.

Many process engineers neglect to consider the effect of dominant variables when assessing capability. Thus, they may have a highly precise machine with respect to its inherent variability, but the differences from station to station or set-up to set-up may make the overall process not capable. Specifications for a machine should include some limits on the differences allowed between stations, set-up changes, and tool changes. Those differences can be quantified in the carefully planned capability study.

Non-normal data. Although industrial data is predominantly normally distributed (in the technical sense of the normal distribution), there are cases in which the output of a process is highly skewed or bimodal (having two peaks). Typical examples of process data that is not normally distributed include torque measurements, weld strengths, and concentricity measurements. Because capability predictions are based on the normal distribution, valid predictions about the process tolerance require the use of special methods.

The control chart will usually not reveal non-normality. To detect it, the data should be viewed graphically in a histogram or plotted on normal probability paper. Analytical tests also exist for testing normality.[21] If the data is found to be non-normal, a two-step response is needed.

First, the process performance itself should be reviewed to see if the distribution is characteristic for that process. If not, the process problem should be corrected, the data discarded, and a new study run. If so, the data can be transformed. The data is analyzed in the transformed state and the natural tolerances are then re-transformed back to the original scale.[22,23] When dealing with non-normal data, an experienced quality engineer or statistician should be consulted.

GAGE AND MEASUREMENT SYSTEM CAPABILITY

Capability of gages and measurement equipment presents some special concerns. When assessing the suitability of a measuring system, four characteristics of the system must be evaluated: (1) accuracy, (2) precision (often called repeatability), (3) reproducibility, and (4) stability. Accuracy is how well the entire set of repeated readings on the gage agrees with a known standard. Precision is the amount of dispersion of repeated readings. Reproducibility reflects how well readings made by one operator agree with those made by another. Stability addresses how constant the accuracy and precision remain with time, and it determines the required calibration and maintenance intervals.

Accuracy can best be assessed by a qualified metrology lab or calibration service. Gages can also be checked for accuracy against ''masters'' of known magnitude, such as gage blocks.

Capability studies for gages address the issues of precision (repeatability) and reproducibility. To isolate the gage variability, repeated measurements must be made on the same sample hardware. The data of interest are the ranges of the repeated readings on the single parts. For measuring systems that depend on operator skill, several operators should make the repeated measurements on the same set of hardware. One procedure recommends using 10 pieces of sample hardware on which each of three operators take three readings.[24] From the ranges of the repeated readings, the standard deviation of the measurement precision is calculated. The difference in average between the operators is an estimate of the reproducibility of the measurement system.

Once the gage precision is quantified, it must be compared to some standard to determine whether it is acceptable. Ideally, the ratio of the measurement standard deviation to the process standard deviation should be 1:10, but a ratio of 1:5 is often considered acceptable.[25] These ratios are based on the following formula:

$$s_t = \sqrt{S_p + S_m} \qquad (3)$$

where:

s_t = total variation
s_p = actual process variation
s_m = measurement error variation

Because the squares are added, a ratio of 1:10 for the standard deviations of measurement to the process translates into a 1% error with respect to precision. The 1:5 ratio translates into a 4% error. The quality engineer must determine if the relative amount of gage error is acceptable in a particular application of the measuring equipment.

QUALITY PROGRAMS

Quality is achieved through the efforts of people. One of the most challenging tasks faced by any manager is to coordinate these efforts in such a way that quality and productivity increase over time. The old-fashioned autocratic approach to management can still be made to work, but more and more managers are adopting modern schemes of participative management. The workforce is becoming more intelligent, better trained, and more demanding of a role in the management of their work. Further, results achieved overseas demonstrate that humane and involved management styles produce favorable long-term results.

There are a variety of successful approaches to organizing human quality efforts. Some of the common ones are quality circles, quality task forces, performance action teams, zero defect programs, quality assurance teams, and "bandwagon" programs; the list goes on and on.

All of these approaches are vehicles for tapping the past resource of employee knowledge and experience. These have proven most effective when working on solving specific problems. A laissez faire approach toward management involvement in these efforts will render them ineffective. Problems or challenges to the team should be clearly defined and all necessary technical assistance provided. Most importantly, employee participation groups must be given adequate training in statistical analysis and other analytical and problem-solving techniques. This is a facet of employee participation programs which, if neglected, precludes success.

This discussion touches only briefly on a vast and vital subject. There is always room for improvement of human attitude, knowledge, and skill in any organization. Human improvement should be a vital, ongoing element of organizational development that leads not only to high-quality products and services, but also to high morale and pride of membership. An organization-wide human involvement program that is managed by specialists and directed by the very top managers should be present in every organization. It should become a part of the organizational culture, not simply a fad that is tacked onto existing efforts in response to persuasion by a consultant or publicity expert.

THE 14-STEP PROGRAM

One of America's leading quality consultants is Phil Crosby. Over a period of years, he developed a 14-step program for building a quality workforce. The 14 steps are as follows:[26]

1. Management commitment. Decide that serious change is going to occur and prepare communications to that effect.
2. Quality improvement team. Bring together representatives of each department, select a chairman, define roles. Explain the program to the team.
3. Quality measurement. Find out just what kind of quality now exists. To accomplish this, a quality maturity grid can be used.
4. Cost of quality evaluation. Make rough preliminary estimates, using the controller's office.
5. Quality awareness. Share the results of quality measurement with all employees. Keep the focus on information *sharing*, not manipulation or force.
6. Corrective action. Bring problems to light, solve them, or formally pass them up to the next level of management.
7. Ad hoc committee for zero defects program. The quality improvement team should develop a strategy to create an awareness of zero defects and how to accomplish it. Beware that this will take a good deal of time and will require firm support by the company's thought leaders.
8. Supervisor training. Formal orientation with all levels of management is essential. All supervisors must be tuned in and able to explain their role to their people.
9. Zero Defects Day. This is a major event where the "new attitude" is put on display. Make it special so people will remember.
10. Goal setting. Each supervisor gets employees to set goals for 30, 60, and 90 days. All goals should be measurable and attainable.
11. Error cause removal. Each person should describe any problem that keeps him or her from performing error-free work. Managers and staff must respond with ways that these problems can be removed.

12. Recognition. Provide rewards and ceremonies that focus attention on effective contributors. The rewards should not emphasize the financial aspect, but should be highly visible.
13. Quality councils. The quality professionals and the team chairpersons should come together to communicate and identify actions to upgrade the program being installed.
14. Do it all over again. A year or more is required to accomplish the first 13 steps. By then it is time to start over with new situations and, often, due to turnover, new people as well.

These steps constitute a sound procedure for upgrading quality attitudes and capabilities throughout an organization. Many variations of this procedure have been developed in the literature.

QUALITY ASSURANCE TEAM

Western Electric Corp. pioneered the use of a three-person team consisting of design engineer, manufacturing engineer, and quality control engineer. These three people ensure that the three corners of the quality triangle are intact and upheld: (1) design intent, (2) manufacturability at low cost, and (3) quality. The use of this quality assurance team is discussed in more detail in the section of this chapter that concerns new product quality.

PERFORMANCE ACTION TEAM

The performance action team is a total participative management process. It obtained its name because managers should be striving for continuously improved performance, not just quality, efficiency, or productivity. The performance of an organizational system (work group, plant, firm, function) is a function of seven interrelated criteria:[27]

1. Effectiveness.
2. Efficiency.
3. Quality.
4. Productivity.
5. Quality of work life.
6. Innovation.
7. Profitability/budgetability.

The performance action team is one way of achieving these seven criteria. It consists of two major components:

1. Performance improvement planning. This focuses on finding high-yield opportunities for improvement, development of consensus, 2-5 year developmental goals, and specific action programs to move to these goals. This component is implemented in the upper and middle levels of the organization.
2. Participative problem solving. This focuses on resolving actual problems facing the work group, with a time horizon of months rather than years.

QUALITY CIRCLES

Quality circles (also called quality control circles) received intense publicity in the United States beginning in 1980 and actually took on many aspects of a fad. Although the principles are sound, flawed implementations have discredited the quality circle process in some people's minds. There is a major professional association (Association for Quality and Participation, formerly known as International Association of Quality Circles) devoted to this technique. It has sponsored conferences and journals devoted to explaining the process and pitfalls.

The major role of the quality circle should be to solve problems associated with achieving product quality, as perceived by the workers in their own areas. The quality circle is a small group of workers who work voluntarily on these problems within their own area. If the program is not voluntary, it should not be called a quality circle program. While there is some dispute about how truly voluntary the circles are in Japan, it is generally felt that without the voluntary aspect, quite a different organizational form is involved. A nonvoluntary group would normally be called a task force or a task team, directed by management rather than by the workers themselves.

Each quality circle has a leader, and the total scheme includes one or more facilitators who assist the groups in meeting their objectives. Typical circles will spend an hour a week in their meeting and will carry the ideas and enthusiasm from this hour on into the other hours of the week.

The facilitator of the circle is key, and the ability of this person is highly influential on the success or failure of the program as a whole. The facilitator normally attends all circle meetings and steering committee meetings. He or she arranges for management presentations, outside visits, publicity, and a host of other details.

The steering committee is intended to provide an organizational link between the management of the organization and the circles program. Typically, the steering committee meets once a month and develops or approves the operating policies for the circles. Before installing the circles, the steering committee must get a firm grip on the philosophy and implementation plan to be followed. If the company is unionized, then a union representative should be on the steering committee, and it should be made clear that quality circles are not intended to interfere with normal union-management collective bargaining.

Although the concept of a quality circle originated in the United States, the major development was in Japan during the period of the 1960s and 1970s. Professor Kaoru Ishikawa was a primary instigator, and one of the tools of quality circles is called the Ishikawa diagram, which he invented and taught to millions of Japanese workers. An Ishikawa diagram is shown in Fig. 23-8. After Japanese success, the circles were imported back into the United States in the early 1980s. Many of the companies that originally thought quality circles could solve all their problems have since learned that this is only one technique among many and should not be expected to work miracles.

Whether the term "quality circles" is used or some other term is used, the concept of voluntary worker participation in problem solving is a powerful aid to achieving quality. However, it should never be forgotten that workers themselves can only address about 15% of the quality problems in a typical workplace. The other 85% are under the control of management and must be addressed by the management team.

ZERO DEFECTS

The zero defects movement began at the Martin Co. of Orlando, FL, in 1961. The concept was very simple: Instead of negotiating for a 1% defective rate, or a 1/2% defective rate, or a 0.01% defective rate, or whatever, the company would simply determine that no defectives at all were acceptable. This was stated as a fundamental goal, with the recognition that it would obviously not be achieved overnight and that transitional procedures would be required.

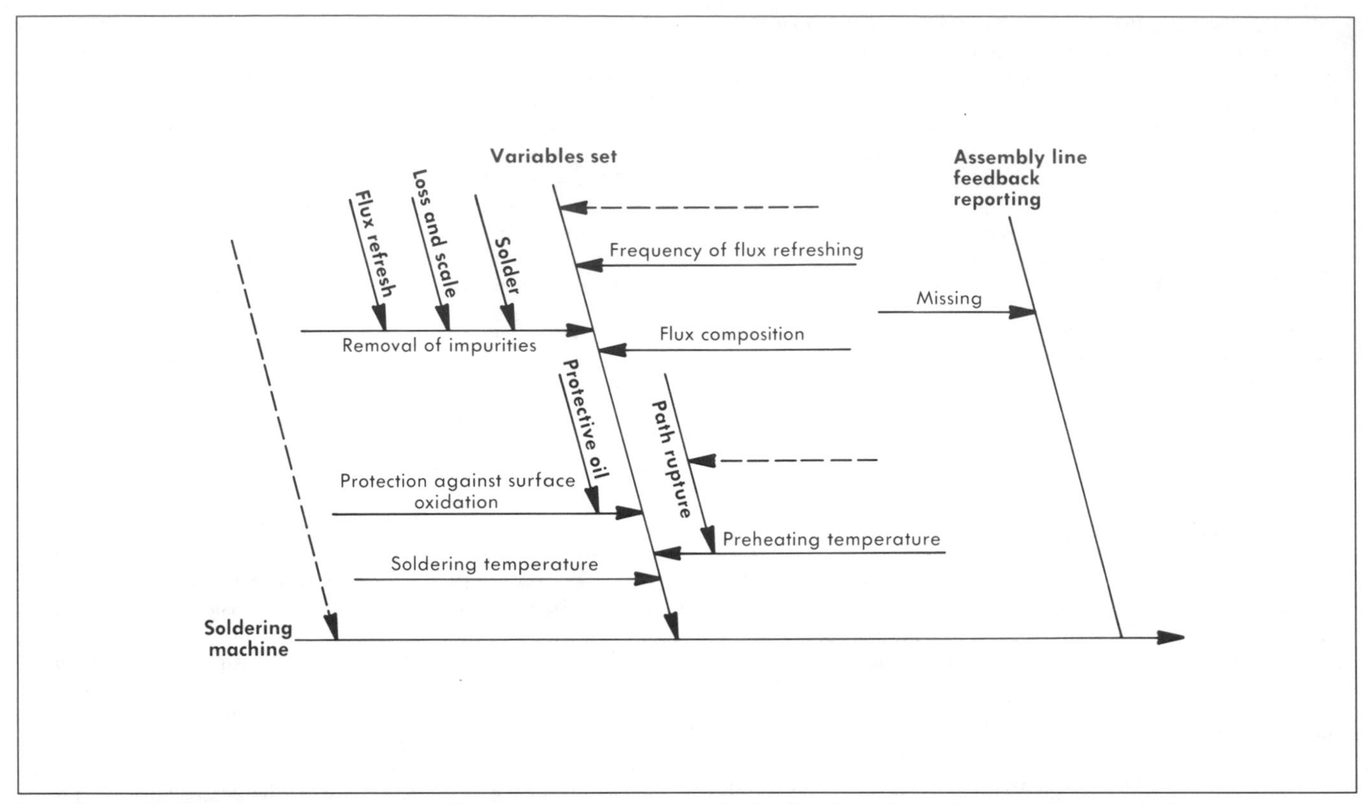

Fig. 23-8 An Ishikawa diagram of a process. (*Copyright American Society for Quality Control, Inc. Reprinted by permission*)

The Martin Co. had good experience with zero defects, and it improved the delivery rate of the Pershing missile that it was then building for the United States Army. Other companies sometimes tried to capture the "flavor" without a full understanding of the underlying processes, and in many cases, zero defects programs were discredited, much the same as quality circles.

The introduction to zero defects is usually accompanied by a high-visibility motivation campaign including rallies, posters, slogans, and perhaps a major picnic or other outing. Each employee is invited to sign a pledge to produce no defects. Coercion is avoided in preference to encouragement and internalization of the concept.

Zero defects is more than a motivational campaign. It is a fundamental philosophy of how to operate a business. It requires not only fired-up employees, but also detection and prevention methods that can give early warning of nonconformances and process problems. In addition, successful achievement requires a high committment from management and corresponding investment to accomplish the goal.

TRAINING PROGRAMS

Widespread and continuing training of the workforce is absolutely essential to maintain and improve quality. A famous example was given by Kaoru Ishikawa as to how important quality control training was in Japan as it built the momentum of worldwide quality leadership:[28]

> Quality control courses through radio began in July 1956. The Nippon Shortwave Broadcasting Co. (NSB) held a 3-month course seven times. Since 1957, NHK (Japan Broadcasting Corp.) held an introductory course called "Management and Quality Control" for 7 weeks (15 minutes per weekday). More than 100,000 copies of this textbook have been distributed. From 1957 to 1962, there were seven NHK radio QC courses. In 1960 and 1961, NHK TV broadcast two courses called "Quality Control" and "Standardization" for 4 months.

Training can take on many different forms. The Japanese emphasis on public radio is unique. Much more common are the training programs incorporated by many colleges, universities, and consulting firms, as well as large companies that provide training for their vendor employees.

In thinking about training, it is important to recognize the three fundamental steps of the learning process:

1. Obtain positive *attitude*.
2. Acquire relevant *knowledge*.
3. Learn necessary *skill*.

If the training program will first develop motivation for the task, then provide knowledge of what the task is all about, and finally develop skill in performing the task, then success is highly likely. Development of proper attitude is often called motivation. This three-step process makes it clear why motivation is not enough. The proper knowledge, skills, and incidentally, tools must also be in place. Additional information on workforce motivation can be found in Chapter 7, "Manufacturing Leadership," and Chapter 10, "Workforce Management," of this volume. Different types of workforce training are discussed in Chapter 9, "Workforce Development," of this Handbook volume.

QUALITY STANDARDS

Standards provide an essential element in the modern industrial system. Without standards, mass production of any kind would be impossible. Standards are to specifications as specifications are to products. The standard spells out how the specifications should perform, just as the specification spells out how the products should perform.

Many different types of standards exist. Many companies have internal standards for drawings, workmanship, scheduling, work quantity, financial reporting, and so on. External standards are imposed on companies through various mechanisms. Much of the standard setting is controlled by national and international voluntary associations, as well as government agencies. There are also mandatory standards spelled out by government agencies. Additional information on internal and external standards can be found in Chapter 14, ''Standards and Certification,'' of this volume.

CLASSIFICATION OF STANDARDS

One author has developed a tripartite classification of standards. This three-faceted classification effort shows the broad scope of standards. The classification is as follows:[29]

1. By subject:
   - Engineering.
   - Transport.
   - Housing/building.
   - Food.
   - Agriculture.
   - Forestry.
   - Textiles.
   - Chemicals.
   - Commerce.
   - Science.
   - Education.
2. By aspect:
   - Forms and contracts.
   - Packaging and labeling.
   - Code of practice, bylaws.
   - Simplification, rationalization.
   - Grading and classification.
   - Tests and analysis.
   - Sampling and inspection.
   - Specification.
   - Nomenclature and symbols.
3. By level:
   - Individual.
   - Company.
   - Association.
   - National.
   - International.

This classification list illustrates the broad scope of standards in modern life. In the context of achieving quality, manufacturers must recognize a simpler classification. One such classification is as follows:

1. Standards for operating a quality program or system, including sampling, auditing, and so on.
2. Product-related standards, including welding, strength, finish, composition, and so on.

This section of the chapter primarily addresses standards for a quality program, with a general discussion of how product standards fit in with the quality program.

The International Organization for Standardization (ISO) defines a standard as follows:[30]

> A technical specification or other document available to the public, drawn up with the cooperation and consensus or general approval of all interests affected by it based on the consolidated results of science, technology, and experience, aimed at the promotion of optimum community benefits and approved by a body recognized on the national, regional, or international level.

An alternate definition is provided by the National Bureau of Standards (NBS):

> A prescribed set of conditions and requirements, of general or broad application, established by authority or agreement, to be satisfied by a material, product, process, procedure, convention, test method; and/or the physical, functional, performance, or conformance characteristic thereof. A physical embodiment of a unit of measurement (an object such as the standard kilogram or an apparatus such as the cesium beam clock).

The NBS standard is somewhat more specific and brings in the important element of a ''physical embodiment.'' Standards need not be reduced to the printed page in all cases, although the vast majority are. The reference standards for metrology, workmanship standards, and surface finish standards are valuable artifacts from which a long chain of day-to-day practice flows.

RELATION OF STANDARDS TO QUALITY

A question that arises in the discussion of standards is: How do standards relate to the attainment or achievement of quality? The standard provides a basis for comparing a measured result against an agreed-on reference. Conceptually, the standard is a communications medium between (1) buyers and sellers, (2) manufacturing and quality control, (3) customers and quality assurance, and (4) preparers and users. It gives the intent of the design, instruction, requirement, criteria, and method.

Another question that may arise is: How does one know what standards are needed? To answer that question, it is necessary to become familiar with the field, talk to the quality control and quality assurance departments or standards coordinator. The American Society for Quality Control (ASQC) has an information service (414/272-8575) that individuals can call for ideas about what standards are available on quality-related topics. The American National Standards Institute (ANSI) also has an overview of all available national standards. In addition, a study of various handbooks reveals many applicable standards for each aspect of the manufacturing process.

For example, drawing specifications should accurately detail limits in size, form, and shape; positional relationships of features; and functional or assembly requirements. One subject that has received increasing interest is geometric dimensioning and tolerancing.[31] It can be described as a means of specifying the geometry or the shape of a piece of hardware on an engineering drawing. It provides the designer with a clear way of expressing design intent and part requirements, which in turn enables the manufacturer to choose the proper method to produce the part. Geometric dimensioning and tolerancing also indicates how the part should be inspected and gaged, thus protecting the design intent.

Geometric dimensioning and tolerancing is rapidly becoming a universal engineering drawing language and technique that manufacturing industries and government agencies are finding essential to their operational well-being. Over the past 30 years, this subject has matured to become an indispensible tool; it assists productivity, quality, and economics in building and marketing products. Geometric dimensioning and tolerancing facilitates the application of computer-aided design, drafting, and manufacturing. The rules of geometric dimensioning and tolerancing become increasingly important as artificial intelligence is applied to product and process design.

The rules and concepts of geometric dimensioning and tolerancing are further discussed in ANSI Y14.5M, "Dimensioning and Tolerancing," as well as other publications listed in the reference and bibliography section of this chapter.

The material should be specified by a generally recognized standard that has been established by one of the professional societies or ANSI. Specifications should include the standards that describe the contents of the designated material, its properties, its working characteristics, and how it should be processed to obtain desired results.

ISO has published a set of international quality standards, the 9000 series; the U.S. versions are the ANSI 90 series. The purpose of the international standards is to improve the flow of international commerce by establishing common quality terminology, and for controlling and assuring the quality of products, parts, and materials on a worldwide basis. The important U.S. and international quality standards are listed in Table 23-3. ANSI is cooperating with ISO and the International Electrotechnical Commission (IEC) to achieve international consensus. When referring to any standards document, it is important to obtain the most current revision.

ADVANTAGES OF STANDARDS

By referring to the applicable standard, a manufacturer can minimize the time and effort of building up specifications from scratch. Of course, this requires that the manufacturer understands what the standard is.

Standards allow the manufacturer to reduce the number of required varieties in stock and help maintain a uniform product. They are invaluable for purchasing agents to secure truly competitive bids and to compare bids.

Standards allow the designer to specify standard materials that are readily available and provide uniform procedures for testing products in different laboratories. Customers can specify a product based on established standards.

A standard specification implies standard methods of testing and standard definitions. In some instances, methods of testing are incorporated within a materials specification, while in other cases some standardizing agencies establish standard methods of testing separately from the materials specifications and make reference to the test methods.

QUALITY AUDITS

Auditing has been an accepted part of financial reporting for many years. An audit is an official verification of records and statements. On the one hand, it would be irresponsible to completely trust the preparer of the financial statement; thus some checking is done. On the other hand, the cost of completely duplicating through independent means every transaction of the accountant would be prohibitive, wasteful, and unnecessary.

Quality audits are necessary because they provide an indication of the health of the unit with respect to quality, the extent to which quality technologies are embedded into the operations of the unit, and the commitment to quality of the people who make up the unit.[32]

In many companies today, quality training is being provided in massive doses—at all levels. But if quality technologies are not used, no amount of training will improve quality, costs, and productivity. The quality audit is one way to identify the need for change in explicit circumstances, to recognize those areas where progress is being made, to assist in setting priorities for improvement projects, to provide assistance in identifying sources of aid for future improvements, and to evaluate internal and external supplier-customer relationships. (Internal supplier-customer relationships exist between such departments as testing, manufacturing, and purchasing.)

Another driving force for quality audits comes from finished goods manufacturers that are reducing the number of suppliers as well as incoming inspection and, in some cases, depending on sole suppliers and just-in-time manufacturers. Increasingly, the trend is for the finished goods manufacturer to require the supplier to conduct internal quality audits. At the same time, the finished goods manufacturer is also conducting audits of the supplier. These requirements are being made part of the contract—for the "privilege of doing business."

Quality audits are becoming more and more popular with companies as they realize the advantages offered. Some of the benefits of a quality audit system are as follows:[33]

- Fosters quality system development.
- Provides information to help management make good decisions.
- Aids in the allocation of resources.
- Spreads technology.
- Reduces product liability costs.
- Reduces overhead.
- Improves morale.
- Increases capacity.
- Generates profit.

TABLE 23-3
Typical Standards for Controlling and Assuring Quality

Standard Number	Standard Title
ANSI/ASQC A1	Definitions, Symbols, Formulas and Tables for Control Charts
ANSI/ASQC A2	Terms, Symbols and Definitions for Acceptance Sampling
ANSI/ASQC A3	Quality Systems Terminology
ANSI/ASQC B1.1	Guide for Quality Control Charts
ANSI/ASQC B1.2	Control Chart Method of Analyzing Data
ANSI/ASQC B1.3	Control Chart Methods of Controlling Quality during Production
ANSI/ASQC C1 (ANSI Z1.8)	Specifications of General Requirements for a Quality Program
ANSI/ASQC E.2	Guide to Inspection Planning
ANSI/ASQC Q1	Generic Guidelines for Auditing of Quality Systems
ANSI/ASQC Z1.4 (MIL-STD-105 D)	Sampling Procedures and Tables for Inspection by Attributes
ANSI/ASQC Z1.9	Sampling Procedures and Tables for Inspection by Variables for Percent Nonconforming
ANSI/ASQC Z1.15	Generic Guidelines for Quality Systems
ANSI/IEEE 730	Software Quality Assurance Plans
ANSI/IEEE 828	Standard for Software Configuration Management Plans
ANSI/IEEE 829	Standard for Software Test Documentation
ANSI/IEEE 830	Guide to Software Requirements Specifications
ANSI/IEEE 983	Guide for Software Quality Assurance Planning
MIL-Q-9858A	Quality Program Requirements
MIL-I-45208A	Inspection System Requirements
MIL-STD-45662	Calibration System Requirements
ISO/8402	Quality Assurance Vocabulary
ISO/9000 (ANSI/ASQC Q90)	Quality Management and Quality Assurance Standards
ISO/9001 (ANSI/ASQC Q91)	Quality Systems: Model for Quality Assurance in Design/Development, Production, Installation and Servicing
ISO/9002 (ANSI/ASQC Q92)	Quality Systems: Model for Quality Assurance in Production and Installation
ISO/9003 (ANSI/ASQC Q93)	Quality Systems: Model for Quality Assurance in Final Inspection and Test
ISO/9004 (ANSI/ASQC Q94)	Quality Management and Quality System Elements—Guidelines

Note: Only the basic specification or standard number has been given. It is necessary to refer to the most current revision given by the specifying body. For example, the latest issue of ANSI/ASQC, ANSI/IEEE, and ISO standards can be obtained from the American National Standards Institute, 1430 Broadway, New York, NY 10018; (212) 354-3300. Standards with "MIL" as part of the standard number can be obtained from the U.S. Government Printing Office, Washington, DC.

TYPES OF AUDITS

Procedures audit, product audit, and process audit are the three major categories of audits. Each has a distinct purpose and should be used only for that purpose.

A procedures audit evaluates the effectiveness of the quality assurance program or system. The standard used during this audit would include the quality assurance policy and procedures manual and the operating manual for each department, including purchasing, receiving, inspection, shipping, material review board, and calibration.

A product audit is a quantitative assessment of conformance to required product characteristics. Product quality audits are generally performed for the following reasons:[34]

- To evaluate the outgoing quality level of the product or group of products.
- To determine if the outgoing quality product meets a predetermined standard level of quality for a product or a group of products.
- To estimate the level of quality originally submitted to inspection.
- To measure the ability of inspection to make valid quality decisions.
- To determine the suitability of the controls.

In effect, the product audit calibrates the inspector. Documents such as assembly drawings, workmanship standards, and test procedures are used for product audits.

A process audit provides an independent assessment of the effectiveness of a quality system through the evaluation of the knowledge of adherence to and adequacy of specific production methods used either in the performance or in the control of the work. The major items for evaluation are:

- Existence of procedures for performing the work as well as for inspecting or testing the work.
- Knowledge that production and quality personnel have about the procedures and specified requirements.
- Conformance of these personnel to the requirements.
- Reasons for deviation from the specified procedures.

The standards used for a process audit are military or commercial process specifications.

A fourth category of audit, which might be part of the product audit or an overall management quality audit, measures customer experience and perception of the organization's products and services. Major items for evaluation include:

- Procedures for acquiring, documenting, and communicating customer feedback.
- Statistical analysis of favorable and unfavorable customer feedback by product, by category (product, service, software, and so on), and by trends.
- Customer perception of quality of product and service vs. competitors.
- Review of episodic input from customers, either good or bad. These reviews made by professionals and independent auditors can be an excellent diagnostic tool for quality performance improvement.

AUDIT ELEMENTS AND PROCEDURES

A quality audit is a documented activity performed by unbiased individuals using written checklists or procedures for the evaluation of compliance with one or more standards. Based on this definition, a quality audit contains four elements:[35]

1. Documented activity. The observations made during the audit must be recorded, and a formal report must be sent to higher management.
2. Unbiased individuals. The auditor must be independent and fair-minded. The auditor should neither supervise nor report to the unit being audited. The auditor must not be responsible for correcting any reported deficiencies.
3. Use of clearly established procedures. Checklists and procedures will provide authority for the audit and ensure that it has been properly planned for both depth and breadth of coverage. A checklist also provides a good framework for reporting of results.
4. Evaluation of compliance with standards. Without standards, the audit is subjective opinion. The standard(s) used should be identified and followed.

As indicated in the third element of a quality audit, it is important to use established procedures to ensure good results. The following guidelines are intended to assist those who want to conduct a quality audit:

1. Locate qualified people and provide the necessary training. Because the auditor must evaluate another's work, sensitivity and integrity are a must. The auditor must be intelligent and discreet to be effective. Proper training is all-important.
2. Find the appropriate standards and then determine what the key elements are in the standards. Mark these to be sure they are checked when the audit occurs.
3. Chart the flow of the activity to be audited. Ideally, there is already a process sheet or procedure that spells this out. It may be included in the reference standard. If not, the audit group should develop the flowchart in cooperation with the audited activity.
4. Build the checklist. This removes much of the subjective element from the audit because the steps to be followed are clearly worked out in advance. A good checklist should have the following parts:
   - Purpose.
   - Scope.
   - Questions.
   - Blank spaces for recording results.
5. Plan and schedule the audit. Surprise audits may occasionally catch malfeasance, but should be used only as a last resort because surprise practically guarantees a lack of cooperation from the audited people and therefore becomes counterproductive if used routinely. The exception is in areas where serious out-of-control situations could be corrected very easily by the audited group. In this case, the group must be told that they are subject to unannounced audits. Scheduling is necessary to get maximum use of a scarce resource (the auditor) and to be sure that different activities are sampled in reasonable quantity. A key element of scheduling for coverage is the selection of an appropriate audit sample. Statistical sampling theory can develop the inevitable tradeoff between excessive audit cost and excessive risk of error.[36,37]
6. Perform the audit. Orient the audited activity, conduct the interview, and document the activity. Look carefully for aberrations and anomalies.
7. Follow up the audit. Do not assume that corrective action will occur without follow-up action. A follow-up report will normally contain the following parts:
   - Summary of findings.
   - Conclusions drawn from findings.
   - Recommendations for further action.

The follow-up report should be reviewed carefully with the audited activity. The audit report should be treated as a confidential document and not released to other elements of the organization. Here is where proper auditor training and attitude is most important. Audit reports often contain bad news. Fair-minded toughness is essential to steer through the narrow channel between destructive criticism and a whitewash.

References

1. Armand V. Feigenbaum, *Total Quality Control*, 3rd ed. (New York: McGraw-Hill Book Co., 1983), p. 65.
2. *Ibid.*, p. 625.
3. "Quality Management and Quality System Elements—Guidelines," ANSI/ASQC Q94 (New York: American National Standards Institute, 1987).
4. Frank M. Gryna, Jr., "How Engineering Can Improve Quality," *Machine Design* (May 8, 1986), pp. 81-85.
5. Ashweni K. Sahni and Stanton D. Myrum, "Fault Tree Analysis of Implantable Leads," *ASQC Quality Congress Transactions* (Milwaukee: American Society for Quality Control, 1987), pp. 491-496.

REFERENCES

6. Willie Hammer, "Designing a Safe System," *Machine Design* (September 3, 1970), pp. 92-97.
7. R.L. Eisner, "Fault Tree Analysis to Anticipate Potential Failure," *ASME Paper* (December 1972).
8. L. P. Sullivan, "Quality Function Deployment," *Quality Progress* (June 1986), pp. 39-50.
9. J. M. Juran, Frank M. Gryna, Jr., and R.S. Bingham, Jr., eds., *Quality Control Handbook*, 3rd ed. (New York: McGraw-Hill Book Co., 1979).
10. R.N. Kacker, "Taguchi's Quality Philosophy: Analysis and Commentary," *Quality Progress*, Vol. XIX, No. 12 (December 1986), pp. 21-24.
11. Juran, Gryna, and Bingham, *op.cit.*, p. 9-16.
12. *Ibid.*, p. 9-17.
13. *Ibid.*
14. A. J. Duncan, *Quality Control and Industrial Statistics*, 4th ed. (Homewood, IL: Richard D. Irwin, Inc., 1974), p. 100.
15. J.M. Juran and F.M. Gryna, Jr., *Quality Planning and Analysis* (New York: McGraw-Hill Book Co., 1980), p. 285.
16. *General Motors Statistical Process Control Manual*, Section 4.1.3, pp. 4-9, available from the GM SPEAR Administrative Staff, GM Technical Center, Warren, MI.
17. Douglas C. Montgomery, *Introduction to Statistical Quality Control* (New York: John Wiley & Sons, Inc., 1985), p. 279.
18. *Ibid.*, p. 174.
19. G.E.P. Box, J.S. Hunter, and W.G. Hunter, *Statistics for Experimenters* (New York: John Wiley & Sons, Inc., 1978). Chapter 6, p. 165, has an excellent discussion on the analysis of variance (ANOVA). Many good programs are also available for personal computers to simplify the calculations required.
20. Juran, Gryna, and Bingham, *op. cit.*, Section 27.
21. Duncan, *op.cit.*, Chapter 27, "Tests for Normality," p. 580.
22. Juran, Gryna, and Bingham, *op. cit.*, pp. 22-65 through 22-67.
23. Box, Hunter, and Hunter, *op. cit.*, pp. 239-241.
24. *General Motors Statistical Process Control Manual*, *op. cit.*, Section 3.
25. Juran, Gryna, and Bingham, *op. cit.*, p. 13-10.
26. Phil Crosby, *Quality is Free* (New York: McGraw-Hill Book Co., 1979).
27. D. Scott Sink and K.J. Kiser, "Performance Action Teams: Update on Continuing Developments," *Proceedings of Annual IIE International Conference* (Norcross, GA: Institute of Industrial Engineers, 1985).
28. Kaoru Ishikawa, *QC Circle Activities*, QC in Japan Series No. 1, 10-11, Sindagaya 5-Chome, Shibayaku (Tokyo: Union of Japanese Scientists and Engineers (JUSE), July 1978).
29. Lal C. Verman, *Standardization—A New Discipline* (Hamden, CT: Archon Books, 1973).
30. *Guide 2: 1978 General Terms and Their Definitions Concerning Standardization and Certification* (Geneva, Switzerland: International Organization for Standardization, 1978).
31. Lowell W. Foster, *Modern Geometric Dimensioning and Tolerance*, 2nd ed. (Fort Washington, MD: National Tooling and Machining Association, 1982), p. 1.
32. John T. Burr, "Overcoming Resistance to Audits," *Quality Progress* (January 1987), pp. 15-18.
33. John H. Farrow, "Quality Audits: An Invitation to Management," *Quality Progress* (January 1987), pp. 11-13.
34. Armando Lopes Pereira, "Quality Audits and International Standards," *Quality Progress* (January 1987), pp. 27-29.
35. Fletcher A. Birmingham, "Audit to Standards for Excellence in Quality," *ASQC Quality Congress Transactions*, 1986, pp. 187-192.
36. *Ibid.*, p. 190.
37. Juran, Gryna, and Bingham, *op. cit.*, p. 21-16.

Bibliography

Barra, Ralph. *Putting Quality Circles to Work: A Practical Strategy for Boosting Productivity and Profits*. Milwaukee: ASQC Quality Press, 1983.

Berger, Roger, and Hart, Thomas H. *Statistical Process Control*. Milwaukee: American Society for Quality Control, 1986.

Berger, Roger W.; Shores, David L.; and Thompson, Mary, eds. *Quality Circles*. Milwaukee: American Society for Quality Control, 1979.

Capability Study Guidelines. Moline, IL: Deere & Company, 1981.

Freund, Richard A. "New International Quality Standards: Management, Assurance Systems and Terminology." *Quality Progress* (May 1986), pp. 18-22.

Gryna, Frank M., Jr. *Quality Circles, A Team Approach to Problem Solving*. New York: AMACOM, 1981.

Johnson, L.M. *Quality Assurance Program Evaluation*. Stockton Trade Press, 1982.

Juran, Joseph M. "International Significance of the QC Circle Movement." *Quality Progress* (November 1980), pp. 18-22.

Koch, W.H. *Products Liability Risk Control*. SME Technical Paper IQ75-538. Dearborn, MI: Society of Manufacturing Engineers, 1975.

MacDonald, B.A. "List of Quality Standards, Specifications and Related Documents." *Quality Progress*, vol. 9, no. 9 (September 1976), pp. 30-35.

Quality Assurance, Quality Control and Inspection Handbook, 3rd ed. Richardson, TX: C.L. Carter Jr. & Associates, 1979.

Quality Engineering Workmanship Standards Manual. Bethesda, MD: Martin Marietta Corp., 1981.

Quality Function Deployment: A Collection of Presentations and QFD Case Studies. Dearborn, MI: American Suppliers Institute, September 1987.

Robinson, Charles B., ed. *How to Plan an Audit*. Milwaukee: American Society for Quality Control, 1987.

Rubinstein, Sidney P. *Participative Systems at Work*. Milwaukee: ASQC Quality Press, 1987.

Sayle, A.J. *Management Audits*. New York: McGraw-Hill Book Co., 1981.

Statistical Quality Control Handbook. Newark, NJ: Western Electric Co. Inc., 1956.

Stephens, Kenneth S. "Standards: A New Frontier for Quality." *Journal of Quality Technology*, vol. 3, no. 4 (December 1977), pp. 125-132.

Sullivan, Charles D. *Standards and Standardization*. New York: Marcel Dekker, Inc., 1983.

Wadsworth, Harrison M.; Stephens, Kenneth S.; Godfrey, A. Blanton. *Modern Methods for Quality Control and Improvement*. New York: John Wiley & Sons, Inc., 1986.

Wilborn, Walter. *Audit Standards: A Comparative Analysis*. Milwaukee: American Society for Quality Control, 1987.

QUALITY COST AND IMPROVEMENT

Quality cost techniques provide a tool for the management of a company to "fine-tune" the quality system. The activities of this quality system exist in all departments of a company: marketing, product design, manufacturing, accounting, service, as well as others. It also extends to the suppliers of raw materials and parts.[1]

Activities of the quality system must be performed together to satisfy the needs of the direct customers and the ultimate users of the company's products. The demands for quality in today's competitive environment can no longer be satisfied by concentrating on inspection, product design, operator education, supplier control, or reliability itself. A systems approach is required for the synergism of all activities working together. Quality cost techniques identify the system in financial terms.[2]

Quality costs are often underestimated, but could represent 35% or more of product costs.[3] Figure 24-1 illustrates the four basic categories of quality cost contained in a quality system: prevention, appraisal, internal failure, and external failure.[4,5,6] Prevention and appraisal costs can be viewed as inputs in the quality system that are normally budgeted and controlled by management. How effectively these inputs, particularly prevention, are applied determines the level to which internal and external failures can occur. The activities contributing to these costs exist in all departments of a company such as marketing, product design, manufacturing, material control, and service. The activities of the quality department, although vital, represent only a minor portion of the effort in the quality system.

Typically the greatest portion of quality cost is spent for failures. In some cases external failure or field service has been reported to be up to 20% of product costs. This amount does not account for customer dissatisfaction that may result in a future loss of sales. Internal failures or scrap and rework have been reported to be up to 15% of labor and material costs.[7]

Appraisal or inspection normally represents the next highest cost in a quality cost system and can be up to 15% of total labor costs. Typically, the smallest portion of quality costs is devoted to preventing problems from occurring.[8]

By using quality cost techniques, the management of a company can examine the relationships between these cost categories in terms of money expended for a particular activity on both a macro and micro level. Relationships among departments and activities within those departments can be evaluated, and changes can then be made to improve preventive efforts, resulting in greater customer satisfaction. With this greater satisfaction, sales should increase in the future. On a near-term basis, failures will decrease along with the need for maintaining an inspection system. Both the short-term and long-range prospects for profitability are therefore improved.

An important question asked by the financial community is why quality cost reports are required in addition to the normal accounting reports currently available. For most companies, the accounting system was not designed to directly evaluate the impact of quality costs on both the business and the customer. Quality for many years was thought to be a subjective measurement and not clearly identifiable with overall operating costs.[9]

For example, it is usually not possible to quantify activities conducted by the marketing department to identify customer needs through normal accounting reports. Also, these reports usually do not provide the means for evaluating the comprehensiveness of the activity in product design that translates these customer needs into standards and requirements for manufacturing. Yet these prevention activities are vital to assure satisfaction.

In manufacturing, failure costs are often understated because rework operations and scrap allowances are considered part of productive or direct labor and are not separately identified. Inspection that is performed by the production operator may likewise be included in productive or direct labor. Without separately identifying these activities, the true potential of improvements may not be known.

The value of quality cost is that it supplements the accounting system by providing a means to identify the effort spent on activities in the quality system. This effort is measured in financial terms. Doing this permits a systems approach to improving quality.

CHAPTER CONTENTS:

APPLICATIONS OF QUALITY COSTS

Quality cost techniques can be used as an effective tool for such things as strategic quality planning, continuous improvement efforts, budgeting purposes, product cost estimating, specific department improvements, and supplier quality relationships. It has rather broad-based applications in the business environment because it uses financial terms that are familiar to all disciplines.

Contributor of this chapter is:* *William O. Winchell*, *Associate Professor, Department of Industrial Engineering, Alfred University.

Reviewers of this chapter are:* *Caroline Bolton*, *Cost of Quality Analyst, B.O.C. Lansing, General Motors Corp.;* *Frank J. Corcoran*, *Quality Assurance Specialist, Kearfott Div., The Singer Co.;* *Edgar W. Dawes*, *Quality Principal, Quality Improvement Associates;* *S. Gopalan*, *Advanced Manufacturing Engineer, General Electric Co.;* *Andrew F. Grimm*, *Director, Quality Control, C & F Stamping Co.;* *J.S. Seward*, *Lawrence Institute of Technology.

CHAPTER 24

APPLICATIONS

STRATEGIC QUALITY PLANNING

Strategic quality planning is the integration of resources in a company to achieve a quality advantage in a marketplace. To achieve the strategic goals, relatively short-term activities must be identified and implemented.

More and more companies recognize that quality must be a fundamental driving strategy to survive in today's marketplace. A planning model used by a company that recognizes the importance of this type of strategy is shown in Fig. 24-2.[10] This company recognizes that quality cannot be degraded to meet other strategic driving forces.

The strategic quality planning process focuses on costs. Because of this, knowledge of the existing quality system defined through quality costs is a vital input to the planning process. Knowledge of these costs can provide the basis for direction leading to the optimum integration of resources in the company. It can also prove invaluable in identifying those short-term activities that lead to the achievement of the strategic goals.

As a result of implementing strategic quality planning, one company identified that inconsistent quality was the result of reworked processes to match changed product designs when it was too late to order new equipment that was needed. To avoid this, product development, manufacturing planning, and supplier commitments were scheduled concurrently instead of in series. Producibility problems in this company are now handled at the same time that laboratory and prototype assessments are evaluated. This minimizes the need to solve manufacturing and supplier problems at the time the product is released for production. The adopted resource allocation plan is shown in Fig. 24-3.[11]

PRODUCT IMPROVEMENT

In the past, quality cost reporting was used in improvement efforts, but it often achieved inadequate results. The main reason for this shortcoming was that the reporting was not accompanied by an improvement action. It is now recognized that a far more realistic objective is for quality cost reporting to support an improvement team whose effort is directed toward improving quality and productivity. Multidiscipline improvement teams can use quality cost reports to point out the strengths and weaknesses of a quality system. These improvement teams can also describe the benefits and ramifications of changes in financial terms. Return on investment (ROI) models and other financial analyses can be constructed directly from quality cost data. These models and analyses can be used to justify improvements to management. Those on the improvement teams can also use this information to rank problems in order of priority, seek out root causes, and implement the most effective, irreversible corrective action. The teams can also track results to ensure that they are headed in the right direction.

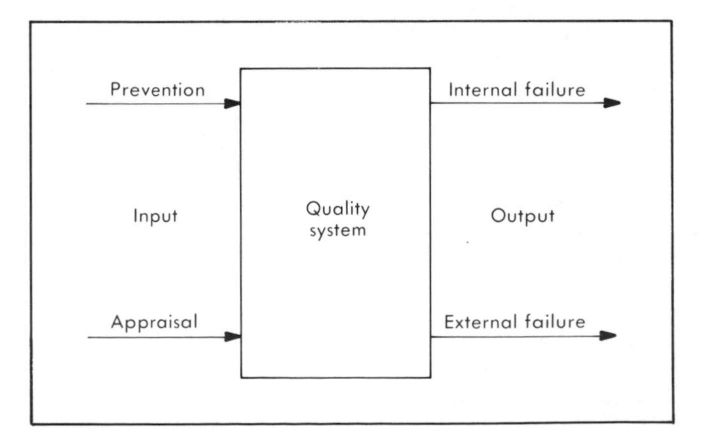

Fig. 24-1 Relationship of the basic quality cost categories with the quality system. *(Copyright American Society for Quality Control, Inc. Reprinted by permission.)*

Companies who collect and report quality costs recognize the value of this information to those making the improvements. They also recognize that reporting will accomplish nothing without someone "making it happen."[12]

BUDGETING

The process of budgeting for many companies consists of adjusting the historical costs of each cost center for changes in economics and expected improvements. With the advent of strategic quality planning, budgeting can now address the changes required to reach goals. Quality cost data and analyses can provide guidance as to the adjustments required.

PRODUCT COST ESTIMATING

For many companies, allocations are used in the cost estimating of new products. These allocations may be based on the

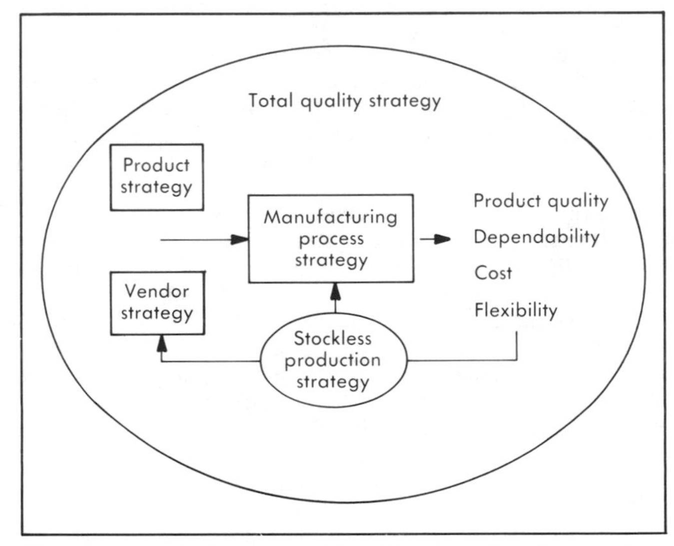

Fig. 24-2 Strategic planning model used by a manufacturing organization. *(Copyright American Society for Quality Control, Inc. Reprinted by permission.)*

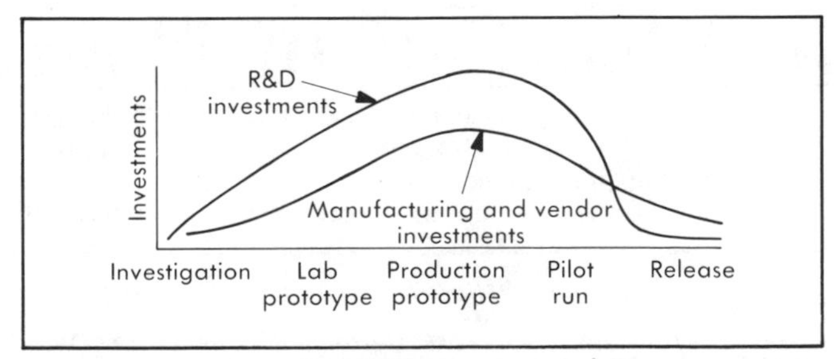

Fig. 24-3 Resource allocation for concurrent strategy. *(Copyright American Society for Quality Control, Inc. Reprinted by permission.)*

"average" experience of products produced in a particular cost center. However, the new product being estimated may have a higher or lower expectancy of, for example, scrap or rework than the average experience. Quality cost data on the parts produced in that cost center can be used to help determine whether the product would be average or not, thus facilitating more accurate estimates.

DEPARTMENTAL IMPROVEMENT

Quality costs have recently been used as a tool for aiding in the improvement of staff departments. In this application, each department is looked on as a separate business with customers usually internal to the company. The department furnishes products or services to these customers. Each department should also have a quality system to ensure that these products or services meet the needs of the customers. As in the traditional use of quality costs, this tool can document the quality system used in each staff department. Except in a general way, it has been found that the definitions used for manufacturing products are not appropriate for the activities in staff departments. Figure 24-4 illustrates a process that may be useful in defining the quality cost categories after the product or service has been identified.[13] The cost categories identified by one company for several staff departments are shown in Table 24-1.[14] The results of applying quality cost techniques for this company in the staff areas is shown in Table 24-2. Besides a reduction in overall quality costs, a shift in effort from correcting failures to prevention and appraisal took place.

SUPPLIER QUALITY COSTS

The quality costs of a supplier can be looked at as either "hidden" or "visible" to the company buying the products.[15]

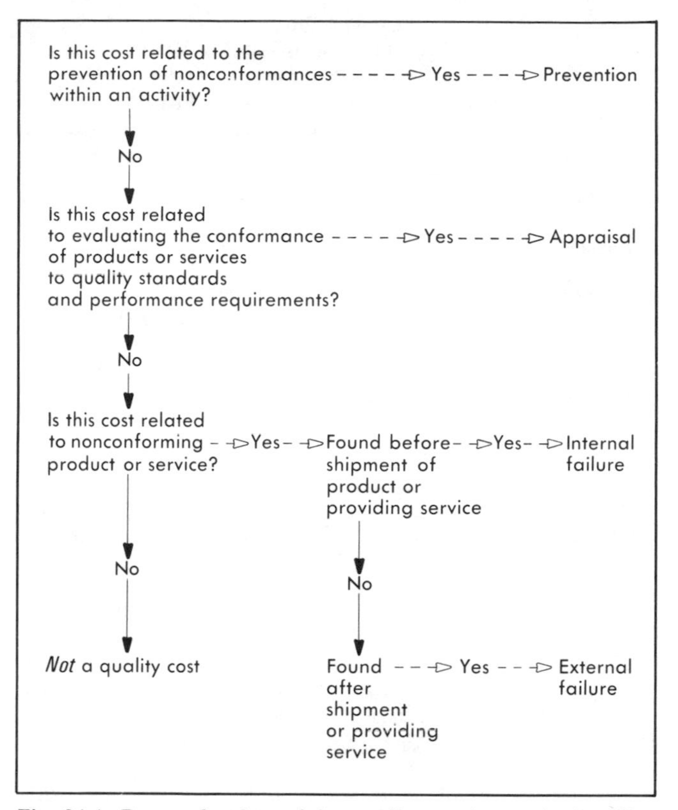

Fig. 24-4 Process for determining quality cost categories in office departments. *(Copyright American Society for Quality Control, Inc. Reprinted by permission.)*

TABLE 24-1
Relationship of Quality Cost Categories to Staff Departments of an Engineering Firm

	Quality Control	Quality Engineering	Supplier Quality	Inspection Engineering
Product	Provide project quality control engineering and inspection service	Provide quality assurance engineering and audit service	Provide supplier surveillance and rating service	Develop inspection documentation
Prevention	Quality plan, quality control cost estimate, quality control instructions	Quality manual audit plan	Supplier survey plan	Establish quality assurance requirement
Appraisals	Proofread inspection plans, inspection audit	Proofread quality auditor evaluation	Source inspection, audit supplier rating audit	Checking quality assurance requirements and inspection instructions
Internal failure	Rewrite inspection plan errors found by quality control supervision	Errors found by quality engineering supervision	Errors found by supervision material review board	Rewrite or update quality assurance requirements and inspection instructions
External failure	Defects or squawks on subsequent operations and/or found on fielded products	Errors reported after publication of audit	Errors found during use or audit of supplier rating system	Required change proposals, update quality assurance requirements and inspection instructions

(with permission from American Society for Quality Control)

TABLE 24-2
Shift in Effort That Occurred as a Result of Applying Quality Costs to Staff Departments of an Engineering Firm

Category	Effort in 1983, %	Effort in 1984, %
Prevention	25	35
Appraisal	34	44
Internal failure	20	14
External failure	21	7

(*with permission from American Society for Quality Control*)

The hidden quality cost is the cost incurred by the supplier prior to shipping; it is much the same as the quality cost of the buying company. It is unlikely that the supplier will reveal these costs for competitive reasons. Although these costs are hidden from the customer, they are included in the purchase price of the products.

There are other supplier quality costs that are visible to the purchaser because they occur in the purchaser's plant. Examples of these costs are rework and scrap on purchased products. The costs of these two examples are the fault of the supplier. Also, some of the warranty costs incurred by the purchaser may be the fault of the supplier. There is a cost incurred by the purchaser if it elects to either absorb these costs or to make the expenditures in a cost recovery debit to the supplier. Some companies have initiated an approach that uses the visible costs to identify improvement opportunities for joint projects between the buying company and its supplier.

MANUFACTURING FOCUS

A key to achieving effective strategy as well as continuous quality improvement within a company is a clear understanding of the quality system. The Quality Cost Technical Committee of the American Society for Quality Control conducted a survey that revealed that some companies define their quality cost report as their scrap report. Most individuals familiar with quality cost concepts would agree that this definition is too narrow. The survey also revealed that there were other companies who include all the categories in their quality system as part of their quality cost report, starting in marketing and extending to product engineering, finance, materials management, manufacturing, and service. For these companies, quality cost is not confined to just the quality department's operations.

At this writing, the quality costs of many companies are somewhere between these two ends of the quality cost reporting spectrum. Typically these companies include 30-40% of the costs associated with the entire quality system. The missing costs, such as the prevention effort in product design, are not as obvious on traditional financial reports. Also missing are rework operations, inspection by the production operator, and process control by the setup person. These activities are all part of standard or direct labor. Warranty or field failure costs typically are not projected to the end of the warranty period and are therefore understated.

Including all activities of the quality system in the initial report may be critical to the success of the improvement team. One study, including the validation effort by product engineering, ultimately led to the conclusion that existing product testing was really based on field problems that occurred as long as 15 years ago, and current problems could be better addressed by determining testing requirements through a design failure mode and effects analysis (DFMEA). Other studies have shown that production operators were getting more money for inspecting products than the quality department. This was a very significant factor in coordinating the transfer of inspection functions from the quality department to manufacturing. After the initial study, which defines the quality system for the improvement team, the team can pick and choose what elements of cost it needs to support its future efforts.

DEFINITIONS

As has been previously mentioned, the costs in a quality system are divided into prevention, appraisal, internal failure, and external failure costs. To assist the user, these four main categories are further divided into different activities applicable to manufacturing and service industries. This subdivision is not meant to be all inclusive, but to provide a general idea of the activities under each cost category. The words used to describe these activities should be changed to reflect the specific language and meanings in use at each company. This will help to ensure a common understanding by those in the company. The description of these categories and activities are abstracted with permission from "Principles of Quality Cost."[16]

Prevention Costs

Prevention is the cost of all activities in a company specifically designed to prevent defects in deliverable products or services. Many of the activities under the prevention costs category are completed before the product is released to production. Activities considered part of the prevention costs category are listed in Table 24-3.

The initial effort in prevention costs is to identify customer or product user needs through customer or user surveys. This work is normally performed by the marketing department. These needs are then translated into quality standards and requirements by the product design department. In addition to the marketing and product design departments, many other departments are also involved at this stage, such as quality, purchasing, manufacturing, and service.

Appraisal Costs

Appraisal costs are the costs of all activities in a company that are associated with measuring, evaluating, or auditing the products or services to ensure conformance with quality standards and performance requirements. Activities under appraisal costs usually occur after the product is in production. Some of the activities in this category include receiving inspection of materials from suppliers, checking performed by production operators, and audits. Other activities included in this category are listed in Table 24-4.

Internal Failure Costs

Whenever quality appraisals are performed, there exists the possibility of discovering products that fail to meet requirements. When this happens, unscheduled and possibly unbudgeted expenses are automatically incurred. For example, when a complete lot of metal parts is rejected for being oversize, the possibility for rework must first be evaluated. Then the cost of rework may be compared to the cost of scrapping the parts and completely replacing them. Finally, a disposition is made and the action is carried out. The total cost of this evaluation, disposition, and subsequent action is an integral part of internal failure costs.

Internal failure costs have been defined to basically include all costs required to evaluate and either correct or replace products or services not conforming to requirements or user needs prior to shipment to the customer. These costs only occur after the product is in production. In general, this includes all the material and labor expenses that are lost or wasted due to defective or otherwise unacceptable work affecting the quality of products or services discovered anywhere during the entire operational sequence. However, corrective action directed

TABLE 24-3
Typical Activities Performed by Various Departments That Affect Prevention Costs in a Quality System

Marketing	Product Design and Development	Purchasing	Operations	Quality Administration
Marketing research	Quality progress reviews	Supplier reviews	Process validation	Administrative salaries
Customer perception surveys	Support activities	Supplier rating	Operations quality planning	Administrative expenses
Contract review	Qualification test	Purchase order technical data reviews	Operations support quality planning	Quality program planning
	Field trials	Supplier quality planning	Operator quality education	Quality performance reporting
			Operator SPC/process control	Quality education
			Design and development	Quality improvement
			Quality related equipment	Quality audits

TABLE 24-4
Typical Activities Performed by Various Departments That Affect Appraisal Costs in a Quality System

Purchasing	Operations	External
Receiving or incoming inspection and tests	Performance of planned inspections, tests and audits • labor operations • quality audits • inspection and test materials	Field performance evaluation
Inspection equipment for purchased material	Set-up and first-piece inspection	Special product evaluations
Supplier product qualification	Special tests (manufacturing)	Field stock and spare parts evaluation
Source inspection and control program	Process control measurements	Test and inspection data review
	Laboratory support	
	Inspection and test equipment • depreciation allowances • measurement equipment expenses • maintenance and calibration labor	
	Endorsement and certification by outside agencies	

toward elimination of the problem in the future is preventive action and may be classified as prevention. Activities performed in various departments as a part of internal failure costs are listed in Table 24-5.

External Failure Costs

External failure costs are those costs due to products or services not meeting customer requirements or needs after they are shipped to the customer. As is obvious, these costs only occur after the product is in production. The elements included in this category are listed in Table 24-6.

BASELINES FOR ANALYSIS

One difficulty encountered in tracking the progress of improvement efforts is that many things vary from one time period to another. This includes such things as the volume produced and wage adjustments. Because of this, quality cost can be best compared when it is a percentage of some appropriate baseline as shown in Fig. 24-5.[17]

For long-term analysis, net sales is the base most often used for presentations to top management. While this measurement may be important from a strategic point of view, individuals doing the analysis and improvement efforts require a baseline that is more related to the amount of work performed, such as operating costs or value-added costs.

Different baselines and their advantages and limitations are given in Table 24-7.[18] Because no one baseline is ideal for all situations, the use of more than one baseline is often required to meet the diverse needs of those reviewing or using the information.

COSTS AFTER DELIVERY

In some instances, there are costs to the company and the consumer that occur after delivery to the customer. These costs depend on the effectiveness of the quality system. Yet they can be easily looked on as being vague by those in the manufacturing environment. This vagueness is often because the costs can only be predicted based on historical records such as warranty. In some cases, such as product liability, cost prediction may not even be possible. Even with warranty that is predictable, only a very small percentage of the quantity shipped usually requires field service. Catching this small percentage with inspection prior to shipping is simply a futile task. Only through prevention—preventing the problems from occurring in the first place—can these costs be minimized to the company and the consumer. The purpose of this section is to explain in a general way the conditions under which these costs may occur so that they may be better understood and dealt with.

TABLE 24-5
Typical Activities Performed by Various Departments That Affect Internal Failure Costs in a Quality System

Product Design	Purchasing	Operations
Corrective action Rework due to design changes Scrap due to design changes Production liaison costs	Purchased material reject disposition costs Purchased material replacement costs Supplier corrective action Rework of supplier rejects Uncontrolled material losses	Disposition Troubleshooting or failure analysis Investigation support Corrective action Rework Repair Reinspection/retest Extra operations Scrap Downgraded end product or service Internal failure labor losses

TABLE 24-6
Typical Elements That Affect External Failure Costs in a Quality System

Customer complaint investigation and resolution
Returned goods evaluation and repair or replace
Retrofit costs
Recall costs
Warranty claims
Liability costs
Penalties
Customer goodwill
Lost sales

TABLE 24-7
Measurement Bases Used for Determining Quality Costs

Base	Advantages	Disadvantages
Direct labor hour	Readily available and understood	Can be drastically influenced by automation
Direct labor dollars	Available and understood; tends to balance any inflation effect	Can be drastically influenced by automation
Standard manufacturing cost dollars	More stability than direct labor dollars	Includes overhead costs both fixed and variable
Value-added dollars	Useful when processing costs are important	Not useful for relating different types of manufacturing departments
Sales dollars	Appeals to higher management	Sales dollars can be influenced by changes in prices, marketing costs, and demand
Product units	Simplicity	Not appropriate when different products are made unless an "equivalent" item can be offered

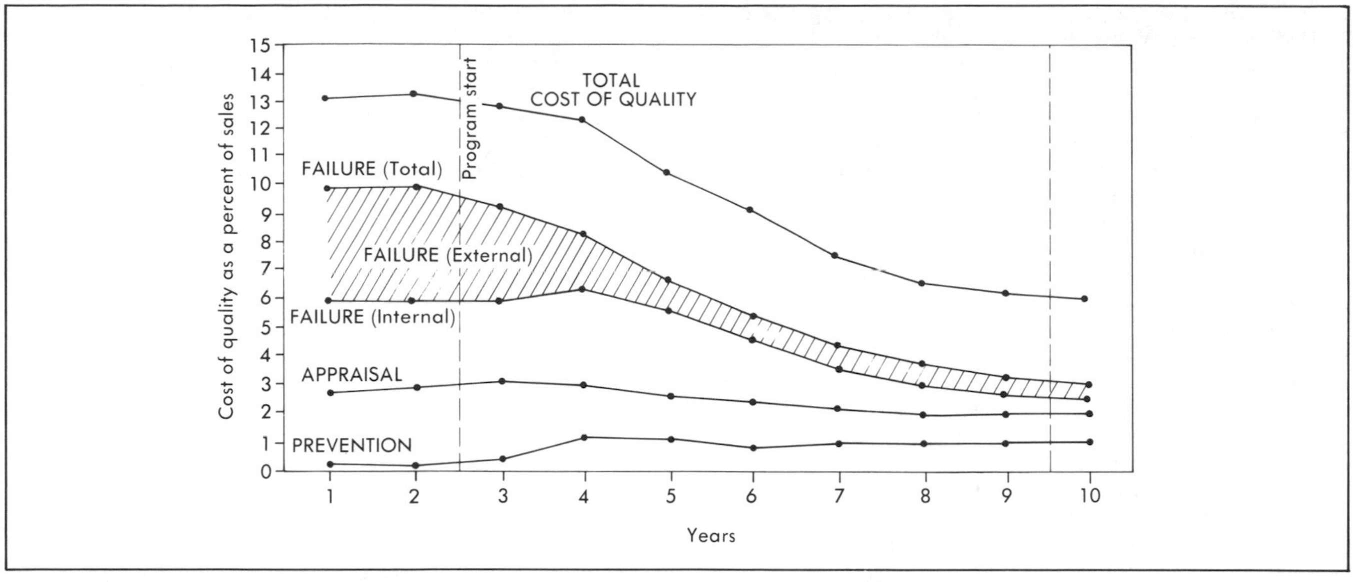

Fig. 24-5 Typical baseline used when comparing quality costs over a period of time. *(Copyright American Society for Quality Control, Inc. Reprinted by permission.)*

Costs to the Company

After the product is delivered to the customer, costs to the company occur because of warranties, negligence claims, product liabilities, violations of government acts, and lost sales. The two primary types of warranties are express warranties and implied warranties.

Express warranties. In general, express warranties mean that the goods furnished must conform to the description given or sample or model shown when the agreement was made. It also applies to any promises or other statements of fact made at this time.[19] As an example of the broadness of this remedy, a breech of express warranty was found based on a representation in an advertisement about the life of the product.[20]

The federal government has enacted the Magnuson-Moss Act to apply to the warranty of consumer products and service contracts. This act does not require a warranty to be given, but regulates the means of disclosing the warranty terms, places limits on disclaimers, promotes informal settlement procedures, and provides some civil remedies if a warranty is given.[21]

Many states have enacted so-called "lemon laws" as an indirect consequence of the Magnuson-Moss Act. These laws set standards between consumers and manufacturers that create a statutory warranty modifying the manufacturer's express warranty.[22]

Implied warranties. Implied warranties differ from express warranties in that they are implicit in any agreement when certain conditions are met.[23] In other words, a manufacturer may be obligated for warranty in this manner even though there was no express warranty given to the buyer. Although there are several implied warranties, only the implied warranty of merchantability and the implied warranty of fitness for particular purpose will be discussed.

The implied warranty of merchantability warrants that the goods sold are of average quality within the industry. Appearance as well as the structural safety and durability of a product are important factors. Products must be fit for any ordinary purposes for which they are used. In addition, they must be adequately contained, packaged, and labeled. The products must also conform to any statements made on the container or label.

The implied warranty of fitness for particular purpose is a much stronger warranty than that of merchantability. The products sold must be fit for the purpose for which they are intended. This warranty is implicit in an agreement that specifies how the products are to be used. The agreement is made known to the seller, and the buyer relies on the skill and expertise of the seller.

In certain circumstances, implied warranties can be excluded from an agreement by disclaimers or inspection by the buyer. But if they are not excluded, remedies for breach may in some situations also include damages for personal injuries.

Negligence. The theory of negligence involves the recovery for injuries suffered when the manufacturer fails to exercise reasonable or prudent care in manufacturing a product.[24] A manufacturer also has a duty to design its product so that it does not present an unreasonable risk of harm to the user. The manufacturer's duty is limited under this theory to consequences reasonably foreseeable. Also, the failure to exercise reasonable and prudent care during production of the product must be the cause of the injury. In some cases, this duty extends to providing warnings to users of the product and installing safety devices to protect them from injury.[25]

Product liability. Product liability is the law imposed on the manufacturer in favor of a customer/user for loss suffered by reason of a defective product.[26] The defect could occur from the manufacture, design, construction, formula, development of standards, preparation, processing, assembly, inspection, testing, listing, certifying, warning, instructing, marketing, advertising, packaging, or the labeling of a product. The customer/user must prove that a defect existed when the product was shipped and that this defect caused the injury. It is not necessary to prove any negligence on the part of the manufacturer. The product can also be proved defective by drawing reasonable inferences from circumstantial evidence.[27]

Safety acts violations. Violations of government safety acts are potential sources of unnecessary cost for the manufacturer.[28,29] Besides the federal statutes concerning OSHA, 29 USC 651, and the motor vehicle and tire safety standards under 15 USC 1390, there are many more enacted by both the federal and state governments. For the Federal Consumer Products Safety Act, under 15 USC 2052, the promulgation of standards may be to prevent unreasonable risk of injury associated with a product. In some cases, such acts have the power to make producers recall products having substantial hazards. Violators in some cases may be subject to civil penalties and also criminal penalties.

Recalls. From time to time, even well-managed companies call back quantities of products from the field to correct problems that they have been unable, for one reason or the other, to anticipate.[30] Recent years have seen voluntary recall of products as diverse as automobiles, adhesives, bicycles, chemical sprays, paint removers, and pacemakers. Mandated recalls initiated by the government are not uncommon. Recent consumer legislation requires that in some cases all parties in the distribution chain be reimbursed by the manufacturer in the event of a recall. The government may also publicize the potential hazard. This publicity may be potentially harmful to the manufacturer's reputation.

Lost sales. Lost sales due to the customer's perception of poor product quality is often an abstract concept for most companies to place in perspective. A company that supplies its product to another firm may find it easier to visualize this concept. It is reflected by the cash flow lost because the customer perceives that better quality could be obtained elsewhere and therefore chooses to cancel or not reward a contract. Another example is the cash flow lost because the customer chooses to reduce the portion of the order "pie" for a similar reason.

For a company that sells directly to consumers, data leading to the identification of lost sales may be obtained through customer surveys. Figure 24-6 illustrates that consumer repurchase intentions vary as the perception of the quality of the product by the consumer changes. If a consumer feels a product is excellent, the chances are high that the consumer will buy that or a similar product again from the same producer. In contrast, a consumer who feels that the quality of a product is poor has a lower likelihood of repurchasing under the same conditions. Customer surveys can determine the estimated repurchase intentions of consumers for each class of perceived quality. This data can be used to calculate the future sales volume and resultant cash flow that will be lost because consumers do not perceive the product to be of the highest quality.

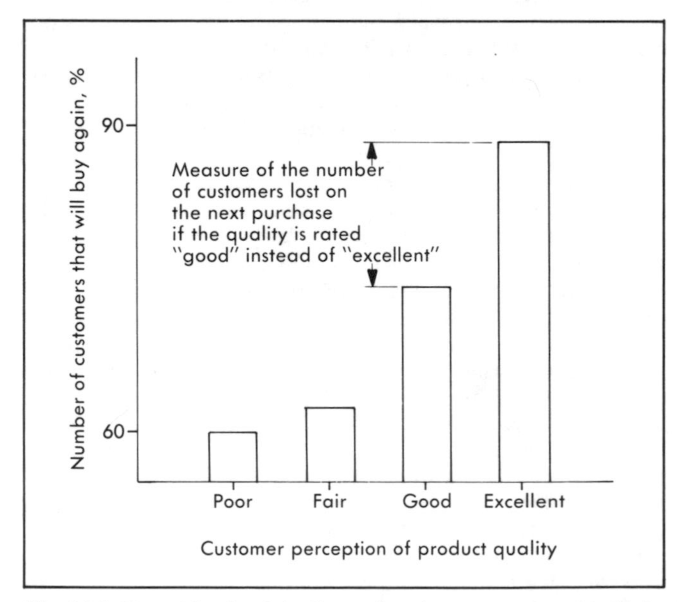

Fig. 24-6 Repeat buying intentions of customers compared to their quality perception.

Costs to the Customer

Lifecycle costs are costs of the product over the entire life of the product from the consumer's viewpoint.[31,32] They include the initial cost of purchasing the product, as well as operating and disposal costs. Operating costs include, among others, maintenance costs, repair costs, and often the time and expense of the consumer in getting these things done. When the perspective is broadened to include the cost to society, disposal cost may also include the effort to restore the materials in the product to raw material that can be recycled.

Figure 24-7 is an example of the lifecycle costs for a computer accessory. Shortly after the initial model was introduced, sales fell off sharply when the consumers found that service cost was forcing the lifecycle cost to be excessive. A new model that had a higher initial cost but lower lifecycle cost was introduced.

The Taguchi loss function is noted because of its similarity to lifecycle cost.[33] This function furnishes the framework around which Taguchi methods are used to gain improvements in products. It is defined as the financial loss imparted to society after a product is shipped (including any internal costs whether the product is shipped or not). The real power of the Taguchi loss function is its impact on changing the way quality is viewed and the methods that are used to justify improvements that do not meet traditional payback guidelines. Interestingly, it has been found that the manufacturer's cost improvements are often less than the loss to society that may be a result of that change. Therefore, the real value to society of making changes can be evaluated by this concept. This value may be a better indication of the long-term effect of the change on the company's future competitive status. Additional information on the loss function and Taguchi methods can be found in Volume IV, *Quality Control and Assembly*, of this Handbook series.

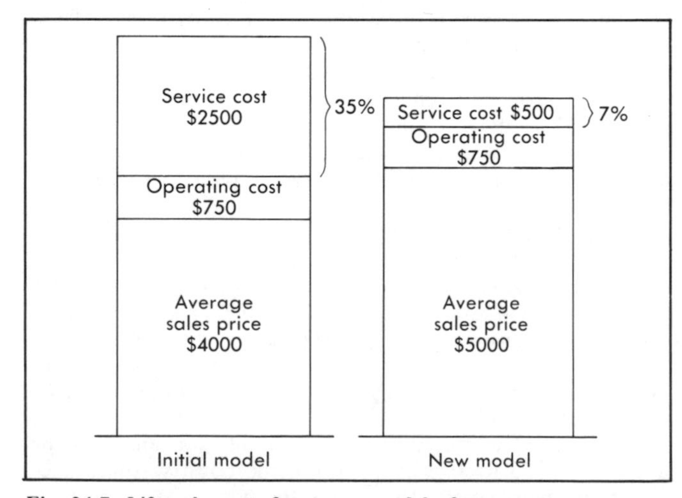

Fig. 24-7 Lifecycle costs for a new model of a computer accessory compared to the initial model. *(Copyright McGraw-Hill Book Co. Reproduced with permission.)*

CONTINUOUS IMPROVEMENT

A systems-oriented approach to quality has been popular for many years in U.S. companies.[34] In some cases the organization of the company has changed to reflect this thinking by adding a systems engineering function. This function was typically assigned the responsibility of pulling together activities associated with the design, manufacture, and assembly of products. Success with this approach has been limited for several reasons.

The first reason concerns a traditional approach widely used in problem solving. For example, a list of the top problems along with corrective action plans is requested. This often leads the whole organization, in a splintered fashion, on a witch hunt to find causal factors and corrective actions. With this approach, the basic problems with systems are often overlooked. Unless these basic problems are solved, a similar list of top problems will be found the next time this information is requested. This approach is often detrimental to continuous quality improvement.

A second reason for the limited success in continuous improvement is related to the way companies are organized. When viewing an organizational chart, a strong vertical relationship typically exists within each department because of the performance objectives. Horizontally among departments a much weaker relationship exists due to many factors, including those same individual department performance objectives. To achieve continuous quality improvement, the horizontal relationship in all departments must be strengthened.[35]

To strengthen the horizontal ties among activities in departments, multidiscipline improvement teams can be used. These teams direct their effort toward improving productivity and quality. As discussed previously, quality cost reports are used as a tool to point out the strengths and weaknesses of the quality system. Those on the teams can also use this information to rank problems in order of priority, seek out root causes, and to implement the most effective, irreversible corrective action. Financial analyses can be constructed directly from cost data. The teams can also track results to ensure that they are headed in the right direction. Companies who collect and report quality cost recognize the value of this information to those directly involved in making the improvements. They also recognize that reporting will accomplish nothing without someone making it happen.[36]

REPORTING

Because there is inherent delay in assembling most data, it is important to recognize that any quality cost report is unlikely to truly represent the exact conditions existing when the report is used.[37] The greatest application for quality cost reporting is in analyzing or justifying improvements or verifying the effectiveness of improvements that have been implemented. The actual improvement process may require the use of data that is more "real time" in nature to fine-tune the approach.

It is important to recognize that quality cost results cannot be compared among operations, plants, or other companies. Also there are really no reliable industry norms for the proper quality cost. Past attempts to make these comparisons have resulted in erroneous conclusions because each organization has unique conditions and quality requirements. It has also been found that it is unlikely that any two organizations will account for quality cost in the same manner. The only proper comparison is to compare the quality cost of an operation with itself over time; this is commonly referred to as trend analysis.

As discussed previously, the initial quality report can be used to allow the improvement team to understand the quality system. A summary page for such a report is depicted in Fig. 24-8.[38] The actual format used should be chosen by the improvement team. In using the report, for example, the improvement team may want to examine various relationships for improvement such as:

- The activity identifying customer perceptions compared to the cost of customer complaints.
- Supplier quality planning activity compared to the cost of required supplier corrective action.
- Operations planning activities compared to operations rework and repair cost.

These are just a few of the comparisons that may lead to improvement efforts in the quality system.

After the initial quality cost study, a decision must be made as to when the study should be repeated. If quality cost reporting is to support the improvement team, then it is logical to assume that specific parts of the study should be repeated when the improvement team needs new or additional information. This should be, for example, when an improvement is in place and a change in results is expected. To generate reports on a regular basis may be wasted effort if no new useful information is provided. Regular reports in the past were perceived as "control" type reports and, in many cases, were not well accepted in the long run by management.

To determine how a situation is doing over time, trend charts can be used. Failure costs, in particular, lend themselves to this type of analysis. Historical data is plotted versus time after normalizing the data by dividing by an appropriate baseline. It is then possible to project trends into the future in view of the improvement projects that have been identified.

There are two types of trend analyses. The first is long range and is principally used for strategic planning and management updates. The second is short range and is usually done in areas of specific interest (see Fig. 24-9).[39]

Failure costs could also be organized in Pareto fashion (the vital few as opposed to the trivial many). This approach provides direction as to the order in which problems should be solved to gain the greatest improvement (see Fig. 24-10).[40]

PROBLEM SOLVING

Although quality costs are valuable tools for identifying improvement areas, they do not solve any problems. Other tools must be used to find the root causes of problems so that irreversible corrective action can be implemented.

The root cause can be defined as the real cause of a problem. This is often quite different than the apparent cause, which appears after a superficial investigation. A frequently asked question is how can a person know when the root cause of the problem has been found. It has been found when the problem can be turned on or off by adding or removing the cause.

Company ______________ Product ______________ Prepared by ______________

Prevention costs	$ (000)	Year-to-Date Current	Year-to-Date Prior year
Marketing/customer			
Product/service development			
Purchasing			
Operations			
Quality administration			
Total			

Appraisal costs	$ (000)	Year-to-Date Current	Year-to-Date Prior year
Product/service development			
Purchasing			
Operations			
External appraisal costs			
Total			

Internal failure costs	$ (000)	Year-to-Date Current	Year-to-Date Prior year
Product/service design			
Purchasing			
Operations (subtotal)			
Material review			
Rework			
Repair			
Reappraisal			
Extra operations			
Scrap			
Total			

External failure costs	$ (000)	Year-to-Date Current	Year-to-Date Prior year
Customer complaints			
Returned goods			
Retrofit costs			
Warranty claims			
Liability costs			
Penalties			
Customer goodwill			
Total			

Baseline data	Year-to-Date Current	Year-to-Date Prior year
Net sales		
Production costs		
Material costs		
Design costs		

Quality cost ratios	Year-to-Date Current	Year-to-Date Prior year
External failure cost/net sales		
Operations failure costs/production costs		
Operations appraisal costs/Production costs		
Purchasing quality costs/material costs		
Design quality costs/design costs		

Fig. 24-8 Typical quality cost summary report for a product. *(Copyright American Society for Quality Control, Inc. Reprinted by permission.)*

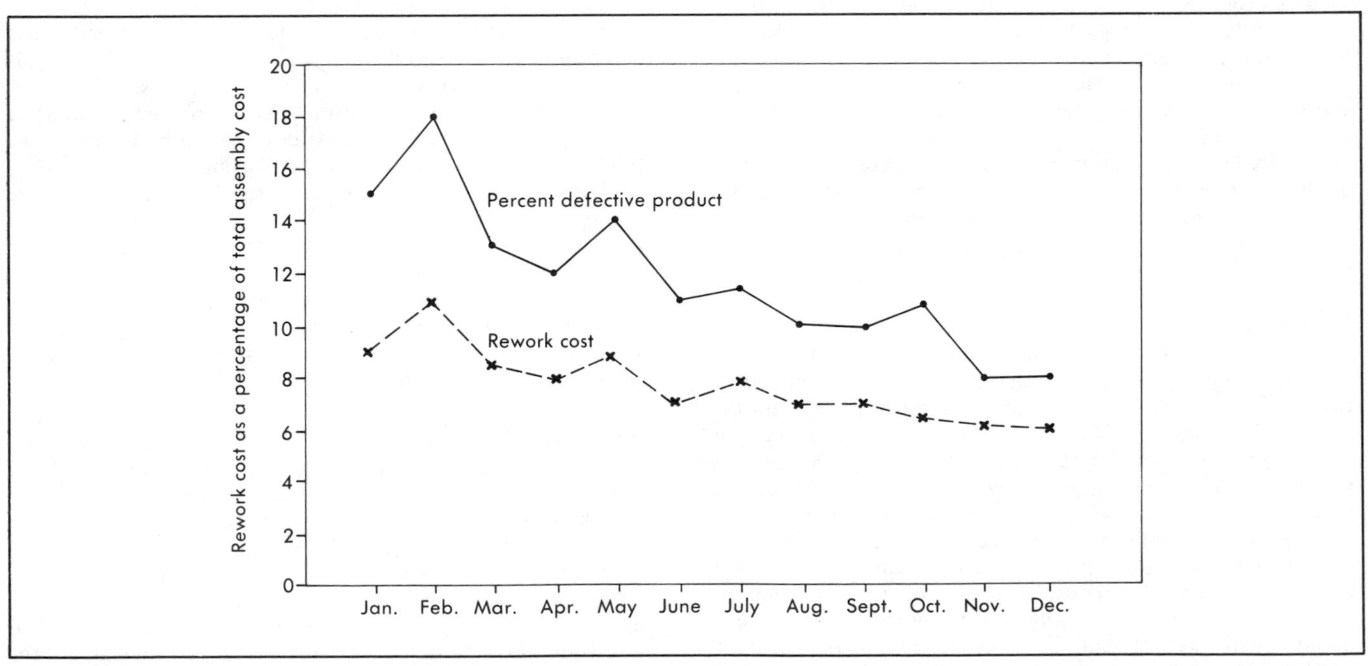

Fig. 24-9 Short-term trend analysis for assembly area quality performance. *(Copyright American Society for Quality Control, Inc. Reprinted by permission.)*

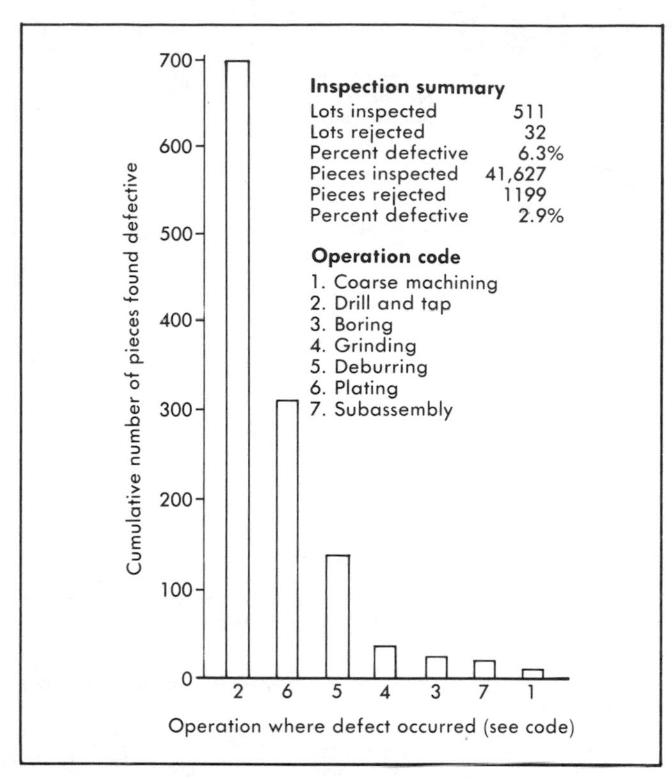

Fig. 24-10 Pareto analysis of eight operations on 11 key part numbers over a two-week period. *(Copyright American Society for Quality Control, Inc. Reprinted by permission.)*

Once the root cause has been found, an irreversible corrective action must be implemented so that there is no foreseeable situation in which the root cause can return. Adhering to this practice ensures permanent improvement.

A problem-solving approach used by one company is illustrated in Fig. 24-11.[41] This system is very effective for several reasons. First, it recognizes that a problem may be solved by a single person when the problem is obvious after some investigation (Phase I). It also recognizes that help may be needed from others to determine potential causes and suggests brainstorming and cause-and-effect analysis by those knowledgeable about the situation (Phase II).

Another important aspect of the problem resolution system is that it recognizes the existence of system or management problems. Quality professionals have consistently maintained that only 15-20% of the problems that occur in manufacturing are within the control of production operators. The remainder of the problems can only be solved by management because they are largely system problems (Phase III and IV).

A commonly used technique for cause-and-effect analysis is called the Ishikawa diagram (see Fig. 24-12).[42] This diagram enables the analysis of an effect or problem for causes by considering the many diverse and complex relationships that exist. The weakness of this approach, as well as other approaches, is that root causes are not distinguishable among all the causes identified. Other methods must be used, such as detailed investigations, comprehensive data analyses, and design of experiments to discover the root causes. Additional information on the subject of design of experiments can be found in Volume IV, *Quality Control and Assembly,* of this Handbook series.

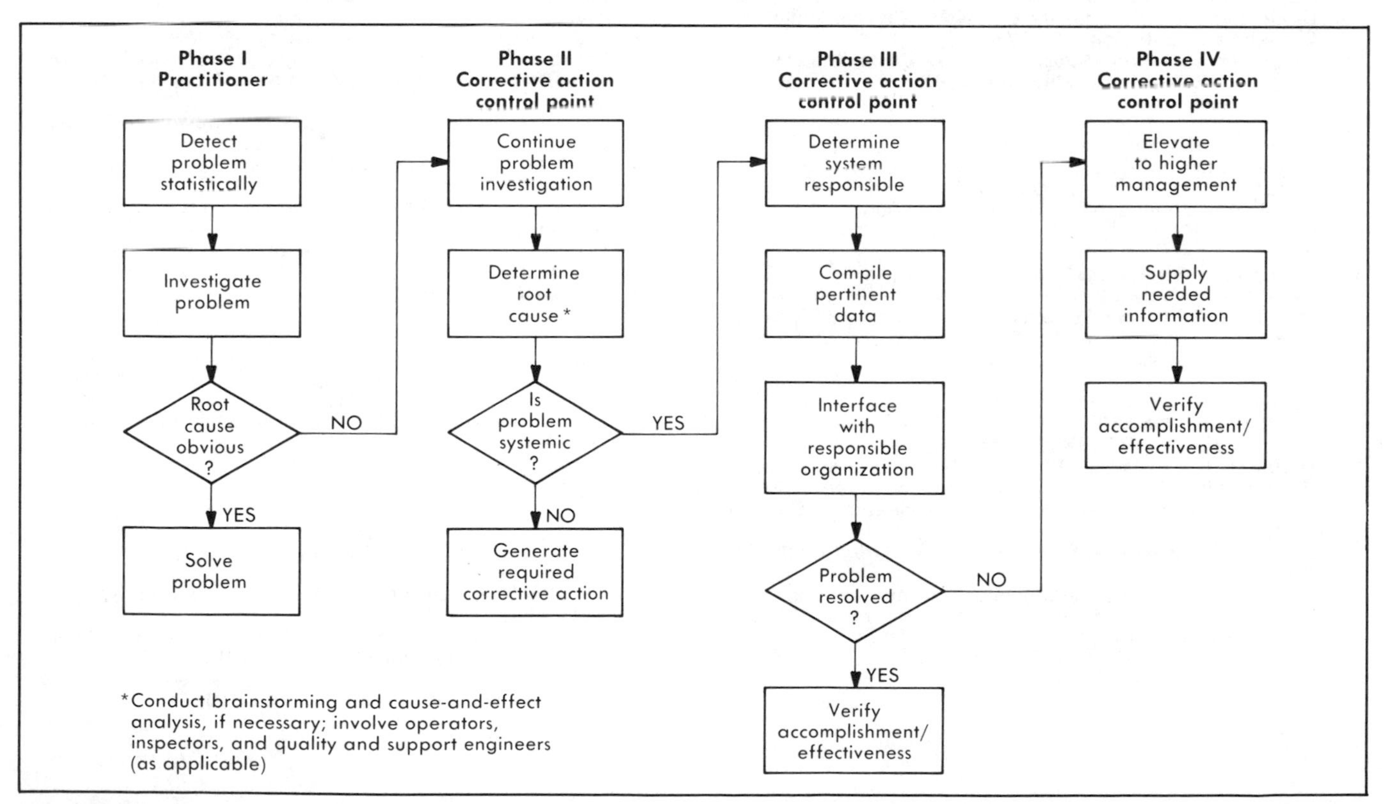

Fig. 24-11 Diagram of a problem-solving approach. *(Copyright American Society for Quality Control, Inc. Reprinted by permission.)*

REFERENCES

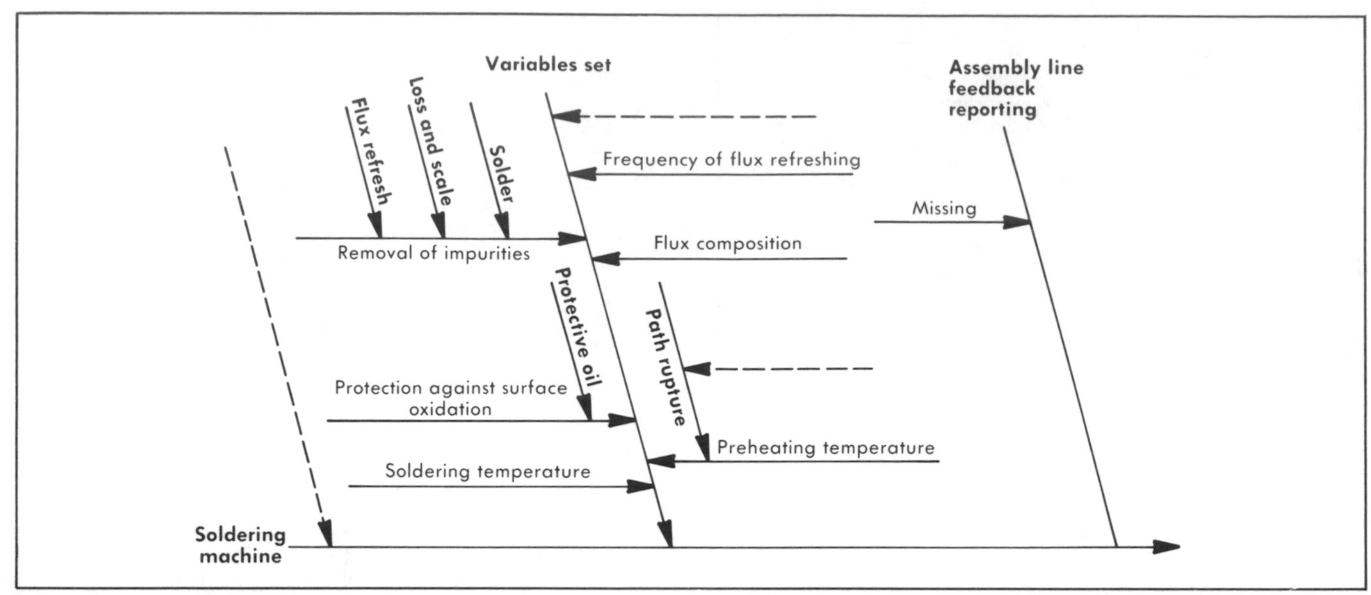

Fig. 24-12 An Ishikawa diagram of a process. *(Copyright American Society for Quality Control, Inc. Reprinted by permission.)*

References

1. *Principles of Quality Cost* (Milwaukee: American Society for Quality Control, 1986).
2. A.V. Feigenbaum, "Quality: Managing the Modern Company," *Quality Progress* (March 1985).
3. Keith E. McKee, "Quality in the 21st Century," *Quality Progress* (June 1983).
4. *Principles of Quality Cost, loc. cit.*
5. *Guide for Reducing Quality Costs* (Milwaukee: American Society for Quality Control, 1977).
6. *Guide for Managing Supplier Quality Costs* (Milwaukee: American Society for Quality Control, 1987).
7. McKee, *loc.cit.*
8. *Ibid.*
9. *Principles of Quality Cost, loc. cit.*
10. Jeffrey P. Kalin, "Quality, Stockless Production and Manufacturing," *38th Annual Quality Congress Transactions* (Milwaukee: American Society for Quality Control, 1984).
11. *Ibid.*
12. *Ibid.*
13. *Principles of Quality Cost, loc. cit.*
14. Lawrence J. Schrader, "An Engineering Organization's Quality Cost Program," *Quality Progress* (January 1986).
15. *Guide for Managing Supplier Quality Costs, loc. cit.*
16. *Principles of Quality Cost, loc. cit.*
17. *Ibid.*
18. Schrader, *loc. cit.*
19. Sales Under the UCC, *Callaghan's Michigan Civil Jurisprudence* (Callaghan & Co., 1986).
20. Herbert E. Greenstone, "A Lawyer's View of Manufacturer's Responsibility," *34th Annual Technical Conference Transactions* (Milwaukee: American Society for Quality Control, 1980).
21. Business Transactions, *Federal Lawyer's Manual* (Callaghan & Co., 1986).
22. Roger D. Billings, Jr., "Automobile Warranty Law: The Quiet Revolution," *Case and Comment* (January-February 1985).
23. Sales Under the UCC, *loc. cit.*
24. Negligence, *Callaghan's Michigan Civil Jurisprudence* (Callaghan & Co., 1986).
25. Greenstone, *loc. cit.*
26. Products Liability, *Callaghan's Michigan Civil Jurisprudence* (Callaghan & Co., 1986).
27. Edward M. Swartz, "The Search for Product Defect," *Case and Comment* (January-February 1986).
28. Business Transactions, *loc. cit.*
29. Consumer Protection, *Callaghan's Michigan Civil Jurisprudence* (Callaghan & Co., 1986).
30. A.V. Feigenbaum, *Total Quality Control*, 3rd ed. (New York: McGraw-Hill Book Co., 1983).
31. *Ibid.*
32. T. David Kiang, "Life Cycle Costing—A New Dimension for Reliability Engineering Challenge," *1976 ASQC Technical Conference Transactions* (Milwaukee: American Society for Quality Control, 1976).
33. L.P. Sullivan, "The Seven Stages of Company-Wide Quality Control," *Quality Progress* (May 1986).
34. *Guide for Reducing Quality Costs, loc. cit.*
35. Sullivan, *loc. cit.*
36. William O. Winchell, "Organizing Quality Cost Efforts to Minimize Difficulties," *39th North East Quality Control Conference Transactions* (Milwaukee: American Society for Quality Control, 1985).
37. *Principles of Quality Cost, loc. cit.*
38. *Ibid.*
39. *Ibid.*
40. *Ibid.*
41. Billie Ruth Marcum, "An Updated Framework for Problem Resolution," *Quality Progress* (July 1985).
42. Edward Kindlarski, "Ishikawa Diagrams for Problem Solving," *Quality Progress* (December 1984).

Bibliography

Ishikawa, Kaoru. *Guide to Quality Control*, 2nd ed. New York: UNIPUB, 1982.

Juran, J.M., and Gryna, Frank M., Jr. *Quality Planning and Analysis*. New York: McGraw-Hill Book Co., 1980.

Juran, J.M., and Gryna, Frank M., Jr. *Quality Control Handbook*, 3rd ed. New York: McGraw-Hill Book Co., 1974.

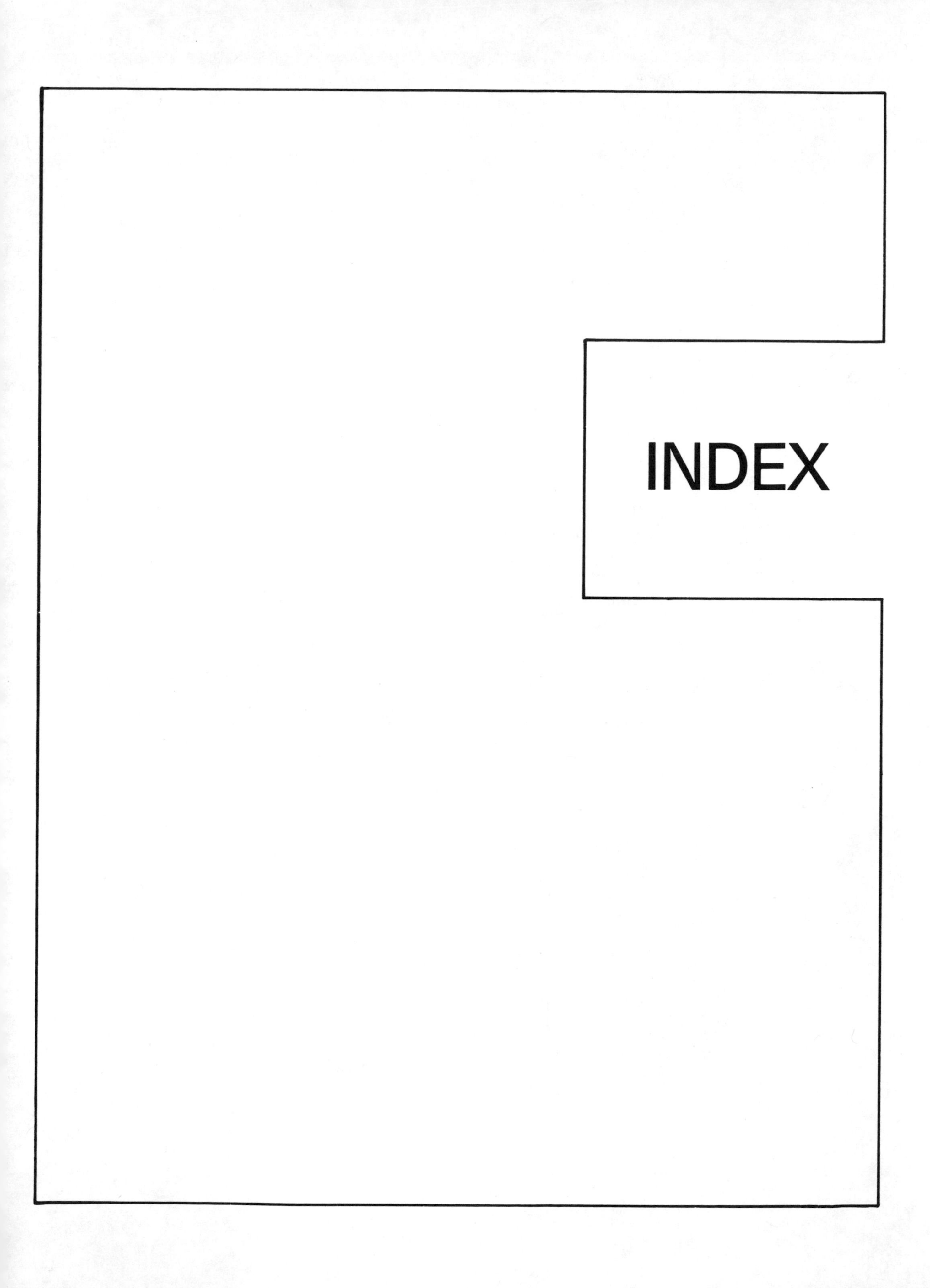
INDEX

A

B

C

D

N

O

P

Q

R

U

V

W

X

Z